Aircraft Anatomy

Aircraft Anatomy

A TECHNICAL GUIDE TO MILITARY AIRCRAFT FROM WORLD WAR II TO THE MODERN DAY

GENERAL EDITORS:
PAUL EDEN &
SOPH MOENG

amber
BOOKS

Previously published in two separate volumes as *Technical Drawings of Aircraft of World War II 1939–1945* and *Modern Military Aircraft Anatomy*

Published by
Amber Books Ltd
United House
North Road
London
N7 9DP
United Kingdom
www.amberbooks.co.uk
Appstore: itunes.com/apps/amberbooksltd
Facebook: www.facebook.com/amberbooks
Twitter: @amberbooks

ISBN: 978-1-78274-655-3

Project Editor: James Bennett
Designer: Neil Rigby

Printed in China

Contents

AIRCRAFT OF WORLD WAR II

Bombers and Dive Bombers

293869

Arado Ar 234

Above: This captured aircraft is an example of the major production model, the Ar 234B-2. The projection above the cockpit is a periscope sight which could serve the two optional 0.79in (20-mm) rear-firing cannon, or give the pilot his only view aft of the aircraft.

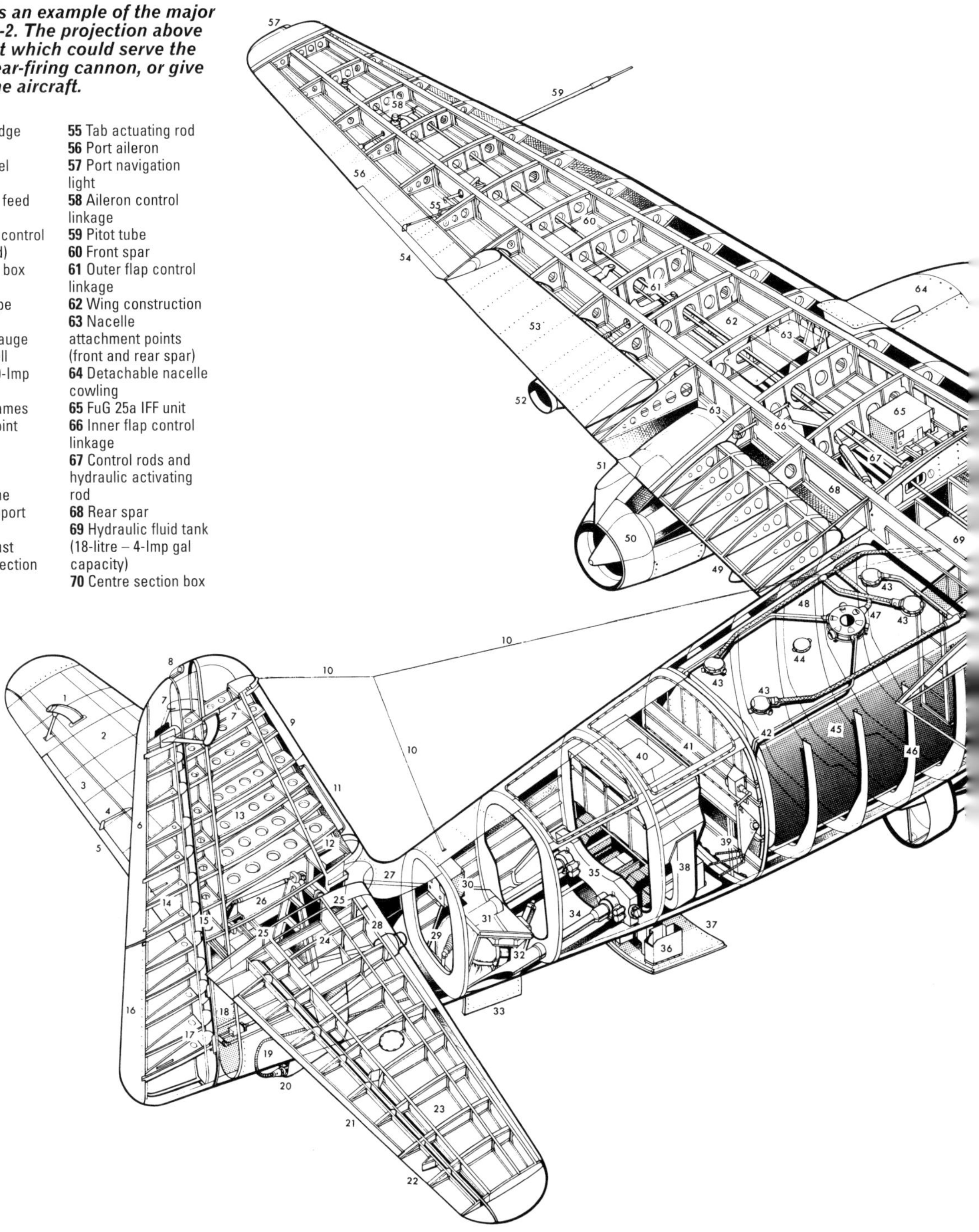

Arado Ar234

Cutaway key
1 Port elevator hinge
2 Tailplane skinning
3 Port elevator
4 Tab actuating rod
5 Elevator trim tab
6 Geared rudder tab (upper)
7 Rudder hinges
8 Tail navigation light
9 Plywood fin leading edge
10 T-aerial
11 Re-transmission aerial
12 Aerial matching unit
13 Tailfin structure
14 Rudder construction
15 Rudder post
16 Rudder tab (lower)
17 Lower rudder hinge
18 Rudder actuating rods
19 Parachute cable
20 Cable anchor point/tailskid
21 Starboard elevator tab
22 Elevator construction
23 Tailplane construction
24 Elevator control linkage
25 Tailplane attachment points
26 Elevator rod
27 Port side control runs
28 Internal mass balance
29 Parachute release mechanism
30 Main FuG 16zy panel (BZA computer)
31 Brake parachute container
32 Starboard MG 151 cannon muzzle
33 Brake chute door (open)
34 Mauser MG 151/20 cannon (rearward firing)
35 Cannon support yoke
36 Spent cartridge chute
37 Access panel (lowered)
38 Ammunition feed chute
39 Tail surface control rods (starboard)
40 Ammunition box
41 Bulkhead
42 Fuel vent pipe
43 Fuel pumps
44 Fuel lever gauge
45 Rear fuel cell (2000-litre –440-Imp gal capacity)
46 Fuselage frames
47 Fuel filler point
48 Fuel lines
49 Inner flap construction
50 Exhaust cone
51 Nacelle support fairing
52 RATO exhaust
53 Outer flap section
54 Aileron tab
55 Tab actuating rod
56 Port aileron
57 Port navigation light
58 Aileron control linkage
59 Pitot tube
60 Front spar
61 Outer flap control linkage
62 Wing construction
63 Nacelle attachment points (front and rear spar)
64 Detachable nacelle cowling
65 FuG 25a IFF unit
66 Inner flap control linkage
67 Control rods and hydraulic activating rod
68 Rear spar
69 Hydraulic fluid tank (18-litre – 4-Imp gal capacity)
70 Centre section box

*The Ar 234**B**-2 was far more versatile than its predecessor, the Ar 234**B**-1, being capable of bombing, pathfinding or reconnaissance missions. This model is equipped with* **Rauchgeräte** *take-off assistance rockets outboard of the engine nacelles.*

SPECIFICATION

Arado Ar 234B-2 Blitz

Type

Single-seat multi-role warplane

Powerplant

Two Junkers Jumo 004B turbojet engines each rated at 1,962 lb st (8.73 kN)

Performance

Maximum speed 460 mph (740 km/h) at 19,685 ft (6000 m); climb to 19,685 ft (6000 m) in 17 minutes 30 seconds with 3,307-lb (1500-kg) bombload; service ceiling 32,810 ft (10000 m); range 1,013 miles (1630 km)

Weights

Empty 11,464 lb (5200 kg); maximum take-off 21,715 lb (9850 kg)

Dimensions

Wing span 46 ft 31/2 in (14.10 m); length 41 ft 51/2 in (12.64 m); height 14 ft 11/2 in (4.30 m); wing area 284.18 sq ft (26.40 m2)

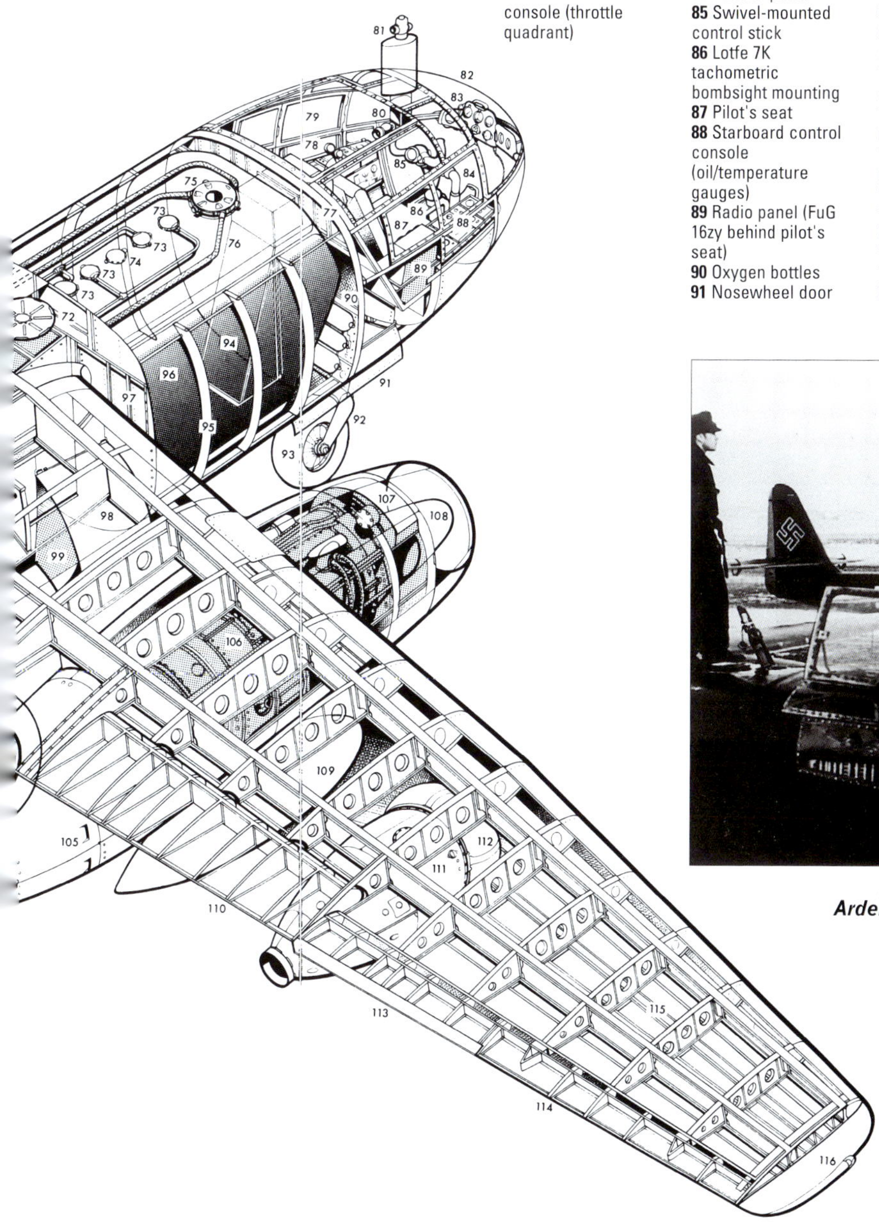

71 FuG 25a ring antenna
72 Suppressed D/F antenna
73 Fuel pumps
74 Fuel level gauge
75 Fuel filler point
76 Fuel lines
77 Bulkhead
78 Port control console (throttle quadrant)
79 Pilot entry hatch (hinged to starboard)
80 Periscopic sight
81 Periscopic head (rearview mirror/gunsight)
82 Clear vision cockpit glazing
83 Instrument panel
84 Rudder pedal
85 Swivel-mounted control stick
86 Lotfe 7K tachometric bombsight mounting
87 Pilot's seat
88 Starboard control console (oil/temperature gauges)
89 Radio panel (FuG 16zy behind pilot's seat)
90 Oxygen bottles
91 Nosewheel door
92 Nosewheel fork
93 Rearward-retracting Nosewheel
94 Nosewheel well centre section
95 Fuselage frames
96 Forward fuel cell (1800-litre – 385-Imp gal capacity)
97 Bulkhead
98 Mainwheel door
99 Starboard mainwheel well
100 Mainwheel leg door
101 Starboard mainwheel leg
102 Forward-retracting mainwheel
103 SC 1000 "Hermann" bomb beneath fuselage
104 Engine exhaust
105 Auxiliary cooling intakes
106 Starboard Jumo 004B turbojet
107 Annular oil tank
108 Riedel starter motor on nose cone
109 Auxiliary tank (300-litre/66-Imp gal) beneath nacelle (not carried with SC 1000 bomb)
110 Flat outer section construction
111 Walter HWK 500A-1 RATO unit
112 RATO recovery parachute pack
113 Aileron tab
114 Starboard aileron construction
115 Wing skin stiffeners
116 Starboard navigation light

Arado 234s line up awaiting another mission during the Ardennes counter-offensive of December 1944–January 1945. The Ar 234s were used for pinpoint attacks on the advancing Allied positions.

Arado Ar 234

This aircraft bears the markings of the 9th Staffel, III Gruppe, Kampfgeschwader (KG) 76. This unit was equipped with Ar 234s in January 1945, and was heavily involved in the attacks on Remagen bridge. KG 76 flew its first sorties with the Ar 234 during the Ardennes offensive in January, although the sortie rate was strictly limited through shortage of fuel. During the second week of February, the Kampfgeschwader was heavily engaged in an attempt to relieve the Allied pressure on Kleve. By the end of March 1945, sorties by KG 76 had virtually ceased.

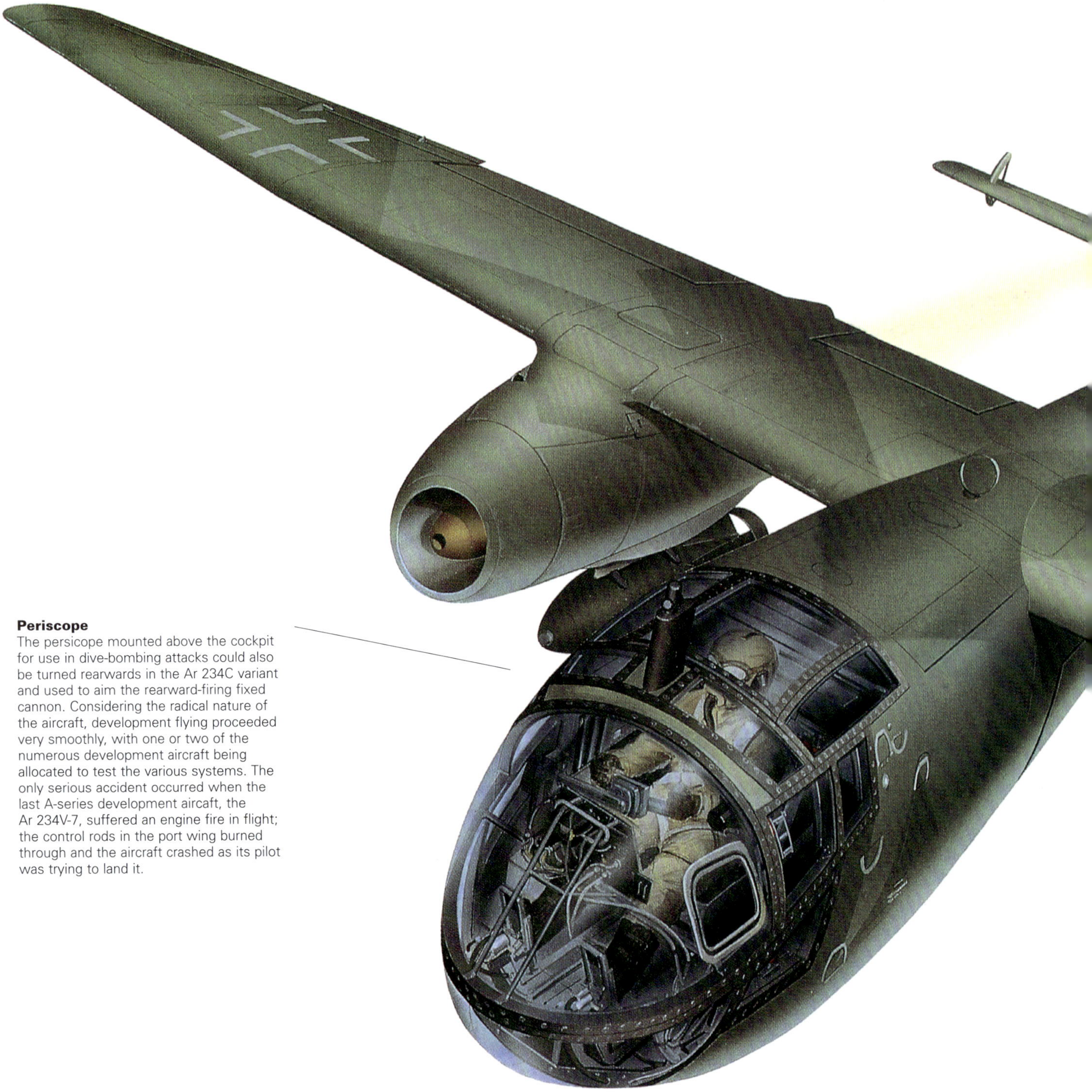

Periscope
The persicope mounted above the cockpit for use in dive-bombing attacks could also be turned rearwards in the Ar 234C variant and used to aim the rearward-firing fixed cannon. Considering the radical nature of the aircraft, development flying proceeded very smoothly, with one or two of the numerous development aircraft being allocated to test the various systems. The only serious accident occurred when the last A-series development aircaft, the Ar 234V-7, suffered an engine fire in flight; the control rods in the port wing burned through and the aircraft crashed as its pilot was trying to land it.

Cockpit
The pilot sat on a primitive ejection seat, with armour platting behind his headrest. The bomber was equipped with a Patin PDS three-axes autopilot with LKS 7D-15 overriding control, enabling the pilot to swing the control column clear so that he could use the Lofte 7K tachometric bomb sight, which was mounted beneath his feet. For shallow dive-bombing a BZA bombing computer was used in conjunction with an RF2C periscopic sight; steep dives were strictly forbidden because of jet surge and the sensitivity of the lateral trim.

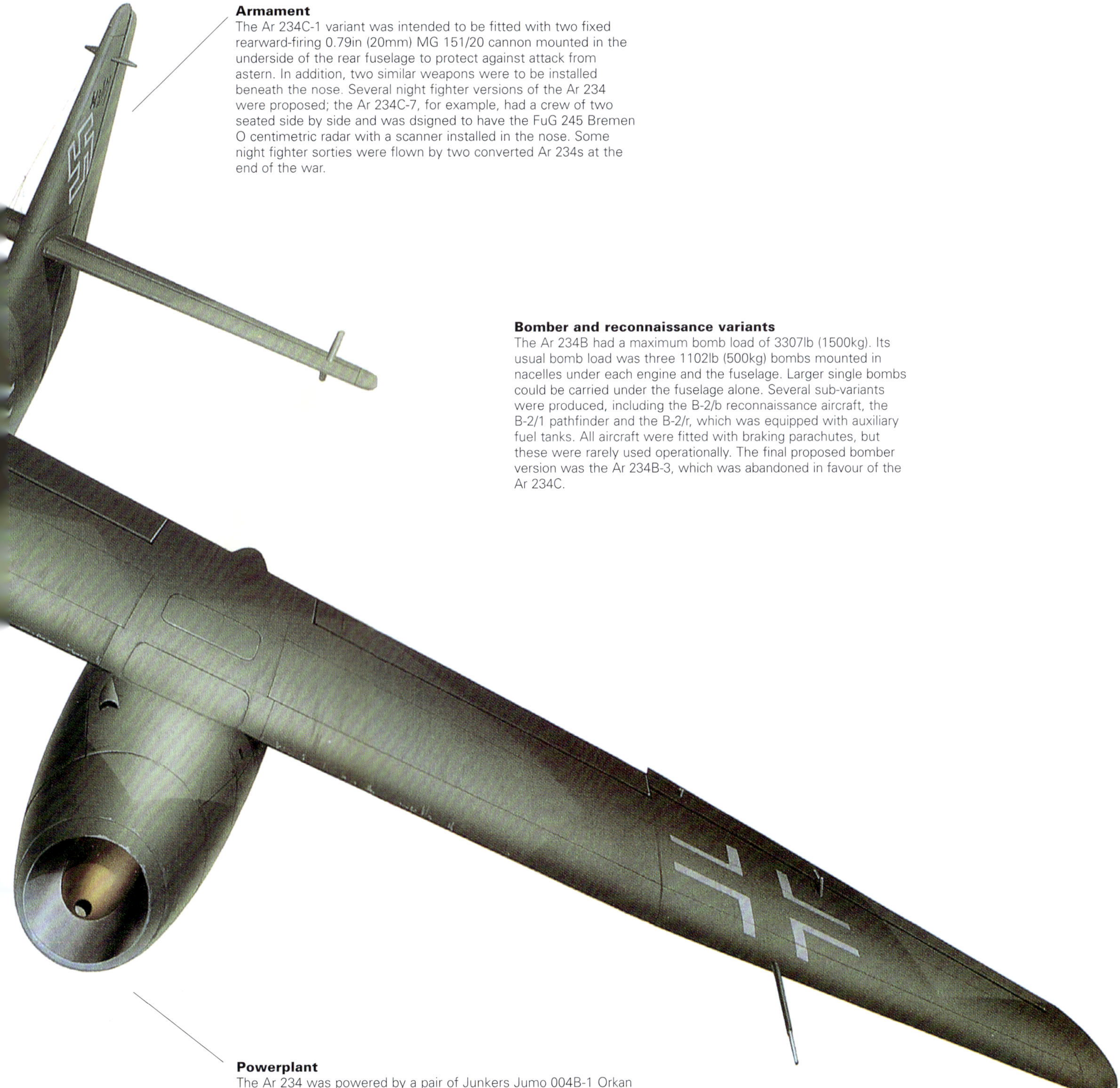

Armament
The Ar 234C-1 variant was intended to be fitted with two fixed rearward-firing 0.79in (20mm) MG 151/20 cannon mounted in the underside of the rear fuselage to protect against attack from astern. In addition, two similar weapons were to be installed beneath the nose. Several night fighter versions of the Ar 234 were proposed; the Ar 234C-7, for example, had a crew of two seated side by side and was dsigned to have the FuG 245 Bremen O centimetric radar with a scanner installed in the nose. Some night fighter sorties were flown by two converted Ar 234s at the end of the war.

Bomber and reconnaissance variants
The Ar 234B had a maximum bomb load of 3307lb (1500kg). Its usual bomb load was three 1102lb (500kg) bombs mounted in nacelles under each engine and the fuselage. Larger single bombs could be carried under the fuselage alone. Several sub-variants were produced, including the B-2/b reconnaissance aircraft, the B-2/1 pathfinder and the B-2/r, which was equipped with auxiliary fuel tanks. All aircraft were fitted with braking parachutes, but these were rarely used operationally. The final proposed bomber version was the Ar 234B-3, which was abandoned in favour of the Ar 234C.

Powerplant
The Ar 234 was powered by a pair of Junkers Jumo 004B-1 Orkan axial flow turbojets, which gave the aircraft a top speed of some 742km/h (461mph) when flying 'clean', without external stores. Some of the development Ar 234s were fitted with the BMW 003A-1 engine, but a great deal of trouble was experienced with the thrust control of this powerplant, which also proved difficult to restart after a flamout during flight. The problem of thrust regulation was solved by the use of the Jumo 004 system. The turbojets had a life of only about 25 hours.

Avro Lancaster

Such was the versatility and load-carrying capability of the Lancaster that many were converted for experimental purposes. This Mk II was used for jet engine testing, with a large dorsal air intake supplying the Metrovick F.2/1, 4 or 4A turbojet mounted in its rear fuselage.

Lancaster B.Mk III

Cutaway key

1 Two 0.303-in (7.7-mm) Browning machine-guns
2 Fraser-Nash power-operated nose turret
3 Nose blister
4 Bomb-aimer's panel (optically flat)
5 Bomb-aimer's control panel
6 Side windows
7 External air temperature thermometer
8 Pitot head
9 Bomb-aimer's chest support
10 Fire extinguisher
11 Parachute emergency exit
12 F24 camera
13 Glycol tank/step
14 Ventilator fairing
15 Bomb-bay doors forward actuating jacks
16 Bomb-bay doors forward bulkhead
17 Control linkage
18 Rudder pedals
19 Instrument panel
20 Windscreen sprays
21 Windscreen
22 Dimmer switches
23 Flight engineer's folding seat
24 Flight engineer's control panel
25 Pilot's seat
26 Flight-deck floor level
27 Elevator and rudder control rods (under floor)
28 Trim tab control cables
29 Main floor/bomb-bay support longeron
30 Fire extinguisher
31 Wireless installation
32 Navigator's seat
33 Canopy rear/down-view blister
34 Pilot's head armour
35 Cockpit canopy emergency escape hatch
36 D/F loop
37 Aerial mast support
38 Electrical services panel
39 Navigator's compartment window
40 Navigator's desk
41 Aircraft and radio compass receiver
42 Wireless operator's desk
43 Wireless operator's seat
44 Wireless operator's compartment window
45 Front spar carry-through/fuselage frame
46 Astrodome
47 Inboard section wing ribs
48 Spar join
49 Aerial mast
50 Starboard inboard engine nacelle
51 Spinner
52 Three-bladed de Havilland constant-speed propellers
53 Oil cooler intake
54 Oil cooler radiator
55 Carburettor air intake
56 Radiator shutter
57 Engine bearer frame
58 Exhaust flame damper shroud
59 Packard-built Rolls-Royce Merlin 28 liquid-cooled engine
60 Nacelle/wing fairing
61 Fuel tank bearer ribs
62 Intermediate ribs
63 Leading-edge structure
64 Wing stringers
65 Wingtip skinning
66 Starboard navigation light
67 Starboard formation light
68 Aileron hinge fairings
69 Wing rear spar
70 Starboard aileron
71 Aileron balance tab
72 Balance tab control rod
73 Aileron trim tab
74 HF aerial
75 Split trailing-edge flap (outboard section)
76 Emergency (ditching) exit
77 Crash axe stowage
78 Fire extinguisher
79 Hydraulic reservoir
80 Signal/flare pistol stowage
81 Parachute stowage box/spar step
82 Rear spar carry-through
83 Bunk backrest
84 Rear spar fuselage frame
85 Emergency packs
86 Roof light
87 Dinghy manual release cable (dinghy stowage in starboard wingroot)
88 Mid-gunner's parachute stowage
89 Tail turret ammunition box
90 Ammunition feed track
91 Emergency (ditching) exit
92 Flame floats stowage
93 Sea markers stowage
94 Roof light
95 Dorsal turret fairing
96 Fraser-Nash power-operated dorsal turret
97 Two 0.303-in (7.7-mm) Browning machine-guns
98 Turret mounting ring
99 Turret mechanism
100 Ammunition track cover plate
101 Turret step

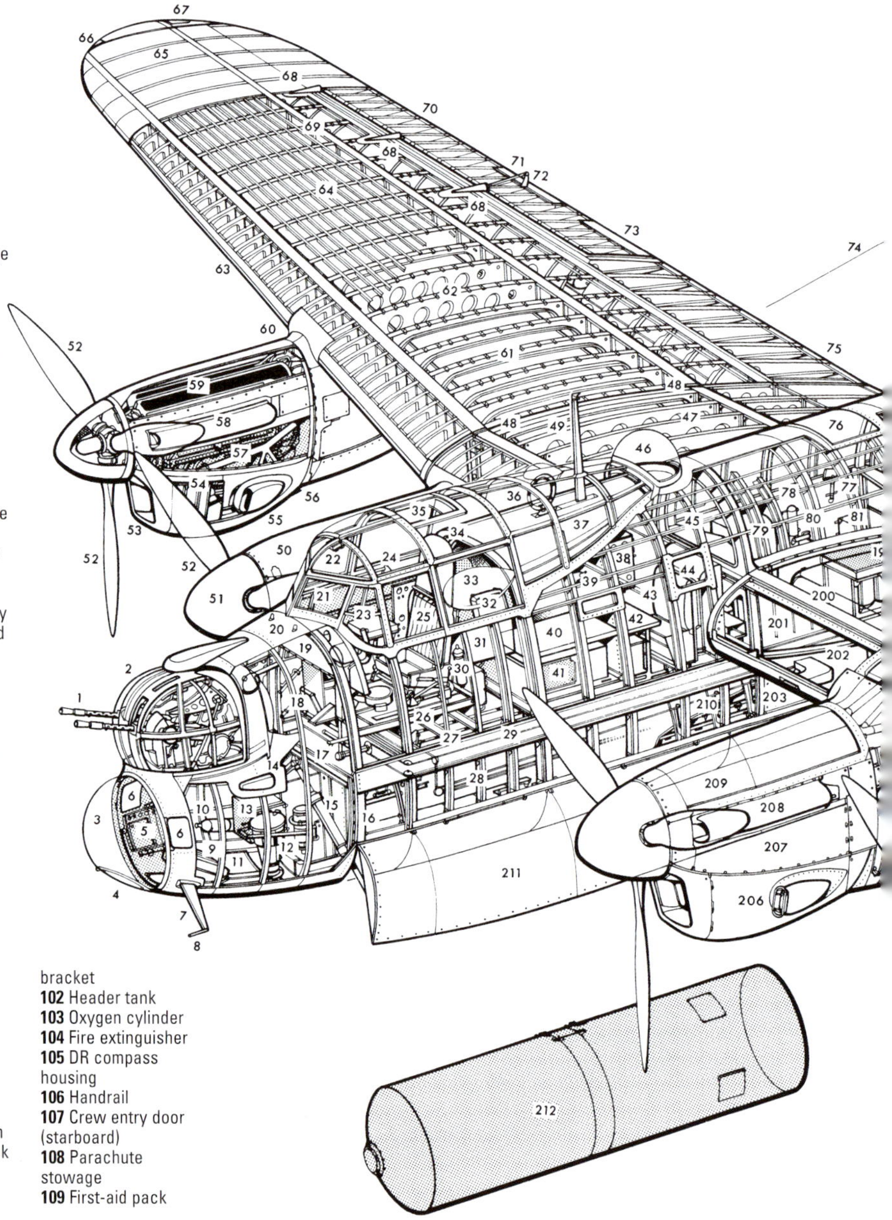

bracket
102 Header tank
103 Oxygen cylinder
104 Fire extinguisher
105 DR compass housing
106 Handrail
107 Crew entry door (starboard)
108 Parachute stowage
109 First-aid pack

A single Lancaster is maintained in flying condition by the Royal Air Force's Battle of Britain Memorial Flight. Based at RAF Coningsby in Lincolnshire, the aircraft wears its true serial number PA474, but has appeared in a variety of colour schemes since it joined the Flight in 1973.

SPECIFICATION

Lancaster Mk I

Dimensions

Length: tail up 69 ft 6 in (21.28 m); tail down 68 ft 10 in (20.98 m)
Wingspan: 102 ft (31.09 m)
Wing area: 1,300 sq ft (120.77 m²)
Total flap area: 146 sq ft (13.56 m²)
Total aileron area: 85.5 sq ft (26.06 m²)
Total fin and rudder area: 111.4 sq ft (10.35 m²)
Rudder area: 20.06 sq ft (1.86 m²)
Tailplane area (including elevators): 237 sq ft (22.02 m²)
Total elevator area: 87.5 sq ft (8.13 m²)
Height: tail up 20 ft 6 in (6.25 m); tail down 20 ft 4 in (6.20 m)
Undercarriage track: 23 ft 9 in (7.24 m)
Wing section: NACA 23018

Powerplant

Four 1,280-hp (955-kW) Rolls-Royce Merlin XX, or 1,460-hp (1089-kW) Merlin 22, or 1,640-hp (1223-kW) Merlin 24 liquid-cooled, 12-cylinder, single-stage supercharged, Vee-type piston engines
Propellers: four de Havilland Type 5140 or Nash Kelvinator A5/138 Hydromatic three-bladed, feathering/constant-speed propellers

Weights

Empty: 37,000 lb (16783 kg)
Empty equipped: 41,000 lb (18614 kg)
Maximum take-off: 68,000 lb (30845 kg)

Fuel and load

Total internal fuel: 2,154 Imp gal (9792 litres)
Maximum bombload: 22,000 lb (9979 kg)

Performance

Maximum speed: 275 mph (442 km/h) at 15,000 ft (4570 m)
Cruising speed: 200 mph (322 km/h) at 15,000 ft (4570 m)
Climb rate: climb to 20,000 ft (6100 m) in 41 minutes 36 seconds
Service ceiling: 20,000 ft (6100 m) at maximum weight
Take-off run (to 50 ft/15 m): 4,649 ft (1417 m)
Landing run (from 50 ft/15 m): 3,002 ft (915 m)

Range

Range (with one auxiliary fuel tank and 7,000-lb/3175-kg payload): 2,530 miles (4072 km)
Range (with standard fuel and 10,000-lb/4540-kg payload): 1,040 miles (1673 km)

Defensive armament

Early production model: nine 0.303-in (7.7-mm) Browning machine-guns (one in F.N.64 ventral, two each in F.N.5 nose and F.N.50 dorsal, and four in F.N.20 tail turrets)

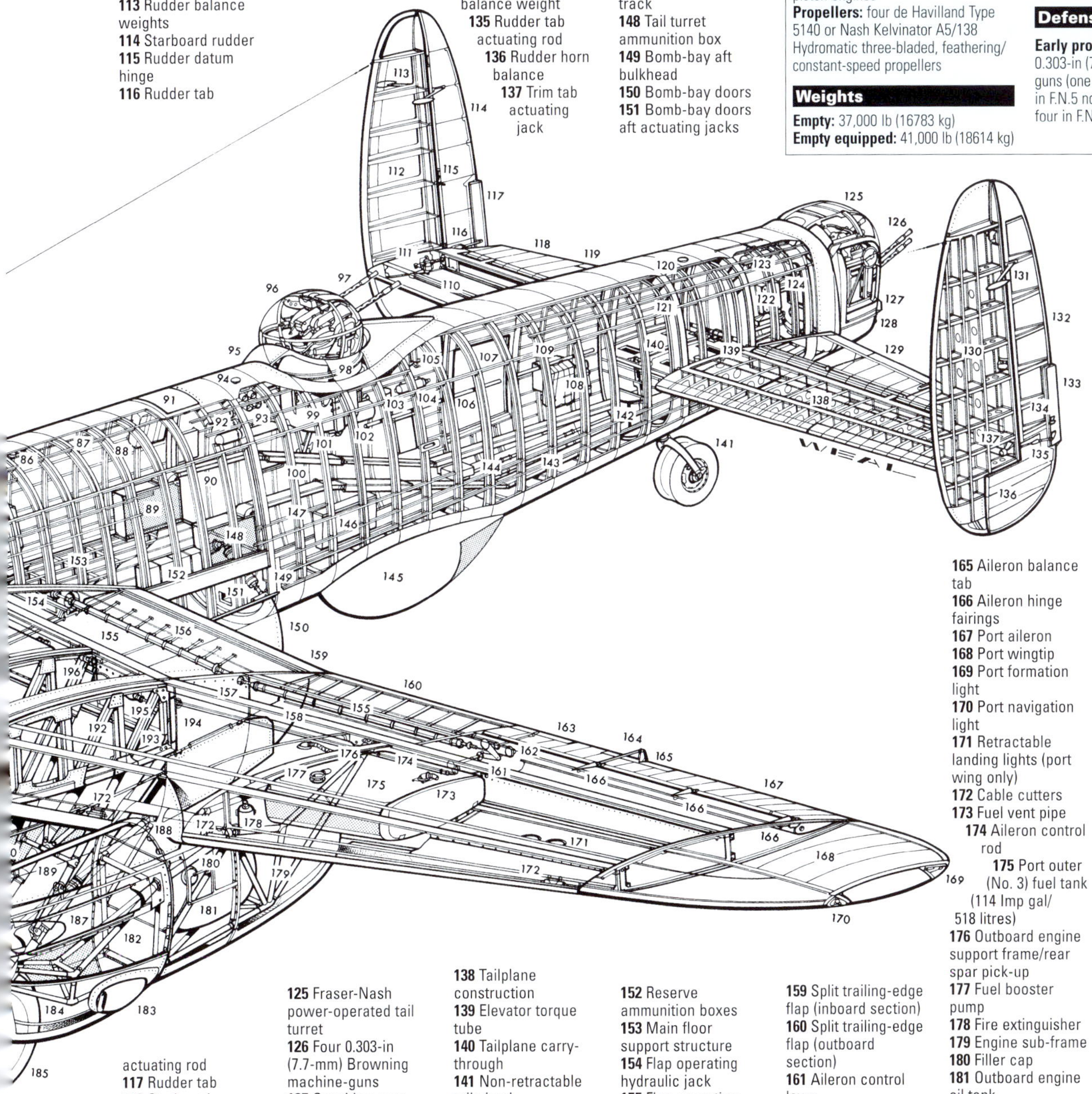

110 Starboard tailplane
111 Rudder control lever
112 Starboard tailfin
113 Rudder balance weights
114 Starboard rudder
115 Rudder datum hinge
116 Rudder tab actuating rod
117 Rudder tab
118 Starboard elevator
119 Elevator balance tab
120 Roof light
121 Tail main frame
122 Parachute stowage
123 Fire extinguisher
124 Tail turret entry door
125 Fraser-Nash power-operated tail turret
126 Four 0.303-in (7.7-mm) Browning machine-guns
127 Cartridge case ejection chutes
128 Rear navigation light
129 Elevator trim tab
130 Fin construction
131 Rudder balance weights
132 Port rudder frame
133 Rudder trim tab
134 Rudder tab balance weight
135 Rudder tab actuating rod
136 Rudder horn balance
137 Trim tab actuating jack
138 Tailplane construction
139 Elevator torque tube
140 Tailplane carry-through
141 Non-retractable tailwheel
142 Elsan closet
143 Ammunition track cover plate
144 Elevator and rudder control rods
145 H_2S (radar-bombing) ventral antenna fairing
146 Dorsal turret step
147 Ammunition feed track
148 Tail turret ammunition box
149 Bomb-bay aft bulkhead
150 Bomb-bay doors
151 Bomb-bay doors aft actuating jacks
152 Reserve ammunition boxes
153 Main floor support structure
154 Flap operating hydraulic jack
155 Flap operating tube
156 Flap toggle links
157 Flap tube connecting link
158 Rear spar
159 Split trailing-edge flap (inboard section)
160 Split trailing-edge flap (outboard section)
161 Aileron control lever
162 Aileron trim tab control cable linkage
163 Aileron trim tab
164 Aileron balance tab control rod
165 Aileron balance tab
166 Aileron hinge fairings
167 Port aileron
168 Port wingtip
169 Port formation light
170 Port navigation light
171 Retractable landing lights (port wing only)
172 Cable cutters
173 Fuel vent pipe
174 Aileron control rod
175 Port outer (No. 3) fuel tank (114 Imp gal/518 litres)
176 Outboard engine support frame/rear spar pick-up
177 Fuel booster pump
178 Fire extinguisher
179 Engine sub-frame
180 Filler cap
181 Outboard engine oil tank
182 Firewall/bulkhead
183 Carburettor air intake
184 Outboard engine support frame
185 Port mainwheel
186 Undercarriage oleo struts
187 Flame-damper shroud
188 Outboard engine support frame/main spar pick-up
189 Undercarriage retraction jacks
190 Oleo strut attachment pin
191 Undercarriage support beam (light-alloy casting)
192 Centre-section outer rib/undercarriage support
193 Location of port intermediate (No. 2) fuel tank (383 Imp gal/1741 litres)
194 Main wheel well
195 Emergency retraction air valve
196 Retraction cylinder attachment
197 Port inner (No. 1) fuel tank (580 Imp gal/2637 litres)
198 Oxygen bottle stowage
199 Rest bunk
200 Main spar
201 Hinged inboard leading edge
202 Cabin heater installation
203 Air intake
204 Inboard engine support frame
205 Inboard engine oil tank
206 Carburettor intake anti-ice guard
207 Port inner nacelle
208 Flame-damper shroud
209 Detachable cowling panels
210 Bomb shackles
211 Bomb-bay doors (open)
212 8,000-lb (3632-kg) bomb

Lancaster B.Mk I (Special)

This specially-modified aircraft spent its entire operational career with the RAF's No. 9 Squadron. It is illustrated here dropping a 12,000-lb (5443-kg) 'Tallboy' bomb on the German navy U-boat pens at Bergen, Norway during a raid on the night of 11/12 January 1945. This particular mission involved a total of 32 Lancasters and one Mosquito, drawn from Nos 9 and 617 Squadrons, both units being experienced in the use of the 'Tallboy'. Contemporary reports record that three 'Tallboys' caused serious damage to the U-boat pens after penetrating their 111/2-ft (3.5-m) thick concrete roof. In addition, two U-boats suffered minor damage, while a cargo ship was seriously damaged and a minesweeper sunk. The accuracy of the raid prevented any repetition of the civilian casualties that had been suffered during two earlier missions, but four Lancasters were lost, one from No. 9 and the others from No. 617 Squadron.

Crew
In regular operational Bomber Command service, the Lancaster was normally crewed by seven men. The flight crew consisted of a pilot, flight engineer, navigator and wireless operator. No second pilot was carried, but most pilots trained the flight engineer in the art of keeping the aircraft straight and level so that the crew could bale out if the pilot was incapacitated. In aircraft fitted with a mid-upper turret, the additional crew members consisted of a tail gunner, mid-upper gunner and a nose gunner who doubled as the bomb-aimer. Aircraft without mid-upper turrets (like that illustrated) still carried two dedicated gunners, one in the tail position and the second in the nose turret. This configuration left the bomb-aimer free to concentrate on just this one task. A crew rest-bunk was provided in the form of a foldaway padded couch aft of the wireless operator's station.

Nose art
In common with many other Bomber Command Lancasters, WS-Y wears colourful nose art. From its mission tally, the aircraft is engaged on its 41st successful mission, 39 of its previous sorties being conventional bombing raids and the 40th an anti-shipping strike. *GETTING YOUNGER EVERY DAY* uses its code letter 'Y' to allude to Youngers brewery and features a character synonymous with that company's beer. Another famous Lancaster of No. 9 Sqn, W4964/WS-J, with the radio call sign 'J-Johnny', received the slogan *Still Going Strong* and the Johnny Walker figure from the whisky brand. 'J-Johnny' accumulated an impressive total of 106 missions, including a 'Tallboy' attack on the *Tirpitz*.

12,000-lb (5443-kg) 'Tallboy'
Having already designed the revolutionary 'bouncing bomb', Barnes Wallis turned his design genius to the creation of the devastating 'Tallboy'. The weapon was first used in action by No. 617 Sqn on the night of 8/9 June 1944 against the Saumur railway tunnel, which was being used by German reinforcements moving towards Normandy.

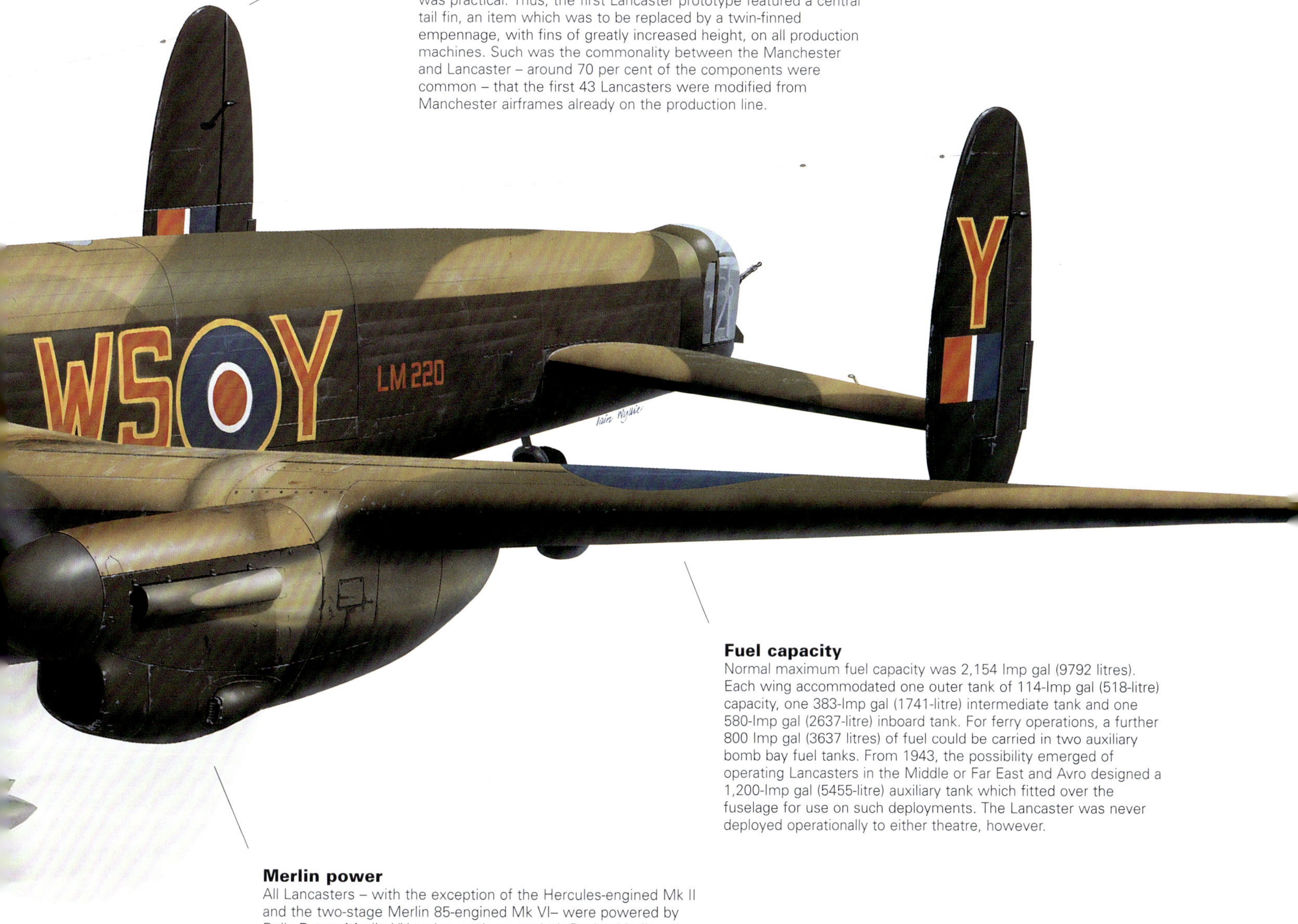

Twin-fins standard
With the need to get the first Manchester Mk III (soon to be known officially as the Lancaster) into the air as soon as possible, Avro decided to use as many standard Manchester components as was practical. Thus, the first Lancaster prototype featured a central tail fin, an item which was to be replaced by a twin-finned empennage, with fins of greatly increased height, on all production machines. Such was the commonality between the Manchester and Lancaster – around 70 per cent of the components were common – that the first 43 Lancasters were modified from Manchester airframes already on the production line.

Fuel capacity
Normal maximum fuel capacity was 2,154 Imp gal (9792 litres). Each wing accommodated one outer tank of 114-Imp gal (518-litre) capacity, one 383-Imp gal (1741-litre) intermediate tank and one 580-Imp gal (2637-litre) inboard tank. For ferry operations, a further 800 Imp gal (3637 litres) of fuel could be carried in two auxiliary bomb bay fuel tanks. From 1943, the possibility emerged of operating Lancasters in the Middle or Far East and Avro designed a 1,200-Imp gal (5455-litre) auxiliary tank which fitted over the fuselage for use on such deployments. The Lancaster was never deployed operationally to either theatre, however.

Merlin power
All Lancasters – with the exception of the Hercules-engined Mk II and the two-stage Merlin 85-engined Mk VI– were powered by Rolls-Royce Merlin XX-series engines, or their Packard-built equivalents. The use of an engine which was also in great demand for both the Halifax, Hurricane and Spitfire, among several other lesser types, placed a great strain on Britain's aero-engine manufacturing industry. The Merlin XX as used in the earliest Lancasters used a single-stage supercharger and was designed for a maximum power output of 1,390 hp (1036 kW). Rolls-Royce built 3,391 Merlin XXs at Crewe, 2,592 at Derby and 9,500 at Glasgow, while Ford built a further 12,538. Total production was 28,021 between 1940 and 1944.

Boeing B-17

Sentimental Journey, N9323Z, an ex-DB-17P and fire-bomber, operates with the Confederate Air Force at Mesa, Arizona. This is one of only 43 B-17s that survive in some form around the world. Of these, 13 are capable of flight and all were original B-17Gs or their derivatives.

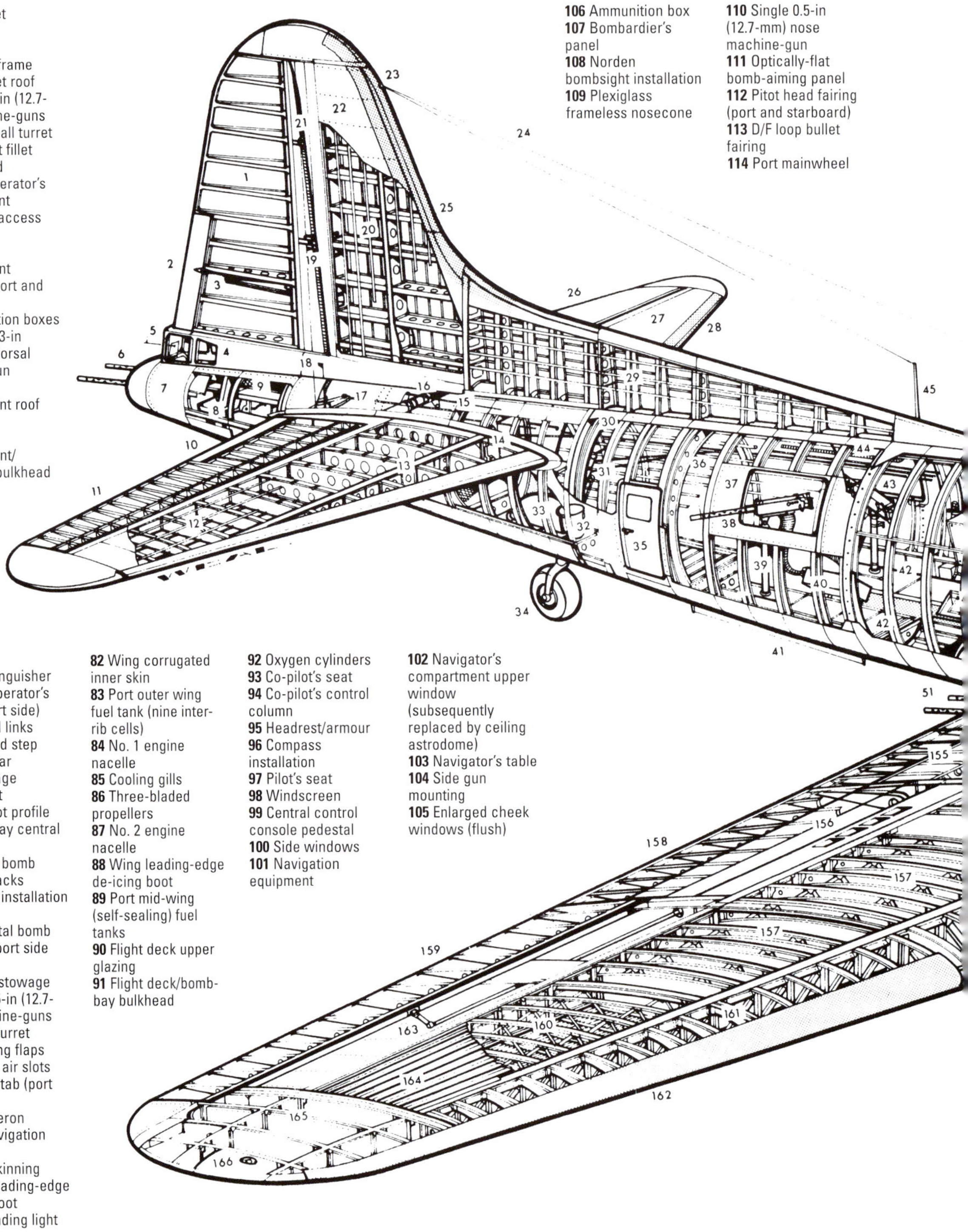

Boeing B-17F

Cutaway key
1 Rudder construction
2 Rudder tab
3 Rudder tab actuation
4 Tail gunner's station
5 Gunsight
6 Twin 0.5-in (12.7-mm) machine-guns
7 Tailcone
8 Tail gunner's seat
9 Ammunition troughs
10 Elevator trim tab
11 Starboard elevator
12 Tailplane structure
13 Tailplane front spar
14 Tailplane/fuselage attachment
15 Control cables
16 Elevator control mechanism
17 Rudder control linkage
18 Rudder post
19 Rudder centre hinge
20 Fin structure
21 Rudder upper hinge
22 Fin skinning
23 Aerial attachment
24 Aerials
25 Fin leading-edge de-icing boot
26 Port elevator
27 Port tailplane
28 Tailplane leading-edge de-icing boot
29 Dorsal fin structure
30 Fuselage frame
31 Tailwheel actuation
32 Toilet
33 Tailwheel (retracted) fairing
34 Fully-swivelling retractable tailwheel
35 Crew entry door
36 Control cables
37 Starboard waist hatch
38 Starboard waist 0.5-in (12.7-mm) machine-gun
39 Gun support frame
40 Ammunition box
41 Ventral aerial
42 Waist gunners' positions
43 Port waist 0.5-in (12.7-mm) machine-gun
44 Ceiling control cable runs
45 Dorsal aerial mast
46 Ball turret stanchion support
47 Ball turret stanchion
48 Ball turret actuation mechanism
49 Support frame
50 Ball turret roof
51 Twin 0.5-in (12.7-mm) machine-guns
52 Ventral ball turret
53 Wingroot fillet
54 Bulkhead
55 Radio operator's compartment
56 Camera access hatch
57 Radio compartment windows (port and starboard)
58 Ammunition boxes
59 Single 0.3-in (7.62-mm) dorsal machine-gun
60 Radio compartment roof glazing
61 Radio compartment/bomb-bay bulkhead
62 Fire extinguisher
63 Radio operator's station (port side)
64 Handrail links
65 Bulkhead step
66 Wing rear spar/fuselage attachment
67 Wingroot profile
68 Bomb-bay central catwalk
69 Vertical bomb stowage racks (starboard installation shown)
70 Horizontal bomb stowage (port side shown)
71 Dinghy stowage
72 Twin 0.5-in (12.7-mm) machine-guns
73 Dorsal turret
74 Port wing flaps
75 Cooling air slots
76 Aileron tab (port only)
77 Port aileron
78 Port navigation light
79 Wing skinning
80 Wing leading-edge de-icing boot
81 Port landing light
82 Wing corrugated inner skin
83 Port outer wing fuel tank (nine inter-rib cells)
84 No. 1 engine nacelle
85 Cooling gills
86 Three-bladed propellers
87 No. 2 engine nacelle
88 Wing leading-edge de-icing boot
89 Port mid-wing (self-sealing) fuel tanks
90 Flight deck upper glazing
91 Flight deck/bomb-bay bulkhead
92 Oxygen cylinders
93 Co-pilot's seat
94 Co-pilot's control column
95 Headrest/armour
96 Compass installation
97 Pilot's seat
98 Windscreen
99 Central control console pedestal
100 Side windows
101 Navigation equipment
102 Navigator's compartment upper window (subsequently replaced by ceiling astrodome)
103 Navigator's table
104 Side gun mounting
105 Enlarged cheek windows (flush)
106 Ammunition box
107 Bombardier's panel
108 Norden bombsight installation
109 Plexiglass frameless nosecone
110 Single 0.5-in (12.7-mm) nose machine-gun
111 Optically-flat bomb-aiming panel
112 Pitot head fairing (port and starboard)
113 D/F loop bullet fairing
114 Port mainwheel

A 1,000-lb (454-kg) bomb is hoisted off its bomb truck ready to be loaded into the bomb bay of a waiting B-17G of the 8th Air Force on the night prior to a mission on 6 June 1944. 2,000-pounders were the largest bombs carried by B-17s; standard weaponload was usually 500-lb (227-kg) or 1,000-lb (908-kg) bombs.

SPECIFICATION

B-17F-25-BO Flying Fortress

Dimensions

Length overall: 74 ft 9 in (22.80 m)
Wingspan: 103 ft 9 in (32.60 m)
Wing area: 1,420 sq ft (131.92 m²)
Height: 19 ft 2 in (5.85 m)
Propeller diameter: 11 ft 7 in (3.54 m)

Powerplant

Four Wright R-1820-97 Cyclone radial piston engines each rated at 1,200 hp (895 kW) at 25,000 ft (7620 m)

Weights

Empty (typical): 34,000 lb (15422 kg)
Loaded (normal): 56,500 lb (25628 kg)
War overload from 1943: 72,000 lb (32660 kg)

Fuel load

Maximum fuel capacity: 1,700 US gal (6435 litres)

Performance

Maximum speed: 299 mph (481 km/h)
Cruising speed: 160 mph (257.5 km/h)
Initial climb rate: 900 ft (274 m) per minute
Service ceiling: 36,000 ft (10975 m)
Combat radius with 5,000-lb (2270-kg) bombload: 800 miles (1287 km)

Crew

Normal complement of nine, but more could be carried. Crew included: bomb-aimer, pilot, co-pilot, upper turret gunner, radio operator, two waist gunners, ball turret gunner and tail gunner.

Armament

Maximum bombload 9,600 lb (4355 kg), later increased to 17,600 lb (7983 kg); defensive firepower normally 10-12 0.5-in (12.7-mm) guns: two cheek-mounted guns, two on top of fuselage, one above radio operator's compartment, two in 'ball' turret below fuselage, two on hand-operated mountings firing through side ports and two in the extreme tail. The G-variant could carry 1-13 guns, the most notable being two chin-mounted .50-cal machine-guns.

115 Flight deck underfloor control linkage
116 Wingroot/fuselage fairing
117 Wing front spar/fuselage attachment
118 Battery access panels (wingroot leading edge)
119 No. 3 engine nacelle spar bulkhead
120 Intercooler pressure duct
121 Mainwheel well
122 Oil tank (nacelle inboard wall)
123 Nacelle structure
124 Exhaust
125 Retracted mainwheel (semi-recessed)
126 Firewall
127 Cooling gills
128 Exhaust collector ring assembly
129 Three-bladed propellers
130 Undercarriage retraction struts
131 Starboard mainwheel
132 Axle
133 Mainwheel oleo leg

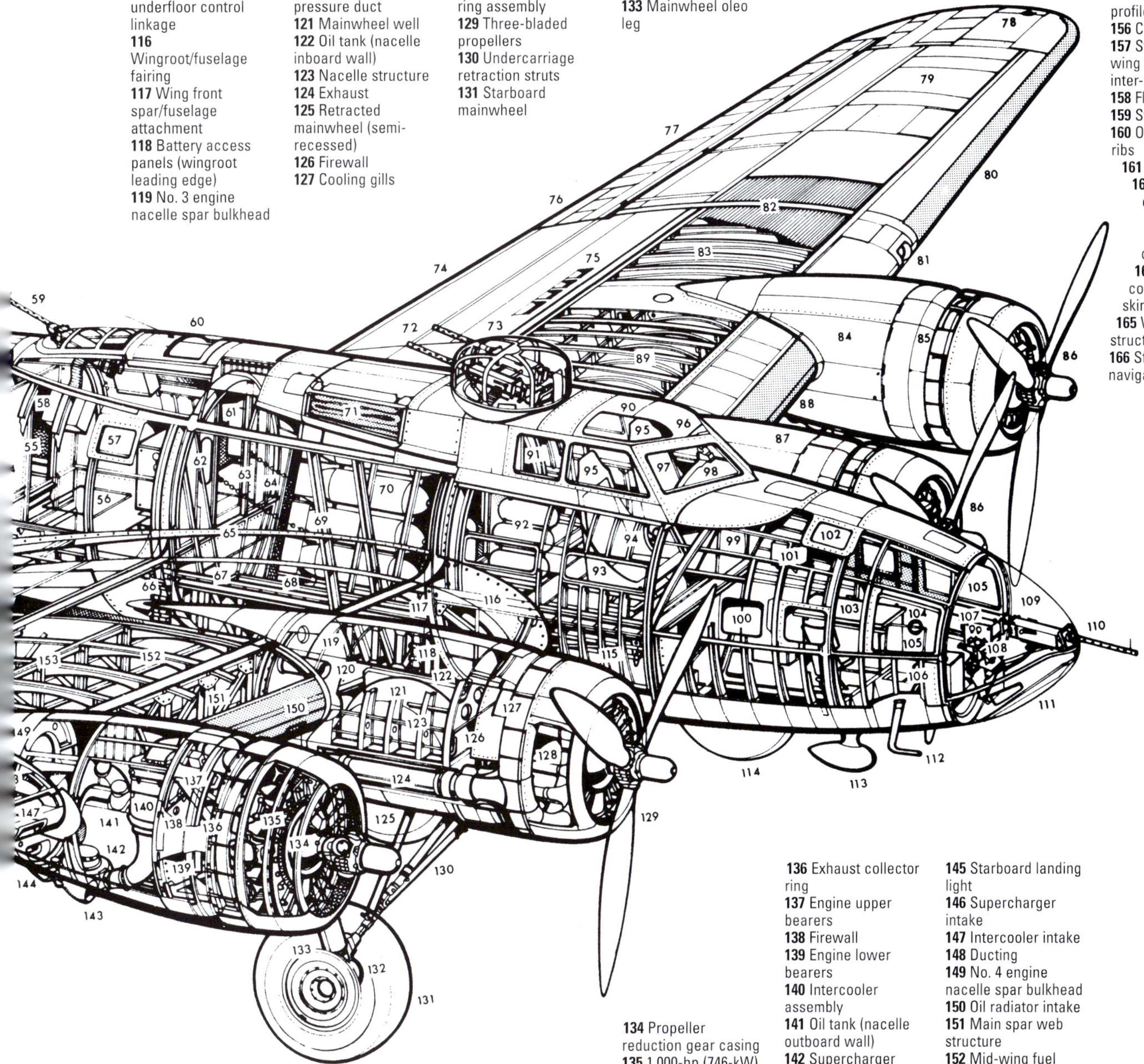

134 Propeller reduction gear casing
135 1,000-hp (746-kW) Wright R-1829-65 radial engine
136 Exhaust collector ring
137 Engine upper bearers
138 Firewall
139 Engine lower bearers
140 Intercooler assembly
141 Oil tank (nacelle outboard wall)
142 Supercharger
143 Intake
144 Supercharger waste-gate
145 Starboard landing light
146 Supercharger intake
147 Intercooler intake
148 Ducting
149 No. 4 engine nacelle spar bulkhead
150 Oil radiator intake
151 Main spar web structure
152 Mid-wing fuel tank rib cut-outs
153 Auxiliary mid spar
154 Rear spar
155 Landing flap profile
156 Cooling air slots
157 Starboard outer wing fuel tank (in inter-rib cut-outs)
158 Flap structure
159 Starboard aileron
160 Outboard wing ribs
161 Spar assembly
162 Wing leading-edge de-icing boot
163 Aileron control linkage
164 Wing corrugated inner skin
165 Wingtip structure
166 Starboard navigation light

B-17G-15-BO

***Chow-hound*, the Boeing-built B-17G 42-31367, was so named by its first crew, entering combat with the 322nd Bomb Squadron of the 91st BG on 29 January 1944. Shortly afterwards, the Group artist, Tony Starcer, added to the nose a painting of Disney's Pluto the dog. By 12 March 1944, *Chow-hound* had completed 15 missions and had been credited with 19 aerial victories. The aircraft went on to fly 26 missions with all of its original engines, one being changed at this point, and the other three 'retired' after 30 'trips'. The last captain of the aircraft was First Lieutenant Jack Thompson, whose crew was flying its 13th mission over Caen on 8 August 1944 when the aircraft was cut in half by flak. Thompson was the only crew member who escaped to become a prisoner of war. *Chow-hound*, like the majority of B-17s delivered to the ETO (European Theatre of Operations), wore the standard Olive Drab (OD) over Neutral Gray scheme introduced at the outbreak of the war. The centre-section of the fin was often painted in medium-green and remained so on most aircraft until the advent of colourful group markings from July 1944 which extended to large areas of the fin. As well as applying colour to the centre of the fin, as in this case, the 91st BG also added red to fixed portions of the horizontal tail surfaces. From February 1944, replacement aircraft arrived with all camouflage paint removed.**

Cockpit
The cockpit of the B-17 was typically spacious and well laid out, as were those of many American warplanes of the period. The pilot was the aircraft commander and sat in the right-hand seat, which was armoured unlike that of the co-pilot. Like all crew positions, the pilots could communicate with each other by interphone, but it usually proved easier for the two men to wear their inboard headphones askew and communicate by shouting. It has been said that one could always differentiate between ex-B-17 pilots and co-pilots by determining in which ear they were deaf! Each pilot shared the flying (which could be of eight hours' duration in close formation), but the co-pilot was responsible for the landing gear, flaps, the starting-up of the engine and power monitoring.

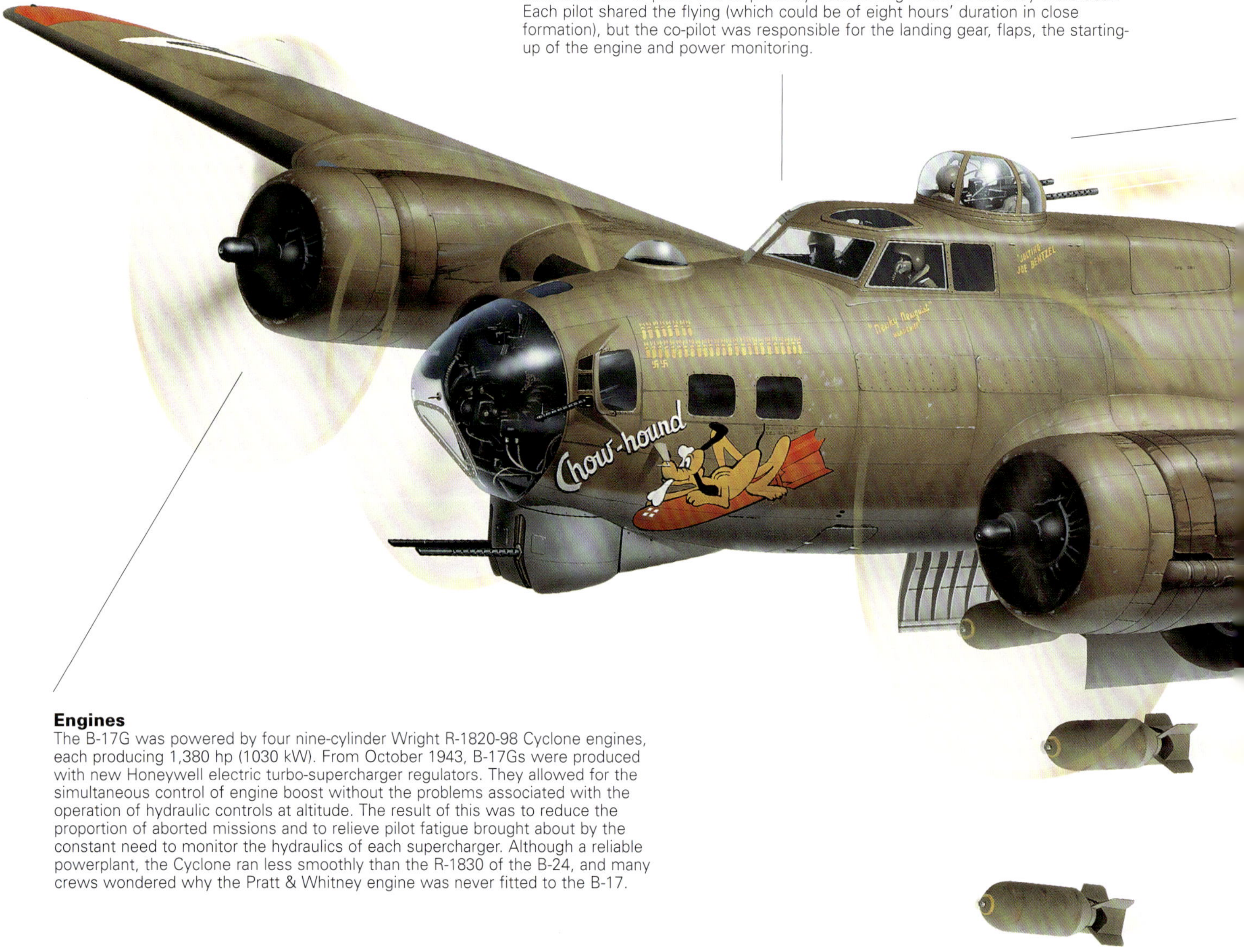

Engines
The B-17G was powered by four nine-cylinder Wright R-1820-98 Cyclone engines, each producing 1,380 hp (1030 kW). From October 1943, B-17Gs were produced with new Honeywell electric turbo-supercharger regulators. They allowed for the simultaneous control of engine boost without the problems associated with the operation of hydraulic controls at altitude. The result of this was to reduce the proportion of aborted missions and to relieve pilot fatigue brought about by the constant need to monitor the hydraulics of each supercharger. Although a reliable powerplant, the Cyclone ran less smoothly than the R-1830 of the B-24, and many crews wondered why the Pratt & Whitney engine was never fitted to the B-17.

Radio room
The radio operator sat in a compartment between the bomb bay and the waist. Unlike the radio operators on RAF heavy bombers, who operated under virtual radio silence, the B-17's radio-man made relatively frequent transmissions, including obtaining 'fixes' to aid the navigator and giving a bombing accuracy report (in the lead aircraft) as the group left the target. The radio operator was always on the alert for recall signals from base or diversions from the primary target that might be called by the group or wing leader in flight. Another duty of the radio operator was to operate the vertical strike camera that recorded bombing results. The single 0.50-in (12.7-mm) gun fitted in the radio room hatch did little more than give the radio operator something to do during fighter attacks, and, with its limited field of fire, sometimes did more damage to the B-17 than to the enemy. The gun was originally fired through the open hatch, but later aircraft had a mount fitted in a closed hatch cover. By the end of the war, most B-17 groups had dispensed with the radio room gun altogether, and the mount was deleted from production.

Top turret
The power-operated top turret was manned by the flight engineer, who was responsible for managing fuel consumption as well as performing a general trouble-shooting role. He had the skills to repair the airframe and engines if away from base, and the weapons, oxygen and radio systems in flight. The engineer sat in a canvas sling (or, if above average height, stood up and used it as a backrest) and operated the turret motion and firing with two levers (twisting the right-hand lever operated the rangefinder for the K-3 computing gunsight). Following trials with a captured B-17, Luftwaffe Me 262 pilots were encouraged to make diving attacks from above because the top turret sight could not calculate quickly enough to track the attacking aircraft. There were two types of Sperry top turret fitted to the B-17G: a low profile early type with large areas of metal sheeting; and the taller (by 6 in/15 cm) later model, as fitted to *Chow-hound*, which had larger glazed areas.

Tail gun position
The tail gunner on the B-17 did not have a proper seat, only a bicycle-type saddle with padded knee rests. With his upper body exposed to the sunlight through the armoured glass sighting screen and Plexiglass side panels, and his lower body in the shade, the gunner could be subjected simultaneously to extremes of temperatures. The wearing of a flak suit as well as a parachute in the confined space was impossible, and gunners had to choose between them, a process made easier by the progressive elimination of armour plating from the turret. On the original type of B-17G tail turret, such as that on *Chowhound*, the sighting of the guns was done with a ring and post system which moved with the guns through a bell crank. The United Airlines Modification Center at Cheyenne, Wyoming developed an improved turret with enlarged glazed areas, and that replaced the original canvas boot from which the guns protruded with a swivelling dome, giving a much improved field of fire and better air-sealing for the gunner. 'Cheyenne' turrets began to arrive in the ETO on new-build Fortresses from June 1944, although some had been fitted at depot level to a number of B-17Gs that had lost their tail turrets in combat. Some B-17s, particularly in the 15th AF, were field-modified with a single 20-mm weapon, although this was not fitted to any production aircraft.

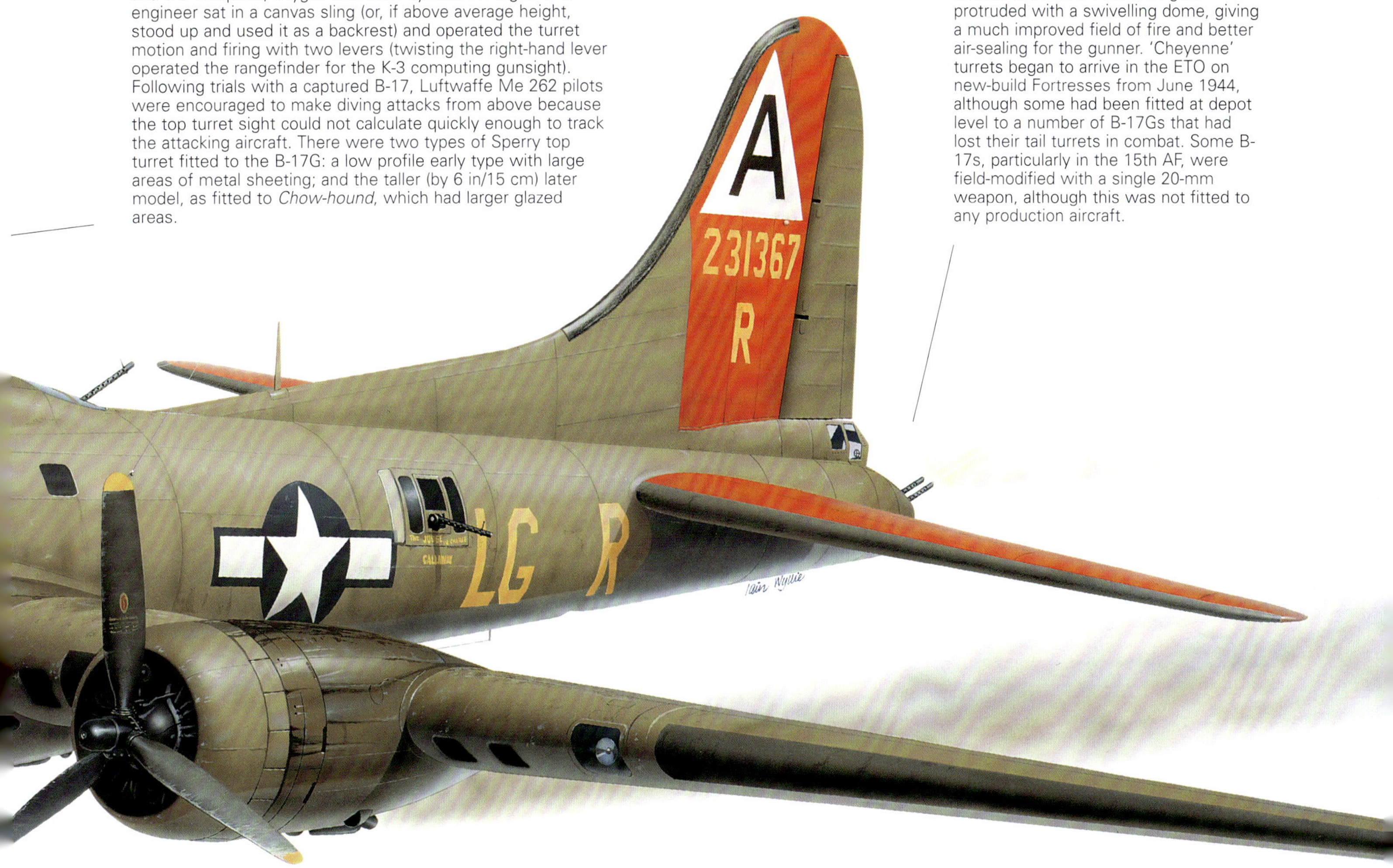

Bombload
The theoretical maximum bombload for a B-17G was 13,600 lb (6170 kg), but for actual operations over Europe, loads of more than 4,000 lb (1815 kg) were seldom carried, and often as little as 2,000 lb (907 kg). The bombs available for use by the 8th Air Force ranged from 2-lb (0.9-kg) incendiaries (bundled in clusters of up to 250) to 2,000-lb (907-kg) demolition bombs, of which only one could practicably be carried in the restricted space of the B-17 bomb bay. The small size of the bomb bay was a limiting factor in the utility of the B-17. A pair of external bomb racks with a capacity of 1,000 lb (454 kg) each was used on occasion, including trials with glide bombs and 'Disney' rocket bombs, although the use of these racks adversely affected aircraft performance.

Waist guns
The waist guns on the earliest B-17Gs were fired through completely open windows, as on the B-17F. In order to reduce the cold and discomfort felt by the gunners, closed Plexiglass windows were added. Initially, this was a framed three-piece unit as seen here but, later, an improved single-piece clear-vision item was introduced, with the gun mounted on the lower sill rather than on a swivel post. It was not until Boeing and Vega -50, and Douglas -25, series aircraft that the problem of gunners obstructing each other in combat was solved by the introduction of staggered waist gun positions. Due to the infrequency of beam attacks in actual combat and the limited field of fire of the guns, the waist position was one of the least effective defensive stations, and also statistically one of the most dangerous due to its exposed nature. Towards the end of the war, many 8th Air Force B-17s carried only one gunner, and the 91st BG flew without either gun in March 1945 as part of a 1st AD (Air Division) experiment.

Boeing B-29 Superfortress

The B-29 was the first of the true 'very heavy' bombers to enter service and its appearance in the Pacific theatre revolutionised the way the war was fought there. Now, American bombers could cross the vast distances of the Pacific and strike the increasingly vulnerable Japanese home islands.

B-29 Superfortress

Cutaway key
1 Temperature probe
2 Nose glazing
3 Optically-flat bomb-aiming panel
4 Bombsight
5 Windscreen panels
6 Forward gun sight
7 Bombardier's seat
8 Pilot's instrument console
9 Control column
10 Co-pilot's seat
11 Pilot's seat
12 Side console panel
13 Cockpit heating duct
14 Nose undercarriage leg strut
15 Steering control
16 Twin nosewheels
17 Retraction struts
18 Nosewheel doors
19 Underfloor control cable runs
20 Pilot's back armour
21 Flight engineer's station
22 Forward upper gun turret, four 0.5-in (12.7-mm) machine-guns, 500 rpg
23 Radio operator's station
24 Chart table
25 Navigator's instrument rack
26 Fire extinguisher bottle
27 Forward lower gun turret, two 0.5-in (12.7-mm) machine-guns, 500 rpg
28 Ventral aerial
29 Navigator's seat
30 Hydraulic system servicing point
31 Access ladder
32 Forward cabin rear pressure bulkhead
33 Armoured bulkhead
34 Pressurised tunnel connecting front and rear cabins
35 Astrodome observation hatch
36 Forward bomb racks
37 Bomb-hoisting winches
38 Catwalk
39 Bomb rack mounting beam
40 Pressurised tunnel internal crawlway
41 D/F loop aerial
42 Radio communications aerials
43 Starboard main undercarriage wheel bay
44 Wing inboard fuel tanks, 1,415 US gal (5356 litres)
45 Starboard inner-engine nacelle
46 Intercooler exhaust flap
47 Engine-cooling air outlet flaps
48 Engine cowling panels
49 Hamilton Standard four-bladed, constant-speed propellers, 16-ft 7-in (5.05-m) diameter
50 Propeller hub pitch change mechanism
51 Starboard outer engine nacelle
52 Exhaust stub
53 Wing outboard fuel tanks, 1,320 US gal (4991 litres), maximum internal fuel load 9,363 US gal (35443 litres) including bomb bay ferry tanks
54 Wing bottom skin stringers
55 Leading-edge de-icing boots
56 Starboard navigation light
57 Fabric-covered aileron
58 Aileron tab
59 Flap guide rails
60 Starboard Fowler-type flap
61 Flap rib construction
62 Inboard nacelle tail fairing
63 Life raft stowage
64 Wing panel centreline joint
65 Wing/fuselage attachment mainframes
66 Pressurisation ducting
67 Heat exchanger
68 Centre-section fuel tank, 1,333 US gal (5046 litres)
69 Cabin heater
70 Pressurisation control valve
71 Fuselage framing
72 Rear bomb bay, 4 x 2000-lb (907-kg) bombs shown
73 Bomb rack
74 Access door
75 Rear cabin front pressure bulkhead
76 Radio aerial mast
77 Upper gun turret sighting hatch
78 Upper gunner's seat
79 Remote gun controller
80 Radio and electronics racks
81 Upper gun turret, two 0.5-in (12.7-mm) machine-guns, 500 rpg
82 Rear pressure bulkhead
83 Finroot fillet
84 Starboard tailplane
85 Starboard elevator
86 Leading-edge de-icing boot
87 Tailfin construction
88 HF aerial cable
89 Fintip fairing
90 Fabric-covered rudder construction

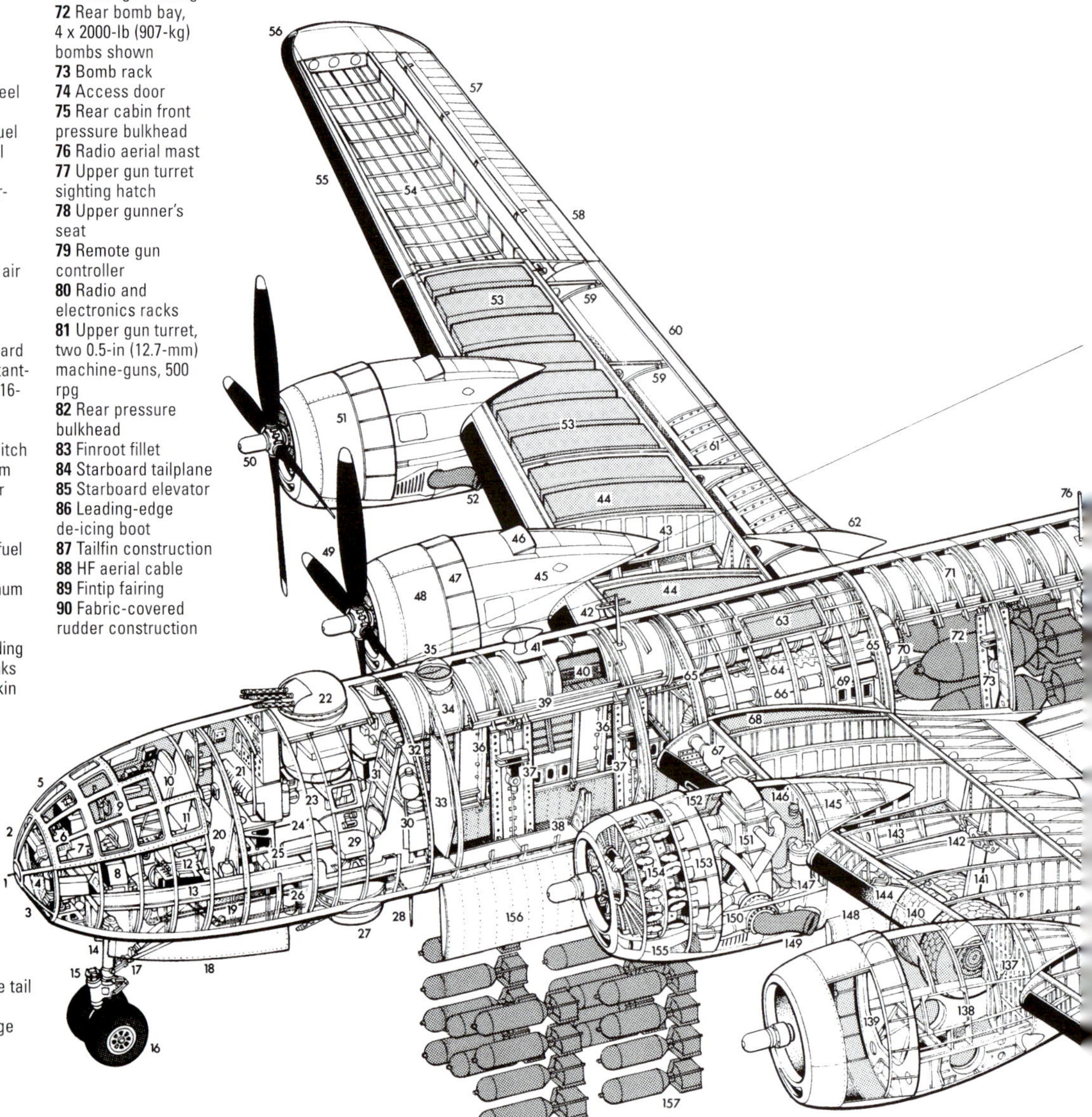

Designed to rectify the problem of the slow rate of fuel transfer with the hose refuelling system, a single KB-29M was converted to a three-hose tanker that could refuel three fighters simultaneously. The aircraft received the new designation YKB-29T and is seen here refuelling three RAF Gloster Meteors. One hose was installed in the tail, while the other two were installed on reels mounted in pods under the wingtips. Since later jet fighters and bombers did not have slow-flight characteristics that were compatible with the obsolete B-29, the triple-hose arrangement was used on jet-assisted B-50 tankers.

SPECIFICATION

B-29 Superfortress

Dimensions

Length: 99 ft (30.18 m)
Height: 29 ft 7 in (9.02 m)
Wingspan: 141 ft 3 in (43.05 m)
Wing area: 1,736 sq ft (161.27 m²)

Powerplant

Four 2,200-hp (1641-kW) Wright R-3350-23 Cyclone 18 turbocharged radial piston engines

Weights

Empty: 70,140 lb (31815 kg)
Maximum take off: 124,000 lb (56245 kg)

Performance

Maximum speed at 25,000 ft (7620 m): 358 mph (576 km/h)
Cruising speed: 230 mph (370 km/h)
Service ceiling: 31,850 ft (9710 m)
Range: 3,250 miles (5230 km)

Armament

Two 0.5-in (12.7-mm) guns each in four remotely-controlled, power-operated turrets, and three 0.50-in (12.7-mm) guns or two 0.50-in (12.7-mm) guns and one 20-mm cannon in the tail turret, plus a bombload of up to 20,000 lb (9072 kg) which could consist of incendiaries, conventional munitions or nuclear weapons

On 14 March 1947, the US Navy took over four B-29-BWs for long-range search missions and assigned them the designation P2B-1S. One aircraft (illustrated) was modified for USN tests of the D-558-II high-speed research aircraft which was carried by the bomber for drop-launching.

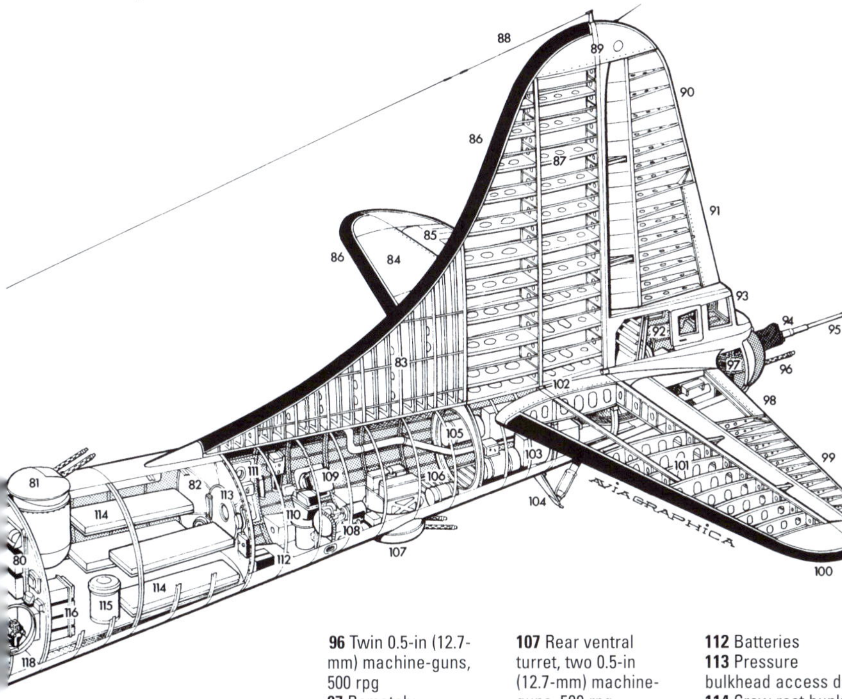

91 Rudder tab
92 Pressurised tail gunner's compartment
93 Armoured-glass window panels
94 Tail gun camera
95 20-mm cannon, 100 rounds
96 Twin 0.5-in (12.7-mm) machine-guns, 500 rpg
97 Remotely-controlled ball turret
98 Elevator tab
99 Port fabric-covered elevator construction
100 Tailplane leading-edge de-icing boot
101 Tailplane construction
102 Fin/tailplane attachment joints
103 Tail turret ammunition boxes
104 Retractable tail bumper
105 Oxygen bottles
106 APU fuel tank
107 Rear ventral turret, two 0.5-in (12.7-mm) machine-guns, 500 rpg
108 Auxiliary power unit (APU)
109 Oblique camera
110 Vertical camera
111 Crew entry door
112 Batteries
113 Pressure bulkhead access door
114 Crew rest bunks
115 Toilet
116 Radio communications tuning units
117 Remote gun sight
118 Gun aiming blister
119 Gunner's seat, port and starboard
120 Voltage regulator
121 Bomb door hydraulic jacks
122 Rear bomb bay doors
123 Port Fowler flap
124 Flap shroud ribs
125 Rear spar
126 Outer wing panel joint
127 Aileron tab
128 Fabric-covered aileron construction
129 Wingtip fairing
130 Port navigation light
131 Wing stringers
132 Outer wing panel ribs
133 Front spar
134 Leading-edge nose ribs
135 Leading-edge de-icing boots
136 Port wing fuel tank bays
137 Engine nacelle firewall
138 Nacelle construction
139 Engine mounting frame
140 Twin mainwheels
141 Main undercarriage leg strut
142 Mainwheel leg pivot mounting
143 Port mainwheel bay
144 Hydraulic retraction jack
145 Nacelle tail fairing
146 Self-sealing oil tank, 85 US gal (322 litres)
147 Hydraulic reservoir
148 Mainwheel doors
149 Exhaust stub
150 Exhaust-driven turbo-supercharger
151 Intercooler
152 Engine-cooling air exit flaps
153 Exhaust collector ring
154 Wright Cyclone R-3350-23 18-cylinder, two-row radial engine
155 Engine intake ducting
156 Forward bomb bay doors
157 20 x 500-lb (227-kg) bombs, maximum bombload 20,000 lb (9072 kg)

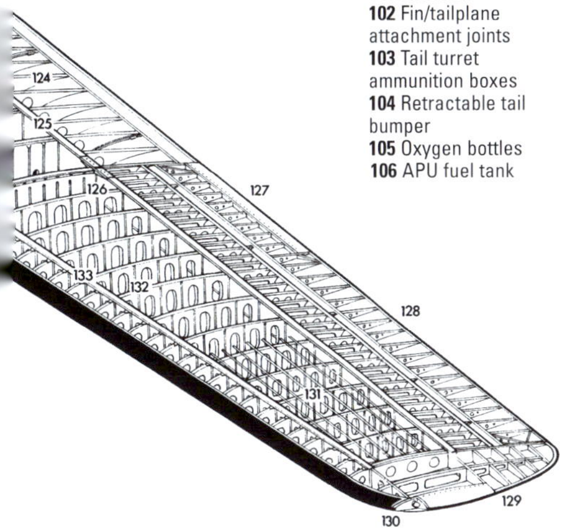

To meet post-war British long-range bomber requirements until the Avro Lincoln could be delivered in sufficient quantity, 87 standard B-29s were loaned to the RAF, which named them Washington. The loan period lasted from March 1950 to 1955.

B-29 Superfortress

The 504th Bombardment Group (Very Heavy) was activated on 11 May 1944. Initially equipped with B-17s, it later received B-29s and was assigned to the 313th Bombardment Wing, 20th Air Force. Combat operations commenced out of the Marianas Islands from January 1945 and the first heavy attacks were made against Japanese airfields and other installations on Maug and Iwo Jima and in the Truk Islands. Throughout 1945, attacks were made against the Japanese home islands, most notably in May 1945 when the Group received a Distinguished Unit Citation for striking the industrial centre at Yokohama. Following this, incendiary raids across Japan and mining operations against enemy shipping resulted in more awards for the Group. After the cessation of hostilities, the B-29s dropped food for POWs and flew over Japan to evaluate damage inflicted by bombardment operations.

Powerplant
Power for the Superfortress came from four Wright R-3350-23 Duplex Cyclones, each with two General Electric turbochargers. Developing 2,200 hp (1641 kW) for take-off, the massive engine drove an equally huge, 16-ft 7-in (5.05-m) diameter four-bladed Hamilton Standard propeller. Throughout the aircraft's early career, the R-3350 proved prone to engine fires, although the powerplant was retained throughout the bomber's life.

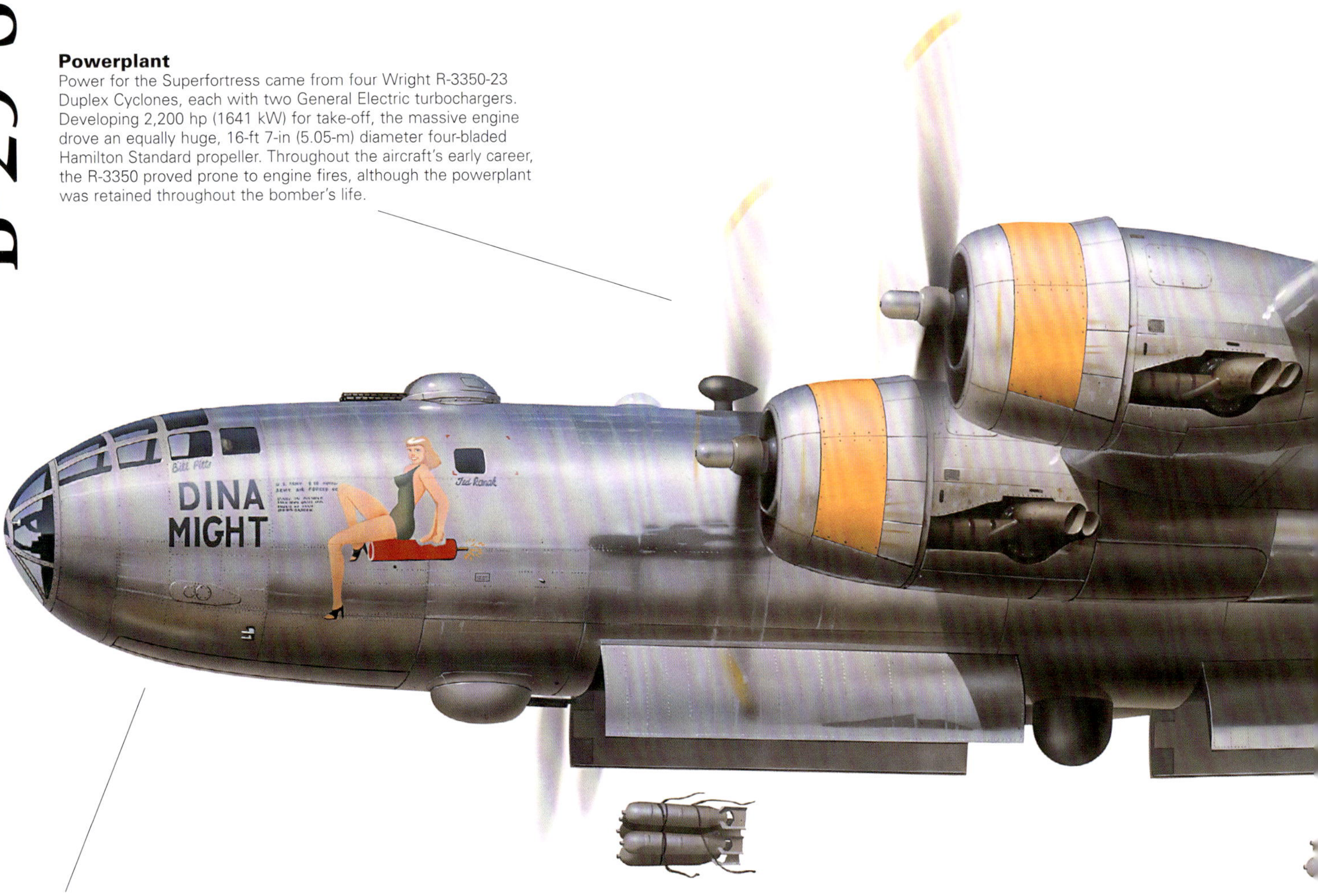

Pressurisation
The B-29 was the world's first pressurised bomber. the aircraft had two separate pressure cabins for the crew, fore and aft, connected by a sealed tunnel that bypassed the unpressurised bomb bays. The large forward cabin provided accommodation for seven of the crew: the pilot, co-pilot, bombardier (who sat in the extreme nose), navigator, flight engineer and two radar operators. The rear fuselage was pressurised between the rear of the bomb bay and a point roughly level with the start of the dorsal fin. In addition to the four gunners' positions, four bunks were provided to enable the crew to rest during long flights, or for relief crews.

Bombs
This aircraft is depicted dropping M47 incendiary bombs 'in train' from both bomb bays. The small incendiaries were tied together in bundles for loading, breaking free from each other in the airstream. Incendiary bombs were widely used against Japanese cities during the last months of the war, causing almost total destruction amid the wooden buildings and enormous casualties. Two vast weapon bays were located either side of the immensely strong wing carry-through structure. Bomb-hoisting winches inside the bomb bays facilitated loading. Up to 20,000 lb (9072 kg) of bombs could be carried internally.

Wing
The slender wing was designed to take a massive loading and bestowed excellent cruise performance, but posed lift problems at low speeds. Powerful Fowler flaps offset this disadvantage, adding 21 per cent to the area of the wing when deployed. Fuel was housed mostly within the wing structure, between the two main spars. Tanks ran through the centre-section out to a point level with the flap/aileron break. For ferry flights, additional fuel could be carried in the bomb bays.

Radar
Often retouched out of wartime-era photographs, the B-29 featured a bombing radar located between the bomb bays. This was the AN/APQ-13 system, developed by Massachusetts Institute of Technology (MIT) and Bell Telephone Labs. It had a 30-in (0.76-m) diameter dish mounted within the radome.

Colour scheme
Although the early B-29s flew with Olive Drab over Neutral Gray camouflage, most served unpainted to reduce weight. The fuselage stripes on this aircraft identified it as a lead-ship, performing the main navigation tasks for the following bomber formation, and supplying bomb-drop information.

Defensive armament
Two turrets were mounted on the forward fuselage, one above and one below. Each had two 0.50-in (12.7-mm) Browning machine-guns, although the upper unit later had four guns. The pressurised cabin made the use of conventional manned turrets impossible. Further twin 50-calibre gun turrets were mounted above and below the rear fuselage, aimed remotely by the three gunners in the rear cabin. Three glazed blisters above and to the sides allowed ample visibility for the gunners. The central gunner was seated on a raised seat which swivelled through 360°. A potent sting in the tail of the Superfortress was provided by a pair of 50-calibre machine-guns and a 20-mm cannon, aimed by a gunner sitting in a pressurised turret with armoured windscreens. Mounted above the cannon was a camera to record firing. A fire control system used an analog computer. The system was normally set up so that the central (upper) gunner controlled both upper turrets, the left and the right gunners shared the lower rear turret, while the bombardier operated the lower forward turret. However, any turret could be overridden to suit individual requirements.

Bristol Blenheim

The two flying 'Blenheims' to survive and all but one of the static exhibits are Bolingbroke Mk IV-Ts. Painstakingly restored from various Bolingbroke Mk IV-T components, this machine flies from Duxford, UK as a fitting tribute to the many aircrew who flew the Blenheim into battle.

Blenheim Mk IV

Cutaway key
1 Starboard navigation light
2 Starboard formation light
3 Wing rib construction
4 Aileron control rod
5 Starboard aileron
6 Aileron tab
7 Starboard outer flap
8 Outboard, long-range fuel tank, capacity 94 Imp gal (427 litres)
9 Fuel tank filler cap
10 Starboard nacelle fairing
11 Main inboard fuel tank, capacity 140 Imp gal (636 litres)
12 Oil tank, capacity 11.5 Imp gal (52 litres)
13 Engine bearers
14 Oil cooler exhaust duct
15 Engine cooling flaps
16 Cowling blister fairings
17 Bristol Mercury XV nine-cylinder radial engine
18 Oil cooler ram air intakes
19 Propeller hub mechanism
20 De Havilland three-bladed propeller
21 Nose compartment glazing
22 Cabin air intake
23 Navigator/ bombardier's instrument panel
24 Bomb-aiming windows
25 Pitot tube
26 Rearward-firing, ventral machine-gun cupola
27 Browning 0.303-in (7.7-mm) machine-gun
28 Fireman's axe
29 Nose compartment escape hatch
30 Fire extinguisher
31 Chart table
32 Fixed foresight
33 Back of instrument panel
34 Foot boards
35 Rudder pedals
36 Compass
37 Control column
38 Windscreen panels
39 Pilot's gunsight
40 Navigator/ bombardier's seat
41 Pilot's seat
42 Engine throttles
43 Venturi tube
44 Pilot's blister observation window
45 Armoured headrest
46 Cockpit roof sliding hatch
47 Parachute stowage
48 Wing centre-section construction
49 Sliding hatch rails
50 Aerial mast
51 Parachute stowage
52 Wing centre-section attachment frame
53 Pneumatic system compressed air bottle
54 Three-man dinghy
55 First-aid box
56 Fuselage double frame
57 Rear gunner's entry/ emergency escape hatch
58 Rear gunner's seat
59 Gun turret
60 Two Browning 0.303-in (7.7-mm) machine-guns
61 Aerial cable
62 Fuselage skin plating
63 Starboard tailplane
64 Starboard elevator
65 Fin construction
66 Rudder balance
67 Fabric-covered rudder construction
68 Rudder tab
69 Tail navigation lights
70 Elevator tab
71 Fabric-covered elevator construction
72 Elevator balance
73 Port tailplane
74 Rudder cables
75 Elevator hinge control
76 Tailwheel shock absorber
77 Tailwheel
78 Control cable cross shaft
79 Tail assembly joint ring
80 Rear fuselage frames
81 Fuselage stringer construction
82 Control cables
83 Access steps
84 Two 4FL flares
85 Trailing-edge flap shroud construction
86 Flap jack
87 Inboard split trailing-edge flap
88 Outer wing spar attachment joint
89 Flap lever mechanism
90 Outboard split trailing-edge flap
91 Rear spar
92 Aileron hinge control
93 Aileron tab
94 Fabric-covered aileron construction
95 Port formation light
96 Wingtip construction
97 Port navigation light
98 Landing and taxiing lamps
99 Wing rib construction
100 Front spar
101 Aileron control rod
102 Leading-edge ribs
103 Ammunition tank
104 Fixed Browning 0.303-in (7.7-mm) machine-gun
105 Outboard, long-range fuel tank, capacity 94 Imp gal (427 litres)
106 Fuel tank filler cap
107 Main wheel well
108 Auxiliary oil tank, capacity 2.5 Imp gal (11 litres)

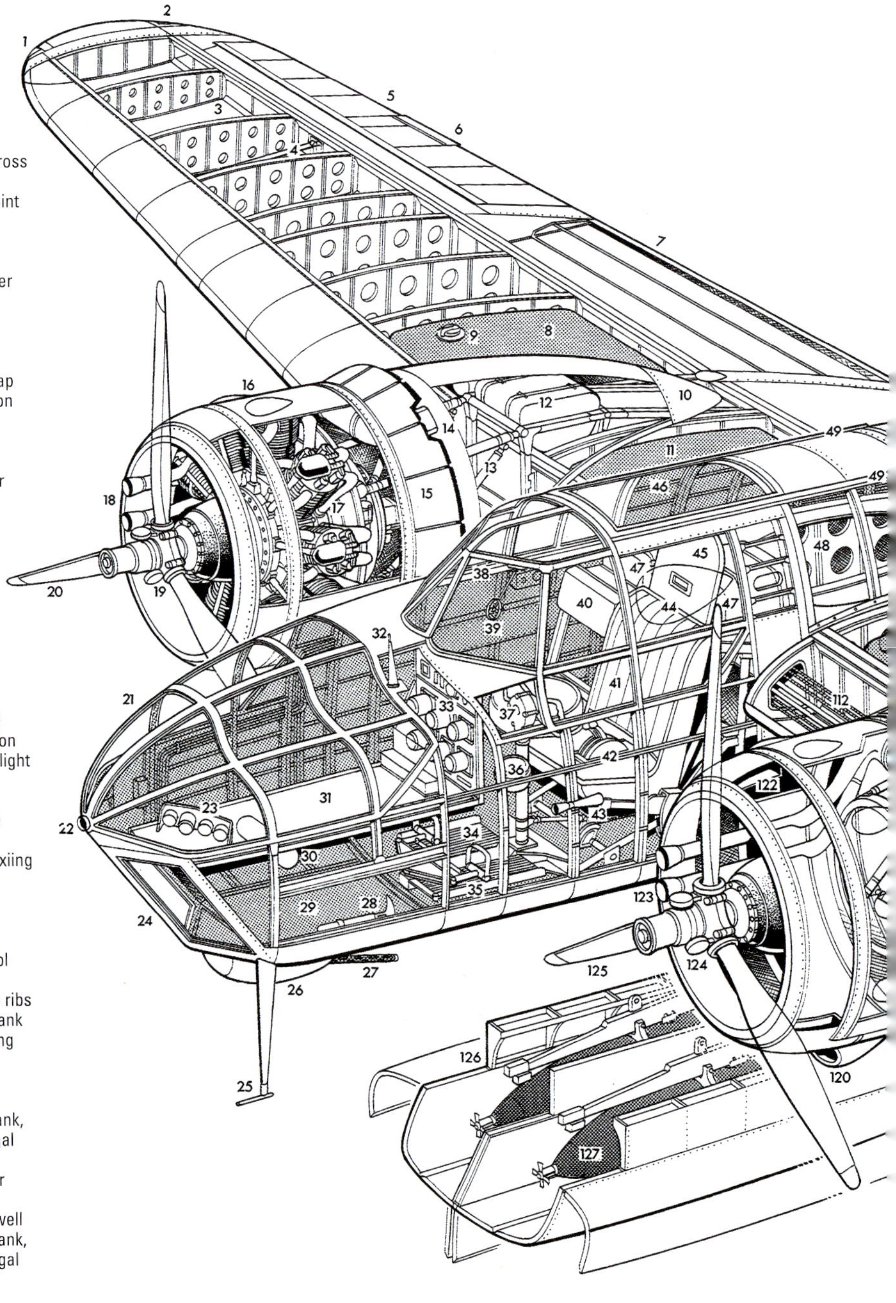

Finland ordered 18 Blenheim Mk Is in October 1936, and these aircraft were modified to carry Swedish-built bombs. The Blenheims were used for bombing and reconnaissance and were supplemented by a further 24 examples. In addition, Valtion Lentokonetehdas of Finland built a total of 55 Mk IIs and Mk IVs. Finnish Blenheims saw action in both the Winter and Continuation Wars against the Soviet Union, and then against German forces in the Lapland War. This Blenheim Mk I is seen in 1944, shortly after transfer to a bomber training unit.

SPECIFICATION

Blenheim Mk IV

Dimensions

Wingspan: 56 ft 4 in (17.17 m)
Length: 42 ft 7 in (12.98 m)
Height: 9 ft 10 in (3.00 m)
Wing area: 469 sq ft (43.57 m^2)

Powerplant

Two 905-hp (675-kW) Bristol Mercury XV radial piston engines

Weights

Empty: 9,790 lb (4441 kg)
Maximum take-off: 14,400 lb (6532 kg)

Performance

Maximum speed at 11,800 ft (3595 m): 266 mph (428 km/h)
Cruising speed: 198 mph (319 km/h)
Service ceiling: 27,260 ft (8310 m)
Maximum range: 1,460 miles (2350 km)

Armament

Four 0.303-in (7.7-mm) machine-guns (one forward-firing in port wing, two in power-operated dorsal turret, and one remotely-controlled in mounting beneath nose and firing aft), plus up to 1,000 lb (434 kg) of bombs internally and 320 lb (145 kg) of bombs externally

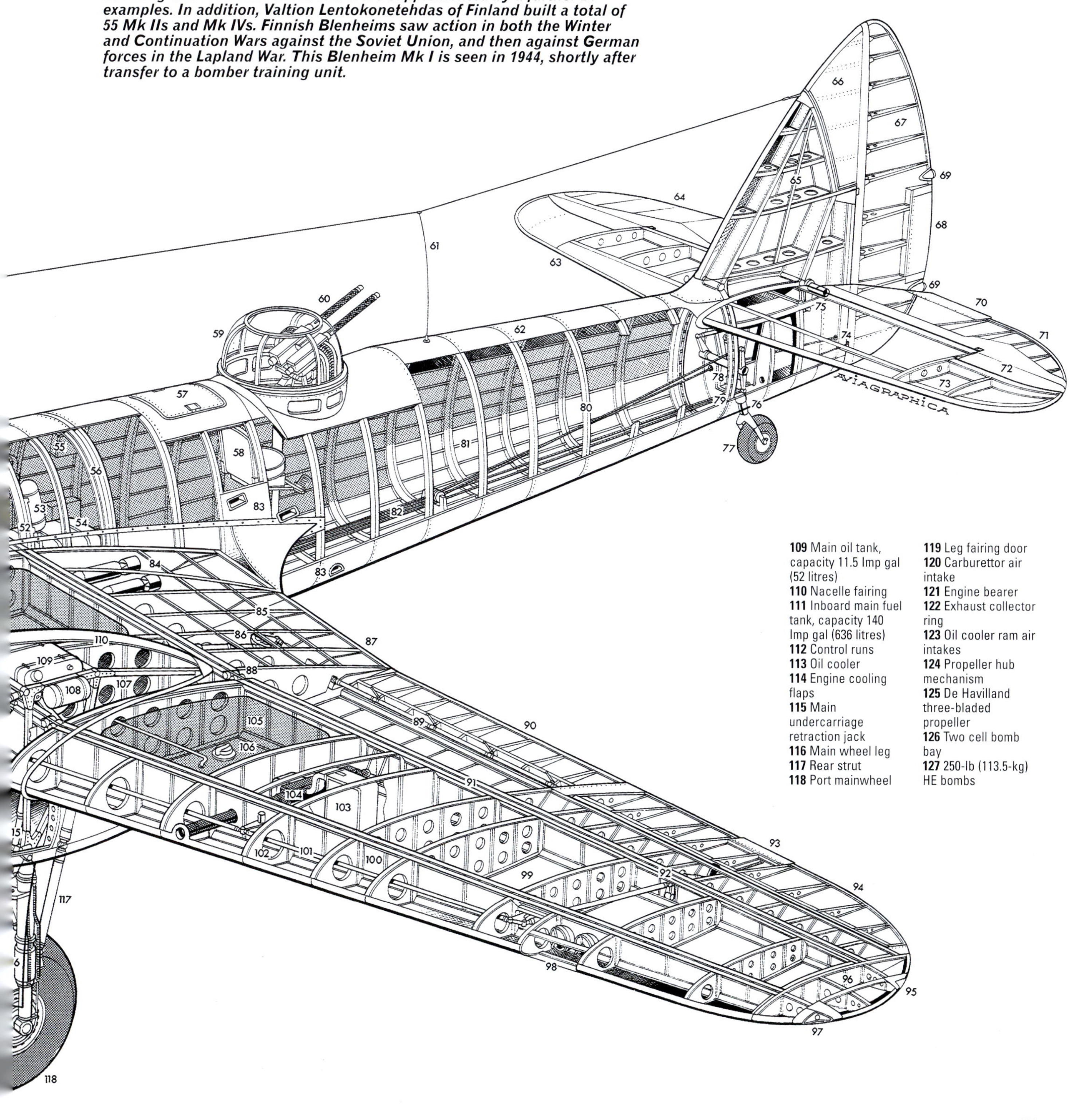

109 Main oil tank, capacity 11.5 Imp gal (52 litres)
110 Nacelle fairing
111 Inboard main fuel tank, capacity 140 Imp gal (636 litres)
112 Control runs
113 Oil cooler
114 Engine cooling flaps
115 Main undercarriage retraction jack
116 Main wheel leg
117 Rear strut
118 Port mainwheel
119 Leg fairing door
120 Carburettor air intake
121 Engine bearer
122 Exhaust collector ring
123 Oil cooler ram air intakes
124 Propeller hub mechanism
125 De Havilland three-bladed propeller
126 Two cell bomb bay
127 250-lb (113.5-kg) HE bombs

Blenheim Mk IV

The RAF's No. 88 Squadron only operated the Blenheim for a short while, with the first Mk IVs arriving in July 1941 and departing in November of that year, to be replaced by Douglas Bostons. During the period that the unit operated the Blenheim, it was involved in Circus operations over northern France and Operation Channel Stop. The Blenheim suffered mixed fortunes in RAF service: while it was able to operate at some risk in daylight raids over Occupied Europe, it never possessed the armour, armament or bombload that would have made it an outstanding aircraft.

Cockpit

The nose of the Blenheim Mk IV was redesigned to bring the windscreen closer to the pilot. The scalloping (on the port side only – where the pilot sat) gave him a better view, but gave the nose a characteristic asymmetric appearance. Several Blenheim units in North Africa fixed 'banshees' (hardwood sirens) under the nose which could be turned across the air stream, causing them to emit a blood-curdling wail. This banshee, like the Ju 87's 'Trumpet of Jericho' siren, had the effect of scaring inexperienced troops at a critical time, but proved to have no long-lasting value.

Fuselage

As a direct derivative of the Type 142, the Blenheim was of all-metal construction, albeit with fabric-covered, cable-operated flying controls. The mid-set cantilever wing (raised by about 16 in/40 cm compared to the Type 142 to allow an internal bomb bay to be provided) had mass-balanced Frise ailerons and trailing-edge split flaps, while the tailplane (with no dihedral) was similarly raised, and was increased in span. Unlike the 142's variable-incidence tailplane, that of the 142M was fixed and provided with longer-chord, trim-tabbed elevators. The raised wing also raised the engine nacelles, which now almost entirely robbed the pilot of any view directly out to the sides. This peculiarity remained a feature of all Blenheim variants, and was the price imposed by the Blenheim's unbroken internal bomb bay, although the later Beaufort and Beaufighter restored the pilot's view by slinging their engine nacelles below the wings.

Armament
Early Blenheims were fitted with the B.I Mks I and II turrets with single Lewis guns, but these were soon superseded by the B.I Mk III with a Vickers 'K' gun. The turret was then modified to mount two 'K' guns, becoming the B.I Mk IIIA, which was then modified again, as supplies permitted, to B.I Mk IV standard, which incorporated two Browning guns with continuous belt feed and an improved rate of fire. The increased firepower of the Mk IV turret enabled Blenheims to better hold their own in many encounters against superior enemy forces and there were several occasions when single Blenheims fought off several Bf 109s and managed to escape into cloudbanks. In general, however, hard manouvering and optimal use of cloud cover were the Blenheim's best chance of survival if attacked by fighters.

Crew access
The gunner accessed his station through a dorsal hatch immediately forward of the turret. The pilot and navigator climbed up over the wing and entered the cockpit through a sliding roof panel. A bail-out hatch was located below the cockpit.

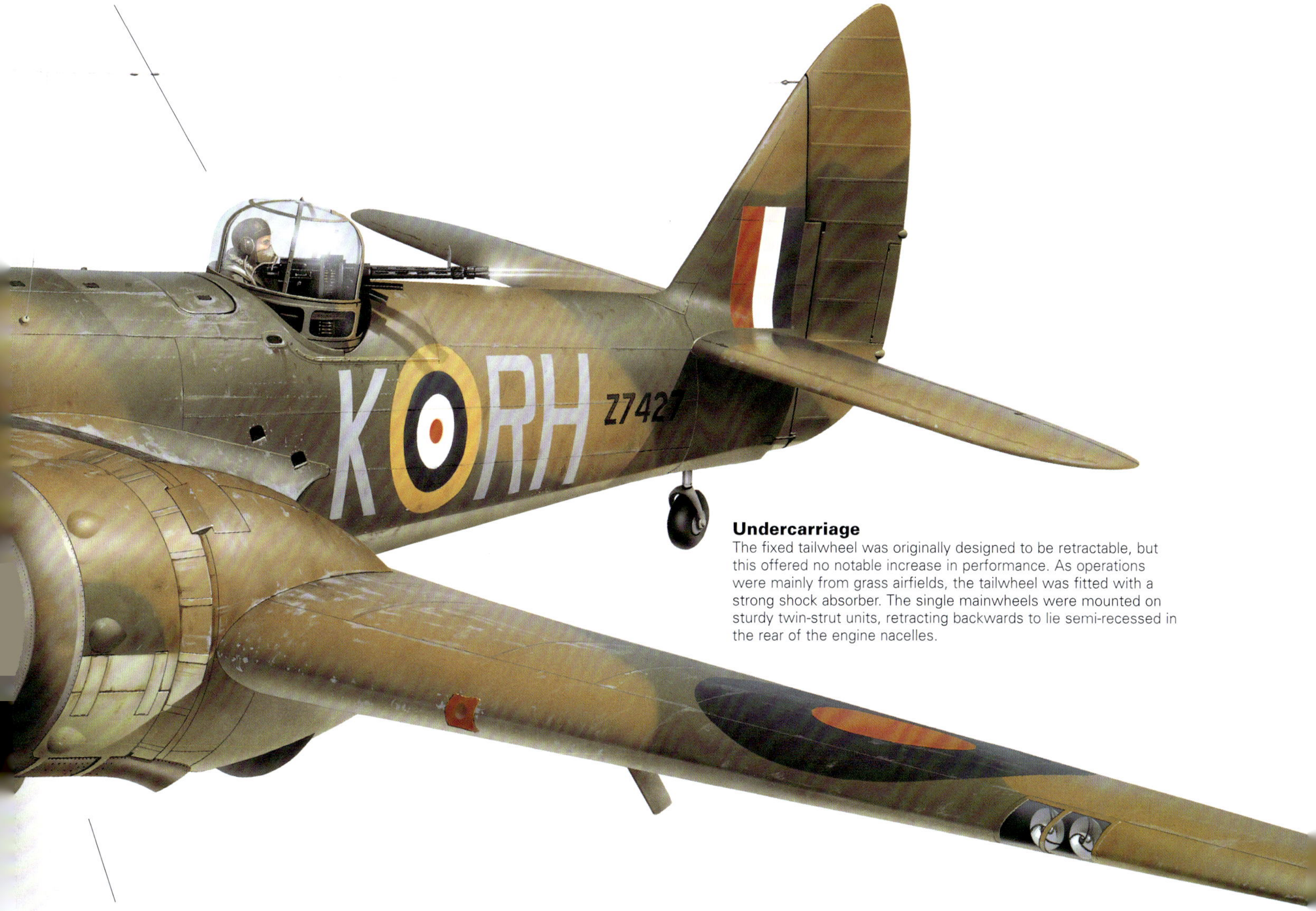

Undercarriage
The fixed tailwheel was originally designed to be retractable, but this offered no notable increase in performance. As operations were mainly from grass airfields, the tailwheel was fitted with a strong shock absorber. The single mainwheels were mounted on sturdy twin-strut units, retracting backwards to lie semi-recessed in the rear of the engine nacelles.

Performance
The Blenheim Mk IV's two 905-hp (675-kW) Bristol Mercury XV engines allowed the aircraft to reach a maximum speed of 266 mph (428 km/h) which made it, at the time, the fastest bomber in the world. However, this speed was rarely attained, especially when the aircraft was bomb-laden; more emphasis had been placed on field performance than on flat-out speed. Aircraft like the Ju 88 and Do 17 – or the Baltimore, Boston, Maryland and Ventura – showed what was possible by using massive two-row radial engines and by accepting longer take-off and landing distances and the need for concrete runways. All these aircraft made the Blenheim look unimpressive and, furthermore, they could all carry much heavier bombloads over larger distances at greater speed.

Consolidated B-24 Liberator

A B-24H of the 453rd BG (part of the Eighth Air Force) returns home after attacking a Nazi air base on 21 February 1944. The contrails that criss-cross the sky belong to the attendant fighters patrolling the skies in search of potential interceptors.

B-24J Liberator

Cutaway key
1 Rudder trim tab
2 Fabric-covered rudder
3 Rudder hinges (metal leading edge)
4 Starboard tailfin
5 Leading-edge de-icing boot
6 Starboard rudder horn
7 Rudder push-pull tube
8 Rear navigation light
9 Tailplane stringers
10 Consolidated (or Motor Products) two-gun electrically-operated tail turret (0.5 in/12.7 mm)
11 Elevator torque tube
12 Elevator trim tab
13 Elevator frame (fabric-covered)
14 Rudder trim tab
15 Tab control linkage
16 Rudder post
17 Light alloy rudder frame
18 HF aerial
19 Tailfin construction
20 Metal-covered fixed surfaces
21 Tailplane front spar
22 Port elevator push/pull tube
23 Elevator drive quadrant
24 Elevator servo unit
25 Rudder servo unit
26 Ammunition feed track (tail turret)
27 Fuselage aft main frame
28 Walkway
29 Signal cartridges
30 Longitudinal 'Z' section stringers
31 Control cables
32 Fuselage intermediate secondary frames
33 Ammunition box
34 Aft fuselage camera installation
35 Lower windows
36 Waist gun support mounting
37 Starboard manually operated waist gun (0.5 in/ 12.7 mm)
38 Waist position (open)
39 Wind deflector plate
40 Waist position hinged cover
41 Port manually-operated 'waist' gun (0.5 in/12.7 mm)
42 Dorsal aerial
43 Ball-turret stanchion support beam
44 Ammunition box
45 Ball-turret stanchion
46 Midships window
47 Turret well
48 Cabin floor
49 Tail-bumper operating jack
50 Tailbumper fairing
51 Briggs-Sperry two-gun electrically-operated ball-turret (0.5 in/12.7 mm)
52 Turret actuation mechanism
53 Bomb-door actuation sprocket (hydraulically-operated)
54 Bomb-door corrugated inner skin
55 Bomb-bay catwalk (box keel)
56 Bomb-bay catwalk vertical channel support members (bomb release solenoids)
57 Bomb-door actuation track and rollers
58 Wing rear spar
59 Bomb-bay access tunnel
60 Fuselage main frame/bulkhead
61 D/F loop housing
62 Whip antenna
63 Oxygen cylinders
64 Aileron cable drum
65 Starboard flap extension cable
66 Wing rib cut-outs
67 Wing centre section carry-through
68 Two 5-man inflatable dinghies
69 Flap hydraulic jack
70 Flap/cable attachments
71 Hydraulically-operated Fowler flap
72 Wing rear spar
73 Port mainwheel well and rear fairing
74 Engine supercharger waste gate
75 Three auxiliary self-sealing fuel cells (port and starboard)
76 Wing outer section
77 Aileron gear boxes
78 Flush riveted smooth metal wing skinning
79 Port statically-balanced aileron (fabric-covered)
80 Port wingtip
81 Port navigation light
82 Wing leading-edge de-icing boot
83 Hopper-type self-sealing oil tank (32.9 US gal/125 litres)
84 Engine nacelle
85 1,200-hp (895-kW) Pratt & Whitney Twin Wasp R-1830-65 14-cylinder two-row radial engine
86 Hamilton Standard Hydromatic constant-speed airscrew (11-ft 7-in/3.53-m diameter)
87 Landing/taxiing light
88 Nacelle structure
89 Supercharger ducting
90 12 self-sealing inter-rib fuel cells (wing centre section)
91 Martin two-gun electrically-operated dorsal turret (0.5 in/12.7 mm)
92 Turret mechanism
93 Fuselage main frame/bulkhead
94 Radio compartment starboard window

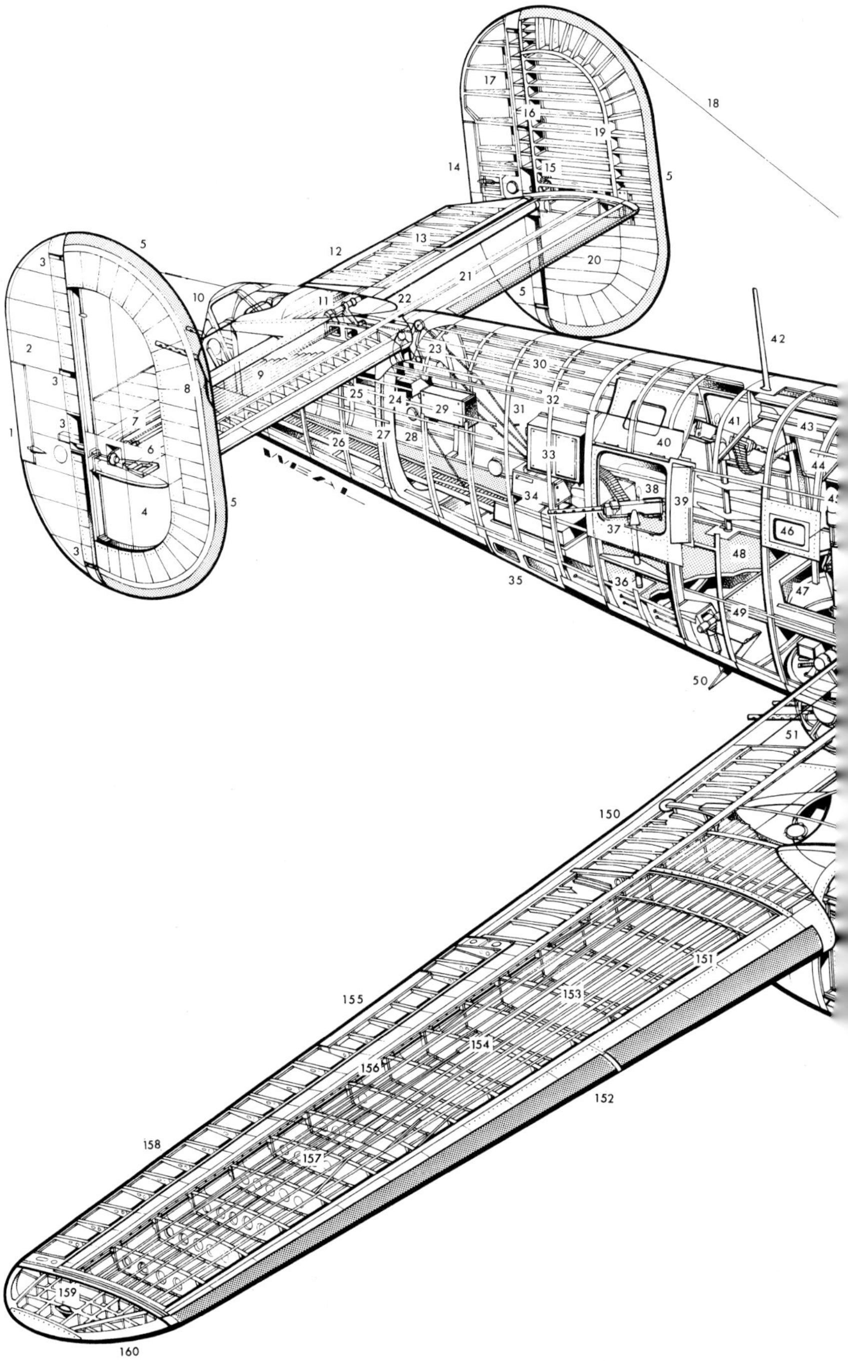

Diamond Lil *can be seen here in 1980 as part of the* **Confederate Air Force.** *The windows under the outboard engine and forward of the insignia stem from the days when the aircraft operated as a* **C-87 Liberator Express.** *Today, it is still flying with the CAF and is based at Midland, TX. Several B-24s still exist in flying condition and many more have been placed in museums and collections around the world.*

SPECIFICATION

B-24J Liberator

Dimensions

Length: 67 ft 2 in (20.47 m)
Height: 18 ft (5.48 m)
Wing span: 110 ft (33.52 m)
Wing area: 1,048 sq ft (319.4m²)

Powerplant

Four Pratt & Whitney R-1830-65 14-cylinder air-cooled radial engines with General Electric B-22 exhaust-driven turbo-chargers, each delivering 1,200 hp (895 kW) at take-off and maintaining this power at a military rating up to 31,800 ft (9692 m)

Weights

Empty: 38,000 lb (17236 kg)
Combat: 56,000 lb (25401 kg)
Maximum overload: 71,200 lb (32295 kg)
Maximum bombload: 12,800 lb (5806 kg)

Performance

Maximum speed at 30,000 ft: 300 mph (483 km/h)
Maximum speed at 20,000 ft: 277 mph (445 km/h)
Maximum continuous speed at 25,000 ft: 278 mph (447 km/h)
Initial climb rate: 1,025 ft (312.42 m) per minute
Service ceiling: 28,000 ft (8534 m)

Range and endurance

Range with 5,000-lb (2268-kg) bombload: 1,700 miles (2735 km) in 7.3 hours at 25,000 ft (7620 m)

Armament

Ten 0.5-in (12.7-mm) Browning machine guns in nose, upper, ventral, ball and tail turrets and in waist positions, with a total of 4,716 rounds. Maximum short-range bomb load was 12,800 lb (5806 kg), while normal offensive load was 5,000 lb (2268 kg)

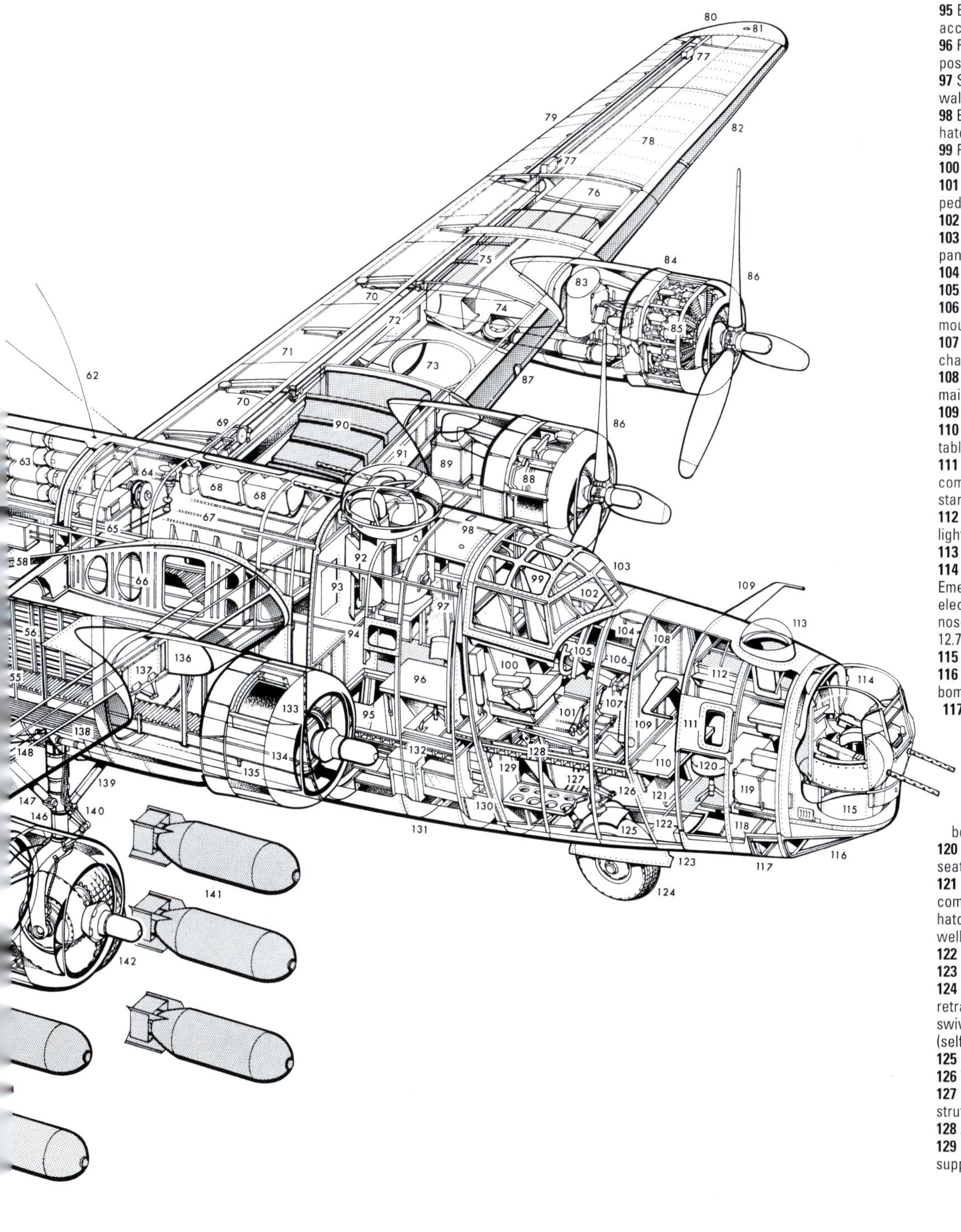

95 Bomb-bay catwalk access trap
96 Radio-operator's position
97 Sound-insulation wall padding
98 Emergency escape hatch
99 Pilot's seat
100 Co-pilot's seat
101 Co-pilot's rudder pedals
102 Instrument panel
103 Windscreen panels
104 Compass housing
105 Control wheel
106 Control wheel mounting
107 Control linkage chain
108 Fuselage forward main frame bulkhead
109 Pitot heads
110 Navigator's chart table
111 Navigator's compartment starboard window
112 Chart table lighting
113 Astro-dome
114 Consolidated (or Emerson) two-gun electrically-operated nose turret (0.5 in/ 12.7 mm)
115 Turret seating
116 Optically-flat bomb-aiming panel
117 Nose side-glazing
118 Bombardier's prone couch
119 Ammunition boxes
120 Navigator's swivel seat
121 Navigator's compartment entry hatch (via nosewheel well)
122 Nosewheel well
123 Nosewheel door
124 Forward-retracting free-swivelling nosewheel (self-aligning)
125 Mudguard
126 Torque links
127 Nosewheel oleo strut
128 Angled bulkhead
129 Cockpit floor support structure
130 Nosewheel retraction jack
131 Smooth-stressed Alclad fuselage skinning
132 Underfloor electrics bay
133 Roll top desk-type bomb-bay doors (four)
134 Supercharger nacelle cheek intakes
135 Ventral aerial (beneath bomb-bay catwalk)
136 Nacelle/wing attachment cut-out
137 Wing front spar nacelle support
138 Undercarriage front-pivoting shaft
139 Drag strut
140 Bendix scissors
141 Internal bomb load (max 8,000 lb/3629 kg)
142 Starboard mainwheel
143 Engine-mounting ring
144 Firewall
145 Monocoque oil tank
146 Mainwheel oleo (Bendix pneudraulic strut)
147 Side brace (jointed)
148 Undercarriage actuating cylinder
149 Starboard mainwheel well and rear fairing
150 Fowler flap structure
151 Wing front spar
152 Wing leading-edge de-icing boot
153 All-metal wing structure
154 Spanwise wing stringers
155 Aileron trim tab (starboard only)
156 Wing rear spar
157 Wing ribs (pressed and built-up former)
158 Statically-balanced aileron (metal frame)
159 Starboard navigation light
160 Wingtip structure

B-24H Liberator

This Liberator, 42-7697, *The Stork*, belongs to the 726th Bombardment Squadron of the 451st Bombardment Group which was part of the 15th Air Force, based in Italy. Within each bomb group, several aircraft had additional patches of colour on the tail and fuselage to identify them as lead ships. In early 1944, B-24s of the 451st carried red discs (as did all those of the 49th BW) as a means of identification.

Crew
The normal crew of a USAAF B-24 was 10 men. This comprised a pilot/aircraft commander and co-pilot seated side by side in a cockpit fitted with dual controls, a bombardier in the lower nose, a nose gunner, a navigator in the forward fuselage, a radio operator in a compartment with a gun turret in the roof, two beam gunners, a ventral 'ball turret' gunner and a tail gunner. The navigator's station had a transparent dome above it to allow him to take sextant readings. Small windows either side gave him some measure of outside visibility. Just aft on either side were mounted the pitot probes.

Bombs and the bomb bay
The central bomb bay was divided into front and rear segments divided by a catwalk which was also the fuselage keel beam. The B-25H's maximum bombload of 12,800 lb (5806 kg) was stowed vertically in racks. These could accommodate bombs from 100 to 1,600 lb (45 to 726 kg) although 2,000-lb (907-kg) bombs could be carried on special rack adaptors. Some Liberators could carry a 4,000-lb (1814-kg) bomb on a rack installed under each wing. The unique roller-type bomb doors retracted upwards into the fuselage sides when opened, offering considerably reduced drag compared to the standard outward-hinging doors. The bombs depicted here are standard high-explosive (HE) weapons, although incendiaries were also carried. A wind-driven vane on the front of each bomb armed the fuse as it fell into the slipstream, a safety measure to prevent bombs from being armed in the aircraft itself.

Camouflage and serial numbers

This Liberator is painted in the standard bomber scheme of Olive Drab upper surfaces and Neutral Gray undersides, the two colours demarcated along the lower fuselage by a wavy line. As weight became more crucial than camouflage, many USAAF bombers were stripped of paint as the war progressed, as a weight-saving measure. Presented in abbreviated form on the fin, the serial number identified this aircraft as one of 1,580 B-24H Liberators built by Ford at Willow Run, Michigan. The factory also built the E, J, L, and M variants.

Powerplant

The prototype XB-24 was originally conceived with Pratt & Whitney R-1830-33 Twin-Wasp 14-cylinder two-row air-cooled radial engines. These were changed during construction for 1,200-hp (895-kW) turbo-supercharged R-1830-43s, with oil coolers mounted on the sides of the engine, giving the aircraft its unmistakable elliptical engine cowlings. The major production variant, the B-24J, was fitted with R-1830-65 engines able to deliver the same power, 1,200 hp (895 kW) at a height of up to 31,800 ft (9692 m). Each engine drove a three-bladed Hamilton Standard Hydromatic constant-speed fully-feathering propeller of 11ft 7-in (3.53-m) diameter.

Armament

Early models of the B-24 had provision for relatively few hand-held (0.3-in/7.62-mm) machine-guns – one installed in the glazed nose together with others which could be fired through apertures in the roof, floor and each side of the fuselage. The fully-developed B-24H sported 10 Browning 0.5-in (12.7-mm) machine-guns installed in electrically-actuated twin-gun nose, upper, ventral and tail turrets, and single-gun waist 'positions' on each side of the fuselage.

Wings

The B-24 was designed around a high-aspect ratio wing employing a 'Davis' high-lift section, so named after an eminent NACA wing designer. This aerofoil was claimed to offer 25 per cent less drag at low speeds and 10 per cent less at higher speeds than other more conventional wing profiles. The use of heavy box spars resulted in a stiff wing structure allowing maximum space for fuel tanks. The wing itself, spanning 110 ft (33.53 m) was shoulder-mounted on the fuselage in order to accommodate a large bomb bay and facilitate bomb-loading.

Most of the trailing edge of the wing was taken up by the sizeable flaps (inboard) and ailerons (outboard), the latter incorporating a trim tab on the starboard side only. The leading edges of the wings, fins and tailplanes were all de-iced by pneumatic rubber boots.

Curtiss SB2C Helldiver

A group of Helldivers heads out on another mission over the Pacific, with F6F Hellcats flying top cover. Though unpopular with pilots, the SB2C was the most successful Allied dive-bomber of World War II.

SB2C Helldiver

Cutaway key
1 Curtiss Electric four-bladed constant-speed propeller
2 Spinner
3 Propeller hub mechanism
4 Spinner backplate
5 Propeller reduction gearbox
6 Carburettor intake
7 Intake ducting
8 Warm air filters
9 Engine cowling ring
10 Oil cooler intake
11 Engine cowlings
12 Wright R-2600-20 Cyclone 14 radial engine
13 Cooling air exit louvres
14 Exhaust collector
15 Exhaust pipe fairing
16 Oil cooler
17 Engine accessories
18 Hydraulic pressure accumulator
19 Boarding step
20 Cabin combustion heater
21 Engine oil tank (25 US gal/94.6 litre capacity)
22 Engine bearer struts
23 Hydraulic fluid tank
24 Fireproof engine compartment bulkhead
25 Aerial mast
26 Starboard wing fold hinges
27 Wing fold hydraulic jack
28 Gun camera
29 Rocket projectiles (4.5-in/11.43-cm)
30 Starboard leading edge slat (open)
31 Slat roller tracks
32 Slat operating cables
33 Starboard navigation light
34 Formation light
35 Starboard aileron
36 Aileron aluminium top skins
37 Aileron control mechanism
38 Starboard dive brake (open position)
39 Windshield
40 Bullet proof internal windscreen
41 Reflector gunsight
42 Instrument panel shroud
43 Cockpit coaming
44 De-icing fluid tank
45 Instrument panel
46 Pilot's pull-out chart board
47 Rudder pedals
48 Control column
49 Cockpit floor level
50 Engine throttle controls
51 Pilot's seat
52 Oxygen bottle
53 Safety harness
54 Armoured seat back
55 Headrest
56 Pilot's sliding cockpit canopy cover
57 Jury strut
58 Wing folded position
59 Fixed bridge section between cockpits
60 Fuel tank filler cap
61 Fuselage fuel tank (110 US gal/416 litre capacity)
62 Fuselage main longeron
63 Handhold
64 Fuselage frame and stringer construction
65 Autopilot controls
66 Sliding canopy rail
67 Aerial lead-in
68 Radio equipment bay
69 Life raft stowage
70 APG-4 low-level bombing radar
71 Gunner's forward sliding canopy cover
72 Gun mounting ring
73 Gunner's seat
74 Footrests
75 Ammunition boxes
76 Armour plate
77 Wind deflector
78 Twin 0.3-in (7.62-mm) machine-guns
79 Retractable turtle decking
80 Gun rest mounting
81 Folding side panels
82 Upper formation light
83 Fin root fillet
84 Starboard tailplane
85 Deck handling handhold
86 Fabric-covered elevator
87 Remote compass transmitter
88 Tailfin construction
89 Aerial cable
90 Sternpost
91 Rudder construction
92 Fabric skin covering
93 Trim tab
94 Balance tab
95 Elevator trim tab
96 Elevator construction
97 Tailplane construction
98 Tailplane spar root fixing
99 Deck arrester hook
100 Arrester hook damper
101 Tail navigation light
102 Tailwheel leg strut
103 Solid tyre tailwheel
104 Leg fairing
105 Rear fuselage frames
106 Tailplane control cables
107 Lifting bar
108 Gunner's floor level
109 Wing root trailing edge fillet
110 Aft end of bomb bay
111 Rear spar centre section fixing
112 Wing walkway
113 Port upper surface flap dive brake
114 Rear spar hinge joint
115 Split trailing edge flaps
116 Balance tab
117 Aileron hinge control
118 Aileron trim tab
119 Lower surface fabric skinning
120 Wing rib construction
121 Wing tip construction
122 Port navigation light
123 Pitot tube
124 Automatic leading edge slat (opens with undercarriage operation)
125 Slat riblets
126 Slat operating cables
127 Main spar
128 Leading edge nose ribs
129 500-lb (226.8-kg) bomb
130 Rocket projectiles (4.5-in/11.43-cm)
131 Drop tank (58 US gal/219.5 litre capacity)
132 Wing fold joint line
133 Main undercarriage leg fairing doors
134 Drag strut
135 Port mainwheel
136 Shock absorber leg strut
137 20-mm wing cannon
138 Cannon barrel fairing

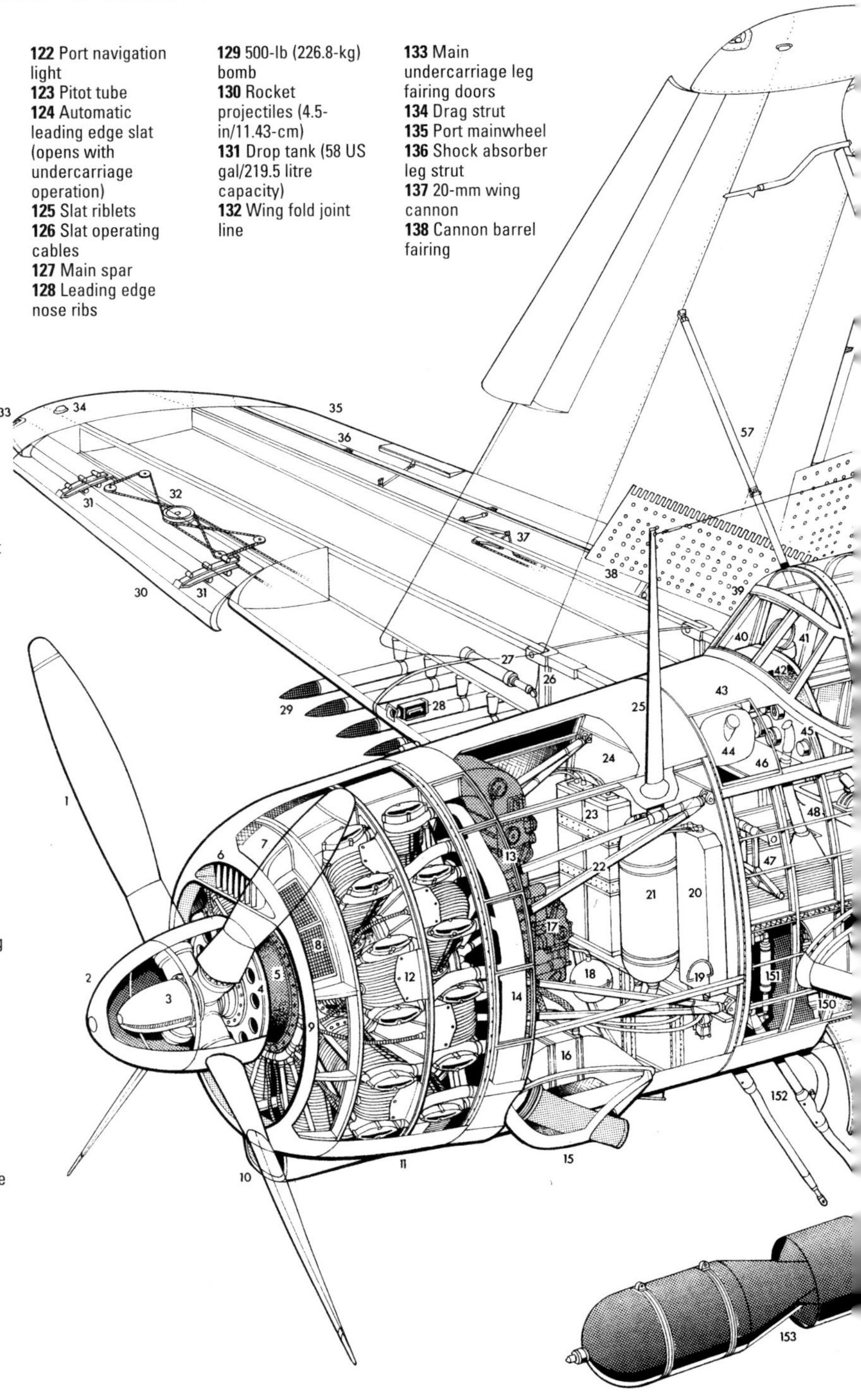

The XSB2C-1 prototype (BuNo. 1758) is seen here having sustained damage in an accident during test-flying in wintry conditions in February 1941. Although it was repaired, the aircraft was never tested by the US Navy before being destroyed after suffering an in-flight wing failure on 21 December 1941.

SPECIFICATION

Curtiss SB2C Helldiver

Type
Two-seat carrierborne and land-based scout-bomber

Powerplant
One Wright R-2600-20 Cyclone 14 radial piston engine rated at 1,900 hp (1417 kW)

Performance
Maximum speed 260 mph (418 km/h) at 16,100 ft (4910 m); cruising speed 148 mph (238 km/h) at optimum altitude; climb to 10,000 ft (3050 m) in 8 minutes 54 seconds; service ceiling 26,400 ft (8045 m); range 1,805 miles (2905 km)

Weights
Empty 10,580 lb (4799 kg); maximum take-off 15,918 lb (7220 kg) in the scout role with maximum fuel

Dimensions
Wingspan 49 ft 8⅝ in (15.15 m); length 36 ft 8 in (11.18 m); height 13 ft 11½ in (4.01 m); wing area 422.00 sq ft (39.20 m2)

Armament
Two 20-mm M2 fixed forward-firing cannon in the leading edges of the wing and two 0.3-in (7.62-mm) Browning trainable rearward-firing machine-guns in the rear of the cockpit, plus up to 2,000 lb (907 kg) of bombs or one torpedo in the lower-fuselage weapons bay and on underwing racks

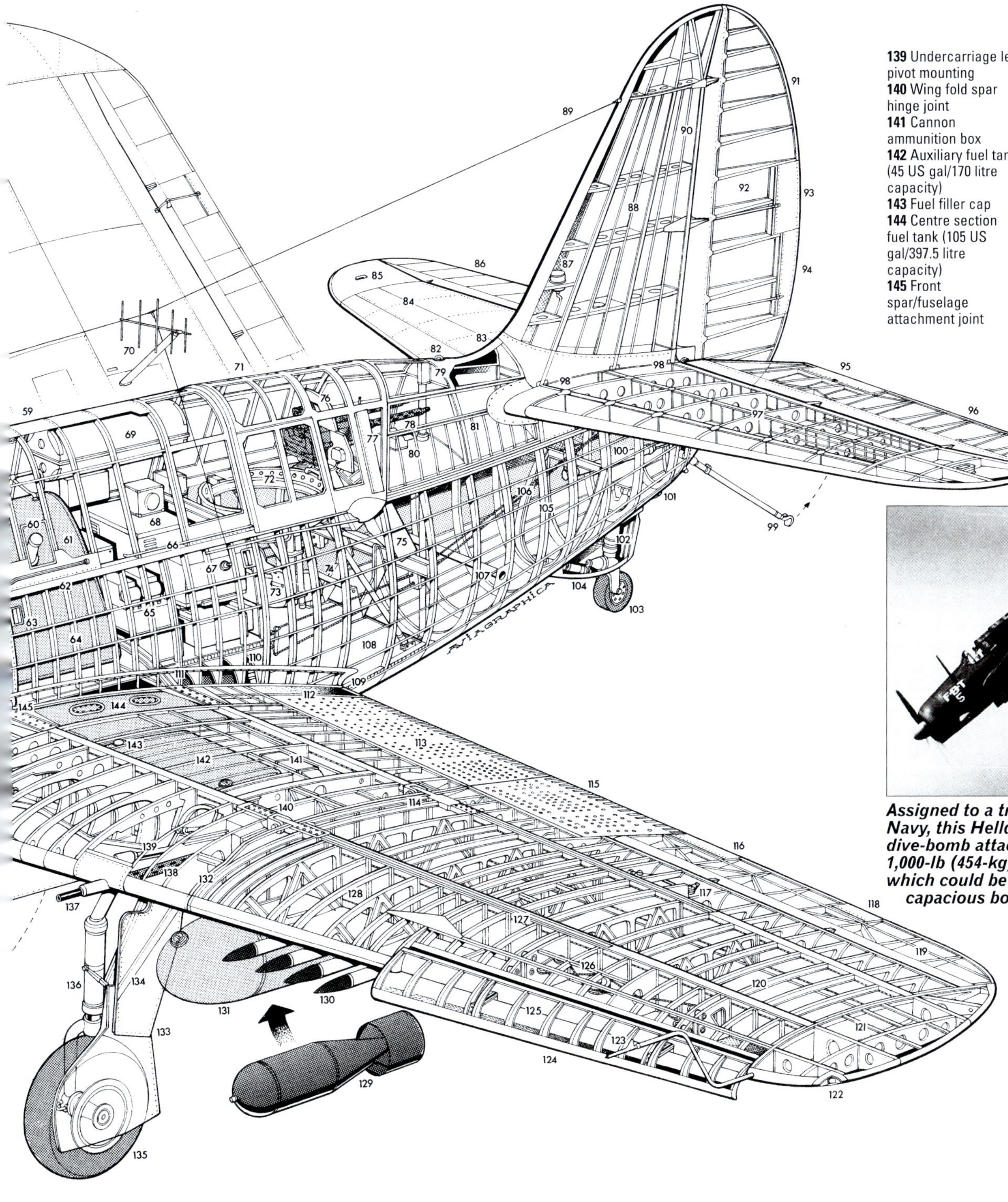

139 Undercarriage leg pivot mounting
140 Wing fold spar hinge joint
141 Cannon ammunition box
142 Auxiliary fuel tank (45 US gal/170 litre capacity)
143 Fuel filler cap
144 Centre section fuel tank (105 US gal/397.5 litre capacity)
145 Front spar/fuselage attachment joint
146 Main undercarriage wheel well
147 Retractable catapult strop
148 Approach light
149 Bomb doors (open)
150 Bomb door hydraulic jack
151 Displacement gear jack
152 H-type bomb displacement arm
153 1,000-lb (453.6-kg) bomb

Assigned to a training unit in the US Navy, this Helldiver engages in a dive-bomb attack. Pictured is a 1,000-lb (454-kg) weapon, two of which could be carried in the capacious bomb bay.

De Havilland Mosquito

Pioneer of unescorted high-speed day bombing raids over Germany, the Mosquito B.Mk IV series II was given a tremendously enthusiastic reception by the squadrons to which it was allocated. First among these was No. 105, to which these aircraft belong.

Mosquito B.XVI

Cutaway key
1 Three-bladed de Havilland type 5000 hydromatic propeller
2 Spinner
3 Starboard engine cowling panels, Merlin 73 engine
4 Exhaust stubs
5 Starboard oil radiator
6 Coolant radiator
7 Radiator air intake
8 Carburettor air intake and guard
9 Fuselage nose skinning
10 Windscreen de-icing fluid nozzle
11 Instrument panel
12 Parachute stowage
13 Junction box
14 Fire axe
15 SYKO apparatus stowage
16 Nose compartment side windows
17 Portable oxygen bottles
18 Mk XIV bombsight
19 Nose glazing
20 Forward navigation/identification light
21 Temperature probe
22 Windscreen de-icing fluid nozzle
23 Optically flat bomb-aiming window
24 Bombsight mounting
25 Bomb selector switches
26 Camera remote control box
27 Bomb aimer's kneeling cushion
28 Signal pistol cartridge racks
29 Rudder pedals
30 Compass
31 Control linkages
32 Oxygen system economiser units
33 Elevator trim handwheel
34 Port radiator ram air intake
35 Oil and coolant radiators
36 Engine throttle levers
37 Ventral entry hatch
38 Control column handwheel
39 Folding chart table
40 Windscreen panels
41 Trailing aerial winch
42 Cockpit roof escape hatch
43 Seat back armour plate
44 Navigator/bombardier's seat
45 Rearward vision blister fairing
46 Pilot's seat
47 Intercom socket
48 Portable fire extinguisher
49 Cabin pressurisation and heating air ducts
50 Non-return air valve
51 Engine control runs
52 Wingroot rib
53 Centre section fuel tanks (two), capacity 68 Imp gal (309 litres) each; 46 Imp gal (209 litres) port and 47.5 Imp gal (216 litres) starboard with 4000-lb (1814-kg) bombload
54 Wing upper surface attachment joint
55 Centre fuel tank filler cap
56 ARI-5083 receiver
57 IFF transmitter/receiver
58 Signal pistol aperture
59 Cockpit aft glazing
60 Rear pressure bulkhead
61 Starboard inboard fuel tanks, capacity 78 Imp gal (355 litres) inner and 66 Imp gal (298 litres) outer
62 Fuel filler cap
63 Nacelle fairing

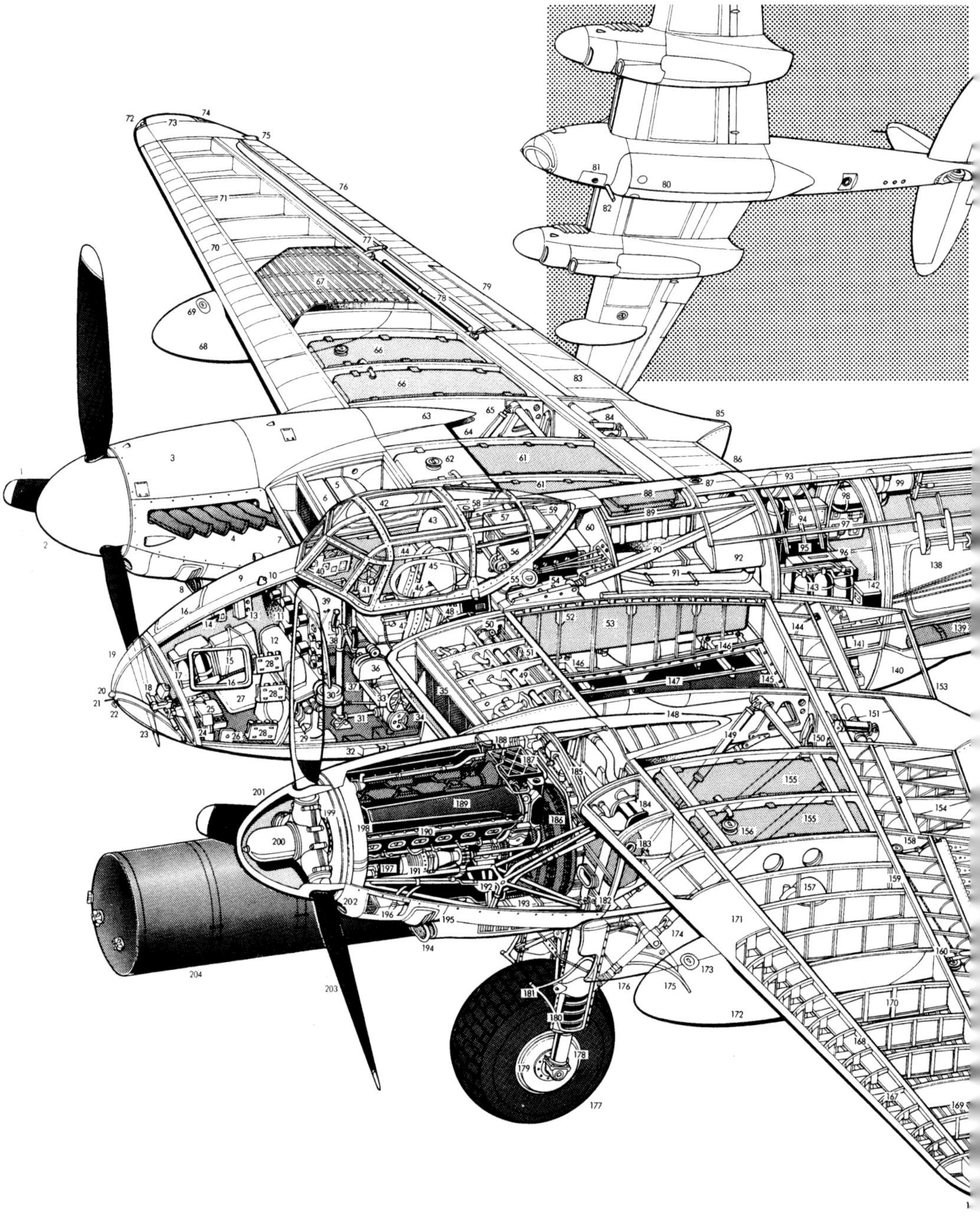

SPECIFICATION

Mosquito B.IV series II	Mosquito PR.Mk 34
Dimensions	**Dimensions**
Length: 40 ft 9½ in (12.43 m) **Height:** 15 ft 3 in (4.65 m) **Wingspan:** 54 ft 2 in (16.51 m) **Wing area:** 454 sq ft (42.18 m^2)	**Length:** 41 ft 6 in (12.65 m) **Height:** 15 ft 3 in (4.65 m) **Wingspan:** 54 ft 2 in (16.51 m) **Wing area:** 454 sq ft (42.18 m^2)
Powerplant	**Powerplant**
Two 1,230-hp (918-kW) Rolls-Royce Merlin 21 inline piston engines	Two 1,710-hp (1276-kW) Rolls-Royce Merlin 113/114 inline piston engines
Weights	**Weights**
Empty: 13,100 lb (5942 kg) **Maximum take-off:** 22,380 lb (10152 kg)	**Empty:** 16,631 lb (7544 kg) **Maximum take-off:** 25,500 lb (11567 kg)
Performance	**Performance**
Maximum speed: 380 mph (612 km/h) at 21,000 ft (6400 m) **Cruising speed:** 265 mph (426 km/h) **Initial climb rate:** 2,500 ft (762 m) per minute **Service ceiling:** 34,000 ft (10363 m) **Range:** 2,040 miles (3283 km)	**Maximum speed:** 425 mph (684 km/h) at 30,500 ft (9295 m) **Cruising speed:** 300 mph (483 km/h) **Initial climb rate:** 2,000 ft (609 m) per minute **Service ceiling:** 43,000 ft (13106 m) **Range:** 3,340 miles (5375 km)
Armament	**Armament**
Normal internal bombload: 2,000 lb (907 kg)	None

Coastal Command wreaked havoc on German shipping throughout the war, attacking vessels in the open sea and in their docks. As many as 34 Mosquitoes would attack a single target. A Mosquito FB.Mk VI of No. 143 Sqn is illustrated, unleashing a cannon and rocket attack on shipping in Sande Fjord, Norway, on 4 April 1945. During this strike, five enemy ships were left burning.

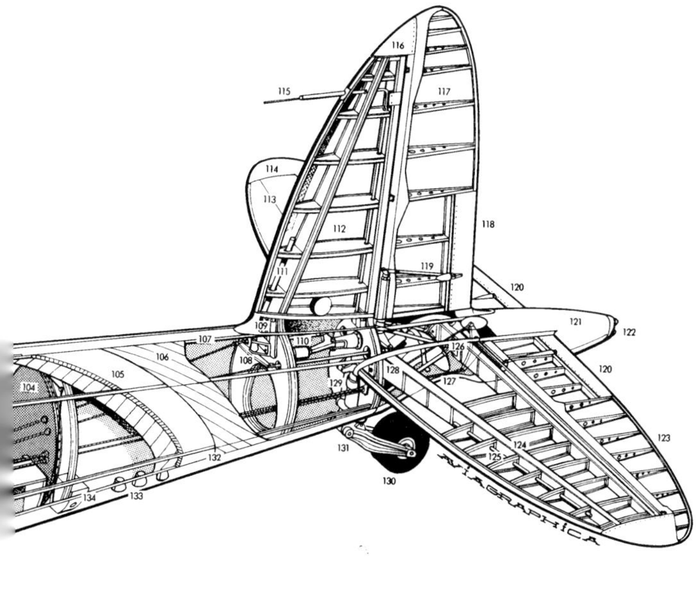

64 Starboard main undercarriage bay
65 Hydraulic retraction jack
66 Outboard fuel tanks, capacity 34 Imp gal (155 litres) inner and 24 Imp gal (109 litres) outer
67 Wing stringers
68 Starboard auxiliary fuel tank, capacity 50 Imp gal (227 litres)
69 Fuel filler cap
70 Plywood leading-edge skinning
71 Wing top skin panelling, double plywood sandwich construction
72 Starboard navigation light
73 Wingtip fairing
74 Formation light
75 Resin light
76 Starboard aileron
77 Aileron hinge control
78 Mass balance weights
79 Aileron tab
80 Underside view showing bulged (increased volume) bomb-bay doors
81 Ventral entry hatch with drift sight aperture
82 Trailing aerial fairing
83 Starboard outer plain flap segment
84 Flap hydraulic jack
85 Nacelle tail fairing
86 Flap inboard segment
87 Oil filler cap
88 Dinghy access panel
89 Two-man dinghy stowage compartment
90 Wing fixing bearer
91 Rear fuselage equipment heater air ducting
92 Long-range oil tank, capacity 10 Imp gal (46 litres)**93** Hydraulic reservoir
94 TR1143 transmitter/ receiver
95 Mk XIV bomb sight computer
96 Batteries
97 Hydraulic and pneumatic systems servicing panel
98 Pneumatic system air bottle
99 De-icing fluid reservoir
100 Picketing equipment stowage
101 Camera motor
102 TR1143 aerial
103 Fuselage stringers, between inner and outer skin laminations
104 Heat-conserving canvas bulkhead cover
105 Fuselage half shell sandwich skin construction (plywood/ balsa/plywood)
106 Diagonal graining pattern
107 Centreline fuselage half shell joint strip
108 Rudder control linkage
109 Fin attachment bulkhead
110 Rudder mass balance weight
111 Ferrite aerial rod
112 Tailfin construction
113 Starboard tailplane
114 Elevator horn balance
115 Pitot tube
116 Rudder horn balance
117 Fabric-covered rudder construction
118 Rudder tab
119 Tab operating rod
120 Elevator tab
121 Tailcone
122 Tail navigation lights
123 Fabric-covered elevator construction
124 Tailplane construction
125 Ferrite aerial rod
126 Elevator operating linkage
127 Tailwheel housing
128 Tailplane spar attachment joint
129 Tailwheel leg strut
130 Retracting tailwheel
131 Levered suspension tailwheel forks
132 Fuselage skin fabric covering
133 Identification code lights, white, amber and green
134 Beam approach aerial
135 Camera mounting
136 F.24 camera
137 Tailplane control cables
138 Rear fuselage entry hatch
139 Crew equipment stowage bag
140 Bulged bomb bay tail fairing
141 Bomb door hydraulic jacks
142 Beam approach receiver
143 Oxygen bottles
144 Flap shroud ribs
145 Inboard fuel tank bay ventral access panel
146 Bomb carriers
147 500-lb (227-kg) short-finned HE bombs (four)
148 Port engine nacelle top fairing
149 Main undercarriage hydraulic retraction jack
150 Undercarriage leg rear strut mounting
151 Flap hydraulic jack
152 Nacelle tail fairing
153 Short plain flap segments
154 All-wooden flap construction
155 Port outer fuel tanks
156 Fuel filler cap
157 Retractable landing lamp
158 Aileron tab control linkage
159 Rear spar
160 Aileron hinge control
161 Aileron tab
162 Aluminium aileron construction
163 Resin lamp
164 Port formation lamp
165 Detachable wingtip fairing
166 Port navigation light
167 Leading-edge nose ribs
168 Front spar, box beam construction
169 Wing lower surface single skin/stringer panel
170 Wingrib construction
171 Plywood leading-edge skinning, fabric-covered
172 Port auxiliary fuel tank, capacity 50 Imp gal (227 litres)
173 Fuel filler cap
174 Main undercarriage rear strut
175 Mudguard
176 Mainwheel doors
177 Port mainwheel
178 Mainwheel leg strut
179 Pneumatic brake disc
180 Rubber compression block shock absorber
181 Spring-loaded door guides
182 Main undercarriage pivot fixing
183 Engine oil tank, capacity 16 Imp gal (73 litres)
184 Cabin heater
185 Fireproof bulkhead
186 Two stage supercharger
187 Intercooler
188 Heywood compressor
189 Rolls-Royce Merlin 72 liquid cooled 12-cylinder Vee engine
190 Exhaust ports
191 Alternator
192 Engine bearers
193 Carburettor air intake duct
194 Intake guard
195 Intercooler radiator exhaust
196 Intercooler radiator
197 Engine mounting block
198 Coolant header tank
199 Spinner armoured backplate
200 Propeller hub pitch change mechanism
201 Spinner
202 Intercooler radiator intake
203 Port three-bladed de Havilland hydromatic propeller
204 4000-lb (1814 kg) HC bomb

Mosquito PR.Mk 34A

The ultimate photo-reconnaissance (PR) version of the Mosquito developed during World War II was the PR.Mk 34/34A. Developed specifically for Far East service with the South East Asia Command, the size of the wing tanks was doubled, and a large overload fuel tank was installed in the bomb bay, which allowed for a range of over 3,500 miles (5632 km). Powered by a Merlin 113 and 114 and equipped with two F.52 vertical cameras and one F.24 oblique camera, the first production PR.Mk 34 flew on 4 December 1944 and 50 were built at Luton by Percival Aircraft Ltd, before contracts were cancelled at the end of the war. Mosquito PR.Mk 34/34As saw post-war service with the RAF – this aircraft served with No. 81 Squadron, operating alongside photo-reconnaissance Spitfires. The squadron was the RAF's Far East Air Force's reconnaissance unit for many years. The Mosquitoes saw action during the Malayan campaign in 1949 and undertook significant mapping surveys of Malaya, Java and Thailand. No. 81 Squadron had the distinction of flying the RAF's last operational Mosquito sortie, with PR.Mk 34A serial RG314, in December 1955. Prior to that it had also flown the last Spitfire mission, and subsequently it would be responsible for the last operational flight of an RAF Meteor.

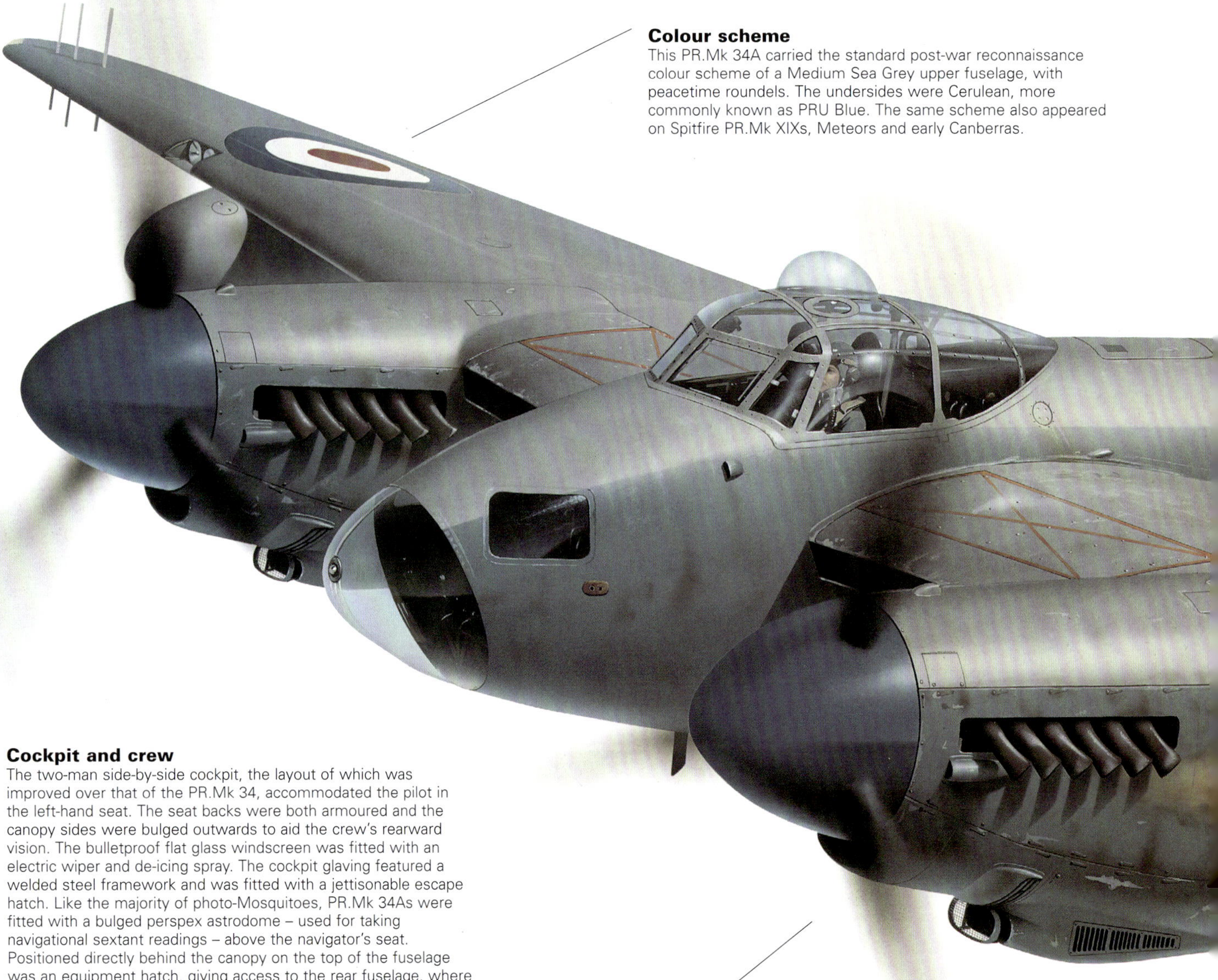

Colour scheme
This PR.Mk 34A carried the standard post-war reconnaissance colour scheme of a Medium Sea Grey upper fuselage, with peacetime roundels. The undersides were Cerulean, more commonly known as PRU Blue. The same scheme also appeared on Spitfire PR.Mk XIXs, Meteors and early Canberras.

Cockpit and crew
The two-man side-by-side cockpit, the layout of which was improved over that of the PR.Mk 34, accommodated the pilot in the left-hand seat. The seat backs were both armoured and the canopy sides were bulged outwards to aid the crew's rearward vision. The bulletproof flat glass windscreen was fitted with an electric wiper and de-icing spray. The cockpit glaving featured a welded steel framework and was fitted with a jettisonable escape hatch. Like the majority of photo-Mosquitoes, PR.Mk 34As were fitted with a bulged perspex astrodome – used for taking navigational sextant readings – above the navigator's seat. Positioned directly behind the canopy on the top of the fuselage was an equipment hatch, giving access to the rear fuselage, where radio equipment was housed. A two-man dinghy was also carried.

Engines and intakes
The PR.Mk 34A differed primarily from the PR.Mk 34 in its engines. The conversion work was undertaken by Marshalls of Cambridge and the aircraft was fitted with Rolls-Royce Merlin 114A two-stage supercharged piston engines, mounted on steel-tube frames attached to the main spar. The PR.Mk 34 was fitted with a Merlin 113 on the starboard side, with a Merlin 114 opposite, the latter driving the cabin supercharger. Each engine was protected by an automatic Graviner fire extinguisher, which could also be operated from the cockpit. The intakes beneath the engines were for the carburettor and were covered with an anti-ice guard. The wingroots housed the intakes for the engine oil and coolant radiators. Coolant temperature was maintained by electro-pneumatic ram-controlled flaps in the radiator duct exits.

Fuselage and fuel load

The Mosquito's unique wooden fuselage had an oval tapering cross-section and was made of balsa, sandwiched between plywood sheeting. This structure was braced internally with several wooden bulkheads. At first, the Air Ministry was unenthusiastic about the de Havilland Company's proposal for such an aircraft and refused to believe that it could be of any value. By the end of the war, 7,781 Mosquitoes of 43 different marks had been built. The Mosquito PR.Mk 34 was fitted with a bulged bomb bay which allowed space for extra fuel tankage. While the bomb bay was still potentially able to accommodate a 4,000-lb (1814-kg) bomb, such a weapon was never used by photo Mosquitoes. A total of 1,524 US gal (5769 litres) of fuel was carried, making the PR.Mk 34/34A the heaviest of all the Mosquitoes, with a loaded weight of 25,500 lb (11567 kg).

Reconnaissance missions

For normal vertical photography, the standard bomb sight was used to align the cameras and was operated by the observer, kneeling in the nose. When the oblique cameras were needed, it was the pilot's responsibility to sight the cameras, so calibration marks were provided on the port-side panels of the cockpit and along the wing.

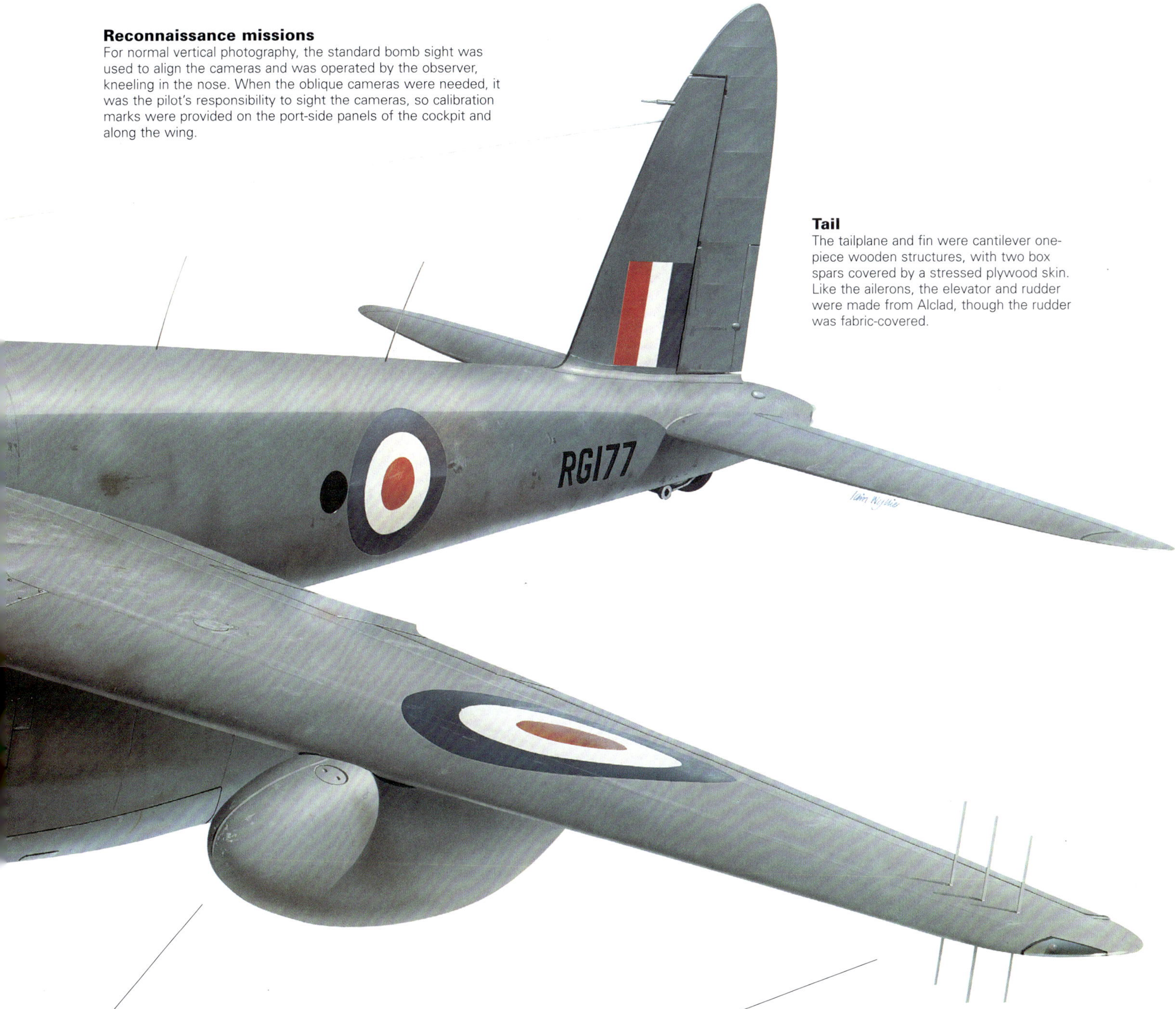

Tail

The tailplane and fin were cantilever one-piece wooden structures, with two box spars covered by a stressed plywood skin. Like the ailerons, the elevator and rudder were made from Alclad, though the rudder was fabric-covered.

Fuel tanks and undercarriage

The PR.Mk 34A was fitted with a pair of underwing fuel tanks which were significantly larger than those fitted to any other version. The two slip tanks housed an extra 400 US gal (1818 litres). Quite often, these would not be carried operationally due to the weight penalties and resultant drag, but they were extremely useful for long-range ferry flights. The main landing gear comprised two interchangeable single wheel units which retracted backwards into their nacelles under the engines. Two Dunlop pneumatic brakes were used per wheel and the shock absorbers were of the rubber-block compression type. An armoured oil tank was also located in each wheel bay.

Wing and antennas

The slender one-piece cantilever wing, with an aspect ratio of 7:1, had a wooden-ply skin stretched over two main spars with inter-connecting spruce stringers. The upper skin was double the thickness of the lower, and was exceptionally strong. Ten self-sealing fuel tanks were housed within. The ailerons were made from aluminium and the leading edge, while still of wooden construction, was covered in fabric. The PR.Mk 34 and PR.Mk 34A had six aerials on each wingtip, dedicated to the Identification Friend or Foe (IFF) transmitter/receiver.

Douglas A-26 Invader

USAF 64-17640 was the first of On Mark Engineering's 40 'production' B-26K Counter Invaders. The aircraft has a solid 'gun nose', eight underwing hardpoints and wingtip fuel tanks.

A-26 Invader

Cutaway key

1 Starboard wing tip
2 Starboard navigation light
3 Water tank
4 Water tank filler cap
5 Aileron hinge control
6 Starboard aileron
7 Aileron tab
8 Landing and taxiing light
9 Control cables
10 Bombardier nose configuration, A-26C
11 Optically flat bomb sight window
12 Bomb bay doors
13 Ventral periscope gunsight
14 Ventral turret
15 Starboard outboard flap
16 Wing access panels
17 Chordwise stiffeners
18 Double slotted flap segments
19 Oil cooler radiator
20 Cooler intake ducting
21 Ram air intake to oil cooler
22 Nacelle fuel tank, capacity 300 US gal (1136 litres)
23 Wing inboard fuel tank, capacity 100 US gal (379 litres)
24 Control runs
25 Oil tank filler
26 Oil tank
27 Carburettor intake ducting
28 Exhaust stubs
29 Cowling air flaps
30 Pratt & Whitney R-2800-27 Double Wasp, two-row 18-cylinder radial engine
31 Carburettor ram air intake
32 Propeller reduction gearbox
33 Propeller hub mechanism
34 Three-bladed propeller
35 Detachable engine cowlings
36 General purpose nose configuration, A-26B
37 Machine-gun barrels
38 Four 0.5-in (12.7-mm) machine-guns, starboard side
39 Spent cartridge case chutes
40 Gun bay bracing strut
41 Two 0.5-in (12.7-mm) machine-guns, port side
42 Ammunition feed chutes
43 Ammunition boxes
44 Pitot tube
45 Nosewheel torque scissors
46 Rearward retracting nosewheel
47 Shock absorber leg strut
48 Nosewheel doors
49 Nosewheel bay/flight deck floor support construction
50 Rudder pedals
51 Interchangeable nose joint bulkhead
52 Autopilot controls
53 Back of instrument panel
54 Fixed foresight
55 Windscreen panels
56 Instrument panel shroud
57 Reflector sight
58 Clear vision panel
59 Control column
60 Pilot's seat
61 Pilot's side window panel/entry hatch
62 Bomb release controls
63 Bombardier/navigator 's seat
64 Canopy hatch handles
65 Bombardier/navigator 's side canopy/entry hatch
66 Oxygen regulator
67 Radio racks
68 Radio receivers and transmitters
69 Bomb-bay armoured roof panel
70 Wing root fillet
71 Armoured wing spar bulkhead
72 Hydraulic accumulators
73 Air filter
74 De-icing valve
75 Aerial mast
76 Double slotted flap inboard section
77 Wing de-icing fluid reservoir
78 De-icing fluid pump
79 Starboard bomb rack, five 100-lb (45-kg) HE bombs
80 Port bomb rack, five 100-lb (45-kg) HE bombs
81 Bomb launcher rails
82 Rear wing spar bulkhead
83 Turret drive motor
84 Upper remotely controlled gun turret
85 Two 0.5-in (12.6-mm) machine-guns
86 Turret mechanism
87 Ammunition boxes
88 Port aft bomb rack, three 100-lb (45-kg) HE bombs
89 Inboard double slotted flap
90 Gunner's bomb bay entry hatch

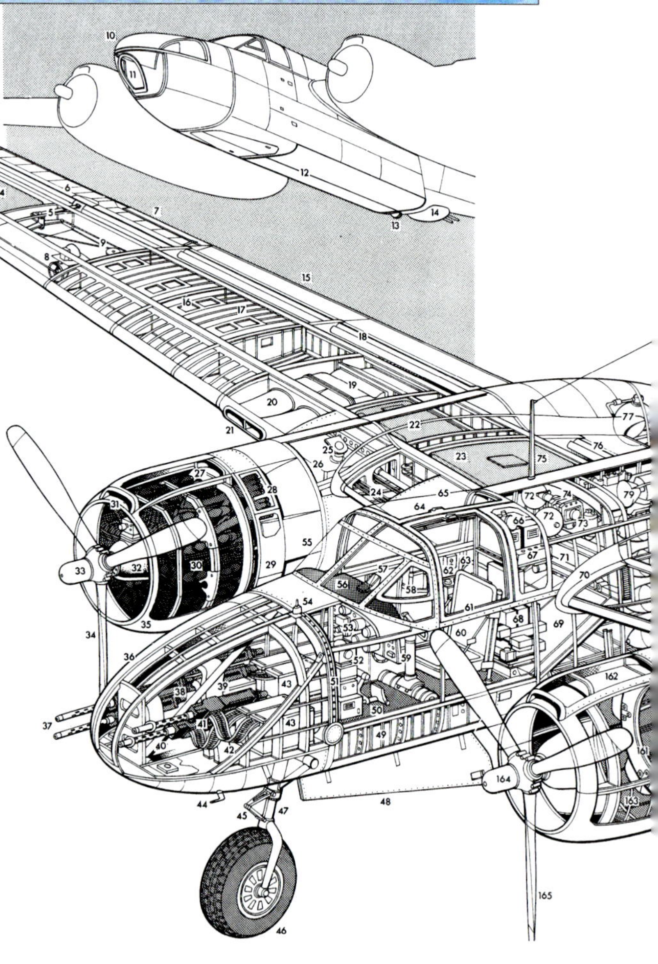

The Invader was first blooded during 1944, the 386th Bomb Group operating A-26Cs (nearest the camera, with gun packs beneath its wings) and A-26Bs against targets on the Continent. 'RU'-coded aircraft belonged to the 554th BS.

SPECIFICATION

Douglas A-26B-15 Invader

Type
Three-seat light attack bomber

Powerplant
Two Pratt & Whitney R-2800-27 or -79 radial piston engines each rated at 2,000 hp (1491 kW)

Performance
Maximum speed 355 mph (571 km/h) at 15,000 ft (4570 m); cruising speed 284 mph (457 km/h) at optimum altitude; climb to 10,000 ft (3050 m) in 8 minutes 6 seconds; service ceiling 22,100 ft (6735 m); range 1,400 miles (2253 km) with standard fuel and warload

Weights
Empty 22,370 lb (10147 kg); maximum take-off 35,000 lb (15876 kg)

Dimensions
Wingspan 70 ft (21.34 m); length 50 ft (15.24 m); height 18 ft 6 in (5.64 m); wing area 540.00 sq ft (50.17 m^2)

Armament
Six 0.5-in (12.7-mm) Browning M2 fixed forward-firing machine-guns in the forward fuselage, two 0.5-in (12.7-mm) Browning M2 trainable machine-guns in the dorsal barbette that could be locked to fire directly forward under pilot control, two 0.5-in (12.7-mm) Browning M2 trainable rearward-firing machine-guns in the optional ventral barbette, and provision for eight 0.5-in (12.7-mm) Browning M2 fixed forward-firing machine-guns installed in four two-gun packs under the outboard wing panels, plus up to 6,000 lb (2722 kg) of disposable stores carried in two lower-fuselage weapons bays and on four underwing hardpoints

91 Oxygen cylinders
92 Life raft
93 Gunner's canopy cover
94 Ditching hatch
95 Upper periscope sight
96 Periscope eyepiece
97 Turret controls
98 Oxygen bottles
99 Gunner's armoured bulkhead
100 Ventral turret ammunition
101 Cabin heater
102 D/F loop antenna fairing
103 Fin root fillet
104 Tailplane control cables
105 Cable pulleys
106 Fin rib construction
107 Starboard tailplane
108 Starboard elevator
109 Fin leading edge
110 Aerial cables
111 Fin tip fairing
112 Fabric covered rudder construction
113 Rudder tab
114 Trim tab control
115 Rudder hinge post
116 Tail navigation lights
117 Elevator tab
118 Port elevator
119 Port tailplane construction

France made good use of the B-26 in colonial wars in Indo-China and Algeria. The firepower housed in the nose of a B-26B – eight 0.5-in (12.7-mm) machine-guns – was ideal for ground strafing.

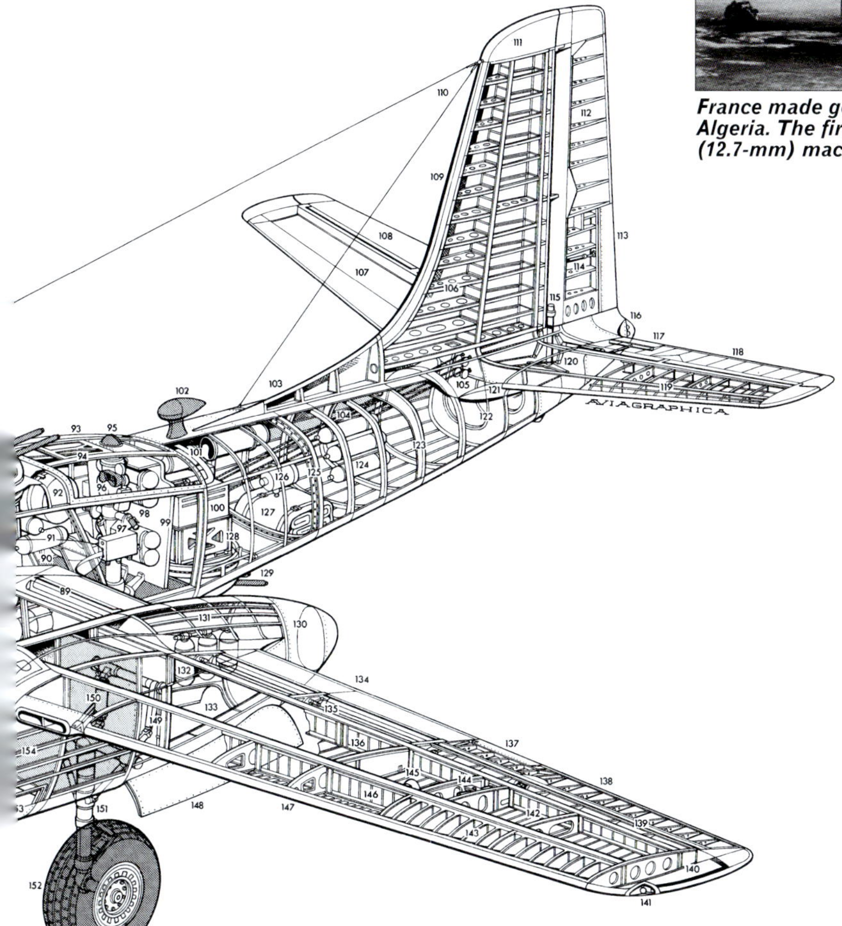

120 Elevator control horns
121 Tailplane root fillet
122 Fin/tailplane fixing frame
123 Rear fuselage construction
124 Oxygen bottles
125 Rear fuselage construction joint bulkhead
126 Turret control amplifier
127 Turret covers
128 Ventral turret control mechanism
129 Two 0.5-in (12.7-mm) machine-guns
130 Port nacelle tailcone
131 Aft nacelle construction
132 Engine fire extinguishers
133 Main undercarriage wheel well
134 Outboard double slotted flaps
135 Flap hinge links
136 Wing rear spar
137 Aileron tab
138 Port aileron
139 Fabric covered aileron construction
140 Port wing tip
141 Port navigation light
142 Wing rib construction
143 Leading edge stiffeners
144 Aileron hinge control
145 Landing and taxiing lamp housing
146 Wing front spar
147 Fluid de-iced leading edge
148 Mainwheel doors
149 Main undercarriage door link mechanism
150 Retraction jack
151 Main undercarriage leg
152 Rearward retracting mainwheel
153 Access panel
154 Nacelle fuel tank, capacity 300 US gal (1136 litres)
155 Oil cooler ram air intake
156 Oil tank filler cap
157 Engine compartment bulkhead/firewall
158 Engine mounting struts
159 Exhaust ducts
160 Cowling cooling air flaps
161 Engine mounting bulkhead
162 Carburettor intake ducting
163 Cowling construction
164 Propeller hub mechanism
165 Three bladed propeller

Douglas DB-7/A-20/P-70 Havoc/Boston

After France's capitulation in 1940, large numbers of DB-7s, -7As and -73s were delivered to the RAF. AL399 is one of the latter, so-designated to avoid confusion with similar DB-7Bs ordered directly by the British. Both the DB-7B and DB-73 were known locally as the Boston Mk III and were the first aircraft in the A-20 family to see service in their intended role.

Boston Mk III

Cutaway key

1 Starboard fabric covered elevator
2 Starboard tailplane
3 Elevator tab
4 Tail navigation and signal lights
5 Tailcone
6 Ruddertab
7 Fabric-covered rudder construction
8 Rudder hinges
9 Pitot tube
10 Fin tip fairing
11 Aerial cable
12 Port elevator
13 Port tailplane
14 Fin leading edge
15 Tail fin construction
16 Elevator hinge control
17 Rudder hinge control
18 Fin attachment joints
19 Tailplane stub attachment
20 Tailplane root fillet
21 Tail bumper
22 Tailcone construction
23 Rear fuselage/tailcone joint frame
24 Fin root fillet
25 Flare launcher tube
26 Reconnaissance flares
27 Ventral hatch cover, open
28 Rear gunner's side window
29 Reconnaissance camera
30 Vickers 0.303-in (7.7-mm) ventral machine-gun
31 Spare ammunition containers
32 Map case
33 Upper identification light
34 Dorsal gun stowage doors
35 Dorsal gun mounting ring
36 Twin Browning 0.303-in (7.7-mm) machine-guns
37 Armour plated screen
38 Rear gunner's cockpit enclosure
39 Gunner's seat
40 Rear emergency control column
41 Trailing aerial reel
42 Wing root trailing-edge fillet
43 Starboard inboard flap
44 Rear spar attachment joint
45 Radio racks
46 Rear gunner's canopy cover, open position
47 Radio receiver
48 Cabin heater pack
49 Propeller de-icing fluid tank
50 D/F loop aerial
51 Aerial mast
52 Radio transmitters
53 Main spar attachment joint
54 Inboard wing panel construction
55 Main undercarriage wheel well housing
56 Hydraulic flap jack
57 Nacelle tail fairing
58 Outer flap construction
59 Main spar
60 Outer wing panel attachment point
61 Wing ribs
62 Aileron tab
63 Fabric covered aerial construction
64 Formation light
65 Starboard navigation light
66 Leading-edge nose ribs
67 Wing stringer construction
68 Main wheel doors
69 Starboard mainwheel
70 Undercarriage leg strut
71 Mainwheel pivot mounting struts
72 Hydraulic retraction jack
73 Engine exhaust
74 Sloping fireproof bulkhead
75 Engine bearer struts
76 Cooling air exit flaps
77 Exhaust collector ring
78 Detachable engine cowlings
79 Hamilton Standard three-bladed, constant speed propeller, 11-ft 3-in (3.43-m) diameter
80 Propeller hub pitch change mechanism
81 Propeller reduction gearbox
82 Wright GR-2600-A5B Cyclone, two-row radial engine
83 Upper cooling air duct
84 Carburettor air intake
85 Starboard oil tank, 19-Imp gal (86-litre) capacity
86 Fuel filler cap
87 Inboard main fuel tank, 110-Imp gal (500-litre) capacity
88 Bomb door central hydraulic jack
89 Wing root fillet
90 Cockpit heater duct

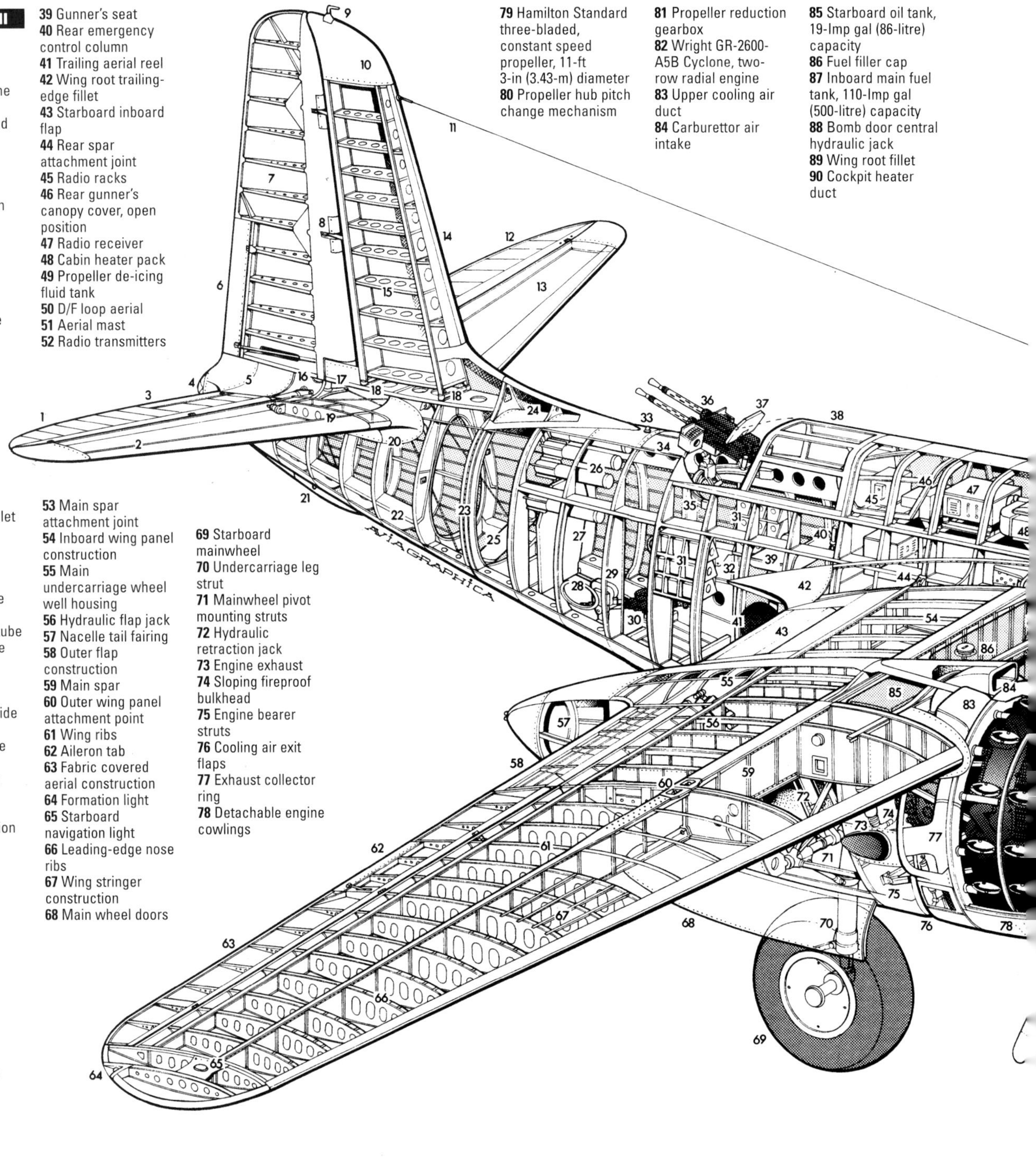

Though the XP-70 was finished in matt black, production P-70 Nighthawks often retained the standard Olive Drab/Neutral Gray scheme applied to A-20s. In general, the P-70 was bereft of markings.

SPECIFICATION

DB-7 (early production)

Dimensions

Length: 46 ft 11¾ in (14.32 m)
Height: 15 ft 10 in (4.83 m)
Wingspan: 61 ft 3 in (18.67 m)
Wing area: 465.00 sq ft (43.20 m²)

Powerplant

Two Pratt & Whitney R-1830-SC3-G 14-cylinder air-cooled radial piston engines each delivering 1,000 hp (746 kW)

Weights

Empty: 11,400 lb (5171 kg)
Gross: 19,040 lb (8636 kg)

Performance

Maximum speed at 13,000 ft (3960 m): 295 mph (475 km/h)
Climb to 12,000 ft (3658 m): 8 minutes **Service ceiling:** 25,800 ft (7835 m)
Combat range: 996 miles (1603 km)

Armament

Four 0.295-in (7.5-mm) fixed forward-firing machine-guns, plus single similar weapons on flexible mount in dorsal and ventral positions; plus up to 1,764 lb (800 kg) of bombs could be carried in the internal bomb bay

A-20G-20 Havoc

Dimensions

Length: 47 ft 11⅛ in (14.63 m)
Height: 17 ft 7 in (5.36 m)
Wingspan: 61 ft 4 in (18.69 m)
Wing area: 465.00 sq ft (43.20 m²)

Powerplant

Two Wright R-2600-23 14-cylinder air-cooled radial piston engines each delivering 1,600 hp (1193 kW)

Weights

Empty: 16,993 lb (7708 kg)
Gross: 24,127 lb (10964 kg)

Performance

Maximum speed at 10,700 ft (3260 m): 317 mph (510 km/h)
Cruising speed: 256 mph (412 km/h)
Climb to 10,000 ft (3048 m): 8 minutes 48 seconds
Service ceiling: 23,700 ft (7225 m)
Combat range: 945 miles (1521 km)

Armament

Six 0.5-in (12.7-mm) Browning M2 fixed forward-firing machine-guns, plus two similar weapons in a power-operated Martin dorsal turret, and one rearward-firing on a flexible mount in the ventral tunnel position; plus up to 4,000 lb (18184 kg) of bombs could be carried in the internal bomb bay, with an extra 2,000 lb (907 kg) on underwing racks

91 Bomb doors
92 Forward pairof 500-lb (227-kg) bombs; maximum bomb load 2,000 lb (907 kg)
93 Lower fuselage box beam construction
94 Bomb carrier
95 Bomb hoist winches
96 Bomb bay top decking
97 Cockpit entry hatch aft extenstion
98 Port inboard main fuel tank, 110-Imp gal (500-litre) capacity

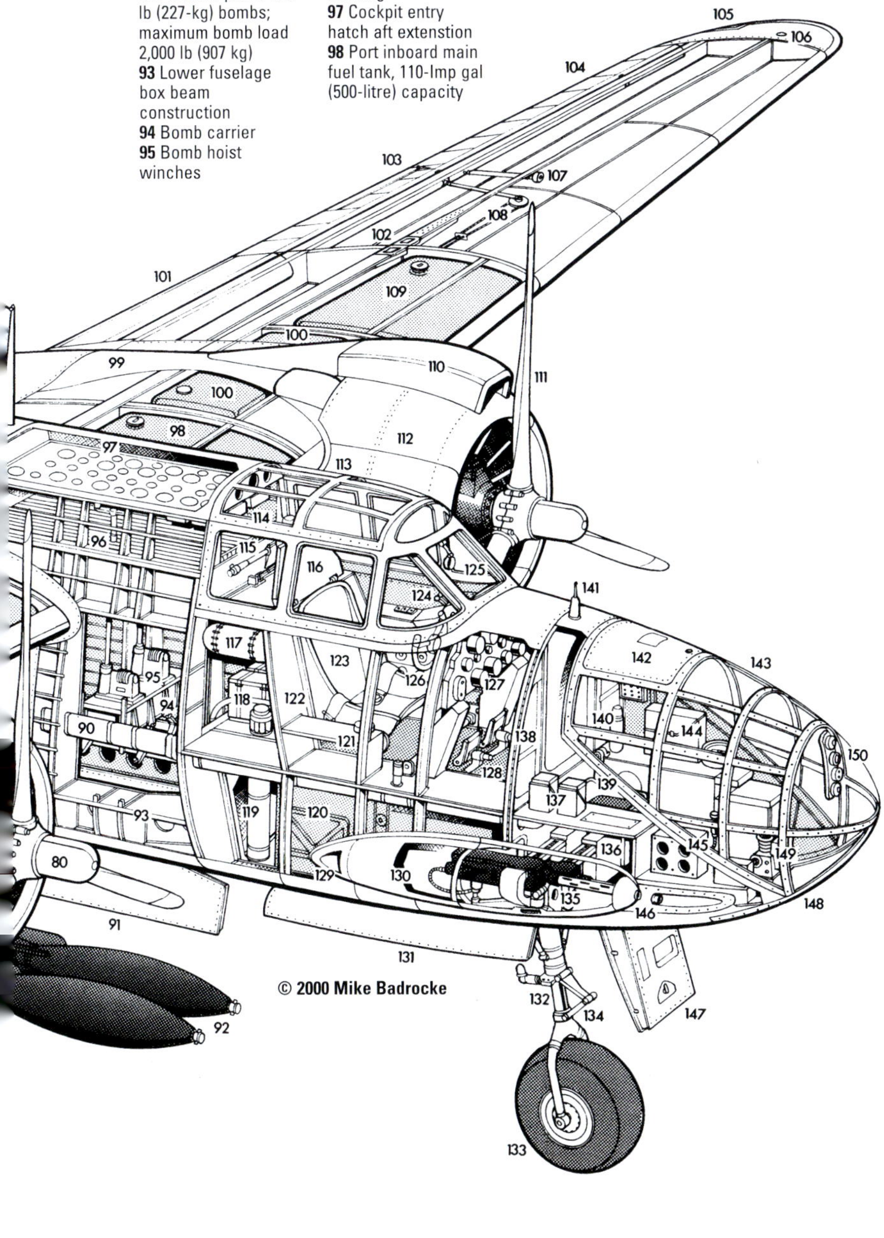

Another new A-20 is run-up prior to a test flight after completion. Douglas built Havocs and Bostons at Long Beach and Santa Monica; small batches were also built by Boeing.

99 Engine nacelle fairing
100 Port oil tank, 19-Imp gal (86-litre) capacity
101 Port outer flap
102 Outer wing panel joint
103 Aileron trim tab
104 Port aileron
105 Formation light
106 Port navigation light
107 Trim tab screw jack
108 Aileron hinge control
109 Port outer auxiliary fuel tank, 51-Imp gal (232-litre) capacity
110 Carburettor intake tropical air filter housing
111 Port propeller
112 Port engine nacelle
113 Cockpit roof entry hatch
114 Emergency equipment packs
115 Crash axe
116 Pilot's folding head armour
117 Hydraulic reservoir
118 Batteries
119 Signal flare chute
120 Nose undercarriage wheel bay
121 Trim tab control handwheels
122 Cockpit sloping bulkhead
123 Pilot's seat
124 Engine throttle and propeller controls
125 Armoured windscreen
126 Control column handwheel
127 Instrument panel
128 Rudder pedals
129 Gun gas exhaust vent
130 Fixed forward gun blister fairing
131 Nosewheel doors
132 Nose undercarriage leg strut
133 Nosewheel
134 Torque scissor links
135 Twin fixed Browning 0.303-in (7.7-mm) machine-guns
136 Ammunition boxes
137 First aid and emergency ration packs
138 Nose compartment joint frame
139 Observer's seat
140 Vacuum flask
141 Pilot's fixed gunsight
142 Observer's ditching hatch
143 Nose compartment glazing
144 Map case
145 Bomb electrical switches and release control
146 Fixed gun muzzles
147 Observer's entry hatch
148 Bomb aiming window
149 Drift sight
150 Observer's instrument panel

A-20G Havoc

A-20G-35-DO 43-10181 *'Joker'* wears the markings of the 647th Bomb Sqn (Light), 410th Bomb Group (Light) around the time of the Allied invasion of Europe in June 1944 – Operation Overlord. The 410th was formed in July 1943 and trained on the A-20 before departing for England, where it joined the 9th Air Force. Entering combat in May 1944, the group's four squadrons of A-20s attacked targets in France in preparation for the invasion, and after D-Day concentrated on lines of communication. The group moved to France in September.

Forward-firing armament

Forward-firing nose-mounted gun armament was a feature of the Havoc's predecessor, the Model 7B, and some form of fixed forward-firing armament was installed in virtually all A-20/Boston variants, with the exception of the F-3A reconnaissance aircraft. French DB-7s and DB-7As carried a pair of 0.295-in (7.5-mm) machine-guns on the fuselage, either side of their nose glazing, while in the DB-7Cs built for the Dutch Koninklijke Marine four 0.303-in (7.7-mm) guns were substituted. Subsequent aircraft in the early batches built for the USAAC (A-20, A-20A, A-20C, A-20E) had four 0.3-in (7.62-mm) guns as standard and it was not until the A-20J and K entered production that 0.5-in (12.7-mm) guns replaced the quartet of '30-calibers'. Nose armament was introduced in USAAF aircraft in the field; A-20As, Bs and Cs were modified with a battery of six 0.5-in (12.7-mm) guns in faired-over glazed noses. The first members of the Havoc/Boston family delivered with factory-installed 'gun noses' were A-20Gs, which were initially fitted with four 20-mm cannon in a solid nose, but were mostly equipped with six 0.5-in (12.7-mm) machine-guns. (The A-20G had no fuselage-mounted forward-firing guns.) The A-20H, which only differed from the A-20G in being powered by different engines, was also equipped with six '50-calibers'. RAF Bostons generally had four 0.303-in (7.7-mm) fuselage-mounted guns; RAF Havoc Mk I night-fighters had an additional four mounted in their noses, while the Havoc Mk II dispensed with the fuselage-mounted guns and boasted 12 nose-mounted 'three-oh-threes'.

Bomb load

An A-20G was able to carry 4,000 lb (1814 kg) of bombs in an internal bay – twice the load of the A-20C; late production aircraft were additionally able to carry a pair of 500-lb (227-kg) bombs on racks beneath each outer wing. Extra armour added further to the variant's overall weight and performance suffered accordingly.

Defensive armament

Dorsal and ventral gun positions were available in most Havocs/Bostons, though the armament fitted varied considerably. French DB-7s were able to carry a single 0.295-in (7.5-mm) machine-gun on a flexible mount in both positions. The DB-7A followed the same pattern, though the original French specifciation called for a single gun to be mounted in the rear of each of the aircraft's engine nacelles, for rearward defence! American aircraft, starting with the A-20, had a pair of flexible 0.3-in (7.62-mm) machine-guns in the open dorsal position and a single 0.3-in (7.62-mm) gun in the ventral position. In the A-20B the twin '30-calibers' were replaced by a single 0.5-in (12.7-mm) gun, this arrangement remaining standard until the 751st A-20G was completed. This and subsequent aircraft had a Martin dorsal turret, with 'twin 50s' installed.

Colours and markings

This A-20G is finished in the Olive Drab upper surfaces that were standard on Ninth Air Force Havocs from their introduction in the ETO. Small patches of disruptive Medium Green were applied to the leading edges of the flying surfaces and the undersides were Neutral Gray. Note the black and white so-called 'invasion stripes' applied for the D-Day landings as an Allied identification feature. The black and white rudder stripes were a 410th Bomb Group marking; the yellow nose cone and propeller hubs signified the 647th Bomb Squadron.

Powerplant

Like the Douglas Model 7B before it, the DB-7 was powered by a pair of Pratt & Whitney R-1830-SC3-G 14-cylinder radials rated at 1,000 hp (746 kW) and equipped with a single-speed supercharger. After 100 of the 270 DB-7s ordered by the French had been completed, a pair of 1,100-hp (820-kW) R-1830-S3C4-Gs, with two-speed superchargers, was substituted. A follow-on order for 100 DB-7As brought a complete change of powerplant, in a further effort to improve performance. The Wright R-2600-A5B 14-cylinder supercharged radials installed were rated at 1,600 hp (1193 kW). The R-2600 remained the powerplant of the US Army's A-20, though in turbocharged form to improve altitude performance. However, the 1,700-hp (1268-kW) R-2600-7 proved troublesome and turbocharging was abandoned. After all, the A-20 was expected to operate at low and medium altitudes, rendering turbocharging unnecessary; all subsequent Havoc variants were powered by supercharged R-2600s. Somewhat ironically, the P-70 interim night-fighters were converted from the original batch of A-20s that should have been powered by turbocharged Wright R-2600-7 engines. Re-engined with R-2600-11s, they were soon found wanting as they were unable to reach the altitudes at which Japanese bombers generally operated.

Douglas SBD Dauntless

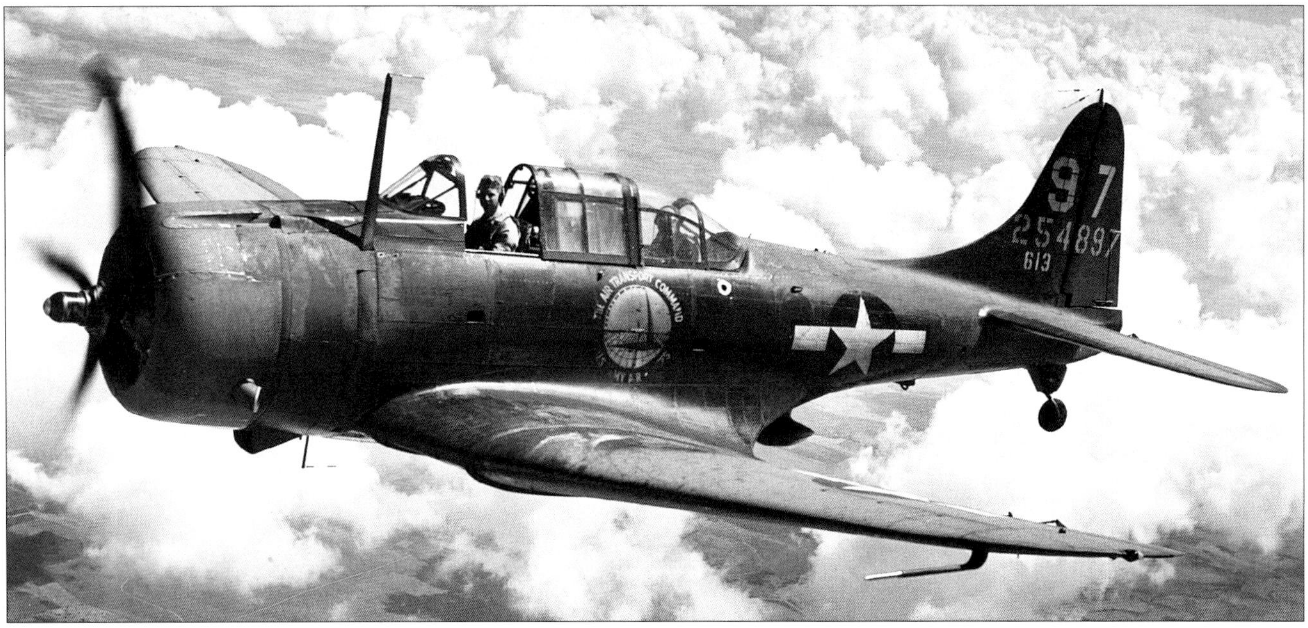

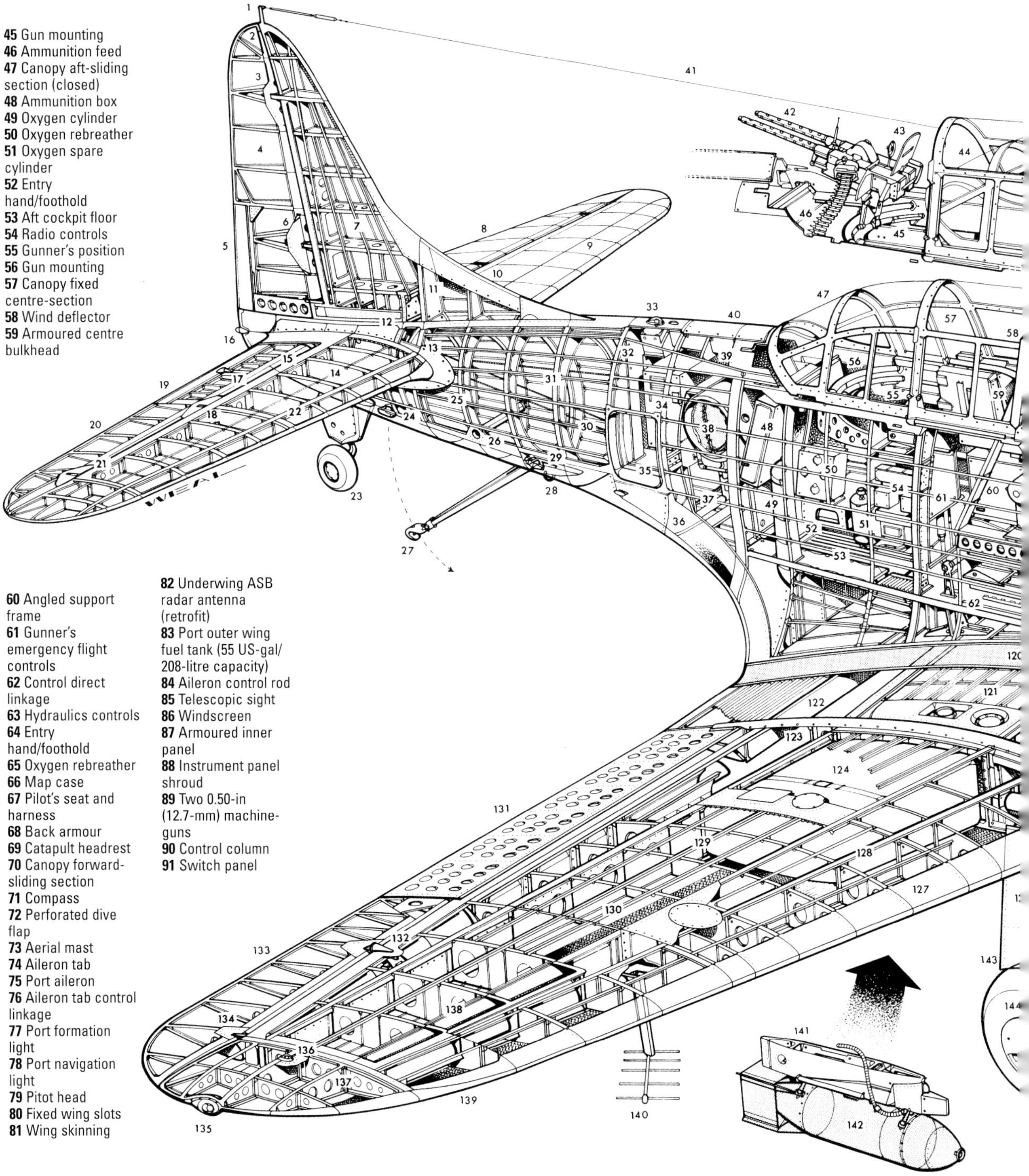

SBD-3 Dauntless

Cutaway key
1 Aerial stub
2 Rudder balance
3 Rudder upper hinge
4 Rudder frame
5 Rudder tab
6 Rudder lower hinge
7 Tailfin structure
8 Port elevator
9 Port tailplane
10 Tailfin root fillet
11 Frame
12 Fuselage frame/ tailfin pick-up
13 Tailplane spar attachment
14 Tailplane structure
15 Elevator torque tube
16 Tail navigation light
17 Elevator tab hinge fairing
18 Elevator hinge
19 Elevator tab
20 Elevator frame
21 Elevator outer hinge
22 Tailplane forward spar
23 Fixed tailwheel (pneumatic tyre on A-24 versions)
24 Arresting hook uplock
25 Fuselage frame
26 Lift point
27 Arresting hook (extended)
28 Tie-down ring
29 Arresting hook pivot
30 Control cables
31 Fuselage structure
32 Bulkhead
33 Section light
34 Radio bay
35 Radio bay access door
36 Wingroot fairing frame
37 Stringers
38 Life-raft cylindrical stowage (access door portside)
39 Dorsal armament stowage
40 Hinged doors
41 Aerial
42 Twin 0.30-in (7.62-mm) Browning machine-guns
43 Gunner's face armour
44 Canopy aft-sliding section (open)
45 Gun mounting
46 Ammunition feed
47 Canopy aft-sliding section (closed)
48 Ammunition box
49 Oxygen cylinder
50 Oxygen rebreather
51 Oxygen spare cylinder
52 Entry hand/foothold
53 Aft cockpit floor
54 Radio controls
55 Gunner's position
56 Gun mounting
57 Canopy fixed centre-section
58 Wind deflector
59 Armoured centre bulkhead
60 Angled support frame
61 Gunner's emergency flight controls
62 Control direct linkage
63 Hydraulics controls
64 Entry hand/foothold
65 Oxygen rebreather
66 Map case
67 Pilot's seat and harness
68 Back armour
69 Catapult headrest
70 Canopy forward-sliding section
71 Compass
72 Perforated dive flap
73 Aerial mast
74 Aileron tab
75 Port aileron
76 Aileron tab control linkage
77 Port formation light
78 Port navigation light
79 Pitot head
80 Fixed wing slots
81 Wing skinning
82 Underwing ASB radar antenna (retrofit)
83 Port outer wing fuel tank (55 US-gal/ 208-litre capacity)
84 Aileron control rod
85 Telescopic sight
86 Windscreen
87 Armoured inner panel
88 Instrument panel shroud
89 Two 0.50-in (12.7-mm) machine-guns
90 Control column
91 Switch panel

SPECIFICATION

SBD-5 Dauntless

Dimensions

Wingspan: 41 ft 7 in (12.67 m)
Length: 33 ft 2 in (10.10 m)
Height: 13 ft 7 in (4.14 m)
Wing area: 325 sq ft (30.19 m²)

Powerplant

One 1,200-hp (895-kW) Wright R-1820-60 Cyclone radial piston engine

Weights

Empty: 6,533 lb (2963 kg)
Loaded: 9,359 lb (4245 kg)
Maximum take-off: 10,700 lb (4854 kg)

Performance

Maximum speed at 10,000 ft (3050 m): 252 mph (406 km/h)
Service ceiling: 25,200 ft (7681 m)
Normal range: 773 miles (1244 km)
Maximum range: 1,370 miles (2205 km)

Armament

Two 0.5-in (12.7-mm) fixed machine-guns in the nose and two 0.3-in (7.62-mm) trainable machine-guns in the rear crewman's position, plus up to 1,600 lb (726 kg) of bombs under the fuselage and 650 lb (295 kg) of bombs under the wings

Above: About 20 Dauntlesses currently exist in museums or private collections. However, only one example is currently airworthy, owned by the Confederate Air Force in Texas. Marked as 2-B-4, the aircraft was built as a USAAF A-24 and then operated by a Mexican photographic service. It was purchased in 1970 and is now painted as a US Navy SBD-3.

Left: The A-24's career was a short one – the USAAF found the Douglas bomber too slow and unpopular with crews and it was quickly relegated to second-line duties in the US. Here, the A-24 found a new lease of life, being used for crew training and ancillary duties. A number of A-24s, or F-24s as they were redesignated, was still in the USAF inventory some two and a half years after the inception of that service in 1947.

Marine Corps Scout Squadron VMS-3 is shown in echelon formation, one favoured for approaching a dive-bombing target. The aircraft are SBD-5s, and the Gull Gray and off-white colour scheme was adopted in early 1944 for the North Atlantic theatre (this unit was based in the Caribbean, however).

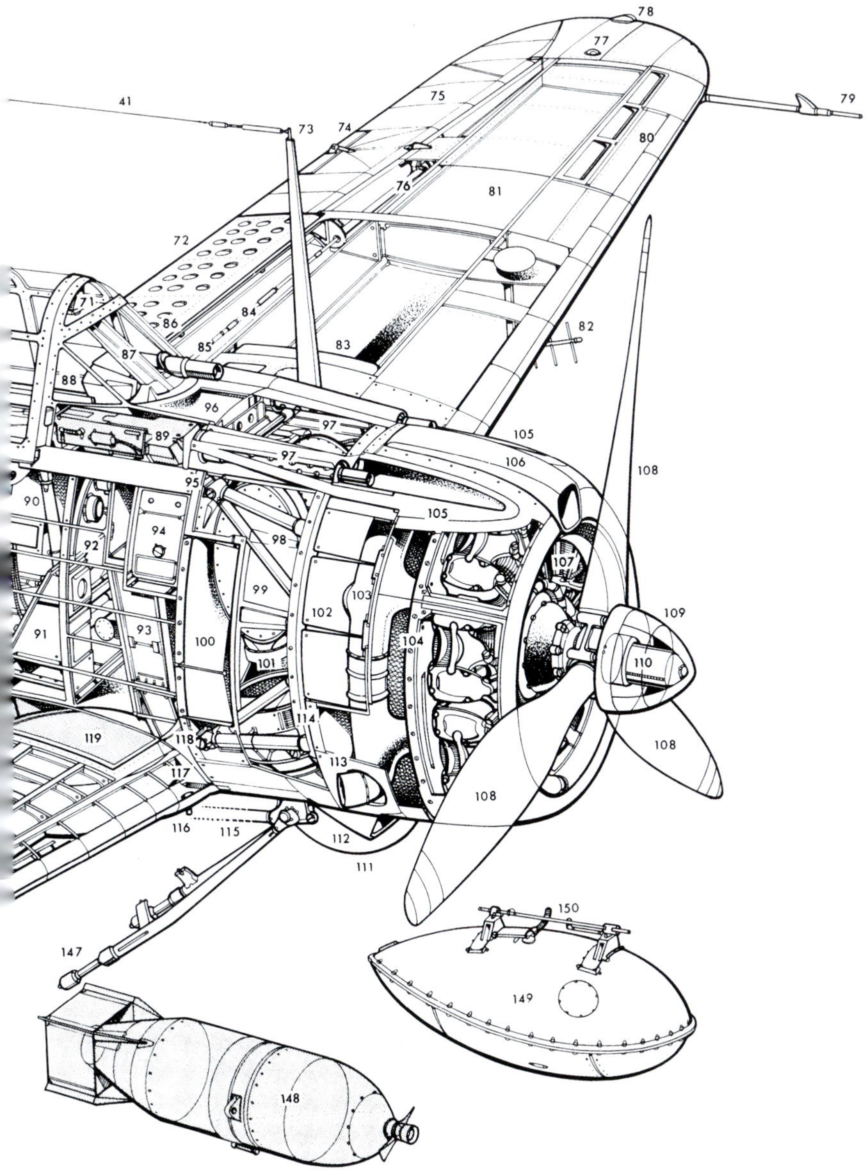

92 Instrument panel
93 Case ejection chute
94 Ammunition box
95 Engine bearer upper attachment
96 Armoured deflection plate
97 Machine-gun barrel shrouds
98 Engine bearers
99 Oil tank
100 Exhaust slot
101 Oil cooler
102 Cooling gills
103 Exhaust manifold
104 Engine cowling ring
105 Machine-gun troughs
106 Carburettor air intake duct
107 Wright R-1820-52 Cyclone radial engine
108 Three-bladed propeller
109 Spinner
110 Propeller hub
111 Port mainwheel
112 Oil cooler intake
113 Exhaust outlet
114 Engine bearers
115 Bomb displacement crutch (in-flight position)
116 Hydraulics vent
117 Case ejection chute outlet
118 Engine bearer lower attachment
119 Starboard mainwheel well
120 Wingroot walkway
121 Starboard inner wing fuel tank (75-US gal/284-litre capacity)
122 Centre-section dive flap (lower)
123 Wing outer-section attachment plate fairing
124 Starboard outer wing fuel tank (55-US gal/208-litre capacity)
125 Mainwheel leg pivot
126 Mainwheel leg door actuation
127 Wing nose ribs
128 Multi-spar wing structure
129 Wing ribs
130 Stiffeners
131 Perforated dive flaps
132 Aileron inner hinge
133 Starboard aileron frame
134 Aileron outer hinge **135** Starboard navigation light
136 Starboard formation light
137 Wingtip structure
138 Fixed wing slots
139 Wing leading edge
140 Underwing radar antenna (retrofit)
141 Underwing stores pylon
142 100-lb (45-kg) bomb
143 Mainwheel leg door
144 Starboard mainwheel
145 Mainwheel axle
146 Mainwheel leg
147 Bomb displacement crutch
148 500-lb (227-kg) bomb
149 Aluminium drop tank (58-US gal/ 220-litre capacity)
150 Underwing shackles/fuel line

SBD-1 Dauntless

Seen during peacetime, this SBD wears the bright colours that typified between-the-wars US Navy and US Marine Corps aircraft. It is the personal aircraft of the commanding officer of VMB-1, the second unit to receive the Dauntless (after VMB-2), and was based at Quantico, Virginia in early 1941 (the unit was renumbered VMSB-132 later that year). It was VMB-2 (later VMSB-232) that received its baptism of fire on the opening day of the war when a number of SBD-2s was surprised during a flight between the USS *Enterprise* and Pearl Harbor, while other SBD-1s of VMB-2 were caught on the ground at Ewa, Hawaii. However, the losses of 7 December 1941 were promptly avenged when Dauntlesses from the USS *Enterprise* contributed to the first sinking of a Japanese submarine.

Powerplant
Power for the SBD-1 came in the form of a Wright R-1820-32 Cyclone. This produced 1,000 hp (746 kW) for take-off, and had a normal rating of 950 hp (709 kW) at 2,300 rpm between sea level and 5,000 ft (1524 m). The large intake on top of the NACA engine cowl fed air to the carburettor.

Bombs
Small pylons underwing could carry a 100-lb (45-kg) bomb each; standard box-fin bombs were the usual ordnance dropped. The main weapon was carried under the centre-section, the maximum weight being 1,600 lb (725 kg). The central bomb was held in a special cradle which swung forward to ensure that the bomb cleared the propeller on release during dive attacks. For level bombing, as here, the cradle was not required.

Pilot

The pilot enjoyed an excellent view from the high-set cockpit, protected to the rear by an armoured backplate, although a bulletproof windscreen was not fitted to this variant. A three-power telescopic sight was provided, which protruded though the windscreen, used for both bomb- and gun-aiming. Immediately ahead of the pilot was a pair of 0.5-in (12.7-mm) Browning machine-guns, each armed with 360 rounds of ammunition, fed from containers behind the fireproof engine bulkhead. The breeches projected into the cockpit, allowing the pilot to clear blockages and recock the guns if there was a stoppage.

Observer/gunner

The observer/gunner sat in the rear cockpit, facing rearwards. The aft section of cockpit slid forwards over him to allow relatively unobstructed firing over the rear hemisphere. A single 0.3-in (7.62-mm) drum-fed machine gun was provided, with 600 rounds. This was stowed in a compartment in the rear fuselage when not in use, covered by folding doors. Two belt-fed guns were fitted to later variants.

Fin

The fin was built integrally with the fuselage and, like the tailplane, employed stressed-skin construction. Both the elevators and rudder were metal-framed with fabric covering. Tabs were built into all the tail control surfaces. Standard radio communications were provided, with a wire aerial running between the fintip and a prominent mast offset to port. The equipment was housed in the rear fuselage, with a rapid-access hatch.

Dive flaps

Much of the trailing edge of the SBD's wing was taken up with perforated split dive flaps. Both upper and lower halves were hinged, the lower operating as conventional flaps during landing and take-off, but both halves operating in unison as dive brakes. To prevent tail buffet, the flaps were perforated with 1.75-in (44.50-mm) holes.

Wing structure

The 'multi-cellular' construction was pioneered by John K. Northrop, the Dauntless's designer. Around two main spars was a Duralumin structure with stressed skin. The centre-section was a rectangular box, while the dihedralled outer panels tapered in both plan and section. Surprisingly, no wing-fold was incorporated. An unusual feature of the SBD was the incorporation of three 'letter-box' slots in the wing, forward of the ailerons, in order to maintain aileron control at slow speeds.

Fairey Swordfish

Swordfish prepare for take-off from the flight deck of HMS Eagle off Mombasa in April 1941. The aircraft are from Nos 813 and 824 Squadrons, which carried out anti-submarine patrols. On 6 June 1941, Swordfish from these squadrons found and sank the U-boat supply ship, Elbe.

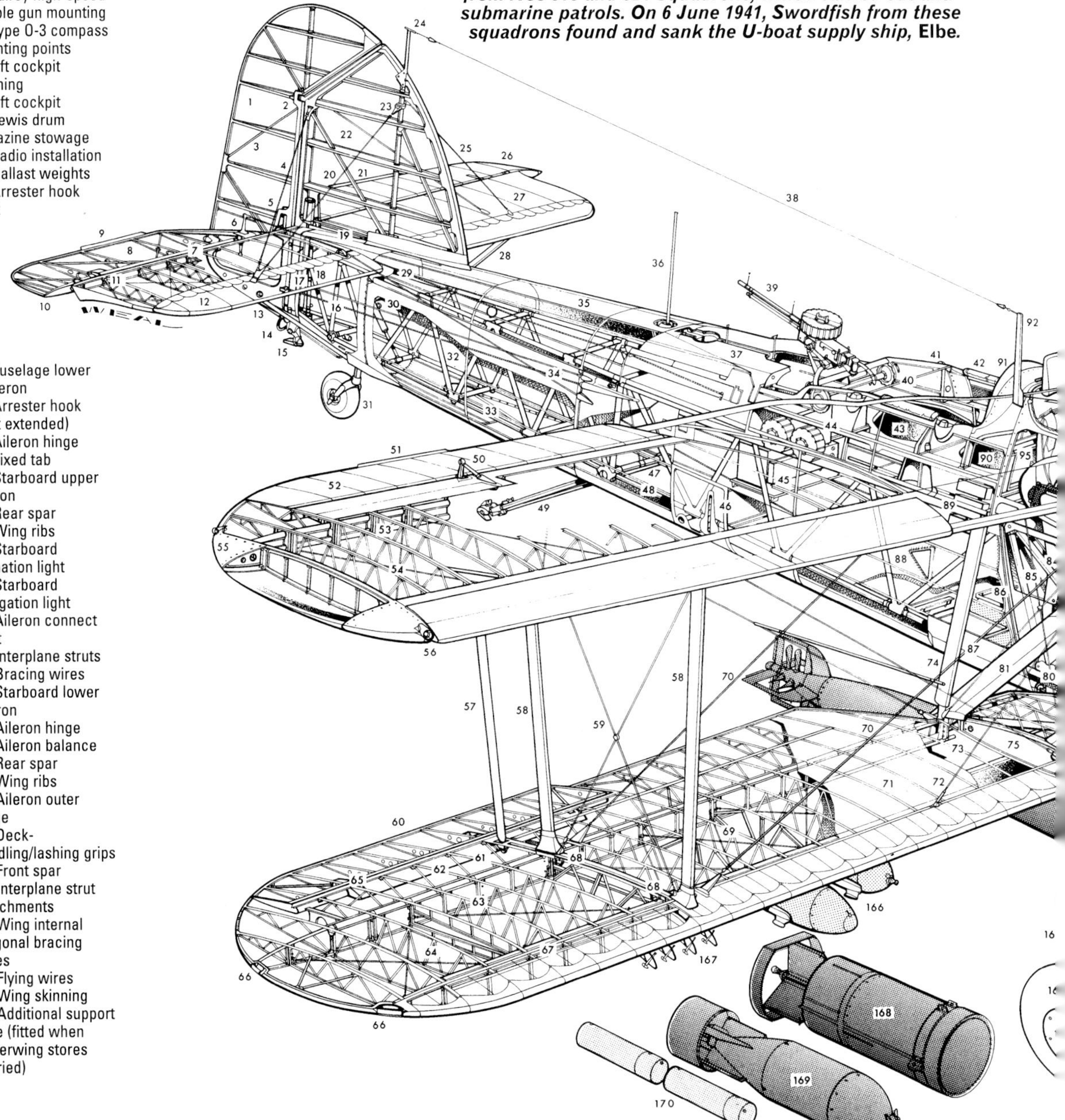

Fairey Swordfish

Cutaway key
1 Rudder structure
2 Rudder upper hinge
3 Diagonal brace
4 External bracing wires
5 Rudder hinge
6 Elevator control horn
7 Tail navigation light
8 Elevator structure
9 Fixed tab
10 Elevator balance
11 Elevator hinge
12 Starboard Tailplane
13 Tailplane struts
14 Lashing down shackle
15 Trestling foot
16 Rear wedge
17 Rudder lower hinge
18 Tailplane adjustment screw
19 Elevator control cable
20 External bracing wires
21 Elevator fixed tab
22 Tailfin structure
23 Bracing wire attachment
24 Aerial stub
25 Bracing wires
26 Port elevator
27 Port Tailplane
28 Tailplane support struts
29 Dinghy external release cord
30 Tailwheel oleo shock absorber
31 Non-retractable Dunlop tailwheel
32 Fuselage framework
33 Arrester hook housing
34 Control cable fairleads
35 Dorsal decking
36 Rod aerial
37 Lewis gun stowage trough
38 Aerial
39 Flexible 0.303-in (7.7-mm) Lewis machine gun
40 Fairey high-speed flexible gun mounting
41 Type O-3 compass mounting points
42 Aft cockpit coaming
43 Aft cockpit
44 Lewis drum magazine stowage
45 Radio installation
46 Ballast weights
47 Arrester hook pivot
48 Fuselage lower longeron
49 Arrester hook (part extended)
50 Aileron hinge
51 Fixed tab
52 Starboard upper aileron
53 Rear spar
54 Wing ribs
55 Starboard formation light
56 Starboard navigation light
57 Aileron connect strut
58 Interplane struts
59 Bracing wires
60 Starboard lower aileron
61 Aileron hinge
62 Aileron balance
63 Rear spar
64 Wing ribs
65 Aileron outer hinge
66 Deck-handling/lashing grips
67 Front spar
68 Interplane strut attachments
69 Wing internal diagonal bracing wires
70 Flying wires
71 Wing skinning
72 Additional support wire (fitted when underwing stores carried)

Seen overflying the newly-commissioned HMS* Ark Royal *in early 1939, this Swordfish Mk I is from No. 820 Squadron. The unit had become the first squadron to deploy aboard the carrier in January of that year.

SPECIFICATION

Type
Two/three-seat carrierborne and land-based torpedo bomber and reconnaissance aircraft

Powerplant
One Bristol Pegasus IIIM radial piston engine rated at 690 hp (515 kW)

Performance
Maximum speed 139 mph (224 km/h) at 4,750 ft (1450 m); cruising speed 128 mph (206 km/h) at 5,000 ft (1525 m); climb to 5,000 ft (1525 m) in 10 minutes 30 seconds; service ceiling 12,400 ft (3780 m); range 1,030 miles (1657 km) with auxiliary fuel and 546 miles (878 km) with a 1,500-lb (680-kg) bombload

Weights
Empty 5,200 lb (2359 kg); maximum take-off 9,250 lb (4196 kg)

Dimensions
Wingspan 45 ft 6 in (13.87 m); length 36 ft 1 in (11.00 m) with the tail down; height 12 ft 101/2 in (3.92 m) with the tail down; wing area 607.00 sq ft (56.39 m2)

Armament
One 0.303-in (7.7-mm) Vickers Mk V fixed forward-firing machine-gun in the starboard side of the forward fuselage and one 0.303-in (7.7-mm) Vickers 'K' or Lewis trainable rearward-firing machine-gun in the rear cockpit, plus up to 1,610 lb (739 kg) of disposable stores carried on one underfuselage and eight underwing hardpoints, and generally comprising one 1,610-lb (739-kg) 18-in (457-mm) torpedo or one 1,500-lb (680-kg) mine carried under the fuselage, or up to 1,500 lb (680 kg) of weapons, including 500-lb (227-kg), 250-lb (113-kg) and 20-lb (9-kg) bombs, carried under the fuselage and lower wing

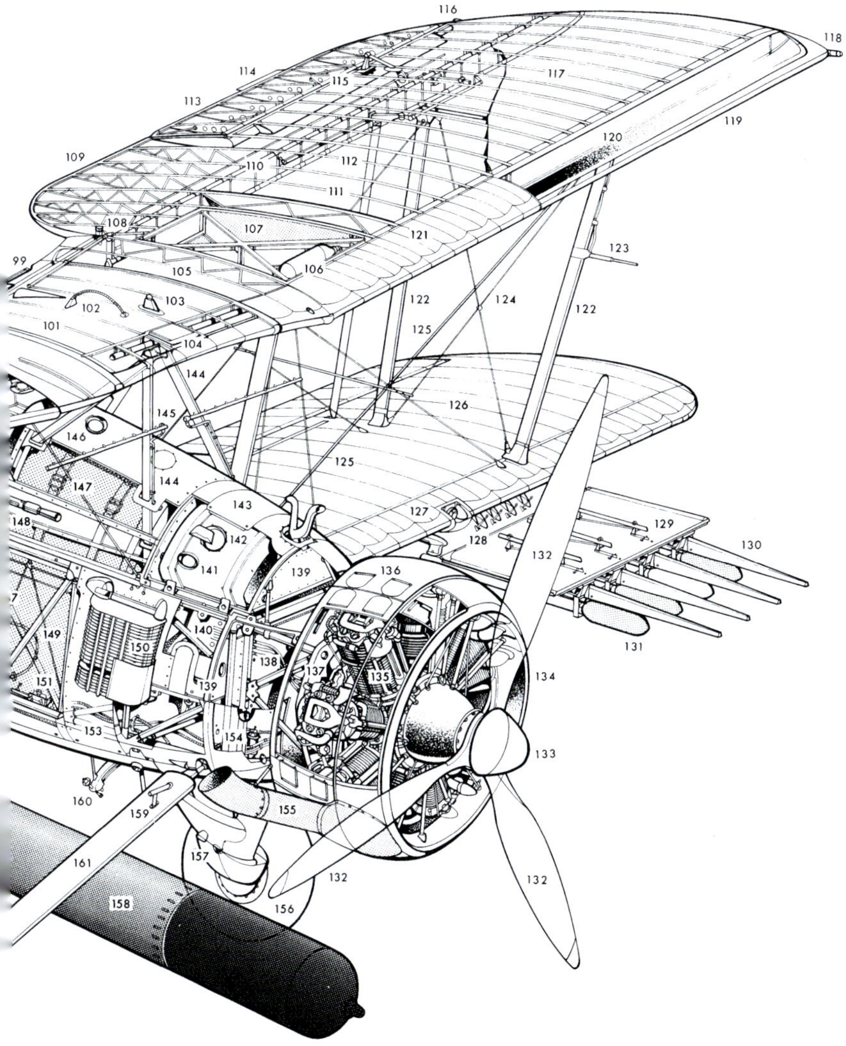

73 Wing fold hinge
74 Inboard interplane struts
75 Stub plane end rib
76 Wing locking handle
77 Stub plane structure
78 Intake slot
79 Side window
80 Catapult spool
81 Drag struts
82 Cockpit sloping floor
83 Fixed 0.303-in (7.7-mm) Vickers gun (deleted from some aircraft)
84 Case ejection chute
85 Access panel
86 Camera mounting bracket
87 Sliding bomb-aiming hatch
88 Zip inspection flap
89 Fuselage upper longeron
90 Centre cockpit
91 Inter-cockpit fairing
92 Upper wing aerial mast
93 Pilot's headrest
94 Pilot's seat and harness
95 Bulkhead
96 Vickers gun fairing
97 Fuel gravity tank (12.5 Imp gal/57 litre capacity)
98 Windscreen
99 Handholds
100 Flap control handwheel and rocking head assembly
101 Wing centre section
102 Dinghy release cord handle
103 Identification light
104 Centre section pyramid strut attachment
105 Diagonal strengtheners
106 Dinghy inflation cylinder
107 Type C dinghy stowage well
108 Aileron control linkage
109 Trailing edge rib sections
110 Rear spar
111 Wing rib stations
112 Aileron connect strut
113 Port upper aileron
114 Fixed tab
115 Aileron hinge
116 Port formation light
117 Wing skinning
118 Port navigation light
119 Leading-edge slot
120 Front spar
121 Nose ribs
122 Interplane struts
123 Pitot head
124 Bracing wires
125 Flying wires
126 Port lower mainplane
127 Landing lamp
128 Underwing bomb shackles
129 Underwing strengthening plate
130 Rocket-launching rails
131 Four 60-lb (27-g) anti-shipping rocket projectiles
132 Three-blade fixed-pitch Fairey-Reed metal propeller
133 Spinner
134 Townend ring
135 Bristol Pegasus IIIM3 (or Mk 30) radial engine
136 Cowling clips
137 Engine mounting ring
138 Engine support bearers
139 Firewall bulkhead
140 Engine controls
141 Oil tank immersion heater socket
142 Filler cap
143 Oil tank (13.75 Imp gal/62.5 litre capacity)
144 Centre section pyramid struts
145 External torpedo sight bars
146 Fuel filler cap
147 Main fuel tank (155 Imp gal/705 litre capacity)
148 Vickers gun trough
149 Fuselage forward frame
150 Oil cooler
151 Fuel filter
152 Stub plane/fuselage attachment
153 Fuel feed lines
154 Dinghy immersion switch
155 Exhaust
156 Port Dunlop mainwheel
157 Jacking foot
158 1,610-lb (730-kg) 18-in (45.7-cm) torpedo
159 Access/servicing footholds
160 Torpedo forward crutch
161 Radius rod fairing
162 Undercarriage axle tube fairing
163 Undercarriage oleo leg fairing
164 Starboard mainwheel
165 Hub cover
166 Underwing bombs
167 Underwing outboard shackles
168 Depth-charge
169 250-lb (113-kg) bomb
170 Anti-shipping flares

Grumman TBF/TBM Avenger

The Confederate Air Force is one of numerous warbird operators to display the Avenger. A dozen or so examples are currently flying in the US, with others in Australia, Canada and Europe while further Avengers await restoration. Many survivors are former fire-fighters.

TBM-1C Avengers

Cutaway key

1 Starboard elevator
2 Fabric-covered aileron construction
3 Elevator trim tab
4 Elevator horn balance
5 Tailplane construction
6 Rudder tab
7 Trim tab control jack
8 Tail navigation light
9 Fabric-covered rudder construction
10 Aerial cable rear mounting
11 Fin construction
12 Port elevator
13 Port tailplane
14 Elevator hinge controls
15 Tailplane support frames
16 Deck arrester hook (lowered)
17 Arrester hook guide rails
18 Rudder hinge control
19 Rear fuselage frames
20 Flush-riveted aluminium skin covering
21 Finroot fairing
22 Tailplane control cables
23 Arrester hook retraction drive motor
24 Lifting tube
25 Rear fuselage frame and stringer construction
26 Tailwheel shock-absorber strut
27 Catapult 'hold-back' shackle
28 Retractable tailwheel
29 Crew compartment rear bulkhead
30 Search flares
31 Parachute flare launch tube
32 Ventral gun turret
33 Ammunition magazine
34 Browning 0.3-in (7.62-mm) machine-gun
35 Machine-gun mounting
36 Gun camera switch box
37 Crew door
38 Parachute stowage
39 Rear fuselage production break point
40 Spare coil stowage rack
41 Bombardier's side window
42 Upper turret spare ammunition magazines
43 Bombardier's folding seat
44 Gun turret mounting ring
45 Gun elevating mechanism
46 Ammunition feed chute
47 Browning 0.5-in (12.7-mm) machine-gun
48 Upper rotating gun turret
49 Bulletproof windscreen
50 Gunner's armoured seat back
51 Aerial cable
52 Port wing, folded position
53 Canopy aft glazing
54 Emergency life raft stowage
55 Hydraulic reservoir
56 Radio communications equipment
57 ASB weapons aiming controller
58 Bomb release levers
59 Cabin heater duct
60 Aft end of bomb bay
61 Fixed wing, root construction
62 Wing fold joint line
63 Browning 0.5-in (12.7-mm)/fixed machine-gun
64 Ammunition feed chute
65 Ammunition magazine (320 rounds)
66 Trailing-edge flap shroud construction
67 Lattice wing ribs
68 Starboard, fabric-covered aileron construction
69 Aileron hinge control
70 Aileron trim tab
71 Starboard wingtip
72 Starboard navigation light
73 Leading-edge ribs
74 Fixed leading-edge slot
75 ASB aerial
76 RT-5/APS-4 search radar pod
77 Radar mounting sway braces
78 Rocket-launching pylons
79 Jettisonable fuel tank
80 Main undercarriage wheel well
81 Sloping main spar
82 Wing fold hinge axis
83 Twin hydraulic folding jacks
84 Machine-gun blast tube
85 Starboard main fuel tank
86 Centre-section main spar
87 Oxygen bottle
88 Autopilot controls
89 Rear cockpit entry hatch

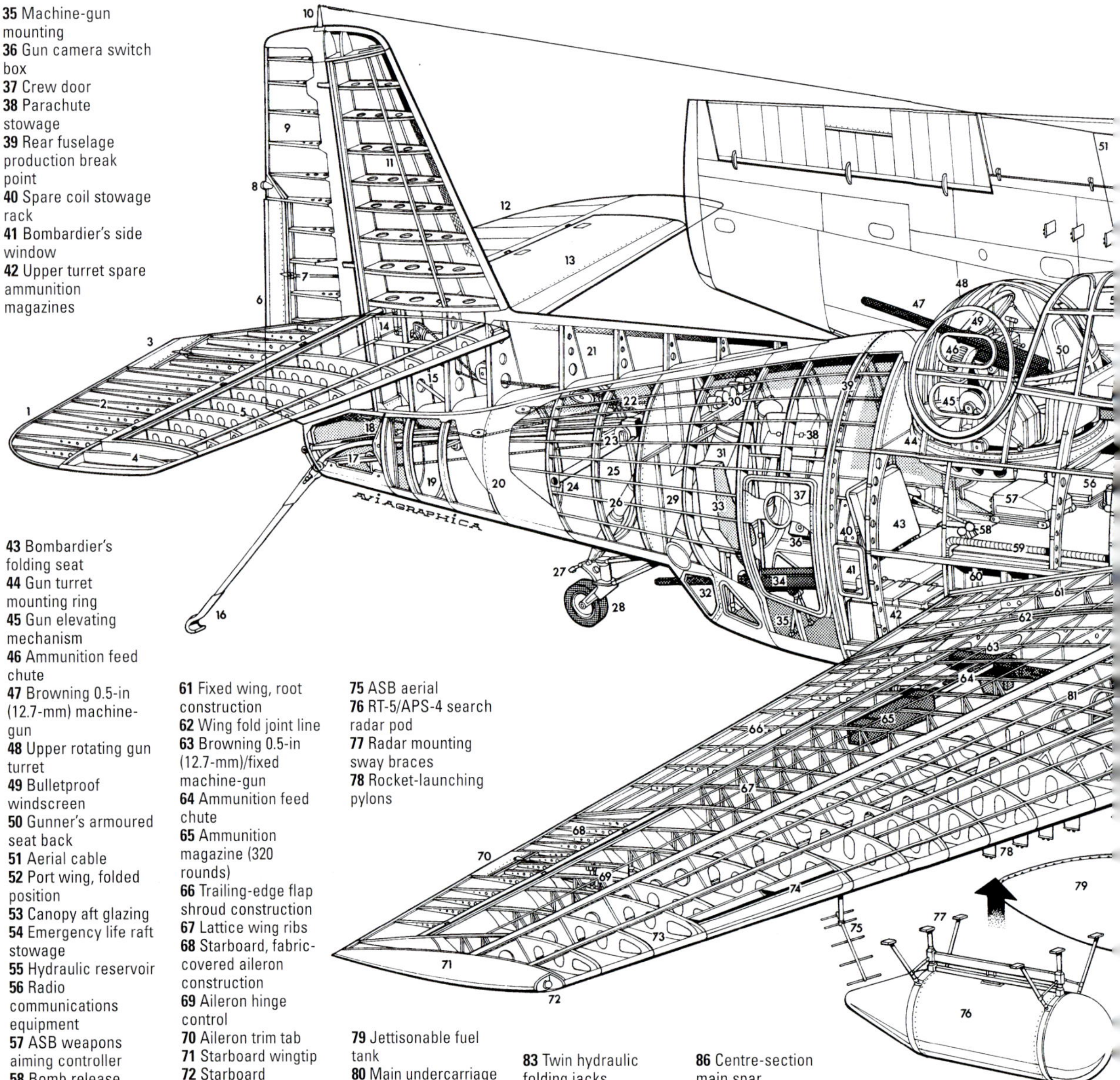

In Fleet Air Arm service the Avenger was initially known as the Tarpon. This flight of Tarpon TR.Mk 1s of No. 846 Squadron was photographed on 10 December 1943. The squadron was then at Macrihanish, and embarked on HMS Tracker in January 1944, transferring to Trumpeter in July 1944 for operations off Norway and in Arctic waters until April 1945. Fleet Air Arm Tarpon crews received their initial training and commissioning at NAS Squantum, Massachusetts in July 1943.

SPECIFICATION

TBF-1C Avenger

Dimensions

Length: 40 ft (12.19 m)
Wingspan: 54 ft 2 in (16.51 m)
Wingspan (folded): 19 ft (5.79 m)
Wing area: 490 sq ft (45.52 m^2)
Height: 16 ft 5 in (5.00 m)

Powerplant

One 1,700-hp (1268-kW) Wright R-2600-8 Cyclone 14-cylinder two-row radial piston engine

Weight

Empty: 10,555 lb (4788 kg)
Normal, loaded: 16,412 lb (7444 kg)
Maximum take-off: 17,364 lb (7876 kg)

Performance

Maximum speed: 257 mph (414 km/h) at 12,000 ft (3660 m)
Cruising speed: 153 mph (246 km/h)
Climb to 10,000 ft (3048 m): 13 minutes
Climb rate: 768 ft (234 m) per minute
Service ceiling: 21,400 ft (6525 m)
Range with torpedo: 1,105 miles (1780 km)
Range as scout: 2,335 miles (3755 km)

Armament

Two 0.5-in/12.7-mm machine-guns firing ahead (in TBF-1 one 0.3-in/7.62-mm forward-firing machine-gun); one 0.5-in (12.7-mm) machine-gun in dorsal turret and one 0.3-in (7.62-mm) machine-gun in lower rear position; internal weapons bay for one 22.4-in (56.9-in) torpedo or up to 2,000 lb (907 kg) of other stores

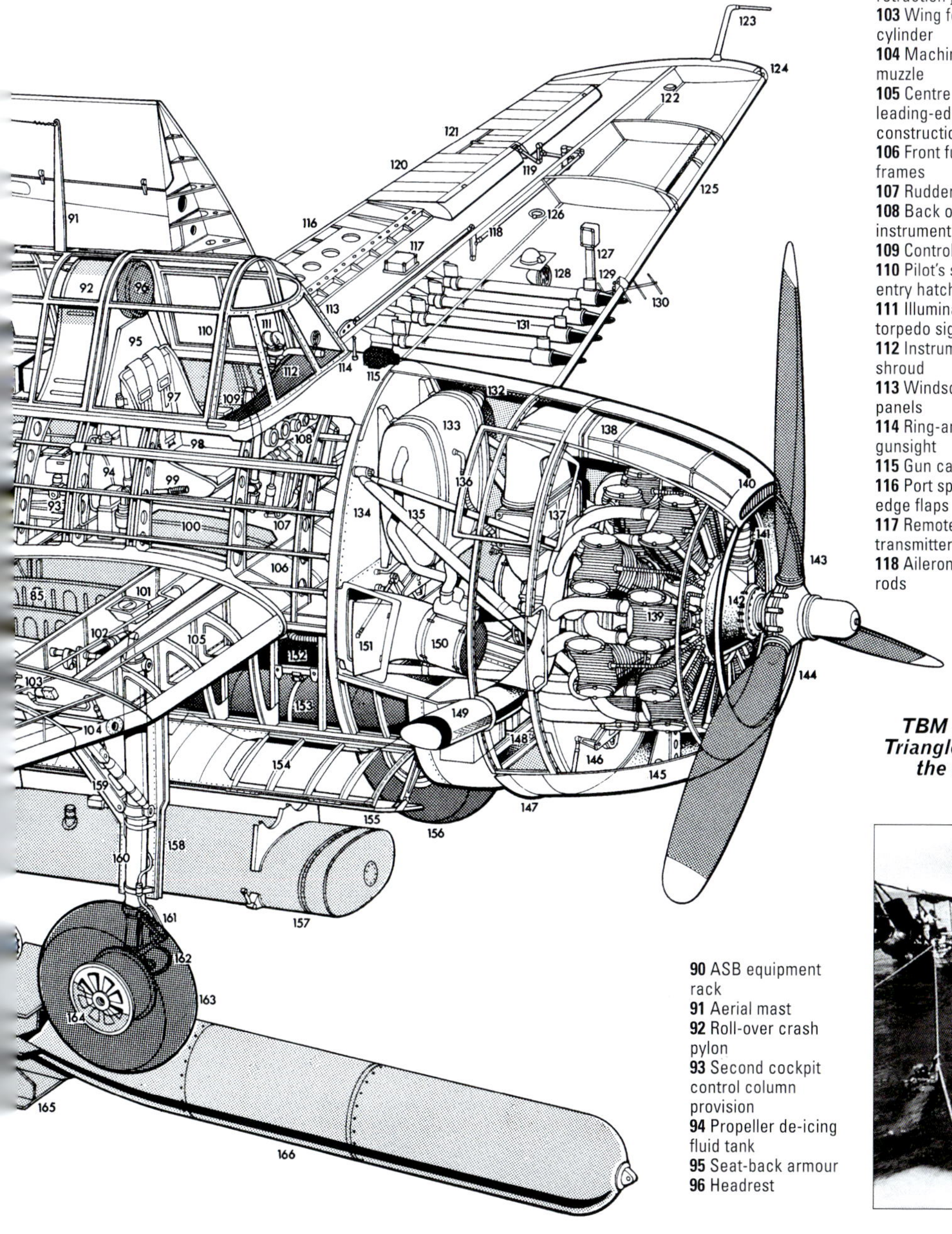

90 ASB equipment rack
91 Aerial mast
92 Roll-over crash pylon
93 Second cockpit control column provision
94 Propeller de-icing fluid tank
95 Seat-back armour
96 Headrest
97 Safety harness
98 Pilot's seat
99 Emergency hydraulic handpump
100 Centre main fuel tank
101 Fuel tank filler cap
102 Main undercarriage retraction jack
103 Wing fold locking cylinder
104 Machine-gun muzzle
105 Centre section leading-edge construction
106 Front fuselage frames
107 Rudder pedals
108 Back of instrument panel
109 Control column
110 Pilot's sliding entry hatch
111 Illuminated torpedo sight
112 Instrument panel shroud
113 Windscreen panels
114 Ring-and-bead gunsight
115 Gun camera
116 Port split trailing-edge flaps
117 Remote compass transmitter
118 Aileron control rods
119 Aileron hinge control
120 Fabric-covered port aileron
121 Aileron trim tab
122 Formation light
123 Pitot tube
124 Port navigation light
125 Fixed leading-edge slot
126 Wing 'tie-down' shackle
127 ASB aerial mounting
128 Retractable landing lamp
129 Red, white and green approach lights
130 Port ASB aerial
131 Ground attack rockets
132 Oil tank filler cap
133 Engine oil tank 13 US gal (49 litres)
134 Engine compartment bulkhead
135 Engine mounting struts
136 Cowling air exit flap
137 Twin carburettors
138 Carburettor air trunking
139 Wright-cyclone, 14-cylinder, two-row radial engine
140 Carburettor air intake
141 Propeller governor
142 Reduction gearbox
143 Hamilton Standard three-bladed propeller
144 Engine cooling intake
145 Engine cowlings
146 Cowling air flap control lever
147 Lower cowling air flap
148 Batteries
149 Starboard exhaust pipe
150 Oil cooler
151 Oil cooler air exit flap
152 Bomb release shackle
153 500-lb (227-kg) bombs
154 Bomb bay door construction
155 Bomb doors (open)
156 Port mainwheel
157 Bomb bay fuel tank (270-US gal/1022-litre capacity)
158 Plain undercarriage leg door
159 Retraction strut
160 Shock absorber leg strut
161 Torque scissor links
162 Hydraulic brake pipe
163 Starboard mainwheel
164 Removable wheel disc cover
165 Torpedo stabilising vanes
166 Mk XIII-2 torpedo

The loss of five Fort Lauderdale-based TBM Avengers and their 14 crewmen in the Bermuda Triangle area remains an enduring aviation mystery. In the mid-1980s some wreckage was recovered, but it was not from the missing flight.

TBM-3 Avenger

A TBM-3 built by Eastern Aircraft, this Avenger served with Torpedo Squadron 4 (VT-4), Air Group 4 (CVG-4), aboard USS *Essex* and is seen as it appeared on 12 January 1945. VT-4 was established on 10 January 1942, aboard the deck of USS *Ranger* (CV-4) while it was in Bermuda. Operations began with the Douglas TBD-1 Devastator, before transition to the TBF-1 Avenger in August 1942. VT-4's Avengers saw action in October 1943 when *Ranger* was deployed as part of Operation Leader, and VT-4 Avengers attacked German shipping along the Norwegian coast in the first US Navy carrier strike in northern Europe. In July 1944 Air Group 4 (including VT-4) moved to the Pacific, entering combat on 11 November 1944 in support of General MacArthur's return to the Philippines, from the deck of *Bunker Hill* (CV-17). In the ensuing days the squadron's Avengers struck targets at Ormoc Bay, Cavite and Clark Field. In late November 1944 the Avengers of VT-4 transferred to USS *Essex* (CV-9) which was promptly damaged by a kamikaze attack off the Philippines before the Avengers had seen action from their new carrier. In January 1945 the repaired carrier received VT-4's TBM-3 Avenger in time for a series of strikes against Western Pacific and Indochinese targets.

Powerplant
Although consideration was given to fitting a 2,000-hp (1491-kW) Pratt & Whitney R-2800 Double Wasp to the TBM-3, Wright's R-2600-20 Cyclone 14, rated at 1,900 hp (1417 kW), was chosen instead, replacing the 1,600-hp (1193-kW) R-2600-8 of the previous production variant, the TBM-1C. A new engine was necessary to counter the marginal performance of the aircraft at normal gross weights, especially when operating from the small decks of escort carriers. The increased cooling requirements of the new variant necessitated some redesign of its cowling: multiple cowl flaps and an oil cooler intake on the lower lip were new features.
The R-2600 – a 14-cylinder, twin-bank derivative of Wright's 1920s-vintage nine-cylinder R-1750/-1820 Cyclone – also powered the Curtiss SB2C Helldiver and, as such, is often described as one of America's 'war-winning' powerplant designs. More than 50,000 were built at Wright's Cincinnati plant.

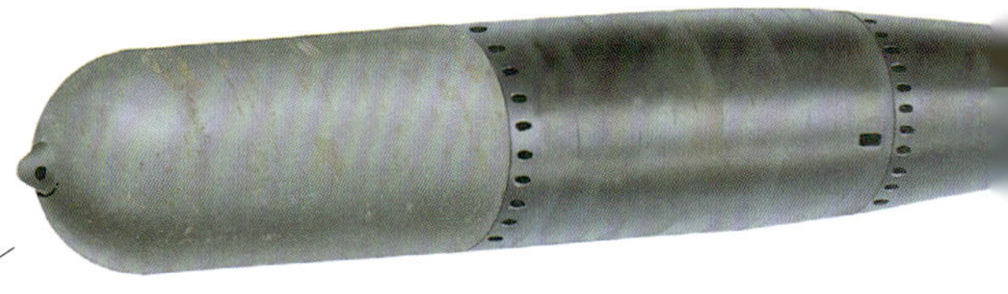

Torpedo
The Mk XIII 'ring-tailed' torpedo was developed by modifying the standard Mk XIII-1A with a 10-in (25.4-cm) steel band welded around the fins. These were first used in August 1944 by VT-13 from aboard USS *Franklin*, west of Iwo Jima. Compared to the unmodified torpedoes, which had to be dropped within a speed range of 115-126 mph (185-204 km/h), and a height of 100 ft (30.50 m), the 'ring tails' could be released from as high as 800 ft (244 m) and as fast as 321 mph (519 km/h). The Mk XIII was designed for aircraft use in about 1938. It had a greater diameter (22.4 in/56.9 cm) than the standard 21-in (53.3-cm) naval torpedo, a steam turbine engine producing 95 bhp (71 kW), a weight of 3,050 lb (1383 kg), and a theoretical maximum speed and range of 53 mph (85 km/h) and 9,000 yd (8230 m), respectively.

Crew

Three crew manned a typical TBM-3: a pilot, radio operator and bombardier. The pilot was housed in a separate cockpit, while the rear portion of the 'glasshouse' contained the radioman and his Grumman 150SE turret, with a 0.50-in (12.7-mm) machine-gun. The rest of the crew compartment consisted of the so-called 'tunnel' in which the bombardier manned a 0.30-in (7.62-mm) machine-gun in the ventral 'stinger' position. The bombardier's main piece of equipment was a Norden bombsight, fitted to early Avengers in an unusual position, sighted through a transparency aft of the weapons bay. He also had a scope for the aircraft's Air-to-Surface Type B (ASB) radar set, the moveable receivers for which (Yagi antennas) were fitted beneath each wing.
The bombsight was deleted in late-production TBM-3s as horizontal bombing was found to be ineffective against manoeuvring ships and less accurate than glide-bombing when attacking smaller land targets and ships at anchor. The radioman then manned the 'stinger' gun, while a dedicated gunner operated the turret. The pilot was able to use a 0.50-in (12.7-mm) machine-gun in each wing to strafe targets.

12 January 1945

On 12 January 1945, 12 TBM-3s, loaded with torpedoes, roared off *Essex*'s flight deck with the mission of attacking shipping on the Saigon River near Cap St Jacques. Among the pilots was Ensign William H. Cannady, a relatively new pilot to VT-4, flying aircraft number '131' (TBM-3 BuNo. 68417). On this occasion, he shared his aircraft with just one other crew member, Aviation Radioman Third Class (ARM3c) J. C. Gerke.

The 'torpeckers' encountered a convoy of merchantmen and their escorts which, along with the shore installations, began filling the sky with intense anti-aircraft fire. Jinking as they approached the release point at an altitude of 250 ft (76 m), the VT-4 Avengers dropped their weapons. Cannady's torpedo combined with those of two of his squadron mates to finish off a merchant vessel.

Bombloads

As a glide- or skip-bomber, the Avenger carried a variety of weapons according to mission, but typical stores included one 2,000-lb (907-kg) general-purpose (GP), one 1,600-lb (726-kg) armour-piercing (AP), two 1,000-lb (454-kg) GP, four 500-lb (227-kg) GP, 12 100-lb (45-kg) GP or four 350-lb (159-kg) depth bombs. The latter was the primary anti-submarine weapon used by the Avenger, along with newly developed aircraft rockets. Rockets were introduced in late 1943, initially in 3.5-in (8.8-cm) form and later with 5-in (12.7-cm) calibre warheads. In early 1944, the first 5-in (12.7-cm) High Velocity Aircraft Rockets (HVARs) were used in combat, nicknamed 'Holy Moses'.

Handley Page Halifax

With dorsal and tail turrets removed and faired over, the Halifax C.Mk VIII was developed from the B.Mk VI as a dedicated transport aircraft. A detachable underfuselage pannier for up to 8,000 lb (3636 kg) of freight could be fitted, and freight, passengers, paratroops or stretcher cases could be carried within the fuselage.

Halifax B.Mk III

Cutaway key
1 Starboard navigation light
2 Formation light
3 Aileron balance weight
4 Wing skinning
5 Starboard aileron
6 Aileron servo tab
7 Trim tab
8 Wing stringer construction
9 Landing/taxiing lamp
10 Carburettor air intake duct
11 Exhaust collector ring
12 Propeller hub pitch change mechanism
13 de Havilland three-bladed propellers
14 Bristol Hercules XVI radial engine
15 Oil cooler intake
16 Cowling air outlet flaps
17 No. 6 fuel tank, capacity 123 Imp gal (559 litres)
18 No. 5 fuel tank, capacity 122 Imp gal (555 litres)
19 Leading-edge oil tank
20 No. 4 fuel tank, capacity 161 Imp gal (732 litres)
21 No. 3 fuel tank, capacity 188 Imp gal (855 litres)
22 Fuel tank breather
23 No. 1 fuel tank, capacity 247 Imp gal (1123 litres)
24 Trailing-edge ribs
25 Starboard flap construction
26 Fuel jettison pipes
27 Starboard main undercarriage wheel bay
28 Inboard wing section bombcells
29 Starboard inner engine cowlings
30 Asymmetric windscreen
31 Nose skinning
32 De-icing fluid tank
33 Nose section frames
34 Spare ammunition drums
35 Bomb aimer's control panel
36 Nose glazing
37 0 303-in (7.7-mm) Vickers 'K' gun
38 Bomb-aiming panels, optically flat
39 Bomb sight
40 Bomb aimer's prone position couch
41 Pitot tube
42 Parachute stowage
43 Navigator's folding seat
44 Chart table
45 Ventral escape hatch
46 Camera
47 Aerial rail
48 Radio transmitters and receivers
49 Radio operator's control panel
50 Rudder pedals
51 Instrument panel
52 Co-pilot's and engineer's folding seats
53 Control column
54 Pilot's seat
55 Cockpit floor level
56 Cabin side windows
57 Radio operator's seat
58 Trailing aerial winch
59 Bomb bay doors (open)
60 Bomb door operating jacks
61 Main floor/bomb bay support longeron
62 Oxygen bottles
63 Parachute stowage
64 Front fuselage diagonal bracing strut
65 Engineer's control panel
66 Astrodome
67 Fuselage skin plating
68 Hydraulic accumulator
69 Batteries
70 D/F loop aerial fairing
71 Nose/centre section joint frame
72 Cabin roof escape hatch
73 Heater duct
74 Rest bunks, port and starboard
75 Hydraulic accumulators
76 Escape ladder
77 Fuselage/rear spar joint
78 Rear escape hatch
79 Fuselage upper longeron

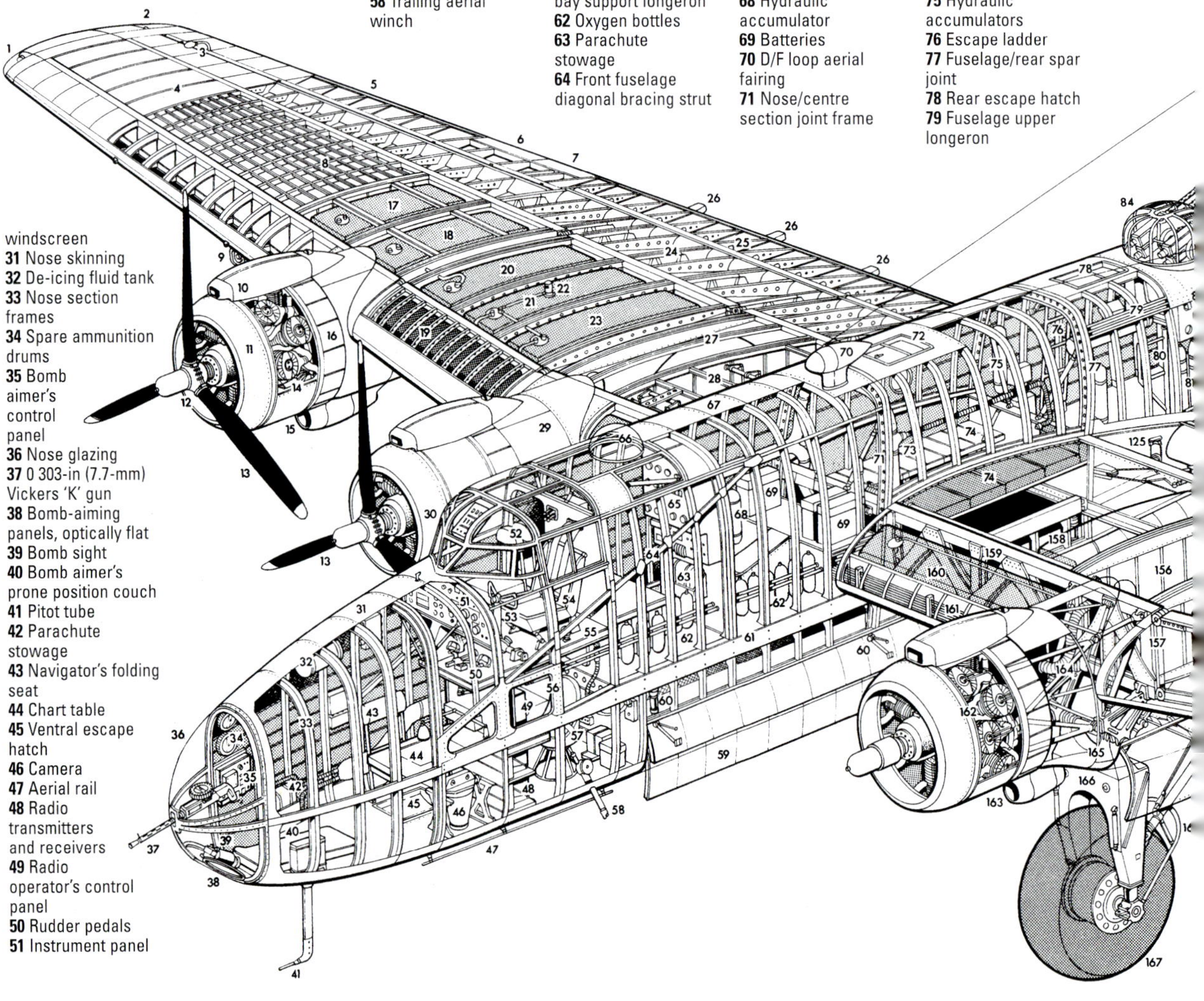

After lying beneath the water of Lake Hoklingen in Norway for 31 years, Halifax Mk II W1048 'S for Sugar' was recovered in 1973 and is now displayed at the RAF Museum, Hendon. The aircraft belonged to No. 35 Squadron and was one of 11 Halifaxes which departed RAF Kinloss for a low-level attack on the German battleship, **Tirpitz**. ***The aircraft was downed by anti-aircraft fire and crash-landed on the frozen lake.***

SPECIFICATION

Halifax B.Mk III

Dimensions

Wingspan: 98 ft 10 in (30.20 m) or later aircraft 104 ft 2 in (31.75 m)
Length: 70 ft 1 in (21.36 m)
Height: 20 ft 9 in (6.32 m)
Wing area: 1,275 sq ft (118.45 m²)

Powerplant

Four 1,615-hp (1204-kW) Bristol Hercules XVI 14-cylinder radial piston engines

Weights

Empty: 38,240 lb (17345 kg)
Maximum take-off: 65,000 lb (29484 kg)
Maximum landing: 55,000 lb (24947 kg)

Fuel capacity

Normal: 1,806 Imp gal (8210 litre) in 12 fuel tanks
Long-range: 2,688 Imp gal (12220 litre) in 17 fuel tanks

Performance

Maximum speed at 13,500 ft (4115 m): 282 mph (454 km/h)
Long-range cruising speed at 20,000 ft (6095 m): 215 mph (346 km/h)
Service ceiling: 24,000 ft (7315 m)
Range with maximum bombload: 1,030 miles (1658 km)
Maximum range: 950 miles (1529 km)

Armament

Four 0.303-in (7.7-mm) Browning machine-guns in Boulton Paul E.Mk I tail turret, four 0.303-in (7.7-mm) machine-guns in Boulton Paul A.Mk VIII dorsal turret and one 0.303-in (7.7-mm) Vickers 'K' machine-gun on trainable mount in nose position plus up to 13,000 lb (5897 kg) of bombs

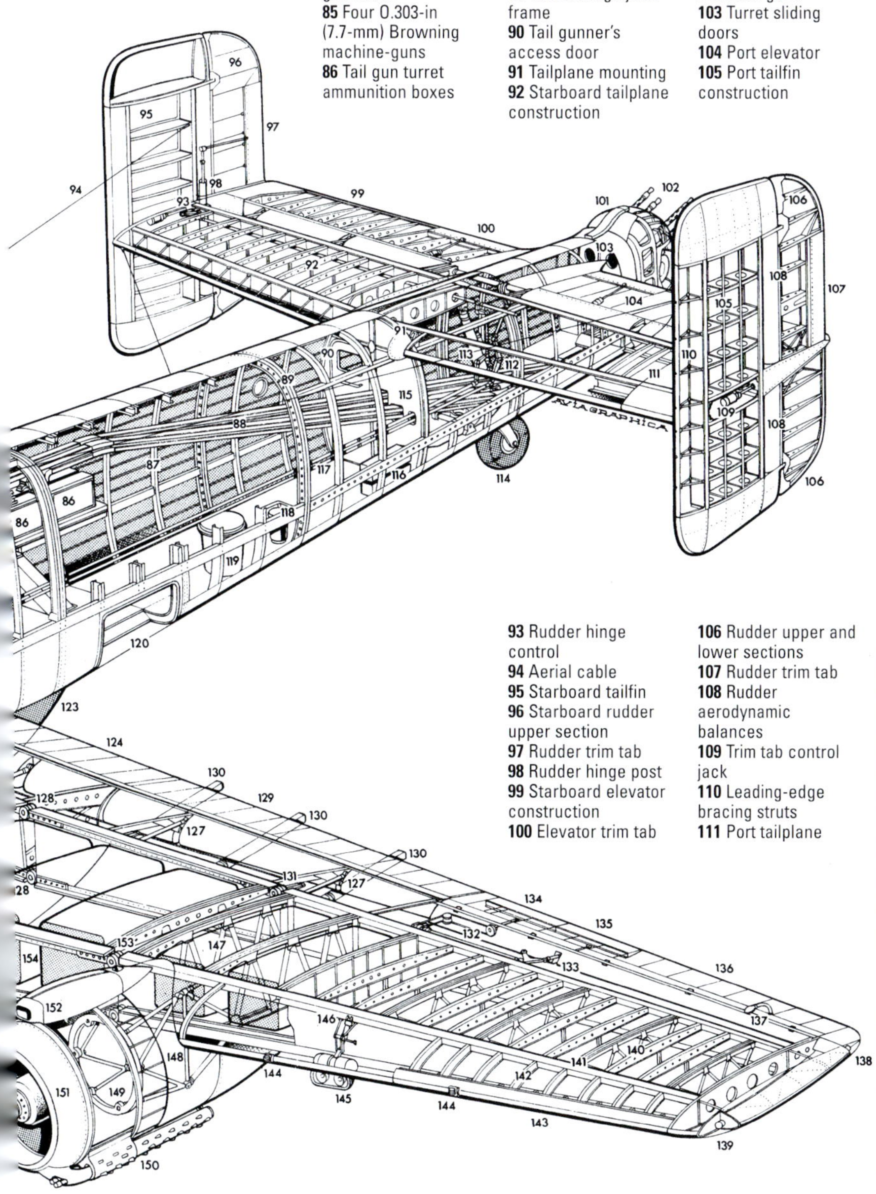

80 Upper turret ladder
81 Flare stowage
82 Sea marker stowage
83 Turret mounting ring
84 Boulton Paul A.Mk.III mid upper gun turret
85 Four 0.303-in (7.7-mm) Browning machine-guns
86 Tail gun turret ammunition boxes
87 Rear fuselage frame construction
88 Ammunition feed tracks
89 Tail fuselage joint frame
90 Tail gunner's access door
91 Tailplane mounting
92 Starboard tailplane construction
93 Rudder hinge control
94 Aerial cable
95 Starboard tailfin
96 Starboard rudder upper section
97 Rudder trim tab
98 Rudder hinge post
99 Starboard elevator construction
100 Elevator trim tab
101 Boulton Paul Type E tail gun turret
102 Four 0.303-in (7.7-mm) Browning machine-guns
103 Turret sliding doors
104 Port elevator
105 Port tailfin construction
106 Rudder upper and lower sections
107 Rudder trim tab
108 Rudder aerodynamic balances
109 Trim tab control jack
110 Leading-edge bracing struts
111 Port tailplane
112 Rudder and elevator control hinges
113 Tailwheel strut
114 Semi-retractable tailwheel
115 Rear fuselage bulkhead
116 ARI 5122 radar bombing control units
117 Tailplane control rods
118 Master compass
119 Toilet
120 Crew entry door, opens inward and upward
121 Flare launch tubes
122 Main fuse
123 H_2S radar bombing antenna fairing
124 Port inner flap
125 Flap jack
126 Dinghy stowage
127 Flap control rods
128 Rear spar inboard section attachment joints
129 Port outer flap
130 Fuel jettison pipes
131 Rear spar outer panel attachment joint
132 Trim tab controls
133 Aileron hinge control
134 Trim tab
135 Aileron servo tab
136 Port aileron
137 Aileron balance weight
138 Formation light
139 Port navigation light
140 Wingrib construction
141 Front spar
142 Leading-edge nose ribs
143 Armoured leading edge
144 Cable-cutters
145 Retractable landing/taxiing lamps
146 Lamp operating jack
147 Outer engine mounting ribs
148 Engine bearer struts
149 Engine mounting ring
150 Flame suppressor exhaust pipe
151 Exhaust collector ring
152 Carburettor intake duct
153 Outer wing panel joint
154 Port wing fuel tanks
155 Main undercarriage jacks
156 Port main wheel bay
157 Inner wing panel front spar joint
158 Wing bomb-cell long-range fuel tank, capacity 96 Imp gal (436 litres)
159 Front spar girder construction
160 Leading-edge No. 2 fuel tank, capacity 62 Imp gal (282 litres)
161 Engine control runs
162 Port inner Bristol Hercules XVI engine
163 Oil cooler air intake
164 Inboard engine bearers
165 Main undercarriage hinge mounting
166 Messier main undercarriage leg
167 Port mainwheel
168 Tyre guard
169 Folding retraction strut
170 Mainwheel door

A Halifax B.Mk VI displaying a relatively subtle example of nose art runs up at an airfield 'somewhere in England'. Piping for the aircraft's Graveley cabin heating system can be seen at the rear of the engine exhaust. As on the Halifax B.Mk III, all of the engines, except the port outer unit, exhausted to starboard. The flame-damping exhausts of the Hercules-engined aircraft are also illustrated to good effect.

Halifax B.Mk VII

This particular B.Mk VII served with No. 408 ('Goose') Squadron of the Royal Canadian Air Force at RAF Linton-on-Ouse, Yorkshire. The unit was formed at RAF Lindholme in June 1941 as one of the many RCAF squadrons operating as part of the RAF's war effort. It initially flew Hampdens before transferring to Halifaxes, then Lancasters, then back to Halifaxes again. The name of the unit came from its badge, which featured a Canada Goose. Graphic nose art such as *Vicky the Vicious Virgin* was very rare on RAF aircraft, but the Canadian squadrons were less strict. The nicknames of the crew were also applied near their positions.

Navigation equipment
The Halifax was equipped with a direction-finding loop in a teardrop radome above the fuselage, which also served as a mast for wire aerials leading back to the fintips. Some Halifaxes were fitted with 'Mandrel' and 'Airborne Cigar' jamming equipment for special radio countermeasures missions.

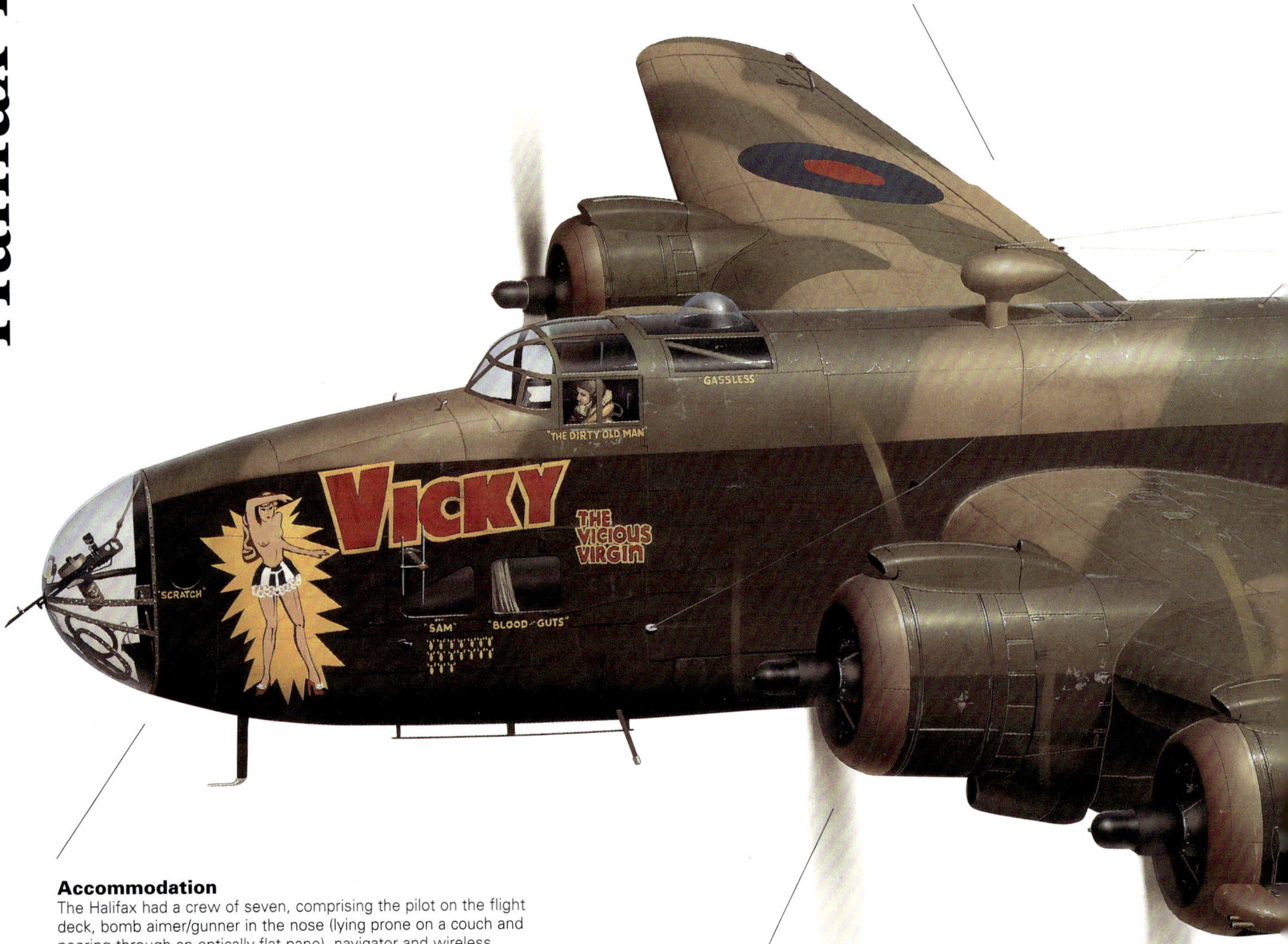

Accommodation
The Halifax had a crew of seven, comprising the pilot on the flight deck, bomb aimer/gunner in the nose (lying prone on a couch and peering through an optically-flat pane), navigator and wireless operator on the lower deck beneath the pilot, flight engineer on the upper deck behind the pilot and the two gunners in the mid-upper and rear turrets. A co-pilot's seat was next to that of the pilot, although no such dedicated crew member was carried. Above the engineer's station was an astrodome for sextant navigation. The crew entered the aircraft through a hatch which opened upwards and inwards.

Bristol power
The B.Mk VII featured the same Bristol Hercules XVI engines as the B.Mk III. These were 14-cylinder two-row sleeve-valve radials, each rated at 1,615 hp (1204 kW) for take-off, 1,675 hp (1249 kW) in 'M' gear at 4,500 ft (1372 m) and 1,455 hp (1085 kW) in 'S' gear at 12,000 ft (3658 m). Each engine drove a de Havilland constant-speed three-bladed propeller. The large intakes above the engines admitted air for the carburettor, while those underneath were for the oil cooler. Engine cooling was by the airflow, which was increased at slow speed by the use of cooling gills around the rear of the cowling. One of the more distinctive features of Hercules-powered Halifax variants was the large flame damper on the exhaust. The cowlings were inherited from the Beaufighter night-fighter.

Tail unit
Throughout its early career, the Halifax suffered from rudder overbalance, which could lead to the aircraft entering a spin in an asymmetric flight condition, such as two unserviceable engines on one side. The Type D fin (out of several different shapes trialled) was introduced to the later variants, curing the problem.

Dorsal and rear turrets
This B.Mk VII is fitted with a Type D tail turret with twin 0.5-in (12.7-mm) machine-guns. Ammunition was stored just aft of the mid-upper turret and fed by tracks to the rear position. Upper protection for the Halifax came courtesy of a Boulton Paul A.Mk VIII mid-upper turret, with a permanent gunner. The four Browning 0.303-in (7.7-mm) machine-guns had 1,160 rounds each. Aircraft which did not have H_2S radar had a Preston-Green ventral gun mounting.

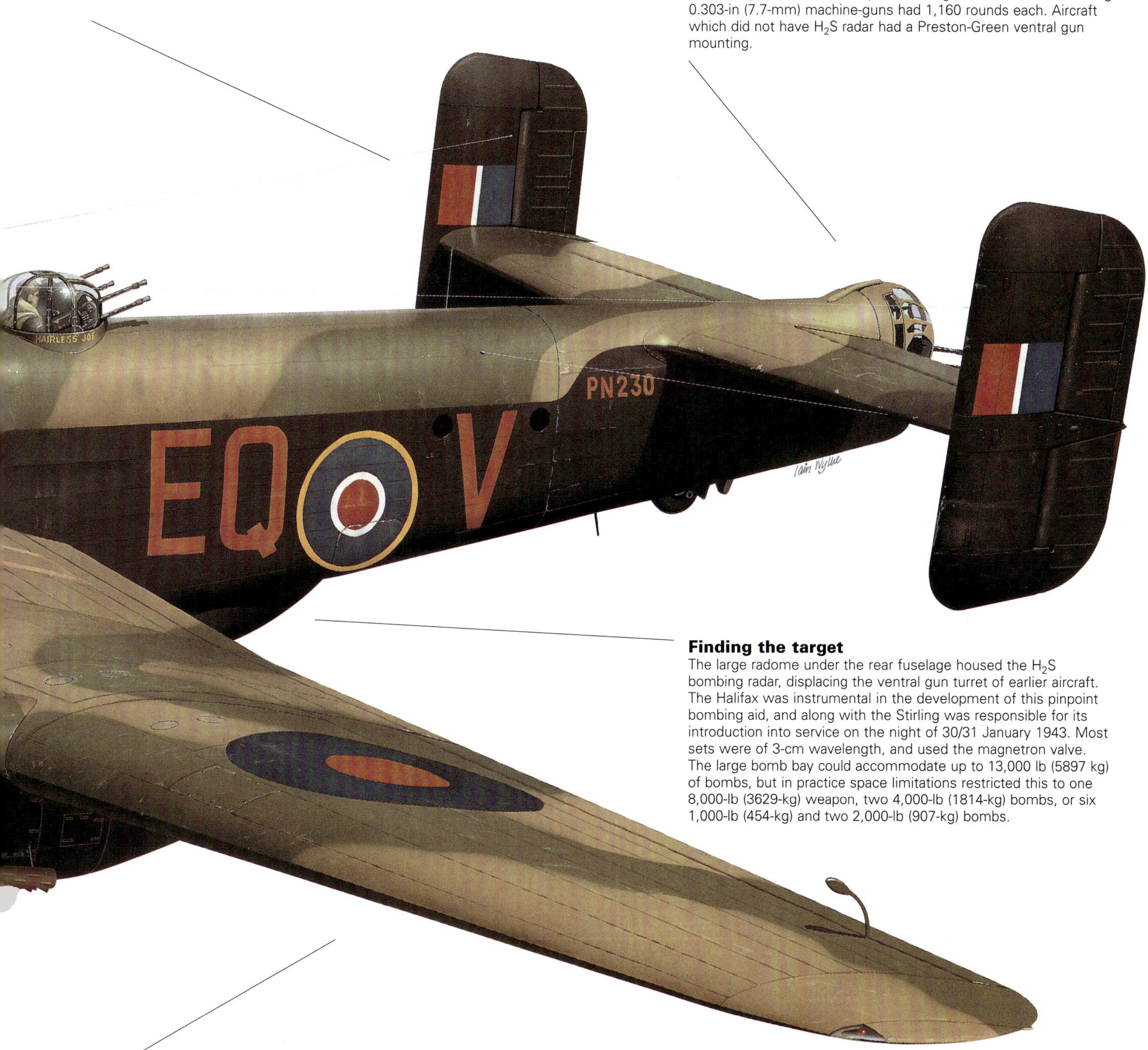

Finding the target
The large radome under the rear fuselage housed the H_2S bombing radar, displacing the ventral gun turret of earlier aircraft. The Halifax was instrumental in the development of this pinpoint bombing aid, and along with the Stirling was responsible for its introduction into service on the night of 30/31 January 1943. Most sets were of 3-cm wavelength, and used the magnetron valve. The large bomb bay could accommodate up to 13,000 lb (5897 kg) of bombs, but in practice space limitations restricted this to one 8,000-lb (3629-kg) weapon, two 4,000-lb (1814-kg) bombs, or six 1,000-lb (454-kg) and two 2,000-lb (907-kg) bombs.

Wing structure
The wing was built around two spars, which carried through the centre-section. The outer wing was made up of two sections, one join being just inboard of the outer engine, and the other being outboard of the inner engine. From the B.Mk III, subsequent Halifax bomber variants had extended rounded wingtips, which had originally been proposed for high-altitude versions. These raised the wing span to 104 ft 2 in (31.75 m) and allowed a substantial increase in maximum take-off weight.

Heinkel He 111

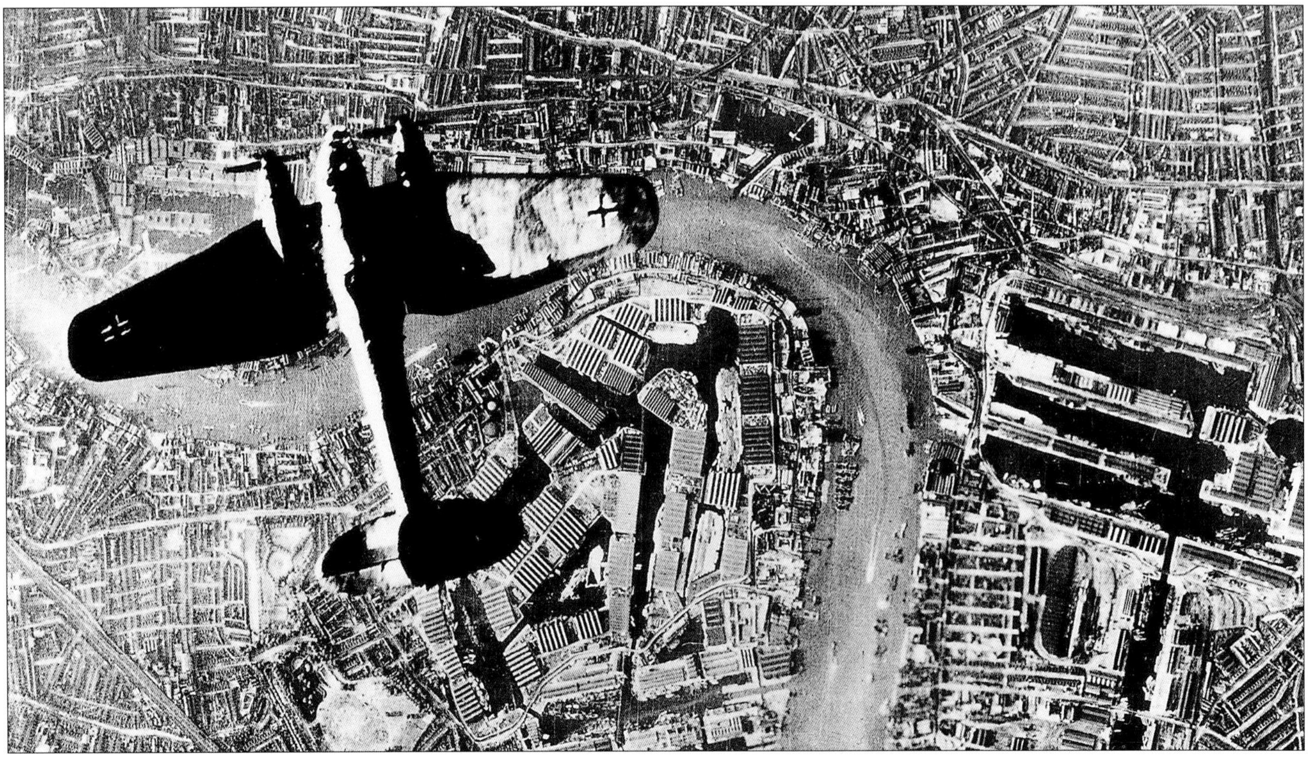

Above: A classic Battle of Britain image of a classic Battle of Britain bomber. This KG 1 He 111 was photographed over West India Docks, London on 7 September 1940.

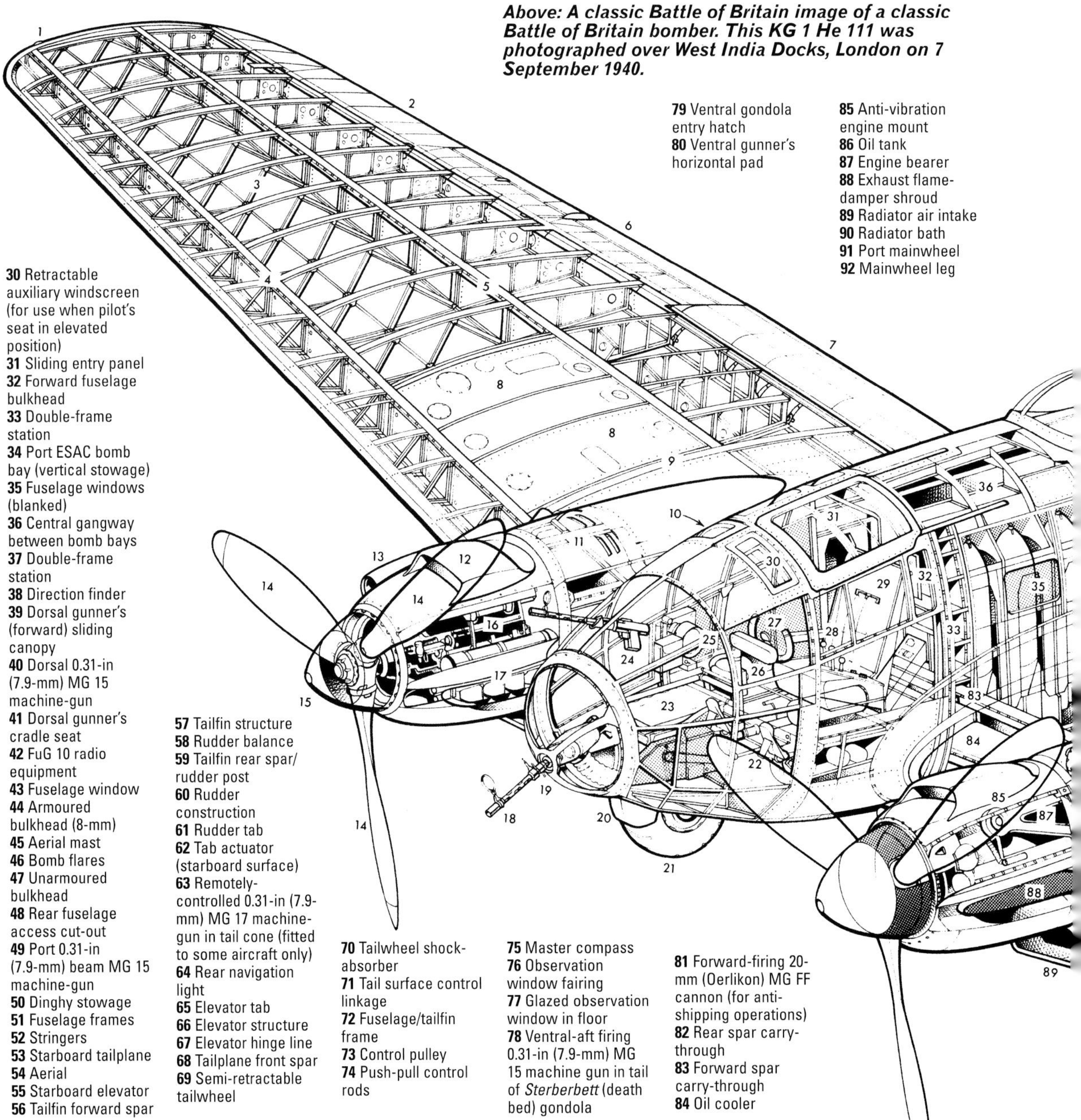

He 111H-3

Cutaway key
1 Starboard navigation light
2 Starboard aileron
3 Wing ribs
4 Forward spar
5 Rear spar
6 Aileron tab
7 Starboard flap
8 Fuel tank access panel
9 Wing centre section/outer panel break line
10 Inboard fuel tank (154-Imp gal/700-litre capacity) position between nacelle and fuselage
11 Oil tank cooling louvres
12 Oil cooler air intake
13 Supercharger air intake
14 Three-bladed VDM airscrew
15 Airscrew pitch-change mechanism
16 Junkers Jumo 211D-1 12-cylinder, inverted-vee, liquid-cooled engine
17 Exhaust manifold
18 Nose-mounted 0.31-in (7.9-mm) MG 15 machine-gun
19 Ikaria ball-and-socket gun mounting (offset to starboard)
20 Bomb sight housing (offset to starboard)
21 Starboard mainwheel
22 Rudder pedals
23 Bomb aimer's horizontal pad
24 Additional 0.31-in (7.9-mm) MG-15 machine-gun (fitted by forward maintenance units)
25 Repeater compass
26 Bomb aimer's folding seat
27 Control column
28 Throttles
29 Pilot's seat
30 Retractable auxiliary windscreen (for use when pilot's seat in elevated position)
31 Sliding entry panel
32 Forward fuselage bulkhead
33 Double-frame station
34 Port ESAC bomb bay (vertical stowage)
35 Fuselage windows (blanked)
36 Central gangway between bomb bays
37 Double-frame station
38 Direction finder
39 Dorsal gunner's (forward) sliding canopy
40 Dorsal 0.31-in (7.9-mm) MG 15 machine-gun
41 Dorsal gunner's cradle seat
42 FuG 10 radio equipment
43 Fuselage window
44 Armoured bulkhead (8-mm)
45 Aerial mast
46 Bomb flares
47 Unarmoured bulkhead
48 Rear fuselage access cut-out
49 Port 0.31-in (7.9-mm) beam MG 15 machine-gun
50 Dinghy stowage
51 Fuselage frames
52 Stringers
53 Starboard tailplane
54 Aerial
55 Starboard elevator
56 Tailfin forward spar
57 Tailfin structure
58 Rudder balance
59 Tailfin rear spar/rudder post
60 Rudder construction
61 Rudder tab
62 Tab actuator (starboard surface)
63 Remotely-controlled 0.31-in (7.9-mm) MG 17 machine-gun in tail cone (fitted to some aircraft only)
64 Rear navigation light
65 Elevator tab
66 Elevator structure
67 Elevator hinge line
68 Tailplane front spar
69 Semi-retractable tailwheel
70 Tailwheel shock-absorber
71 Tail surface control linkage
72 Fuselage/tailfin frame
73 Control pulley
74 Push-pull control rods
75 Master compass
76 Observation window fairing
77 Glazed observation window in floor
78 Ventral-aft firing 0.31-in (7.9-mm) MG 15 machine gun in tail of *Sterberbett* (death bed) gondola
79 Ventral gondola entry hatch
80 Ventral gunner's horizontal pad
81 Forward-firing 20-mm (Oerlikon) MG FF cannon (for anti-shipping operations)
82 Rear spar carry-through
83 Forward spar carry-through
84 Oil cooler
85 Anti-vibration engine mount
86 Oil tank
87 Engine bearer
88 Exhaust flame-damper shroud
89 Radiator air intake
90 Radiator bath
91 Port mainwheel
92 Mainwheel leg

***Above:** In the aftermath of **Operation Husky**, the Allied action to liberate **Sicily** in July 1943, some 600 Luftwaffe aircraft were left derelict on the island. This **He 111H**, left astride a bomb crater at **Catania**, was one of them.*

SPECIFICATION

He 111H-16

Dimensions

Length: 53 ft 9½ in
Height: 13 ft 1¼ in
Wingspan: 74 ft 1¾ in
Wing area: 931.10 sq ft (16.40 m²)

Powerplant

Two Junkers Jumo 211F-2 inline piston engines, each rated at 1,350 hp (1006 kW) for take off

Weights

Empty: 19,136 lb (8680 kg)
Maximum take-off: 30,864 lb (14000 kg)

Performance

Never-exceed speed: 298 mph (480 km/h)
Maximum speed at maximum take-off weight at sea level: 217 mph (349 km/h)
Maximum speed at maximum take-off weight at 13,120 ft (4000 m): 252 mph (406 km/h)
Normal range with maximum bomb load at sea level: 1,212 miles (1950 km)
Climb to 13,120 ft (4000 m) at maximum take-off weight: 23 minutes 30 seconds
Service ceiling at maximum take-off weight: 21,980 ft (6700 m)
***g* limits at 24,251 lb (11000 kg):** +2.7

Armament

One 20-mm MG FF cannon in the nose, one 0.51-in (13-mm) MG 131 machine-gun in the dorsal position, two 0.31-in (7.9-mm) MG 81 machine-guns in the rear of the ventral gondola and one or two 0.31-in (7.9-mm) MG 15 or MG 81 machine-guns in each of the beam positions, plus up to 5,512 lb (2500 kg) of bombs, torpedoes or other stores

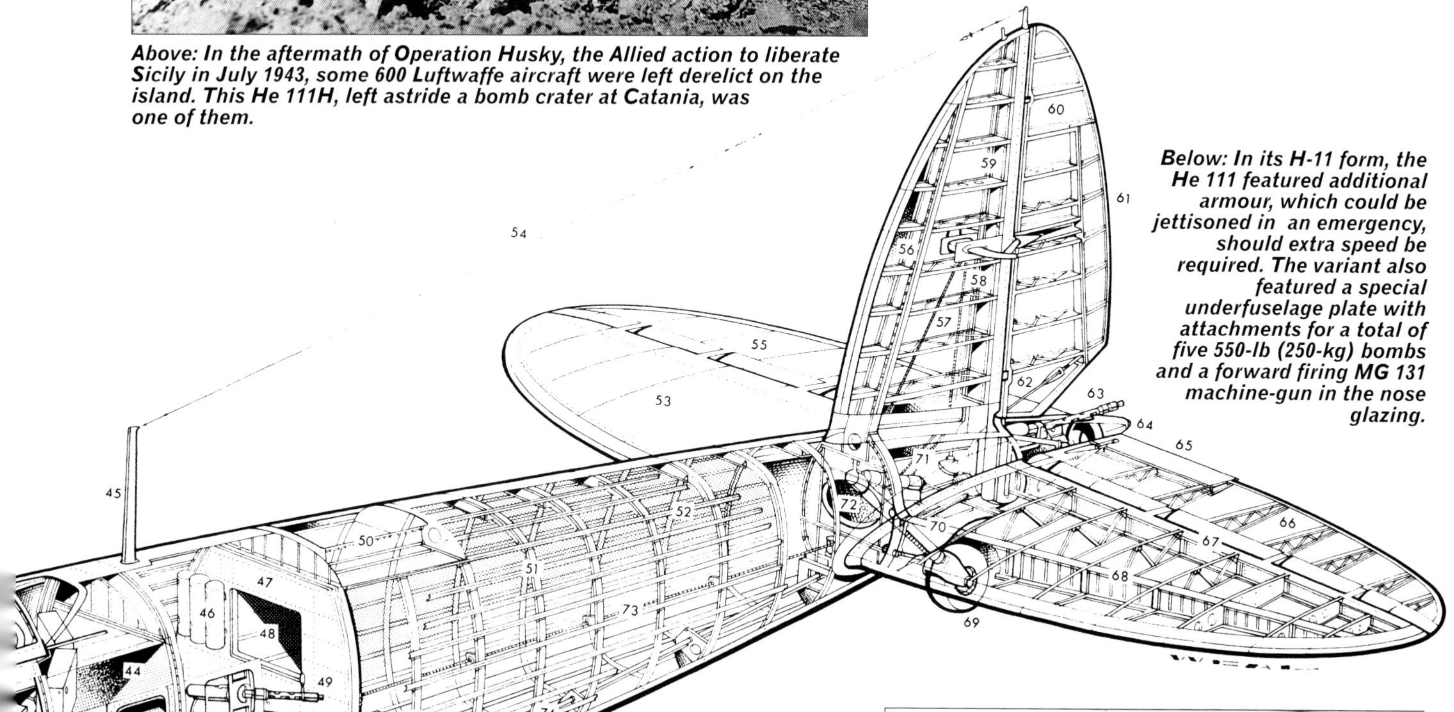
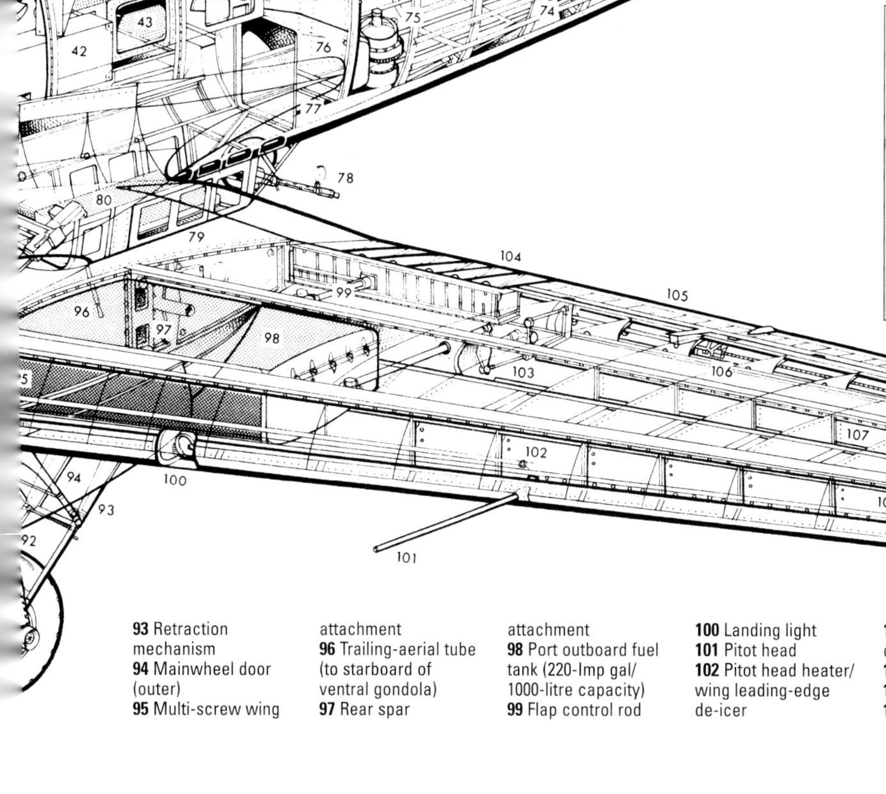
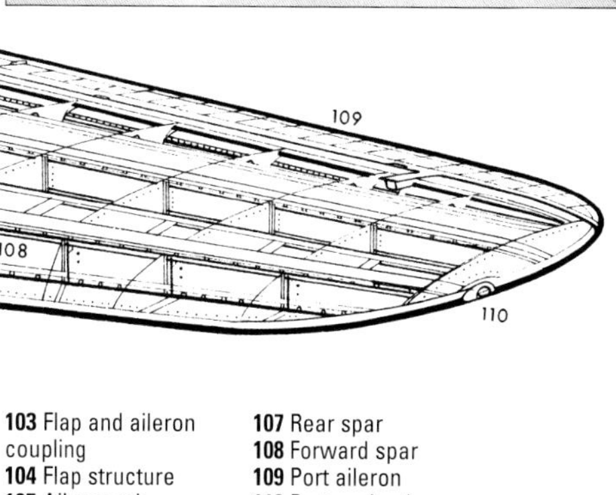

***Below:** In its **H-11** form, the **He 111** featured additional armour, which could be jettisoned in an emergency, should extra speed be required. The variant also featured a special underfuselage plate with attachments for a total of five 550-lb (250-kg) bombs and a forward firing **MG 131** machine-gun in the nose glazing.*

93 Retraction mechanism
94 Mainwheel door (outer)
95 Multi-screw wing attachment
96 Trailing-aerial tube (to starboard of ventral gondola)
97 Rear spar attachment
98 Port outboard fuel tank (220-Imp gal/ 1000-litre capacity)
99 Flap control rod
100 Landing light
101 Pitot head
102 Pitot head heater/ wing leading-edge de-icer
103 Flap and aileron coupling
104 Flap structure
105 Aileron tab
106 Tab actuator
107 Rear spar
108 Forward spar
109 Port aileron
110 Port navigation light

He 111H-22

This aircraft is illustrated as it appeared on the strength of III Kampfgeschwader 3 based at Venlo in Holland in the late summer of 1944. The unit was tasked with the continuation of Operation Rumpelkammer, the Fi 103 (V1) missile offensive against the UK. III./KG 3 employed He 111H-22s alongside converted H-16s and H-20s to air-launch the missiles. In September 1944, KG 53 was withdrawn from the Eastern Front, the crews of I./KG 53 then being absorbed into II. and III./KG 53 and III./KG 3 being redesignated as an all-new I./KG 53. By November 1944 the II. Gruppe of KG 53 had joined I. Gruppe in making Fi 103 attacks, but operating from the Oldenburg and Bremen areas of Germany. Meanwhile, I./KG 53 itself had been forced out of its Dutch base by the advancing Allied armies and had also been forced to withdraw to Germany. KG 27 probably provided veteran crews to KG 53, although itself disbanded between January and April 1945, at which point its III. Gruppe was probably training for the Fi 103 mission.

Wing, tail and rear fuselage
The all-metal wing was constructed around a two-spar structure, with the spars continuing through the fuselage forward and aft of the bomb bay. At its inboard trailing edges on both sides, the wing carried hydraulically-operated slotted flaps, while outboard of these were ailerons equipped with two-section trim tabs. In the landing sequence, the selection of flaps automatically deflected the ailerons downwards, providing greater lift at this critical stage than the flaps alone would have achieved and apparently not affecting the aircraft's trim. The tailplane could be adjusted on the ground through a range of 3° 12' into three separate positions for trimming purposes, while the rudder was the only control surface activated by the autopilot. With the automatic pilot engaged, three different rates of turn could be selected via a switch on the control column. To the rear of the crew compartment, aft of the cabin bulkhead, the fuselage was more-or-less empty, although it did provide accommodation for the emergency dinghy, master compass and tail control cable runs. A hatch in the bulkhead allowed the crew to access the rear fuselage for reasons of maintenance.

Fi 103 launches
With the withdrawal of conventional V-1 launch sites into central Holland, the UK, specifically London, fell out of the missile's range, hence the switch to airborne launching. The air-launched Fi 103 was normally sent on its way by an He 111 flying at around 1,500 ft (457 m) at night over the North Sea. The missiles were carried on special pylons under the wing root and could be launched from either port or starboard, although only one missile was ever carried at a time. The first air-launched Fi 103s were sent against Southampton on 7 July 1944, with minimal results. Nevertheless, by the end of August some 90 missiles had been fired at the town, with another 300 being aimed at London and 20 at Gloucester. By mid-December a further 865 missiles had been expended, but the cost in terms of aircraft and crews was high. When operations ceased on 14 January 1945, KG 53 had lost 77 He 111s, 12 of them when their missiles detonated immediately after take off.

Defensive armament

On the He 111H-22 defensive weapons were usually reduced to just two (as here) or three guns. The dorsal turret mounted a 0.51-in (13-mm) MG 131 machine-gun with 1,000 rounds of ammunition. A vane protruding from the rear of the turret, inline with the gun, offset the aerodynamic drag caused by the weapon's barrel as the turret rotated. A 20-mm MG FF cannon with 180 rounds was mounted in the extreme nose, while a 0.31-in (7.9-mm) MG 15 could be fixed on an Ikaria universal mounting to fire through the foremost upper glazed panel to the right of the cockpit area.

Fuel system

Fuel was contained in two tanks in each wing, positioned between the wing spars. A 154-Imp gal (700-litre) capacity main fuel tank was fitted between each engine nacelle and the fuselage, while 220-Imp gal (1000-litre) reserve tanks were fitted immediately outboard of the nacelles. The engines took fuel from the main tanks first, fuel being automatically moved from the reserve to the main tanks by electric transfer pumps when the main tanks were down to 44 Imp gal (200 litres) of fuel remaining.

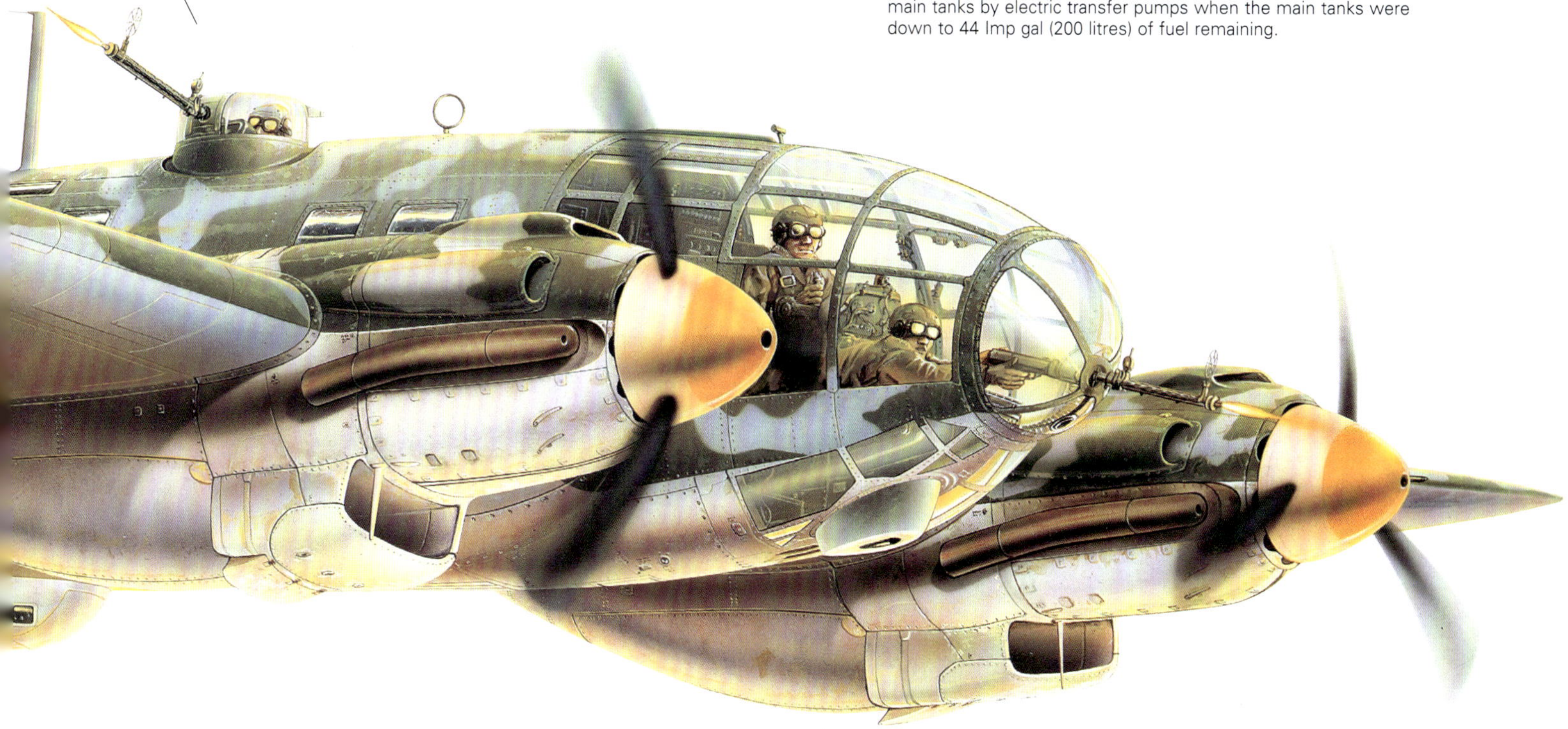

Jumo powerplant

Since the He 111H-22s were modified from H-21 airframes on the production line, they retained the Jumo 213E-1 engines of the night-bomber variant. These delivered 1,750 hp (1305 kW) for take-off and 1,320 hp (985 kW) at 32,000 ft (9753 m). V1-carriers converted from H-16 bomber and H-20 multi-role glider tug/transport airframes retained the Jumo 211F-2 engines of these variants, producing just 1,350 hp (1007 kW) on take off and as little as 1,060 hp (791 kW) at 17,390 ft (5300 m).

Fieseler Fi 103

With a typical launch weight of 4,806 lb (2180 kg), the Fi 103 was better known as the V-1 (Vergeltungswaffe Eins – Reprisal Weapon One). At launch, it was simply aimed in the direction of its target, the chances of hitting a city-sized area in this manner being fairly high. The weapon was powered by a 661-lb st (2.94-kN) Argus-Schmidt As 014 pulse-jet which operated at 47 Hz to produce the distinctive sound of the 'buzz-bomb'. An autopilot kept the weapon on course during its 32-minute flight and maximum range was in excess of 150 miles (240 km). The Amatol warhead weighed 1,874 lb (850 kg) and was detonated by an impact fuse in the missile's nose.

Heinkel He 115

The He 115 was fitted with a permanent ladder to allow crew access via the floats. The ladder is clearly visible in this wartime colour photograph.

He 115

Cutaway key
1 7.9-mm MG 15 machine-gun
2 Gunsights
3 Ikaria nose mounting
4 Cartridge collector chute
5 Nose ring
6 Entry/escape hatch
7 Nose glazing
8 Bomb/torpedo-sight
9 Selector panel
10 Handhold
11 Bombardier's kneeling-pad
12 Ventral glazing
13 Bombardier's/navigator's hinged seat
14 Duplicate throttle controls
15 Duplicate control column
16 Instrument panel
17 Nose compartment windscreen
18 Fixed glazing
19 Electrics panel
20 Batteries
21 Cockpit/nose access
22 Smoke floats
24 Fuselage frame
25 Cockpit floor
26 Rudder pedals
27 Throttles
28 Control column
29 Instrument panel
30 Windscreen
31 Starboard nacelle oil tank location
32 Engine bearer supports
33 Cooling gills
34 Starboard BMW 132K nine-cylinder radial engine
35 Nacelle nose ring
36 Propeller hub
37 Spinner
38 VDM three-bladed metal propeller of 10.83-ft (3.30-m) diameter
39 Nacelle hinged access/maintenance panels
40 Leading-edge hinged access/servicing panel
41 Starboard outer main fuel tank
42 Leading-edge tank (provision)
43 Wing structure
44 Front spar
45 Starboard navigation light
46 Starboard outer rib
47 Aileron outer hinge
48 Starboard aileron
49 Aileron tabs
50 Rear spar
51 Aileron tab hinge fairing
52 Control linkage
53 Flap outer section
54 Aileron profile
55 Starboard flap
56 Canopy hinged section
57 Fixed section
58 Cockpit rear-sliding canopy
59 Pilot's seat
60 Leading-edge inboard hinged access/servicing panel
61 Front spar carry-through
62 Fuselage/spar main frame
63 Front spar
64 Port inner main fuel tank
65 Filler cap
66 Fuselage centre bay
67 Wireless installation
68 Aerial mast
69 Dorsal identification light
70 Rear spar carry-through
71 Wireless operator's position
72 Flare stowage
73 Pistol flare port
74 Wireless operator/gunner's swivel seat
75 Cockpit coaming
76 Canopy fixed section
77 Gunner's hinged canopy section
78 Dorsal 7.9-mm MG 15 machine-gun
79 Ammunition magazine stowage (1,500 rounds)
80 Cockpit warm air
81 Wing upper surface walkway
82 Rescue dinghy stowage
83 Port flap inner section
84 Trailing-edge flap
85 Crew entry ladder (port and starboard)
86 Ladder attachments
87 Handholds
88 Fuselage frame
89 Dorsal skinning
90 Semi-monocoque fuselage structure
91 Control runs
92 Compass installation
93 Stringers
94 Fuselage aft frame
95 Port tailplane forward attachment
96 Tailfin leading edge
97 Starboard tailplane
98 Starboard elevator mass balances
99 Tailplane spar
100 Aerial
101 Elevator outer hinge
102 Starboard elevator tab
103 Aerial attachment
104 Rudder upper hinge
105 Tailfin structure
106 Front spar
107 Rudder mass balances
108 Starboard tailplane lower brace strut
109 Port tailplane aft attachment
110 Rudder tab hinge fairing
111 Rudder
112 Rudder tab upper section
113 Rudder tab lower section
114 Elevator tab
115 Port elevator upper mass balance
116 Tab hinge fairing
117 Port elevator
118 Elevator outer hinge
119 Port elevator lower mass balance
120 Port tailplane lower brace strut
121 Tailplane front spar
122 Mooring attachment
123 Brace strut/fuselage fairings
124 Fuselage aft main frame
125 Ventral skinning
126 Wing construction break-point
127 Rib strap joint
128 Port flap outer section
129 Aileron control linkage
130 Rear spar
131 Aileron tab hinge fairing
132 Aileron tab
133 Outer hinge
134 Port aileron
135 Port wingtip
136 Port navigation light
137 Outer rib
138 Wing structure
139 Front spar
140 Pitot head
141 Wing leading edge
142 Landing lamp
143 Port float aft section
144 Float sternpost
145 Mooring bollard
146 Spar section
147 Ladder/float attachment
148 Port outer main fuel tank
149 Filler cap

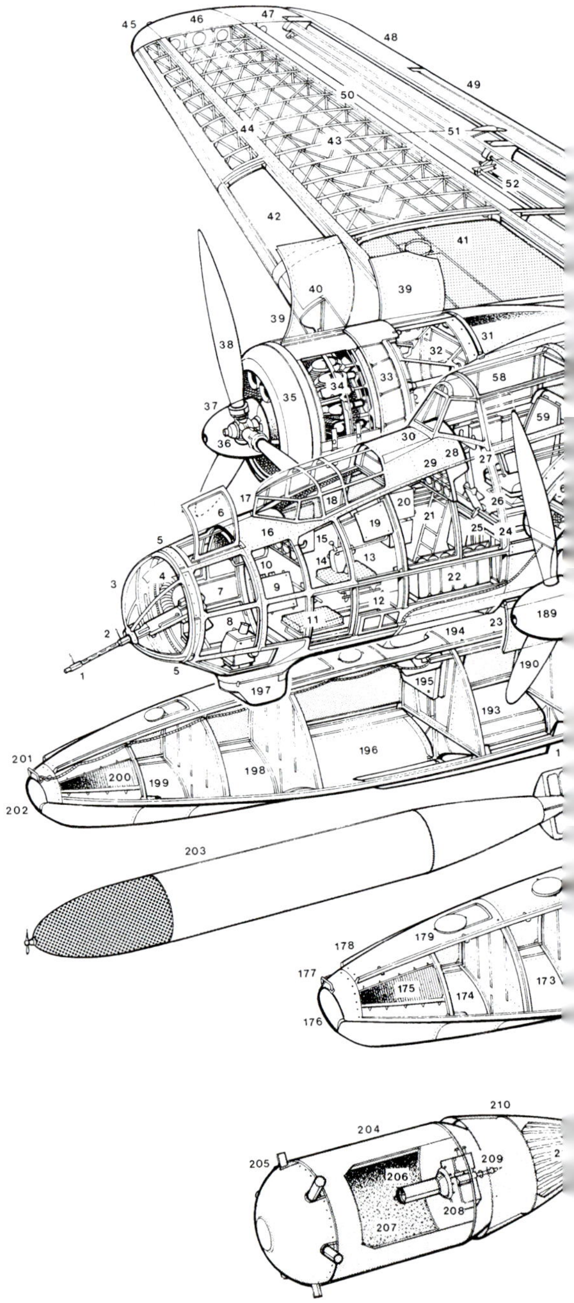

Norway became an important area of operations for He 115 seaplanes during World War II. Ironically, Norway's Marinens Flyvevåben also operated the type, and a few enemy He 115s were captured by both sides during the fighting. Six Norwegian-flown He 115s bombed German positions during the battle for Narvik.

SPECIFICATION

Type
Three-seat coastal general-purpose and torpedo bomber floatplane

Powerplant
Two BMW 132K radial piston engines each rated at 960 hp (716 kW)

Performance
Maximum speed 186 mph (300 km/h) at 3,280 ft (1000 m); cruising speed 180 mph (290 km/h) at 6,560 ft (2000 m); climb to 9,845 ft (3000 m) in 22 minutes 18 seconds; service ceiling 16,950 ft (5165 m); range 1,740 miles (2800 km) with maximum fuel

Weights
Empty 15,146 lb (6870 kg); normal take-off 23,545 lb (10680 kg)

Dimensions
Wingspan 73 ft 1 in (22.28 m); length 56 ft 9 in (17.30 m); height 21 ft 73⁄4 in (6.60 m); wing area 933.23 sq ft (86.70 m2)

Armament
One 15-mm MG 151 fixed forward-firing cannon on the lower port side of the nose, one 0.31-in (7.92-mm) MG 17 fixed rearward-firing machine-gun in the rear of each engine nacelle, one 0.31-in (7.92-mm) MG 15 trainable forward-firing machine gun in the nose position and one 0.31-in (7.92-mm) MG 15 trainable rearward-firing machine-gun in the dorsal position, plus up to 2,205 lb (1000 kg) of disposable stores carried in a lower-fuselage weapons bay and on two underwing hardpoints

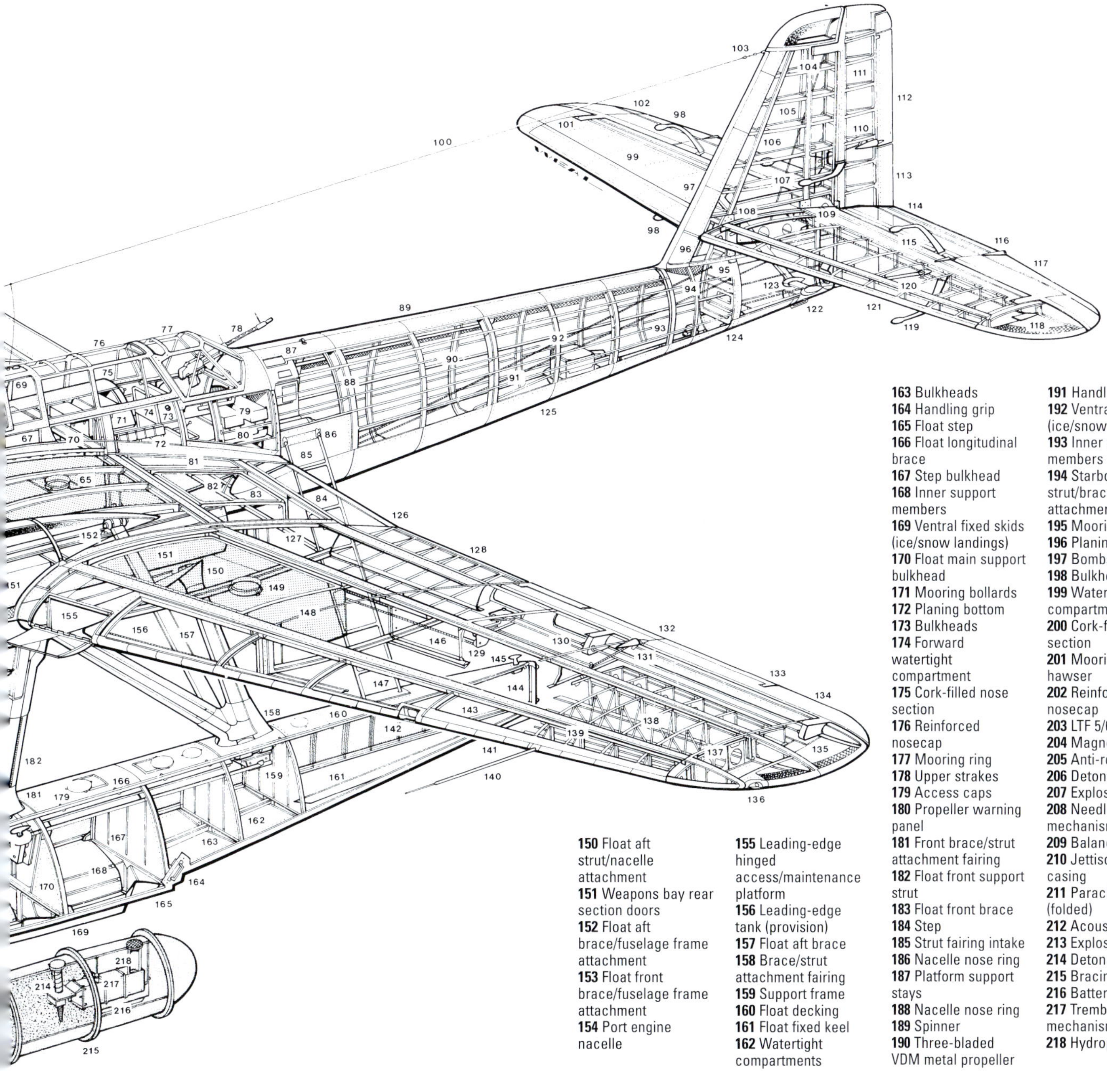

150 Float aft strut/nacelle attachment
151 Weapons bay rear section doors
152 Float aft brace/fuselage frame attachment
153 Float front brace/fuselage frame attachment
154 Port engine nacelle
155 Leading-edge hinged access/maintenance platform
156 Leading-edge tank (provision)
157 Float aft brace
158 Brace/strut attachment fairing
159 Support frame
160 Float decking
161 Float fixed keel
162 Watertight compartments
163 Bulkheads
164 Handling grip
165 Float step
166 Float longitudinal brace
167 Step bulkhead
168 Inner support members
169 Ventral fixed skids (ice/snow landings)
170 Float main support bulkhead
171 Mooring bollards
172 Planing bottom
173 Bulkheads
174 Forward watertight compartment
175 Cork-filled nose section
176 Reinforced nosecap
177 Mooring ring
178 Upper strakes
179 Access caps
180 Propeller warning panel
181 Front brace/strut attachment fairing
182 Float front support strut
183 Float front brace
184 Step
185 Strut fairing intake
186 Nacelle nose ring
187 Platform support stays
188 Nacelle nose ring
189 Spinner
190 Three-bladed VDM metal propeller
191 Handling grip
192 Ventral fixed skids (ice/snow landings)
193 Inner support members
194 Starboard float strut/brace attachment fairing
195 Mooring bollards
196 Planing bottom
197 Bombsight fairing
198 Bulkheads
199 Watertight compartment
200 Cork-filled nose section
201 Mooring ring/steel hawser
202 Reinforced nosecap
203 LTF 5/6 torpedo
204 Magnetic mine
205 Anti-rolling horns
206 Detonator
207 Explosive charge
208 Needle/contact mechanism
209 Balance magnet
210 Jettisonable aft casing
211 Parachute (folded)
212 Acoustic mine
213 Explosive charge
214 Detonator
215 Bracing straps
216 Battery
217 Trembler/contact mechanism
218 Hydrophone

Heinkel He 177

An audacious British operation saw this He 177A-5/R6 being taken from the Luftwaffe airfield at Blagnac. The machine was tested in autumn 1944 and into 1945. Officially named Grief (Griffon), the He 177 proved almost as dangerous to its crews as it did to the enemy, and it was generally considered a poor machine by its British test pilots.

Ju 87D-3

Cutaway key

1 Starboard navigation light
2 Detachable wingtip
3 FuG 101 radio altimeter (FM)
4 Aileron control runs
5 Starboard aileron
6 Aileron trim tab
7 Spring-loaded geared tab
8 Aileron counter-balance
9 FuG 102 radio altimeter (pulsed)
10 Tab mechanism
11 Fowler flap outboard track
12 Fowler flap position (extended)
13 Aileron tab control linkage
14 Flap actuating cylinder (hydraulic)
15 Control cables
16 Main spar (outboard section)
17 Wing ribs
18 Auxiliary front spar
19 Heated leading edge
20 Oil radiator intake
21 Starboard Hs 293 radio-controlled glide-bomb
22 Starboard outer mainwheel door (open position)
23 Starboard outer mainwheel well
24 Balloon cable-cutter in leading edge
25 Starboard ETC weapons rack
26 Radiator outlet flap
27 Radiator outlet flap
28 Hot-air ducting
29 Mainwheel door actuating cylinder
30 No. 8 (starboard outer) fuel tank of 1,120 litre/246.5 Imp gal capacity (flexible bag)
31 Fuel filler cap
32 Fowler flap outer section
33 auxiliary rear spar
34 Wing dihedral break point
35 Fowler flap track
36 Starboard fuel starting tank (9 litre/2 gal capacity)
37 Starboard oil tanks
38 Main hydraulic tank (starboard only) (32 litre/7 gal capacity)
39 Fuel filler cap
40 No. 3 (starboard inner) fuel tank of 621 litre/136.5 Imp gal capacity (metal/self sealing)
41 Fowler flap inner section
42 Main spar (inboard section)
43 Starboard inner mainwheel well
44 Engine supercharger
45 Nacelle fairing
46 Wing spar attachment point and fairing
47 Engine accessories
48 Daimler-Benz DB 610A-1 24-cylinder liquid-cooled engine
49 Anti-vibration side-mounting pad
50 Supercharger and wing de-icing intakes
51 Nacelle former
52 Coolant vents
53 Engine forward mounting
54 Cooling gills
55 Double-gear crank casing
56 Single propeller shaft
57 Propeller de-icing saddle tank
58 Nacelle cooling profile
59 Propeller variable-pitch mechanism
60 Propeller boss
61 Blade cuffs
62 VDM four-bladed propeller (right-handed)
63 Chin intake
64 Flame damper exhaust
65 Starboard outer mainwheel leg
66 Starboard inner mainwheel leg
67 Starboard outer mainwheel
68 D/F loop in dorsal blister
69 Emergency hydraulic tank (25 litre/5.5 Imp gal)
70 No. 7 fuselage frame
71 C-Stand ammunition tank (1,000 rounds)
72 Dorsal barbette remote drive motor
73 Revi gunsight with slotted 10-mm armour protection
74 Remote control sighting cupola
75 Barbette traverse control handle
76 Barbette elevation control handle
77 Main radio panel (FuG 10P; general-purpose set) (FuG17Z: VHF communication and homing) (FuG BL 2F: Blind-approach)
78 First-aid pack
79 Navigator's take-off/landing station
80 Window
81 Gunner's seat
82 Emergency jettison panels (port and starboard)
83 Bomb aimer's seat (raised)
84 External rear-view mirror
85 Engine control panel (starboard)
86 Internal rear-view mirror
87 Offset ring-and-bead gunsight
88 MG 81 7.9-mm machine-gun (A1-Stand)
89 Circular gun mounting
90 Balloon cable-cutters in nose horizontal frames
91 Ammunition feed
92 A1-Stand ammunition tank (1,000 rounds)
93 Hinged window panel (port and starboard)
94 Pilot's seat (armour plate: 9-mm back, 6-mm seat)
95 Rudder pedals
96 Cockpit hot-air
97 Lower glazed section often overpainted/armoured
98 Lotfe 7D bombsight fairing
99 'Boxed' gunsight
100 MG 151 20-mm cannon (A2-Stand)
101 Bullet-proof glass in nose of 'bola'
102 De-icing intake
103 Ventral crew entry hatch
104 Telescopic ladder
105 Actuating arm
106 MG 161 20-mm cannon ammunition feed
107 De-icing air heater/blower
108 A2-Stand ammunition tank (300 rounds)
109 Toilet installation
110 C-Stand ammunition feed
111 Thermos flasks
112 Circular vision port
113 MG 131 13-mm machine gun (C-Stand) at rear of 'bola'
114 'Fritz X' (Kramer X-1) radio-controlled bomb
115 Cruciform main fins
116 SAP warhead
117 Tail fin structure
118 Air-brake attachment
119 Ventral bomb rack (only fitted if forward bomb bay blanked off)
120 Forward-bomb bay (often blanked off)
121 Fuel tank retaining strap lugs
122 Internal bomb shackle
123 Bomb bay central partition
124 No. 4 (fuselage) fuel tank (1520 litre/334 Imp gal)
125 Fuel filler cap
126 Barbette remote drive cooling duct and linkage
127 Remote control dorsal barbette (B1-Stand)
128 Twin 10-mm MG 131 guns
129 No. 13 fuselage frame
130 Barbette structure
131 B1-Stand double ammunition tank (1,000 rounds per gun)
132 Central bomb bay (often blanked off)
133 Bomb bay door (outer section)
134 Port inner mainwheel well
135 No. 5 (fuselage) fuel tank (1520 litre/334 Imp gal) (Replaced by 3450 litre/759 Imp gal tank if bomb bay blanked off) (metal/self sealing)
136 Fuel filler cap
137 No. 19 fuselage frame

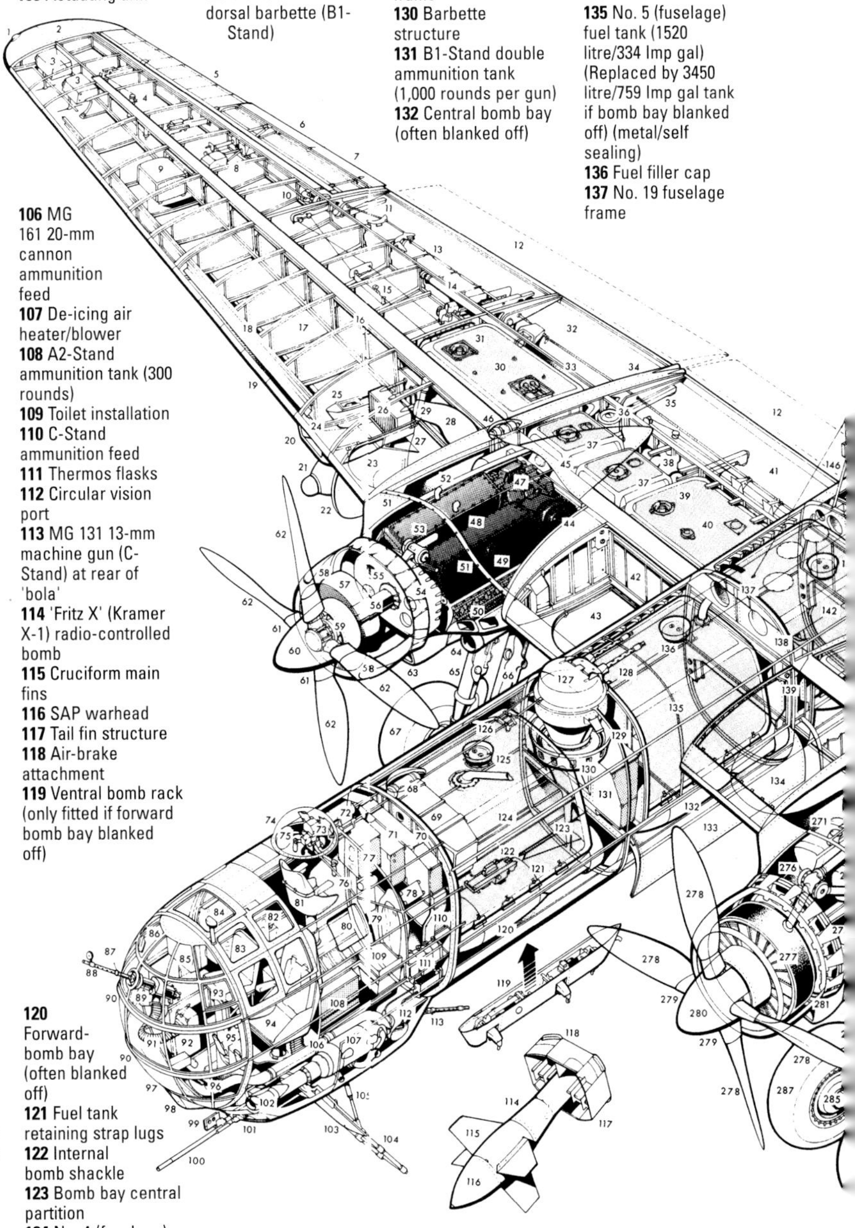

During Operation Steinbock, experienced He 177 crews found that by entering enemy airspace at 29,527 ft (9000 m), and attacking at full power and in a shallow dive at about 435 mph (700 km/h), they stood a chance of avoiding interception.

SPECIFICATION

Type
Five-seat heavy bomber and missile-carrier

Powerplant
Two Daimler-Benz DB 610A-1/B-1 24-cylinder inverted-Vee piston engines each rated at 2,950 hp (2200 kW)

Performance
Maximum speed 304 mph (490 km/h) at 19,685 ft (6000 m); cruising speed 258 mph (415 km/h) at 19,685 ft (6000 m); initial climb rate 623 ft (190 m) per minute; service ceiling 26,245 ft (8000 m); range 3,417 miles (5500 km) with two Hs 293 missiles

Weights
Empty 37,038 lb (16800 kg); max take-off 68,343 lb (31000 kg)

Dimensions
Wingspan 103 ft 1¾ in (31.44 m); length 66 ft 11¼ in (20.40 m); height 20 ft 11¾ in (6.39 m); wing area 1,097.95 sq ft (102.00 m2)

Armament
One 0.31-in (7.92-mm) MG 81J trainable forward-firing machine-gun in the nose position, one 20-mm MG 151/20 trainable forward-firing cannon in the ventral gondola, two 0.31-in (7.92-mm) MG 81 trainable rearward-firing machine-guns in the ventral gondola, two 0.51-in (13-mm) MG 131 trainable machine-guns in the remotely controlled power-operated dorsal barbette, one 0.51-in (13-mm) MG 131 trainable machine-gun in the power-operated dorsal turret and one 20-mm MG 151/20 trainable rearward-firing cannon in the tail position, plus up to 13,228 lb (6000 kg) of disposable stores carried in a lower-fuselage weapons bay and on underwing hardpoints

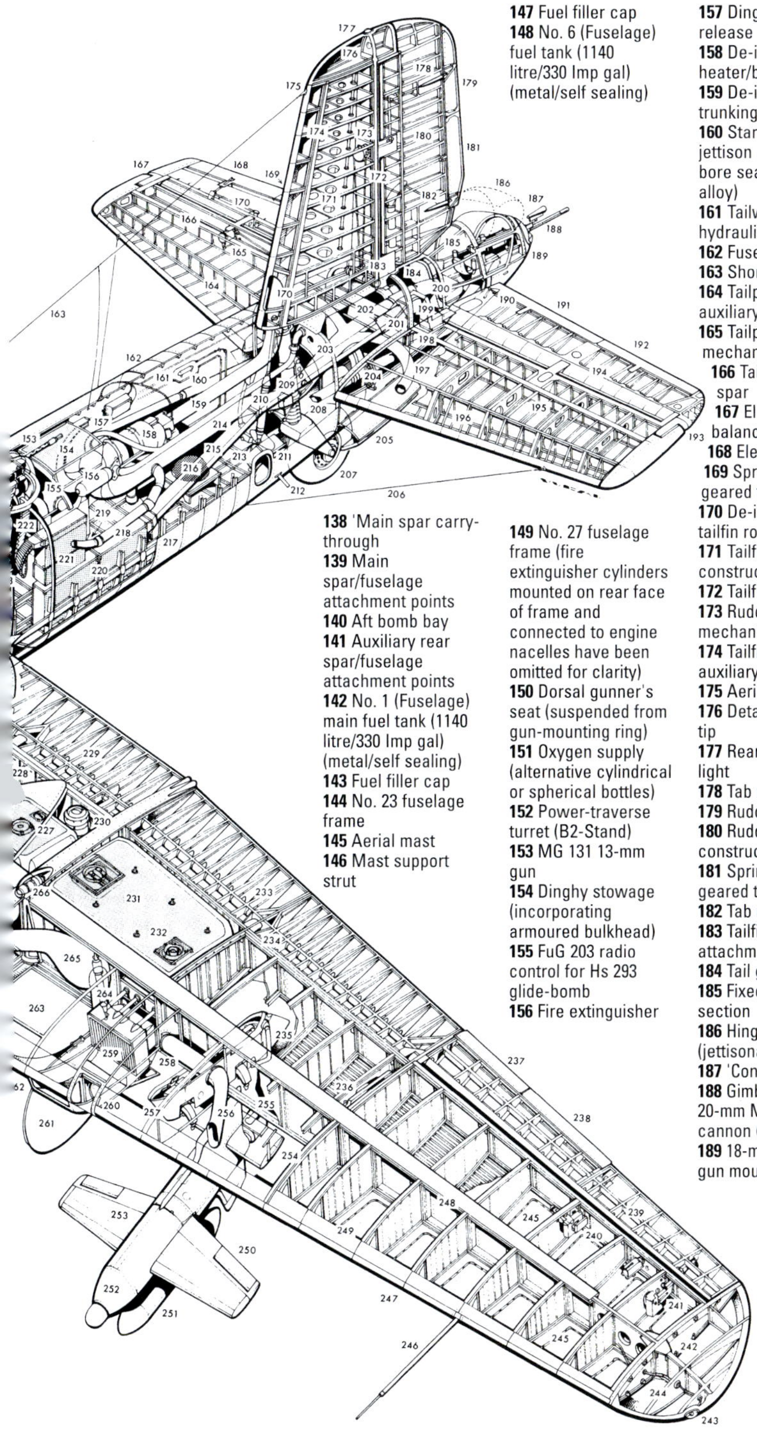

138 'Main spar carry-through
139 Main spar/fuselage attachment points
140 Aft bomb bay
141 Auxiliary rear spar/fuselage attachment points
142 No. 1 (Fuselage) main fuel tank (1140 litre/330 Imp gal) (metal/self sealing)
143 Fuel filler cap
144 No. 23 fuselage frame
145 Aerial mast
146 Mast support strut
147 Fuel filler cap
148 No. 6 (Fuselage) fuel tank (1140 litre/330 Imp gal) (metal/self sealing)
149 No. 27 fuselage frame (fire extinguisher cylinders mounted on rear face of frame and connected to engine nacelles have been omitted for clarity)
150 Dorsal gunner's seat (suspended from gun-mounting ring)
151 Oxygen supply (alternative cylindrical or spherical bottles)
152 Power-traverse turret (B2-Stand)
153 MG 131 13-mm gun
154 Dinghy stowage (incorporating armoured bulkhead)
155 FuG 203 radio control for Hs 293 glide-bomb
156 Fire extinguisher
157 Dinghy manual release
158 De-icing air heater/blower
159 De-icing intake trunking
160 Starboard fuel jettison pipe (large-bore seamed light alloy)
161 Tailwheel hydraulic lines
162 Fuselage skinning
163 Short-wave aerial
164 Tailplane forward auxiliary spar
165 Tailplane tab mechanism
166 Tailplane main spar
167 Elevator balance
168 Elevator trim tab
169 Spring-loaded geared tab
170 De-icing intake in tailfin root
171 Tailfin construction
172 Tailfin main spar
173 Rudder hinge mechanism
174 Tailfin forward auxiliary spar
175 Aerial attachment
176 Detachable tailfin tip
177 Rear navigation light
178 Tab mechanism
179 Rudder trim tab
180 Rudder construction
181 Spring-loaded geared tab
182 Tab mechanism
183 Tailfin/fuselage attachment point
184 Tail gunner's seat
185 Fixed canopy section
186 Hinged (jettisonable) hood
187 'Coned' gunsight
188 Gimbal-mounted 20-mm MG 151 cannon (H-Stand)
189 18-mm armoured gun mounting
190 Tab hinge
191 Spring-loaded geared tab
192 Elevator trim tab
193 Elevator balance
194 Elevator construction
195 Tailplane construction
196 Heated leading edge
197 Hot-air ducting
198 Tailplane/fuselage attachment points
199 H-Stand ammunition feed motor
200 Tail position hot-air
201 First-aid pack
202 Continuous main spar carry-through
203 No. 44 fuselage frame
204 Tailplane auxiliary spar/fuselage attachment points
205 Hinged tailwheel doors
206 FuG 203 aerial (Hs 293 control)
207 Tailwheel
208 Port fuel jettisonable pipe (large-bore seamed light alloy)
209 Tailwheel leg
210 Retraction mechanism
211 Rectangular vision port
212 Trailing aerial lead-in and matching unit
213 Trailing aerial winch
214 Main hot-air duct
215 H-Stand ammunition feed
216 Master compass
217 Semi-monocoque fuselage construction
218 Dorsal turret hot-air
219 Jettisonable floor/entry escape hatch
220 H-Stand ammunition tank (800 rounds)
221 B2-Stand ammunition tank (1,000 rounds)
222 Ammunition feed
223 Flexible chute
224 Empty belt link and cartridge collector box
225 Aft bomb bay door (outer section)
226 No. 2 (port inner) fuel tank (621 litre/136.5 Imp gal) (metal/self-sealing)
227 Port oil tanks
228 Auxiliary rear spar
229 Fowler flap construction (inner section)
230 Port fuel starting tank (9 litre/2 gal)
231 No. 7 (port outer) fuel tank 1120 litre/246 Imp gal) (flexible bag)
232 Fuel filler cap
233 Fowler flap construction (outer section)
234 Flap hinge fairing
235 ETC rack hot air
236 Fowler flap track attachment
237 Spring-loaded geared tab
238 Aileron trim tab
239 Port aileron construction
240 Tab mechanism
241 Aileron mechanism
242 Wingtip attachment bolts
243 Port navigation light
244 Detachable wingtip
245 Wing undersurface access/inspection panels
246 Pitot head
247 Heated leading edge
248 Main spar (outboard section)
249 Auxiliary front spar
250 Hs 293 radio-controlled glider-bomb
251 590-kg (1,300-lb) thrust rocket motor housing
252 500-kg (1,100-lb) warhead
253 Wing control surfaces
254 Tail-mounted aerial masts (radio signal receivers)
255 Tracking flare installation
256 Outboard leading-edge hot air
257 Port ETC weapons rack
258 Oil radiator outlet flap
259 Twin oil radiators (port engines)
260 Searchlight/landing light
261 Port outer mainwheel door (open position)
262 Oil radiator intake
263 Port outer mainwheel well
264 Mainwheel door actuating cylinder
265 Hot air ducting
266 Wing spar attachment point and fairing
267 Individual undercarriage/main spar attachment
268 Engine bearer ball socket
269 Hydraulic retracting jack attachment
270 Upper engine bearer
271 Coolant tanks
272 Engine support strut
273 Mainwheel oleo leg pivot points
274 Supercharger and wing de-icing intakes
275 Cooling gills
276 Engine forward mounting
277 Segmented annular radiator
278 VDM four-bladed propeller (left-handed)
279 Blade cuffs
280 Propeller boss
281 Chin intake
282 Flame damper exhaust
283 Port inboard mainwheel oleo leg (inward retracting)
284 Port outboard mainwheel oleo leg (outward retracting)
285 Mainwheel axle
286 Port outer mainwheel
287 Port inner mainwheel

Junkers Ju 87

No aircraft in history was ever as effective (when unopposed) as the infamous 'Stuka', nor so vulnerable when it encountered opposition. Its devastating effects in the early months of the war were only equalled by its dismal failure when it met the RAF over England a few weeks later.

Ju 87D-3

Cutaway key
1 Spinner
2 Pitch-change mechanism housing
3 Blade hub
4 Junkers VS 11 constant speed airscrew
5 Anti-vibration engine mounting attachments
6 Oil filler point and marker
7 Auxiliary oil tank (5.9-Imp gal/26.8-litre capacity)
8 Junkers Jumo 211J-1 12-cylinder inverted-Vee liquid-cooled engine
9 Magnesium alloy forged engine mount
10 Coolant (Glysantin-water) header tank
11 Ejector exhaust stubs
12 Fuel injection unit housing
13 Induction air cooler
14 Armoured radiator
15 Inertia starter cranking point
16 Ball joint bulkhead fixing (lower)
17 Tubular steel mount support strut
18 Ventral armour (0.315 in/8 mm)
19 Main oil tank (9.9-Imp gal/45-litre capacity)
20 Oil filling point
21 Transverse support frame
22 Rudder pedals
23 Control column
24 Heating point
25 Auxiliary air intake
26 Ball joint bulkhead fixing (upper)
27 Bulkhead
28 Oil tank (6.8-Imp gal/ 31-litre capacity)
29 Oil filler point and marker (Intava 100)

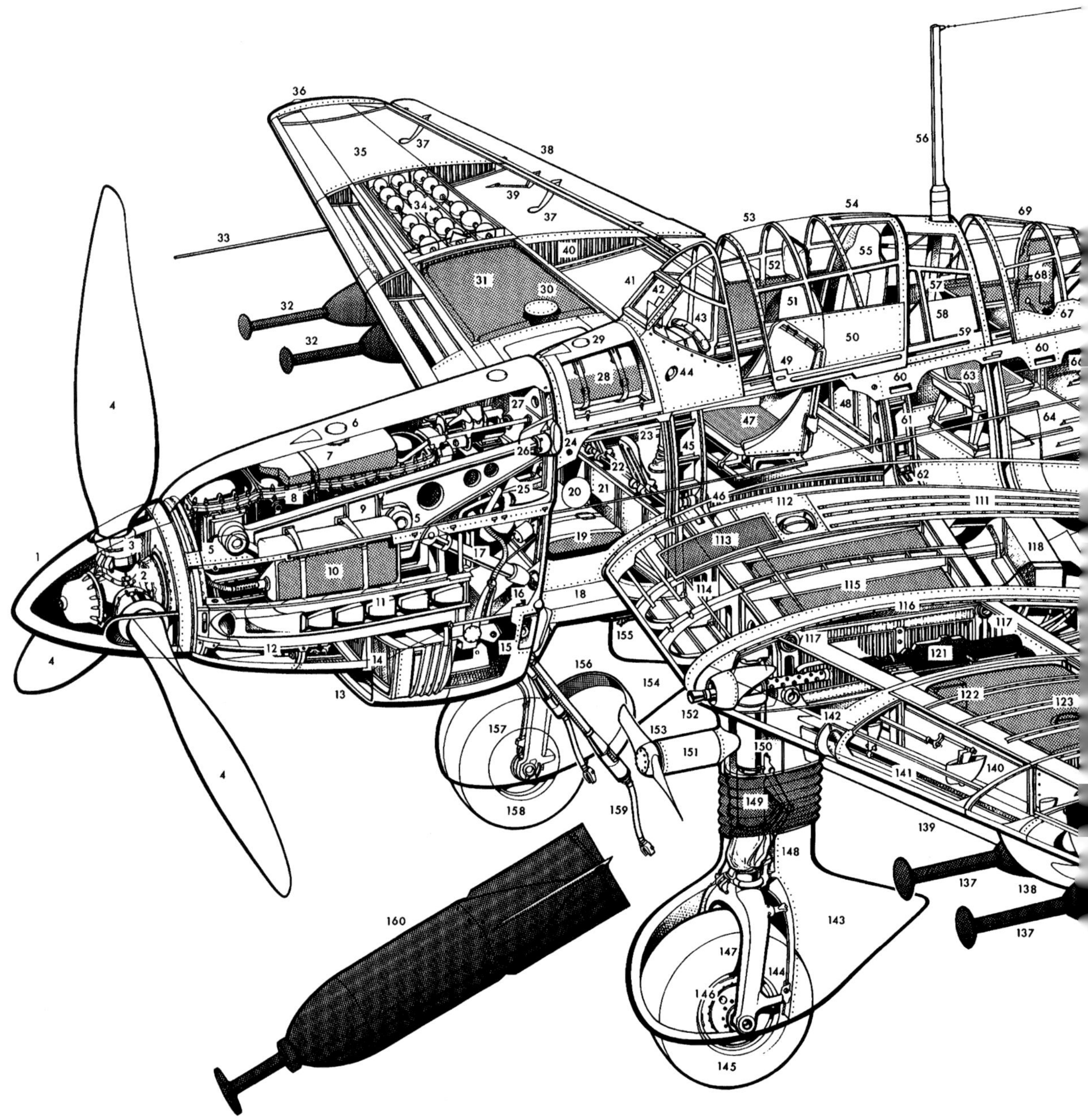

The Battle of Kursk saw the Stuka reach the end of its career as a successful dive-bomber – it was at this battle that Germany mounted its last, ultimately unsuccessful, great armoured offensive. The Ju 87 was no longer regarded as a 'terror weapon' and ground commanders had learned how to keep the number of casualties to a minimum. Evidence of this altered perspective of the Stuka can be seen in the omission of the propeller sirens from the aircraft.

SPECIFICATION

Ju 87G-1

Dimensions

Length: 37 ft 8 in (11.50 m)
Wingspan: 49 ft 2 in (15 m)
Wing area: 362.6 sq ft (33.69 m^2)
Height: 12 ft 9 in (3.9 m)

Powerplant

One Junkers Jumo 211J-1 12 cylinder inline piston engine rated at 1,400 hp (1044 kW)

Weight

Empty: 9,700 lb (4400 kg)
Normal loaded: 12,880 lb (5842 kg)
Maximum take-off: 14,550 lb (6600 kg)

Performance

Maximum speed: 195 mph (314 km/h)
Cruising speed: 118 mph (190 km/h)
Ceiling: 24,248 ft (7390 m)
Combat radius: 199 miles (320 km)

Armament

Two 37-mm BK 3,7 cannon and one flexible 0.312-in (7.92-mm) MG 81 machine-gun. In the event of no armoured targets being available, bombs could be carried instead of cannon. One 3,968-lb (1800-kg), 2,205-lb (1000-kg), 1,102-lb (500-kg) or 551-lb (250-kg) bomb could be carried beneath the fuselage and four 110-lb (50-kg), two 551-lb (250-kg) or two packs of 92 4.4-lb (2-kg) anti-personnel bombs under the wings

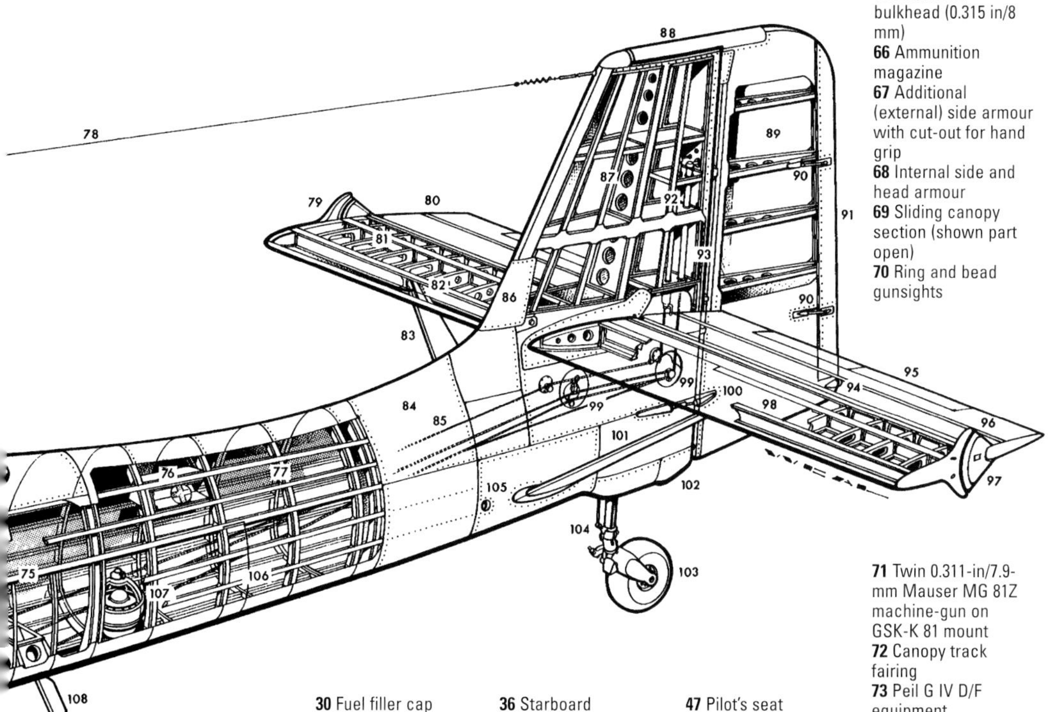

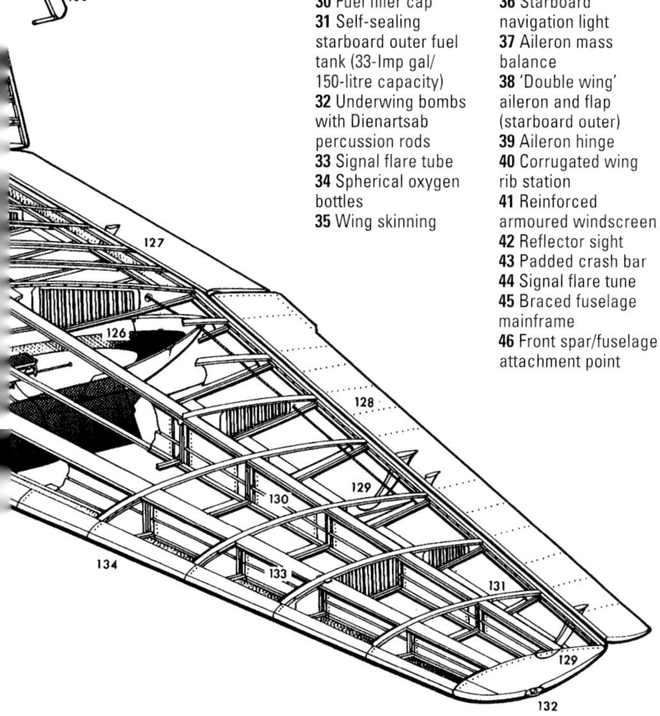

30 Fuel filler cap
31 Self-sealing starboard outer fuel tank (33-Imp gal/ 150-litre capacity)
32 Underwing bombs with Dienartsab percussion rods
33 Signal flare tube
34 Spherical oxygen bottles
35 Wing skinning
36 Starboard navigation light
37 Aileron mass balance
38 'Double wing' aileron and flap (starboard outer)
39 Aileron hinge
40 Corrugated wing rib station
41 Reinforced armoured windscreen
42 Reflector sight
43 Padded crash bar
44 Signal flare tune
45 Braced fuselage mainframe
46 Front spar/fuselage attachment point
47 Pilot's seat (reinforced with 0.158-in/ 4-mm side armour and 0.315-in/8-mm rear armour)
48 Inter-cockpit bulkhead
49 Sliding canopy handgrip
50 External side armour
51 Pilot's back armour (0.315 in/8 mm)
52 Headrest
53 Aft-sliding cockpit canopy (shown part open)
54 Radio mast cut-out
55 Anti-crash hoop (magnesium casting)
56 Radio mast
57 Radio equipment (FuGe 16) compartment
58 Additional (internal) side armour
59 Canopy track
60 Handholds/footrests
61 Braced fuselage mainframe
62 Rear spar/fuselage attachment point
63 Radio operator/ gunner's seat (folding)
64 Floor armour (0.2 in/ 5 mm)
65 Armoured bulkhead (0.315 in/8 mm)
66 Ammunition magazine
67 Additional (external) side armour with cut-out for hand grip
68 Internal side and head armour
69 Sliding canopy section (shown part open)
70 Ring and bead gunsights
71 Twin 0.311-in/7.9-mm Mauser MG 81Z machine-gun on GSK-K 81 mount
72 Canopy track fairing
73 Peil G IV D/F equipment
74 Circular plexiglass access panel
75 Back-to-back L-section stringers (fuselage horizontal break)
76 First-aid stowage
77 Z-section fuselage frames
78 Radio aerial
79 Faired elevator mass balance
80 Starboard elevator
81 Tailplane structure
82 Tailplane brace/spar attachment point
83 Tailplane bracing strut
84 Fuselage skinning
85 Control runs
86 Tailfin attachment fairing
87 Tailfin structure
88 Rudder horn balance
89 Rudder
90 Rudder trim tab controls
91 Rudder trim tab
92 Rudder control linkage
93 Rudder post
94 Rear navigation light
95 Elevator tab
96 Port elevator
97 Faired elevator mass balance
98 Tailplane front spar
99 Control pulley circular access panels
100 Rudder lower hinge fairing
101 Tailplane bracing strut
102 Emergency tailskid
103 Tailwheel
104 Tailwheel leg
105 Jacking point
106 Fuselage stringers
107 Master compass
108 Crew entry step (port and starboard)
109 Entry step support (with control run cut-outs)
110 Wingroot fairing
111 Non-slip walkway (aft section external metal strakes)
112 Fuel filler point
113 Non-slip walkway (forward section composite surface)
114 Leading-edge structure
115 Self-sealing port inner wing fuel tank (52.8-Imp gal/240-litre capacity)
116 Wing-joint external cover strip
117 Ball-and-socket wing attachment points
118 Armoured coolant radiator (port and starboard)
119 Inboard flap structure
120 Flap hinge
121 Rheinmetall-Borsig MG 17 machine-gun of 0.312-in/7.92-mm calibre (port and starboard)
122 Ammunition tank (1,000 rounds capacity) inboard of rib
123 Port outer self-sealing fuel tank (33-Imp gal/150-litre capacity)
124 Corrugated wing rib
125 ETC bomb rack support bar
126 ETC bomb rack underwing fairing
127 Port outboard flap
128 Port aileron
129 Aileron mass balance
130 Rear spar
131 Wing rib
132 Port navigation light
133 Front spar
134 Wing leading edge
135 Underwing bombload (two 110-lb/ 50-kg bombs) on multi-purpose carrier
136 Bomb shackles
137 Dienartstab percussion rod attachments
138 ETC 50/VIII fairing
139 Air brake (extended)
140 Air brake activating mechanism
141 Airbrake (retracted)
142 Landing lamp
143 Wheel spat
144 Fork/spat attachment
145 Port mainwheel
146 Brake reservoir filler port
147 Cantilever fork
148 Leather shroud
149 Oleo-pneumatic shock absorber
150 Mainwheel leg
151 Siren fairing
152 Barrel of MG 17 machine-gun
153 Wind-driven siren
154 Starboard wheel spat
155 PVC ventral bomb rack
156 Bomb cradle
157 Starboard wheelfork
158 Starboard mainwheel
159 Bomb release trapeze
160 551-lb (250-kg) bomb with Dienartstab attachment

Ju 87B-2

Stuka-(Sturzkampfflugzeug)-geschwader 77 was involved in the Ju 87's main operations throughout the war, elements of its complement taking part in the Polish campaign. Following the successful march through northern Europe, the unit then faced the fighters of the RAF in the Battle of Britain, and suffered accordingly. A successful campaign in Greece and the Balkans followed, before the Stukageschwader turned its attentions to the Eastern Front, where it fought until late 1943. By this time, the unit had been redesignated Schlachtgeschwader and was transitioning to the more powerful Fw 190.

Powerplant

The Ju 87B was powered by a Junkers Jumo 211Da 12-cylinder liquid-cooled engine. This unit was rated at 1,200 hp (900 kW) for take-off (2,400 rpm) and 1,100 hp (825 kW) at 4,920 ft (1500 m). The increase in power offered by this engine over the earlier Jumo 210 of the A-series enabled a greater bombload to be carried. The radiator was housed in an armoured 'bath' beneath the engine. Hydraulically-operated cooling gills immediately behind the radiator increased airflow through the engine at low speeds. The Ju 87B-1 model featured simple port exhausts, but the B-2 introduced ejector-type stubs behind an aerodynamic fairing. Angled back, these provided a small but useful amount of thrust. The Ju 87D offered more power, with a Jumo 211J-1 rated at up to 1,410 hp (1050 kW) with an induction air cooler and a strengthened crankshaft.

Fuel

Fuel was carried in two large tanks mounted in the inboard (anhedral) wing sections. The Ju 87R and later variants introduced additional fuel in the outboard wing sections.

Undercarriage

The immensely sturdy main undercarriage was shrouded by large trouser fairings around the legs, with spats around the wheels. This arrangement replaced the braced and trousered main gear of the Ju 87A. On the Eastern Front in the winter, many Ju 87s operated with their spats removed, since mud quickly accumulated and clogged the wheels.

Entry steps

On either side of the rear fuselage were mounted permanent steps, just aft of the wing trailing edge. Both crew entered the aircraft by climbing up these steps on to the wing and thence to their respective cockpits via hand/footholds.

Cockpit
The Ju 87 featured a crew of two, although Bs were often flown as single-seaters. Both pilot and radio operator/gunner sat under separate sliding canopies, the latter facing to the rear. Armour was provided where possible.

Control balances
A pair of large weights projecting from the underside of the wing balanced each aileron. The elevators had faired mass balances, resulting in the distinctive surfaces at the tips, while the rudder had a slim horn balance at the fintip. Two sections of trim tabs were incorporated in the elevators, while the rudder had a one-piece tab running virtually the full height of the fin.

Rear gun
To provide a limited measure of protection for the woefully vulnerable Ju 87, a single MG 15 0.312-in (7.92-mm) machine-gun was placed on a flexible mount in the rear of the cockpit, aimed by the radio operator. Later variants introduced the more potent Mauser MG 81Z twin-gun mount.

Wings
The characteristic inverted gull wings of the Ju 87 were built around a two-spar structure with closely-spaced ribs. The centre-section was integral with the fuselage. The cranked wing proved immensely strong, and kept the length of the fixed undercarriage short; the classic Junkers 'double wing' arrangement was used. The inboard surfaces (two sections) acted as flaps, while the outboard surfaces provided roll control.

Tailplane
The strong tailplane was a two-spar structure. On the Ju 87B it was externally braced by two struts; on the refined Ju 87D these struts were formed into one aerodynamic strut. The elevators were not large, but provided enough authority to pull the aircraft out of a 90° dive with ease.

Junkers Ju 88

Ju 88A-5 (M2-MK), of Küstenfliegergruppe 106, tasked with attacking shipping in the Irish Sea on 26 November 1941, landed by miscalculation at RAF Chivenor. After a wheels-down landing, the aircraft was flown to Duxford, with a Hudson escort, to join No. 1426 (Enemy Aircraft) Flight. The machine later starred in the film 'In Which We Serve'.

Ju 88G-1

Cutaway key
1 Starboard navigation light
2 Wingtip profile
3 FuG 227 Flensburg radar receiver antenna
4 Starboard aileron
5 Aileron control runs
6 Starboard flaps
7 Flap-fairing strip
8 Wing ribs
9 Starboard outer fuel tank (91-Imp gal/ 415-litre capacity)
10 Fuel filler cap
11 Leading-edge structure
12 Annular exhaust slot
13 Cylinder head fairings
14 Adjustable nacelle nose ring
15 Twelve-bladed cooling fan
16 Propeller boss
17 Variable-pitch VS 111 wooden propeller
18 Leading-edge radar array
19 FuG 220 Lichtenstein SN-2 intercept radar array
20 Nose cone
21 Forward armoured bulkhead
22 Gyrocompass
23 Instrument panel
24 Armoured glass windscreen
25 Folding seat
26 Control column
27 Rudder pedal/brake cylinder
28 Control runs
29 Pilot's armoured seat
30 Sliding window section
31 Headrest
32 Jettisonable canopy roof section
33 Gun restraint
34 Wireless-operator/ gunner's seat
35 Rheinmetall Borsig MG 131 machine-gun (0.51-in/13-mm calibre)
36 Radio equipment (FuG 10P HF, FuG 16ZY VHF, FuG 25 IFF)
37 Ammunition box (500 rounds of 0.51-in/ 13-mm)
38 FuG 220 Lichtenstein SN-2 indicator box
39 FuG 227 Flensburg indicator box
40 Control linkage
41 Bulkhead
42 Armoured gunmount
43 Aerial post traverse check
44 Fuel filler cap
45 Whip aerial
46 Forward fuselage fuel tank (105-Imp gal/ 480-litre capacity)
47 Fuselage horizontal construction joint
48 Bulkhead
49 Fuel filler cap
50 Aft fuselage fuel tank (230-Imp gal/1045-litre capacity)

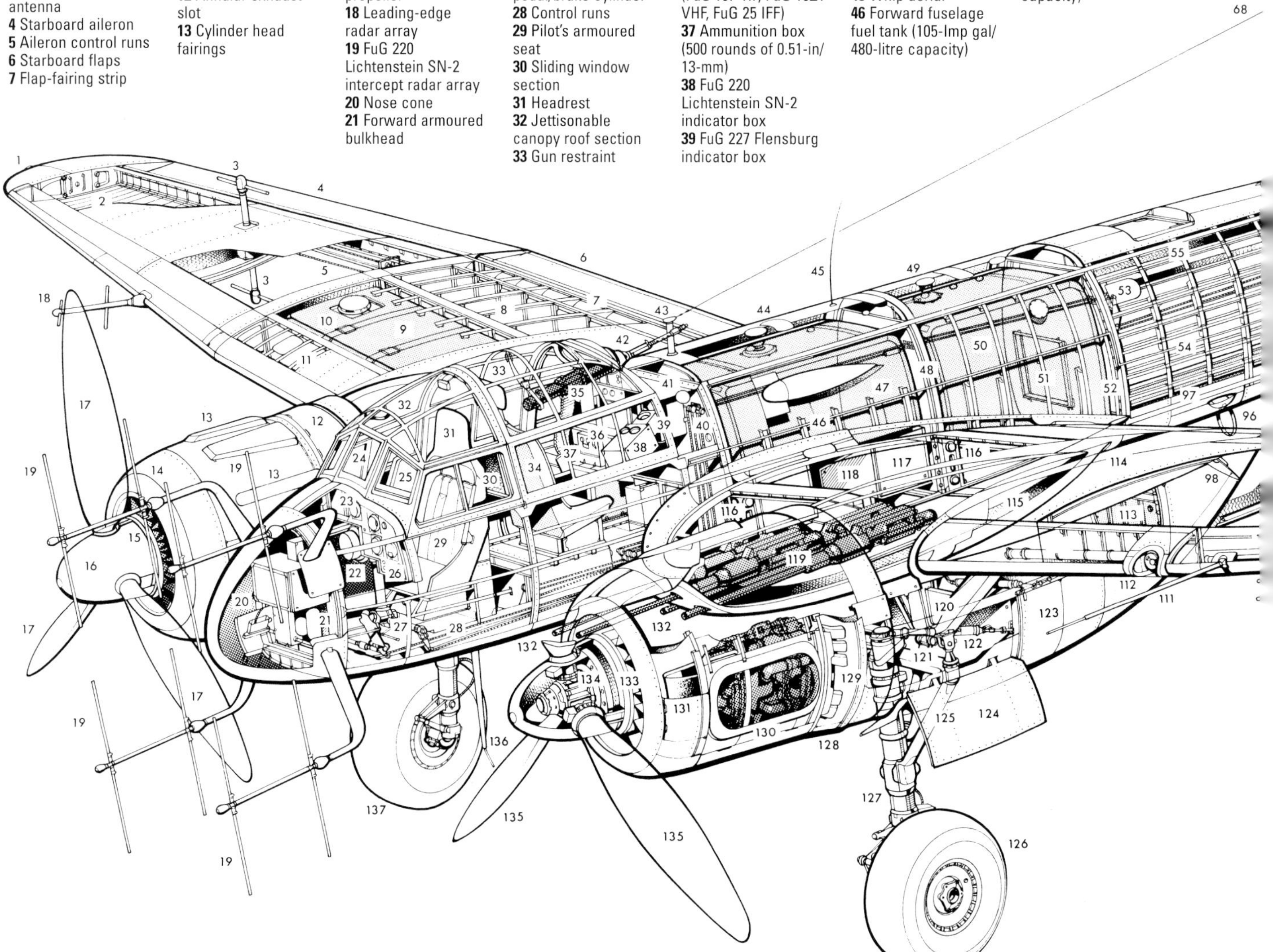

Illustrated is a Ju 88A-1 (or perhaps an A-5) of KG 30, the first operational Ju 88 unit. Operating initially in the anti-shipping role from bases around Scandinavia, Germany, and the Netherlands, the Gruppe was initially known as I./KG 25, prior to a change of designation in September 1939. KG 30's first major mission came on 26 September 1939, when aircraft struck British shipping, although damage was limited due to the fact that the SC 500 bombs of the Ju 88A-1s and A-0s failed to detonate. On 9 October two KG 30 Ju 88A-1s were shot down – the first combat losses of the type.

SPECIFICATION

Ju 88A-4

Dimensions

Length: 47 ft 2¾ in (14.40 m)
Height: 15 ft 11 in (4.85 m)
Wingspan: 65 ft 7½ in (20.00 m)
Wing area: 586.63 sq ft (54.50 m²)
Wing loading: 41.8-46.9 lb/sq ft (204-229 kg/m²)

Powerplant

Two liquid-cooled Junkers Jumo 211J-1 or J-2 12-cylinder inline piston engines each rated at 1,340-hp (999-kW)

Weights

Empty: 21,737 lb (9860 kg)
Normal loaded: 26,686 lb (12105 kg)
Maximum loaded: 30,865 lb (14000 kg)

Performance

Maximum speed at 17,390 ft (5300 m): 292 mph (470 km/h) at 27,557 lb (12500 kg)
Climb to 16,405 ft (5000 m): 18 mins
Climb to 17,716 ft (5400 m): 23 mins
Economical cruising speed: 230 mph (370 km/h) at 17,390 ft (5300 m)
Maximum cruising speed: 248 mph (399 km/h) at 16,405 ft (5000 m)
Service ceiling: 26,900 ft (8199 m)
Range: 1,112 miles (1789 km) with 637 Imp gal (2896 litres) of fuel
Maximum range: 1,696 miles (2730 km) with 886 Imp gal (4028 litres) of fuel
Take-off speed: 115 mph (185 km/h)
Approach speed: 155 mph (250 km/h)

Armament

Defensive: one fixed or free-mounted forward-firing 0.31-in (7.92-mm) MG 81 machine-gun and one free-mounted forward-firing 0.51-in (13-mm) MG 131 or two 0.31-in (7.92-mm) MG 81 machine-guns, plus two 0.31-in (7.92-mm) MG 81 machine-guns firing aft above the fuselage, and one 0.51-in (13-mm) MG 131 machine-gun or two 0.31-in (7.92-mm) MG 81 machine-guns firing aft below the fuselage
Offensive: ten 50-kg (110-lb) SC 50 bombs internally, and four 250-kg (551-lb) SC 250 or two 500-kg (1,105-lb) SC 500 bombs externally, or four 500-kg (1,105-lb) SC 500 bombs externally

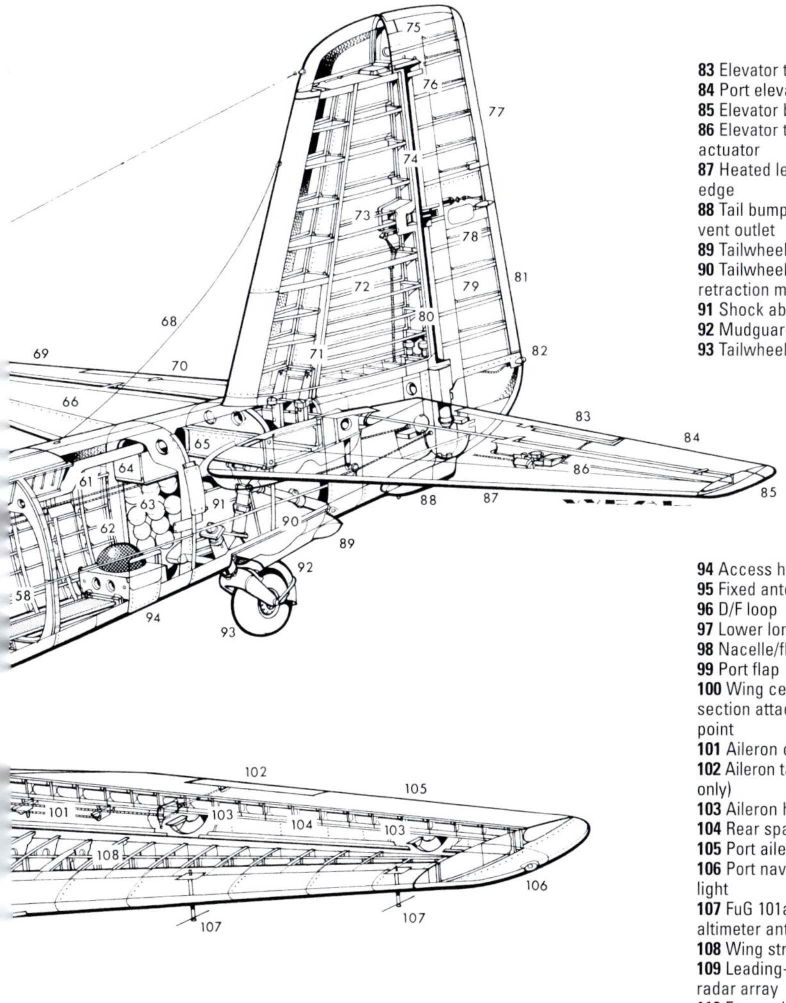

51 Access hatch
52 Bulkhead
53 Control linkage access plate
54 Fuselage stringers
55 Upper longeron
56 Maintenance walkway
57 Control linkage
58 Horizontal construction joint
59 Z-section fuselage frames
60 Dinghy stowage
61 Fuel vent pipe
62 Master compass
63 Spherical oxygen bottles
64 Accumulator
65 Tailplane centre section carry-through
66 Starboard tailplane
67 Elevator balance
68 Aerial
69 Starboard elevator
70 Elevator tab
71 Tailfin forward spar/ fuselage attachment
72 Tailfin structure
73 Rudder actuator
74 Rudder post
75 Rudder mass balance
76 Rudder upper hinge
77 Rudder tab (upper section)
78 Inspection/ maintenance handhold
79 Rudder structure
80 Tailfin aft spar/ fuselage attachment
81 Rudder tab (lower section)
82 Rear navigation light
83 Elevator tab
84 Port elevator
85 Elevator balance
86 Elevator tab actuator
87 Heated leading edge
88 Tail bumper/fuel vent outlet
89 Tailwheel doors
90 Tailwheel retraction mechanism
91 Shock absorber leg
92 Mudguard
93 Tailwheel
94 Access hatch
95 Fixed antenna
96 D/F loop
97 Lower longeron
98 Nacelle/flap fairing
99 Port flap
100 Wing centre/outer section attachment point
101 Aileron controls
102 Aileron tab (port only)
103 Aileron hinges
104 Rear spar
105 Port aileron
106 Port navigation light
107 FuG 101a radio altimeter antenna
108 Wing structure
109 Leading-edge radar array
110 Forward spar
111 Pitot head
112 Landing lamp
113 Mainwheel well rear bulkhead
114 Port outer fuel tank location (91-Imp gal/ 415-litre capacity)
115 Ventral gunpack (offset to port)
116 Ball-and-socket fuselage/wing attachment points
117 Port inner fuel tank location (934-Imp gal/ 425-litre capacity)
118 Ammunition boxes for MG 151 cannon (200 rpg)
119 Mauser MG 151/20 cannon (four) of 20-mm calibre
120 Mainwheel leg retraction yoke
121 Leg pivot member
122 Mainwheel door actuating jack
123 Mainwheel door (rear section)
124 Mainwheel door (forward section)
125 Leg support strut
126 Port mainwheel
127 Mainwheel leg
128 Annular exhaust slot
129 Exhaust stubs (internal)
130 BMW 801D air-cooled radial engine (partly omitted for clarity)
131 Annular oil tank
132 Cannon muzzles (depressed 5°)
133 Twelve-bladed cooling fan
134 Propeller mechanism
135 Variable-pitch wooden VS 111 propeller
136 FuG 167 antenna
137 Starboard mainwheel

Lt Johannes Geismann of I. Gruppe/KG 77 poses in front of the impressive kill tally on the rudder of his Ju 88. KG 77 transferred from the Russian Front to Sicily in the early summer of 1942, from where it was primarily engaged in attacks against Allied shipping involved in supplying Malta and North Africa.

Ju 88A-5

This Junkers Ju 88 wears the codes of KG 30, whose I. Gruppe was the first Luftwaffe unit to employ the Ju 88 in combat when, on 26 September 1939, it undertook a mission attacking British warships in the Firth of Forth. Based initially at Westerland-Sylt with Ju 88A-1s, the geschwader quickly re-located to Denmark, Norway and, as depicted here, the Netherlands, by which time the unit had re-equipped with the improved Ju 88A-5 model. During the Battle of Britain, KG 30 was heavily involved in raids over the British Isles, suffering heavy losses. For much of the remainder of the conflict, the unit was preoccupied with anti-shipping missions in the North Sea, ranging from convoy attacks off the Norwegian coast to high-speed raids on naval vessels based in British harbours.

Cockpit
The four-man crew sat close together in the forward fuselage. The pilot sat in the front of the upper cockpit, offset to port, while below him and to starboard sat the bombardier/second pilot, who had access to the glazed nose. In the rear of the upper cabin sat the flight engineer, facing to the rear to operate the gun in the back of the canopy. Alongside him, slightly below, sat the radio operator, who had access to the ventral gondola to operate the rear-facing lower gun. A prominent mast from the cockpit roof held a single wire aerial running back to the tail.

Bombload
Two fuselage bomb bays were provided, carrying a maximum of 28 50-kg (110-lb) SC 50 bombs. Four external racks were fitted under the inner wings which could carry a 500-kg (1,102-lb) SC 500 bomb or a similar weapon, while the A-4/A-5 introduced the provision for a further bomb rack under each outer wing. As illustrated, the Ju 88 could also carry the Luftminen Type B (LM-B) mine. Due to the low level from which the weapon was dropped, parachute retarding was required either to break the fall of the mine when used at sea, or to delay the detonation when used as a bomb over land. The rear of the mine was released on dropping to allow the parachute to deploy.

Wing structure
The dihedral wing was of stressed-skin construction around a strong two-spar box. The A-4/A-5 introduced a new wing with extended tips which allowed operations at higher weights. For divebombing, slatted divebrakes were fitted underneath the wing leading edge, swinging forward into the airflow when deployed. Metal-skinned ailerons replaced the fabric-covered units of earlier variants and drooped in concert with the flaps to provide additional lift at low speeds. The leading edges of the wings could be de-iced by hot air bled from the engines.

Powerplant
The Junkers Ju 88A-5 preceded the A-4 on the production lines due to development delays with the latter's Jumo 211J engines. The A-5 had the A-4's new wing but retained the Jumo 211G-1 engines of the A-1/A-2 models. These were 12-cylinder liquid-cooled units which produced 1,200 hp (894 kW) each at take-off settings. The circular engine cowlings of the Ju 88 were deceiving, for the engines carried within were actually inline units. The circular cross-section section allowed the use of annular radiators and oil coolers.

Rear fuselage
Oxygen bottles and the radio equipment were mounted in the rear fuselage just forward of the tail. A hatch in the rear wall of the aft bomb bay allowed maintenance personnel access to the rear fuselage, with a walkway provided back to the equipment. At the very rear of the fuselage, fuel could be rapidly jettisoned in flight through a fairing situated beneath the tail. Forward of this fairing was the fully-retracting tailwheel which was enclosed, when retracted, by bulged doors.

Armament
Early Ju 88 variants had the standard defensive armament of three 0.311-in (7.9-mm) MG 15 machine-guns. One was fired by the pilot from the windscreen while two faced aft, one from the rear of the canopy (fired by the flight engineer) and one from the rear of the ventral gondola (fired by the radio operator). Cockpit weapons were all operated through separate ball/socket mountings. This armament generally proved to be insufficient, and aircraft were steadily improved with additional lateral-firing weapons and guns in the nose panel. Other alterations included the substitution of the MG 15 machine-gun for the far superior MG 81 and 0.51-in (13-mm) MG 131. The bombardier was provided with a bombsight in the extreme nose for conventional bombing. For dive-bombing attacks, the pilot had a sight mounted on the cockpit ceiling, which could be swung sideways when not in use. In all, there were at least 40 different armament schemes employed by the Ju 88.

Junkers Ju188

A line-up of Ju 188D-2s of 1.(F)/FAGr 124 at Kirkenes, Norway. The Ju 188D-2 was intended primarily for the maritime strike and reconnaissance roles, and usually carried FuG 200 Hohentwiel radar.

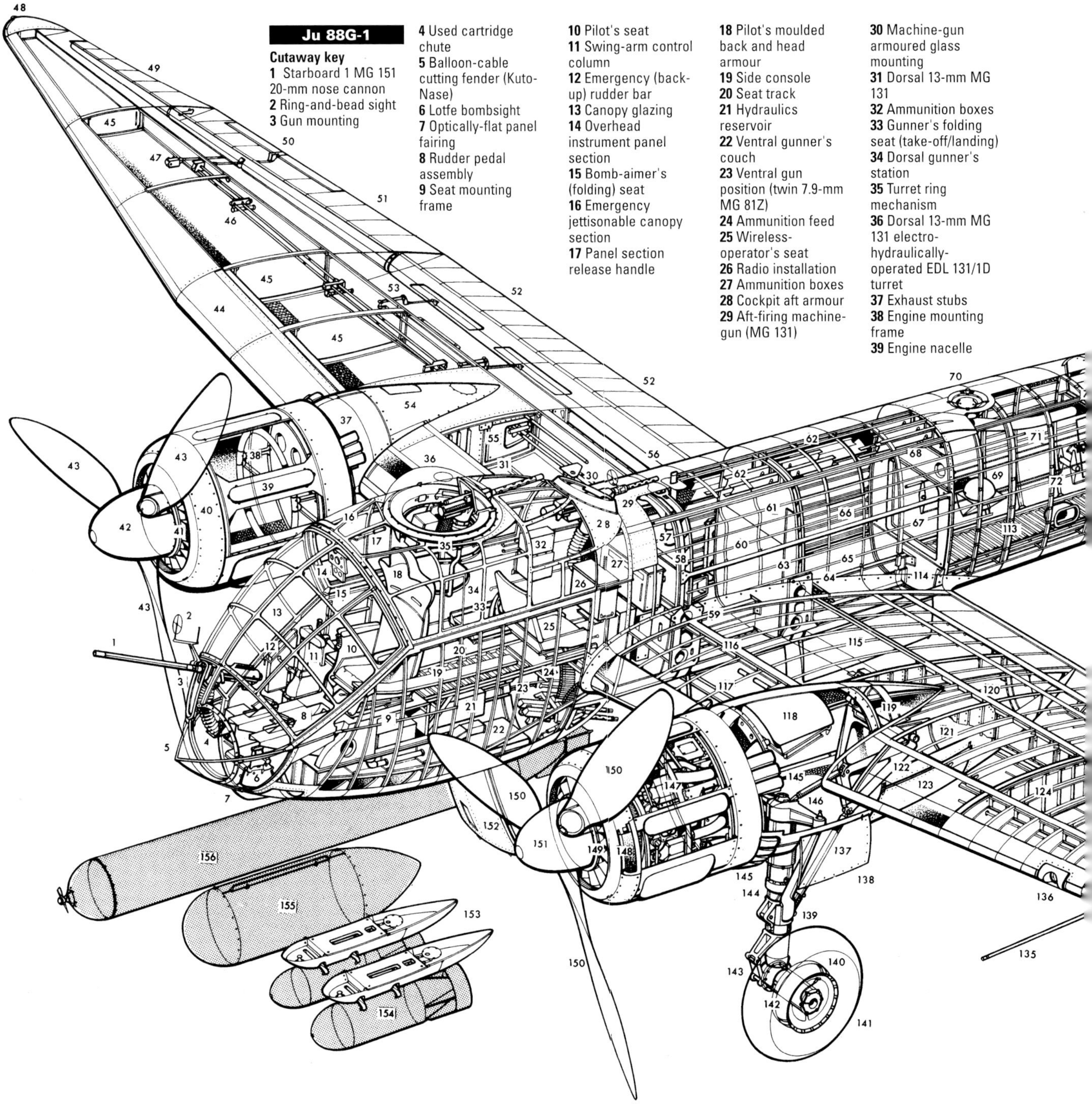

Ju 88G-1

Cutaway key

1 Starboard 1 MG 151 20-mm nose cannon
2 Ring-and-bead sight
3 Gun mounting
4 Used cartridge chute
5 Balloon-cable cutting fender (Kuto-Nase)
6 Lotfe bombsight
7 Optically-flat panel fairing
8 Rudder pedal assembly
9 Seat mounting frame
10 Pilot's seat
11 Swing-arm control column
12 Emergency (back-up) rudder bar
13 Canopy glazing
14 Overhead instrument panel section
15 Bomb-aimer's (folding) seat
16 Emergency jettisonable canopy section
17 Panel section release handle
18 Pilot's moulded back and head armour
19 Side console
20 Seat track
21 Hydraulics reservoir
22 Ventral gunner's couch
23 Ventral gun position (twin 7.9-mm MG 81Z)
24 Ammunition feed
25 Wireless-operator's seat
26 Radio installation
27 Ammunition boxes
28 Cockpit aft armour
29 Aft-firing machine-gun (MG 131)
30 Machine-gun armoured glass mounting
31 Dorsal 13-mm MG 131
32 Ammunition boxes
33 Gunner's folding seat (take-off/landing)
34 Dorsal gunner's station
35 Turret ring mechanism
36 Dorsal 13-mm MG 131 electro-hydraulically-operated EDL 131/1D turret
37 Exhaust stubs
38 Engine mounting frame
39 Engine nacelle

The Ju 88 V44 was the second of the Ju 188 development vehicles and introduced the enlarged tail surfaces. As such, it was redesignated as the Ju 188 V1 during mid-1942, and was joined on the flight test programme by another aircraft to hasten development.

SPECIFICATION

Junkers Ju118

Type
Four-seat medium bomber

Powerplant
Two 1,700-hp (1268-kW) BMW 801D-2 radial piston engines

Performance
Maximum speed 310 mph (500 km/h) at 19,685 ft (6000 m); service ceiling 30,660 ft (9345 m); range 1,209 miles (1945 km)

Weights
Empty equipped 21,737 lb (9860 kg); maximum take-off 31,989 lb (9750 kg)

Dimensions
Wingspan 72 ft 2 in (22.00 m); length 49 ft 1/2 in (14.95 m); height 14 ft 7 in (4.44 m); wing area 602.80 sq ft (56 m2)

Armament
One forward-firing 20-mm MG 151 cannon in nose, a single 0.51-in (13-mm) MG 131 in dorsal turret and at the rear of the cockpit canopy, and one 0.31-in (7.92-mm) MG 18 machine-gun in lower front fuselage firing aft, plus a maximum bombload of 6,614 lb (3000 kg)

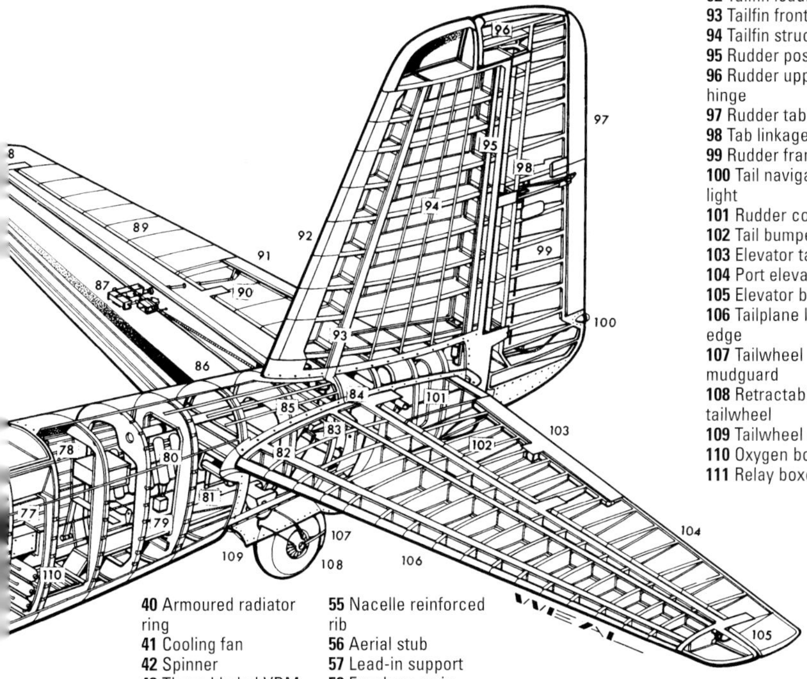

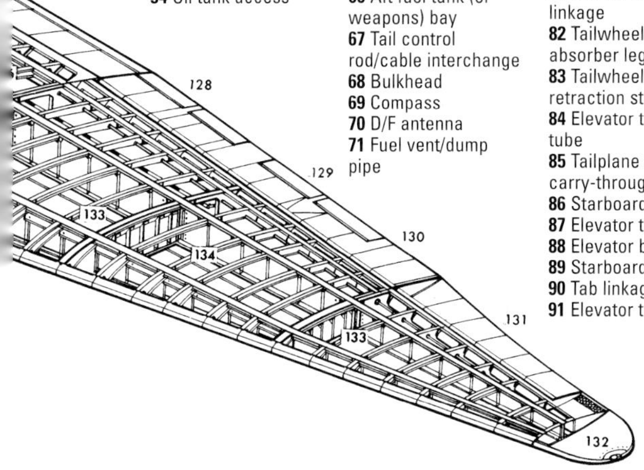

40 Armoured radiator ring
41 Cooling fan
42 Spinner
43 Three-bladed VDM propeller
44 Wing leading edge
45 Main rib stations
46 Control rod linkage
47 Tab servo
48 Starboard navigation light
49 Outer aileron
50 Aileron servo tab
51 Inner aileron
52 Starboard flaps
53 Flap mechanism
54 Oil tank access
55 Nacelle reinforced rib
56 Aerial stub
57 Lead-in support
58 Fuselage main frame
59 Front spar/fuselage attachment
60 Forward fuel tank bay
61 Bulkhead
62 Fuel tank support beams
63 Aft spar/fuselage attachment
64 Flaps motor
65 Centre keel
66 Aft fuel tank (or weapons) bay
67 Tail control rod/cable interchange
68 Bulkhead
69 Compass
70 D/F antenna
71 Fuel vent/dump pipe
72 Elevator control cables
73 Water filter
74 Dinghy release cord spool
75 Dinghy stowage
76 Electrics panel
77 First-aid kit
78 Fuselage frames
79 Rudder internal mass balance
80 Oxygen cylinders
81 Rudder control linkage
82 Tailwheel shock-absorber leg
83 Tailwheel retraction strut
84 Elevator torque tube
85 Tailplane spar carry-through
86 Starboard tailplane
87 Elevator tab motor
88 Elevator balance
89 Starboard elevator
90 Tab linkage
91 Elevator tab
92 Tailfin leading edge
93 Tailfin front spar
94 Tailfin structure
95 Rudder post
96 Rudder upper hinge
97 Rudder tabs
98 Tab linkage
99 Rudder frame
100 Tail navigation light
101 Rudder controls
102 Tail bumper
103 Elevator tab
104 Port elevator
105 Elevator balance
106 Tailplane leading edge
107 Tailwheel mudguard
108 Retractable tailwheel
109 Tailwheel doors
110 Oxygen bottles
111 Relay boxes
112 Auto-pilot control
113 Aft fuselage crawlway
114 Wingroot fillet
115 Aft (bomb) bay doors
116 Front spar
117 Front bay doors
118 Oil tank
119 Nacelle aft structure
120 Rear spar
121 Mainwheel well
122 Undercarriage attachment
123 Mainwheel doors
124 Rib station
125 Intermediate ribs
126 Control rods
127 Port flaps
128 Aileron trim tab
129 Aileron servo tab
130 Inner port aileron
131 Outer port aileron
132 Port navigation light
133 Outer rib stations
134 Wing structure
135 Pitot head
136 Landing light
137 Undercarriage retraction strut
138 Mainwheel leg door
139 Strut/leg join
140 Brake drum
141 Port mainwheel
142 Axle
143 Torque links
144 Mainwheel leg
145 Exhaust stubs
146 Undercarriage pivot
147 BMW 801 C radial air-cooled engine
148 Oil cooler circular radiator
149 Cooling fan
150 Three-bladed VDM propeller
151 Spinner
152 Crew ventral entry hatch
153 Wingroot ETC weapon racks for
154 Two bombs, or alternatively
155 Auxiliary fuel tank (Ju 188F) or
156 Aerial torpedo (Ju 188E-2)

The BMW 801-powered Ju 188E series was delivered ahead of the Jumo 213-powered Ju 188A. This pre-production Ju 188E-0 was modified to act as a fast staff transport for General-Luftzeugmeister Erhard Milch.

Lockheed A-29 Hudson

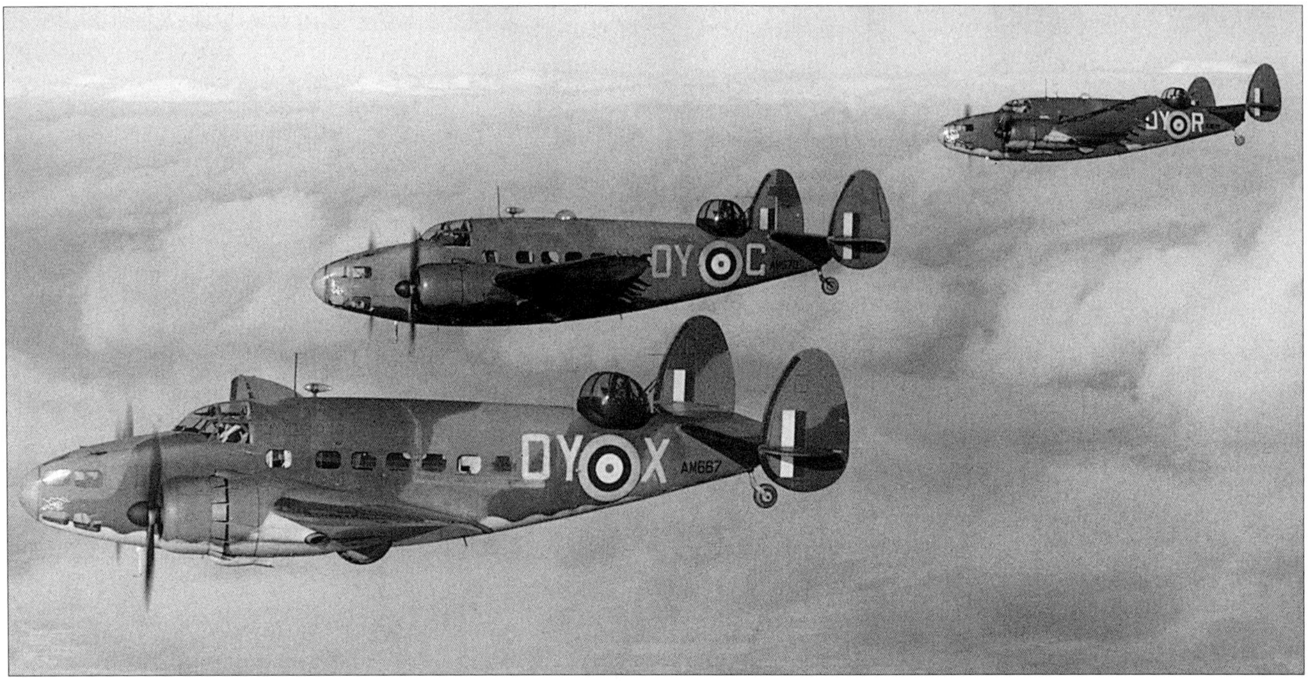

By the beginning of 1942, Coastal Command possessed 12 squadrons of Hudsons (these examples are from No. 48 Sqn), but only five of these were engaged on anti-shipping duties, while the remainder carried out anti-submarine work or air-sea rescue. The Hudsons conducted daylight sorties over the North Sea, along the coasts of Norway and the Netherlands, hunting singly or in small groups.

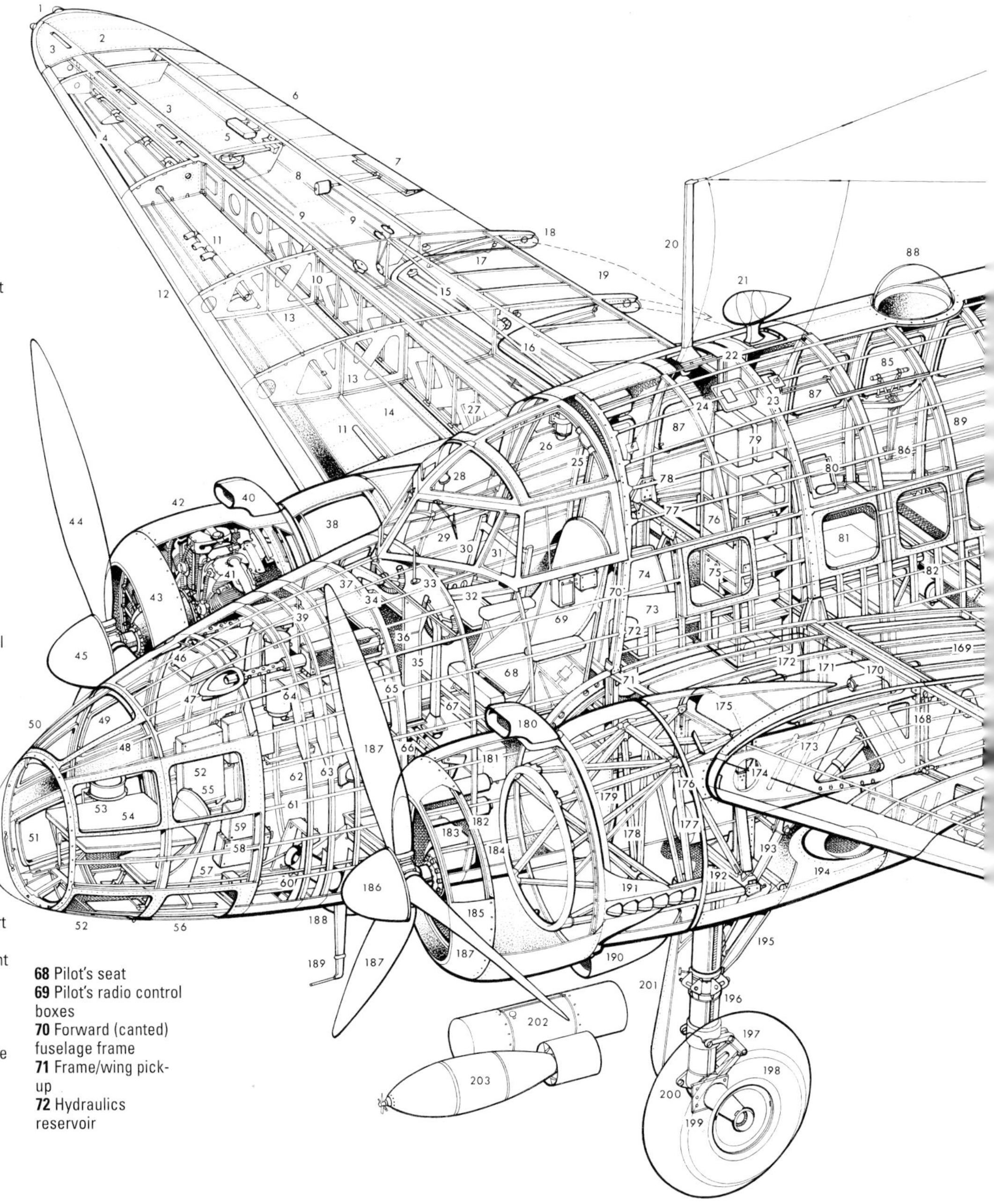

A-29 Hudson

Cutaway key
1 Starboard navigation/identification lights
2 Starboard wingtip
3 De-icing slots
4 Internal vanes
5 Aileron internal mass balance
6 Starboard aileron
7 Aileron tab
8 Tab mechanism
9 Control cables
10 Wing main spar structure
11 De-icing tubes
12 Leading-edge de-icing boot
13 Main wing rib stations
14 Wing skinning
15 Flap control cables
16 Flap tracks
17 Flap cables/pulleys
18 Track fairings
19 Port flap (extended)
20 Aerial mast
21 D/F loop fairing
22 Supported structure
23 Aerial lead-in
24 Cockpit cold air
25 Flight deck sun-blind frames
26 Windscreen wiper motor
27 Jettisonable canopy hatch
28 Console light
29 Windscreen wipers
30 Second pilot's jump seat
31 Adjustable quarterlight
32 Windscreen frame support member
33 External gunsight
34 Second-pilot's (back-up) centre column (cantilevered)
35 Central instrument console
36 Starboard nose compartment entry tunnel
37 Bulkhead
38 Starboard engine oil tank
39 Fixed forward firing 0.303-in (7.7-mm) Browning machine-guns (two)
40 Carburettor intake
41 Wright R-1820-GI-02A radial engine
42 Starboard nacelle
43 Cowling nose ring
44 Three-bladed propeller
45 Spinner
46 Nose compartment cold air
47 Machine-gun muzzles
48 Nose structure
49 Roof glazing
50 Window frames
51 Nosecone
52 Navigator's side windows
53 Compass
54 Navigator's table
55 Navigator's (sliding) seat
56 Bomb-aimer's flat panels
57 Bomb-aimer's prone position
58 Bomb selector/switch panel
59 Navigator's instrument panel
60 Forward flare chute
61 Bombsight support
62 Nose frames
63 Nose compartment warm air
64 Windscreen de-icing tank
65 Machine-gun ammunition magazine
66 Rudder pedal assembly
67 Pilot's control column
68 Pilot's seat
69 Pilot's radio control boxes
70 Forward (canted) fuselage frame
71 Frame/wing pick-up
72 Hydraulics reservoir

One Hudson Mk IV was built for, and delivered to, Sperry Gyroscope Co. Registered NX21771, this aircraft was used by Sperry for the testing of aircraft instruments. The aircraft was still being used in 1999 for research and development activities.

SPECIFICATION

Hudson Mk VI

Dimensions

Length: 44 ft 4 in (13.51 m)
Wingspan: 65 ft 6 in (19.96 m)
Height: 11 ft 11 in (3.63 m)
Wing area: 551 sq ft (51.19 m²)

Powerplant

Two 1,200-hp (895-kW) Pratt & Whitney R-1830-S3C4-G Twin Wasp radial piston engines

Weights

Empty: 12,929 lb (5864 kg)
Loaded: 18,500 lb (8391 kg)
Maximum take-off: 22,360 lb (10,142 kg)

Performance

Maximum speed at 15,000 ft (4572 m): 261 mph (420 km/h)
Cruising speed: 224 mph (360 km/h)
Service ceiling: 27,000 ft (8230 m)
Range: 2,160 miles (3476 km)
Endurance: 6 hours 55 minutes

Armament

Twin 0.303-in (7.7-mm) machine-guns in fixed forward and dorsal turret installations and one 0.303-in (7.7-mm) machine-gun in a ventral position, plus two optional 0.303-in (7.7-mm) machine-guns in beam positions, plus up to 1,000 lb (454 kg) of bombs or depth charges

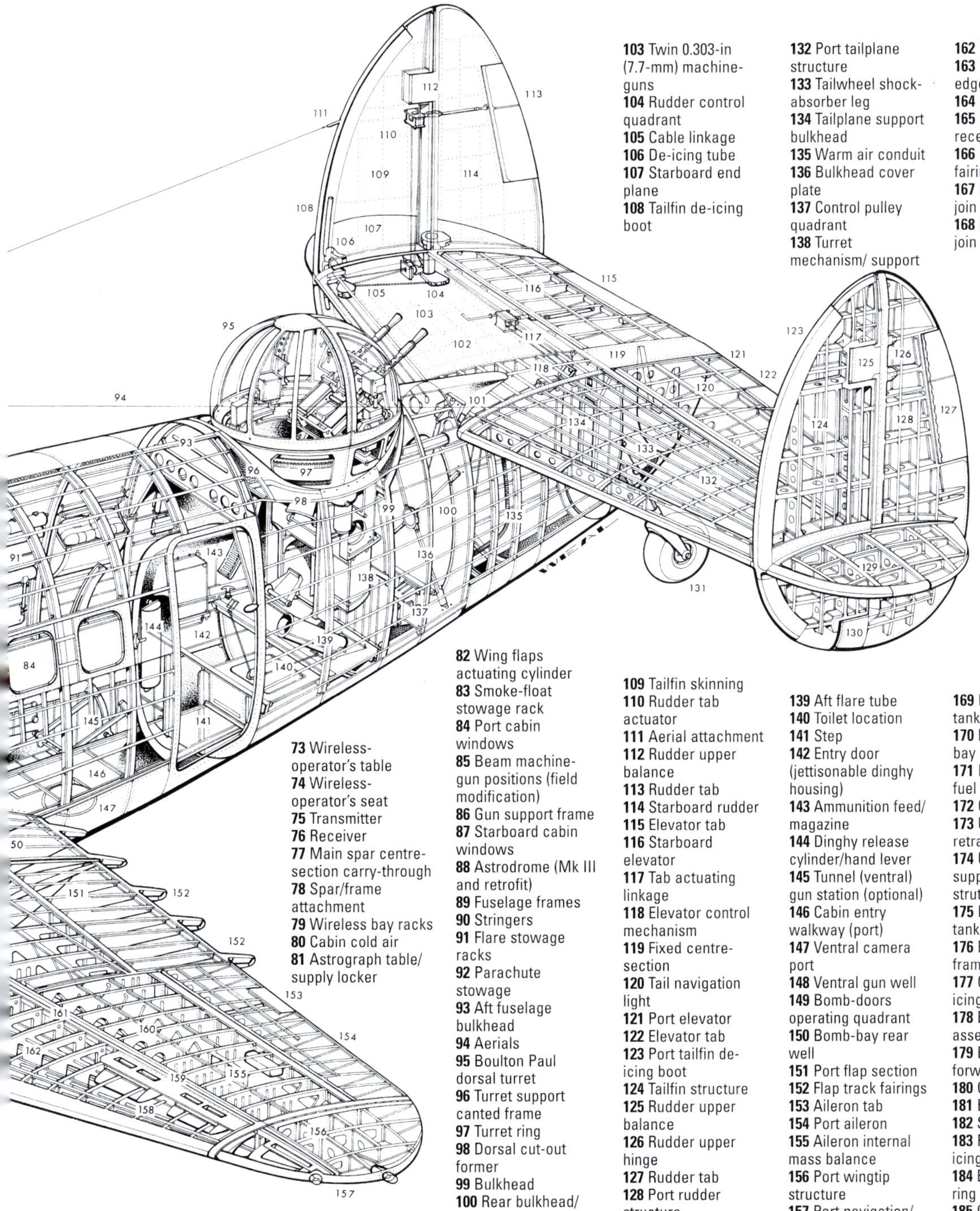

73 Wireless-operator's table
74 Wireless-operator's seat
75 Transmitter
76 Receiver
77 Main spar centre-section carry-through
78 Spar/frame attachment
79 Wireless bay racks
80 Cabin cold air
81 Astrograph table/supply locker
82 Wing flaps actuating cylinder
83 Smoke-float stowage rack
84 Port cabin windows
85 Beam machine-gun positions (field modification)
86 Gun support frame
87 Starboard cabin windows
88 Astrodrome (Mk III and retrofit)
89 Fuselage frames
90 Stringers
91 Flare stowage racks
92 Parachute stowage
93 Aft fuselage bulkhead
94 Aerials
95 Boulton Paul dorsal turret
96 Turret support canted frame
97 Turret ring
98 Dorsal cut-out former
99 Bulkhead
100 Rear bulkhead/tailplane support
101 Tail surface control linkage
102 Starboard tailplane
103 Twin 0.303-in (7.7-mm) machine-guns
104 Rudder control quadrant
105 Cable linkage
106 De-icing tube
107 Starboard end plane
108 Tailfin de-icing boot
109 Tailfin skinning
110 Rudder tab actuator
111 Aerial attachment
112 Rudder upper balance
113 Rudder tab
114 Starboard rudder
115 Elevator tab
116 Starboard elevator
117 Tab actuating linkage
118 Elevator control mechanism
119 Fixed centre-section
120 Tail navigation light
121 Port elevator
122 Elevator tab
123 Port tailfin de-icing boot
124 Tailfin structure
125 Rudder upper balance
126 Rudder upper hinge
127 Rudder tab
128 Port rudder structure
129 Port end plane
130 Rudder lower balance
131 Fixed tailwheel
132 Port tailplane structure
133 Tailwheel shock-absorber leg
134 Tailplane support bulkhead
135 Warm air conduit
136 Bulkhead cover plate
137 Control pulley quadrant
138 Turret mechanism/ support
139 Aft flare tube
140 Toilet location
141 Step
142 Entry door (jettisonable dinghy housing)
143 Ammunition feed/ magazine
144 Dinghy release cylinder/hand lever
145 Tunnel (ventral) gun station (optional)
146 Cabin entry walkway (port)
147 Ventral camera port
148 Ventral gun well
149 Bomb-doors operating quadrant
150 Bomb-bay rear well
151 Port flap section
152 Flap track fairings
153 Aileron tab
154 Port aileron
155 Aileron internal mass balance
156 Port wingtip structure
157 Port navigation/ identification lights
158 Internal vanes
159 Wing slots
160 Wing structure
161 Main spar
162 Nose ribs
163 Port wing leading-edge de-icing boot
164 Rib assembly
165 Mainwheel recess
166 Port nacelle fairing
167 Rear spar wing join
168 Main spar wing join
169 Port wing aft fuel tank
170 Fuselage bomb-bay actuating cylinder
171 Port wing forward fuel tank
172 Control servos
173 Undercarriage retraction cylinder
174 Undercarriage support/attachment strut
175 Port engine oil tank bay
176 Engine support frame
177 Carburettor anti-icing tank
178 Engine bearer assembly
179 Bomb-bay forward wall
180 Carburettor intake
181 Battery
182 Smoke floats
183 Propeller anti-icing tank (fuselage)
184 Engine bearer ring
185 Cowling nose ring
186 Spinner
187 Three-bladed propeller
188 Starboard mainwheel
189 Pitot head
190 Oil cooler intake
191 Exhaust louvres
192 Landing gear fulcrum
193 Drag strut
194 Exhaust stub
195 Side strut
196 Mainwheel oleo leg
197 Torque links
198 Port mainwheel
199 Axle hub
200 Towing lug
201 Undercarriage door
202 Float marker
203 250-lb (113.5-kg) A/S bomb

Hudson Mk VI

No. 48 Sqn was the first RAF unit to receive the Avro Anson, with which it undertook coastal patrols until October 1941. Duties included convoy and anti-submarine flights and Armed Rover reconnaissance patrols. Shortly after receipt of the Hudson, it moved to Gibraltar to cover the Torch landings in North Africa, and remained on the 'Rock' until February 1944, flying patrols over the western Mediterranean and Atlantic, occasionally tangling with Fw 200 Condors. On return to the UK, the squadron was assigned to the transport role with Dakotas, in time for the invasion of France.

Defensive armament
The principal defence for the Hudson lay within a Boulton Paul 'C' Mk II turret mounted on a ring frame. This turret contained a pair of Browning 0.303-in (7.7-mm) machine-guns, fed from magazines beneath the guns. Forward-firing armament was provided by a pair of 0.303-in (7.7-mm) machine-guns in the upper forward fuselage, aimed by the pilot using an external sight and fed from a magazine underneath the weapon. Although not fitted to this aircraft, 0.303-in (7.7-mm) machine-guns could be located on flexible mounts, firing from the beam position. Another option was a single gun, mounted in the lower rear fuselage.

Structure
The single-spar tailplane structure was mounted on top of the rear fuselage, and continued beyond the endplate fins. Large elevators were fitted either side of a small fixed centre-section, each incorporating tabs. Endplate fins were carried by the tailplane, each fitted with de-icing boots on the leading edges. The rudders were large, tabbed and fitted with aerodynamic balances near the top and at the bottom.

Camouflage
During the early part of the war, the RAF's Hudson force wore standard Dark Green/Dark Earth camouflage, but later adopted the Coastal Command colour scheme of Dark Green/Ocean Grey upper surfaces and white sides/undersurfaces. Full-colour roundels were carried, but the squadron code letters were not normally worn.

Cabin windows
Surprisingly, the Hudson was fitted with airliner-style cabin windows, inherited from the Lockheed Model 14 design from which it was hastily derived. The Mk VI could be reconfigured for passenger and light freight transport if required and many spent the last part of the war in this role.

Equipment
A prominent direction-finding loop antenna was fitted to the Hudson (enclosed in a teardrop faring in earlier versions). Behind this was an astrodome for use with a sextant while undertaking celestial navigation. The radio aerials were wires strung between a dorsal mast and the two fintips.

Cockpit
The flight deck had a fixed station for the pilot and a folding jump-seat for the co-pilot, this seat allowing inflight access to the front section. The wireless operator occupied a station immediately behind the pilot on the port side, facing outwards. The glazed nose portion was occupied by the navigator, who was provided with a sliding seat and chart table behind the glazed nosecone. Behind and beneath his seat was a flat pane for bomb-aiming. An important addition to the Coastal Command Hudson fleet from January 1940 was ASV radar, which allowed the detection of surfaced submarines in poor weather or at night. The antennas for the system were positioned under the wings and nose.

Mitsubishi G4M 'Betty'

The 'Flying Cigar' appellation is fully justified in this view of a G4M1. Built in larger numbers than any other Japanese bomber, the type saw considerable success in long-range bombing duties.

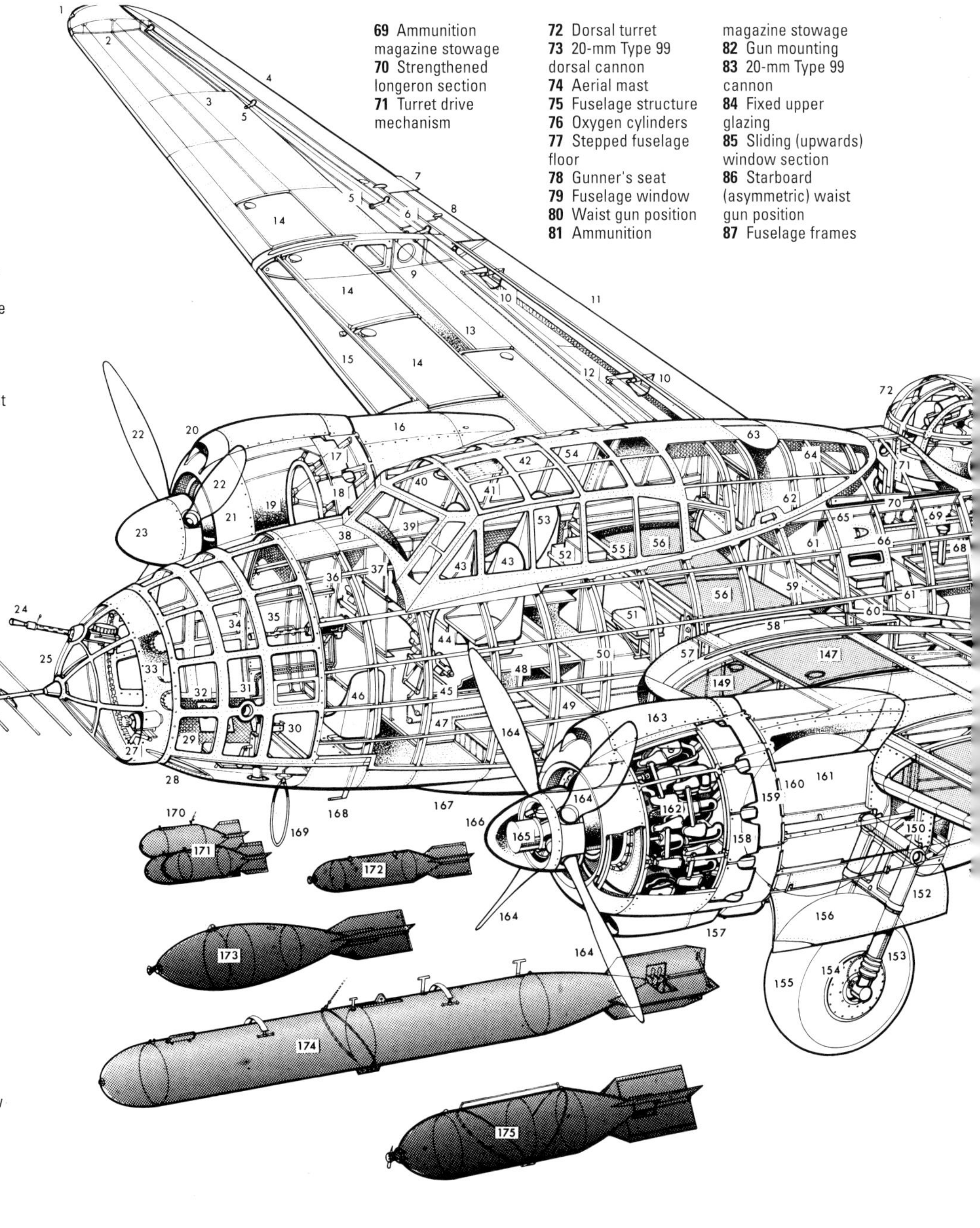

G4M 'Betty'

Cutaway key
1 Starboard navigation light
2 Starboard wingtip
3 Wing outboard spars
4 Starboard aileron
5 Aileron hinges
6 Aileron tab linkage
7 Fixed tab
8 Aileron trim tab
9 Wing join station
10 Flap hinge fairings
11 Starboard flap
12 Flap controls
13 Wing spar
14 Starboard wing fuel tanks
15 Starboard outer oil tank
16 Engine nacelle fairing
17 Cooling gills
18 Individual exhaust stubs
19 Engine bearer
20 Intake
21 Cowling ring
22 Four-blade propeller
23 Spinner
24 7.7-mm Type 92 machine-gun
25 Nose turret
26 Nose radar antenna
27 Nose turret drive mechanism
28 Bomb-aimer's flat panel
29 Bomb-aimer's couch
30 Type 90 bombsight
31 Additional cheek gun/drift sight mounting (port and starboard)
32 Bomb panel
33 Ammunition magazine stowage
34 Nose glazing
35 Additional machine-gun (stowed)
36 Rudder pedal assembly
37 Control console
38 Coaming
39 Flat windscreen panels
40 Overhead controls
41 Sun blinds
42 Flight deck emergency escape hatch
43 Pilot's seats
44 Control column
45 Rudder pedal assembly
46 Bomb-aimer's seat
47 Control linkage
48 Flight deck floor level
49 Nose compartment access walkway
50 Fuselage structure
51 Navigation/wireless-operator's station
52 Equipment racks
53 Commander's seat
54 Cockpit roof glazing
55 Front spar carry-through
56 Fuselage centre-section fuel tanks
57 Front spar/fuselage attachment
58 Over-spar centre step section
59 Rear spar carry-through
60 Rear spar/fuselage attachment
61 Gunner's take-off/land jump seats (two)
62 Emergency handhold (down to wing upper surface)
63 Emergency escape hatch
64 Dorsal frames
65 Intake scoop
66 Fuselage window
67 Dorsal gunner's step
68 Dorsal turret mount
69 Ammunition magazine stowage
70 Strengthened longeron section
71 Turret drive mechanism
72 Dorsal turret
73 20-mm Type 99 dorsal cannon
74 Aerial mast
75 Fuselage structure
76 Oxygen cylinders
77 Stepped fuselage floor
78 Gunner's seat
79 Fuselage window
80 Waist gun position
81 Ammunition magazine stowage
82 Gun mounting
83 20-mm Type 99 cannon
84 Fixed upper glazing
85 Sliding (upwards) window section
86 Starboard (asymmetric) waist gun position
87 Fuselage frames

Unbeknown to the Royal Navy at the time, the G4M made its debut against Allied forces when a number joined G3Ms in attacking and sinking Royal Navy battleships on 10 December 1941. Here IJN personnel load a torpedo aboard a G4M1.

SPECIFICATION

Mitsubishi G4M 'Betty'

Type
Seven-crew long-range bomber

Powerplant
Two 1,825-hp (1361-kW) Mitsubishi MK4T Kasei 25 radial piston engines

Performance
Maximum speed 292 mph (470 km/h) at 16,895 ft (5150 m); service ceiling 30,250 ft (9220 m); maximum range 2,694 miles (4335 km)

Weights
Empty 18,049 lb (8350 kg); maximum take-off 27,558 lb (12500 kg)

Dimensions
Wingspan 82 ft 1⁄4 in (25.00 m); length 63 ft 113⁄4 in (19.50 m); height 19 ft 81⁄4 in (6.00 m); wing area 841.01 sq ft (78.13 m2)

Armament
Four 20-mm cannon and two 7.7-mm (0.303-in) machine-guns, plus one 1,764-lb (800-kg) torpedo or 2,205 lb (1000 kg) of bombs

88 Longerons
89 Cannon muzzle trough
90 Crew circular entry hatch
91 Latch
92 Walkway to tail turret
93 Fuselage window
94 Starboard radar aerial internal support
95 Aft fuselage structure
96 Fuselage frame/tailfin support
97 Tailfin join
98 Starboard tailplane skinning
103 Tailfin leading edge
104 Tailfin structure
105 Aerial attachment
106 Rudder balance
107 Rudder frame
108 Rudder post
109 Access panels
116 Open tail turret (glazed side segments)
117 Tail 20-mm Type 99 cannon
118 Elevator tab
119 Port elevator
120 Elevator balance
130 Non-retractable tailwheel
131 Lower longeron
132 Waist station floor level
133 Bulged bomb-bay aft contour
134 Port flap section
135 Wing structure
136 Rear main spar
137 Wing inboard/outboard join
138 Aileron trim tab
139 Fixed tab
140 Port aileron
141 Wing ribs
142 Port wingtip
143 Port navigation light
144 Front main spar
145 Panel joins
146 Nose ribs
147 Port wing fuel tanks (four)
148 Spar join
149 Port wing oil tanks (two)
150 Undercarriage attachment
151 Nacelle fairing
152 Mainwheel leg
159 Individual exhaust stubs
160 Cooling/exhaust stubs
161 Mainwheel bay
162 Mitsubishi Kasei 25 (MK4T) engine
163 Engine upper intake
164 Four-blade Sumitomo VDM propeller
165 Propeller hub
166 Spinner
167 Bulged bomb-bay forward contour
168 Pitot tube (offset/angled to starboard)
169 D/F loop
170 Weapons load, inc:
171 Twelve 110-lb (50-kg) bombs (4 x 3),
172 Four 551-lb (250-kg) bombs (2 x 2),
173 Two 1,102-lb (500-kg) bombs
174 One naval torpedo, or
175 One 1,764-lb (800-kg) bomb

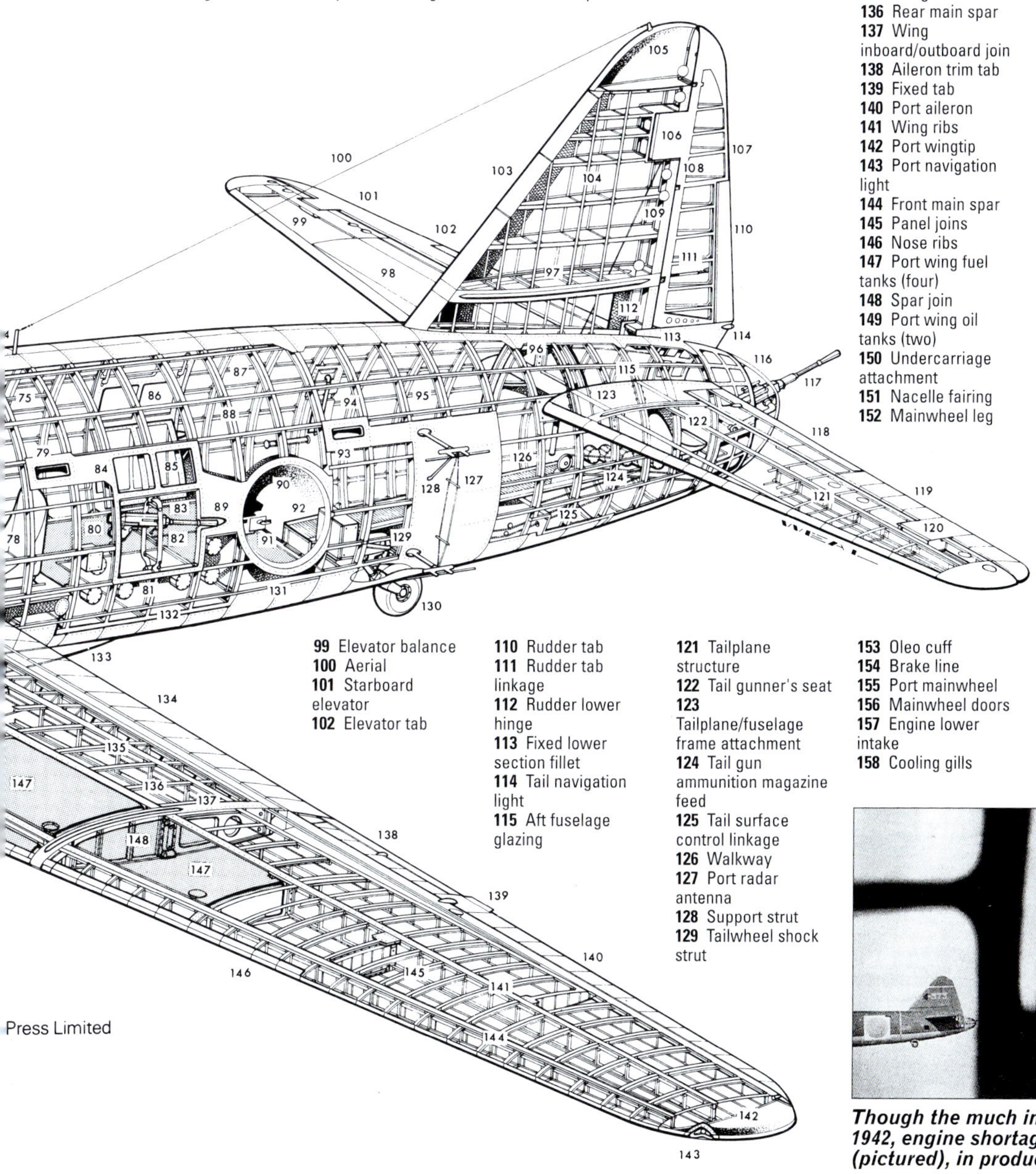

99 Elevator balance
100 Aerial
101 Starboard elevator
102 Elevator tab
110 Rudder tab
111 Rudder tab linkage
112 Rudder lower hinge
113 Fixed lower section fillet
114 Tail navigation light
115 Aft fuselage glazing
121 Tailplane structure
122 Tail gunner's seat
123 Tailplane/fuselage frame attachment
124 Tail gun ammunition magazine feed
125 Tail surface control linkage
126 Walkway
127 Port radar antenna
128 Support strut
129 Tailwheel shock strut
153 Oleo cuff
154 Brake line
155 Port mainwheel
156 Mainwheel doors
157 Engine lower intake
158 Cooling gills

Press Limited

Though the much improved G4M2 was available from 1942, engine shortages kept its predecessor, the G4M1 (pictured), in production until early 1944.

North American B-25 Mitchell

In the Pacific, the B-25 came to be used for missions for which it had never been intended, shining as a low-level strafer and ship-bomber. Illustrated is an attack on Hong Kong harbour.

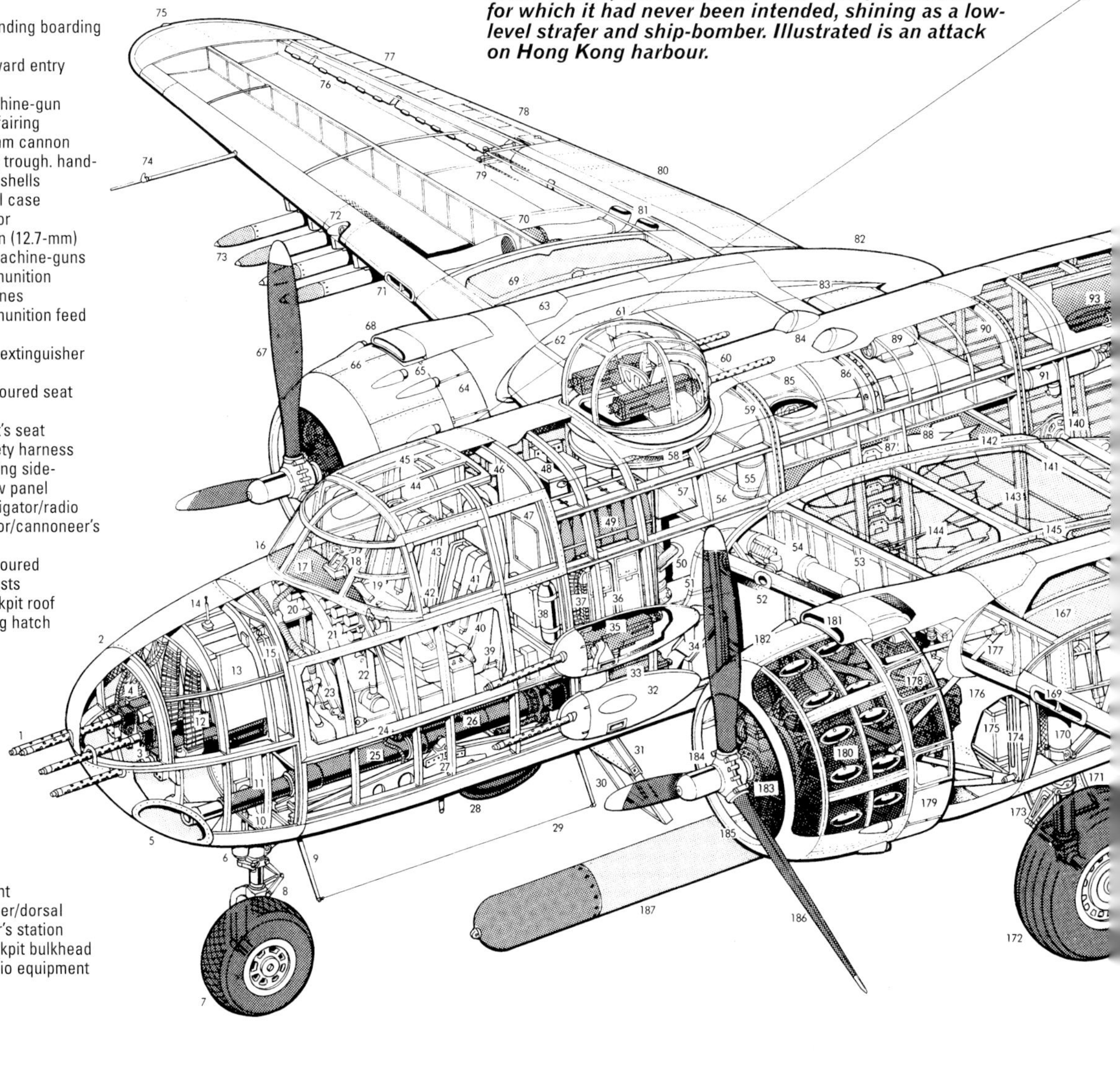

B-25H Mitchell

Cutaway key
1 Nose machine-gun barrels
2 Hinged nose compartment access door
3 4 x 0.5-in (12.7-mm) machine-guns
4 Ammunition feed chutes
5 Cannon muzzle aperture
6 Nosewheel steering control
7 Aft-retracting nosewheel
8 Torque scissor links
9 Aerial mast
10 Nosewheel leg pivot mounting
11 Cannon barrel
12 Ammunition feed chutes
13 Machine-gun ammunition magazines, 400 rpg
14 Fixed bead sight
15 Armoured bulkhead
16 Windscreen panels
17 Instrument panel shroud
18 Pilot's gunsight
19 Direct vision opening window panel
20 Windscreen de-misting air ducts
21 Instrument panel
22 Control column
23 Rudder pedals
24 Cockpit armoured skin-plating
25 T13E1 75-mm cannon
26 Recoil mechanism
27 Cannon mounting subframe
28 D/F loop aerial
29 HF aerial cable
30 Extending boarding ladder
31 Forward entry hatch
32 Machine-gun blister fairing
33 75-mm cannon loading trough. hand-loaded shells
34 Shell case collector
35 0.5-in (12.7-mm) fixed machine-guns
36 Ammunition magazines
37 Ammunition feed chutes
38 Fire extinguisher bottle
39 Armoured seat backs
40 Pilot's seat
41 Safety harness
42 Sliding side-window panel
43 Navigator/radio operator/cannoneer's seat
44 Armoured headrests
45 Cockpit roof ditching hatch
46 Flight engineer/dorsal gunner's station
47 Cockpit bulkhead
48 Radio equipment racks

SPECIFICATION

B-25H Mitchell

Dimensions

Wingspan: 67 ft 7 in (20.60 m)
Length: 51 ft (15.54 m)
Height: 15 ft 9 in (4.80 m)
Wing area: 610 sq ft (56.67 m²)

Powerplant

Two 1,700-hp (1268-kW) Wright R-2600-13 14-cylinder air-cooled radial engines

Weights

Empty: 19,975 lb (9061 kg)
Maximum take-off: 36,047 lb (16351 kg)

Performance

Maximum speed: 275 mph (443 km/h) at 13,000 ft (3960 m)
Climb to 15,000 ft (4570 m): 19 min
Service ceiling: 23,800 ft (7255 m)
Normal range: 1,350 miles (2173 km)

Armament

One 75-mm T13E1 gun with 21 shells in nose, four 0.5-in (12.7-mm) machine-guns in extreme nose, four in 'blisters' on side of nose, two in dorsal turret, two in extreme tail and one in each waist position of rear fuselage, plus up to eight 5-in (127-mm) rocket projectiles under the outer wings and up to 3,000 lb (1361 kg) of bombs carried internally

During World War II, Britain received B-25s as part of the Lend-Lease agreement. This began in 1942 with the initial allocation of 23 B-25Bs that took the RAF name Mitchell Mk I. However, it was to be the Mitchell Mk II – the equivalent of the USAAC's B-25C and B-25D – that would prove to be the RAF's first operational B-25, and the first of an eventual total of 543 joined the RAF from mid-1942. Deliveries to Britain concluded with 316 Mitchell Mk IIIs (B-25J), although some of these aircraft were diverted to the RCAF.

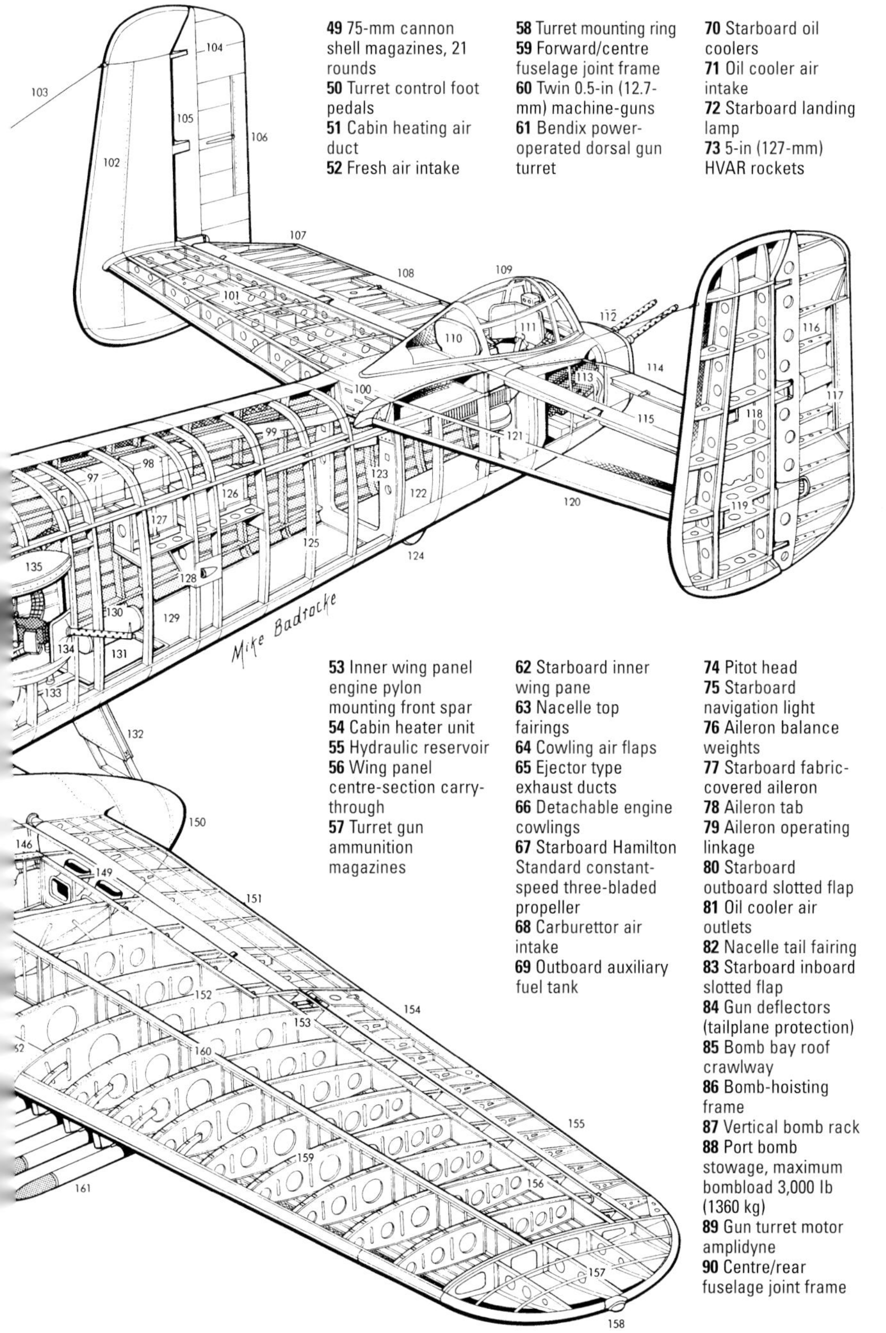

49 75-mm cannon shell magazines, 21 rounds
50 Turret control foot pedals
51 Cabin heating air duct
52 Fresh air intake
53 Inner wing panel engine pylon mounting front spar
54 Cabin heater unit
55 Hydraulic reservoir
56 Wing panel centre-section carry-through
57 Turret gun ammunition magazines
58 Turret mounting ring
59 Forward/centre fuselage joint frame
60 Twin 0.5-in (12.7-mm) machine-guns
61 Bendix power-operated dorsal gun turret
62 Starboard inner wing pane
63 Nacelle top fairings
64 Cowling air flaps
65 Ejector type exhaust ducts
66 Detachable engine cowlings
67 Starboard Hamilton Standard constant-speed three-bladed propeller
68 Carburettor air intake
69 Outboard auxiliary fuel tank
70 Starboard oil coolers
71 Oil cooler air intake
72 Starboard landing lamp
73 5-in (127-mm) HVAR rockets
74 Pitot head
75 Starboard navigation light
76 Aileron balance weights
77 Starboard fabric-covered aileron
78 Aileron tab
79 Aileron operating linkage
80 Starboard outboard slotted flap
81 Oil cooler air outlets
82 Nacelle tail fairing
83 Starboard inboard slotted flap
84 Gun deflectors (tailplane protection)
85 Bomb bay roof crawlway
86 Bomb-hoisting frame
87 Vertical bomb rack
88 Port bomb stowage, maximum bombload 3,000 lb (1360 kg)
89 Gun turret motor amplidyne
90 Centre/rear fuselage joint frame
91 Rear fuselage heater unit
92 Starboard 0.5-in (12.7-mm) waist machine-gun
93 Dinghy stowage
94 Dinghy hatch
95 Fuselage skin panelling
96 Ammunition feed chutes
97 Starboard waist gun ammunition box
98 Starboard tailgun ammunition box
99 Tailgun feed chute
100 Tailplane centre section
101 Tailplane rib and spar construction
102 Starboard tailfin
103 HF aerial cable
104 Fabric-covered rudder
105 Rudder horn balance
106 Rudder tab
107 Fabric-covered elevator construction
108 Elevator tab
109 Tail gunner's seat
110 Head armour
111 Rearward armoured panel
112 Tail barbette
113 Twin 0.5-in (12.7-mm) machine-guns
114 Elevator tab
115 Port elevator
116 Port rudder rib construction
117 Rudder tab
118 Fin rib construction
119 Fin/tailplane attachment joint
120 Port tailplane
121 Tail gunner's seat
122 Tail compartment access
123 Rear fuselage/ tailplane joint frame
124 Tail bumper
125 Fuselage frame construction
126 Port tail gun ammunition box
127 Port waist gun ammunition box
128 Air scoop
129 Fuselage walkway
130 Emergency stores pack
131 Rear entry hatchway
132 Extending boarding ladder
133 Gun pintle mounting
134 Flexible canvas seal
135 Port waist gun cupola
136 Port 0.5-in (12.7-mm) waist machine-gun
137 Cartridge case collector
138 Port inboard slotted flap segment
139 Flap rib construction
140 Emergency stores pack
141 Inner wing rear spar
142 Fuselage/wing skin joint strap
143 Rear main fuel tank, 164 US gal (621 litres)
144 Forward main fuel tank, 151 US gal (572 litres)
145 Auxiliary fuel tanks, 152 US gal (575 litres) in three fuel cells per wing
146 Flap actuator links
147 Flap hydraulic jack
148 Port oil coolers
149 Oil cooler exhaust ducts
150 Nacelle tail fairing
151 Port outer-slotted flap segment
152 Outer wing panel rib construction
153 Aileron spar
154 Aileron tab
155 Port aileron rib construction
156 Aileron spar
157 Wingtip rib construction
158 Port navigation light
159 Outerwing panel leading edge ribs
160 Main spar
161 5-in (127-mm) HVAR rockets
162 Port landing lamp
163 Mainwheel doors
164 Main undercarriage wheel bay
165 Outer wing panel joint rib
166 Mainwheel hydraulic retraction jack
167 Nacelle auxiliary fuel tank
168 Mainwheel mounting subframe
169 Oil cooler air intake
170 Mainwheel shock absorber leg strut
171 Mainwheel leg door
172 Port mainwheel
173 Torque scissor links
174 Engine mounting subframe
175 Battery stowage
176 Engine bay firewall
177 Engine bearer struts
178 Accessory equipment bay
179 Cowling air flaps
180 Wright R-2600-13 14-cylinder two-row radial engine
181 Carburettor air intake
182 Detachable cowling panels
183 Propeller reduction gearbox
184 Propeller hub pitch change mechanism
185 Cowling nose ring/ cooling air intake
186 Port three-bladed propeller
187 2,000-lb (907-kg) torpedo

B-25D Mitchell

41-29896 was a B-25D-1-NC Mitchell. The B-25D had the company designation NA-82A, and was essentially similar to the B-25C (NA-82) except that the 2,290 built were constructed at a new North American factory established in Kansas, instead of at the Inglewood, California plant. B-25s saw service in the North African theatre from August 1942, when four bomber groups were brought in to help counter Rommel's *Afrika Korps*. The following year, the Axis forces were on the retreat and some B-25s were modified as strafers, these aircraft going on to participate in every major campaign from Tunisia to the German surrender in northern Italy.

Cockpit
The B-25 was flown by a crew of two, comprising aircraft commander in the left-hand seat and co-pilot/navigator in the right-hand seat. The extensively glazed cockpit gave the crew good visibility. Later versions with fuselage guns had a reflector gunsight added above the dashboard. The B-25C and the B-25D were the first Mitchell models to introduce an autopilot, greatly aiding long-range flying and reducing crew fatigue.

Nose section
The glazed nose section housed the bombardier, equipped with a Norden bombsight. His escape hatch was located at the rear of the compartment on the port side. Two machine-guns were incorporated in the nose section, one fixed to fire forward and the other on a flexible mount, aimed by the bombardier.

Rear fuselage
The empty rear fuselage provided space for the carriage of equipment, as well as the dorsal gunner. Observation panels were incorporated to provide lateral visibility, and in later models were replaced by waist guns. Rear protection was provided by two 0.5-in (12.7-mm) machine-guns located in a Bendix dorsal turret. (The later B-25J had the turret moved forward and a separate tail turret added.) The characteristic twin-fin tail unit was built up around a two-spar central structure, with leading-edge sections and large control surfaces attached.

Payload
The single bomb bay was located between the forward and aft wing carry-through structures. Although short and narrow, it was nearly the height of the fuselage. Bombs were held in side-by-side vertical racks and the maximum payload was 5,200 lb (2359 kg), comprising 3,200 lb (1452 kg) carried internally, plus eight 250-lb (113-kg) bombs on wing racks. B-25Cs and B-25Ds occasionally carried a 2,000-lb (907-kg) torpedo externally for shipping attacks.

Petlyakov Pe-2

The popular 'Peshka' was built in enormous numbers in order to satisfy air force demand: regular series production ended in early 1945, by which date 11,427 examples had been completed. These VVS aircraft are pictured engaging in a high-speed level bomb-run during the war. The foreground Pe-2, an early production example, has been provided with a makeshift upper surface winter camouflage scheme.

Pe-2FT

Cutaway key
1 Glazed nose cone
2 Muzzle of starboard 7.62mm ShKAS machine-gun
3 Muzzle of port 7.62-mm ShKAS machine-gun
4 Nose compartment
5 Lower side glazing
6 Bomb-aimer's optically flat glazing
7 Three-blade VISh-61 constant-speed metal propeller
8 Propeller hub
9 Spinner
10 Starter clog
11 Oil cooler intake
12 Oil cooler installation
13 Adjustable outlet flap
14 Engine bearing lower cross frame
15 Engine bearer assembly
16 Klimov M-105RA 12-cylinder Vee engine
17 Engine cool air
18 Carburettors
19 Ventral D/F loop
20 Rudder pedal assembly
21 Cartridge case collector chute
22 Fixed nose armament (port 7.62-mm ShKAS optional)
23 Ammunition box (500 rpg)
24 Nose panelling
25 Main instrument panel
26 Split windscreen
27 Fixed quarter light
28 Instrument side panel
29 Starter dog
30 Control column
31 Pilot's seat with 9-mm armoured back
32 Control run linkage
33 Navigator's (starboard) seat support frame
34 Three-blade VISh-61 propeller
35 Crew entry hatch
36 Oil cooler intake
37 Propeller hub
38 Pilot's adjustable armoured headrest (9-mm)
39 Cockpit canopy
40 Aerial mast
41 Pitot head
42 Aerials
43 Dorsal flexible 7.62-mm ShKAS machine-gun
44 Canopy hinged alt section
45 Exhaust collector shroud
46 Engine cool air intake
47 Dorsal armament ammunition box (750 rounds)
48 Dorsal gunner's seat **49** Port engine nacelle fasteners
50 Fuselage main fuel tank (114 Imp gal/518 litre capacity)
51 Fuselage port saddle tank (11.65 Imp gal/53 litre capacity)
52 Port engine bearer assembly
53 Cooling pipe
54 Oil cooler
55 Adjustable outlet flap
56 Flap actuating servo motor
57 Carburettor air intake
58 Port undercarriage/wing spar attachment
59 Port engine oil tank
60 End fib profile
61 Port outer radiator intake
62 Radiator ducting
63 Front spar/nacelle fixing
64 Port wing root fuel tank (39.6 Imp gal/80 litre capacity)
65 Undercarriage radius rod/wing spar attachment
66 Part outer radiator assembly
67 Underwing dive-brake (extended)
68 Dive-brake hinge fairings
69 Landing lamp
70 Port outer wing inboard fuel tank (31.5 Imp gal/143 litre capacity)
71 Port outer wing outboard fuel tank (23.5 Imp gal/107 litre capacity)
72 Wing leading-edge strip
73 Wing stiffeners
74 Nose rib stations
75 Wing front spar
76 Wing fibs
77 Wing rear spar
78 Wing skinning
79 Port navigation light
80 Port wingtip
81 Aileron attachment/hinge points

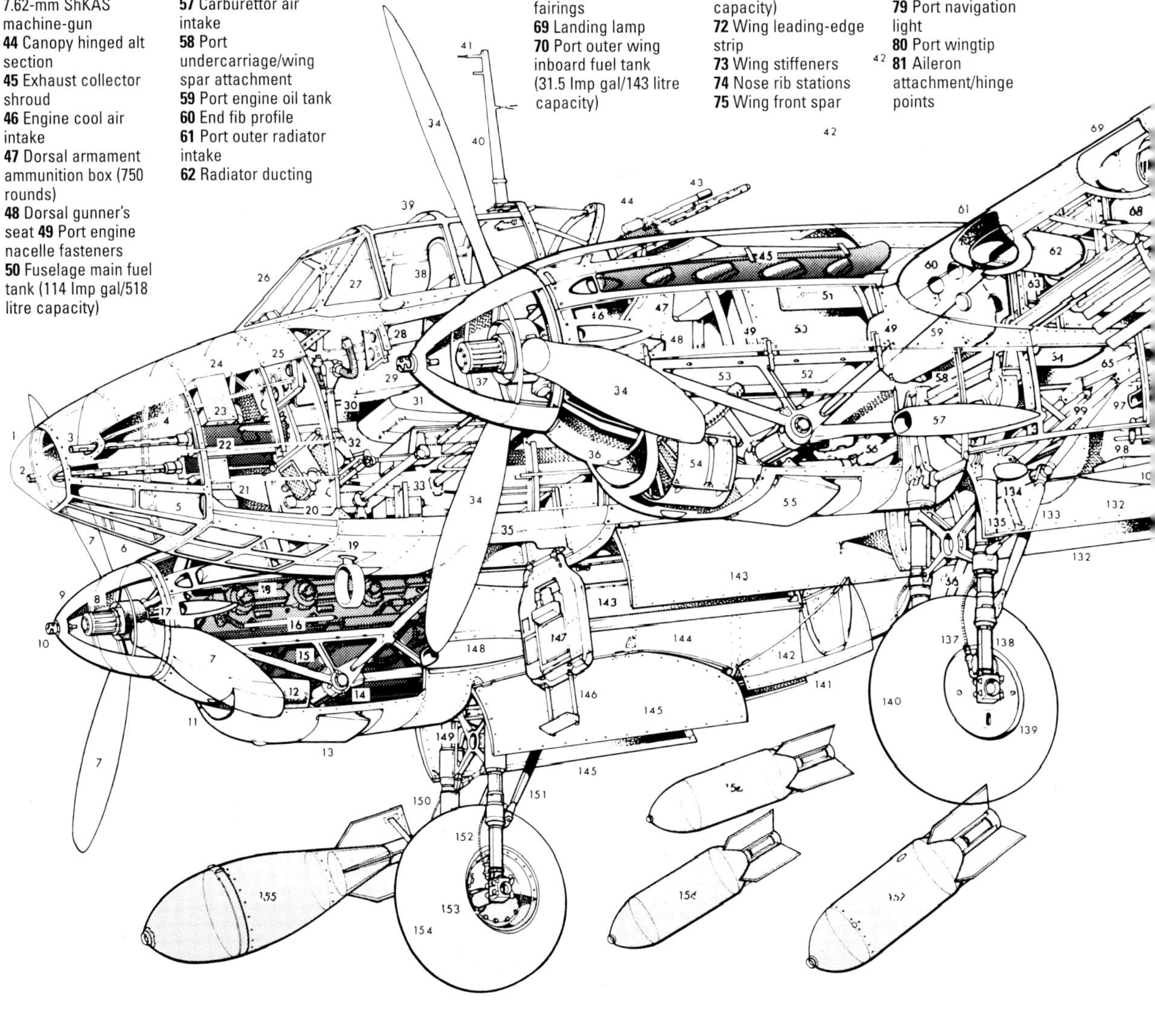

The 82-mm calibre RS-82* (Reaktivnyi Snaryad, *reaction missile) was the most numerous Soviet air-to-ground rocket of the Great Patriotic War. The weapon, a simple projectile with an ogival nose with a fuze windmill coupled to a drum-shaped central portion and four-fin conical tail, came as a surprise to Nazi and Allied forces alike when first used against German Panzers by Il-2s in June 1941. This Pe-2 is armed with a total of ten RS-82 rockets on underwing rails.

SPECIFICATION

Pe-2FT (unless stated)

Dimensions

Length: 41 ft 11½ in (12.78 m)
Wingspan: 56 ft 2 in (17.11 m)
Wing area: 435.95 sq ft (40.5 m²)
Height: 11 ft 2⅔ in (3.42 m)

Powerplant

Two Klimov M-105RA (VK-105PF) 12-cylinder liquid-cooled piston engines, rated at 1,260-hp (939-kW), with two-speed superchargers

Weights

Empty weight: 13,117 lb (5950 kg)
Loaded weight (normal): 17,130 lb (7770 kg) (Pe-2)
Loaded weight (maximum): 18,783 lb (8520 kg) (Pe-2)
Wing loading: 38.9-43 lb/sq ft (190-210 kg/m²) (Pe-2)

Performance

Maximum speed at sea level: 279 mph (449 km/h)
Maximum speed at 16,400 ft (5000 m): 361 mph (581 km/h)
Maximum speed at 6,560 ft (2000 m): 314 mph (506 km/h) (Pe-2)
Range: 1,100 miles (1770 km)
Climb to 9,840 ft (3000 m): 3.5 mins (Pe-2)
Climb to 16,400 ft (5000 m): 7 mins (Pe-2)
Service ceiling: 29,530 ft (9000 m)

Armament

(Fixed) Typically two 0.3-in (7.62-mm) ShKAS machine-guns firing ahead aimed by pilot, MV-3 dorsal turret with single 0.5-in (12.7-mm) UBT, one ShKAS in rear ventral position, and one ShKAS in left or right rear beam position. (Disposable) See text

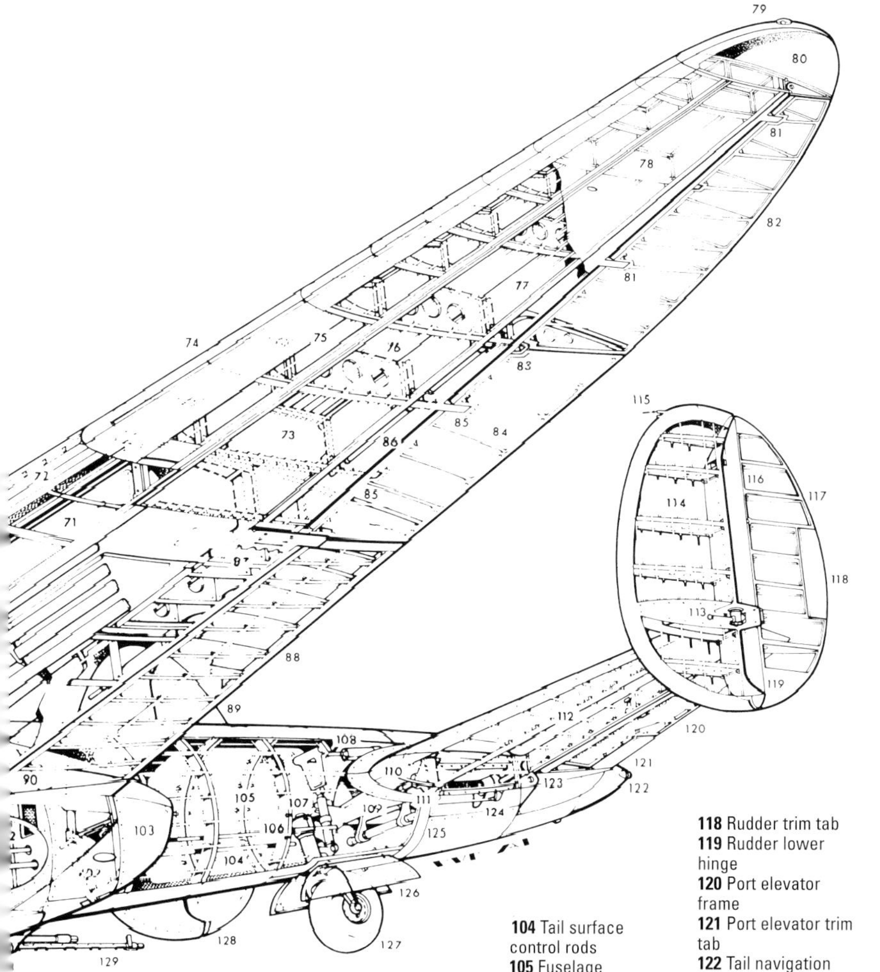

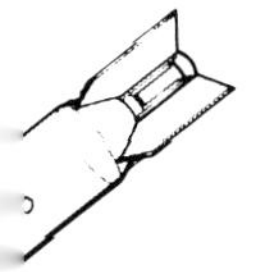

Above: The Pe-2UT multirole trainer had the former No. 1 (fuselage) fuel tank replaced by an extra pilot station at the rear, and was equipped with dual controls.

During TsAGI tests, this Pe-2 had its standard MV-3 rear-facing gun turret replaced by a VUB-1 installation, similarly equipped with a single-blade compensator (weathercock). The 12.7-mm UB gun had the same calibre as the equivalent 0.5-in Browning, but was lighter and faster-firing.

82 Port aileron outer section
83 Aileron actuating control linkage
84 Port aileron inner section
85 Aileron attachment/hinge points
86 Aileron control rod
87 Stiffening plate
88 Port flap outboard section
89 Starboard rudder
90 Flap inboard rib fillet
91 Rear spar/nacelle fixing
92 Fuselage beam window
93 Tail surface control rods
94 Cooling louvres
95 Ventral armament ammunition box 1750 rounds)
96 Ventral machine gun mounting/retraction frame
97 Periscopic sight
98 Control grips
99 Undercarriage radius rods
100 Ventral gunner's (prone) pad
101 Ventral retractable 7.62 mm ShKAS machine-gun
102 Wort nacelle bomb-bay
103 Port nacelle tail cone
104 Tail surface control rods
105 Fuselage structure
106 Tailwheel leg assembly
107 Shock-absorber strut
108 Tailwheel retraction mechanism and cylinder
109 Tailplane centre-section brace
110 Rudder control push-pull rod link
111 Tailplane attachment fillet
112 Tailplane structure
113 Rudder control linkage
114 Tailfin structure
115 Aerial attachment pick-up
116 Rudder upper hinge
117 Rudder frame
118 Rudder trim tab
119 Rudder lower hinge
120 Port elevator frame
121 Port elevator trim tab
122 Tail navigation light
123 Elevator attachment/hinge bracket
124 Elevator actuating rod and internal counterweight
125 Aft fuselage frame/ tailplane front spar join
126 Tailwheel doors
127 Retractable tailwheel
128 Starboard tailfin/rudder
129 Ventral machine gun (deployed)
130 110-lb (50-kg) engine nacelle bay bomb-load
131 Ventral armament wind deflector fairing
132 Port undercarriage doors
133 Undercarriage lower brace strut
134 Retraction jack
135 Door actuating link
136 Mainwheel leg cross-brace
137 Brake lines
138 Mainwheel oleo legs
139 Hub plate
140 Port mainwheel
141 Starboard nacelle bomb bay doors (open)
142 Starboard nacelle bomb bay
143 Fuselage bomb-bay doors
144 Starboard engine nacelle mainwheel well
145 Starboard undercarriage doors
146 Telescopic entry step
147 Crew entry hatch door/integral step
148 Carburettor air intake
149 Mainwheel leg cross-brace
150 Brake lines
151 Undercarriage lower brace strut
152 Mainwheel oleo legs
153 Hub assembly
154 Starboard mainwheel
155 551-lb (250-kg) (optional) underwing bomb load
156 441-lb (200-kg) (4x 110-lb/50-kg) fuselage internal bomb load
157 882-lb (400-kg) (4x 220-lb/100-kg) underwing external bomb load

Pe-2FT

Operated by the 12th Guards Dive-bomber Air Regiment (12 GvBAP) of the Baltic Fleet, Aviatsiya Voenno-Morskovo Flota (AV-MF), flying sorties over the Gulf of Finland in 1944, this Series 205 Pe-2FT (*Frontovii trebovanii*, front-line requirement) wears standard pattern camouflage, consisting of a two-tone green and tan upper surface scheme and pale undersides. The prized *'Guardiya'* badge on the nose signifies that the 12 BAP had been recognised as a Guards unit. Note also the 'Kremlin star' national insignia. The pilot of aircraft '01' was Lt Col Vasili I. Rakov, HSU (Hero of the Soviet Union, awarded 7 February 1940 whilst serving with the 57 BAP), who flew 170 missions, and became commander of the 12 GvBAP after leading a successful attack by 28 Pe-2s on the German anti-aircraft cruiser *Niobe* at the port of Kotka, southern Finland, on 16 July 1944 (an action for which he received a second HSU, on 22 July).

Notable *'Peshka'* operators

As the standard VVS tactical bomber of the Great Patriotic War, the Pe-2 equipped the majority of Soviet bomber regiments by the end of the war. However, among those with particularly distinguished service records were the 5 SBAP (Fast Bomber Aviation Regiment) commanded by Col F. P. Kotlyar (Commander of the 4 GvBAD, Guards Bomber Aviation Division, at the end of the war) and becoming the 8 GvBAP (Guards Bomber Aviation Regiment) in spring 1942, the 6 BAP (Bomber Aviation Regiment) of the 202 BAD (Bomber Aviation Division), the 24th Orlov Red Banner Regiment, the 24 BAP, the 35 GvBAP, the 80 BAP, the 86th Guards Stalingrad Regiment, the 150 BAP commanded by Lt Col I. S. Polbin in early 1942, the 779 BAP and the 82 GvBAP.

Bombload

The Pe-2's offensive stores capability included a standard fuselage bomb bay which could accommodate up to six 100-kg (220-lb) FAB-100 general-purpose HE bombs or four 250-kg (551-lb) FAB-250 bombs, whilst additional 100-kg bombs could be carried singly in the rear of each engine nacelle, with a further two more such weapons carried on racks under each wing centre section. Later in the war, after the power of the Pe-2 had been increased, the four external weapons carried underwing could be increased to 250-kg weapons. Among important roles bestowed upon the Pe-2 was that of dive-bomber. To reduce speed during the dive and stabilise the aircraft, large and powerful slatted 'Venetian-blind' hydraulically-actuated dive-brakes hinged down from under the wing. Carrying a 2,205-lb (1000-kg) bomb load, the Pe-2FT had a range of approximately 817 miles (1315 km).

High-speed capability

In 1941, when Nos. 8 and 134 Sqns, RAF, were sent to Murmansk in order to familiarise Soviet pilots with the Hurricane, the RAF fighters sometimes flew escort to a Pe-2 regiment. Hurricane pilots discovered that 'in an operation that lasted as long as an hour they had to go all-out to keep station', whilst the Pe-2s 'climbed and flew at a rate that that astounded our boys considerably' (Flt Lt H. Griffiths). Maximum permissible speed of the Pe-2 was 450 mph (725 km/h) or 373 mph (600 km/h) with dive-brakes extended. After the introduction of the Bf 109G-2 in late 1942, measures were taken to further improve performance, with the M-105RA engine replaced by the M-105PF/RF rated at 1,260 hp (939-kW) at 2,560 ft (780 m) and 1,180 hp (880-kW) at 8,860 ft (2,700 m), improving take-off performance and increasing maximum speed.

Pe-2FT modifications

Most Pe-2FTs were updated to this standard 'in field' by factory teams who visited operational units. The Pe-2FT offered improved defensive armament in response to requests from crews, who were facing Bf 109Fs by early 1942. The most obvious modification was the replacement of the bomb aimer/navigator's ShKAS with a UBT in the rear turret. A feature of the new lightweight manually operated MV-3 turret was the weathercock. When swung to the side the barrel of the gun caused considerable drag, so the weathercock was fitted in order to offset the large forces required by the bomb aimer/navigator to swing the gun under combat conditions.

Crew positions and duties

The Pe-2 was flown by a crew of three comprising pilot, bomb aimer/navigator and radio operator. The bomb aimer/navigator normally occupied the rear cockpit turret, but just prior to the bombing run would slide forward to the bomb aiming position, lying prone in the nose. The radio operator, who doubled as the ventral gunner, was enclosed in the rear cabin. Although rather spartan by Western standards, Soviet cockpits were provided with all necessary equipment with which the pilot could perform his task. With his head and back protected by 9-mm armour, the pilot also had an adjustable headrest. The distinguishing feature of the Pe-2FT was the bomb aimer/navigator's rear-facing Mozharovsky-Venyevidov MV-3 turret. This mounted a single 0.5-in (12.7-mm) UBT gun, replacing the 0.3-in (7.62 mm) ShKAS weapon of previous variants. The radio operator was responsible for the ventral rear-facing gun aimed from the rear fuselage position. This was aimed using both a periscope and oval side windows. Initially a ShKAS, the weapon was subsequently replaced with a 0.5-in (12.7-mm) calibre UBT machine-gun in some later aircraft, including the example illustrated. However, the one or two beam ShKAS weapons of this Pe-2FT, fired through port and/or starboard hatches, are not visible.

Tupolev Tu-2

Later production Tu-2s (throughout the war the Tu-2S designation was applied to production machines) featured numerous modifications compared to earlier aircraft. Early in production the cowling diameter was reduced (Block 20); small blisters added over the valve gear and a redesigned metal nose. By Block 50, the Tu-2 had been progressively equipped with windows for the ventral gunner, a new VUB-68 gun mount for the radio operator, an additional retractable landing light (all Block 44), improved Lu-68 ventral gun installation (Block 46), extended nose glazing, new straight-top canopy and new VUS-1 navigator's gun mount (Block 48).

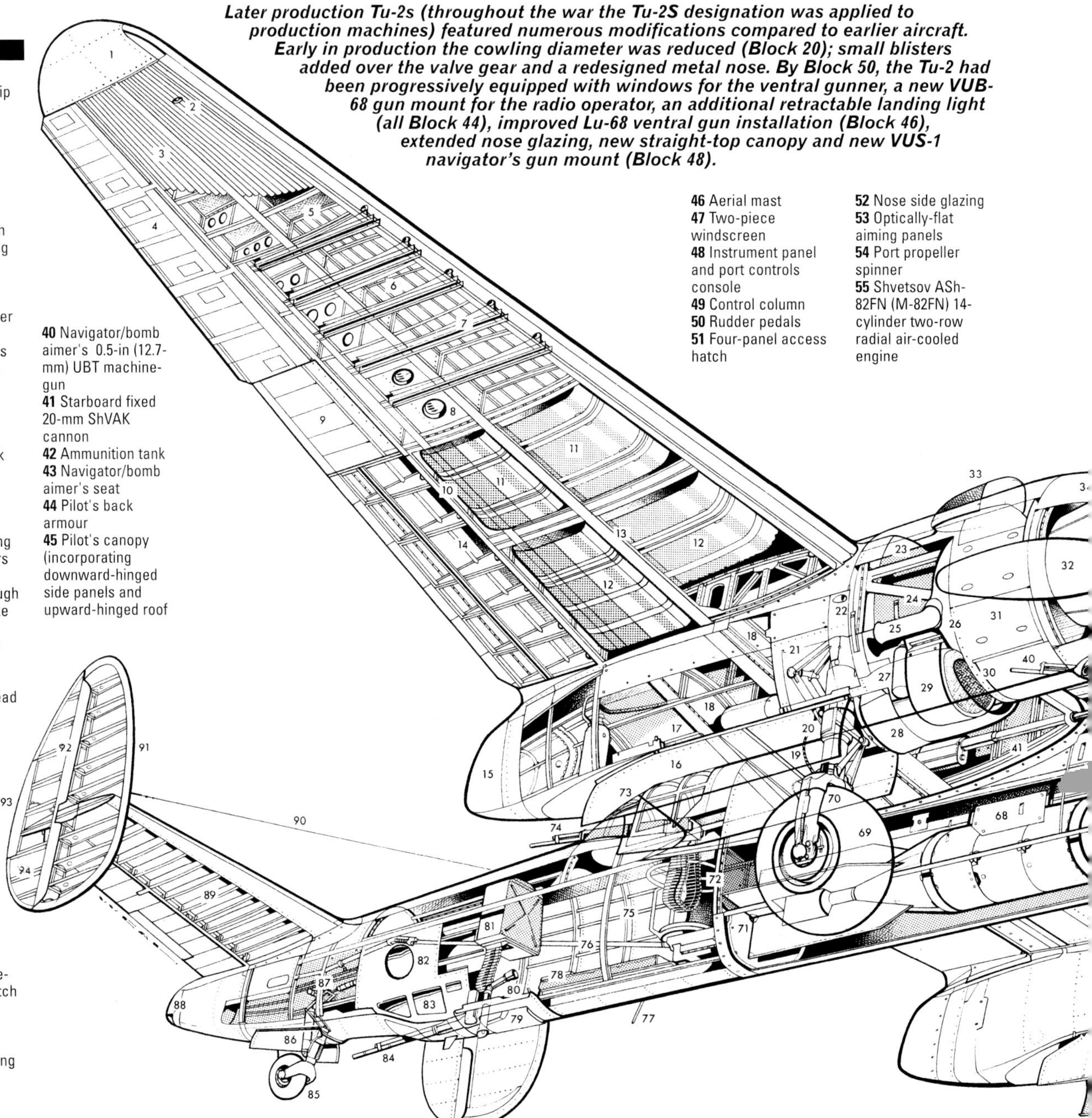

Tu-2S

Cutaway key

1 Detachable wingtip
2 Starboard navigation light
3 Corrugated inner skin (upper and lower)
4 Starboard aileron (outer section)
5 Wing construction
6 Skin-strengthening stringers
7 Light bomb or rocket-launching attachments (five per side)
8 Twin landing lights
9 Starboard aileron (inboard section)
10 Rear spar
11 Outboard fuel tank set
12 Inboard fuel tank set
13 Main spar
14 Landing flap assembly
15 Nacelle aft fairing
16 Mainwheel doors
17 Mainwheel well
18 Spar carry-through
19 Mainwheel brake cable
20 Mainwheel oleo
21 Mainwheel retraction scissors
22 Fireproof bulkhead
23 Oil tank
24 Engine bearers
25 Exhaust stub
26 Exhaust collector ring
27 Fire-suppression bottle
28 Radiator outlet shutter
29 Oil radiator
30 Oil cooler intake
31 Detachable cowling panels
32 AV-5-157A three-bladed variable-pitch metal propeller
33 Carburettor air intake
34 Low-drag cowling ring
35 Cooling louvres
36 Propeller boss
37 Hucks-type starter dog
38 Propeller shaft
39 Gear housing
40 Navigator/bomb aimer's 0.5-in (12.7-mm) UBT machine-gun
41 Starboard fixed 20-mm ShVAK cannon
42 Ammunition tank
43 Navigator/bomb aimer's seat
44 Pilot's back armour
45 Pilot's canopy (incorporating downward-hinged side panels and upward-hinged roof
46 Aerial mast
47 Two-piece windscreen
48 Instrument panel and port controls console
49 Control column
50 Rudder pedals
51 Four-panel access hatch
52 Nose side glazing
53 Optically-flat aiming panels
54 Port propeller spinner
55 Shvetsov ASh-82FN (M-82FN) 14-cylinder two-row radial air-cooled engine

Andrei Nikolaevich Tupolev (in white hat), pictured visiting a front-line Tu-2 unit, developed the Tu-2, along with his design team, under prison conditions. Tupolev had been imprisoned in a Moscow jail in 1937 but was transferred to Bolshevo, where he worked at the Central Design Bureau N29 of the NKVD (TsKB-29) with other aviation specialists the following year. The full-scale mock-up of Samolyet 103 (the future Tu-2) was constructed from timber in a forest close to the prison.

SPECIFICATION

Tu-2S (Block 10)

Dimensions

Fuselage length: 45 ft 3⅓ in (13.80 m)
Wing span: 61 ft 10½ in (18.86 m)
Wing area: 525 ft^2 (48.80 m^2)
Height: 14 ¾ ft (4.50 m)

Powerplant:

Two 14-cylinder Shvetsov ASh-82FN (M-82FN) radial piston engines each normally rated at 1,523 hp (1136 kW)

Weights:

Empty weight: 16,477 lb (7474 kg)
Loaded weight: 25,044 lb (11,360 kg)
Maximum take-off weight: 28,219 lb (12,800 kg)

Normal fuel capacity: 615 Imp gal (2800 litre)

Performance:

Maximum speed: 342 mph (550 km/h) at 17,716 ft (5400 m)
Cruising speed: 275 mph (442 km/h) at 19,030 ft (5800 m)
Service ceiling: 31,170 ft (9500 m)
Maximum range: 1,305 miles (2,100 km)
Range with 3,307 lb (1500 kg) bombs: 1,553 miles (2500 km)
Range with 4,960 lb (2500 kg) bombs: 870 miles (1400 km)
Climb to 16,400 ft (5000 m): 9.5 min
Landing speed: 98 mph (158 km/h))

Right: Initial work on Samolyet 103 (the name derived from that of Tupolev's design team at the TsKB-29: KB-103) began on 1 March 1940, by which time TsKB-29 had been re-located to Tupolev's Moscow offices in Radio Street, which then became a prison. Rolled out on 3 October 1940 after construction of the prototype at factory N156, the first aircraft was then transferred to Chkalovskaya airbase, where it began NII-VVS testing on 1 December. At this stage, Tupolev's continued status as 'enemy of the people' meant that the aircraft had to be known as Samolyet 103, rather than by an ANT-series designation. Series production began and ended in 1941 in favour of fighter production, however, it was restarted in 1943, by which time the aircraft was officially known as the Tu-2.

56 Exhaust pipe cluster
57 Quick-release cowling clips
58 Ejector exhaust pipe
59 Oil cooler intake
60 25-kg (55-lb) OFAB fragmentation bombs
61 Port mainwheel
62 Mainwheel shock-absorber scissors
63 Mainwheel doors
64 Exhaust gill
65 Centre-hinged main bomb bay doors
66 Forward bomb shackle
67 Single FAB-1000 2,205-lb (1000-kg) bomb
68 Access panel
69 Starboard mainwheel
70 Mainwheel fork
71 Aft bomb bay bulkhead
72 Radio operator's position
73 Dorsal glazing
74 Dorsal 0.5-in (12.7-mm) UBT machine-gun
75 Fuselage construction
76 Control cable shroud
77 Stub aerial
78 Ventral gunner's couch
79 Aft crew entry hatch
80 Ventral gunner's aiming periscope
81 Ammunition tank
82 Beam observation window
83 Ventral glazing
84 Ventral 0.5-in (12.7-mm) UBT machine-gun
85 Retractable tailwheel
86 Tailwheel doors
87 Tailwheel retraction mechanisms
88 Aft navigation light
89 Tailplane construction
90 Aerial
91 Starboard tailfin
92 Rudder post
93 Rudder tab
94 Starboard rudder

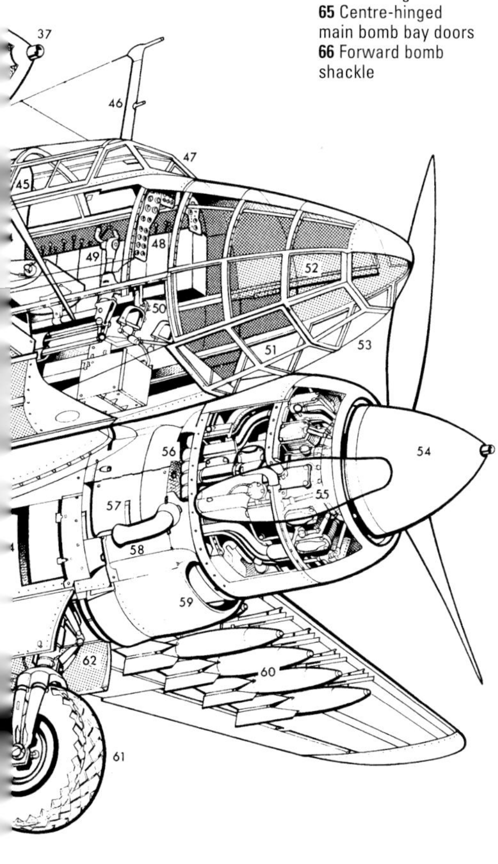

Above: The five-seat Tu-2DB (ANT-65) long-range bomber introduced a long-span wing, supercharged (exhaust-driven) liquid-cooled engines and a twin-pilot cockpit, whilst retaining the Tu-2's weapons and bombload.

A torpedo-carrying capability had been envisaged during the original Samolyet 103 scheme, and was realised in the Tu-2T (Torpedonosyets), with two prototypes built in 1945-46. Changes included the introduction of a pylon under each wingroot for the carriage of a 45-36-AN aerial torpedo strengthened landing gear, and increased fuel capacity in a sealed bomb bay. The Tu-2T, which had a 2,360 mile (3800 km) range, was built in series for the AV-MF in 1947, it replaced the Il-4T, and served until the mid-1950s with the Baltic, Black Sea and Northern Fleets.

Tu-2S

This Tu-2S (ANT-61) wears a similar camouflage scheme as those aircraft which participated in the Aviation Day parade over Moscow on 18 August 1945. There were several camouflage variations involving two-tone upper surfaces, but many Tu-2s were completed with simple dark green upper surfaces and pale blue undersides. Red star national markings were carried prominently, even during wartime. The first unit to receive the Tu-2 was the 3rd Air Army on the Kalinin Front, which received its first three 103VS aircraft in late April 1942. Following the war, the Tu-2 remained in front-line Soviet air force service until 1950, and was exported to China, North Korea and Warsaw Pact clients, flying Korean War sorties with the Chinese.

Early operational history

The first operational evaluation unit to fly the Tu-2 was Maj. Gen. M. M. Gromov's 3rd Air Army, which received the 103VS (ANT-60) first series production variant in April 1942. From May, the definitive production aircraft (now known as the Tu-2S) was being delivered to the 132 BAP (*Bombardirovochnaya Aviatsionnyi Polk*, Bomber Aviation Regiment), which converted from the SB-2 with assistance of instructors from the Omsk flying school, and a GAZ 166 test pilot. This initial regiment left for the front in September, and a second regiment had become operational by the end of the year.

Design and construction, post-war service

From the outset, Samolyet 103 was envisaged as an all-metal aircraft, utilising smooth duraluminium for the airframe, wings and skin, and steel for the engine mounts and undercarriage. The large bomb compartment beginning just behind the cockpit took up most of the fuselage's volume. Later (post-Block 50) modifications to the construction of the Tu-2S comprised an improved form fixed canopy for the radio operator/gunner and addition of a starboard wingroot walkway and hand-hold (Block 50); metal wingtips (from the 21st aircraft of Block 52) and the introduction of pulsating rubber de-icers for the wing leading edges and tail surfaces, increased-area fins and rudders and new AV-9VF-21K propellers with four square-tipped reduced diameter 'paddle' blades (Block 59). Block 61 saw the enlargement of the engine air inlets in order to incorporate dust filters.

Total Soviet production amounted to 2,257, whilst further aircraft were license-built in China (some of these later being fitted with Chinese HS8 engines and surviving in service until the 1960s). Tu-2s exported to Bulgaria, China, Hungary, North Korea, Poland and Romania tended to be taken from surplus Soviet stocks. The Tu-2 received the ASCC (Allied Standards Co-ordinating Committee) reporting name 'Bat'.

Shvetsov ASh-82FN powerplant
The first powerplant for the Tu-2 family was Mikulin's unreliable 12-cylinder liquid-cooled AM-37, developing 1,400 hp (1044 kW), which was used by the ANT-58, the first prototype Samolyet 103. The ASh-82FN (or M-82FN) was installed on the production Tu 2S (ANT-61), replacing the earlier air-cooled ASh-82 radial piston engine of the Samolyet 103V (ANT-60) production prototype. Both units were developments of the Mikulin M-82, however, the 14-cylinder FN produced between 1,450 hp (1082 kW) at 15,255 ft (4650 m) and 1,850 hp (1380 kW) at 2,500 rpm (for take-off), compared to the 1,330 hp (992 kW) output of the basic ASh-82. Normal rated output for the engine was 1,523-hp (1136-kW). Each powerplant drove a three-bladed 12-ft 6-in (3.8-m) AV-5-157A propeller, with fully feathering constant-speed blades. These metal units replaced the VISh-61T propellers of the Samolyet 103, and the VISh-61E 12-ft 5-in (37.8-m) diameter units of the Samolyet 103U.

Combat record
Despite the fact that the Tu-2 took a long time to appear in numbers within front-line units (due to materials shortages, evacuation of factories, and production schedules favouring the Yak-1 fighter), the Tu-2 was veteran of many battles during the Great Patriotic War. The first major duty was during the Battle of Kursk in June 1944, when 18 aircraft of the 285 BAP bombed enemy forces and attacked positions behind the lines. Over 600 Tu-2s, lead by Col I. Piskok of the 334 BAP took part in the Battle of Vyborg, beginning in 9 June 1944. These aircraft also bombed positions around the pre-Baltic, including the Kiviniemi, Valkiarv and Kivennopa regions. On 7 April 1945, a force of 516 Tu-2s and Pe-8s attacked defences around Koningsberg (Kaliningrad) before the city was captured by the Soviets. During the first day alone of the Battle of Berlin, Tu-2s dropped a total of 97 tonnes of bombs on the German capital.

Gun armament
Forward-firing armament for the Tu-2S consisted of two fixed 20-mm ShVAK cannon, one mounted in each wingroot, and provided with 200 rounds of ammunition per gun. Further defensive armament comprised one manually-operated 0.5-in (12.7-mm) UBT heavy machine-gun (with 250 rounds) behind the pilot, for operation by the navigator/bomb-aimer. Two further manually-operated UBT machine-guns were located in the rear dorsal and ventral positions, operated by the radio operator/gunner and a dedicated gunner, respectively. Sitting on the centreline, the pilot was provided with a PBP-1A reflector sight in order to aim the pair of wingroot ShVAKs. As well as operating the ShKAS mounted on a TSS installation, and firing to the rear of the bulged Plexiglass canopy, the navigator/bomb-aimer could fly the aircraft in an emergency using a folding control wheel.

Tu-2S experimental modifications
Two Tu-2s, unofficially designated Tu-2K (*Katapult*) were utilised for ejection seat tests in 1944-45: two different configurations were trialled, one with the ejection seat in the navigator's position, the other with an open cockpit on the radio operator's station. Another aircraft, the Tu-2N was a series aircraft used to test the Rolls-Royce Nene turbojet in an underslung installation in July 1947. At least two further Tu-2Ls were used as testbeds for other jet engines (at least five types, including the RD-10, RD-20, RD-45, RD-500 and TR-1) at the LII. Other aircraft were used for testing radar systems, bomb sights and guns, whilst the aircraft was also employed as a tanker in early looped-hose refuelling trials with jet aircraft, including the Yak-15. A number of de-militarised Tu-2s were used by Aeroflot as freighters (Tu-2G) with a 4,410-lb (2000-kg) payload; and in 1948-49, a single Block 60 aircraft was modified to carry a GAZ-6TB scout car, which could be air-dropped from a semi-recessed underfuselage position.

Vickers Wellington

Equipped with underwing rockets and a Leigh Light (shown here retracted under the rear fuselage), this Wellington Mk XIV also carries ASV.Mk III radar in a 'chin' radome. These ASW Wellingtons served with 10 RAF Coastal Command squadrons.

Wellington Mk XVI

Cutaway key
1 Nash and Thomson power-operated tail gun turret
2 Four Browning 0.303-in (7.7-mm) machine-guns
3 Cartridge case ejection chute
4 Elevator tab
5 Elevator rib construction
6 Elevator horn balance
7 Tailplane tip construction
8 R.3003 aerial cable
9 Tailplane leading-edge de-icing boot
10 Tailplane geodetic construction
11 Elevator torque shaft
12 Fin/tailplane attachment main frames
13 Gun turret entry doors
14 Rudder tab
15 Tail navigation and formation lights
16 Rudder rib construction
17 Rudder mass-balance weights
18 HF aerial cable
19 Fin tip construction
20 Tailfin geodetic construction
21 leading-edge de-icing boot
22 Starboard fabric-covered elevator
23 Aluminium alloy-skinned tailplane tip segment
24 Port fabric-covered tailplane
25 Tailplane control rods
26 Ammunition feed chutes
27 Tailwheel pivot fixing
28 Castoring tailwheel
29 Ventral aerial mast
30 Tailwheel retraction jack
31 Vacuum flask stowage
32 Tail turret ammunition boxes
33 Boarding ladder stowage
34 Life jacket container
35 Engine turning crank handle
36 Radio altimeter aerial
37 Fuselage fabric covering
38 Lower longeron
39 Footboards
40 Browning 0 303-in (7.7-mm) machine-gun
41 Beam gunner's window, port and starboard
42 Ammunition box
43 Beam gunner's swivelling seat
44 Reconnaissance flares
45 Leigh-Light mounting frames
46 Hydraulic actuator
47 Light extension and retraction mechanism
48 Leigh-Light control panel
49 Fuselage upper longeron
50 Cabin roof geodetic frame construction
51 Beam approach aerial
52 Fabric support stringers
53 Mid-cabin window panels, port and starboard
54 Flare/marker launch tube
55 Rear spar attachment main frame
56 Marine marker stowage
57 Toilet
58 TR.9J transmitter/receiver
59 Dinghy emergency equipment pack
60 ASV radar operator's seat
61 Intercom socket
62 Parachute stowage
63 Astrodome observation hatch
64 ASV Mk III radar receiver
65 Radar equipment rack
66 Wing main spar cut-out
67 Pneumatic system CO_2 bottles
68 Oil filler cap
69 Starboard engine oil tank, 16-Imp gal (73-litre) capacity
70 Nacelle fuel tank, 58-Imp gal (264-litre) capacity
71 Main undercarriage hydraulic retraction jack
72 Dinghy inflation bottle
73 Dinghy stowage
74 Leigh-Light, extended
75 Searchlight cooling air scoop
76 Rear spar
77 Flap shroud ribs
78 Fuel jettison pipe
79 Starboard split trailing-edge flap
80 Aileron trim tab
81 Starboard aileron
82 Aileron rib construction
83 Starboard formation light
84 Wing-tip fairing
85 Starboard navigation light
86 Armoured leading-edge panel
87 Geodetic wing panel construction

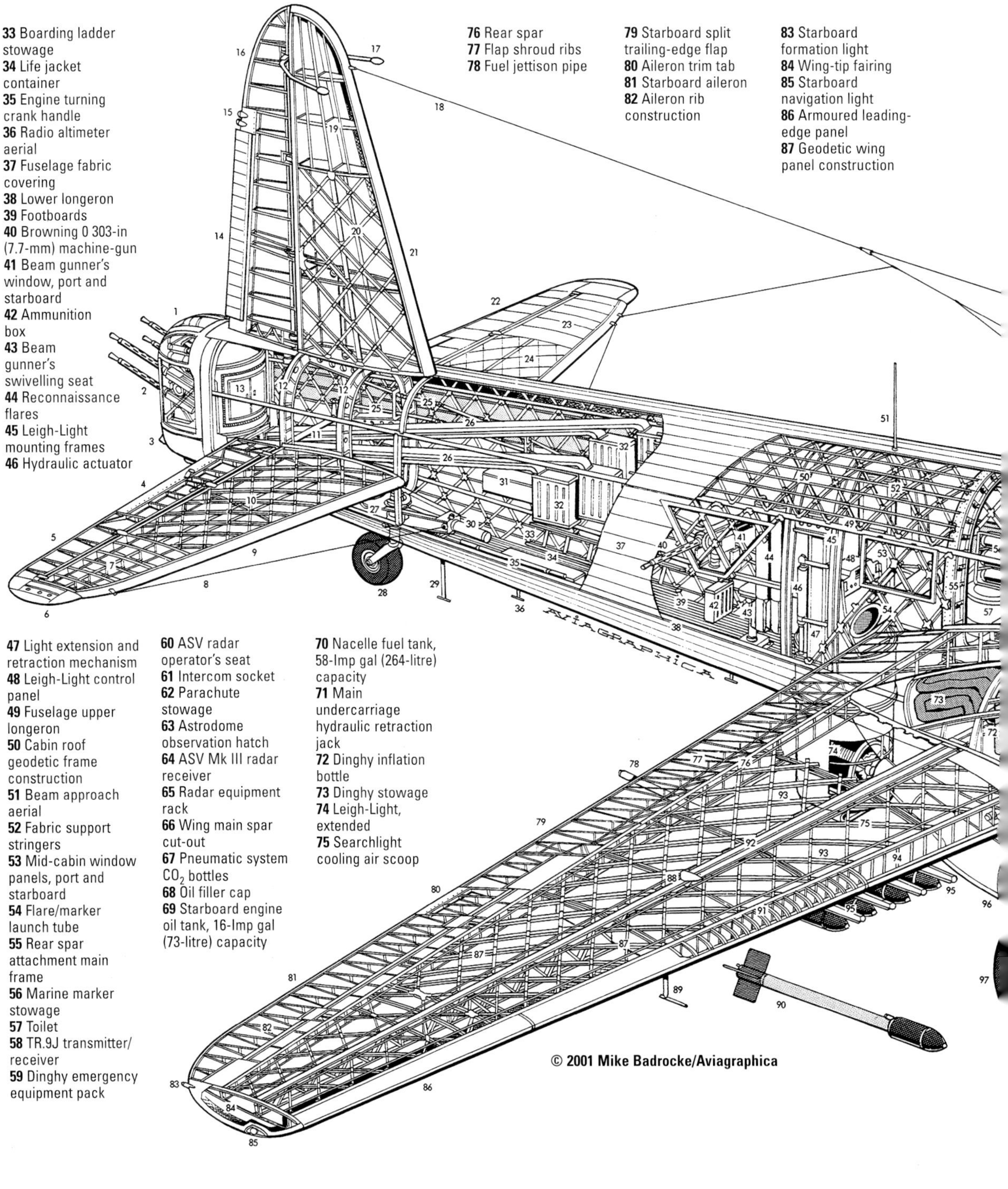

Above: The first production Wellingtons were 180 Mk I aircraft ordered in August 1936, all but the first of which were powered by Bristol Pegasus XVIII radials. The first examples were delivered to No. 99 Sqn in October 1938; L4280 (pictured) went to No. 148 Sqn in March 1939. Also among these first production aircraft were six machines ordered by the RNZAF though, with war looming all remained with the RAF.

SPECIFICATION

Wellington Mk I

Dimensions

Length: 61 ft 3 in (18.67 m)
Height: 17 ft 5 in (5.31 m)
Wingspan: 86 ft (26.2 m)
Wing area: 840.00 sq ft (78.04 m^2)

Powerplant

Two Bristol Pegasus XVIII radial piston engines, each rated at 1,050 hp (783 kW)

Weights

Empty: 18,000 lb (8165 kg)
Gross: 24,850 lb (11272 kg)

Performance

Maximum speed at 15,000 ft (4572 m): 245 mph (394 km/h)
Initial climb rate: 1,120 ft/min (341 m/min)
Climb to: 15,000 ft (4572 m) in 18 minutes
Service ceiling: 21,600 ft (6584 m)

Range

3,200 miles (5150 km) at 180 mph (290 km/h)/15,000 ft (4572 m)

Armament

A single 0.303-in (7.7-mm) Browning machine-gun in a Vickers nose turret, plus a pair of similar weapons in a Vickers tail turret and ventral Fraser Nash FN.9 'dust bin' turret. Offensive load was up to 4,500 lb (2041 kg) of bombs

Wellington B.Mk X

Dimensions

Length: 64 ft 7 in (19.68 m)
Height: 17 ft 6 in (5.33 m)
Wingspan: 86 ft 2 in (26.3 m)
Wing area: 840.00 sq ft (78.04 m^2)

Powerplant

Two Bristol Hercules VI/XVI radial piston engines, each rated at 1,675 hp (1249 kW)

Weights

Empty: 22,474 lb (10194 kg)
Gross: 36,500 lb (16556 kg)

Performance

Maximum speed: 255 mph (410 km/h)
Climb to: 15,000 ft (4572 m) in 27.7 minutes
Service ceiling: 22,000 ft (6706 m)

Range

1,885 miles (3034 km) at 180 mph (290 km/h) with a 1,500 lb (680 kg) bomb load

Armament

Two 0.303-in (7.7-mm) Browning machine-guns in an FN.5 nose turret, plus four similar weapons in an FN.20A tail turret and a single beam gun either side of the rear fuselage. Offensive load was up to 4,000 lb (1814 kg) of bombs

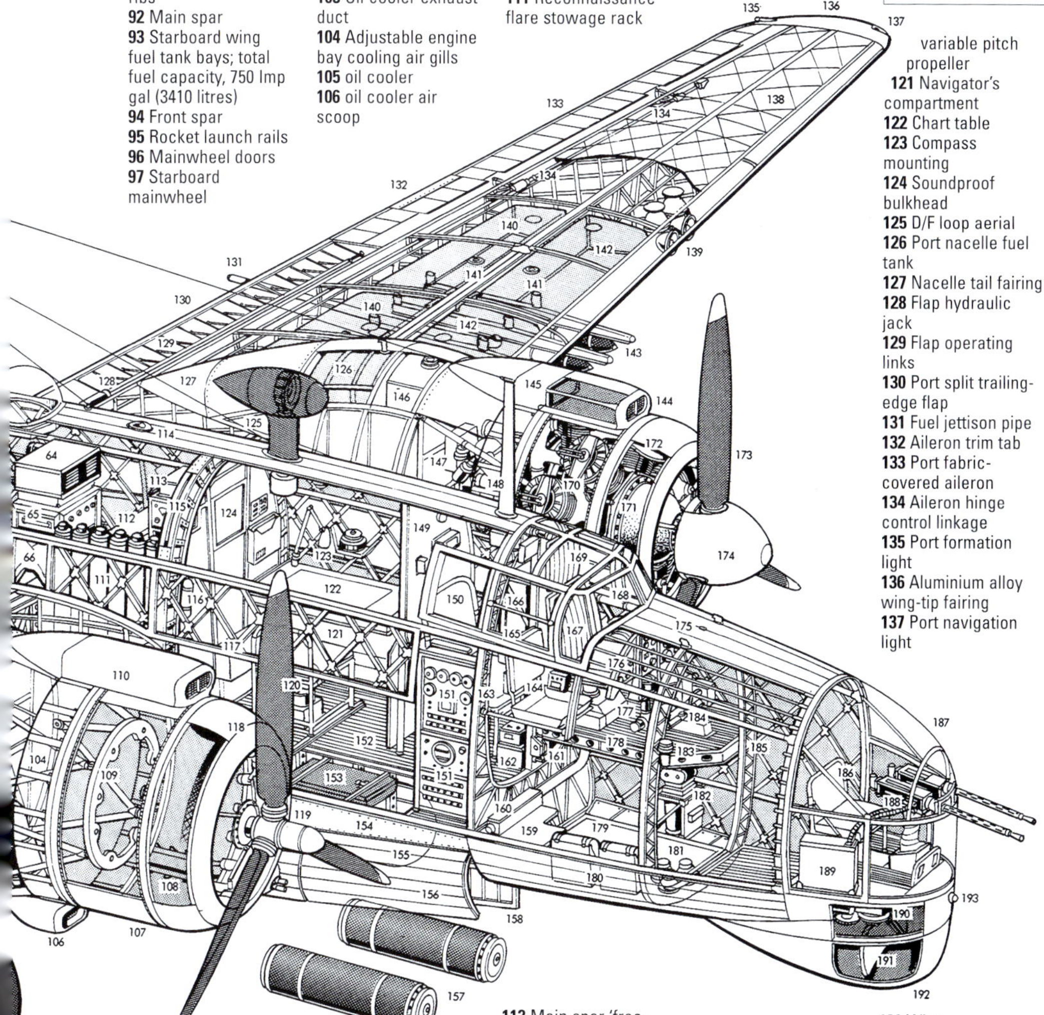

88 Outer wing panel spar joint
89 Pitot head
90 60-lb (27-kg) air-to-surface rocket projectile
91 Leading-edge nose ribs
92 Main spar
93 Starboard wing fuel tank bays; total fuel capacity, 750 Imp gal (3410 litres)
94 Front spar
95 Rocket launch rails
96 Mainwheel doors
97 Starboard mainwheel
98 Hydraulic brake pipe
99 Oleo-pneumatic shock absorber leg
100 Main undercarriage pivot fixing
101 Engine bay fireproof bulkhead
102 Engine bearer struts
103 Oil cooler exhaust duct
104 Adjustable engine bay cooling air gills
105 oil cooler
106 oil cooler air scoop
107 Three-segment detachable engine cowling panels
108 Flame suppressing exhaust pipe on inboard side
109 Engine mounting ring frame
110 Carburettor air intake duct
111 Reconnaissance flare stowage rack
112 Main spar 'free-floating' centre-section carry-through
113 Beam approach equipment rack
114 Upper identification light
115 Front spar attachment main frame
116 Fire extinguisher bottle
117 Forward cabin window panel
118 Engine cowling nose ring
119 Propeller hub pitch change mechanism
120 Starboard variable pitch propeller
121 Navigator's compartment
122 Chart table
123 Compass mounting
124 Soundproof bulkhead
125 D/F loop aerial
126 Port nacelle fuel tank
127 Nacelle tail fairing
128 Flap hydraulic jack
129 Flap operating links
130 Port split trailing-edge flap
131 Fuel jettison pipe
132 Aileron trim tab
133 Port fabric-covered aileron
134 Aileron hinge control linkage
135 Port formation light
136 Aluminium alloy wing-tip fairing
137 Port navigation light
138 Wing panel fabric skinning
139 Retractable landing lamps
140 Wing aft fuel tank train (60 Imp gal/273 litre inboard, 57 Imp gal/259 litre centre and 50 Imp gal/227 litre outboard)
141 Fuel filler caps
142 Wing forward fuel tank train (52 Imp gal/236 litre inboard, 55 Imp gal/250 litre centre and 43 Imp gal/195 litre outboard)
143 Port rocket launch rails
144 Carburettor filtered air intake
145 Intake trunking
146 Port engine oil tank
147 Engine accessory equipment compartment
148 Aerial mast
149 Armoured bulkhead
150 Radio operator's compartment
151 Electrical distribution panels
152 Main cabin floor level
153 Bomb-bay emergency flotation bags (14)
154 Bomb-bay lateral support beam
155 Flotation bag inflated position
156 Outer bomb doors, open
157 250-lb (113-kg) depth charges; 5,000-lb (2268-kg) maximum internal load
158 Triple cell bomb-bay doors
159 Internal step
160 Cabin heater air duct
161 Hydraulic system hand pump
162 HT battery
163 Cockpit section main frame
164 Second pilot's folding seat
165 Sliding cockpit side window panel
166 Radio equipment racks
167 Pilot's seat
168 Windscreen panels
169 Cockpit roof glazing
170 Bristol Hercules XVII 14-cylinder sleeve-valve two-row radial engine
171 Propeller reduction gearbox
172 Townend ring exhaust collector
173 de Havilland three-bladed variable pitch propeller
174 Spinner
175 Windscreen washer ducts
176 Instrument panel
177 Control column
178 Cockpit floor level
179 Ventral entry hatch
180 Parachute stowage
181 Downward identification lights
182 Reconnaissance camera
183 Hand bearing compass holder
184 Rudder pedals
185 Nose compartment construction
186 Nose gunner's seat
187 Glazed nose compartment
188 Manually-operated 0.303-in (7.7-mm) Browning machine-guns
189 Ammunition box
190 Radar scanner drive mechanism
191 ASV Mk III radar scanner
192 Nose radome
193 Forward navigation light

X3662 was among the aircraft equipping No. 115 Squadron, RAF Bomber Command at RAF Marham, Norfolk between late 1941 and March 1943, when the unit re-equipped with Avro Lancaster B.Mk IIs. A Blackpool-built aircraft, X3662 was one of 500 Mk IC aircraft ordered in 1940, 50 of which were completed as such before the order was amended to one for Mk IIIs. No. 115 Sqn was one of the best-known bomber units in the RAF. Having operated the Handley Page Harrow, it acquired its first Wellingtons (Mk Is) in April 1939. During its time equipped with the 'Wimpy', the squadron became the first RAF unit to bomb a mainland enemy target, attacking the airfield at Stavanger/Sola in April 1940 while 'on loan' to Coastal Command.

Wellington B.Mk III

Crew positions

The Wellington Mk III generally flew with a crew of five, comprising a pilot, bomb-aimer/nose gunner, navigator, radio-operator and rear gunner. The pilot occupied the lefthand seat on the flight deck, a folding seat on the starboard side allowing access to the nose compartment in flight. Just aft of the nose turret was a prone bomb-aiming position. Behind the flight deck were positions for the wireless operator, who faced forward immediately behind the pilot, but seperated from the pilot by a bulkhead, and the navigator, who was further aft facing outwards on the port side. The rear crew positions were defensive in nature, comprising the beam gunner's station in the rear fuselage (manned by the navigator and wireless operator), and the tail turret, which was accessed from under the tail.

Structure and fuel capacity

The Wellington was built on the 'geodetic' principal developed by Dr Barnes Wallis, involving a lattice work of diagonals which provided great strength under both bending and twisting loads. With the various loads borne by this structure, a fabric covering was possible, thus saving weight. Four conventional longerons ran the length of the fuselage to form the basis of the framework. Three bomb bays made up the entire centre fuselage beneath the wing. In the top of the bays were flotation bags which were inflated in the event of ditching to give the crew time to escape. Fuel was carried in the wings and engine nacelles. Tanks in the upper nacelles each had a capacity of 58 Imp gal (284 litres), while those ahead of the main spar held 52 Imp gal (236 litres) inboard, 55 Imp gal (250 litres) in the centre and 43 Imp gal (195 litres) outboard. Another three tanks were fitted aft of the spar, these containing 167 Imp gal (758 litres) in all.

Powerplant
The Wellington Mk I was powered by a pair of Bristol Pegasus XVIII radials, though concerns over supplies of this engine and the continuing need for more power prompted development of the Mk II (with Rolls-Royce Merlin Xs), Mk III and Mk IV (Pratt & Whitney Twin Wasps). The Wellington B.Mk III employed a pair of 14-cylinder, two-row Bristol Hercules XIs, each rated at 1,500 hp (1119 kW) and it was the Hercules that was to be the most widely used of the Wellington's varied powerplants. The most numerous variant, the Mk X (of which 3,803 were completed), employed a pair of 1,675-hp (1249-kW) Hercules XVIs.

Bomb load
Typically the Wellington was able to carry a bomb load of up to 4,500 lb (2041 kg), comprising nine 500-lb (227-kg) bombs. Alternatively, a pair of 2,000 lb (907 kg) bombs could be carried, or (in later machines and earlier aircraft so converted) a single 4,000-lb (1814-kg) 'Cookie' HC device.

Defensive armament
In original Mk I guise the Wellington was lightly armed with Vickers nose and tail turrets, the former with mounting a single Browning 0.303-in (7.7-mm) machine-gun and the latter a pair of similar weapons. A retractable two-gun Fraser Nash FN.9 'dust bin' turret was also fitted. In the Mk IA FN.5 nose and FN.10 tail turrets were fitted in place of the original units, each sporting a pair of 'three-oh-threes'. From the Mk IC the ventral FN.9 turret was deleted and replaced by a single beam gun either side of the rear fuselage. The Merlin-engined Mk II introduced the FN.20A four-gun tail turret; this change was carried forward into the Mk III and Mk X. Concern about the Wellington's defensive capabilities prompted the development of a new ventral turret installation, but this was rejected by Bomber Command on the grounds that its weight would have resulted in a reduced bomb load.

Fighter and Ground Attack Aircraft

PZ86
e Last
f the Many!

Bell P-39 Airacobra

When the fuselage side intakes and turbo supercharger were removed, the XP-39 prototype became the XP-39B. The removal of these features, however, turned the P-39 from a potentially great fighter into something of an 'also ran'.

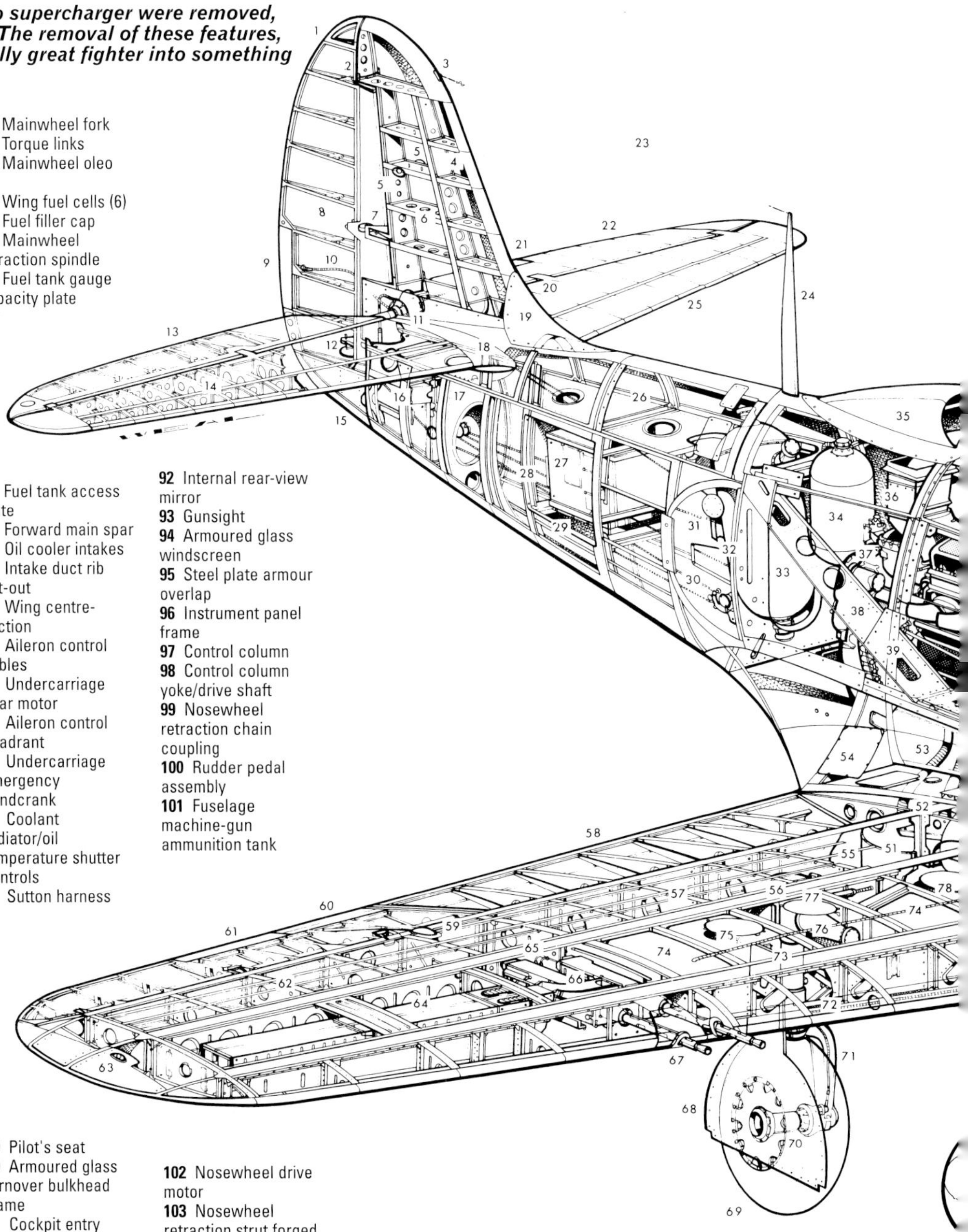

Airacobra

Cutaway key

1 Aluminium sheet rudder tip
2 Rudder upper hinge
3 Aerial attachment
4 Fin forward spar
5 Tall navigation lights
6 Fin structure
7 Rudder middle hinge
8 Rudder
9 Rudder tab
10 Rudder tab flexible shaft
11 Elevator control quadrant
12 Rudder control quadrant
13 Starboard elevator
14 Starboard tailplane
15 Rudder lower hinge
16 Control cables
17 Fuselage aft frame
18 Diagonal brace
19 Fin root fillet
20 Elevator hinge fairing
21 Elevator tab (port only)
22 Port elevator
23 Aerial
24 Aerial mast
25 Port tailplane
26 Aft fuselage semi-monocoque
27 Radio installation
28 Access panel
29 Radio equipment tray
30 Control quadrant
31 Oil tank armour plate
32 Aft fuselage/central chassis bulkhead
33 Engine oil tank
34 Prestone (cooler) expansion tank
35 Carburettor intake fairing
36 Carburettor intake shutter housing
37 Engine accessories
38 Central chassis web
39 Frame
40 Starboard longitudinal fuselage beam
41 Exhaust stubs
42 Allison V-1710-35 Vee 12-cylinder engine
43 Engine compartment decking
44 Aft-vision glazing
45 Crash turnover bulkhead
46 Turnover bulkhead armour plate
47 Auxiliary air intake
48 Ventral Prestone (coolant) radiator
49 Rear main spar/centre section attachment
50 Cylindrical oil radiator
51 Ventral controllable shutters
52 Auxiliary spar/centre section attachment
53 Hoses
54 Shutter control rod access doors
55 Starboard mainwheel well
56 Mainwheel leg/rear main spar attachment point
57 Wing structure
58 Port flap structure
59 Aileron tab control link fairing
60 Aileron trim tab
61 Aileron servo tab
62 Wing rib
63 Starboard navigation light
64 Ammunition tanks
65 Two 0.3-in (7.62-mm) machine-guns
66 Inboard gun ammunition feed chute
67 Machine-gun barrels
68 Mainwheel door fairing
69 Starboard mainwheel
70 Axle
71 Mainwheel fork
72 Torque links
73 Mainwheel oleo leg
74 Wing fuel cells (6)
75 Fuel filler cap
76 Mainwheel retraction spindle
77 Fuel tank gauge capacity plate
78 Fuel tank access plate
79 Forward main spar
80 Oil cooler intakes
81 Intake duct rib cut-out
82 Wing centre-section
83 Aileron control cables
84 Undercarriage gear motor
85 Aileron control quadrant
86 Undercarriage emergency handcrank
87 Coolant radiator/oil temperature shutter controls
88 Sutton harness
89 Pilot's seat
90 Armoured glass turnover bulkhead frame
91 Cockpit entry doors
92 Internal rear-view mirror
93 Gunsight
94 Armoured glass windscreen
95 Steel plate armour overlap
96 Instrument panel frame
97 Control column
98 Control column yoke/drive shaft
99 Nosewheel retraction chain coupling
100 Rudder pedal assembly
101 Fuselage machine-gun ammunition tank
102 Nosewheel drive motor
103 Nosewheel retraction strut forged 'A'-frame attachments

While the P-39 saw only limited in Europe and Africa, it was in the Pacific theatre that the Airacobra, along with the P-40, was the dominant fighter until 1944. These P-39s can be seen escorting a C-47 over New Guinea en route to Wau.

SPECIFICATION

Type
Single-seat fighter and fighter-bomber

Powerplant
One Allison V-1710-85 Vee engine rated at 1,200 hp (895 kW)

Performance
Maximum speed 376 mph (605 km/h) at 15,000 ft (4570 m); cruising speed 200 mph (322 km/h) at optimum altitude; climb to 15,000 ft (4570 m) in 6 minutes 6 seconds; service ceiling 38,270 ft (11665 m); range 975 miles (1569 km)

Weights
Empty 6,400 lb (2903 kg); maximum take-off 8,800 lb (3992 kg)

Dimensions
Wingspan 34 ft (10.36 m); length 30 ft 2 in (9.19 m); height 12 ft 5 in (3.75 m); wing area 213.00 sq ft (19.79 m2)

Armament
One 37-mm T9 fixed forward-firing cannon, two 0.5-in (12.7-mm) fixed forward-firing machine-guns and four 0.3-in (7.62-mm) fixed forward-firing machine-guns, plus one 500-lb (227-kg) bomb carried externally

The P-39Q represented the final attempt to mould the Airacobra into a world-class fighter. This is a P-39Q-20, most of which were built for the USSR and delivered without guns, although this example was destined for the USAAF and retained its armament.

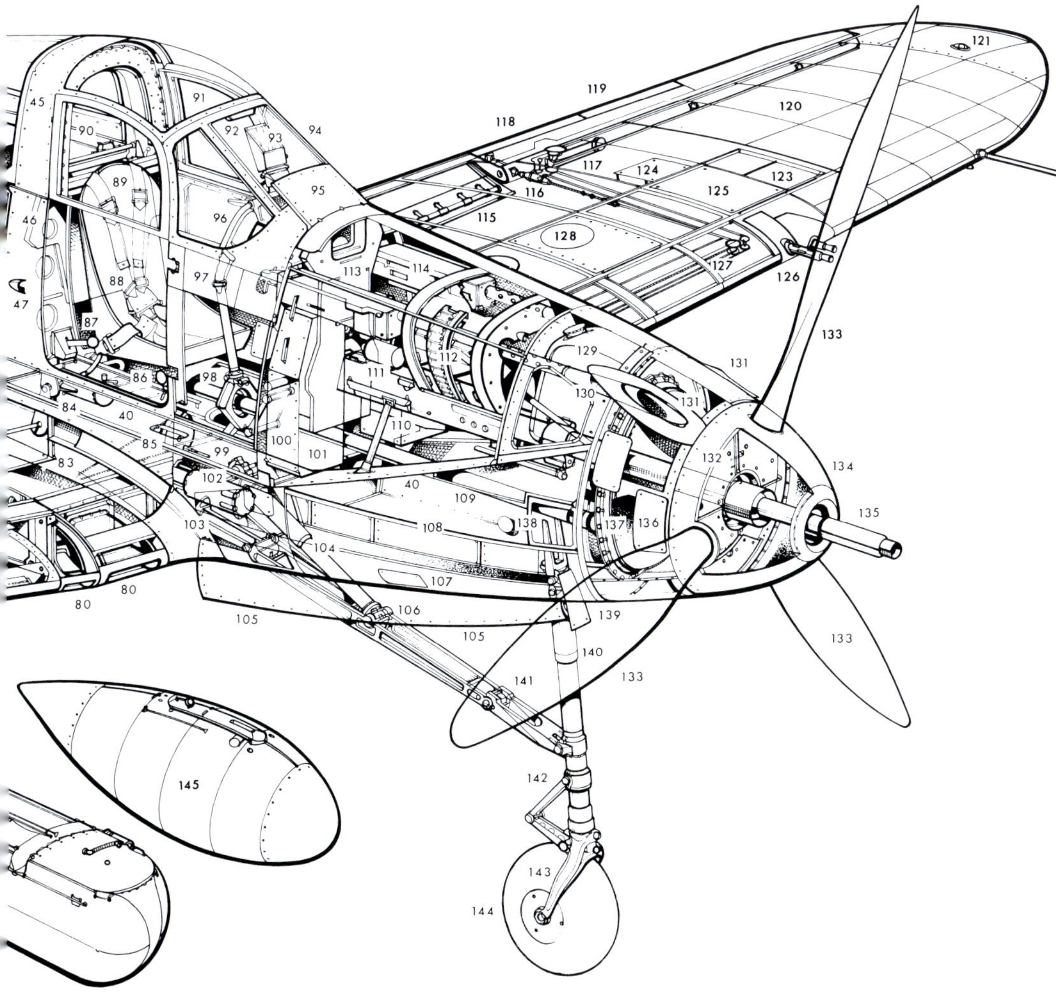

104 Retraction screw
105 Nosewheel doors
106 Link assembly
107 Access plate
108 Nosewheel well
109 Drive shaft
110 Cannon aft support frame
111 37-mm M4 cannon breech
112 Circular endless belt-type cannon magazine (30 rounds)
113 Cockpit forward armoured plate
114 Two 0.5-in (12.7-mm) fuselage machine-guns
115 Flap links
116 Aileron tab actuating link
117 Aileron control
118 Aileron trim tab
119 Aileron servo tab
120 Wing skinning
121 Port navigation light
122 Pitot tube
123 Ammunition feed chute access
124 Gun charge cable access
125 Wing gun service access
126 Machine-gun barrels
127 Aileron and tab control pulleys
128 Fuel tank filler cap
129 Reduction gear oil tank
130 Machine-gun blast tubes
131 Machine-gun ports
132 Reduction gear box frontal armour
133 Three-bladed Curtiss Electric constant speed propeller
134 Spinner
135 Cannon muzzle
136 Blast tube access
137 Reduction gear casing
138 Nosewheel link
139 Nosewheel door forward fairing
140 Nosewheel oleo
141 Link assembly
142 Torque links
143 Axle fork
144 Rearward-retracing nosewheel
145 Ventral stores, options including auxiliary fuel tank, or;
146 Two-man life raft

Bristol Beaufighter

No. 404 Squadron (RCAF) was equipped with Beaufighter TF.Mk Xs. The aircraft, four of which are seen here in June 1944 while based at RAF Davidstow Moor, Cornwall, covered the western Channel during the D-Day invasion of Normandy.

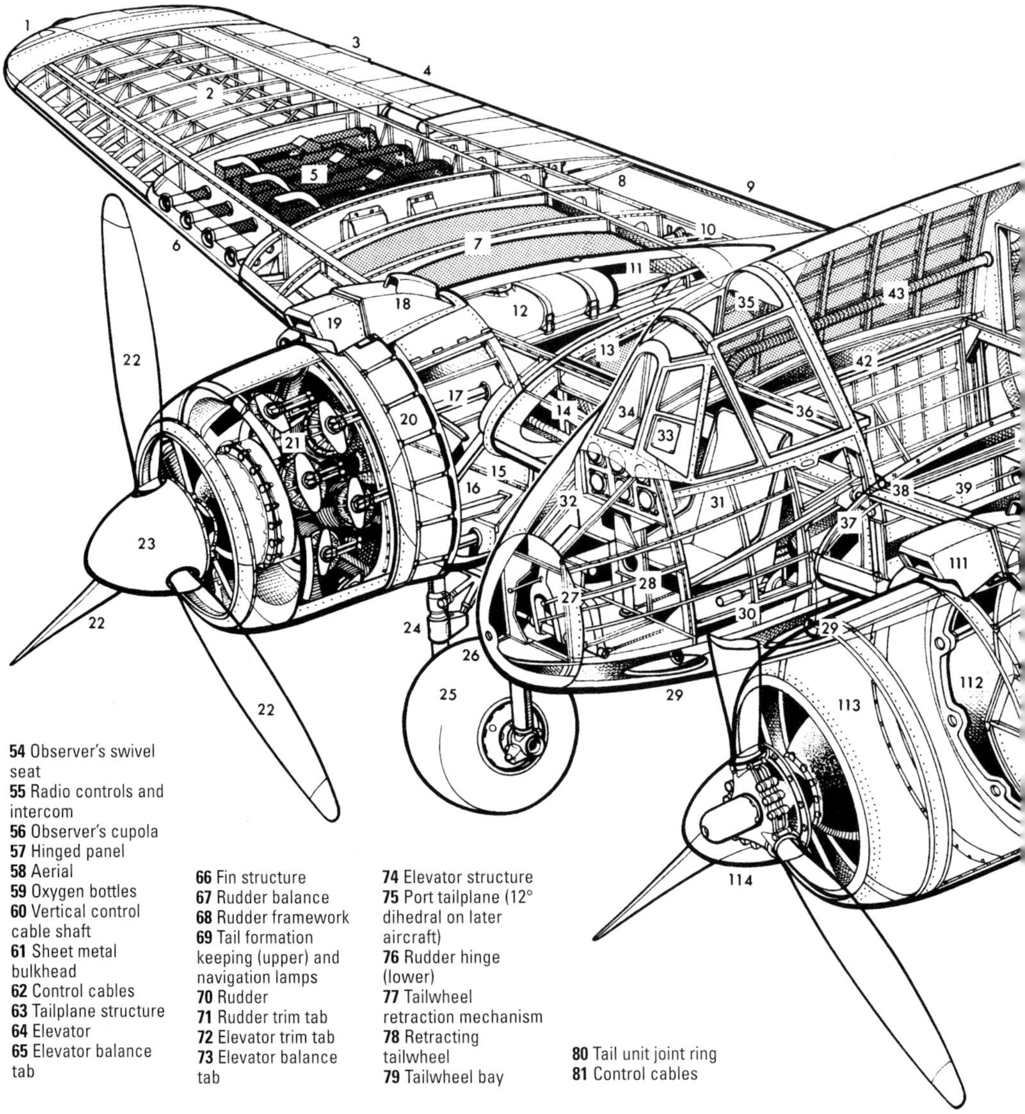

Beaufighter Mk 1

Cutaway key

1 Starboard navigation light (fore) and formation-keeping light (aft)
2 Wing structure
3 Aileron adjustable tab
4 Starboard aileron
5 Four Browning 0.303-in (7.7-mm) machine-guns
6 Machine-gun ports
7 Starboard outer wing fuel tank, capacity 87 Imp gal (395 litres)
8 Split trailing-edge flaps, hydraulically-actuated
9 Starboard flap
10 Flap-operating jack
11 Starboard nacelle tail fairing
12 Oil tank capacity 17 Imp gal (77 litres)
13 Starboard inner wing fuel tank, capacity 188 Imp gal (855 litres)
14 Cabin air duct
15 Hinged leading-edge sections
16 Engine bulkhead
17 Engine bearers
18 Auxiliary intake
19 Supercharger air intake
20 Engine cooling flaps
21 1,560-hp (1164-kW) Bristol Hercules III radial engine
22 de Havilland Hydromatic propeller
23 Propeller spinner
24 Lockheed oleo-pneumatic shock absorber
25 Starboard mainwheel, with Dunlop brakes
26 Forward identification lamp in nose cap
27 Rudder pedals
28 Control column
29 Cannon ports
30 Seat adjusting lever
31 Pilot's seat
32 Instrument panel
33 Clear vision panel
34 Flat bulletproof windscreen
35 Fixed canopy (sideways-hinged on later aircraft)
36 Spar carry-through step
37 Nose centre-section attachment point
38 Fuselage/centre-section attachment point
39 Pilot's entry/emergency escape hatch
40 Underfloor cannon blast tubes
41 Fuselage/centre-section attachment points
42 Centre-section attachment longeron reinforcement
43 Cabin air duct
44 Cannon heating duct
45 Rear spar carry-through
46 Bulkhead cutout (observer access to front hatch)
47 Bulkhead
48 Hydraulic header tank
49 Aerial mast
50 Monocoque fuselage construction
51 Starboard cannon (two 20-mm)
52 Floor level
53 Steps
54 Observer's swivel seat
55 Radio controls and intercom
56 Observer's cupola
57 Hinged panel
58 Aerial
59 Oxygen bottles
60 Vertical control cable shaft
61 Sheet metal bulkhead
62 Control cables
63 Tailplane structure
64 Elevator
65 Elevator balance tab
66 Fin structure
67 Rudder balance
68 Rudder framework
69 Tail formation keeping (upper) and navigation lamps
70 Rudder
71 Rudder trim tab
72 Elevator trim tab
73 Elevator balance tab
74 Elevator structure
75 Port tailplane (12° dihedral on later aircraft)
76 Rudder hinge (lower)
77 Tailwheel retraction mechanism
78 Retracting tailwheel
79 Tailwheel bay
80 Tail unit joint ring
81 Control cables

A potent addition to RAF Coastal Command's strike capability was provided by the Beaufighter Mk VIC (illustrated) and TF.Mk X torpedo-bombers, which carried a single Mk XII torpedo with the usual battery of four 20-mm cannon. The first Beaufighter Strike Wing was formed at RAF North Coates in November 1942. The Perspex bulge immediately behind the cockpit covered the D/F loop for the radio compass.

SPECIFICATION

Beaufighter TF.Mk X

Dimensions

Length: 41 ft 8 in (12.70 m)
Wingspan: 57 ft 10 in (17.63 m)
Height: 15 ft 10 in (4.83 m)
Wing area: 503 sq ft (46.73 m²)

Powerplant

Two 1,735-hp (1295-kW) Bristol Hercules XVII sleeve-valve radial engines

Weights

Empty: 15,600 lb (7076 kg)
Maximum take-off: 25,200 lb (11431 kg)

Performance

Maximum speed at 1,300 ft (395 m): 303 mph (488 km/h)
Cruising speed: 249 mph (401 km/h)
Climb to 5,000 ft (1525 m): 3 minutes 30 seconds
Service ceiling: 15,000 ft (4570 m)
Range: 1,470 miles (2366 km)

Armament

Four forward-firing 20-mm cannon under the nose, six forward-firing 0.303-in (7.7-mm) machine-guns in the wings and one 0.303-in (7.7-mm) Vickers 'K' gun in the dorsal position, plus one torpedo and two 250-lb (113-kg) bombs or eight 60-lb (27-kg) rocket projectiles

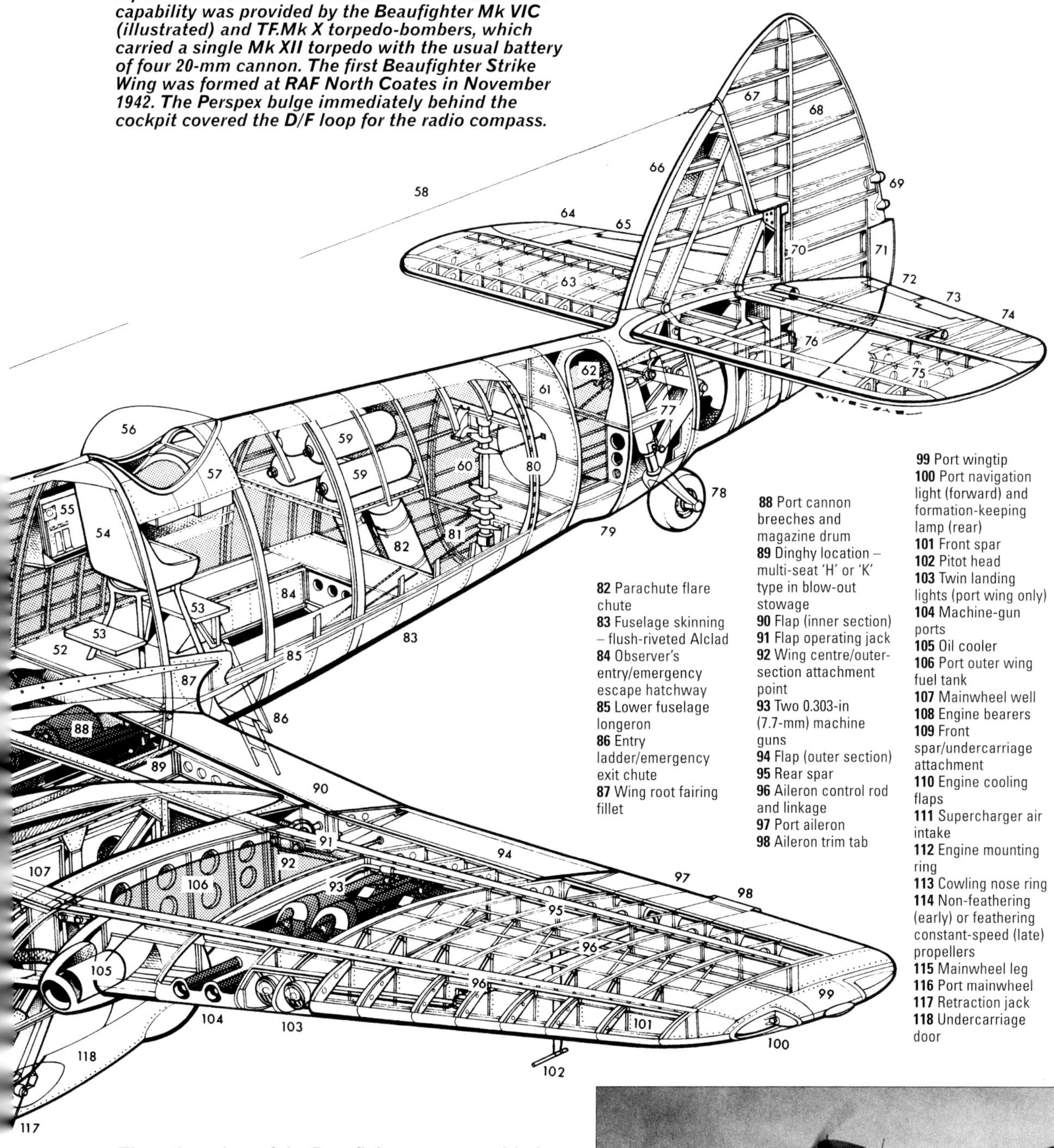

82 Parachute flare chute
83 Fuselage skinning – flush-riveted Alclad
84 Observer's entry/emergency escape hatchway
85 Lower fuselage longeron
86 Entry ladder/emergency exit chute
87 Wing root fairing fillet
88 Port cannon breeches and magazine drum
89 Dinghy location – multi-seat 'H' or 'K' type in blow-out stowage
90 Flap (inner section)
91 Flap operating jack
92 Wing centre/outer-section attachment point
93 Two 0.303-in (7.7-mm) machine guns
94 Flap (outer section)
95 Rear spar
96 Aileron control rod and linkage
97 Port aileron
98 Aileron trim tab
99 Port wingtip
100 Port navigation light (forward) and formation-keeping lamp (rear)
101 Front spar
102 Pitot head
103 Twin landing lights (port wing only)
104 Machine-gun ports
105 Oil cooler
106 Port outer wing fuel tank
107 Mainwheel well
108 Engine bearers
109 Front spar/undercarriage attachment
110 Engine cooling flaps
111 Supercharger air intake
112 Engine mounting ring
113 Cowling nose ring
114 Non-feathering (early) or feathering constant-speed (late) propellers
115 Mainwheel leg
116 Port mainwheel
117 Retraction jack
118 Undercarriage door

The only variant of the Beaufighter to serve with the USAAF was the Mk VIF. During 1942-43, the USAAF desperately needed a night-fighter aircraft for the Mediterranean theatre, so the RAF provided sufficient aircraft to arm four squadrons of the 12th Air Force. These aircraft saw extensive service during the German withdrawal of North Africa, especially in providing night air cover for the army landings at Anzio and Salerno.

Beaufighter Mk 21

A8-186 was an Australian-built Beaufighter Mk 21 which served with No. 22 Squadron, RAAF during the first half of 1945. The name *'Beau-gunsville'* on the nose of the aircraft refers to the island of Bougainville in the Solomon Islands which saw heavy fighting by Australian forces in late 1944 and early 1945. No. 22 Sqn was part of the RAAF's No. 77 (Attack) Wing which flew strike missions against the Celebes and surrounding Japanese-held islands from its base on the island of Morotai. In the last months of the war, the unit moved to newly-captured islands and was mainly involved in harrying the retreating Japanese forces. This particular aircraft is one of the few Australian Beaufighters to survive intact and is currently on display at the Camden Museum of Aviation, New South Wales.

Cockpits

The single-place cockpit was dominated by the large reflector sight mounted above the dashboard. Flight controls were effected by a yoke-style control column which incorporated the firing button. Engine controls were located to the port side, while the starboard had navigation and general function controls. A distinguishing feature of the Beaufighter Mk 21 was the bulge ahead of the cockpit, intended to cover a Sperry autopilot although this was not, in the end, fitted on operational aircraft. The rear cockpit enclosed a rearward-facing observer/radio operator beneath a 'bubble' canopy. In some variants a machine-gun was provided giving a measure of self defence from the rear. The second crew member was also tasked, when necessary, with reloading the four 20-mm cannon located in the rear of the front fuselage.

Powerplant

The Australian-built Beaufighter Mk 21 was powered by a pair of Bristol Hercules Mk XVIII radial engines. Each had 14 cylinders in two rows with sleeve valves, and developed 1,735 hp (1295 kW). Large air inlets on the upper surface of each engine nacelle provided air for the engine's supercharger. Further forward, a ring of cooling flaps around the rear of the cowling admitted cooling air during taxiing and low-speed flight. They were closed shut at high speeds when ram air on the front of the engine provided adequate cooling.
A 17-Imp gal (77-litre) oil tank was mounted in the top of each nacelle, aft of the engine.

Gun armament

The Beaufighter Mk 21's main gun armament was four 20-mm cannon mounted in the lower fuselage, firing from beneath the cockpit. The ammunition drum and firing mechanisms for these reached back to a point level with the trailing edge of the wing. Unlike British Beaufighters, which were mostly equipped with six 0.303-in (7.7-mm) machine-guns in the wings (two in the port and four in the starboard), the Australian Mk 21 introduced four 0.5-in (12.7-mm) machine-guns (two in each wing). The barrels of the guns projected forward of the leading edge of the wing.

Australian markings
Royal Australian Air Force aircraft adopted RAF-style squadron codes during World War II. 'DU' was the code for No. 22 Squadron. The standard RAAF serial consisted of a type designator (in this case A8) followed by a serial number. British-built examples of the Beaufighter serving with the RAAF were given the prefix 'A19'. This aircraft was the 186th from the Fisherman's Bend factory in Australia. The roundels of all RAF and RAAF aircraft serving in the Far East theatre had the red centre omitted to avoid confusion with Japanese aircraft. The fin flash also only featured blue and white segments.

Fuselage
The exceptional strength of the Beaufighter was based on an enormously strong centre-section, which also included the engine bearers. To this all-metal structure was added the forward and rear fuselage, engines and outer wing panels. The rear fuselage behind the observer accommodated oxygen bottles in the upper section. Parachute flares could be carried in the lower section for illuminating targets at night. The lower fuselage of the rear centre section housed a dinghy to be deployed in the event of a ditching. A rotating shaft aft of the rear cockpit transferred the flight control run from the lower to the upper fuselage.

Wing structure
The two-spar wing was built in four major sections: the centre-section, two outer sections (one containing the wing guns) and a detachable tip. Fuel was contained in large tanks between the engines and the guns, and between the engines and the fuselage. The pitot sensor for the air speed indicator was held on a long mast beneath the port wing, which ensured that it was situated in undisturbed airflow. On the same wing, large twin landing lights were fitted behind a flush transparent cover. Australian Mk 21s also had provision for underwing armament which normally consisted of up to two 250-lb (113-kg) bombs and eight rocket projectiles.

Curtiss P-40 Warhawk

Encouraged by the performance being demonstrated by European interceptors powered by liquid-cooled inline engines, Curtiss decided, in 1938, to install a 1,160-hp (865-kW) supercharged Allison V-1710-19 in its radial-powered P-36A. Designated XP-40, the aircraft was the first American fighter to enter mass production. This early P-40C wears pre-war markings possibly belonging to the 20th Pursuit Group, stationed at Hamilton Field, California.

P-40E Warhawk

Cutaway key

1 Rudder aerodynamic balance
2 Rudder upper hinge (port external)
3 Radio aerial bracket/ insulator
4 Rear navigation light (port and starboard)
5 Tailfin structure
6 Rudder post/support tube
7 Rudder structure
8 Rudder trim tab
9 Rudder trim tab push-rod (starboard external)
10 Elevator tab
11 Elevator structure
12 Elevator aerodynamic balance
13 Tailplane structure
14 Rudder lower hinge
15 Rudder control horn
16 Tab actuator flexible drive shafts
17 Tailplane attachment lugs
18 Elevator control horn
19 Tab control rear sprocket housing/chain drive
20 Tailwheel retraction mechanism
21 Access panel
22 Tailwheel door
23 Retractable tailwheel
24 Tailwheel leg
25 Lifting point
26 Tailwheel lower attachment
27 Trim control cable turnbuckles
28 Elevator control cables
29 Tailwheel upper attachment
30 Access panel
31 Port tailplane
32 Port elevator
33 Radio aerials
34 Monocoque fuselage structure
35 Hydraulic reserve tank
36 Automatic recognition device
37 Aerial lead-in
38 Radio aerial mast
39 Hand starter crank stowage
40 Radio bay access door (port)
41 Radio receiver/ transmitter
42 Support frame
43 Battery stowage
44 Ventral aerial (optional)
45 Hydraulic system vent and drain
46 Rudder control cable turnbuckle
47 Oxygen bottles
48 Radio equipment installation (optional)
49 Hydraulic tank
50 Hydraulic pump
51 Wingroot fillet
52 Streamline ventral cowl
53 Wing centreline splice
54 Fuselage fuel tank, capacity 51.5 Imp gal (234 litres)
55 Canopy track
56 Fuel lines
57 Rear-vision panels
58 Pilot's headrest
59 Rearward-sliding cockpit canopy
60 Rear view mirror (external)
61 Bulletproof windshield
62 Instrument panel coaming
63 Electric gunsight
64 Throttle control quadrant
65 Trim tab control wheels
66 Flap control lever
67 Pilot's seat
68 Elevator control cable horn
69 Seat support (wing upper surface)
70 Hydraulic pump handle
71 Control column
72 Rudder pedal/brake cylinder assembly
73 Bulkhead
74 Oil tank, capacity 108 Imp gal (491 litres)
75 Ring sight
76 Flap control push rod rollers
77 Aileron control cables
78 Aileron cable drum
79 Aileron trim tab drive motor
80 Aileron trim tab
81 Port aileron
82 Port navigation light
83 Pitot head
84 Wing skinning
85 Ammunition loading panels
86 Bead sight
87 Coolant expansion tank, capacity 29 Imp gal (132 litres)
88 Carburettor intake
89 Engine bearer support attachment
90 Air vapour eliminator
91 Hydraulic emergency reserve tank
92 Junction box
93 Engine support tubes
94 Engine mounting vibration absorbers
95 Exhaust stacks
96 Cowling panel lines
97 Allison V-1710-39 engine
98 Carburettor Intake fairing
99 Propeller reduction gear casing
100 Coolant thermometer
101 Propeller hub shaft
102 Spinner
103 Curtiss Electric propeller
104 Radiator (divided) intakes
105 Intake trunking
106 Oil cooler radiator (centreline)
107 Glycol radiators (port and starboard)
108 Radiator mounting brackets
109 Glycol radiator intake pipe
110 Port mainwheel
111 Controllable cooling gills
112 Access panel (oil drain)
113 Engine bearer support truss
114 Fresh air intake
115 Wingroot fairing
116 Fuselage frame/wing attachment
117 Walkway
118 Wing/fuselage splice plate
119 Split flap structure
120 Aileron fixed tab
121 Starboard aileron
122 Starboard wingtip construction
123 Starboard navigation light
124 Wing rib

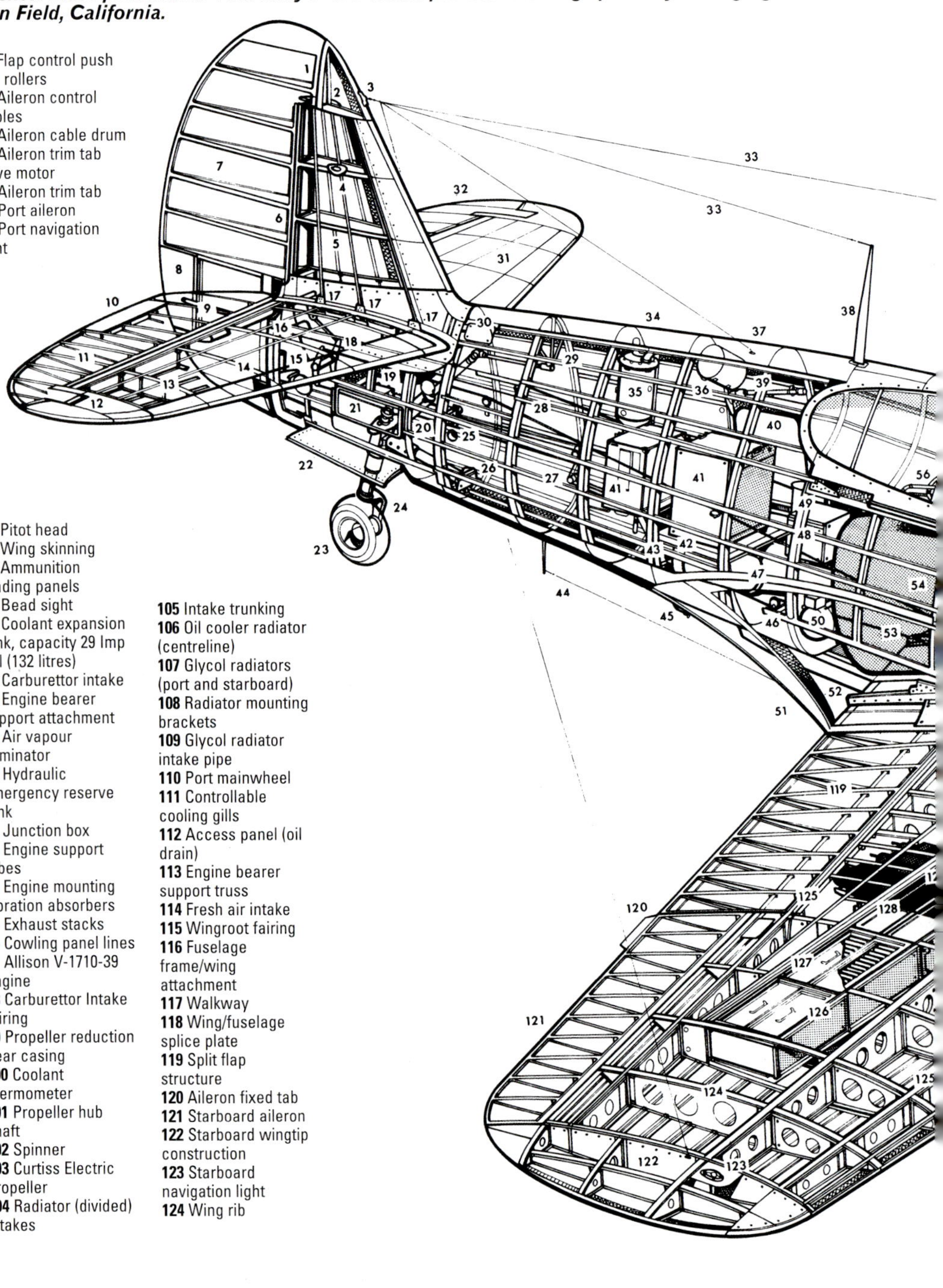

A Randolph Field-based P-40E illustrates the type's standard olive Drab paint scheme. The first of Warhawk variant to serve with the USAAF in Europe in 1942, the P-40E flew with a number of American squadrons in the Mediterranean theatre, but proved generally inferior to most other Allied fighters.

SPECIFICATION

P-40E (Model H87-B2) Warhawk

Dimensions

Length: 31 ft 2 in (9.49 m)
Height: 10 ft 7 in (3.22 m)
Wingspan: 37 ft 3½ in (11.35 m)
Wing area: 236 sq ft (21.92 m²)

Powerplant

One 1,150-hp (857-kW) Allison V-1710 inline piston engine

Weights

Empty: 6,350 lb (2880 kg)
Gross: 8,280 lb (3756 kg)
Maximum take-off: 9,200 lb (4173 kg)

Fuel

Normal: 147 Imp gal (558 litre)
Maximum: 200 Imp gal (756 litre)

Performance

Maximum speed (clean) at 15,000 ft (4572 m): 366 mph (589 km/h)
Climb to 15,000 ft (4572 m): 7 min 36 seconds
Climb in one minute: 2,050 ft (625 m)
Cruising speed: 308 mph (496 km/h)
Speed for maximum range: 190 mph (306 km/h)
Landing speed: 82 mph (132 km/h)
Service ceiling: 29,000 ft (8839 m)
Range: 350 miles (563 km)
Maximum range: 950 miles (1529 km)

Armament

Six 0.5-in (12.7-mm) machine-guns in the wings; plus provision for six 20-lb (9 kg) bombs or one 500-lb (227-kg) bomb mounted under the fuselage

P-40N (Model H87-V) Warhawk

Dimensions

Length: 33 ft 4 in (10.16 m)
Height: 12 ft 4 in (3.75 m)
Wingspan: 37 ft 3½ in (11.35 m)
Wing area: 236 sq ft (21.92 m²)

Powerplant

One 1,360-hp (1014-kW) supercharged Allison V-1710 inline piston engine

Weights

Empty: 6,000 lb (2722 kg)
Gross: 4,400 lb (2903 kg)
Maximum take-off: 8,850 lb (4014 kg)

Fuel

Normal: 122 Imp gal (462 litre)
Maximum: 292 Imp gal (1105 litre)

Performance

Maximum speed (clean) at 15,000 ft (4572 m): 343 mph (552 km/h)
Climb to 15,000 ft (4572 m) : 6 min 42 seconds
Climb in one minute: 2,120 ft (646 m)
Cruising speed: 263 mph (423 km/h)
Speed for maximum range: 198 mph (319 km/h)
Landing speed: 85 mph (137 km/h)
Service ceiling: 31,000 ft (9449 m)
Range: 750 miles (1207 km)
Maximum range: 1,400 miles (2253 km)

Armament

Six 0.5-in (12.7-mm) machine-guns in the wings; plus up to 700-lb (318-kg) of bombs

125 Multi (7)-spar wing structure
126 Inboard gun ammunition box (235 rounds)
127 Centre gun ammunition box (235 rounds)
128 Outboard gun ammunition box (235 rounds)
129 Three 0.50-in (12.7-mm) M-2 Browning machine-guns
130 Ammunition feed chute
131 Starboard wheel well
132 Wing centre section main fuel tank, capacity 42.1 Imp gal (191 litres)
133 Undercarriage attachment
134 Wing centre section reserve fuel tank, capacity 29.2 Imp gal (133 litres)
135 Retraction cylinder
136 Retraction arm/links
137 Machine-gun barrel forward support collars
138 Blast tubes
139 Bevel gear
140 Undercarriage side support strut
141 Gun warm air
142 500-lb (227-kg) bomb (ventral stores)
143 Undercarriage oleo leg fairing
144 Undercarriage fairing door
145 Machine/gun ports
146 Hydraulic brake line
147 One (or two) underwing 40-lb (18-kg) bomb(s)
148 Oleo leg
149 Torque links
150 Axle
151 30-in (762-mm) diameter smooth contour mainwheel tyre
152 Tow ring/jack point **153** Ventral auxiliary tank, capacity 43.3 Imp gal (195 litres)
154 Vent line
155 Sway brace pads
156 External fuel line
157 Shackle assembly
158 Filler neck
159 Alternative ventral 250 lb (114 kg) bomb with:
160 Extended percussion fuse

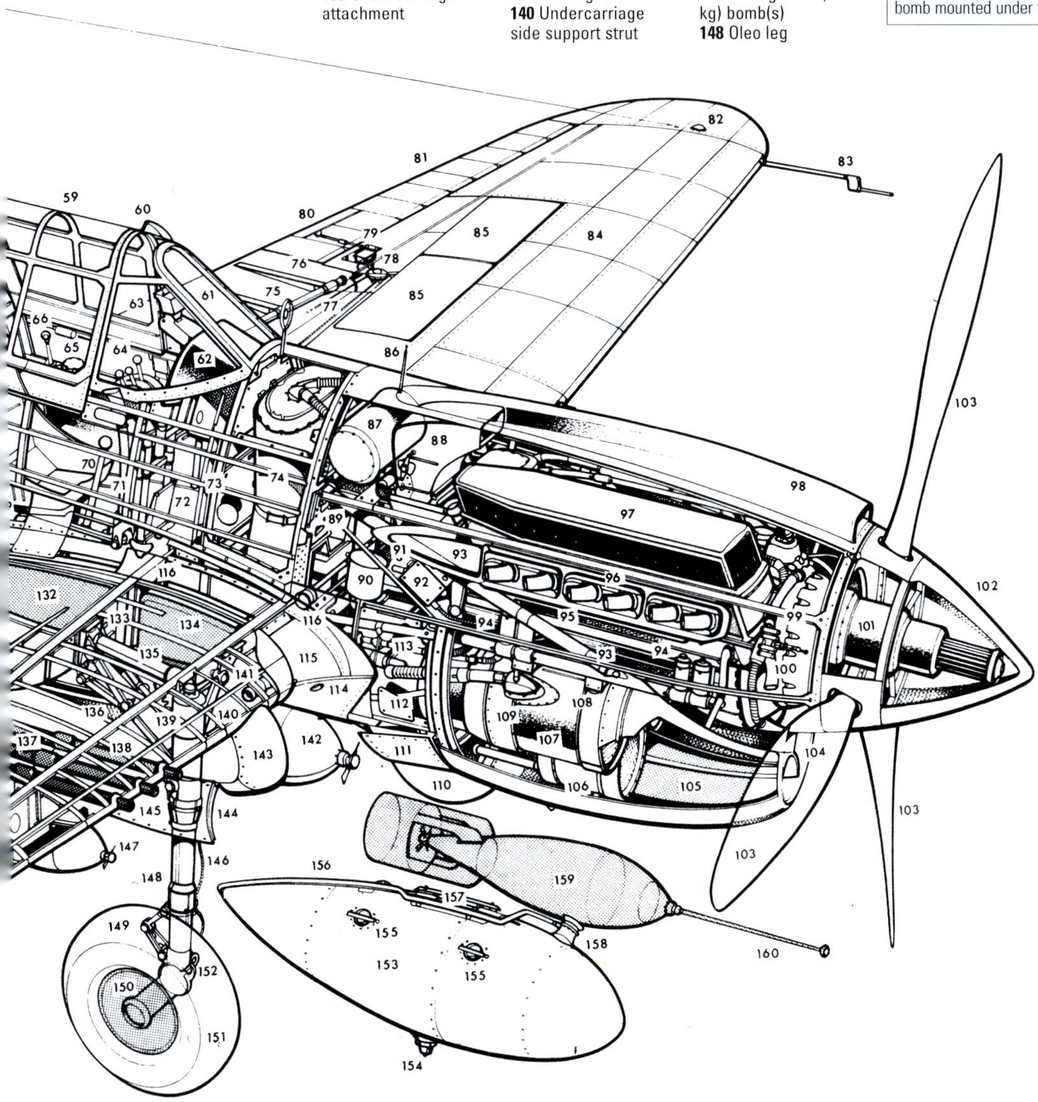

A quartet of P-40F Warhawks performs a classic fighter peel-off. The 'F' was identified by the lack of an intake above the engine, denoting the fitment of a Packard-built V-1650 Merlin engine. In total, 1,311 of this model were completed, many with a lengthened rear fuselage.

P-40E Warhawk

In 1941 Madam Chiang Kai-shek, the National Secretary of Aviation (regarded by many as the real power in Nationalist China) gave Brig. Gen. Claire L. Chennault – founder of the American Volunteer Group in China (AVG) – nearly $9 million via the China Defence Supplies Corporation to obtain the fighters he wanted. At the time, the standard fighter used by the AVG (later called the 'Flying Tigers') in China was the Curtiss P-36 Hawk. A batch of 100 P-40Cs had originally been intended for supply to Great Britain under Lend-Lease, but as they were not of a sufficiently high performance to take on German fighters over Western Europe, they were transferred to the Chinese air force, which in turn forwarded them to the AVG in the southern China/Burma area. The P-40s went into action for the first time during December 1941. Among these aircraft was this example flown by Lt Dallas Clinger. Flying with the 76th Fighter Squadron, 23rd Fighter Group, he achieved 'ace' status in the Chinese theatre, scoring five kills over Japanese aircraft.

Powerplant

The P-40E was powered by an Allison V-1710-39 12-cylinder inline engine, with an output of 1,150 hp (857 kW). Most P-40s were Allison-powered, although the P-40F (1,311 built) featured the Packard V-1650 licence-built Merlin. A long inlet duct which admitted air for the carburettor positioned at the rear of the engine was mounted prominently above the cowling. Just forward of the cockpit and the engine bulkhead was an oil tank, containing 10.8 Imp gal (49 litres). Immediately forward of this was an expansion tank for the coolant system, containing 2.9 Imp gal (13 litres). Cooling air for the three radiators mounted underneath the engine was admitted by a characteristic chin intake. The central unit cooled the oil, while the two outer radiators cooled glycol for the main cooling system.

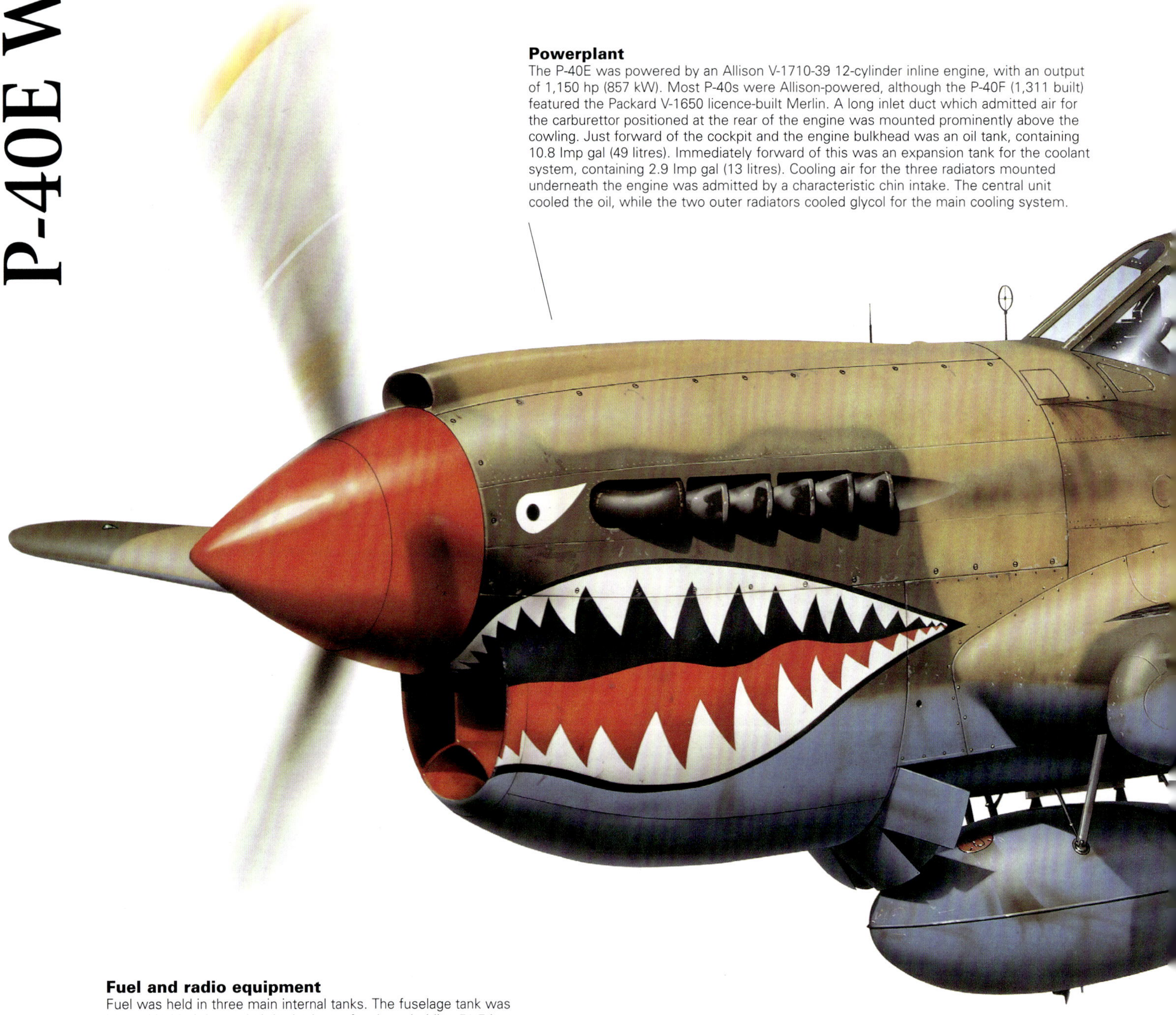

Fuel and radio equipment

Fuel was held in three main internal tanks. The fuselage tank was located behind the cockpit in the lower fuselage, holding 51.5 Imp gal (234 litres). Two tanks were located in the centre section, the main one holding 42.1 Imp gal (191 litres), and a reserve holding 29.2 Imp gal (133 litres). From a point on the fin ran a cable antenna for the radio set, extending to the rear fuselage, wingtips and the prominent mast behind the cockpit. Equipment for the radio set was carried in the rear fuselage.

Tail surfaces and insignia
The large rudder and elevators were both balanced aerodynamically by horns. Trimming was by tabs, that on the rudder actuated by an external rod on the starboard side. 'Hold'n My Own' was a typically derisory tail marking which left no doubt as to the pilot's intentions. In reality, the P-40 was barely adequate in combat against Japanese fighters such as the Nakajima Ki-43 'Oscar' and Mitsubishi A6M 'Zeke', both of which possessed a much higher degree of manoeuvrability. However, a high level of training and superior tactics enabled AVG air crew to gain limited air superiority.

Cockpit
Despite the glazed rear-vision panels incorporated into the decking behind the pilot's headrest, the P-40 did not have good vision in the all-important rear quadrant. A mirror mounted on the canopy rail, offset to port, allowed the pilot to check his 'six'. The P-40E was provided with an electric reflector gunsight on top of the dashboard, providing range and offset information. If this unit failed, a back-up ring and bead sight was fitted to the upper fuselage decking for emergency use.

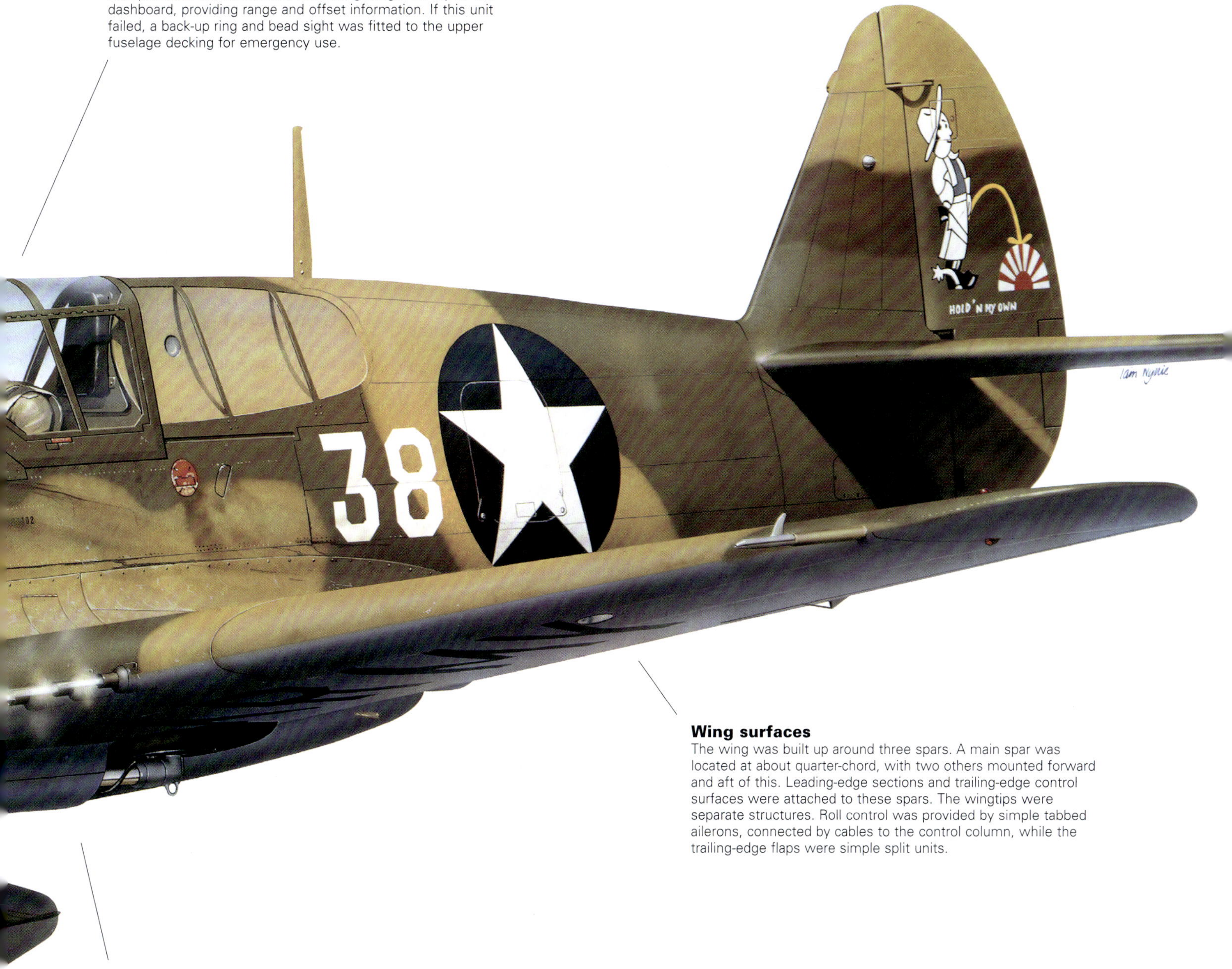

Wing surfaces
The wing was built up around three spars. A main spar was located at about quarter-chord, with two others mounted forward and aft of this. Leading-edge sections and trailing-edge control surfaces were attached to these spars. The wingtips were separate structures. Roll control was provided by simple tabbed ailerons, connected by cables to the control column, while the trailing-edge flaps were simple split units.

Undercarriage and wing guns
Fairings under the wing held the main undercarriage struts. The main wheels were held flat within the wing section, deploying forwards. The wheel had to rotate through 90° during the deployment process to be correctly aligned for landing. The P-40E introduced a six 0.50-in (12.7-mm) calibre machine-gun armament. Each gun had 235 rounds available, all armament held in quick-change boxes outboard of the guns. Many P-40s carried a centrally-mounted drop tank, holding 43.3 Imp gal (197 litres), to augment their internal fuel load. The normal range of the aircraft was in the region of 350 miles (563 km).

Dewoitine D.520

After a heroic, yet ultimately futile, effort against the superior Messerschmitt Bf 109 during the Battle of France, the D.520 found itself fighting alongside its former enemy against Allied Forces in North Africa.

Dewoitine D.520

Cutaway key
1 Cannon port
2 Spinner
3 Three-blade Ratier Electric propeller
4 Cannon barrel blast tube
5 Coolant water tank
6 Safety vent
7 Cowling forward frame
8 Auxiliary intake
9 Chin intake
10 Coolant piping
11 Oil cooler intake
12 Intake duct
13 Oil radiator
14 Engine bearer frames
15 Engine accessories
16 Exhaust stubs
17 Hispano-Suiza 12Y45 engine
18 Cowling rear frame
19 Cannon ammunition drum (60 rounds)
20 Oil tank
21 Starboard wing fuel tank
22 Wing skinning
23 Starboard navigation light
24 Starboard aileron
25 Aileron hinge
26 Emergency ring and bead gunsight
27 Fuselage main fuel tank
28 Fuselage main frame upper member
29 Engine bearer upper attachment
30 Bulkhead
31 20-mm HS 404 cannon breech
32 Compressor outlet
33 Extinguisher
34 Szydlowski compressor
35 Engine bearer support frame
36 Wing root fairing
37 Starboard mainwheel
38 Port mainwheel well
39 Ventral radiator bath intake
40 Undercarriage retraction mechanism
41 Mainwheel leg pivot
42 Wing machine-gun blast tubes
43 Machine-gun ports
44 Mainwheel leg
45 Port mainwheel
46 Mainwheel cover
47 Mainwheel leg door
48 Port wing fuel tank
49 Wing nose ribs
50 Pitot head
51 Port navigation light
52 Wingtip
53 Port aileron frame
54 Aileron hinge
55 Wing rear false spar
56 Wing skinning
57 Wing ribs
58 Two 7.5-mm MAC 1934 machine-guns
59 Ammunition feed
60 Wing main spar
61 Ammunition boxes (675 rpg)
62 Gun hot air
63 Radiator bath
64 Wing flap inboard profile
65 Radiator outlet flap
66 Port wing flap
67 Retractable radio aerial
68 Wing root fairing
69 Fuselage main frame lower member
70 Wing flap control linkage
71 Rudder pedal bar
72 Instrument panel
73 Command radio receiver
74 Control column grip
75 HF receiver
76 Windscreen
77 OPL RX 39 gunsight
78 Canopy track
79 Pilot's seat
80 Seat adjustment lever
81 Seat mounting frame

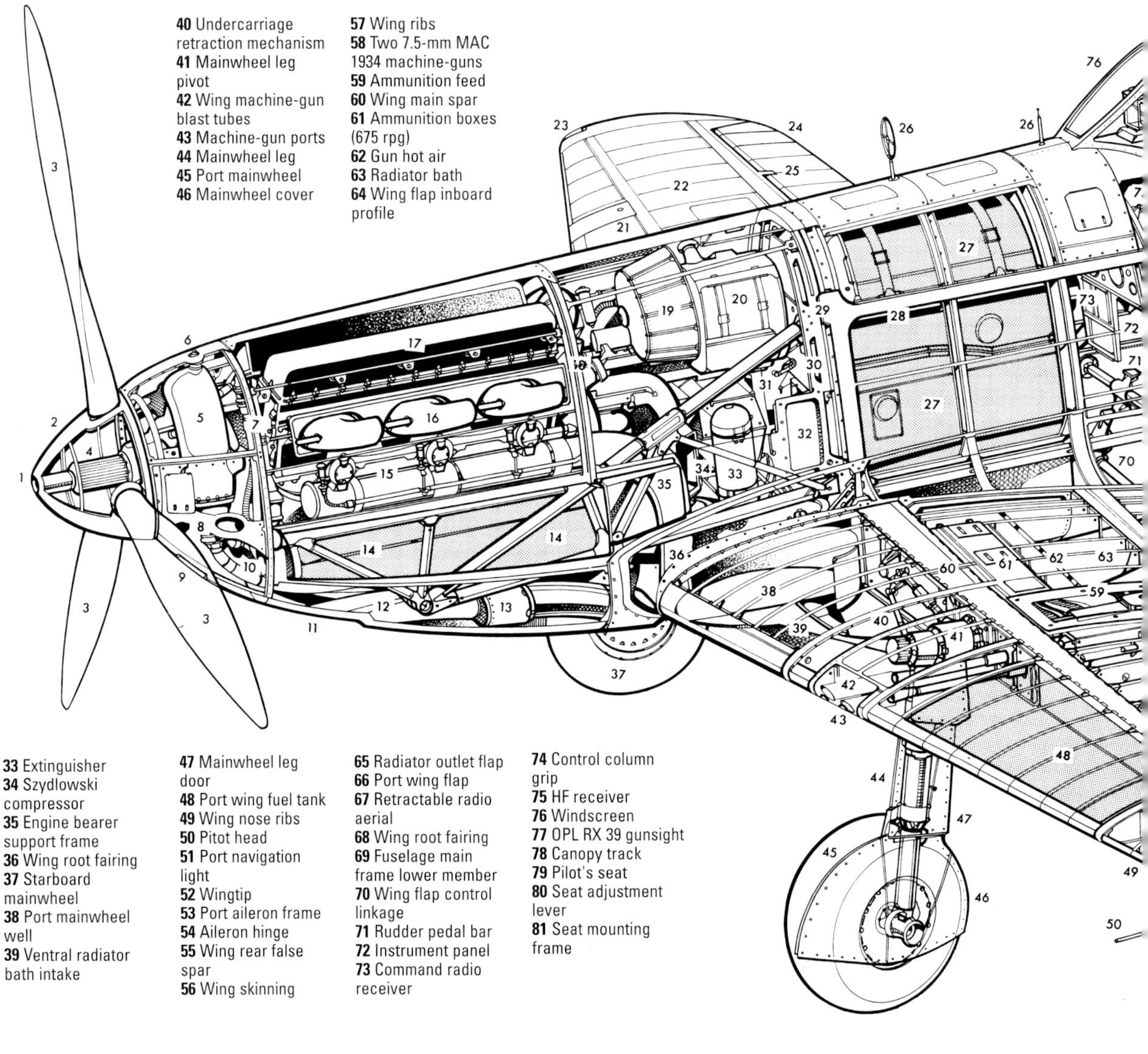

*The extensive use of the **D.520** by Luftwaffe fighter schools was an expression of the desperation faced by that service late in the war. Accident rates among the young students were high.*

SPECIFICATION

Dewoitine D.520

Type

Single-seat fighter

Powerplant

One Hispano-Suiza 12Y-45 Vee piston engine rated at 935 hp (967 kW)

Performance

Maximum speed 332 mph (534 km/h) at 18,045 ft (5500 m); cruising speed 230 mph (370 km/h) at optimum altitude; climb to 13,125 ft (4000 m) in 5 minutes 48 seconds; service ceiling 34,450 ft (10500 m); range 950 miles (1530 km)

Weights

Empty 4,449 lb (2036 kg); maximum take off 5,902 lb (2677 kg)

Dimensions

wingspan 33 ft 51⁄2 in (10.20 m); length 28 ft 21⁄2 in (8.60 m); height 8 ft 51⁄4 in (2.57 m); wing area 171.91 sq ft (15.97 m2)

Armament

One 20-mm Hispano-Suiza HS-404 fixed forward-firing cannon between the engine's cylinder banks, and four 0.295-in (7.5-mm) MAC 34 M39 fixed forward-firing machine-guns in the leading edges of the wing

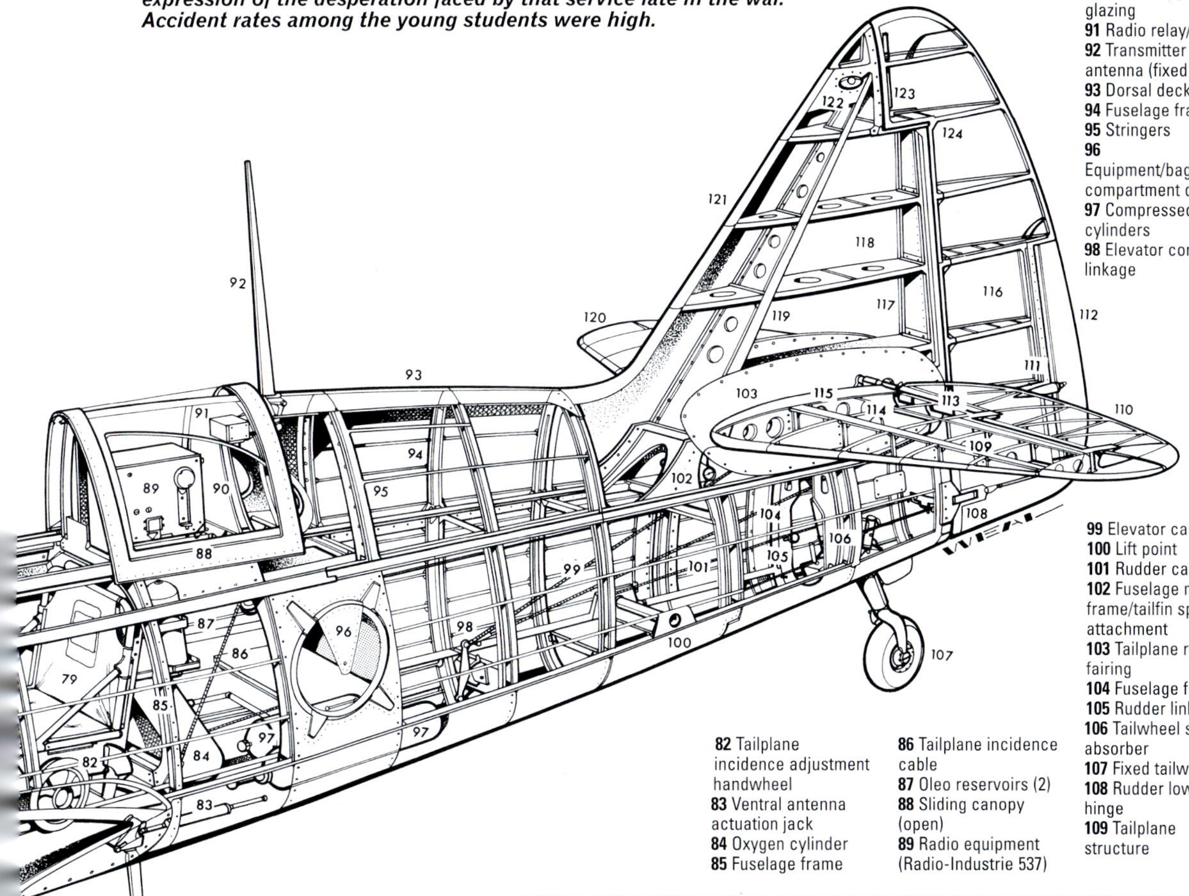

82 Tailplane incidence adjustment handwheel
83 Ventral antenna actuation jack
84 Oxygen cylinder
85 Fuselage frame
86 Tailplane incidence cable
87 Oleo reservoirs (2)
88 Sliding canopy (open)
89 Radio equipment (Radio-Industrie 537)
90 Aft canopy fixed glazing
91 Radio relay/lead-in
92 Transmitter antenna (fixed)
93 Dorsal decking
94 Fuselage frames
95 Stringers
96 Equipment/baggage compartment door
97 Compressed air cylinders
98 Elevator control linkage
99 Elevator cables
100 Lift point
101 Rudder cables
102 Fuselage main frame/tailfin spar attachment
103 Tailplane root fairing
104 Fuselage frame
105 Rudder linkage
106 Tailwheel shock absorber
107 Fixed tailwheel
108 Rudder lower hinge
109 Tailplane structure
110 Port elevator frame
111 Rudder tab hinge fairing
112 Rudder tab
113 Elevator control horn
114 Elevator torque tube
115 Tailplane attachment
116 Rudder frame
117 Rudder post
118 Tailfin structure
119 Tailfin front spar
120 Starboard tailplane
121 Tailfin leading-edge
122 Tail navigation light
123 Rudder internal balance
124 Rudder upper hinge

*Marcel Doret is seen at the controls of the first **D.520** on its second flight on 8 October 1938. The short fin gave the type the appearance of a racer rather than a fighting machine.*

Dornier Do335

During initial tests at Oberpfaffenhofen and Rechlin, the Do 335 V1 demonstrated superb acceleration and generally good handling.

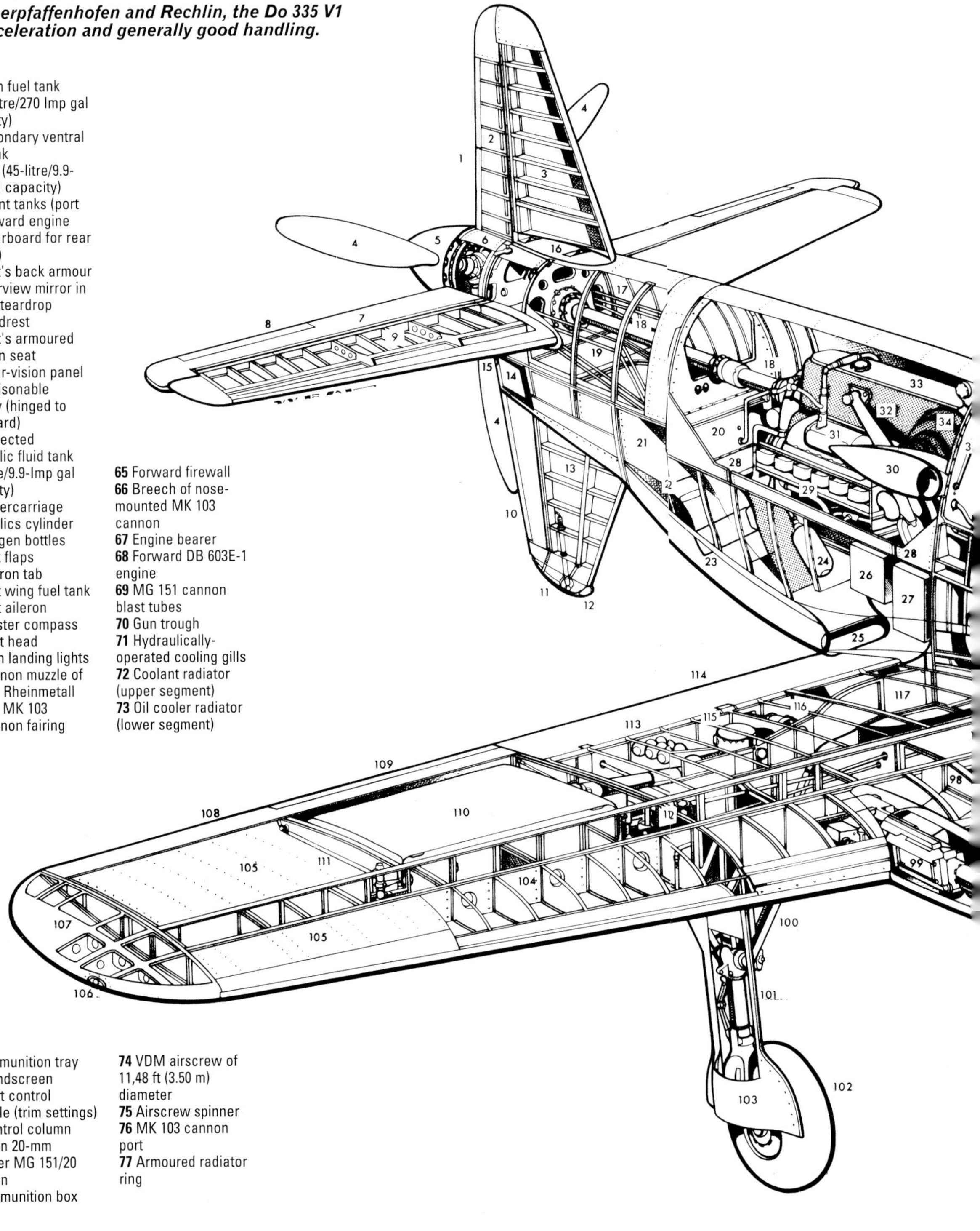

Dornier Do335

Cutaway key

1 Upper rudder trim tab
2 Upper rudder
3 Upper tailfin (jettisonable by means of explosive bolts)
4 VDM airscrew of 3.30 m (10.83 ft) diameter
5 Airscrew spinner
6 Airscrew pitch mechanism
7 Starboard elevator
8 Elevator tab
9 Metal stressed-skin tailplane
10 Ventral rudder
11 Tail bumper
12 Tail bumper oleo shock-absorber
13 Ventral tailfin (jettisonable for belly landing)
14 Coolant outlet
15 Rear navigation light
16 Explosive bolt seatings
17 Rudder and elevator tab controls
18 Hollow airscrew extension shaft
19 Rear airscrew lubricant feeds
20 Aft bulkhead
21 Coolant trunking
22 Oil cooler radiator
23 Coolant radiator
24 Fire extinguisher
25 Ventral air intake
26 FuG 25a IFF
27 FuG 125a blind landing receiver
28 Rear engine access cover latches
29 Exhaust stubs
30 Supercharger intake
31 Coolant tank
32 Engine bearer
33 Aft Daimler-Benz DB 603E-1 12-cylinder inverted-Vee liquid-cooled engine rated at 1340 kW (1,800 hp) for take-off and 1415 kW (1,900 hp) at 1800 m (5,905 ft)
34 Supercharger
35 Aft firewall
36 FuG 25a ring antenna
37 Fuel filler cap
38 Main fuel tank (1230-litre/270 Imp gal capacity)
39 Secondary ventral fuel tank
40 Two (45-litre/9.9-Imp gal capacity) lubricant tanks (port for forward engine and starboard for rear engine)
41 Pilot's back armour
42 Rearview mirror in glazed teardrop
43 Headrest
44 Pilot's armoured ejection seat
45 Clear-vision panel
46 Jettisonable canopy (hinged to starboard)
47 Protected hydraulic fluid tank (45-litre/9.9-Imp gal capacity)
48 Undercarriage hydraulics cylinder
49 Oxygen bottles
50 Port flaps
51 Aileron tab
52 Port wing fuel tank
53 Port aileron
54 Master compass
55 Pitot head
56 Twin landing lights
57 Cannon muzzle of 30-mm Rheinmetall Borsig MK 103
58 Cannon fairing
59 Ammunition tray
60 Windscreen
61 Port control console (trim settings)
62 Control column
63 Twin 20-mm Mauser MG 151/20 cannon
64 Ammunition box
65 Forward firewall
66 Breech of nose-mounted MK 103 cannon
67 Engine bearer
68 Forward DB 603E-1 engine
69 MG 151 cannon blast tubes
70 Gun trough
71 Hydraulically-operated cooling gills
72 Coolant radiator (upper segment)
73 Oil cooler radiator (lower segment)
74 VDM airscrew of 11,48 ft (3.50 m) diameter
75 Airscrew spinner
76 MK 103 cannon port
77 Armoured radiator ring

Ten pre-production Do 335A-0 fighter-bombers were completed, the second (illustrated) being evaluated in the US post-war.

SPECIFICATION

Dornier Do335

Type
Single-seat fighter-bomber

Powerplant
Two Daimler-Benz DB 603A-2 inverted Vee piston engines each rated at 1,750 hp (1305 kW)

Performance
Maximum speed 478 mph (770 km/h) at 21,000 ft (6400 m); cruising speed 426 mph (685 km/h) at 23,295 ft (7100 m); climb to 26,245 ft (8000 m) in 14 minutes 30 seconds; service ceiling 37,400 ft (11400 m); range 857 miles (1380 km)

Weights
Empty 16,314 lb (7400 kg); maximum take-off 21,164 lb (9600 kg)

Dimensions
Wingspan 45 ft 31⁄4 in (13.80 m); length 45 ft 51⁄4 in (13.85 m); height 16 ft 43⁄4 in (5.00 m); wing area 414.42 sq ft (38.50 m2)

Armament
One 30-mm MK 103 fixed forward-firing cannon between the cylinder banks of the forward engine and two 15-mm MG 151/15 fixed forward-firing cannon in the upper part of the forward fuselage, plus one 500-kg (1,102-lb) SC500 or two 250-kg (551-lb) SC250 bombs carried in a lower-fuselage weapons bay and two 250-kg (551-lb) SC250 bombs carried under the wing

78 Coolant tank (15-litre/3.3-Imp gal capacity)
79 Exhaust stubs
80 Nosewheel oleo leg
81 Nosewheel scissors
82 Damper
83 Nosewheel
84 Mudguard
85 Retraction strut
86 Nosewheel door
87 MK 103 cannon ammunition tray
88 Collector tray
89 Accumulator
90 Electric systems panel
91 Ejector seat compressed air bottles
92 Rudder pedals
93 Ammunition tray
94 Armour
95 Cannon fairing
96 MK 103 barrel
97 Muzzle brake
98 Ammunition feed chute
99 Starboard MK 103 wing cannon
100 Mainwheel retraction strut
101 Oleo leg
102 Starboard mainwheel
103 Mainwheel door
104 Forward face of box spar
105 Stressed wing skinning
106 Starboard navigation light
107 Wingtip structure
108 Starboard aileron
109 Aileron trim tab
110 Starboard wing fuel tank
111 Aileron control rod
112 Trim tab linkage
113 Oxygen bottles
114 Starboard flaps
115 Starter fuel tank
116 Flap hydraulic motor
117 Starboard mainwheel well
118 Boxspar
119 Compressed air bottles (emergency undercarriage actuation)
120 Mainspar/fuselage attachment points

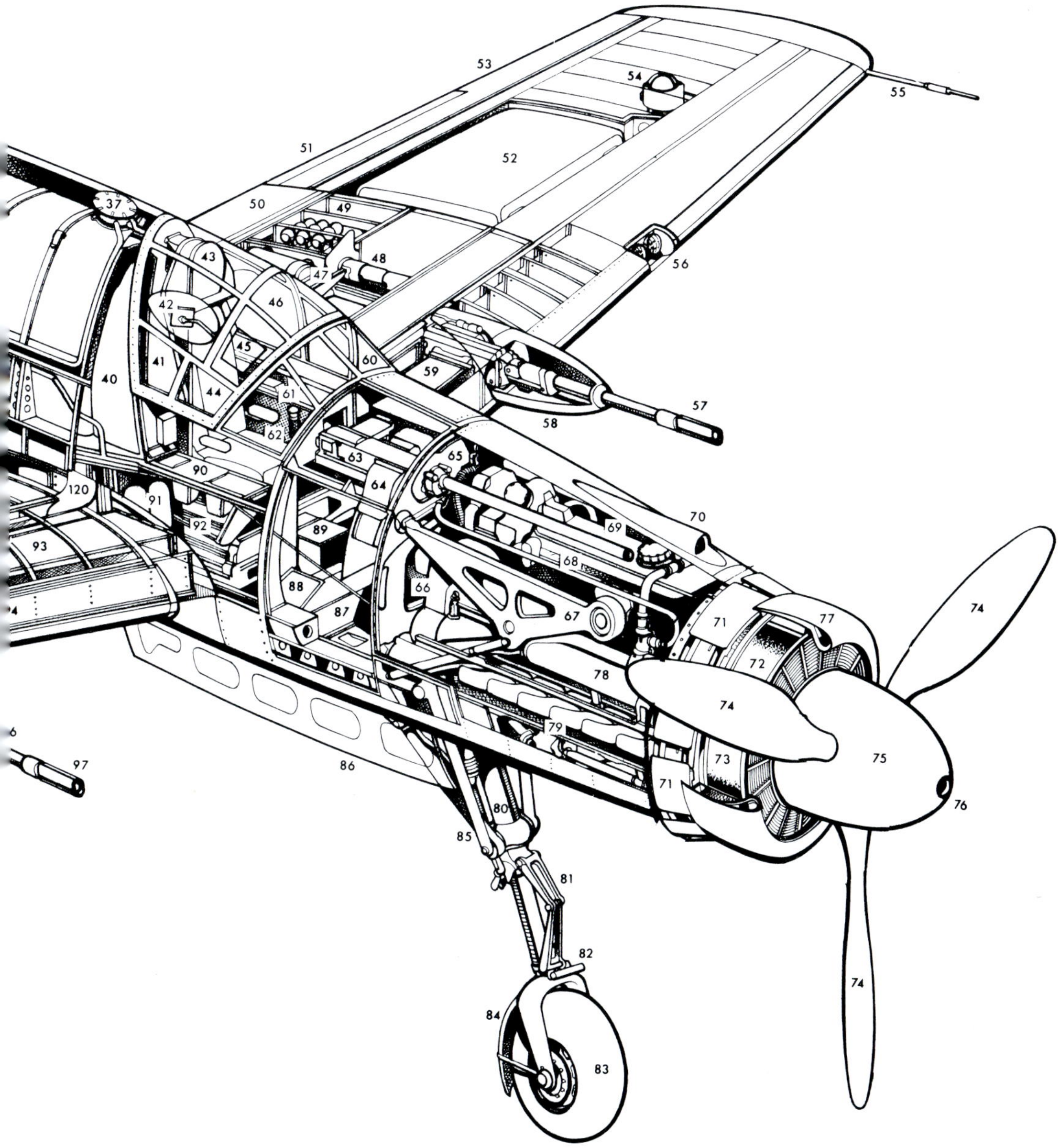

Focke Wulf Fw 190

Restored in the markings of JG 3 'Udet', this Fw 190D-9 belongs to the US National Air and Space Museum. The 'Dora-9' was in most respects an excellent fighter at low level (acquiring the nickname 'Downstairs Maid') but by the time it appeared the Luftwaffe's fortunes were in a spiral descent from which there could be no recovery.

Fw 190A-8

Cutaway key
1 Pitot head
2 Starboard navigation light
3 Detachable wingtip
4 Pitot tube heater line
5 Wing lower shell floating rib
6 Aileron hinge points
7 Wing lower shell stringers
8 Leading-edge ribs
9 Front spar
10 Outermost solid rib
11 Wing upper shell stringers
12 Aileron trim tab
13 Aileron structure
14 Aileron activation/ control linkage
15 Ammunition box (125 rpg)
16 Starboard 20-mm MG 151/20E wing cannon (sideways mounted)
17 Ammunition box rear suspension arm
18 Flap structure
19 Wing flap under skinning
20 Flap setting indicator peephole
21 Rear spar
22 Inboard wing construction
23 Undercarriage indicator
24 Wing rib strengthening
25 Ammunition feed chute
26 Static and dynamic air pressure lines
27 Cannon barrel
28 Launch tube bracing struts
29 Launch tube carrier strut
30 Mortar launch tube (auxiliary underwing armament)
31 Launch tube internal guide rails
32 21-cm (WfrGr 21) spin-stabilised Type 42 mortar shell
33 VDM three-bladed adjustable-pitch constant-speed propeller
34 Propeller boss
35 Propeller hub
36 Starboard undercarriage fairing
37 Starboard mainwheel
38 Oil warming chamber
39 Thermostat
40 Cooler armoured ring (0.25-in/6.5-mm)
41 Oil tank drain valve
42 Annular oil tank (12.1 Imp gal/55 litres)
43 Oil cooler
44 12-bladed engine cooling fan
45 Hydraulic-electric pitch control unit
46 Primer fuel line
47 Bosch magneto
48 Oil tank armour (0.22-in/5.5-mm)
49 Supercharger air pressure pipes
50 BMW 801D-2 14-cylinder radial engine
51 Cowling support ring
52 Cowling quick-release fasteners
53 Oil pump
54 Fuel pump (engine rear face)
55 Oil filter (starboard)
56 Wingroot cannon synchronisation gear
57 Gun troughs/cowling upper panel attachment
58 Engine mounting ring
59 Cockpit heating pipe
60 Exhaust pipes (cylinders 11-14)
61 MG 131 link and casing discard chute
62 Engine bearer assembly
63 MG 131 ammunition boxes (400 rpg)
64 Fuel filter recess housing
65 MG 131 ammunition cooling pipes
66 MG 131 synchronisation gear
67 Ammunition feed chute
68 Twin fuselage 13-mm MG 131 machine-guns
69 Windscreen mounting frame
70 Emergency power fuse and distributor box
71 Rear hinged gun access panel
72 Engine bearer/ bulkhead attachment
73 Control column
74 Transformer
75 Aileron control torsion bar
76 Rudder pedals (EC pedal unit with hydraulic wheelbrake operation)
77 Fuselage/wing spar attachment
78 Adjustable rudder push rod
79 Fuel filler head
80 Cockpit floor support frame
81 Throttle lever
82 Pilot's seat back plate armour (0.31-in/ 8-mm)
83 Seat guide rails
84 Side-section back armour (0.19 in/5-mm)
85 Shoulder armour (0.19-in/5-mm)
86 Oxygen supply valve
87 Steel frame turnover pylon
88 Windscreen spray pipes
89 Instrument panel shroud
90 1.18-in/30-mm armoured glass quarterlights
91 1.96-in/50-mm armoured glass windscreen
92 Revi 16B reflector gunsight
93 Canopy
94 Aerial attachment
95 Headrest
96 Head armour (0.47-in/12 mm)
97 Head armour support strut
98 Explosive charge canopy emergency jettison unit
99 Canopy channel side
100 Auxiliary tank: fuel (25.2 Imp gal/115 litres) or GM-1 (18.7 Imp gal/ 85 litres)
101 FuG 16ZY transmitter-receiver unit
102 Handhold cover
103 Primer fuel filler cap
104 Autopilot steering unit (PKS 12)
105 FuG 16ZY power transformer
106 Entry step cover plate
107 Two tri-spherical oxygen bottles (starboard fuselage wall)
108 Auxiliary fuel tank filler point
109 FuG 25a transponder unit
110 Autopilot position integration unit
111 FuG 16ZY homer bearing converter
112 Elevator control cables
113 Rudder control DUZ flexible rods
114 Fabric panel (bulkhead 12)
115 Rudder differential unit
116 Aerial lead-in
117 Rear fuselage lift tube
118 Triangular stress frame
119 Tailplane trim unit
120 Tailplane attachment fitting
121 Tailwheel retraction guide tube
122 Retraction cable lower pulley
123 Starboard tailplane
124 Aerial
125 Starboard elevator
126 Elevator trim tab
127 Tailwheel shock strut guide
128 Fin construction
129 Retraction cable under pulley
130 Aerial attachment strut
131 Rudder upper hinge
132 Rudder structure

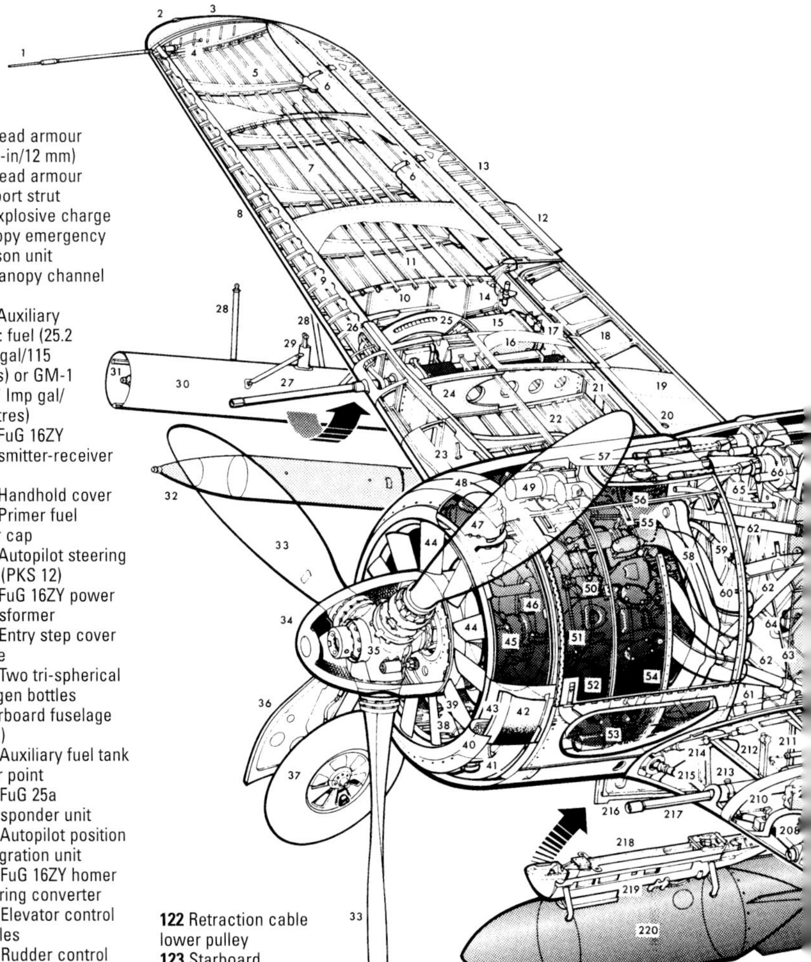

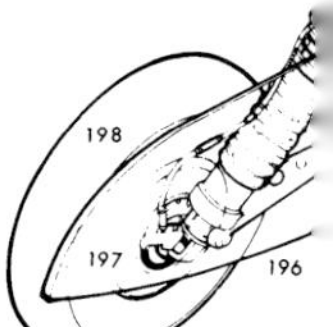

The first operational Fw 190s were sent to France to equip JG 2 and JG 26, which faced the RAF across the English Channel. These Fw 190As are from 7. Staffel/Jagdgeschwader 2 'Richthofen'. A feature of many Fw 190s was a painted design aft of the engine cowling to mask the prolific exhaust stains.

SPECIFICATION

Fw 190A-8

Powerplant

one BMW 801D-2 14-cylinder radial piston engine rated at 2,100-hp (1567-kW)

Performance

Maximum speed: (clean) 408 mph (654 km/h)
Initial climb rate: 2,363 ft (720 m) per minute
Normal range: 500 miles (805 km)
Service ceiling: 37,400 ft (11400 m)

Weights

Empty: 7,000 lb (3170 kg)
Maximum loaded: 10,800 lb (4900 kg)

Dimensions

Span: 34 ft 5 in (10.50 m)
Length: 29 ft (8.84 m)
Height: 13 ft (3.96 m)
Wing area: 197 sq ft (18.30 m^2)

Armament

(A-8/R2) two 0.31-in (7.9-mm) MG 17 machine-guns, four 20-mm MG 151/20 cannon, one 500-kg (1,100-lb) and two 250-kg (550-lb) bombs, or one 66-Imp gal (300-litre) drop tank

Fw 190D-9

Powerplant

one Junkers Jumo 213A-1 12-cylinder inverted-Vee piston engine developing 2,242 hp (1670 kW) at sea level with MW-50 methanol boosting

Performance

Maximum speed: 426 mph (686 km/h) at 21,654 ft (6600 m), 357 mph (575 km/h) at sea level
Maximum range: 520 miles (837 km) on internal fuel
Climb: 2.1 minutes to 6560 ft (2000 m); 7.1 minutes to 32,800 ft (10000 m)

Weights

Empty: 7,694 lb (3590 kg)
Maximum loaded: 10,670 lb (4850 kg)

Dimensions

Span: 34 ft 5 in (10.50 m)
Length: 33 ft 5 in (10.19 m
Height: 11 ft 0¼ in (3.36 m)
Wing area: 197 sq ft (18.30 m^2)

Armament

two 20-mm MG 151 cannon with 250 rounds per gun, two 13-mm MG 131 machine-guns with 475 rounds per gun, ETC 504 fuselage rack for one 1,100-lb (500-kg) SC 500 bomb

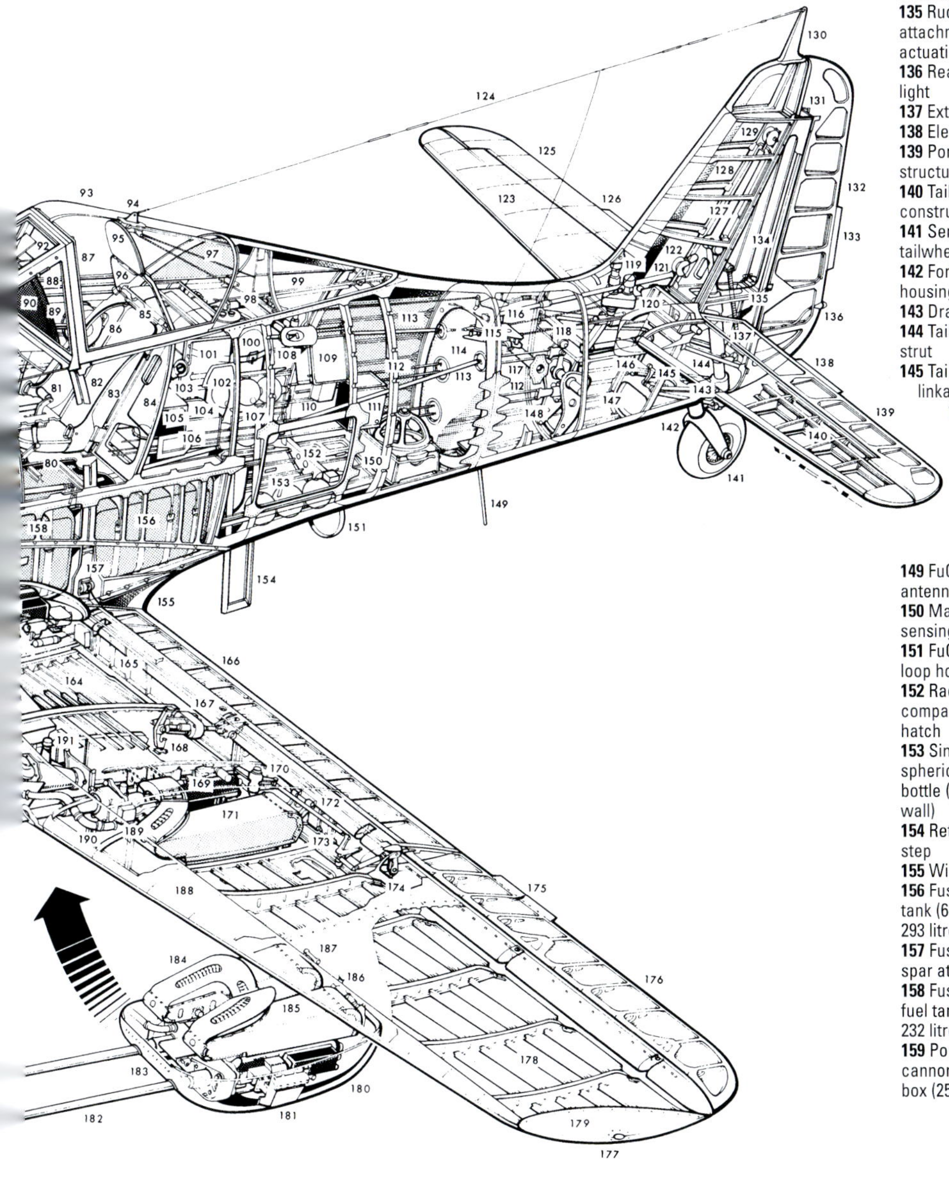

133 Rudder trim tab
134 Tailwheel retraction mechanism access panel
135 Rudder attachment/ actuation fittings
136 Rear navigation light
137 Extension spring
138 Elevator trim tab
139 Port elevator structure
140 Tailplane construction
141 Semi-retracting tailwheel
142 Forked wheel housing
143 Drag yoke
144 Tailwheel shock strut
145 Tailwheel locking linkage
146 Elevator actuation lever linkage
147 Angled frame spar
148 Elevator differential bellcrank
149 FuG 25a ventral antenna
150 Master compass sensing unit
151 FuG 16ZY fixed loop homing antenna
152 Radio compartment access hatch
153 Single tri-spherical oxygen bottle (port fuselage wall)
154 Retractable entry step
155 Wingroot fairing
156 Fuselage rear fuel tank (64.5 Imp gal/ 293 litres)
157 Fuselage/rear spar attachment
158 Fuselage forward fuel tank (51 Imp gal/ 232 litres)
159 Port wingroot cannon ammunition box (250 rpg)
160 Ammunition feed chute
161 Port wingroot MG 151/20E cannon
162 Link and casing discard chute
163 Cannon rear mount support bracket
164 Upper and lower wing shell stringers
165 Rear spar
166 Spar construction
167 Flap position indicator scale and peephole
168 Flap actuating electric motor
169 Port 20-mm MG 151/20E wing cannon (sideways mounted)
170 Aileron transverse linkage
171 Ammunition box (125 rpg)
172 Ammunition box rear suspension arm
173 Aileron control linkage
174 Aileron control unit
175 Aileron trim tab
176 Port aileron structure
177 Port navigation light
178 Outboard wing stringers
179 Detachable wingtip
180 A-8/R1 variant underwing gun pack (in place of outboard cannon)
181 Link and casing discard chute
182 Twin unsynchronised 20-mm MG 151/20E cannon
183 Light metal fairing (gondola)
184 Ammunition feed chutes
185 Ammunition boxes (125 rpg)
186 Carrier frame restraining cord
187 Ammunition box rear suspension arms
188 Leading-edge skinning
189 Ammunition feed chute
190 Ammunition warming pipe
191 Aileron bellcrank
192 Mainwheel strut mounting assembly
193 EC-oleo shock strut
194 Mainwheel leg fairing
195 Scissors unit
196 Mainwheel fairing
197 Axle housing
198 Port mainwheel
199 Brake lines
200 Cannon barrel
201 FuG 16ZY Morane antenna
202 Radius rods
203 Rotating drive unit
204 Mainwheel retraction electric motor housing
205 Undercarriage indicator
206 Sealed air jack
207 BSK 16 gun camera
208 Retraction locking hooks
209 Undercarriage locking unit
210 Armament collimation tube
211 Camera wiring conduits
212 Wheel well
213 Cannon barrel blast tube
214 Wheel cover actuation strut
215 Ammunition hot air
216 Port inboard wheel cover
217 Wingroot cannon barrel
218 ETC 501 carrier unit
219 ETC 501 bomb rack
220 SC 500 bomb (500 kg/1,102 lb)

Fw 190F-2

First appearing in combat in the skies over France in September 1941, the Focke-Wulf Fw 190A was an unpleasant shock to the RAF, for this pugnacious-looking radial-engined fighter was clearly superior to the RAF's Spitfire Mk Vs. Enjoying a considerable period of dominance in the Western Front fighter v. fighter war, the type also came to be increasingly used as a fighter-bomber. This role was developed further with the Fw 190F series, which became dedicated close support aircraft and served with distinction on the Russian Front, where the type rapidly superseded the Junkers Ju 87 as the main Luftwaffe ground attack aircraft. Around 550 Fw 190F-1s, F-2s and F-3s were built between late 1942 and mid-1943. They were so successful that the F model was reinstated in production as the Fw 190F-8 in spring 1944, this version being based on the Fw 190A-8 and upgunned with 13-mm MG 131s in the upper fuselage. This aircraft is an Fw 190F-2 of 5. Staffel, II. Gruppe of Schlachtgeschwader 1, and is shown as it would have appeared in 1943 on the Russian Front.

Powerplant
The Fw 190 smashed the theory that only sleek inline-engined fighters could achieve good performance. The Fw 190F-2 was based on the Fw 190A-5 airframe, and featured the BMW 801D-2 powerplant. This engine developed 1,700 hp (1268 kW) for take-off and 1,440 hp (1074 kW) at 18,700 ft (5700 m). MW-50 water-methanol boosting was fitted as standard. Around the front of the engine was an annular oil cooler, and a 12-bladed cooling fan which built up pressure in the engine compartment. Cooling air was ejected at the rear of the cowling, downstream of the flush exhaust outlets.

Armament
Gun armament consisted of a pair of o.31-in (7.9-mm) MG 17 machine-guns in the upper fuselage decking, featuring a characteristic bulge over the breech block and each armed with 1,000 rounds. In each wingroot was a 20-mm MG 151/20 cannon with 200 rounds. A centreline ETC 250 bomb rack was used for a single 250-kg (551-lb) weapon or four 50-kg (110-lb) bombs on an ER 4 adaptor. Two ETC 50 racks (for 50-kg/110-lb) bombs could be fitted under each wing. In addition to standard SC 250 and SC 50 bombs, the Fw 190F regularly carried the AB 250 *splitterbombe* (cluster bomb) which could dispense a variety of sub-munitions. These included 224 SD 1, 144 SD 2, 30 SD 4 or 17 SD 10 anti-personnel/armour minelets, 184 B 1 incendiaries or 116 B 2 steel-nosed incendiaries. The widely-used SD 2 was the feared 'butterfly' bomb, which deployed small wings to slow its descent to earth.

Radio
The Fw 190F-2 used the FuG 16Z radio equipment, which had first been introduced to the family by the Fw 190A-4. The wire aerial stretched from a short mast on top of the fin to a spool mounted on the canopy. This was spring-loaded to take up the wire antenna when the canopy was slid backwards and to maintain tension in flight.

Cockpit
Well laid out and effectively armoured, the cockpit provided the pilot with good visibility in flight, but the broad nose and tail-down stance made ground visibility poor. Introduced on late Fw 190Fs was a bulged canopy which dramatically improved the pilot's view of the world.

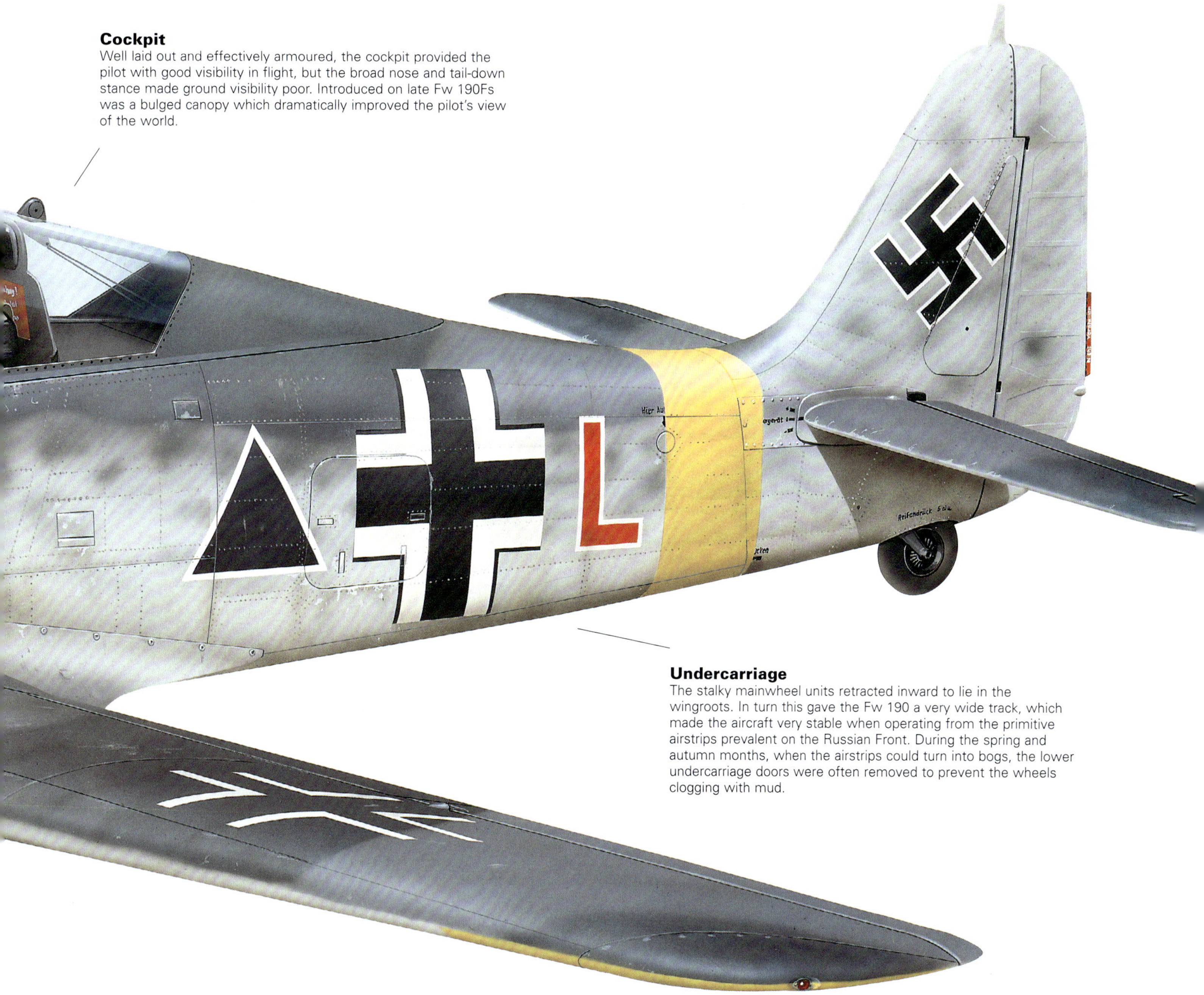

Undercarriage
The stalky mainwheel units retracted inward to lie in the wingroots. In turn this gave the Fw 190 a very wide track, which made the aircraft very stable when operating from the primitive airstrips prevalent on the Russian Front. During the spring and autumn months, when the airstrips could turn into bogs, the lower undercarriage doors were often removed to prevent the wheels clogging with mud.

Markings
This Fw 190F-2 is finished in typical camouflage with a light grey base, two-tone splinter top surfaces and mottling along the fuselage. The Russian Front theatre markings consisted of a yellow fuselage band and yellow panels under the wingtips and cowling. The Schlachtgeschwadern originally used symbols as unit identifications, Schlachtgeschwader 1 employing a black triangle. The red colour of the individual aircraft code signified 5. Staffel, as did the red background to the unit badge and similarly-coloured spinner. The gun-toting Mickey Mouse badge was applied by all II./SchG 1 staffeln, and had been inherited from one of the Gruppe's predecessors – IV.(Schlacht)/Lehrgeschwader 2.

Fuel
Fuel was accommodated in two self-sealing tanks located beneath the pilot's seat. They were separated by the rear spar tie-through member (the Fw 190 was built using a very strong through-spar construction). The forward tank held 51 Imp gal (232 litres) while the rear tank held 64.5 Imp gal (293 litres).

Gloster Meteor

As the operational life of the Meteor began to draw to a close, a large number of Meteor F.Mk 8s was modified to become target-towing tugs and advanced trainers. The target tugs were unofficially designated 'F(TT).Mk 8' and the trainers 'T.Mk 8'. Both types received high-visibility markings and a 'T.Mk 8' of No. 85 Sqn, with bright orange Dayglo markings, is seen here accompanying a Lightning of No. 5 Sqn.

Meteor F.Mk III

Cutaway key

1 Starboard detachable wingtip
2 Starboard navigation light
3 Starboard recognition light
4 Starboard aileron
5 Aileron balance tab
6 Aileron mass balance weights
7 Aileron control coupling
8 Aileron torque shaft
9 Chain sprocket
10 Cross-over control runs
11 Front spar
12 Rear spar
13 Aileron (inboard) mass balance
14 Nacelle detachable tail section
15 Jet pipe exhaust
16 Internal stabilising struts
17 Rear spar 'spectacle' frame
18 Fire extinguisher spray ring
19 Main engine mounting frame
20 Engine access panel(s)
21 Nacelle nose structure
22 Intake internal leading-edge shroud
23 Starboard engine intake
24 Windscreen de-icing spray tube
25 Reflector gunsight
26 Cellular glass bulletproof windscreen
27 Aft-sliding cockpit canopy
28 Demolition incendiary (cockpit starboard wall)
29 RPM indicators (left and right of gunsight)
30 Pilot's seat
31 Forward fuselage top deflector skin
32 Gun wobble button
33 Control column grip
34 Main instrument panel
35 Nosewheel armoured bulkhead
36 Nose release catches (10)
37 Nosewheel jack bulkhead
38 Nose ballast weight location
39 Nosewheel mounting frames
40 Radius rod (link and jack omitted)
41 Nosewheel pivot bearings
42 Shimmy-damper/ self-centring strut
43 Gun camera
44 Camera access
45 Aperture
46 Nose cone
47 Cabin cold-air intake
48 Nosewheel leg door
49 Picketing rings
50 Tension shock absorber
51 Pivot bracket
52 Mudguard
53 Torque strut
54 Doorhoop
55 Wheel fork
56 Retractable nosewheel
57 Nosewheel doors
58 Port cannon trough fairings
59 Nosewheel cover
60 Intermediate diaphragm
61 Blast tubes
62 Gun front mount rails
63 Pilot's seat pan
64 Emergency crowbar
65 Canopy de-misting silica gel cylinder
66 Bulletproof glass rear view cut-outs
67 Canopy track
68 Sea bulkhead
69 Entry step
70 Link ejection chutes
71 Case ejection chutes
72 20-mm Hispano Mk III cannon
73 Belt feed mechanism
74 Ammunition feed necks
75 Ammunition tanks
76 Aft glazing (magazine bay top door)
77 Leading ramp
78 Front spar bulkhead
79 Oxygen bottles (2)
80 Front spar carry-through
81 Tank bearer frames
82 Rear spar carry-through
83 Self-sealing (twin compartment) main fuel tank, capacity 165 Imp gal (750 litres) in each half
84 Fuel connector pipe
85 Return pipe
86 Drain pipes
87 Fuel filler caps
88 Tank doors (2)
89 T.R.1143 aerial mast
90 Rear spar bulkhead (plywood face)
91 Aerial support frame

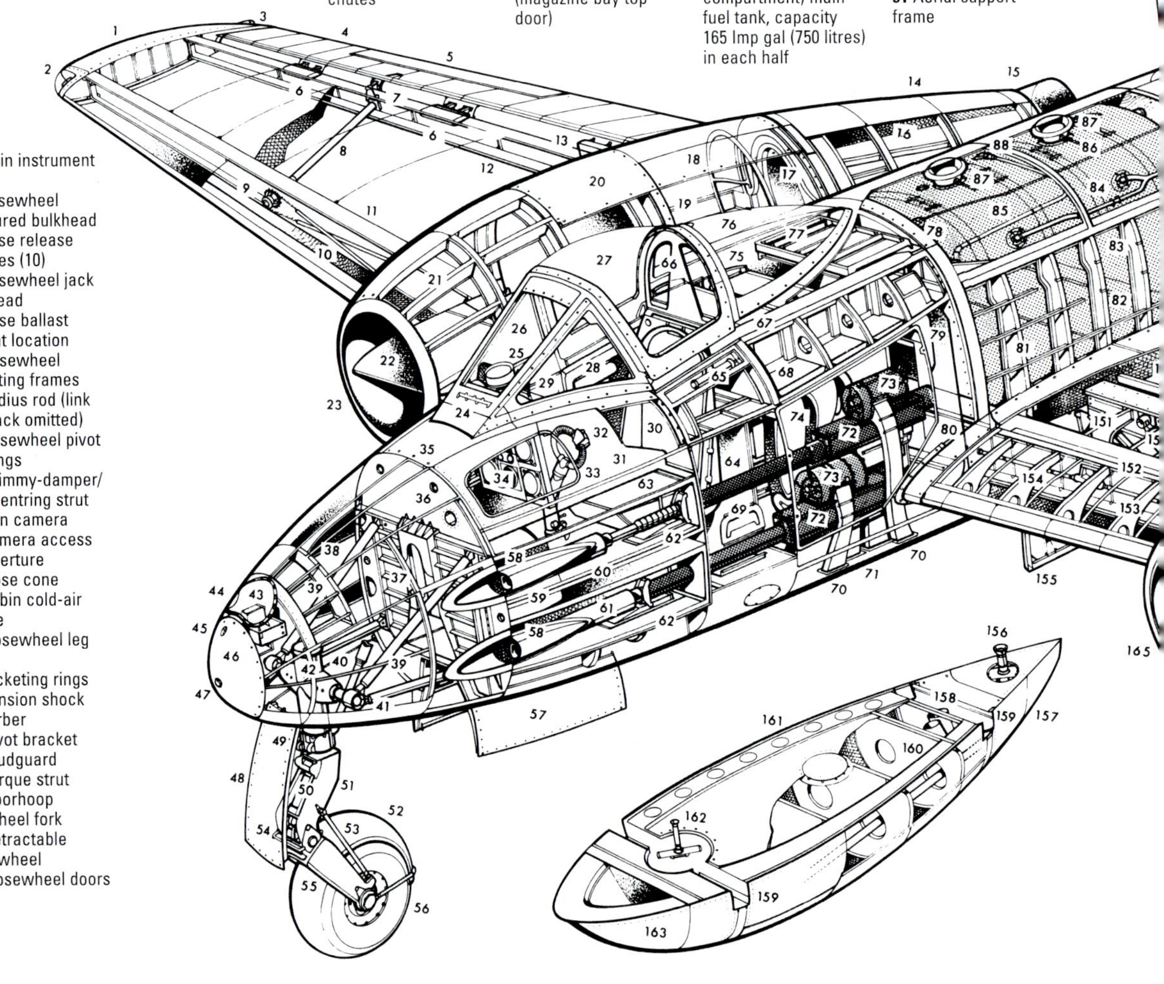

EE389 was the first Meteor involved in inflight-refuelling trials, in April 1949. Here it is seen, with its airbrakes deployed and probe clearly evident, about to refuel from a Lancaster Mk III tanker. The neat probe installation in the nose of the aircraft served as the basis of a similar fitting applied to a number of Mk 4s and Mk 8s.

SPECIFICATION

Meteor F.Mk 8

Dimensions

Length: 44 ft 7 in (13.59 m)
Wingspan: 37 ft 2 in (11.33 m)
Height: 13 ft (3.96 m)
Wing area: 350 sq ft (32.515 m²)
Aspect ratio: 3.9
Root chord: 11 ft 9 in (3.6 m)

Powerplant

Two 3,500-lb (15.5-kN) thrust Rolls-Royce Derwent 8 turbojets

Weights

Empty: 10,684 lb (4846 kg)
Maximum overload: 15,700 lb (7122 kg)

Performance

Maximum speed at sea level: 592 mph (953 km/h)
Maximum speed at 30,000 ft (9144 m): 550 mph (885 km/h)
Climb to 30,000 ft (9144 m): 6 minutes 30 seconds
Service ceiling: 44,000 ft (13410 m)
Range without wing drop tanks: 690 miles (1111 km)
Endurance at 40,000 ft (12192 m) with 420 Imp gal (1909 litres) of fuel: 592 mph (953 km/h)

Armament

Four fixed 20-mm British Hispano cannon in the nose with 195 rounds per gun

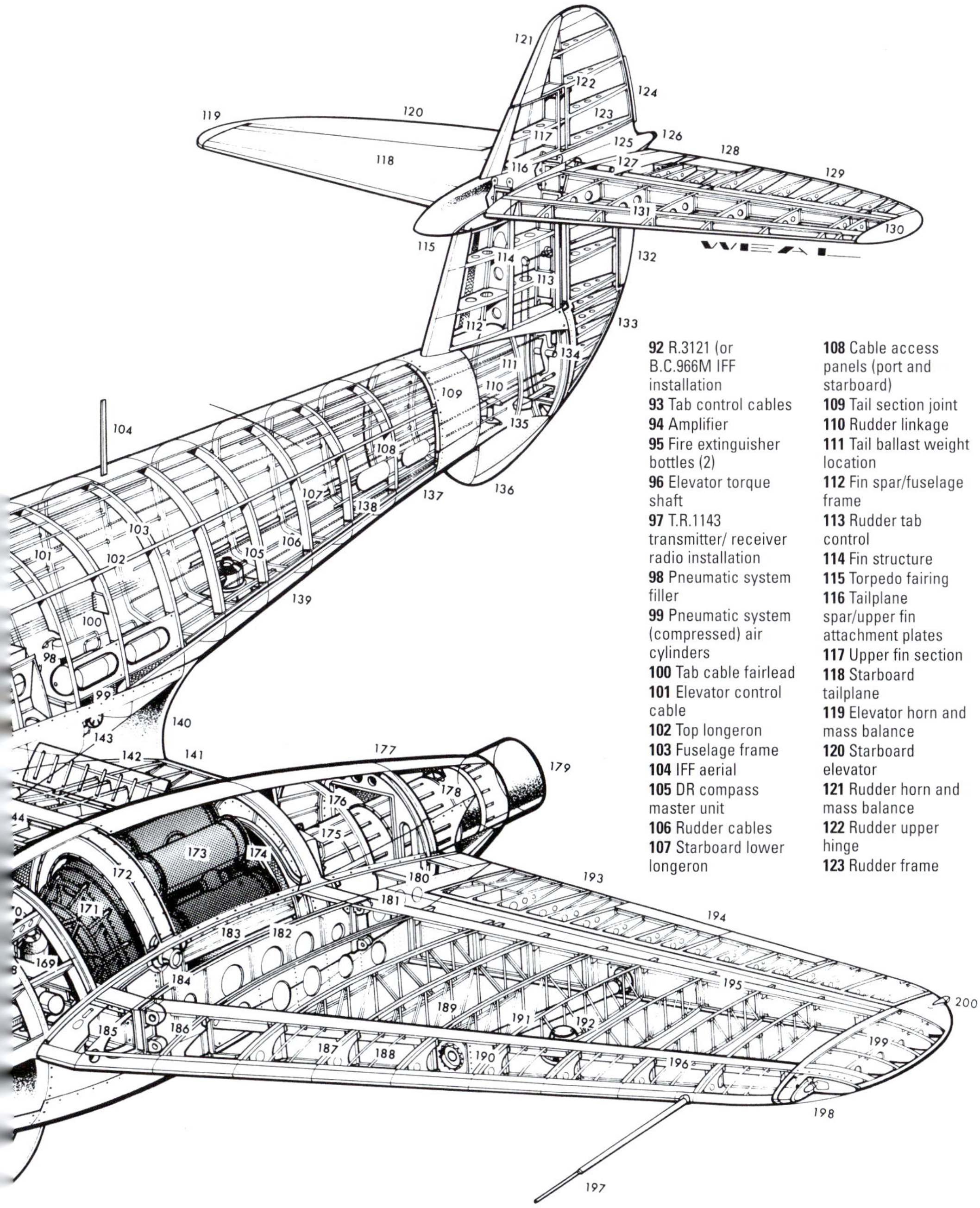

92 R.3121 (or B.C.966M IFF installation
93 Tab control cables
94 Amplifier
95 Fire extinguisher bottles (2)
96 Elevator torque shaft
97 T.R.1143 transmitter/ receiver radio installation
98 Pneumatic system filler
99 Pneumatic system (compressed) air cylinders
100 Tab cable fairlead
101 Elevator control cable
102 Top longeron
103 Fuselage frame
104 IFF aerial
105 DR compass master unit
106 Rudder cables
107 Starboard lower longeron
108 Cable access panels (port and starboard)
109 Tail section joint
110 Rudder linkage
111 Tail ballast weight location
112 Fin spar/fuselage frame
113 Rudder tab control
114 Fin structure
115 Torpedo fairing
116 Tailplane spar/upper fin attachment plates
117 Upper fin section
118 Starboard tailplane
119 Elevator horn and mass balance
120 Starboard elevator
121 Rudder horn and mass balance
122 Rudder upper hinge
123 Rudder frame
124 Fixed tab
125 Rear fairing
126 Tail navigation light
127 Elevator torque shaft
128 Elevator trim tab
129 Elevator frame
130 Elevator horn and mass balance
131 Tailplane structure
132 Rudder combined balance trim tab
133 Rudder lower section
134 Elevator push-rod linkage
135 Rudder internal/ lower mass balance weight
136 Emergency landing tailskid
137 Tail section riveted joint
138 Port lower longeron
139 Fuselage stressed skin
140 Wingroot fairing
141 Inboard split flap
142 Airbrake (upper and lower surfaces)
143 Flap indicator transmitter
144 Rear spar
145 Inter-coupler cables (airbrake/airbrake and flap/flap)
146 Port mainwheel well
147 Roof rib station
148 Front diaphragm
149 Undercarriage beam
150 Undercarriage retraction jack
151 Undercarriage sidestay/downlock
152 Front spar
153 Nose ribs
154 Aileron control runs
155 Mainwheel door inner section
156 Ventral tank transfer pipe
157 Tank rear fairing
158 Filler stack pipes
159 Ventral tank attachment strap access doors
160 Anti-surge baffles
161 Fixed ventral fuel tank, capacity 105 Imp gal (477 litres)
162 Air pressure inlet
163 Tank front fairing
164 Port mainwheel
165 Starboard engine intake
166 Intake internal leading edge shroud
167 Auxiliary gearbox drives (vacuum pump/generator)
168 Nacelle nose structure
169 Starter motor
170 Oil tank
171 Rolls-Royce W.2B/23C Derwent I
172 Main engine mounting frame
173 Combustion chambers
174 Rear spar spectacle frame
175 Jet pipe thermo-coupling
176 Nacelle aft frames
177 Nacelle detachable tail section
178 Jet pipe suspension link
179 Jet pipe exhaust
180 Gap fairing tail section
181 Rear-spar outer wing fixing
182 Outer wing rib No. 1
183 Engine end rib
184 Engine mounting/ removal trunnion
185 Gap fairing nose section
186 Front-spar outer wing fixing
187 Nose ribs
188 Intermediate riblets
189 Wing ribs
190 Aileron drive chain sprocket
191 Aileron torque shaft
192 Retractable landing lamp
193 Port aileron
194 Aileron balance tab
195 Rear spar
196 Front spar
197 Pitot head
198 Port navigation light
199 Outer wing rib No. 10/wingtip attachment
200 Port recognition light

Meteor NF.Mk 12

RAF Fighter Command aircraft adopted colourful squadron markings during the 1950s, these replacing the two-letter codes which dated from the war years. A stylised adaptation of the traditional squadron badge was usually worn on the fin, with colourful bars flanking the roundel. No. 153 Squadron, of No. 11 Group based at RAF West Malling, Kent, used black-edged white bars, with inward-pointing red chevrons, and carried a version of the squadron's official bat and six pointed star on the fin. The flight markings of No. 153 Squadron came in the shape of coloured intake leading edges and fin bullet fairings. 'A' Flight used red as its identifying colour.

Cockpit

Like the NF.Mk 11, the Meteor NF.Mk 12 retained the same heavily framed, sideways-opening cockpit as the T.Mk 7 trainer. The port windscreen side panel was split diagonally and one portion could be opened to act as a direct vision panel if the canopy iced up. Although a handful of Meteor single-seaters was fitted with ejection seats, the two-seaters (including the night-fighters) were never thus equipped. Abandoning the aircraft was fraught with difficulties, and was impossible at low level.

Radome

With the arrival of the NF.Mk 12 the Meteor night-fighter lost its distinctive undernose bulge. This had housed the scanner mounting-bracket, which was redesigned on the APS.21. The Mk 12 used this American-built radar rather than the British AI Mk 10 used in the NF.Mk 11. This necessitated a slight increase in fuselage length (17 in/43 cm), and to balance this, an increase in fin area. This increase was obtained by adding fillets to the fin leading edge which faired towards into the leading-edge tailplane bullet fairing.

Armament

The Meteor night-fighters retained the same four British Hispano Mk V 20-mm cannon armament as the day-fighter and fighter-reconnaissance Meteors, but these were relocated to the wings, outboard of the engine nacelles. This was a major modification, as the cannon access doors had to be part of the stressed structure of the wings, which had to be strengthened to withstand this variant's higher IAS (Indicated Air Speed) limits. One hundred and sixty rounds of ammunition were carried for each gun in ammunition tanks further outboard.

Powerplant

The Meteor NF.Mk 12 was powered by a pair of 3,800-lb (16.9-kN) thrust Rolls-Royce Derwent 9s. These replaced the 3,600-lb (16-kN) Derwent 8s of the slightly lighter NF.Mk 11, and raised the limiting Mach number from 0.79 to 0.81, necessitating some airframe strengthening. The turbojets were fed by simple, round, pitot intakes, inside which the front spar continued unbroken, immediately ahead of the engine, enclosed in a fairing approximating in shape to the wing leading edge outboard of the engine nacelle. The NF.Mk 12 had the enlarged, so-called 'deep breather' intakes fitted to late F.Mk 8s.

Tail
Although they used the round-tipped wings of the early Meteors, the Meteor night-fighters had the square-cropped horizontal tailplanes of the F.Mk 8. These were fixed, with conventional trailing-edge elevators, and trim tabs inboard. The two-piece rudder, above and below the tailplane, was joined together by a torque tube, and acted as a single control surface. The lower section incorporated a manually adjustable trim tab.

Outer wing panels
Day-fighter Meteors after the F.Mk III had their outer wings redesigned with square-cropped wingtips and reduced span to improve roll-rate. The original outer wing was re-introduced on the photo-reconnaissance PR.Mk 10 and retained on the night-fighters. The night fighters were originally to have had swept outer-wing panels, but this idea was soon abandoned.

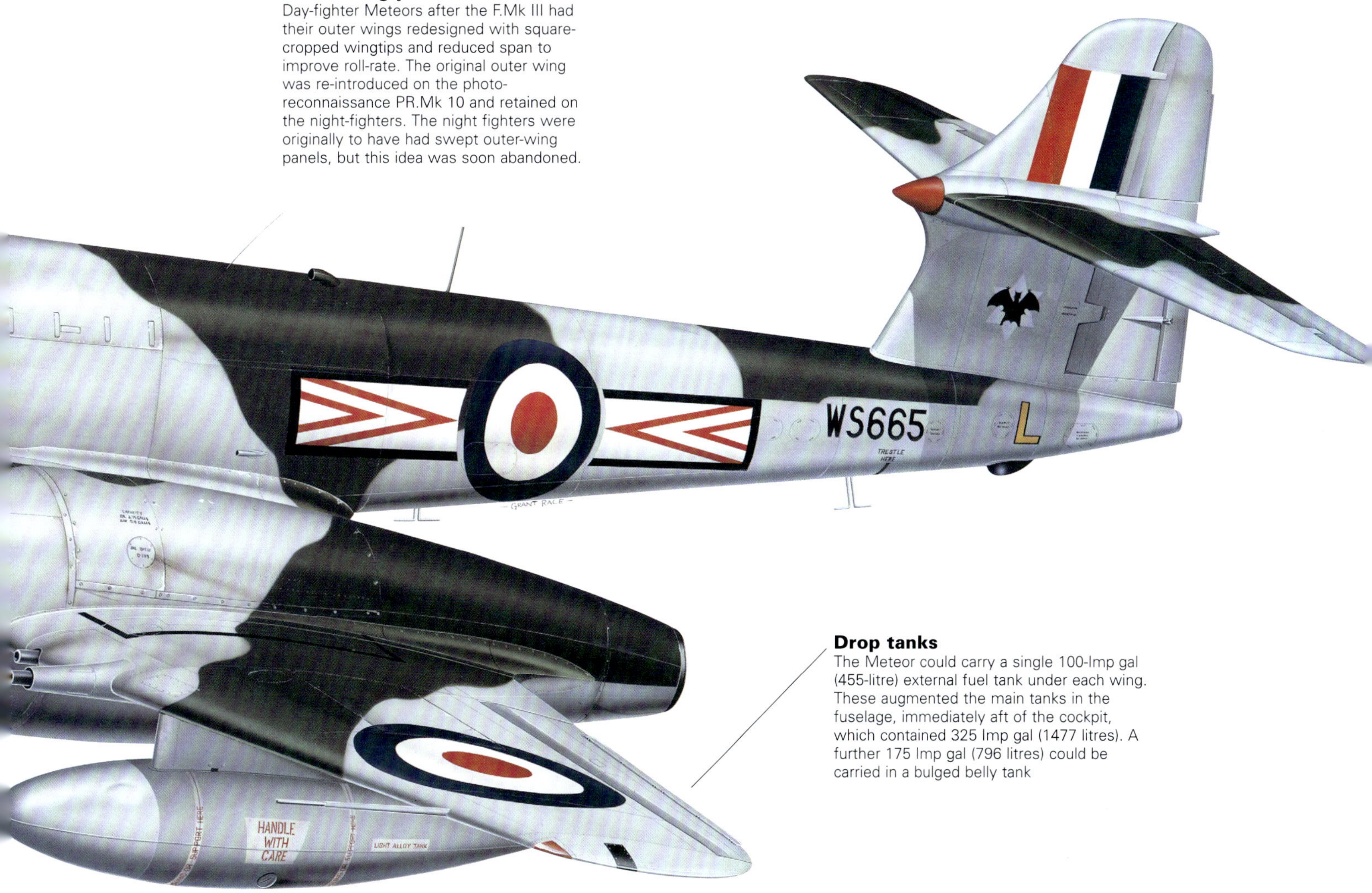

Drop tanks
The Meteor could carry a single 100-Imp gal (455-litre) external fuel tank under each wing. These augmented the main tanks in the fuselage, immediately aft of the cockpit, which contained 325 Imp gal (1477 litres). A further 175 Imp gal (796 litres) could be carried in a bulged belly tank

Grumman F4F Wildcat

***Operation Torch, the Allied invasion of North Africa in November 1942, was the earliest of the small number of major actions in which the Wildcat was involved in the European theatre. Here an F4F-4 of VF-41 is waved on its way from the wooden deck of USS* Ranger.**

F4F-4 Wildcat

Cutaway key
1 Starboard navigation light
2 Wingtip
3 Starboard formation light
4 Rear spar
5 Aileron construction
6 Fixed aileron tab
7 All riveted wing construction
8 Lateral stiffeners
9 Forward canted main spar
10 'Crimped' leading-edge ribs
11 Solid web forward ribs
12 Starboard outer gun blast tube
13 Carburettor air duct
14 Intake
15 Curtiss three-bladed, constant-speed propeller
16 Propeller cuffs
17 Propeller hub
18 Engine front face
19 Pressure baffle
20 Forward cowling ring
21 Cooler intake
22 Cooler air duct
23 Pratt & Whitney R-1830-86 radial engine
24 Rear cowling ring/ flap support
25 Controllable cowling flaps
26 Downdraft ram air duct
27 Engine mounting ring
28 Anti-detonant regulator unit
29 Cartridge starter
30 Generator
31 Intercooler
32 Engine accessories
33 Bearer assembly welded cluster joint
34 Main beam
35 Lower cowl flap
36 Exhaust stub
37 Starboard mainwheel
38 Undercarriage fairing
39 Lower drag link
40 Hydraulic brake
41 Port mainwheel
42 Detachable hub cover
43 Low-pressure tyre
44 Axle forging
45 Upper drag link
46 Oleo shock strut
47 Ventral fairing
48 Wheel well
49 Pivot point
50 Landing light
51 Main forging
52 Compression link
53 Gun camera port
54 Counter balance
55 Anti-detonant tank
56 Retraction sprocket
57 Gear box
58 Stainless steel firewall
59 Engine bearers
60 Actuation chain (undercarriage)
61 Engine oil tank
62 Oil filler
63 Hoisting sling installation
64 Bullet resistant windscreen
65 Reflector gunsight
66 Panoramic rear-view mirror

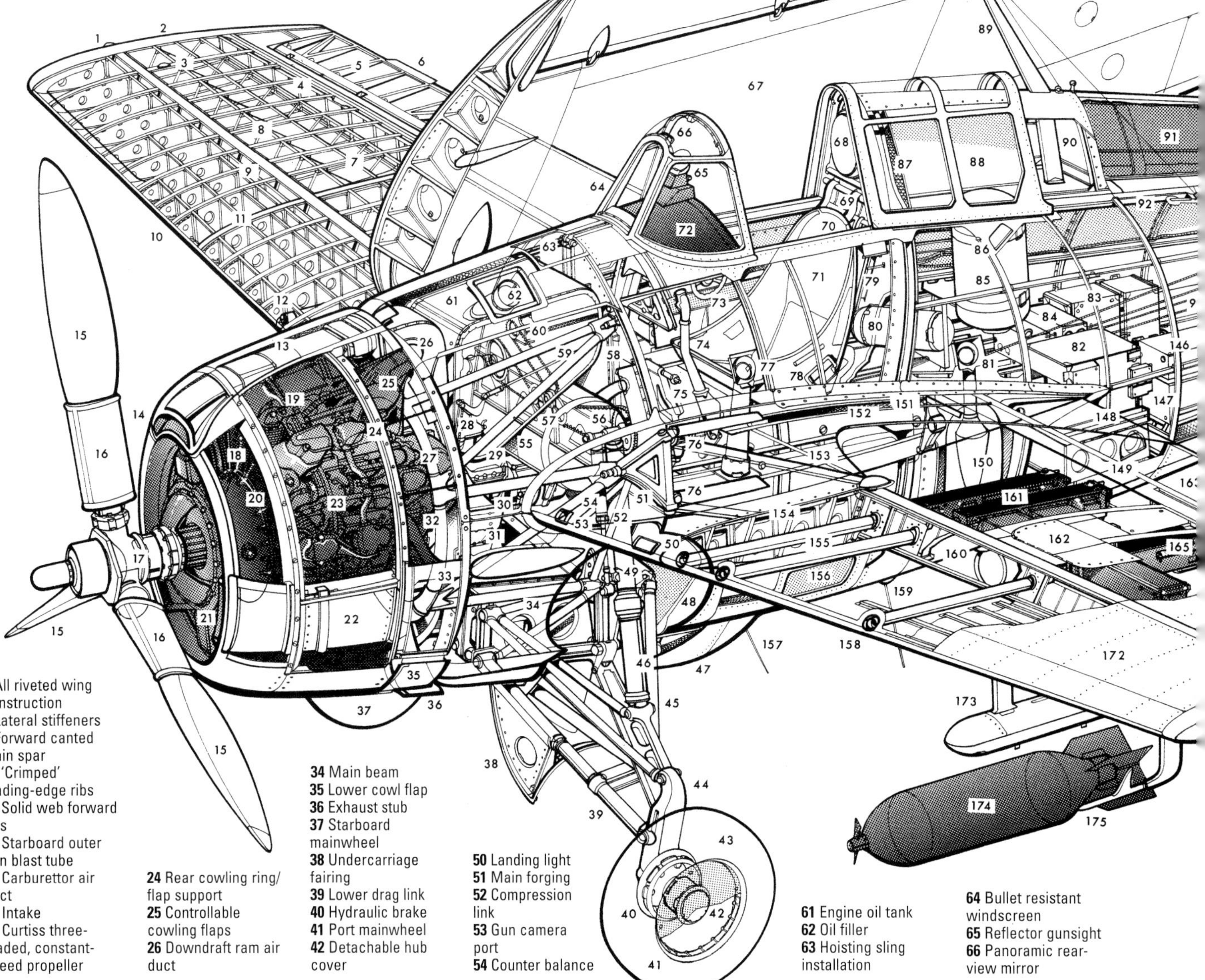

The F4F-3S Wildcatfish was an attempt to produce a floatplane fighter in the same mould as the Japanese A6M2 'Rufe'. One hundred were ordered, but the aircraft's poor performance (including a top speed of 266 mph/428 km/h) led to their completion as conventional F4F-3 training aircraft.

SPECIFICATION

FM-2 Wildcat

Dimensions

Length: 28 ft 11 in (8.81 m)
Height: 9 ft 11 in (3.02 m)
Wingspan: 38 ft 0 in (11.58 m)
Wing area: 260 sq ft (24.15 m²)

Powerplant

One Wright R-1820-56/-56A or -56W/-56WA (with water injection) Cyclone 9 radial piston engine rated at 1,350 hp (1007 kW)

Weights

Empty: 5,448 lb (2471 kg)
Loaded: 7,487 lb (3396 kg)
Maximum take-off: 8,271 lb (3752 kg)

Fuel

Internal fuel: 117 US gal (443 litres)

Maximum external load: two 58-US gal (220-litre) drop tanks

Performance

Maximum speed at 28,800 ft (8780 m): 332 mph (534 km/h)
Cruising speed: 164 mph (264 km/h)
Climb rate: 3,650 ft (1113 m) per minute
Service ceiling: 34,700 ft (10577 m)

Range

Normal: 900 miles (1448 km)
Maximum: 1,310 miles (2108 km) with external fuel

Armament

Four wing-mounted 0.5-in (12.7-mm) Browning M2 machine-guns each with 430 rounds of ammunition, plus six underwing rocket projectiles or up to 500 lb (227 kg) of bombs

The Fleet Air Arm trialled the installation of rocket projectiles on a number of its fighter types, including the Wildcat. Experiments conducted at A&AEE Boscombe Down involved several machines, including this Wildcat Mk IV, with three 25-lb (11-kg) rockets beneath each wing. British Mk I rails (seen here) and American Mk V zero-length launchers were tested, with some success, between 1942 and 1944, but the weapon was never used operationally from FAA Wildcats.

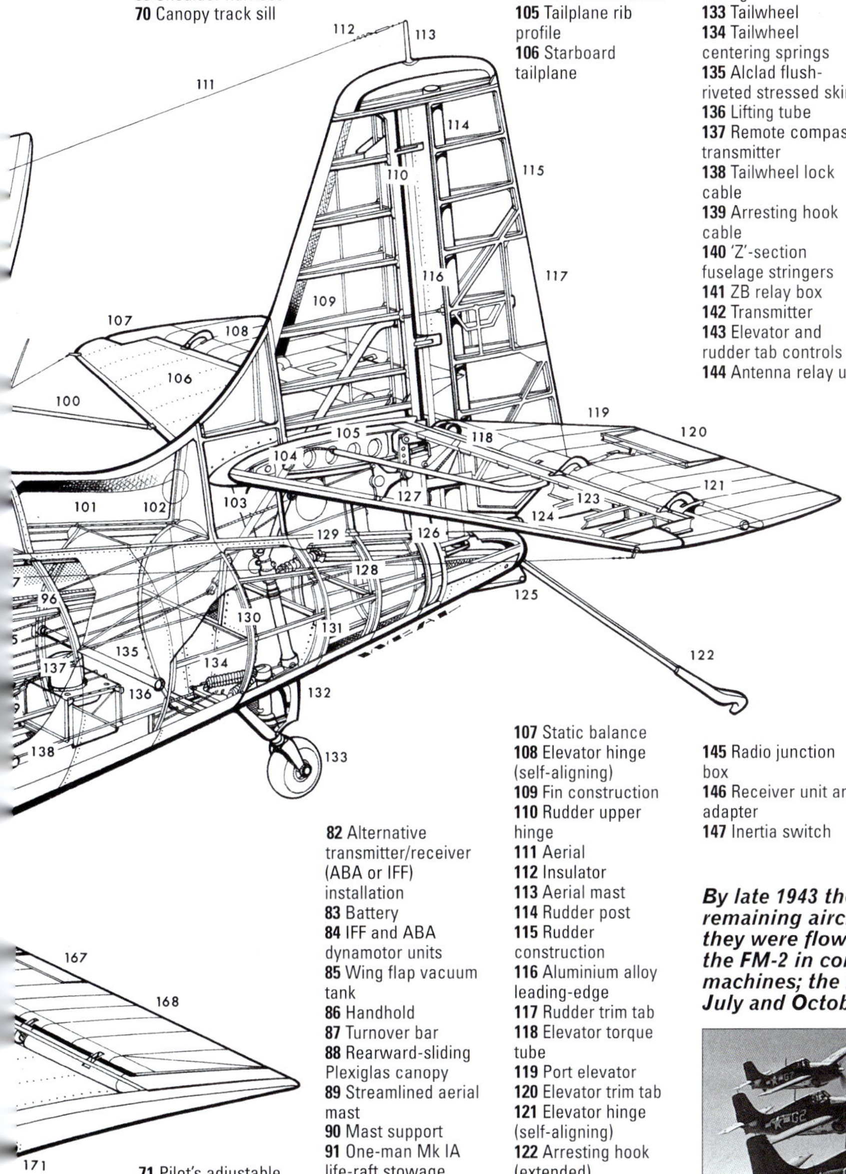

67 Wing fold position
68 Adjustable headrest
69 Shoulder harness
70 Canopy track sill
71 Pilot's adjustable seat
72 Instrument panel shroud
73 Undercarriage manual crank
74 Control column
75 Rudder pedals
76 Fuselage/front spar attachment
77 Main fuel filler cap
78 Seat harness attachment
79 Back armour
80 Oxygen cylinder
81 Reserve fuel filler cap
82 Alternative transmitter/receiver (ABA or IFF) installation
83 Battery
84 IFF and ABA dynamotor units
85 Wing flap vacuum tank
86 Handhold
87 Turnover bar
88 Rearward-sliding Plexiglas canopy
89 Streamlined aerial mast
90 Mast support
91 One-man Mk IA life-raft stowage
92 Upper longeron
93 Toolkit
94 Aerial lead-in
95 Elevator and rudder control runs
96 'L'-section fuselage frames
97 IFF aerial
98 Dorsal lights
99 Whip aerial
100 Wing-fold jury strut
101 Fin fairing
102 Access panel
103 Tailwheel strut extension arm
104 Rudder trim tab control flexible shaft
105 Tailplane rib profile
106 Starboard tailplane
107 Static balance
108 Elevator hinge (self-aligning)
109 Fin construction
110 Rudder upper hinge
111 Aerial
112 Insulator
113 Aerial mast
114 Rudder post
115 Rudder construction
116 Aluminium alloy leading-edge
117 Rudder trim tab
118 Elevator torque tube
119 Port elevator
120 Elevator trim tab
121 Elevator hinge (self-aligning)
122 Arresting hook (extended)
123 Tailplane spar
124 Rear navigation light
125 Towing lug
126 Rudder torque tube support
127 Elevator control linkage
128 Rudder control cable
129 Arresting hook spring
130 Tailwheel shock strut
131 Rear fuselage frame/bulkhead
132 Forged castor fairing
133 Tailwheel
134 Tailwheel centering springs
135 Alclad flush-riveted stressed skin
136 Lifting tube
137 Remote compass transmitter
138 Tailwheel lock cable
139 Arresting hook cable
140 'Z'-section fuselage stringers
141 ZB relay box
142 Transmitter
143 Elevator and rudder tab controls
144 Antenna relay unit
145 Radio junction box
146 Receiver unit and adapter
147 Inertia switch
148 Radio equipment support rack
149 Entry foothold
150 Reserve fuel tank, capacity 27 US gal (102 litres)
151 Fuselage/rear spar attachment
152 Wing hinge line
153 Main (underfloor) fuel tank, capacity 117 US gal (443 litres)
154 Stub wing end rib and fairing
155 Inboard gun blast tubes
156 Plexiglas observation panel
157 Ventral antenna
158 Outboard gun port
159 ZB antenna
160 Fixed D/F loop
161 Two 0.50-in (12.7-mm) Browning M2 machine-guns
162 Outboard gun access/loading panels
163 ABA antenna
164 Flap profile
165 Outboard 0.50-in (12.7-mm) Browning M2 machine-gun
166 Aileron control linkage
167 Aileron trim tab
168 Port aileron
169 Aileron hinges (self-aligning)
170 Port formation light
171 Port navigation light
172 Wing skinning
173 Bomb rack (optional)
174 Fragmentation bomb
175 Pitot head

By late 1943 the F4F-4 had disappeared from front-line service, the few remaining aircraft having been issued to training units in the US where they were flown by the next generation of Wildcat pilots, destined to fly the FM-2 in combat. Large fuselage codes identify these aircraft as training machines; the red 'star and bar' surrounds date the picture at between July and October 1943.

F4F-3 Wildcat

This F4F-3 carries the markings of Marine Corps Fighting Squadron 121 (VMF-121) and was flown by Bruce Porter during his time with the unit between September 1941 and February 1942, based at Tafuna, Samoa. Porter also flew Wildcats with VMF-441 and VMF-111 before returning to VMF-121 at Guadalcanal in March 1943. By then the squadron was equipped with F4U-1s and it was in a Corsair that he scored his first three Zero kills, plus a pair of 'probables'. Towards the end of the war in the Pacific, Porter achieved ace status with two bomber kills gained while flying F6F-5N Hellcat night-fighters with VMF(N)-542 from Okinawa. VMF-121 was the top-scoring Wildcat unit of the war, claiming 160 kills in all.

Powerplant
The XF4F-2 Wildcat prototype was first flown in 1937 powered by a 1,050-hp (783-kW) Pratt & Whitney R-1830-66 Twin Wasp (with single-stage, single-speed supecharging), but this engine suffered major teething problems (including crankshaft failures) and was replaced during the 1938 redesign of the aircraft with a 1,200-hp (895-kW) R-1830-76, with two-stage, two-speed supercharging. This more reliable powerplant was subsequently fitted to the first 100 F4F-3s, later aircraft being powered by the similarly-rated R-1830-86, with twin magnetos. Concerns that supplies of this engine could not be maintained led to a batch of 95 F4F-3As employing an R-1830-90 derivative, with a single-stage supercharger, but in the F4F-4 the R-1830-86 was reinstated.

Fuel stowage
Most Wildcat variants carried fuel in a 117-US gal (443-litre) main tank situated beneath the cockpit and in a 27-US gal (102-litre) emergency tank in the rear fuselage. The F4F-3 was able to carry a non-jettisonable 42-US gal (159-litre) external tank, flush-mounted under each wing, though this installation appears not to have been used in service. For ferrying purposes a number of USMC aircraft were modified in Hawaii, during 1942, to carry a single, jettisonable version of this tank under the centreline. A proper drop tank installation was designed for the F4F-4; a 58-US gal (220-litre) teardrop-shaped tank could be carried by this and subsequent variants. The F4F-7 long-range reconnaissance variant (of which just 21 were completed) had a rigid wing containing additional fuel tanks. With a total capacity of no less than 555 US gal (2101 litres), the F4F-7 had an endurance of 25 hours. As a safety measure, a special fuel dump system was also installed.

Wing folding

F4F-3s and G-36As for the Aéronavale did not have a folding mainplane. Known by Grumman as the 'sto-wing' this was introduced in the F4F-4 and was an important innovation, especially where the Wildcat was operated from small escort aircraft-carriers with limited (or even non-existent) below-deck space. In the XF4F-4 prototype wing folding was accomplished hydraulically, though this equipment was omitted in production aircraft, presumably to save weight. Aircraft-carrier capacity was improved by 150 per cent when aircraft with folding wings were embarked, the Wildcat's span being just 14 ft 4 in (4.37 m) in its folded state.

Offensive armament

In its original guise the XF4F-2 carried a pair of 0.5-in (12.7-mm) machine-guns in the upper engine cowling and had the option of another pair mounted in the wings. The first production F4F-3s had 0.3-in (7.62-mm) cowl-mounted guns and a pair of '50-calibers' in the wings, but most of these machines (including the aircraft depicted here) dispensed with the cowl guns and were fitted with four wing-mounted '50s' instead. In the F4F-4 armament was improved, each wing carrying three 0.5-in (12.7-mm) machine-guns, a pair near the wing fold and another further outboard. The General Motors-built FM-1 derivative had four guns (with more rounds per weapon) and this configuration was continued in the XF4F-8 (FM-2) as a weight-saving measure. The F4F-7 reconnaissance variant was unarmed.

Undercarriage

The Wildcat's undercarriage was very similar in configuration to that of Grumman's earlier amphibian and FF/SF, F2F and F3F fighter designs, and retracted into the sides of the lower forward fuselage. Gear retraction was a manual operation, requiring 30 full turns of a hand crank in the cockpit. This operation required the F4F's pilot to change hands on the stick during take-off, often resulting in 'porpoising' as the pilot attempted to maintain control.

Grumman F6F Hellcat

The Hellcat remained in USN service following the conclusion of the Pacific war, in particular with units from the Naval Air Reserve. Additional war-surplus airframes were converted to unmanned F6F-5K status, with wingtip camera pods for the tracking of air-to-air missiles (AAMs) and surface-to-air missiles (SAMs).

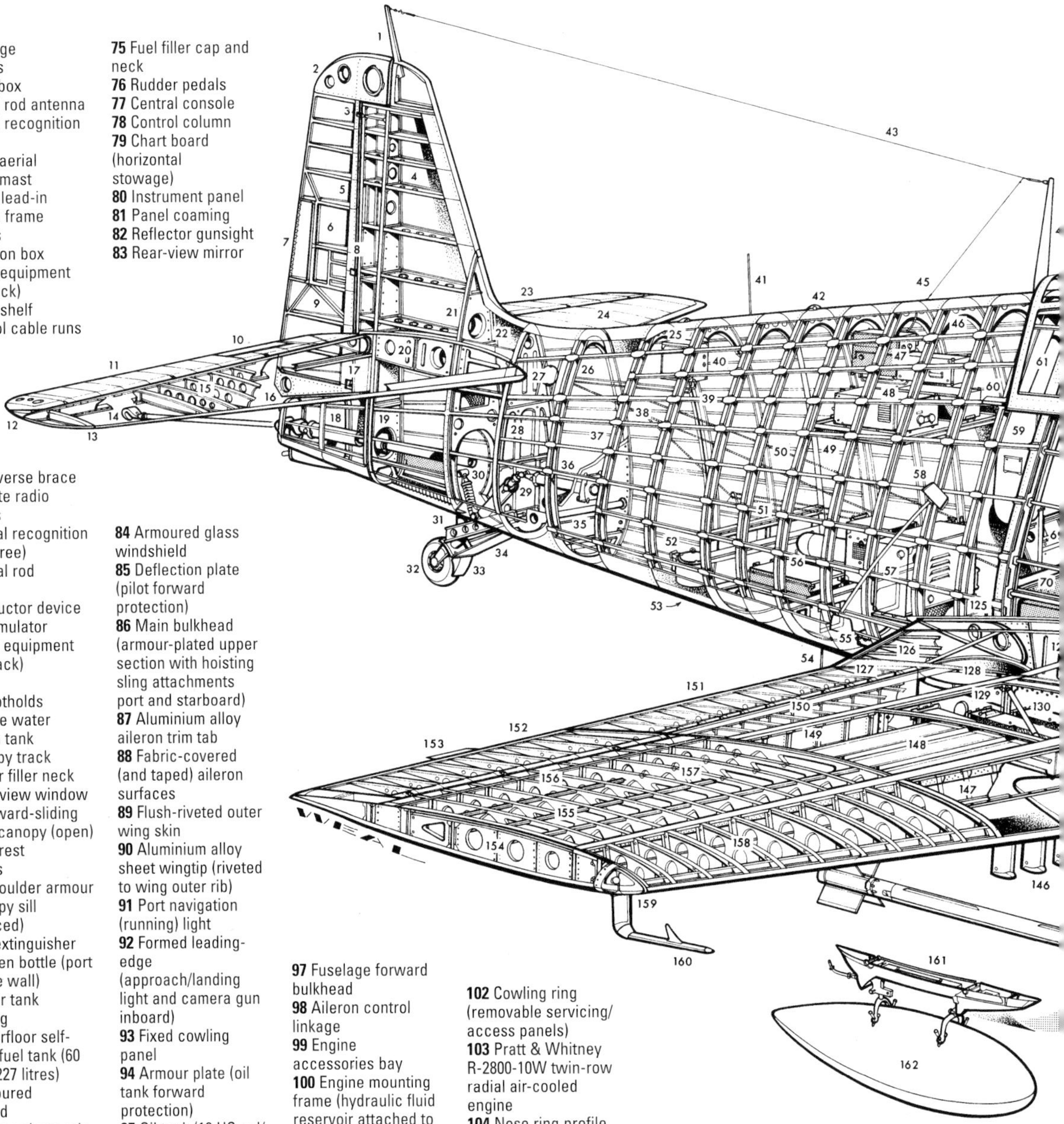

F6F-5 Hellcat

Cutaway key
1 Radio mast
2 Rudder balance
3 Rudder upper hinge
4 Aluminium alloy fin ribs
5 Rudder post
6 Rudder structure
7 Rudder trim tab
8 Rudder middle hinge
9 Diagonal stiffeners
10 Aluminium alloy elevator trim tab
11 Fabric-covered (and taped) elevator surfaces
12 Elevator balance
13 Flush-riveted leading-edge strip
14 Arrester hook (extended)
15 Tailplane ribs
16 Tail navigation (running) light
17 Rudder lower hinge
18 Arrester hook (stowed)
19 Fin main spar lower cut-out
20 Tailplane end rib
21 Fin forward spar
22 Fuselage/finroot fairing
23 Port elevator
24 Aluminium alloy-skinned tailplane
25 Section light
26 Fuselage aft frame
27 Control access
28 Bulkhead
29 Tailwheel hydraulic shock-absorber
30 Tailwheel centring mechanism
31 Tailwheel steel mounting arm
32 Rearward-retracting tailwheel (hard rubber tyre)
33 Fairing
34 Steel plate door fairing
35 Tricing sling support tube
36 Hydraulic actuating cylinder
37 Flanged ring fuselage frames
38 Control cable runs
39 Fuselage longerons
40 Relay box
41 Dorsal rod antenna
42 Dorsal recognition light
43 Radio aerial
44 Radio mast
45 Aerial lead-in
46 Dorsal frame stiffeners
47 Junction box
48 Radio equipment (upper rack)
49 Radio shelf
50 Control cable runs
51 Transverse brace
52 Remote radio compass
53 Ventral recognition lights (three)
54 Ventral rod antenna
55 Destructor device
56 Accumulator
57 Radio equipment (lower rack)
58 Entry hand/footholds
59 Engine water injection tank
60 Canopy track
61 Water filler neck
62 Rear-view window
63 Rearward-sliding cockpit canopy (open)
64 Headrest
65 Pilot's head/shoulder armour
66 Canopy sill (reinforced)
67 Fire extinguisher
68 Oxygen bottle (port fuselage wall)
69 Water tank mounting
70 Underfloor self-sealing fuel tank (60 US gal/227 litres)
71 Armoured bulkhead
72 Starboard console
73 Pilot's seat
74 Hydraulic handpump
75 Fuel filler cap and neck
76 Rudder pedals
77 Central console
78 Control column
79 Chart board (horizontal stowage)
80 Instrument panel
81 Panel coaming
82 Reflector gunsight
83 Rear-view mirror
84 Armoured glass windshield
85 Deflection plate (pilot forward protection)
86 Main bulkhead (armour-plated upper section with hoisting sling attachments port and starboard)
87 Aluminium alloy aileron trim tab
88 Fabric-covered (and taped) aileron surfaces
89 Flush-riveted outer wing skin
90 Aluminium alloy sheet wingtip (riveted to wing outer rib)
91 Port navigation (running) light
92 Formed leading-edge (approach/landing light and camera gun inboard)
93 Fixed cowling panel
94 Armour plate (oil tank forward protection)
95 Oil tank (19 US gal/72 litres)
96 Welded engine mount fittings
97 Fuselage forward bulkhead
98 Aileron control linkage
99 Engine accessories bay
100 Engine mounting frame (hydraulic fluid reservoir attached to port frames)
101 Controllable cooling gills
102 Cowling ring (removable servicing/access panels)
103 Pratt & Whitney R-2800-10W twin-row radial air-cooled engine
104 Nose ring profile
105 Reduction gear housing

Five VF-6 F6F-3 Hellcats are seen on the flight deck of the USS* Intrepid *(CV-11). Pictured during February 1944, off the Marshall Islands, the unit claimed a total of 14.66 kills during that year alone. VF-6 Hellcat aces included F. J. Blair with five kills before being killed in action; C. J. Chambers with 5.3; T. T. Coleman Jr with 10 kills (he also saw service with VF-83); H. G. Odenbrett with seven victories; J. D. Robbins with five victories and the legendary Alex Vraciu with 19 aerial victories.

SPECIFICATION

F6F-5 Hellcat

Dimensions

Span: 42 ft 10 in (13.08 m)
Span (wings folded): 16 ft 2 in (4.93 m)
Length: 33 ft 7 in (10.23 m)
Height: 13 ft 1 in (3.99 m)
Wing area: 334 sq ft (31.03 m²)
Wing loading: 38.17lb/sq ft (410.87 kg/m²

Powerplant

One 2,200-hp (1641-kW) Pratt & Whitney R-2800-10W Double Wasp 18-cylinder radial piston engine

Weights

Empty: 9,153 to 9,239 lb (4152 to 4191 kg)
Normal take-off: 12,500 lb (5670 kg)
Maximum take-off: 15,413 lb (6991 kg)

Fuel

Internal fuel: 250 US gal (66 litres)
External fuel: 150 US gal (39 litres)

Performance

Maximum speed at medium altitude: 386 mph (621 km/h)
Initial climb rate (clean): 3,410 ft (1039 m) per minute
Service ceiling: 37,300 ft (11369 m)
Cruising speed: 168 mph (270 km/h)
Time to altitude: 7 min 30 secs to 20,000 ft (6096 m)

Range

Range on internal fuel: 1,040 miles (1674 km)

Armament

Six 0.5-in (12.7-mm) Browning machine-guns each with 400 rounds, plus provision for two or three bombs up to a maximum total of 2,000 lb (907 kg), plus up to six 5-in (127-mm) High Velocity Aircraft Rockets (HVAR)

106 Three-bladed Hamilton Standard Hydromatic controllable-pitch propeller
107 Propeller hub
108 Engine oil cooler (centre) and supercharger intercooler
109 Oil cooler deflection plate under-protection
110 Oil cooler duct
111 Intercooler intake duct
112 Mainwheel fairing
113 Port mainwheel
114 Auxiliary tank support/attachment arms
115 Cooler outlet and fairing
116 Exhaust cluster
117 Supercharger housing
118 Exhaust outlet scoop
119 Wing front spar web
120 Wing front spar/fuselage attachment bolts
121 Undercarriage mounting/pivot point on front spar
122 Inter-spar self-sealing fuel tanks (port and starboard: 87.5 US gal (133 litres) each)
123 Wing rear spar/fuselage attachment bolts
124 Structural end rib
125 Slotted wing flap profile
126 Wing flap centre-section
127 Wing fold line
128 Starboard wheel well (doubler-plate reinforced edges)
129 Gun bay
130 Removable diagonal brace strut
131 Three 0.5-in (12.7-mm) Colt Browning machine-guns
132 Auxiliary tank aft support
133 Blast tubes
134 Folding wing joint (upper surface)
135 Machine-gun barrels
136 Fairing
137 Undercarriage actuating strut
138 Mainwheel leg oleo hydraulic shock strut
139 Auxiliary tank sling/brace
140 Long-range auxiliary fuel tank (jettisonable)
141 Mainwheel aluminium alloy fairing
142 Forged steel torque link
143 Low pressure balloon tyre
144 Cast magnesium wheel
145 Underwing 5-in (12.7-cm) air-to-ground RPS
146 Mark V zero-length rocket launcher installation
147 Canted wing front spar
148 Inter-spar ammunition box bay (lower surface access)
149 Wing rear spar (normal to plane of wing)
150 Rear sub spar
151 Wing flap outer-section
152 Frise-type aileron
153 Aileron balance tab
154 Wing outer rib
155 Wing lateral stiffeners
156 Aileron spar
157 Wing outer-section ribs
158 Leading-edge rib cut-outs
159 Starboard navigation (running) light
160 Pitot head
161 Underwing stores pylon (mounted on fixed centre-section inboard of mainwheel leg)
162 Auxiliary fuel tank

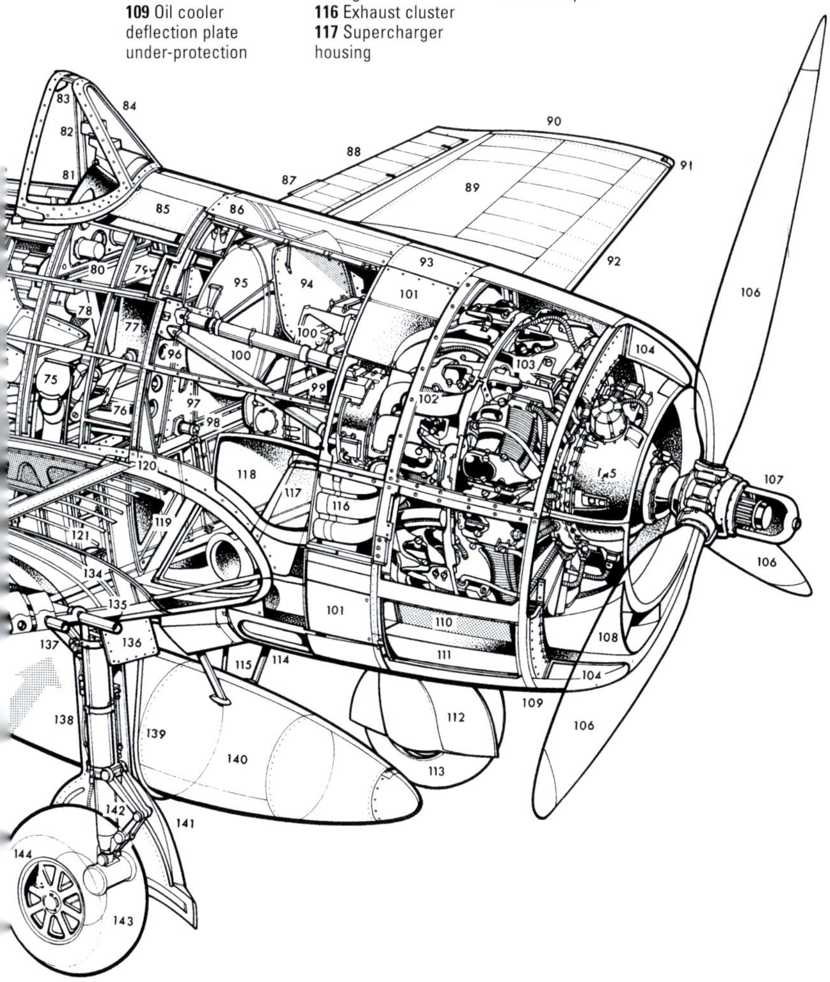

***VF-16 Hellcats are pictured aboard USS* Randolph *(CV-15) in March 1945. Later that year the unit would prove to be the second most successful F6F squadron operator, with 18 kills between July and August 1945, putting it just one behind the leading VF-31, aboard the* Belleau Wood.**

F6F-5P Hellcat

Incorporating a rear fuselage camera installation, the Grumman F6F-5P was unique among Hellcat variants in possessing a limited tactical reconnaissance capability. This example served with Fighter Squadron VF-84, aboard the carrier USS *Bunker Hill*, during February 1945. The aircraft is depicted as it appeared during raids on Tokyo, when the Hellcats involved wore a distinctive yellow cowling, repainted in midnight blue following the campaign. A pre-dawn launch on 16 February began the series of strikes on Tokyo, led by 16 aircraft-carriers, whose often inexperienced Hellcat pilots met the opposition in great numbers. USS *Bunker Hill* operated a mixture of VF-84's Hellcats, as well as two Marine F4U Corsair units, making its major combat debut over Tokyo. The Hellcat unit that saw most action over the Japanese capital during the first days of the operation was the 'Vipers' of VF-80, who claimed 24 aircraft on the morning of the 16th, when one of the unit's pilots, Lt A. L. Anderson, destroyed four different aircraft types – two 'Oscars', one 'Tojo', one 'Tony' and a Zero.

Wing and tail structure
The Hellcat's robust wing was comprised of two strong central spars, the outer panels of which folded backwards, skewing through 90° in order to lie flat against the rear fuselage. Outboard of the wing-fold mechanism for carrier stowage, the Hellcat wing featured a slight dihedral. The fin and tailplane were both built around a single central spar for structural strength, the part of the spar that comprised the fin running from the fintip to the bottom of the fuselage. The rudder carried a small balance near the top, and a central tab. The fixed section of the fin carried a radio mast. Full-span elevators on the tailplane had tabs, and were slotted to allow the uninterrupted movement of the rudder.

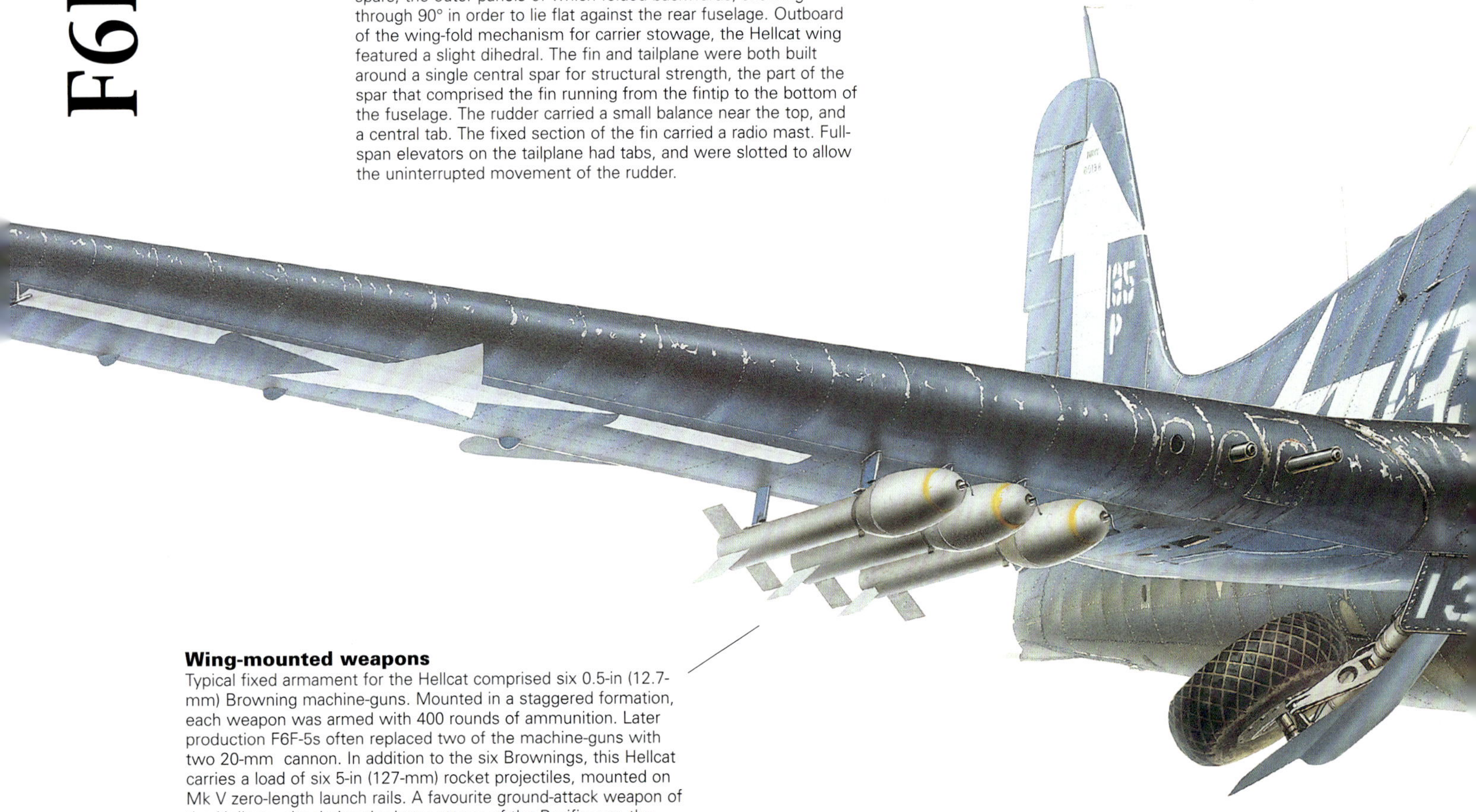

Wing-mounted weapons
Typical fixed armament for the Hellcat comprised six 0.5-in (12.7-mm) Browning machine-guns. Mounted in a staggered formation, each weapon was armed with 400 rounds of ammunition. Later production F6F-5s often replaced two of the machine-guns with two 20-mm cannon. In addition to the six Brownings, this Hellcat carries a load of six 5-in (127-mm) rocket projectiles, mounted on Mk V zero-length launch rails. A favourite ground-attack weapon of the Hellcat units during the latter stages of the Pacific war, the unguided rockets were heavily utilised during the assaults on Iwo Jima and Okinawa.

Reconnaissance equipment and undercarriage
Ahead of the 'sting'-type arrester hook that projected from the extreme rear of the fuselage was the camera installation, unique to the F6F-5P. This sloped downwards on the port side of the rear fuselage underside. Around 200 of the 8,100 F6F-5 airframes built were equipped with this reconnaissance package. Other equipment situated in the rear fuselage included radio and navigation systems aft of the cockpit, and a radio compass further back on the floor of the rear fuselage. The Hellcat's undercarriage was developed to withstand the punishment imposed by carrier operations, and comprised a single main oleo that rotated through 90° to lie flat within the wing structure. The tailwheel unit was also fully retractable.

Cockpit and systems
The Hellcat pilot was seated high in the fuselage under a sliding canopy. Well protected by armour plating, particularly at his rear, the pilot did, however, suffer from poor rear quadrant visibility. A reflector gunsight ahead of him enabled the pilot to aim his weapons. Distinctive external systems included the starboard wingtip air speed indicator pitot head, mounted in the undisturbed airflow beneath the wingtip, just by the starboard wingtip navigation light. The radio aerial, stretched between the fintip mast and dorsal mast, led to the rear fuselage equipment bay. The F6F-5 introduced an upright dorsal mast, as opposed to the forward-sloping mast of the F6F-3.

Powerplant and fuel
The F6F-5 Hellcat was powered by a single Pratt & Whitney R-2800-10W Double Wasp radial engine, with two rows of nine cylinders each. Developing 2,200 hp (1641 kW), the unit was angled down by three degrees to allow the fitment of a zero-incidence wing. The engine drove a three-bladed Hamilton Standard Hydromatic controllable-pitch propeller, ahead of a reduction gear housing. Internal self-sealing fuel tanks under the cockpit and in the inner wing structure carried a total of 235 US gal (889 litres). Tanks were filled through fuselage side and wingroot caps.

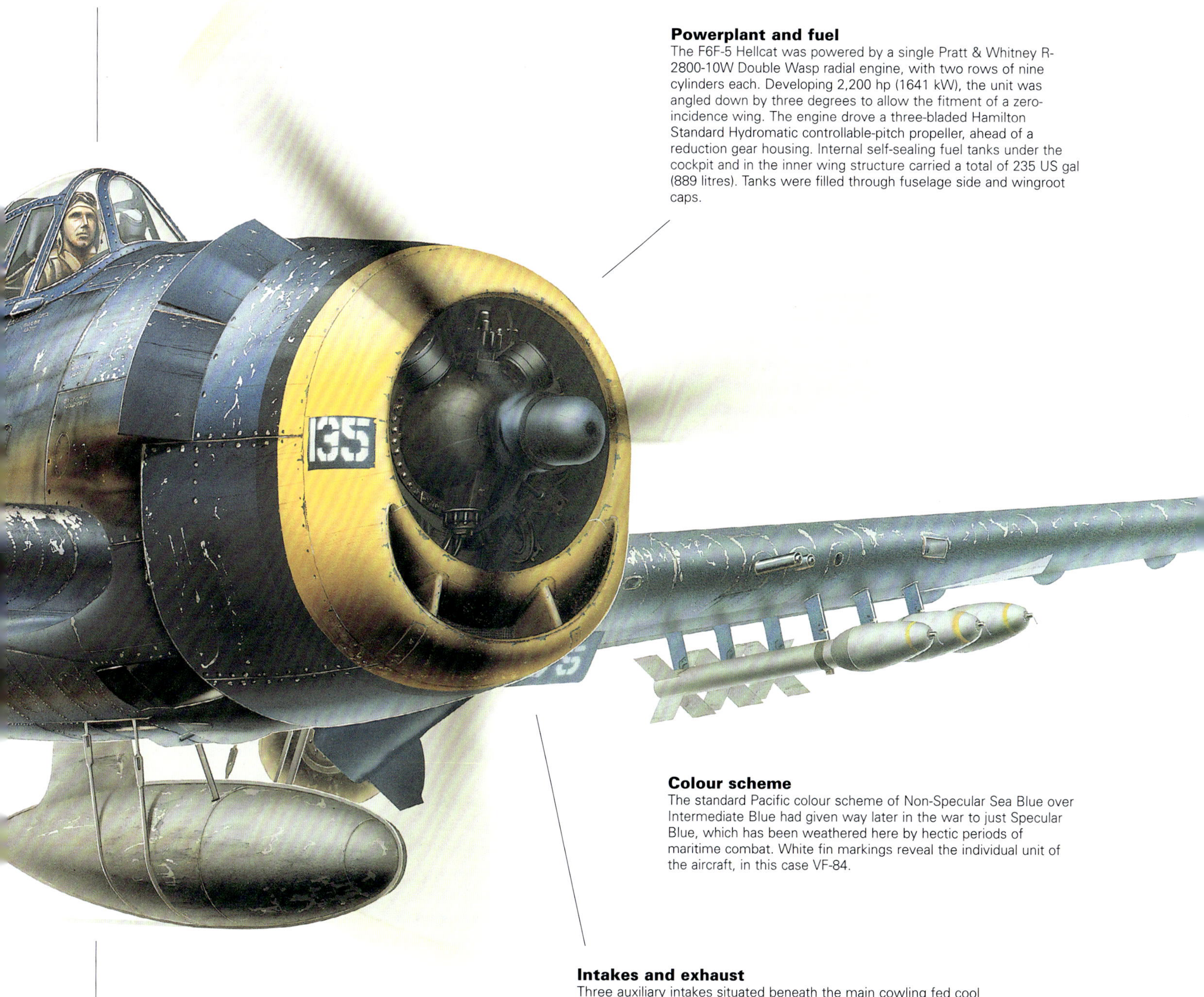

Colour scheme
The standard Pacific colour scheme of Non-Specular Sea Blue over Intermediate Blue had given way later in the war to just Specular Blue, which has been weathered here by hectic periods of maritime combat. White fin markings reveal the individual unit of the aircraft, in this case VF-84.

Intakes and exhaust
Three auxiliary intakes situated beneath the main cowling fed cool air to the engine oil supply (centre section), and the supercharger (two side sections). The main aperture cooled the engine, aided by a series of cooling gills on the rear of the cowling, which promoted a greater flow of air through the engine at low speeds, particularly for the rear row of cylinders. Cylinder exhaust gases were collected in ducts prior to passage through a cluster of exhausts on either side of the engine. These ejected the exhaust fumes from under the rear of the engine cowling and into shallow troughs.

External loads
The later F6F variants were capable of carrying up to 2,000 lb (907 kg) of external stores, although all models were equipped to carry a single 150-US gal (500-litre) auxiliary fuel tank, slung under the fuselage centreline. Another later variant, the F6F-5N, was equipped with night-interception radar, carried in a starboard wing-mounted pod. The faired-in housing contained the AN/APS-6 centrimetric radar, the scope of which was mounted on the centre of the pilot's instrument panel. Pilot radar controls allowed alteration of the radome calibration, signal level adjustment, and the changing of the focus and intensity of the received image.

Hawker Hurricane

Having served their time with operational squadrons, early-production Hurricane Mk Is often found their way into training units, in this case the Empire Central Flying School, Hullavington. This unarmed Hurricane Mk IA (seen with a pair of Spitfire Mk IIAs) was on charge with the ECFS in 1942.

Hurricane Mk I

Cutaway key
1 Starboard navigation light
2 Wingtip fairing
3 Fabric-covered aileron
4 Aluminium alloy wing skin panelling
5 Aileron hinge control
6 Starboard outer wing panel
7 Inboard torsion box heavy-gauge skin panel
8 Starboard landing lamp
9 Rotol three-bladed propeller
10 Spinner
11 Propeller hub pitch change mechanism
12 Spinner back plate
13 Propeller reduction gearbox
14 Cowling fairing
15 Starboard machine-gun muzzles
16 Upper engine cowling
17 Coolant pipes
18 Rolls-Royce Merlin III 12-cylinder liquid-cooled Vee engine
19 Exhaust stubs
20 Engine-driven generator
21 Forward engine mounting
22 Ignition control unit
23 Engine bearer struts
24 Lower engine cowlings
25 Starboard mainwheel
26 Manual-type inertia starter
27 Hydraulic pumps
28 Carburettor air intake
29 Cooling air scoop
30 Rear engine mounting
31 Single-stage supercharger
32 Port magneto
33 Coolant system header tank
34 External bead sight
35 Coolant filler cap
36 Starboard wing gun bay
37 Ammunition magazines
38 Starboard Browning 0.303-in (7.7-mm) machine-guns (4)
39 Fuel filler cap
40 Engine bay canted bulkhead
41 Rear engine mounting struts
42 Pneumatic system air bottle (gun firing)
43 Wing spar centre-section carry-through
44 Lower longeron/wing spar joint
45 Rudder pedals
46 Pilot's foot boards
47 Control column linkage
48 Fuselage (reserve) fuel tank, capacity 28 Imp gal (127 litre)
49 Fuel tank bulkhead
50 Control column hand grip
51 Instrument panel
52 Reflector gunsight
53 Starboard split trailing-edge flap
54 Bulletproof windscreen panel
55 Canopy internal handle
56 Rear view mirror
57 Sliding cockpit canopy cover
58 Plexiglass canopy panels
59 Canopy framework
60 Canopy external handle
61 Starboard side 'break-out' emergency exit panel
62 Safety harness
63 Seat height adjustment lever
64 Oxygen supply cock
65 Engine throttle lever
66 Elevator trim tab control handwheel
67 Oil pipes to radiator
68 Radiator flap control lever
69 Cockpit section tubular fuselage framework
70 Coolant system piping
71 Pilot's oxygen cylinder
72 Boarding step
73 Seat back armour
74 Pilot's seat
75 Armoured headrest
76 Turn-over crash pylon struts
77 Canopy rear fairing construction
78 Sliding canopy rail
79 Battery
80 TR 9D radio transmitter/receiver
81 Radio shelf
82 Downward identification light
83 Flare launch tube
84 Handgrip
85 Plywood skin panel
86 Dorsal fairing stringers
87 Upper identification light
88 Aerial mast
89 Aerial lead-in
90 Wooden dorsal section formers
91 Fuselage upper longeron
92 Rear fuselage fabric covering
93 Aluminium alloy tailplane leading edge
94 Starboard fabric-covered tailplane
95 Fabric-covered elevator
96 Aluminium alloy fin leading edge
97 Forward fin mounting post
98 Tailplane spar attachment joint
99 Elevator hinge control
100 Fin rib construction
101 Tailfin fabric covering
102 Diagonal bracing strut

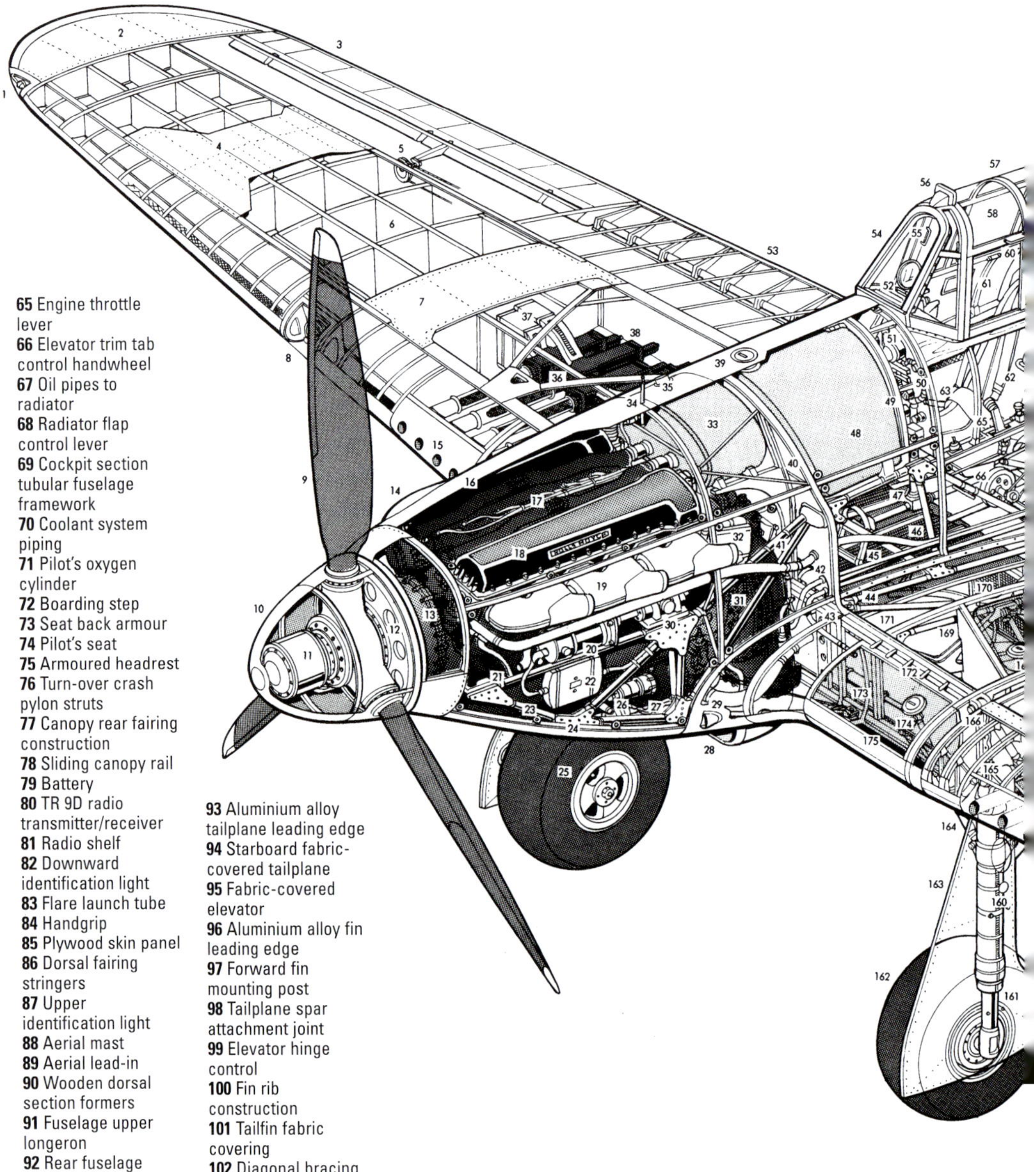

103 Stern post
104 Rudder mass balance weight
105 Aileron cable
106 Rear aerial mast
107 Fabric-covered rudder
108 Aluminium alloy rudder framework
109 Tail navigation light
110 Rudder tab
111 Elevator trim tab
112 Port elevator rib construction
113 Elevator horn balance
114 Port tailplane rib construction
115 Diagonal spar bracing struts
116 Rudder control horn
117 Tail control access panel
118 Ventral tailwheel fairing
119 Fixed, castoring tailwheel
120 Dowty shock absorber tailwheel strut
121 Ventral fin framework
122 Lifting bar socket
123 Aluminium alloy lateral formers
124 Tail control cables
125 Rear fuselage tubular framework
126 Diagonal wire bracing
127 Lateral stringers
128 Fuselage lower longeron
129 Pull-out boarding step
130 Wing root trailing-edge fillet
131 Ventral access hatch
132 Walkway
133 Flap hydraulic jack
134 Inner wing panel rear spar
135 Outer wing panel spar attachment joint
136 Gun heater air duct
137 Wing panel joint cover strip
138 Flap shroud ribs
139 Port split trailing-edge flap
140 Aluminium alloy aileron rib construction
141 Port fabric-covered aileron
142 Aileron hinges
143 Wingtip fairing construction
144 Port navigation light
145 Leading-edge nose ribs
146 Front spar
147 Intermediate spars
148 Ventral pitot head
149 Rear spar
150 Aluminium alloy wing rib construction
151 Wing stringers
152 Port landing lamp
153 Inboard double-web strengthened spar section
154 Outboard ammunition magazines, 338 rounds each
155 Port Browning 0.303-in (7.7-mm) machine-guns (4)
156 Inboard ammunition magazines, 324 and 338 rounds
157 Diagonal gun bay ribs
158 Gun barrel blast tubes
159 Machine-gun muzzles
160 Main undercarriage leg strut
161 Oleo-pneumatic shock absorber strut
162 Port mainwheel
163 Mainwheel leg fairing
164 Side-locking strut
165 Main undercarriage leg pivot fixing
166 Outer wing panel front spar bolted joint
167 Fuel filler cap
168 Port wing main fuel tank, capacity 34.5 Imp gal (157 litre)
169 Centre-section strut framework
170 Ventral oil and coolant radiator
171 Main undercarriage wheel bay
172 Oil tank attachments
173 Mainwheel hydraulic retraction jack
174 Oil filler cap
175 Leading-edge oil tank, capacity 9 Imp gal (41 litre), port side only

Twelve-gun Hurricane Mk IIBs of No. 601 (County of London) Squadron are seen over the English countryside, some time in 1941. This Royal Auxiliary Air Force unit operated Mk IIBs for most of 1941, on sweeps over the English Channel and escort sorties. Towards the end of the year, it began its short stint as the only operational RAF Bell P-39 Airacobra unit. Later, the unit was equipped with Spitfires and served with distinction in the Mediterranean.

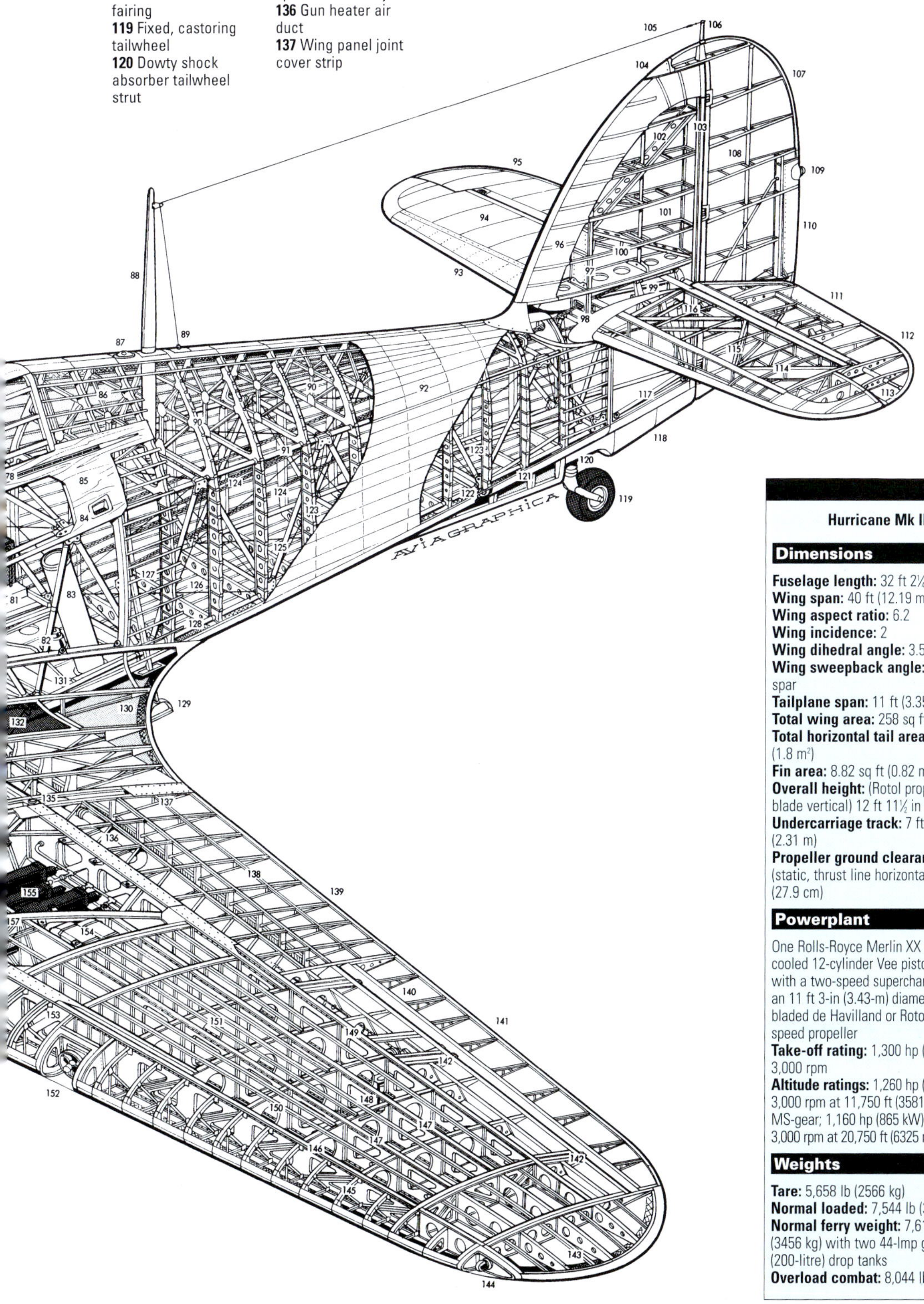

SPECIFICATION

Hurricane Mk IIC

Dimensions

Fuselage length: 32 ft 2¼ in (9.81 m)
Wing span: 40 ft (12.19 m)
Wing aspect ratio: 6.2
Wing incidence: 2
Wing dihedral angle: 3.5°
Wing sweepback angle: 3° on front spar
Tailplane span: 11 ft (3.35 m)
Total wing area: 258 sq ft (23.97 m²)
Total horizontal tail area: 19.6 sq ft (1.8 m²)
Fin area: 8.82 sq ft (0.82 m²)
Overall height: (Rotol propeller, top blade vertical) 12 ft 11½ in (3.95 m)
Undercarriage track: 7 ft 7 in (2.31 m)
Propeller ground clearance: (static, thrust line horizontal) 11 in (27.9 cm)

Powerplant

One Rolls-Royce Merlin XX liquid-cooled 12-cylinder Vee piston engine with a two-speed supercharger, driving an 11 ft 3-in (3.43-m) diameter three-bladed de Havilland or Rotol constant-speed propeller
Take-off rating: 1,300 hp (969 kW) at 3,000 rpm
Altitude ratings: 1,260 hp (940 kW) at 3,000 rpm at 11,750 ft (3581 m) in MS-gear; 1,160 hp (865 kW) at 3,000 rpm at 20,750 ft (6325 m) in S-gear

Weights

Tare: 5,658 lb (2566 kg)
Normal loaded: 7,544 lb (3422 kg)
Normal ferry weight: 7,619 lb (3456 kg) with two 44-Imp gal (200-litre) drop tanks
Overload combat: 8,044 lb (3648 kg)

Fuel and load

Total internal fuel: 97 Imp gal (441 litre), comprising two wing fuel tanks each of 34.5-Imp gal (157-litre) capacity, plus one reserve tank in rear fuselage of 28-Imp gal (127-litre) capacity
External fuel: provision for two 44-Imp gal (200-litre) drop tanks or two 88-Imp gal (400-litre) ferry tanks
Maximum external load: 1,000 lb (454 kg)

Performance

Maximum level speeds: (clean) 260 mph (418 km/h) at sea level; 329 mph (529 km/h) at 17,800 ft (5425 m)
Rate of climb: 2,760 ft (841 m) per minute at sea level; 1,670 ft (509 m) per minute at 15,000 ft (4572 m)
Service ceiling: 32,400 ft (9875 m)
Absolute ceiling: 35,600 ft (10850 m)
Time to climb: 15,000 ft (4572 m) in 5.7 minutes; 20,000 ft (6096 m) in 7.7 minutes

Range

460 miles (740 km) clean at optimum range speed (178 mph; 286 km/h); 920 miles (1481 km) with two 44-Imp gal (200-litre) drop tanks at optimum range speed

Armament

Fixed: four wing-mounted 20-mm Hispano cannon, with a total of 364 rounds (91 per gun)
External stores: provision for two underwing racks, each able to carry a 250-lb (113-kg) or 500-lb (227-kg) bomb

Hurricane Mk IID

HV663/'U' was a Hurricane Mk IID completed by Hawker Aircraft Ltd towards the end of 1943 and shipped to North Africa. Initially allocated to No. 71 Operational Training Unit at Carthago, Sudan in February 1943, the aircraft would have been utilised in the training of fighter pilots in desert conditions. Later in 1943, it was transferred to No. 6 Squadron, RAF which had operated this Hurricane variant since June 1942, in an anti-tank role against Rommel's Afrika Korps. After a period based in Egypt on shipping protection duties, No. 6 Squadron returned to the desert the following February, again in the anti-tank role, eventually moving westwards into Tunisia. The Mk IID's effectiveness as a 'tank buster' earned the unit the nickname 'The Flying Can-Openers' – a motif (a can-opener with a set of wings) to be found on the squadron's aircraft to this day. In common with a number of RAF units in the Western Desert, No. 6's aircraft dispensed with unit codes (though they carried 'JV' for a time), but retained an individual aircraft letter.

Powerplant and propeller
Early-production Hurricane Mk Is were fitted with a Rolls-Royce Merlin II rated at 1,030 hp (768 kW) at 16,250 ft (4953 m) and driving a Watts two-bladed, fixed-pitch wooden propeller with a diameter of 11 ft 6 in (3.5 m). Later-production aircraft benefited from a de Havilland three-bladed, constant-speed propeller which much improved take-off performance, climb rate and maximum speed at altitude. This and a similar Rotol design appeared from late 1938 and necessitated modifications to the aircraft's engine, which became the Merlin III. The Hurricane Mk II introduced the Merlin XX engine with two-stage supercharging and a rating of 1,260 hp (940 kW) at 11,750 ft (3580 m). Early Mk IIs were equipped with an enlarged coolant header tank which lengthened the nose of these aircraft by some 6.5 in (16.5 cm). However, as this change presented handling problems, it was abandoned after a handful of aircraft were completed. The illusion that most other Hurricane Mk IIs had longer noses than Mk Is was a result of the longer Rotol propeller spinner fitted to these machines. Hurricane Mk IVs benefited from a Merlin 27, modified to produce peak power (1,620 hp; 1208 kW) on take-off and operate more efficiently at high temperatures. Canadian-built Hurricane Mk Xs were fitted with Packard-built Merlin 28s (equivalent to the Merlin XX), while Mk XIIs had an improved Packard Merlin 29, rated at 1,300 hp (970 kW). Vokes tropical filters were fitted to aircraft intended for operation in the Mediterranean and Far East, including most Mk IIDs and Mk IVs.

Gun armament

Eight Browning 0.303-in (7.7-mm) machine-guns equipped Hurricane Mk IAs and Mk IIAs. The Mk IIB introduced an additional pair of Brownings in each wing, making this variant popular with pilots. However, with the recognition that rifle-calibre guns had their limitations against the armour and self-sealing fuel tanks installed in Luftwaffe aircraft, cannon were soon introduced. After trials with Oerlikon and Hispano 20-mm designs, the latter was chosen, four being fitted to the Hurricane Mk IIC (two of which were sometimes removed in service, especially in the Middle East, to save weight). Mk IIDs featured two Vickers 40-mm 'S' guns but retained a pair of Browning guns for sighting purposes, firing tracer ammunition. The 'E' wing, as fitted to the Mk IV (originally known as the Mk IIE), was a 'universal' design intended to be reconfigured according to the role of the squadron to which the aircraft was issued. Four Hispano cannon could be installed, but the most common armament fits were external – rockets (four per wing), a pair of 'S' guns or two 250-lb (113-kg) or 500-lb (227-kg) bombs.

Armour

Successive Hurricane variants were fitted with increased amounts of armour, largely for protection from ground fire. Later-production Mk IIDs were fitted with an additional 368 lb (167 kg) of armour plate over early-build examples and Mk IICs. The Mk IV had still more armour fitted, mainly around the nose and radiator intake under the fuselage. This variant had a considerably reduced top speed and an all-up weight of 8,462 lb (3838 kg), compared to the 6,218 lb (2820 kg) of the earliest Mk I aircraft.

Structure and markings

Among the changes made to the Hurricane's airframe during early production, in light of service experience, were the introduction of locally strengthened, metal-skinned wings, a slightly enlarged rudder and a ventral underfin on later-build Hurricane Mk Is. The rear fuselage remained fabric-covered throughout the life of the design, this ironically contributing to the machine's ability to sustain battle damage without loss of the aircraft (and its pilot). Repairs were also more straightforward and thus often easier to accomplish at squadron level, compared to the all-metal monocoque airframe of the Spitfire. This Mk IID is finished in the standard RAF Dark Earth/Middle Stone/Azure Blue applied to aircraft in the Mediterranean theatre.

Hawker Tempest

Seen fresh from the Hawker factory at Langley, this Tempest Mk II was fitted with bomb racks and was delivered to the A&AEE at Boscombe Down in September 1945 for weapons trials with 1,000-lb (454-kg) bombs.

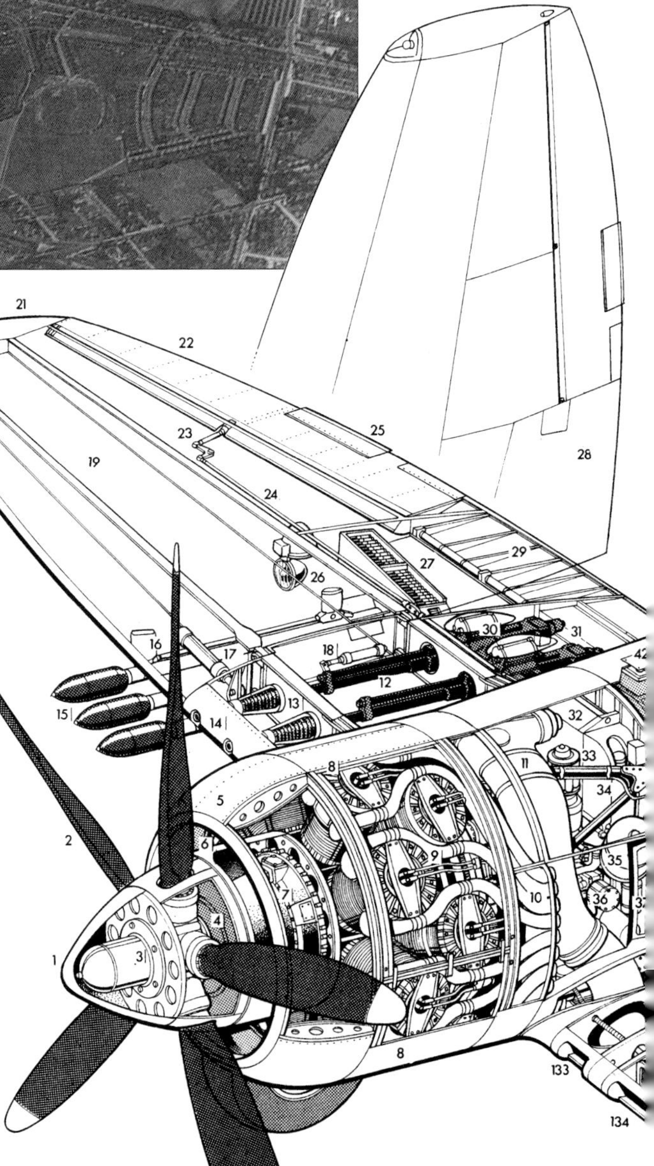

Hawker Sea Fury

Cutaway key
1 Spinner
2 Rotol five-bladed constant-speed propeller of 12 ft 9 in (3.89 m) diameter
3 Propeller hub pitch change mechanism
4 Spinner backplate
5 Engine cowling ring
6 Cooling air intake
7 Propeller reduction gear casing
8 Detachable engine cowlings
9 Bristol Centaurus Mk 18 18-cylinder two-row radial sleeve valve engine
10 Exhaust stubs
11 Carburettor intake ducting
12 Starboard British Hispano Mk 5 20-mm cannon
13 Recoil springs
14 Cannon muzzles
15 60-lb (27.22 kg) ground attack rocket projectiles
16 Zero-length rocket launcher rails
17 Wing folding jack
18 Wing fold latching mechanism
19 Starboard outer wing panel
20 Starboard navigation light
21 Wing tip fairing
22 Starboard aileron
23 Aileron hinge control
24 Push-button control rod
25 Aileron spring tab
26 Retractable landing/taxiing lamp
27 Ammunition box (290 rounds port and starboard)
28 Starboard wing folded position
29 Outer split trailing edge flap
30 Ammunition feed drum blister fairings
31 Cannon breeches
32 Oil tank (14 Imp gal/63.65 litre capacity)
33 Engine cartridge starter
34 Engine bearer struts
35 Hydraulic reservoir
36 Accessory drive gearbox
37 Engine cooling air outlet
38 Wing front spar attachment joint
39 Fireproof engine compartment bulkhead
40 Fuselage double frame
41 Main fuel tank (97 Imp gal/441 litre)
42 Fuel tank vent
43 Filler cap
44 Fuselage top longeron
45 Rudder pedals
46 Auxiliary fuselage fuel tank (30 Imp gal/136 litre)
47 Fuselage bottom longeron
48 Rear wing spar attachment joint
49 Oxygen bottle
50 Control column
51 Instrument panel
52 Bullet proof windscreen
53 Mk 4B reflector sight
54 Windscreen framing
55 Pilot's starboard side console
56 Pilot's seat
57 Engine throttle and propeller controls
58 Radio equipment
59 Port side console
60 Seat back armour plate
61 Safety harness
62 Headrest
63 Armoured headrest support
64 Sliding cockpit canopy cover
65 Canopy rails
66 Tailplane control rod
67 Rear fuselage joint frame
68 Whip aerial
69 Fuselage skin plating
70 Elevator push-pull control rod
71 Tailplane attachment joint frame
72 Fin root fillet
73 Starboard tailplane
74 Starboard elevator
75 Tailfin construction
76 Curved fin leading edge
77 Sternpost
78 Rudder construction
79 Mass balance weight
80 Rudder tab
81 Deck arrester hook
82 Elevator trim tab
83 Port elevator
84 Tailplane construction
85 Tailplane spar joints
86 Rudder hinge control
87 Tail navigation light
88 Arresting hook attachment link
89 Tailwheel hydraulic retraction jack
90 Tailwheel
91 Tailwheel doors
92 Rear fuselage double bulkhead
93 Tailwheel bay
94 Tailwheel bay bulkhead
95 Fuselage frame and stringer construction
96 Rudder push-pull control rod
97 Remote compass transmitter
98 Ventral aerial
99 Handgrip
100 Radio transmitter/receiver
101 Trailing edge wing root fillet
102 Retractable 'stirrup-type' step
103 Inboard split trailing edge flap
104 Flap shroud structure
105 Gun heater duct
106 Inboard ammunition box
107 Ammunition guide track
108 Port British Hispano Mk 5 20-mm cannon
109 Ammunition feed drums
110 Outer ammunition box (145 rounds)
111 Outer split trailing edge flap
112 Port retractable landing/taxiing lamp
113 Aileron spring tab
114 Aileron construction
115 Wing tip fairing
116 Port navigation light
117 Pitot tube
118 Rear spar
119 Wing rib construction
120 Main spar
121 Leading edge nose ribs
122 1,000-lb (453.6-kg) HE bomb
123 60-lb (27.22-kg) ground attack rockets
124 Port drop tank (45 or 90 Imp gal/204.5 or 409 litres)
125 Tank pylon
126 Wing fold hydraulic jack
127 Wing fold hinge joints
128 Cannon barrel mountings
129 Port interspar fuel tank (28 Imp gal/127 litres)
130 Main undercarriage wheel bay
131 Mainwheel door
132 Hydraulic retraction jack
133 Port carburettor air intake
134 Oil cooler ram air intake
135 Oil radiator (starboard leading edge has 17 Imp gal/77 litre fuel tank)
136 Port cannon muzzles
137 Pivoted main undercarriage shock absorber leg strut
138 Undercarriage leg fairing door
139 Port mainwheel

SPECIFICATION

Hawker Tempest Mk V Series 2

Type
single-seat fighter and fighter-bomber

Powerplant
one Napier Sabre IIA H-type piston engine rated at 2,180 hp (1626 kW)

Performance
maximum speed 426 mph (686 km/h) at 18,500 ft (5640 m); cruising speed 391 mph (629 km/h) at 18,800 ft (5730 m); initial climb rate 4,700 ft (1433 m) per minute; climb to 20,000 ft (6095 m) in 6 minutes 6 seconds; service ceiling 36,000 ft (10975 m); range 1,300 miles (2092 km) with drop tanks or 820 miles (1320 km) with standard fuel

Weights
Empty 9,000 lb (4082 kg); maximum take-off 13,540 lb (6142 kg)

Dimensions
Wingspan 41 ft (12.50 m); length 33 ft 8 in (10.26 m); height 16 ft 1 in (4.90 m); wing area 302.00 sq ft (28.06 m2)

Armament
Four 20-mm Hispano Mk V fixed forward-firing cannon in the leading edges of the wing, plus up to 2,000 lb (907 kg) of disposable stores carried on two underwing hardpoints, and generally comprising two 1,000- or 500-lb (454- or 227-kg) bombs or eight 60-lb (27-kg) air-to-surface unguided rockets.

British piston-engined fighter design reached a pinnacle in the Hawker Fury and Sea Fury, which had its origins in a lightweight derivative of the Tempest. In this view of the second prototype, the Fury's raised cockpit and reduced wingspan are evident. Though it failed to see RAF service during World War II, a naval version – the Sea Fury – served the post-war Fleet Air Arm well.

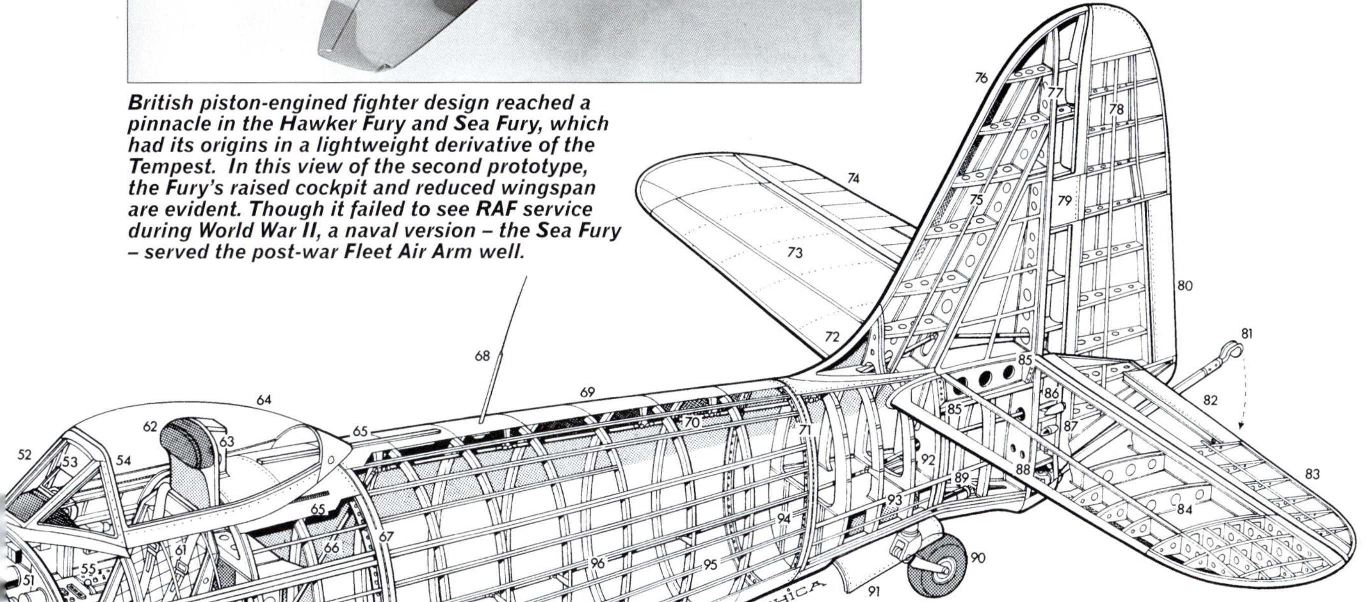

Seen at Newchurch in the spring of 1944, No. 486 Squadron, RNZAF was to become, along with No. 3 Squadron, one of the RAF's most successful units against the V-1 flying-bombs.

Hawker Tempest Mk II

The Tempest Mk II was powered by the Bristol Centaurus V 18-cylinder two row radial engine. The origins of the Tempest II can be traced to a proposal by Hawker Aircraft for a Typhoon Mk II with a radial engine. The prototype of the Centaurs-Typhoon Mk II was abandoned without being flown because sufficient experience had already been amassed with the Centaurus-powered Tornado, another Hawker design. However, delays in the engine's development meant that the Mk II entered service after the Mk V, which had a Napier Sabre engine.

Construction
The Tempest pilot enjoyed an excellent all-round view, which was a considerable improvement over that of its predecessor, the Hawker Typhoon. Production of the Tempest F. Mk.II was originally to be undertaken jointly by Hawker and Gloster Aircraft, but by the time the second prototype flew on 18 September, 1943, the number of aircraft on order had increased to 600 and some of the production work was switched to the Bristol Aeroplane Company because of Gloster's other commitment, the Meteor jet fighter

Markings
This aircraft has the markings of No 16 Squadron RAF, based in northern Germany in 1946-47 as part of the 2nd Tactical Air Force. The squadron only flew Tempests for about two years before converting to the de Havilland Vampire Mk 5 jet fighter-bomber. The squadron originally formed part of the Fassberg Wing, the other two squadrons being Nos 33 and 26. In November 1947 the Wing moved to Gütersloh, where it spent the remainder of its Tempest days. The first Tempest II squadron, No 33, later redeployed to Malaya.

Cockpit
A reflector gun sight was mounted on a bar above the instrument panel. A dimmer switch was also mounted on the instrument panel and had three positions, marked 'off', 'night' and 'day'. A supply of warm air to the cockpit was controlled by a lever on the starboard cockpit wall, the lever being moved downwards from 'off' to 'on'. Two ventilators were provided, one on each side of the instrument panel. The radio installation, fitted behind the pilot, comprised a TR1143 transceiver and A1271 beam approach equipment. Provision was also made for IFF (Identification Friend or Foe) equipment.

Armament
The Tempest two carried a fixed armament of four 0.79in (20mm) Hispano V cannon, specially designed to fit inside the aircraft's very thin wing section, and 800 rounds of ammunition. The type could also carry up 907kg (2000lb) of underwing stores, including eight rocket projectiles, as seen here. The Tempest Mk II gave excellent postwar service in the fighter-bomber role with the RAF in the Far East, carrying out many attacks on communist terrorist hideouts during the early stages of Operation Firedog, the counter-insurgency operations in Malaya.

Fuel capacity
The Tempest had less internal fuel tankage than the Typhoon, as its new, thin wing prevented the installation of wing fuel tanks. To compensate for this a new tank was installed behind the engine. Although the Tempest had a shorter endurance than the Typhoon, its higher speed meant that it had a similar range. Hawker designed a 45 gal streamlined long-range fuel tank for the Tempest, replacing the cylindrical tank that had been used on the Hurricane and Typhoon. Napalm tanks were also developed, but were detrimental to handling.

Hawker Typhoon/Tempest

This rare wartime colour photograph illustrates some of the modifications that the Typhoon underwent, as shown by the light coloured panels. The aircraft's serial number on the rear of the fuselage has been painted out.

Typhoon Mk IB

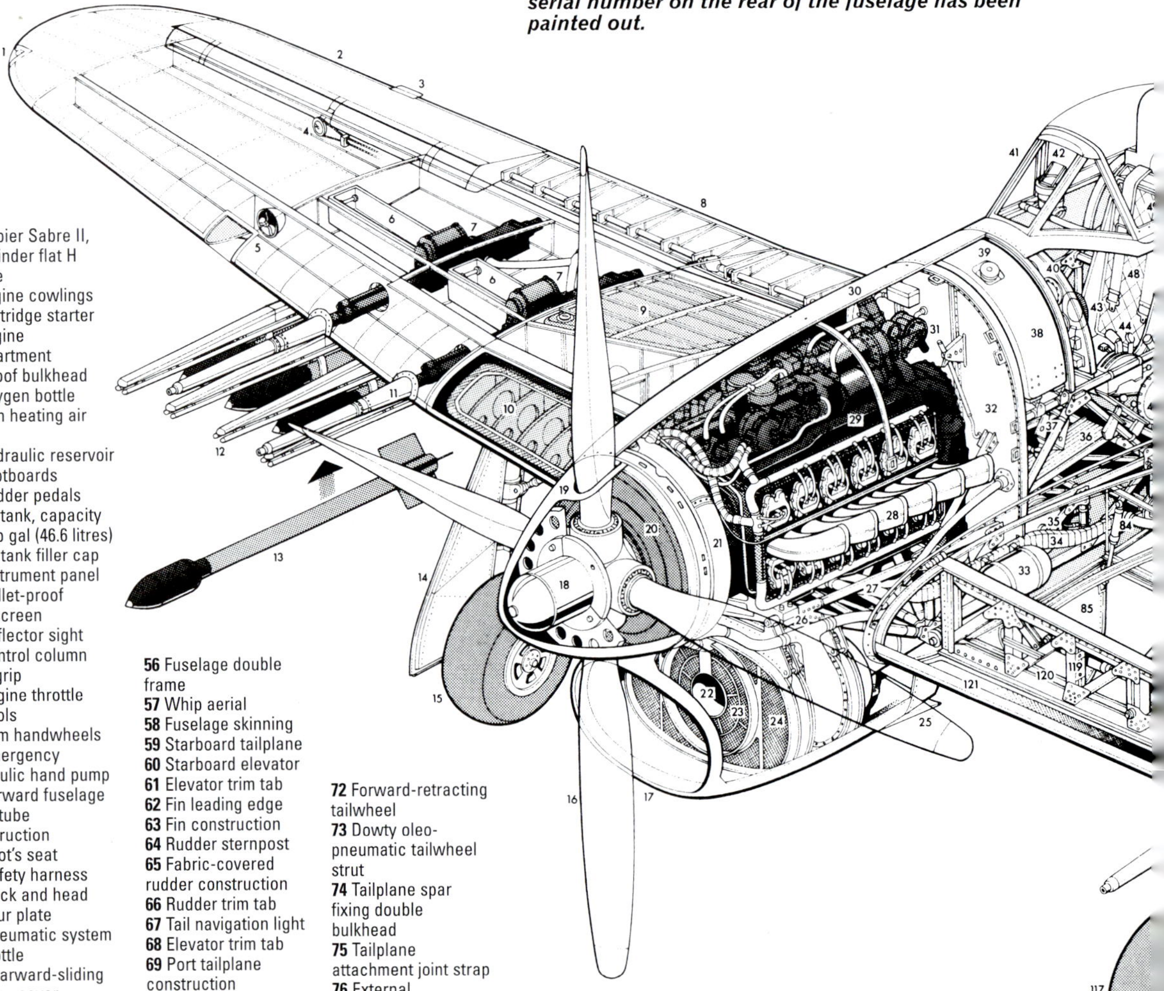

Cutaway key
1 Starboard navigation light
2 Starboard aileron
3 Fixed trim tab
4 Aileron hinge control
5 Landing lamp
6 Ammunition boxes
7 Starboard Hispano Mk II 20-mm cannon
8 Split trailing-edge flaps
9 Starboard main fuel tank, capacity 40 Imp gal (182 litres)
10 Self-sealing leading-edge fuel tank, capacity 35 Imp gal (159 litres)
11 Cannon barrel fairings
12 Rocket launcher rails **13** 60-lb (27-kg) ground attack rockets
14 Main undercarriage leg fairing
15 Starboard mainwheel
16 de Havilland four-bladed propeller
17 Air intake
18 Propeller pitch change mechanism
19 Spinner
20 Armoured spinner backplate
21 Coolant tank, capacity 7¼ Imp gal (33 litres)
22 Supercharger ram air intake
23 Oil radiator
24 Coolant radiator
25 Radiator shutter
26 Engine mounting block
27 Tubular steel engine support framework
28 Exhaust stubs
29 Napier Sabre II, 24-cylinder flat H engine
30 Engine cowlings
31 Cartridge starter
32 Engine compartment fireproof bulkhead
33 Oxygen bottle
34 Gun heating air duct
35 Hydraulic reservoir
36 Footboards
37 Rudder pedals
38 Oil tank, capacity 18 Imp gal (46.6 litres)
39 Oil tank filler cap
40 Instrument panel
41 Bullet-proof windscreen
42 Reflector sight
43 Control column handgrip
44 Engine throttle controls
45 Trim handwheels
46 Emergency hydraulic hand pump
47 Forward fuselage steel tube construction
48 Pilot's seat
49 Safety harness
50 Back and head armour plate
51 Pneumatic system air bottle
52 Rearward-sliding canopy cover
53 Aft fuselage joint
54 Canopy rails
55 Radio transmitter/receiver
56 Fuselage double frame
57 Whip aerial
58 Fuselage skinning
59 Starboard tailplane
60 Starboard elevator
61 Elevator trim tab
62 Fin leading edge
63 Fin construction
64 Rudder sternpost
65 Fabric-covered rudder construction
66 Rudder trim tab
67 Tail navigation light
68 Elevator trim tab
69 Port tailplane construction
70 Tailplane spar attachments
71 Tailwheel hydraulic jack
72 Forward-retracting tailwheel
73 Dowty oleo-pneumatic tailwheel strut
74 Tailplane spar fixing double bulkhead
75 Tailplane attachment joint strap
76 External strengthening fishplates
77 Elevator mass balance

A fine 1945 photograph of an echelon of Tempests from the CFE, the busy RAF Central Fighter Establishment at West Raynham, where most of the RAF's trials and evaluations on fighter-type aircraft were conducted. All are Mk V Series 2: SN328, SN108 and EJ884. SN328 has eight of the lengthened rocket rails fitted.

SPECIFICATION

Typhoon Mk IB (early production)

Dimensions

Length: 31 ft 11 in (9.73 m)
Height: 14 ft 10 in (4.52 m)
Wing span: 41 ft 7 in (12.67 m)
Wing area: 279 sq ft (25.90 m²)

Powerplant

One 2,180-hp (1626-kW) Napier Sabre IIA inline piston engine

Weights

Empty: 8,800 lb (3992 kg)
Maximum take-off: 13,250 lb (6010 kg)

Performance

Maximum speed (clean) at 18,000 ft (5486 m): 405 mph (652 km/h)
Climb to 15,000 ft (4572 m) from sea level: 5 minutes 55 seconds
Service ceiling: 34,000 ft (10670 m)
Range: (clean) 610 miles (982 km), and with 2,000-lb (910-kg) bombload 510 miles (821 km)

Armament

four 20-mm Hispano cannon each with 140 rounds, plus two bombs of up to 1,000-lb (454-kg) each, or eight 60-lb (27-kg) rockets or other stores such as 45-Imp gal (205-litre) drop tanks.

Tempest Mk V

Dimensions

Length: 33 ft 8 in (10.26 m)
Height: 16 ft 1 in (4.90 m)
Wing span: 41 ft (12.50 m)
Wing area: 302 sq ft (28.06 m²)

Powerplant

One 2,180-hp (1626-kW) Napier Sabre IIA inline piston engine

Weights

Empty: 9,000 lb (4082 kg)
Maximum take-off: 13,540 lb (6142 kg)

Performance

Maximum speed (clean) at 18,500 ft (5639 m): 426 mph (686 km/h)
Service ceiling: 36,500 ft (11125 m)
Maximum Range: 1,530 miles (2462 km)

Armament

four 20-mm Hispano cannon, plus two bombs of up to 1,000-lb (454-kg) each, or rocket projectiles

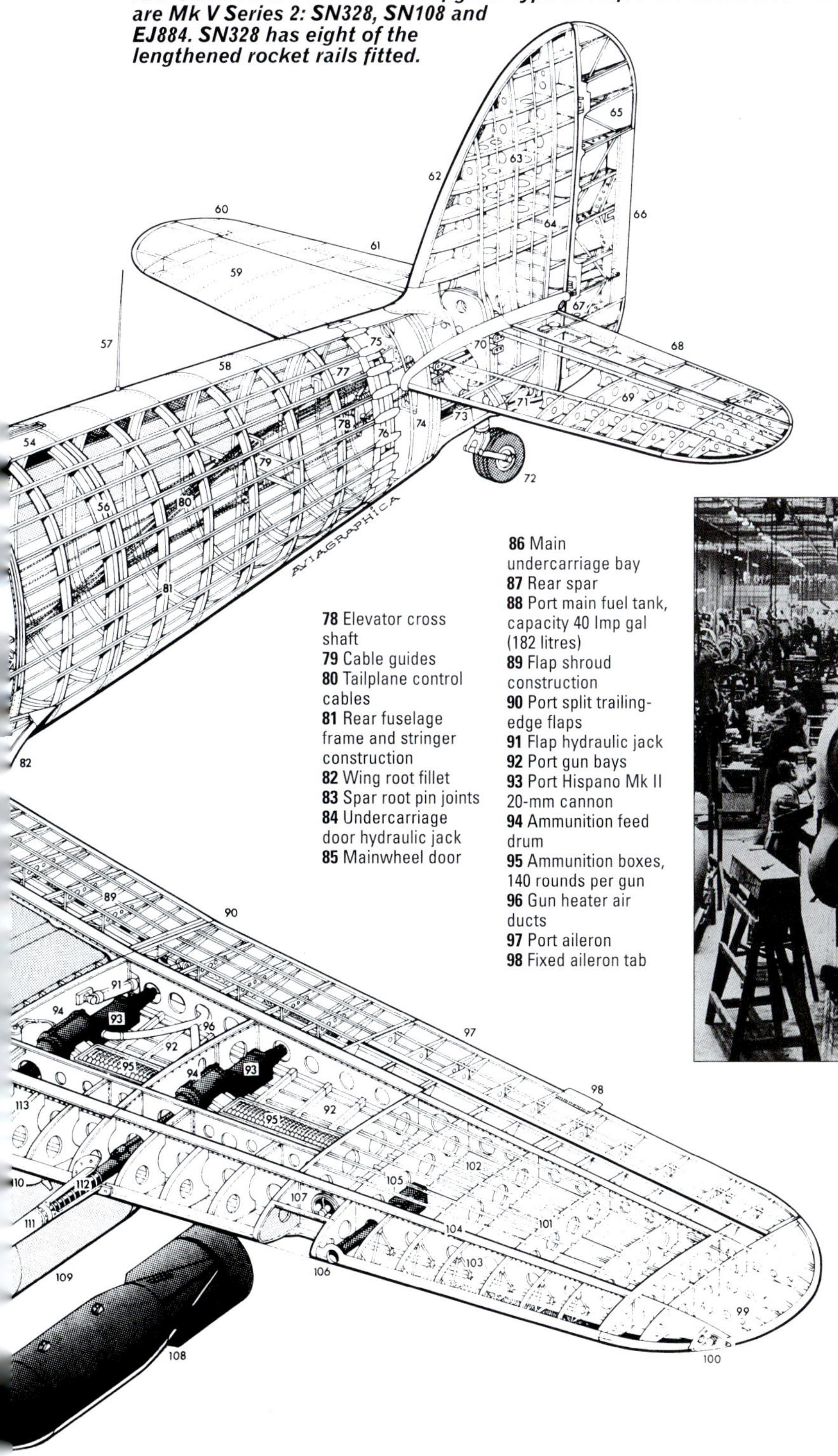

78 Elevator cross shaft
79 Cable guides
80 Tailplane control cables
81 Rear fuselage frame and stringer construction
82 Wing root fillet
83 Spar root pin joints
84 Undercarriage door hydraulic jack
85 Mainwheel door
86 Main undercarriage bay
87 Rear spar
88 Port main fuel tank, capacity 40 Imp gal (182 litres)
89 Flap shroud construction
90 Port split trailing-edge flaps
91 Flap hydraulic jack
92 Port gun bays
93 Port Hispano Mk II 20-mm cannon
94 Ammunition feed drum
95 Ammunition boxes, 140 rounds per gun
96 Gun heater air ducts
97 Port aileron
98 Fixed aileron tab
99 Wingtip construction
100 Port navigation light
101 Wing rib construction
102 Wing stringers
103 Front spar
104 Leading-edge nose ribs
105 Gun camera
106 Camera port
107 Landing lamp
108 1,000-lb (454-kg) bomb
109 Long-range tank, capacity 90 Imp gal (409 litres)
110 Underwing stores pylon
111 Cannon barrel fairings
112 Recoil spring
113 Leading-edge construction
114 Main undercarriage leg
115 Undercarriage leg fairing door
116 Oleo-pneumatic shock absorber strut
117 Port mainwheel
118 Undercarriage locking mechanism
119 Mainwheel hydraulic jack
120 Wing spar inboard girder construction
121 Port leading-edge fuel tank, capacity 35 Imp gal (159 litres)

Hawker could not handle the mass-production of the Typhoon on top of its continuing commitment to making the Hurricane, so the Typhoon was assigned to Gloster, a sister-firm in the Hawker-Siddeley group. Clearly visible in this view are the armoured backplates to the spinner. The delivery of Typhoons to the RAF was hindered by strikes when the Gloster workers tried to obtain a pay rise because of the complex construction methods used in making the new aircraft.

Typhoon Mk IB

Pictured unleashing a pair of rockets against a ground target in the Normandy area during August 1944, this Typhoon operated with No. 175 Squadron as part of the 2nd Tactical Air Force. Known as 'The Seven-Ton Brute', the Typhoon excelled in the ground-attack role, not only striking enemy airfields, but also lines of communication, river and canal barges and trains and locomotives. Often operating in pairs on missions known as 'Rhubarbs', Typhoon pilots were simply given the order to attack targets of opportunity, if not required by ground forces to provide specific close-air support missions. During these sorties pilots often showed great initiative in the bid to inflict maximum damage on the Third Reich. Perhaps one of the most sensational attacks was by No. 609 and 198 Squadrons on Rommel's headquarters in a château on the Cherbourg peninsula on the eve of D-Day. Rommel was in charge of the German defences against the coming invasion, but had left the château shortly before the Typhoons swept in. Had he been in residence at the time, his chances of survival would have been slight. However, a few weeks later in July Allied aircraft, believed to be Typhoons of No. 193 Squadron, attacked Rommel's staff car, wounding him and ending his military career.

Powerplant
Bias against radial engines ensured that the best engine for the Typhoon, the Bristol Centaurus, was not selected. Instead prototypes were ordered with the Rolls-Royce Vulture (to be known as the Hawker Tornado) and the Napier Sabre (to be known as the Hawker Typhoon). The Vulture engine proved unreliable beyond redemption and was cancelled, and the Sabre-engined Typhoon entered service almost by default. The Napier Sabre IIA was a 24-cylinder, inline water/glycol cooled monster, with sleeve valves. Power output was 2,180 hp (1624 kW) at sea level and 1,830 hp (1363 kW) at 11,500 ft (3500 m). Power from the Sabre was transmitted by a three-bladed de Havilland Hydromatic constant-speed propeller, although later a four-bladed unit was occasionally fitted. Both units had a diameter of 14 ft (4.27 m).

Canopy, oil tank and aerial
Early Typhoons featured a heavily-framed canopy with a car-type outward-hinging door manufactured by Rover. Complaints from pilots about the lack of visibility particularly to the rear, led to the development of a one-piece teardrop sliding canopy, with much improved vision. In addition the solid radio mast of early versions was replaced by a whip aerial further back on the spine. By the end of 1943 most Typhoons had this important modification, as did all the Tempests. Housed in the upper fuselage slightly ahead of the cockpit was the oil tank for the Sabre engine. Since it had a capacity of 18 Imp gal (46.6 litres) many pilots were initially wary that ground-fire would set the tank alight. However, additional armour plating fitted around this area relieved any fears pilots held.

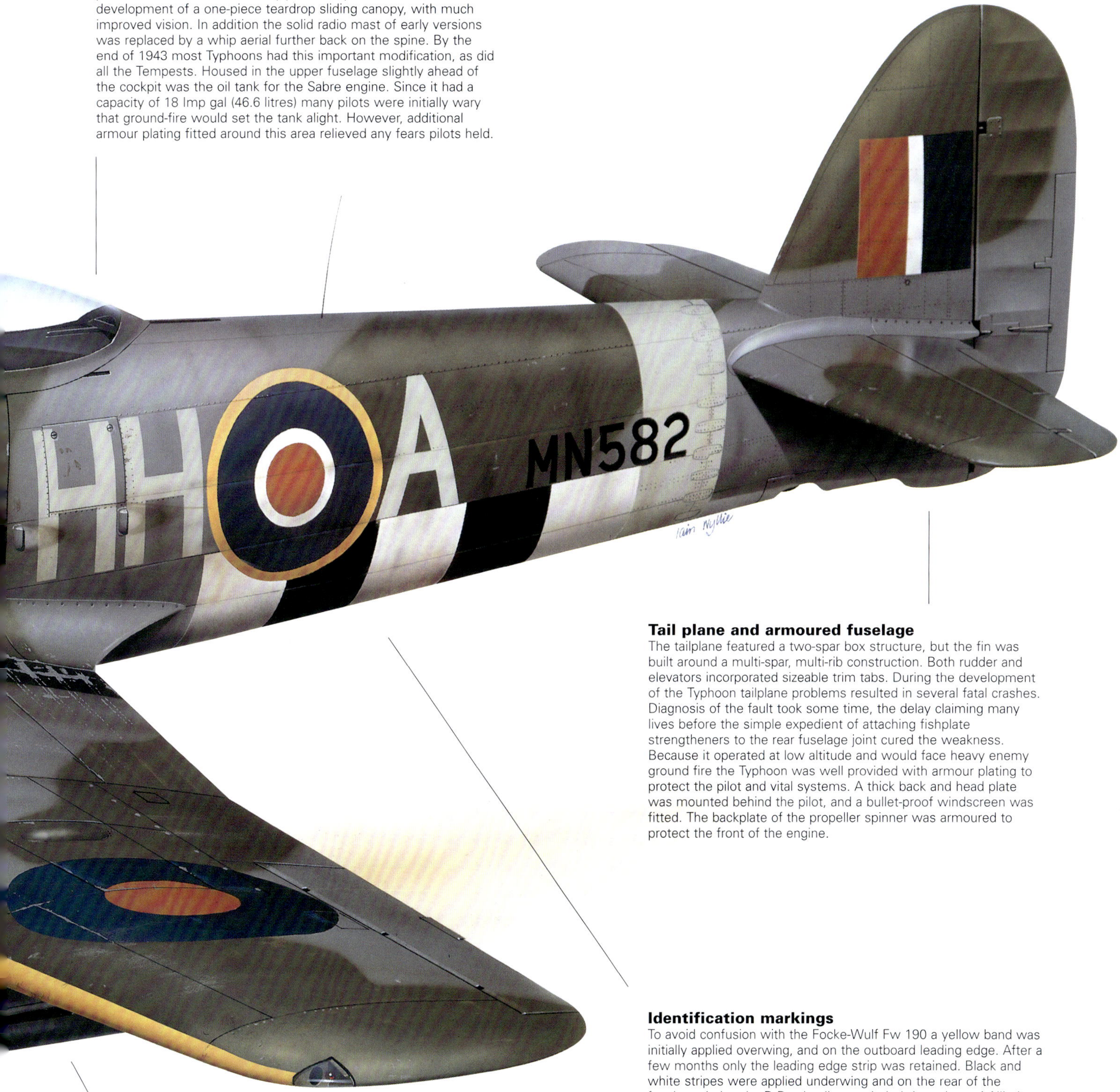

Tail plane and armoured fuselage
The tailplane featured a two-spar box structure, but the fin was built around a multi-spar, multi-rib construction. Both rudder and elevators incorporated sizeable trim tabs. During the development of the Typhoon tailplane problems resulted in several fatal crashes. Diagnosis of the fault took some time, the delay claiming many lives before the simple expedient of attaching fishplate strengtheners to the rear fuselage joint cured the weakness. Because it operated at low altitude and would face heavy enemy ground fire the Typhoon was well provided with armour plating to protect the pilot and vital systems. A thick back and head plate was mounted behind the pilot, and a bullet-proof windscreen was fitted. The backplate of the propeller spinner was armoured to protect the front of the engine.

Identification markings
To avoid confusion with the Focke-Wulf Fw 190 a yellow band was initially applied overwing, and on the outboard leading edge. After a few months only the leading edge strip was retained. Black and white stripes were applied underwing and on the rear of the fuselage during the D-Day landings to help 'trigger-happy' Allied infantry units to identify friendly aircraft.

Underwing pylons
For bombing missions, the Typhoon could lift a pair of 1,000-lb (454-kg) bombs on underwing racks. Alternatively, the racks could carry 45- or 90-Imp gal (205- or 410-litre) drop tanks for extended range, although from June 1944 onwards most missions were short-range attacks close in front of friendly lines. For these sorts of missions rockets were used. Depicted here is a Typhoon from No. 175 Squadron, which worked up with rockets during the spring of 1944 and proved its expertise with this devastating weapon throughout the final year of the war. Eight Rocket Projectiles (RPs) could be carried by the Typhoon, each a 3-in (76-mm) rocket with a 60-lb (27-kg) warhead. These proved more than enough to penetrate the armour of German battle tanks.

Heinkel He 162 Salamander

This He 162A-2 was allocated to 3. Staffel/JG 1 at its Leck base in May 1945. By this time, the 50 aircraft had been reorganised into one single Gruppe, Einsatz-Gruppe I./JG 1; many pilots from other fragmented units at Leck were absorbed by this new Gruppe.

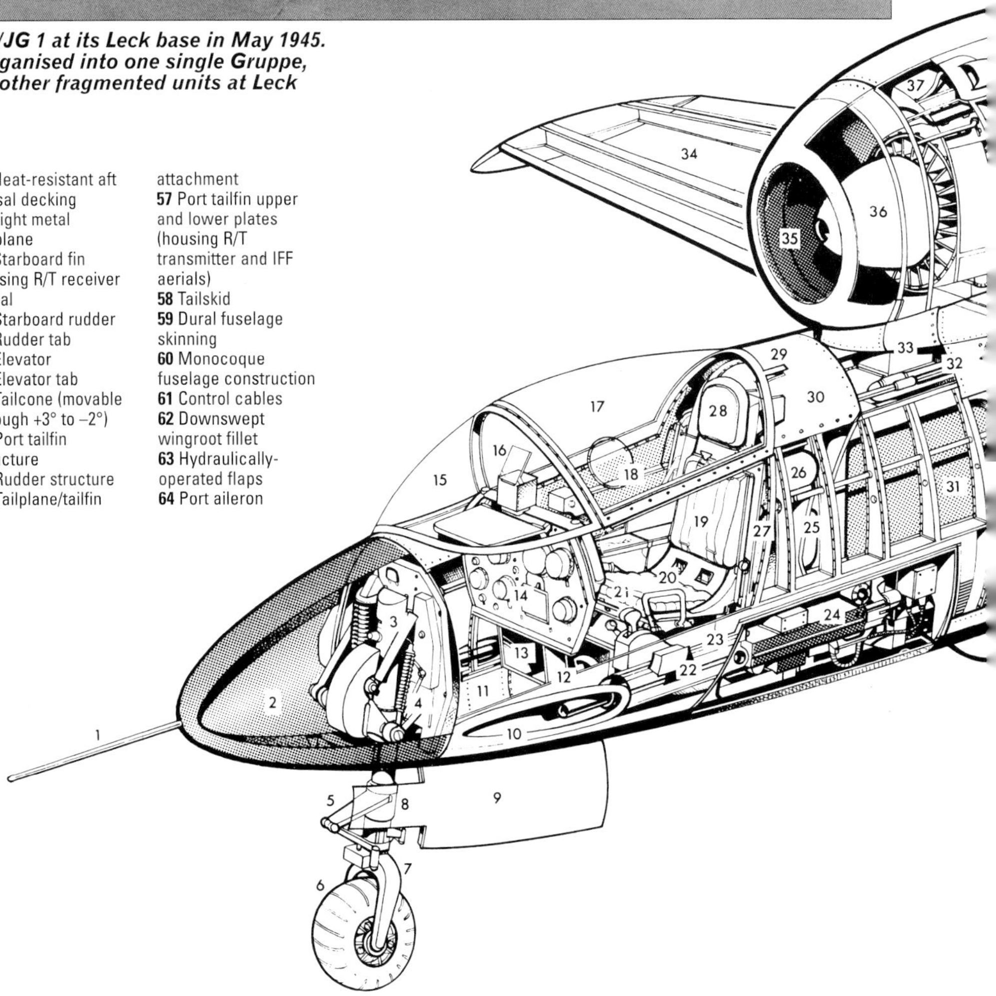

He 162

Cutaway key
1 Pitot tube
2 Moulded plywood nose cap
3 Nosewheel retraction mechanism
4 Spring-loaded nosewheel extension assembly
5 Shock absorber scissor
6 Nosewheel
7 Nosewheel fork
8 Nosewheel leg
9 Nosewheel door
10 Gun trough
11 Nosewheel well
12 Rudder pedal
13 Window panel (visual nosewheel retraction check)
14 Wooden instrument panel
15 One-piece moulded windscreen
16 Revi 16G gunsight (interchangeable with the Revi 16B)
17 Jettisonable hinged clear-vision canopy
18 Ventilation disc
19 Heinkel cartridge-operated ejection seat
20 Ejection seat handle grip
21 Throttle control quadrant
22 Retractable entry step
23 Gun barrel shroud in cockpit wall
24 Port 20-mm MG 151 cannon
25 Ammunition chute
26 Main oxygen supply bottle (3.5-pint/2-litre capacity)
27 Explosive charge ejector rail
28 Pilot's headrest
29 Canopy hinge
30 Ammunition box behind cockpit (120 rounds per gun)
31 Flexible main tank (153-Imp gal/695 litre capacity)
32 Fuel lines
33 FuG 25a IFF radio compartment
34 Beech plywood wing skinning
35 Jet intake
36 Riedel two-stroke starter motor bullet
37 Oil tank
38 BMW 003E-1 Sturm axial-flow turbojet
39 Auxiliary intake
40 Seven-stage axial compressor casing
41 FuG 24 R/T homing loop
42 Annular combustion chamber
43 Exhaust centre body
44 Exhaust outlet
45 Jet efflux fairing
46 Heat-resistant aft dorsal decking
47 Light metal tailplane
48 Starboard fin housing R/T receiver aerial
49 Starboard rudder
50 Rudder tab
51 Elevator
52 Elevator tab
53 Tailcone (movable through +3° to –2°)
54 Port tailfin structure
55 Rudder structure
56 Tailplane/tailfin attachment
57 Port tailfin upper and lower plates (housing R/T transmitter and IFF aerials)
58 Tailskid
59 Dural fuselage skinning
60 Monocoque fuselage construction
61 Control cables
62 Downswept wingroot fillet
63 Hydraulically-operated flaps
64 Port aileron

Considerable numbers of Heinkel He 162s were captured by the Allies in varying degrees of repair and some still exist in museums. This aircraft was flown on 26 flights by the British RAE at Farnborough. It was restored for display at RAF St Athan.

SPECIFICATION

Heinkel He 162A-2 Salamander

Type

Single-seat fighter

Powerplant

One BMW 003A-1 turbojet engine rated at 1,764 lb st (7.85 kN)

Performance

Maximum speed 522 mph (840 km/h) at 19,685 ft (6000 m); initial climb rate 3,780 ft (1152 m) per minute; service ceiling 39,370 ft (12000 m); endurance 57 minutes at 35,990 ft (10970 m)

Weights

Empty 4,520 lb (2050 kg); maximum take-off 5,941 lb (2695 kg)

Dimensions

Wingspan 23 ft 7½ in (7.20 m); length 29 ft 8⅓ in (9.05 m); height 8 ft 6⅓ in (2.60 m); wing area 120.56 sq ft (11.20 m2)
Armament: two 20-mm MG 151/20 fixed forward-firing cannon in the underside of the nose

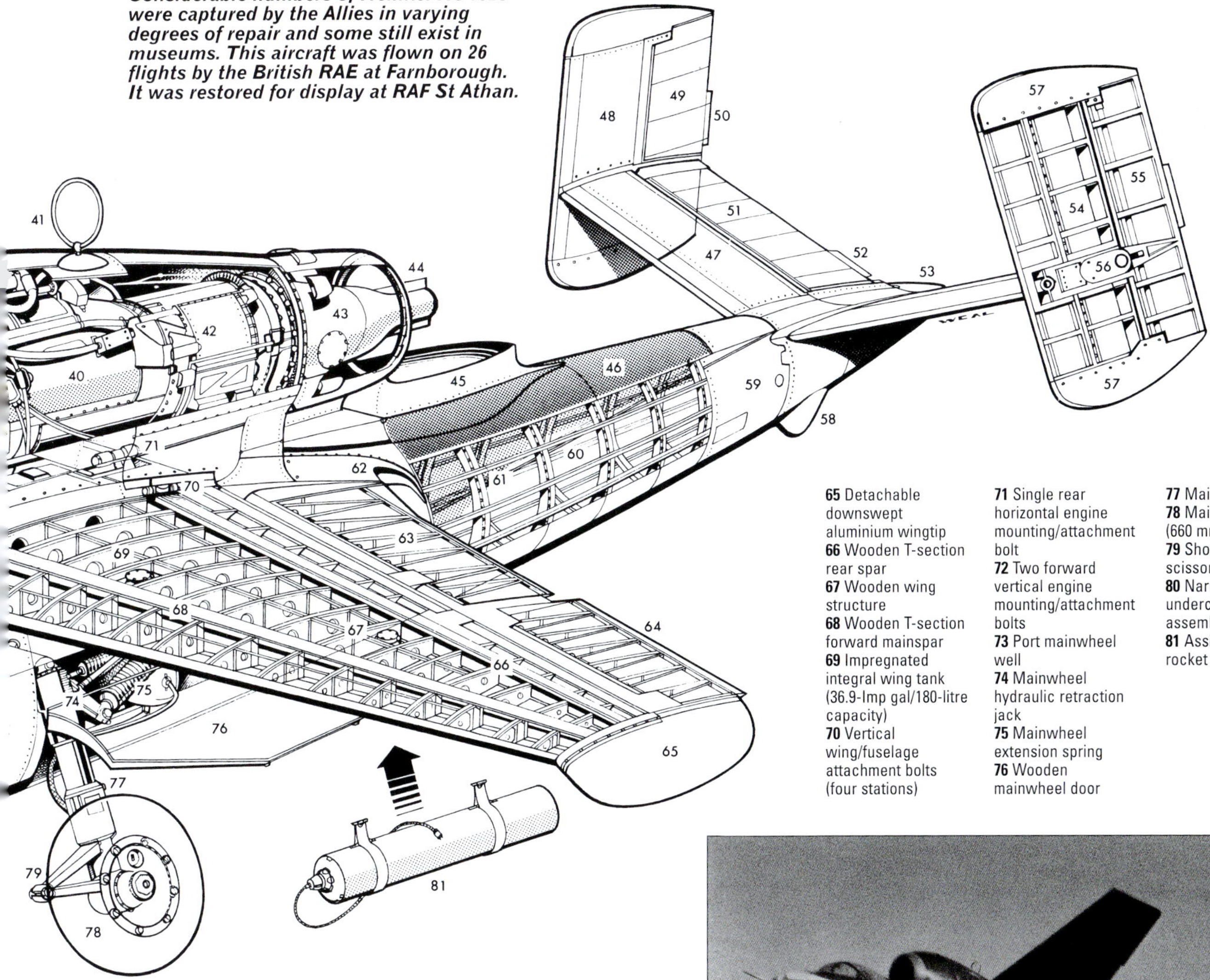

65 Detachable downswept aluminium wingtip
66 Wooden T-section rear spar
67 Wooden wing structure
68 Wooden T-section forward mainspar
69 Impregnated integral wing tank (36.9-Imp gal/180-litre capacity)
70 Vertical wing/fuselage attachment bolts (four stations)
71 Single rear horizontal engine mounting/attachment bolt
72 Two forward vertical engine mounting/attachment bolts
73 Port mainwheel well
74 Mainwheel hydraulic retraction jack
75 Mainwheel extension spring
76 Wooden mainwheel door
77 Mainwheel leg
78 Mainwheel tyre (660 mm x 190 mm)
79 Shock absorber scissor
80 Narrow-track main undercarriage assembly
81 Assisted-take-off rocket unit

An He 162A-1 shows the turned-down wingtips first introduced on the V3 to overcome excessive dihedral problems. The A-1s were manufactured in parallel with the prototypes, the latter being regarded as A-0 pre-production airframes.

Ilyushin Il-2/10 'Bark'/'Beast'

As an attack aircraft more feared even than the German Stuka or the British Typhoon, the Il-2 was one of the world's all-time great military aircraft. In terms of numbers built, it is unsurpassed. An Il-2M3 is illustrated.

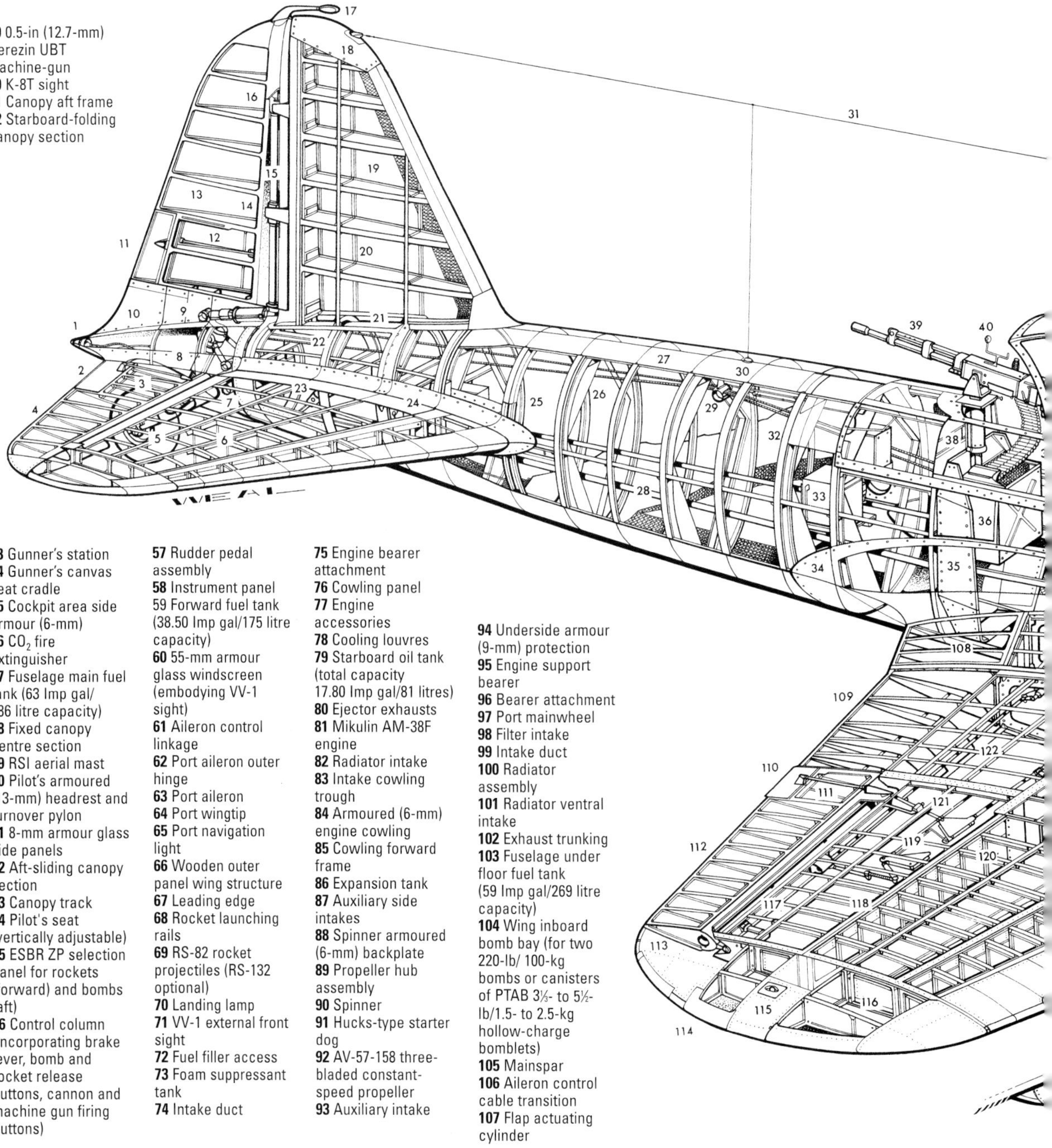

Il-2M3

Cutaway key
1 Rear navigation light
2 Elevator tab
3 Elevator tab hinge
4 Starboard elevator frame
5 Non-retractable tailwheel
6 Wooden tailplane structure
7 Tailwheel leg assembly
8 Elevator torque tube
9 Tailwheel shock-absorber
10 Tail cone fairing
11 Rudder tab
12 Rudder tab actuating rod
13 Rudder frame
14 Rudder hinge
15 Rudder post
16 Rudder upper hinge
17 Rudder mass balance
18 Aerial attachment
19 Wooden tail fin structure
20 Tail fin spars
21 Rudder control cables
22 Fuselage aft frames
23 Elevator internal mass balance
24 Tailplane spar/ fuselage attachment
25 Fuselage intermediate frame
26 Strengthened wooden frame
27 Plywood skinning
28 Elevator control rods
29 Rudder control cable pulleys
30 Aerial lead-in
31 Aerial
32 Fuselage frame
33 Provision for camera (portside window) or DAG-10 grenade launcher
34 Wingroot fairing
35 Armour protection (13-mm) for rear gunner
36 Ammunition tank (150 rounds)
37 Ammunition feed
38 Gun mounting post
39 0.5-in (12.7-mm) Berezin UBT machine-gun
40 K-8T sight
41 Canopy aft frame
42 Starboard-folding canopy section
43 Gunner's station
44 Gunner's canvas seat cradle
45 Cockpit area side armour (6-mm)
46 CO_2 fire extinguisher
47 Fuselage main fuel tank (63 Imp gal/ 286 litre capacity)
48 Fixed canopy centre section
49 RSI aerial mast
50 Pilot's armoured (13-mm) headrest and turnover pylon
51 8-mm armour glass side panels
52 Aft-sliding canopy section
53 Canopy track
54 Pilot's seat (vertically adjustable)
55 ESBR ZP selection panel for rockets (forward) and bombs (aft)
56 Control column (incorporating brake lever, bomb and rocket release buttons, cannon and machine gun firing buttons)
57 Rudder pedal assembly
58 Instrument panel
59 Forward fuel tank (38.50 Imp gal/175 litre capacity)
60 55-mm armour glass windscreen (embodying VV-1 sight)
61 Aileron control linkage
62 Port aileron outer hinge
63 Port aileron
64 Port wingtip
65 Port navigation light
66 Wooden outer panel wing structure
67 Leading edge
68 Rocket launching rails
69 RS-82 rocket projectiles (RS-132 optional)
70 Landing lamp
71 VV-1 external front sight
72 Fuel filler access
73 Foam suppressant tank
74 Intake duct
75 Engine bearer attachment
76 Cowling panel
77 Engine accessories
78 Cooling louvres
79 Starboard oil tank (total capacity 17.80 Imp gal/81 litres)
80 Ejector exhausts
81 Mikulin AM-38F engine
82 Radiator intake
83 Intake cowling trough
84 Armoured (6-mm) engine cowling
85 Cowling forward frame
86 Expansion tank
87 Auxiliary side intakes
88 Spinner armoured (6-mm) backplate
89 Propeller hub assembly
90 Spinner
91 Hucks-type starter dog
92 AV-57-158 three-bladed constant-speed propeller
93 Auxiliary intake
94 Underside armour (9-mm) protection
95 Engine support bearer
96 Bearer attachment
97 Port mainwheel
98 Filter intake
99 Intake duct
100 Radiator assembly
101 Radiator ventral intake
102 Exhaust trunking
103 Fuselage under floor fuel tank (59 Imp gal/269 litre capacity)
104 Wing inboard bomb bay (for two 220-lb/ 100-kg bombs or canisters of PTAB 3½- to 5½-lb/1.5- to 2.5-kg hollow-charge bomblets)
105 Mainspar
106 Aileron control cable transition
107 Flap actuating cylinder

This trio of Shturmoviks was photographed on the Voronezh Front during the Kursk campaign in July 1943. During this campaign, as with all the others in which the type was engaged, the Il-2 proved devastating against German armour. Note the 37-mm cannon being fired in this view of an attack.

SPECIFICATION

Il-2M3

Dimensions

Length: 38 ft 2½ in (11.65 m)
Height: 13 ft 8 in (4.17 m)
Wingspan: 47 ft 10¾ in (14.60 m)
Wing area: 414.42 sq ft (38.50 m²)

Powerplant

One 1,720-hp (1282-kW) Mikulin AM-38F piston engine

Weights

Empty: 9,976 lb (4525 kg)
Empty (with NS-37 cannon): 10,196 lb (4625 kg)
Loaded: 14,021 lb (6360 kg)
Loaded (with NS-37 cannon): 13,580 lb (6160 kg)

Fuel

Internal: 1,179 lb (535 kg)

Range

Normal: 475 miles (765 km)

Performance

Maximum speed at sea level: 242 mph (390 km/h)
Maximum speed at 4,920 ft (1500 m): 255 mph (410 km/h)
Climb to 3,280 ft (1000 m): 2 minutes 12 seconds
Climb to 9,843 ft (3000 m): 6 minutes 54 seconds
Service ceiling: 19,685 ft (6000 m)
Take-off run: 1,296 ft (395 m)
Landing run at 90 mph (145 km/h): 1,755 ft (535 m)

Armament

Two 23-mm VYa cannon and two 0.3-in (7.62-mm) ShKAS machine-guns (all wing-mounted) and one 0.5-in (12.7-mm) UBT machine-gun for the gunner, plus 220-lb (100-kg) bombs (four carried internally and two under the fuselage), or two 551-lb (250-kg) bombs (under the fuselage), eight RS-82 rockets or four RS-132 rockets under the outer wing panels

108 Starboard flap inboard section
109 Flap outboard section profile
110 Aileron tab
111 Tab hinge
112 Starboard aileron
113 Aileron profile
114 Starboard wingtip assembly
115 Starboard navigation light
116 Wing outer rib
117 Aileron outer hinge
118 Wooden wing structure
119 Flap control linkage
120 Wing ribs
121 Aileron tab control
122 Flap rods
123 Wing centre/outer section joint/capping strake
124 Wing attachment points
125 Undercarriage retraction cylinder
126 Forward extremity of weapons bay
127 ShKAS 0.3-in (7.62-mm) machine gun
128 Machine-gun barrel support
129 Cannon barrel fairing
130 Volkov-Yartsev VYa 23 23-mm cannon
131 Ammunition access panel
132 Magazine for starboard VYa-23 cannon (150 rounds)

These aircraft represent the Il-2M3 in its ultimate form. All have 37-mm cannon, and the second aircraft at least, has yet to fire its underwing RS-82 rockets. All four of these machines have similar canopy shapes, but at least five main variations on canopy design have been noted, along with a range of minor changes. Such differences resulted from variations during manufacture and from local field modifications.

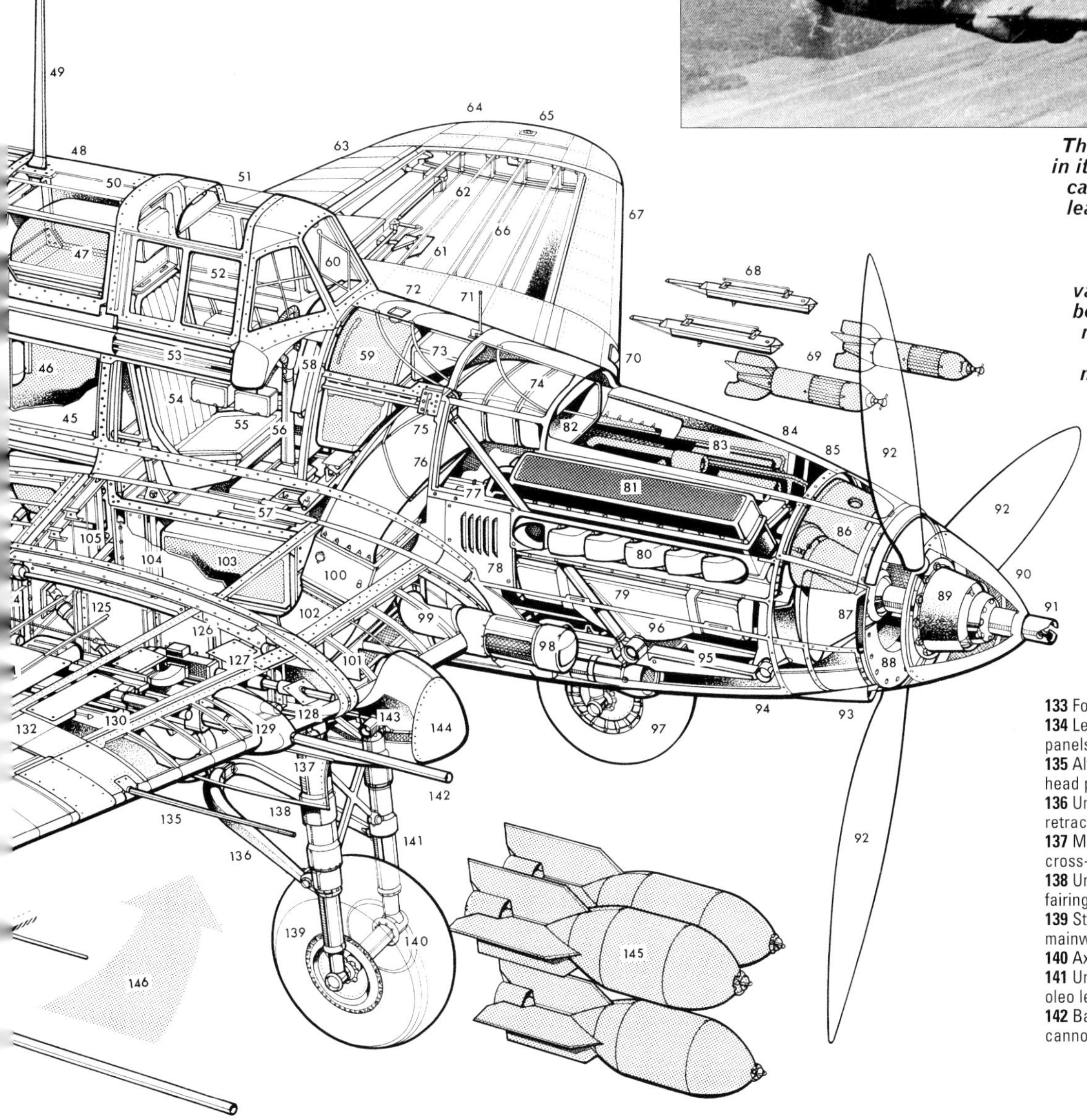

133 Forward spar
134 Leading edge panels
135 Alternative pitot head positions
136 Undercarriage retraction linkage
137 Mainwheel leg cross-brace
138 Undercarriage fairing doors
139 Starboard mainwheel
140 Axle
141 Undercarriage oleo legs
142 Barrel of VYa-23 cannon
143 Undercarriage pivot point
144 Fairing nose section
145 Starboard bomb load comprising three (two internal and one external) 220-lb (100-kg) bombs which could be released singly or in salvo
146 Alternative main armament (Il-2 Type M3) of 37-mm Nudelman-Suranov NS-11-37 (NSK-OKB-16) cannon

Il-2M3

This aircraft flew with the 566th Shturmovik Aviation Squadron of the 277th Shturmovik Aviation Division, Frontal Aviation, Soviet Air Forces. It is illustrated as it appeared while in action on the Leningrad Front in the Summer of 1944, while being flown by the squadron commander Lt V. I. Mykhlik. The slogan to the rear of the fuselage star read 'Revenge for Khristenko', while that in front stated 'For Leningrad'. Khristenko had been a pilot on the squadron, but was killed earlier in 1944, while Mykhlik was awarded two Hero of the Soviet Union awards and survived to finish the war as a Captain.

Rear fuselage structure

From 1944, the Il-2M3 was produced with an all-Dural fuselage structure. Prior to this however, the rear fuselage of earlier models, including the Il-2 and Il-2M, had been a wooden monocoque structure. This created many problems early in the type's service career, with glue often failing where sections were jointed and, indeed, the whole rear fuselage being prone to failure. These problems were cured by A. K. Belyenkov, a repair engineer who devised a fix employing four steel strengthening pieces and which later became standard in the composite-structure production aircraft. The rear fuselage contained a battery, the flap actuation equipment, oxygen bottles, a pneumatic bottle and the radio system.

Flying surfaces

On the earliest production Il-2s, the fin was of wooden construction, while the tailplanes, wings and all control surfaces were generally of all-metal construction. There were exceptions however, with a number of basic Il-2 aircraft having wooden outer wing and tailplane structures. The leading edges of the outer wing panels of the Il-2 and Il-2M were swept at 5° in order to improve longitudinal stability. In the Il-2/AM-38F (which preceded the ultimate late-production Il-2m3), this sweep angle was increased to 15° with a view to decreasing drag, and aircraft with this revised wing were first in action over Stalingrad in January 1943. The wing centre section had two spars, with a box in between used to accommodate bombs. Two bomb cells were created in each wing, each covered by a pair of doors that were opened manually by the pilot for bomb release, and typically carrying a 100-kg (220-lb) FAB-100 high-explosive general purpose bomb.

Armour protection

Armour plate of 4-, 5-, 6-, 8- and 12-mm thickness was used to protect the Il-2's crew and vital systems. The engine was protected by 4-mm thick armour at the front and 5-mm material covered the lower and under sides. Other armour included a 7-mm thick section protecting the bulkhead behind the rear crew member. In order to minimise the weight penalties inherent in adding armour to an aircraft, Ilyushin, along with Mikoyan-Gurevich and Sukhoi, was working on methods of making the armour part of the load-bearing structure of the airframe. In the case of the Il-2, this was only possible around the engine and all three manufacturers suffered grevious problems in making accurate sections of armour. Bolt and rivet holes frequently failed to match up on armour sections that were supposed to be joined together and this problem persisted until an armour-welding process was perfected for use on the Il-2M and subsequent models.

Passengers
It was not uncommon for the Il-2 to carry passengers. These were normally carried only in times of urgent need, where a normal transport might be too vulnerable, and were normally accommodated lying on the floor of the gunners cockpit. Many downed crews were rescued from crash sites by their fellow Il-2 flyers, sometimes by using the lying in the cockpit method, but also by securing themselves to the main undercarriage units, which were then kept extended for the return to base.

Tactics
The basic Il-2 operating group consisted of 4-12 aircraft and an attack might involve several groups. Standard echelon, line abreast, line astern and vic formations were employed, although the favourite attack formation was the 'circle of death'. Approached in echelon right, the 'circle of death' was flown with the aircraft in line astern and spaced at 1,970-ft (600-m) intervals. The circle was off-set to one side of the target, the aircraft attacking from behind and making attack runs in turn until all ammunition had been expended. This tactic frequently kept the enemy under constant attack by at least one aircraft for up to 30 minutes. Bombs were often delivered in steep diving attacks, while for attacks against emplacements, infantry and vehicles in the open, Il-2s would often run in to the target at an altitude between 15 and 30 ft (4.57 and 9.00 m), firing their fixed armament and rockets horizontally.

Weapons
The rockets leaving the underwing racks of this machine are twin-finned, 51-lb (23.10-kg) RS-132 weapons. These rockets were 36⅘ in (93.50-cm) long and were common Il-2 rockets along with the four-finned RS-82, although RS-132 compatibility was not introduced until mid-1941. Early Il-2s could only carry six such rockets, while this was later upgraded to eight. The Il-2/AM-38 introduced a new two-tier launch rack for the RS-82, which allowed up to 32 RS-82s to be carried. These weapons were normally aimed using either the PBP-1 optical sight or the less-complicated VV-1 sight, although crews also made markings on the windscreen and canopy to provide references for gun and rocket firing and for bombing. As well as conventional high-explosive bombs, the Il-2 frequently employed cluster weapons, in the form of canisters containing 200 3⅓- to 5½-lb (1.5- to 2.5-kg) hollow-charge PTAB bomblets for use against armour. Other options included the unusual DAG-10 grenade launcher, which released infantry grenades on small parachutes, hopefully into the path of any pursuing fighters. Fixed armament originally consisted of four forward-firing machine-guns in the form of two ShVAK and two ShKAS weapons in the wings. From mid-1941, the 20-mm ShVAK weapons were replaced by a pair of 23-mm calibre VYa-23 guns and with the introduction of the second crew member in the Il-2M, a 0.5-in (12.7-mm) UBT machine-gun was added on a flexible mount in the rear cockpit. The ultimate Il-2 gun configuration was displayed by a number of Il-2M3s, which featured two ShVAKs, the UBT and a pair of 37-mm NS-OKB-16 cannon. Being of such a large calibre, the NS-OKB-16s were restricted to just 50 rounds each, but these were capable of penetrating the armour of the formidable PzKW VI Tiger tank.

Kawasaki Ki-61 Hien

This Ki-61-I of the 37th Sentai was among those that fought in the last stages of the defence of the Philippines, before being forced to redeploy to Formosa and Okinawa in the last year of the war.

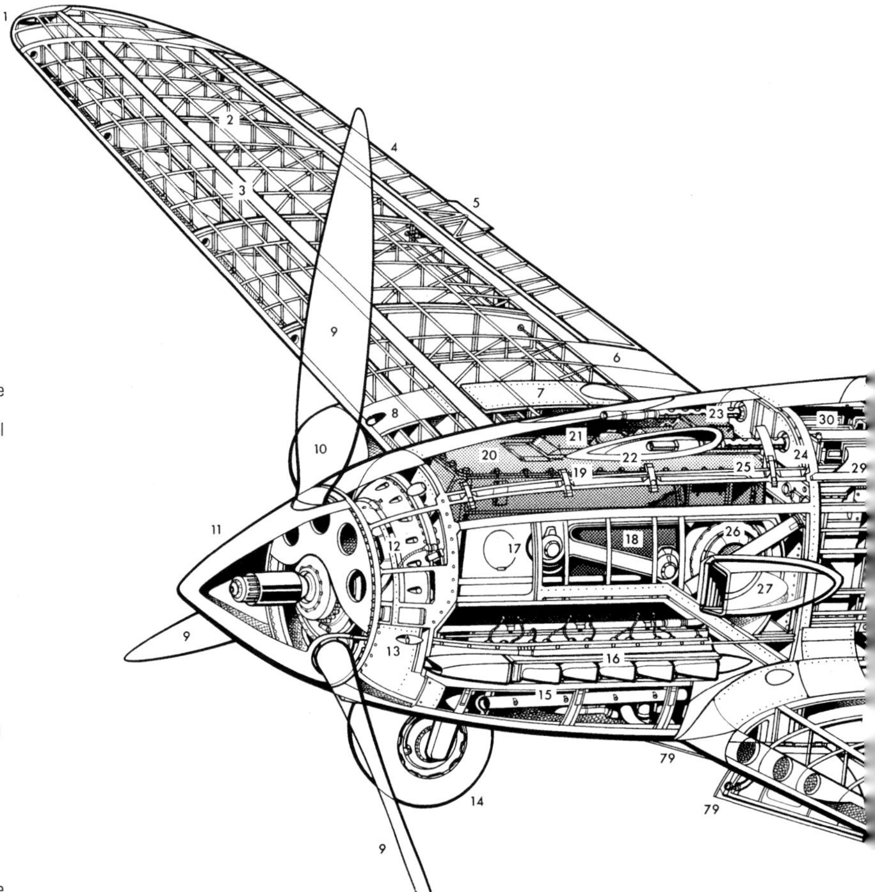

Kawasaki Ki-61

Cutaway key
1 Starboard navigation light
2 Wing rib bracing
3 Wing spar
4 Starboard aileron
5 Aileron tab
6 Starboard flap
7 Wing gun access panel
8 Gun port
9 Three-blade constant-speed propeller
10 Auxiliary drop-tank (43.9 Imp gal/200 litres)
11 Propeller boss
12 Propeller reduction gear housing
13 Air intake duct
14 Starboard mainwheel
15 Lower cowling quick-release catches
16 Exhaust stubs
17 Anti-vibration mounting pad
18 Engine bearer
19 Upper cowling quick-release catches
20 Kawasaki Ha-40 (Army Type 2) engine
21 Engine accessories
22 Gun port
23 Cannon barrels
24 Firewall
25 Cowling panel line
26 Supercharger
27 Supercharger intake
28 Ammunition tanks
29 Ammunition feed chute
30 Two 20-mm Ho-5 cannon
31 Sloping windscreen
32 Gunsight
33 Control column
34 Pilot's seat (armoured)
35 Fuselage frame
36 Rearward-sliding cockpit
37 Pilot's headrest
38 Rear-vision cut-out
39 Aft glazing
40 Canopy track
41 Spring-loaded handhold
42 Fuselage fuel tank (36.2 Imp gal/165 litres)
43 Fuselage equipment access door (upward hinged)
44 Radio pack (Type 99-111)
45 Aerial mast
46 Aerial lead-in
47 Aerial
48 Elevator control cables
49 Upper longeron
50 Rudder cable
51 Fuselage join
52 Starboard tailplane
53 Starboard elevator
54 Tailfin root fairing
55 Tailfin structure
56 Rear navigation light (port and starboard)
57 Aerial stub mast
58 Rudder balance
59 Rudder fixed trim tab
60 Rudder post
61 Rudder framework
62 Elevator tab
63 Elevator fixed trim tab
64 Port elevator
65 Elevator control cable
66 Rudder hinge
67 Rear fuselage frame/tailplane attachment
68 Tailwheel retraction jack
69 Tailwheel doors
70 Retractable tailwheel
71 Tailwheel shock absorber oleo
72 Lower longeron
73 Radiation bath air outlet
74 Adjustable gill
75 Radiator
76 Radiator intake ducting
77 Intake
78 Main spar/fuselage attachment point
79 Inboard mainwheel doors
80 Mainwheel well
81 Landing light
82 Mainwheel pivot point
83 Mainwheel leg
84 Oleo shock-absorber section (leather-sleeved)
85 Mainwheel single fork
86 Port mainwheel
87 Mainwheel door
88 Separate mainwheel leg fairing
89 Gun port
90 Machine gun barrel
91 Wing-mounted 12.7-mm Ho-103 machine-gun
92 Gun access panel
93 Bomb/tank shackle
94 Port flap
95 Main spar
96 Wing ribs
97 Auxiliary drop-tank (43.9 Imp gal/200 litres)
98 Pitot head
99 Metal wing skin
100 Aileron tab
101 Port aileron
102 Wingtip structure
103 Port navigation light

By Western standards the Japanese 53-US gal (200-litre) drop tank carried by all operational versions of the Ki-61 was a crude store, which reduced its maximum speed by around 50 mph (80 km/h) but increased the range of the Ki-61-II KAIa from 684 miles (1100 km) to 995 miles (1600 km).

SPECIFICATION

Kawasaki Ki-61-Ic

Type
Single-seat fighter

Powerplant
One 1,175-hp (876-kW) Kawasaki Ha-40 V-12 piston engine

Performance:
Maximum speed 348 mph (560 km/h); service ceiling 32,810 ft (10000 m); maximum range 1,181 miles (1900 km)

Weights
Empty 5,798 lb (2630 kg); maximum take-off 7,650 lb (3470 kg)

Dimensions
Wingspan 39 ft 4 1/2 in (12.00 m); length 29 ft 4 1/4 in (8.95 m); height 12 ft 11 in (3.70 m); wing area 215.29 sq ft (20.00 m2)

Armament
As detailed under variants, plus provision for two drop tanks or two 551-lb (250-kg) bombs

Ki-61-KAIs of the Akeno Flying Training School, the main home-based Hien training unit, are run up before another training sortie. KAI variants had strengthened wings and cannon armament.

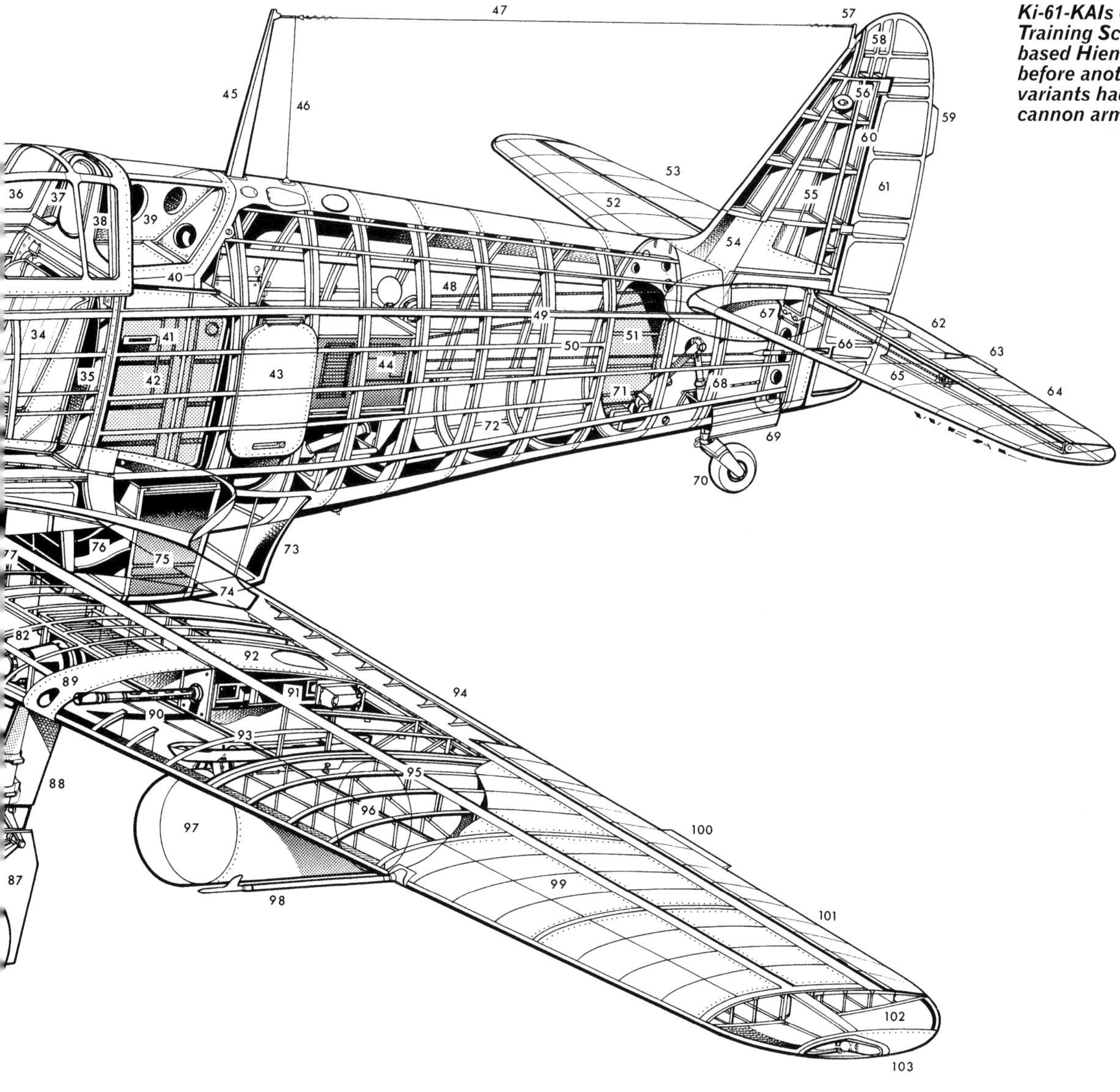

Lavochkin La-5/7

With the La-5 and La-7, Lavochkin created one of the most capable fighter families of World War II. Operating from a typically rough airstrip, this La-5 was engaged in combat against the Luftwaffe's Luftflotte 4 during the Battle of Kursk in the summer of 1943.

La-5FN

Cutaway key

1 Hucks-type starter dog
2 Spinner
3 Propeller balance
4 Controllable frontal intake louvres
5 VISh-105V metal propeller
6 Nose ring profile
7 Intake centrebody
8 ShVAK cannon port
9 Supercharger air intake
10 Supercharger intake trunk fairing
11 Blast tube
12 Shvetsov ASh-82FN 14-cylinder two-row radial
13 Cowling ring
14 Cowling panel hinge line
15 Exhaust pipes
16 Exhaust outlet cluster (seven per side)
17 Outlet cover panel
18 Engine accessories
19 Mainspar/fuselage attachment
20 Ammunition tanks (200 rpg)
21 Link and cartridge ejection chutes
22 Engine bearer upper support bracket
23 Cannon breech fairing
24 Paired 20-mm ShVAK cannon
25 Supercharger intake trunking
26 Stressed Bakelite-ply skinning
27 Automatic leading-edge slat (obliquely-operated)
28 Pitot head
29 Starboard navigation light
30 Wingtip
31 Dural-framed fabric-covered aileron
32 Aileron trim tab
33 Armoured glass windscreen (2 1⁄10 in/ 55 mm thick)
34 PBP-1A reflector gunsight
35 Cockpit air intake
36 Control column
37 Outlet louvres
38 Rudder pedal assembly
39 Underfloor control linkage
40 Rear spar/fuselage attachment
41 Rudder and elevator trim hand wheels
42 Seat height adjustment
43 Boost controls
44 Seat harness
45 Pilot's seat
46 Throttle quadrant
47 Hydraulics main valve
48 Aft-sliding cockpit canopy
49 Fixed aft transparent cockpit fairing
50 Armoured glass screen (2 3⁄5 in/66 mm thick)
51 Canopy track
52 RSI-4KhF HF R/T installation
53 Radio equipment shelf
54 Dural fuselage side panels
55 Control cables
56 Plywood-sheathed birch frames with triangular-section wooden stringers
57 Stressed Bakelite-ply skinning
58 Accumulator
59 Accumulator access panel
60 Tailfin front spar attachment
61 Aerial mast
62 Radio aerials
63 Starboard tailplane
64 Elevator hinge
65 Dural-framed fabric-covered elevator
66 Tailfin leading edge
67 Tailfin wooden structure (plywood skinning)
68 Aerial stub
69 Rudder balance
70 Rudder upper hinge
71 Dural-framed fabric-covered rudder
72 Rudder trim tab
73 Rear navigation light
74 Rudder centre hinge
75 Elevator control lever
76 Tailplane/fuselage attachment
77 Rudder control lever
78 Elevator trim tab
79 Dural-framed fabric-covered elevator
80 Wooden two-spar tailplane structure (plywood skinning)
81 Tailwheel doors
82 Aft-retracting tailwheel (sometimes locked in extended position)
83 Tailwheel leg
84 Tailwheel shock strut
85 Retraction mechanism
86 Stressed Bakelite-ply skinning

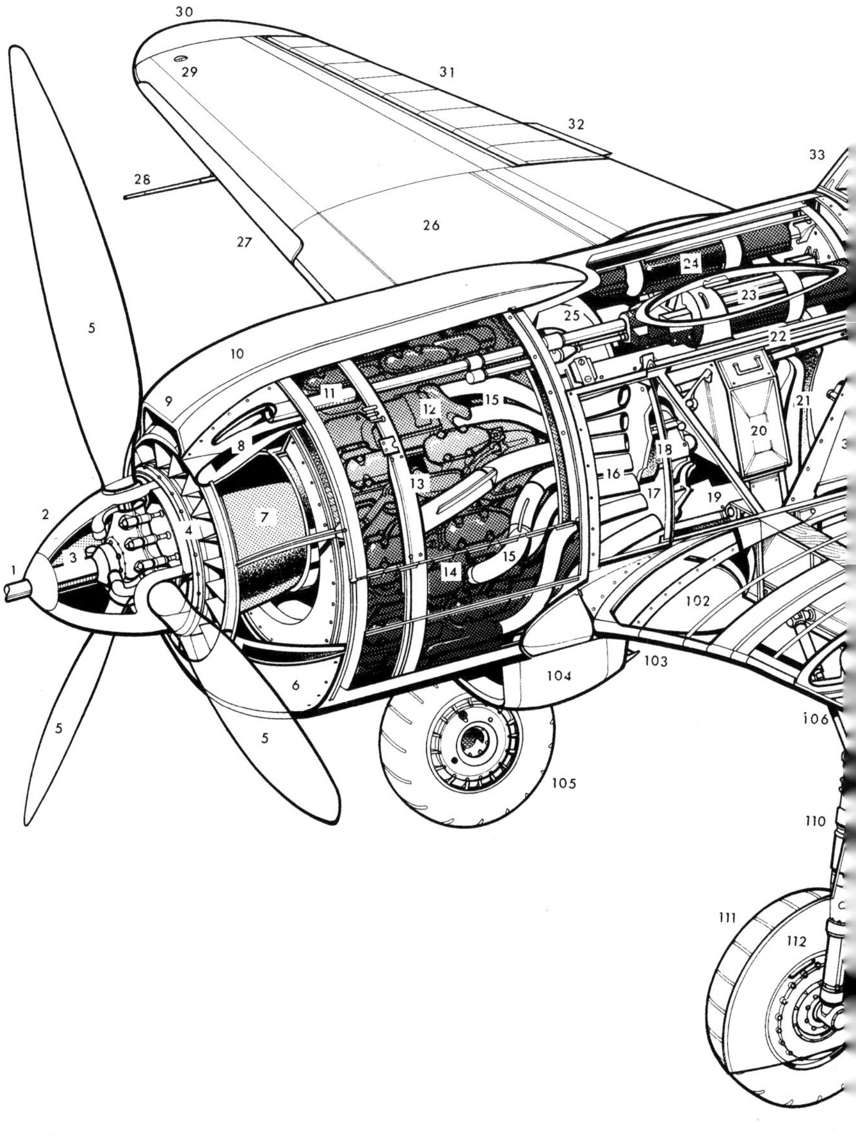

Semyon Lavochkin (second from the right) posed for this photograph with Soviet naval pilots, beside an La-7. The Naval Air Forces also contributed to the Soviet victory, acting as four distinct air forces – the Red Banner Baltic Fleet, the Northern Fleet, the Black Sea Fleet and the Pacific Fleet. None of the constituent units appears to have operated the La-7 during the war, although both the 3rd GvIAP-KBF and the 4th GvIAP-KBF flew La-5 variants.

SPECIFICATION

La-5FN

Dimensions

Length: 28 ft 5⅛ in (8.67 m)
Height: 8 ft 4 in (2.54 m)
Wingspan: 32 ft 1¾ in (9.80 m)
Wing area: 189 sq ft (17.59 m²)

Powerplant

One Shvetsov ASh-82FN 14-cylinder, twin-row, air-cooled, direct fuel injection radial piston engine rated at 1,850 hp (1380 kW) or 1,470 hp (1096 kW) at altitude

Weights

Empty: 5,743 lb (2605 kg)
Gross: 7,323 lb (3322 kg)

Fuel and load

Internal fuel: 102 Imp gal (464 litres)
Maximum external load: 880 lb (400 kg)

Performance

Maximum speed at 21,325 ft (6500 m): 403 mph (648 km/h)
Maximum speed at sea level: 356 mph (573 km/h)
Climb to 16,400 ft (5000 m): 5 minutes
Service ceiling: 36,090 ft (11000 m)
Take-off run: 951 ft (290 m)
Landing run: 1,673 ft (510 m)
Time for a 360° turn: 18 seconds

Range

Operational: 360 miles (580 km)

Armament

Two ShVAK 20-mm cannon with 200 rounds per gun, plus provision for two 50-kg (110-lb) AO-25M, FAB-50 or FAB-50M bombs on two underwing D3-40 racks or up to six RS-82 82-mm (3¼-in) unguided rockets

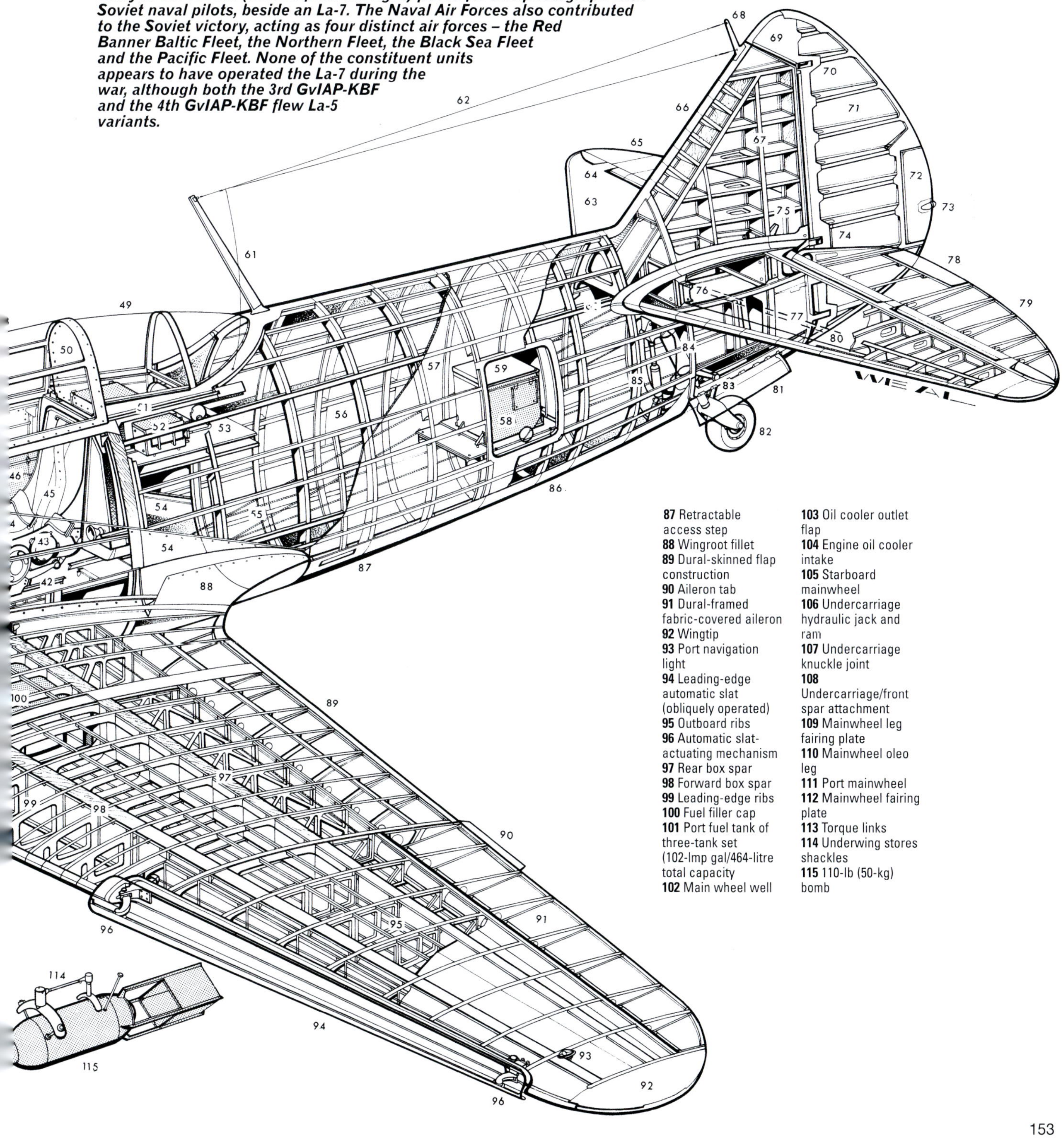

87 Retractable access step
88 Wingroot fillet
89 Dural-skinned flap construction
90 Aileron tab
91 Dural-framed fabric-covered aileron
92 Wingtip
93 Port navigation light
94 Leading-edge automatic slat (obliquely operated)
95 Outboard ribs
96 Automatic slat-actuating mechanism
97 Rear box spar
98 Forward box spar
99 Leading-edge ribs
100 Fuel filler cap
101 Port fuel tank of three-tank set (102-Imp gal/464-litre total capacity
102 Main wheel well
103 Oil cooler outlet flap
104 Engine oil cooler intake
105 Starboard mainwheel
106 Undercarriage hydraulic jack and ram
107 Undercarriage knuckle joint
108 Undercarriage/front spar attachment
109 Mainwheel leg fairing plate
110 Mainwheel oleo leg
111 Port mainwheel
112 Mainwheel fairing plate
113 Torque links
114 Underwing stores shackles
115 110-lb (50-kg) bomb

La-5FN

Kapitan Petr Yakovlevich Likholetov served with the 159th IAP for the entirety of his operational career. From 1941 to summer 1942, the unit was equipped with Yak-7 fighters, Likholetov probably gaining his first 12 victories on this type. From summer 1942 onwards, the 159th flew the P-40, La-5 and La-5FN in succession. Some confusion surrounds the allocation of Likholetov's final score, although he notched up at least 30 individual/group kills, of which certainly one was scored by ramming a Bf 109 with his Yak-7. He finished his flying career on the La-5FN, since at the end of 1944 he was seriously injured in a car accident and died of his injuries on 13 July 1945.

M-82 engine

Shvetsov's M-82 (later ASh-82) was a 14-cylinder, twin-row radial piston engine displacing 41.20 litres. In its ASh-82FN form for the La-5FN, the engine carried the *forsirovanny neprosredstvenno*, or directly boosted, designation, the letters 'FN' stencilled on the cowling of this aircraft in Cyrillic referring to this. In fact, the engine used a direct fuel injection system, which finally solved the inverted flight/engine cut-out problem that had previously dogged the La-5. A change associated with the introduction of the -82FN engine, and a key La-5FN recognition feature, was the addition of a new supercharger air intake stretching across the entire chord of the upper cowling. The intake beneath the centre forward fuselage in line with the wing leading edge served the oil cooler. The location of this item was moved to be in line with the wing trailing edge on the La-7. A flap at the rear of the intake controlled the through-flow of air.

Canopy design and gunsight

Early production La-5s, in common with the prototype, used the canopy and curved windscreen design of the earlier LaGG-3. Later aircraft used a redesigned windscreen with a flat forward panel which enabled the use of a PBP-1A gunsight. This modified arrangement also allowed the installation of a 2⅒-in (55-mm) thick armoured glass panel behind the windscreen, while a 2⅗-in (66-mm) thick armoured glass panel was fitted behind the pilot's head. The La-5F featured entirely revised rear glazing, which was combined with a cut-down rear fuselage to improve visibility to the rear. The La-5FN featured at least three distinctly different canopy arrangements. The first FNs shared the canopy design of the La-5F, while later aircraft added a small external grab handle to the port-side lower canopy frame and a vertical frame was added up the centre of the canopy at a later date. The aircraft illustrated demonstrates only the former canopy modification. Early La-7s employed a similar canopy/windscreen arrangement to the early La-5FN. The La-7 also introduced the much improved PBP-1B(V) reflector gunsight. This device overcame the limitations of the simple PBP-1A gunsight, whose inherent lack of accuracy led to pilots using three bursts of fire for each target – one before, one during and one after the target had passed through the sight.

Cannon armament

All the La-5 variants were armed with a pair of 20-mm ShVAK (*Shpital'nyi/Vladimirov Aviatsionnyi Krupnokaliber*, or Shpital'nyi/Vladimirov large-calibre cannon) weapons mounted in the upper part of the engine cowling to either side of the fuselage centreline. Boris G. Shpital'nyi and S. V. Vladimirov developed the weapon from the earlier ShKAS 0.3-in (7.62-mm) machine-gun. Each cannon weighed 92½ lb (45 kg), with a barrel 49 in (1.25 m) long and an overall length of 69⅕ in (1.76 m). Some 200 rounds per gun were carried in ammunition tanks between the cockpit and the engine, while the cannon were cable of firing at between 750 and 800 rounds per minute. The majority of La-7s was delivered with a three-gun ShVAK installation, although aircraft delivered from 1945 introduced the 20-mm calibre Berezin B-20 cannon. As with the ShVAKs, these weapons were installed in the form of two to the left of the centreline, with the third to the right, but in the same general position as those of the La-5. However, supplies of this weapon could not keep up with demand – it was also used in the PV-20 gun-turret system for Soviet bombers such as the Tu-4 'Bull' – and many La-7s flew with just two B-20s installed.

Disposable stores

All La-5/7-family aircraft were equipped to mount a single D3-40 bomb rack beneath each wing for bombs of up to 110 lb (50 kg) in weight. Though rarely used, these could include the FAB-50 GP bomb or AO-25M or FAB-50M fragmentation bombs. With the La-5FN already established in service, the capability to carry two or three RS-82 rockets under each wing was introduced. At least 2.5 million of these rockets were built, each 34 in (86.40 cm) long and weighing 15 lb (6.82 kg). The La-5's rocket capability was also little used.

Structure

Aft of the engine mounting bulkhead, the La-5 airframe was more or less identical to the desperately inadequate LaGG-3. As such, the fuselage was a semi-monocoque wooden structure utilising plywood-sheathed birch frames and wooden stringers as its basis with Bakelite-ply skinning. The wooden components were made of a specially developed material known as DSP-10 delta wood in the initial production La-5s. This material used between five and eight birch strips glued together across their grain and impregnated with a substance known as VIAM-B3. The latter consisted of boric acid, borax and phenol-formaldehyde resin. Predictably, supplies of the German-supplied resin soon dried up and the majority of production machines used pine rather than delta wood components. Bakelite ply was made by gluing layers of birch strip together and then combining them with Bakelite film at a temperature of 302°F (150°C). As La-5 development continued, metal was increasingly introduced in to the structure, starting with the wing spars in later La-5FNs.

Antennas and radios

Many changes of aerial mast and wire antenna arrangement occurred throughout the service life of the La-5/7 series, this aircraft exhibiting the standard La-5FN installation. RSI-4 radio was standard in early-build FNs, but later machines featured the RSI-4KhF unit. The main controls for the radio, including the knobs used to change frequencies, were mounted beneath the instrument panel. Changing frequencies could be a little awkward, but pilots praised the simple transmit switch on the throttle. This was pressed when the pilot wished to talk, and released at all other times.

Pressurised fuel tanks

In an effort to prevent fuel tank fires, especially in combat, the La-5/7 series featured a gas pressurisation system for the fuel tanks. Gas from the left-side exhaust manifold was piped to a filter in the rear fuselage. It was then cooled and filtered before passing to the fuel tanks in the wing/fuselage centre section. The system was operated by a cock in the cockpit, the pilot having to remember to switch the system on when entering combat and to deactivate it afterwards. If the system was used for too long, the fuel tanks were prone to deformation due to the excess pressure.

Camouflage and markings

It was commonplace for Soviet fighters to carry slogans on their fuselage sides, proclaiming or urging victory over Germany or loyalty to the Soviet Motherland. The inscription on Likholetov's aircraft translates as 'For Vasek and Zhora' and while it appears here in white, it may have been applied in yellow. A variety of camouflage colours and schemes was used, this aircraft being illustrated in what appears to have become a standard La-5FN pattern of two shades of blue grey over light grey.

Lockheed P-38 Lightning

An F-4 Lightning of the 3rd Photographic Reconnaissance Group banks towards the photographer, displaying the type's unique lines. The 3rd Group arrived at La Senia, Algeria, in mid-December 1942 and flew countless photographic missions in support of the Allied armies.

P-38 Lightning

Cutaway key
1 Starboard navigation light
2 Wingtip trailing edge strake
3 Landing light (underwing)
4 Starboard aileron
5 Aileron control rod/quadrant
6 Wing outer spar
7 Aileron tab drum
8 Aileron tab control pulleys
9 Aileron tab control rod
10 Aileron trim tab
11 Fixed tab
12 Tab cable access
13 Flap extension/retraction cables
14 Control pulleys
15 Flap outer carriage
16 Fowler-type flap (extended)
17 Control access panel
18 Wing spar transition
19 Outer section leading-edge fuel tanks (P-38J-5 and subsequent) capacity 46 Imp gal (208 litres) each
20 Engine bearer/bulkhead upper attachment
21 Firewall
22 Triangulated tubular engine bearer supports
23 Polished mirror surface panel (undercarriage visual check)
24 Cantilever engine bearer
25 Intake fairing
26 Accessories cooling intake
27 Oil radiator (outer sections) and intercooler (centre section) tripleintake
28 Spinner
29 Curtiss-Electric three blade (left) handed propeller
30 Four machine gun barrels
31 Cannon barrel
32 Camera-gun aperture
33 Nose panel
34 Bulkhead
35 Machine gun blast tubes
36 Four 0.5-in (12.7-mm) machine guns
37 Cannon flexible hose hydraulic charger
38 Chatellerault-feed cannon magazine (150 rounds)
39 Machine gun firing solenoid
40 Cannon ammunition feed chute
41 Nose armament cowling clips
42 Case ejection chute (port lower machine gun)
43 Ammunition box and feed chute (port lower machine gun)
44 Case ejection chute (port upper machine gun)
45 Ammunition box and feed chute (port upper machine gun)
46 Radio antenna
47 Ejection chute exit (shrouded when item 52 attached)
48 Nosewheel door
49 Nosewheel shimmy damper assembly and reservoir
50 Torque links
51 Towing eye
52 Type M10 triple-tube 4.5-in (11.4-cm) rocket-launcher
53 Rearward-retracting nosewheel
54 Alloy spokes cover plate
55 Fork
56 Rocket-launcher forward attachment (to 63)
57 Nosewheel lower drag struts
58 Nosewheel oleo leg
59 Nosewheel pin access
60 Side struts and fulcrum
61 Actuating cylinder
62 Upper drag strut
63 Rocket-launcher forward attachment bracket
64 Rudder pedal assembly
65 Engine controls quadrant
66 Instrument panel
67 Spectacle grip cantilevered control wheel
68 Non-reflective shroud
69 Lynn-3 reflector sight mounting
70 Optically-flat bullet-proof windscreen
71 External rear-view mirror
72 Armoured headrest
73 Rearward-hinged canopy
74 Pilot's armoured seat back
75 Canopy bracing
76 Downward-winding side windows
77 Wing root fillets
78 Nosewheel well
79 Port reserve fuel tank, capacity 50 Imp gal (227 litres)
80 Fuel filler cap
81 Main (double I-beam) spar
82 Fuel filler cap
83 Flap inner carriage
84 Port main fuel tank, capacity 75 Imp gal (341 litres)
85 Flap control access
86 Flap structure
87 Entry ladder release
88 Flap drive motor
89 Fuel surge tank
90 Radio equipment compartment
91 Turnover support pylon

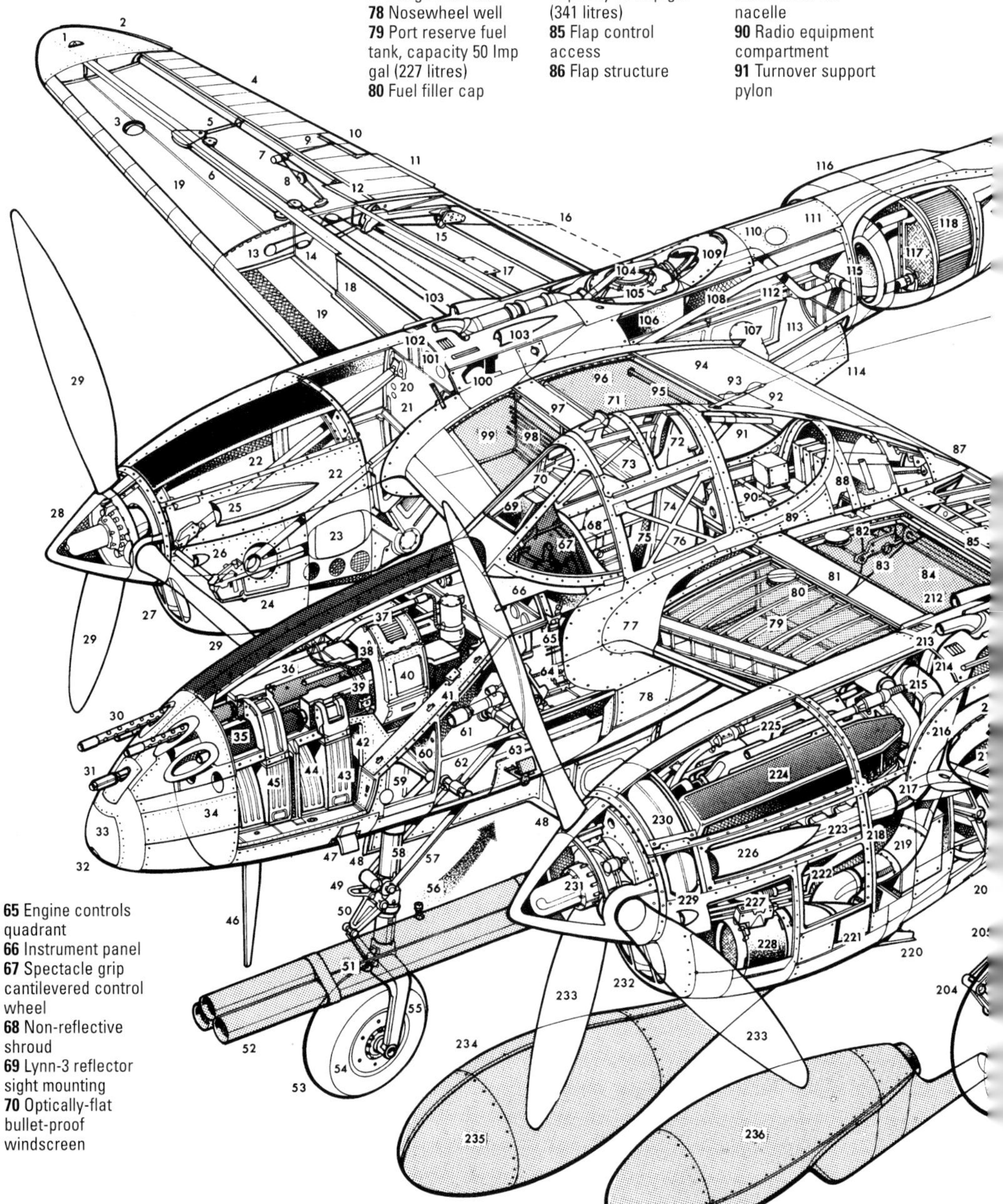

P-38Hs were the first to have the bar added to the national insignia (illustrated). This factory-fresh example is seen on a test flight from Lockheed's Burbank facility in California, prior to delivery to the USAAF.

SPECIFICATION

Lockheed P-38L Lightning

Type
Single-seat fighter

Powerplant
Two 1,475-hp (1100-kW) Allison V-1710-111/-113 V-12 piston engines

Performance
Maximum speed 414 mph (666 km/h) at 25,000 ft (7620 m); service ceiling 44,000 ft (13410 m); normal range 450 miles (724 km)

Weights
Empty 12,800 lb (5806 kg); maximum take-off 21,600 lb (9798 kg)

Dimensions
Wingspan 52 ft (15.85 m); length 37 ft 10 in (11.53 m); height 12 ft 10 in (3.91 m); wing area 328 sq ft (30.47 m2)

Armament
Four 0.5-in (12.7-mm) machine-guns and one 20-mm cannon, plus up to 3,200 lb (1451 kg) of bombs

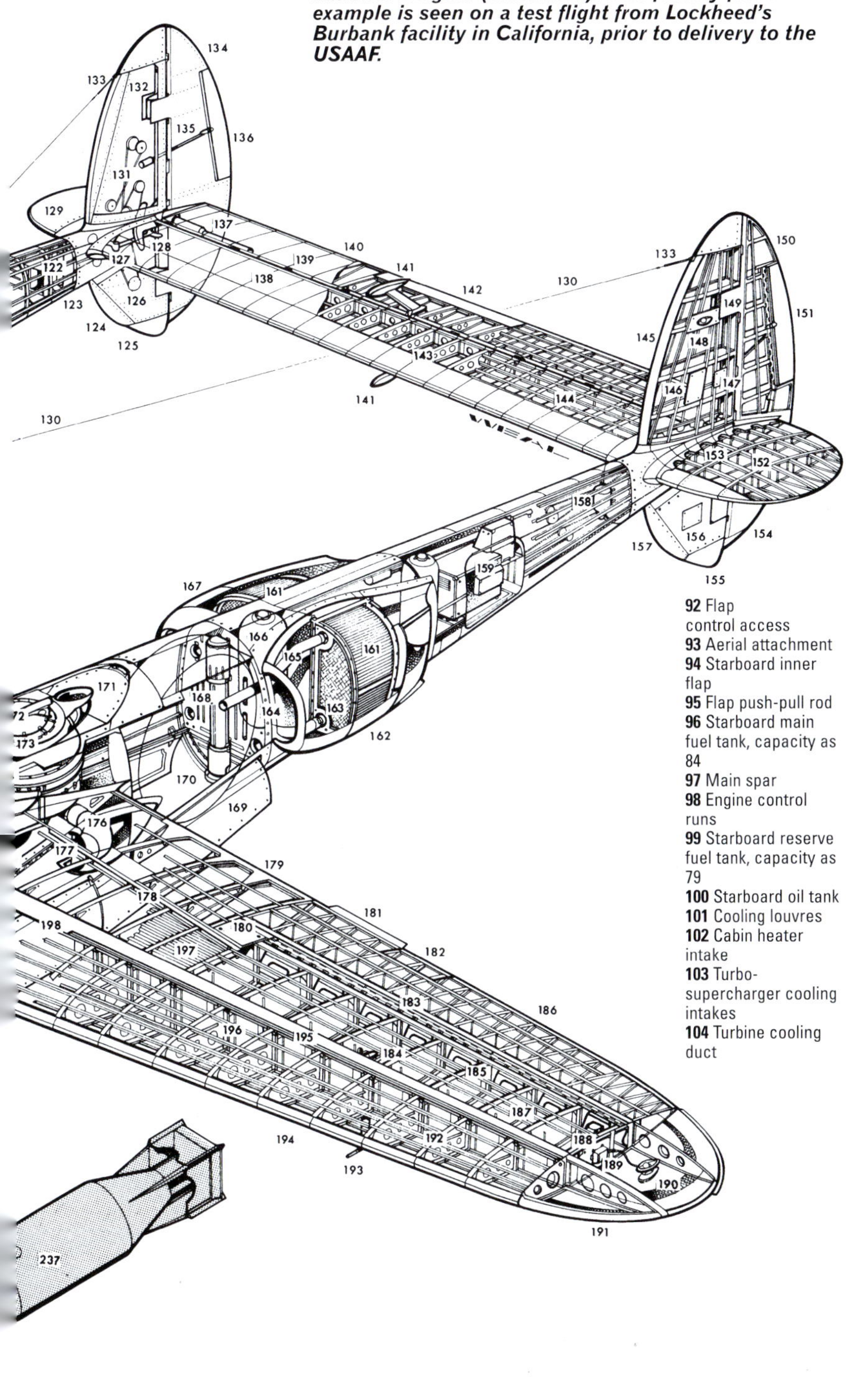

92 Flap control access
93 Aerial attachment
94 Starboard inner flap
95 Flap push-pull rod
96 Starboard main fuel tank, capacity as 84
97 Main spar
98 Engine control runs
99 Starboard reserve fuel tank, capacity as 79
100 Starboard oil tank
101 Cooling louvres
102 Cabin heater intake
103 Turbo-supercharger cooling intakes
104 Turbine cooling duct
105 Exhaust turbine
106 Supercharger housing
107 Wingroot/boom fillet
108 Coolant/radiator return pipe (left and right)
109 Exhaust waste gate outlet
110 Access panel
111 Boom Joint (Station 265)
112 Radiator/coolant supply pipe
113 Mainwheel well
114 Mainwheel doors
115 Radiator intake
116 Starboard outer radiator fairing
117 Radiator grille
118 Engine coolant radiator assembly
119 Exit flap
120 Tool and baggage compartment
121 Boom structure
122 D/R master compass housing
123 Boom/tail attachment joint (Station 393)
124 Starboard lower fin
125 Tail bumper skid shoe
126 Elevator control pulley
127 Rudder stop
128 Elevator control horn
129 Fixed tip
130 Radio aerials
131 Tail surface control pulleys
132 Aerodynamic mass balance
133 Aerial attachments
134 Starboard rudder
135 Tab control rod and drum
136 Rudder trim tab
137 Elevator abbreviated torque tube
138 Tailplane stressed skin
139 Elevator pin hinges (eight off)
140 Elevator
141 Upper and lower mass balances
142 Elevator trim tab
143 Tailplane structure
144 Stiffeners
145 Port fin structure
146 Elevator pulley access
147 Rudder tab drum access
148 Tail running light (port)
149 Aerodynamic mass balance
150 Rudder framework
151 Rudder trim tab
152 Fixed tip structure
153 Tail surfaces/boom (quatrefoil bulkhead) attachment flanges
154 Rudder lower section
155 Tail bumper skid shoe
156 Elevator pulley access
157 Port lower fin
158 Elevator, rudder, and table cables
159 Battery compartment
160 Radiator exit flap
161 Engine coolant radiator assembly
162 Radiator housing
163 Radiator/coolant supply pipe
164 Radiator intake
165 Coolant/radiator return pipe
166 Oxygen cylinder
167 Port inner radiator fairing
168 Flare tube (port and starboard booms)
169 Mainwheel doors
170 Mainwheel well
171 Exhaust waste gate outlet
172 Turbine cooling duct
173 Exhaust turbine
174 Supercharger assembly
175 Supercharger/intercooler duct
176 Carburettor intake duct
177 Carburettor air intake
178 Abbreviated rear spar
179 Flap outer section
180 Tab cable access
181 Fixed tab
182 Aileron trim tab
183 Aileron full-span piano-wire hinge
184 Underwing pitot attachment
185 Raked web stiffener (outboard of rear spar)
186 Aileron structure
187 Outer wing pressed sheet ribs
188 Aileron counterweight
189 Junction box
190 Port navigation lights
191 Port wingtip structure
192 Leading-edge ribs
193 Pitot head
194 Wing leading-edge skin join (fabric-covered piano-wire hinge)
195 Wing outer section I-beam box spar
196 Leading-edge stringers (no fuel tanks in early P-38 Js)
197 Wing inner surface corrugation
198 Spar single/double I-beam box spar transition
199 Mainwheel leg doors
200 Rearward-retracting mainwheel
201 Mainwheel oleo leg
202 Alloy spoked hub
203 Cantilever axle
204 Torque links
205 Hydraulic brake cable
206 Drag strut
207 Side strut
208 Drag links
209 Fulcrum
210 Actuating cylinder
211 Multi-bolt outer wing fixings
212 Turbo-supercharger cooling intakes
213 Cabin heater intake
214 Cooling louvres
215 Carburettor duct
216 Outer section wing fillet
217 Insulated exhaust shroud duct
218 Intercooler/carburettor duct
219 Supercharger/intercooler duct
220 Outlet
221 Oil radiator shutter
222 Intercooler
223 Exhausts
224 Allison V-1710-89/91 twelve-cylinder Vee engine
225 Magnetos/distributors
226 Intake fairing
227 Header feed pipes
228 Port outer oil radiator
229 Spark-plug and magneto cooling intake
230 Coolant header tank
231 Propeller hub
232 Oil radiator (outer sections) and intercooler (centre section tripleintake
233 Curtiss-Electric three-blade (right) handed propeller
234 Inner section underwing stores including
235 Jettisonable auxiliary fuel tank, or
236 Smoke generator, or
237 1,000-lb (454-kg) bomb

Macchi MC.200/202/205

The Cucaracham (cockroach) device on the rear fuselage of these early MC.200s identifies the aircraft as belonging to 22° Gruppo, 52° Stormo, based at Ciampino for the defence of Rome in 1940. In fact, these machines are from 371ª Squadriglia.

MC.200 Saetta (Serie XIX)

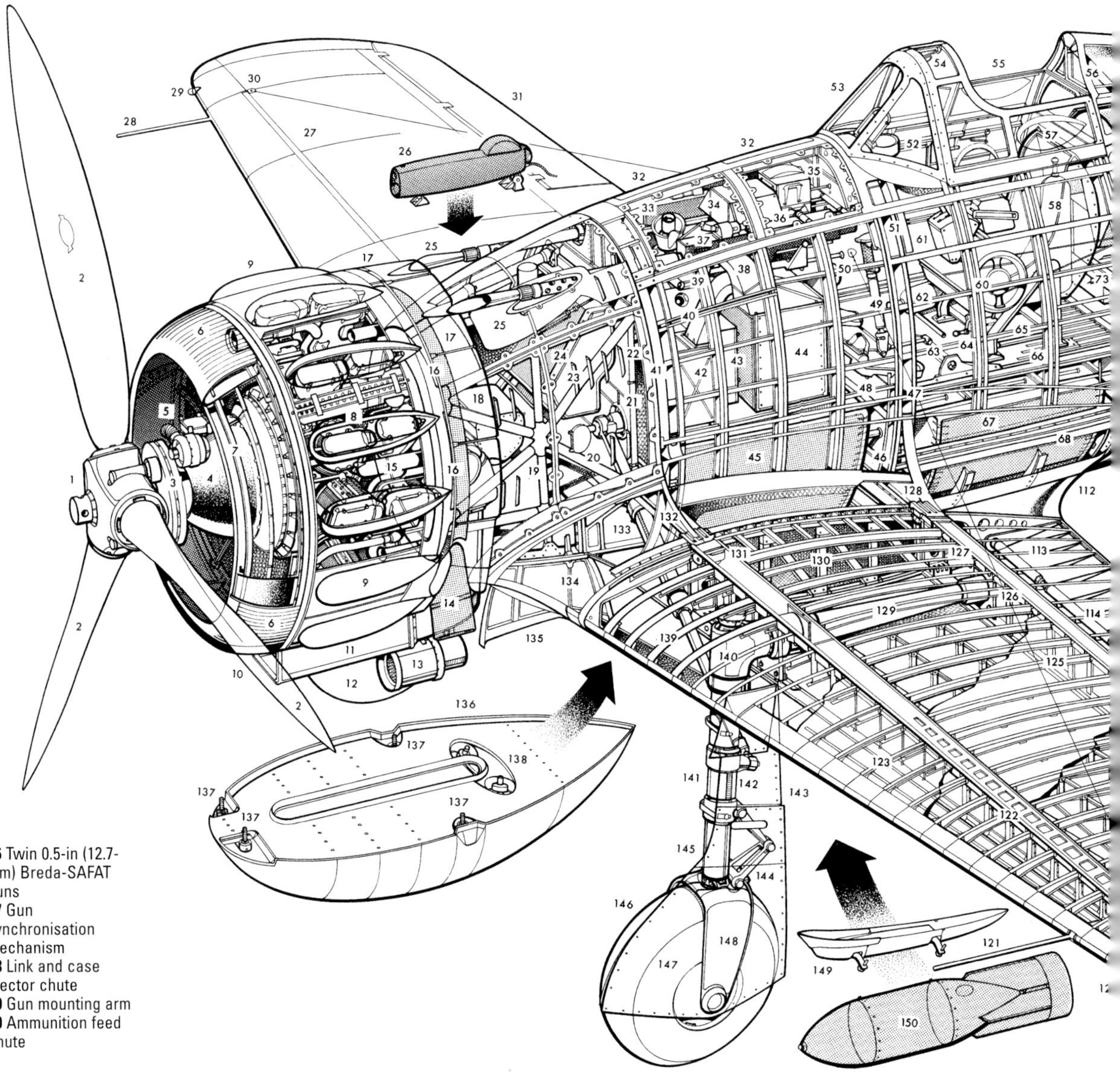

Cutaway key
1 Propeller hub
2 Variable-pitch propeller
3 Hub plate
4 Casing
5 Pitch control mechanism
6 Oil radiator
7 Cowling ring
8 Fiat A.74 RC.38 14-cylinder radial air-cooled engine
9 Cowling rocker arm fairings
10 Carburettor intake
11 Intake housing
12 Starboard mainwheel
13 Intake filter
14 Exhaust outlet
15 Engine mounting ring
16 Exhaust collector ring
17 Adjustable cowling gills
18 Zenith compressor
19 Engine ring bearer frames
20 Oil filler access
21 Undercarriage retraction jack attachment
22 Firewall bulkhead
23 Cooling louvres
24 Oil tank (9-Imp gal/ 42-litre capacity)
25 Machine-gun muzzle ports
26 F.M.62 gun camera (mounted mid-chord starboard wing join)
27 Starboard mainplane **28** Starboard pitot tube (heated)
29 Starboard navigation light
30 Aerial attachment
31 Starboard aileron
32 Cowling access panels
33 Fuel filler cap
34 Allocchio Bacchini B.30 R/T set
35 Battery
36 Twin 0.5-in (12.7-mm) Breda-SAFAT guns
37 Gun synchronisation mechanism
38 Link and case ejector chute
39 Gun mounting arm
40 Ammunition feed chute

Late-production MC.200s such as these of 81ª Squadriglia, 6º Gruppo, 1º Stormo (based at Catania, Sicily late in 1940), took part in many sorties over Malta during this period. Deployed to North Africa in April 1941, the Saetta soon proved no match for the RAF's Hurricane Mk II and P-40 in air combat. So it was that, the following year, the MC.200 was used for the first time as a fighter-bomber, usually escorted by more modern fighters. By the time of the Allied invasion of Sicily the type was wholly outdated, though 42 remained in the front-line order of battle, of which 23 were still in use in September 1943. The Saetta also saw some use on the Eastern Front, 22º and, later, 21º Gruppi deploying to Russia in April 1941. These two units flew 6,361 sorties in all, claiming 85 Soviet kills for 15 losses.

SPECIFICATION

MC.200 Saetta	MC.205V Veltro
Dimensions	**Dimensions**
Length: 26 ft 10¼ in (8.19 m) **Height:** 11 ft 5¾ (3.50 m) **Wingspan:** 34 ft 8½ in (10.58 m) **Wing area:** 180.84 sq ft (16.80 m²)	**Length:** 29 ft 0½ in (8.58 m) **Height:** 9 ft 11½ (3.04 m) **Wingspan:** 34 ft 8½ in (10.58 m) **Wing area:** 180.84 sq ft (16.80 m²)
Powerplant	**Powerplant**
One Fiat A.74 RC.38 14-cylinder air-cooled radial piston engine delivering 870 hp (649 kW)	One Fiat RA.1050 RC.58 Tifone 12-cylinder inverted Vee piston engine delivering 1,475 hp (1100 kW)
Weights and loads	**Weights and loads**
Empty: 4,178 lb (1895 kg) **Maximum take-off:** 5,710 lb (2590 kg)	**Empty:** 5,691 lb (2581 kg) **Maximum take-off:** 7,514 lb (3408 kg)
Performance	**Performance**
Maximum speed at 14,765 ft (4500 m): 312 mph (502 km/h) **Cruising speed:** 283 mph (455 km/h) **Service ceiling:** 29,200 ft (8900 m) **Range with auxiliary fuel:** 540 miles (870 km)	**Maximum speed at 23,620 ft (7200 m):** 399 mph (642 km/h) **Cruising speed:** 310 mph (500 km/h) **Service ceiling:** 37,090 ft (16370 m) **Range:** 646 miles (1040 km)
Armament	**Armament**
Two 0.5-in (12.7-mm) Breda-SAFAT machine-guns in upper cowling; later aircraft had two additional 0.303-in (7.7-mm) guns mounted in the wings. Late-production aircraft were also able to carry up 882 lb (400 kg) of bombs on underwing pylons	Two 0.5-in (12.7-mm) Breda-SAFAT machine-guns in upper cowling, plus two wing-mounted 0.303-in (7.7-mm) machine-guns; late series aircraft had the wing-mounted guns replaced by two 20-mm cannon

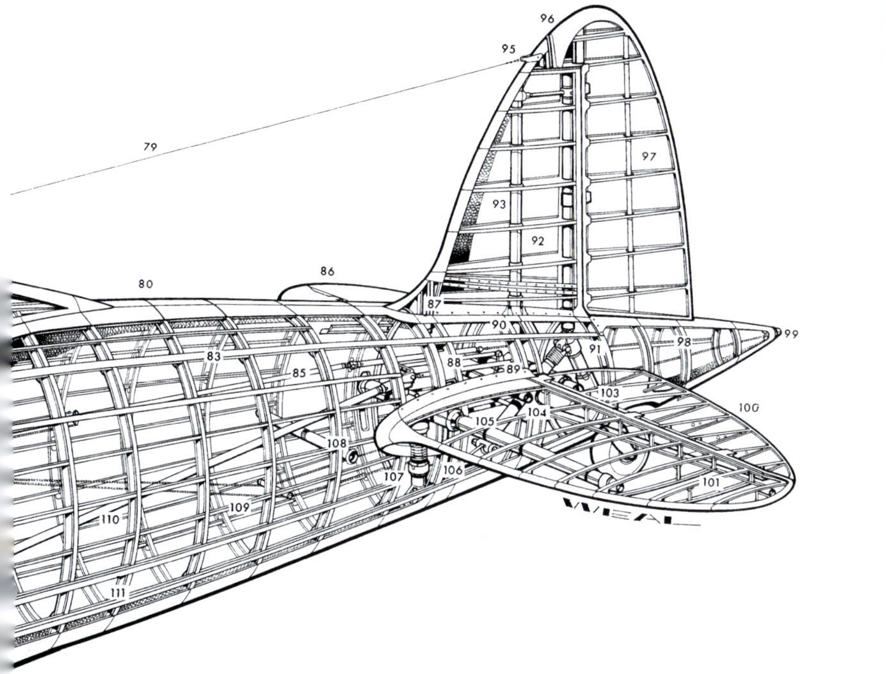

MC.201 – a one-off prototype

Answering the Regia Aeronautica's call in 1938 for an 'economical interceptor' to follow on from the Fiat G.50 and Macchi MC.200, a number of proposals were received. Macchi's was the MC.201, effectively a Saetta fitted with the new 1,000-hp (746-kW) Fiat A.76 RC.40 radial. A single prototype (MM436) was constructed, but development of the A.76 was abandoned so this first flew with the Saetta's A.74 engine installed. By early 1940 interest in the 'economical interceptor' had waned and a second MC.201 was completed with a DB 601 engine, as a forerunner to the MC.202.

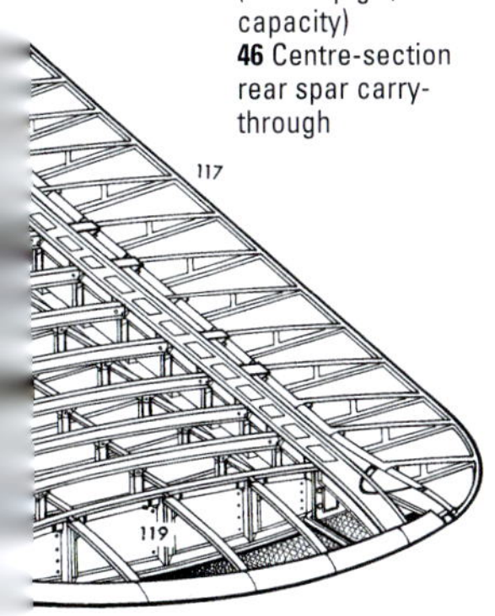

41 Fuselage forward frame (Frame 0)
42 Supplementary magazine
43 Ammunition magazine
44 Link/spent case collector
45 Main fuel tank (52.3-Imp gal/238-litre capacity)
46 Centre-section rear spar carry-through
47 Fuselage frame (Frame 4)
48 Rudder pedal/heel rest assembly
49 Control column
50 Aerial attachment
51 Instrument panel
52 San Giorgio reflector gunsight
53 Windscreen
54 Canopy side-panel lock/release
55 Cutaway canopy side-panels
56 Cutaway canopy side-panels
57 Turnover pylon structure
58 SILMA CO_2 fire extinguisher bottle (fuselage starboard wall)
59 Pilot's seat
60 Adjustable tailplane trim wheel
61 Throttle quadrant
62 Pilot's oxygen cylinder (to right of seat)
63 Control linkage
64 Seat adjustment handle
65 Seat main frame
66 Cockpit floor
67 Underfloor fuel tank (16.5-Imp gal/75-litre capacity)
68 Lower longeron
69 Entry foothold
70 Cylinder support frame
71 Compressed air cylinder (2.2-Imp gal/10-litre capacity)
72 Hydraulic reservoir (flap actuation)
73 Garelli compressor (fuselage starboard wall)
74 Hydraulic reservoir (undercarriage actuation)
75 Auxiliary fuel tank (18-Imp gal/83-litre capacity)
76 Fuel filler access cut-out
77 Fairing formers
78 Stub aerial mast
79 Aerial
80 Fuselage skin
81 Fuselage structure
82 Frame
83 Upper longeron
84 Stringer
85 Rudder control rod
86 Starboard horizontal tail surfaces
87 Tailfin front attachment
88 Fuselage frame (Frame 16)
89 Elevator control horns
90 Tailplane attachment (Frame 17)
91 Fuselage aft frame (Frame 18)
92 Tailfin structure
93 Support tube
94 Rudder post
95 Aerial attachment
96 Rudder balance
97 Rudder frame
98 Tail cone
99 Tail navigation light
100 Port elevator
101 Port tailplane structure
102 Non-retractable tailwheel
103 Tailwheel shock strut
104 Tailplane incidence torque tube (+ 1° 45' to -5° 30')
105 Tailplane support tube
106 Tailwheel strut attachment
107 Tailplane incidence screw
108 Lifting tube
109 Tailplane incidence control cables
110 Elevator control rod
111 Lower longeron
112 Wingroot fillet
113 Flap profile
114 Flap-operating rod
115 Flap structure
116 Wing rear spar
117 Port aileron structure
118 Wing outer section ribs
119 Port wingtip structure
120 Port navigation light
121 Port pitot tube (unheated)
122 Wing front spar
123 Leading-edge rib sections
124 Wing skin
125 Aerial
126 Undercarriage/rear spar attachment
127 Wing outer/inner section front spar join
128 Wing root fairing former
129 Undercarriage rotation spindle
130 Centre-section outer rib
131 Wing outer/inner section front spar join
132 Frame 0 carry-through
133 Undercarriage retraction strut
134 Port mainwheel well
135 Mainwheel door inner section
136 Auxiliary jettisonable fuel tank (33-Imp gal/ 150-litre capacity)
137 Attachment lugs
138 Fuel connections
139 Mainwheel leg well
140 Undercarriage pivot
141 Mainwheel leg
142 Retraction strut attachment
143 Leg doors (hinged)
144 Torque links
145 Shock strut
146 Port mainwheel
147 Mainwheel door outer section
148 Axle fork
149 Underwing stores pylon
150 Bomb

MC.202 Folgore

MM9066 is an MC.202 Serie VII of 151ª Squadriglia, 20° Gruppo, 51° Stormo of the Regia Aeronautica, based in Sicily towards the end of 1942. Named *Dai Banana!* (Go, banana!) the aircraft was flown by Italian ace Sergeant Ennio Tarantola, who had been a banana importer before the war. Seven of his eventual eight kills are marked on the Folgore's tail. As well as Macchi, both Breda and SAI-Ambrosini were given contracts to build Folgores, the seventh production batch (Serie VII) of 100 aircraft being completed by Macchi. In all, 1,300 aircraft in 15 batches were authorised, though only about 1,150 were completed before the Italian surrender.

National insignia
In addition to the white cross on the tail, Regia Aeronautica aircraft carried the *fasces* national insignia of the Italian Fascist party on the fuselage. The symbol was of Roman origin, consisting of a bundle of rods with an axe-head protruding, and symbolised the power of the magistrates. Wing roundels comprised three black *fasces* devices in a black circle.

Control surfaces
The statically- and aerodynamically-balanced ailerons were of high aspect ratio and reached right to the wingtip. The MC.202 was considered to be one of the most manouevrable and, above all, one of the most beautifully harmonised fighters of the war.

Wing and tail structure
The wing was built around a two-spar structure, which carried through the lower fuselage. Apart from the wing root fuel tank and the gun/ammunition installation, there was nothing contained within the structure itself. The tail surfaces exhibited classical curves, the moveable surfaces being metal-framed but fabric-skinned. The large rudder had an auxiliary spar and featured an aerodynamic horn balance at the tip.

Armament
Though comparatively fast and agile the Folgore was inadequately armed. Aircraft in the first six production batches carried a pair of 0.5-in (12.7-mm) Breda-SAFAT machine-guns mounted in the upper fuselage decking ahead of the cockpit. These were armed with 400 rounds each. From Serie VII the MC.202 featured a gun in each wing in an attempt to redress this problem. These were 0.303-in (7.7-mm) Breda-SAFAT guns, each with 500 rounds. This armament arrangement was repeated in early-production MC.205Vs, though a pair of 20-mm MG 151 cannon replaced the wing machine-guns in later aircraft.

Powerplant
Alfa Romeo acquired a licence for the Daimler-Benz DB 601A inverted vee-12 engine, which was built for the MC.202 as the RA.1000 RC.41 Monsoni. The engine developed 1,075 hp (802 kW) for take-off at 2,500 rpm and had a normal maximum rating of 1,040 hp (776 kW) at 2,400 rpm. The first production MC.202s were fitted with German-built engines, 400 of which were supplied. Alfa Romeo began DB 601 production at its Pomigliano d'Arco plant, initially using components supplied from Germany. The first wholly Italian-made engines came off the line in the summer of 1941. Eventually 2,000 were completed. The Monsoni engine drove a Piaggio P.1001 three-bladed, constant-speed propeller of 10 ft (3.05 m) diameter.

Cockpit
In the Folgore's cockpit the pilot was provided with a San Giorgio reflector sight, offset to starboard, and secondary ring-and-bead sight, offset to port. The framed canopy hinged to starboard for access. The characteristic fairing behind the cockpit improved aerodynamics and incorporated a cut-out on either side so that rearward visibility was impaired as little as possible. An armoured seat was provided and, from Serie VII onward, an armoured glass windscreen was fitted.

Camouflage and markings
Depending on the theatre, MC.202s were mostly finished in greens and browns. This aircraft has the classic 'sand-and-spinach' finish, the basic desert scheme being overpainted with green patterns for a disruptive effect. The prominent marking applied over the white fuselage band is that of 51° Stormo, depicting 'Sorci Verdi' (Green Mice, representing bomber aircraft) being chased by a black cat (a 51° Stormo fighter). This insignia was retained by the Gruppo post-war and is displayed on its AMX aircraft in 2000. The small 'AS' marking under the aircraft's serial number stands for Africa Settentrionale (North Africa) and indicated that the aircraft had been tropicalised for service in the Mediterranean threatre.

Fuel
The Folgore's main fuel tank was located in the fuselage between the cockpit and engine, this holding 60 Imp gal (270 litres). Either side of this was a pair of wing root tanks, each holding 9 Imp gal (40 litres), while an overload fuel tank was available in the rear fuselage behind the pilot's seat for a further 18 Imp gal (80 litres). The Serie XI aircraft introduced plumbing for drop tanks.

Messerschmitt Bf 109E

A Bf 109E-7/Trop of JG 5 over the northern sector of the Eastern Front in 1942. This example was fitted with dust and sand filters and a survival kit. This was standard for tropical versions of the Bf 109.

Bf 109E-4

Cutaway key
1 Hollow propeller hub
2 Spinner
3 Three-bladed VDM variable-pitch propeller
4 Propeller pitch-change mechanism
5 Spinner back plate
6 Glycol coolant header tank
7 Glycol filler cap
8 Cowling fastener
9 Chin intake
10 Coolant pipe fairing
11 Exhaust forward fairing
12 Additional (long-range) oil tank
13 Daimler-Benz DB 601A engine
14 Supplementary intakes
15 Fuselage machine-gun troughs
16 Anti-vibration engine mounting pads
17 Exhaust ejector stubs
18 Coolant pipes (to underwing radiators)
19 Oil cooler intake
20 Coolant radiator
21 Radiator outlet flap
22 Cowling frame
23 Engine mounting support strut
24 Spent cartridge collector compartment
25 Ammunition boxes (starboard loading)
26 Engine supercharger
27 Supercharger air intake fairing
28 Forged magnesium alloy cantilever engine mounting
29 Engine mounting/forward bulkhead attachment
30 Ammunition feed chutes
31 Engine accessories
32 Two fuselage-mounted MG17 machine-guns
33 Blast tube muzzles
34 Wing skinning
35 Starboard cannon access
36 20-mm MG FF wing cannon
37 Leading-edge automatic slot
38 Slot tracks
39 Slot actuating linkage
40 Wing main spar
41 Intermediate rib station
42 Wing end rib
43 Starboard navigation light
44 Aileron outer hinge
45 Aileron metal trim tab
46 Starboard aileron
47 Aileron/flap link connection
48 Combined control linkage
49 Starboard flap frame
50 Cannon ammunition drum access
51 Fuselage machine-gun cooling slots
52 Gun mounting frame
53 Firewall/bulkhead
54 Instrument panel near face (fabric covered)
55 Oil dipstick cover
56 Control column
57 Oil filler cap (tank omitted for clarity)
58 Rudder pedal assembly
59 Aircraft identity data plate (external)
60 Mainspar centre-section carry-through
61 Underfloor control linkage

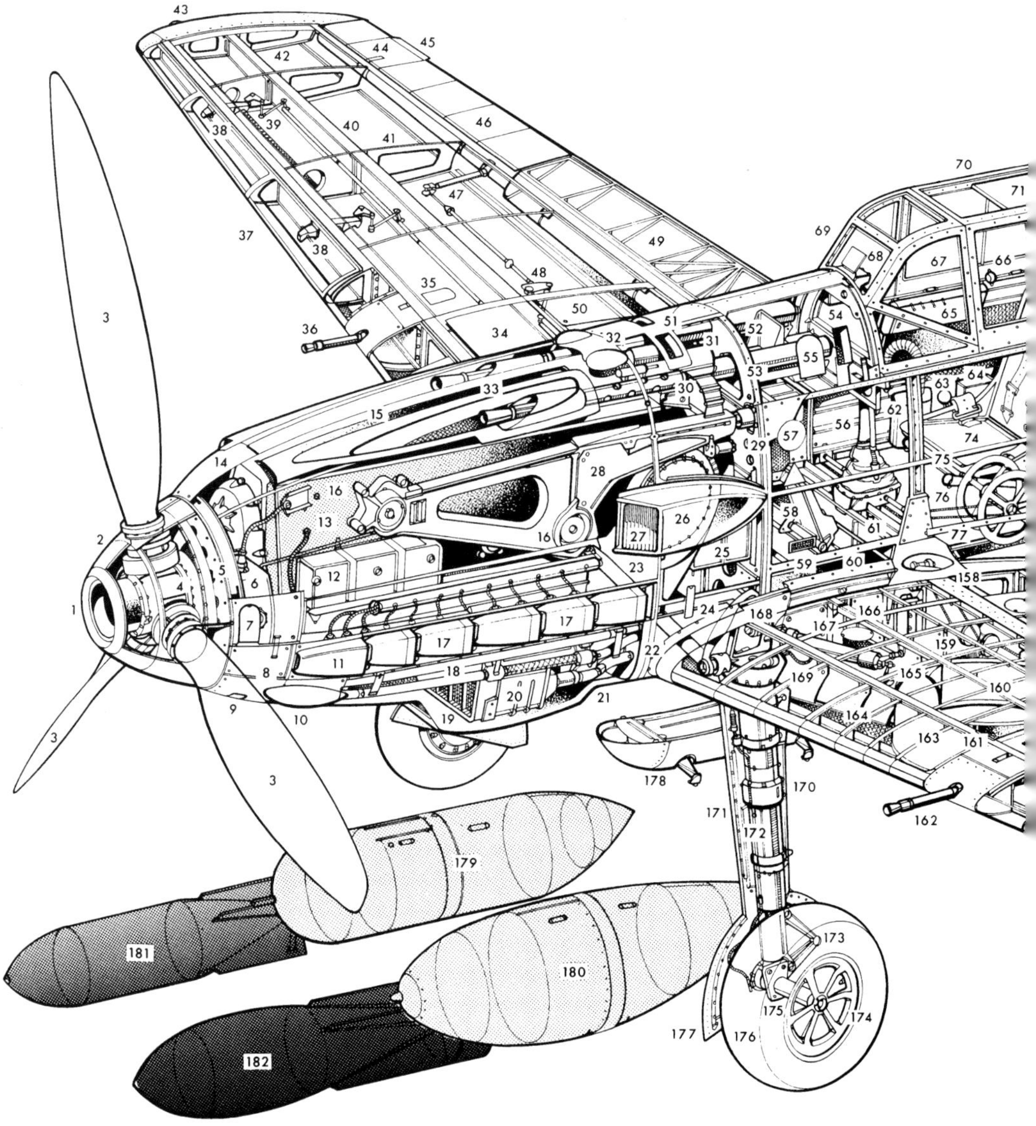

This Bf 109E-1 of I/JG 20 is seen in August/September 1939, i.e. upon the outbreak of World War II. This Gruppe did not participate in the Polish campaign, remaining in eastern Germany on metropolitan defence duties around Dresden. In July 1940 I./JG 20 became III./JG 51 and the 'black cat' emblem was used by 8/JG 51.

SPECIFICATION

Bf 109 E-4

Dimensions

Length overall: 28 ft 6 in (8.7 m)
Wing span: 32 ft 6 in (9.9 m)
Wing area: 176 sq ft (16.4m²)

Powerplant

One Daimler Benz DB 601Aa 12 cylinder inverted-Vee liquid cooled engine with direct fuel injection. Rated at 1,020 hp (761 kW) at 14,7675 ft (4500 m).

Weights

Empty: 4,440 lb (2018 kg)
Maximum take-off: 5,520 lb (2509 kg)

Performance

Maximum speed: 357 mph (575 km/h) at 19,700 ft (3200 m)
Service ceiling: 36,000 ft (10972 m)

Range

434 miles (700 km) on internal fuel

Armament

Two 20-mm (Oerlikon) MG FF cannon with 60 r.p.g in wings and two fuselage-mounted 0.312-in (7.9-mm) Rheinmetall Borsig MG 17 machine guns with 500 r.p.g. E-4/B fighter-bombers were capable of carrying either four 110-lb (50-kg) bombs or one 551-lb (250-kg) bomb.

Bf 109 T-2

generally similar to Bf 109 E-4 except

Dimensions

Length overall: 28 ft 9 in (8.76 m)
Wing span: 36 ft 4¼ in (11.08 m)
Wing area: 188,368 sq ft (17.5m²)

Powerplant

One Daimler Benz DB 601N cylinder inverted-Vee liquid cooled engine with direct fuel injection. Rated at 1,270 hp (947 kW) at 16,400 ft (5000 m).

Weights

Empty: 4,409 lb (2000 kg)
Maximum take-off: 6,786 lb (3078 kg)

Performance

Service ceiling: 34,450 ft (10500 m)

Range

568 miles (914 km) (with 66-Imp gal/300-litre drop tank)

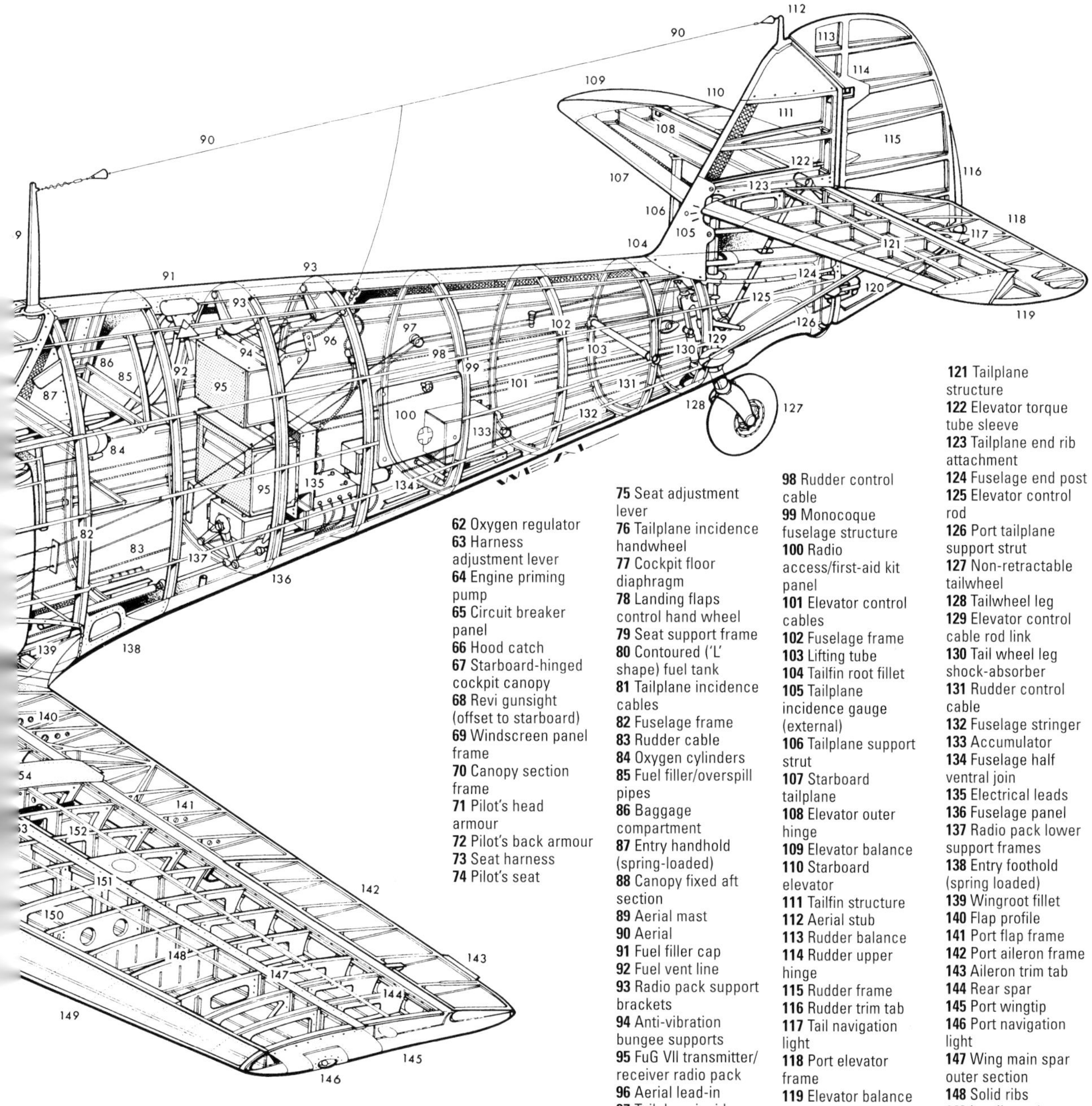

62 Oxygen regulator
63 Harness adjustment lever
64 Engine priming pump
65 Circuit breaker panel
66 Hood catch
67 Starboard-hinged cockpit canopy
68 Revi gunsight (offset to starboard)
69 Windscreen panel frame
70 Canopy section frame
71 Pilot's head armour
72 Pilot's back armour
73 Seat harness
74 Pilot's seat
75 Seat adjustment lever
76 Tailplane incidence handwheel
77 Cockpit floor diaphragm
78 Landing flaps control hand wheel
79 Seat support frame
80 Contoured ('L' shape) fuel tank
81 Tailplane incidence cables
82 Fuselage frame
83 Rudder cable
84 Oxygen cylinders
85 Fuel filler/overspill pipes
86 Baggage compartment
87 Entry handhold (spring-loaded)
88 Canopy fixed aft section
89 Aerial mast
90 Aerial
91 Fuel filler cap
92 Fuel vent line
93 Radio pack support brackets
94 Anti-vibration bungee supports
95 FuG VII transmitter/ receiver radio pack
96 Aerial lead-in
97 Tailplane incidence cable pulley
98 Rudder control cable
99 Monocoque fuselage structure
100 Radio access/first-aid kit panel
101 Elevator control cables
102 Fuselage frame
103 Lifting tube
104 Tailfin root fillet
105 Tailplane incidence gauge (external)
106 Tailplane support strut
107 Starboard tailplane
108 Elevator outer hinge
109 Elevator balance
110 Starboard elevator
111 Tailfin structure
112 Aerial stub
113 Rudder balance
114 Rudder upper hinge
115 Rudder frame
116 Rudder trim tab
117 Tail navigation light
118 Port elevator frame
119 Elevator balance
120 Rudder control quadrant
121 Tailplane structure
122 Elevator torque tube sleeve
123 Tailplane end rib attachment
124 Fuselage end post
125 Elevator control rod
126 Port tailplane support strut
127 Non-retractable tailwheel
128 Tailwheel leg
129 Elevator control cable rod link
130 Tail wheel leg shock-absorber
131 Rudder control cable
132 Fuselage stringer
133 Accumulator
134 Fuselage half ventral join
135 Electrical leads
136 Fuselage panel
137 Radio pack lower support frames
138 Entry foothold (spring loaded)
139 Wingroot fillet
140 Flap profile
141 Port flap frame
142 Port aileron frame
143 Aileron trim tab
144 Rear spar
145 Port wingtip
146 Port navigation light
147 Wing main spar outer section
148 Solid ribs
149 Leading-edge automatic slot
150 Rib cut-outs
151 Control link access plate
152 Wing rib stations
153 Port wing 20-mm MG FF cannon installation
154 Ammunition drum access panel
155 Inboard rib cut-outs
156 Flap visual position indicator
157 Control access panel
158 Main spar/fuselage attachment fairing
159 Wing control surface cable pulleys
160 Port mainwheel well
161 Wheel well (zipped) fabric shield
162 20-mm MG FF wing cannon
163 Wing front spar
164 Undercarriage leg tunnel rib cut-outs
165 Undercarriage lock mechanism
166 Wing/fuselage end rib
167 Undercarriage actuating cylinder
168 Mainwheel leg/fuselage attachment bracket
169 Leg pivot point
170 Mainwheel oleo leg
171 Mainwheel leg door
172 Brake lines
173 Torque links
174 Mainwheel hub
175 Axle
176 Port main wheel
177 Mainwheel half-door
178 Ventral ETC centre-line stores pylon, possible loads include:
179 Early-type (wooden) drop tank
180 66-Imp gal (300-litre) (Junkers) metal drop tank
181 551-lb (250-kg) HE bomb, or
182 551-lb (250-kg) SAP bomb

Bf 109E-4

This Messerschmitt Bf 109E-4 is part of I. Gruppe, Jagdgeschwader 3 and was based at Grandvillier, France in August 1940. Gruppenkommandeur Hans von Hahn's aircraft, it displays the *Tatzelwurm* (a dragon resembling a crested or spiked worm) on the nose, a symbol carried by all the aircraft in I. Gruppe of JG 3. The emblem was applied in green to Stab (staff) aircraft, in white to aircraft of 1. Staffel, in red to aircraft of 2. Staffel, and in yellow to aircraft of 3. Staffel. The foremost part of the spinner was painted in the same colour.

Up-engined 'Emils'

The Bf 109E-4/N was powered by a DB 601N engine, whose flattened (instead of concave) piston heads gave a higher compression ratio, and which used higher octane fuel. The Bf 109E-7 was similarly powered, and the Bf 109E-7/Z introduced a nitrous oxide (GM1) supercharger injection system, which boosted altitude performance and was known as the 'Ha-Ha' system. A similar performance increase was provided both in the Bf 109E-8 and the recce-roled Bf 109E-9 by the new DB 601E engine, with increased maximum rpm, improved supercharging, and a take-off power of 1,350 hp (1006 kW). The downside was that the new engine was considerably less economical and, even with a 22-Imp gal (100-litre) increase in fuel capacity (to 88 Imp gal/400 litres), the Bf 109E's already inadequate reach was not greatly improved.

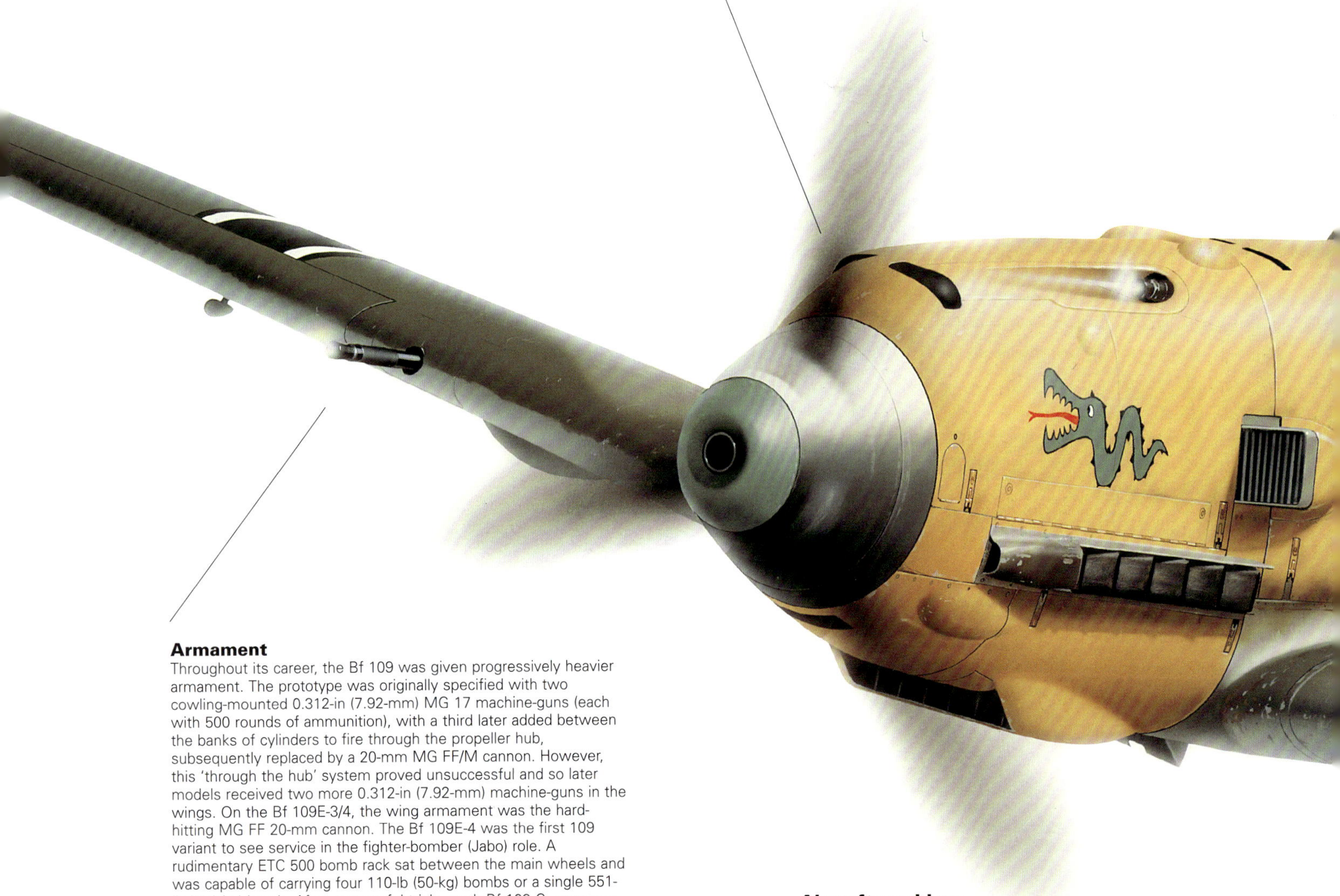

Armament

Throughout its career, the Bf 109 was given progressively heavier armament. The prototype was originally specified with two cowling-mounted 0.312-in (7.92-mm) MG 17 machine-guns (each with 500 rounds of ammunition), with a third later added between the banks of cylinders to fire through the propeller hub, subsequently replaced by a 20-mm MG FF/M cannon. However, this 'through the hub' system proved unsuccessful and so later models received two more 0.312-in (7.92-mm) machine-guns in the wings. On the Bf 109E-3/4, the wing armament was the hard-hitting MG FF 20-mm cannon. The Bf 109E-4 was the first 109 variant to see service in the fighter-bomber (Jabo) role. A rudimentary ETC 500 bomb rack sat between the main wheels and was capable of carrying four 110-lb (50-kg) bombs or a single 551-lb (250-kg) bomb. After successful trials, each Bf 109 Gruppe was ordered to convert some of its Staffeln to the Jabo role. These aircraft received the designation BF 109E-4/B to illustrate their new role.

Aircraft markings

In the confused cut-and-thrust of fighter-versus-fighter combat, a means of easy recognition was regarded as essential and, from mid-August, the Bf 109Es had wing and tail tips painted in white or yellow, together with a patch at the top of the rudder. Whole rudders were soon painted in yellow, and the tips of noses (and sometimes whole cowlings) followed. Some aircraft had tail units that were painted yellow in their entirety, but this was unusual.

Camouflage
The Luftwaffe's Messerschmitt Bf 109s originally wore dark-green upper surfaces, with light blue-grey (Hellblau) undersides, but these colours were replaced by a disruptive two-tone green camouflage on the top surfaces. The two shades of green were so similar that they looked almost like a single colour, and weathering further reduced the differentiation between the shades. In the winter of 1939, the top surface colours were limited to the upper surfaces of the wings and tailplanes, plus the top decking of the fuselage. The fuselage sides and fin were painted Hellblau. On most aircraft, the top surface camouflage was repainted, with RLM 71 Dunkelgrün (the lighter of the two tones used previously) and RLM 02 Grau (a pale grey-green). These had a greater tonal variation than the previous colours. The basic new camouflage scheme was then modified at unit level during the Battle of Britain. Some aircraft had a neutral grey area painted along the wing leading-edge to soften the demarcation between the dark topsides and light undersurfaces. Fuselage sides were often overpainted with a mottled RLM 02 Grau, as seen on this aircraft.

Personal markings
The aircraft belonging to the Gruppenkommandeur of JG 3's I. Gruppe, Hans 'Vadder' ('Daddy') von Hahn, carried his own distinctive version of the standard double chevron (Winkel) or 'triangle and chevron' Gruppenkommandeur's marking. Under the cockpit was a cockerel's head, a play on his name (Hahn meaning cockerel in German), an insignia also used by his better-known namesake, Hans 'Assi' Hahn. Von Hahn began the war as the adjutant of JG 53, later becoming the Staffelkapitän of 8./JG 53 under Werner Mölders during the Battle of France. At the height of the Battle of Britain, when Herman Göring replaced his older fighter leaders (many of whom, like Theo Osterkamp and Dr Mix, had fought and even become aces in World War I), von Hahn was promoted to become the Gruppenkommandeur of I./JG 3, replacing Gunther Lützow who was, in turn, promoted to become Geschwaderkommodore. Von Hahn took the *Tatzelwurm* Gruppe to Russia in June 1941 and returned with the Gruppe in January 1942, having added 17 kills to his tally, thus bringing his total to 34 in 300 missions. The Gruppe was redesignated II./JG 1 and started converting to the Fw 190, and von Hahn began the first of a succession of staff and training appointments, ending the war as Jafü (fighter leader) Oberitalien (Upper Italy). He died in Frankfurt in 1957, aged 53.

Anatomy
The Bf 109 was built as a semi-monocoque fuselage, constructed in two halves and joined at the centreline. The undercarriage was attached directly to the fuselage, allowing the wing to be easily removed without the need for assembling jacks. The wing itself was a single-spar construction, the cannon-armed variants having a strengthened cut-out in the spar to accommodate the weapon. Similarly, the tail and tailplanes were simple single-spar structures, the tailplanes being externally braced by a single strut. The incidence of the tailplane could be altered via a torque tube in landing configuration to offset the effects of deploying both the flaps and the drooping ailerons. The wing slots, flaps and ailerons were actuated by rods, but the tail surfaces had cable runs through the rear fuselage, the rudder having a simple left/right pivoting arm. A tube was located transversely across the fuselage just forward of the tail unit, through which a rod could be inserted for lifting the aircraft's tail. The rear fuselage was largely empty but did provide housing for the two oxygen bottles and the FuG 7 radio equipment, which was supported on brackets attached to both the upper and lower fuselage frames. In the forward fuselage, the engine was cantilevered on A-frame engine bearers, above which were mounted the guns. The ammunition tanks were located below and behind the engine, feeding upwards into the weapons. The glycol coolant was held in a header tank just aft of the propeller.

Messerschimitt Bf 109G/K

It was not until the development of the G model that the Messerschmitt Bf 109 went into mass-scale production and with this new variant came the hitherto unseen practice of producing a myriad of versions. By the middle of the war, the Bf 109 had lost its supremacy, yet continued to remain the backbone of the Luftwaffe until the end of the war. Even during the final days of the Reich, the late-model Bf 109Gs and Ks managed to put up fierce resistance to the invading Allies.

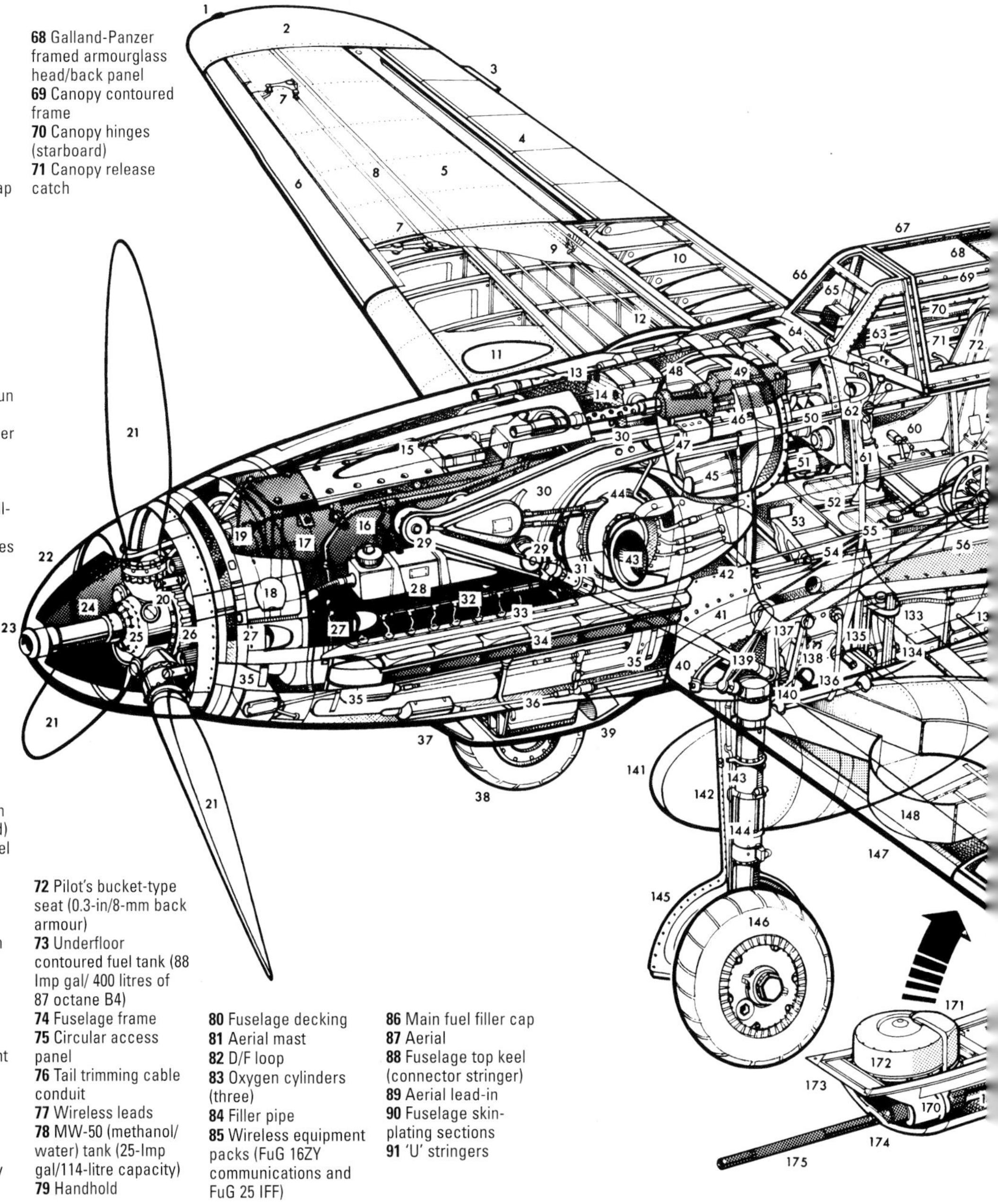

Bf 109G-14

Cutaway key
1 Starboard navigation light
2 Starboard wingtip
3 Fixed trim tab
4 Starboard Frise-type aileron
5 Flush-rivetted stressed wing skinning
6 Handley Page automatic leading-edge slot
7 Slot control linkage
8 Slot equaliser rod
9 Aileron control linkage
10 Fabric-covered flap section
11 Wheel fairing
12 Port fuselage machine-gun ammunition feed fairing
13 Port 13-mm Rheinmetall-Borsig MG 131 machine-gun
14 Engine accessories
15 Starboard machine-gun trough
16 Daimler-Benz DB 605AM 12-cylinder inverted-Vee liquid-cooled engine
17 Detachable cowling panel
18 Oil filter access
19 Oil tank
20 Propeller pitch-change mechanism
21 VDM electrically-operated constant-speed propeller
22 Spinner
23 Engine-mounted cannon muzzle
24 Blast tube
25 Propeller hub
26 Spinner back plate
27 Auxiliary cooling intakes
28 Coolant header tank
29 Anti-vibration rubber engine mounting pads
30 Elektron forged engine bearer
31 Engine bearer support strut attachment
32 Plug leads
33 Exhaust manifold fairing strip
34 Ejector exhausts
35 Cowling fasteners
36 Oil cooler
37 Oil cooler intake
38 Starboard mainwheel
39 Oil cooler outlet flap
40 Wingroot fillet
41 Wing/fuselage fairing
42 Firewall/bulkhead
43 Supercharger air intake
44 Supercharger assembly
45 20-mm cannon magazine drum
46 13-mm machine-gun ammunition feed
47 Engine bearer upper attachment
48 Ammunition feed fairing
49 13-mm Rheinmetall-Borsig MG 131 machine-gun breeches
50 Instrument panel
51 20-mm Mauser MG 151/20 cannon breech
52 Heelrests
53 Rudder pedals
54 Undercarriage emergency retraction cables
55 Fuselage frame
56 Wing/fuselage fairing
57 Undercarriage emergency retraction handwheel (outboard)
58 Tail trim handwheel (inboard)
59 Seat harness
60 Throttle lever
61 Control column
62 Cockpit ventilation inlet
63 Revi 16B reflector gunsight (folding)
64 Armoured windshield frame
65 Anti-glare gunsight screen
66 3½-in (90-mm) armoured glass windscreen
67 Erla-Haube clear-vision hinged canopy
68 Galland-Panzer framed armourglass head/back panel
69 Canopy contoured frame
70 Canopy hinges (starboard)
71 Canopy release catch
72 Pilot's bucket-type seat (0.3-in/8-mm back armour)
73 Underfloor contoured fuel tank (88 Imp gal/ 400 litres of 87 octane B4)
74 Fuselage frame
75 Circular access panel
76 Tail trimming cable conduit
77 Wireless leads
78 MW-50 (methanol/ water) tank (25-Imp gal/114-litre capacity)
79 Handhold
80 Fuselage decking
81 Aerial mast
82 D/F loop
83 Oxygen cylinders (three)
84 Filler pipe
85 Wireless equipment packs (FuG 16ZY communications and FuG 25 IFF)
86 Main fuel filler cap
87 Aerial
88 Fuselage top keel (connector stringer)
89 Aerial lead-in
90 Fuselage skin-plating sections
91 'U' stringers

As the tide of war turned in favour of the Allies, the retreating Germans left a number of aircraft in their wake. This Bf 109G-2/Trop was captured at Gambut in Libya and made airworthy by an RAAF crew. It was initially transported around the Middle East for testing and eventually made it to England, where it was used in comparative trials against Allied fighters. It survived the war and remained in storage for many years until it was restored to an airworthy condition. The aircraft's first post-rebuild flight took place on 17 March 1991 and the '109 received its old identification of 'Black 6'. However, in 1998 the aircraft made a forced landing and was damaged to such an extent that it will spend the rest of its days in the static line at Duxford Air Museum.

SPECIFICATION

Bf 109G-6

Dimensions

Length: 29 ft 7 in (9.02 m)
Wingspan: 32 ft 6½ in (9.92 m)
Height: 8 ft 2½ in (2.50 m)
Wing area: 172.75 sq ft (16.05 m²)

Powerplant

One Daimler Benz DB 605AM 12-cylinder inverted-Vee liquid-cooled engine rated at 1,475 hp (1100 kW) for take-off and 1,355 hp (1010 kW) at 18,700 ft (5700 m)

Weights

Empty: 5,953 lb (2700 kg)
Loaded: 6,495 lb (3150 kg)

Performance

Maximum speed at 22,967 ft (7000 m): 387 mph (623 km/h)
Maximum speed at sea level: 338 mph (544 km/h)
Climb to 19,868 ft (6000 m): 6 minutes
Service ceiling: 38,551 ft (11750 m)
Absolute ceiling: 39,700 ft (12100 m)

Range

Normal: 450 miles (725 km)
Maximum: 615 miles (990 km)

Armament

One 30-mm Rheinmentall Borsig MK 108 engine-mounted cannon with 60 rounds or two 20-mm Mauser MG 151/20 cannon with 150 rounds, and two 13-mm Rheinmetall-Borsig MG 131 fuselage-mounted machine-guns with 300 rounds per gun

Bf 109K-4

Same as Bf 109G except for:

Powerplant

One Daimler Benz DB 605DCM 12-cylinder inverted-Vee liquid-cooled engine rated at 2,000 hp (1492 kW) for take-off and 1,800 hp (1343 kW) at 16,400 ft (5000 m)

Weights

Loaded: 7,438 lb (3370 kg)

Performance

Maximum speed at 24,610 ft (7500 m): 387 mph (623 km/h)
Maximum speed at sea level: 378 mph (608 km/h)

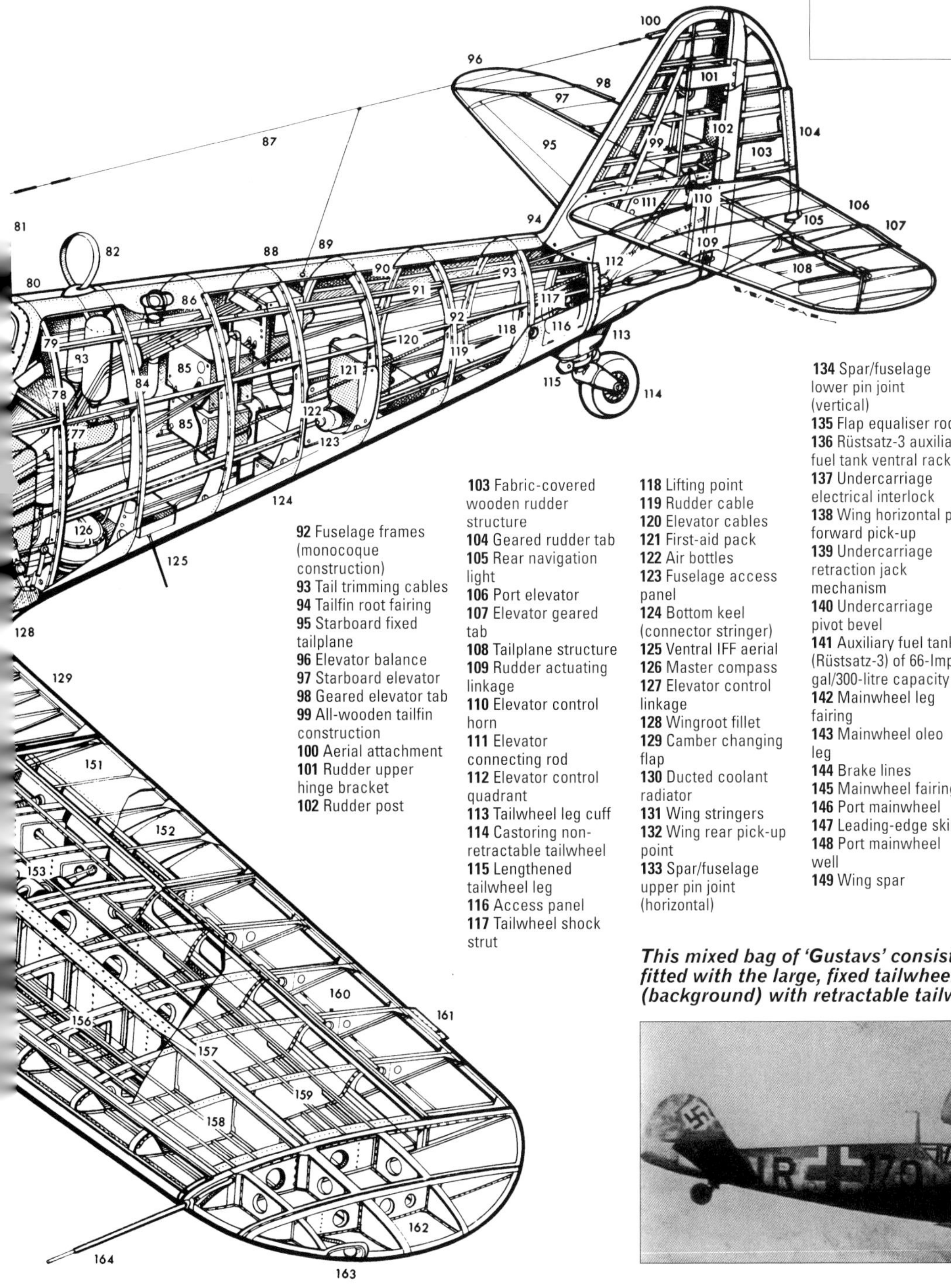

92 Fuselage frames (monocoque construction)
93 Tail trimming cables
94 Tailfin root fairing
95 Starboard fixed tailplane
96 Elevator balance
97 Starboard elevator
98 Geared elevator tab
99 All-wooden tailfin construction
100 Aerial attachment
101 Rudder upper hinge bracket
102 Rudder post
103 Fabric-covered wooden rudder structure
104 Geared rudder tab
105 Rear navigation light
106 Port elevator
107 Elevator geared tab
108 Tailplane structure
109 Rudder actuating linkage
110 Elevator control horn
111 Elevator connecting rod
112 Elevator control quadrant
113 Tailwheel leg cuff
114 Castoring non-retractable tailwheel
115 Lengthened tailwheel leg
116 Access panel
117 Tailwheel shock strut
118 Lifting point
119 Rudder cable
120 Elevator cables
121 First-aid pack
122 Air bottles
123 Fuselage access panel
124 Bottom keel (connector stringer)
125 Ventral IFF aerial
126 Master compass
127 Elevator control linkage
128 Wingroot fillet
129 Camber changing flap
130 Ducted coolant radiator
131 Wing stringers
132 Wing rear pick-up point
133 Spar/fuselage upper pin joint (horizontal)
134 Spar/fuselage lower pin joint (vertical)
135 Flap equaliser rod
136 Rüstsatz-3 auxiliary fuel tank ventral rack
137 Undercarriage electrical interlock
138 Wing horizontal pin forward pick-up
139 Undercarriage retraction jack mechanism
140 Undercarriage pivot bevel
141 Auxiliary fuel tank (Rüstsatz-3) of 66-Imp gal/300-litre capacity
142 Mainwheel leg fairing
143 Mainwheel oleo leg
144 Brake lines
145 Mainwheel fairing
146 Port mainwheel
147 Leading-edge skin
148 Port mainwheel well
149 Wing spar
150 Flap actuating linkage
151 Fabric-covered control surfaces
152 Slotted flap structure
153 Leading-edge slot-actuating mechanism
154 Slot equaliser rod
155 Handley Page automatic leading-edge slot
156 Wing stringers
157 Spar flange decrease
158 Wing ribs
159 Flush-rivetted stressed wing skinning
160 Metal framed Frise-type aileron
161 Fixed trim tab
162 Wingtip construction
163 Port navigation light
164 Angled pitot head
165 Rüstsatz-6 optional underwing cannon gondola
166 14-point plug connection
167 Electrical junction box
168 Cannon rear mounting bracket
169 20-mm Mauser MG 151/20 cannon
170 Cannon front mounting bracket
171 Ammunition feed chute
172 Ammunition magazine drum
173 Underwing panel
174 Gondola fairing
175 Cannon barrel

This mixed bag of 'Gustavs' consists of a Bf 109G-3 (nearest the camera) fitted with the large, fixed tailwheel, a G-6 (centre), and an early machine (background) with retractable tailwheel, probably a G-2.

Bf 109K-4

III./JG 77 and III./JG 27 were the first units to re-equip with the Bf 109K-4, the first of which came off the production line in August 1944. The first recorded combat engagement involving K-4s came on 2 November, when JG 27 engaged Allied fighters escorting a large bomber force near Leipzig. In the ensuing fight, JG 27 suffered its worst one-day losses of the war to date – 27 pilots were killed and 11 wounded. Seven Allied 'kills' were claimed by the unit – all Mustangs. Later in the month other units received K-4s, including III./JG 3, III./JG 4, III./JG 26 and I./JG 27, and by 1 January 1945, the day of the Luftwaffe's ill-fated Operation Bodenplatte against Allied airfields, the following Gruppen had also received the variant: II./JG 2, I./JG 4, IV./JG 4 and II./JG 11. It is believed that no unit was exclusively equipped with the Bf 109K; a mix of Gs and Ks was the norm. During January 1945, a number of units equipped with K-4s transferred to the Eastern Front, including III./JG 27, and during late January and February a number of eastern Gruppen was issued with the type.

Engine and propeller
Bf 109K-0 pre-production aircraft and the first K-4s were powered by the DB 605DB variant of Daimler-Benz's inverted-Vee, 12-cylinder, liquid-cooled DB 605. (The DB 605D series was derived from the 605A, featuring a larger supercharger derived from that designed for the DB 603 engine.) Later machines utilised the DB 605DC, differing from the 605DB in having a higher maximum supercharger boost pressure and being strengthened accordingly. The higher boost pressure allowed maximum power output to be attained while using lower-octane fuel, a valuable feature as shortages of first-rate fuel worsened as the war continued. Like its predecessors, the Bf 109K-4 also used MW-50 (methanol-water) injection to boost power, but only when the engine was running on lower-octane fuel. The DB 605 drove a wide-bladed VDM9-12159A propeller with a 10-ft (3-m) diameter, and used fuel that was held in an 88-Imp gal (400-litre) tank located behind the pilot. A Rüstsatz kit (R3) was applied to some aircraft to improve range, with a 66-Imp gal (300-litre) drop tank suspended under the fuselage.

Reconnaissance variants
Among the Rüstzustände kits planned for the K-4 were three which gave the type a reconnaissance capability: the R2, R5 and R6. The R2 was to carry an Rb 50/30 or Rb 70/30 vertical camera behind the cockpit, the R5 an Rb 32/7x9 or Rb 12/7x9 and the R6 a BSK 16 gun camera in the port wing leading edge for bomber and fighter-bomber units.

Identifying features

Several features of the Bf 109K-4's airframe had been carried over from late-production 'Gustavs'. These included rectangular upper wing surface bulges to accommodate wider, low-pressure main gear tyres (fitted to improve ground-handling), a bulged DB 605D series sump cover (below the leading exhaust nozzle) and the FuG 16ZY radio aerial under the port wing. The rudder usually had a Flettner tab and two fixed trim tabs (to obviate the need for rudder applications when diving and climbing), although some aircraft did not have the fixed tabs. Unique to the K-4 were a relocated DF loop, a refined lower engine cowling design, a relocated fuselage access hatch, a taller, retractable tailwheel and additional main gear wheel well doors (often removed in the field). Fixed aileron trim tabs were also a feature of the K-4; a few aircraft were additionally equipped with hinged Flettner aileron tabs.

Armament

Bf 109K-4s carried the standard Bf 109G-6 armament of a single, engine-mounted, Rheinmetall-Borsig 30-mm MK 108 cannon and a pair of Rheinmetall-Borsig MG 131 13-mm machine-guns above the engine, each with 300 rounds. The MK 108 was a highly effective weapon, but was prone to jamming during hard manoeuvring, leaving the aircraft with just two machine-guns. As an alternative to the MK 108, the earlier Mauser MG 151/20 20-mm cannon, as fitted to early Bf 109Gs, could be installed, but this seems not to have been deemed necessary in production K-4s. The wing of the Bf 109K was built to accommodate not only Mauser MG 151/20 cannon but, alternatively, two MK 108s using an internal mount. An anti-bomber variant, the K-6, was to be equipped to the same standard as the late K-4, but with the addition of an MK 108 cannon in each wing. The weight penalty was considerable and handling suffered accordingly. Although prototype service trials, probably with III./JG 3, were carried out from February 1945, the K-6 is thought not to have entered production. Various Rüstsätze kits of the type developed for the Bf 109G could be applied to the K-4. The R1 was a fighter-bomber modification allowing the carriage of a 551-lb (250-kg) or 1,102-lb (500-kg) bomb under the fuselage, while the R4 allowed the installation of an additional MG 151/20 under each wing.

Messerschmitt Me 110

The Luftwaffe used Bf 110Cs during the lightning war against Poland during September 1939 where they enjoyed relative immunity. Indeed, even when German attention was turned westwards, the high-flying Bf 110s could outperform the Hurricane Mk I and outclimb the Spitfire Mk I. It was only when the Bf 110 was pushed into the bomber escort role, a task for which it was never intended, that it found itself outclassed.

Bf110G-4b/R3

Cutaway key

1 *Hirschgeweih* (stag's antlers) array for the FuG 220b Lichtenstein SN-2 radar
2 Quad di-pole type antenna for the FuG 212 Lichtenstein C-1 radar
3 Camera gun
4 Cannon muzzles
5 Cannon ports
6 Blast tubes
7 Starboard mainwheel
8 Armour plate (0.4 in; 10 mm)
9 Twin 30-mm Rheinmetall Borsig Mk 103 (Rüstsatz/ Field Conversion Set 3) with 135 rpg
10 Armoured bulkhead
11 Supercharger intake
12 Position of nacelle-mounted instruments on day-fighter model
13 Exhaust flame damper
14 Auxiliary tank
15 Three-bladed VDM airscrew
16 Leading-edge automatic slat
17 Pitot tube
18 FuG 227/1 Flensburg homing aerial fitted to some aircraft by forward maintenance units (to home onto Monica tailwarning radar emissions)
19 Stressed wing skinning
20 Starboard aileron
21 Trim tab
22 Slotted flap
23 Hinged canopy roof
24 Armoured glass windscreen (2.4 in; 60 mm)
25 Instrument panel
26 Cockpit floor armour (0.16 in; 4 mm)
27 Twin 20-mm Mauser MG 151 cannon with 300 rounds (port) and 350 rounds (starboard)
28 Pilot's seat
29 Control column
30 Pilot's back and head armour (0.315 in; 8 mm)
31 Cannon magazine
32 Centre section carrythrough
33 Radar operator's swivel seat
34 D/F loop
35 Aerial mast
36 Upward-firing cannon muzzles
37 Two 30-mm MK 108 cannon in *schräge Musik* (oblique music) installation firing obliquely upward (optional installation supplied as an Umrust-Bausatz/Factory Conversion Set)
38 Ammunition drums
39 Aft cockpit bulkhead
40 FuG 10P HF R/T set
41 FuB 12F airfield blind-approach receiver

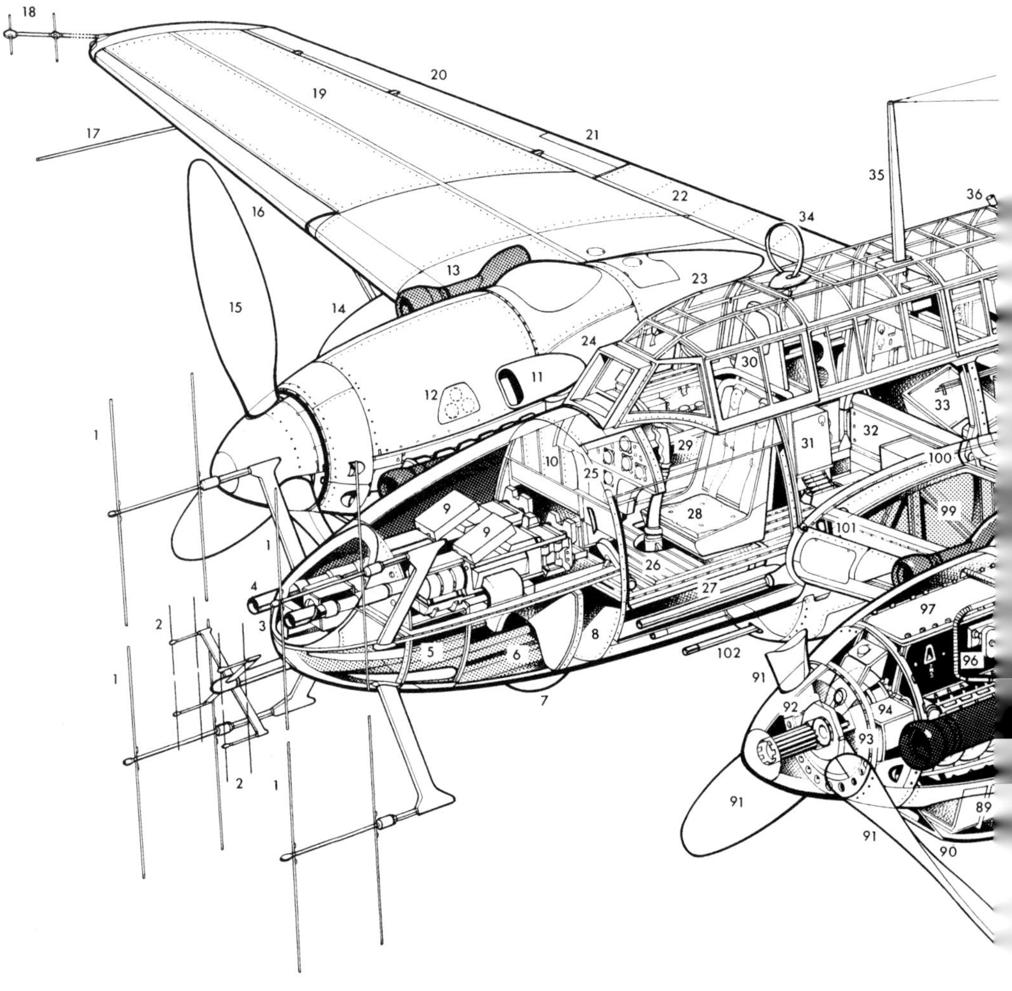

Sharkmouths adorned numerous wartime aircraft, few as strikingly as the gaping **Haifisch** *jaws on the* ***Bf 110s*** *of* ***II./ZG 76****, the source of many of the crews and aircraft that would go on to equip* ***NJG 1****. The sharkmouth markings were retained on individual* ***ZG 76*** *aircraft transferred to the night-fighter arm, though in some cases the white teeth were oversprayed in black, this being the standard colour for night-fighters before serious investigation into night camouflage began. Aircraft adorned like this were often photographed, more frequently than their plain counterparts, so as so to generate morale for the Luftwaffe's new night-fighting unit. Below the cockpit, the first of eight flags (representing the nations against which* ***ZG 76*** *fought) can just be seen.*

SPECIFICATION

Bf 110C-4

Dimensions

Length: 41 ft 6 in (12.65 m)
Height: 11 ft 6 in (3.50 m)
Wingspan: 53 ft 3 in (16.27 m)
Wing area: 413.3 sq ft (38.40 m²)

Powerplant

Two 1,100-hp (821-kW) Daimler Benz DB 601A inverted-Vee 12-cylinder piston engines

Weights

Empty: 11,454 lb (5200 kg)
Maximum take-off: 14,881 lb (6750 kg)

Performance

Maximum speed at 22,965 ft (7000 m): 349 mph (560 km/h)
Initial climb rate: 2165 ft (660 m) per minute
Cruising speed 22,965 ft (7000 m): 301 mph (480 km/h)
Service ceiling: 32,810 ft (10000 m)
Normal range: 482 miles (775 km)
Maximum range: 565 miles (910 km)

Armament

Two 20-mm MG 151 cannon and four 0.311-in (7.92-mm) MG 17 guns in the nose firing forward, and one 0.311-in (7.92-mm) MG 15 gun in the rear cockpit firing aft. Provision was also made for the carriage of two 551-lb (250-kg) bombs for the *Jabo* role

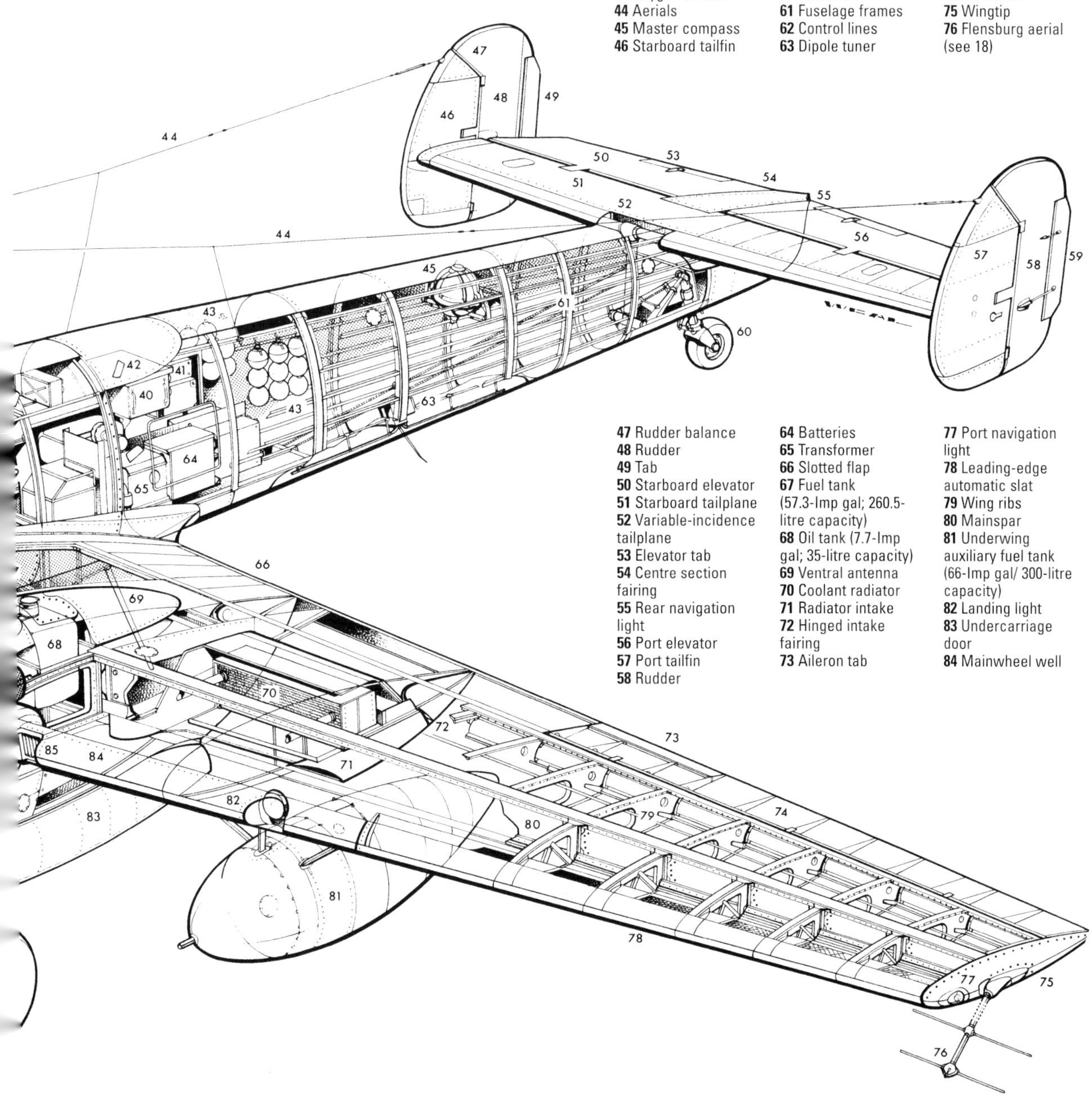

42 Handhold
43 Oxygen bottles
44 Aerials
45 Master compass
46 Starboard tailfin
47 Rudder balance
48 Rudder
49 Tab
50 Starboard elevator
51 Starboard tailplane
52 Variable-incidence tailplane
53 Elevator tab
54 Centre section fairing
55 Rear navigation light
56 Port elevator
57 Port tailfin
58 Rudder
59 Hinged tab
60 Tailwheel
61 Fuselage frames
62 Control lines
63 Dipole tuner
64 Batteries
65 Transformer
66 Slotted flap
67 Fuel tank (57.3-Imp gal; 260.5-litre capacity)
68 Oil tank (7.7-Imp gal; 35-litre capacity)
69 Ventral antenna
70 Coolant radiator
71 Radiator intake
72 Hinged intake fairing
73 Aileron tab
74 Aileron construction
75 Wingtip
76 Flensburg aerial (see 18)
77 Port navigation light
78 Leading-edge automatic slat
79 Wing ribs
80 Mainspar
81 Underwing auxiliary fuel tank (66-Imp gal/ 300-litre capacity)
82 Landing light
83 Undercarriage door
84 Mainwheel well
85 Supercharger intake
86 Undercarriage pivot point
87 Mainwheel leg
88 Mainwheel
89 Oil cooler
90 Oil cooler intake
91 VDM propeller
92 Pitch-change mechanism
93 Armoured ring (0.2 in; 5 mm)
94 Coolant tank
95 Exhaust flame damper
96 Anti-vibration engine mounting pad
97 Daimler-Benz DB 605B-1 12-cylinder inverted-Vee engine (rated at 1,475 hp/ 1100 kW for take-off and 1,355 hp/1011 kW at 18,700 ft/5700 m)
98 Forged engine bearer
99 Fuel tank (82.5-Imp gal; 375-litre capacity)
100 Fuselage/mainspar attachment point
101 Fuselage/forward auxiliary spar attachment point
102 Waffenwanne 151Z, a ventral tray housing a pair of 20-mm MG 151 cannon (optional)

Bf 110C-4/B

Bf 110s sported many camouflage schemes depending on season and theatre. This aircraft bears what was to become the standard scheme – light-grey fuselage and undersurfaces with darker mottling on the fuselage sides, and disruptive camouflage on the upper surfaces. The most famous marking applied to a Bf 110 was the *Wespen* (wasp) of Zerstörergeschwader 1. The nucleus of II./SKG (schnelles Kampfgeschwader) 210 was provided by 1. Staffel of ZG 1, so the wasp markings stayed with the unit. In 1942 II./SKG 210 returned to ZG 1 control, reuniting the markings with their original operator. The Erprobungsgruppe 210 was established with ZG 1 personnel as an experimental unit for the evaluation of *Jabo* (fighter-bomber) tactics, and it was assigned the first Bf 110C-4/Bs. It commenced cross-Channel raids in July 1940, and became effective at surprise low-level attacks. After being renamed as a Schnellkampfgeschwader (fast bomber wing), the unit dispatched its II. Gruppe to the Eastern Front. The unit fought through the southern sector to Stalingrad, by which time it had been redesignated II./ZG 1 and had been all but decimated.

Cockpit
The Bf 110's cockpit was designed for a crew of three comprising, from front to rear, a pilot, radio operator and gunner. This three-man configuration, which was to prove so unwieldy as the war progressed, had also been adopted by British and French aircraft of the time. In operational conditions, however, the roles of rear gunner and radio operator were usually combined. The view forward and above was extremely good for both crew members, whereas little could be seen downward with this overall layout. The Bf 110C-4 was the first model to introduce some armour protection for the crew.

Powerplant
In deference to its newly-found Jabo role, the Bf 110C-4/B featured the uprated Daimler Benz DB 601N engines (as opposed to the DB 601A of the regular Bf 110C-4). Although standard power outputs were similar, the DB 601N offered 1,200 hp (895 kW) with full boost for one minute at take-off, and an emergency power rating of 1,270 hp (947 kW) for a similar period at 16,400 ft (5000 m). The extra power was achieved by the use of 96 octane C3 fuel, increased compression ratio and redesigned piston heads. Early models of the Bf 110 featured a deep radiator bath under each engine, but the Bf 110C introduced shallow glycol radiators under the wings, outboard of each engine. The radiator intake led back up into the wing, air flow being controlled by electrically-operated exit flaps. The oil cooler for each engine was mounted underneath, fed with air by a small chin inlet. The oil tank was situated behind the engine block and had a capacity of 9 Imp gal (43 litres).

Armament
The nose of the Bf 110C-4B carried four 0.311-in (7.9-mm) MG 17 machine-guns, staggered so that they fitted into the slim fuselage. Each weapon was provided with 1,000 rounds, held in magazines beneath the guns. Two MG FF 20-mm cannon were mounted in the lower fuselage beneath the pilot's seat, but firing through blast troughs underneath the nose. Each had 180 rounds, manually loaded from tanks in the gunner's compartment. To protect the Bf 110 from rear attack, the gunner/radio operator was provided with a single 0.311-in (7.9-mm) MG 15 machine-gun with 750 rounds on a flexible Arado mount. His station was provided with an upward-hinging hood which provided a better field of fire and some protection from the airflow when firing. The Bf 110C-4/B was the first version to be dedicated to the *Jabo* role, for which it was provided with a pair of ETC 250 bomb racks, nestling under the fuselage centre-section. These each carried a 551-lb (250 kg) bomb.

Tail/wing units
The Bf 110's tail was mounted simply on the top of the rear fuselage. The tailplane was small, as the long rear fuselage provided a large moment. Two elevators were fitted, each tabbed and balanced with a notch to allow full rudder deflection. The endplate fins had aerodynamically-balanced rudders. As with that of the Bf 109, the Bf 110's wing featured a single spar, joined to a carry-through member in the fuselage, running behind the pilot's seat. The trailing edge of the wing had large single-slotted flaps inboard, unbroken even behind the engine nacelle. Outboard were slotted ailerons with external mass balances. On the outer section of the leading edge were automatic Handley Page slots.

Fuel
Fuel was provided in four tanks, located in the inner wings either side of the main spar. The forward tanks each held 82 Imp gal (373 litres), while the rear tanks each held 58 Imp gal (264 litres). Later versions could carry drop tanks to increase range.

Undercarriage
The main undercarriage was a sturdy and simple single-strut, single-wheel construction, retracting to the rear into bays in the rear of the engine nacelles, where it was fully enclosed by two doors. The tailwheel was not retractable. The extended tailcone shown here housed a dinghy and survival equipment.

Armament
The nose of the Bf 110C-4B carried four 0.311-in (7.9-mm) MG 17 machine-guns, staggered so that they fitted into the slim fuselage. Each weapon was provided with 1,000 rounds, held in magazines beneath the guns. Two MG FF 20-mm cannon were mounted in the lower fuselage beneath the pilot's seat, but firing through blast troughs underneath the nose. Each had 180 rounds, manually loaded from tanks in the gunner's compartment. To protect the Bf 110 from rear attack, the gunner/radio operator was provided with a single 0.311-in (7.9-mm) MG 15 machine-gun with 750 rounds on a flexible Arado mount. His station was provided with an upward-hinging hood which provided a better field of fire and some protection from the airflow when firing. The Bf 110C-4/B was the first version to be dedicated to the *Jabo* role, for which it was provided with a pair of ETC 250 bomb racks, nestling under the fuselage centre-section. These each carried a 551-lb (250 kg) bomb.

Messerschmitt Me 163

Above: An Me 163B-1a launching at Bad Zwischenahn, home of the trials unit Erprobungskommando 16, which accepted its first Me 163B during May 1944.

Me 163B Komet

Cutaway key

1 Generator drive propeller
2 Generator
3 Compressed air bottle
4 Battery and electronics packs
5 Cockpit ventilation intake
6 Solid armour (⅗-in; 15-mm) nosecone
7 Accumulator pressuriser
8 Direct cockpit air intake
9 FuG 25a radio pack
10 Rudder control assembly
11 Hydraulic and compressed air points
12 Elevon control rocker-bar
13 Control relay
14 Flying controls assembly box
15 Plastic rudder pedals
16 Radio tuning controls
17 Torque shaft
18 Port T-Stoff cockpit tank (13 Imp gal/60 litre capacity)
19 Control column
20 Hinged instrument panel
21 Armourglass windscreen brace
22 Revi 16B gunsight
23 Armourglass internal windscreen (3½-in; 90-mm)
24 Armament and radio switches (starboard console)
25 Pilot's seat
26 Back armour (⅓-in; 8-mm)
27 Head and shoulder armour ½-in/13-mm)
28 Radio frequency selector pack
29 Headrest
30 Mechanically jettisonable hinged canopy
31 Ventilation panel
32 Fixed leading-edge wing slot
33 Trim tab
34 Fabric-covered starboard elevon
35 Position of underwing landing flap
36 Inboard trim flap
37 FuG 16yz radio receiving aerial
38 T-Stoff filler cap
39 Main unprotected T-Stoff fuselage tank (229 Imp gal/1040 litre capacity)
40 Aft cockpit glazing
41 Port cannon ammunition box (60 rounds)
42 Starboard cannon ammunition box (60 rounds
43 Ammunition feed chute
44 T-Stoff starter tank
45 Rudder control upper bell crank
46 C-Stoff filler cap
47 HWK 509A-1 motor turbine housing
48 Main rocket motor mounting frame
49 Rudder control rod
50 Disconnect point
51 Aerial matching unit
52 Fin front spar/ fusetage attachment point
53 Tailfin construction
54 Rudder horn balance
55 Rudder upper hinge
56 Rudder frame
57 Rudder trim tab
58 Rudder control rocker-bar
59 Linkage fairing
60 Fin rear spar/ fuselage attachment point
61 Rocket motor combustion chamber
62 Tailpipe
63 Rudder root fairing
64 Rocket thrust orifice
65 Vent pipe outlet
66 Hydraulic cylinder
67 Lifting point
68 Tailwheel fairing
69 Steerable tailwheel
70 Tailwheel axle fork
71 Tailwheel oleo
72 Tailwheel steering linkage
73 Coupling piece/ vertical lever
74 Wing root fillet
75 Combustion chamber support brace
76 Gun-cocking mechanism
77 Trim flap control angle gear (bulkhead mounted)
78 Worm gear
79 Trim flap mounting
80 Port inboard trim flap
81 Elevon mounting
82 Rocker-bar
83 Elevon actuation push rod
84 Port elevon
85 Wing rear spar
86 Trim tab
87 Elevon outboard hinge
88 Wingtip bumper
89 Wing construction
90 Fixed leading-edge wing slot
91 Elevon control bell crank
92 Position of port underwing landing flap
93 Push-rod in front spar
94 Front spar
95 FuG 25a aerial
96 Pitot head
97 Wing tank connecting pipe fairing

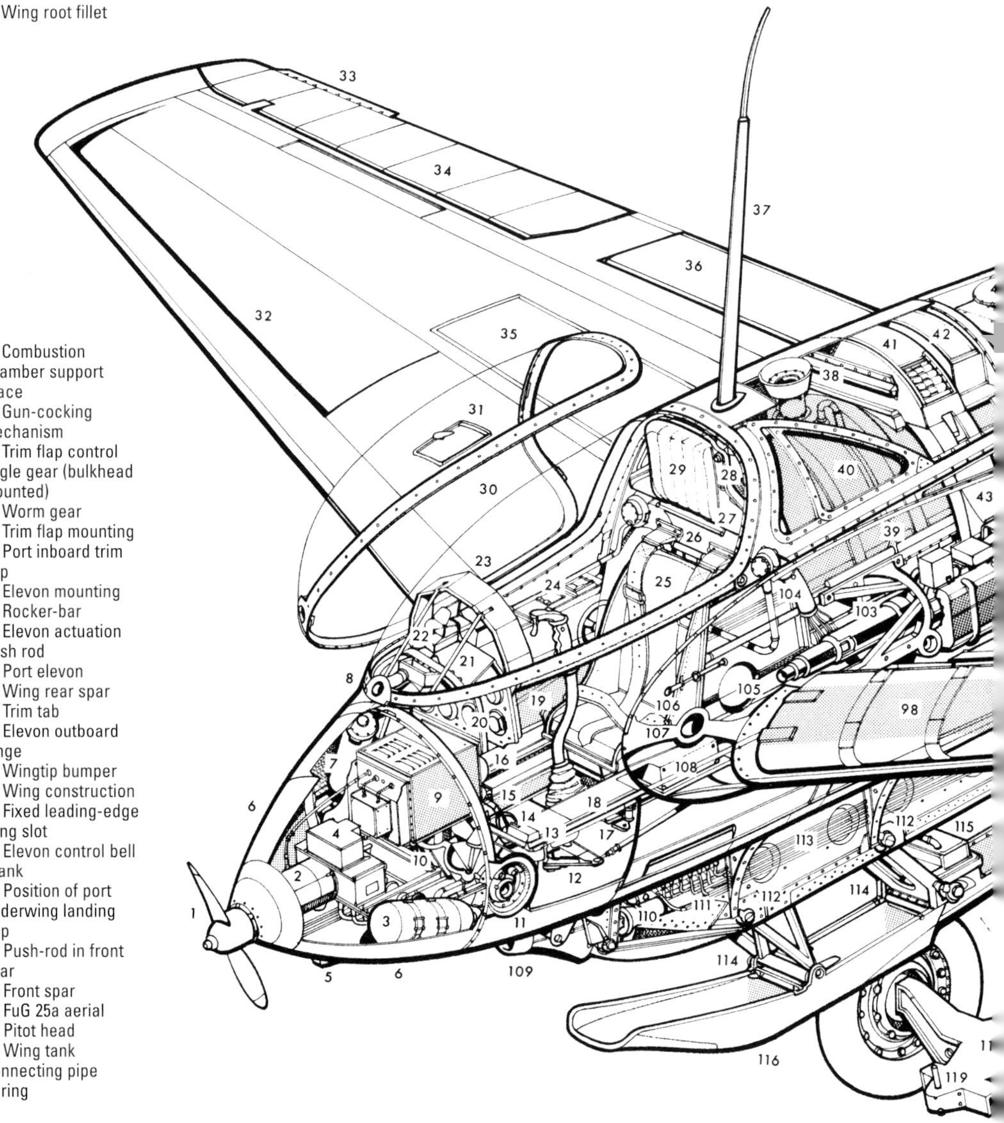

SPECIFICATION

Me 163B-1a Komet

Dimensions

Length: 19 ft 2 in (5.85 m)
Height on take-off dolly: 9 ft (2.76 m)
Wingspan: 30 ft 7 in (9.40 m)
Wing area: 199.10 sq ft (18.50 m²)

Powerplant

One Walter HWK 509A-1 or A-2 rocket motor pump-fed with hypergolic (spontaneously reacting) T-stoff and C-stoff, with high-altitude thrust of 3,748 lb (16.67 kN)

Weights

Empty: 4,190 lb (1900 kg)
Maximum take-off: 9,502 lb (4310 kg)

Performance

Maximum speed at low level: about 510-520 mph (830 km/h)
Maximum speed above 9,845 ft (3000 m): 597 mph (960 km/h)
Initial climb rate: 16,080 ft (4900 m) per minute
Practical range (not allowing for combat): 80 miles (130 km)
Maximum rocket endurance (allowing for periods at reduced thrust): 7 minutes 30 seconds
Service ceiling: 39,370 ft (12000 m)

Armament

Two 30-mm Rheinmetall MK 108 cannon each with 60 rounds

Above: Two of the Me 163B prototypes, V6 and V18, were modified with prototypes of the HWK 509C-1 motor with main and cruising thrust chambers, to give much better flight endurance. Here the V6 blows steam through its propellant lines in the summer of 1944. Note the repositioned retractable tailwheel.

First unit

In early 1943 a special Me 163B test squadron was formed at Karlshagen under Hauptmann Wolfgang Späte, but while this was still in its early stages Peenemunde was raided by the RAF and the unit, Erprobungskommando 16, was moved to Bad Zwischenahn. This was the centre for most Komet flying for the next year, and the aircraft became known to the Allies from high-flying reconnaissance photographs taken here in December 1943. By this time the programme had been further delayed by a raid of the very kind the Komet had been invented to prevent. The Messerschmitt factory at Regensburg was heavily hit by Boeing B-17s on 17 August 1943, many of the pre-production batch being destroyed. The main production, however, was to be dispersed throughout Germany under the control of Klemm Technik, with final assembly at a secret Schwarzwald (Black Forest) centre and then guarded rail shipment to the flight-test base at Lechfeld.

This giant production plan suffered many further problems, and the flow did not begin to arrive at Lechfeld until February 1944. The production interceptor was designated Me 163B-1a, and although in many ways seemingly crude it was actually a very refined aircraft as a result of the prolonged experience with earlier variants. The first operational flight of the new machine came on 14 May 1944, when Späte, now a Major, took the red-painted V41 into action.

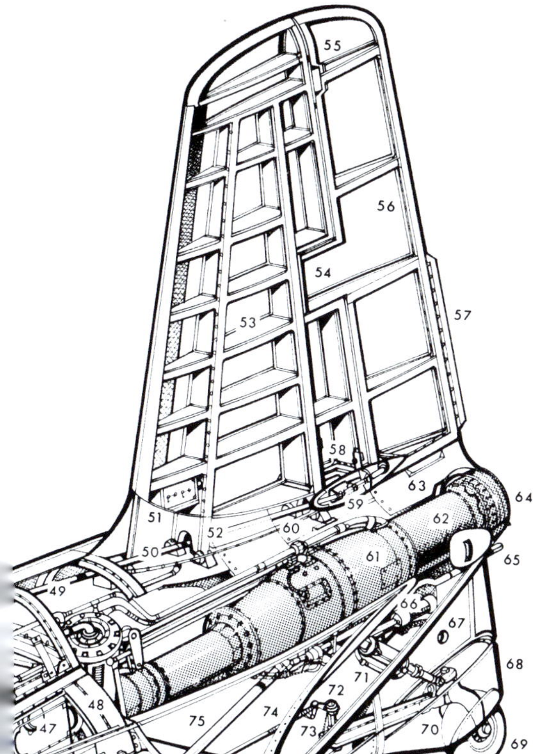

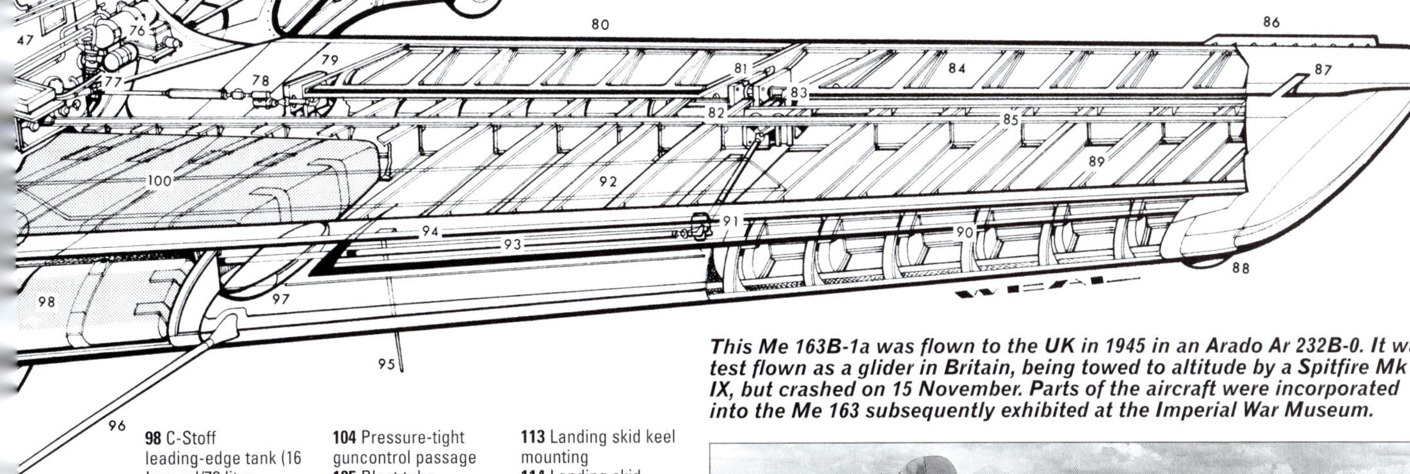

This Me 163B-1a was flown to the UK in 1945 in an Arado Ar 232B-0. It was test flown as a glider in Britain, being towed to altitude by a Spitfire Mk IX, but crashed on 15 November. Parts of the aircraft were incorporated into the Me 163 subsequently exhibited at the Imperial War Museum.

98 C-Stoff leading-edge tank (16 Imp gal/73 litre capacity)
99 Gun-cocking compressed air bottle
100 Main C-Stoff wing tank (38 Imp gal/173 litre capacity)
101 Port 30-mm MK 108 short-barrel cannon
102 Expanding shell and link chute
103 Gun forward mounting frame
104 Pressure-tight guncontrol passage
105 Blast tube
106 Gun alignment mechanism
107 Cannon port
108 FuG 23a FF pack
109 Tow-bar attachment point
110 Compressed-air ram for landing skid
111 Hydraulics and compressed-air pipes
112 Landing skid pivots
113 Landing skid keel mounting
114 Landing skid mounting brackets
115 Trolley jettison mechanism
116 Landing skid
117 Take-off trolley frame
118 Take-off trolley retaining lugs
119 Take-off trolley alignment pins
120 Low-pressure tyre

Me 163B Komet

This aircraft carried the Baron von Munchhausen badge of JG 400's I. Gruppe when flying from Brandis, near Leipzig, in late 1944. JG 400 was activated at Bad Zwischenahn in February 1944 and was soon transferred to Wittmundhaven, where a second staffel was formed. The unit then moved on to Venlo, Holland in June, before moving on to Brandis in August. It conducted operations against USAAF heavy bombers from the latter base and a second Gruppe was formed for a similar role in November 1944. I. Gruppe was subsequently dissolved in April 1945, but II. Gruppe continued to the end, retreating eventually to Nordholz. I. Gruppe claimed nine bombers destroyed for 14 Me 163s shot down and many more lost in accidents, while II. Gruppe made no claims.

Cockpit
The cockpit was comfortable, although there was no system available for pressurisation other than a plain ram inlet at the front. The canopy was a flimsy Plexiglas moulding, hinged on the right side and with little ability to resist hail or birds at the speeds the Komet could attain. There was a hinged ventilation window on the left side of the hood, and another air inlet on the underside of the nose. Nose and back armour was provided, but the seat was not of the new ejection type and it was impossible to get out at high airspeeds.

Powerplant
The Me 163B was powered by a single Hellmuth Walter Werke R II-211 rocket motor, with fuel for six minutes at full throttle. Derived from Von Braun's 650-lb st (2.89-kN) A 1 rocket engine of 1935, the engine was closely based on Walter's TP-1 and TP-2 'Cold' rockets using hydrogen peroxide (T-stoff) with an aqueous solution of sodium or calcium permanganate (Z-stoff) as a catalyst. Essentially the engine consisted of a steam generator into which the two fuels were sprayed using compressed air. This drove a turbine, which powered the pump that delivered T-stoff to the combustion chamber. The TP-2 was redesignated as the HWK (Hellmuth Walter Kiel) R I-203, and was developed progressively into the R II-203, redesignated HWK 509A in production form, which powered early Me 163 prototypes. Substitution of a solution of 30 per cent hydrazine hydrate, 57 per cent methyl alcohol, 13 per cent water and 17 per cent cupracyanide (C-stoff) for the Z-stoff resulted in a hot rocket engine with more thrust and greater reliability, which did not generate a white vapour trail. Before each flight the entire system had to be drained and flushed through with scrupulous care, using vast amounts of water. The motor was started with T-stoff fed from a separate starter tank in the top of the rear fuselage, while an electric motor cranked up the turbopumps. The tanks were pressurised, and once the feed reached the turbopumps the liquids were supplied under high pressure at the rate of 17.64 lb (8 kg) per second, combusting spontaneously on contact in the chamber. Sea-level thrust was about 3,307 lb (14.71 kN), rising with reducing atmospheric pressure to 3,748 lb (16.61 kN) at high altitude. The Type 509A could be throttled back to 220 lb st (0.98 kN) idling rating, but it was inefficient at this level and could often stop entirely. The entire rear fuselage and motor could readily be detached. Although crude compared with later units, the Type 509A was a remarkable achievement and, although over 7 ft (2.13 m) long, weighed little over 220 lb (100 kg).

Wing

The Me 163's wing was smaller and simpler than those of the precursor aircraft, and although it appeared swept it was mainly its taper that gave a quarter-chord sweep angle of 23° 20'. The wooden structure was simple, with two widely spaced spars and a skin of fabric-covered ply usually 0.31 in (8 mm) thick. Outboard on the trailing edge were the only control surfaces, other than the rudder: large manual fabric-covered elevons used for both pitch and roll. The trim tabs were plain metal bent on the ground with pliers to give the required behaviour. Inboard were large, plain, hinged flaps which were lowered hydraulically by screwjacks before landing, in unison with main landing flaps ahead of them on the underside of the wing. The landing flaps caused strong nose-up trim, and the trailing-edge flaps cancelled this out with equal nose-down trim. Trials with the Me 163 V1 and V4 during October 1941 saw the aircraft reaching speeds of up to 550 mph (885 km/h), with maximum speed limited by fuel capacity. To get around this, Heini Dittmar had the Me 163 V4 fully tanked and then towed into the air behind a Bf 110C tug. Lighting the rocket, Dittmar accelerated to 623.85 mph (1003.96 km/h - equivalent to Mach 0.84) when compressibility effects forced the aircraft into a steep dive, from which Dittmar recovered by cutting the engine. The sudden change in pitch stability was due to the fact that the Me 163 V4 had essentially retained the wing of the DFS 194, with considerable washout which caused wingtip compressibility stalls. On the Me 163B the wing was considerably redesigned, with reduced sweep on the trailing edge, constant sweep on the leading edge and with low drag fixed slots on the outer 40 per cent of the wing leading edge. These removed the danger of tip stalling, and also made the Me 163 unspinnable. Even with fully crossed controls, the aircraft would only sideslip.

Fuselage and armament

The small fuselage was of light alloy, covered mainly with detachable panels to gain access to the densely packed interior. The largest item was the T-stoff tank of 229-Imp gal (1040-litre) capacity, which filled the space between the cockpit and the motor. Smaller T-stoff tanks filled each side of the cockpit. The C-stoff was housed in two 38-Imp gal (173-litre) tanks between the wing spars and two 16-Imp gal (73-litre) tanks in the leading edges. The Z-stoff originally used as a catalyst in the Me 163A was prone to clogging feed pipes, but the T-stoff fuel had even worse characteristics. Highly unstable and prone to spontaneous combustion when exposed to organic material (such as human flesh), T-stoff was also highly corrosive. The Me 163 pilot was surrounded by T-stoff tanks in flight, and had to wear a non-organic flying suit made of asbestos-Mipolamfibre. The C-stoff catalyst used in the Me 163 was also highly reactive, and had to be stored in glass or enamelled containers. The Me 163B was initially armed with a pair of Mauser 20-mm MG 151 cannon, but from the 47th pre-production aircraft these were replaced by 30-mm Rheinmetall-Borsig MK 108 cannon, each with 60 rounds of ammunition.

Messerschmitt Me 262 Schwalbe

The Me 262 was among the fastest aircraft operating in any theatre during World War II, capable of overhauling the Mosquito and the Mustang, but its lack of manoeuvrability made it vulnerable, and like all aircraft, it was a sitting duck when in the circuit to land, a fact used to full advantage by the Allies. This machine, III./EJG 2's 'White 10', is seen flown by Leutnant Kurt Bell during the making of a Luftwaffe training film.

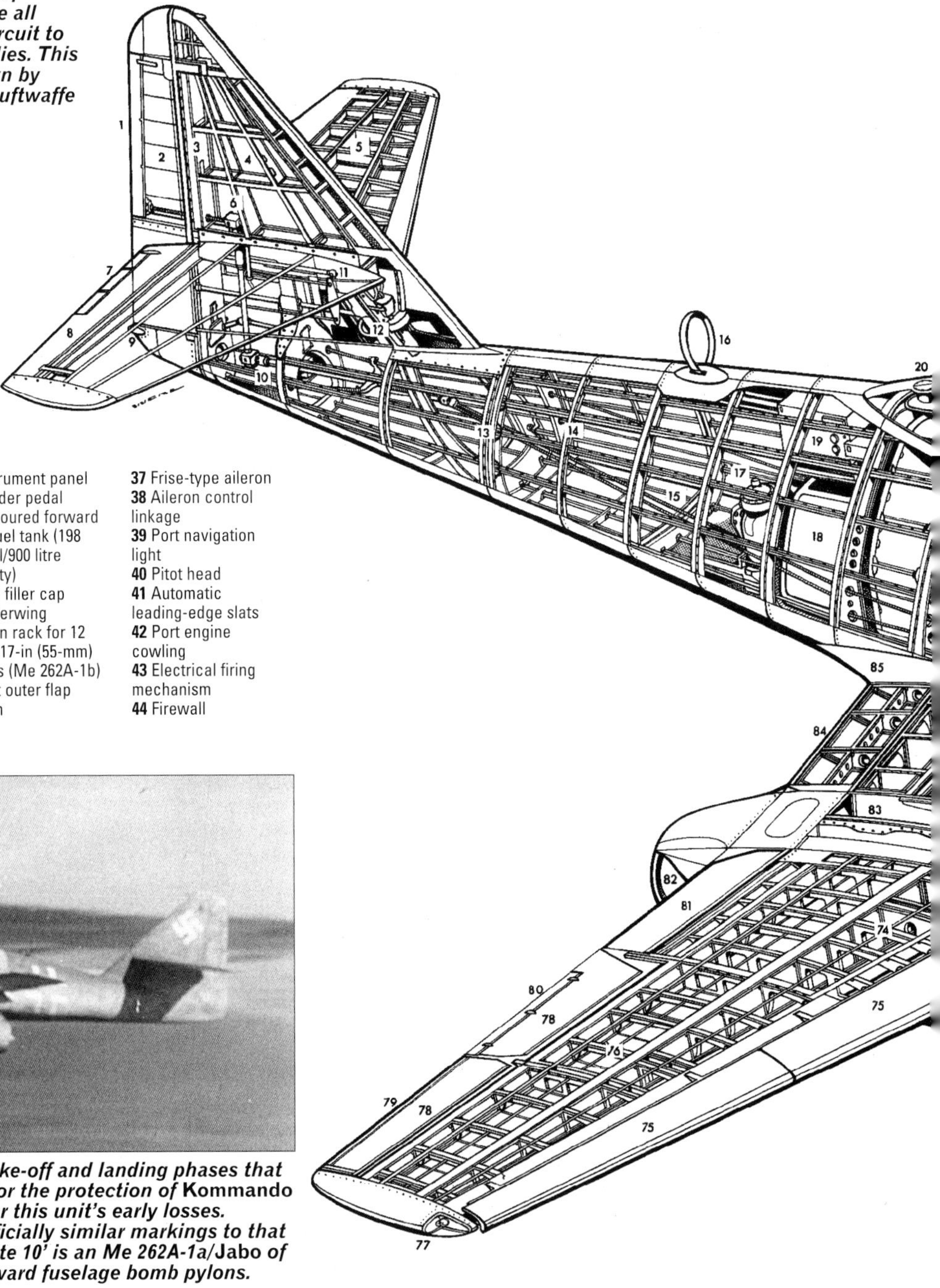

Me 262A-1a/b

Cutaway key

1 Flettner-type geared trim tab
2 Mass-balanced rudder
3 Rudder post
4 Tail fin structure
5 Tailplane structure
6 Rudder tab mechanism
7 Flettner-type servo tab
8 Starboard elevator
9 Rear navigation light
10 Rudder linkage
11 Elevator linkage
12 Tailplane adjustment mechanism
13 Fuselage break point
14 Fuselage construction
15 Control runs
16 FuG 25a loop antenna (IFF)
17 Automatic compass
18 Aft auxiliary self-sealing fuel tank (132 Imp gal/600 litre capacity)
19 FuG 16zy R/T
20 Fuel filler cap
21 Aft cockpit glazing
22 Armoured aft main fuel tank (198 Imp gal/900 litre capacity)
23 Inner cockpit shell
24 Pilot's seat
25 Canopy jettison lever
26 Armoured 0.59-in (15-mm) head rest
27 Canopy (hinged to starboard)
28 Canopy lock
29 Bar-mounted Revi 16B sight (for both cannon and R4M missiles)
30 Armourglass windscreen 3.54-in (90-mm)
31 Instrument panel
32 Rudder pedal
33 Armoured forward main fuel tank (198 Imp gal/900 litre capacity)
34 Fuel filler cap
35 Underwing wooden rack for 12 R4M 2.17-in (55-mm) rockets (Me 262A-1b)
36 Port outer flap section
37 Frise-type aileron
38 Aileron control linkage
39 Port navigation light
40 Pitot head
41 Automatic leading-edge slats
42 Port engine cowling
43 Electrical firing mechanism
44 Firewall

So vulnerable was the Me 262 during the take-off and landing phases that two Staffeln of Fw 190D-9s were assigned for the protection of **Kommando Nowotny** *based at Achmer and Hesepe after this unit's early losses. Although the pictured aircraft wears superficially similar markings to that flown for the camera by Kurt Bell, this 'White 10' is an Me 262A-1a/*Jabo *of EKdo 262, identified by the presence of forward fuselage bomb pylons.*

Kommando Nowotny line-up, probably at Achmer in late September 1944. Whilst 'White 19' and 'White 4' are serviced, in the foreground 'Green 3' is towed out towards the runway by a Kettenkrad motorcycle half-track. Austrian-born Major Walter Nowotny, already a Russian front expert, was chosen by Adolf Galland to head the new Kommando, of which he assumed control on 20 September 1944. Kommando Nowotny's first combat sorties were undertaken on 7 October, when the two-Staffel strong jet unit flew against bombers returning from raids on oil targets within Germany. Flying from Achmer, two of the unit's pilots, Franz Schall and his wingman Helmut Lennartz, each claimed a B-24D Liberator destroyed.

SPECIFICATION

bMe 262A-1a/b

Dimensions:

Length: 34 ft 9 in (10.58 m
Height: 12 ft 7 in (3.83 m)
Span: 40 ft 11 in (12.5 m)
Wing area: 234 sq ft (21.73 m²)
Wing leading edge sweepback: 18° 32'

Powerplant:

Two Junkers Jumo 004B-1, -2 or -3 axial-flow turbojets each rated at 1,984 lb st (8.83 kN)

Weights:

Empty: 3,778 lb (3795 kg)
Empty equipped: 9,742 lb (4413 kg)
Maximum take-off: 14,080 lb (6387 kg)

Performance:

Maximum speed at sea level: 514 mph (827 km/h)
Maximum speed at 9,845 ft (3000 m): 530 mph (852 km/h)
Maximum speed at 19,685 ft (6000 m): 540 mph (869 km/h)
Maximum speed at 26,245 ft (8000 m): 532 mph (856 km/h)
Initial climb rate: 3,937 ft (1200 m) per minute
Service ceiling: over 40,000 ft (12190 m)
Range: 652 miles (1050 km) at 29,530 ft (9000 m)
Landing speed: 109 mph (175 km/h)

Armament:

Four 30-mm Rheinmetall-Borsig Mk 108A-3 cannon with 100 rounds per gun for the upper pair and 80 rounds per gun for the lower pair, and aimed with a Revi 16.B gunsight or EZ.42 gyro-stabilised sight. Provision for 12 R4M air-to-air rockets under each wing (Me 262A-1b)

'White 4' wears a typical **Kommando Nowotny** *camouflage scheme, comprising a principally solid fuselage colour applied together with distinctive mottled vertical tail surfaces. Following the death of Nowotny on 8 November 1944, control of the unit was passed to Oberleutnant Georg-Peter Eder, however, the unit flew its final mission from Achmer on 17 November. Former pilots went on to form the nucleus of JG 7, the first and only Luftwaffe jet fighter* **Geschwader**.

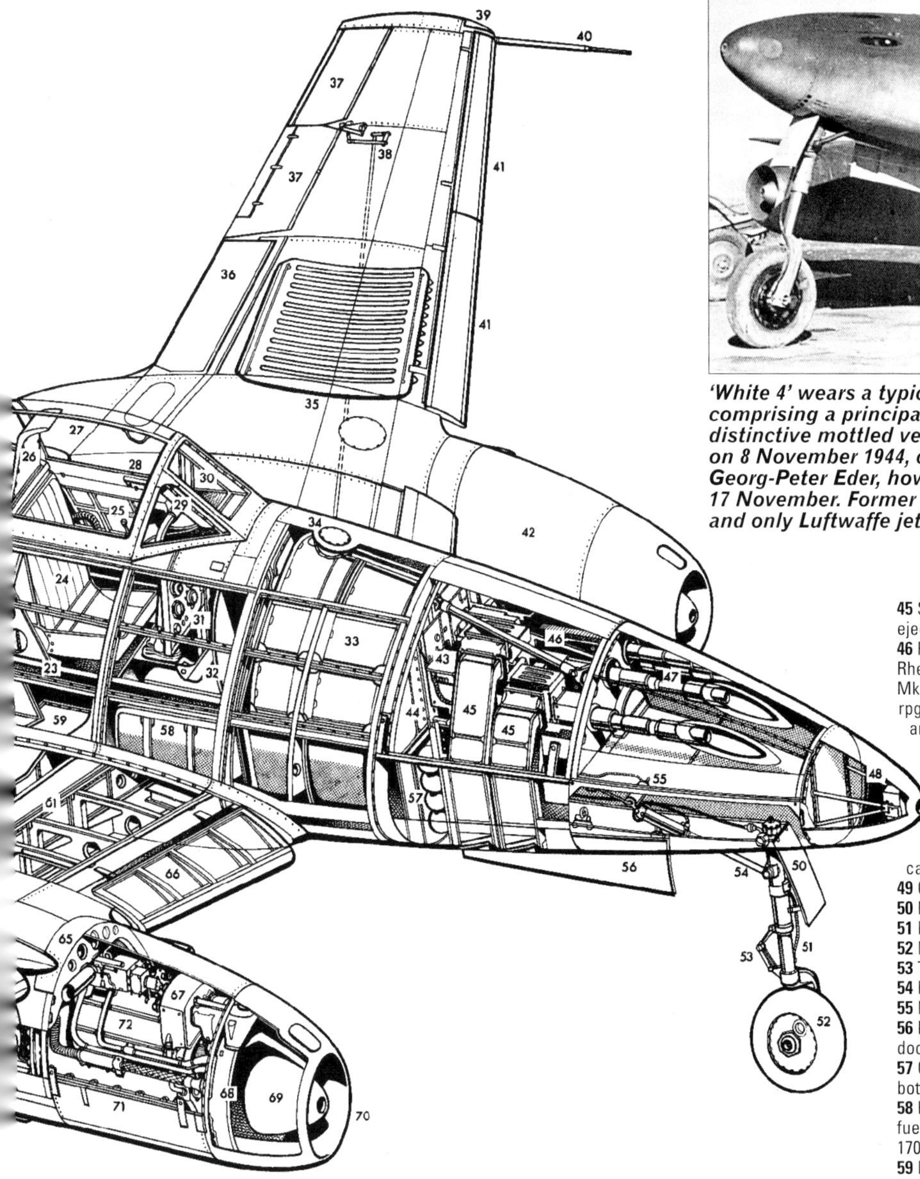

45 Spent cartridge ejector chutes
46 Four 30-mm Rheinmetall Borsig Mk 108 cannon (100 rpg belt-fed ammunition for upper pair and 80 rpg for lower pair)
47 Cannon muzzles
48 Combat camera
49 Camera aperture
50 Nosewheel fairing
51 Nosewheel leg
52 Nosewheel
53 Torque scissors
54 Retraction jack
55 Hydraulic lines
56 Main nosewheel door (starboard)
57 Compressed air bottles
58 Forward auxiliary fuel tank (37 Imp gal/ 170 litre capacity)
59 Mainwheel well
60 Torque box
61 Main spar
62 Mainwheel leg pivot point
63 Mainwheel door
64 Mainwheel retraction rod
65 Engine support arch
66 Leading-edge slat structure
67 Auxiliaries gearbox
68 Annular oil tank
69 Riedel starter motor housing
70 Engine air intake
71 Hinged cowling section
72 Junkers Jumo 004B-2 axial-flow turbojet
73 Starboard mainwheel
74 Wing structure
75 Automatic leading-edge slats
76 Mainspar
77 Starboard navigation light
78 Frise-type ailerons
79 Trim-tab
80 Flettner-type geared tab
81 Starboard outer flap section
82 Engine exhaust orifice
83 Engine support bearer
84 Starboard inner flap structure
85 Faired wing root

Me 262A-1a Schwalbe

This Me 262A-1a operated from Parchim in north east Germany with 9 Staffel, Jagdgeschwader III, during the early months of 1945. JG III was part of III Gruppe, and was responsible to 1 Jagddivision of I Jagdkorps. Blue and red 'Defence of the Reich' bands were worn on the rear fuselage, and the Gruppe's leaping greyhound emblem was painted on the fuselage forward of the cockpit. This aircraft was captured at Lechfeld, shipped to the United States from France, and evaluated by the USAF at Freeman Field in Indiana before being passed on to the National Air and Space Museum in Washington DC where it still resides.

Wings
The Me 262 was equipped with leading edge slots which opened automatically in a turn or in a climb if the airspeed fell below 280 mph (451 km/h), or to below 185 mph (298 km/h) in a gliding attitude. Fighter versions of the type were able to carry up to 12 R4M unguided 2.16-in (55-mm) rocket projectiles under each wing, fired from a wooden rack. The trajectory of the rockets was similar to that of the cannon, so the same sight could be used. The rockets scattered to fill the same space as that occupied by a four-engined bomber at a range of 600 yards. Each rocket had a 16-oz (4.54-kg) warhead, with a considerable blast effect.

Powerplants
The Me 262 was powered by a pair of Junkers Jumo 004B-1 axial flow turbojets, each rated at about 1,984 lb (8.82 kN) static thrust. This gave a maximum speed of approximately 540 mph (869 km/h) at 20,000 ft (6096 m). The aircraft could exceed this speed in a dive, soon reaching its limiting Mach number. The Germans experienced extreme difficulties in maintaining engine production, the factories producing the jet engines being primary targets for Allied bombers. This led to poor reliability of the powerplants, and shortages of chrome and nickel prevented the turbine blades from being manufactured with sufficient strength to withstand the extreme temperatures encountered, thus making the life of the engine very short.

Elevators and ailerons
Pre-production Me 262s were fitted with fabric-covered elevators, but these proved highly susceptible to ballooning in high-speed dives, and on several occasions the fabric actually tore. Thereafter all production aircraft were fitted with metal-skinned elevators. The Me 262 had good aileron control at all altitudes, and the aircraft showed very little tendency to spin following a stall. Directional stability, however, was not so good, although the powerful rudder enabled the pilot to keep the aircraft under control.

Armament
Fighter Me 262s were armed with four 30-mm Rheinmetall Borsig Mk 108A-3 cannon, with 100 rounds for each of the upper guns and 80 for the lower guns. The guns were clustered closely together on the centreline, packing a substantial punch without tricky harmonisation problems. The fighter-bomber version of the Me 262 was dubbed *Sturmvogel* (Stormbird) to distinguish it from the standard fighter. It differed from the latter only in having bomb fusing equipment and provision for two undernose pylons, which could accomodate a pair of 551-lb (250-kg) bombs, or a single 1,102-lb (500-kg) bomb. These aircraft were referred to as 'Super Speed Bombers', because Hitler hated the use of the term 'fighter', even in the phrase 'fighter-bomber'.

Messerschmitt Me 410 Hornisse

With its neatly cowled engines and purposeful nose contours, the Me 210 looked the part, but it was plagued with vicious and unpredictable handling qualities. This aircraft is one of those fitted with a longer rear fuselage, which largely cured the design's major faults.

Me 410 Hornisse

Cutaway key
1 Starboard navigation light
2 Starboard detachable wingtip
3 Main spar
4 Wing leading-edge slat
5 Aileron control rods
6 External balance (underwing)
7 Starboard aileron
8 Tab (ground-adjustable only)
9 Aileron trim tab
10 Trim tab control
11 Slatted airbrakes (above and below wing)
12 Wing centre/outer section join
13 Starboard underwing radiator
14 Boundary layer bleed
15 Radiator flap section
16 Radiator flap motor (in flap section)
17 Starboard oil filter
18 Cowling panelling
19 Starboard engine supercharger intake
20 Starboard nacelle
21 Exhaust stub cover
22 Oil cooler intake (adjustable flap)
23 Auxiliary intake
24 Coolant filter access
25 Spinner
26 Three-bladed constant-speed VDM propeller
27 Starboard mainwheel
28 Bomb-bay doors (open)
29 Two 7.9-mm MG 17 machine-gun ports
30 Two 20-mm MG 151 cannon ports
31 Cabin air intake
32 Cabin air heater
33 Nose glazing
34 Rudder pedals
35 Instrument panel side sections
36 Instrument panel lower section
37 Control column
38 Pilot's heelboards
39 MG 151 cannon blast tube
40 Bomb bay
41 Bomb winch cable hoist
42 Port instrument console
43 Throttle quadrant
44 Pilot's seat
45 Starboard instrument console (weapons/oxygen)
46 Revi C/12D weapons sight
47 Armoured windscreen
48 Hinged cockpit canopy
49 Pilot's armoured head/backrest
50 Canopy internal bracing
51 Ammunition magazines (1,000 rounds 7.9-mm/350 rounds 20-mm)
52 Pilot's entry handhold
53 Ammunition feed chutes
54 Port weapons breeches
55 Mainspar centre-section carry-through
56 Observer's seat
57 Electrical main distribution panel
58 Beam armament master switch and ammunition counter
59 Sighting head for FDSL beam barbettes
60 Hinged cockpit canopy section
61 Aerial mast (angled to starboard)
62 D/F loop aerial housing
63 Optically flat side windows
64 Barbette elevation input shaft
65 Barbette traverse input shaft
66 Observer's entry handhold
67 EZ2 D/F receiver remote control unit
68 FuG 10 radio receiver
69 EZ2 D/F receiver
70 FuG 10 radio transmitter
71 Rear spar centre-section carry-through
72 Wingroot fairing
73 Barbette electrics junction box
74 Access panel/handhold
75 Barbette torque amplifier
76 Barbette ring gears
77 Barbette centre rotating drum
78 Ammunition around drum (500 rpg)
79 Port beam gun fairing
80 10-mm MG 131 beam gun
81 Aerial unit
82 Rear fuselage access panel
83 FuG 25 IFF transformer
84 FuG 25 transponder
85 Aerial lead-in
86 Master compass
87 Fuselage frames
88 Course control drive
89 Skin panelling dorsal join
90 Rear fuselage structure
91 Control rods
92 Tailwheel support frame

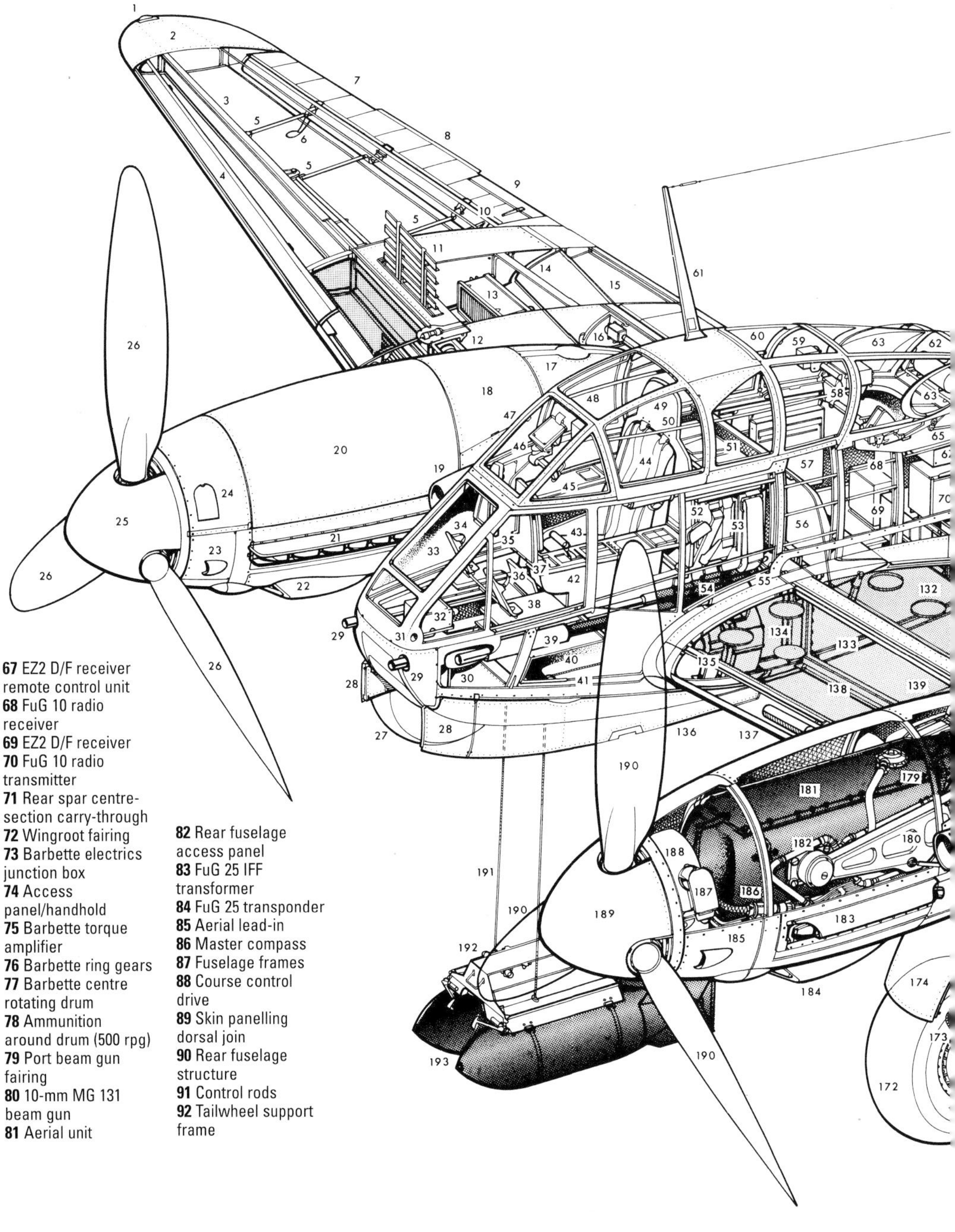

Me 210A-2s from the Stabsschwarm, III./ZG 1 fly low over the Tunisian countryside in 1943. Only a small number of units used Me 210s before the improved Me 410 entered service.

SPECIFICATION

Messerschmitt Me 410A-1/U2

Type

Two-seat heavy fighter

Powerplant

Two 1,850-hp (1380-kW) Daimler-Benz 603A inverted V-12 inline piston engines

Performance

Maximum speed 364 mph (625 km/h) at 21,980 ft (6700 m); service ceiling 32,180 ft (10000 m); maximum range 1,050 miles (1690 km)

Weights

Empty equipped 16,574 lb (7518 kg); maximum take-off 21,276 lb (9650 kg)

Dimensions

Wingspan 53 ft 7¾ in (16.35 m); length 40 ft 11½ in (12.48 m); height 14 ft ½ in (4.28 m); wing area 389.67 sq ft (36.20 m2)

Armament

Four 20-mm MG 151 cannon and two 0.31-in (7.92-mm) MG 17 machine-guns firing forward, plus two 0.51-in (13-mm) MG 131 machine-guns in remotely-controlled rear-firing barbettes

The Me 210 V1 first prototype clearly showed its Bf 110 genes, although the twin tail was swiftly changed after the first few flights.

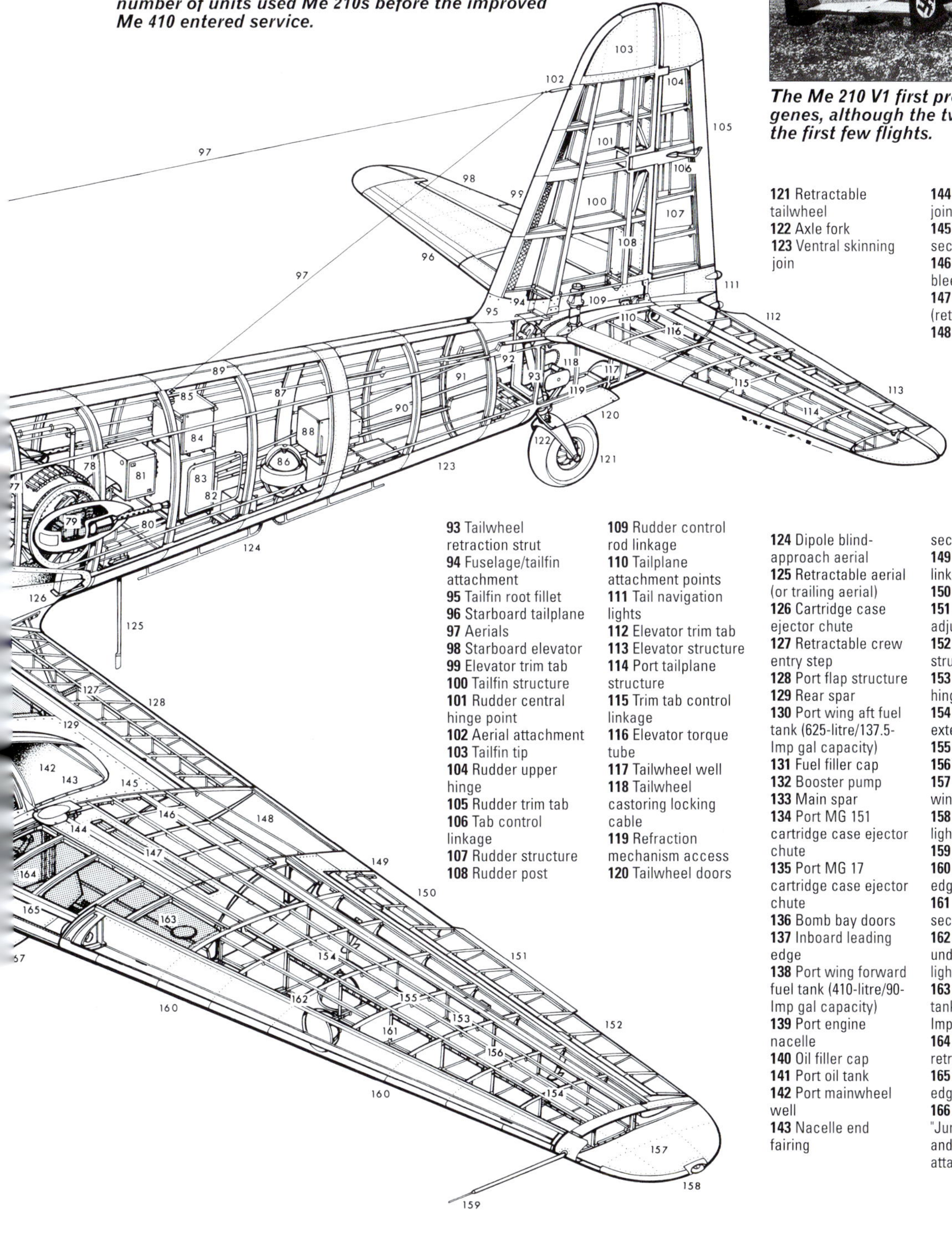

93 Tailwheel retraction strut
94 Fuselage/tailfin attachment
95 Tailfin root fillet
96 Starboard tailplane
97 Aerials
98 Starboard elevator
99 Elevator trim tab
100 Tailfin structure
101 Rudder central hinge point
102 Aerial attachment
103 Tailfin tip
104 Rudder upper hinge
105 Rudder trim tab
106 Tab control linkage
107 Rudder structure
108 Rudder post
109 Rudder control rod linkage
110 Tailplane attachment points
111 Tail navigation lights
112 Elevator trim tab
113 Elevator structure
114 Port tailplane structure
115 Trim tab control linkage
116 Elevator torque tube
117 Tailwheel well
118 Tailwheel castoring locking cable
119 Refraction mechanism access
120 Tailwheel doors
121 Retractable tailwheel
122 Axle fork
123 Ventral skinning join
124 Dipole blind-approach aerial
125 Retractable aerial (or trailing aerial)
126 Cartridge case ejector chute
127 Retractable crew entry step
128 Port flap structure
129 Rear spar
130 Port wing aft fuel tank (625-litre/137.5-Imp gal capacity)
131 Fuel filler cap
132 Booster pump
133 Main spar
134 Port MG 151 cartridge case ejector chute
135 Port MG 17 cartridge case ejector chute
136 Bomb bay doors
137 Inboard leading edge
138 Port wing forward fuel tank (410-litre/90-Imp gal capacity)
139 Port engine nacelle
140 Oil filler cap
141 Port oil tank
142 Port mainwheel well
143 Nacelle end fairing
144 Wing main spar join cover
145 Wing centre/outer section join
146 Boundary layer bleed
147 Slatted airbrakes (retracted in slot)
148 Radiator flap
section
149 Trim tab control linkage
150 Aileron trim tab
151 Tab (ground-adjustable only)
152 Port aileron structure
153 Aileron main hinge point
154 Underwing external balances
155 Wing ribs
156 Stringers
157 Port detachable wingtip
158 Port navigation light
159 Pitot head
160 Wing leading-edge slat
161 Main spar outer section
162 Retractable underwing landing light
163 Wing outer fuel tank (170-litre/37.4-Imp gal capacity)
164 Undercarriage retraction strut
165 Wing leading-edge reinforcing tube
166 Wing join "Junkers" type ball-and-socket attachment
167 Mainwheel well door
168 Undercarriage brace strut
169 Mainwheel leg
170 Oleo shock absorbers
171 Hub brakes
172 Port mainwheel
173 Torque links
174 Mainwheel leg door
175 Drag strut
176 Supercharger air intake
177 Firewall
178 Engine accessories/magneto
179 Oil breather pipes
180 Engine bearer mounting
181 Daimler-Benz DB 603A 12-cylinder liquid-cooled engine
182 Anti-vibration mounting pad
183 Exhaust stub cover
184 Oil cooler intake (adjustable flap)
185 Auxiliary intake
186 Coolant pipes
187 Coolant filter access
188 Coolant header tank
189 Spinner
190 Three-bladed constant-speed VDM propeller
191 Bomb hoist cables
192 Bomb rack
193 Two SC 500 bombs navigation light

Mitsubishi A6M Reisen 'Zeke'

Many Zeros were captured by the Allies and used for test and evaluation purposes. The first was captured in 1942, following an inconclusive Japanese attack on Dutch Harbor in the Aleutians. An A6M2 flown by Petty Officer Koga was forced to make an emergency landing at Akutan Island and, upon touching down, the aircraft flipped over and the pilot was killed. The machine was transported to San Diego, repaired, and flown for evaluation against Allied aircraft.

A6M2 Reisen

Cutaway key
1 Tail navigation light
2 Tail cone
3 Tailfin fixed section
4 Rudder lower brace
5 Rudder tab (ground adjustable)
6 Fabric-covered rudder
7 Rudder hinge
8 Rudder post
9 Rudder upper hinge
10 Rudder control horn (welded to torque tube)
11 Aerial attachment
12 Tailfin leading edge
13 Forward spar
14 Tailfin structure
15 Tailfin nose ribs
16 Port elevator
17 Port tailplane
18 Piano-hinge join
19 Fuselage dorsal skinning
20 Control turnbuckles
21 Arrester hook release/retract steel cable runs
22 Fuselage frame/tailplane centre brace
23 Tailplane attachments
24 Elevator cables
25 Elevator control horns/torque tube
26 Rudder control horns
27 Tailwheel combined retraction/shock strut
28 Elevator trim tab
29 Tailwheel leg fairing
30 Castored tailwheel
31 Elevator frame (fabric-covered)
32 Elevator outer hinge
33 Tailplane structure
34 Forward spar
35 Elevator trim tab control rod (chain-driven)
36 Fuselage flotation bag rear wall
37 Arrester hook (extended)
38 Arrester hook pivot mounting
39 Elevator trim tab cable guide
40 Fuselage skinning
41 Fuselage frame stations
42 Arrester hook position indicator cable (duralumin tube)
43 Elevator cables
44 Rudder cables
45 Trim tab cable runs
46 Arrester hook pulley guide
47 Fuselage stringers
48 Fuselage flotation bag front
49 Fuselage construction join
50 Wingroot fillet formers
51 Compressed air cylinder (wing gun charging)
52 Transformer
53 'Ku' type radio receiver
54 Oxygen cylinder (starboard) carbon dioxide fire extinguisher cylinder (port)
55 Battery
56 Radio tray support
57 Radio transmitter
58 Canopy/fuselage fairing
59 Aerial mast support/lead-in
60 Aerial
61 Aerial mast (forward raked)
62 Canopy aft fixed section
63 Aluminium and plywood canopy frame
64 Crash bulkhead/headrest support
65 ' Ku' type D/F frame antenna mounting (late models)
66 Canopy track
67 Turnover truss
68 Pilot's seat support frame
69 Starboard elevator control bell crank
70 Aileron control push-pull rod
71 Wing rear spar/fuselage attachment
72 Fuselage aft main double frame
73 Aileron linkage
74 Landing-gear selector lever
75 Flap selector lever
76 Seat adjustment lever
77 Pilot's seat
78 Cockpit canopy rail
79 Seat support rail
80 Elevator tab trim handwheel
81 Fuel gauge controls
82 Throttle quadrant
83 Reflector gunsight mounting (offset to starboard)
84 Sliding canopy
85 Plexiglas panels
86 Canopy lock/release
87 Windscreen
88 Fuselage starboard 0.303-in (7.7-mm) machine-gun
89 Control column
90 Radio control box
91 Radio tuner
92 Elevator control linkage
93 Rudder pedal bar assembly
94 Cockpit underfloor fuel
95 Wing front spar/fuselage attachment
96 Fuselage forward main double frame
97 Ammunition magazine

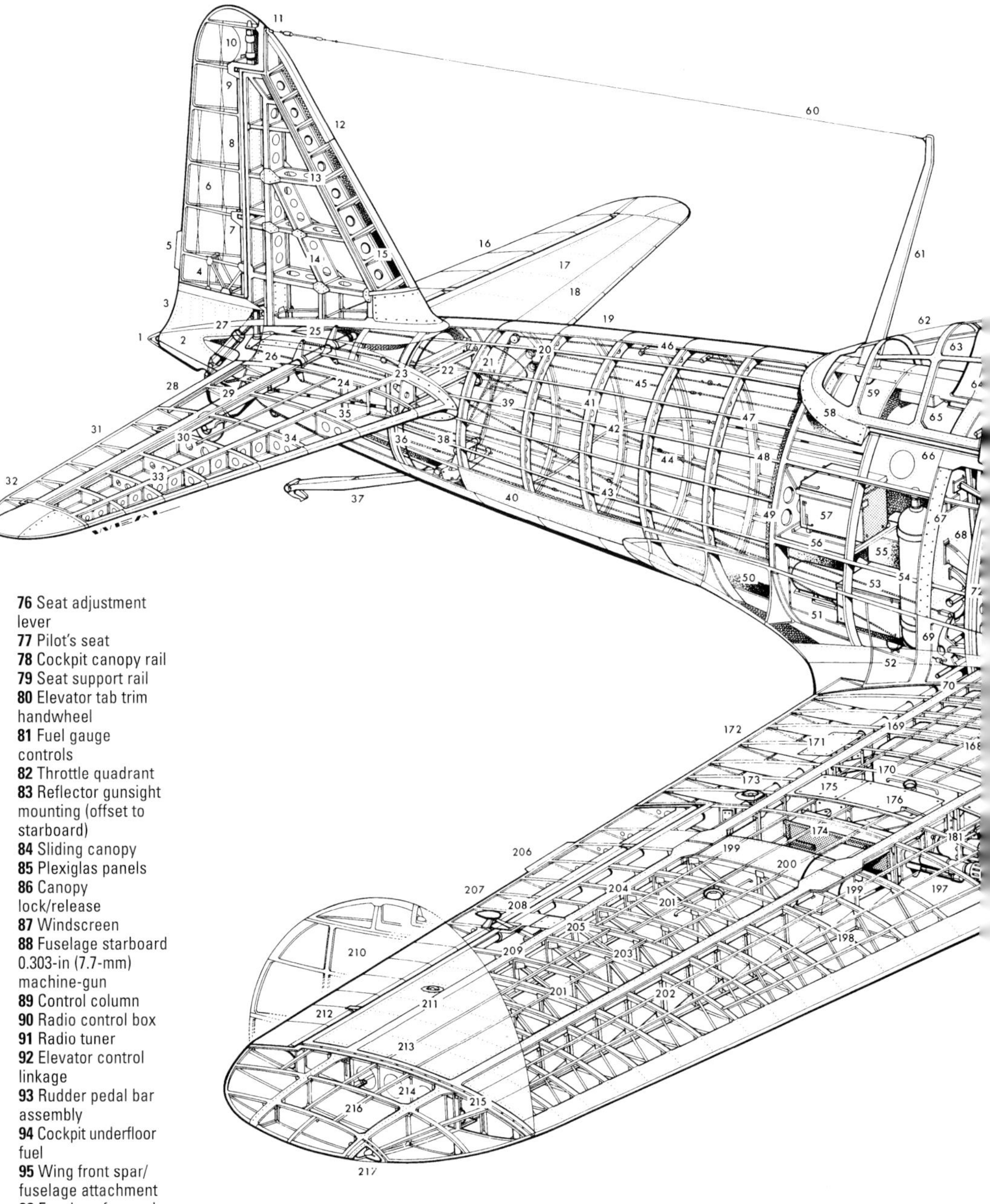

As the A6M2 began to reach service units, it gradually replaced the IJN's previous single-seat fighter, the ultra-manoeuvrable Mitsubishi A5M. The latter served on into the early war years as is proved by this collection of A6M2s and A5Ms lined up on an airfield at the beginning of the war in the Pacific. The US Navy received its first taste of Japanese airpower during the attack on Pearl Harbor, which was planned in great secrecy and took months of preparation. The 'Zekes' which escorted the attacking bombers destroyed four US aircraft in the air without a single loss. Subsequent operations in the Java Sea, against Wake Island and against Darwin proved equally successful.

SPECIFICATION

Mitsubishi Zero A6M2 Model 21

(unless otherwise noted)

Dimensions

Wingspan: 39 ft $4\frac{7}{16}$ in (12 m)
A6M3, 5, 8: 36 ft $1\frac{1}{16}$ in (11 m)
Length (including A6M3): 29ft $8\frac{11}{16}$ in (12 m)
A6M5: 29 ft $11\frac{3}{32}$ in (9.12 m)
A6M8: 30 ft $3\frac{21}{32}$ in (9.24 m)
Height: 10 ft $\frac{1}{16}$ in (3.05 m)
A6M3, 5: 11 ft $6\frac{5}{32}$ in (3.51 m)
Wing area: 241.541 sq ft (22.44 m^2)
A6M3: 231.746 sq ft (21.53 m^2)
A6M5, 8: 229.27 sq ft (21.30 m^2)

Powerplant

One Nakajima NK1C Sakae 12 14-cylinder air cooled radial engine, rated at 940 hp (700 kW) at take-off and 950 hp (708 kW) at 13,780 ft (4200 m), driving a three-bladed metal propeller.
A6M3, 5a, b, c: One Nakajima NK1F Sakae 21 14-cylinder air cooled radial engine, rated at 1,130 hp (843 kW) at take-off and 1,100 hp (820 kW) at 9,350 ft (2850 m) and 980 hp (731 kW) at 19,685 ft (6000 m), driving a three-bladed metal propeller.
A6M8: One Mitsubishi MK8P Kinsei 62 14-cylinder radial, rated at 1,560 hp (1163 kW) at take-off, 1,340 hp (999 kW) at 2,100 ft (6890 m) and 1,180 hp (880 kW) at 5,800 ft (19030 m) driving a three-bladed metal propeller.

Weights

Empty: 3,704 lb (1680 kg)
A6M3: 3,984 lb (1807 kg)
A6M5: 4,136 lb (1876 kg)
A6M8: 4,740 lb (2150 kg)
Loaded: 5,313 lb (2410 kg)
A6M3: 5,609 lb (2544 kg)
A6M5: 6,025 lb (2733 kg)
A6M8: 6,945 lb (3150 kg)
Maximum: 6,164 lb (2796 kg)
Wing loading: 22 lb/sq ft (107.4 kg/m^2)
A6M3: 24.2 lb/sq ft (118.1 kg/m^2)
A6M5: 26.3 lb/sq ft (128.3 kg/m^2)
A6M8: 30.3 lb/sq ft (147.9 kg/m^2)

Fuel and load

Normal fuel: 130 Imp gal (590.98 litres)
A6M3: 134 Imp gal (609.16 litres)
Maximum fuel: 202 Imp gal (918.29 litres)
External fuel: 72.6 Imp gal (330 litres)
Maximum weapon load: 1,790 lb (811.92 kg)

Performance

Maximum speed: 288 kt (332 mph; 534 km/h) at 14,930 ft (4550 m)
A6M3: 294 kt (338 mph; 545 km/h) at 19,685 ft (6000 m)
A6M5: 305 kt (351 mph; 565 km/h) at 19,685 ft (6000 m)
A6M8: 309 kt (365 mph; 573 km/h) at 19,685 ft (6000 m)
Cruising speed: 180 kt (207 mph; 334 km/h)
A6M3, 5: 200 kt (230 mph; 370 km/h)
Service ceiling: 32,810 ft (10000 m)
A6M3: 36,250 ft (11050 m)
A6M5: 38,520 ft (11740 m)
A6M8: 37,075 ft (11200 m)

Range

Normal range: 1,010 nm (1,162 miles; 1870 km)
Maximum range: 1,675 nm (1930 miles; 1162 km)

Armament

Fixed: Two 0.303-in (7.7-mm) Type 97 machine-guns in the upper fuselage decking and two wing-mounted 20-mm Type 99 cannon. The A6M8 carried two wing mounted 0.6-in (13.2-mm) Type 3 machine-guns and two wing-mounted 20-mm Type 99 cannon.
External: Normal load was two 132-lb (60-kg) bombs. For suicide missions, one 551-lb (250-kg) bomb was carried. Maximum external load for A6M7 and A6M8 was 1,102 lb (500 kg). Eight 22-lb (10-kg) or two 132-lb (60-kg) air-to-air rockets could be carried by the A6M6c and A6M8. There were also hardpoints for drop tanks; usually a 72.6-Imp gal (330-litre) tank was carried, but the A6M7 and A6M8 carried a 77-Imp gal (350-litre) tank.

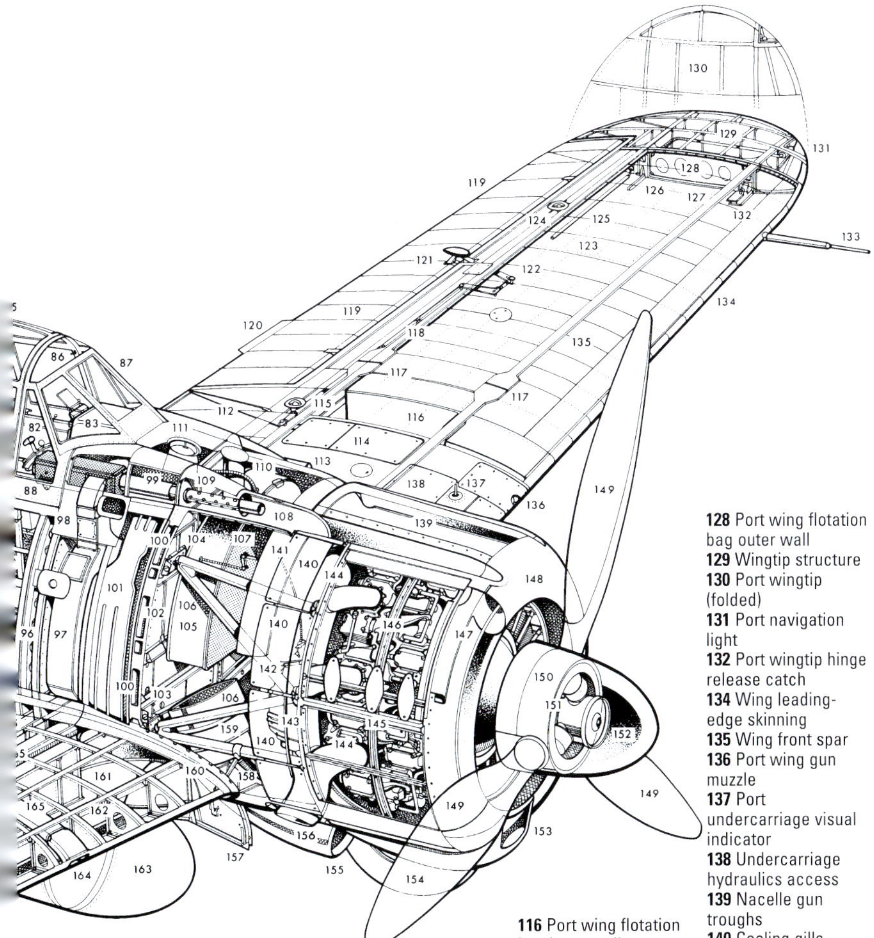

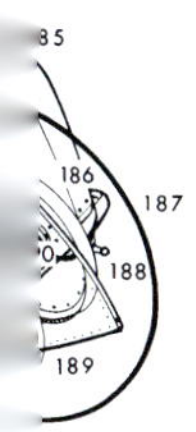

98 Ammunition feed
99 Blast tube
100 Cooling louvres
101 Fuselage fuel tank capacity 34 Imp gal (155 litres)
102 Firewall bulkhead
103 Engine bearer lower attachment
104 Engine bearer upper attachment
105 Oil tank capacity 12.7 Imp gal (58 litres)
106 Bearer support struts
107 Cooling gill adjustment control
108 Machine-gun muzzle trough
109 Barrel fairing
110 Oil filler cap
111 Fuselage fuel tank filler cap
112 Port flap profile
113 Port fuselage machine-gun
114 Port wing gun access panels
115 Port inner wing identification light
116 Port wing flotation bag inner wall
117 Wing spar joins
118 Aileron control rods
119 Port aileron (fabric covered)
120 Aileron tab (ground adjustable)
121 Aileron external counter-balance
122 Control linkage
123 Wing skinning
124 Port inner wing identification light
125 Port navigation light lead conduit
126 Wingtip hinge
127 Wing end rib
128 Port wing flotation bag outer wall
129 Wingtip structure
130 Port wingtip (folded)
131 Port navigation light
132 Port wingtip hinge release catch
134 Wing leading-edge skinning
135 Wing front spar
136 Port wing gun muzzle
137 Port undercarriage visual indicator
138 Undercarriage hydraulics access
139 Nacelle gun troughs
140 Cooling gills
141 Fuselage gun synchronisation cable
142 Bearer support strut assembly
143 Carburettor
144 Exhaust manifold
145 Cowling panel fastener clips
146 950 hp Nakajima Sakae 12 radial engine
147 Cowling inner ring profile
148 Cowling nose ring
149 Three-bladed propeller
150 Spinner
151 Propeller gears
152 Hub
153 Carburettor intake
154 Port main wheel
155 Oil cooler intake
156 Exhaust outlet
157 Starboard mainwheel inner door fairing
158 Engine bearer support brace
159 Oil cooler
160 Wingroot fasteners
161 Starboard mainwheel well
162 Front auxiliary spar cutouts
163 Auxiliary fuel tank
165 Intake trunking
166 Front main spar
167 Starboard wing fuel tank capacity 43 Imp gal (195 litres)
168 Fuel filler cap
169 Rear main spar
170 Flap actuating cylinder
171 Access cover
172 Starboard flap structure
173 Starboard inner wing identification light
174 Starboard wing 20-mm cannon
175 Access panels
176 Ammunition magazine (underwing loading)
177 Landing gear hydraulic retraction jack
178 Hydraulic lines
179 Starboard undercarriage visual indicator
180 Landing gear pivot axis
181 Undercarriage/spar mounting
182 Starboard wing gun muzzle
183 Starboard undercarriage leg
184 Oleo travel
185 Welded steel wheel fork
186 Wheel uplock latch
187 Starboard mainwheel
188 Wheel door fairing ball and swivel closure
189 Mainwheel door fairing
190 Axle hub
191 Access plate
192 Hinge
193 Left fairing attachments
194 Brake line
195 Leg fairing
196 Leg fairing upperflap
197 Wing gun barrel support collar
198 Wing nose ribs
199 Cartridge ejection chute
200 Wing spar joins
201 Wing outer structure
202 Front spar outer section
203 Inter-spar ribs
204 Rear spar outer section
205 Aileron control access
206 Aileron (ground adjustable)
207 Starboard aileron frame
208 Aileron external counter balance
209 Control linkage
210 Starboard wingtip (folded)
211 Starboard outerwing identification light
212 Aileron outer hinge
213 Starboard wing flotation bag outer wall
214 Wing end rib
215 Starboard wingtip hinge release catch
216 Wingtip structure
217 Starboard navigation light

A6M2 Reisen 'Zeke'

This A6M2 Reisen belonged to the 2nd Sentai, 1st Koku Kentai (air fleet) flying from the carrier *Hiryu* during the Battle of Midway in June 1942. Wearing typical markings for the period, the aircraft is painted overall light grey and features a black cowling and a borderless Hinomaru (national insignia). The Imperial Japanese Navy concentrated its Reisens in large formations for maximum effect although, with four Japanese carriers being sunk at Midway by the US Navy Pacific Fleet, the Imperial Fleet was unable to mount further mass air attacks.

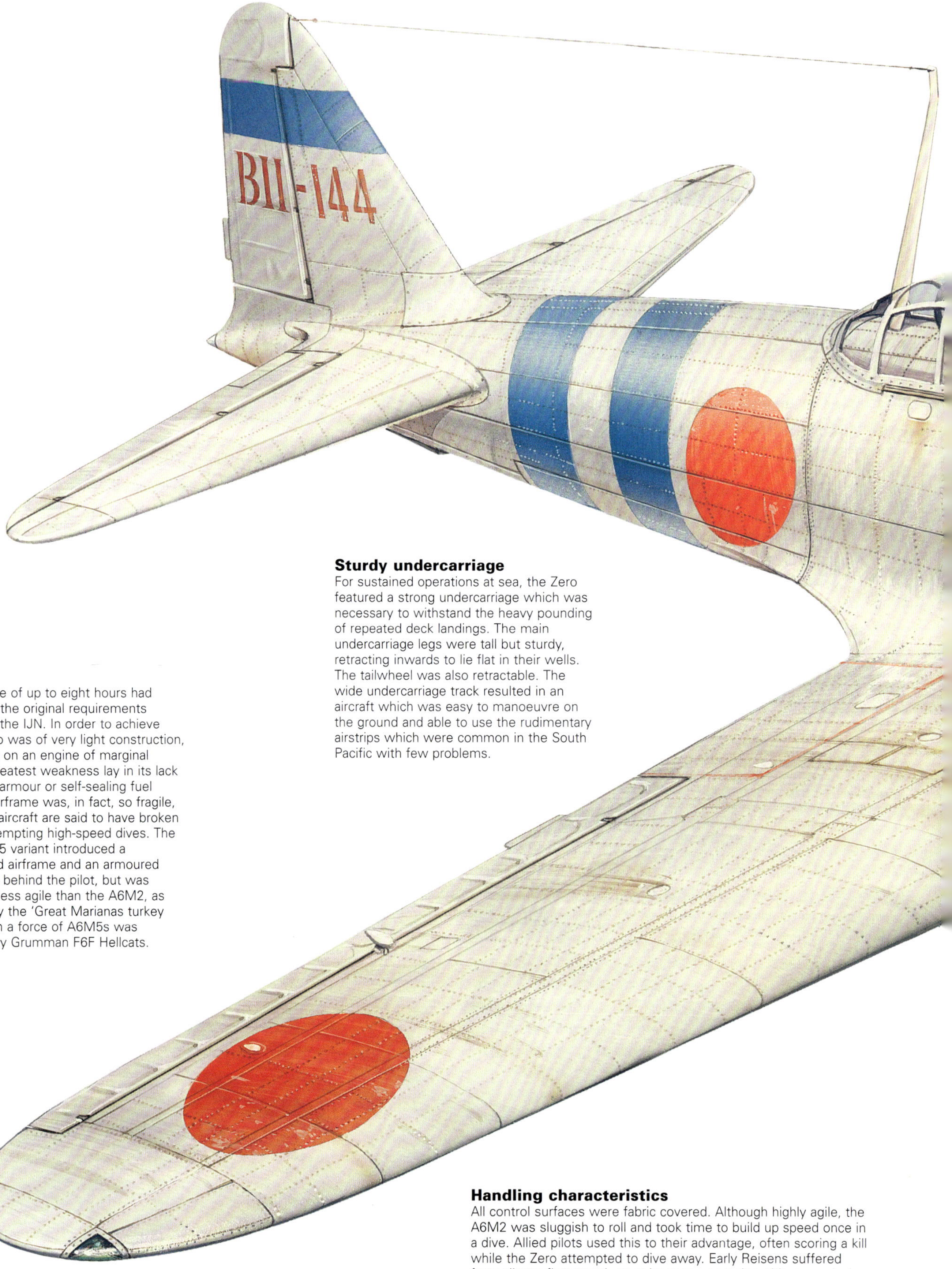

Sturdy undercarriage
For sustained operations at sea, the Zero featured a strong undercarriage which was necessary to withstand the heavy pounding of repeated deck landings. The main undercarriage legs were tall but sturdy, retracting inwards to lie flat in their wells. The tailwheel was also retractable. The wide undercarriage track resulted in an aircraft which was easy to manoeuvre on the ground and able to use the rudimentary airstrips which were common in the South Pacific with few problems.

Armour
An endurance of up to eight hours had been one of the original requirements specified by the IJN. In order to achieve this, the Zero was of very light construction, while relying on an engine of marginal power. Its greatest weakness lay in its lack of adequate armour or self-sealing fuel tanks. The airframe was, in fact, so fragile, that several aircraft are said to have broken up while attempting high-speed dives. The interim A6M5 variant introduced a strengthened airframe and an armoured glass screen behind the pilot, but was heavier and less agile than the A6M2, as evidenced by the 'Great Marianas turkey shoot', when a force of A6M5s was annihilated by Grumman F6F Hellcats.

Handling characteristics
All control surfaces were fabric covered. Although highly agile, the A6M2 was sluggish to roll and took time to build up speed once in a dive. Allied pilots used this to their advantage, often scoring a kill while the Zero attempted to dive away. Early Reisens suffered from aileron flutter and, on at least one occasion, this prevented the pilot from pulling out of a dive, his aircraft crashing into the sea. The ailerons on subsequent aircraft were modified as a result.

Fuselage and canopy
With the exception of the control surfaces, the Zero was of all-metal construction. There was no armour for the pilot, no bulletproof windshield and no jettisonable hood. The cockpit arrangement was typical for fighters of the era except for its size. While big enough for a Japanese pilot, a typical Western pilot in full flying gear would find the cockpit cramped. The canopy used ordinary glass, which was all flat plate except for the segments in the curved top.

Engine
The standard engine of the A6M2 variant was a 14-cylinder Nakajima NK1C Sakae 12 radial piston engine. Fitted with a single-stage supercharger and driving a three-bladed airscrew, it produced 950 hp (708 kW). Even by 1940 standards, this was not a particularly powerful engine, and the excellent agility and performance of the Zero could only be maintained in later, heavier models by the substitution of a more powerful engine. The A6M3 introduced an improved Sakae 21 with a two-stage supercharger, boosting power output to 1,130 hp (843 kW). Throughout the war, the chief engineer on the Zero, Jiro Horikoshi, had favoured the 1,560-hp (1163-kW) Mitsubishi Kinsei 62 radial as the original powerplant and was finally granted his wish with the construction of the A6M8. Approximately 6,300 of this variant were ordered but Allied bombing hindered its development and only two prototypes were built.

Fuel
A6M2s had three fuel tanks, one in each wingroot and a fuselage tank directly in front of the pilot. During operations in the Pacific, most Reisens also carried a single drop tank under the fuselage, extending maximum range out to 1,864 miles (3000 km) and endurance to eight hours on intermediate engine power.

Propeller
The A6M1 prototype was completed in March 1939 and first flew in April. One of the key changes during the flight test programme was the change from a two-bladed variable-pitch propeller to a Sumitomo-Hamilton three-bladed constant-speed unit. The three-bladed metal propeller became standard on all subsequent variants.

Armament
Although the Zero's designers sacrificed protection to save weight, they decided from the very beginning to equip their nimble fighter with heavy 20-mm cannon. Although slow-firing, these hard-hitting weapons gave the Zero a decided advantage over US fighters armed only with machine-guns. Later versions of the Zero carried a number of different weapon configurations, one of the most radical being that of the night-fighter version of the A6M5. It had two 0.303-in (7.7-mm) Type 97 machine-guns in the upper fuselage decking, two wing-mounted 20-mm Type 99 cannon and one fuselage-mounted oblique-firing Type 99 cannon for engaging bombers from relative safety.

Mitsubishi Ki-46 'Dinah'

This 16 Dokuritsu Hiko-tai Ki-46-III has the upward/oblique-firing 1.45-in (37-mm) cannon that was typical of the variant.

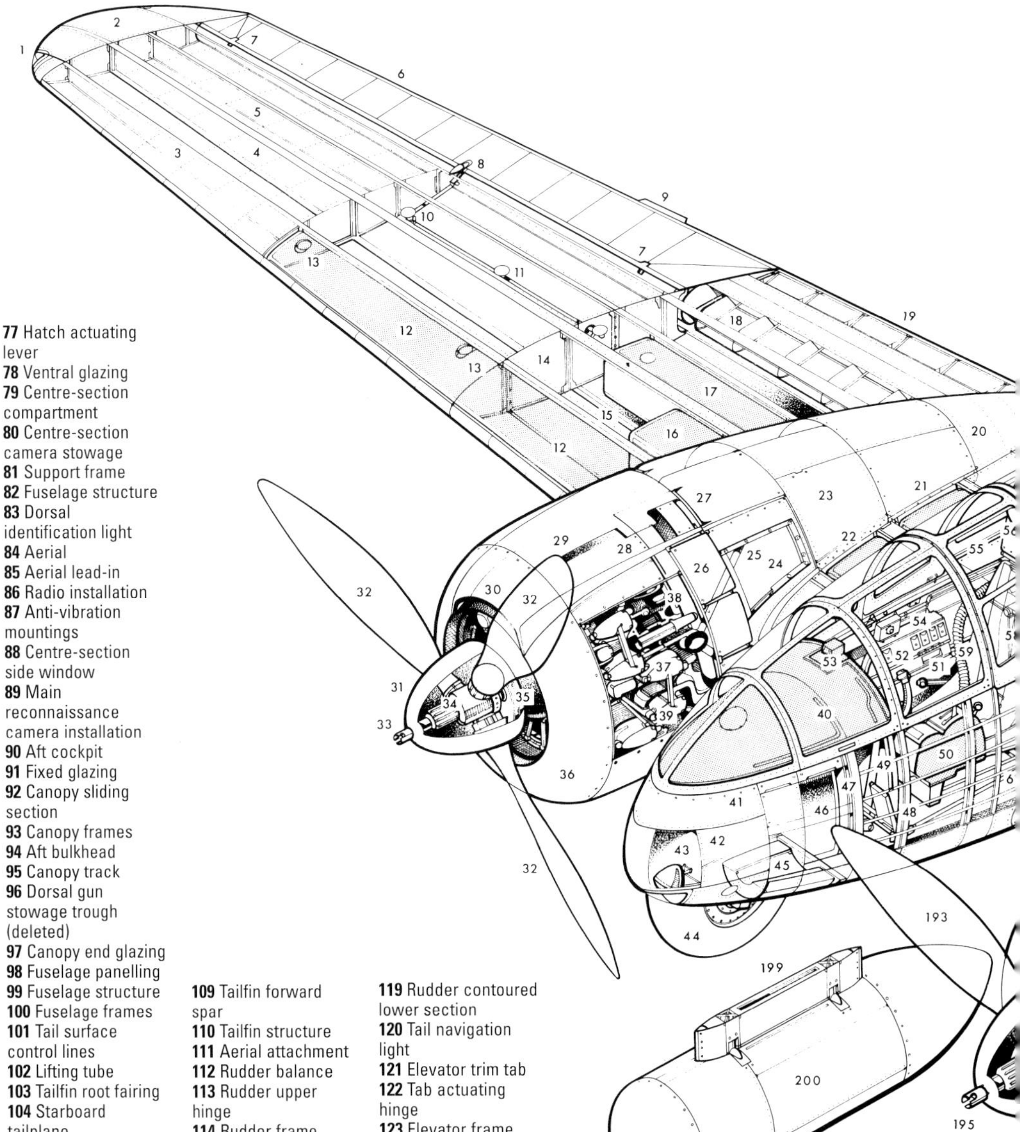

Ki-46 'Dinah'

Cutaway key
1 Starboard navigation light
2 Starboard wingtip
3 Wing front spar
4 Main spar
5 Auxiliary rear spar
6 Starboard aileron
7 Aileron hinges
8 Aileron actuating hinge fairing
9 Aileron fixed tab
10 Access plates
11 Control rods
12 Leading-edge fuel tank
13 Filler/access points
14 Rib station
15 Centre spar
16 Centre fuel tank
17 Aft fuel tank
18 Flap profile
19 Starboard flap outer section
20 Starboard nacelle aft fairing
21 Wing inner aft fuel tank
22 Wing inner centre fuel tank
23 Nacelle panels
24 Access
25 Engine bearer ring support
26 Cooling gills
27 Exhaust slots
28 Cowling inner ring
29 Inner trunking
30 Intake slot
31 Spinner
32 Three-blade propeller
33 Starter dog
34 Propeller hub
35 Reduction gear housing
36 Cowling nose ring
37 Mitsubishi Ha-112 Otsu radial engine
38 Exhaust manifold
39 Unstepped nose glazing
40 Inner coaming
41 Fixed frame
42 Nose panels
43 Nose landing lamp
44 Starboard mainwheel
45 Nose access/(optional) camera hatch
46 Nose (optional) fuel tank
47 Fuselage forward frame
48 Rudder pedal assembly
49 Control column
50 Throttle quadrant
51 Seat adjustment lever
52 Control horn
53 Compass housing
54 Starboard electrics panel
55 Canopy sliding section
56 Pilot's headrest
57 Pilot's 13-mm back armour
58 Pilot's seat and harness
59 Oxygen hose
60 Seat support frame
61 Control rod linkage
62 Wing root fillet
63 Wing front spar/fuselage frame
64 Main spar centre-section carry-through
65 Wing control surface actuating rods
66 Canopy track
67 Canopy fixed aft glazing
68 Armoured headrest support
69 Aerial mast
70 Dorsal decking
71 Fuselage main (contoured cut-out) fuel tank
72 Spring-loaded hand/entry grips
73 Cockpit former longeron
74 Fuel feed lines
75 Centre-section camera mounting rings
76 Ventral sliding hatch
77 Hatch actuating lever
78 Ventral glazing
79 Centre-section compartment
80 Centre-section camera stowage
81 Support frame
82 Fuselage structure
83 Dorsal identification light
84 Aerial
85 Aerial lead-in
86 Radio installation
87 Anti-vibration mountings
88 Centre-section side window
89 Main reconnaissance camera installation
90 Aft cockpit
91 Fixed glazing
92 Canopy sliding section
93 Canopy frames
94 Aft bulkhead
95 Canopy track
96 Dorsal gun stowage trough (deleted)
97 Canopy end glazing
98 Fuselage panelling
99 Fuselage structure
100 Fuselage frames
101 Tail surface control lines
102 Lifting tube
103 Tailfin root fairing
104 Starboard tailplane
105 Elevator balance
106 Starboard elevator
107 Elevator hinge
108 Tailfin leading-edge
109 Tailfin forward spar
110 Tailfin structure
111 Aerial attachment
112 Rudder balance
113 Rudder upper hinge
114 Rudder frame
115 Rudder trim tab
116 Rudder actuating hinge
117 Rudder tab hinge fairing
118 Rudder post
119 Rudder contoured lower section
120 Tail navigation light
121 Elevator trim tab
122 Tab actuating hinge
123 Elevator frame
124 Elevator hinge
125 Elevator balance
126 Tailplane structure
127 Tailplane front spar
128 Control cables

The most obvious external change in the Ki-46-III was the distinctive redesign of the forward fuselage with a new canopy design.

SPECIFICATION

Mitsubishi Ki-46-III

Type

Two-seat reconnaissance aircraft

Powerplant

Two 1,500-hp (1119-kW) Mitsubishi Ha-112-II radial piston engines

Performance

Maximum speed 391 mph (630 km/h) at 19,685 ft (6000 m); service ceiling 34,450 ft (10500 m); range 2,485 miles (4000 km)

Weights

Empty 8,444 lb (3830 kg); maximum take-off 14,330 lb (6500 kg)

Dimensions

Wingspan 48 ft 23⁄4 in (14.70 m); length 36 ft 1 in (11.00 m); height 12 ft 83⁄4 in (3.88 m); wing area 344.46 sq ft (32.00 m2)

Armament

Ki-46-I and Ki-46-II had a single 7.7-mm (0.303-in) rear-firing machine-gun on a trainable mount; III Kai two 20-mm Ho-5 and oblique 37 Ho-203

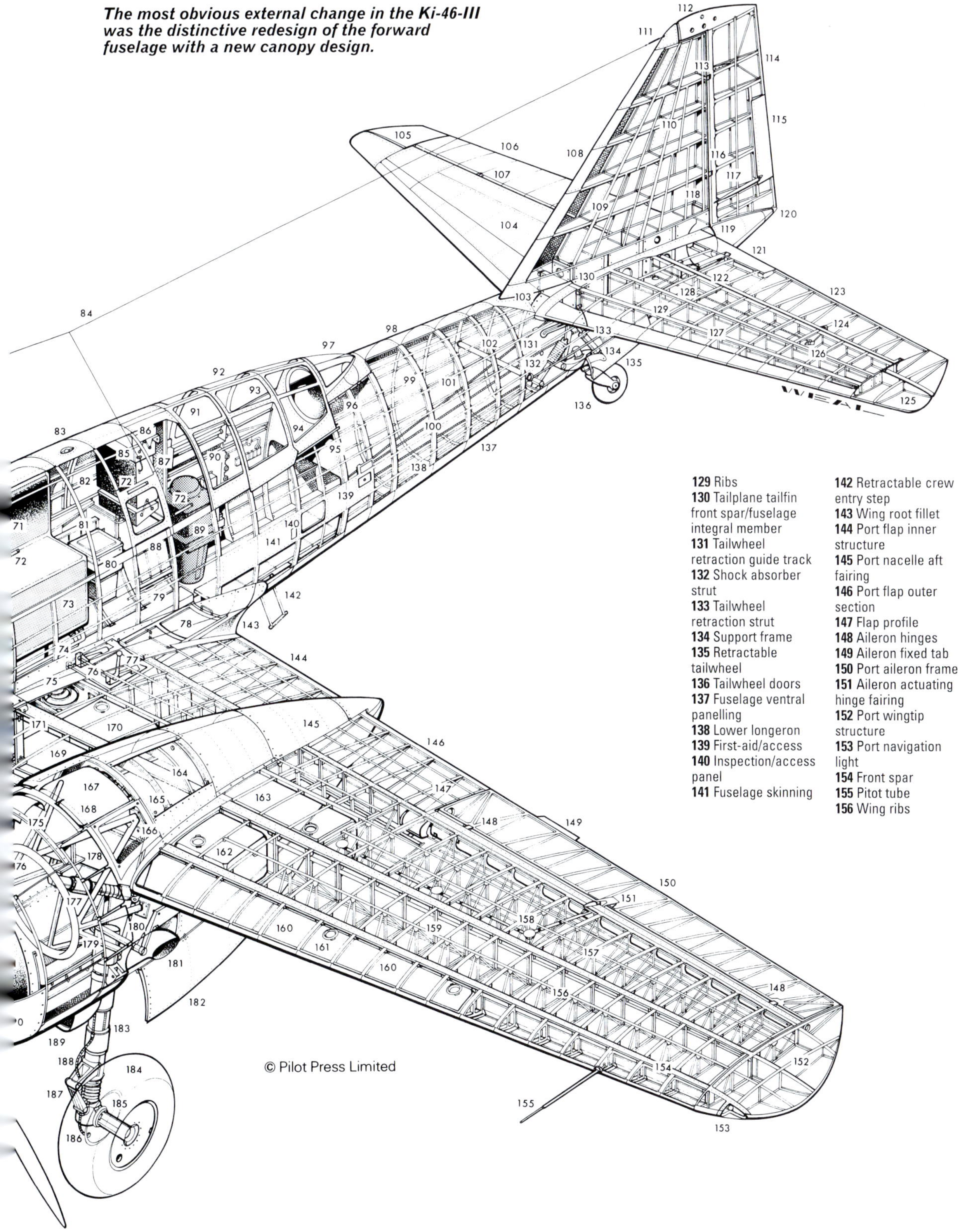

129 Ribs
130 Tailplane tailfin front spar/fuselage integral member
131 Tailwheel retraction guide track
132 Shock absorber strut
133 Tailwheel retraction strut
134 Support frame
135 Retractable tailwheel
136 Tailwheel doors
137 Fuselage ventral panelling
138 Lower longeron
139 First-aid/access
140 Inspection/access panel
141 Fuselage skinning
142 Retractable crew entry step
143 Wing root fillet
144 Port flap inner structure
145 Port nacelle aft fairing
146 Port flap outer section
147 Flap profile
148 Aileron hinges
149 Aileron fixed tab
150 Port aileron frame
151 Aileron actuating hinge fairing
152 Port wingtip structure
153 Port navigation light
154 Front spar
155 Pitot tube
156 Wing ribs
157 Wing structure
158 Access panels
159 Wing main spar
160 Leading-edge fuel tank
161 Filler/access
162 Centre fuel tank
163 Aft fuel tank
164 Nacelle formers
165 Bulkhead frame
166 Engine bearer ring support attachment
167 Port nacelle oil tank
168 Exhaust slots
169 Wing inner centre fuel tank
170 Wing inner aft fuel tank
171 Wing main spar attachment
172 Wing front spar attachment
173 Leading-edge ribs
174 Cowling frame
175 Cowling inner frame
176 Engine bearer ring
177 Undercarriage retraction strut
178 Cooling gills
179 Mainwheel leg pivot
180 Engine bearer ring lower support strut
181 Exhaust
182 Mainwheel door
183 Mainwheel leg
184 Port mainwheel
185 Axle
186 Brake line
187 Torque links
188 Shock strut
189 Lower intake
190 Engine cowling nose ring
191 Inner ring
192 Gear housing
193 Three-blade Sumitomo propeller
194 Spinner
195 Starter dog
196 Propeller hub
197 Intake trunking
198 Intake slot
199 Ventral (centre-line) tank pylon
200 Auxiliary ventral fuel tank (101 Imp gal/460 litre capacity)

Nakajima Ki-43

The Japanese puppet state of Manchukuo was supplied with Ki-43s for home defence. Here the Japanese equivalent of the British 'Hucks starter' is seen being employed to start a section of Hayabusas.

Nakajima Ki-43

Cutaway key

1 Starboard navigation light
2 Wingtip
3 Starboard fabric-covered aileron
4 Aileron actuating linkage
5 Aileron control rod
6 Control rod connecting fittings
7 Aileron tab
8 Flap outer cable drum
9 Flap travel
10 Flap control cables
11 Radio mast
12 Light alloy wing skinning
13 Starboard undercarriage fairing
14 Gun port fairings
15 Nose ring
16 Annular oil cooler
17 Two-blade two-pitch metal propeller
18 Spinner
19 Starter dog
20 Supercharger air intake
21 Intake fairing
22 Nakajima Ha-25 (Type 99) 14-cylinder two-row radial engine
23 Cowling gills
24 Exhaust collector ring
25 Exhaust outlet
26 Engine lower bearers
27 Oil regulator valve
28 Oil pressure tank
29 Engine accessories
30 Engine upper bearers
31 Cowling gill controls
32 Two 0.5-in (12.7-mm) Type 89 machine-guns
33 Gun gas outlet
34 Cartridge link ejection chute
35 Fireproof (No. 1) bulkhead
36 Ammunition magazine (500 rpg)
37 Cartridge ejection chute
38 Gun breech fairing
39 Telescopic gun sight
40 One-piece curved windscreen
41 Radio aerial
42 Aft-sliding cockpit canopy
43 Turnover structure
44 Seat back
45 Seat adjustment rails
46 Seat pan
47 Throttle quadrant
48 Instrument panel
49 Control column
50 Rudder pedals
51 Underfloor control linkage
52 Seat support frame
53 Control cable and rod bearings
54 Oxygen cylinders
55 Rudder cable pulleys
56 Transceiver
57 Type 96 Hi-3 radio installation
58 Receiver unit
59 Transmitter unit
60 Anti-vibration mounting slings
61 Fuselage construction break
62 Inspection/access panel
63 Fuselage stringers
64 Fuselage structure
65 Frame
66 Fuselage upper longeron
67 Elevator control cables
68 Fuselage skinning
69 Tailwheel shock strut
70 Tail unit attachment
71 Tailfin root fairing
72 Starboard tailplane
73 Elevator balance
74 Starboard elevator
75 Tailfin leading edge
76 Tailfin structure
77 Rear navigation light
78 Aerial attachment
79 Rudder upper hinge
80 Rudder post
81 Rudder frame
82 Rudder trim tab
83 Rudder middle hinge
84 Elevator control lever
85 Elevator trim tab
86 Elevator frame
87 Elevator balance
88 Tailplane structure
89 Rudder control lever
90 Non-retractable tailwheel
91 Cantilever tailwheel leg
92 Tailwheel leg/bulkhead attachment
93 Rudder cables
94 Fuselage skinning
95 Wing fillet
96 Flap inboard profiles
97 Flap actuating cylinder

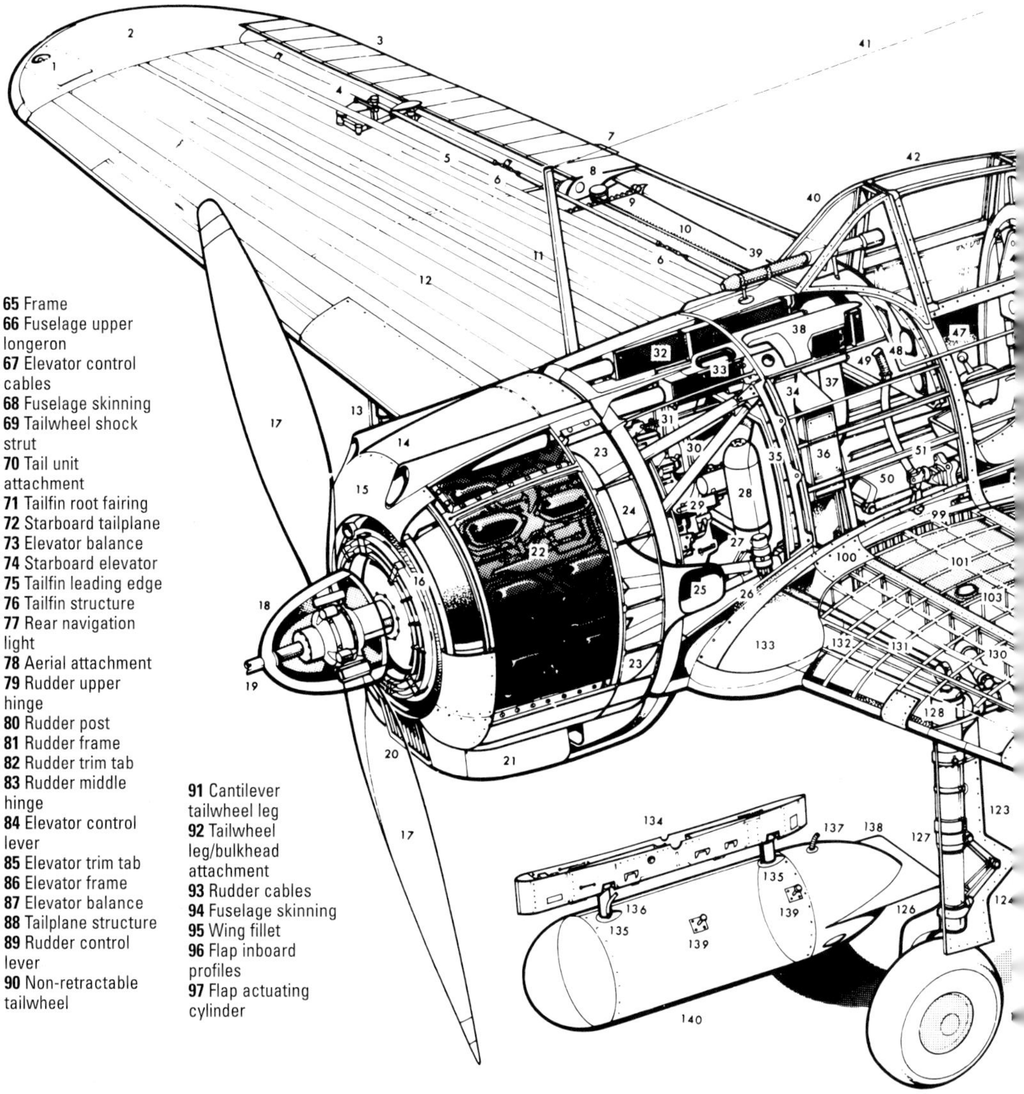

After participating in the Philippines campaign, the 50th Sentai returned to Japan for conversion to Ki-43-Is for subsequent operations in Burma. Photographed at Tokorozawa in June 1942, these Ki-43-I-Hei belonged to either the unit's 1st Chutai (with white lightning markings) or 3rd Chutai (yellow lightning).

SPECIFICATION

Nakajima Ki-43-IIb

Type

Single-seat fighter/fighter-bomber
Powerplant
one 1,150-hp (858-kW) Nakajima Ha-115 radial piston engine

Performance

Maximum speed 329 mph (530 km/h) at 13,125 ft (4000 m); service ceiling 36,745 ft (11200 m); maximum range 1,988 miles (3200 km)

Weights

Empty 4,211 lb (1910 kg); maximum take-off 5,710 lb (2590 kg)

Dimensions

Wingspan 35 ft 6¾ in (10.84 m); length 29 ft 3¼ in (8.92 m); height 10 ft 8¾ in (3.27 m); wing area 230.36 sq ft (21.40 m2)

Armament

Two 0.50-in (12.5-mm) forward-firing machine-guns, plus two underwing racks each able to carry a 551-lb (250-kg) bomb

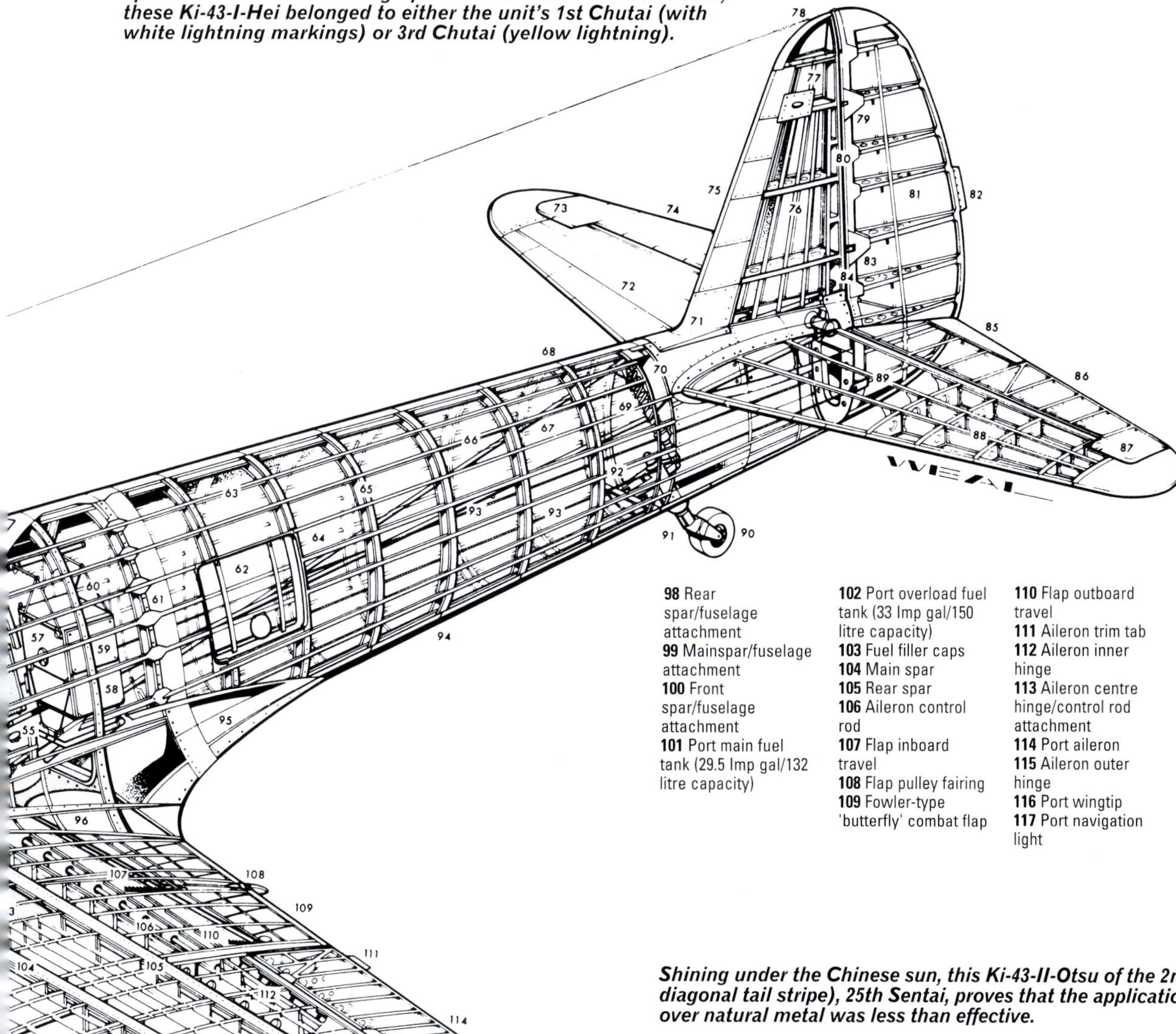

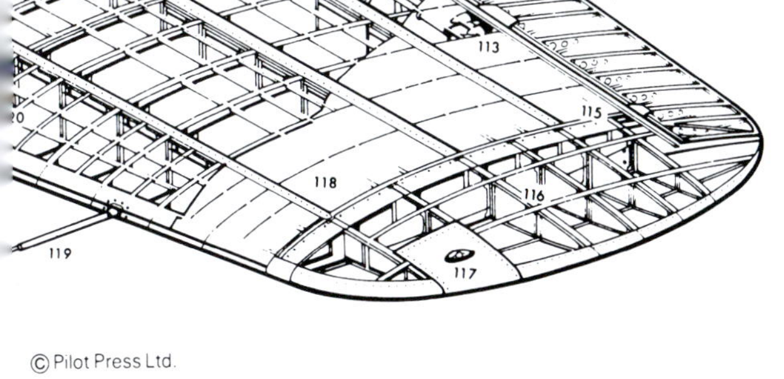

98 Rear spar/fuselage attachment
99 Mainspar/fuselage attachment
100 Front spar/fuselage attachment
101 Port main fuel tank (29.5 Imp gal/132 litre capacity)
102 Port overload fuel tank (33 Imp gal/150 litre capacity)
103 Fuel filler caps
104 Main spar
105 Rear spar
106 Aileron control rod
107 Flap inboard travel
108 Flap pulley fairing
109 Fowler-type 'butterfly' combat flap
110 Flap outboard travel
111 Aileron trim tab
112 Aileron inner hinge
113 Aileron centre hinge/control rod attachment
114 Port aileron
115 Aileron outer hinge
116 Port wingtip
117 Port navigation light
118 Wing skinning
119 Pitot head
120 Leading edge ribs
121 Front spar
122 Landing light
123 Mainwheel leg fairing
124 Torque links
125 Port mainwheel
126 Axle fork
127 Mainwheel oleo
128 Mainwheel leg pivot
129 Gear support bearer
130 Gear actuating cylinder
131 Emergency actuation cables
132 Leading edge rib cut-outs
133 Mainwheel well
134 Underwing drop tank pylon (mounted aft and just inboard of the main undercarriage attachment point)
135 Tank suspension lugs
136 Air vent
137 Fuel pipe connection
138 Tank fin
139 Sway brace attachment points
140 Jettisonable 44-Imp gal (200-litre) tank

Shining under the Chinese sun, this Ki-43-II-Otsu of the 2nd Chutai (red diagonal tail stripe), 25th Sentai, proves that the application of green mottle over natural metal was less than effective.

Nakajima Ki-84 Hayate

This is one of the initial pre-production batch of 83 Ki-84s, seen at Tachikawa in August 1943 from where it was flown by the Army Air Arsenal as a service trials aircraft.

Nakajima Ki-84

1 Starter dog
2 Spinner
3 Constant-speed electrically-operated Pe-32 propeller
4 Propeller reduction gear housing
5 Carburettor air intake
6 Starboard 20-mm Ho-5 cannon muzzle
7 Gun camera port
8 Starboard leading-edge fuel tank (14.7 Imp gal/67 litre capacity)
9 Mainspar
10 Starboard navigation light
11 Starboard wingtip
12 Fabric-covered aileron
13 Aileron control link fairing
14 Aileron trim tab
15 Flap track extension fairing
16 Starboard Fowler-type flap
17 Wing cannon ammunition box access
18 Wing cannon access covers
19 Carburettor intake trunking
20 Machine-gun blast tube
21 Machine-gun trough
22 Army Type 4 Model 21 (Nakajima Ha-45-21) 18-cylinder radial air-cooled engine
23 Cowling fasteners
24 Aluminium cylinder fans
25 Oil cooler intake
26 Starboard mainwheel
27 Oil cooler housing
28 Ejector exhaust stubs
29 Cowling gills
30 Engine bearers
31 Oil tank (11 Imp gal/50 litre capacity)
32 Vent
33 Gun cooling muffle
34 Firewall/bulkhead
35 Ho-103 machine-gun (two) of 13-mm calibre
36 Main fuel tank (47.7 Imp gal/217 litre capacity)
37 Port ammunition tank (350 rounds)
38 Fuel filler cap
39 Rudder pedals
40 Control column
41 Instrument panel
42 Fuselage flush-riveted stressed-skin panels
43 Reflector sight (offset to starboard)
44 Armourglass (65-mm) windscreen
45 Aft-sliding cockpit canopy
46 Canopy lock/release
47 Pilot's headrest
48 Pilot's head armour/turnover support
49 Canopy fixed aft glazing
50 Canopy track
51 Entry handgrip
52 Pilot's 13-mm back armour
53 Elevator trim handwheel
54 Pilot's seat (adjustable vertically)
55 Throttle quadrant
56 Flap setting lever
57 Undercarriage selector lever
58 Underfloor control runs
59 Flap-rod linkage
60 Water-methanol tank
61 Mid-fuselage construction break
62 Radio equipment tray
63 Type 4 Hi no.3 radio communications pack
64 Aerial lead-in
65 Aerial mast
66 Aerials
67 Light alloy semi-monocoque fuselage structure
68 Fuselage upper longeron
69 Oval section fuselage aft frames
70 Aft fuselage construction break
71 Starboard tailplane
72 Elevator balance
73 Starboard elevator (fabric covered)
74 Elevator trim tab
75 Tailfin leading edge
76 Tailfin structure
77 Rear navigation/formation light
78 Aerial stub attachment

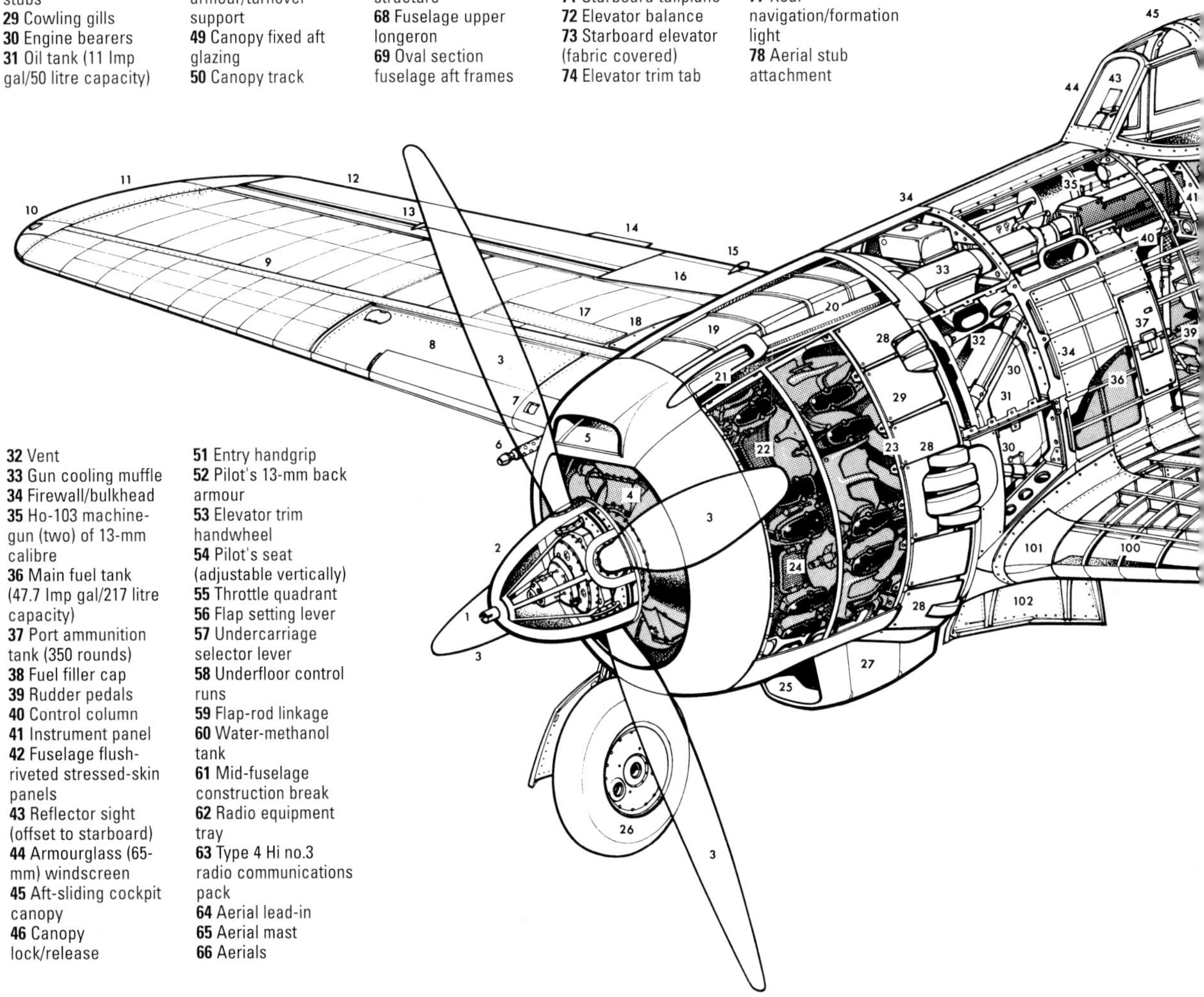

This aircraft is one of the three Ki-106 prototypes built by Tachikawa late in the war. Although it looked exactly like a Ki-84, it was built entirely of wood in an attempt to save strategic materials.

SPECIFICATION

Nakajima Ki-84-Ia

Type
Single-seat interceptor fighter/fighter-bomber

Powerplant
One 1,900-hp (1416-kW) Nakajima Ha-45 radial piston engine

Performance
Maximum speed 392 mph (631 km/h) at 20,080 ft (6120 m); service ceiling 34,350 ft (10500 m); maximum range 1,347 miles (2168 km)

Weights
Empty 5,864 lb (2660 kg); maximum take-off 8,576 lb (3890 kg)

Dimensions
Wingspan 36 ft 10 1/2 in (11.24 m); length 32 ft 6 1/2 in (9.92 m); height 11 ft 1 1/2 in (3.39 m); wing area 226.05 sq ft (21.00 m2)

Armament
Two 0.5-in (12.7-mm) machine-guns and two 20-mm cannon, plus underwing racks for two 551-lb (250-kg) bombs

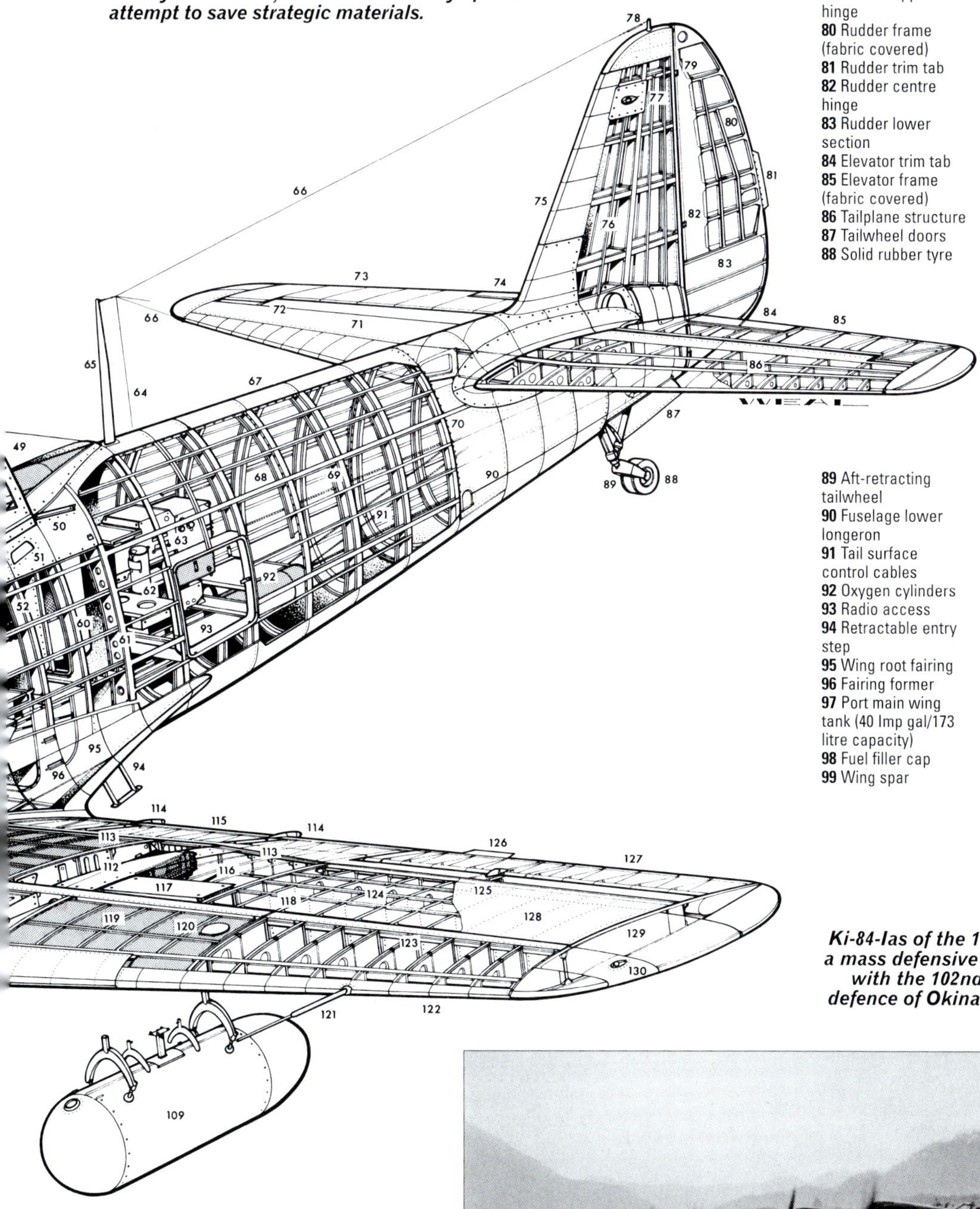

79 Rudder upper hinge
80 Rudder frame (fabric covered)
81 Rudder trim tab
82 Rudder centre hinge
83 Rudder lower section
84 Elevator trim tab
85 Elevator frame (fabric covered)
86 Tailplane structure
87 Tailwheel doors
88 Solid rubber tyre
89 Aft-retracting tailwheel
90 Fuselage lower longeron
91 Tail surface control cables
92 Oxygen cylinders
93 Radio access
94 Retractable entry step
95 Wing root fairing
96 Fairing former
97 Port main wing tank (40 Imp gal/173 litre capacity)
98 Fuel filler cap
99 Wing spar
100 Undercarriage leg cut-outs
101 Mainwheel wells
102 Mainwheel doors
103 Port 20-mm Ho-5 cannon muzzle
104 Wheel brake hydraulic lines
105 Shock-absorber links
106 Port mainwheel
107 Axle
108 Mainwheel leg fairing
109 Underwing auxiliary fuel tank (44 Imp gal/200 litres capacity)
110 Landing light
111 Cannon blast tube
113 Flap tracks
114 Flap track extension fairings
115 Fowler-type flap structure
116 Rear auxiliary spar
117 Cannon ammunition tank (150 rounds)
118 Spar join
119 Port auxiliary leading-edge tank (14.7 Imp gal/67 litre capacity)
120 Fuel filler cap
121 Pitot tube
123 Main spar outer section
124 Wing ribs
125 Aileron control rod link fairing
126 Aileron trim tab
127 Aileron frame (fabric covered)
128 Wing skinning
129 Port wingtip
130 Port navigation light

Ki-84-Ias of the 101st Sentai start their engines prior to a mass defensive action in the latter part of 1944. Along with the 102nd Sentai, the unit was embroiled in the defence of Okinawa, and enjoyed conspicuous success in raids on US airfields.

North American P-51 Mustang

It is a tribute to the Mustang's popularity that today there are almost 150 airworthy examples. This aircraft represents 'Chuck' Yeager's **Glamorous Glen III.**

P-51B Mustang

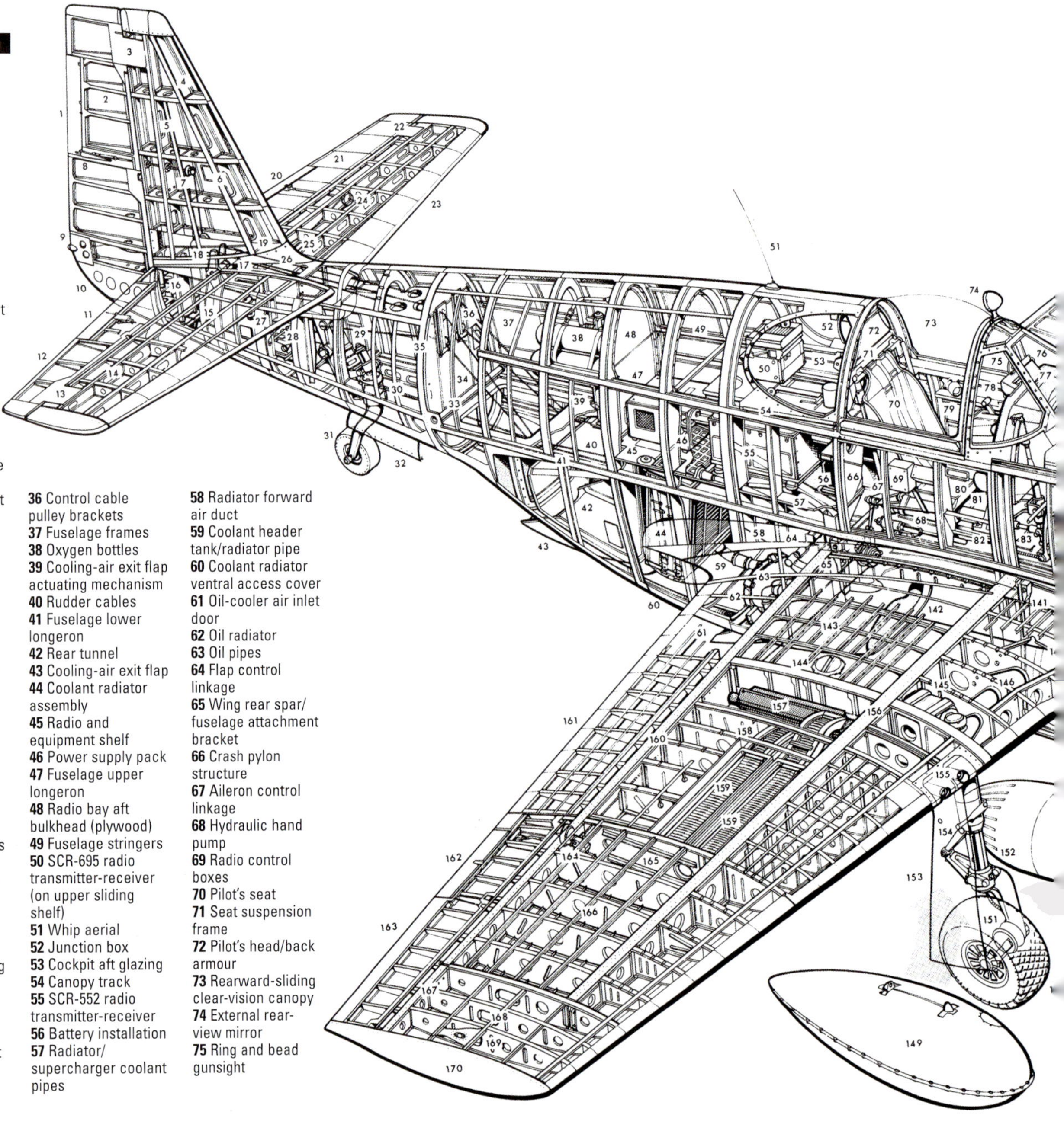

Cutaway key

1 Plastic (Phenol fibre) rudder trim tab
2 Rudder frame (fabric covered)
3 Rudder balance
4 Fin front spar
5 Fin structure
6 Access panel
7 Rudder trim-tab actuating drum
8 Rudder trim-tab control link
9 Rear navigation light
10 Rudder metal bottom section
11 Elevator plywood trim tab
12 Starboard elevator frame
13 Elevator balance weight
14 Starboard tailplane structure
15 Reinforced bracket (rear steering stresses)
16 Rudder operating horn forging
17 Elevator operating horns
18 Tab control turnbuckles
19 Fin front spar/ fuselage attachment
20 Port elevator tab
21 Fabric-covered elevator
22 Elevator balance weight
23 Port tailplane
24 Tab control drum
25 Fin root fairing
26 Elevator cables
27 Tab control access panels
28 Tailwheel steering mechanism
29 Tailwheel mount
30 Tailwheel leg assembly
31 Forward-retracting steerable tailwheel
32 Tailwheel doors
33 Lifting tube
34 Fuselage aft bulkhead/break point
35 Fuselage break point
36 Control cable pulley brackets
37 Fuselage frames
38 Oxygen bottles
39 Cooling-air exit flap actuating mechanism
40 Rudder cables
41 Fuselage lower longeron
42 Rear tunnel
43 Cooling-air exit flap
44 Coolant radiator assembly
45 Radio and equipment shelf
46 Power supply pack
47 Fuselage upper longeron
48 Radio bay aft bulkhead (plywood)
49 Fuselage stringers
50 SCR-695 radio transmitter-receiver (on upper sliding shelf)
51 Whip aerial
52 Junction box
53 Cockpit aft glazing
54 Canopy track
55 SCR-552 radio transmitter-receiver
56 Battery installation
57 Radiator/ supercharger coolant pipes
58 Radiator forward air duct
59 Coolant header tank/radiator pipe
60 Coolant radiator ventral access cover
61 Oil-cooler air inlet door
62 Oil radiator
63 Oil pipes
64 Flap control linkage
65 Wing rear spar/ fuselage attachment bracket
66 Crash pylon structure
67 Aileron control linkage
68 Hydraulic hand pump
69 Radio control boxes
70 Pilot's seat
71 Seat suspension frame
72 Pilot's head/back armour
73 Rearward-sliding clear-vision canopy
74 External rear-view mirror
75 Ring and bead gunsight

76 Bullet-proof windshield
77 Gyro gunsight
78 Engine controls
79 Signal-pistol discharge tube
80 Circuit-breaker panel
81 Oxygen regulator
82 Pilot's footrest and seat mounting bracket
83 Control linkage
84 Rudder pedal
85 Tailwheel lock control
86 Wing centre-section
87 Hydraulic reservoir
88 Port wing fuel tank filler point
89 Port Browning 0.5-in (12.7-mm) guns
90 Ammunition feed chutes
91 Gun-bay access door (raised)
92 Ammunition box troughs
93 Aileron control cables
94 Flap lower skin (Alclad)
95 Aileron profile (internal aerodynamic balance diaphragm)
96 Aileron control drum and mounting bracket
97 Aileron trim-tab control drum
98 Aileron plastic (Phenol fibre) trim tab
99 Port aileron assembly
100 Wing skinning
101 Outer section sub-assembly
102 Port navigation light
103 Port wingtip
104 Leading-edge skin
105 Landing lamp
106 Weapons/stores pylon
107 500-lb (227-kg) bomb
108 Gun ports
109 Gun barrels
110 Detachable cowling panels
111 Firewall/integral armour
112 Oil tank
113 Oil pipes
114 Upper longeron/engine mount attachment
115 Oil-tank metal retaining straps
116 Carburettor
117 Engine bearer assembly
118 Cowling panel frames
119 Engine aftercooler
120 Engine leads
121 Packard V-1650 (R-R Merlin) 12-cylinder liquid-cooled engine
122 Exhaust fairing panel
123 Stub exhausts
124 Magneto
125 Coolant pipes
126 Cowling forward frame
127 Coolant header tank
128 Armour plate
129 Propeller hub
130 Spinner
131 Hamilton Standard Hydromatic propeller
132 Carburettor air intake, integral with (133)
133 Engine-mount front-frame assembly
134 Intake trunk
135 Engine-mount reinforcing tie
136 Hand-crank starter
137 Carburettor trunk vibration-absorbing connection
138 Wing centre-section front bulkhead
139 Wing centre-section end rib
140 Starboard mainwheel well
141 Wing front spar/fuselage attachment bracket
142 Ventral air intake (radiator and oil cooler)
143 Starboard wing fuel tank
144 Fuel filler point
145 Mainwheel leg mount/pivot
146 Mainwheel leg rib cut-outs
147 Main gear fairing doors
148 Auxiliary fuel tank (plastic/pressed-paper composition, 108 US gal/409 litres)
149 Auxiliary fuel tank (metal 75 US gal/284 litres)
150 27-in (0.69-m) smooth-contour mainwheel
151 Axle fork
152 Towing lugs
153 Landing-gear fairing
154 Main-gear shock strut
155 Blast tubes
156 Wing front spar
157 Gun bay
158 Ammunition feed chutes
159 Ammunition boxes
160 Wing rear spar
161 Flap structure
162 Starboard aileron tab
163 Starboard aileron
164 Starboard aileron tab adjustment mechanism (ground setting)
165 Wing rib strengthening
166 Outboard section structure
167 Outer section single spar
168 Wingtip sub-assembly
169 Starboard navigation light
170 Detachable wingtip

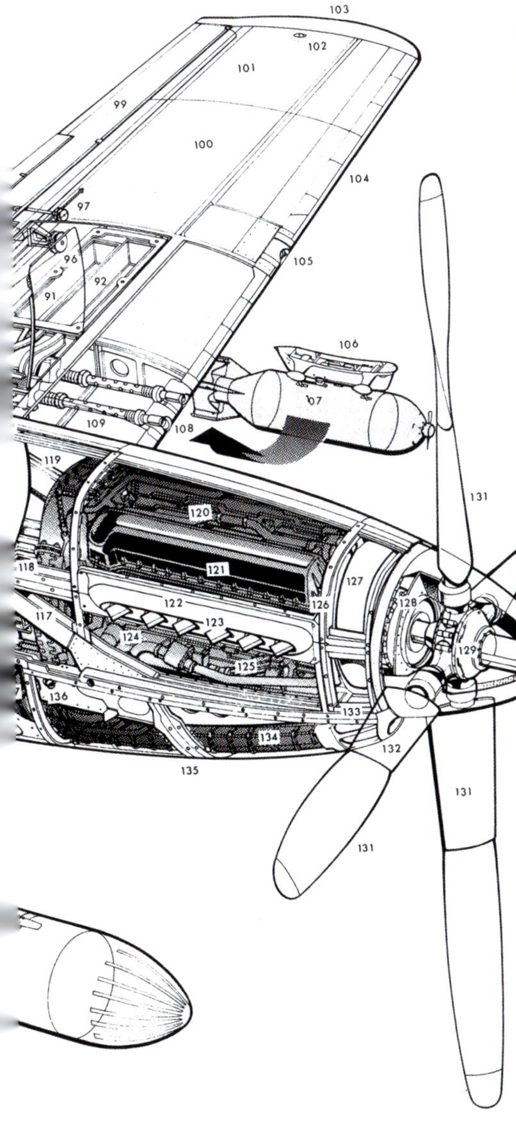

SPECIFICATION

P-51D-5-NA Mustang

Dimensions

Fuselage length: 32 ft 3$\frac{5}{16}$ in (9.84 m)
Wing span: 37 ft 0$\frac{5}{16}$ in (11.27 m)
Wing aspect ratio: 4.46
Wing incidence: approx 1°
Wing dihedral angle: 5°
Wing sweepback angle: 3° 3′ 32″
Tailplane span: 13 ft 2$\frac{1}{8}$ in (3.83 m)
Total wing area: 233.19 sq ft (21.66 m^2)
Total aileron area: 12.64 sq ft (1.17 m^2)
Total flap area: 32.60 sq ft (3.03 m^2)
Total horizontal tail area: 27.85 sq ft (2.59 m^2)
Total elevator area: 13.05 sq ft (1.21 m^2)
Fin area: 8.83 sq ft (0.82 m^2)
Rudder area: 10.25 sq ft (0.95 m^2)
Overall height: 13 ft 8 in (4.16 m) (taxiing attitude, statically loaded)
Undercarriage track: 11 ft 10 in (3.61 m)
Maximum wing loading: 49.7 lb/sq ft (242.9 kg/m^2)

Powerplant

One Packard-built Merlin V-1650-7 liquid-cooled 12-cylinder Vee inline piston engine with two-speed two-stage supercharger, driving an 11 ft 2-in (3.40-m) diameter Hamilton Standard constant-speed hydromatic four-bladed propeller
Take-off rating: 1,450 hp (1094 kW)
War emergency ratings: 1,695 hp (1264 kW) at 10,300 ft (3139 m) and 1,390 hp (1036 kW) at 24,000 ft (7315 m)
Military ratings: 1,550 hp (1156 kW) at 13,000 ft (3962 m) and 1,260 hp (940 kW) at 26,500 ft (8077 m)
Continuous ratings: 1,150 hp (858 kW) at 17,000 ft (5182 m) and 980 hp (731 kW) at 29,000 ft (8839 m)
Propeller pitch range: 23° - 65°

Weights

Empty operating: 7,125 lb (3232 kg)
Basic: 7,635 lb (3466 kg)
Combat weight (clean): 10,100 lb (4582 kg)
Maximum take-off: 11,600 lb (5262 kg) with two 108-US gal (409-litre) drop tanks

Fuel and load

Total internal fuel: 269 US gal (1018 litres), comprising total of 184 US gal (696 litres) in self-sealing wing fuel cells, plus 85 US gal (322 litres) in self-sealing fuselage tank
External fuel: provision for two 75-US gal (284-litre), 110-US gal (416-litre) or 108-US gal (409-litre) fuel tanks
Maximum weapon load: 2,000 lb (908 kg)

Performance

Maximum level speeds: (clean, no racks fitted) 395 mph (636 km/h) at 5,000 ft (1524 m), 413 mph (665 km/h) at 15,000 ft (4572 m), 437 mph (703 km/h) at 25,000 ft (7620 m) and 433 mph (697 km/h) at 30,000 ft (9144 m)
Maximum cruising speed: 362 mph (583 km/h)
Rate of climb: 3,475 ft (1059 m) per minute at 5,000 ft (1524 m)
Service ceiling: 41,900 ft (12771 m)
Time to climb: 20,000 ft (6096 m) in 7 minutes 20 seconds, 30,000 ft (9144 m) in 12 minutes 36 seconds
Landing speed: 100 mph (161 km/h)

Range

Range (clean): 825 nm (950 miles; 1529 km) at 25,000 ft (7620 m) at maximum cruise power
Maximum range (absolute): 1998 nm (2,304 miles; 3703 km) with two 108-US gal (409-litre) drop tanks, 1433 nm (1650 miles; 2655 km) at 25,000 ft (7620 m) at maximum cruise power
Operational radius with maximum fuel: 1129 nm (1,300 miles; 2092 km)

Armament

Fixed: six wing-mounted 0.50-in (12.7-mm) Colt-Browning M2 machine guns, with total 1880 rounds, comprising 400 rounds for each inboard gun and 270 rounds each for outboard pairs
External stores: two underwing hardpoints for two 500-lb (227-kg) or 1,000-lb (454-kg) free-fall general-purpose bombs, plus bazooka-type tubes for six 4.5-in (114-mm) M-8 explosive rockets or (from P-51D-25-NA) six 5-in (127-mm) high-velocity aircraft rockets (HVARs) carried on zero-length launchers with tanks or bombs, or 10 HVARS without tanks or bombs

A pair of P-51Ds of the 458th Fighter Squadron, 506th Fighter Group, Seventh Air Force, are seen through the aft gun sight of a B-29 bomber while on an escort mission to Japan. This was only made possible by the taking of Iwo Jima in February 1945 which was achieved at bloody cost to the US Marines. No fewer than five airfields were built on the island which eventually became home for three Mustang fighter groups. These undertook escort missions as well as fighter-bomber sweeps over Japan.

P-51D Mustang

Li'l Butch **was an Inglewood-built P-51D-20-NA assigned to the 47th Fighter Squadron, 15th Fighter Group, of the USAAF's Seventh Air Force in the Pacific. It wears the 47th Fighter Squadron's distinctive yellow-edged black wing and fuselage bands and fin triangle, and a black-banded yellow spinner. Other squadrons in the 15th FG used diagonal green stripes edged in black on wings, tail and fuselage, along with black and yellow fin, wing and tailplane tips. These large and prominent markings allowed easy differentiation between P-51s and Japanese fighters. Although it had Mustangs for only a relatively brief time (its aircraft arrived at Iwo Jima from 6 March 1945), the 15th FG notched up an impressive record. The 47th FS alone scored 36 kills during its brief Mustang era.**

Engine supercharger and exhaust
The Merlin's two-speed, two-stage supercharger (reportedly designed by Wright) maintained pressures within the induction system equal to the pressures experienced at sea level. When any gas is compressed, its temperature rises. An intercooler is therefore used between the two supercharger compressors, giving a denser mixture of fuel and air in the cylinder. The supercharged Merlin produced more power at 26,000 ft (7925 m) than the Allison did on take-off. The Merlin's twelve cylinders exhausted through individual stub exhausts projecting from the sides of the cowling. These exhausts were often faired in, with a shroud, as seen here. They were usually of circular cross-section, unlike the flattened fish-tail exhausts fitted to many of the earlier Allison-engined aircraft. The later Twin Mustang (which was also Allison-powered) introduced flame-damping exhausts.

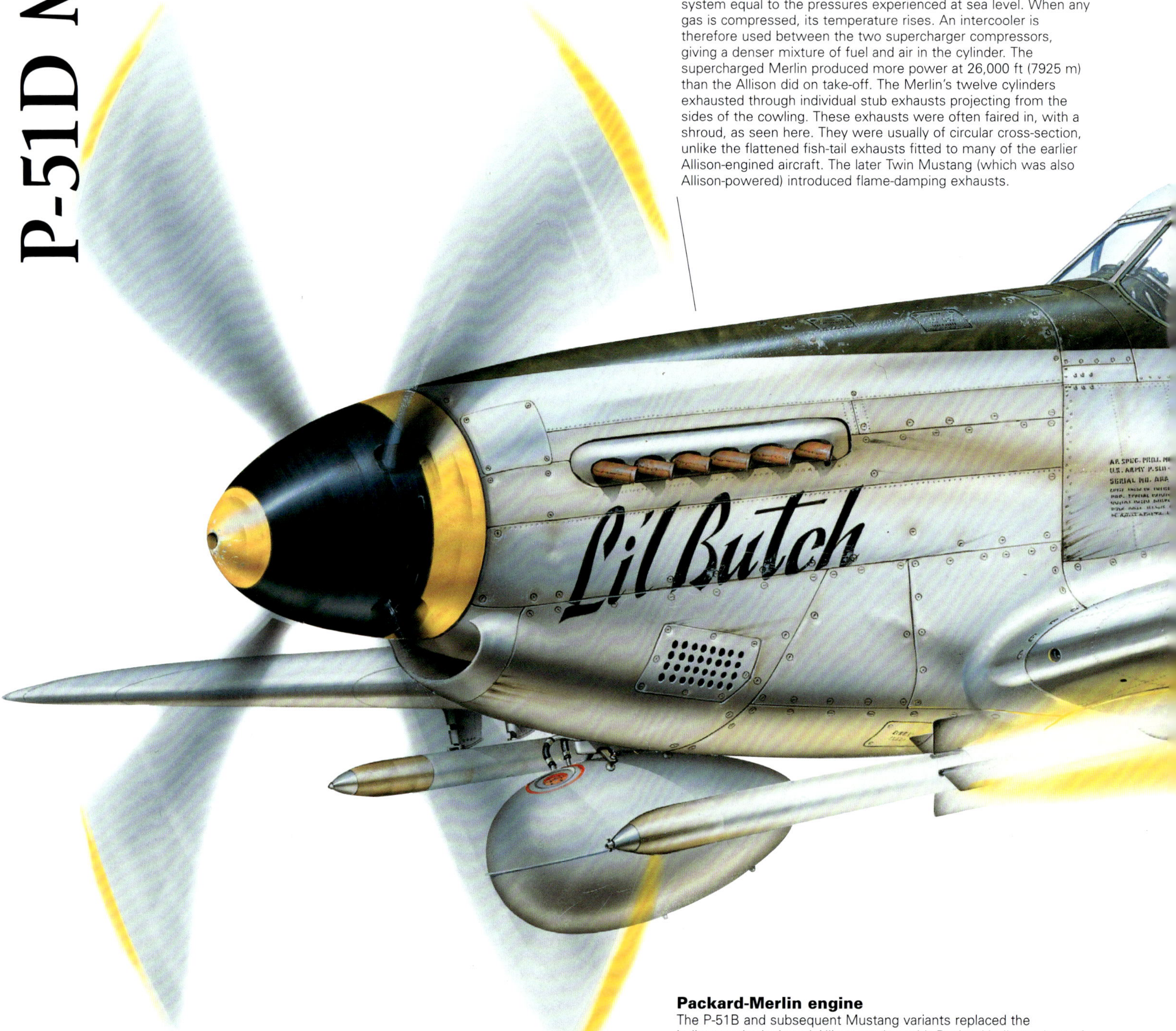

Propeller
Most examples of the P-51D were fitted with a J-6523A-24 or K-6523A-24 Hamilton Standard propeller, with drag reducing 'cuffs' at the roots. Such cuffs were often removed in service, especially after the war. The distinctive paddle-shaped blades were highly effective in the thin air found at high altitude. Some later aircraft used a square-tipped 'paddle-bladed' Model 6547A-6 Hamilton Standard prop, while P-51Ks used a four-bladed A20-156-24M Aeroproducts propeller.

Packard-Merlin engine
The P-51B and subsequent Mustang variants replaced the indigenously designed Allison engine with Packard-built versions of the Rolls-Royce Merlin. The Merlin (famously the powerplant for the Spitfire, Hurricane, Lancaster and Mosquito) was by 1942 a mature and reliable powerplant. A V-12 engine, the Merlin was, like the Allison, cooled by a mix of water and glycol, and was of similar overall dimensions. Compared to the American engine, it developed greater power at high altitudes, and promised superior range and endurance. It was a natural choice of engine for the Mustang. Rolls-Royce and North American started examining Merlin-engined Mustang versions virtually simultaneously, both opting to use the Packard-built V-1650-3 (equivalent to the Rolls-Royce Merlin 68) with two-stage supercharger and intercooler. The two programmes were soon merged, although two NAA and five Rolls-Royce prototypes were flown.

Colour scheme
Most USAAF P-51Ds wore an overall natural metal finish. European-based aircraft had camouflaged upper surfaces during 1944, but reverted to overall natural metal. Many P-51s had their wings painted silver, while the fuselages were left in bare metal finish. The top decking forward of the cockpit was painted matt olive drab to serve as an anti-dazzle panel. National markings were without red centre discs or stripes, in order to avoid any confusion with the Japanese 'meatball'. Various schemes were used to avoid misidentification – P-51s in Europe wore white (or black on silver aircraft) noses and stripes on wings and tails, while some aircraft in the Far East had equally prominent black bands.

Mustang canopies
The spacious cockpit of the P-51D was covered by an aft-sliding, blown, bubble canopy, which offered some improvement in all round visibility, by comparison with the standard canopy on the high-backed P-51B/C and earlier versions. The bubble canopy was produced in several versions, with a lower, reduced-drag canopy introduced from the middle of the P-51D-5-NA production run. The Dallas-built P-51Ks had a larger canopy, with a kinked 'trailing edge' just above the back of the canopy rail. The TF-51D conversions and Cavalier's post-war Mustangs each used yet another canopy shape. Various racing Mustangs use tiny 'pimple' canopies for the ultimate in drag reduction.

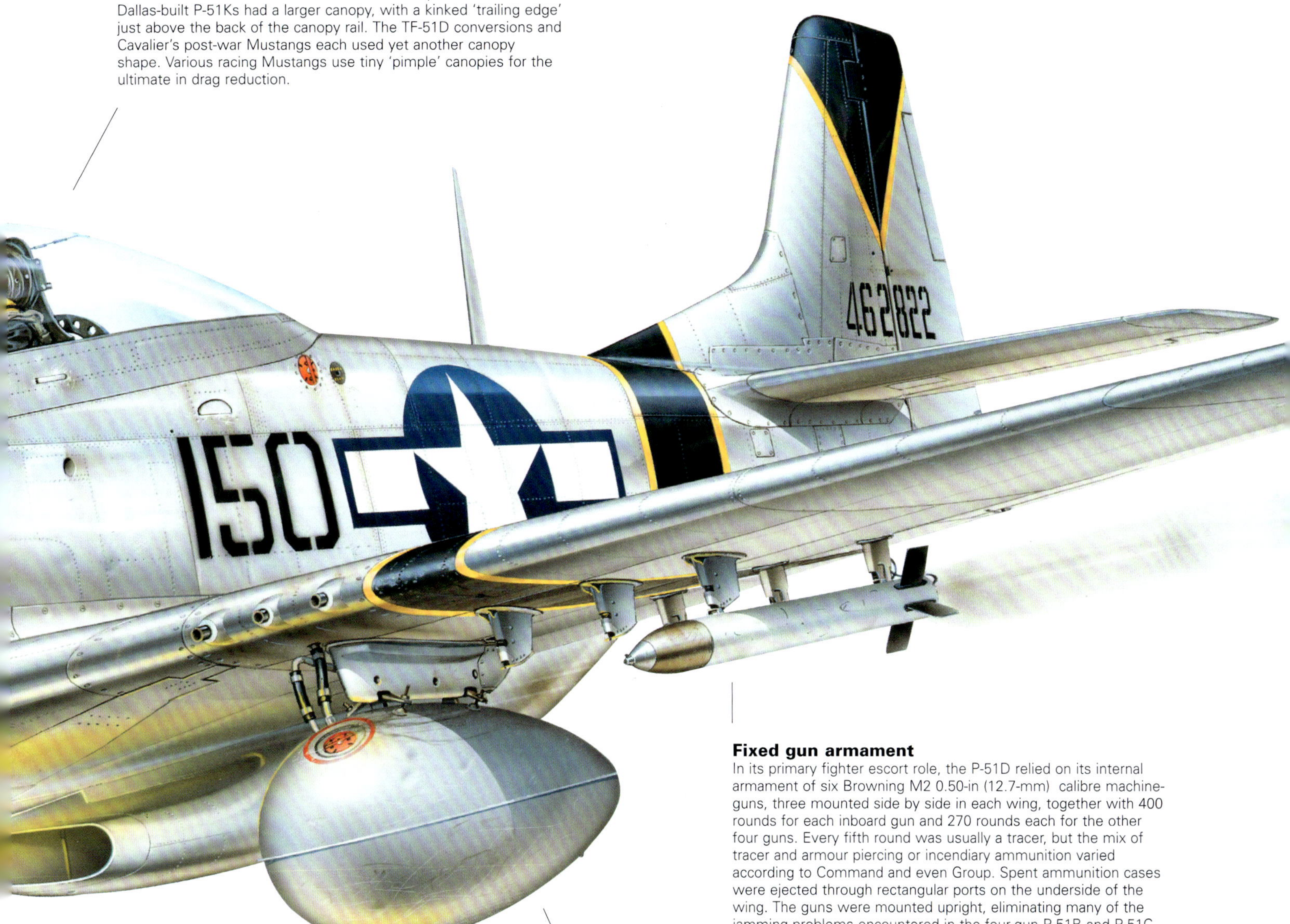

Fixed gun armament
In its primary fighter escort role, the P-51D relied on its internal armament of six Browning M2 0.50-in (12.7-mm) calibre machine-guns, three mounted side by side in each wing, together with 400 rounds for each inboard gun and 270 rounds each for the other four guns. Every fifth round was usually a tracer, but the mix of tracer and armour piercing or incendiary ammunition varied according to Command and even Group. Spent ammunition cases were ejected through rectangular ports on the underside of the wing. The guns were mounted upright, eliminating many of the jamming problems encountered in the four-gun P-51B and P-51C.

Intake
The liquid-cooled Merlin engine relied on a complex belly-mounted air-cooled radiator. Aspirated by a massive ventral airscoop, the radiator's location made it vulnerable to ground fire. The intake of the Merlin Mustang, like that of the A-36, was fixed, while the P-51 and P-51A had articulated intakes, with opening lips to admit more airflow under certain conditions, especially on the ground. All Mustangs had an opening rear vent, downstream of the radiator itself. In certain circumstances, the expulsion of hot air from this duct actually generated a small amount of thrust and thus gave a small improvement in performance.

Underwing stores
The Mustang also had underwing hardpoints for an external fuel tank or bomb of 500 lb (227-kg), or from the P-51D, 1,000 lb (454-kg) weight. This pylon could even mount a 'bazooka' type three-tube M10 launcher for 4.5-in (114-mm) rockets. The standard P-51D tank (seen here) was made of metal, and contained 75 US gal (284 litres), a larger tank of similar shape contained 110 US gal (416 litres). In an effort to save metal, and to avoid littering Germany with useful aluminium which could be recycled for its own war machine, a 108-US gal (409-litre) tank of impregnated paper construction was used by Eighth Air Force Mustang units. Redundant fuel tanks were often converted to serve as napalm containers. The P-51 could also be fitted with zero-length launchers for unguided 5-in (127-mm) rocket projectiles (seen here). These launchers were simple streamlined mini-pylons, to which rockets were clipped directly.

Northrop P-61 Black Widow

Above: This Saipan-based P-61A is seen cruising over the Pacific, some time during January 1945. The gun turrets' straight trailing edge indicates a fixed installation, although other P-61s, without the troublesome rotating unit, are known to have used the factory-produced 'teardrop'-shaped housing.

P-61B Black

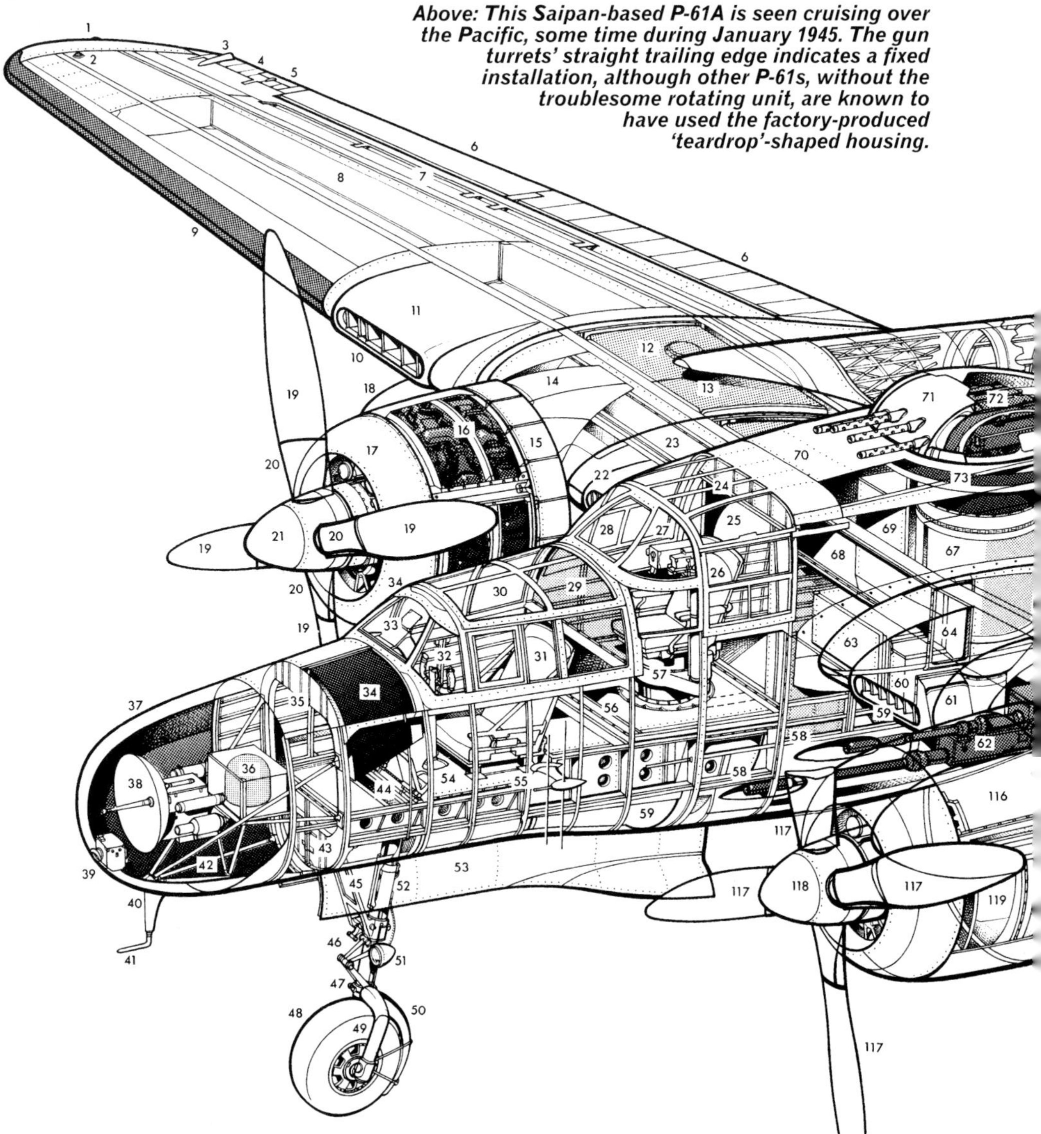

Cutaway key
1 Starboard navigation light
2 Starboard formation light
3 Aileron hinge fairing
4 Conventional aileron
5 Aileron tab
6 Full-span flaps (Zapp type)
7 Retractable aileron (operable as spoiler)
8 Wing skinning
9 De-icer boot
10 Intercooler controllable shutters
11 Intercooler and supercharger induction
12 Fuel filler cap
13 Starboard outer wing fuel tank
14 Nacelle fairing
15 Cooling gills
16 Pratt & Whitney R-2800-65 engine
17 Nacelle ring
18 Starboard outer auxiliary tank
19 Four-bladed Curtiss Electric propeller
20 Propeller cuffs
21 Propeller boss
22 Heater air induction
23 Front spar
24 Plexiglas canopy
25 Cannon access bulkhead cut-out
26 Front gunner's compartment
27 Sighting station
28 Bullet-resistant windshield
29 Inter-cockpit/ compartment armour (shaded)
30 Pilot's canopy
31 Pilot's seat
32 Control column
33 Gunsight (fixed cannon)
34 Bullet-resistant windshield
35 Fuselage structural joint (armour plate deleted for clarity)
36 Radar modulator
37 Dielectric nosecone
38 SCR-720 radar scanner
39 Gun camera (gunsight aiming point)
40 Mast
41 Pitot head
42 Radar equipment steel support tube
43 Bulkhead (centre joint)
44 Rudder pedals
45 Drag strut
46 Torque link
47 Towing eye
48 Nosewheel
49 Cantilever steel strut
50 Mud guard (often deleted)
51 Taxi lamp
52 Air-oil shock strut (shimmy damper on forward face)
53 Nosewheel door
54 Cockpit floor
55 Radar aerials
56 Gunner's compartment floor (stepped)
57 Gunner's seat-swivel mechanism
58 Cannon ports
59 Heater air induction
60 Cannon ammunition magazines
61 Ammunition feed chute
62 Four 20-mm cannon in ventral compartment
63 Magazine forward armour plate
64 Front spar fuselage cut-out
65 Magazine rear armour plate
66 Rear spar fuselage
67 Dorsal turret support/drive motor
68 Front spar carry-through
69 Turret support forward armour plate
70 Flush-riveted aluminium alloy skin

Four P-61Cs and an F-15 move out for a thunderstorm penetration mission in 1947 from the All-Weather Flying Center in Ohio. The aircraft would penetrate the storm at different intervals, recording temperatures and pressures and photographing cloud patterns. The nosecone of the lead Black Widow shows the battering caused by rain, hail and lightning.

SPECIFICATION

P-61B Black Widow

Dimensions

Length: 49 ft 7 in (15.11 m)
Wingspan: 66 ft $^3/_4$ in (20.11 m)
Height: 14 ft 8 in (4.47 m)
Wing area: 662.36 sq ft (61.53 m^2)

Powerplant

Two Pratt & Whitney R-2800-65 Double Wasp 18-cylinder radial engines with a 2,000-hp (1491-kW) military rating and a 2,250-hp (1678-kW) war emergency rating

Weights

Empty: 23,450 lb (10637 kg)
Maximum overload: 36,200 lb (16420 kg)

Performance

Maximum speed at war emergency engine rating: 366 mph (589 km/h)
Initial climb rate at military power: 2,090 ft (637 m) per minute
Range at long-range cruise power: 1,350 miles (2172 km)

Armament

Four 20-mm M2 cannon each with 200 rounds; (on some aircraft) dorsal turret with four 0.5-in (12.7-mm) Colt-Browning machine-guns each with 560 rounds; four external pylons each rated at up to 1,600 lb (726 kg) and able to carry bombs or other stores

F-15A Reporter

Dimensions

Same as P-61B except for
Length: 50 ft 4 in (15.34 m)

Powerplant

Two Pratt & Whitney R-2800-73 Double Wasp supercharged 18-cylinder radial engines with a 2,100-hp (1566-kW) military rating and 2,800-hp (2088-kW) maximum rating with water injection

Weights

Maximum take-off: 32,190 lb (16420 kg)

Performance

Maximum speed: 400 mph (708 km/h)
Service ceiling: 41,000 ft (12497 m)
Range: 4,000 miles (6437 km)

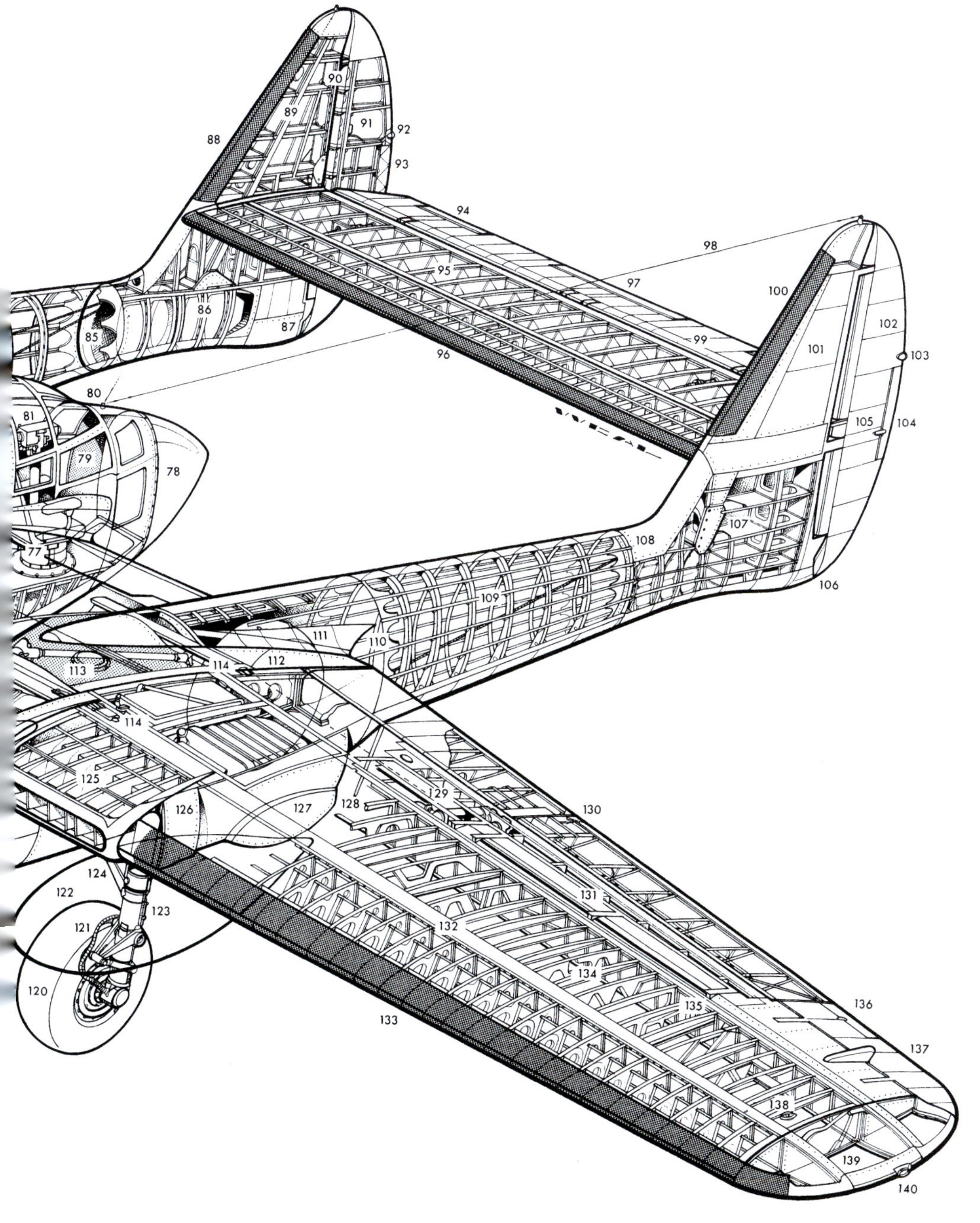

71 Gun mantle (four 0.50-in/12.7-mm machine- guns)
72 General Electric remote control power turret
73 Turret drive ring
74 Rear spar carry-through
75 Turret support rear armour plate
76 Radio operator/rear gunner's compartment
77 Gunner's seat-swivel mechanism
78 Plexiglas tailcone
79 Rear compartment glazing
80 Aerial attachment
81 Sighting station
82 Anti-collision beacon
83 Tailboom structure (inner stringers deleted for clarity)
84 Control runs
85 Tailboom/fin attachment
86 Fin spar attachment (inner face)
87 Rudder lower hinge
88 De-icer boot
89 Fin structure
90 Rudder upper hinge
91 Rudder
92 Rear navigation light
93 Rudder tab
94 Balance tab
95 Horizontal stabiliser structure
96 De-icer boot
97 Trim tab
98 Aerials
99 Elevator
100 De-icer boot
101 Port fin
102 Rudder
103 Rear navigation light
104 Rudder tab
105 Tab hinge fairing
106 Rudder lower hinge
107 Fin spar attachment (outer face)
108 Tailboom/fin attachment butt
109 Tailboom structure
110 Tailboom joint
111 Wing/boom fairing fillet
112 Mainwheel well
113 Port outer wing fuel tank
114 Spar dihedral-break attachment bolts
115 Cooling gills
116 Port inner auxiliary tank
117 Four-bladed Curtiss Electric propeller
118 Propeller boss
119 Nacelle construction
120 Port mainwheel
121 Hydraulic and airbrake pressure lines
122 Port outer auxiliary tank
123 Mainwheel leg (hydraulic shock strut)
124 Drag strut
125 Intercooler and supercharger induction trunking
126 Mainwheel flap
127 Mainwheel door
128 Radio antenna (port and starboard booms)
129 Wing flap lock
130 Full-span flaps (Zapp type)
131 Retractable aileron (operable as spoiler)
132 Front spar
133 De-icer boot
134 Wing structure
135 Rear spar
136 Aileron tab
137 Port aileron
138 Port formation light
139 Port wingtip
140 Port navigation light

P-61B Black Widow

The aircraft depicted here, 42-39404 *Midnite Madness II*, was the second of two P-61s with this name. Referred to as *Midnite Madness I* and *II*, their nose-art appears to have been identical, although the 'II' was added under the name on 239404 at some point. This aircraft was assigned to Captain James W. Bradford of the 548th NFS. Flying from le Shima on the night of 24 June 1945, Bradford, together with radar operator 1st Lieutenant Lawrence K. Lunt and gunner Master Sergeant Reno H. Sukow, destroyed a Mitsubishi G4M 'Betty' – one of only five victories credited to the squadron. That morning, a P-61 of the 548th's sister squadron, the 549th NFS, had destroyed another 'Betty' off Iwo Jima, for its only kill of the war. The fate of *Midnite Madness II* is unclear: it eventually either ended up in the 'boneyard' at Clark AB or was lost in an accident immediately after the war. James Bradford went on to become a group commander of the Instrument School (All Weather) at Moody AFB, did a tour with the Republic of Korea AF in F-86s as a senior advisor and was involved in an exchange tour with the RAF, flying Meteor T.Mk 7s. He became wing commander at the Training Base at Laredo, Texas and finished his career with a tour at the Pentagon, retiring as a full colonel.

Powerplant
The engines used on the P-61B were Pratt & Whitney R-2800-65 Double Wasp 18-cylinder radials rated at 2,000 hp (1491 kW). Late P-61s and the P-61B were fitted with water injection which provided war emergency power (also called combat emergency power). The application of water injection increased manifold pressure and therefore power. Two water tanks, each with a capacity of 26 US gal (98 litres), allowed war emergency power to be applied for up to 15 minutes.

Radio equipment
In addition to the SCR-720 radar set, nearly 30 other 'black boxes' were installed in the Black Widow. In reality, this number was mainly made up of control units and junction boxes for the main radar and communication sets. The principal items found in an early P-61B included SCR-522 VHF command radio (two sets), SCR-695 IFF, SCR-718 radar altimeter, AN/APS-13 tail-warning radar and SCR-729 IFF/beacon locater. By sending out pulses and 'interrogating' ground beacons, the SCR-729 could be used as a navigation aid.

Cannon armament
The four 20-mm Hispano M2 cannon were mounted in the P-61's belly in a staggered installation, with the two outboard guns located much further back, the breeches being mounted just forward of the centre-section trailing-edge flaps. Being located beneath the aircraft, the muzzle flash was not a threat to the crew's night vision. Containers for 200 rounds per gun were mounted fore and aft of the turret installation. The forward pair was accessible to the gunner in flight. To save on strategic materials, these boxes were initially made from a pressed fibre material that unfortunately broke down in humid conditions. Locally-made aluminium replacements had to be devised, and this material was later adopted for production. The pilot fired the cannon with a button on the right side of the control wheel and aimed with an LY-3N gunsight (L-1 on the P-61A). Microswitches prevented the activation of the guns with the nose gear down or the ladder in place. The nosewheel had to be raised and the fuselage supported to allow boresighting of the guns.

548th Night Fighter Squadron

The 548th NFS was assigned to the 301st Fighter Wing at this time. Unlike day-fighter units, the NFS squadrons did not belong to groups. Activated in Hawaii in September 1944, they were assigned to the 7th Fighter Wing of the 7th Air Force from October. The 548th was attached to the 301st Wing during May and June 1945 and was back under the 7th AF from July until December. Its sole commanding officer was Major Robert D. Curtis. As well as night-fighter patrols and the escort of night-bombing B-29s, the 548th engaged in a number of long-range interdiction missions, particularly against the Bonin Islands, for which the P-61s were fitted with 5-in (127-mm) HVAR rockets.

Dorsal turret

Due to the buffeting experienced when the General Electric A-4 turret was rotated or elevated, the bulk of P-61As and many P-61Bs, including the P-61B-1, were delivered from the factory without the unit. The need for additional firepower, and the crew composition of units already in the field, led some Pacific squadrons to accept aircraft with fixed installations, allowing forward-firing only. *Midnite Madness II* had such a fixed turret, allowing for the retention of a three-man crew who had been together since training. The original *Midnite Madness*, a P-61A-11, had been equipped with the fully-rotating turret, which weighed 1,600 lb (726 kg) and was the same model as that fitted to the B-29, which had a higher production priority – another reason why turrets were dropped from the P-61 line. The turret was fitted with four 0.50-in (12.7-mm) machine-guns, firing 800 rounds per minute. Each gun had 560 rounds of ammunition.

Plastics

The P-61 made extensive use of non-metallic components. The nosecone was originally frosted Plexiglas, later painted to reduce its visibility; the black coat caused the Plexiglas to sag in the heat. Eventually, a new nosecone made of resin-impregnated fibreglass was introduced on the production line; this did not suffer as much from the heat and could be left uncovered on the ground. Canopy sections were made of Lucite, manufactured by Dupont. The tailcone was made of two hemispheres of Lucite, cemented together along the vertical axis. After a number of these cones imploded at high speed, an aluminium reinforcing bar was added.

Polikarpov I-16

Created at a time when other manufacturers were still producing biplane fighters with fixed landing gear, the Polikarpov I-16 was a landmark in fighter design.

I-16 Type 10

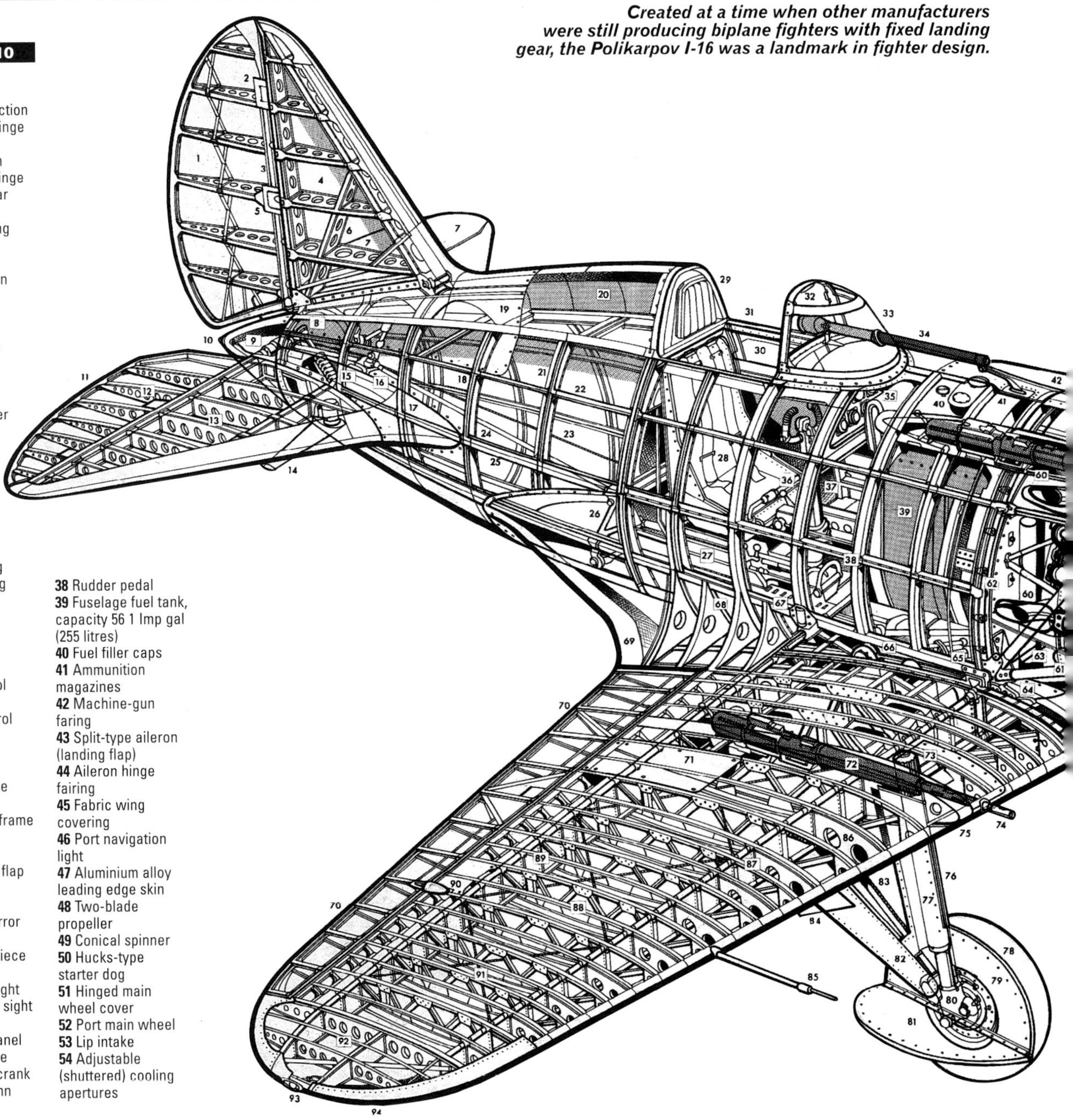

Cutaway key
1 Rudder construction
2 Rudder upper hinge
3 Rudderpost
4 Fin construction
5 Rudder lower hinge
6 Fin auxiliary spar
7 Port tailplane
8 Rudder actuating mechanism
9 Tail cone
10 Rear navigation light
11 Elevator construction
12 Elevator hinge
13 Tailplane construction
14 Tailskid
15 Tailskid damper
16 Control linkage (elevator and rudder)
17 Tailplane filet
18 Fuselage half frames
19 Fin root fairing
20 Dorsal decking
21 Fuselage monocoque construction
22 Main upper longeron
23 Rudder control cable
24 Elevator control rigid rod
25 Main lower longeron
26 Control linkage crank
27 Seat support frame
28 Pilot's seat
29 Headrest
30 Cockpit entry flap (port)
31 Open cockpit
32 Rear-view mirror (optional)
33 Curved one-piece windshield
34 Tubular gunsight (PBP-1 reflector sight optional)
35 Instrument panel
36 Undercarriage retraction handcrank
37 Control column
38 Rudder pedal
39 Fuselage fuel tank, capacity 56 1 Imp gal (255 litres)
40 Fuel filler caps
41 Ammunition magazines
42 Machine-gun faring
43 Split-type aileron (landing flap)
44 Aileron hinge fairing
45 Fabric wing covering
46 Port navigation light
47 Aluminium alloy leading edge skin
48 Two-blade propeller
49 Conical spinner
50 Hucks-type starter dog
51 Hinged main wheel cover
52 Port main wheel
53 Lip intake
54 Adjustable (shuttered) cooling apertures

SPECIFICATION

Polikarpov I-16 Type 24

Dimensions
Span: 29 ft 1.6 in (8.88 m)
Length: 19 ft 8.125 in (6.04 m)
Height: 7 ft 9.75 in (2.41 m)
Wing area: 160.06 sq ft (14.87 m²)

Powerplant
One Shvetsov M-62 nine-cylinder radial air-cooled engine rated at 1,000 hp (746 kW) at 2,000 rpm for take-off and 800 hp (597 kW) at 2,100 rpm at 13,780 ft (4200 m), driving an AV-2 two-bladed two-pitch propeller

Weights
Empty equipped: 3,252 lb (1475 kg)
Normal loaded: 4,215 lb (1912 kg)
Maximum take-off: 4,541 lb (2060 kg)

Performance
Maximum speed at sea level: 273 mph (440 km/h)
Maximum speed at 9,840 ft (3000 m): 304 mph (489 km/h)
Range (clean): 373 miles (600 km)
Range with external tanks: (600 miles (1100 km)
Rate of climb: 5.8 minutes to 16,405 ft (5000 m)
Service ceiling: 31,070 ft (9470 m)

Armament
Four 0.3in (7.62-mm) Shpital'ny-komaritsky ShKAS machine-guns (two synchronised in fuselage and two unsynchronised in wings) with 650 rounds per gun;
Alternatively, two fuselage-mounted ShKAS with 650 rpg and two wing-mounted 20-mm Shpital'ny-Viadimirov ShVAK cannon with 180 rpg
A Single 0.47-in (12.7-mm) Berezin UBK heavy machine-gun with 300 rounds, firing through the lower cowling, could be added, usually in place of the wing-mounted armament.
Later model I-16s could carry two-220-lb (100-kg) bombs or six RS-82 rockets.

Above: The Soviets failed to capitalise on the lead they gained in fighter design with the I-16, a fact which saw the type committed to actions in 1941 in which it was at a considerable disadvantage. Even so, the I-16's contribution to Russia's defence cannot be underestimated.

Right: The first UTI tandem trainers were produced in 1935. Later versions were fitted with a simple fixed undercarriage. The motto of the Soviet Air Force training schools was "If you can fly the I-16, you can fly anything."

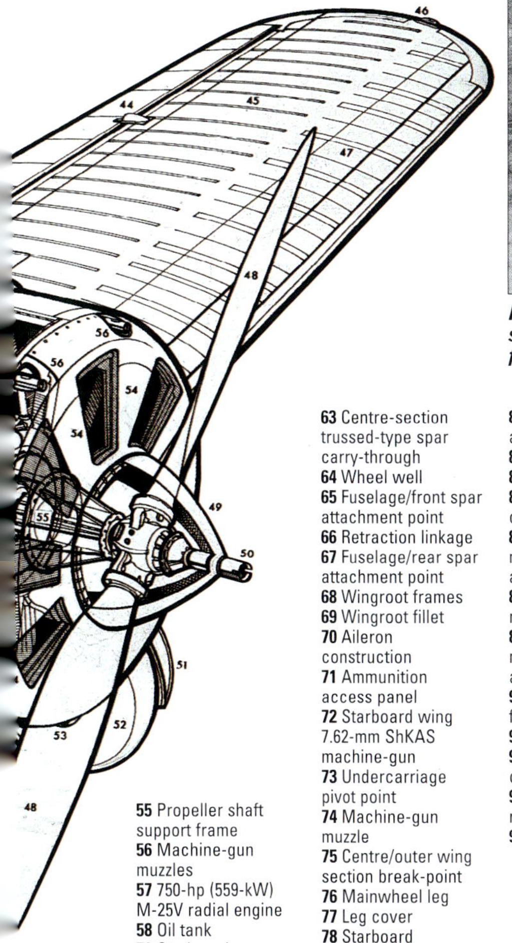

55 Propeller shaft support frame
56 Machine-gun muzzles
57 750-hp (559-kW) M-25V radial engine
58 Oil tank
59 Starboard synchronized 0.3-in (7.62-mm) ShKAS machine gun
60 Exhaust exit ports
61 Engine bearers
62 Firewall/bulkhead
63 Centre-section trussed-type spar carry-through
64 Wheel well
65 Fuselage/front spar attachment point
66 Retraction linkage
67 Fuselage/rear spar attachment point
68 Wingroot frames
69 Wingroot fillet
70 Aileron construction
71 Ammunition access panel
72 Starboard wing 7.62-mm ShKAS machine-gun
73 Undercarriage pivot point
74 Machine-gun muzzle
75 Centre/outer wing section break-point
76 Mainwheel leg
77 Leg cover
78 Starboard mainwheel
79 Mainwheel cover
80 Axle
81 Hinged cover flap
82 Actuating rod cover
83 Retraction actuating rod
84 Cover flap
85 Pitot head
86 Leading-edge construction
87 KhMA chrome-molybdenum steel alloy front spar
88 Alternate dural ribs/frames
89 KhMA chrome-molybdenum steel alloy rear spar
90 Aileron hinge fairing
91 Wire cross-bracing
92 Wingtip construction
93 Starboard navigation light
94 Wingtip edging

Above: One of the most unusual variants of the I-16 was the Sostavnoi pikiruyushchy bombardirovshchik – SPB or composite dive bomber – in which two bomb-armed I-16 Type 5s were hung beneath the wings of an ANT-6 carrier aircraft. In August 1941 the combination was tested in action against oil and communications targets in Romania

I-16 Type 5

One of the great warplanes in history, the Polikarpov I-16 was the first low-winged monoplane with retractable landing gear to enter service with any air force. It saw its combat debut with republican forces in Spain, who knew the diminutive aircraft as the *Mosca*, or 'Fly'. Its nationalist opponents had a different name for the stubby fighter – they called it the *Rata* or 'Rat'. But to the majority of pilots who flew the I-16, the men of the Soviet air force who had to wrestle with its unruly handling, it was known as *Ishak* or 'Mule'. However, as they got to know its quirks, that was often contracted with back-handed affection as *Ishachka*, or 'Little Mule'. The example depicted here is a Polikarpov I-16 Type 5 Mosca, serving with the Aviación Militar Republicana during the Spanish Civil War.

Tail surfaces
The tail surfaces were necessarily large to counter the lack of stability caused by the short rear fuselage. In spite of the designers' best efforts, the I-16 had only limited stability longitudinally, and needed concentration from the pilot at all times. However, this instability brought great dividends in manoeuvrability at high speeds, where the rod-actuated elevators were noticeably effective. The rudder was actuated by pulley and cable.

Canopy
The canopy was another innovation, the whole unit, including windscreen, sliding forward on rails for ingress/egress. In practice the pilots were fearful of not being able to get out quickly in an emergency and the canopy was often locked in the open position. This led to later I-16s featuring a truncated canopy and partially open cockpit.

Markings
The entire rudder was taken up with the red, yellow and purple markings of the Republican air force. The 'Popeye' emblem was that of the 4a Escuadrilla de Moscas. Other squadron markings were the cartoon character Betty Boop (1a Escuadrilla de Moscas), a pelican on a crutch (2a Escuadrilla) and a double-six domino (3a Escuadrilla).

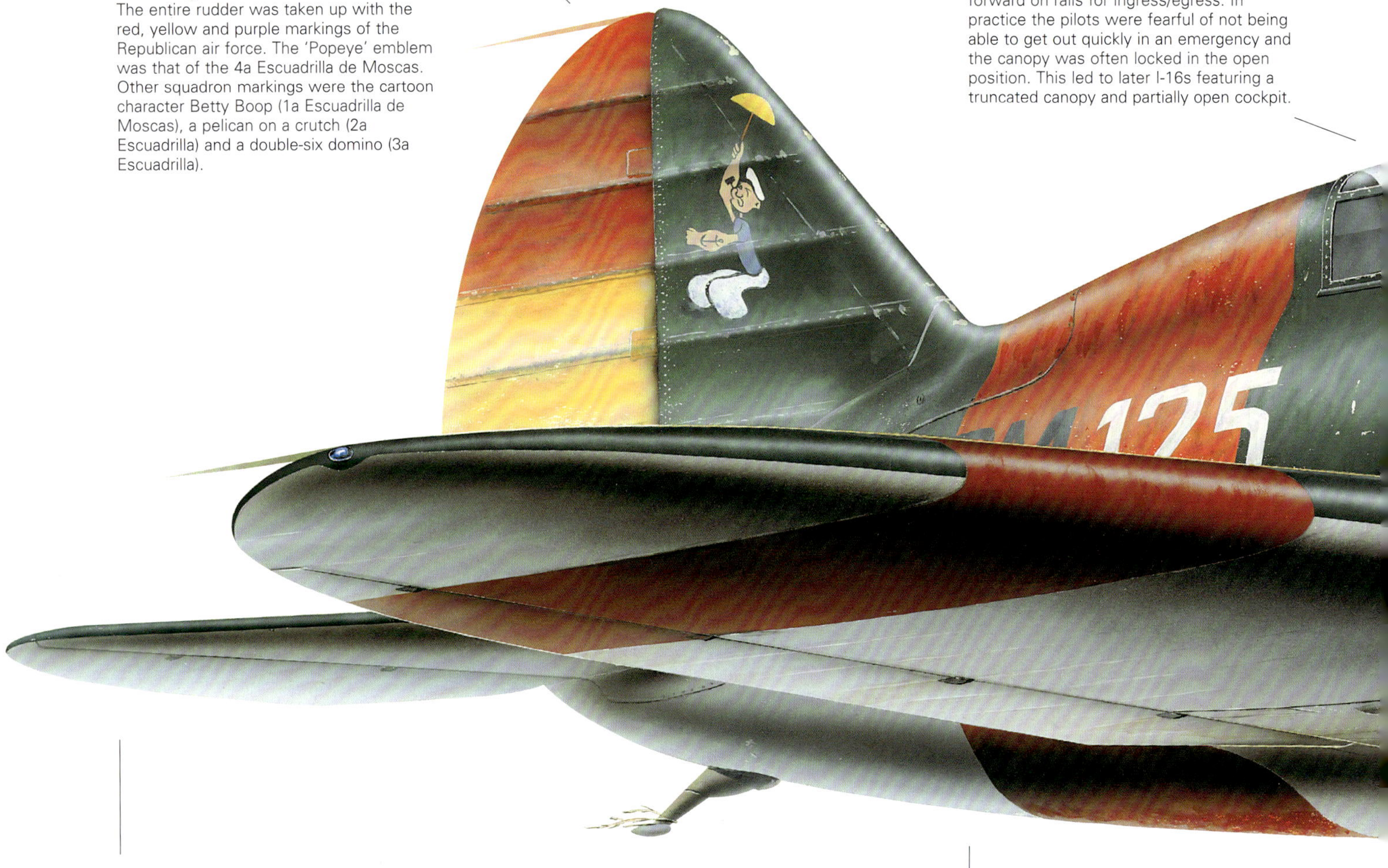

Wing structure
The I-16 had a metal two-spar wing structure, with trussed KhMA chrome-molybdenum steel alloy centre-section spars and tubular outer spars. Wing ribs were made of dural and skinning was aluminium inboard and fabric outboard The long ailerons were operated by rods and bell cranks. They could be drooped by 15° to act as flaps on landing, although this did little to reduce the high stalling speed of the I-16.

Fuel
The Mosca's fuel was housed in a single tank located in the central fuselage between the cockpit and engine installation. Total capacity was 56 Imp gal (255 litres). No fuel gauge was fitted in the cockpit, the pilot having to listen to the engine note to determine when fuel was low while keeping a close eye on his watch!

Fuselage structure
The fuselage structure was built in two halves, divided vertically. The main structure consisted of four longerons and eleven half-frames built of pine, to which was bonded a *shpon* skin – a kind of plywood made from long strips of birch glued together.

Cockpit
The cramped cockpit was equipped with only rudimentary instruments. No radio or oxygen equipment was fitted, and there was no indicator for the undercarriage. The pilot was provided with a control column with a yoke-type grip, and a cable-cutter for severing the undercarriage retraction cables if they became stuck partially open.

Gunsight
The Mosca was provided with a simple Aldis telescope sight which projected into the cockpit through the windscreen. Four magnification powers were provided, and the sight featured simple cross-hairs.

Powerplant
Power for the I-16 Type 5 was provided by a Shvetsov M-25 nine-cylinder radial, which was a 'Sovietised' version of the Wright Cyclone SR-1820-F-3. The heavy cowling completely enclosed the engine, with large apertures in the front to admit cooling air. These were shuttered to control the amount of air entering the engine compartments. The cylinders exhausted through individual ports arranged around the rear of the cowling.

Armament
The I-16Type 5 featured a ShKAS 0.3-in (7.62-mm) machine-gun in each wing, known to the Spanish as the Rusa E rápida. The weapon weighed 22 lb (10 kg) and fired at a rate of 1,800 rpm. Each gun had 900 rounds, the 0.34-oz (9.6g) bullets being fired at a muzzle velocity of 2,706 ft/sec (825 m/sec). In spite of a tendency to jam, the highly accurate ShKAS was regarded in the early 1930s as the best aircraft gun in the world.

Undercarriage
The I-16 was innovative in the introduction of retractable main wheels. The pilot operated a crank in the cockpit which raised the undercarriage by cables, the wheels retracting inwards to lie in wells between the centre-section spars, covered by a hinged flap. The effort of retracting the undercarriage inevitably resulted in an erratic flight path after take-off as the crank was operated.

Propeller
The two-bladed AV-1 propeller was of metal construction, and had a diameter of 9 ft 2 in (2.80 m). It had two pitch options, and the large spinner was fitted with a Hucks starter dog.

Republic P-47 Thunderbolt

Ironically, the P-47, which developed as a light fighter for bomber interception, was first posted to Europe in the escort role supporting the USAAF's heavy bomber fleet, like this B-17.

P-47D

Cutaway key

1 Rudder upper hinge
2 Aerial attachment
3 Fin flanged ribs
4 Rudder post/fin aft spar
5 Fin front spar
6 Rudder trim tab worm and screw actuating mechanism (chain-driven)
7 Rudder centre hinge
8 Rudder trim tab
9 Rudder structure
10 Tall navigation light
11 Elevator fixed tab
12 Elevator trim tab
13 Starboard elevator structure
14 Elevator outboard hinge
15 Elevator torque tube
16 Elevator trim tab worm and screw actuating mechanism
17 Chain drive
18 Starboard tailplane
19 Tail jacking point
20 Rudder control cables
21 Elevator control rod and linkage
22 Fin spar/fuselage attachment points
23 Elevator
24 Aerial
25 Port tailplane structure (two spars and flanged ribs)
26 Tailwheel retraction worm gear
27 Tailwheel anti-shimmy damper
28 Tailwheel oleo
29 Tailwheel doors
30 Retractable and steerable tailwheel
31 Tailwheel fork
32 Tailwheel mount and pivot
33 Rudder cables
34 Rudder and elevator trim control cables
35 Lifting tube
36 Elevator rod linkage
37 Semi-monocoque all-metal fuselage construction
38 Fuselage dorsal 'razorback' profile
39 Aerial lead-in
40 Fuselage stringers
41 Supercharger air filter
42 Supercharger
43 Turbine casing
44 Turbosupercharger compartment air vent
45 Turbo-supercharger exhaust hood fairing (stainless steel)
46 Outlet louvres
47 Intercooler exhaust doors (port and starboard)
48 Exhaust pipes
49 Cooling air ducts
50 Intercooler unit (cooling and supercharged air)
51 Radio transmitter and receiver packs (Detrola)
52 Canopy track
53 Elevator rod linkage
54 Aerial mast
55 Formation light
56 Rearward-vision frame cut-out and glazing
57 Oxygen bottles
58 Supercharger and cooling air pipe (supercharger to carburettor) port
59 Elevator linkage
60 Supercharger and cooling air pipe (supercharger to carburettor) starboard
61 Central duct (to intercooler unit)
62 Wingroot air louvres
63 Wingroot fillet
64 Auxiliary fuel tank (100 US gal/379 litres)
65 Auxiliary fuel filler point
66 Rudder cable turnbuckle
67 Cockpit floor support
68 Seat adjustment lever
69 Pilot's seat
70 Canopy emergency release (port and starboard)
71 Trim tab controls
72 Back and head armour
73 Headrest
74 Rearward-sliding canopy
75 Rear view mirror fairing
76 'Vee' windshields with central pillar
77 Internal bulletproof glass screen
78 Gunsight
79 Engine control quadrant (cockpit port wall)
80 Control column
81 Rudder pedals
82 Oxygen regulator
83 Underfloor elevator control quadrant
84 Rudder cable linkage
85 Wing rear spar/fuselage attachment (tapered bolts/bushings)
86 Wing-supporting lower bulkhead section
87 Main fuel tank (205 US gal/776 litres)
88 Fuselage forward structure
89 Stainless steel/Alclad firewall bulkhead
90 Cowl flap valve
91 Main fuel filler point
92 Anti-freeze fluid tank
93 Hydraulic reservoir
94 Aileron control rod
95 Aileron trim tab control cables
96 Aileron hinge access panels
97 Aileron and tab control linkage
98 Aileron trim tab (port wing only)
99 Frise type aileron
100 Wing rear (No. 2) spar
101 Port navigation light
102 Pitot head
103 Wing front (No. 1) spar
104 Wing stressed skin
105 Four gun ammunition troughs (individual bays)
106 Staggered gun barrels
107 Removable panel
108 Inter spar gun bay access panel
109 Forward gunsight bead
110 Oil feed pipes
111 Oil tank (28.6 US gal/108 litres)
112 Hydraulic pressure line
113 Engine upper bearers
114 Engine control correlating cam
115 Eclipse pump (anti-icing)
116 Fuel level transmitter
117 Generator
118 Battery junction box
119 Storage battery
120 Exhaust collector ring
121 Cowl flap actuating
122 Exhaust outlets to collector ring
123 Cowl flaps
124 Supercharged and cooling air ducts to carburettor (port and starboard)
125 Exhaust upper outlets
126 Cowling frame
127 Pratt & Whitney R-2800-59 18-cylinder twin-row engine
128 Cowling nose panel
129 Magnetos
130 Propeller governor
131 Propeller hub

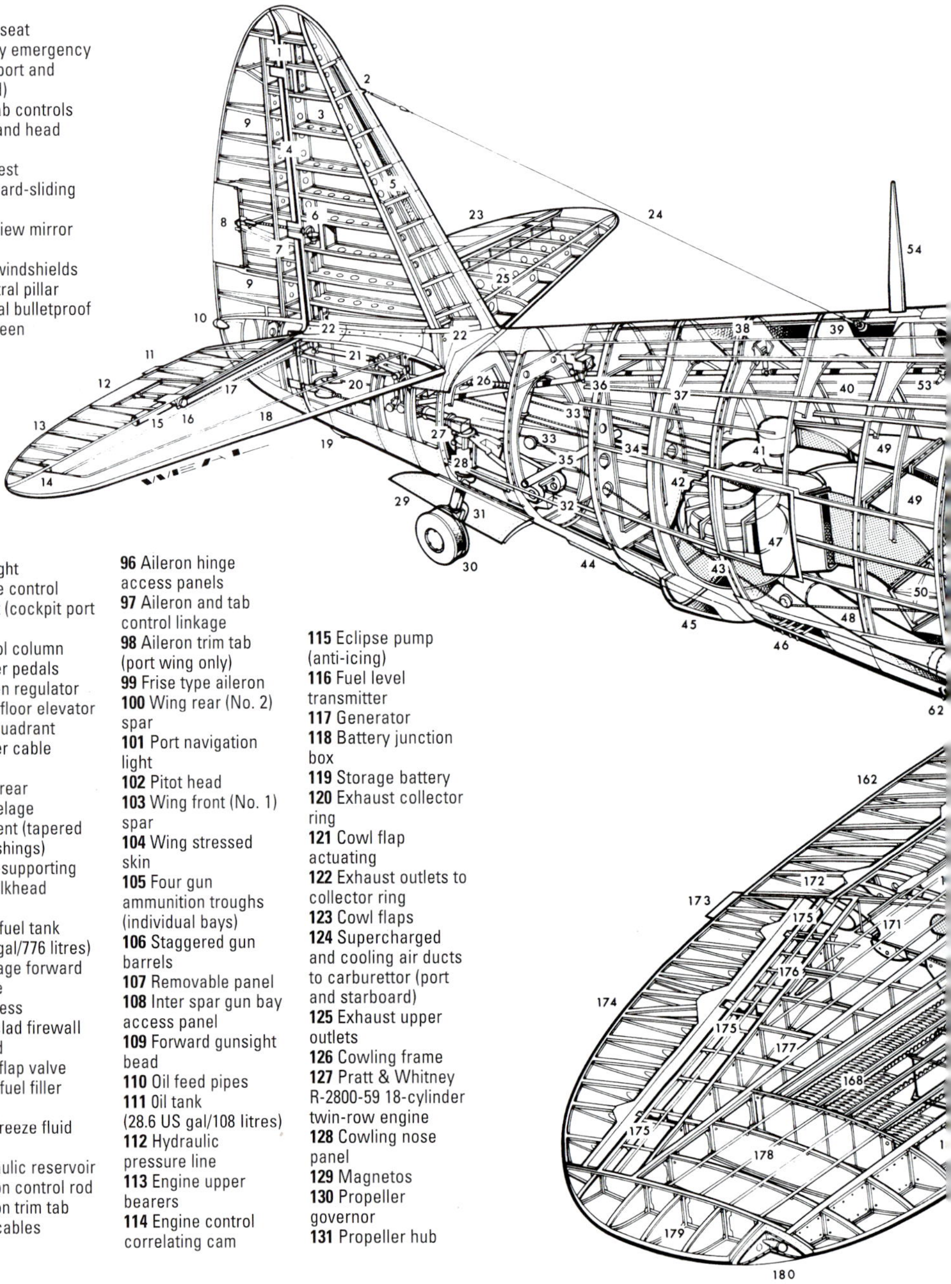

The first examples of the P-47 went into combat in April 1943. Designed as an interceptor, the P-47, as exemplified by this P-47D model, quickly found its niche as a long-range, hard-hitting ground attack platform. The long-legged P-47 was a credible escort aircraft too, drop tanks enabling it to fly from the UK as far as Berlin.

SPECIFICATION

Republic P-47N-1-RE Thunderbolt (unless otherwise stated)

Dimensions

Length: 36 ft 1¾ in (11.02 m)
Wingspan: 42 ft 6½ in (12.82 m)
Wing area: 322.2 sq ft (65.99 m²)
Height overall (unloaded): 14 ft 6 in (4.42 m)

Powerplant

One Pratt & Whitney R-2800-57 rated at 2,800 hp (2088 kW) with water injection
P-47B: one Pratt & Whitney R-2800-21 rated at 2,000 hp (1492 kW)
P-47C-5-RE: one Pratt & Whitney R-2800-21 rated at 2,000 hp (1492 kW)
P-47D-30-RA: one Pratt & Whitney R-2800-59 rated at 2,430 hp (1812 kW) with water injection
P-47M-1-RE: one Pratt & Whitney R-2800-57 rated at 2,800 hp (2088 kW) with water injection
Propeller: Curtiss Electric C642S-B40
Propeller diameter: 13 ft (3.96 m)

Weights

Empty weight: 10,988 lb (4984 kg)
Gross weight: 13,823 lb (6270 kg)
Maximum take-off weight: 21,200 lb (9616 kg)

Fuel load

Total fuel capacity: 1,270 US gal (4807 litres)
Total internal fuel capacity: 570 US gal (2157 litres)
Total external fuel capacity: 700 US gal (2650 litres)
Total wing fuel capacity: 186 US gal (704 litres)
Fuel consumption: 300 US gal (1136 litres) per hour

Performance

Maximum speed at altitude: 467 mph (752 km/h) at 32,000 ft (9754 m)
Cruising speed: 200 mph (322 km/h) with 300-US gal (1134-litre) underwing fuel tanks
Climb rate: 14.2 min to 25,000 ft (7620 m)
Landing speed: 98 mph (158 km/h)
Take-off run: 3,800 ft (1158 m) at gross weight
Maximum service ceiling: 43,000 ft (13106 m)

Range

Range with maximum fuel load: 2,000 miles (3219 km) at 25,000 ft (7620 m)
Normal operational range: 800 miles (1288 km) at 25,000 ft (7620 m)

Armament

Six or eight 0.5-in (12.7-mm) machine-guns in wing with 267- 500 rounds per gun; up to three 1,000-lb (454-kg) bombs, or up to eight rockets on underwing rails, or a combination of bombs and rockets

132 Reduction gear casing
133 Spinner
134 Propeller cuffs
135 Four blade Curtiss constant-speed electric propeller
136 Oil cooler intakes (port and starboard)
137 Supercharger intercooler (central) air intake
138 Ducting
139 Oil cooler feed pipes
140 Starboard oil cooler
141 Engine lower bearers
142 Oil cooler exhaust variable shutter
143 Fixed deflector
144 Excess exhaust gas gate
145 Belly stores/weapons shackles
146 Metal auxiliary drop tank (75 US gal/284 litres)
147 Inboard mainwheel well door
148 Mainwheel well door actuating cylinder
149 Camera gun port
150 Cabin air-conditioning intake (starboard wing only)
151 Wingroot fairing
152 Wing front spar/fuselage attachment (tapered bolts/bushings)
153 Wing inboard rib mainwheel well recess

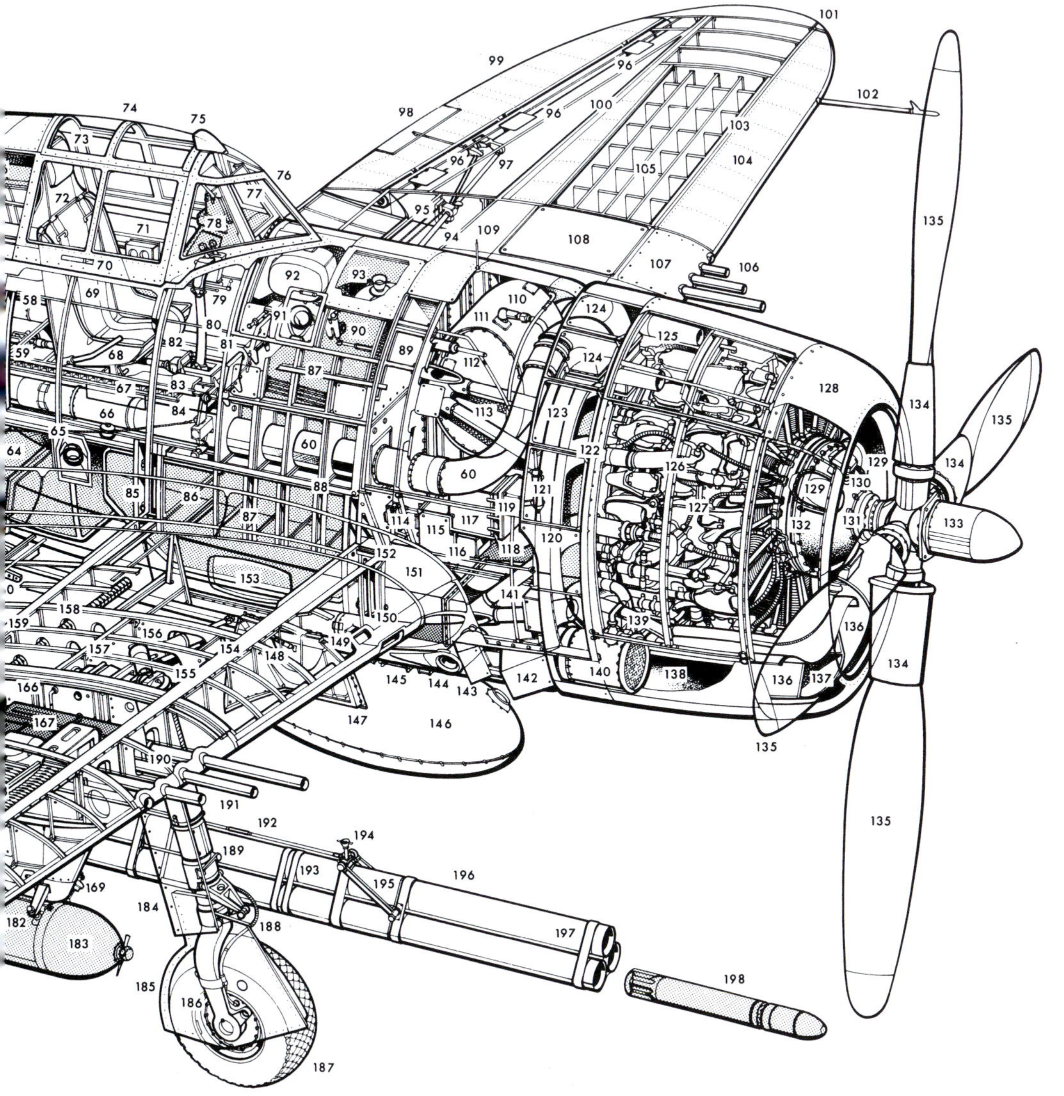

154 Wing front (No. 1) spar
155 Undercarriage pivot point
156 Hydraulic retraction cylinder
157 Auxiliary (undercarriage mounting) wing spar
158 Gun bay warm air flexible duct
159 Wing rear (No. 2) spar
160 Landing flap inboard hinge
161 Auxiliary (No. 3) wing spar inboard section (flap mounting)
162 NACA slotted trailing-edge landing flaps
163 Landing flaps centre hinge
164 Landing flap hydraulic cylinder
165 Four 0.5-in (12.7-mm) Browning machine-guns
166 Inter-spar gun bay inboard rib
167 Ammunition feed chutes
168 Individual ammunition troughs
169 Underwing stores/weapons pylon
170 Landing flap profile
171 Flap door
172 Landing flap profile
173 Aileron fixed tab (starboard wing only)
174 Frise-type aileron structure
175 Aileron hinge/steel forging spar attachments
176 Auxiliary (No. 3) wing spar outboard section (aileron mounting)
177 Multi-cellular wing construction
178 Wing outboard ribs
179 Wingtip structure
180 Starboard navigation light
181 Leading-edge rib sections
182 Bomb shackles
183 500-lb (227 kg) M43 demolition bomb
184 Undercarriage leg fairing (overlapping upper section)
185 Mainwheel fairing (lower section)
186 Wheel fork
187 Starboard mainwheel
188 Brake lines
189 Landing gear air-oil shock strut
190 Machine-gun barrel blast tubes
191 Staggered gun barrels
192 Rocket launcher slide bar
193 Centre strap
194 Front mount (attached below front spar between inboard pair of guns)
195 Deflector arms
196 Triple tube 4.5-in (11.5-cm) rocket launcher (Type M10)
197 Front retaining band
198 4.5-in (11.5-cm) M8 rocket projectile

P-47D-25-RE Thunderbolt

David Carl Schilling flew 132 combat missions with the 56th Fighter Group, scoring 22½ kills. In the aircraft pictured, during 1943, Schilling shot down three Fw 190s (and one shared) and damaged a single Me 210. Based with the 8th Air Force in England, he was made CO of his Group during 1944. Post-war, he converted to the P-80 and, in 1948 in the F-84, became the first pilot to make a non-stop jet fighter flight across the Atlantic. Following his death in 1956, Smoky Hill AFB was renamed Schilling AFB in his honour.

Powerplant
The Thunderbolt was powered by the Pratt & Whitney R-2800 series Double Wasp. A powerful and durable 18-cylinder radial air-cooled unit, the Double Wasp in the P-47 produced over 2,430 hp (1813 kW) at full power. Already immensely strong and fast, the P-47D-25 series was further advantaged by the addition of a rakish cut-down rear fuselage and streamlined teardrop canopy. Additional power was provided by a turbo-supercharger. The P-47D model illustrated introduced not only an improved system of turbocharger exhaust, but a new water injection system as standard for both the R-2800-21 and -59 engines. Later P-47D production models carried the improved R-2800-63 powerplant.

Propeller
Most models in the P-47 series carried a characteristic large diameter four-bladed propeller, and the P-47D-22 introduced this giant 12.2-ft (3.71-m) diameter Hamilton Standard Hydromatic model, with paddle-blades. This was fitted to the aircraft along with other modifications and improvements, including a small dorsal fin in order to offset the loss of keel surface entailed after the introduction of the bubble canopy. These changes came as a result of extensive flight testing of the experimental XP-47K in July 1943, the first Thunderbolt to feature a cut-down rear fuselage and all-round vision canopy, as opposed to the framed aft-sliding canopy of earlier aircraft.

Rocket armament
A favourite weapon of the Thunderbolt pilots in Europe, particularly when engaging enemy armour, was the M-8 unguided rocket. Launched from triple-round M110 tube launchers underwing, the 4.5-in (11.5-cm) rockets could destroy the heavily-armoured Tiger or Panther tanks when used against them in low-level attacks. When the cloud base permitted, Thunderbolts would also knock out bridges, tanks and vehicles by dive-bombing from altitudes between 9,843 ft (3000 m) and 13,123 ft (4000 m), dropping high-explosive bombs, and pulling out at no lower than 820 ft (250 m). This was necessary in order to protect themselves from fragments from their own bombs. Known as *'jabos'* (short for *jagdbombers*, or fighter-bombers), the P-47 attacks struck terror into the Axis troops below them.

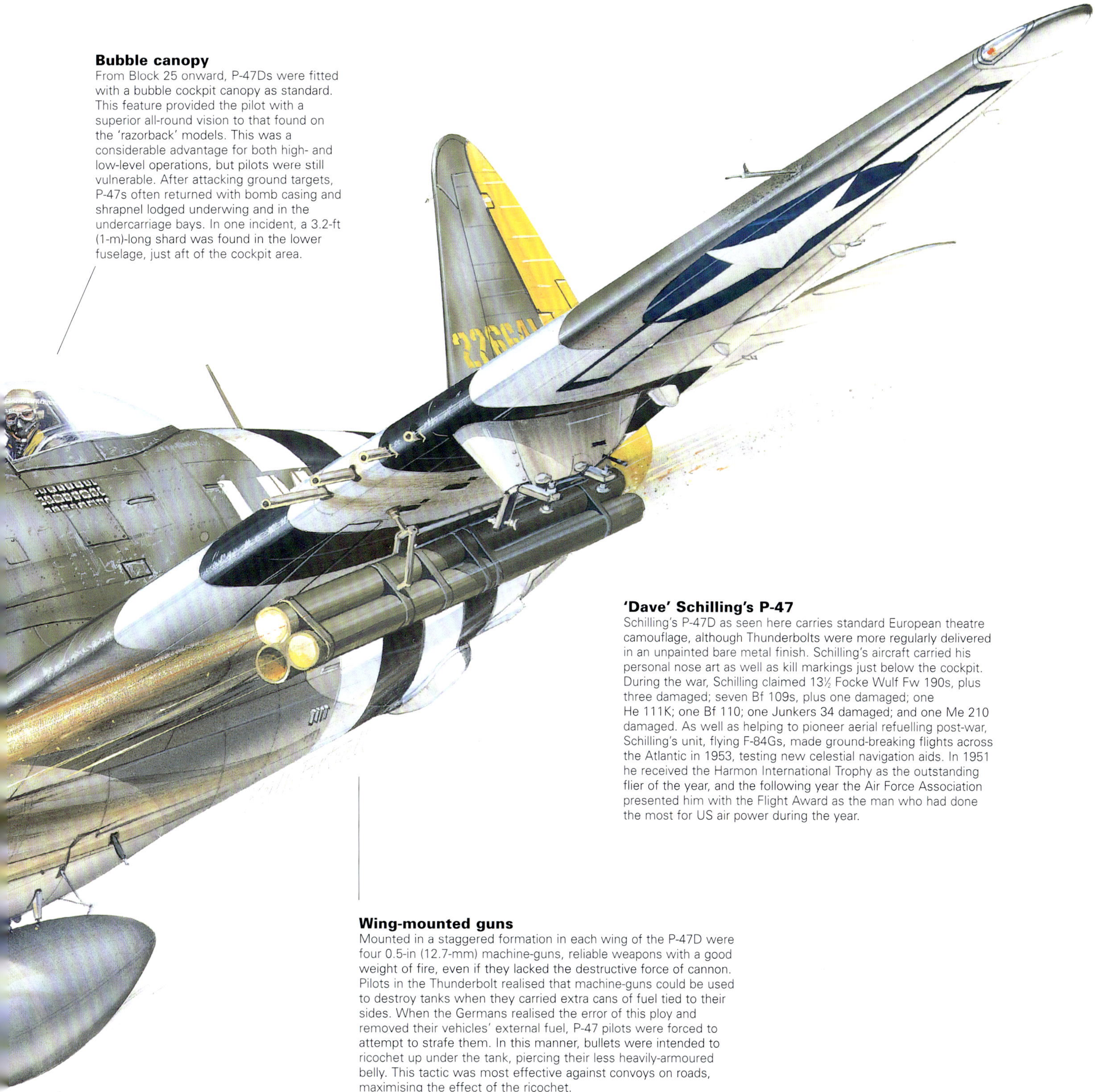

Bubble canopy
From Block 25 onward, P-47Ds were fitted with a bubble cockpit canopy as standard. This feature provided the pilot with a superior all-round vision to that found on the 'razorback' models. This was a considerable advantage for both high- and low-level operations, but pilots were still vulnerable. After attacking ground targets, P-47s often returned with bomb casing and shrapnel lodged underwing and in the undercarriage bays. In one incident, a 3.2-ft (1-m)-long shard was found in the lower fuselage, just aft of the cockpit area.

'Dave' Schilling's P-47
Schilling's P-47D as seen here carries standard European theatre camouflage, although Thunderbolts were more regularly delivered in an unpainted bare metal finish. Schilling's aircraft carried his personal nose art as well as kill markings just below the cockpit. During the war, Schilling claimed 13½ Focke Wulf Fw 190s, plus three damaged; seven Bf 109s, plus one damaged; one He 111K; one Bf 110; one Junkers 34 damaged; and one Me 210 damaged. As well as helping to pioneer aerial refuelling post-war, Schilling's unit, flying F-84Gs, made ground-breaking flights across the Atlantic in 1953, testing new celestial navigation aids. In 1951 he received the Harmon International Trophy as the outstanding flier of the year, and the following year the Air Force Association presented him with the Flight Award as the man who had done the most for US air power during the year.

Wing-mounted guns
Mounted in a staggered formation in each wing of the P-47D were four 0.5-in (12.7-mm) machine-guns, reliable weapons with a good weight of fire, even if they lacked the destructive force of cannon. Pilots in the Thunderbolt realised that machine-guns could be used to destroy tanks when they carried extra cans of fuel tied to their sides. When the Germans realised the error of this ploy and removed their vehicles' external fuel, P-47 pilots were forced to attempt to strafe them. In this manner, bullets were intended to ricochet up under the tank, piercing their less heavily-armoured belly. This tactic was most effective against convoys on roads, maximising the effect of the ricochet.

Fuel capacity
The main fuel tank in the fuselage centre section was armoured, and augmented in this instance by a single 101-US gal (384-litre) drop-tank mounted under the centreline. When fully loaded, the bulky Thunderbolt could be as much as three times heavier than an early model Supermarine Spitfire, and in terms of range the P-47 was far superior to the Luftwaffe's Bf 109G fighter. The range of the P-47 made it a highly capable escort fighter for waves of bombers both over Europe, and later the Pacific. The endurance of the Thunderbolt originated from the fact that the aircraft had filled an initial requirement for a long-range bomber interceptor. By the time the P-47 entered the war, it was a long-range bomber escort as well as a capable ground-attack fighter.

Supermarine Spitfire Mk V

Built as a Spitfire Mk I and later modified to Mk VB standard, R6923 was shot down by a Bf 109 on 22 June 1941. The Spitfire was closely matched to the Bf 109 for much of World War II, with pilot skill often the deciding factor in a dogfight.

Spitfire Mk VB

Cutaway key
1 Aerial stub attachment
2 Rudder upper hinge
3 Fabric-covered rudder
4 Rudder tab
5 Sternpost
6 Rudder tab hinge
7 Rear navigation light
8 Starboard elevator tab
9 Rear navigation light
10 Elevator balance
11 Tailplane front spar
12 IFF aerial
13 Castoring non-retractable tailwheel
14 Tailwheel strut
15 Fuselage double frame
16 Elevator control lever
17 Tailplane spar/fuselage attachment
18 Fin rear spar (fuselage frame extension)
19 Fin front spar (fuselage frame extension)
20 Port elevator tab hinge
21 Port elevator
22 IFF aerial
23 Port tailplane
24 Rudder control lever
25 Cross shaft
26 Tailwheel oleo access plate
27 Tailwheel oleo shock-absorber
28 Fuselage angled frame
29 Battery compartment
30 Lower longeron
31 Elevator control cables
32 Fuselage construction
33 Rudder control cables
34 Radio compartment
35 Radio support tray
36 Flare chute
37 Oxygen bottle
38 Auxiliary long-range fuel tank
39 Dorsal formation light
40 Aerial lead-in
41 HF aerial
42 Aerial mast
43 Cockpit aft glazing
44 Voltage regulator
45 Canopy track
46 Structural bulkhead
47 Headrest
48 Plexiglass canopy
49 Rear-view mirror
50 Entry flap (port)
51 Air bottles (alternative rear fuselage stowage)
52 Sutton harness
53 Pilot's seat (moulded Bakelite)
54 Datum longeron
55 Seat support frame
56 Wingroot fillet
57 Seat adjustment lever
58 Rudder pedal frame
59 Elevator control connecting tube
60 Control column spade grip
61 Trim wheel
62 Reflector gunsight

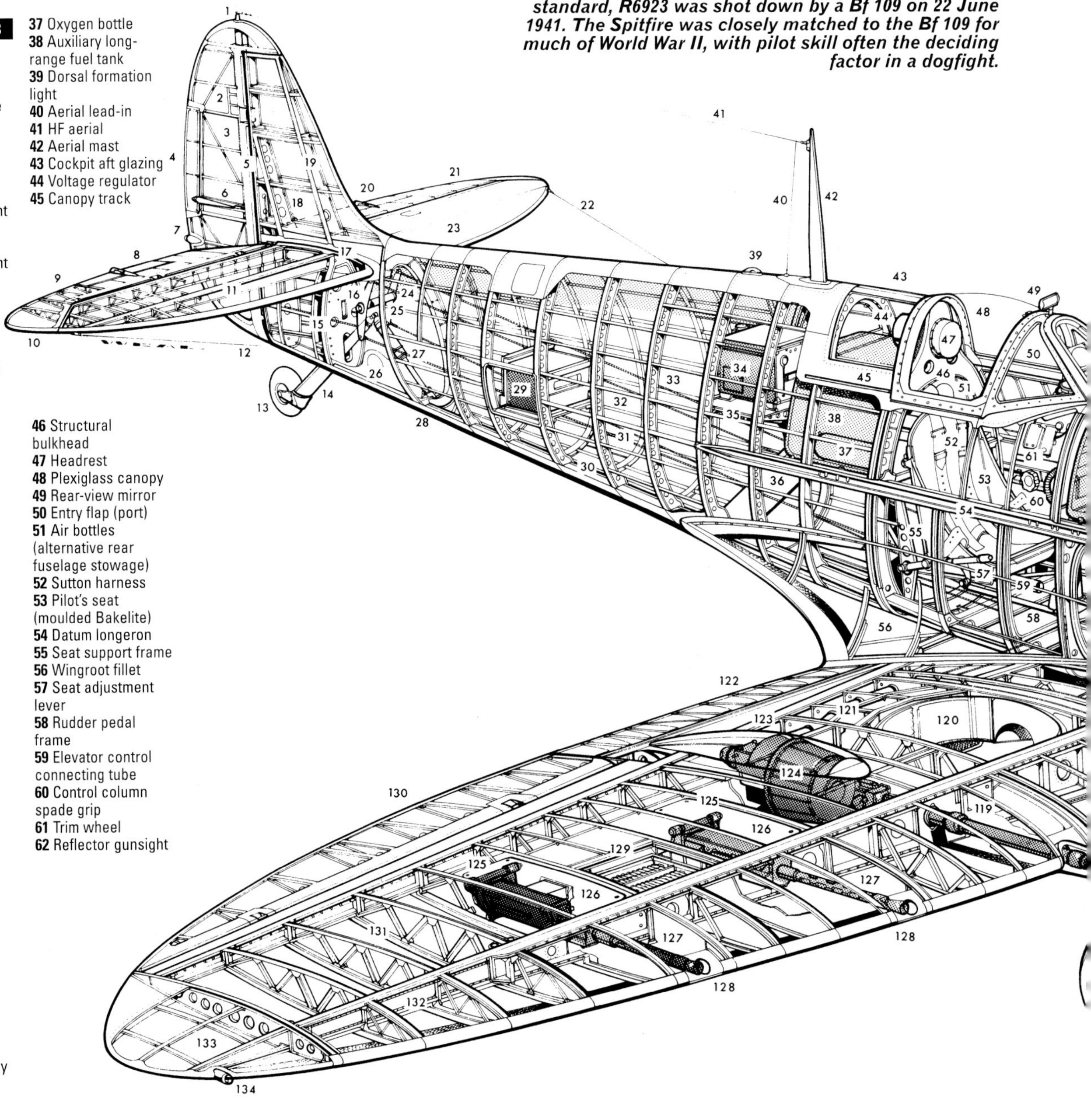

***Above:** With its underfuselage ferry tank and an additional 29 Imp gal (132 litres) of fuel in the rear fuselage tank, **BR202** had a range of 1,450 miles (2334 km) at 15,000 ft (4572 m). The success of this installation led to 17 specially stripped-down aircraft being ferried from the UK to Gibraltar non-stop, and then on to Malta. Although one aircraft failed to complete the journey, the remaining 16 were soon rearmed and ready for action, long before any ship-delivered machines had arrived.*

63 External windscreen armour
64 Instrument panel
65 Main fuselage fuel tank (48 Imp gal/ 218 litre)
66 Fuel tank/longeron attachment fittings
67 Rudder pedals
68 Rudder bar
69 King post
70 Fuselage lower fuel tank (57 Imp gal/ 168 litre)
71 Firewall/bulkhead
72 Engine bearer attachment
73 Steel tube bearers
74 Magneto
75 'Fishtail'/exhaust manifold
76 Gun heating intensifier
77 Hydraulic tank
78 Fuel filler cap
79 Air compressor intake
80 Air compressor
81 Rolls-Royce Merlin 45 engine
82 Coolant piping
85 Port cannon wing fairing
84 Flaps
85 Aileron control cables
86 Aileron push tube
87 Bellcrank
88 Aileron hinge
89 Port aileron
90 Machine-gun access panels
91 Port wingtip
92 Port navigation light
93 Leading-edge skinning
94 Machine-gun ports (protected)
95 20-mm cannon muzzle
96 Three-blade constant-speed propeller
97 Spinner
98 Propeller hub
99 Coolant tank
100 Cowling fastening
101 Engine anti-vibration mounting pad
102 Engine accessories
103 Engine bearers
104 Main engine support member
105 Coolant pipe
106 Exposed oil tank

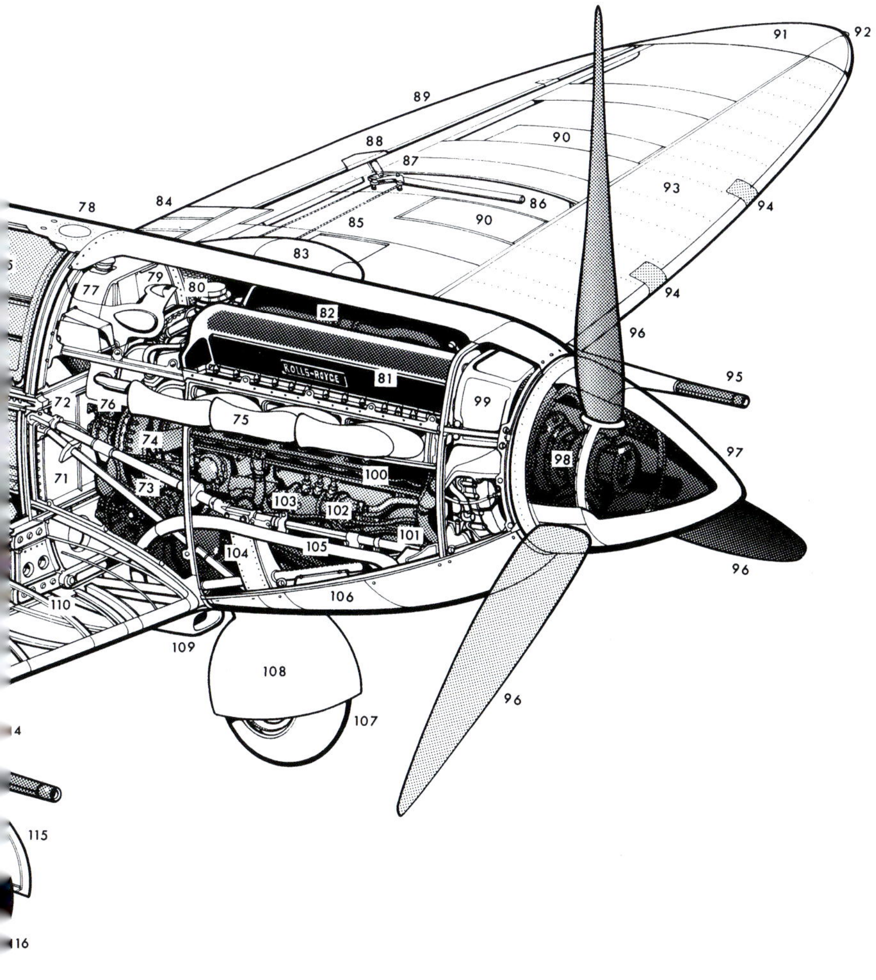

SPECIFICATION

Spitfire Mk V

Dimensions

Length: 29 ft 11 in (9.13 m); some late production aircraft 30 ft 2½ in (9.21 m)
Wing span: 36 ft 10 in (11.23 m); with clipped wings 32 ft 2 in (9.80 m)
Wing aspect ratio: 5.68
Wing area: 242 sq ft (22.45 m²); with clipped wings 231 sq ft (21.46 m²)
Total flap area: 15.6 sq ft (1.45 m²)
Total aileron area: 18.90 sq ft (1.76 m²)
Fin area: 4.61 sq ft (0.43 m²)
Rudder area: 8.23 sq ft (0.76 m²)
Tailplane area: 31.46 sq ft (2.92 m²)
Elevator area: 13.26 sq ft (1.23 m²)
Height: tail down with propeller vertical 11 ft 5½ in (3.49 m)
Wing loading: 36.0 lb/sq ft (175.8 kg/m²)
Wing section: root NACA 2213

Powerplant

Mk VA, VB and VC: one Rolls-Royce Merlin 45 liquid-cooled, 12-cylinder, single-stage supercharged, Vee-type piston engine rated at 1,230 hp (972 kW) for continuous operation or 1,515 hp (1130 kW) for five minutes of emergency combat power at 11,000 ft (3353 m); or one Merlin 46 rated at 1,190 hp (887 kW) continuous
Mk VC: one Merlin 50 rated at* 1,230 hp (917 kW) continuous; or one Merlin 50A or 55 rated at 1,100 hp (820 kW) continuous; or one Merlin 56 rated at 1,580 hp (1178 kW) continuous
LF.Mk V: one Merlin 45M rated at 1,580 hp (1178 kW) at 2,600 ft (792 m); or one Merlin 50M rated at 1,230 hp (917 kW) continuous; or one Merlin 55M rated at 1,580 hp (1178 kW) continuous
Propeller: one de Havilland Type 5/29A or 5/39, or 45/1 or 45/4, Hydromatic three-bladed, constant-speed propeller of 10 ft 9 in (3.28 m) diameter; or one Rotol RX5/10 Jablo three-bladed, constant-speed propeller of 10 ft 3 in (3.12 m) diameter; or one RX5/14 or RS5/24 of 10 ft 9 in (3.28 m) diameter

Weights

Mk VA: empty 4,960 lb (2658 kg), normal loaded 6,237 lb (2829 kg), maximum loaded 6,700 lb (3039 kg)
Mk VB (with 30-Imp gal/136-litre auxiliary tank): empty 5,065 lb (2297 kg), normal loaded 6,630 lb (3007 kg), maximum loaded 6,700 lb (3039 kg)
Mk VC (with 30-Imp gal/136-litre auxiliary tank): empty 5,100 lb (2313 kg), normal loaded 6,785 lb (3078 kg), maximum loaded (with 'B' wing armament) 7,300 lb (3311 kg)
LF.Mk VB (with 30-Imp gal/136-litre auxiliary tank): empty 5,050 lb (2291 kg), normal loaded 6,615 lb (3000 kg)

Fuel and load

Total internal fuel: 85 Imp gal (386 litres), held as 48 Imp gal (218 litre) in upper fuselage tank and 37 Imp gal (168 litre) in lower fuselage tank (additional 29-Imp gal (132-litre) rear fuselage tank for ferry flights only)
External fuel: 30-Imp gal (136-litre), 90-Imp gal (409-litre) or 170-Imp gal (773-litre) underfuselage auxiliary tanks
Maximum external load (Mk VB and VC): 500 lb (227 kg)

Performance

Maximum speed: VA 376 mph (605 km/h) at 19,500 ft (5944 m); **VB** 369 mph (594 km/h) at 19,500 ft (5944 m); **VC** 374 mph (601 km/h) at 13,000 ft (3962 m); **LF.Mk VB** 357 mph (574 km/h) at 6,000 ft (1829 m)
Maximum diving speed: 450 mph (724 km/h)
Initial rate of climb: 4,740 ft (1445 m) per minute; 3240 ft (988 m) per minute at 5,000 ft (1524 m); 3,250 ft (1474 m) per minute at 15,000 ft (4572 m)
Maximum rate of climb at sea level (LF.Mk VB): 4,750 ft (2155 m) per minute
Service ceiling: 36,500 ft (11125 m)

Range

Maximum range: 470 miles (756 km) on internal fuel; 1,135 miles (1827 km) with 90-Imp gal (409-litre) external tank

Armament

VA: eight 0.303-in (7.7-mm) Browning machine-guns with 350 rpg (rounds per gun)
VB: four 0.303-in (7.7-mm) Browning machine-guns with 350 rpg, plus two 20-mm Hispano cannon with 60 rpg
VC: VA or VB armament, with the option of four 20-mm Hispano cannon with 120 rpg, plus one 500-lb (227-kg) bomb or two 250-lb (113-kg) bombs

107 Port mainwheel
108 Mainwheel fairing
109 Carburettor air intake
110 Stub/spar attachment
111 Mainwheel leg pivot point
112 Main spar
113 Leading-edge ribs (diagonals deleted for clarity)
114 Mainwheel leg shock-absorber
115 Mainwheel fairing
116 Starboard mainwheel
117 Angled axle
118 Cannon barrel support fairing
119 Spar cut-out
120 Mainwheel well
121 Gun heating pipe
122 Flap structure
123 Cannon wing fairing
124 Cannon magazine drum (120 rounds)
125 Machine-gun support brackets
126 Gun access panels
127 0.303-in (7.7-mm) machine-gun barrels
128 Machine-gun ports
129 Ammunition boxes (350 rounds per gun)
130 Starboard aileron construction
131 Wing ribs
132 Single-tube outer spar section
133 Wingtip structure
134 Starboard navigation light

Spitfire Mk VC

In July 1943, No. 2 Squadron of No. 7 Wing, South African Air Force (SAAF), Desert Air Force, began relinquishing its Kittyhawks for the Spitfire Mk VC. Declared operational with its new machines in Sicily on 23 August, the unit moved to a newly captured airfield in Italy. Flying from a succession of Italian bases, No. 2 retained its Mk VCs until March 1944.

Merlin power

With the introduction of the 50-series Merlin to the Spitfire Mk V, a longstanding problem was eliminated. Previously, pilots flying negative-*g* manoeuvres found the engines of their Spitfires cutting out. Rolls-Royce devised a diaphragm-operated carburettor which eliminated the fuel starvation problem. The new carburettor was introduced on the 50-series powerplants, and transformed the fighting qualities of the Spitfire. Pilots were warned, however, that prolonged periods of negative-*g* or inverted flight could result in a catastrophic drop in engine oil pressure.

Engine and propeller

Engine efficiency was improved by the installation of a constant-speed propeller as standard. The pitch of the propeller blades was automatically adjusted according to the flight regime, ensuring that the engine always operating at its optimum level of about 3,000 rpm for maximum power delivery. Most Spitfire Mk Vs used Rotol three-bladed propellers.

Cockpit access

Entry to the Spitfire was gained via a downward-hinged door on the port side of the fuselage. The cockpit canopy slid backward along prominent rails running along the sides of the fuselage behind the cockpit. Extensive testing was carried out to ensure that the canopy would separate cleanly from the aircraft in the event of the pilot jettisoning it in flight. Tests revealed that the canopy could show an alarming tendency to rotate very slowly with minimum lift after separation if the aircraft's rate of descent were too high. This slow rotation caused the canopy to fly toward the fin, a problem which was solved by adjusting the centre of gravity of the canopy itself.

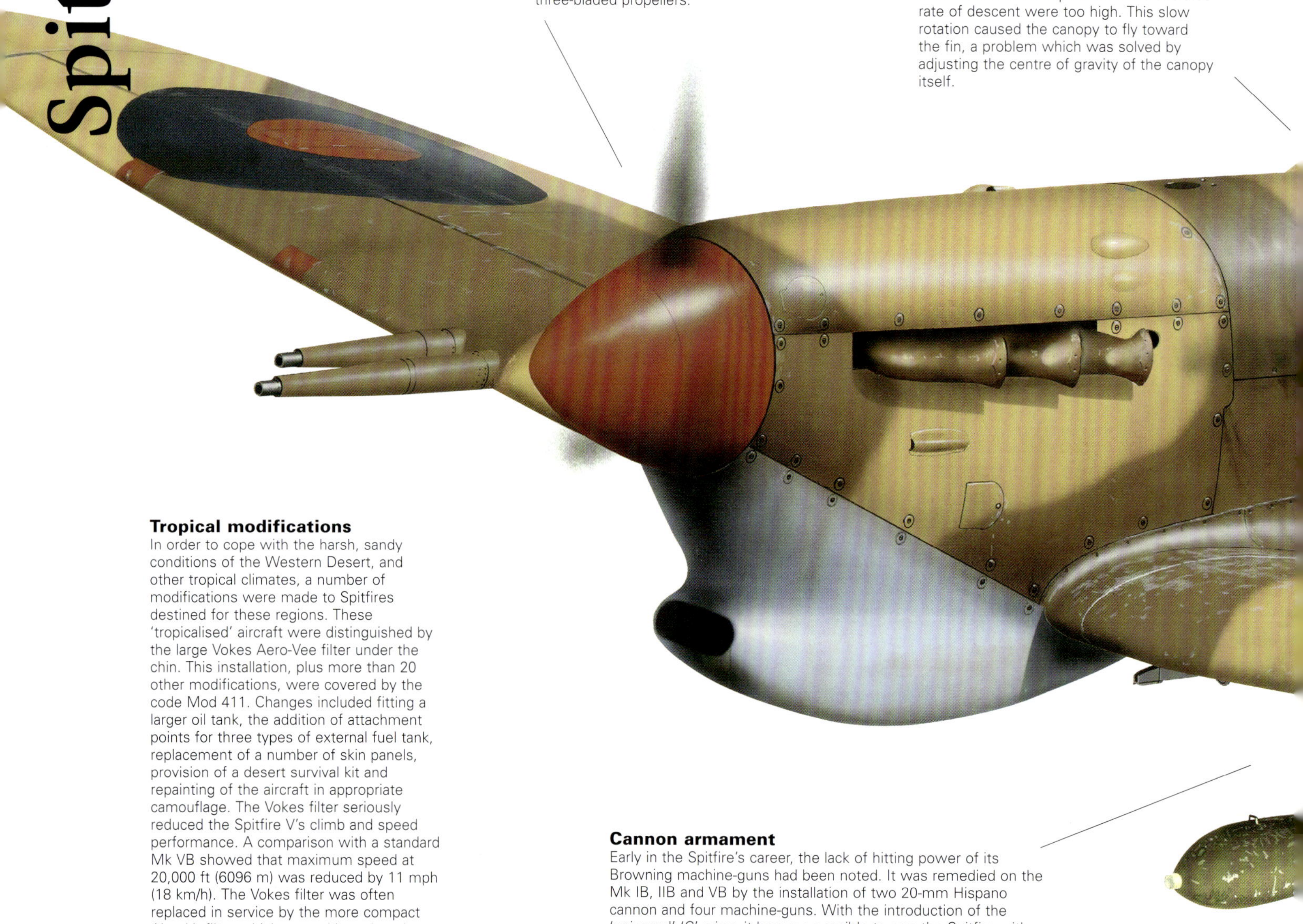

Tropical modifications

In order to cope with the harsh, sandy conditions of the Western Desert, and other tropical climates, a number of modifications were made to Spitfires destined for these regions. These 'tropicalised' aircraft were distinguished by the large Vokes Aero-Vee filter under the chin. This installation, plus more than 20 other modifications, were covered by the code Mod 411. Changes included fitting a larger oil tank, the addition of attachment points for three types of external fuel tank, replacement of a number of skin panels, provision of a desert survival kit and repainting of the aircraft in appropriate camouflage. The Vokes filter seriously reduced the Spitfire V's climb and speed performance. A comparison with a standard Mk VB showed that maximum speed at 20,000 ft (6096 m) was reduced by 11 mph (18 km/h). The Vokes filter was often replaced in service by the more compact Aboukir filter which caused less drag; it was designed by the RAF's No. 103 Maintenance Unit at Aboukir, Egypt.

Cannon armament

Early in the Spitfire's career, the lack of hitting power of its Browning machine-guns had been noted. It was remedied on the Mk IB, IIB and VB by the installation of two 20-mm Hispano cannon and four machine-guns. With the introduction of the 'universal' 'C' wing, it became possible to arm the Spitfire with four of the Hispano weapons, each supplied with 120 rounds of ammunition. This powerful internal armament, coupled with a respectable bombload, made the Spitfire Mk VC a potent fighter-bomber, especially when combined with the clipped wingtips which gave improved low-level manoeuvrability. Some Mk VBs and VCs were delivered to the Middle East with clipped wings, but many were modified in-theatre by No. 103 MU. Aircraft so-altered featured wooden wingtip fairings, which were of a slightly more rounded shape than those fitted to UK-based machines. In addition to its modification activities, which included various non-standard armament configurations, the MU was also responsible for reconditioning 140 aircraft per month.

No. 2 Squadron, SAAF

By mid-1944, No. 7 Wing, SAAF, whose springbok marking appears on the tail of this aircraft, formed part of the Desert Air Force with No. 3 Wing, SAAF and three RAF wings. As a component of No. 7 Wing, No. 2 Sqn received its Spitfire Mk Vs in July 1943, a month before it moved to Sicily. The squadron relocated again during September, moving onto the Italian mainland and occupying newly-captured Axis airfields. Though flown intensively during the squadron's time in Italy, the Spitfire Mk Vs were on charge with No. 2 for just eight months. Spitfire Mk IXs arrived in February 1944, and a few Mustang Mk IVs were taken 'on charge' in June 1945. On 12 July the squadron's personnel left Tissano airfield en route for South Africa.

Airframe variations

Apart from the three wing types fitted to the Spitfire Mk V, a host of other variations existed between individual aircraft. Many were the result of factory-fitted items, but many more resulted from field modifications. Tropicalisation resulted in reduced performance regardless of the filter type installed, but, beyond this, minor airframe modifications could amount to significant performance variations between airframes. In 1943 a thorough investigation into the effects of such minor changes was conducted at Farnborough, with startling results. A Spitfire Mk VB was subjected to modifications which ranged from minor engineering changes to gap-filling and simple polishing. Speed gained by each modification combined to give a 28.5-mph (45.9-km/h) increase in maximum speed.

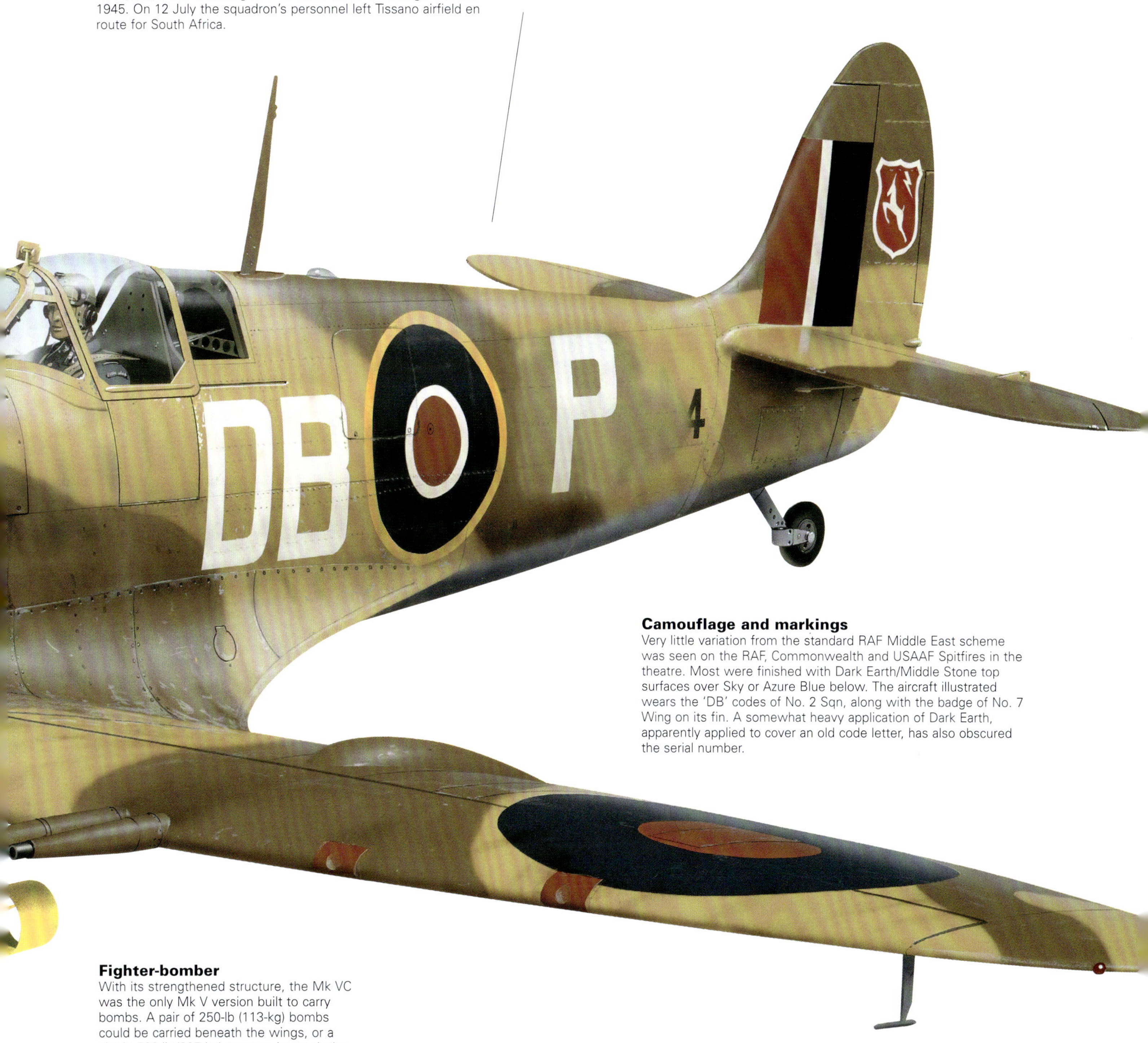

Camouflage and markings

Very little variation from the standard RAF Middle East scheme was seen on the RAF, Commonwealth and USAAF Spitfires in the theatre. Most were finished with Dark Earth/Middle Stone top surfaces over Sky or Azure Blue below. The aircraft illustrated wears the 'DB' codes of No. 2 Sqn, along with the badge of No. 7 Wing on its fin. A somewhat heavy application of Dark Earth, apparently applied to cover an old code letter, has also obscured the serial number.

Fighter-bomber

With its strengthened structure, the Mk VC was the only Mk V version built to carry bombs. A pair of 250-lb (113-kg) bombs could be carried beneath the wings, or a single 500-lb (227-kg) weapon beneath the fuselage. On its arrival in Italy, No. 2 Sqn, SAAF provided top cover for Allied operations, but, as enemy air activity dwindled, the unit dedicated more of its time to fighter-bomber sorties. The stronger wing of the VC also freed pilots from the restrictions they had faced when flying earlier Mk Vs. Installing a more powerful engine into the basic Spitfire Mk II airframe had given pilots the potential for overstressing the aircraft's structure during high-*g* manoeuvres.

Undercarriage and wheels

Compared to the Spitfire Mk I, the Mk V's weight had increased considerably. With the wheels remaining unchanged in size, Dunlop was obliged to supply tyres of increased ply-rating. The tailwheel was a fixed castoring unit, Supermarine having retained the simple unit of the Mk II. A retractable tailwheel, as fitted to the Mk IV, did not appear on a production Spitfire until the Mk VII was introduced. Pilots always treated the Spitfire with respect on the ground, since the aircraft possessed less-than-ideal ground-handling characteristics, owing to the narrow track of its main landing gear. Aircraft operating in strong crosswinds, especially over rough ground or in slippery, muddy conditions, were easily ground looped (spun horizontally around their main wheels), a problem which also made Spitfires and Seafires tricky to operate from aircraft-carriers.

Several Spitfire Mk Vs were flown from carriers as the final leg of their delivery from the UK to Malta. In April and May 1942, the USS *Wasp* and HMS *Eagle* between them delivered 126 tropicalised Mk Vs to the island.

Supermarine Griffon Spitfire

Re-engining the Spitfire with the more powerful Griffon engine gave the aircraft a new lease of life. As well as increasing outright performance, the Griffon also allowed a considerable increase in weights, opening the door to a new period of Spitfire development. The aircraft illustrated is a Mk XIV.

Spitfire F.Mk 21

Cutaway key
1 Starboard elevator construction
2 Elevator tab
3 Tail navigation light
4 Rudder trim tab
5 Fabric-covered rudder construction
6 Sternpost
7 Rudder balance weight
8 Fin main spar
9 Tailfin construction
10 Tail ballast weights
11 Fin secondary spar
12 Rudder trim jack
13 Tailplane trim jack
14 Tailplane construction
15 Tailwheel doors
16 Mudguard
17 Tailwheel retraction jack
18 Elevator control rods
19 Tailwheel
20 Fuselage double bulkhead
21 Port elevator
22 Port tailplane
23 Finroot fillet fairing
24 Tail assembly joint frame
25 Oxygen cylinder
26 Six-cartridge signal flare launcher
27 Tailplane control cables
28 Access door
29 Fuselage ballast weights
30 Battery
31 R.3067 radio receiver
32 Radio access door
33 Whip aerial
34 Harness release
35 TR.1143 radio transmitter
36 Radio track
37 Fuselage frame and stringer construction
38 Wingroot trailing edge fillet
39 Control cable runs
40 Fuselage main longeron
41 Port side access door
42 Canopy aft glazing
43 Sliding canopy rail
44 Voltage regulator
45 Fuselage double frame
46 Seat support frame work
47 Back armour
48 Pilot's seat
49 Sutton harness
50 Head armour
51 Sliding cockpit canopy cover
52 Rear-view mirror
53 Windscreen framing
54 Bulletproof windscreen
55 Reflector gunsight
56 Port side entry hatch
57 Instrument panel
58 Control column
59 Compass mounting
60 Undercarriage control lever
61 Seat adjusting handle
62 Seat pan armour plate
63 Wingroot rib
64 Radiator shutter jack
65 Coolant radiator, oil cooler on port side
66 Gun heating duct
67 Wing rear spar
68 Flap hydraulic jack
69 Flap shroud ribs
70 Tubular flap spar
71 Starboard split trailing-edge flap
72 Aileron control bellcrank
73 Aileron hinge
74 Aileron tab
75 Aluminium skinned aileron construction
76 Wingtip fairing
77 Starboard navigation light
78 Wingtip construction
79 Aileron outer hinge rib
80 Wing rib construction
81 Main spar
82 Leading-edge nose ribs
83 Ammunition feed drums
84 Mainwheel fairing door
85 Ammunition feed drums
86 Blister fairings
87 Ammunition belt feed

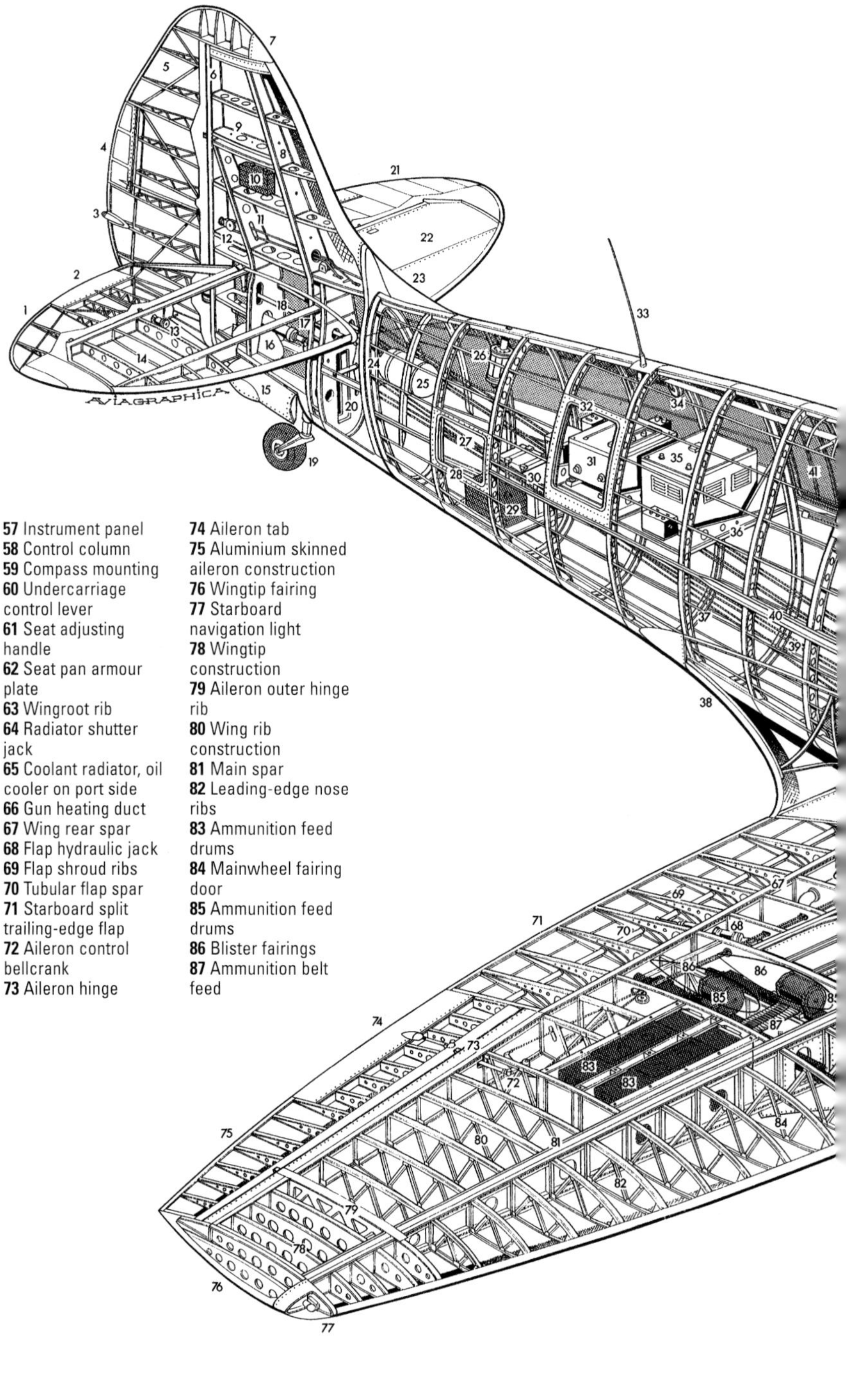

Clipped wings improved the Mk XII's speed and rate of roll at low altitudes. Up to 20,000 ft (6100 m), the Spitfire Mk XII was easily superior to the Mk IX, but above this altitude the earlier mark was faster.

Large numbers of surplus Griffon Spitfires became available in the immediate post-war period and the aircraft played an important role in the rebuilding of many European air arms. Such were the qualities of the Griffon Spitfires that many served on into the early jet era. These FR.Mk 14Es of the Force Aérienne Belge/Belgische Luchtmacht carry the 'IQ' codes of the Ecole de Chasse at Coxyde, which operated the variant between 1948 and 1954.

88 20-mm British Hispano, Mk II cannon barrels
89 Cannon barrel support fairing
90 Recoil springs
91 Fuel filler cap
92 Leading-edge fuel tank, capacity 17 Imp gal (77 litres)
93 Main undercarriage wheel well
94 Mainwheel blister fairing
95 Undercarriage retraction link
96 Undercarriage leg pivot
97 Shock absorber leg strut
98 Hydraulic brake pipe
99 Starboard main wheel
100 Main wheel leg fairing door
101 Undercarriage torque scissors
102 Fuel pipe runs
103 Main spar stub attachment
104 Lower main fuel tank, capacity 48 Imp gal (218 litres)
105 Upper main fuel tank, capacity 36 Imp gal (164 litres)
106 Fuel filler cap
107 Oil tank vent
108 Oil tank, capacity 9 Imp gal (41 litres)
109 Oil tank access door
110 Engine compartment fire proof bulkhead
111 Port split trailing-edge flap
112 Flap hydraulic jack
113 Flap synchronising jack
114 Port twin 20-mm British Hispano cannon
115 Spent cartridge case ejector chute

SPECIFICATION

Spitfire F.Mk XII

Dimensions

Length: 30 ft 9 in (9.37 m)
Wingspan, clipped: 32 ft 7 in (9.93 m)
Wing area, clipped: 242 sq ft (22.45 m²)
Height: 11 ft (3.35 m)

Powerplant

One 1,735-hp (1294-kW) Rolls-Royce Griffon IIB, III or IV liquid-cooled, 12-cylinder inline piston engine with single-stage supercharging, driving a four-bladed Rotol propeller

Weight

Maximum loaded: 7,400 lb (3356 kg)

Performance

Maximum speed at 12,500 ft (3810 m): 389 mph (626 km/h)
Service ceiling: 37,350 ft (11387 m)

Armament

Two 20-mm British Hispano cannon and four 0.303-in (7.7-mm) Browning machine-guns

Spitfire F.Mk XIV

Dimensions

Length: 32 ft 8 in (9.96 m)
Wingspan: 36 ft 10 in (10.98 m)

Powerplant

One 2,050-hp (1529-kW) Rolls-Royce Griffon 65 liquid-cooled, 12-cylinder inline piston engine with two-stage supercharging, driving a five-bladed Rotol propeller

Weight

Empty: 6,376 lb (2892 kg)
Maximum loaded: 10,065 lb (4565 kg)

Performance

Maximum speed at 24,500 ft (7468 m): 439 mph (707 km/h)
Service ceiling: 43,000 ft (13110 m)

Armament

Two 20-mm British Hispano cannon and four 0.5-in (12.7-mm) Browning machine-guns plus (post-war only) one 500-lb (227-kg) bomb under the fuselage and two 250-lb (114-kg) bombs underwing, or up to 12 60-lb (27-kg) rocket projectiles underwing

Spitfire F.Mk 22

Dimensions

Length: 32 ft 11 in (10.03 m)
Wingspan: 36 ft 11 in (11.25 m)

Powerplant

As F.Mk XIV

Weight

Maximum loaded: 10,086 lb (4574 kg)

Performance

Maximum speed at 25,000 ft (7622 m): 449 mph (723 km/h)
Service ceiling: 45,500 ft (13872 m)

Armament

Four 20-mm Hispano Mk 5 cannon plus one 500-lb (227-kg) bomb under the fuselage and two 500-lb (227-kg) bombs underwing, or up to four 300-lb (136-kg) rocket projectiles underwing

116 Ammunition feed drums
117 Ammunition belt feeds
118 Ammunition boxes, 150 rounds per gun
119 Aileron control bellcrank
120 Aileron tab
121 Port aileron
122 Wingtip fairing
123 Port navigation light
124 Pitot tube
125 Cannon barrel fairings
126 Cannon barrels
127 Port leading edge fuel tank, capacity 17 Imp gal (77 litres)
128 Upper engine cowling
129 Hydraulic fluid tank
130 Intercooler
131 Compressor intake
132 Generator
133 Heywood compressor
134 Engine bearer attachment
135 Hydraulic pump
136 Coolant pipes
137 Gun camera
138 Camera port
139 Engine air intake duct
140 Port mainwheel
141 Engine bearer
142 Cartridge starter
143 Exhaust stubs
144 2,035-hp (1517-kW) Rolls Royce Griffon 61 engine
145 Engine magnetos
146 Coolant header tank
147 Front engine mounting
148 Lower engine cowling
149 Spinner backplate
150 Propeller hub pitch change mechanism
151 Spinner
152 Rotol five-bladed constant speed propeller

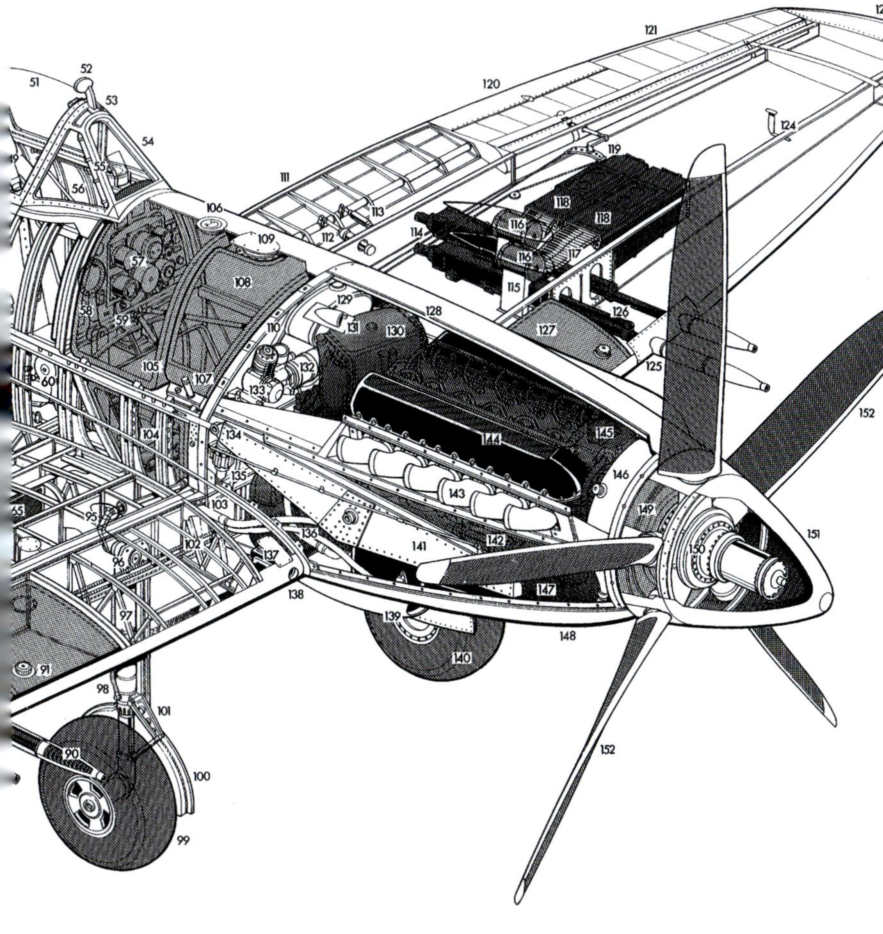

For many years Spitfire PR.Mk 19 PS853 flew with the RAF's Battle of Britain Memorial Flight. The aircraft joined the RAF in 1945 for service with the Temperature and Humidity Flight (THUM) at Woodvale. PS853 was sold to Rolls-Royce in 1996, in order to supply funds for the rebuilding of BBMF Hurricane LF363.

Spitfire FR.Mk 18E

This aircraft, 'H' (serial unknown) of No. 28 Sqn, of the RAF's Far East Air Force, was based at Kai Tak, Hong Kong in late 1950. It is finished in the standard RAF fighter scheme of the day – Dark Green/Dark Sea Grey upper surfaces over Medium Sea Grey undersurfaces. National insignia consisted of Type D roundels of the type introduced in 1947. Squadron marks included a red spinner and a unit badge on the fin. The latter featured a demi-Pegasus, representing the white horse on the Downs near Yatesbury, Wiltshire where the squadron had received its fighters (Sopwith Camels) in 1917, after a period as a training unit. For the duration of the Korean War, Spitfires based in Hong Kong carried black and white recognition stripes on the rear fuselage (covering the serial number on the port side of this aircraft) and on both the upper and lower surfaces of each wing.

No. 28 in Hong Kong
Civil war in China and an increase in Nationalist movements in the Far East (often supported by Communist elements) brought a need for extra fighter reinforcement in the colony, and in May 1949 Spitfire FR.Mk 18s of No. 28 Sqn were deployed to Kai Tak from Malaya. At the same time, No. 80 Sqn with Spitfire F.Mk 24s left Germany for Hong Kong aboard HMS *Ocean*, arriving in August. Operational from September, No. 80 Sqn assumed daytime ground-attack and high- and low-level interception roles, with No. 28 Sqn covering medium-level interceptions. Other tasks included navy and army co-operation and anti-piracy patrols.

Development of the Griffon engine
The Rolls-Royce Griffon was a 12-cylinder, 60° Vee engine, with a bore and stroke of 6 in (15.24 cm) and 6⅔ in (16.764 cm), respectively. This equated to a swept volume of 2,240 cu in (36.7 litres), some 35 per cent greater than that of the Merlin. Developed from the Buzzard via the 'R' racing engine (as fitted to the S.6 Schneider Trophy floatplanes), the Griffon was otherwise very much like an enlarged Merlin, incorporating many of the latter's features, but rotating in the opposite direction (anti-clockwise, from the pilot's point of view). A derated 'R' engine, known as the Griffon I, first ran in 1933 but was destined never to fly. Under pressure to develop a smaller engine (the Merlin), Rolls suspended Griffon development until 1939 and then carried out an extensive redesign to make the Griffon II a far more compact powerplant. As it was necessary to keep the engine's size and weight length within limits imposed by the configurations of existing fighter aircraft, much work went into keeping the Griffon's length to a minimum (centring on a redesign of the engine's supercharger drive system). Such was the success of these measures that the overall length of the Griffon was in the range 72-81 in (183-206 cm). For comparison, a Merlin with a single-speed supercharger measured 69 in (175 cm) in length, while a two-speed, two-stage Merlin was 88.7 in (225 cm) long.

Camera fit
Unlike that of its predecessor, the Mk 14, Spitfire Mk 18 production was almost entirely of the FR.Mk 18 fighter-reconnaissance variant, characterised by its cut-down rear fuselage, bubble canopy and oblique camera ports behind the cockpit, either side of the fuselage. The standard camera fit for the type consisted of three F.24 cameras with lenses of varying focal lengths – usually one 14-in (35.6-cm) and two 20-in (50.8-cm). One of these was fitted to look obliquely through one of the two oblique ports, while the other pair was positioned to look downwards through the lower fuselage. An alternate load was a single F.52 vertical camera in the same position.

Air-to-ground armament
Of the Griffon-engined Spitfire variants, the Mk XII was confined to an interception role, as was the Mk XIV in wartime RAF service. However, the latter variant was able to carry a 500-lb (227-kg) bomb under the fuselage and two 250-lb (114-kg) bombs underwing, as well as six rocket projectiles (RPs) (or 12 in pairs). The FR.Mk 18, effectively a strengthened late-production Mk XIV, was able to carry similar loads. Maximum loads quoted for the Mk 21, 22 and 24 appear to have been three 500-lb (227-kg) bombs or four 300-lb (136-kg) RPs, though in RAF service, the Mk 21 and 22 were confined to a fighter role. However, RPs were commonly seen fitted to Mk 24s, rocket-firing forming part of the training programme for No. 80 Sqn at Kai Tak.

The Spitfire Mk XVIII in RAF service
The first Spitfire Mk XVIII (NH872) flew in June 1945, the new mark being externally identical to a late production Mk XIV, with its bubble canopy and 'E' wing armament. Internally, the Mk XVIII had a stronger wing spar structure and two 31-Imp gal (141-litre) fuel tanks in the rear fuselage. Like the Mk XIV, there was a fighter-reconnaissance variant, the FR.Mk XVIII, with a single fuselage fuel tank and three cameras behind the cockpit. The Mk XVIII (known as the Mk 18 from 1947) joined RAF squadrons from August 1946, when No. 208 Sqn re-equipped in Palestine. Deliveries began in earnest in 1947; in January No. 60 Sqn re-equipped at Seletar, Singapore, followed by Nos 11 (with whom Mk 18s operated alongside Mk 14s), 28 (from February 1947) and 81 Sqns (from August, to augment the unit's PR.Mk 19s and Mosquito PR.Mk 34s). In the Middle East, No. 32 Sqn operated the new variant from April 1947.

Vought F4U Corsair

By far the most important operators of the Corsair in World War II were the fighter squadrons of the US Marine Corps, based on island airstrips in the Pacific. These aircraft were photographed leaving parking areas on Majuro Atoll, in the Marshalls, on 29 August 1944. The squadron belongs to the 4th Marine Air Wing.

F4U-1A Corsair

Cutaway key
1 Spinner
2 Three-bladed Hamilton Standard constant-speed propeller
3 Reduction gear housing
4 Nose ring
5 Pratt & Whitney R-2800-8W Double Wasp 18-cylinder two-row engine
6 Exhaust pipes
7 Hydraulically-operated cowling
8 Fixed cowling panels
9 Wing leading-edge unprotected integral fuel tank, capacity 62 US gal (235 litres)
10 Truss-type main spar
11 Leading-edge rib structure
12 Starboard navigation light
13 Wingtip
14 Wing structure
15 Wing ribs
16 Wing outer-section (fabric skinning aft of main spar)
17 Starboard aileron
18 Ammunition boxes (maximum total capacity 2,350 rounds)
19 Aileron trim tab
20 Aerial mast
21 Forward bulkhead
22 Oil tank, capacity 28 US gal (106 litres)
23 Oil tank forward armour plate
24 Fire suppressor cylinder
25 Supercharger housing
26 Exhaust trunking
27 Blower assembly
28 Engine support frame
29 Engine control runs
30 Wing main spar carry-through structure
31 Engine support attachment
32 Upper cowling deflection plate (0.1-in/ 0.25-cm aluminium)
33 Fuel filler cap
34 Fuselage main fuel tank, capacity 237 US gal (897 litres)
35 Upper longeron
36 Fuselage forward frames
37 Rudder pedals
38 Heelboards
39 Control column
40 Instrument panel
41 Reflector sight
42 Armoured-glass windshield
43 Rear-view mirror
44 Rearward-sliding cockpit canopy
45 Handgrip
46 Headrest
47 Pilot's head and back armour
48 Canopy frame
49 Pilot's seat
50 Engine control quadrant
51 Trim tab control wheels
52 Wing folding lever
53 Centre/aft fuselage bulkhead
54 Radio shelf
55 Radio installation
56 Canopy track
57 Bulkhead
58 Aerial lead-in
59 Aerial mast
60 Aerials
61 Heavy sheet skin plating
62 Dorsal identification light
63 Longeron
64 Control runs
65 Aft fuselage structure
66 Compass installation
67 Lifting tube
68 Access/inspection panels
69 Fin/fuselage forward attachment

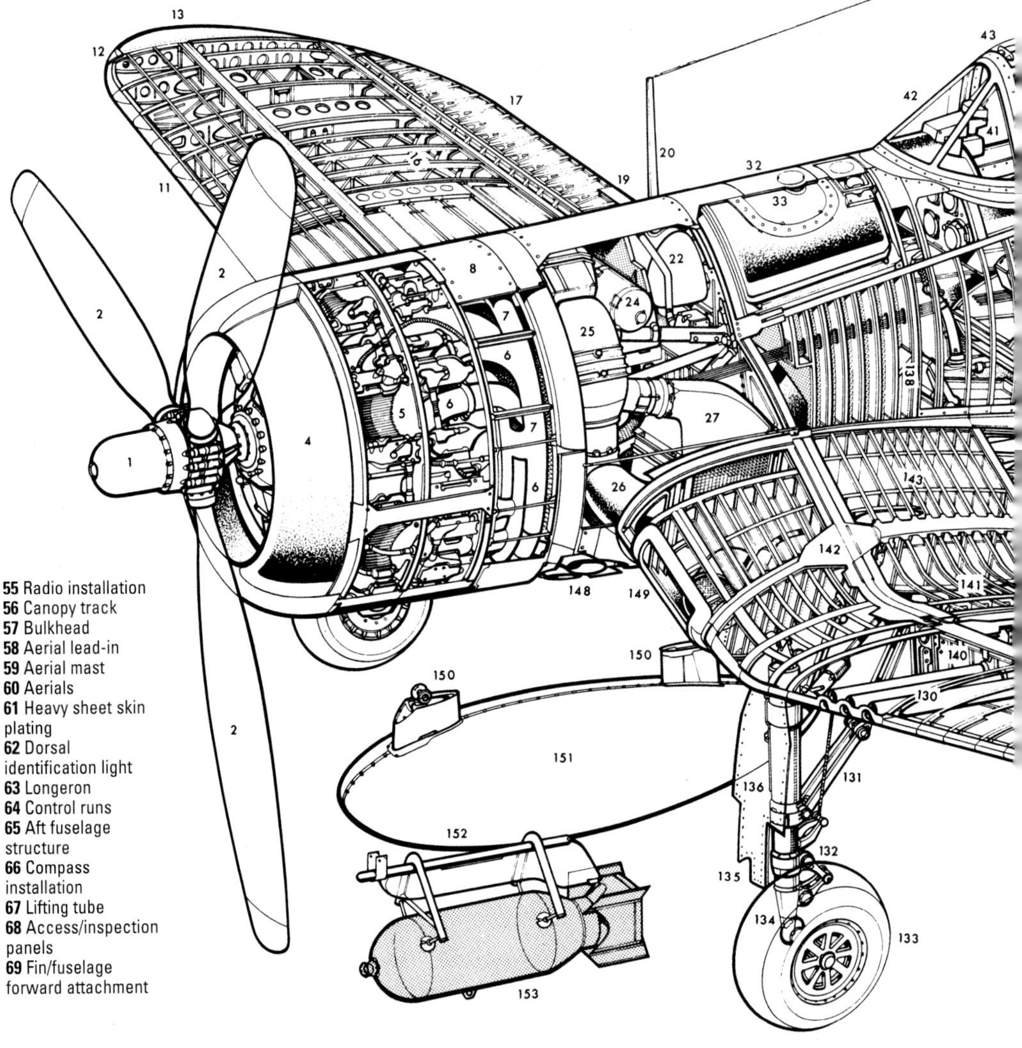

During the Korean War, aircraft from both VF-92 (based aboard CVA-45 USS *Valley Forge* from November 1952 to June 1953) and VF-94 (operating from CVA-47 USS *Philippine Sea* between December 1952 and August 1953) carried the tail code 'N'. The units' Corsairs shared their carrier decks with more modern F9F-2s and AD Skyraiders.

SPECIFICATION

F4U-1A Corsair

Dimensions

Length: 33 ft 4 in (10.16 m)
Height: 16 ft 1 in (4.9 m)
Wingspan: 41 ft (12.5 m)
Wing area: 314 sq ft (29.17 m^2)

Powerplant

One 2,000-hp (1492-kW) Pratt & Whitney R-2800-8W Double Wasp 18-cylinder radial piston engine

Weights

Empty: 8,982 lb (4074 kg)
Loaded: 14,000 lb (6350 kg)

Fuel

External fuel: provision for one 175-US gal (662-litre) centreline tank

Performance

Maximum speed at 19,900 ft (6065 m): 417 mph (671 km/h)
Maximum speed at sea level: 316 mph (509 km/h)
Initial climb rate: 2,890 ft (881 m) per minute
Service ceiling: 36,900 ft (11247 m)

Range

Maximum range: 1,015 miles (1633 km)

Armament

Six 0.50-in (12.7-mm) Browning MG53-2 machine-guns in the folding outer wing section (replaced in the F4U-1C by four 20-mm cannon), and (F4U-1D/Corsair II and later models) provision for a bombload of 2,000 lb (908 kg), comprising bombs up to the weight of 1,000-lb (454-kg) or up to eight 5-in (127-mm) HVARs

106 Tailwheel/arrester hook cylinder
107 Tailwheel retraction strut
108 Bulkhead attachment points
109 Fuselage skinning
110 Bulkhead frame
111 Elevator/rudder control runs
112 Entry hand/foothold
113 Hydraulically-operated flap inboard section
114 Wing fold line
115 'Flap gap' closure plate
116 Hydraulically-operated flap outboard section
117 Aileron balance tab (port only)
118 Aileron trim tab
119 Port aileron
120 Deck landing grip
121 Port wingtip
122 Port navigation light
123 Pitot head
124 Leading-edge ribs
125 Wing outer section structure
126 Ammunition boxes
127 Three 0.5-in (12.7-mm) Colt-Browning MG53-2 wing machine-guns with 400 rpg (inboard pair) and 375 rpg (outboard)
128 Wing fold outboard cylinder
129 Wing leading-edge unprotected integral fuel tank, capacity 62 US gal (235 litres), deleted from final 150 Corsair IIs
130 Machine-gun blast tubes
131 Mainwheel retraction strut
132 Torque links
133 Port mainwheel
134 Axle
135 Mainwheel leg fairing
136 Mainwheel oleo leg
137 Mainwheel leg pivot point
138 Undercarriage main spar attachment
139 Undercarriage actuating cylinder
140 Main spar fold point
141 Mainwheel well
142 Contoured main spar inboard section
143 All-aluminium wing centre section
144 Main spar/fuselage attachment
145 Blower radiator
146 Oil cooler
147 Engine supercharger intake duct
148 Exhaust stacks
149 Engine supercharger air intake
150 Auxiliary fuel tank centre-line attachment points
151 'Duramold' auxiliary drop tank capacity 175 US gal (662 litres)
152 Bomb attachment shackle (underwing inner section, (F4U-1D and Corsair II only))
153 Bombload, up to 1,000 lb (454 kg) each side (F4U-1D and Corsair II only)

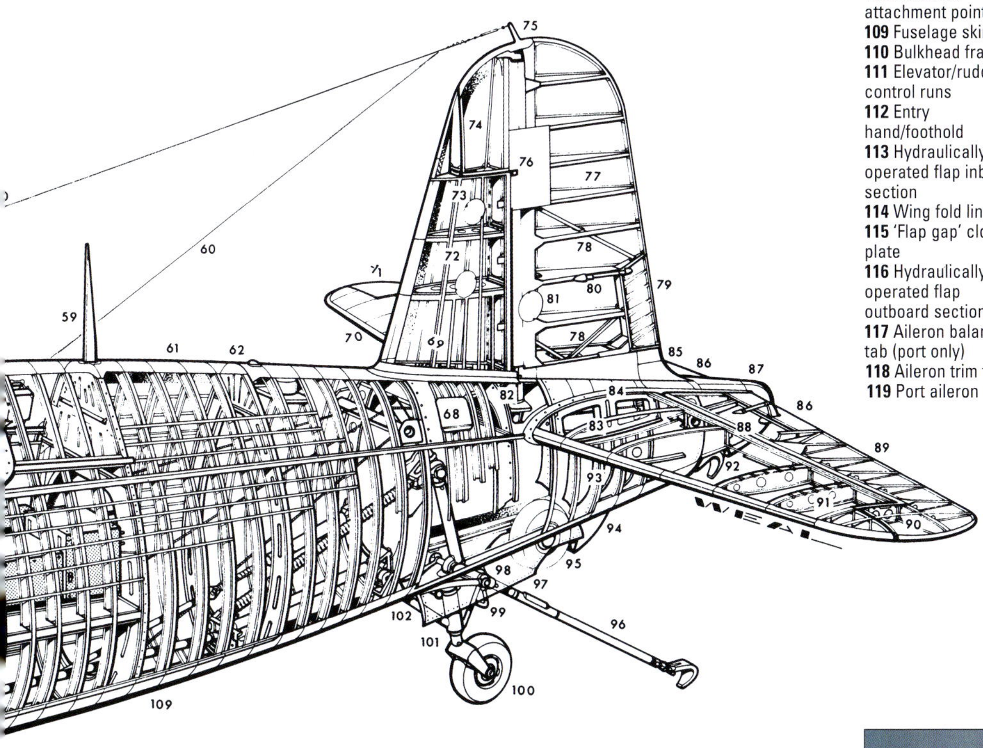

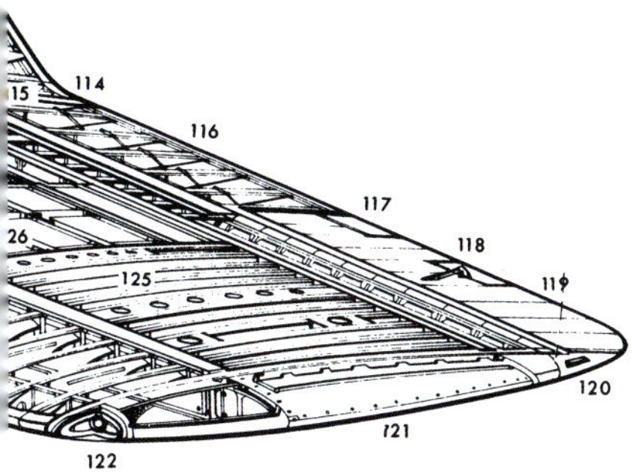

70 Starboard tailplane
71 Elevator balance
72 Fin structure
73 Inspection panels
74 Rudder balance
75 Aerial stub
76 Rudder upper hinge
77 Rudder structure
78 Diagonal bracing
79 Rudder trim tab
80 Trim tab actuating rod
81 Access panel
82 Rudder post
83 Tailplane end rib
84 Elevator control runs
85 Fixed fairing root
86 Elevator trim tabs (port and starboard)
87 Tail cone
88 Rear navigation light
89 Port elevator
90 Elevator balance
91 Port tailplane structure
92 Arrester hook (stowed)
93 Tail section frames
94 Fairing
95 Tailwheel (retracted)
96 Arrester hook (lowered)
97 Tailwheel/hook doors
98 Tailwheel/hook attachment/ pivot
99 Mooring/tie-down lug
100 Rearward-retracting tailwheel
101 Tailwheel oleo
102 Support strut
103 Arrester hook actuating strut
104 Aft/tail section bulkhead
105 Arrester hook shock absorber

A total of 312 USN F4Us was lost during the Korean conflict. The Corsair in its AU-1 guise soldiered on in US Navy Reserve service until 1957, and remained in Marines Reserve service a year longer. A small number of F4U-7 Corsairs remained in service with the Aéronavale until 1964.

F4U-5N Corsair

Wearing the 'WF' tailcode of VMF(N)-513 (nicknamed the 'Flying Nightmares') in white on top of overall midnight blue, and coded '4', this F4U-5N night-fighter carries colours and markings typical of fighters in the Korean theatre. The 'U-birds' of VMF(N)-513 played a vital role in the Korean War from an early stage. While the Pusan Perimeter was surrounded, the squadron's Corsairs helped other close support types to force the North Korean forces into resupplying their army by night. Then, augmenting a small number of F-82G Twin Mustangs, the Corsairs maintained their vigil under the cover of darkness, harassing enemy forces on the ground, and maintaining air superiority. The F4U-5N Corsairs were shorter-ranged than the first F-82 night-fighters to operate over Korea and, with no bases in Korea itself, the aircraft were initially forced to fly from Itazuke in Japan, their radius of action limited to just north of the Pusan Perimeter.

'Flying Nightmares' mission
A typical Korean operation for the F4U-5N pilot of VMF(N)-513 involved flying at tree-top height under the cover of darkness, following roads and other landmarks lit by moonlight. When a re-supply truck or convoy coming from the Chinese border was spotted, a quick wing-over would be followed by a strafing pass, attacking the target at close range due to the reduced visibility. The Corsairs often ran the gauntlet of heavy-calibre automatic AAA weapons, although tracer from these weapons on the ground also served to reveal targets to them.

Pusan Perimeter crossing
When the North Koreans crossed the 38th Parallel on the morning of Sunday 25 June 1950, the South Korean army quickly fell apart, with the hastily-assembled American reinforcements unable to stop the advancing troops. Falling back to the south coast, opposite southern Japan, the US and UN forces organised the defensive Pusan Perimeter. Aerial bombardment of the North Korean People's Army forced the enemy to operate during the night and in bad weather, when their activities began to come under attack from night-fighters, including Corsairs, behind the Perimeter.

Radar
The wing leading edge-mounted APS-19A radar superseded the crude APS-4 radar carried by the first F4U-2 night fighter. The APS-19A had an excellent range of some 80 miles (130 km) for ground-mapping, and an even longer range for picking up a ground beacon. In the interception role, the radar could pick up a fighter-sized target at ranges of up to three miles (4.8 km), although a detection range of two miles (3.2 km) was more usual. Initial problems were experienced when the radar sets were knocked out by the recoil from the wing cannon.

Cockpit and armament

The cockpit of the F4U-5N was covered by a bulged clear-view canopy, affording much better visibility than earlier Corsair variants, the first of which featured a lower 'birdcage' canopy. The raised clear-view canopy was first introduced on the Brewster-built F3A-1 and the Goodyear-built FG-1, and from the 689th F4U-1, although the latter retained two thin horizontal frames. Over Korea, the gloss midnight blue night-fighter finish with white codes worn by VMF(N)-513's Corsairs and Tigercats was later replaced with matt black and dull red codes in an effort to reduce conspicuity. The four wing-mounted 20-mm cannon carried a mixed load of 800 incendiary, high-explosive and armour-piercing rounds. Particularly effective weapons for attacks on resupply convoys were fragmentation bombs, six of which were typically carried on underwing racks. This example, however, carries a full load of eight 5-in (127-mm) HVARs. After this weapon was found to be too light to damage T-34 tanks, it was modified to 'Ram rocket' status, with 6.5-in (165-mm) anti-tank aerial rocket (ATAR) warheads.

Post-war improvements

Following the conclusion of the war in the Pacific theatre, development of the Corsair continued, with the first XF4U-5 (featuring uprated R-2800-32W powerplant and low-drag metal-skinned wings and tail) flying on 4 April 1946. Production F4U-5s and subsequent sub-variants also incorporated ailerons with spring tabs, and heated windshield, gun bays, and pitot head. A similar low-altitude version with the single row R-2800-83W engine and increased armour was the F4U-6, later designated AU-1.

Powerplant

The F4U-5 and F4U-5N were powered by a 2,850-hp (2126-kW) Pratt & Whitney R-2800-32W Double Wasp two-row 18-cylinder radial piston engine, with two-stage supercharger with twin supercharger inlets in the lower half of the cowling. Combined with aerodynamic improvements, the new engine gave the F4U-5 and F4U-5N dramatically improved performance, with a maximum speed of 470 mph (756 km/h) and a range of 1,120 miles (1802 km). In order to improve the pilot's visibility over the nose, the engine was drooped by 2.75°.

Yakovlev Yak-1/3/7/9

Yak-3

Cutaway key
1 Rudder trim tab
2 Rudder structure
3 Rudder post
4 Tail fin structure
5 Aerial attachment
6 Tail fin leading edge spar
7 Spar attachment points
8 Tail fin root fairing
9 Elevator control horns
10 Rudder lower hinge
11 Elevator torque tube
12 Rear navigation light
13 Elevator trim tab
14 Elevator structure
15 Tailplane construction
16 Tailwheel doors
17 Retractable tailwheel
18 Tailwheel oleo
19 Tailwheel well
20 Wheel-impact door-closure struts
21 Tailwheel retraction jack
22 Lifting tube
23 Tubular steel fuselage framework
24 Ventral former
25 Elevator control cables
26 Diagonal brace wires
27 Dorsal former
28 Decking
29 Aerial
30 Aerial attachment/ lead-in
31 Canopy fixed aft glazing
32 Armourglass screen
33 Canopy track
34 HF (RSI-6M) radio equipment
35 Accumulator
36 Equipment rack
37 Hydraulic reservoir
38 Ventral coolant radiator housing
39 Control rod linkage
40 Radiator bath aft fairing
41 Radiator
42 Seat support frame
43 Pilot's seat pan
44 Trim tab control console (port)
45 Padded (armoured) seat back
46 Switchbox
47 Aft-sliding cockpit canopy
48 Reflector sight
49 One-piece moulded armourglass windscreen
50 Instrument panel coaming
51 Control column
52 Instrument panel starboard console
53 Control linkage
54 Rudder pedal bar
55 Bulkhead
56 Frame
57 Gun support tray
58 Bracket
59 0.5-in (12.7-mm) ShVAK machine-gun (port and starboard)
60 Port flap
61 Guide rollers
62 Aileron push-rod control linkage
63 Aileron trim tab
64 Port aileron
65 Port wingtip
66 Port navigation light
67 Pitot tube
68 Forward spar
69 Port outboard fuel tank
70 Fuel filler cap

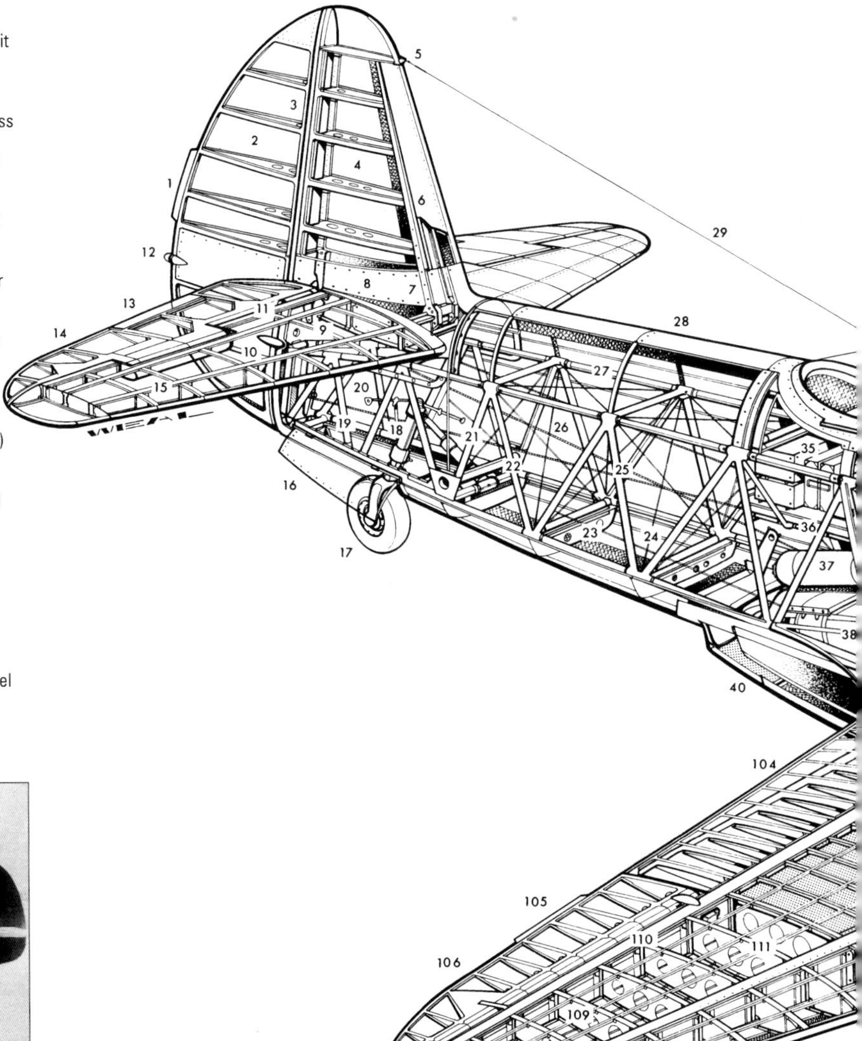

Although it introduced a number of refinements compared to the original Yak-7A fighter, the Yak-7B was still very much an interim machine. Problems afflicting the type ranged from the minor, including a slow climb rate; long take-off and landing runs and poor cockpit canopy; to the potentially lethal. In the latter category were the wing skins, which caused many crashes on separating from the wing.

***Above:** Yakovlev built and tested a single example of the **Yak-9P** (P – pushechnyi or cannon) in March 1945. The aircraft was lighter than the standard Yak-9, even though it had a 20-mm **ShVAK** cannon with 175 rounds of ammunition in place of the standard **UBS** machine-gun. The aircraft is shown here as it appeared during spin tests in the post-war period.*

***Left:** Having entered production late in 1940, the Yak-1 was already struggling to match the Messerschmitt **Bf 109F** by June 1941. It formed the basis of suceeding designs, however, including the Yak-9 which matured as a superb multi-role fighter and the superlative Yak-3 dogfighter, which was able to meet the late-mark **Bf 109** and formidable **Fw 190A** on more than equal terms.*

***Right:** A distinctive feature of the Yak-1 and the Yak-1M was their overlapping two-part main undercarriage doors, those of the -1M being considerably larger. A further distinguishing feature between the two subvariants concerned their canopies, that of the Yak-1 fairing directly into the fuselage spine and that of the -1M being raised above the cut down rear fusleage decking to provide the pilot with an improved all-round view. The aircraft illustrated was the first Yak-1 to emerge from the **GAZ-292** factory at Saratov.*

SPECIFICATION

Yak-9

Dimensions

Length: 28 ft 5 in (8.66 m)
Height: 8 ft 7 in (2.60 m)
Wingspan: 31 ft 11½ in (9.74 m)
Wing area: 184.61 sq ft (17.15 m²)

Powerplant

One M-105PF-1 V-12 liquid-cooled piston engine rated at 1,260 hp (940 kW)

Weights

Empty: 5,066 lb (2298 kg)
Loaded: 6,669 lb (3025 kg)

Performance

Maximum speed at sea level: 331 mph (533 km/h)
Maximum speed at 13,123 ft (4000 m): 371 mph (597 km/h)
Range: 457 miles (735 km)
Climb to 16,404 ft (5000 m): 5 minutes 30 seconds
Service ceiling: 32,808 ft (10000 m)
Take-off run: 1,247 ft (380 m)
Landing run: 1,640 ft (500 m)

Armament

One 20-mm ShVAK cannon mounted to fire through tthe propeller hub and provided with 120 rounds and one 0.5-in (12.7-mm) UBS fixed forward-firing machine-gun mounted on the port side of the upper rear fuselage and provided with 200 rounds. Later aircraft were fitted with a similar UBS installation to starboard

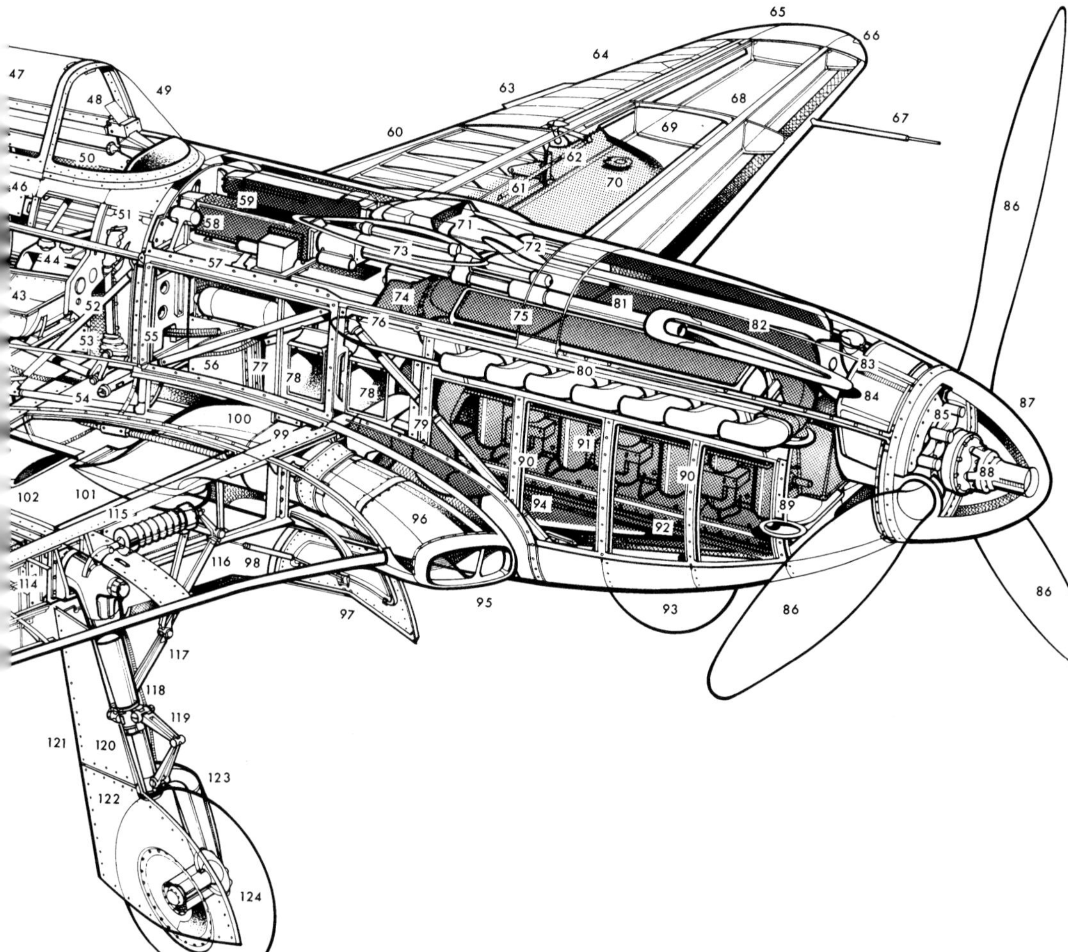

71 Supercharger intake scoop
72 Intake ducting
73 Gun cocking mechanism fairings
74 Supercharger housing
75 Cowling frame
76 Engine bearer/ firewall attachment
77 Oil tank
78 Ammunition boxes
79 Cowling aft frame
80 Exhaust stubs
81 Blast tubes
82 Gun muzzle troughs
83 Filler cap
84 Coolant header tank
85 Propeller pitch mechanism
86 VISh-105 variable-pitch metal propeller
87 Propeller spinner
88 Propeller hub
89 Auxiliary intake
90 Cowling attachment frames
91 VK-105 12-cylinder liquid-cooled Vee engine
92 Coolant ducting
93 Port mainwheel
94 Engine bearer
95 Oil cooler intake
96 Ducting
97 Mainwheel well door inboard section
98 Wheel-impact door-closure struts
99 Main spar cut-out
100 Oil cooler housing
101 Oil cooler outlet fairing
102 Radiator intake
103 Radiator grill
104 inset flap structure
105 Aileron trim tab
106 Aileron frame
107 Starboard wingtip
108 Starboard navigation light
109 Outboard wing ribs
110 Rear spar
111 Stringers
112 Starboard outboard fuel tank
113 Front spar
114 Undercarriage/spar attachment plate
115 Undercarriage retraction cylinder
116 Mainwheel leg well
117 Undercarriage downlock strut
118 Brake lines
119 Torque links
120 Mainwheel oleo leg
121 Mainwheel leg fairing plate
122 Mainwheel fairing plate
123 Axle fork
124 Starboard mainwheel

Yak-3

General Major Georgii Nefedovich flew this Yak-3 while leading the 303rd IAD. He fought over Kursk and the 3rd Ukrainian Front with the unit and in 1945 moved into combat over East Prussia. Nefedovich began his combat career over Spain, having volunteered to fight on the Republican side during the Civil War. Flying an I-15, he scored six kills before returning to the Soviet Union. Nefedovich scored a further three victories over China in 1938 and by the end of the Great Patriotic War his personal total stood at 23 kills.

Cockpit and handling

The Yak-3 cockpit was a primitive affair compared to that of contemporary Allied or, indeed, Axis types. Blind flying had to be done on primary instruments alone, there being no gyroscopic instruments provided for this purpose. In general, the Yak-3 was a difficult aircraft to fly, especially for the novice pilot. The aircraft had a high stalling speed, which meant constant attention during slow-speed flight, and would willingly drop a wing on the approach if speed was allowed to drop. The tendancy to swing on landing and take-off was also a constant problem and ground-loops were relatively frequent. Nevertheless, in the hands of a competent pilot the Yak-3 was probably the most effective air combat fighter in the world at the time, having less than half the weight of the majority of its western counterparts and easily being able to out manoeuvre the best the Luftwaffe could offer in close-in, high-*g* dogfighting.

Powerplant

Although the Yak-3 was built in 18 versions, these can be broken down into two basic forms, the 'frontal' variants and the 'high-altitude' machines. The former, like that illustrated, were powered by variants of the VK-105PF liquid-cooled inline piston engine, typically driving a VISh-105SV three-bladed propeller. For high-altitude operations the Yak-3 was equipped with variants of the VK-105PD engine, which featured two-stage supercharging and, in its Yak-3 application, usually drove a three-bladed VISh-105L-2 propeller. All production variants of the Yak-3 dispensed with the distinctive undernose oil cooler air intake of the Yak-1, the radiator for this purpose having been moved to an intake in the port wing root. In air-to-air combat the Yak-3 was the most effective of the Yak-1/3/7/9 family, thanks to its light weight, small size and considerable agility, and the repositioned oil cooler intake provided a useful recognition aid for Luftwaffe fighter pilots, allowing them to judge the calibre of the opposition before engaging.

Armament

Since the Yak-3 was designed as a light-weight fighter, it was used exclusively in air-to-air roles and was never given provision for air-to-ground weapons. The earliest production Yak-3s were armed with a single engine-mounted 20-mm UBS cannon with 200 rounds and a single 0.5-in (12-7-mm) ShVAK machine-gun with 120 rounds synchronised to fire through the propeller disc, but pilot criticism soon led to a second ShVAK being added. Subsequent variants introduced a range of different gun combinations, the most powerful being the one engine-mounted 37-mm cannon and twin synchronised 20-mm cannon arrangement of the Yak-3T.

Stucture
In producing the Yak-3, Yakovlev combined a fuselage based closely on that of the Yak-1, with the composite wing of the Yak-9. Compared to the Yak-1, the fuselage of the Yak-3 had a lower rear deck and the cockpit was covered by a canopy similar to that of the Yak-1M. The fuselage retained the plywood skinning over steel-tube framework construction of the Yak-1. The wing used a considerable amount of steel and light alloy, replacing much of the wood used in the Yak-1's wing, largely thanks to supplies of steel becoming available from the US. This wing structure caused many problems when first introduced into combat on the Yak-9, with the plywood wing skins becoming loose and, on occasion ripping off the airframe inflight. With much 'encouragement' from Stalin, Yakovlev eventually traced the problem to the nitro-cellulose dope used to protect the plywood from the elements: that being applied by one of the factories was not waterproof and after just a short time in the field the plywood had been weakened to the point where it was no longer able to remain attached to the wing.

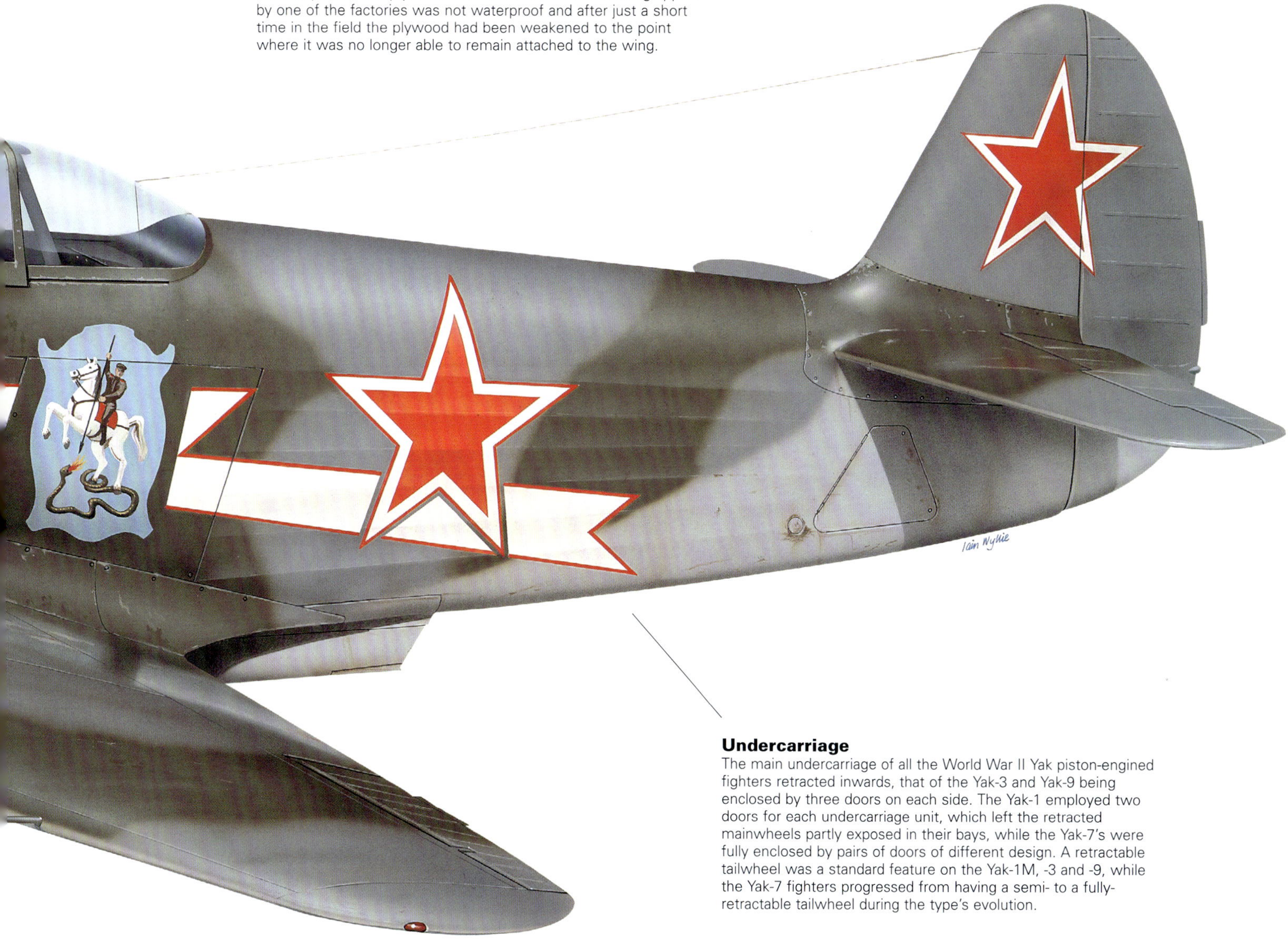

Undercarriage
The main undercarriage of all the World War II Yak piston-engined fighters retracted inwards, that of the Yak-3 and Yak-9 being enclosed by three doors on each side. The Yak-1 employed two doors for each undercarriage unit, which left the retracted mainwheels partly exposed in their bays, while the Yak-7's were fully enclosed by pairs of doors of different design. A retractable tailwheel was a standard feature on the Yak-1M, -3 and -9, while the Yak-7 fighters progressed from having a semi- to a fully-retractable tailwheel during the type's evolution.

Transport, Reconnaissance and Maritime Patrol Aircraft

4152

Consolidated PBY Catalina

This is one of a small number of conversions of PBY-4 and PBY-5 Catalinas to the mine warfare role. A special electricity-generating plant supplied a large current to cables inside the ring, generating a magnetic field to explode magnetic mines. The mine warfare task was a dangerous one, the aircraft having to fly over the minefield at low level.

PBY-5A Catalina

Cutaway key

1 Starboard tailplane
2 Tailplane leading-edge de-icing
3 Tail navigation light
4 Starboard fabric-covered elevator
5 Elevator tab
6 Rudder trim tab
7 Fabric-covered rudder construction
8 Tailcone
9 Elevator push-pull control rod
10 Rudder control horn
11 Tail mooring point
12 Lower fin structure integral with tail fuselage
13 Tailplane centre-section attachment
14 Upper fin construction
15 Aerial cables
16 Fin leading-edge de-icing
17 Port tailplane
18 Cooling air intake
19 Rear fuselage frame and stringer construction
20 Ventral tunnel gun hatch
21 0.3-in (7.62-mm) machine-gun
22 Fuselage skin plating
23 Target-towing reel
24 Flare launch tube
25 Rear fuselage bulkhead
26 Bulkhead door
27 0.5-in (12.7-mm) beam machine-gun
28 Starboard beam gun cupola
29 Cupola opening side window
30 Flexible gun mounting
31 Port beam gun cupola
32 Gunner's folding seat
33 Semi-circular gun platform
34 Walkway
35 Hull bottom V-frames
36 Wardroom bulkhead
37 Crew rest bunks
38 Wardroom
39 Starboard mainwheel
40 Hull planing bottom step
41 Planing bottom construction
42 Fuselage skin planing
43 Mainwheel housing
44 Hydraulic retraction jack
45 Telescopic leg strut
46 Fore and aft wing support struts
47 Wing mounting centre pylon construction
48 Pylon tail fairing
49 Starboard wing integral fuel tank, capacity 875 US gal (3312 litres)
50 Fuel jettison pipe
51 1,000-lb (454-kg) bomb
52 Smoke generator tank
53 Trailing-edge ribs
54 Fabric-covered trailing edge
55 Rear spar
56 Aileron trim tab
57 Starboard retractable wingtip float
58 Float support struts
59 Retraction linkage
60 Fabric-covered starboard aileron
61 Static discharge wicks
62 Wingtip aerial mast
63 Float up-lock
64 Float leg housing
65 Starboard navigation light
66 Leading-edge de-icing boot
67 Float retracting gear
68 Front spar
69 Wing rib/stringer construction
70 ASV radar aerial
71 Outer wing panel attachment joint
72 Wing lattice ribs
73 Bomb-carrier and release unit
74 Two 500-lb (227-kg) bombs
75 Leading-edge nose ribs
76 Position of pitot tube on port wing
77 Landing lamp
78 Landing lamp glare shield
79 Starboard engine nacelle fairing
80 Hydraulic accumulator
81 Engine oil tank
82 Fireproof bulkhead
83 Exhaust stub
84 Engine bearer struts
85 Detachable engine cowlings
86 Curtiss Electric three-bladed constant-speed propeller, 12-ft (3.66-m) diameter
87 Propeller hub pitch-change mechanism
88 Pratt & Whitney R-1830-92 Twin Wasp two-row radial engine
89 Aerial cable lead-in
90 D/F loop aerial
91 Oil cooler
92 Control runs through pylon front fairing
93 Pylon step
94 Engineer's control panel
95 Flight engineer's seat
96 Wing mounting fuselage main frame
97 Radio and radar control units
98 Cabin heater
99 Front cabin walkway
100 Port main undercarriage leg strut

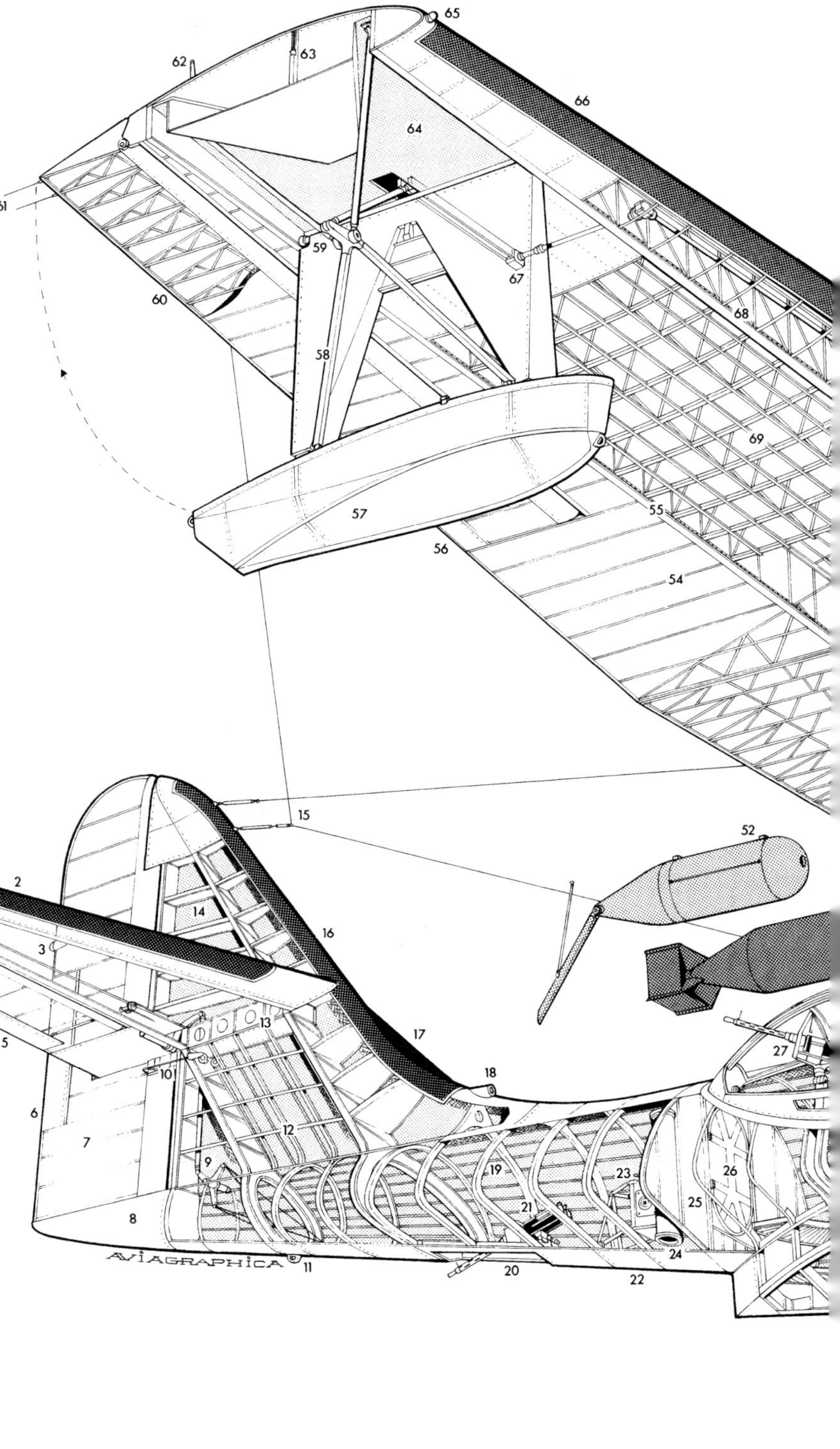

An original colour photograph taken early in the Aleutians campaign of 1942 shows a US Navy PBY-5 taking on supplies before slipping its moorings for a patrol. A pair of bombs can be seen under the wing (maximum load was four, each of 1,000 lb/ 454 kg). Defensive armament would be provided by two 0.3-in (7.62-mm) and two 0.5-in (12.7-mm) guns.

SPECIFICATION

Catalina Mk IB

Dimensions

Length: 65 ft 2 in (19.86 m)
Height: 17 ft 11 in (5.46 m)
Wingspan: 104 ft (31.70 m)
Wing area: 1,400 sq ft (130.06 m²)

Powerplant

Two 1,200-hp (895-kW) Pratt & Whitney R-1830 radial piston engines

Weights

Empty: 14,240 lb (6459 kg)
Maximum take-off: 27,080 lb (12283 kg)

Performance

Maximum speed at 10,500 ft (3200 m): 190 mph (306 km/h)
Cruising speed: 179 mph (288 km/h)
Service ceiling: 24,000 ft (7315 m)
Range: 4,000 miles (6437 km)

Armament

One 0.303-in (7.7-mm) machine-gun in bow and two 0.303-in (7.7-mm) machine-guns in each side blister and in ventral position, plus up to 4,000 lb (1814 kg) of bombs, torpedoes, smoke generators or depth charges

Painted black and tasked with harrying Japanese forces at night, the 'Black Cats' were specially selected Catalina squadrons. They were responsible for the sinking of thousands of tons of Japanese shipping during the Pacific War.

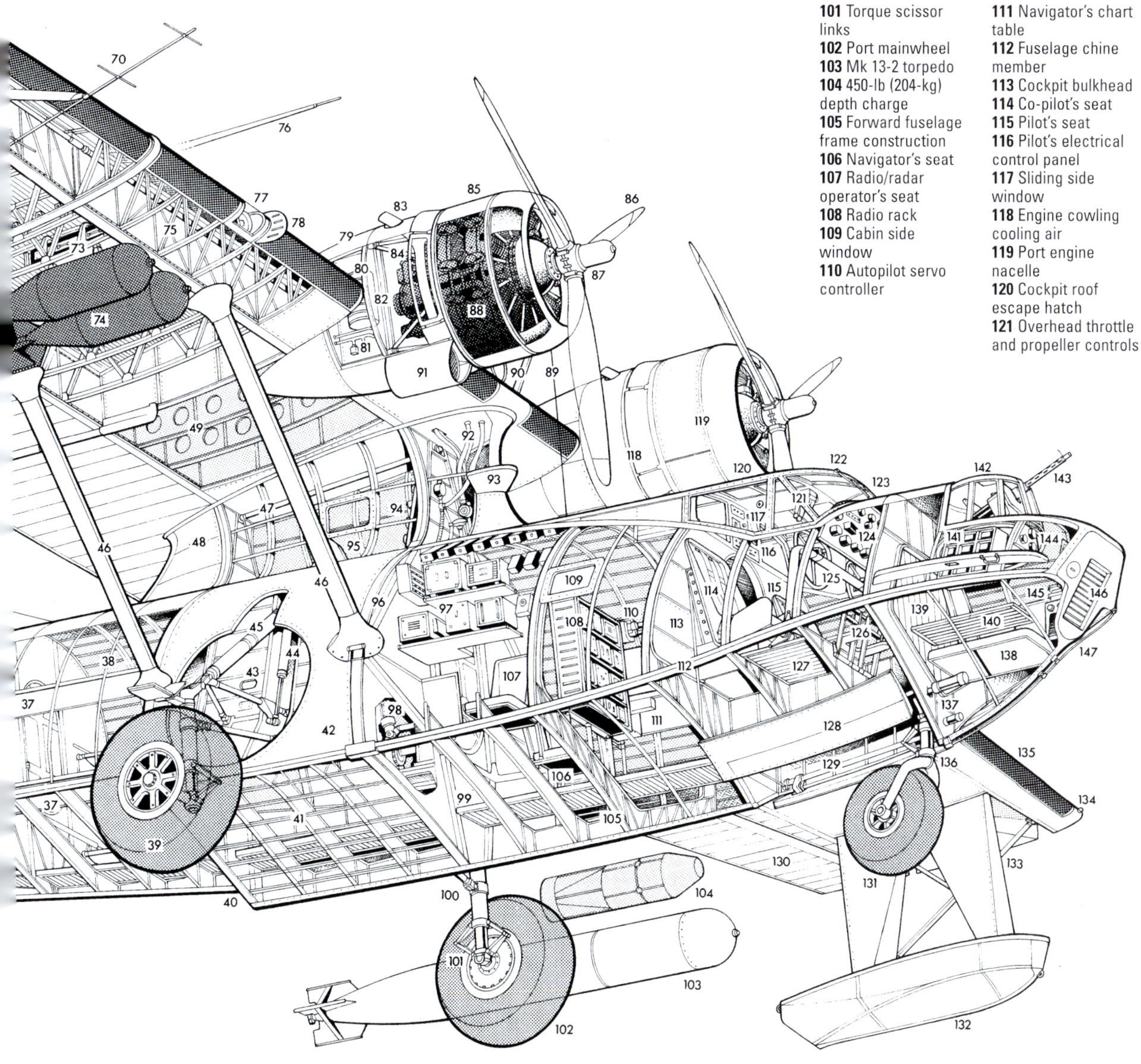

101 Torque scissor links
102 Port mainwheel
103 Mk 13-2 torpedo
104 450-lb (204-kg) depth charge
105 Forward fuselage frame construction
106 Navigator's seat
107 Radio/radar operator's seat
108 Radio rack
109 Cabin side window
110 Autopilot servo controller
111 Navigator's chart table
112 Fuselage chine member
113 Cockpit bulkhead
114 Co-pilot's seat
115 Pilot's seat
116 Pilot's electrical control panel
117 Sliding side window
118 Engine cowling cooling air
119 Port engine nacelle
120 Cockpit roof escape hatch
121 Overhead throttle and propeller controls
122 Windscreen wipers
123 Curved windscreens
124 Instrument panel
125 Control column yoke and handwheels
126 Rudder pedals
127 Cockpit flooring
128 Nose undercarriage hatch doors
129 Nosewheel bay
130 Port aileron
131 Nosewheel
132 Port retractable wingtip float
133 Float support struts
134 Port navigation light
135 Leading-edge de-icing boot
136 Nosewheel forks
137 Nose undercarriage retraction jack
138 Front gunner/bomb aimer's station
139 Curtained bulkhead
140 Gunner's footboards
141 Spare ammunition containers
142 Front rotating gun turret
143 0.3-in (7.62-mm) machine-gun
144 Bomb aimer's instrument panel
145 Drift sight
146 Bomb-aiming window with protective blind
147 Anchor cable

Catalina Mk IVA

No. 202 Sqn was formed in October 1914 as No. 2 Sqn, RNAS. For most of World War II, it was based at Gibraltar to secure the Strait and for patrols in the western Mediterranean and Atlantic. It moved to Ireland in September 1944 for anti-submarine patrols at night, maintaining a detachment at Sullom Voe in the Shetlands. Wartime equipment included the Saro London, Fairey Swordfish, Short Sunderland and the Catalina which, in Mk I, II and IV forms, served from May 1941 to June 1945. Today, the squadron flies Westland Sea King HAR.Mk 3As on search and rescue duties around the British coastline.

Leigh Light
An important aid for the Catalina's patrol duties was the high-illumination Leigh Light carried under the starboard wing – this denied surfacing submarines the cover of darkness. One notable attack took place in July 1944 when a Catalina of No. 210 Squadron operating out of Sullom Voe discovered U-boat *U-347*. As the Catalina made its first run over the U-boat, the depth charges failed to release. Flight Officer Cruickshank made a second pass, under intensive anti-aircraft fire, this time dropping the depth charges directly onto the U-boat, destroying it. However, the attack had been costly, with the navigator killed and Cruickshank badly injured. Nevertheless, he managed to return to base, where doctors treated him for 72 wounds. He was later awarded the VC for valour.

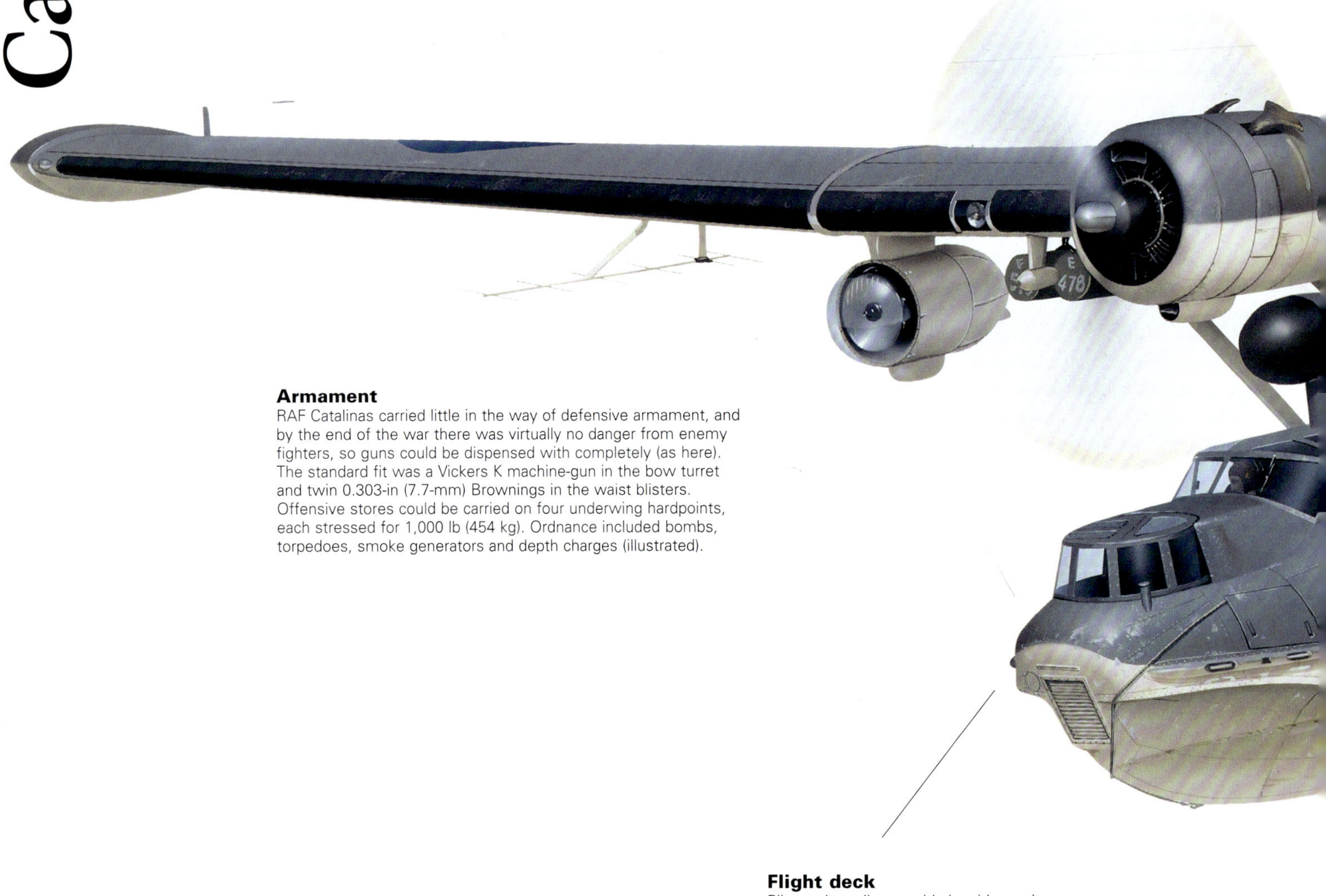

Armament
RAF Catalinas carried little in the way of defensive armament, and by the end of the war there was virtually no danger from enemy fighters, so guns could be dispensed with completely (as here). The standard fit was a Vickers K machine-gun in the bow turret and twin 0.303-in (7.7-mm) Brownings in the waist blisters. Offensive stores could be carried on four underwing hardpoints, each stressed for 1,000 lb (454 kg). Ordnance included bombs, torpedoes, smoke generators and depth charges (illustrated).

Flight deck
Pilot and co-pilot sat side-by-side on the flight deck and were provided with a roof escape hatch for emergency egress. The throttles and propeller controls were mounted in the overhead position.

Fuel
Fuel was held in integral tanks between the wing spars of the centre-section. This gave a range of around 2,350 miles (3780 km) with a full military load. Catalinas were renowned for their incredible endurance, which could be augmented in the Mk IVA with additional fuel tanks. Missions of up to 27 hours were undertaken.

Radar
The Catalina Mk IVA was fitted with ASV Mk II radar to spot surfaced U-boats. The antennas for this system were mounted under the wings. A centimetric surveillance radar was mounted in a large teardrop radome above the flight deck.

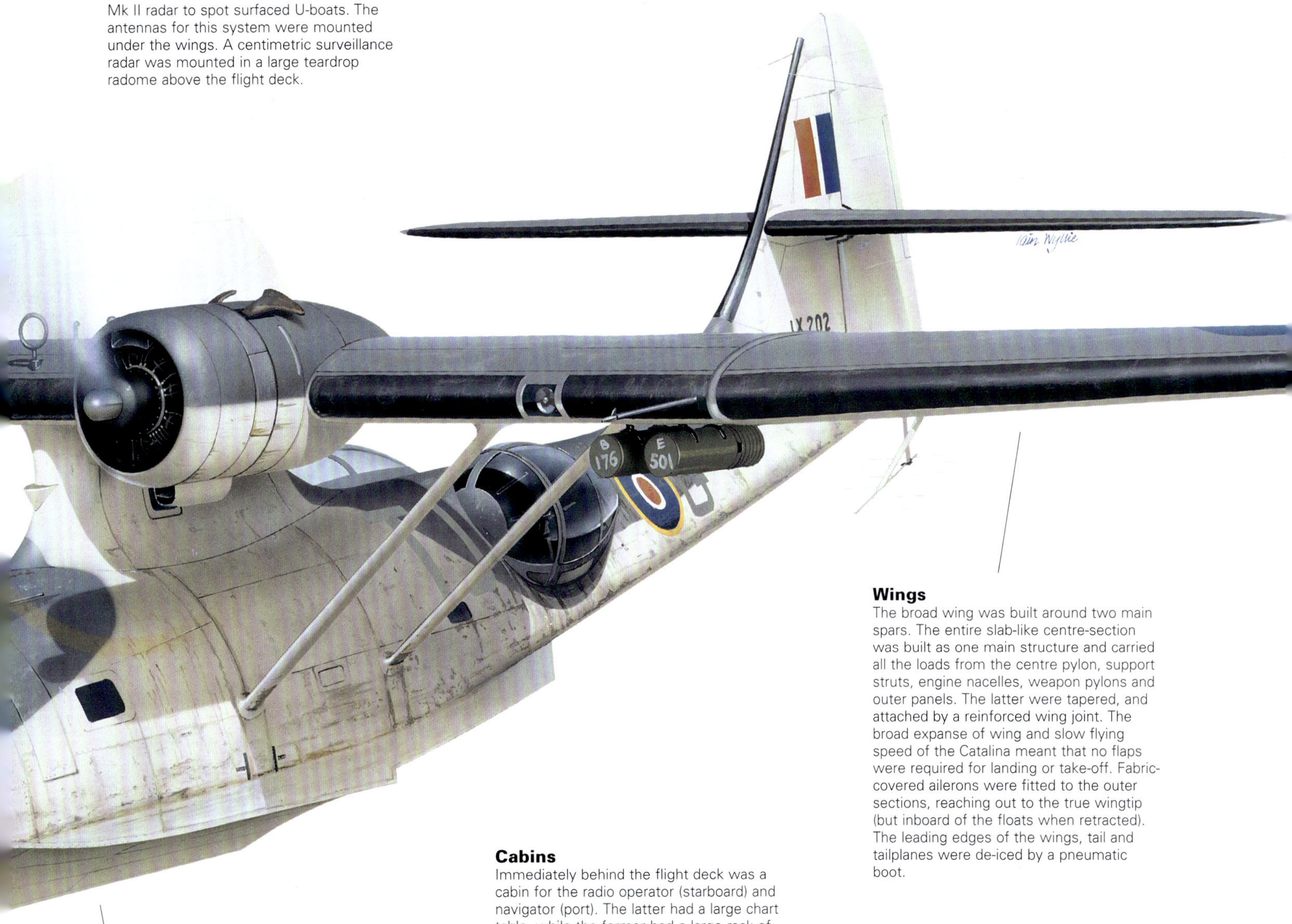

Wings
The broad wing was built around two main spars. The entire slab-like centre-section was built as one main structure and carried all the loads from the centre pylon, support struts, engine nacelles, weapon pylons and outer panels. The latter were tapered, and attached by a reinforced wing joint. The broad expanse of wing and slow flying speed of the Catalina meant that no flaps were required for landing or take-off. Fabric-covered ailerons were fitted to the outer sections, reaching out to the true wingtip (but inboard of the floats when retracted). The leading edges of the wings, tail and tailplanes were de-iced by a pneumatic boot.

Cabins
Immediately behind the flight deck was a cabin for the radio operator (starboard) and navigator (port). The latter had a large chart table, while the former had a large rack of radio equipment. Small windows gave some view of the outside world. A central walkway led through the entire hull from the flight deck to the rear, finally entering the aft cabin, where the gunners/observers worked. A semi-circular platform allowed these crewmen to swing guns through a wide arc, while the bulged waist blisters enabled them to carry out surveillance over a large area during patrols. The flight engineer's station was situated in the centre of the aircraft, projecting up into the wing support pylon. Small windows were incorporated into the pylon. Behind the flight engineer, under the wing trailing edge, was a wardroom, complete with crew rest bunks.

Hull
The stepped hull was all-metal, with one main deck and an unusual semi-circular upper section. The Catalina Mk IVA was the same as the US Navy's PBY-5, and was strictly a flying-boat. The Mk IIIA was a PBY-5A with amphibian gear, which retracted to a semi-enclosed position on the fuselage sides.

Douglas C-47 Skytrain/Dakota

South Africa's isolation during the years of apartheid meant that the SADF had to run its fleet of Dakotas long after other first line air forces had retired the type. But the tough and reliable Douglas design gave excellent service in the bush wars in Namibia and Angola.

Dakota Mk IV

Cutaway key

1 Hinged nose cone, access to instruments and controls
2 Rudder pedals
3 Instrument panel
4 Windscreen de-icing fluid spray nozzle
5 Starboard propeller
6 Windscreen panels
7 Co-pilot's seat
8 Engine throttles
9 Control column
10 Cockpit floor level
11 Access panels to control cable runs
12 Pilot static tubes
13 Aerial cables
14 Propeller de-icing fluid tank
15 Pilot's seat
16 Cockpit bulkhead
17 Cockpit roof escape hatch
18 Whip aerial
19 Starboard landing/taxiing lamp
20 Windscreen de-icing fluid tank
21 Starboard baggage compartment
22 Electrical fuse panel
23 Crew entry door
24 ADF loop aerial housing
25 Life raft stowage
26 Port baggage compartment
27 Main cabin bulkhead
28 Radio operator's seat
29 Air scoop
30 Heating and ventilating system heat exchangers
31 Astrodome observation hatch
32 Starboard outer wing panel
33 Pneumatic leading-edge de-icing boot
34 Starboard navigation light
35 Starboard aileron
36 Aileron cable controls
37 Trim tab
38 Trim tab control gear
39 Flap control shaft
40 Starboard outer flap
41 Fuselage frame and stringer construction
42 Centre fuselage main frames
43 Centre wing section corrugated inner skin
44 Port main fuel tank, capacity 210 US gal (794 litres)
45 Port auxiliary fuel tank capacity 201 US gal (761 litres)
46 Wing spar attachments
47 Flap hydraulic jack
48 Centre section flap
49 Floor beam construction
50 Cabin window panels
51 Window panel grommets for small arms attachments
52 Paratroop seating, 28 paratroops
53 Starboard emergency exit window
54 Port emergency exit window
55 Cabin lining panels
56 Overhead heating and ventilating duct
57 Rear cabin frames
58 Fuselage skin plating
59 Rear cabin bulkhead
60 First aid kit
61 Access door to tail controls
62 Fin root fillet
63 Starboard tailplane
64 Starboard elevator
65 Fin leading-edge pneumatic de-icing boot
66 Fin construction
67 Aerial cables
68 Rudder aerodynamic balance
69 Hinge post
70 Rudder construction
71 Fabric covering
72 Rudder trim tab
73 Trim tab control gear
74 Rudder and elevator control horns
75 Fuselage tail fairing
76 Elevator trim tab
77 Port elevator construction
78 Fabric covered elevator
79 Leading-edge pneumatic de-icing boot
80 Tailplane construction
81 Tailplane attachment joint
82 Rudder stop cables
83 Tailplane centre section
84 Tailwheel
85 Shock absorber leg strut
86 Tailwheel mounting plate
87 Tailwheel strut
88 Rudder and elevator control cables
89 Tail fuselage joint frame
90 Toilet
91 Rear freight door
92 Forward freight door
93 Paratroop/passenger door
94 Fuselage stringer construction
95 Freight floor
96 Wing root trailing-edge fillet
97 Inboard split trailing-edge flap
98 Flap shroud construction
99 Fuel filler caps

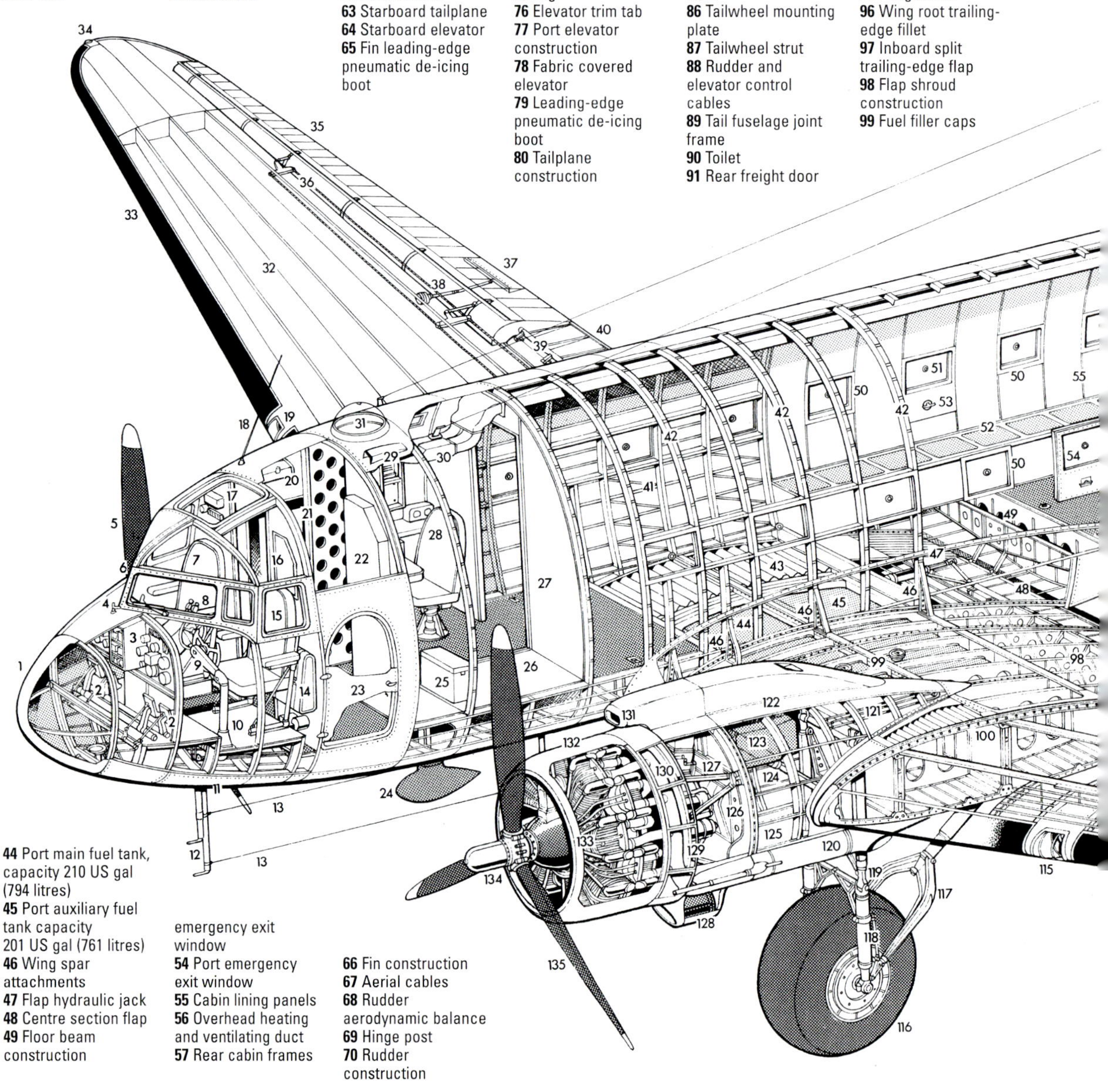

Central America is one of the last bastions of the C-47 in military service. Still capable of providing air forces like that of Honduras with effective service, it will soldier on well into the 21st Century, after more than 60 years in the front line.

SPECIFICATION

Douglas C-47A Skytrain

Dimensions

Wingspan: 95 ft 6 in (29.11 m)
Wing area: 987 sq ft (91.69 m²)
Length: 63 ft 9 in (19.43 m)
Height: 17 ft (5.18 m)

Powerplant

Two Pratt & Whitney R-1830-92 Twin Wasp 14-cylinder air-cooled two-row radial piston engines, each rated at 1,200 hp (895 kW)

Typical Weights

Empty: 17,865 lb (8103 kg)
Normal loaded: 26,000 lb (11793 kg)
Maximum take-off: 31,000 lb (14061 kg)
Maximum internal load: 10,000 lb (4536 kg)
Maximum fuel load: 4,824 lb (2188 kg)

Performance

Maximum speed: 230 mph (370 km/h) at 8,800 feet (2680 m)
Cruising speed: 185 mph (298 km/h) at 10,000 feet (3050 m)
Service ceiling: 24,000 ft (7315 m)
Maximum range: 1,600 miles (2575 km)
Rate of climb: 1,170 ft (357 m) per minute

Accommodation/Payload

Crew: three (pilot, co-pilot, radio operator/navigator)
Accommodation: 14 passengers as sleeper transport/ambulance; 21 passengers as standard airliner; 28 fully equipped troops; 31 passengers as high-density airliner

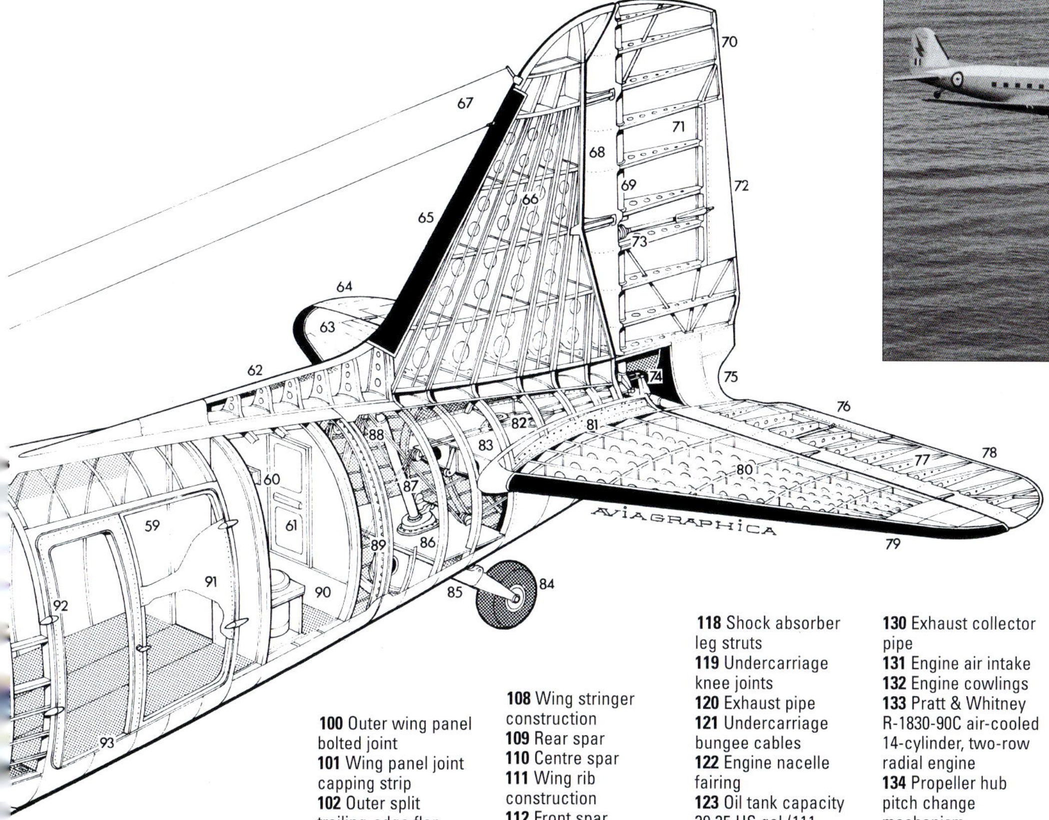

An Australian Dakota flies over the Straits of Malacca as it heads towards the RAAF base at Butterworth. The RAAF used Dakotas into the 1970s, until they were replaced by Lockheed C-130 Hercules and DHC-4 Caribou.

100 Outer wing panel bolted joint
101 Wing panel joint capping strip
102 Outer split trailing-edge flap
103 Port aileron
104 Aileron fabric covering
105 Detachable wingtip joint rib
106 Port navigation light
107 Leading-edge pneumatic de-icing boot
108 Wing stringer construction
109 Rear spar
110 Centre spar
111 Wing rib construction
112 Front spar
113 Leading-edge nose ribs
114 Leading-edge stringers
115 Port landing/taxiing lamp
116 Port mainwheel
117 Main undercarriage rear strut
118 Shock absorber leg struts
119 Undercarriage knee joints
120 Exhaust pipe
121 Undercarriage bungee cables
122 Engine nacelle fairing
123 Oil tank capacity 29.25 US gal (111 litres)
124 Undercarriage retraction jack
125 Mainwheel well
126 Engine fireproof bulkhead
127 Engine bearer struts
128 Oil cooler
129 Cooling air exit flaps
130 Exhaust collector pipe
131 Engine air intake
132 Engine cowlings
133 Pratt & Whitney R-1830-90C air-cooled 14-cylinder, two-row radial engine
134 Propeller hub pitch change mechanism
135 Hamilton Standard constant speed three-bladed propeller

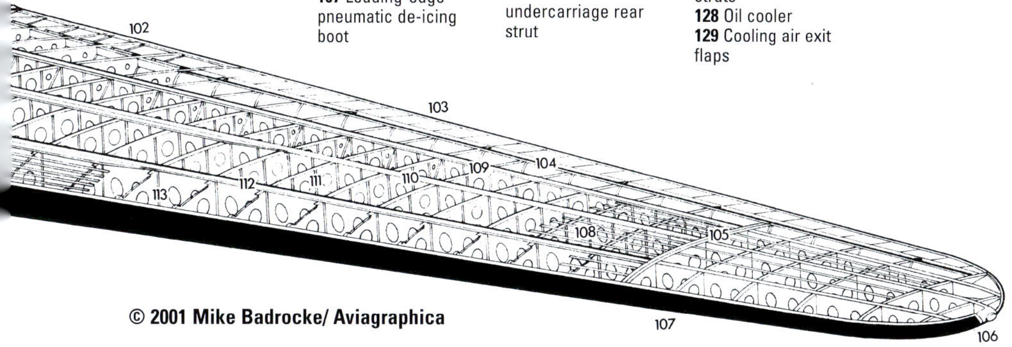

South African 'Parabats' drop from a C-47. The SADF used its Dakotas in combat as late as 1987, dropping troops in support of operations against SWAPO guerrilla bases in southern Angola.

C-47A Skytrain

The Douglas DC-3 has had a longer service history than any other aircraft. In its military form as the C-47, it contributed mightily to the Allied victory during World War II, and when large numbers of the ten thousand or more C-47s built came on to the market after the war they were quickly snapped up by civil and military operators alike. Simple, rugged and dependable, the C-47 has been in front-line service for more than fifty years: this example was flown by the Grupo de Transportes Aéreos Militares of the Fuerza Aérea Colombiana in the 1990s.

Cockpit and crew
The C-47 has a flight crew of two, though in military service it also carried a radio operator/ navigator in a compartment behind the cockpit. The sharply-angled windscreens contrast with the gentle curves elsewhere on the aircraft.

Powerplant and nacelle
This C-47 Skytrain is fitted with two Pratt & Whitney Twin Wasp R-1830-90C air-cooled, 14-cylinder two-row radial engines, developing 1,050 hp (783 kW) at 2,550 rpm. Some variants were fitted with Wright Cyclone nine-cylinder radials of similar size and power. The engines are mounted into streamlined nacelles, each of which also contains a 33.25 US gal (126-litre) oil tank. The wheels are retracted hydraulically into the rear of the nacelle, but are not fully enclosed when retracted. They can be raised or lowered in 15 seconds.

Colour scheme
Most Colombian C-47s flew in a natural metal finish, but the demands of the endless struggle against leftist guerrillas and the powerful drug lords of the cocaine cartels saw at least two aircraft painted in this two-tone camouflage scheme.

Rudder
The DC-3/C-47 has an unpowered rudder of extremely broad chord, which gives very good control authority at low speeds. This is essential when delivering troops or supplies by parachute, which is part of the reason that the C-47 remained a useful military aircraft for more than half a century. The rudder itself has a simple trailing-edge trim tab for fine control.

Cabin and exits
Military variants of the long-serving Douglas transport have stronger cabin floors and rear fuselages than their civil DC-3 counterparts, primarily to take heavier loads. Similarly, they are usually fitted with a large two-part cargo door at the port rear of the fuselage, with a smaller door in one leaf to allow the deployment of airborne troops. Airliner versions of the DC-3 were designed to carry between 18 and 21 passengers: the military C-47 could carry four and a half tonnes of cargo, or 28 fully-equipped paratroops in canvas or aluminium folding bucket seats mounted along the sides of the cabin.

Tail fairing
This Colombian C-47 would have been fitted with a glider towing hook in its earlier life as a World War II aircraft, but as with most such aircraft postwar the hook was removed and the space was faired over.

Horizontal tail
The C-47 has a large horizontal tail, fitted with simple fabric-covered trimming tabs on the ailerons. The leading edge of the tailplane is fitted with a de-icer boot – a rubber strip which can be inflated and deflated to dislodge any potentially dangerous build-up of ice.

Static line
There are two types of military parachutist. Special forces often jump from high altitude, pulling their own ripcords, but most paratroopers are dropped at very low level (under 975 ft/300 m). Their parachutes are opened automatically by a static line clipped to the aircraft, which is left streaming out of the door after the trooper has jumped.

Fieseler Fi 156 Storch

Shortly after the fall of Paris to the German Blitzkrieg in 1940, this Storch landed in the Place de la Concorde, demonstrating its excellent STOL characteristics.

Fi 156 Storch

Cutaway key
1 Fixed tab
2 Rudder construction
3 Rudder balance
4 Navigation light
5 Tailplane bracing strut
6 Tailplane tab
7 Elevator construction
8 Wooden tailplane construction
9 Lower surface elevator slat
10 Elevator bell crank
11 Tailplane pivot mounting
12 Fin construction
13 Fin leading edge
14 Elevator fabric covering
15 Port tailplane
16 Elevator balance
17 Tailplane trim jack
18 Tailskid strut cuff
19 Tailskid
20 Tailskid support strut
21 Welded steel-tube fuselage framework
22 Tailplane trim cables
23 Elevator push-pull control rods
24 Rudder push-pull control rod
25 Fuselage fabric covering
26 Zip-fastened access panel
27 Stowage locker door
28 Gun sight
29 7.9-mm MG 15 machine-gun (provision for three 50-round magazines)
30 LL-K machine-gun swivel mounting
31 Cartridge case collector box
32 Rear cabin bulkhead
33 Cabin roof construction
34 Radio aerial (for FuG XVII)
35 Stub wing spar attachment
36 Flap operating rod
37 Port flap
38 Wing root fuel tank (16.28 Imp gal/74 litre capacity)
39 Rear wing bracing strut
40 Strut attachment rib
41 Wing fabric covering
42 Aileron balance tab
43 Port aileron
44 Port wing tip
45 Navigation light
46 Leading edge fixed slat
47 Aileron control rod
48 Searchlight
49 Pitot head
50 Forward wing bracing strut
51 Flap operating jack
52 Port undercarriage framework
53 Access step
54 Windscreen
55 Compass
56 Downward vision windows
57 Trim control
58 Control column
59 Instrument panel shroud
60 Instrument access panel
61 Engine cowlings, detachable
62 Oil tank filler

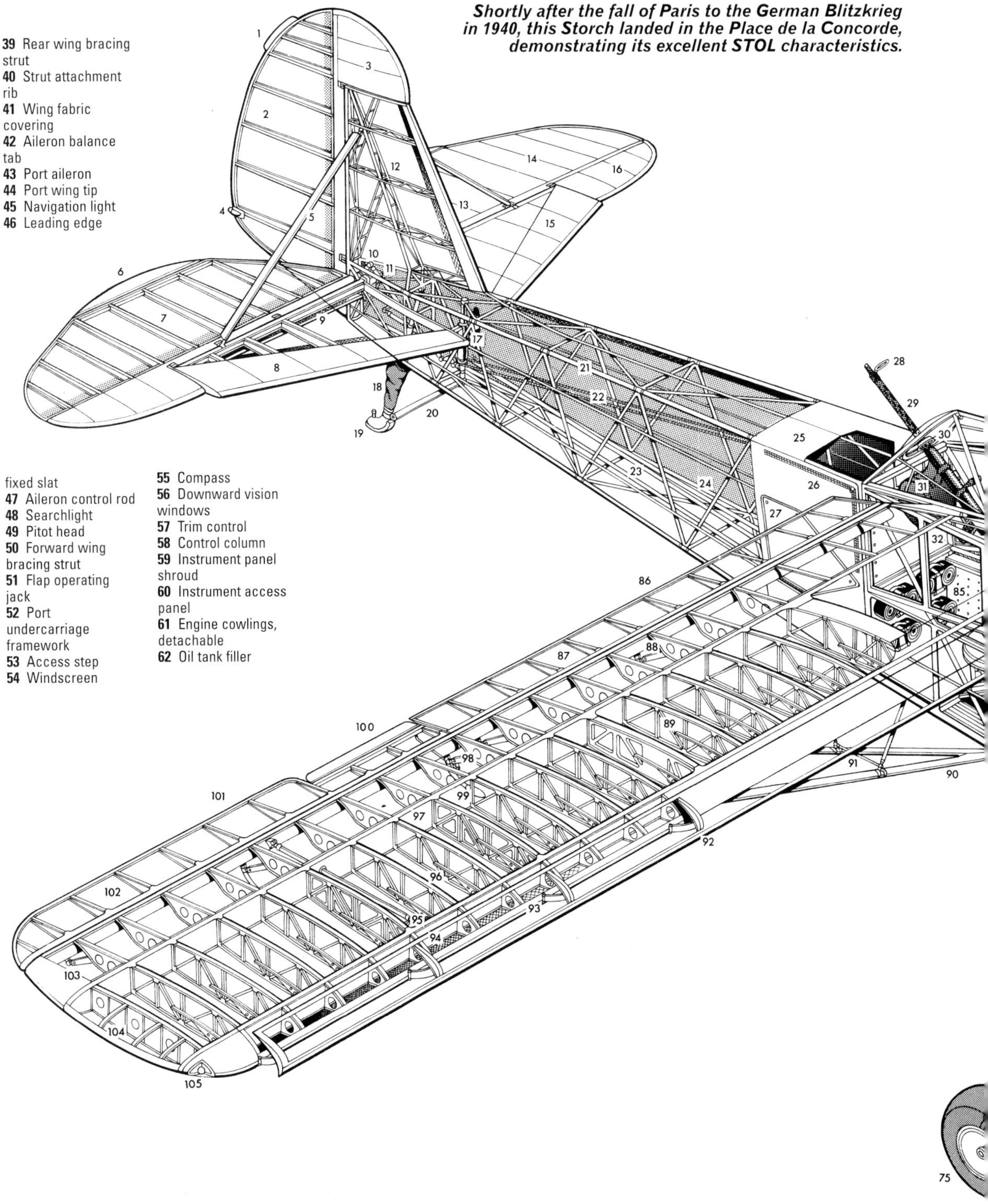

Fi 156 V4 (the fourth prototype), D-IFMR, had ski undercarriage and a drop tank fitted for test purposes.

SPECIFICATION

Fieseler Fi 156C-2

Type

Two-seat army co-operation/ reconnaissance aircraft

Powerplant

One 240-hp (179-kW) Argus As 10C-3 8-cylinder inverted Vee piston engine

Performance

Maximum speed 109 mph (175 km/h) at sea level, economical cruising speed 81 mph (130 km/h), service ceiling 15,090 ft (4600 m), range 239 miles (385 km)

Weights

Empty 2,050 lb (930 kg); maximum take-off 2,921 lb (1325 kg)

Dimensions

Wingspan 46 ft 9 in (14.25 m), length 32 ft 5¾ in (9.90 m), height 10 ft (3.05 m), wing area 279.87 sq ft (26.00 m2) Armament: one rear-firing 0.31-in (7.92-mm) machine-gun on pivoted mount

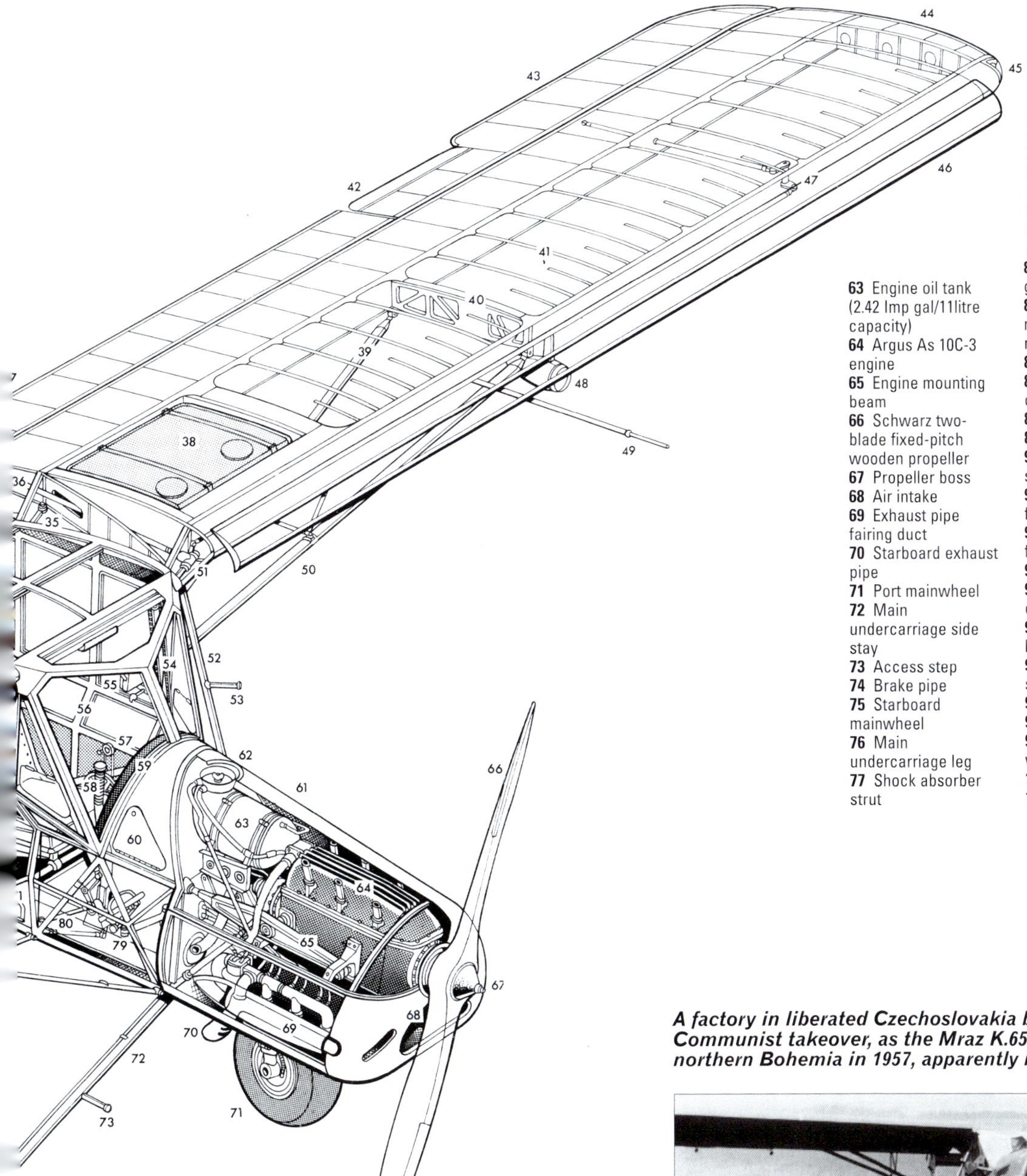

63 Engine oil tank (2.42 Imp gal/11litre capacity)
64 Argus As 10C-3 engine
65 Engine mounting beam
66 Schwarz two-blade fixed-pitch wooden propeller
67 Propeller boss
68 Air intake
69 Exhaust pipe fairing duct
70 Starboard exhaust pipe
71 Port mainwheel
72 Main undercarriage side stay
73 Access step
74 Brake pipe
75 Starboard mainwheel
76 Main undercarriage leg
77 Shock absorber strut
78 Undercarriage mounting framework
79 Rudder pedal
80 Control rod linkage
81 Entry step
82 Cabin door
83 Pilot's seat
84 Observer's/ gunner's seat
85 Ammunition magazines (two of 50-round capacity)
86 Starboard flap
87 Plywood flap construction
88 Flap hinge
89 Lattice ribs
90 Wing bracing Vee struts
91 Strut supporting framework
92 Leading edge fixed slat
93 Slat attachment
94 Leading edge construction
95 Aileron control rod linkage
96 Fabric bracing strips
97 Wooden main spar
98 Aileron hinge
99 Aileron balance weight
100 Balance tab
101 Starboard aileron
102 Plywood aileron construction
103 Aileron outer hinge
104 Wing tip construction
105 Navigation light

A factory in liberated Czechoslovakia briefly built the Fi 156 prior to the Communist takeover, as the Mraz K.65 Cap. These aircraft are seen in northern Bohemia in 1957, apparently in use as glider tugs.

Focke Wulf Fw 200 Condor

Chocks strain under the power of four BMW-Bramo radials as this Fw 200C-1 is run up prior to a test flight. Note the bomb racks outboard of the outer engines – a feature of all but the first two C-1s.

Fw 200C-4/U3

Cutaway key
1 Starboard navigation light
2 Wing skinning
3 Starboard aileron
4 Aileron trim tabs
5 Outboard mainspar
6 Aileron control run
7 Wing ribs (centre section)
8 Wing ribs (forward section)
9 Wing dihedral break point
10 Starboard flap (outer section)
11 Starboard flap (centre section)
12 Starboard flap (inner section)
13 Wing fuel tank covers
14 Inboard mainspar structure
15 Starboard outer oil tank
16 Multiple exhaust stubs
17 Cooling gills
18 Starboard outer nacelle (angled)
19 Three-blade VDM controllable-pitch metal-bladed propeller
20 Propeler boss
21 Carburettor air intake
22 Auxiliary fuel tank (66 Imp gal/300 litre capacity)
23 Starboard inner nacelle
24 FuG 200 Hohentwiel search radar array (port antenna omitted for clarity)
25 Nose D/F loop
26 Nose bulkhead
27 Rudder pedals
28 Hand-held 0.51-in (13-mm) MG 131 machine-gun (D-Stand)
29 Lotfe 7D bomb sight fairing
30 Ventral gondola side windows (gondola offset to starboard)
31 Rear dorsal gunner's take- offseat
32 Pilot's circular vision port
33 First pilot's seat
34 Sliding windscreen panel
35 Co-pilot's seat (co-pilot also served as bomb- aimer)
36 Flight deck entry
37 Arc-of-fire interrupter gate
38 Cabin air inlet (starboard side only)
39 Hydraulically-operated Fw 19 turret mounting single 0.31-in (7.9-mm) MG 15 machine-gun (A-Stand)
40 Gunner's seat
41 Ammunition racks (A-Stand)
42 Bulkhead
43 Radio operator's rectangular vision port
44 Ventral gondola entry hatch
45 Radio operator's station (A-Stand gunner's station)
46 Ammunition racks (D-Stand)
47 Ammunition racks (D-Stand)
48 Ventral gondola centre section (with maximum capacity of one 198- Imp gal/900-litre armoured fuel tank or 12 110-lb/50-kg bombs)
49 Underfloor control runs
50 Cabin window stations (staggered two to port and three to starboard)
51 Underfloor structure
52 Fuselage oil tank
53 De-icing fluid reservoir
54 Aerial mast
55 Five main fuselage fuel tanks (canted)
56 Mainspar fuselage carry-through structure
57 Rear ventral gunner's take-off seat
58 Upper fuselage longeron
59 Mainframe
60 Cabin ventilators/air extractors
61 Fuselage side walls
62 Ammunition racks (C-Stand)
63 Second radio operator's take-off seat
64 Strengthened fuselage frame
65 Dorsal D/F loop
66 Starboard 0.31-in (7.9-mm) MG 15 machine-gun (F-Stand)
67 Beam gunners' take-off seats
68 Bulkhead
69 Dorsal aft gunner's position (B-Stand)
70 Dorsal glazing
71 Ammunition racks (B-Stand)
72 Hinged canopy section
73 MG 15 machine-gun 0.31-in (7.9-mm calibre)
74 Rear fuselage frames
75 Starboard tailplane
76 Endplate-fin balance
77 Starboard elevator
78 Elevator hinge
79 Elevator tab
80 Tailfin front spar structure
81 Tailfin structure
82 Rudder balance
83 Rudder construction
84 Electrically-operated rudder trim tab (upper section)

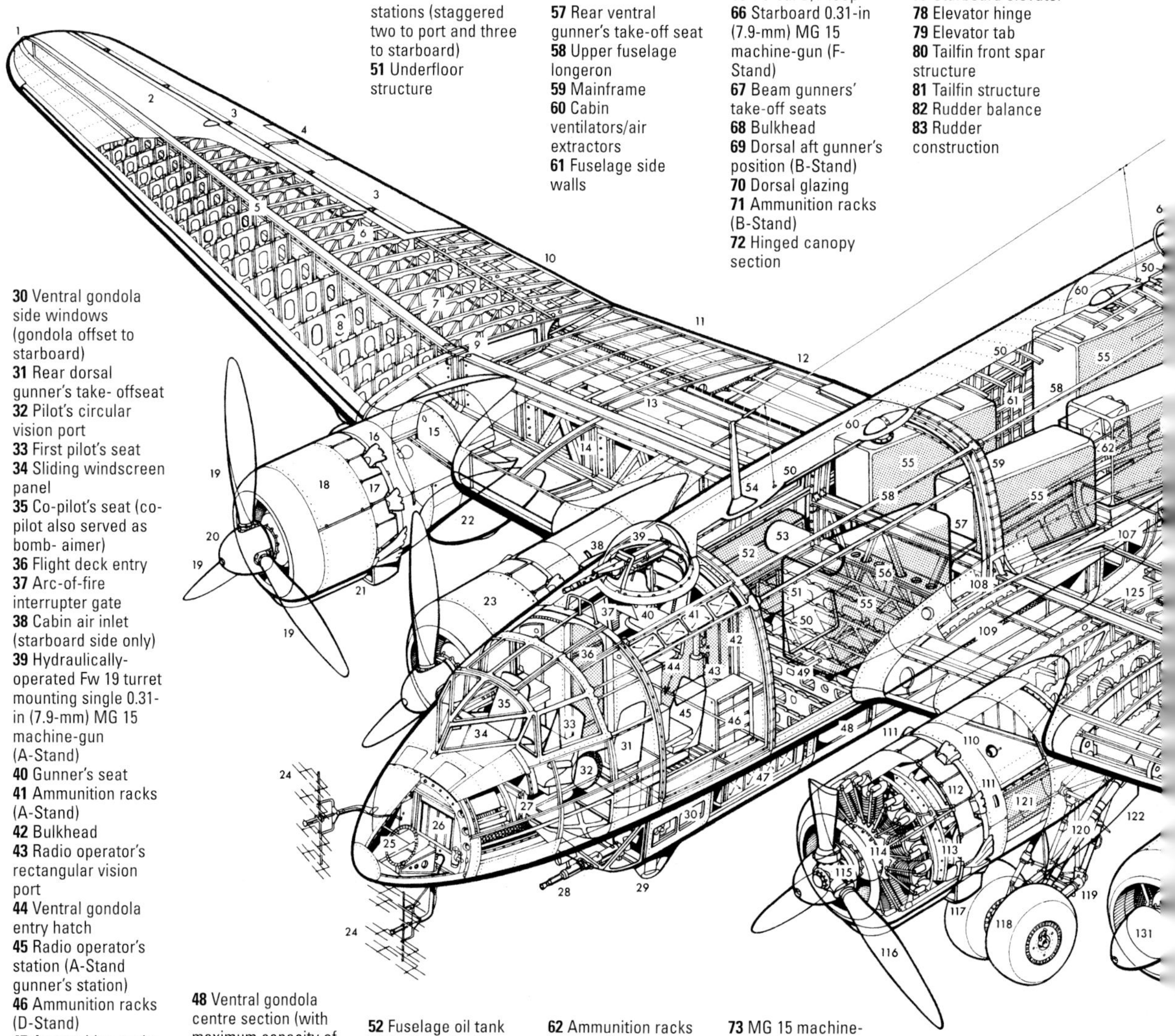

A close-up view of an Fw 200C-3/U2 shows details of the redesigned engine nacelles made necessary by the variant's new BMW-Bramo Fafnir engines, the Fw 19 forward turret reintroduced in this variant and the Condor's complex main undercarriage. Also visible is the ventral gondola which, in the C-3/U2, housed a Lotfe 7D bomb sight and (though omitted from this aircraft) an MG 131 0.51-in (13-mm) machine-gun.

SPECIFICATION

Fw 200C-3/U4 Condor

Dimensions

Wing span: 107 ft 9 in (32.85 m)
Wing area: 1,290.10 sq ft (119.85 m²)
Length: 76 ft 11 in (23.45 m)
Height: 20 ft 8 in (6.30 m)

Powerplant

four BMW-Bramo 323R-2 Fafnir nine-cylinder radial piston engines, each rated at 1,200 hp (895 kW)

Typical weights

Empty: 37,490 lb (17005 kg)
Maximum take-off: 50,057 lb (24520 kg)

Performance

Maximum speed: 224 mph (360 km/h)
Cruising speed: 208 mph (335 km/h)
Service ceiling: 19,685 ft (6000 m)
Range: 2,212 miles (3560 km)
Endurance: 14 hours

Armament

four 13-mm (0.51-in) MG 131 machine-guns in dorsal and beam positions, and one MG 131 or one 20-mm MG 151 cannon in forward ventral gondola; maximum bomb load of 2100 kg (4,630 lb) comprising two 500-kg (1,102-lb), two 250-kg (551-lb) and 12 50-kg (110-lb) bombs

This was the fate which befell a number of Condors after months of punishing low-level flying and violent manoeuvres when avoiding Flak and/or fighters. The type's airframe was never sufficiently strengthened for the military role, an 'achilles heel' becoming apparent in the rear spar of the mainplane. This proved liable to catastrophic failure, resulting in wing failure (above) or a broken back on landing (below).

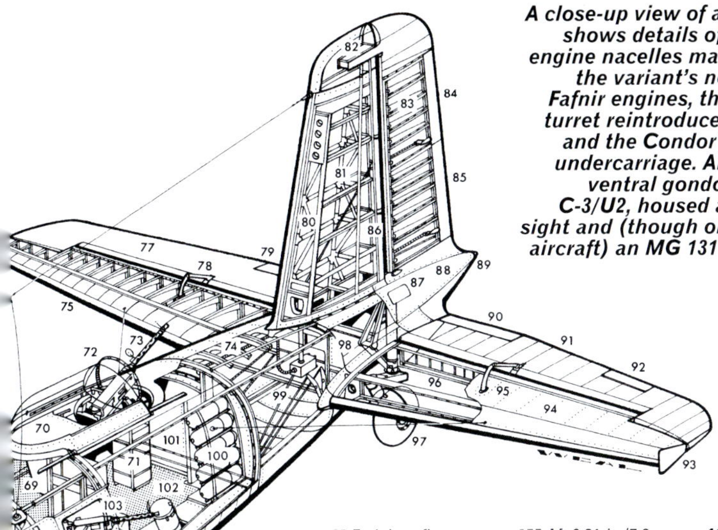

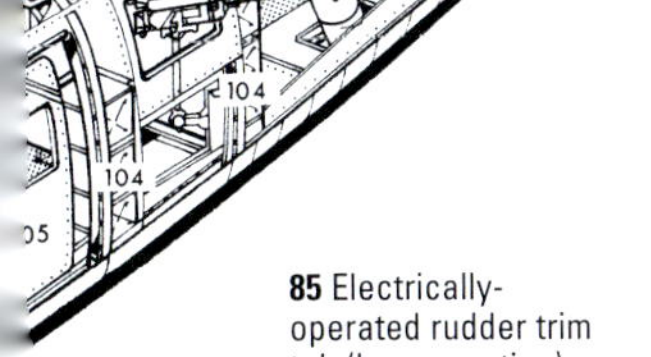

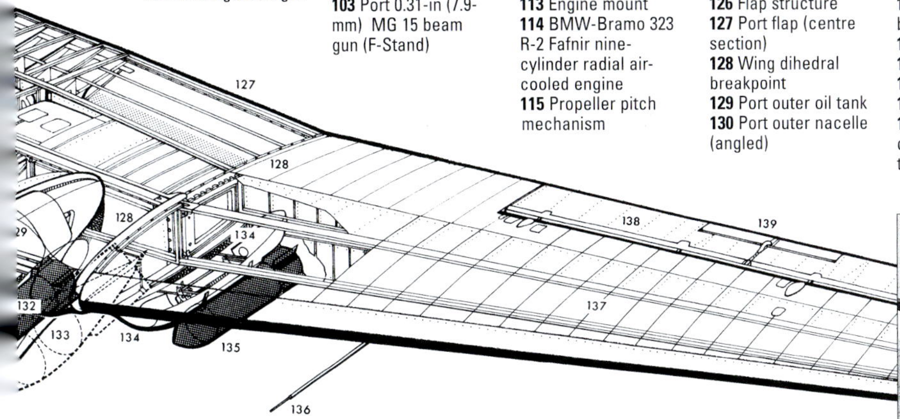

85 Electrically-operated rudder trim tab (lower section)
86 Rudderpost
87 Tail wheel mechanism access panel
88 Tail cone
89 Aft navigation light
90 Elevator tab
91 Port elevator
92 Electrically-operated elevator tab (port only)
93 Endplate-fin balance
94 Port tailplane
95 Elevator hinge
96 Tailplane
97 Forward-retracting tailwheel
98 Tailwheel retraction mechanism
99 Control runs
100 Oxygen bottles
101 Aft bulkhead
102 Chute for Schwan D/F buoys, Lux light-buoys or flares
103 Port 0.31-in (7.9-mm) MG 15 beam gun (F-Stand)
104 Ammunition racks (F-Stand) – starboard racks identical
105 Entry door
106 Aft 0.31-in (7.9-mm) MG 15 ventral gun (C-Stand)
107 Ventral gondola side windows
108 Main fuselage/wing attachment points
109 Ventral weapons/overload fuel bay
110 Port inner nacelle
111 Multiple exhaust stubs
112 Cooling gills
113 Engine mount
114 BMW-Bramo 323 R-2 Fafnir nine-cylinder radial air-cooled engine
115 Propeller pitch mechanism
116 Three-blade VDM controllable-pitch metal-bladed propeller
117 Carburettor air intake
118 Twin mainwheels
119 Forward-retracting hydraulically-operated main undercarriage member
120 Retraction jack
121 Mainwheel well
122 Mainwheel door
123 Wing structure
124 Main spar
125 Wing fuel tanks
126 Flap structure
127 Port flap (centre section)
128 Wing dihedral breakpoint
129 Port outer oil tank
130 Port outer nacelle (angled)
131 Propeller boss
132 Semi-recessed 551-lb (250-kg) bomb beneath outboard nacelle
133 Position of 1102-lb (500-kg) bomb on outboard nacelle rack (external)
134 Port underwing bomb rack
135 551-lb (250-kg) bomb
136 Pitot head
137 Wing skinning
138 Port aileron
139 Aileron trim tabs
140 Electrically-operated aileron trim tab (port only)

The Fw 200C-4/U1 was a one-off armed VIP transport version employed by SS commander Heinrich Himmler. Equipped to accommodate 11 passengers, the aircraft had a private compartment for Himmler with an armour-plated seat and an escape hatch in the floor. The aircraft is seen here in RAF markings during post-war evaluation at RAE Farnborough.

Fw 200C-1 Condor

F8+AH of I./KG 40, IV Fliegerkorps, Luftflotte 3, was based at Bordeaux-Mérignac, France, in 1940. In addition to KG 40s' badge, this Condor carries a name in white on its nose cone and kill markings on its fin. This particular aircraft was flown by Obertleutnant Edgar Petersen, the commander of KG 40 and architect of much of the Condor's considerable success over British shipping during its first month of maritime operations.

Radar
Early Fw 200C-4s had FuG Rostock search radar, served by antennas on the nose and outer wings, but this was soon replaced by FuG 200, with the nose antenna arrays. A few aircraft had both Rostock and Hohentwiel, the Rostock having greater range and wider search angle but longer minimum range.

Crew
The addition of an extra gunner in the Fw 200C-3/U4 and subsequent versions brought the crew complement to seven, having been raised from five in the Fw 200C–1 and C-2 to six in the C-3. The basic five-man crew consisted of a pilot and co-pilot, with a flight engineer/gunner, a navigator/bombardier (who also doubled as radio operator/gunner) and a rear dorsal gunner

Bombload
The Fw 200C-1 carried four 250-kg (551-lb) SC 250 bombs on its armed reconnaissance missions. These were carried externally, two under the outboard engine nacelles and two on racks under the wings. The gondola accommodated a similar-sized cement bomb, which was used for calibrating the bombsight or assessing ballistics immediately before the main weapons were released.

Defensive armament
The first four basic Fw 200C-0s were unarmed transports, but the next six had defensive armament and bomb racks for the maritime reconnaissance role. The defensive armament consisted of a single 0.31-in (7.9-mm) MG 15 machine-gun in the vestigial turret above and behind the flight deck, with two similar weapons firing from a downward hatch and from the glazed fairing above the rear fuselage. The C-1 replaced the ventral MG 15 with an offset gondola, in the nose of which was a 20-mm MG FF on a flexible mounting, and with an MG 15 in its tail. The turret above the fuselage was replaced by a fixed cupola with an MG 15 on a flexible mounting. The C-3 replaced the cupola with a powered turret, and introduced two MG 15s behind sliding beam panels, while the C-3/U1 introduced an HDL 151 turret with a 15-mm MG 151 cannon. The gondola's MG FF was replaced by another MG 151. The C-3/U2 and U4 reintroduced the Fw 19 forward upper turret, while the C-3/U3 had an EDL 131 turret with a 0.51-in (13-mm) MG 131, and with another MG 131 in the aft dorsal position. The C-4, C-6 and C-8 reintroduced the high drag HDL 151 turret and had MG 131s in the aft dorsal and beam positions.

Powerplants
The Fw 200C was powered by the same 830-hp (620-kW) BMW 132H air-cooled, nine-cylinder piston engines as its airline progenitor, the Fw 200B-2, although the nacelles were lengthened and the aircraft received long-chord cowlings. The Fw 200C-3 and subsequent versions were powered by the BMW-Bramo 323R-2 Fafnir, rated at 1,000 hp (745 kW) for take off, or 1,200 hp (894 kW) with water-methanol injection. The Fw 200C-2 introduced low drag, cut-down outboard engine nacelles, although C-6 and C-8 missile carriers had deeper outboard nacelles.

Focke Wulf Fw 189 Uhu

Everything about the Fw 189 was slender, especially the wings and tail booms. Despite this, it was an immensely strong aircraft, able to take large amounts of battle damage. The latter was a vital asset in the low-level, over the battlefield environment in which the Uhu operated.

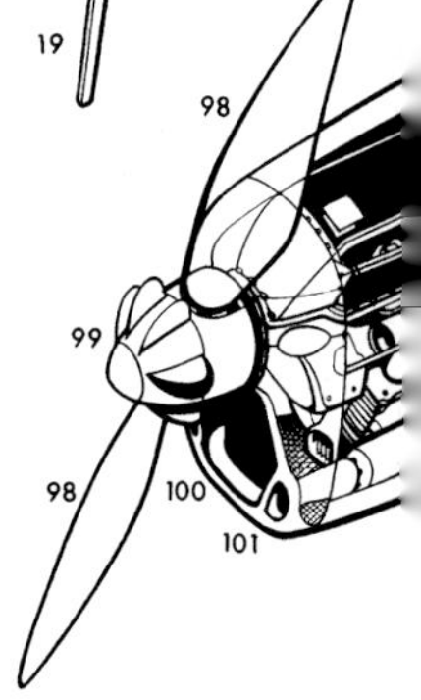

Fw-189 Uhu

Cutaway key
1 Starboard navigation light
2 Aileron control linkage (outer and inner)
3 Starboard aileron
4 Aileron tab
5 Starboard outer flap control linkage
6 Pitot tube
7 ETC 50/VIIId underwing rack fairings
8 Two 50-kg (110-lb) SC 50 bombs
9 Papier-maché 'screamers' attached to bomb fins
10 Wing centre/outer section join
11 Starboard engine nacelle
12 Air intake
13 Argus two-bladed controllable-pitch propeller
14 Pitch control vanes
15 Oil cooler intake
16 Engine air intake
17 FuG 212 Lichtenstein C-1 radar array (fitted to night fighter adaptation)
18 Starboard mainwheel
19 Ventral radio mast
20 Optically flat nose panels
21 Rudder pedals
22 GV 219d bomb sight
23 Control column
24 Bomb switch panel
25 Pilot's ring-and-bead sight (for fixed wing-root machine-guns)
26 Padded overhead instrument panel
27 Navigator's swivel seat
28 Throttle levers
29 Pilot's seat
30 Mainspar carry-through
31 Centre-hinged two-piece canopy hatch
32 Turnover bar with attached plasticised anti-glare curtain
33 Radio equipment
34 Shell collector box
35 Centre-section camera well (one RB 20/30, RB 50/30, RB 21/18 or RB 15/18 camera)
36 Canvas shell collection chute
37 Dorsal turret
38 MG 81Z twin 7.9-mm machine-gun
39 MG 151 (15-mm) fixed cannon in 'schräge Musik' installation (fitted to night fighter adaptation)
40 Starboard tailboom
41 Rudder and elevator control cables
42 Ammunition stowage (dorsal position)
43 Entry handholds
44 Centre-section flap below crew nacelle
45 Wing-root gun access panel (raised)
46 Rear turret-cone drive motor
47 Rear gunner's two-piece quilted pad
48 Ammunition stowage (rear position)
49 Rear canopy opening
50 MG 81Z twins 7.9-mm machine-guns (trunnion mounted)
51 Revolving Ikarin powered cone turret
52 Field or fire cut out
53 Aft glazing
54 Tailboom mid-section strengthening frame
55 Starboard tailfin
56 Starboard rudder
57 Rudder tab
58 Elevator construction
59 Tailplane forward spar
60 Elevator tab
61 Tailplane construction
62 Tailwheel hinged (two piece) door
63 Tailwheel (swivelling)
64 Tailwheel retraction mechanism
65 Tailwheel well (offset to port)
66 Tailfin construction
67 Rudder tab
68 Rear navigation light
69 Tail bumper
70 Tailboom frames
71 Tailboom upper longeron
72 Mid-section strengthening frame
73 Tail surface control cables
74 External stiffening strake (upper and lower)
75 Master compass
76 Wing-root fairing
77 Port outer flap construction
78 Aileron tab
79 Aileron construction
80 Port navigation light
81 Wing stringers (upper shell)
82 Lower shell wing inner skin stringers
83 Two-piece shaped wing ribs
84 Mainspar structure
85 Mainspar/boom attachment point
86 Rear spar/boom attachment point
87 Port fuel tank (24.2 Imp gal/110 litres)
88 Centre section one-piece flap
89 Wing walkway
90 Fixed 7.9-mm MG 17 machine-gun

With a completely redesigned fuselage nacelle, the Fw 189B was intended as a five-seat trainer. Ten of the Fw 189B-1 production aircraft were delivered before Fw 189A production began.

The first V1 prototype took to the air in July 1938, with Kurt Tank himself at the controls. The aircraft, registered D-OPVN, differed little from the production aircraft which followed.

SPECIFICATION

Focke-Wulf Fw 189A-1

Type

Two-seat short-range reconnaissance aircraft

Powerplant

Two 465-hp (347-kW) Argus As 410A-1 12-cylinder inverted Vee piston engines

Performance

Maximum speed 208 mph (335 km/h); cruising speed 196 mph (315 km/h); service ceiling 22,965 ft (7000 m); range 416 miles (670 km)

Weights

Empty 6,185 lb (2805 kg); maximum take-off 8,708 lb (3950 kg)

Dimensions

Wingspan 60 ft 41/2 in (18.40 m); length 39 ft 51/2 in (12.03 m); height 10 ft 2 in (3.10 m); wing area 409.04 sq ft (38.00 m2)

Armament: two flexible 0.31-in (7.92-mm) MG 15 machine-guns, two 0.31-in (7.92-mm) MG 17 machine-guns and four 110-lb (50-kg) bombs

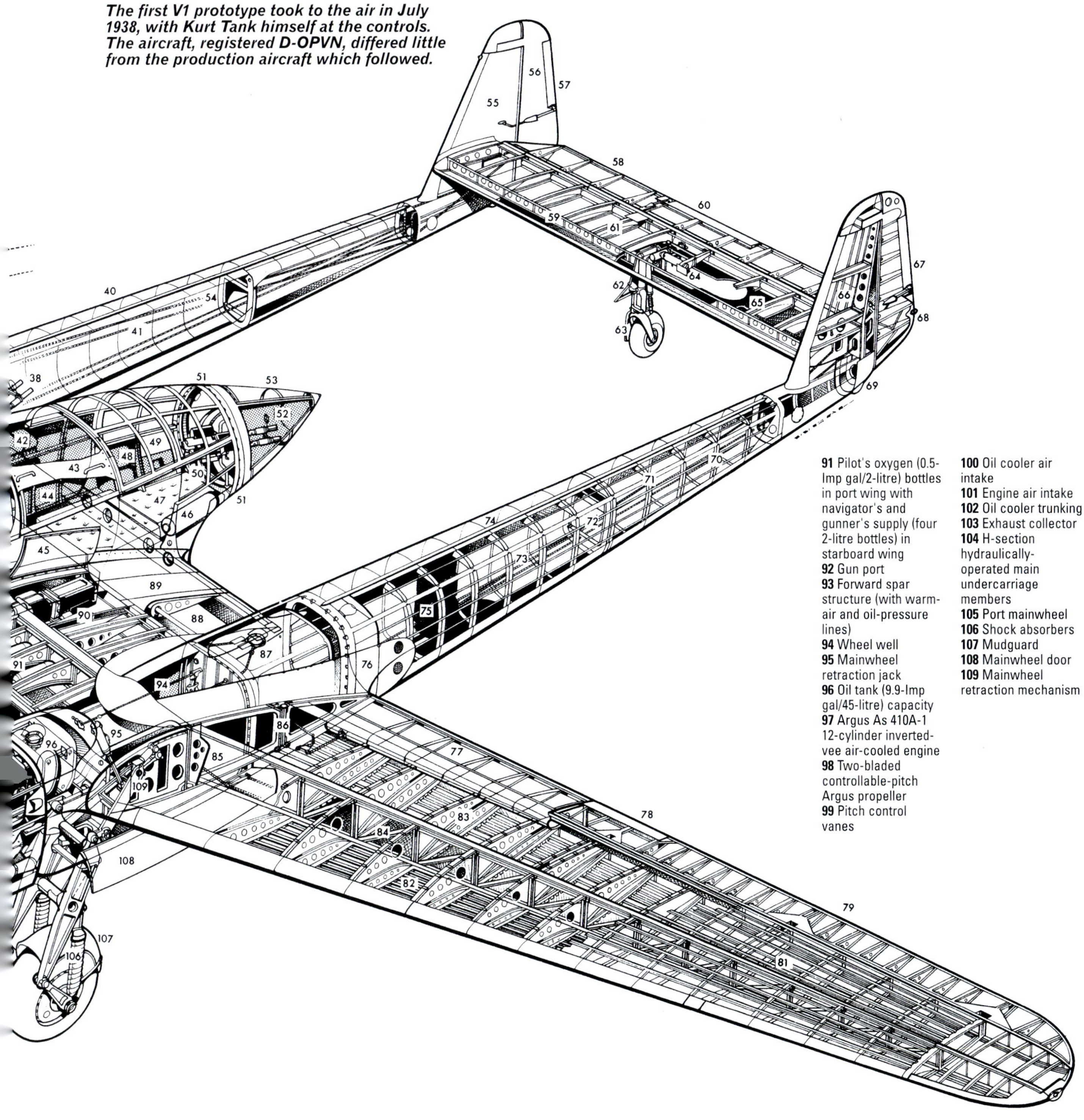

91 Pilot's oxygen (0.5-Imp gal/2-litre) bottles in port wing with navigator's and gunner's supply (four 2-litre bottles) in starboard wing
92 Gun port
93 Forward spar structure (with warm-air and oil-pressure lines)
94 Wheel well
95 Mainwheel retraction jack
96 Oil tank (9.9-Imp gal/45-litre) capacity
97 Argus As 410A-1 12-cylinder inverted-vee air-cooled engine
98 Two-bladed controllable-pitch Argus propeller
99 Pitch control vanes
100 Oil cooler air intake
101 Engine air intake
102 Oil cooler trunking
103 Exhaust collector
104 H-section hydraulically-operated main undercarriage members
105 Port mainwheel
106 Shock absorbers
107 Mudguard
108 Mainwheel door
109 Mainwheel retraction mechanism

Junkers Ju 52

Although it was the best transport type the Luftwaffe possessed, the Ju 52/3m was slow, lacked manoeuvrability and defensive armament and was an easy target for marauding Allied fighters.

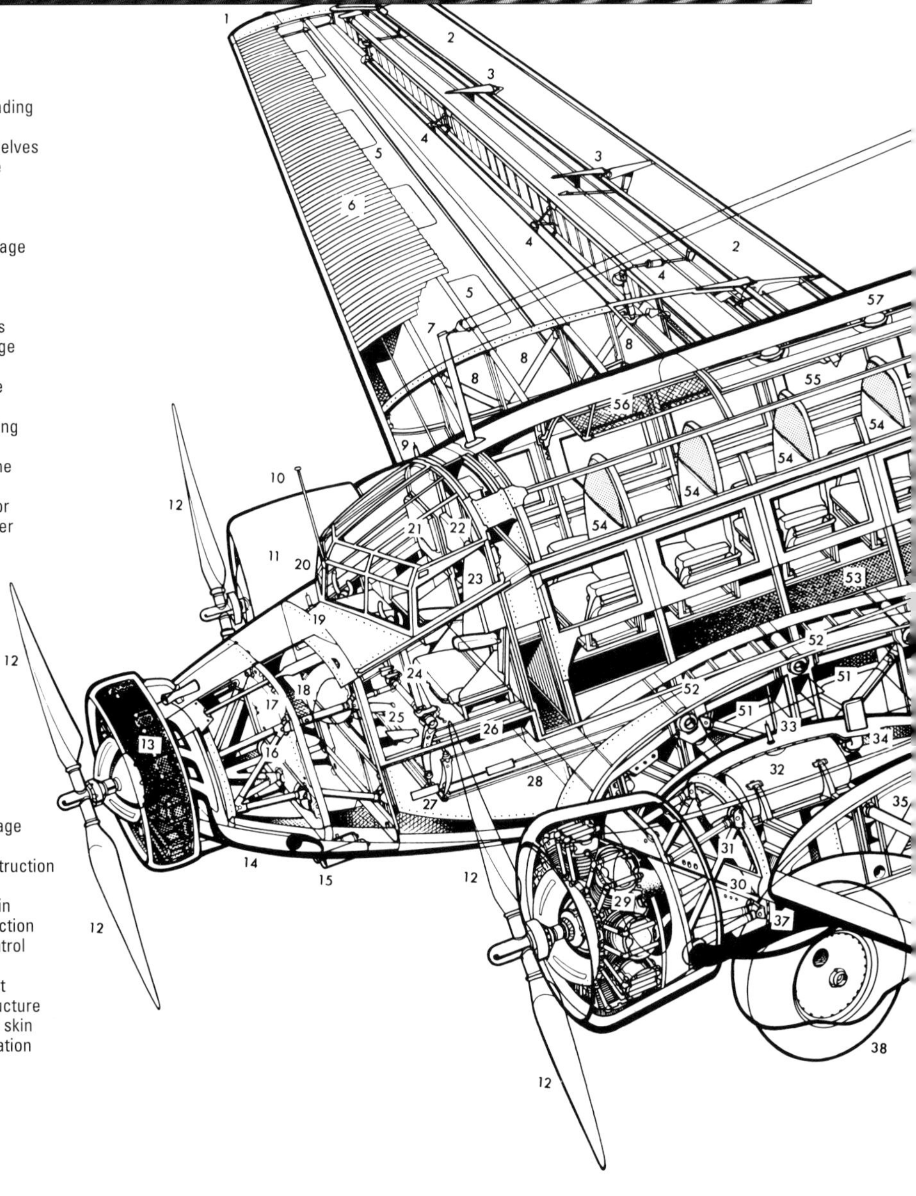

Junkers Ju-52

Cutaway key
1 Starboard navigation light
2 Drooping aileron section of Junkers 'double wing'
3 Aileron hinge fairings
4 Control linkage
5 Underwing inspection panels
6 Corrugated wing skin
7 Aerial mast
8 Wing strut diagonal bracing
9 Starboard oil filler cap
10 House-flag mast
11 Starboard engine cowling (NACA cowling)
12 Junkers metal two-blade propeller
13 Centre BMW 132A radial engine (in Townend ring)
14 Exhaust
15 Filter intakes
16 Engine bearers
17 Bulkhead
18 Centre oil tank
19 Oil filler cap
20 Flat windscreen panels
21 Co-pilot's seat
22 Radio-operator's jump-seat
23 Pilot's seat
24 Control column
25 Rudder pedals
26 Raised cockpit floor level
27 Control linkage
28 Control lines
29 Port BMW 132A radial engine (in NACA cowling)
30 Bulkhead
31 Engine bearers
32 Engine oil tank
33 Oil filler cap
34 Fuel filler cap
35 Mainwheel support strut
36 Mainwheel spat
37 Exhaust
38 Port mainwheel
39 Corrugated wing skin
40 Multi-spar wing structure
41 Diagonal cross-brace members
42 Pitot head
43 Port navigation light
44 Drooping aileron section of Junkers 'double wing'
45 Aileron hinge fairings
46 Trim tab
47 Tab control
48 Control runs
49 Inner section trailing-edge flap
50 Control linkage
51 Port wing fuel tanks
52 Fuselage/wing ball-and-socket attachment points
53 Centre aisle
54 Single-seat cabin arrangement (17 maximum)
55 Upper fuselage longeron
56 Luggage rack
57 Ceiling lights
58 Aerial
59 Two-seat rear bench
60 Passenger entry door
61 Underfloor control lines
62 Entry vestibule
63 Entry steps attachment
64 Toilet compartment
65 Rear cabin bulkhead
66 Cabin ventilation
67 Cargo compartment
68 Luggage loading hatch
69 Baggage shelves
70 Aft fuselage frames
71 Inspection walkway
72 Lower fuselage longeron
73 Fuselage construction
74 Control lines
75 Rear fuselage frame
76 Fin/fuselage attachment
77 Tailskid spring
78 Tailskid
79 Port tailplane structure
80 Port elevator
81 Lower rudder hinge
82 Control linkage
83 Multi-spar tailplane construction
84 Elevator corrugated skin
85 Fin construction
86 Rudder control linkage
87 Rudder post
88 Rudder structure
89 Corrugated skin
90 Rear navigation light

German forces in North Africa relied heavily on supplies and reinforcements delivered by Ju 52/3ms. These aircraft are seen in April 1941 shortly before the German invasion of Crete which saw the Ju 52/3m in its other major role – as a paratroop assault transport.

SPECIFICATION

Junkers Ju 52/3mg3e

Type

Medium bomber and troop transport

Powerplant

Three 725-hp (541-kW) BMW 132A-3 radial piston engines

Performance

Maximum speed 171 mph (275 km/h) at 2,955 ft (900 m); service ceiling 19,360 ft (5900 m); range with auxiliary fuel 808 miles (1300 km)

Weights

Empty 12,610 lb (5720 kg); maximum take-off 23,149 lb (10500 kg)

Dimensions

Wingspan 95 ft 11½ in (29.25 m); length 62 ft (18.90 m); height 18 ft 2½ in (5.55 m); wing area 1,189.45 sq ft (110.50 m2)

Armament

Two 0.31-in (7.92-mm) MG15 machine-guns, plus up to 1,102 lb (500 kg) of bombs

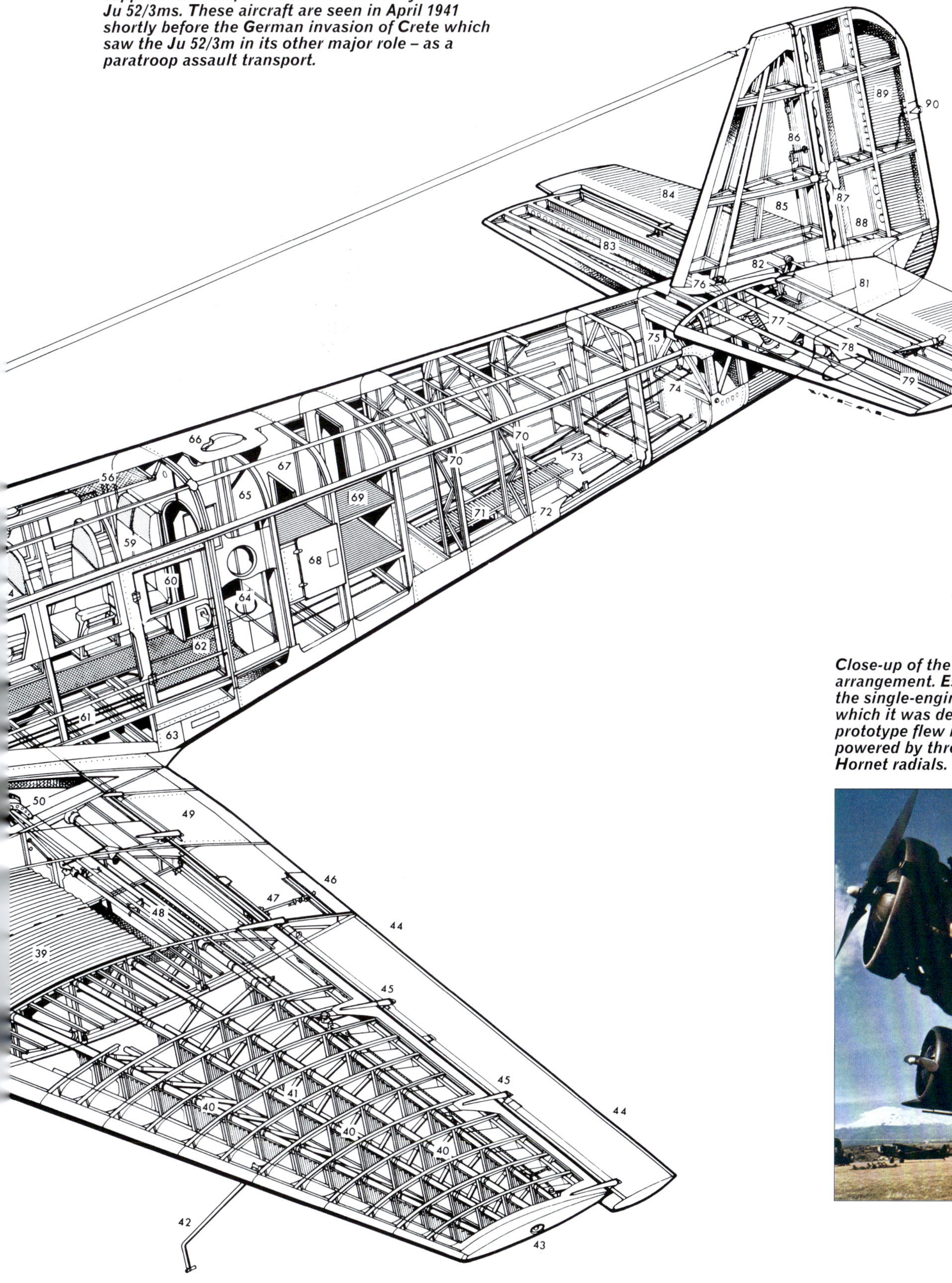

Close-up of the Ju 52/3m's engine arrangement. Essentially similar to the single-engined variant from which it was developed, the prototype flew in April 1932, powered by three Pratt & Whitney Hornet radials.

Junkers Ju 52

Development of the basic single-engined Ju 52 design led the Junkers firm to produce a three-engined version, the Ju 52/3m. The prototype, coded 4007, flew in April 1932, powered by three licence-built Pratt & Whitney Hornet nine-cylinder radials. The Ju 52/3mg5e, seen here, was powered by three BMW 132T-2 radials. The two outboard engines faced slightly outwards to reduce yaw should one of them fail. The exhaust gases were collected by annular ducts, which gave the aircraft its distinctive appearance.

Radio equipment
The Ju 52 had a loop antenna for direction-finding equjipment, and an aerial mast mounted behind the cockpit. Radio equipment was constantly improved, and several Ju 52s were fitted with specialist equipment of one kind or another. This included a large-diameter duralumin hoop which, when energized by a small auxiliary motor, was used for detonating magnetic mines. Aircraft carrying this device were used by the Minensuchgruppe based at Cognac, Le Leu and Biarritz. Later versions of the Ju 52 wre fitted with an automatic pilot.

Undercarriage
The Ju 52's undercarriage had to be very robust to cope with repeated landings on rough fields. Spats were issued to reduce the drag created by the wheels, but on operations these would clog up with mud and debris, and were rarely fitted. Many aircraft were fitted with floats in place of the wheeled undercarriage. The first Ju 52/3m floatplanes were operated by Aero O/Y, a Finnish company, and Sweden's AB Aerotransport. One Lufthansa Ju 52 was used to fly the Olympic Torch from Greece to Berlin for the 1936 Olympic Games.

Armament
The dorsal hatch had a mounting for a 0.31in (7.92mm) MG 15 machine gun. A transparent fairing was fixed in front of it to give the gunner some protection from the slipstream in flight. The first Ju 52 variant for the Luftwaffe, the Ju 52mg3e, was designed as a heavy bomber with a crew of four and armed with two MG 15 machine guns, one mounted in the dorsal position and the other carried in a retractable 'dustbin' suspended under the fuselage. The Ju 52 was soon superseded by more modern types of bomber aircraft.

Construction
The corrugated fuselage was a common feature of many early Junkers designs. The metal skin was load-bearing, and the corrugation gave it immense strength for little weight penalty. The first aircraft to use this type of skin was the Junkers J4 of 1917, and the Ju 52 was the last. The first Junkers all-metal monoplane was the J1, completed in 1915; it was developed via the Junkers D1 monoplane fighter into the F13 four-passenger transport, which was one of the most advanced aircraft if its time. The prototype flew on 25 June, 1919.

Operations
When fitted with seats the Ju 52 could carry up to 18 passengers, sitting in single rows separated by an aisle. The aircraft's value as a military transport was first demonstrated during the Spanish Civil War, when 20 Ju 52/3ms were used to ferry 10,000 colonial troops from Morocco to Spain in support of General Franco's nationalist uprising. The Ju 52s then carried out many bombing operations in support of the land battle around Madrid in November 1936. The Spanish nicknamed the Ju 52 Pava (Turkey).

Savoia Marchetti SM.79 Sparviero

One of the highly successful torpedo-bomber units equipped with the Sparviero, 278ª Squadriglia, 132º Gruppo was based in Sicily and Sardinia for much of 1942/43, operating against Allied shipping in the Mediterranean.

SM.79-I Sparviero

Cutaway key

1 Starboard pitot tube
2 Starboard navigation light
3 Three-spar wing structure
4 Aileron mass balance
5 Starboard aileron
6 Aileron control rods
7 Flap linkage
8 Starboard slotted flap
9 Starboard outer fuel tank, capacity 33 Imp gal (150 litres)
10 Flap rod
11 Wing skinning
12 Leading-edge slot
13 Starboard engine nacelle
14 Propeller
15 Exhaust collector ring
16 Three-bladed propeller
17 Spinner
18 Engine cowling ring
19 Alfa Romeo 126 radial engine
20 Exhaust
21 Engine bearers
22 Firewall bulkhead
23 Oil filter
24 Starboard mainwheel
25 Ventral landing lamp
26 Undercarriage cylinders
27 Nose compartment access panel
28 Intake
29 Centre engine oil tank
30 Undercarriage warning horns
31 Fire extinguisher
32 Instrument panel
33 Control column
34 Rudder pedals
35 Main fuel filter
36 Flap control linkage
37 Oil cooler intakes
38 Front spar/fuselage frame pick-up
39 Pilot's seat
40 Central control console
41 Pilot's seat
42 Windscreen panels
43 Roof panels
44 Fixed forward-firing 12.7-mm (0.5-in) Breda-SAFAT machine-gun
45 Dorsal fairing frame
46 Ammunition tank
47 Link collector box
48 Radio transmitter/ receiver
49 Radio operator's position
50 Main spar carry-through
51 Central forward fuel tank, capacity 46 Imp gal (210 litres)
52 Port forward fuel tank, capacity 44 Imp gal (200 litres)
53 Port aft fuel tank, capacity 111 Imp gal (505 litres)
54 Main spar/fuselage frame pick-up
55 Rear spar/fuselage frame pick-up
56 Central aft fuel tank, capacity 128 Imp gal (580 litres)
57 Oxygen cylinder
58 Crew compartment entry steps
59 Flight mechanic's seat
60 Bulkhead partition
61 Ammunition tanks
62 Link collector box
63 Radio operator's seat
64 Fixed window
65 Flexible link chute
66 Dorsal blister
67 Dorsal glazing
68 Dorsal flexible 12.7-mm (0.50-in) Breda-SAFAT machine-gun
69 Gun support bracket
70 Sliding fairing
71 Bomb bay support frame
72 Bomb vertical stowage attachment lugs
73 Twelve 100-kg (220-lb) bombs (alternatives: five 250-kg (550-lb) or two 500 kg (1,100 lb) bombs
74 Bomb magazine (offset to starboard)
75 Bomb bay doors
76 Fuselage frames
77 Crew entry catwalk
78 OMI vertical camera installation
79 Ventral gondola
80 Jozza bombsight
81 Bomb-aimer's rudder control handwheel
82 Starboard side windows (three)
83 Radio receiver
84 Fuel header tank, capacity 8.8 Imp gal (40 litres)
85 Dorsal fairing fixed aft section
86 Plywood dorsal skinning
87 D/F loop
88 Starboard waist position
89 Waist gun mounting bar
90 Verey cartridge stowage boxes
91 Crew entry doorway
92 Bomb-aimer's sliding knee supports
93 Flexible link chute
94 Ventral gondola fairing
95 Ventral flexible 12.7-mm (0.50-in) Breda-SAFAT machine-gun
96 Gondola hinged aftersection
97 Port side windows (two)
98 Ammunition feed
99 Waist machine gun
100 Ammunition tank
101 Mist window frame
102 Elevator control linkage
103 Control rods
104 Fuselage structure
105 Fabric side covering
106 Fin/fuselage attachment
107 Fin spar
108 Starboard tailplane
109 Elevator balance
110 Starboard elevator
111 Fin structure
112 Rudder upper hinge
113 Rudder torque tube
114 Rudder frame
115 Rudder tab
116 Rudder lower hinge
117 Tailplane brace strut
118 Rudder tab control link
119 Port elevator

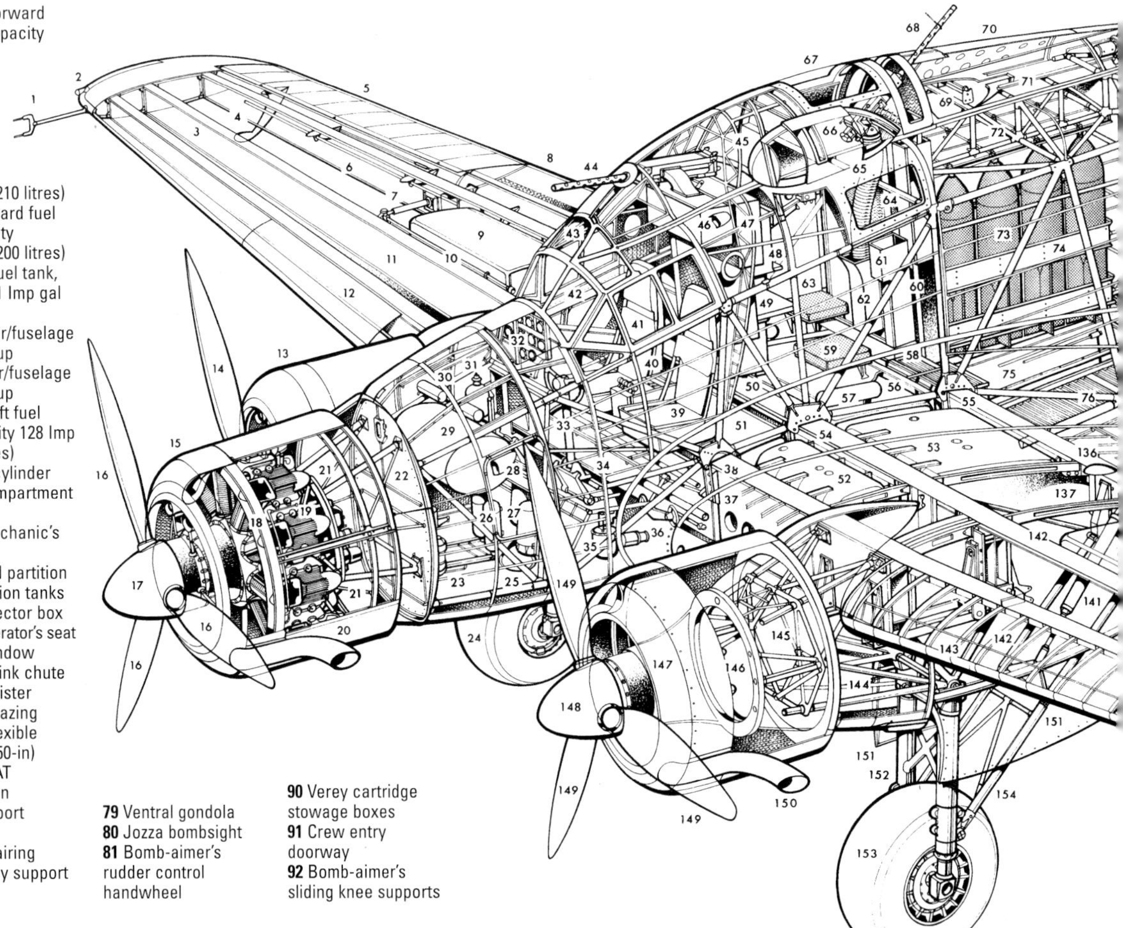

An experimental twin-torpedo installation was tested at Gorizia in the spring of 1938. As it caused a marked deterioration in aircraft performance, the modification was abandoned.

SPECIFICATION

SM.79-I Sparviero

Dimensions

Length: 51 ft 2 in (15.60 m)
Height: 15 ft 1 in (4.60 m)
Wingspan: 69 ft 6¾ in (21.20 m)
Wing area: 664.2 sq ft (61.70 m²)

Powerplant

Three Alfa Romeo 126 RC.34 nine-cylinder radial piston engines, each rated at 780 hp (582 kW) for take off

Weights

Empty: 15,322 lb (6950 kg)
Maximum take-off: 23,655 lb (10730 kg)

Performance

Maximum speed at 13,125 ft (4000 m): 267 mph (430 km/h)
Climb to 13,125 ft (4000 m): 13 minutes 15 seconds
Maximum range at 211 mph (340 km/h): 2,050 miles (3300 km)
Service ceiling at maximum take-off weight: 21,325 ft (6500 m)

Armament

One fixed 12.7-mm (0.5-in) machine-gun firing forward over cabin roof, guns of the same calibre in dorsal position and in rear of ventral position, one 7.7-mm (0.303-in) machine-gun for beam defence, plus a maximum bomb load of five 250-kg (551-lb) or one 45-cm (17.70-in) torpedo

This SM.79-II was seized by the Luftwaffe after the Armistice and employed as a transport. Note that it retains its defensive armament.

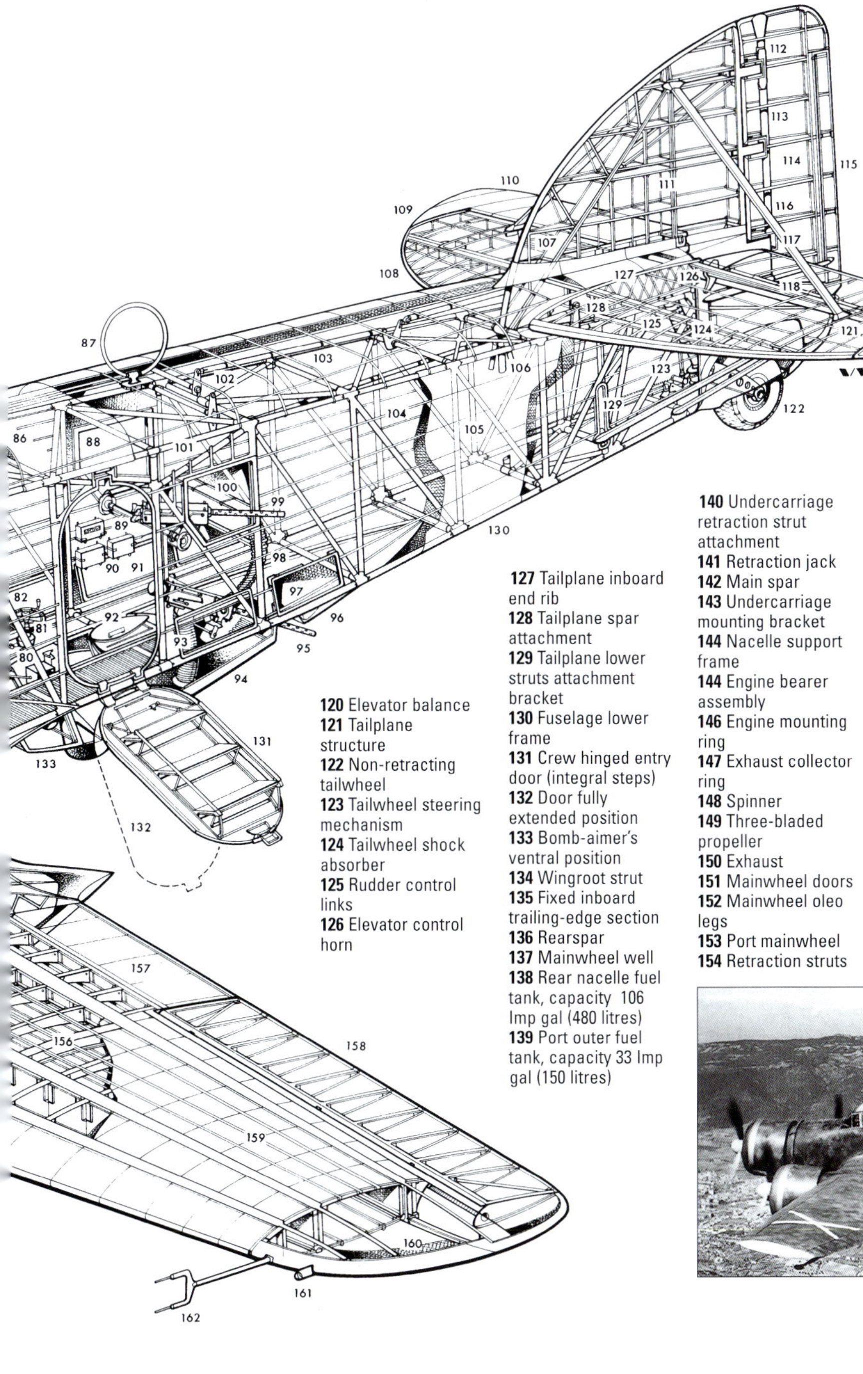

120 Elevator balance
121 Tailplane structure
122 Non-retracting tailwheel
123 Tailwheel steering mechanism
124 Tailwheel shock absorber
125 Rudder control links
126 Elevator control horn
127 Tailplane inboard end rib
128 Tailplane spar attachment
129 Tailplane lower struts attachment bracket
130 Fuselage lower frame
131 Crew hinged entry door (integral steps)
132 Door fully extended position
133 Bomb-aimer's ventral position
134 Wingroot strut
135 Fixed inboard trailing-edge section
136 Rearspar
137 Mainwheel well
138 Rear nacelle fuel tank, capacity 106 Imp gal (480 litres)
139 Port outer fuel tank, capacity 33 Imp gal (150 litres)
140 Undercarriage retraction strut attachment
141 Retraction jack
142 Main spar
143 Undercarriage mounting bracket
144 Nacelle support frame
144 Engine bearer assembly
146 Engine mounting ring
147 Exhaust collector ring
148 Spinner
149 Three-bladed propeller
150 Exhaust
151 Mainwheel doors
152 Mainwheel oleo legs
153 Port mainwheel
154 Retraction struts
155 Leading-edge slot
156 Wooden wing structure
157 Slotted flap
158 Aileron frame
159 Wing skinning
160 Port wingtip structure
161 Port navigation light
162 Port pitot tube

Sparvieros in Spain

From the outset the Regia Aeronautica test pilots expressed enthusiasm for the SM.79, and production orders were placed, before the end of 1935. Early aircraft, SM.79-Is, with three 780-hp (582-kW) Alfa Romeo 126 RC.34 radials, entered service with the 8' and 11P Stormi Bombardamento Veloce (fast bomber groups) in 1936. In 1937 these units were sent to Spain to serve with the Aviacion del Tercio, flying during the Spanish Civil War as the 27° and 28° Gruppi (Falchi delle Baleari, Hawks of Balearics, based in the Balearic Islands), and the 29° and 30° Gruppi (Sparvieri, or Sparrows). These units, together with two of SM.81s, flew 5,318 sorties, dropped 11,850 tons of bombs and scored 224 direct hits on government vessels. At the end of the Civil War the new Spanish government took over 80 SM.79s and these came to provide a major portion of the Spanish air force's bombing arm for many years to come. Pictured are SM.79-I bombers of XXVII Gruppo Bombardamento Terrestre, Aviacion del Tercio during the Spanish Civil War. Note the green blotching on sand camouflage of the nearest aircraft, which contrasts with the brown and sand schemes of the other aircraft.

SM.79-II Sparviero

This 'sparrowhawk' of 283ª Squadriglia, 130° Gruppo Autonomo Aerosiluranti was based at Gerbini, Sicily during 1942. During this period the 280° and 283° Squadriglie concentrated on attacking the Malta convoys, most notably 'Harpoon' during June. By the beginning of January 1943 the Gruppo had just nine serviceable aircraft and withdrew to Italy to reform and retrain.

Powerplant
Early versions of the SM.79 featured only two engines, but the major production for the Regia Aeronautica and Aerosiluranti were three-engined. The SM.79-II was the major production version, and featured the Piaggio P.XI RC.40 radial, rated at 1,000 hp (746 kW). The central engine was mounted externally on bearers attached to a firewall bulkhead. The gap between the cowling and the fuselage was partially filled with a secondary cowl which still allowed the passage of cooling air through the engine itself. The oil tank was mounted behind the bulkhead.

Armament
Above the cockpit was a single Breda-SAFAT 12.7-mm (0.5-in) machine-gun, fixed in the roofing and fired by the pilot. A similar weapon on a flexible mount was situated in the rear of the blister, covered by a panel when not in use. Protection to the rear and sides was provided by a 7.7-mm (0.303-in) machine-gun for the waist position, and a Breda-SAFAT 12.7-mm (0.5-in) gun in the rear of the ventral gondola. The waist gun was mounted on a lateral bar so that it could fire out of either beam window.

Camera
To provide a limited reconnaissance capability, the SM.79 could be fitted with an OMI vertical camera, situated in the rear fuselage forward of the ventral gondola and peering through a flat pane in the cabin floor.

Bomb load
The SM.79 carried its bombs vertically, stowed in a bay offset to starboard aft of the rear spar. A maximum load was five 250-kg (551-lb) weapons, with two 500-kg (1,102-lb) bombs or 12 100-kg (220-lb) bombs as common alternatives. The SM.79 scored its greatest successes as a torpedo-bomber with the Gruppi Aerosiluranti (specialist torpedo-bombing units). These carried one or two weapons under the fuselage, usually a 45-cm (17.70-in) weapon.

Fuel
The S.M.79 carried its fuel in 11 tanks. Six were inboard of the nacelles located between the spars, comprising port and starboard forward (44 Imp gal/200 litres each), port and starboard aft (111 Imp gal/505 litres each), central forward (46 Imp gal/210 litres) and central aft (128 Imp gal/580 litres). A further tank of 33 Imp gal (150 litres) was located between the central and rear spars outboard of each engine nacelle, while the rear of each nacelle had a 106-Imp gal (480-litre) tank. A small (8.8-Imp gal/40-litre) header tank in the upper rear fuselage brought the internal capacity to 770 Imp gal (3500 litres).

Fuselage and wing structure
The fuselage was built on a traditional frame structure, the main members running along the four corners of the fuselage and forming a rectangular frame work. The curved upper decking was a secondary structure, skinned in plywood. The fuselage sides were fabric-covered. The Sparviero had a three-spar wing, the spars carrying through the lower fuselage and the top of each engine nacelle. The structure was of wood.

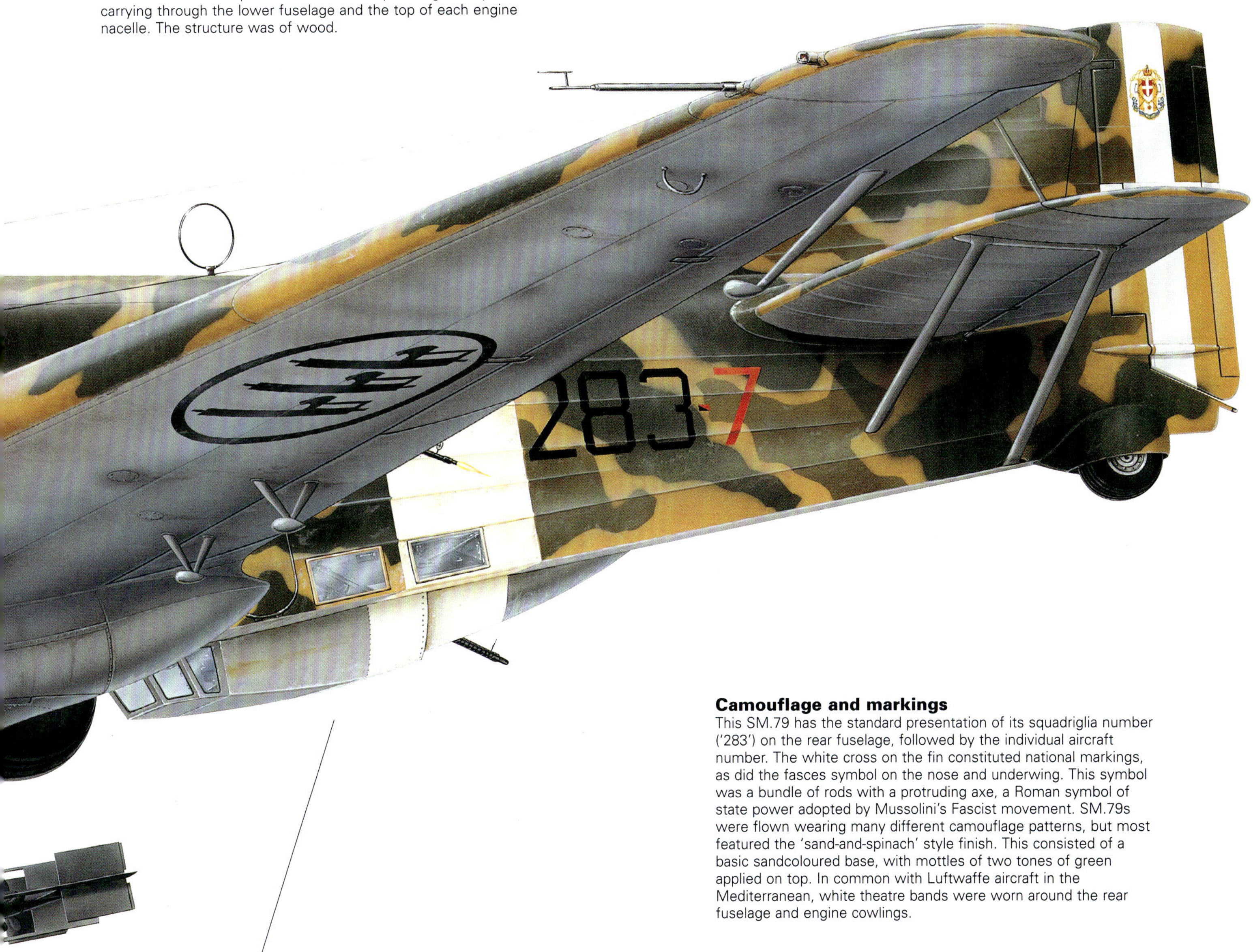

Camouflage and markings
This SM.79 has the standard presentation of its squadriglia number ('283') on the rear fuselage, followed by the individual aircraft number. The white cross on the fin constituted national markings, as did the fasces symbol on the nose and underwing. This symbol was a bundle of rods with a protruding axe, a Roman symbol of state power adopted by Mussolini's Fascist movement. SM.79s were flown wearing many different camouflage patterns, but most featured the 'sand-and-spinach' style finish. This consisted of a basic sandcoloured base, with mottles of two tones of green applied on top. In common with Luftwaffe aircraft in the Mediterranean, white theatre bands were worn around the rear fuselage and engine cowlings.

Ventral gondola
In addition to housing the rear-facing gun, the gondola provided a bomb-aiming station, the bomb-aimer being provided with forward-and downward-facing flat panes. At his side was a small wheel with which he made minute corrections to the aircraft's path by controlling the rudder. The rear of the gondola was a hinged fairing which retracted to allow the gun to fire.

Short Sunderland

BOAC Sunderland Mk III G-AGIA started life as RAF aircraft ML728 and joined the airline in July 1943 for use on priority passenger and mail services to West Africa, India and (after VJ-Day) Burma. Twenty-four aircraft were converted for use on these routes, all military equipment and armament being removed and basic bench-and-mattress seating being fitted.

Sunderland Mk

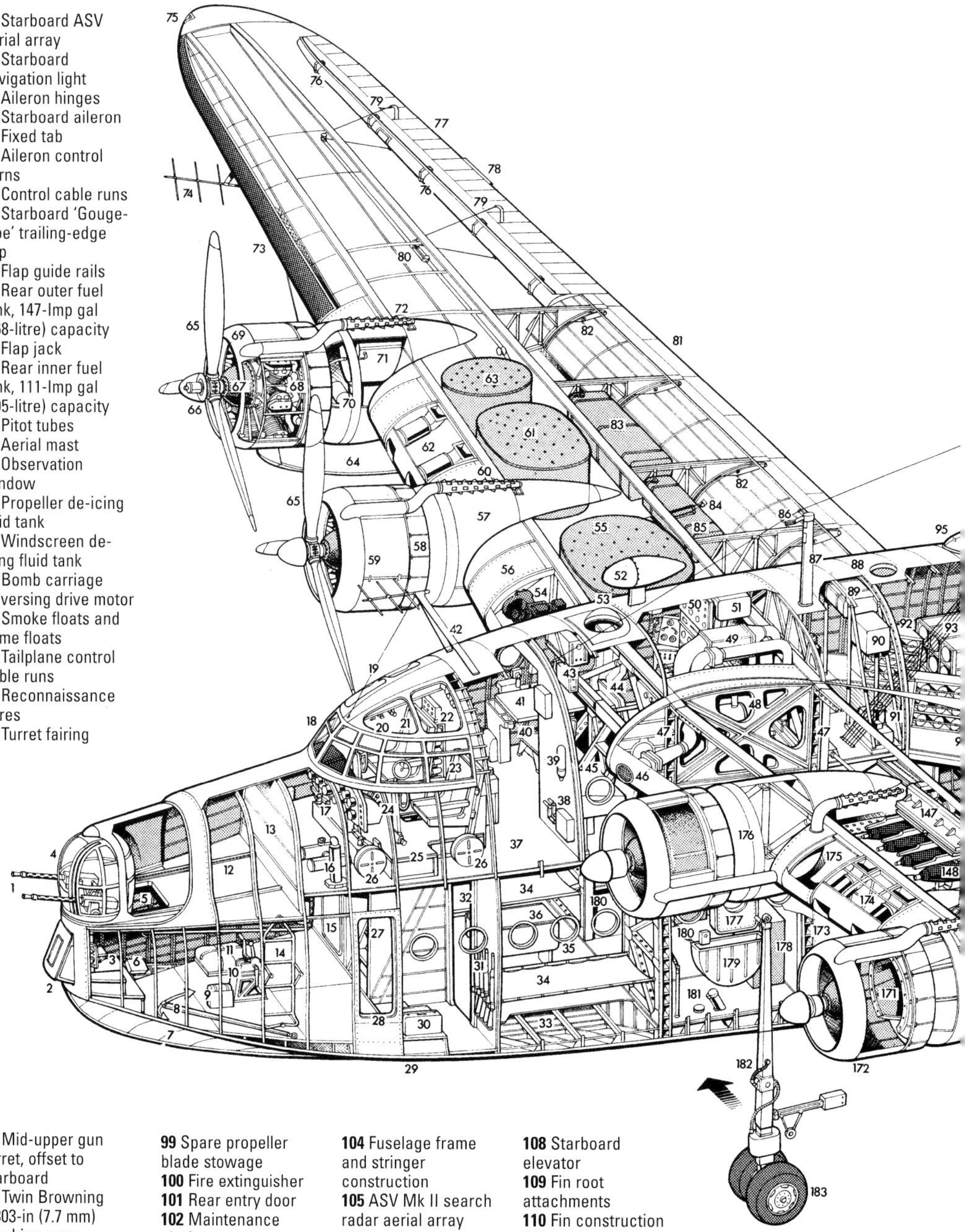

Cutaway key

1 Twin Vickers 0.303-in (7.7-mm) machine-guns
2 Bomb aiming window, retractable
3 Bomb aimers station
4 Retractable nose turret
5 Front entry/mooring hatch
6 Mooring cable stowage
7 Hull planing bottom
8 Anchor
9 Parachute stowage
10 Anchor winch
11 Dinghy
12 Front turret rails
13 Cockpit bulkhead
14 Mooring ladder
15 Toilet compartment door, starboard side
16 Nose gun turret hydraulic reservoir
17 Instrument panel
18 Windscreens
19 Cockpit roof glazing
20 Overhead control panels
21 Co-pilot's seat
22 Signal cartridge rack
23 Pilot's seat
24 Control column
25 Raised cockpit floor level
26 Autopilot controllers
27 Stairway between upper and lower decks
28 Front entry door
29 Fuselage chine member
30 Crew luggage locker
31 Rifle rack
32 Wardroom door
33 Planing bottom hull construction
34 Wardroom bunks
35 Window panes
36 Folding table
37 Upper deck floor level
38 Parachute stowage
39 Fire extinguisher
40 Navigator's seat
41 Chart table
42 Forward ASV radar aerial mast
43 Navigator's instrument panel
44 Flight engineer's aft facing seat
45 Radio operator's station
46 Air intake duct
47 Wing/fuselage attachment main frames
48 Wing root rib cut-outs
49 Air conditioning plant
50 Engineer's control panels
51 Carburettor de-icing fluid tank
52 D/F loop aerial
53 Astrodome observation hatch
54 Auxiliary Power Unit
55 Forward inner fuel tank, 529-Imp gal (2405-litre) capacity
56 Fold-down, leading-edge maintenance platform
57 Starboard inner engine nacelle
58 Cowling air flaps
59 Detachable engine cowlings
60 Flame suppressor exhaust pipe
61 Forward inner fuel tank, 325-Imp gal (1477-litre) capacity
62 Oil coolers
63 Forward outer fuel tank, 132-Imp gal (600-litre) capacity
64 Starboard wing tip float
65 de Havilland three-bladed, constant-speed propeller, 12ft 9in (3.89 m) diameter
66 Propeller hub pitch change mechanism
67 Engine reduction gearbox
68 Bristol Pegasus XVIII, nine-cylinder radial engine, 1065 hp (794 kW)
69 Exhaust collector ring
70 Oil filter
71 Oil tank, 32-Imp gal (145-litre) capacity
72 Flame suppressor exhaust pipe
73 Leading edge de-icing
74 Starboard ASV aerial array
75 Starboard navigation light
76 Aileron hinges
77 Starboard aileron
78 Fixed tab
79 Aileron control horns
80 Control cable runs
81 Starboard 'Gouge-type' trailing-edge flap
82 Flap guide rails
83 Rear outer fuel tank, 147-Imp gal (668-litre) capacity
84 Flap jack
85 Rear inner fuel tank, 111-Imp gal (505-litre) capacity
86 Pitot tubes
87 Aerial mast
88 Observation window
89 Propeller de-icing fluid tank
90 Windscreen de-icing fluid tank
91 Bomb carriage traversing drive motor
92 Smoke floats and flame floats
93 Tailplane control cable runs
94 Reconnaissance flares
95 Turret fairing
96 Mid-upper gun turret, offset to starboard
97 Twin Browning 0.303-in (7.7 mm) machine guns
98 Fuselage skin plating
99 Spare propeller blade stowage
100 Fire extinguisher
101 Rear entry door
102 Maintenance platform stowage
103 Observation window
104 Fuselage frame and stringer construction
105 ASV Mk II search radar aerial array
106 Leading edge de-icing
107 Starboard tailplane
108 Starboard elevator
109 Fin root attachments
110 Fin construction
111 Leading edge de-icing
112 Fin tip construction

The Sunderland's flight deck was relatively spacious and provided an excellent all-round view for its pilots. The covered buttons on the pilot's control wheel (left) are for bomb release and firing the four forward-firing, nose-mounted guns.

SPECIFICATION

Sunderland Mk III

Dimensions

Wing span: 112 ft 9½ in (34.38 m)
Wing area: 1,487 sq ft (119.85 m²)
Length: 85 ft 4 in (26.01 m)
Height (on beaching chassis): 32 ft 2 in (9.79 m)

Powerplant

Four Bristol Pegasus XVIII nine-cylinder radial piston engines, each rated at 1,065 hp (794 kW)

Typical weights

Empty: 33,000 lb (14969 kg)
All-up: 58,000 lb (26308 kg)

Performance

Maximum speed: 212 mph (341 km/h)
Initial climb rate: 790 ft (241 m) per minute
Service ceiling: 15,000 ft (4570 m)
Range: 3,000 miles (4828 km) at 145 mph (233 km/h)
Endurance: 20 hours

Armament

One 0.303-in (7.7-mm) Vickers GO machine-gun in nose turret, two 0.303-in (7.7-mm) Browning machine-guns in mid-upper turret, four similar Brownings in tail turret; optional second nose turret gun, four fixed Brownings firing ahead and twin 0.5-in (12.7-mm) Brownings fired from waist hatches; assorted ordnance to total weight of 4,960 lb (2250 kg) housed in hull and cranked out under wings prior to attack

Above: This view of a Sunderland Mk I's interior, looking forward towards the cockpit, gives an impression of the spaciousness of the aircraft's fuselage, with its double-deck layout.

Below: WAAFs service a Sunderland's Bristol 'Peggies' between sorties from a Coastal Command base. The Sunderland was always considered underpowered until the Mk V entered service.

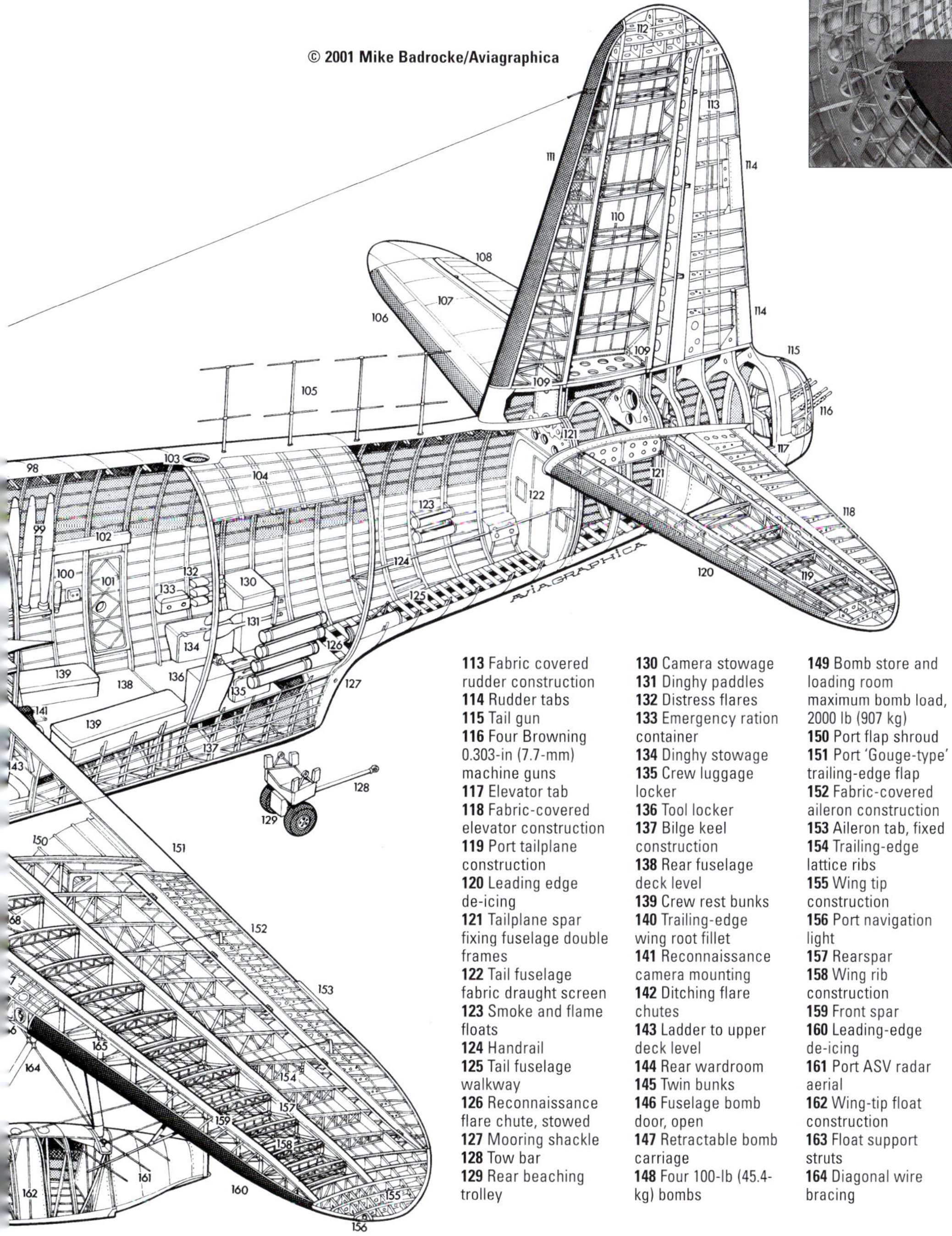

113 Fabric covered rudder construction
114 Rudder tabs
115 Tail gun
116 Four Browning 0.303-in (7.7-mm) machine guns
117 Elevator tab
118 Fabric-covered elevator construction
119 Port tailplane construction
120 Leading edge de-icing
121 Tailplane spar fixing fuselage double frames
122 Tail fuselage fabric draught screen
123 Smoke and flame floats
124 Handrail
125 Tail fuselage walkway
126 Reconnaissance flare chute, stowed
127 Mooring shackle
128 Tow bar
129 Rear beaching trolley
130 Camera stowage
131 Dinghy paddles
132 Distress flares
133 Emergency ration container
134 Dinghy stowage
135 Crew luggage locker
136 Tool locker
137 Bilge keel construction
138 Rear fuselage deck level
139 Crew rest bunks
140 Trailing-edge wing root fillet
141 Reconnaissance camera mounting
142 Ditching flare chutes
143 Ladder to upper deck level
144 Rear wardroom
145 Twin bunks
146 Fuselage bomb door, open
147 Retractable bomb carriage
148 Four 100-lb (45.4-kg) bombs
149 Bomb store and loading room maximum bomb load, 2000 lb (907 kg)
150 Port flap shroud
151 Port 'Gouge-type' trailing-edge flap
152 Fabric-covered aileron construction
153 Aileron tab, fixed
154 Trailing-edge lattice ribs
155 Wing tip construction
156 Port navigation light
157 Rearspar
158 Wing rib construction
159 Front spar
160 Leading-edge de-icing
161 Port ASV radar aerial
162 Wing-tip float construction
163 Float support struts
164 Diagonal wire bracing

165 Wing spar girder construction
166 Landing lamps
167 Leading-edge rib construction
168 Diagonal wire-braced wing ribs
169 Fold-down, leading-edge maintenance platform
170 Engine nacelle construction
171 Engine mounting ring
172 Porrt outer engine nacelle
173 Oil cooler intakes
174 Oil coolers
175 Exhaust shroud heat exchangers
176 Port inner engine nacelle
177 Emergency escape hatch
178 Ice chest
179 Drogue container
180 Galley compartments, port and starboard
181 Watertight trailing aerial socket
182 Main beaching gear leg strut
183 Twin beaching wheels

Sunderland Mk II

Sunderland Mk II T9087 was an aircraft of No. 201 Squadron, Coastal Command, based on Lough Erne, Northern Ireland from September 1941 until March 1944. One of 20 Sunderland IIs built by Blackburn, in Dumbarton, it is depicted as it would have appeared during 1942, with the unit's 'ZM' codes and typical 'early war' Coastal Command camouflage of Dark Slate Grey, Extra Dark Sea Grey extending well down the sides of the hull and with Sky lower surfaces.

Powerplant
Early Sunderlands, like this aircraft, had 1,010-hp (753-kW) Bristol Pegasus XXII engines, but after the 35th airframe had been completed, 1,065-hp (794-kW) Pegasus XVIIIs, with two-speed superchargers, had been substituted. It was not until the Mk V the 1,200-hp (865-kW) Pratt & Whitney Twin Wasps were adopted, the last giving Short's airframe sufficient power to compensate for its increased weight.

Flight deck
The two pilots sat side-by-side with full dual controls, with the navigator and wireless operator behind them facing to starboard and port respectively. The flight engineer sat further back, facing aft. Because the Air Ministry distrusted the optical qualities of the curved windscreen of the Empire, the windscreen and canopies consisted of flat Triplex panels.

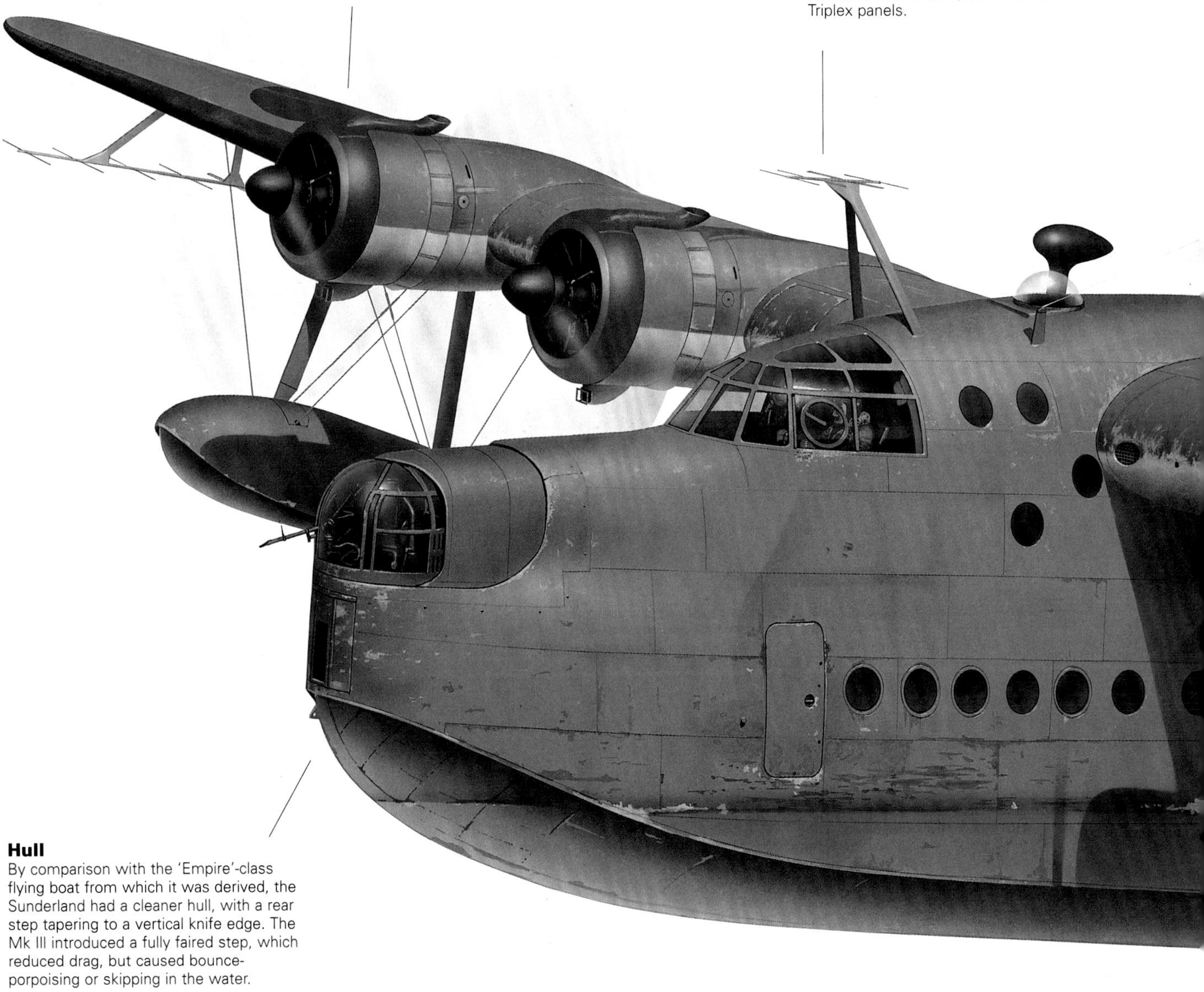

Hull
By comparison with the 'Empire'-class flying boat from which it was derived, the Sunderland had a cleaner hull, with a rear step tapering to a vertical knife edge. The Mk III introduced a fully faired step, which reduced drag, but caused bounce-porpoising or skipping in the water.

Twin-deck layout
Forward of the entrance door on the lower deck was the dinghy and anchor stowage (and the bomb-aimers compartment with its hinging bomb-aimers window), while the wardroom with its folding bunks and tables was situated aft. Behind this was a small workshop, and aft of the rear step was a walkway to the tail gunner's compartment. The upper deck began with the spacious flight deck, with the galley between the main frames under the wing and the bomb room aft.

Fuel tanks
The Sunderland had three drum-like fuel tanks in the inner part of each wing, aft of the leading edge. The innermost of these contained 529 Imp gal (2405 litres), the next 325 Imp gal (1477 litres) and the outermost 132 Imp gal (600 litres). Aft of these were box-like tanks containing 147 Imp gal (668 litres, outer) and 111 Imp gal (505 litres, inner).

D/F loop aerial and aerial mast
A streamlined fairing contained the direction finding (D/F) loop antenna, and was located adjacent to the navigator's astrodome, which itself was used for taking sun and star shots with an old-fashioned sextant. A huge aerial mast above the centre fuselage also served as a mounting point for twin pitot tubes.

Radar
Most Mk II Sunderlands were fitted with ASV Mk II radar, which necessitated the fitting of four vertical dipole mast antennas above the fuselage and 16 transmitter loops on masts projecting horizontally from the rear fuselage in two fore-and-aft rows of four masts. Yagi homing aerials were carried below the outer wings and above the flight deck. The Germans eventually developed radar warning gear capable of detecting the ASV Mk II's 4.92-ft (1.5-m) transmissions. This was countered by ASV Mk III, which incorporated Bomber Command's H_2S.

Mid-upper turret
In all but the first few Mk IIs, a pair of open K-gun mountings (with streamlined metal slipstream protectors) in the upper fuselage were replaced by a power-operated Frazer Nash FN.7 mid-upper turret offset to starboard. This was similar to the unit used in the Blackburn Botha, and contained a pair of 0.303-in (7.7-mm) machine-guns.

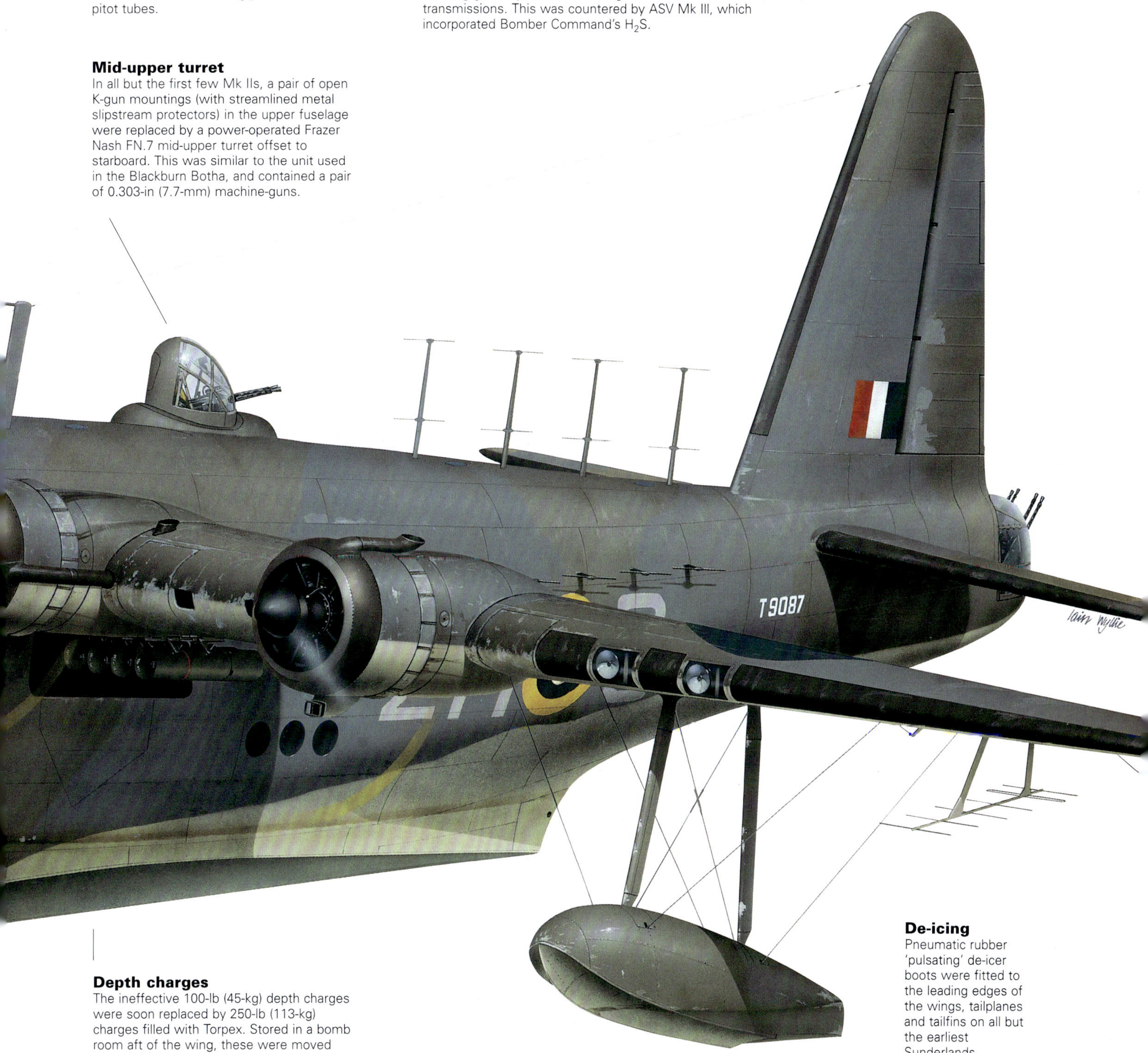

Depth charges
The ineffective 100-lb (45-kg) depth charges were soon replaced by 250-lb (113-kg) charges filled with Torpex. Stored in a bomb room aft of the wing, these were moved electrically out below the wing through hinged drop panels in the fuselage.

De-icing
Pneumatic rubber 'pulsating' de-icer boots were fitted to the leading edges of the wings, tailplanes and tailfins on all but the earliest Sunderlands.

Defensive armament
The Sunderland's generous array of defensive armament, which the Luftwaffe believed to include 20-mm cannon, resulted in it being nicknamed the *Fliegende Stachelschwein* ('Flying Porcupine'). While it is true that in several incidents, lone Sunderlands shot down several attacking fighters, the aircraft's 0.303-in (7.7-mm) machine-gun armament often proved insufficiently light to inflict decisive damage on an attacking aircraft.

15
DANGER-FAN INTAKE-DANGER
RESCUE
VMFA-121

MODERN AIRCRAFT

Attack Aircraft

Dassault Etendard/Super Etendard

Although the spotlight has been focused on the Super Etendard's precision attack operations, the type has always had a respectable conventional attack capability. This Flottille 14F aircraft totes four MATRA 155 rocket pods, each able to carry 18 2.70-in (68-mm) SNEB rockets.

Super Etendard

Cutaway key
1 Radome
2 Scanner housing
3 Flat plate radar scanner
4 Scanner tracking mechanism
5 Thomson-CSF/ESD Agave multi-mode radar equipment package
6 Refuelling probe housing
7 Retractable inflight refuelling probe
8 Nav/attack avionics equipment
9 UHF aerial
10 Pitot head
11 Temperature probe
12 Refuelling probe retraction link and jack
13 Cockpit front pressure bulkhead
14 Instrument panel shroud
15 Windscreen panels
16 Head-up display
17 Control column
18 Rudder pedals
19 Cockpit section framing
20 Pressure floor level
21 Side console panel
22 Engine throttle lever
23 Radar hand controller
24 Nose undercarriage pivot fixing
25 Carrier deck approach lights
26 Nosewheel leg doors
27 Hydraulic steering jacks
28 Nosewheel forks
29 Nosewheel, aft retracting
30 Nose undercarriage leg strut
31 Rear breaker strut
32 Hydraulic retraction jack
33 Port engine air intake
34 Boundary layer splitter plate
35 Air-conditioning system ram air intake
36 Cockpit sloping rear pressure bulkhead
37 Boundary layer spill duct
38 Hispano-SEMMB built Martin-Baker CM4A ejection seat
39 Starboard engine air intake
40 Ejection seat headrest
41 Face blind firing handle
42 Cockpit canopy cover, upward hinging
43 Canopy hinge point
44 Canopy emergency release
45 Air-conditioning plant
46 Intake duct framing
47 Ventral cannon blast trough
48 Cannon barrel
49 Oxygen bottles (two)
50 Navigation and communications avionics equipment racks
51 Martin Pescador air-to-surface missile (Argentine aircraft only)
52 Martin Pescador guidance pod
53 Starboard external fuel tank
54 Equipment bay dorsal access panels
55 Fuel system inverted flight accumulator
56 Intake suction relief door
57 DEFA 552A 30-mm cannon (two)
58 Ground power and intercom sockets
59 Ventral cannon pack access door
60 Ammunition magazine, 125 rounds per gun
61 Air system pre-cooler, avionics cooling air
62 Forward fuselage bag-type fuel tanks, total internal capacity 719 Imp gal (3270 litres)
63 Fuel tank access panels
64 Wing spar attachment main frames
65 Fuselage dorsal systems ducting
66 Avionics cooling air exit louvres
67 IFF aerial
68 Starboard wing integral fuel tank
69 Pylon attachment points
70 MATRA 155 rocket launcher pack, 18 x 2.70-in (68-mm) rockets
71 Leading-edge dog-tooth
72 Leading-edge flap control rod and links
73 Starboard leading-edge flap, lowered
74 Aileron hydraulic actuator
75 Wing fold hydraulic jack
76 Outboard, folding wing tip panel
77 Strobe identification light
78 Starboard navigation light
79 Starboard wingtip folded position
80 Fixed portion of trailing edge
81 Starboard aileron
82 Aileron hinge control
83 Aileron/spoiler interconnecting link
84 Spoiler hydraulic actuator
85 Starboard spoiler, open
86 Double-slotted Fowler-type flap, down position
87 Rear fuselage bag-type fuel tanks
88 Rudder control cables
89 Engine starter housing
90 Compressor intake
91 Forward engine mounting bulkhead
92 Accessory gearbox drive shaft
93 Gearbox driven generators (two)
94 Engine accessory equipment
95 SNECMA Atar 8K50 non-afterburning turbojet engine
96 Engine bleed air duct to air conditioning system
97 Rudder control cable quadrant
98 Fin spar attachment joint
99 Leading-edge access panel to control runs
100 All-moving tailplane pitch trim control electric motor
101 Tailplane root leading-edge aerodynamic notch
102 Elevator hydraulic actuator
103 Upper/lower fin segment joint
104 Tailplane sealing plate
105 Rudder hydraulic actuator
106 Tailfin construction
107 Starboard all-moving tailplane
108 Forward radar-warning antenna
109 VOR aerial (Argentine aircraft only)
110 VHF aerial

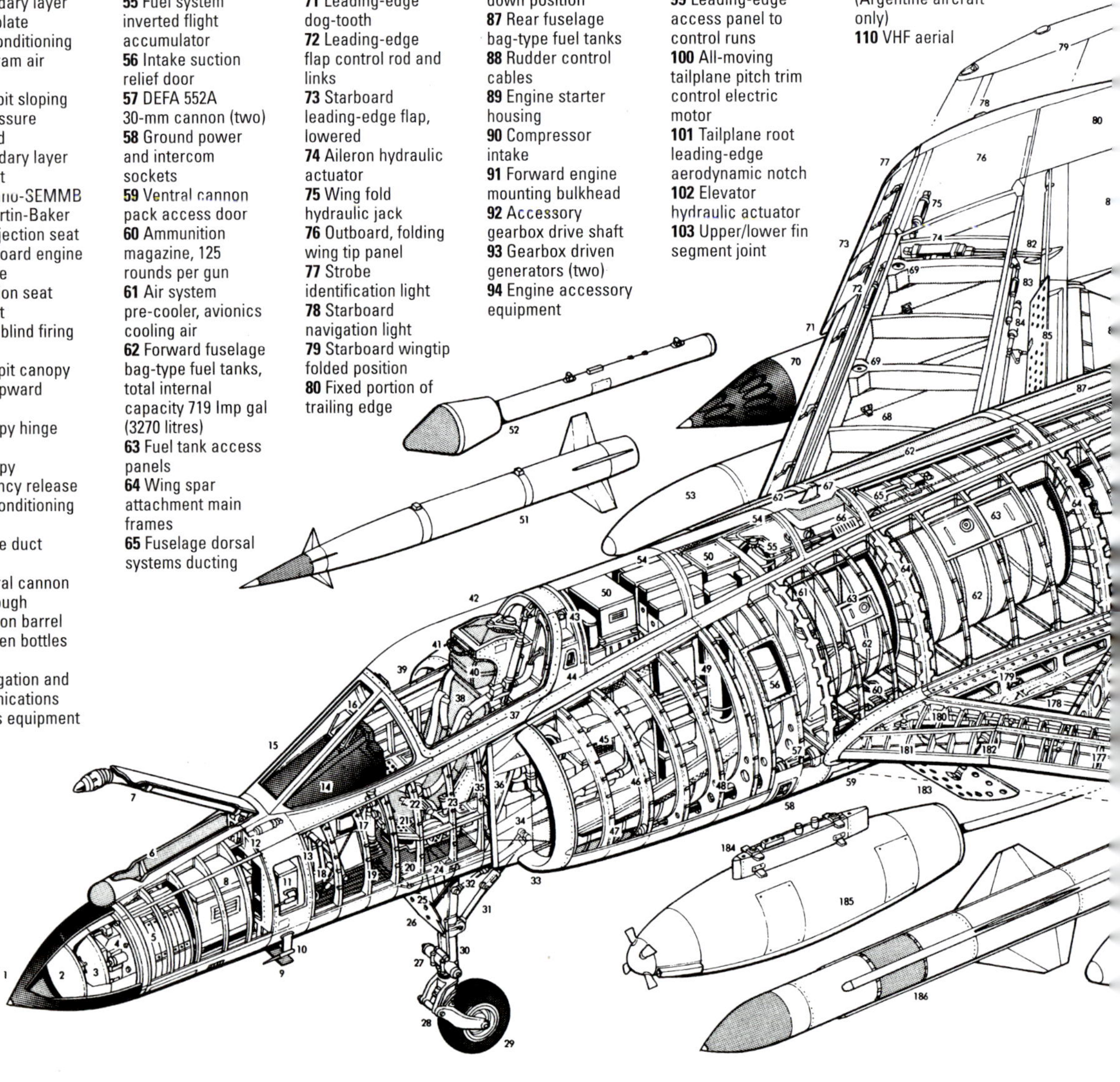

With the F-8P Crusaders of Flottille 12F having retired in December 1999, and the Etendard IVPMs of 16F and Alizés of 6F in 2000, the Super Etendard is the last survivor of the 'old' generation of Aéronavale aircraft. However, it will serve for the first decade of the 21st century alongside the Aéronavale's 'new' generation, in the shape of the Northrop Grumman Hawkeye and Dassault Rafale M. This particular 'SEM', seen during one of the last cruises of the F-8P 'Crouze', is a Standard 3 aircraft carrying an ATLIS II designator pod.

SPECIFICATION

Super Etendard

Dimensions

Length: 46 ft 11½ in (14.31 m)
Height: 12 ft 8 in (3.86 m)
Wingspan: 31 ft 6 in (9.60 m)
Wingspan (folded): 25 ft 7 in (7.80 m)
Wing area: 305.71 sq ft (28.40 m²)
Wing aspect ratio: 3.23
Wheel track: 11 ft 6 in (3.50 m)
Wheelbase: 15 ft 9 in (4.80 m)

Powerplant

One SNECMA Atar 8K50 turbojet rated at 11,023 lb (49.05 kN) thrust

Weights

Empty equipped: 14,330 lb (6500 kg)
Maximum take-off: 26,455 lb (12000 kg)

Fuel and load

Internal fuel: 719 Imp gal (3270 litres)
External fuel: 616 Imp gal (2800 litres)
Maximum ordnance: 4,630 lb (2100 kg)

Performance

Maximum level speed at 36,090 ft (11000 m): 744 kt (857 mph; 1380 km/h)
Maximum level speed at sea level: 637 kt (733 mph; 1180 km/h)
Maximum rate of climb at sea level: 19,685 ft (6000 m) per minute
Service ceiling: 44,950 ft (13700 m)
Operational radius: 459 nm (528 miles; 850 km) on hi-lo-hi anti-ship mission carrying two drop tanks and one AM39 Exocet

Armament

Two 30-mm DEFA 552A cannon each with 125 rounds. Four underwing and one centreline hardpoint for drop tanks and various weapons including bombs up to 881 lb (400-kg), rocket pods, BAP100 anti-runway bombs, BAT120 area-denial weapons, EU4/Paveway II LGBs, AS30L laser-guided missile, AM39 Exocet, ASMP tactical nuclear missile or Magic 2 short-range AAMs

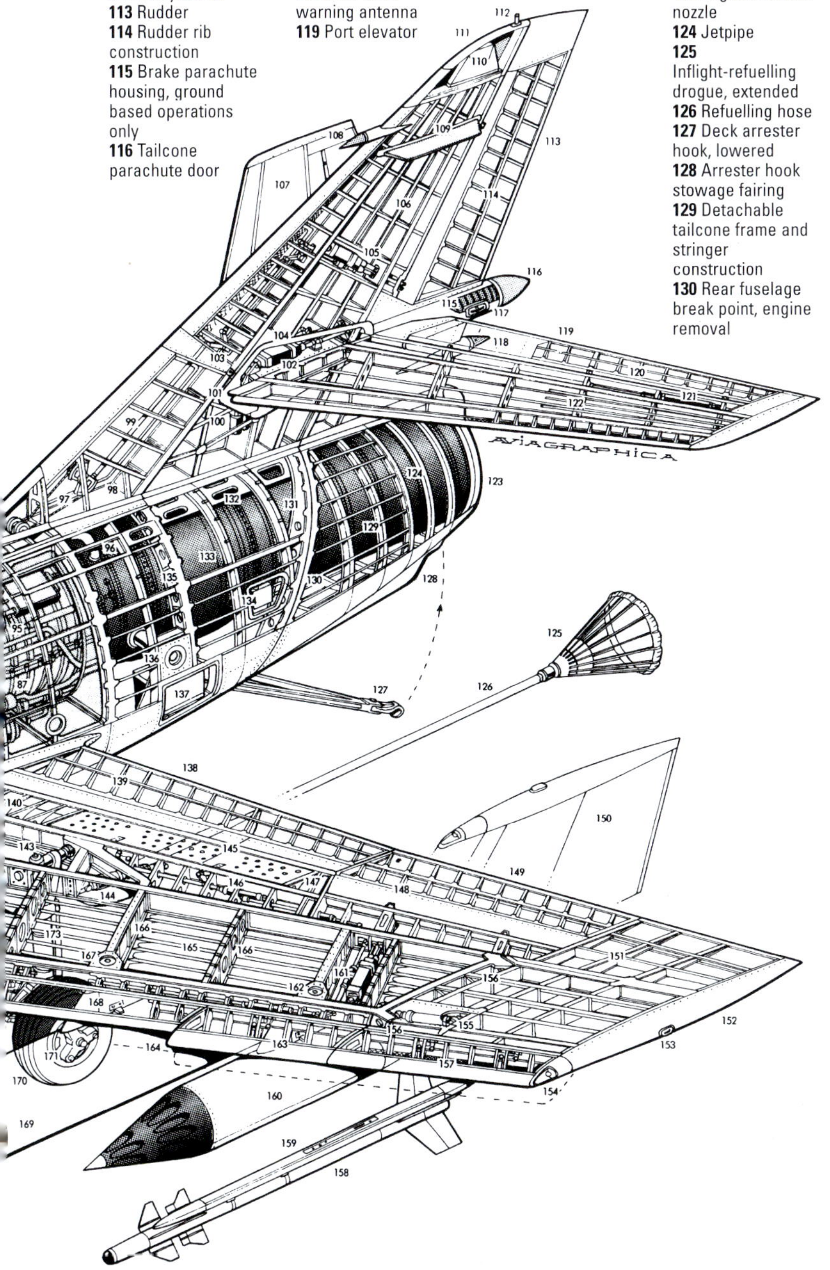

111 Fintip aerial fairing
112 Command telemetry aerial
113 Rudder
114 Rudder rib construction
115 Brake parachute housing, ground based operations only
116 Tailcone parachute door
117 Tail navigation and anti-collision lights
118 Rear radar warning antenna
119 Port elevator
120 Elevator rib construction
121 Elevator damper
122 Port all-moving tailplane construction
123 Engine exhaust nozzle
124 Jetpipe
125 Inflight-refuelling drogue, extended
126 Refuelling hose
127 Deck arrester hook, lowered
128 Arrester hook stowage fairing
129 Detachable tailcone frame and stringer construction
130 Rear fuselage break point, engine removal

The ATLIS II laser designation pod, introduced by the Super Etendard Modernisé Standard 3, is normally carried on the centreline pylon. It provides daylight-only capability.

131 Sloping fin spar attachment bulkhead
132 Engine bay heat shroud
133 Engine turbine section
134 Radar warning power amplifier
135 Fin spar and engine mounting bulkhead
136 Main engine mounting spigot
137 Aft avionics equipment bays, port and starboard
138 Port double-slotted Fowler-type flap
139 Flap rib construction
140 Flap shroud ribs
141 Inboard flap guide rail
142 Main undercarriage wheel bay
143 Main undercarriage leg pivot fixing
144 Flap hydraulic jack
145 Port spoiler
146 Spoiler hydraulic jack and control links
147 Outboard flap guide rail
148 Aileron rib construction
149 Port aileron
150 Port wingtip, folded position
151 Wingtip panel construction
152 Wingtip fairing
153 Port navigation light
154 Strobe identification light
155 Wing fold hydraulic jack
156 Wing fold hinge joints
157 Outboard leading-edge flap segment
158 MATRA 550 Magic air-to-air missile
159 Missile launch rail
160 MATRA 155 18 x 2.70-in (68-mm) rocket pod
161 Aileron hydraulic actuator
162 Outboard pylon attachment joint
163 Outboard stores pylon
164 Leading-edge dog-tooth
165 Machined wing skin/stringer panel
166 Wing rib construction
167 Inboard pylon attachment joint
168 Inboard stores pylon
169 External fuel tank, 242 Imp gal (1100 litres)
170 Port mainwheel
171 Hydraulic multi-plate disc brake
172 Torque scissor links
173 Main undercarriage leg strut
174 Hydraulic retraction jack
175 Port wing integral fuel tank bays
176 Inboard leading-edge flap segment
177 Leading-edge flap rib construction
178 Ventral catapult strop hook
179 Wingroot bolted attachment joint
180 Leading-edge flap hydraulic jack
181 Extended chord wingroot leading-edge
182 Airbrake hydraulic jack
183 Ventral airbrake, port and starboard
184 Fuselage centreline pylon
185 Inflight refuelling 'buddy' pack
186 AM39 Exocet AShM

Dassault Mirage III

The most far-reaching upgrade proposed for the Mirage III was conversion to relaxed stability configuration with the installation • an FBW system borrowed from the Mirage 2000. The Mirage 3N (Nouvelle Génération) first flew on 21 December 1982 wearing a sma blue and white colour scheme and was fitted with foreplane Marketed as a 'poor man's' Mirage 2000, the 3NG was demonstrated . Farnborough and Paris, but export orders did not materialis

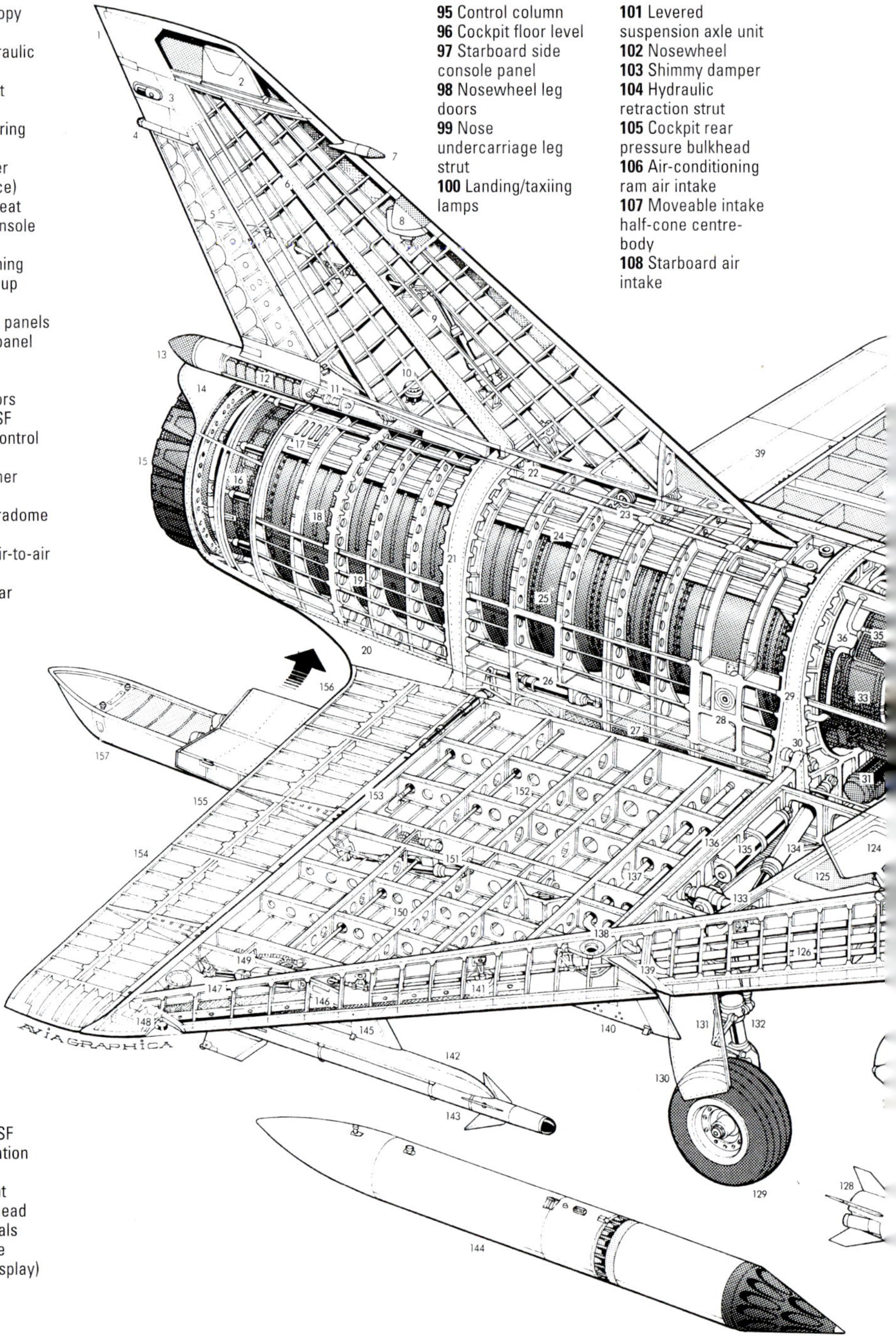

Mirage IIIE

Cutaway key

1 Glass-fibre fintip aerial fairing
2 VHF aerial
3 Tail navigation and anti-collision lights
4 Tail radar warning antenna
5 Rudder construction
6 Fin main spar
7 Passive radar antenna
8 UHF aerial
9 Rudder hydraulic actuator
10 Magnetic detector
11 Parachute release link
12 Brake parachute housing
13 Parachute fairing
14 Exhaust nozzle shroud
15 Variable-area exhaust nozzle flaps
16 Nozzle jacks
17 Cooling air louvres
18 Jet pipe
19 Rear fuselage frame and stringer construction
20 Wingroot trailing-edge fillet
21 Fin attachment main frame
22 Fin spar attachment joint
23 Control cable runs
24 Engine bay/jet pipe thermal lining
25 Afterburner duct
26 Elevon compensator hydraulic jack
27 Ventral fuel tank
28 Main engine mounting
29 Wing spar/fuselage main frame
30 Main spar joint
31 Engine gearbox driven generator
32 Engine accessory compartment
33 SNECMA Atar 9C afterburning turbojet
34 Cooling system air intakes
35 Heat exchanger
36 Engine oil tank
37 IFF aerial
38 Port wing integral fuel tank, total internal capacity 733 Imp gal (3330 litres)
39 Inboard elevon
40 Outboard elevon
41 Port navigation light
42 Cambered leading-edge ribs
43 Port wing pylon fixing
44 Leading-edge notch
45 Port leading-edge fuel tank
46 Main undercarriage pivot fixing
47 Fuselage dorsal systems ducting
48 Air system piping
49 Turbojet intake
50 Engine starter housing
51 Fuselage fuel tanks
52 Equipment cooling system air filter
53 Computer system voltage regulator
54 Oxygen bottles
55 Inverted flight fuel system accumulator
56 Intake ducting
57 Matra 530 missile computer
58 VHF radio transmitter/receiver
59 Gyro platform multiplier
60 Doppler transceiver
61 Navigation system computer
62 Air data computer
63 Nord missile encoding supply
64 Radio altimeter transceiver
65 Heading and inertial correction computer
66 Armament junction box
67 Radar programme controller
68 Canopy external release
69 Canopy hinge
70 Radio and electronics bay access fairing
71 Fuel tank stabilising fins
72 286-Imp gal (1300-litre) auxiliary fuel tank (374-Imp gal/1700-litre alternative)
73 137-Imp gal (625-litre) drop tank
74 Cockpit canopy cover
75 Canopy hydraulic jack
76 Ejection seat headrest
77 Face blind firing handle
78 Martin-Baker (Hispano licence) RM4 ejection seat
79 Port side console panel
80 Canopy framing
81 Pilot's head-up display
82 Windscreen panels
83 Instrument panel shroud
84 Instrument pressure sensors
85 Thomson-CSF Cyrano II fire control radar
86 Radar scanner dish
87 Glass-fibre radome
88 Pitot tube
89 Matra 530 air-to-air missile
90 Doppler radar fairing
91 Thomson-CSF Doppler navigation radar antenna
92 Cockpit front pressure bulkhead
93 Rudder pedals
94 Radar scope (head-down display)
95 Control column
96 Cockpit floor level
97 Starboard side console panel
98 Nosewheel leg doors
99 Nose undercarriage leg strut
100 Landing/taxiing lamps
101 Levered suspension axle unit
102 Nosewheel
103 Shimmy damper
104 Hydraulic retraction strut
105 Cockpit rear pressure bulkhead
106 Air-conditioning ram air intake
107 Moveable intake half-cone centre-body
108 Starboard air intake

During the Six-Day war, the air forces of the Arab nations surrounding Israel continued their attacks, but were met with swift resistance from the IDF/AF's fighter units. The Israeli Shahaks (local name for the Mirage III) were at the forefront of the defensive operations. The top-scoring Shahak was 59 (above), which was credited with scoring 13 kills in the hands of a number of different pilots during its Israeli service career. During the Six-Day War the aircraft downed an Egyptian Il-14 on 5 June 1967 and a MiG-19 the next day. Its finest hour was on 10 July 1970 when, piloted by Israeli Baharav, it downed a pair of Egyptian MiG-21s.

SPECIFICATION

Mirage IIIE

Dimensions

Length: 49 ft 3½ in (15.03 m)
Height: 14 ft 9 in (4.50 m)
Wingspan: 26 ft 11⅔ in (8.22 m)
Wing area: 376.75 sq ft (35.00 m^2)
Aspect ratio: 1.94
Wheel track: 10 ft 4 in (3.15 m)
Wheel base: 15 ft 11¾ in (4.87 m)

Powerplant

One SNECMA Atar 9C-3 rated at 9,436 lb st (41.97 kN) dry and 13,668 lb st (60.80 kN) with afterburning, and provision for one jettisonable SEPR 84 rocket booster rated at 3,307 lb st (14.71 kN)

Weights

Empty: 15,542 lb (7050 kg)
Normal take-off: 21,164 lb (9600 kg)
Maximum take-off: 30,203 lb (13700 kg)

Fuel and load

Internal fuel: 631.4 US gal (2390 litres)
External fuel: Up to two 449-, 343-, 291- or 165-US gal (1700-, 1300-, 1100- or 625-litre) drop tanks, or two 132-US gal (500-litre) non-jettisonable supersonic tanks, or two 66-US gal (250-litre) JL-100 combined drop tanks/rocket launchers, or two 291-US gal (1100-litre) fuel/electronic equipment tanks
Maximum ordnance: 8,818 lb (4000 kg)

Performance

Maximum level speed 'clean' at 39,370 ft (12000 m): 1,268 kt (1,460 mph; 2350 km/h)
Cruising speed at 36,090 ft (11000 m): 516 kt (594 mph; 956 km/h)
Ferry range with three drop tanks: 2,152 nm (2,486 miles; 4000 km)
Combat radius: 647 nm (746 miles; 1200 km)
Maximum rate of climb at sea level: More than 16,405 ft (5000 m) per minute
Climb to 36,090 ft (11000 m): 3 minutes
Service ceiling: 55,775 ft (17000 m) or 75,460 ft (23000 m) with rocket pack
Take-off run: Between 2,297 and 5,249 ft (700 and 1600 m), depending on mission-related maximum weight
Landing run: 2,297 ft (700 m) with brake chute
***g* limits:** +4.83 in a sustained turn at Mach 0.9 at 16,405 ft (5000 m)

Armament

Cannon armament of two 30-mm DEFA 552 cannon with 125 rounds per gun. Basic IIIC interceptor version with centreline pylon for one radar-guided missile, initially Nord 5103 or MATRA R.511, subsequently MATRA R530 (Hughes AIM-26 Falcon on Swiss aircraft). Two wing pylons for infra-red guided missile, either AIM-9B/P Sidewinder or MATRA R550 Magic. Attack capability in form of JL-100 fuel tank/rocket pod. Mirage IIIE multi-role aircraft introduced a maximum of five pylons with a maximum weaponload of 8,818 lb (4000 kg) including most free-fall bombs and rocket pods. Attack missiles include Aérospatiale AS30 and MATRA AS37 Martel. French aircraft wired for 15-kT yield AN52 tactical nuclear free-fall bomb

109 Nosewheel well door (open position)
110 Intake centre-body screw jack
111 Air-conditioning plant
112 Boundary layer bleed air duct
113 Centre fuselage bomb rack
114 882-lb (400-kg) HE bombs
115 Cannon barrels
116 30-mm DEFA cannon (2), 250 rounds per gun
117 Ventral gun pack
118 Auxiliary air intake door
119 Electrical system servicing panel
120 Starboard 30-mm DEFA cannon
121 Front spar attachment joint
122 Fuel system piping
123 Airbrake hydraulic jack
124 Starboard airbrake, upper and lower surfaces (open position)
125 Airbrake housing
126 Starboard leading-edge fuel tank
127 AS37 Martel, radar-guided air-to-ground missile
128 Nord AS30 air-to-air missile
129 Starboard mainwheel
130 Mainwheel leg door
131 Torque scissor links
132 Shock absorber leg strut
133 Starboard main undercarriage pivot fixing
134 Hydraulic retraction jack
135 Main undercarriage hydraulic accumulator
136 Wing main spar
137 Fuel system piping
138 Inboard pylon fixing
139 Leading-edge notch
140 Starboard inner stores pylon
141 Control rod runs
142 Missile launch rail
143 AIM-9 Sidewinder air-to-air missile
144 JL-100 fuel and rocket pack, 55 Imp gal (250 litres) of fuel plus 18 x 68-mm unguided rockets
145 Outboard wing pylon
146 Outboard pylon fixing
147 Front spar
148 Starboard navigation light
149 Outboard elevon hydraulic jack
150 Starboard wing integral fuel tank
151 Inboard elevon hydraulic actuator
152 Wing multi-spar and rib construction
153 Rear spar
154 Outboard elevon construction
155 Inboard elevon construction
156 Elevon compensator
157 110-Imp gal (500-litre) auxiliary fuel tanks

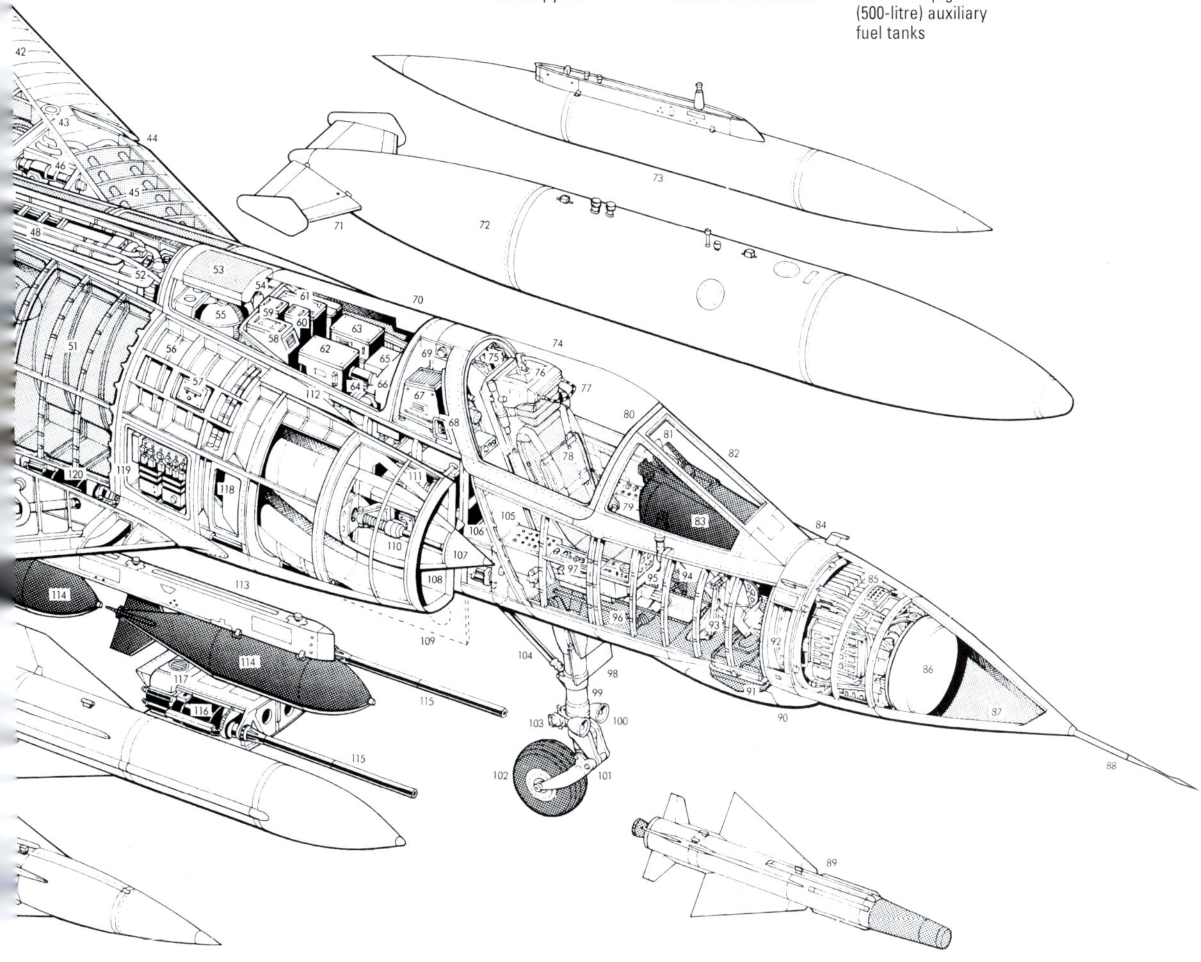

Dassault Mirage 5/50

Seen here accompanying an RAF Bucca (wearing Desert Storm 'pink') is a Mirage from the Belgian air force's No. 8 Escadrille. nearest the Belgian Mirage came to combat on 6 January 1991 when 15 aircraft were dep to Erhac, Turkey to boost offensive forces i event that Iraq attacked Turkey (durin imminent UN operation to liberate Kuv

Mirage 50

Cutaway key
1 Pitot head
2 Radome
3 Scanner housing
4 Flat plate radar scanner
5 Scanner tracking mechanism
6 Thomson-CSF/EMID Agave lightweight multi-mode radar (Cyrano IV alternative fit)
7 Nose compartment construction
8 Fixed inflight-refuelling probe (optional)
9 Nose compartment access door
10 Avionics equipment compartment
11 Ventral radar altimeter aerials
12 IFF aerial
13 Instrument system total pressure head
14 Front pressure bulkhead
15 Incidence probe
16 Ventral Doppler aerial (optional equipment)
17 Cockpit pressure floor level
18 Rudder pedals
19 Radar scope (head-down display)
20 Control column
21 Instrument panel shroud
22 Windscreen panels
23 Pilot's head-up display
24 CC 420 30-mm cannon pod
25 Reconnaissance variant nose profile
26 Forward oblique OMERA 53 long-range camera
27 Lateral OMRAF 100 cameras, port and starboard
28 Camera mounting frame
29 Lateral oblique OMERA F 200 cameras, port and starboard
30 Cannon pod ammunition magazine, 250 rounds
31 Cockpit canopy cover, upward-hinging
32 Ejector seat headrest
33 Face blind-firing handle
34 Martin-Baker (Hispano licence-built) RM.4 ejector seat
35 Safety harness
36 Starboard side console panel
37 Engine throttle lever
38 Cockpit section framing
39 Port side console panel
40 Nose landing gear
41 Nosewheel leg doors
42 Nose landing-gear leg strut
43 Landing/taxiing lamps
44 Levered suspension axle beam
45 Nosewheel, aft-retracting
46 Hydraulic retraction jack/lock strut
47 Cockpit rear pressure bulkhead
48 Elevon artificial feel unit
49 Port engine air inlet
50 Movable inlet half-cone centrebody
51 Air-conditioning ram air intake
52 Battery
53 Canopy emergency release
54 Canopy hydraulic jack
55 Radar programme controller
56 Canopy hinge point
57 374-Imp gal (1700-litre) external fuel tank; 110-Imp gal (500-litre) alternative
58 Dorsal spine fairing
59 Forward fuselage fuel lank, total internal capacity 750 Imp gal (3410 litres)
60 Boundary layer spill duct
61 Air-conditioning plant
62 Inlet centrebody screw jack
63 Inlet duct
64 Ventral cannon muzzle blast trough
65 Cannon barrel
66 Intake suction relief door, spring-loaded
67 Electrical system equipment
68 Weapons system control units
69 Oxygen bottles (two)
70 Inverted flight fuel system accumulator
71 Voltage regulator
72 Leading-edge fuel tank
73 Starboard main landing-gear pivot fixing
74 Equipment cooling system air filter
75 Fuselage dorsal systems ducting
76 Bleed air and fuel system piping
77 Fuselage fuel tanks
78 Bifurcated intake trunking
79 Port 30-mm DEFA cannon
80 Front spar attachment joint
81 Engine accessory equipment drive shaft
82 Accessory equipment cooling air duct
83 Engine starter housing
84 Turbojet intake
85 Hydraulic fluid reservoir
86 Gearbox-driven generator
87 Main spar attachment joint
88 Wing spar/fuselage main frame
89 Engine compressor section
90 Bleed air pre-cooler ram air intake
91 Engine oil tank
92 VHF aerial
93 Starboard wing integral fuel tank
94 Elevon control rods
95 Leading-edge notch
96 MATRA R.550 Magic air-to-air missile
97 Missile launch rail
98 Cambered leading-edge ribs
99 Starboard navigation light
100 Outboard elevon

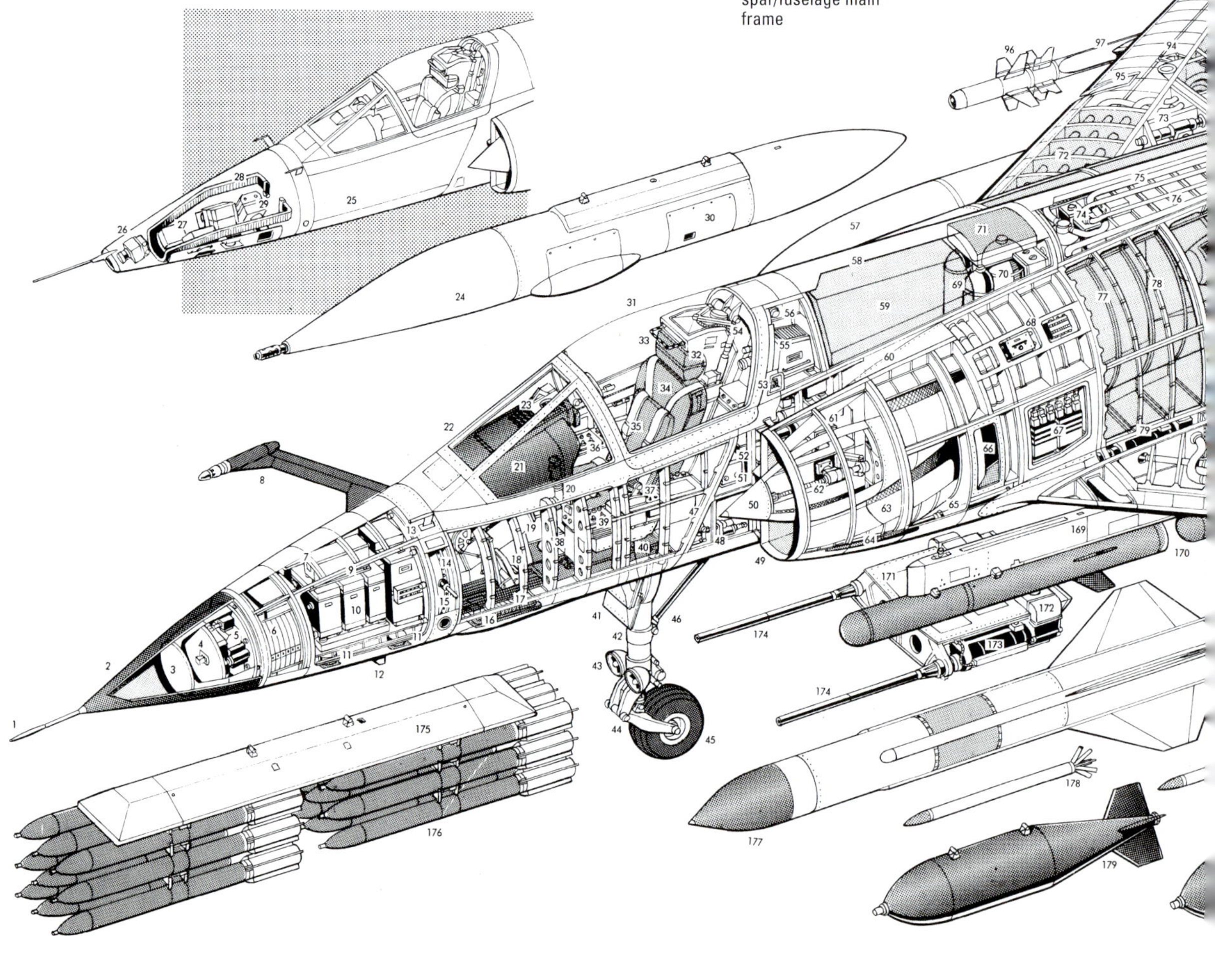

Improvements in low-speed handling and short field performance were sought through the Milan (Kite) programme undertaken in collaboration with Switzerland. Actual production did not go ahead but the enhanced flying characteristics led to the fitment of canards to Israeli Kfirs.

SPECIFICATION

Mirage 50M

Dimensions

Length: 51 ft ½ in (15.56 m)
Height: 14 ft 9 in (4.50 m)
Wingspan: 26 ft 11½ in (8.22 m)
Aspect ratio: 1.94
Area: 376.75 sq ft (35.00 m^2)
Optional canard foreplane area: 10.76 sq ft (1.00 m^2)
Wheel track: 10 ft 4 in (3.15 m)
Wheel base: 15 ft 11¾ in (4.87 m)

Powerplant

One SNECMA Atar 09K-50 rated at 11,055 lb st (49.20 kN) dry and 15,873 lb st (70.60 kN) with afterburning

Weights

Empty equipped: 15,763 lb (7150 kg)
Normal take-off: 22,046 lb (10000 kg)
Maximum take-off: 32,407 lb (14700 kg)

Fuel and load

Internal fuel for Mirage III conversions: 5,044 lb (2288 kg)
Internal fuel for Mirage 5 conversions: 5,974 lb (2710 kg)
External fuel: Up to two 449-, 343-, 291- or 165-US gal (1700-, 1300-, 1100- or 625-litre) drop tanks, or two 132-US gal (500-litre) non-jettisonable supersonic tanks, or two 66-US gal (250-litre) JL-100 combined drop tanks/rocket launchers, or two 291-US gal (1100-litre) non-jettisonable Bidon Cyclope fuel/electronic equipment tanks
Maximum ordnance: 8,818 lb (4000 kg)

Performance

Maximum level speed 'clean' at 39,370 ft (12000 m): 1,262 kt (1,453 mph; 2338 km/h)
Cruising speed at 36,090 ft (11000 m): 516 kt (594 mph; 956 km/h)
Combat radius: 817 miles (1315 km)
Maximum rate of climb at sea level: 36,614 ft (11160 m) per minute
Climb to 45,000 ft (13715 m): 4 minutes 42 seconds
Service ceiling: 59,055 ft (18000 m)
Take-off run at maximum take-off weight: 2,625 ft (800 m)

Armament

Similar to that of Mirage III family, although not nuclear-capable. Some aircraft designated Mirage 5 have Cyrano fire-control radar and are, in effect, Mirage IIIs. These can launch radar-guided air-to-air missiles, but other Mirage 5 family members cannot. Egypt's Mirage 5s employ US weapons, including the Rockeye cluster bomb. Venezuelan and some Pakistani aircraft (5PA3) have Cyrano IVM3 or Agave radar and the ability to launch the AM39 Exocet anti-ship missile

101 Elevon hydraulic actuators
102 Inboard elevon
103 Rudder control cable runs
104 SNECMA Atar 09K-50 afterburning turbojet engine
105 Main engine mounting trunnion
106 Fuselage ventral fuel tank
107 Elevon compensator hydraulic actuator
108 Engine turbine section
109 Engine bay internal heat shield
110 Fin spar attachment joints
111 Rudder control quadrant
112 Rudder hydraulic actuator
113 Remote compass transmitter
114 Fin main spar
115 UHF aerial

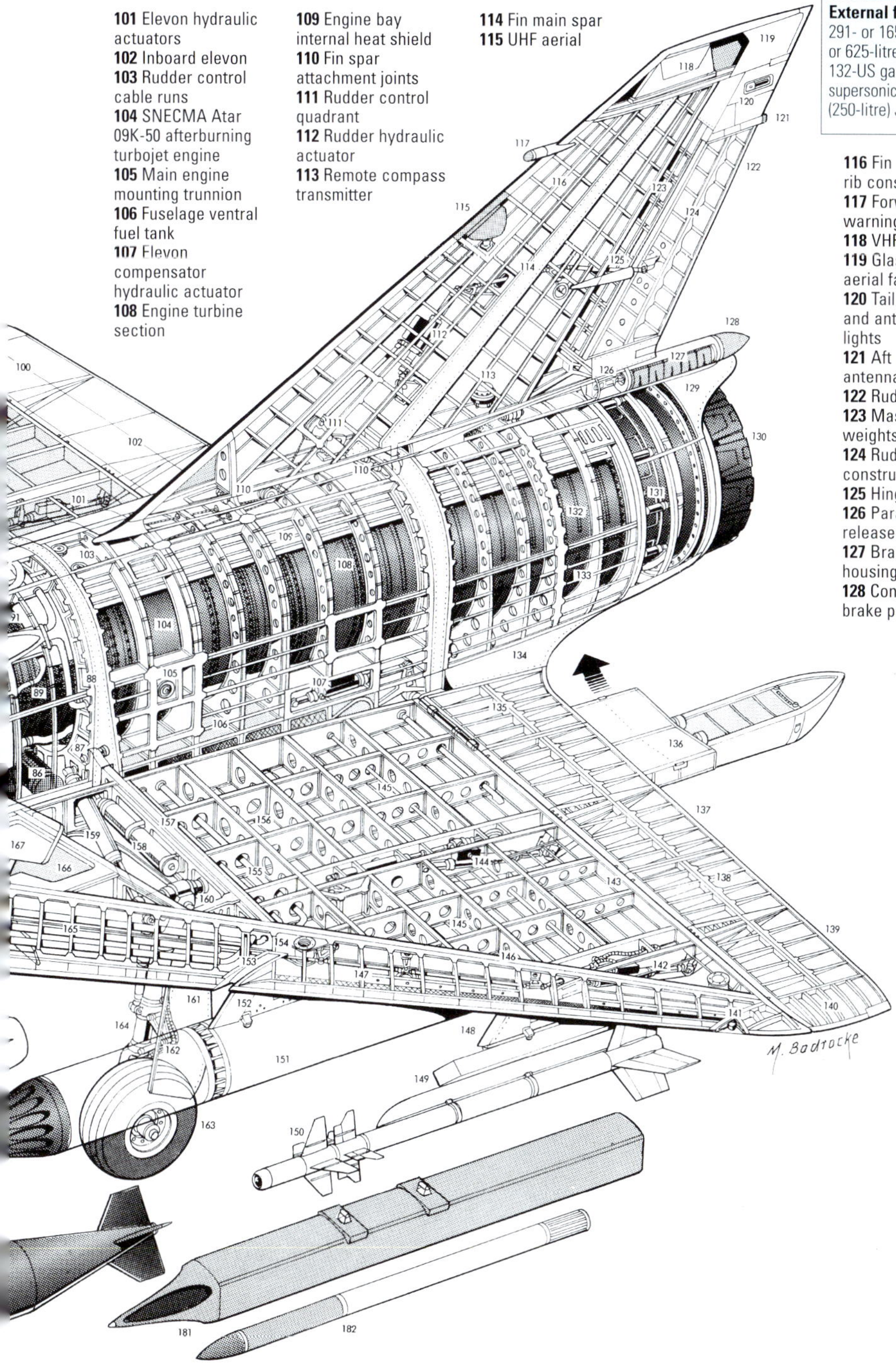

116 Fin leading-edge rib construction
117 Forward radar warning antenna
118 VHF aerial
119 Glassfibre fintip aerial fairing
120 Tail navigation and anti-collision lights
121 Aft radar warning antenna
122 Rudder
123 Mass balance weights
124 Rudder rib construction
125 Hinge control link
126 Parachute release link
127 Brake parachute housing
128 Conic fairing brake parachute door
129 Exhaust nozzle shroud
130 Variable-area exhaust nozzle flaps
131 Nozzle control jacks
132 Afterburner ducting
133 Rear fuselage frame and stringer construction
134 Wingroot trailing-edge fillet
135 Elevon inboard compensator panel
136 110-Imp gal (500-litre) auxiliary ventral fuel tank
137 Inboard elevon
138 Elevon rib construction
139 Outboard elevon
140 Cambered wingtip
141 Port navigation light
142 Outboard elevon hydraulic actuator
143 Rearspar
144 Inboard elevon hydraulic actuator
145 Port wing integral fuel tank
146 Front spar
147 Cambered leading-edge ribs
148 Outboard missile pylon
149 Missile launch rail
150 MATRA R.550 Magic air-to-air missile
151 JL.100 combined fuel and rocket pack, 56 Imp gal (250 litres) of fuel plus 18 x 68-mm FFAR
152 Inboard stores pylon
153 Leading-edge notch
154 Wing pylon attachment joint
155 Fuel system piping
156 Wing rib and multi-spar construction
157 Main spar
158 Landing-gear hydraulic accumulator
159 Hydraulic retraction jack/lock strut
160 Main landing-gear leg pivot fixing
161 Mainwheel leg door
162 Torque scissor links
163 Port mainwheel
164 Shock absorber leg strut
165 Port leading-edge fuel tank
166 Airbrake housing
167 Port airbrake, upper and lower surfaces, open position
168 Airbrake hydraulic jack
169 Fuselage centreline stores pylon
170 MATRA Durandal penetration bombs (10)
171 Ventral cannon pack, lowered
172 Ammunition feed chute, 125-rounds per gun
173 DEFA 30-mm cannon (two)
174 Cannon barrels
175 Centreline pylon bomb adaptor
176 BA-100 runway-cratering retarded bombs (18)
177 AM39 Exocet air-to-surface missile
178 68-mm folding fin aircraft rockets (FFAR)
179 551-lb (250-kg) general-purpose HE bomb
180 882-lb (400-kg) general-purpose HE bomb
181 C4 rocket launcher
182 100-mm unguided FFAR

Dassault Mirage F1

Possibly one of the most aesthetically pleasing combat aircraft yet designed, th Mirage F1 has demonstrated its considerable combat capability in conflicts ranging from anti-guerrilla strikes in South Africa to Desert Storm.

Mirage F1CT

Cutaway key
1 Pitot head
2 Glass-fibre radome
3 Radar scanner housing
4 Inflight-refuelling probe
5 Dynamic pressure sensor
6 Thomson-CSF Cyrano IVMR radar equipment module
7 Incidence probe
8 TMV 630A laser rangefinder
9 Rudder pedals
10 Control column
11 Instrument panel shroud
12 Windscreen panels
13 Thomson VE120 head-up display
14 Upward-hinging cockpit canopy cover
15 Martin-Baker F10M zero-zero ejection seat
16 Engine throttle lever
17 Side console panel
18 Nose undercarriage hydraulic retraction jack
19 Twin nosewheels, aft-retracting
20 Hydraulic steering mechanism
21 TACAN aerial
22 Cockpit sloping rear pressure bulkhead
23 Canopy jack
24 Canopy emergency release
25 Central intake control actuator
26 Moveable half-cone intake centre-body
27 Port air intake
28 Air-conditioning equipment bay
29 Intake centre-body screw jack
30 Intake suction relief door
31 Pressure refuelling connection
32 Port airbrake panel
33 Airbrake hydraulic jack
34 Retractable landing lamp
35 Forward fuselage integral fuel tank
36 Boundary layer spill duct
37 Avionics equipment bay
38 Power amplifier
39 Strobe light (white) and anti-collision beacon (red)
40 Fuel system inverted flight accumulator
41 30-mm DEFA cannon, starboard side only
42 Ammunition magazine, 135 rounds
43 External fuel tank
44 Starboard wing integral fuel tank
45 Forged steel wing attachment fitting
46 Inboard pylon attachment hardpoint
47 MATRA-Philips Phimat chaff/flare pod
48 Leading-edge flap
49 Starboard navigation light
50 Wingtip missile launch rail
51 MATRA Magic air-to-air missile
52 Starboard aileron
53 Two-segment double-slotted flaps
54 Spoiler panel (open)
55 Wing panel attachment machined fuselage main frame
56 Fuel system filters
57 Engine intake centre-body/starter housing
58 Wing panel attachment pin joints
59 Engine accessory equipment gearbox
60 SNECMA Atar 9K-50 afterburning engine
61 Engine bleed air pre-cooler
62 Rear spar attachment joint
63 Rear fuselage integral fuel tank
64 Engine turbine section
65 Engine bay thermal lining
66 Fin spar attachment joint
67 Starboard all-moving tailplane
68 Forward SHERLOC ECM antenna fairing
69 UHF antenna
70 VOR aerial
71 Fin-tip aerial fairing
72 IFF/VHF 1 aerial
73 Rear navigation light and anti-collision beacon
74 Aft SHERLOC ECM antenna
75 Rudder
76 Rudder hydraulic actuator
77 Rudder trim actuator
78 VHF 2 aerial
79 Brake parachute housing
80 Variable-area afterburner nozzle
81 Nozzle control jacks
82 Port all-moving tailplane
83 Honeycomb trailing-edge panel
84 Multi-spar tailplane construction
85 Tailplane pivot fitting
86 Tailplane hydraulic actuator
87 Autopilot controller
88 Port ventral fin
89 Inboard double-slotted flap segment
90 Flap hydraulic jack
91 Spoiler hydraulic jack
92 Port spoiler housing and actuating linkage
93 Port aileron hydraulic actuator
94 Outboard double-slotted flap segment
95 Port aileron
96 Wingtip missile interface unit
97 Port navigation light
98 Leading-edge flap
99 Port MATRA Magic air-to-air missile
100 68-mm rocket projectile
101 MATRA 18-round rocket launcher
102 Thomson-CSF ECM pod
103 Outer pylon attachment hardpoint
104 Wing panel multi-spar construction
105 Port wing integral fuel tank
106 Main undercarriage hydraulic retraction jack
107 Shock absorber strut
108 Twin mainwheels
109 Levered suspension axle
110 Mainwheel leg strut and leg rotating linkage
111 Leading-edge flap hydraulic jack
112 Main undercarriage wheel bay
113 Port ammunition bay, unused
114 Centre fuselage weapon pylon
115 881-lb (400-kg) HE bombs
116 Underwing MATRA-Corral conformal chaff/flare dispenser
117 Multiple bomb-carrier
118 Thomson-Brandt BAP-100 runway-cratering bomb or BAT-120 area denial/anti-armour munition
119 MATRA Belouga submunition dispenser
120 MATRA Durandal retarded concrete-piercing bomb

Libya ordered 38 F1s, including 16 F1EDs. The aircraft are tasked mainly with air defence, but have a limited ground-attack capability. This was practised against Chad in the 1980s, the Libyan Mirages declining to engage French F1s based in the area.

SPECIFICATION

Mirage F1C
(unless otherwise noted)

Dimensions

Wingspan without tip stores: 27 ft 6¾ in (8.40 m)
Mirage F1B: 27 ft 8⅓ in (8.44 m)
Wingspan with tip-mounted Magic AAMs: 30 ft 6¾ in (9.32 m)
Wing aspect ratio: 2.82
Wing area: 269.11 sq ft (25.00 m^2)
Length: 50 ft 2½ in (15.30 m)
Mirage F1B: 51 ft ⅕ in (15.55 m)
Wheel track: 8 ft 2½ in (2.50 m)
Wheel base: 16 ft 4¾ in (5.00 m)
Height: 14 ft 9 in (4.50 m)
Mirage F1B: 14 ft 8⅘ in (4.49 m)

Powerplant

One SNECMA Atar 9K-50 turbojet rated at 11,023 lb st (49.03 kN) dry and 15,785 lb st (70.21 kN) with afterburning

Weights

Empty: 16,314 lb (7400 kg)
Operating empty (Mirage F1B including pilots): 18,078 lb (8200 kg)
Normal take-off: 24,030 lb (10900 kg)
Mirage F1B: 24,691 lb (11200 kg)
Maximum take-off: 35,715 lb (16200 kg)

Fuel and load

Internal fuel capacity: 1,134 Imp gal (4300 litres)
Mirage F1B: 1,017 Imp gal (3850 litres).
External fuel capacity: Provision for one 581-Imp gal (2200-litre) tank on centreline and two 299-Imp gal (1130-litre) tanks under the wings
Maximum weaponload: 13,889 lb (6300 kg)

Performance

Maximum level speed 'clean' at 36,090 ft (11000 m): 1,453 mph (2338 km/h)
Maximum rate of climb at sea level: 41,930 ft (12780 m) per minute
Mirage F1B (without afterburning): 13,780 ft (4200 m) per minute
Service ceiling: 65,615 ft (20000 m)
Mirage F1B (stabilised supersonic ceiling): 52,495 ft (16000 m)
Take-off run at 25,353-lb (11500-kg) weight: 1,969 ft (600 m)
Landing run at 18,739-lb (8500-kg) weight: 2,198 ft (670 m)

Range

Combat radius: 264 miles (425 km) on a hi-lo-hi attack mission with 14 551-lb (250-kg) bombs, or 373 miles (600 km) on a lo-lo-lo attack mission with six 551-lb (250-kg) bombs and two drop tanks, or 863 miles (1390 km) on a hi-lo-hi attack mission with two 551-lb (250-kg) bombs and three drop tanks
Endurance: 2 hours 15 minutes on a CAP with two Super 530 AAMs and one drop tank
Mirage F1B normal training mission endurance: 2 hours

Armament

Two fixed internal DEFA 553 30-mm cannon with 135 rounds per gun; standard air-to-air load of two MATRA Magic or AIM-9 Sidewinder missiles on wingtip rails and either one MATRA R.530 on the centreline station or two Super 530Fs underwing. A limited ground-attack capability is available using various unguided bombs, cluster munitions and rockets.

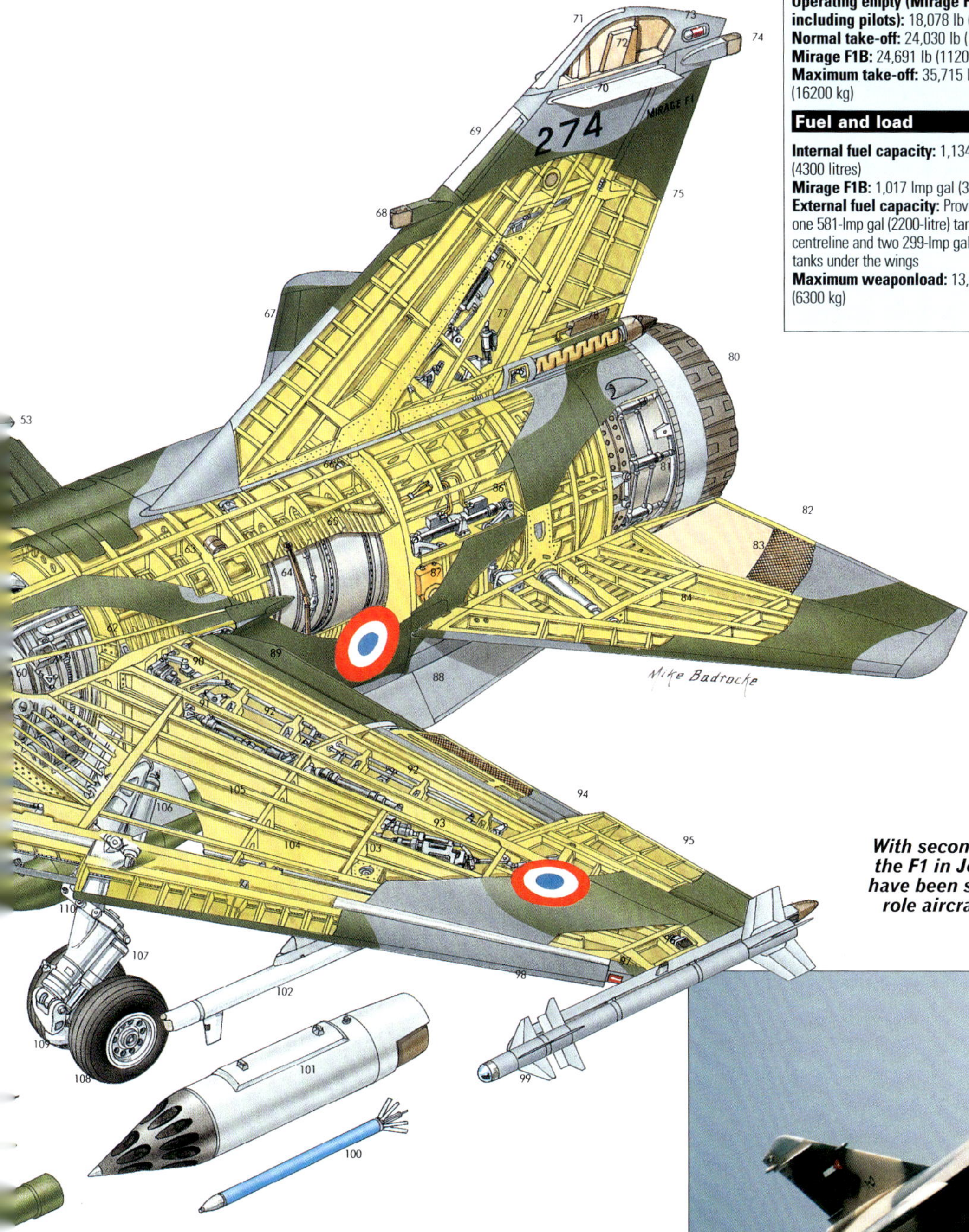

With second-hand ex-USAF F-16s complementing the F1 in Jordanian service, a number of Mirages have been sold to Spain. The F1 is a potent multi-role aircraft and it is unlikely that Jordan would wish to relinquish its entire fleet.

Dassault Mirage 2000

The various models within the Mirage 2000 family are, between them, capable of carrying a wide range of contemporary French weaponry. Early production aircraft were optimised for air-to-air operations, but did feature a limited ground-attack capability. This ability was enhanced in the Mirage 2000-5 which, while primarily a long-range interceptor has an expanded air-to-ground capability; the 2000N nuclear attack variant; and the 2000D, which is a dedicated conventional attack variant of the 2000N.

Mirage 2000C

Cutaway key
1 Pitot tube
2 Glass fibre radome
3 Flat-plate radar scanner
4 Thomson-CSF RDM multi-role radar unit (initial production aircraft)
5 Cassegrain monopulse planar antenna
6 Thomson-CSF RDI pulse-Doppler radar unit (later production aircraft)
7 Radar altimeter aerial
8 Angle-of-attack probe
9 Front pressure bulkhead
10 Instrument pitot heads
11 Temperature probe
12 Fixed inflight refuelling probe
13 Frameless windscreen panel
14 Instrument panel shroud
15 Static ports
16 Rudder pedals
17 Low-voltage formation light strip
18 VHF aerial
19 Nosewheel jack door
20 Hydraulic retraction jack
21 Nose landing gear leg strut
22 Twin nosewheels
23 Towing bracket
24 Torque scissor links
25 Landing/taxiing lamps
26 Nosewheel steering jacks
27 Nose landing gear leg doors
28 Cockpit flooring
29 Centre instrument console
30 Control column
31 Pilot's head-up display (HUD)
32 Canopy arch
33 Cockpit canopy cover
34 Starboard air intake
35 Ejection seat headrest
36 Safety harness
37 Martin-Baker Mk 10 zero-zero ejection seat
38 Engine throttle control and airbrake switch
39 Port side console panel
40 Nosewheel bay
41 Cannon muzzle blast trough
42 Electrical equipment bay
43 Port air intake
44 Intake half-cone centre body
45 Air-conditioning system ram air intake
46 Cockpit rear pressure bulkhead
47 Canopy emergency release handle
48 Hydraulic canopy jack
49 Canopy hinge point
50 Starboard intake strake
51 IFF aerial
52 Radio and electronics bay
53 Boundary layer bleed air duct
54 Air-conditioning plant
55 Intake centre-body screw jack
56 Cannon muzzle
57 Pressure refuelling connection
58 Port intake strake
59 Intake suction relief doors (above and below)
60 DEFA 554 30-mm cannon
61 Cannon ammunition box
62 Forward fuselage integral fuel tanks
63 Radio and electronics equipment
64 Fuel system equipment
65 Anti-collision light
66 Air system pre-cooler
67 Air exit louvres
68 Starboard wing integral fuel tank, total internal fuel capacity 836 Imp gal (3800 litres)
69 Wing pylon attachment hardpoints
70 Leading-edge slat hydraulic drive motor and control shaft
71 Slat screw jacks
72 Slat guide rails
73 Starboard wing automatic leading-edge slats
74 Matra 550 Magic 'dogfight' AAM
75 Missile launch rail
76 Outboard wing pylon
77 Radar warning antenna
78 Starboard navigation light

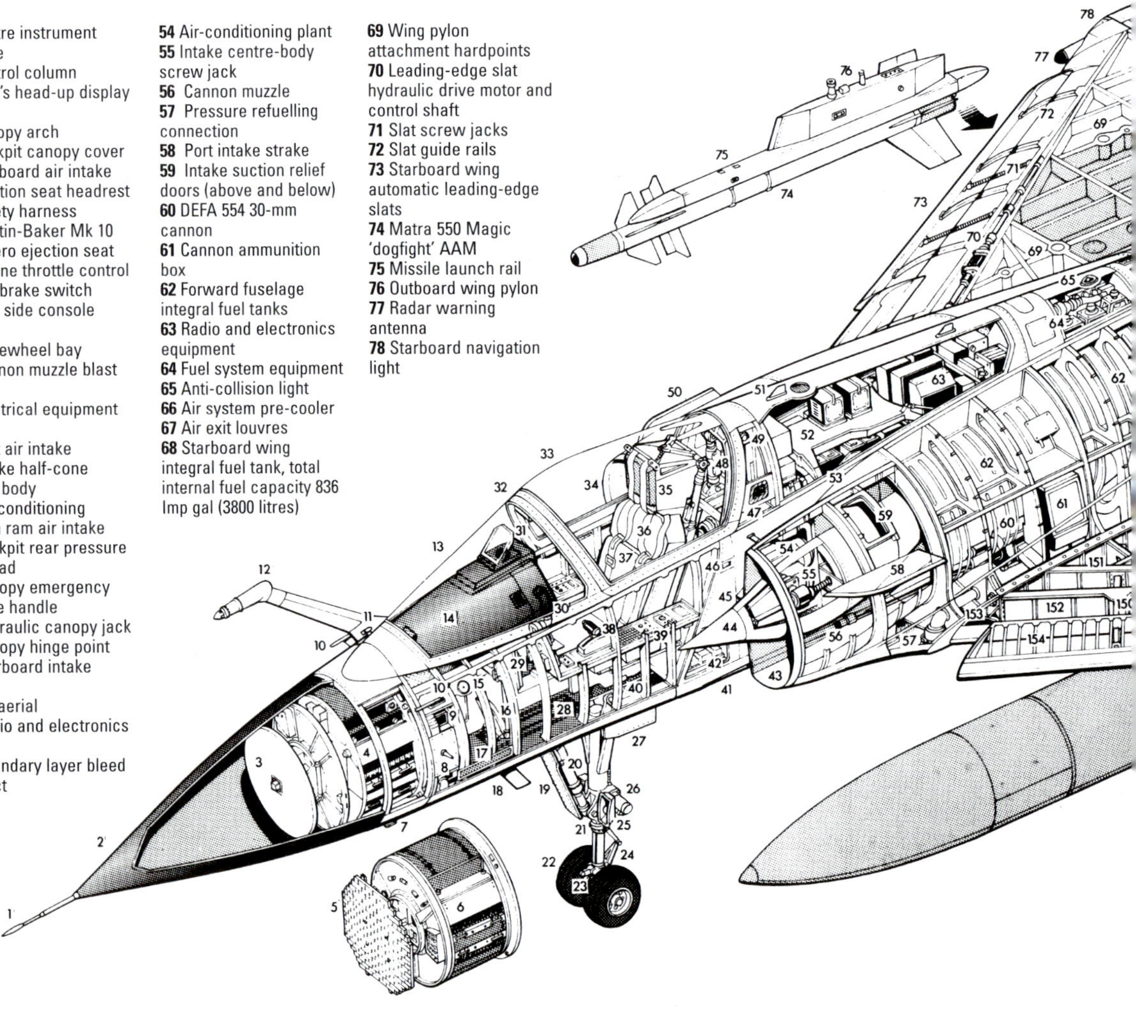

The Mirage 2000-01 first flew on 10 March 1978 from Istres in France and was piloted by Jean Coureau. During its first flight, basic handling and performance of the new fighter were validated, and the aircraft was pushed to Mach 1.3 in afterburner. While the Mirage 2000 is externally similar to its predecessor, the Mirage III, it is the adoption of technologies under the skin, such as its 'fly-by-wire' controls, that made it such an advancement over previous Mirages.

SPECIFICATION

Mirage 2000C

Dimensions

Fuselage length: 47 ft 1¼ in (14.36 m)
Wingspan: 29 ft 11½ in (9.13 m)
Wing area: 441.33 sq ft (41.00 m²)
Wing aspect ratio: 2.03
Height: 17 ft ¾ in (5.20 m)
Wheel track: 11 ft 1¾ in (3.40 m)
Wheel base: 16 ft 4¾ in (5.00 m)

Powerplant

One SNECMA M53-P2 turbofan rated at 14,462 lb st (64.33 kN) dry and 21,384 lb st (95.12 kN) with afterburning

Weights

Empty: 16,534 lb (7500 kg)
Normal take-off: 23,534 lb (10680 kg)
Maximum take-off: 37,478 lb (17000 kg)

Fuel and load

Internal fuel: 6,966 lb (3160 kg)
External fuel: 8,201 lb (3720 kg) in one 343-US gal (1300-litre) drop tank and two 449 US-gal (1700-litre) drop tanks.
Maximum ordnance: 13,889 lb (6300 kg)

Performance

Maximum speed at high level: Mach 2.2
Maximum speed at sea level: Mach 1.2
Minimum speed in stable flight: 100 kts (115 mph; 185 km/h)
Range: Over 850 nm (979 miles; 1575 km) with 4,409 lb (2000 kg) of underwing ordnance and external fuel tanks
Service ceiling: 54,000 ft (16460 m)
Reaction time: Under five minutes, from brakes-off to interception of Mach 3 target at 80,000 ft (24400 m)

Armament

Two internal DEFA 554 30-mm cannon with 125 rounds per gun. Total of 13,889 lb (6300 kg) of stores carried on five underfuselage and four underwing hardpoints. Standard air defence load is two MATRA Magic 2 infra-red missiles and two Super 530D radar-guided missiles. Early aircraft were only equipped to fire the Super 530F. In the ground-attack role, up to 18 551-lb (250-kg) bombs or BAP 100 anti-runway bombs, two 1,984-lb (900-kg) BGL 1000 laser-guided bombs, six Belouga cluster bombs, two AS30L laser-guided air-to-surface missiles, two ARMAT anti-radiation missiles or two AM39 Exocet anti-ship missiles are options.
Mirage 2000D/N/S: The N is dedicated to the carriage of the 1,874-lb (850-kg) ASMP stand-off nuclear missile (150 kT or 300 kT yield). The 2000D/S has provision for the MATRA APACHE, Durandal, F4 rocket pod or Dassault CC630 gun pod.

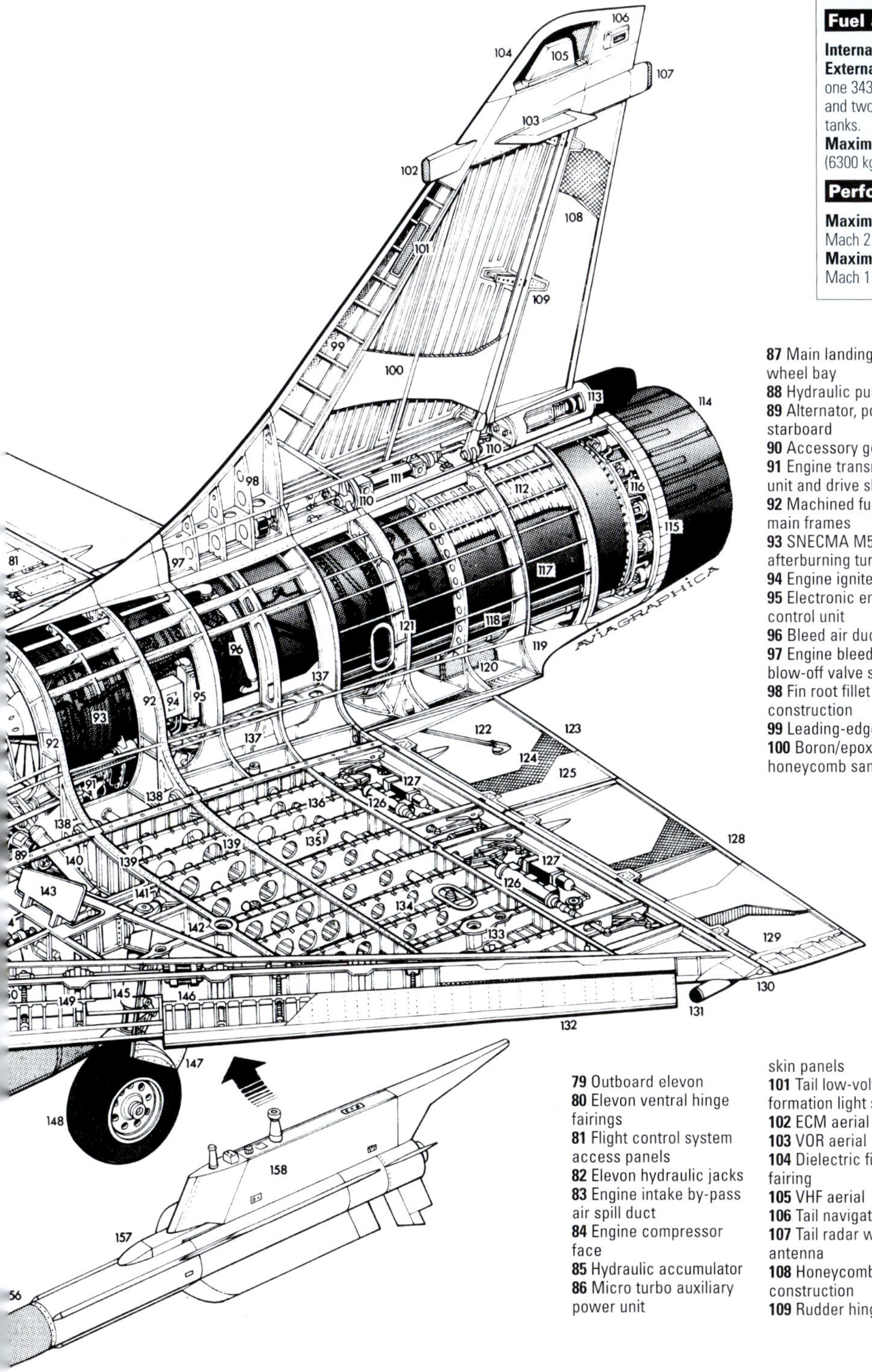

79 Outboard elevon
80 Elevon ventral hinge fairings
81 Flight control system access panels
82 Elevon hydraulic jacks
83 Engine intake by-pass air spill duct
84 Engine compressor face
85 Hydraulic accumulator
86 Micro turbo auxiliary power unit
87 Main landing gear wheel bay
88 Hydraulic pump
89 Alternator, port and starboard
90 Accessory gearbox
91 Engine transmission unit and drive shaft
92 Machined fuselage main frames
93 SNECMA M53-5 afterburning turbofan
94 Engine igniter unit
95 Electronic engine control unit
96 Bleed air ducting
97 Engine bleed air blow-off valve spill duct
98 Fin root fillet construction
99 Leading-edge ribs
100 Boron/epoxy/ carbon honeycomb sandwich fin skin panels
101 Tail low-voltage formation light strip
102 ECM aerial fairing
103 VOR aerial
104 Dielectric fin tip fairing
105 VHF aerial
106 Tail navigation light
107 Tail radar warning antenna
108 Honeycomb rudder construction
109 Rudder hinge
110 Fin spar attachment joints
111 Rudder hydraulic jack
112 Engine bay thermal lining
113 ECM equipment housing
114 Variable-area afterburner exhaust nozzle
115 Tailpipe sealing flaps
116 Fueldraulic nozzle control jacks
117 Afterburner tailpipe
118 Engine withdrawal rail
119 Wing root extended trailing-edge fillet
120 Ventral brake parachute housing
121 Rear engine mounting main frame
122 Runway emergency arrestor hook
123 Port inboard elevon
124 Elevon honeycomb construction
125 CArbon fibre skin panels
126 Elevon hydraulic control jacks
127 Fly-by wire electronic system command units
128 Outboard elevon
129 Elevon tip construction
130 Port navigation light
131 Radar warning antenna
132 Outboard automatic leading-edge slat
133 Outboard wing pylon attachment hardpoints
134 Machined upper-and lower-wing skin/stringer panels
135 Port wing integral fuel tank
136 Wing rib construction
137 Rear fuselage/ wingroot fairing integral fuel tank
138 Wing spar attachment joints
139 Main spars
140 Landing gear hydraulic retraction jack
141 Main landing gear leg pivot fixing
142 Inboard pylon attachment hardpoints
143 Port airbrakes (open) above and beneath wing
144 Airbrake hydraulic jack
145 Main landing gear leg strut
146 Leading-edge slat hydraulic drive motor
147 Mainwheel leg door
148 Port mainwheel
149 Slat guide rails
150 Screw jacks
151 Auxiliary spar
152 Wing front spar
153 Front spar attachment joint
154 Inboard automatic leading edge slat rib construction
155 3,741-Imp gal (1700-litre) auxiliary fuel tank (fuselage centreline or wing inboard stations)
156 MATRA Super 530 medium-range AAM
157 Missile launch rail
158 Inboard wing pylon

Dassault Ouragan/Mystère

Sharing several important features with the F-100 Super Sabre (45° swept wing, powered slab tailplane and flattened oval intake), the Super Mystère holds the distinction of being Europe's first truly supersonic fighter, following the F-100 and the Soviet MiG-19. This aircraft carries the codes and red lightning flash (outlined in black) of EC 2/10 'Seine'. Note the crude two-'eyelid' afterburner nozzle.

Mystère IVA

Cutaway key
1 Engine air intake
2 Radar rangefinder antenna
3 Intake divider
4 Gun camera
5 Nose electronics compartment access door
6 Radar transmitter
7 Radar receiver
8 Nose undercarriage wheel bay
9 Hydraulic retraction jack
10 Battery
11 Cockpit front pressure bulkhead
12 Rudder pedals
13 Cockpit floor level
14 Nose undercarriage pivot fixing
15 Cannon blast trough
16 Cannon muzzle
17 Nosewheel leg door
18 Landing/taxiing lamp
19 Nose undercarriage leg strut
20 Nosewheel
21 Torque scissor links
22 Bifurcated intake duct framing
23 Control column
24 Cockpit coaming
25 Instrument panel shroud
26 Gyro gunsight
27 Windscreen panels
28 Cockpit canopy cover
29 Ejection seat face blind firing handle
30 Headrest
31 Pilot's ejection seat
32 Canopy emergency release lever
33 Engine throttle lever
34 Port side console panel
35 Cockpit pressurised enclosure
36 Cannon mounting
37 DEFA 30-mm cannon
38 Spent cartridge case collector box
39 Gun bay access panel
40 Ammunition feed chute
41 Ammunition box (150 rounds per gun)
42 Control rod runs
43 Cockpit armoured rear pressure bulkhead
44 Sliding canopy rail
45 Oxygen bottle
46 Forward fuselage fuel tank (total internal capacity 396 Imp gal/ 1800 litres)
47 Wing root fillet
48 Aileron hydraulic booster
49 Intake duct framing
50 Radio and electronics equipment bay
51 UHF aerial
52 Starboard wing fuel cells
53 Aileron push-pull control rods
54 Pitot tube
55 Starboard navigation light
56 Wing tip fairing
57 Starboard aileron
58 Aileron hinge control
59 Split trailing-edge flap
60 Flap torque shaft actuator
61 Fuel tank access door
62 Fuel filler cap
63 Control rod duct
64 Centre fuselage fuel tank
65 Wing front spar/ fuselage main frame
66 Wing centre-section carry-through
67 Wing skin bolted root joint
68 Rear spar/fuselage main frame
69 Main undercarriage wheel bay
70 Hydraulic reservoir
71 Engine accessory compartment
72 Fuel system piping
73 Control rod runs
74 Dorsal spine fairing
75 Engine bay access door
76 Generator
77 Compressor intake filter screens
78 Intake plenum chamber
79 Main engine mounting
80 Hispano Suiza Verdon 350 centrifugal-flow turbojet
81 Rear fuselage break point (engine removal)
82 Engine flame cans
83 Fin root fillet
84 Engine turbine section
85 Tailplane control rods
86 Trimming tailplane electric screwjack
87 Tailplane sealing plate
88 Elevator control linkage
89 Rudder push-pull control rod
90 Starboard tailplane

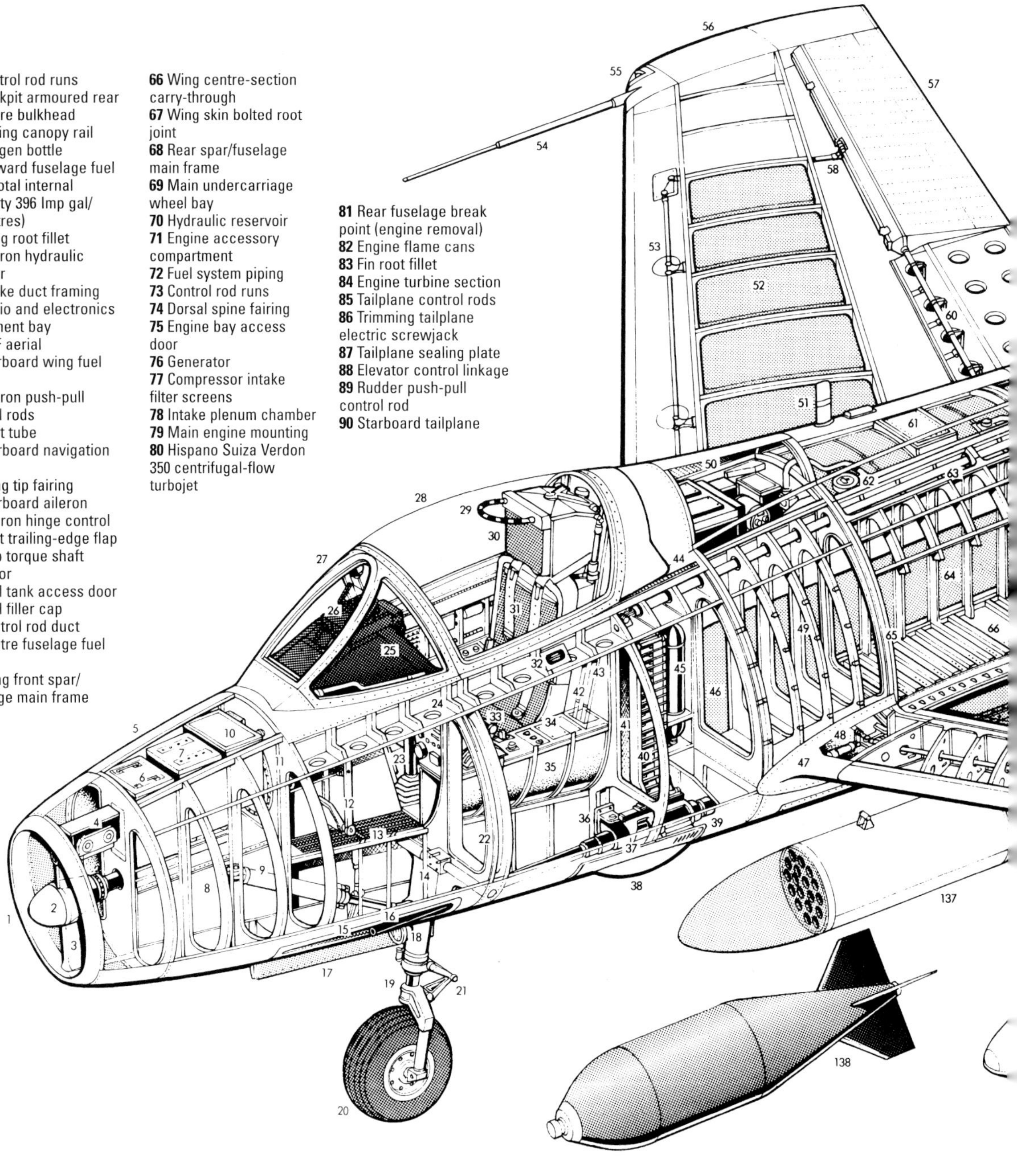

Israel's Super Mystères were re-engined to provide greater dry thrust (the new J52 did not have an afterburner) and better fuel consumption. This Sa'ar, as the upgraded aircraft were known, carries two Shafrir air-to-air missiles

SPECIFICATION

Super Mystère B2

Dimensions

Length: 46 ft 4¼ in (14.13 m)
Height: 14 ft 11 in (4.55 m)
Wingspan: 34 ft 6 in (10.52 m)
Wing area: 376.75 sq ft (35.00 m²)
Wheelbase: 14 ft 11½ in (4.56 m)

Powerplant

One SNECMA Atar 101-G-2/-3 turbojet rated at 9,833 lb (43.76 kN) thrust with afterburning

Weights

Empty equipped: 15,282 lb (6932 kg)
Maximum take-off: 22,046 lb (10000 kg)

Performance

Maximum speed: 646 mph (1040 km/h) at sea level and 743 mph (1195 km/h) at 39,370 ft (12000 m)
Initial climb rate: 17,505 ft (5335 m) per minute
Service ceiling: 55,775 ft (17000 m)
Normal range: 540 miles (870 km)

Armament

Two 30-mm DEFA 551 cannon plus 35 68-mm SNEB rockets in retractable pack; plus up to 2,205 lb (1000 kg) of stores on four underwing hardpoints, including bombs up to 1,102-lb (500-kg); some aircraft later upgraded to carry AIM-9 Sidewinder or Shafrir AAMs

Among this array of Super Mystère weaponry is the 35-round SNEB rocket pack (centre), and the Nord AS20 guided missile.

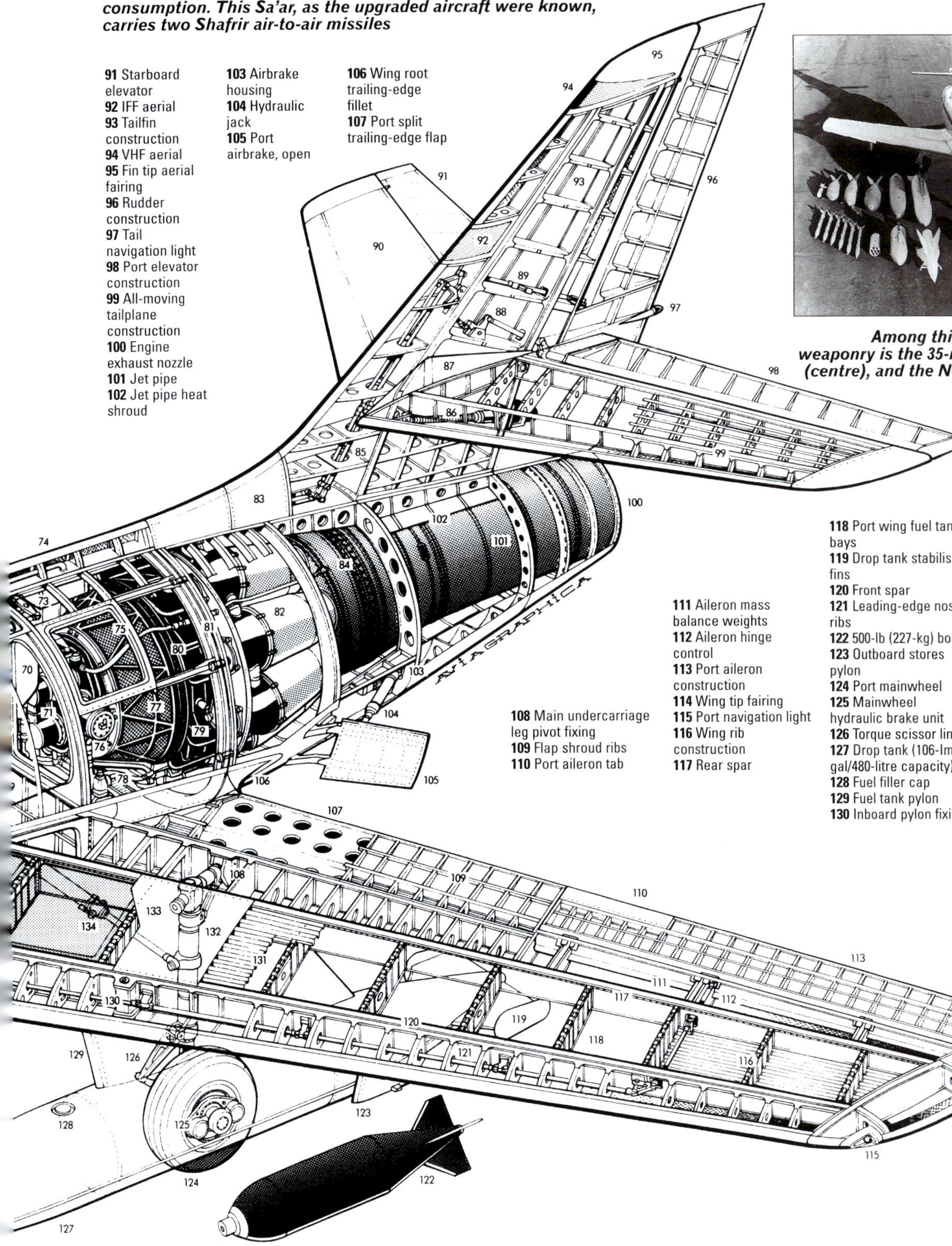

91 Starboard elevator
92 IFF aerial
93 Tailfin construction
94 VHF aerial
95 Fin tip aerial fairing
96 Rudder construction
97 Tail navigation light
98 Port elevator construction
99 All-moving tailplane construction
100 Engine exhaust nozzle
101 Jet pipe
102 Jet pipe heat shroud
103 Airbrake housing
104 Hydraulic jack
105 Port airbrake, open
106 Wing root trailing-edge fillet
107 Port split trailing-edge flap
108 Main undercarriage leg pivot fixing
109 Flap shroud ribs
110 Port aileron tab
111 Aileron mass balance weights
112 Aileron hinge control
113 Port aileron construction
114 Wing tip fairing
115 Port navigation light
116 Wing rib construction
117 Rear spar
118 Port wing fuel tank bays
119 Drop tank stabilising fins
120 Front spar
121 Leading-edge nose ribs
122 500-lb (227-kg) bomb
123 Outboard stores pylon
124 Port mainwheel
125 Mainwheel hydraulic brake unit
126 Torque scissor links
127 Drop tank (106-Imp gal/480-litre capacity)
128 Fuel filler cap
129 Fuel tank pylon
130 Inboard pylon fixing
131 Corrugated wing skin sandwich panel
132 Main undercarriage leg strut
133 Wing skin panelling
134 Main undercarriage hydraulic retraction jack
135 Inboard fuel cells
136 Aileron push-pull control rod
137 18 x 68-mm (2.68-in) rocket launcher
138 1,000-lb (454-kg) HE bomb

de Havilland DH.100 Vampire

The first pre-production Vampire trainer for the Royal Navy, WW458 was delivered for evaluation in January 1952. Following trials with Nos 759 and 781 Squadrons, Fleet Air Arm it was withdrawn from flying duties and, in early 1954, became an instructional airframe.

Vampire FB.Mk 5

Cutaway key

1 Ciné camera port
2 Cockpit fresh air intake
3 Nosewheel leg door
4 Pivoted axle nosewheel suspension
5 Anti-shimmy nosewheel tyre
6 Nose undercarriage leg strut
7 Nosewheel door
8 Cannon muzzle blast trough
9 Nosewheel hydraulic jack
10 Nose undercarriage pivot fixing
11 Radio
12 Gun camera
13 Windscreen fluid de-icing reservoir
14 Armoured instrument access panel
15 Cockpit front bulkhead
16 Rudder pedals
17 Cockpit floor level
18 Nosewheel housing
19 Instrument panel
20 Reflector gunsight
21 Windscreen panels
22 Side console switch panel
23 Control column
24 Engine throttle
25 Tailplane trim handwheel
26 Undercarriage and flap selector levers
27 Control linkage
28 Cannon barrels beneath cockpit floor
29 Pull-out boarding step
30 Control system cable compensator
31 Emergency hydraulic handpump
32 Pilot's seat
33 Safety harness
34 Sliding canopy rails
35 Cockpit heater
36 Cockpit canopy cover
37 Pilot's head and back armour
38 Hydraulic system reservoir
39 Radio equipment bay
40 Ammunition tanks (150 rounds per gun)
41 Plywood/balsa/plywood fuselage skinning
42 Boundary layer splitter
43 Port engine air intake
44 Ventral gun bay (4 x 20-mm Hispano cannon)
45 Spent cartridge case and link ejector chute
46 Cannon bay access panel
47 Cockpit heating and pressurising intake
48 Intake ducting
49 Fuselage/front spar attachment joint
50 Fuselage/main spar attachment joint
51 Engine bay fire wall
52 Fuselage fuel tank (total internal system capacity 400 Imp gal/1818 litres)
53 Fuel filler cap
54 Wooden skin section fabric covering
55 Cockpit air heat exchanger
56 Engine bearer struts
57 de Havilland Goblin DGn 2 centrifugal-flow turbojet
58 Cabin blower
59 Engine accessories
60 Engine bay access panels
61 Starboard wingroot fuel tank
62 Starboard main undercarriage, retracted position
63 Leading-edge fuel tank
64 Starboard drop tank (112 Imp gal/509 litres)
65 Drop tank pylon
66 Starboard wing fuel tanks
67 Fuel filler cap
68 Gyrosyn compass remote transmitter
69 Starboard navigation light
70 Wingtip fairing
71 Starboard aileron
72 Aileron mass balance weights
73 Trim tab
74 Aileron hinge control
75 Starboard trailing-edge airbrake segment (open)
76 Airbrake hydraulic jack
77 Starboard outer split trailing-edge flap

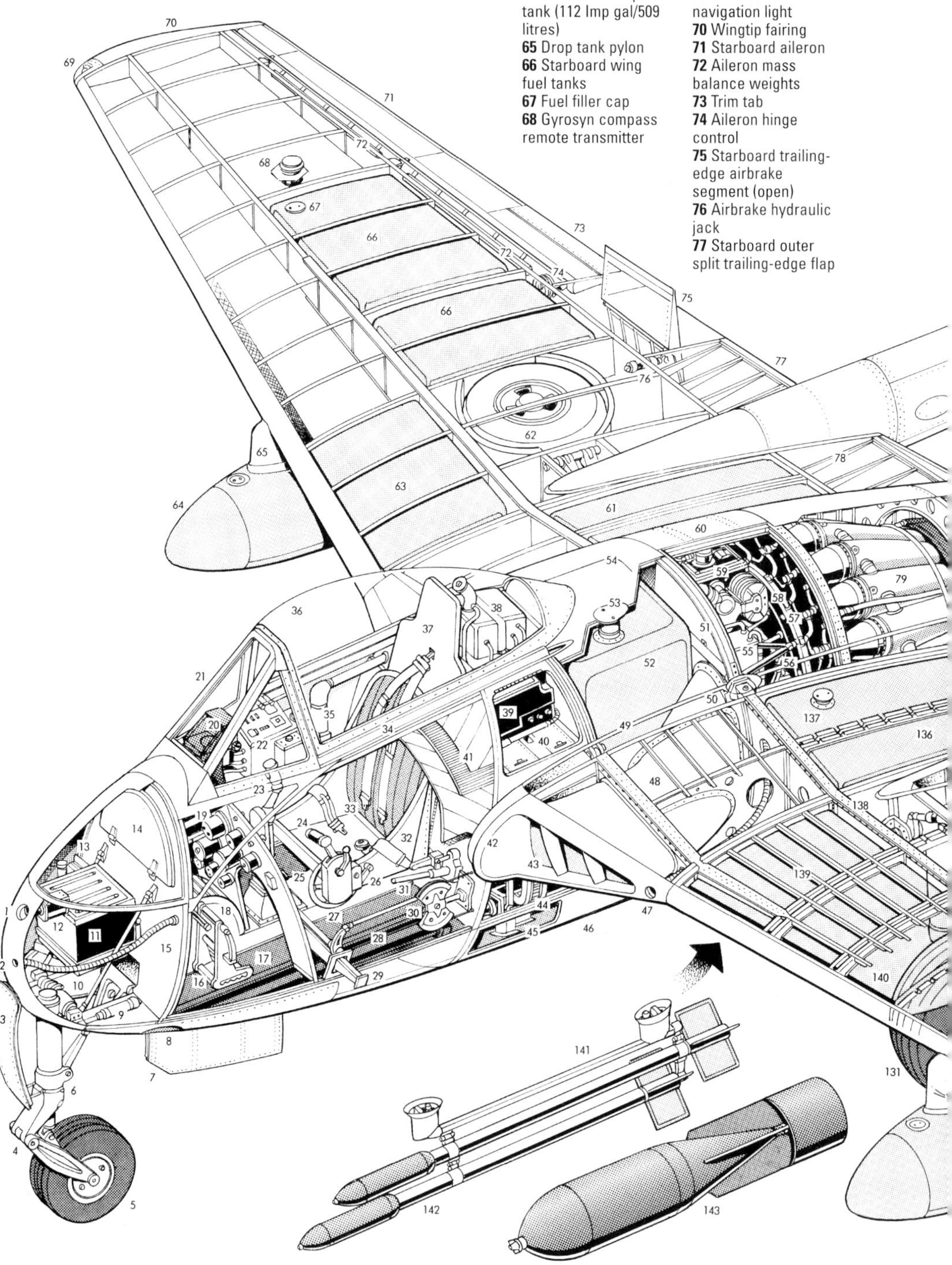

No. 601 (County of London) Squadron, Royal Auxiliary Air Force, received the Vampire F.Mk 3 in December 1949. Based at North Weald, the squadron took its Vampires on summer armament training camps from 1950-52. Here, the Vampires bask in the hot Mediterranean sun on the island of Malta during the summer camp of 1951. No. 601 displayed the unit's red/black triangular insignia on the tailbooms and the squadron's winged sword badge on the nose.

SPECIFICATION

Vampire F.Mk 3

Dimensions

Length: 30 ft 9 in (9.37 m)
Height: 8 ft 10 in (2.69 m)
Wingspan: 40 ft (12.19 m)
Wing area: 266 sq ft (24.71 m²)
Wing loading: 39.4 lb/sq ft (192 kg/m²)

Powerplant

One de Havilland Goblin 2 centrifugal-flow turbojet rated at 3,100 lb st (14 kN)

Weights

Empty: 7,134 lb (3236 kg)
Maximum take-off: 12,170 lb (5520 kg)

Fuel

Internal fuel: 530 Imp gal (2409 litres)
External fuel: 200 Imp gal (909 litres) in drop tanks

Performance

Maximum level speed at sea level: 531 mph (855 km/h)
Maximum level speed at 17,500 ft (5334 m): 525 mph (845 km/h)
Maximum level speed at 30,000 ft (9144 m): 505 mph (813 km/h)
Rate of climb at sea level: 4,375 ft (1334 m) per minute
Rate of climb at 20,000 ft (6096 m): 2,500 ft (762 m) per minute
Rate of climb at 40,000 ft (12192 m): 990 ft (302 m) per minute
Service ceiling: 43,500 ft (13259 m)
Take-off run to 50 ft (15.24 m) at maximum weight: 3,540 ft (1079 m)
Landing run from 50 ft (15.24 m): 3,300 ft (1006 m)

Range and endurance

Range at sea level: 590 miles (949 km) at 350 mph (463 km/h)
Range at 30,000 ft (9144 m): 1,145 miles (1843 km) at 350 mph (463 km/h)
Patrol duration at sea level: 2 hours at 220 mph (354 km/h)
Patrol duration at 30,000 ft (9144 m): 2 hours 35 mins at 220 mph (354 km/h)

Armament

Four 20-mm Hispano cannon mounted in the front of the lower fuselage. Ammunition of 150 rounds per gun, giving a total of 600 rounds

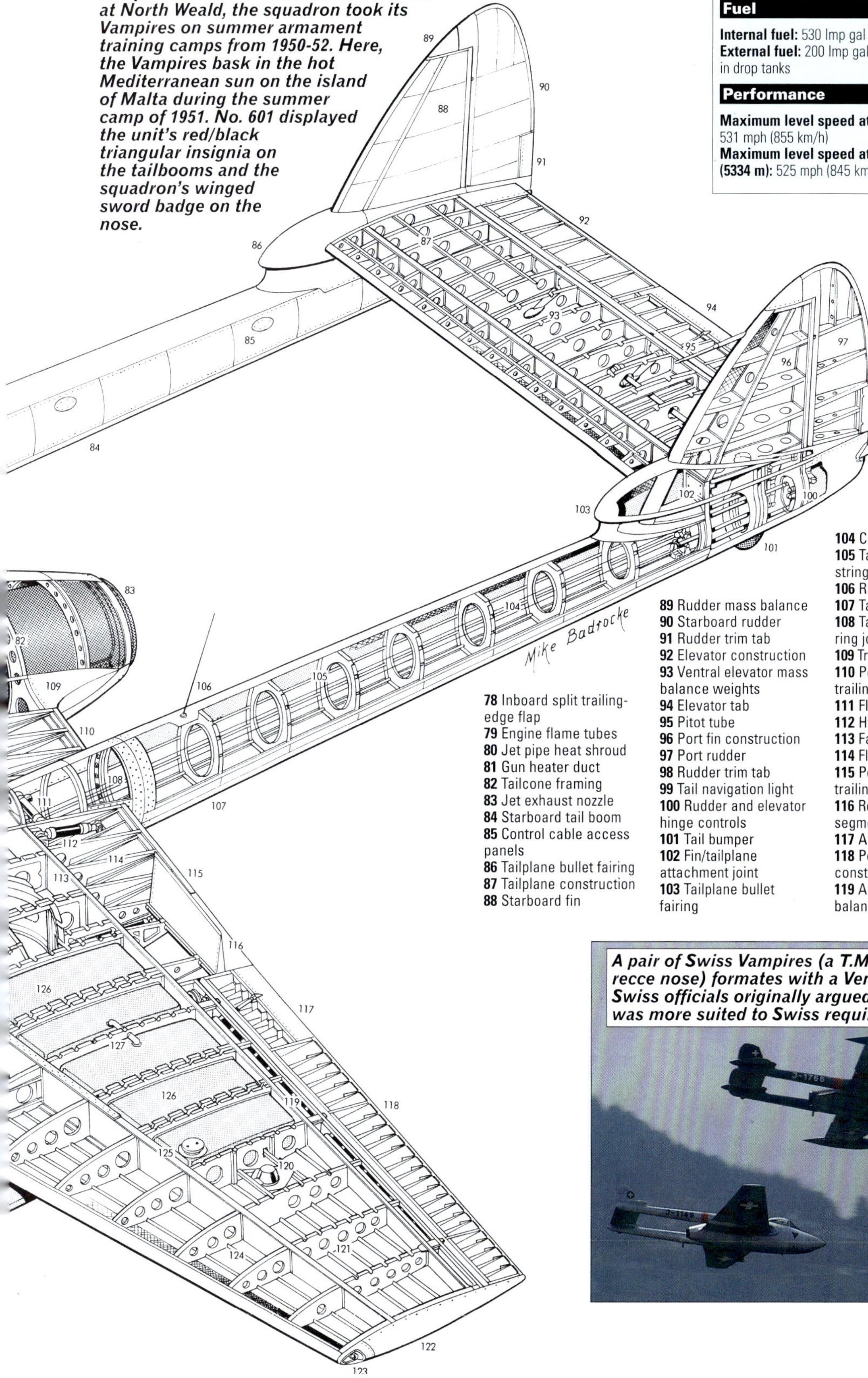

78 Inboard split trailing-edge flap
79 Engine flame tubes
80 Jet pipe heat shroud
81 Gun heater duct
82 Tailcone framing
83 Jet exhaust nozzle
84 Starboard tail boom
85 Control cable access panels
86 Tailplane bullet fairing
87 Tailplane construction
88 Starboard fin
89 Rudder mass balance
90 Starboard rudder
91 Rudder trim tab
92 Elevator construction
93 Ventral elevator mass balance weights
94 Elevator tab
95 Pitot tube
96 Port fin construction
97 Port rudder
98 Rudder trim tab
99 Tail navigation light
100 Rudder and elevator hinge controls
101 Tail bumper
102 Fin/tailplane attachment joint
103 Tailplane bullet fairing
104 Control cable runs
105 Tailboom frame and stringer construction
106 Radio aerial mast
107 Tailboom skinning
108 Tailboom attachment ring joint
109 Trailing-edge root fillet
110 Port inboard split trailing-edge flap
111 Flap interconnection
112 Hydraulic flap jack
113 False rear spar
114 Flap shroud ribs
115 Port outboard split trailing-edge flap
116 Rotating trailing-edge segment airbrake, open
117 Aileron tab
118 Port aileron construction
119 Aileron mass balance weights
120 Retractable landing/taxiing lamp
121 Wingrib and stringer construction
122 Wingtip fairing
123 Port navigation light
124 Leading-edge nose ribs
125 Fuel filler cap
126 Port wing main fuel tanks
127 Fuel tank interconnection
128 Pylon attachment rib
129 Port 112-Imp gal (509-litre) drop tank
130 Drop tank pylon
131 Port mainwheel
132 Mainwheel door actuating linkage
133 Port mainwheel bay
134 Retraction linkages and locks
135 Main undercarriage leg strut pivot fixing
136 Wingroot fuel tank
137 Fuel filler cap
138 Main spar
139 Wing stringers
140 Leading-edge fuel tank
141 Rocket launcher rail
142 60-lb (27-kg) unguided ground attack rocket
143 500-lb (227-kg) HE bomb

A pair of Swiss Vampires (a T.Mk 55 and an FB.Mk 6 with a recce nose) formates with a Venom (foreground). Some Swiss officials originally argued that the Hawker Sea Fury was more suited to Swiss requirements than the Vampire.

English Electric Lightning

Designers and engineers from English Electric (and later from its successor BAC) can be justifiably proud of the Lightning. Wholly conceived and developed in Britain, it was the first British aircraft to exceed Mach 2, and was also an unofficial world speed record holder.

Lightning F.Mk 6

Cutaway key

1 Pitot head boom
2 Intake bullet fairing
3 Ferranti AIRPASS radar antenna/scanner
4 Engine air intake lip
5 Hot-air de-icing
6 Bullet lower spacer
7 G 90 camera
8 Radar pack
9 Bullet upper spacer (electrical leads)
10 Forward equipment bay
11 Forward fuse box
12 Capacitor box
13 LOX container
14 Light fighter sight control unit
15 De-icing/demister air
16 Radar ground cooling air coupling
17 Nosewheel door mechanism torque shaft and operating rods
18 Nosewheel bay
19 Nosewheel doors
20 Nosewheel strut
21 Roller guide bracket
22 Forward-retracting nosewheel
23 Castor auto-disconnect
24 Shimmy damper and centring unit
25 Aft door (linked to leg)
26 Flight refuelling probe (detachable)
27 Nosewheel strut pivot pin
28 Heat exchanger
29 Nosewheel hydraulic jack
30 Intake ducting
31 Cockpit canted floor
32 Engine power control panel
33 Control column
34 Instrument panel shroud
35 Rudder pedal assembly
36 Canopy column
37 Rain dispersal duct
38 Windscreen (electro- thermal)
39 CRT display unit (starboard)
40 Airpass (light fighter) attack sight
41 Standby magnetic compass
42 Canopy top panel demisting ducts
43 Magnesium-forged canopy top frame
44 IFF aerial
45 Chemical air driers
46 Starboard (armaments) console
47 Ejection seat face-blind/firing-handle
48 Air conditioning duct
49 Rear pressure bulkhead
50 Martin-Baker ejection seat
51 Port instrument panels
52 Cockpit ladder attachment
53 Cockpit emergency ram air intake
54 Lower (No. 1) engine intake duct frames
55 Firestreak weapons pack
56 Launch sequence units
57 Control units
58 Port missile pylon
59 Firestreak missile
60 Fuse 'windows'
61 Armament safety break panel
62 Aileron accumulator pressure gauges
63 Accumulator group bay
64 Plessey LTSA starter in lower (No. 1) engine nosecone
65 Lower (No. 1) engine intake
66 Wingrooot inboard fairing
67 Main equipment bay
68 Selector address unit
69 Electronic unit
70 Air data computer
71 Converter signal unit (datalink)
72 Communications T/R (two)
73 Canopy hinge
74 Dorsal spine bays
75 AC fuse and relay box (cold-air unit and water boiler to starboard)
76 28-Volt battery
77 Upper (No. 2) engine intake duct
78 Fuselage frames
79 Water heater tank and extractor
80 Wing/fuselage main attachment point
81 Aileron idler lever
82 Aileron control push-pull tubes
83 Tube attachment brackets
84 Fuselage multi-bolt forward/centre-section join
85 ADEN gun muzzle
86 Leading-edge integral fuel
87 Muzzle blast tube
88 Aileron tube triple-roller guides
89 Access
90 Fuel lines
91 Non-return valve
92 Detachable leading-edge sections
93 Shuttle valve
94 Undercarriage strut fixed fairing
95 Shock-absorber strut
96 Port mainwheel
97 Brake unit
98 Tubeless tyre
99 Torque links
100 Red Top missile
101 Aft fairing flap
102 Undercarriage pivot
103 Radius rod (inward- breaking)
104 Undercarriage retraction jack
105 Door jack sequence valve
106 Door master locking mechanism
107 Collector tank and booster pumps (two)
108 Aerodynamic leading-edge slot
109 Tank pressurising intake/vent (in slot)
110 Mainwheel door
111 Undercarriage jack sequence valve
112 Door latch linkage
113 Port mainwheel well
114 Aileron control push-pull tubes
115 Aileron movement restrictor
116 Aileron autostabiliser actuator
117 Aileron control linkage
118 Aileron hydraulic runs
119 Cambered leading-edge extension
120 Localiser aerial
121 Port navigation light
122 Port wingtip
123 Port aileron
124 Aileron powered flying-control units
125 Control linkage
126 Wing outer structure
127 Aileron mass balance
128 Wing outer fixed section
129 Flap outer actuator jack
130 Flap sections
131 Flap integral tank
132 Angled aft spar
133 Undercarriage attachment
134 Refuelling/defuelling valve
135 Flap inner actuator jack
136 Three-way cock (manual)
137 DC transfer pump
138 Gate valves
139 Wing/fuselage rear main attachment point
140 Lower (No. 1) engine intermediate jet pipe forward face
141 Wing inboard structure
142 Wing integral fuel
143 Intermediate spar booms (T-section)
144 Port ADEN cannon (forward ventral pack)
145 Wing rib stations
146 Fuel vent pipe
147 Multi-bolt wing attachment plate
148 Access panels
149 Upper (No. 2) engine duct frames
150 Fuselage break frame
151 Voltage regulators
152 Start tank
153 Engine pump units
154 Solenoid valves
155 Communications antenna
156 Starter control unit
157 HF igniter units
158 Fuselage frame
159 Main wing box upper skin
160 Forged centre rib (multi-bolt attachment)
161 Starter exhaust
162 Upper (No. 2) engine nosecone
163 Generator cooling ram-air intake

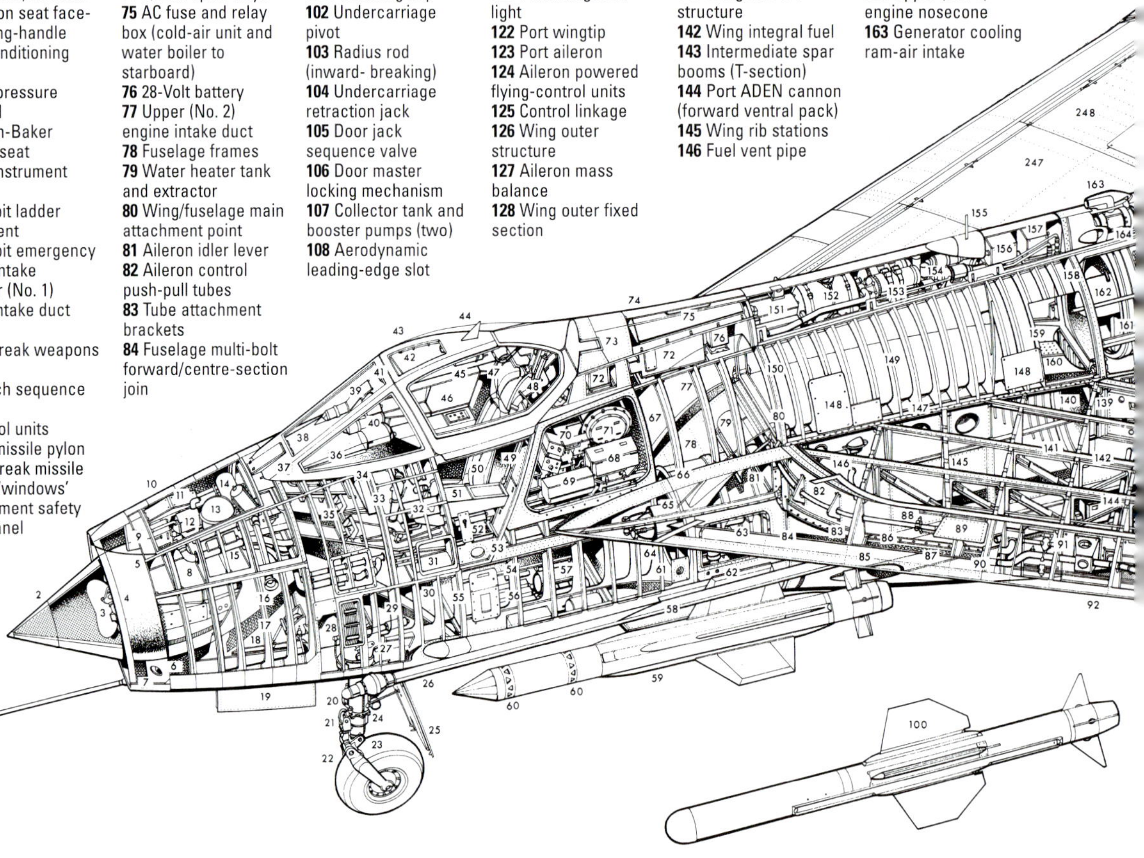

Lack of funds led to the protracted development of a multi-role Lightning targeted for export, and this prevented the fighter from achieving its full potential on the export market. Sales were made to just one major overseas operator, the Royal Saudi Air Force (RSAF). Shown lifting off smartly at the 1966 Farnborough air show, XR770 was an RAF F.Mk 6 painted in RSAF markings for demonstration purposes.

164 Stand-by generator
165 Anti-icing bleed air
166 Upper (No. 2) Avon 301 turbojet engine and reheat units
167 Airpass recorder unit
168 Engine front mounting point
169 Engine accessories
170 No. 2 engine bleed-air turbopump (reheat fuel)
171 Engine bay firewalls
172 Integral pumps (two)
173 HE ignition units
174 Voltage regulator
175 Current sensing unit
176 Rubber spring feel mechanism
177 Auxiliary intake
178 Main mounting trunnion
179 Aft (port) equipment bays
180 Electronic unit
181 IFF coder
182 Tailplane controls
183 Tailplane trim actuator and feel unit
184 Ventral fuel tank (aft section)
185 Fin
186 Reheat cooling lower intake
187 Tailplane autostabiliser actuator
188 Gearbox oil filler
189 AC generator
190 Glide-path receiver
191 IFF transmit/ receive unit
192 Outlet
193 No. 2 engine intermediate jet pipe
194 Refrasil heat shrouds
195 Stress-bearing upper (No. 2) engine hatch
196 Port airbrake
197 Airbrake hydraulic actuator jack
198 DC generator
199 Main accessory-drive unit
200 Airbrake lower frame
201 Turbine exhaust (from 199)
202 Tailplane accumulator and nitrogen bottle
203 Reheat 'hotshot' igniter box
204 Tailplane drive triangular unit
205 Tailplane powered flying-control unit
206 Tailplane spigot
207 Pivot spar
208 All-moving tailplane
209 Light alloy honeycomb structure

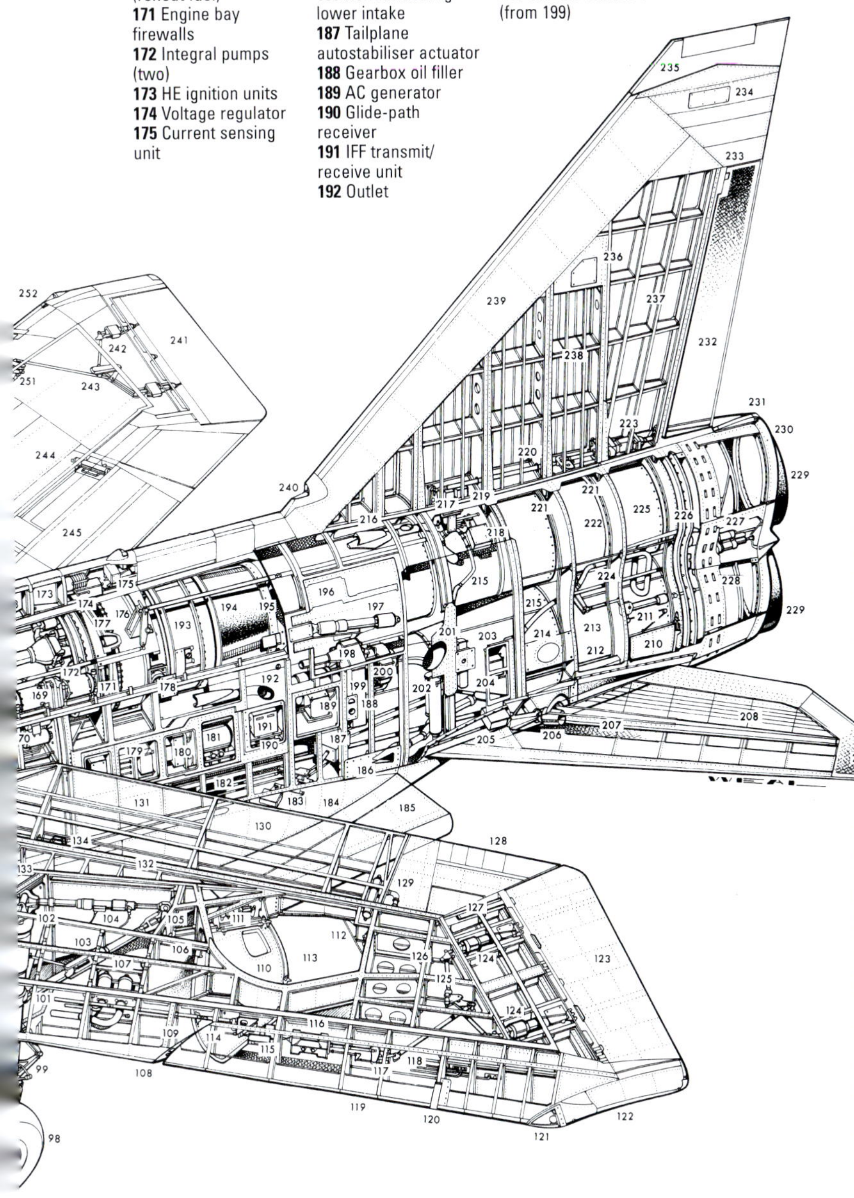

SPECIFICATION

Lightning F.Mk 53

Dimensions

Length overall (including probe): 55 ft 3 in (16.84 m)
Wing span: 34 ft 10 in (10.62 m)
Wing aspect ratio: 2.65
Wing sweepback angle: 60° on leading edge and 52° on trailing edge
Tailplane span: 14 ft 6 in (4.42 m)
Wing area: 458.50 sq ft (42.60 m²)
Overall height: 19 ft 7 in (5.97 m)
Undercarriage track: 12 ft 9 in (3.89 m)
Wheel base: 18 ft 1½ in (5.52 m)
Maximum wing loading: 90.9 lb/sq ft (444.0 kg/m²)

Powerplant

Two Rolls-Royce Avon RA.24 Mk 302C turbojets each rated at 11,000 lb (48.92kN) static military thrust and 16,300 lb (72.49kN) with afterburning

Weights

Empty operating: 28,040 lb (12719 kg), 29,600 lb (13426 kg) with gun pack and missiles
Maximum take-off (fully armed): 41,700 lb (18915 kg)

Fuel and load

Maximum internal fuel: 10,608 lb (4812 kg); pressure refuelling of wing tanks via adaptor under port wing trailing edge, provision for detachable fixed inflight-refuelling probe projecting forward from port wing
Maximum internal fuel capacity: 1277 Imp gal (5805 litres)
Wing tank capacity: 690 Imp gal (3137 litres)
Ventral tank capacity: 587 Imp gal (2668 litres), or 515 Imp gal (2342 litres) with 30-mm cannon pack fitted
Total external fuel capacity: 520 Imp gal (2364 litres) in two overwing tanks
Maximum weapon load: 6,000 lb (2722 kg)

Performance

Maximum level speed at high altitude: 1,320 kt (1,520 mph; 2446 km/h) at 40,000 ft (12190 m) or Mach 2.3
Maximum level speed at low altitude: approximately 723 kt (834 mph; 1340 km/h) or Mach 1.1
Cruising speed for optimum range: 517 kt (595 mph; 958 km/h)
Initial rate of climb: 50,000 ft (15240 m) per minute
Actual ceiling: in excess of 60,000 ft (18290 m)
Service ceiling: 60,000 ft (18290 m)
Time to climb: 150 seconds from brakes release to Mach 0.9 at 40,000 ft (12190 m)
Acceleration: 210 seconds from Mach 1 to over Mach 2
Landing speed: 160 kt (183mph; 295km/h)
Crosswind limits: 25 kts (29 mph; 46 km/h) in dry conditions and 15 kts (17 mph; 28 km/h) in wet conditions
Take-off run: 3,300 ft (1006 m) at 38,500 lb (17464 kg) with afterburner
Landing run: 3,600 ft (1097 m) with parachute at 29,000 lb (17237 kg), 4,500 ft (1371 m) at 38,000 lb (17237 kg)

Range

Maximum range on internal fuel: 800 miles (1287 km)
Combat radius on internal fuel: 373 miles (600 km)

Armament

Fixed: two 30-mm ADEN Mk 4 cannon, each with 120 rounds, in gun pack forward of ventral fuel tank
Forward external stores: standard intercept load of two Red Top or Firestreak infra-red-homing short-range air-to-air missiles; in place of missile pack, alternative loads of pack containing twin retractable launchers for total of 44 2-in (51-mm) MicroCell spin-stabilised rockets, reconnaissance pack containing five Vinten Type 360 70-mm cameras, or night reconnaissance pack containing optical cameras and linescan equipment
Wing-mounted stores: single hardpoint under outer wing capable of carrying two 1,000-lb (454-kg) high-explosive, retarded or fire bombs, two MATRA Type 155 launchers containing total of 36 2.68-in (68-mm) SNEB rockets, two flare pods or two machine-gun pods; single hardpoint above inner wing capable of carrying single 1,000-lb (454-kg) bomb, one MATRA Type 155 rocket launcher, two MATRA Type 100 combined fuel/rocket launchers each containing 18 2.68-in (68-mm) SNEB rockets and 50 Imp gal (227 litre) of fuel, or one 260-Imp gal (1182-litre) fuel tank.

210 Braking parachute box internally-retracting doors
211 Cable operating assembly
212 Fuselage aft frame
213 Lower (No. 1) engine reheat jet pipe
214 Trunnion access panel**215** AMCU air pipes
216 Reheat cooling upper intake
217 Rudder feel unit
218 Rudder trim actuator
219 Rudder autostabiliser actuator
220 Rudder linkage
221 Fin spar/fuselage bolts
222 Fuselage frame formers
223 Rudder powered flying-control unit
224 Reheat jet pipe mounting rail
225 Upper (No. 2) engine reheat jet pipe
226 Rear rollers
227 Air-driven nozzle actuator
228 Jet pipe trunnion access panel
229 Variable propelling nozzles
230 Streamer cable around rear lip (spring-clipped)
231 Parachute streaming anchor and jettison unit
232 Rudder light-alloy honeycomb structure
233 Flutter damper
234 Communications antenna
235 Dielectric tip
236 Compass unit
237 Angled aft spars
238 Main fin structure
239 Fin leading-edge panels
240 Accessory drive cooling air
241 Starboard aileron
242 Aileron powered flying-control units
243 Control linkage
244 Starboard flap outer actuator jack
245 Starboard flap
246 Wing panels
247 Wing skinning
248 Wing integral fuel
249 Aileron control push-pull tubes
250 Aileron movement restrictor
251 Aileron autostabiliser actuator
252 Starboard navigation light
253 Glide-slope aerial

Fairchild A-10 Thunderbolt II

The A-10 at its peak formed the backbone of the USAF's close air support force, a position it has now lost to the ubiquitous F-16. Nevertheless, the type continues to provide excellent service, although in reduced numbers, notably in the forward air control role. This 'decorated' example served with the now-deactivated 91st TFS based at RAF Woodbridge, England.

A-10A Thunderbolt II

Cutaway key
1 Cannon muzzles
2 Nose cap
3 ILS aerial
4 Air-to-air refuelling receptacle (open)
5 Nosewheel bay (offset to starboard)
6 Cannon barrels
7 Rotary cannon barrel bearing
8 Gun compartment ventilating intake
9 L-band radar warning aerial
10 Electrical system relay switches
11 Windscreen rain dispersal airduct
12 Pave Penny laser receiver and tracking pod
13 Windscreen panel
14 Head-up display symbol generator
15 Pilot's head-up display screen
16 Instrument panel shroud
17 Air-to air refuelling pipe
18 Titanium armour cockpit enclosure
19 Rudder pedals
20 Battery
21 General Electric GAU-8/A 30-mm seven-barrelled rotary cannon
22 Ammunition feed ducts
23 Steering cylinder
24 Nose undercarriage leg strut
25 Nosewheel
26 Nosewheel scissor links
27 Retractable boarding ladder
28 Ventilating air outlets
29 Ladder stowage box
30 Pilot's side console panel
31 Engine throttles
32 Control column
33 McDonnell Douglas ACES 2 ejection seat
34 Headrest canopy breakers
35 Cockpit canopy cover
36 Canopy hinge mechanism
37 Space provision for additional avionics
38 Angle of attack probe
39 Emergency canopy release handle
40 Ventral access panels to gun compartment
41 Ammunition drum (1,174 rounds)
42 Ammunition drum armour plating
43 Electrical system servicing panel
44 Ventral fin
45 Spent cartridge-case return chute
46 Control cable runs
47 Avionics compartments
48 Forward/centre fuselage joint bulkhead
49 Aerial selector switches
50 FF aerial
51 Anti-collision light
52 UHF TACAN aerial
53 Starboard wing integral fuel tank
54 Wing skin plating
55 Outerwing panel attachment joint strap
56 Starboard fixed-wing pylons
57 ALE-37A chaff dispenser pod
58 ALQ-119 electronic countermeasures pod
59 Pitot tube
60 Starboard drooped wingtip fairing
61 Split aileron/ deceleron mass balance
62 Deceleron open position
63 Starboard aileron/ deceleron
64 Deceleron hydraulic jack
65 Aileron hydraulic jack
66 Control linkages
67 Aileron tab
68 Tab balance weight
69 Slotted trailing-edge flaps
70 Outboard flap jack
71 Flap synchronising shafts
72 Fuselage self-sealing fuel cells (maximum internal fuel capacity 10,700 lb/4853 kg)
73 Fuselage main longeron
74 Longitudinal control and services duct
75 Air-conditioning supply duct
76 Wing attachment fuselage main frames
77 Gravity fuel fillercaps
78 Engine pylon fairing
79 Pylon attachment joint
80 Starboard intake
81 Intake centre cone
82 Engine fan blades
83 Night/adverse weather two-seater variant
84 Radar pod (forward-looking infra-red in starboard pod)
85 Engine mounting struts
86 Nacelle construction
87 Oil tank
88 General Electric TF34-GE-100 turbofan
89 Rear engine mounting
90 Pylon trailing-edge fillet
91 Engine exhaust duct
92 Fan air duct
93 Rudder hydraulic jack
94 Starboard tailfin
95 X-band aerial
96 Rudder mass balance weight
97 Starboard rudder
98 Elevator tab
99 Tab control rod
100 Starboard elevator
101 Starboard tailplane
102 Tailplane attachment frames
103 Elevator hydraulic jacks
104 Tailcone
105 Tail navigation light
106 Rear radar warning receiver aerial
107 Honeycomb elevator construction
108 Port vertical tailfin construction
109 Honeycomb rudder panel
110 Rudder hydraulic jack
111 Formation light
112 Vertical fin ventral fairing
113 Tailplane construction

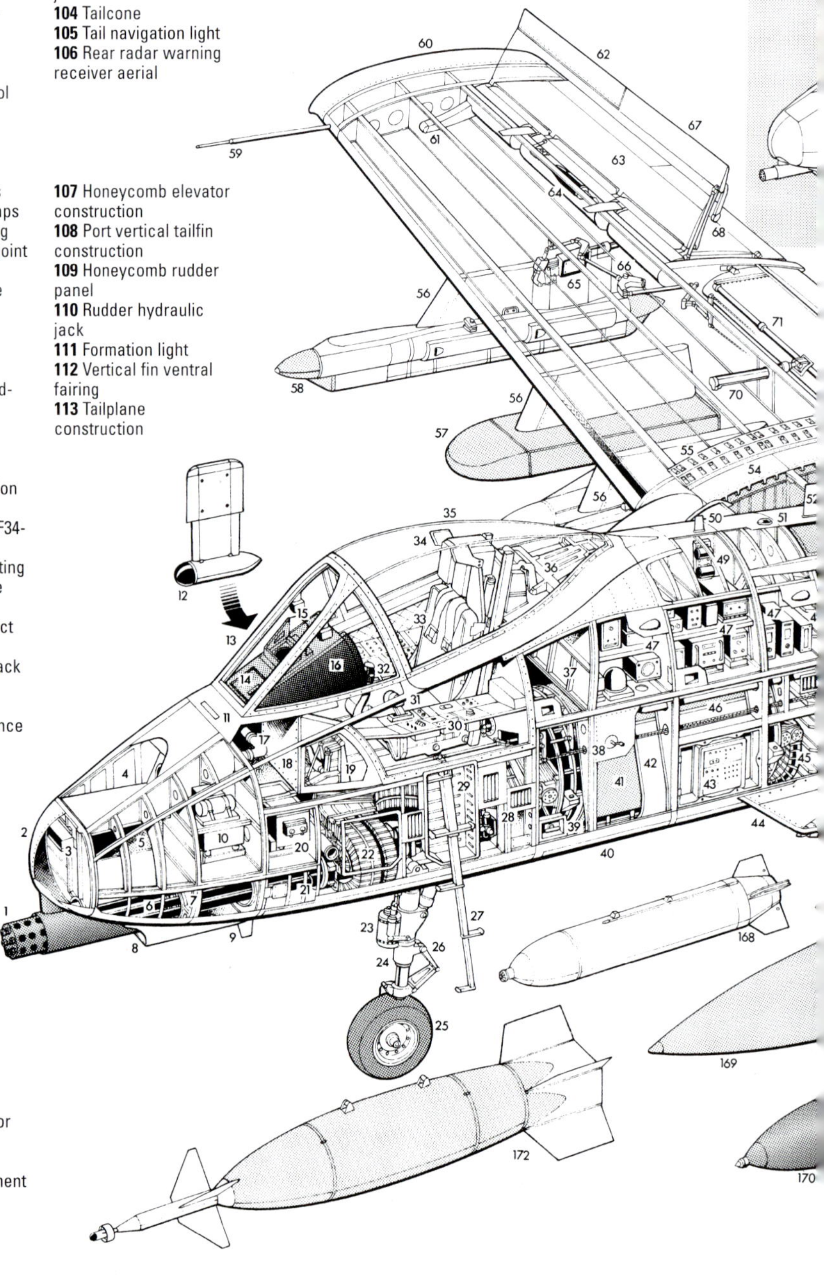

Placing the refuelling receptacle immediately in front of the pilot allowed easy alignment for rendezvous, as seen here with a KC-135 Stratotanker. During Desert Storm, A-10s were able to loiter over the battlefield for far longer than expected, allowing pilots to remain on station until the best target became available.

SPECIFICATION

A-10 Thunderbolt II

Dimensions

Length overall: 53 ft 6 in (17.53 m)
Height: 14 ft 8 in (4.47 m)
Wingspan: 57 ft 6 in (17.53 m)
Aspect ratio: 6.54
Wing area: 506.00 sq ft (47.01 m^2)
Tailplane span: 18 ft 10 in (5.74 m)
Wheel track: 17 ft 2½ in (5.25 m)

Powerplant

Two General Electric TF34-GE-100 turbofans each rated at 9,065 lb st (40.32 kN) dry

Weights

Basic empty: 21,541 lb (9771 kg)
Operating empty: 24,959 lb (11321 kg)
Maximum take-off: 50,000 lb (22680 kg)

Fuel and load

Internal fuel: 10, 700 lb (4853 kg)
External fuel: three 600-US gal (2271-litre) drop tanks
Maximum ordnance: 16,000 lb (7258 kg)

Performance

Maximum level speed 'clean' at sea level: 381 kt (439 mph; 706 km/h)
Maximum cruising speed at 5,000 ft (1525 m): 337 kt (387 mph; 623 km/h)
Maximum rate of climb at sea level: about 6,000 ft (1828 m) per minute
Take-off run: 4,000 ft (1220 m) at maximum take-off weight or 1,450 ft (442 m) at forward strip weight
Landing run: 2,000 ft (610 m) at maximum landing weight or 1,300 ft (396 m) at forward strip weight

Range

Ferry range with two drop tanks: 2,131 nm (2,454 miles; 3949 km)
Combat radius: 540 nm (620 miles; 1000 km) on a deep strike mission or 250 nm (288 miles; 463 km) on a close air support mission with a 1.7-hour loiter

Armament

One GAU-8/A 30-mm cannon with 1,350 rounds, plus up to 16,000 lb (7258 kg) of mixed ordnance (including laser-guided bombs, rockets, CBUs and Maverick missiles) on 11 external store stations

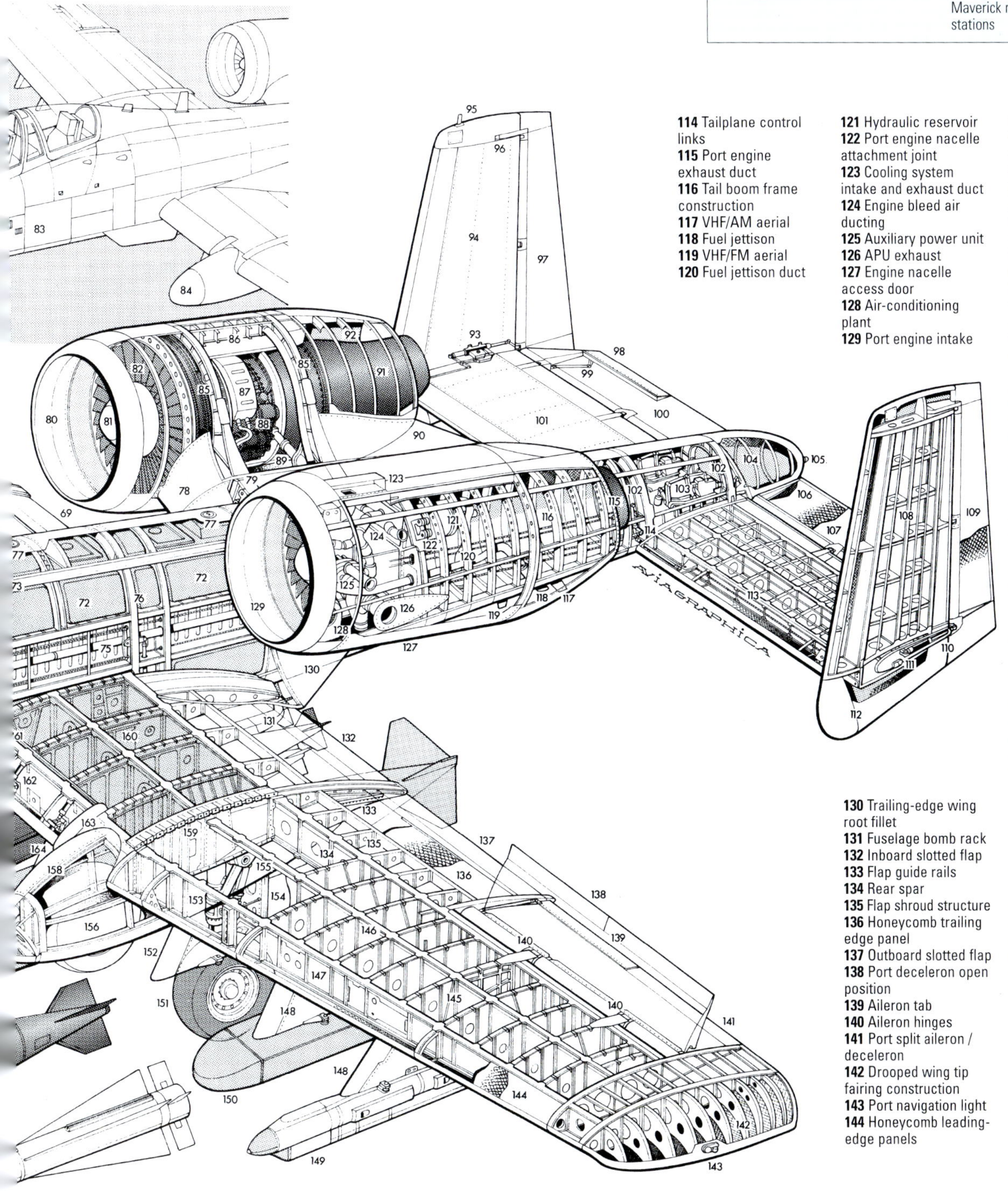

114 Tailplane control links
115 Port engine exhaust duct
116 Tail boom frame construction
117 VHF/AM aerial
118 Fuel jettison
119 VHF/FM aerial
120 Fuel jettison duct
121 Hydraulic reservoir
122 Port engine nacelle attachment joint
123 Cooling system intake and exhaust duct
124 Engine bleed air ducting
125 Auxiliary power unit
126 APU exhaust
127 Engine nacelle access door
128 Air-conditioning plant
129 Port engine intake
130 Trailing-edge wing root fillet
131 Fuselage bomb rack
132 Inboard slotted flap
133 Flap guide rails
134 Rear spar
135 Flap shroud structure
136 Honeycomb trailing edge panel
137 Outboard slotted flap
138 Port deceleron open position
139 Aileron tab
140 Aileron hinges
141 Port split aileron / deceleron
142 Drooped wing tip fairing construction
143 Port navigation light
144 Honeycomb leading-edge panels
145 Wing rib construction
146 Centre spar
147 Leading-edge spar
148 Two outer fixed pylons (1,000-lb/454-kg capacity)
149 ALQ-119 electronic countermeasures pod
150 ALE-37A chaff dispenser
151 Port mainwheel
152 2,500-lb (1134-kg) capacity stores pylon
153 Main undercarriage leg strut
154 Undercarriage leg doors
155 Main undercarriage leg pivot fixing
156 Port mainwheel semi-recessed housing
157 Pressure refuelling point
158 Undercarriage pod fairing
159 Outer wing panel attachment joint
160 Port wing integral fuel tank
161 Inboard leading-edge slat
162 Slat hydraulic jacks
163 Slat end plate
164 2,500-lb (1134-kg) stores pylon
165 3,500-lb (1588-kg) capacity fuselage pylon
166 Bomb ejector rack
167 Mk 82 500-lb (227-kg) bombs
168 Rockeye anti-armour cluster bomb
169 600-US gal (2271-litre) long-range ferry tank
170 Mk 84 2,000-lb (907-kg) bomb
171 Maverick air-to-ground missile
172 Paveway 3,000-lb (1360-kg) laser-guided bomb

General Dynamics F-111

The three-tone T.O. 1-1-4 Southeast Asia camouflage, as applied to this F-111, proved highly effective. On daylight missions in Vietnam, the F-111 was used for hitting tactical or strategic targets, but was occasionally used as a bomb leader, taking less sophisticated aircraft on bombing missions and supplying drop commands to increase their accuracy.

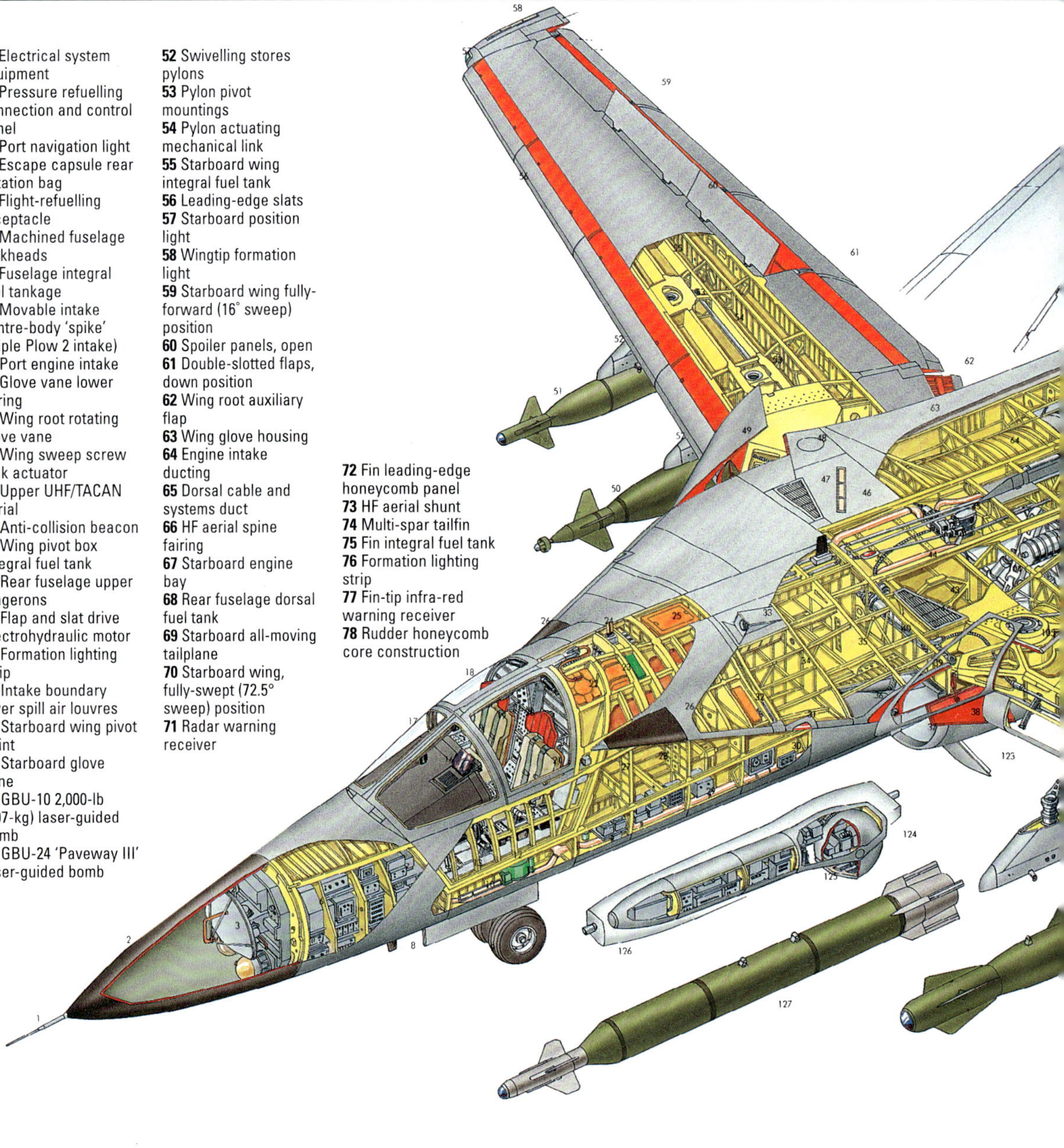

F-111F Aardvark

Cutaway key

1 Pitot head
2 Glass-fibre radome
3 AN/APQ-161 navigation and attack radar
4 AN/APQ-146 terrain-following radar
5 Radar equipment module
6 Avionics equipment bay
7 Flight control computers
8 Lower UHF/TACAN antenna
9 Twin nosewheels, forward-retracting
10 Electro-luminescent formation lighting strip
11 Liquid oxygen converter
12 Pressurised escape capsule joint frame
13 Rudder pedals
14 Control column
15 Engine throttle levers and wing sweep control
16 Pilot's head-up display
17 Single curvature windscreen panels
18 Upward-hinged cockpit canopy covers
19 Tactical navigator's seat
20 Rear bulkhead consoles
21 Pilot's seat
22 Escape capsule recovery parachute stowage
23 Self-righting air bag (two)
24 UHF/IFF aerial
25 Stabilising and brake parachute stowage
26 Leading-edge flush ECM antennas
27 Forward fuselage fuel tank
28 Tank floor/weapons bay roof
29 Electrical system equipment
30 Pressure refuelling connection and control panel
31 Port navigation light
32 Escape capsule rear flotation bag
33 Flight-refuelling receptacle
34 Machined fuselage bulkheads
35 Fuselage integral fuel tankage
36 Movable intake centre-body 'spike' (Triple Plow 2 intake)
37 Port engine intake
38 Glove vane lower fairing
39 Wing root rotating glove vane
40 Wing sweep screw jack actuator
41 Upper UHF/TACAN aerial
42 Anti-collision beacon
43 Wing pivot box integral fuel tank
44 Rear fuselage upper longerons
45 Flap and slat drive electrohydraulic motor
46 Formation lighting strip
47 Intake boundary layer spill air louvres
48 Starboard wing pivot point
49 Starboard glove vane
50 GBU-10 2,000-lb (907-kg) laser-guided bomb
51 GBU-24 'Paveway III' laser-guided bomb
52 Swivelling stores pylons
53 Pylon pivot mountings
54 Pylon actuating mechanical link
55 Starboard wing integral fuel tank
56 Leading-edge slats
57 Starboard position light
58 Wingtip formation light
59 Starboard wing fully-forward (16° sweep) position
60 Spoiler panels, open
61 Double-slotted flaps, down position
62 Wing root auxiliary flap
63 Wing glove housing
64 Engine intake ducting
65 Dorsal cable and systems duct
66 HF aerial spine fairing
67 Starboard engine bay
68 Rear fuselage dorsal fuel tank
69 Starboard all-moving tailplane
70 Starboard wing, fully-swept (72.5° sweep) position
71 Radar warning receiver
72 Fin leading-edge honeycomb panel
73 HF aerial shunt
74 Multi-spar tailfin
75 Fin integral fuel tank
76 Formation lighting strip
77 Fin-tip infra-red warning receiver
78 Rudder honeycomb core construction

F-111s can carry a formidable and varied amount of offensive weaponry, though the picture above shows practice bombs. Weapons are placed on a trolley which can be lowered or raised accordingly, and greatly increases the speed and ease of reloading. F-111Fs equipped with GBU-12D/Bs were used to attack Iraqi armoured and motorised formations during the Gulf War and scored 10 times as many tank kills as the F-16 during the conflict.

SPECIFICATION

F-111F Aardvark

Dimensions

Length overall: 73 ft 6 in (22.40 m)
Height: 17 ft 1½ in (5.22 m)
Wingspan (spread): 63 ft (19.20 m)
Wingspan (swept): 31 ft 11½ in (9.74 m)
Aspect ratio (spread): 7.56
Aspect ratio (swept): 1.55
Wing area (spread): 657.07 sq ft (61.07m^2)
Wing area (swept): 525 sq ft (48.77m^2)

Powerplant

Two Pratt & Whitney TF30-P-100 turbofans, each rated at 25,000 lb st (111.65 kN) with afterburning

Weights

Operating empty: 47,481 lb (21537 kg)
Maximum take-off: 100,000 lb (45360 kg)

Fuel and load

Internal fuel: 5,025 US gal (19021 litres)
External fuel: up to four 600-US gal (2271-litre) drop tanks
Maximum ordnance: 31,500 lb (14228 kg)

Performance

Maximum level clean speed at 36,000 ft (10975 m): 1,433 kt (1,650 mph; 2655 km/h)
Cruising speed at high altitude: 496 kt (571 mph; 919 km/h)
Service ceiling: 60,000 ft (18290 m)
Take-off distance to 50 ft (15 m): 3,120 ft (951 m)
Landing run: less than 3,000 ft (915 m) at normal landing weight

Range

Range with maximum internal fuel: 2,540 nm (2,925 miles; 4707 km)

Armament

One M61 multi-barreled 20-mm cannon and two B43 bombs in the internal weapon bay. The external stores were carried on three attachments under each wing. Primary weapons included the 500-lb (227-kg) GBU-12 Paveway II, 2,000-lb (907-kg) GBU-10 Paveway II and 2,000-lb (907-kg) GBU-24 Paveway III. Both 2,000-lb (907-kg) weapons were available with either a standard Mk 84 warhead or BLU-109 penetration warhead. The GBU-28 'Deep Throat' was a 4,800-lb (2177-kg) Paveway III hastily developed for Desert Storm. 'Dumb' ordnance such as iron bombs, cluster weapons and the BLU-107 Durandal runway-cratering munition cuold also be carried. An F-111F speciality was the GBU-15 2,000-lb (907-kg) EO-guided bomb. This had either a Mk 84 or BLU-109 warhead, and a TV- or IR-seeker adapted from those fitted to Maverick missiles. For defence, the F-111F routinely carried the AIM-9P-3 Sidewinder

79 Rudder hydraulic actuator
80 Afterburner nozzle control jacks
81 Translating primary iris afterburner nozzle
82 Afterburner duct air mixing intakes
83 Chaff dispenser
84 Variable area exhaust nozzle
85 Rear ECM antenna fairing
86 Port all-moving tailplane
87 Static dischargers
88 Port wing, fully-swept position
89 Control surface honeycomb core leading and trailing edges
90 Tailplane pivot fixing
91 Tailplane hydraulic actuator
92 Port engine bay
93 Port ventral fin

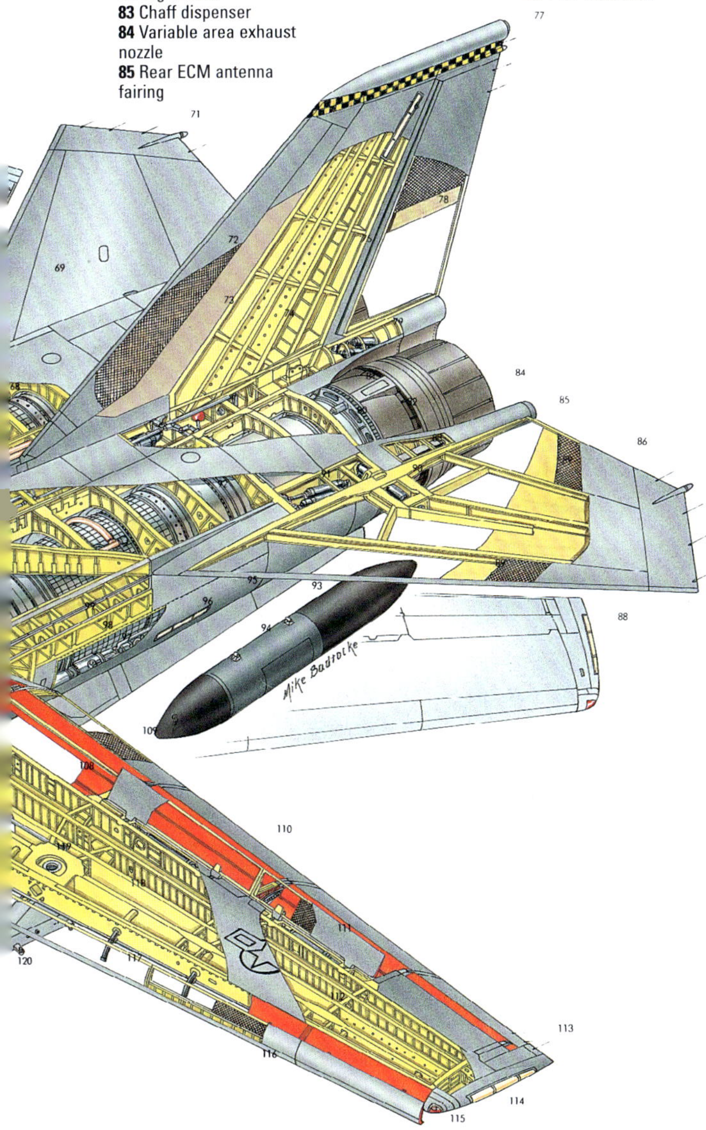

94 AN/AXQ-14 two-way datalink weapon control and guidance pod, carried beneath rear fuselage
95 Engine bay access doors
96 Formation lighting strip
97 Engine accessory equipment compartment
98 Fuselage flank fuel tank
99 Wing root pneumatic seal
100 Pratt & Whitney TF30- P-100 afterburner turbofan
101 Conical engine intake centrebody
102 Hydraulic reservoir
103 Main undercarriage wheel bay
104 Retraction breaker strut and shock absorbers
105 Wing pivot bearing
106 Flap drive shaft and screw jacks
107 Wing root auxiliary flap
108 Ventral flap sealing plate
109 Flap honeycomb core construction
110 Port double-slotted flap
111 Port spoiler panels
112 Multi-spar wing panel construction
113 Static dischargers
114 Wingtip formation light
115 Port position light
116 Port leading-edge slat
117 Slat guide rails
118 Port wing integral fuel tank
119 Pylon pivot mountings
120 Port swivelling stores pylons
121 Leading-edge slat drive shaft and transfer gearbox
122 Port mainwheel
123 Ventral airbrake/mainwheel door
124 AN/AVQ-26 Pave Tack laser designator pod, carried in weapons bay
125 Swivelling and rotating sighting unit
126 Pave Tack rotary housing
127 GBU-28 'Deep Throat' laser-guided bomb
128 GBU-15 2,000-lb (907-kg) EO-guided bomb

Gloster Javelin

Although plagued by serviceability and structural problem Javelin was a highly capable bomber-interceptor, equipped as i with four 30-mm ADEN cannon and, later, with the addition o IR-seeking missiles. XH881 was an example of the ultimate va the FAW.Mk 9. It was built as an FAW.Mk 7, but retained by G for conversion to Mk 9 standard before delivery to No. 25

Javelin FAW.Mk 9R

Cutaway key

1 Detachable flight refuelling probe, used for overseas deployment
2 Glass-fibre radome
3 AI.Mk 22 radar scanner dish (American AN/APQ-43)
4 Scanner tracking mechanism
5 Radar transmitter/receiver
6 Radar mounting bulkhead
7 Instrument venturi
8 Aft-retracting nosewheel
9 Mudguard
10 Torque scissor links
11 Lower IFF antenna
12 Nose equipment bay access door
13 Additional (long-range) oxygen bottles
14 Radar modulator
15 Upper IFF antenna
16 Front pressure bulkhead
17 Rudder pedals
18 Standard oxygen bottle stowage, port and starboard
19 Side console panel
20 Engine throttle levers
21 Control column
22 Pilot's instrument panel
23 Instrument panel shroud
24 Windscreen rain dispersal air duct
25 Starboard engine intake
26 Windscreen panels
27 Pilot's gyro gunsight
28 Rearward-sliding cockpit canopy
29 Ejection seat faceblind firing handle
30 Pilot's Martin-Baker Mk 4 ejection seat
31 Seat mounting rails
32 Port engine air intake
33 Intake lip bleed air de-icing
34 Intake duct framing
35 Rebecca homing antenna
36 Radar altimeter transmitting antenna
37 Missile cooling system heat exchanger
38 Cold air unit and compressor
39 Port intake duct
40 Radar operator's instrument console
41 Radar indicator
42 Fixed canopy centre section
43 Missile cooling system air bottles
44 Radar operator's rearward-sliding canopy
45 Radar operator's Martin-Baker ejection seat
46 Cockpit pressure shell framing
47 Missile control system equipment
48 Engine compressor intake
49 IPN engine starter fuel tank
50 Central equipment bay
51 Engine-driven gearbox with generators and hydraulic pumps
52 Cabin air system heat exchanger
53 Flight control rods
54 Gee antenna
55 Canopy tail fairing with heat exchanger outlet duct
56 Wing spar attachment fuselage main frame
57 Starboard main undercarriage wheel bay
58 Gun heating system air reservoirs
59 Starboard leading edge fuel tanks Nos 1, 2 and 3. Total internal capacity 950 Imp gal (4319 litres)
60 100-Imp gal (454-litre) external pylon tanks
61 Starboard wing pylons
62 Pylon aerodynamic fairings
63 Cannon muzzle blast fairings with frangible caps
64 Cannon barrel blast tubes
65 30-mm ADEN cannon, four carried for Far Eastern deployment, two only for European operations
66 Link collector boxes
67 Gun camera
68 Aileron control rod and pitch stabiliser
69 Aileron servodyne
70 Vortex generators, three rows
71 Starboard pitot head
72 Starboard navigation light
73 Formation light
74 Starboard aileron
75 Aileron spar
76 Fixed portion of trailing edge
77 Starboard airbrake, upper and lower surfaces
78 Airbrake hydraulic jack (2)
79 Flap hydraulic jack (2)
80 Ventral flap panel
81 Ammunition magazines, 100 rounds per gun
82 Rear fuel tanks, Nos 4 and 5
83 Engine exhaust, zone-3, cooling air intake
84 Artificial feel simulator pressure heads
85 Starboard engine bay
86 Fuselage centre keel structure
87 Rudder feel simulator
88 Port Armstrong Siddeley Sapphire Sa.7R turbojet with 12 per cent limited reheat
89 Engine bay firewall
90 Turbine section
91 Central fuel system collector tanks
92 Engine exhaust duct
93 Fin-mounted bulkhead
94 Fin spar attachment joint
95 Servomotor
96 Rudder servodyne
97 Fin rib structure
98 Leading-edge ribs and control runs
99 Hydraulic accumulators
100 Tailplane hydraulic power control unit
101 Tailplane operating beam
102 Fixed tailplane centre section
103 Tailplane spar bearing
104 Tubular tailplane spar
105 Starboard trimming tailplane
106 Starboard elevator
107 UHF antenna
108 Tail navigation light
109 Tail warning radar antenna
110 Elevator operating linkage
111 Port elevator rib structure
112 Tailplane single spar and rib structure
113 Rudder rib structure
114 Afterburner nozzles
115 Detachable fuselage tail section, engine removal
116 Afterburner duct
117 Tail section joint frame
118 Wing rear spar attachment joint
119 Port flap housing
120 Flap hydraulic jacks
121 Airbrake hydraulic jacks
122 Semi-span rear spar
123 Port airbrake panel, upper and lower surfaces
124 Fixed trailing-edge rib structure
125 Cartridge case ejection chutes
126 Port aileron
127 Aileron rib structure
128 Aileron servodyne
129 Formation light
130 Wingtip member structure
131 Port navigation light

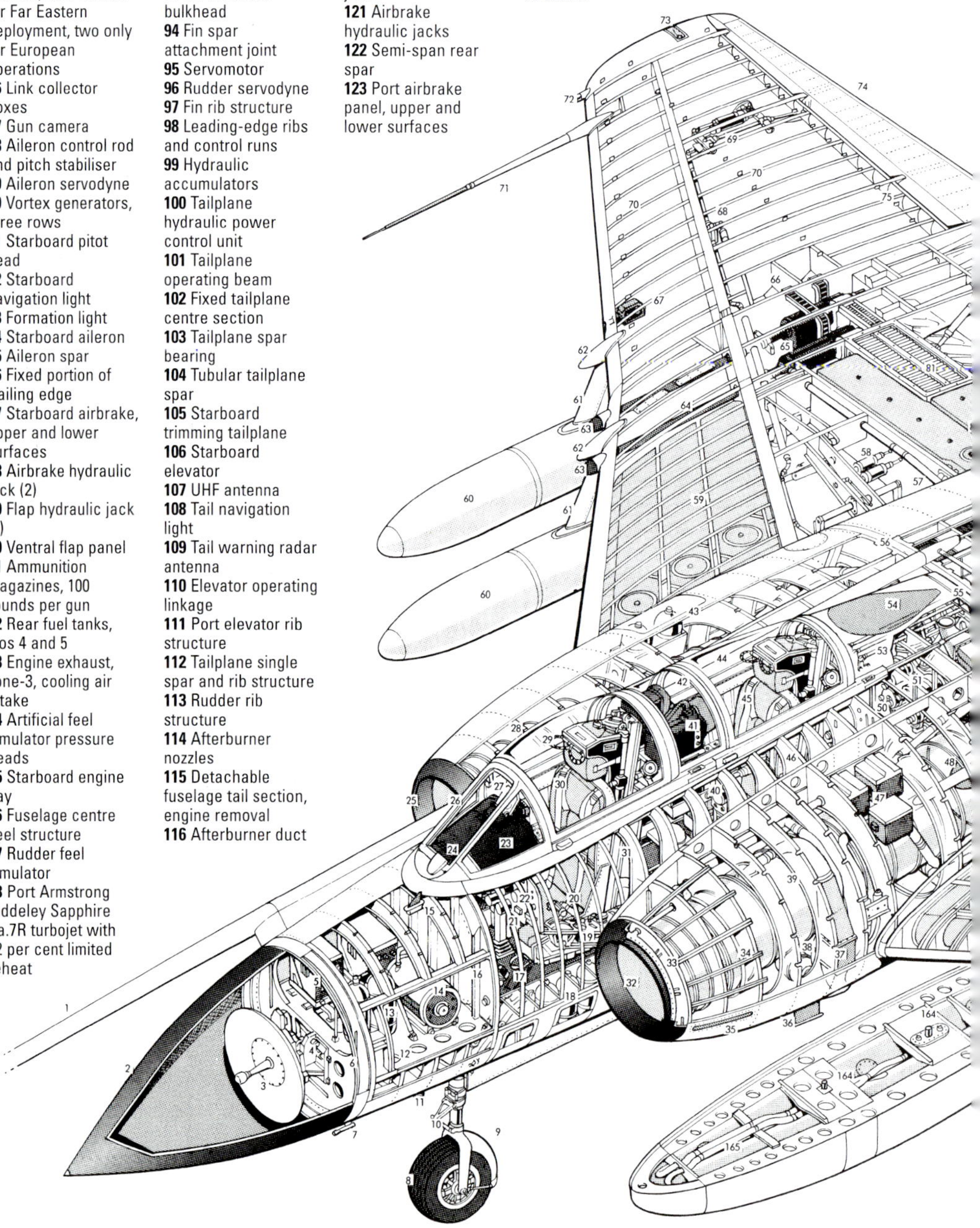

Javelin FAW.Mk 7 XH712 never entered squadron service in its original mark, spending its early career as a trials airframe. It was initially engaged in handling trials with the Aircraft and Armament Experimental Establishment, before passing to de Havilland as a testbed for both Firestreak (illustrated above) and Red Top AAM trials. In 1959 it was converted to FAW.Mk 9 standard and passed into storage. It was once more modified, this time to Mk 9R standard, before passing to No. 23 Sqn late in 1962. It saw out its service with No. 29 Sqn, before being scrapped on 30 June 1967.

SPECIFICATION

Javelin FAW.Mk 7/8/9/9R

Dimensions

Length, Mk 7/9/9R: 56 ft 4 in (17.17 m)
Length, Mk 8: 55 ft 2½ in (16.83 m)
Height: 16 ft (4.88 m)
Wingspan: 52 ft (15.85 m)
Wing area: 927 sq ft (86.12 m²)
Wheel track: 23 ft 4 in (7.11 m)

Powerplant

Mk 7: Two Armstrong Siddeley Sapphire Sa.7 turbojets rated at 11,000 lb st (48.92 kN) dry
Mk 8/9/9R: Two Armstrong Siddeley Sapphire Sa.7R turbojets rated at 11,000 lb st (48.92 kN) dry and 12,300 lb st (54.70 kN) with 12 per cent afterburning at 20,000 ft (6096 m)

Weights

Take-off, 'clean', Mk 7: 35,690 lb (16188 kg)
Take-off, 'clean', Mk 8: 37,410 lb (16968 kg)
Take-off, 'clean', Mk 9: 38,100 lb (17272 kg)
Overload, with two ventral tanks, Mk 7: 40,270 lb (18266 kg)
Overload, with two ventral tanks Mk 8: 42,510 lb (19282 kg)
Overload, with two ventral tanks Mk 9: 43,165 lb (19578 kg)

Fuel and load

Internal fuel, Mk 7: 915 Imp gal (4158 litres)
Internal fuel, Mk 8/9/9R: 950 Imp gal (4319 litres)
External fuel: All variants could carry up two 250-Imp gal (1137-litre) conformal ventral tanks
Drop tanks, Mk 7/8/9: ventral tanks plus up to four 100-Imp gal (454-litre) tanks
Drop tanks, Mk 9R: ventral tanks plus up to four 230-Imp gal (1046-litre) tanks

Performance

Maximum level speed 'clean' at sea level, Mk 7: 708 mph (1141 km/h)
Maximum level speed 'clean' at sea level, Mk 8/9: 702 mph (1130 km/h)
Climb to 45,000 ft (13716 m), Mk 7: 6 minutes 36 seconds
Climb to 50,000 ft (15240 m), Mk 8/9: 9 minutes 15 seconds
Service ceiling, Mk 7: 52,800 ft (16039 m)
Service ceiling, Mk 8/9: 52,000 ft (15849 m)
Absolute ceiling, Mk 7: 54,100 ft (16489 m)
Absolute ceiling, Mk 8/9: 54,000 ft (16459 m)

Armament

Up to four 30-mm ADEN cannon in the outer wing panels, each with 100 rounds, plus up to four de Havilland Propellers Firestreak IR-homing air-to-air missiles

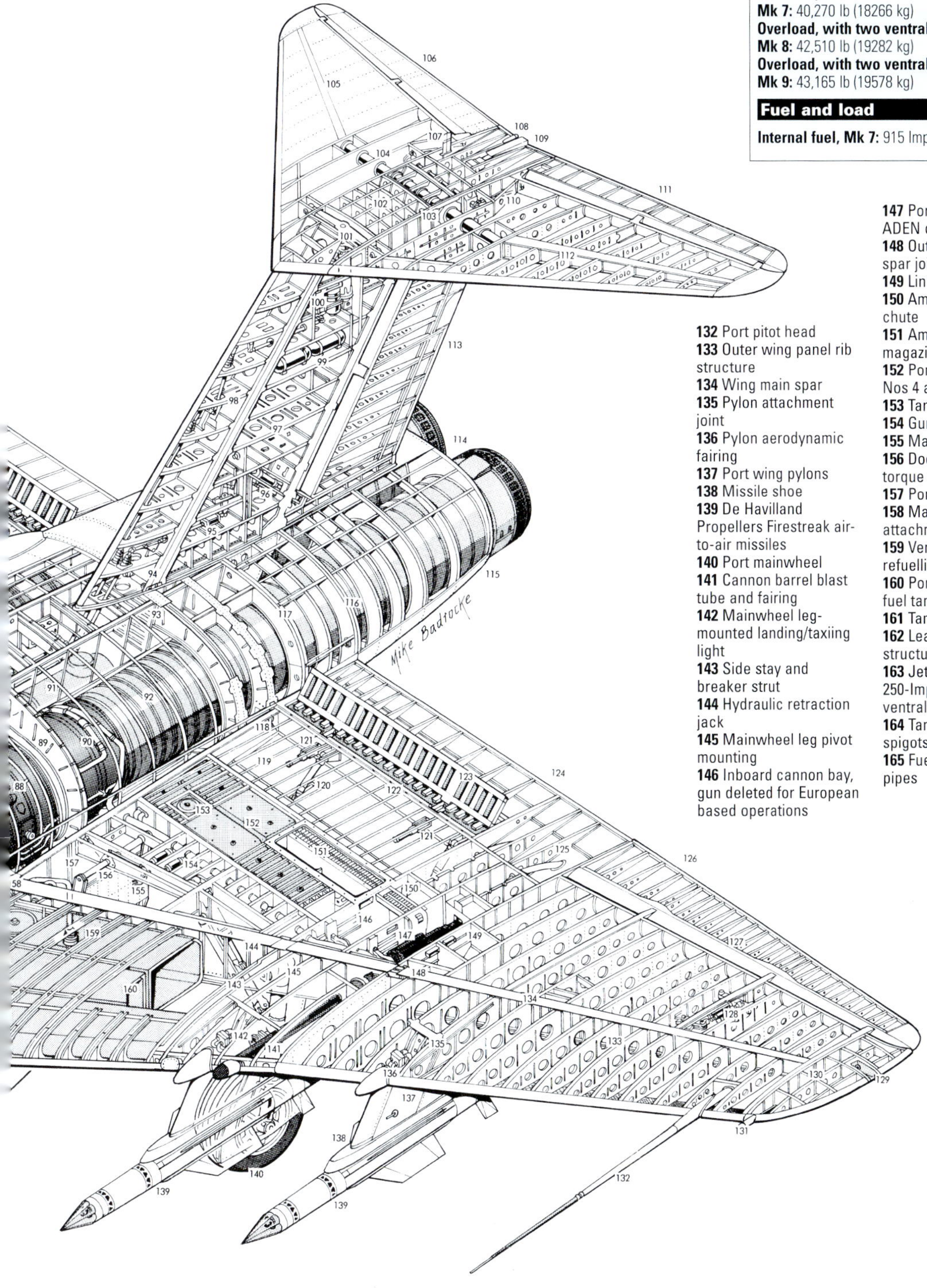

132 Port pitot head
133 Outer wing panel rib structure
134 Wing main spar
135 Pylon attachment joint
136 Pylon aerodynamic fairing
137 Port wing pylons
138 Missile shoe
139 De Havilland Propellers Firestreak air-to-air missiles
140 Port mainwheel
141 Cannon barrel blast tube and fairing
142 Mainwheel leg-mounted landing/taxiing light
143 Side stay and breaker strut
144 Hydraulic retraction jack
145 Mainwheel leg pivot mounting
146 Inboard cannon bay, gun deleted for European based operations
147 Port outboard 30-mm ADEN cannon
148 Outer wing panel spar joint
149 Link collector boxes
150 Ammunition feed chute
151 Ammunition magazines
152 Port rear fuel tanks, Nos 4 and 5
153 Tank access panel
154 Gun heating air duct
155 Mainwheel door
156 Door operating torque shaft and linkage
157 Port mainwheel bay
158 Main spar attachment bolted joint
159 Ventral pressure refuelling connection
160 Port leading-edge fuel tanks, Nos 1, 2 and 3
161 Tank access panels
162 Leading-edge rib structure
163 Jettisonable 250-Imp gal (1137-litre) ventral fuel tank (2)
164 Tank mounting spigots
165 Fuel vent and feed pipes

Hawker Hunter

India was one of the most important export customers for the Hunter in both single-and two-seat form. A total of 252 new-build and refurbished examples were delivered between 1957 and 1973.

Hunter FGA.Mk 9

Cutaway key
1 Radome
2 Radar scanner dish
3 Ram air intake
4 Camera port
5 Radar ranging equipment
6 Camera access panel
7 Gun camera
8 Ground pressurisation connection
9 Nosewheel door
10 Oxygen bottles
11 IFF aerial
12 Electronics equipment
13 Nosewheel bay
14 De-icing fluid tank
15 Pressurisation control valves
16 Cockpit front bulkhead
17 Nose landing gear leg
18 Nosewheel forks
19 Forward retracting nosewheel
20 Nosewheel leg door
21 Cannon muzzle port
22 Gun blast cascade deflectors
23 Rudder pedals
24 Bullet proof windscreen
25 Cockpit canopy framing
26 Reflector gunsight
27 Instrument panel shroud
28 Control column
29 Cockpit section fuselage frames
30 Rearward sliding cockpit canopy cover
31 Pilot's starboard side console
32 Martin Baker Mk 3H ejector seat
33 Throttle control
34 Pilot's port side console
35 Cannon barrel tubes
36 Pneumatic system airbottles
37 Cockpit canopy emergency release
38 Cockpit rear pressure bulkhead
39 Air conditioning valve
40 Ejector seat headrest
41 Firing handle
42 Air louvres
43 Ammunition tanks
44 Ammunition link collector box
45 Cartridge case ejectors
46 Batteries
47 Port air inlet
48 Boundary layer splitter plate
49 Inlet lip construction
50 Radio and electronics equipment bay
51 Sliding canopy rail
52 Air conditioning supply pipes
53 Control rod linkages
54 Communications aerial
55 Fuselage double frame bulkhead
56 Boundary layer air outlet
57 Secondary air inlet door spring loaded
58 Inlet duct construction
59 Forward fuselage fuel tank
60 Starboard inlet duct
61 Starboard wing fuel tank
62 230 Imp gal (1046 litre) drop tank
63 Inboard pylon mounting
64 Leading edge dog tooth
65 100-Imp gal (455-litre) drop tank
66 Outboard pylon mounting
67 Wing fence
68 Leading edge extension
69 Starboard navigation light
70 Starboard wingtip
71 Whip aerial
72 Fairey hydraulic aileron booster jack
73 Starboard aileron
74 Aileron control rod linkage
75 Flap cut out section for drop tank clearance
76 Starboard flap construction
77 Flap hydraulic jack
78 Flap synchronising jack
79 Starboard main landing gear mounting
80 Retraction jack
81 Starboard landing gear bay
82 Dorsal spine fairing
83 Main wing attachment frames
84 Main spar attachment joint
85 Engine starter fuel tank
86 Air conditioning system
87 Engine inlet compressor face
88 Air conditioning pre-cooler
89 Cooling air outlet louvres
90 Rear spar attachment
91 Aileron control rods
92 Front engine mountings
93 Rolls Royce Avon 207 engine
94 Bleed air duct
95 Engine bay cooling flush air Intake
96 Rear engine mounting
97 Rear fuselage joint ring
98 Joint ring attachment bolts
99 Tailplane control rods
100 Fuel piping from rear tank

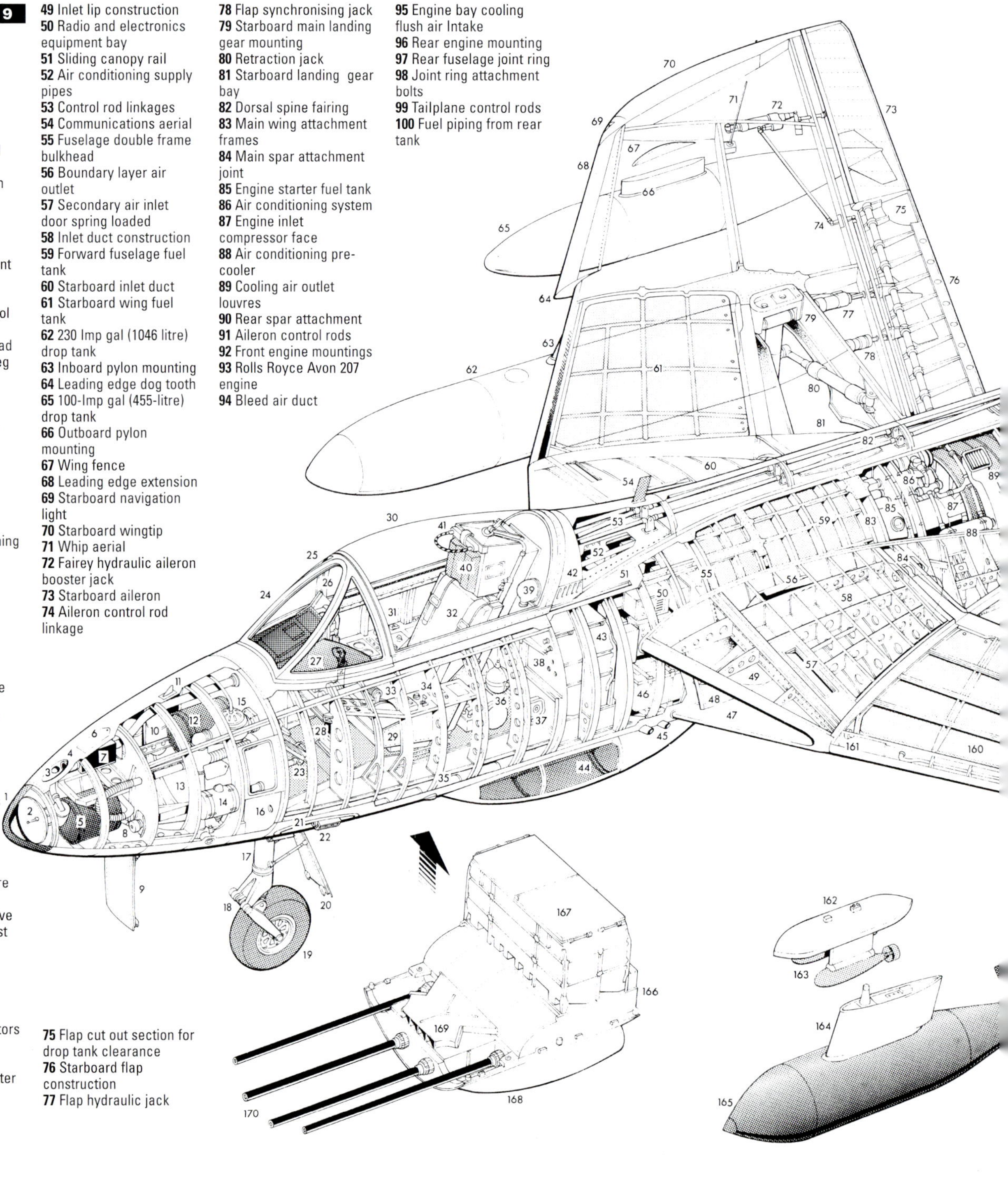

One of ten new-build T.Mk 8s constructed for the Royal Navy at Hawker's Kingston-upon-Thames site, XL584 was delivered in October 1958 for service with No. 764 Sqn, FAA. This aircraft, along with two other examples, subsequently returned to Hawker for preparation in high-gloss epoxy dark blue and white paint for use by the Flag Officer (Flying Training) at Yeovilton, then becoming known as 'the Admiral's barges'. By 1975, like most FAA Hunters, XL584 had been reallocated to the Fleet Requirement and Air Direction Unit (FRADU) and had adopted this high visibility colour scheme.

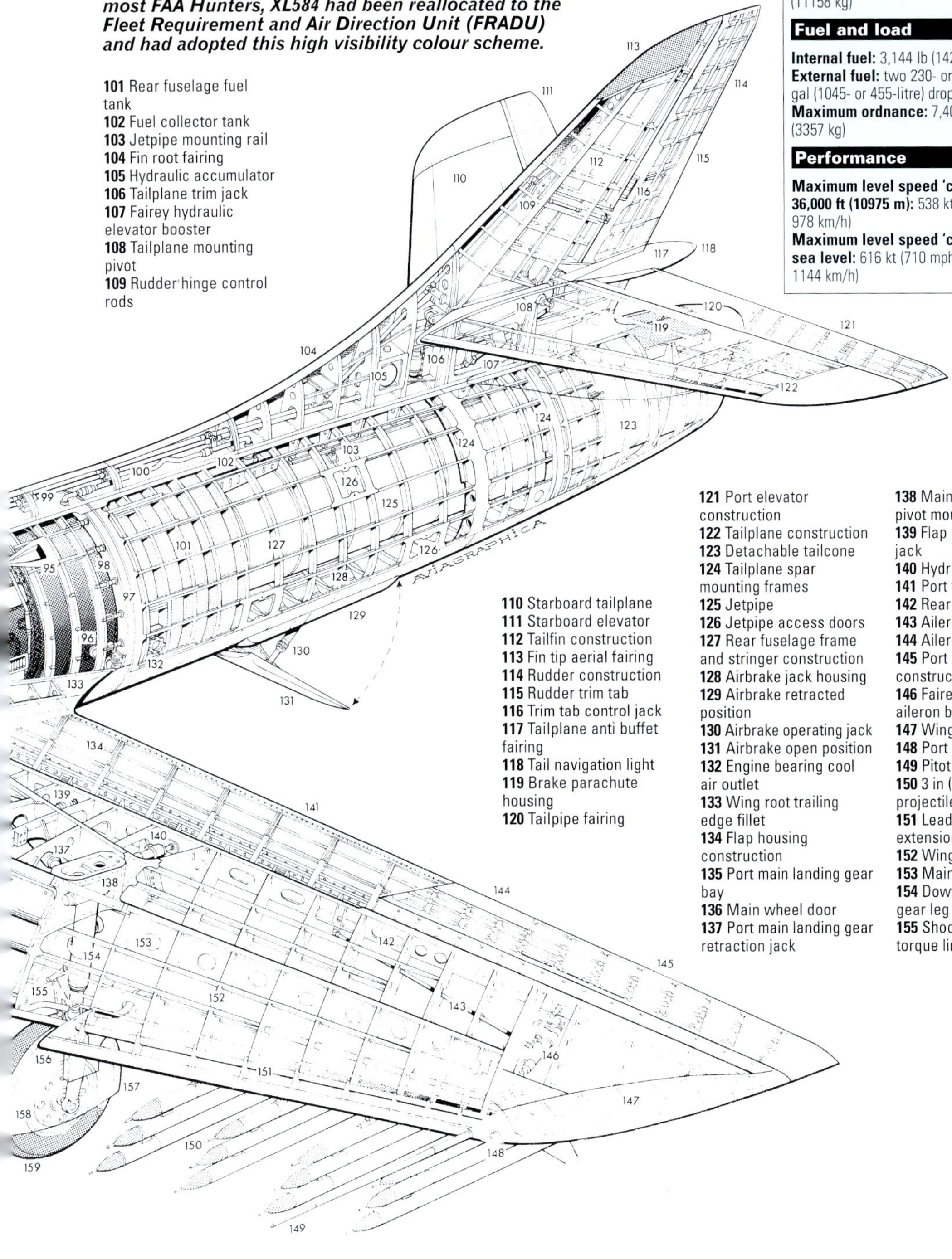

101 Rear fuselage fuel tank
102 Fuel collector tank
103 Jetpipe mounting rail
104 Fin root fairing
105 Hydraulic accumulator
106 Tailplane trim jack
107 Fairey hydraulic elevator booster
108 Tailplane mounting pivot
109 Rudder hinge control rods
110 Starboard tailplane
111 Starboard elevator
112 Tailfin construction
113 Fin tip aerial fairing
114 Rudder construction
115 Rudder trim tab
116 Trim tab control jack
117 Tailplane anti buffet fairing
118 Tail navigation light
119 Brake parachute housing
120 Tailpipe fairing
121 Port elevator construction
122 Tailplane construction
123 Detachable tailcone
124 Tailplane spar mounting frames
125 Jetpipe
126 Jetpipe access doors
127 Rear fuselage frame and stringer construction
128 Airbrake jack housing
129 Airbrake retracted position
130 Airbrake operating jack
131 Airbrake open position
132 Engine bearing cool air outlet
133 Wing root trailing edge fillet
134 Flap housing construction
135 Port main landing gear bay
136 Main wheel door
137 Port main landing gear retraction jack
138 Main landing gear leg pivot mounting
139 Flap synchronising jack
140 Hydraulic flap jack
141 Port flap
142 Rear spar
143 Aileron control rods
144 Aileron trim tab
145 Port aileron construction
146 Fairey hydraulic aileron booster
147 Wing tip construction
148 Port navigation light
149 Pitot tube
150 3 in (7.62 cm) rocket projectiles
151 Leading edge extension ribs
152 Wing rib construction
153 Main spar
154 Dowty main landing gear leg
155 Shock absorber torque links
156 Leading-edge dog-tooth
157 Mainwheel doors
158 Dunlop-Maxaret anti-skid wheel brakes
159 Port mainwheel
160 Port wing fuel tank: total internal fuel capacity 392 Imp gal (1782 litres)
161 Leading edge pin joint
162 ML twin stores carrier
163 9-kg (20-lb) practice bombs
164 Inboard wing pylon
165 1000-lb (454-kg) bomb
166 Four 30-mm ADEN gun pack
167 Ammunition boxes, 150 rounds per gun
168 Link collector box
169 Gun gas purging air duct
170 Cannon barrels remaining in aircraft when gun pack is withdrawn

SPECIFICATION

Hunter FGA.Mk 9

Dimensions

Length overall: 45 ft 10½ in (13.98 m)
Height: 13 ft 2 in (4.01 m)
Wingspan: 33 ft 8 in (10.26 m)
Aspect ratio: 3.25
Wing area: 349 sq ft (32.42 m²)
Tailplane span: 11 ft 10 in (3.61 m)
Wheel track: 14 ft 9 in (4.50 m)
Wheel base: 15 ft 9 in (4.80 m)

Powerplant

One Rolls-Royce Avon RA.28 Mk 207 turbojet rated at 10,150 lb st (45.15 kN)

Weights

Empty equipped: 14,400 lb (6532 kg)
Normal take-off: 18,000 lb (8165 kg)
Maximum take-off: 24,600 lb (11158 kg)

Fuel and load

Internal fuel: 3,144 lb (1426 kg)
External fuel: two 230- or 100-Imp gal (1045- or 455-litre) drop tanks
Maximum ordnance: 7,400 lb (3357 kg)

Performance

Maximum level speed 'clean' at 36,000 ft (10975 m): 538 kt (620 mph; 978 km/h)
Maximum level speed 'clean' at sea level: 616 kt (710 mph; 1144 km/h)
Maximum cruising speed at 36,000 ft (10975 m): 481 kt (554 mph; 892 km/h)
Economical cruising speed at optimum altitude: 399 kt (460 mph; 740 km/h)
Service ceiling: 50,000 ft (15240 m)
Maximum rate of climb at sea level: about 8,000 ft (2438 m) per minute
Take-off run: 2,100 ft (640 m) at normal take-off weight
Take-off distance to 50 ft (15 m): 3,450 ft (1052 m) at normal take-off weight
Landing run: 3,150 ft (960 m) at normal landing weight

Range

Ferry range with two drop tanks: 1,595 nm (1,840 miles; 2961 km)
Combat radius: 385 nm (443 miles; 713 km) on hi-lo-hi attack mission with typical warload and two drop tanks

Armament

Four 30-mm ADEN cannon mounted in a pack beneath the forward fuselage with up to 150 rounds of ammunition per gun. Inboard pylons could carry either British or foreign bombs of up to 1000 lb (454 kg), 2-in (5.08-cm) multiple rocket batteries, 100-Imp gal (455-litre) Napalm bombs, practice bomb carriers and a variety of other stores. Outboard pylons could carry launchers for 24 3-in (7.62-cm) rocket projectiles with various warheads or other types of rocket projectile

Hawker Siddeley Harrier

Several of No. IV (AC) Squadron's Harrier GR.Mk 3s gained a colourful fin during their last couple of years of service. Initially, the colours were applied to the commanding officer's aircraft for the squadron's 75th anniversary celebrations. Coloured fins were also tested in a bid to avoid collisions during low-level training by increasing conspicuity.

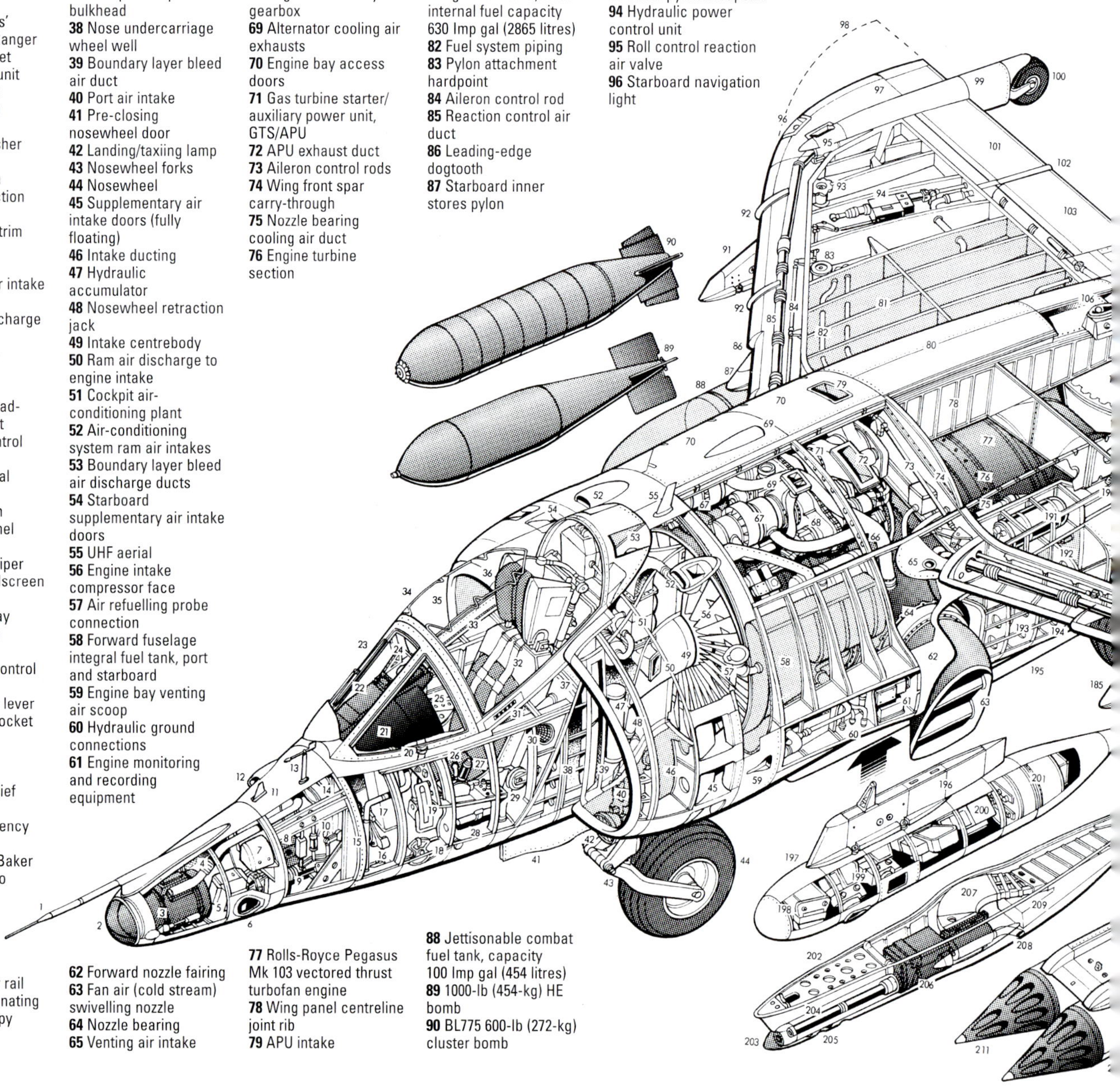

Harrier GR.Mk 3

Cutaway key

1 Pitot tube
2 Laser window protective 'eyelids'
3 Ferranti Laser Ranger and Marked-Target Seeker (LRMTS) unit
4 Cooling air duct
5 Oblique camera
6 Camera port
7 Windshield washer reservoir
8 Inertial platform
9 Nose pitch reaction control air duct
10 Pitch feel and trim actuator
11 IFF aerial
12 Cockpit ram air intake
13 Yaw vane
14 Cockpit air discharge valve
15 Front pressure bulkhead
16 Rudder pedals
17 Nav/attack 'head-down' display unit
18 Underfloor control linkages
19 Canopy external handle
20 Control column
21 Instrument panel shroud
22 Windscreen wiper
23 Birdproof windscreen panels
24 Head-up display
25 Starboard side console panel
26 Nozzle angle control lever
27 Engine throttle lever
28 Ejection seat rocket pack
29 Fuel cock
30 Cockpit pressurisation relief valve
31 Canopy emergency release
32 Pilot's Martin-Baker Type 9D, zero-zero ejection seat
33 Sliding canopy rail
34 Miniature detonating cord (MDC) canopy breaker
35 Starboard air intake
36 Ejection seat headrest
37 Cockpit rear pressure bulkhead
38 Nose undercarriage wheel well
39 Boundary layer bleed air duct
40 Port air intake
41 Pre-closing nosewheel door
42 Landing/taxiing lamp
43 Nosewheel forks
44 Nosewheel
45 Supplementary air intake doors (fully floating)
46 Intake ducting
47 Hydraulic accumulator
48 Nosewheel retraction jack
49 Intake centrebody
50 Ram air discharge to engine intake
51 Cockpit air-conditioning plant
52 Air-conditioning system ram air intakes
53 Boundary layer bleed air discharge ducts
54 Starboard supplementary air intake doors
55 UHF aerial
56 Engine intake compressor face
57 Air refuelling probe connection
58 Forward fuselage integral fuel tank, port and starboard
59 Engine bay venting air scoop
60 Hydraulic ground connections
61 Engine monitoring and recording equipment
62 Forward nozzle fairing
63 Fan air (cold stream) swivelling nozzle
64 Nozzle bearing
65 Venting air intake
66 Alternator cooling air ducts
67 Twin alternators
68 Engine accessory gearbox
69 Alternator cooling air exhausts
70 Engine bay access doors
71 Gas turbine starter/auxiliary power unit, GTS/APU
72 APU exhaust duct
73 Aileron control rods
74 Wing front spar carry-through
75 Nozzle bearing cooling air duct
76 Engine turbine section
77 Rolls-Royce Pegasus Mk 103 vectored thrust turbofan engine
78 Wing panel centreline joint rib
79 APU intake
80 Centre-section fairing panels
81 Starboard wing integral fuel tank, total internal fuel capacity 630 Imp gal (2865 litres)
82 Fuel system piping
83 Pylon attachment hardpoint
84 Aileron control rod
85 Reaction control air duct
86 Leading-edge dogtooth
87 Starboard inner stores pylon
88 Jettisonable combat fuel tank, capacity 100 Imp gal (454 litres)
89 1000-lb (454-kg) HE bomb
90 BL775 600-lb (272-kg) cluster bomb
91 Starboard outer stores pylon
92 Wing fences
93 Outer pylon hardpoint
94 Hydraulic power control unit
95 Roll control reaction air valve
96 Starboard navigation light

RAF Harrier GR.Mk 3s and Royal Navy Sea Harrier FRS.Mk 1s share the flight deck of HMS* Hermes *during the Falklands conflict of 1982. With the Sea Harrier's adoption of the air defence role, the GR.Mk 3s were heavily involved in the close air support (CAS) tasking and suffered a number of casualties.

SPECIFICATION

Harrier GR.Mk 3

Dimensions

Length: 46 ft 10 in (14.27 m)
Height: 11 ft 11 in (3.63 m)
Wingspan: 25 ft 3 in (7.70 m) with combat tips or 29 ft 8 in (9.04 m) with ferry tips
Wing area: 201.10 sq ft (18.68 m^2) with combat tips or 216.00 sq ft (20.07 m^2) with ferry tips
Wing aspect ratio: 3.18 with combat tips or 4.08 with ferry tips
Tailplane span: 13 ft 11 in (4.24 m)
Outrigger track: 22 ft 2 in (6.76 m)

Powerplant

One Rolls-Royce Pegasus Mk 103 rated at 21,500 lb st (95.94 kN)

Weights

Empty equipped: 12,300 lb (5579 kg)
Operating empty: 13,535 lb (6139 kg)
Normal take-off: 23,500 lb (10660 kg)
Maximum take-off: 25,200 lb (11431 kg)

Fuel and loadd

Internal fuel: 5,060 lb (2295 kg)
External fuel: up to two 330-Imp gal (1500-litre) ferry tanks or two 190- or 100-Imp gal (864- or 455-litre) drop tanks
Maximum ordnance: 5,000 lb (2268 kg) authorised or 8,000 lb (3269 kg) demonstrated

Performance

Maximum level speed 'clean' at sea level: 635 kt (730 mph; 1176 km/h)
Maximum rate of climb at sea level: 29,000 ft (8840 m) per minute
Climb to 40,000 ft (12190 m) after VTO: 2 minutes 23 seconds
Service ceiling: 51,200 ft (15605 m)
Take-off run at maximum take-off weight: 1,000 ft (305 m)
Landing run at normal landing weight: 0 ft (0 m)

Armament

One or two 30-mm ADEN cannon plus a wide variety of laser-guided and dumb bombs, cluster bombs and rocket pods plus two AIM-9L air-to-air missiles

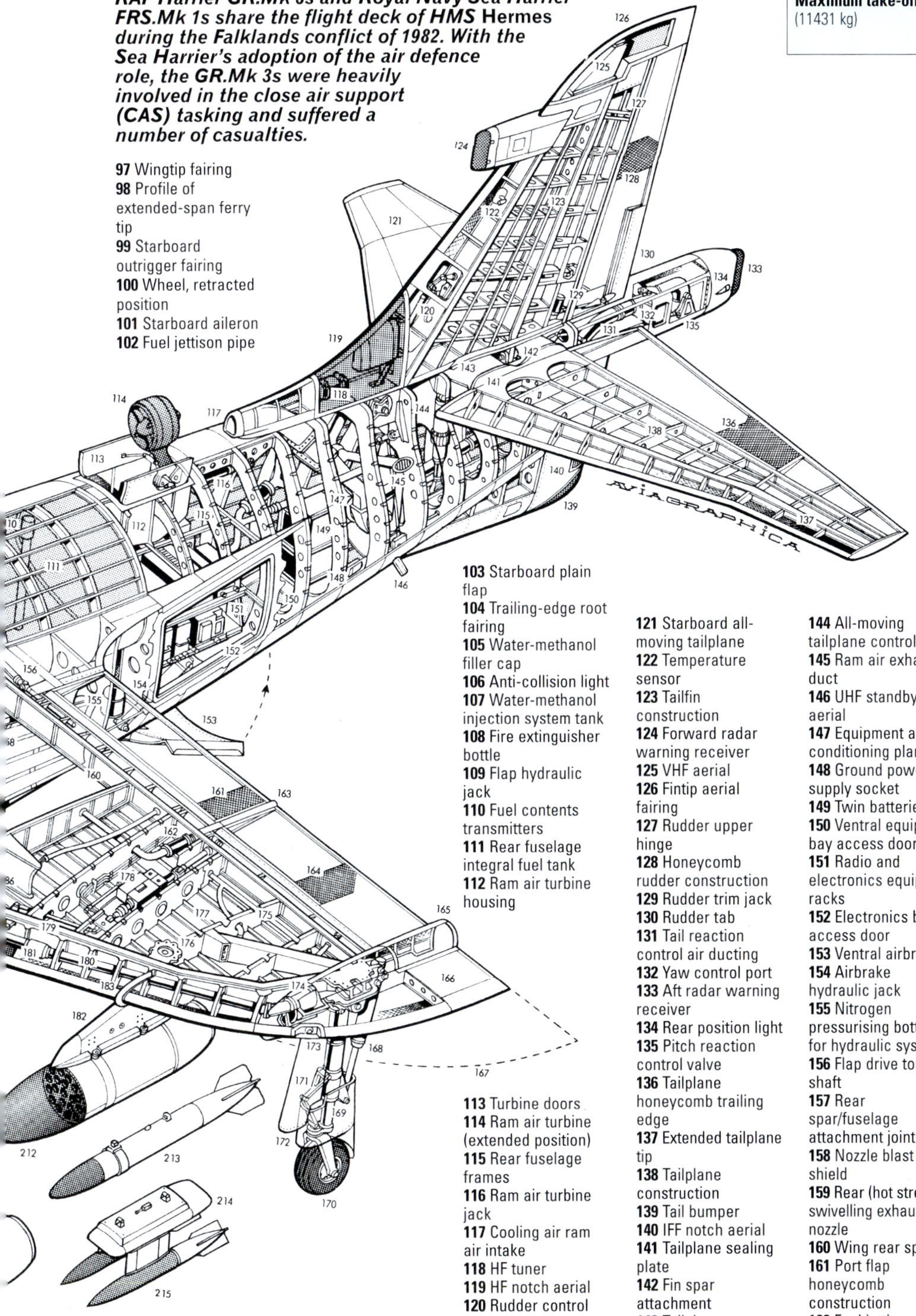

97 Wingtip fairing
98 Profile of extended-span ferry tip
99 Starboard outrigger fairing
100 Wheel, retracted position
101 Starboard aileron
102 Fuel jettison pipe
103 Starboard plain flap
104 Trailing-edge root fairing
105 Water-methanol filler cap
106 Anti-collision light
107 Water-methanol injection system tank
108 Fire extinguisher bottle
109 Flap hydraulic jack
110 Fuel contents transmitters
111 Rear fuselage integral fuel tank
112 Ram air turbine housing
113 Turbine doors
114 Ram air turbine (extended position)
115 Rear fuselage frames
116 Ram air turbine jack
117 Cooling air ram air intake
118 HF tuner
119 HF notch aerial
120 Rudder control rod linkages
121 Starboard all-moving tailplane
122 Temperature sensor
123 Tailfin construction
124 Forward radar warning receiver
125 VHF aerial
126 Fintip aerial fairing
127 Rudder upper hinge
128 Honeycomb rudder construction
129 Rudder trim jack
130 Rudder tab
131 Tail reaction control air ducting
132 Yaw control port
133 Aft radar warning receiver
134 Rear position light
135 Pitch reaction control valve
136 Tailplane honeycomb trailing edge
137 Extended tailplane tip
138 Tailplane construction
139 Tail bumper
140 IFF notch aerial
141 Tailplane sealing plate
142 Fin spar attachment
143 Tailplane centre section/carry-through
144 All-moving tailplane control jack
145 Ram air exhaust duct
146 UHF standby aerial
147 Equipment air-conditioning plant
148 Ground power supply socket
149 Twin batteries
150 Ventral equipment bay access door
151 Radio and electronics equipment racks
152 Electronics bay access door
153 Ventral airbrake
154 Airbrake hydraulic jack
155 Nitrogen pressurising bottles for hydraulic system
156 Flap drive torque shaft
157 Rear spar/fuselage attachment joint
158 Nozzle blast shield
159 Rear (hot stream) swivelling exhaust nozzle
160 Wing rear spar
161 Port flap honeycomb construction
162 Fuel jettison valve
163 Fuel jettison pipe
164 Aileron honeycomb construction
165 Outrigger wheel fairing
166 Wingtip fairing
167 Profile of extended ferry tip
168 Hydraulic retraction jack
169 Shock absorber leg strut
170 Port outrigger wheel
171 Torque scissor links
172 Outrigger wheel leg fairings
173 Port navigation light
174 Roll control reaction valve
175 Wing rib construction
176 Outer pylon hardpoint
177 Machined wing skin/stringer panel
178 Aileron power control unit
179 Front spar
180 Leading-edge nose ribs
181 Reaction control air ducting
182 Port outer stores pylon
183 Leading-edge fences
184 Twin mainwheels
185 Port inner stores pylon
186 Fuel and air connections to pylon
187 Inboard pylon hardpoint
188 Port wing fuel tank end rib
189 Pressure refuelling connection
190 Wing bottom skin panel/fuselage attachment joint
191 No. 1 hydraulic system reservoir (No. 12 to starboard)
192 Centre fuselage integral fuel tank, port and starboard
193 Nozzle fairing construction
194 Leading-edge dogtooth
195 Cushion augmentation strake (fitted in place of gun pod)
196 Centreline stores pylon
197 Reconnaissance pod
198 Forward F.135 camera
199 Port F.95 Mk 7 oblique cameras
200 Starboard F.95 Mk 7 oblique cameras
201 Signal data converter (SDC) unit
202 Cannon pod
203 Frangible nose cap
204 Cannon barrel
205 Blast suppression ports
206 ADEN 30-mm revolver-type cannon
207 Ammunition feed chute
208 Link ejector chute
209 Ammunition box, 130 rounds
210 ML twin stores carrier
211 Matra 155 rocket launchers, 18 x 2.68-in (68-mm) rockets
212 MATRA 116M rocket launcher, 19 x 2.68-in (68-mm) rockets
213 LEPUS flare
214 Twin light stores carrier
215 28-lb (13-kg) practice bomb

IAI Kfir

Although its has now passed from regular front-line Israeli service, the Kfir remains a potent warplane with three operators, including Ecuador. The type proved supremely capable in combat over the Lebanon, scoring air-to-air kills as well as flying its primary attack mission.

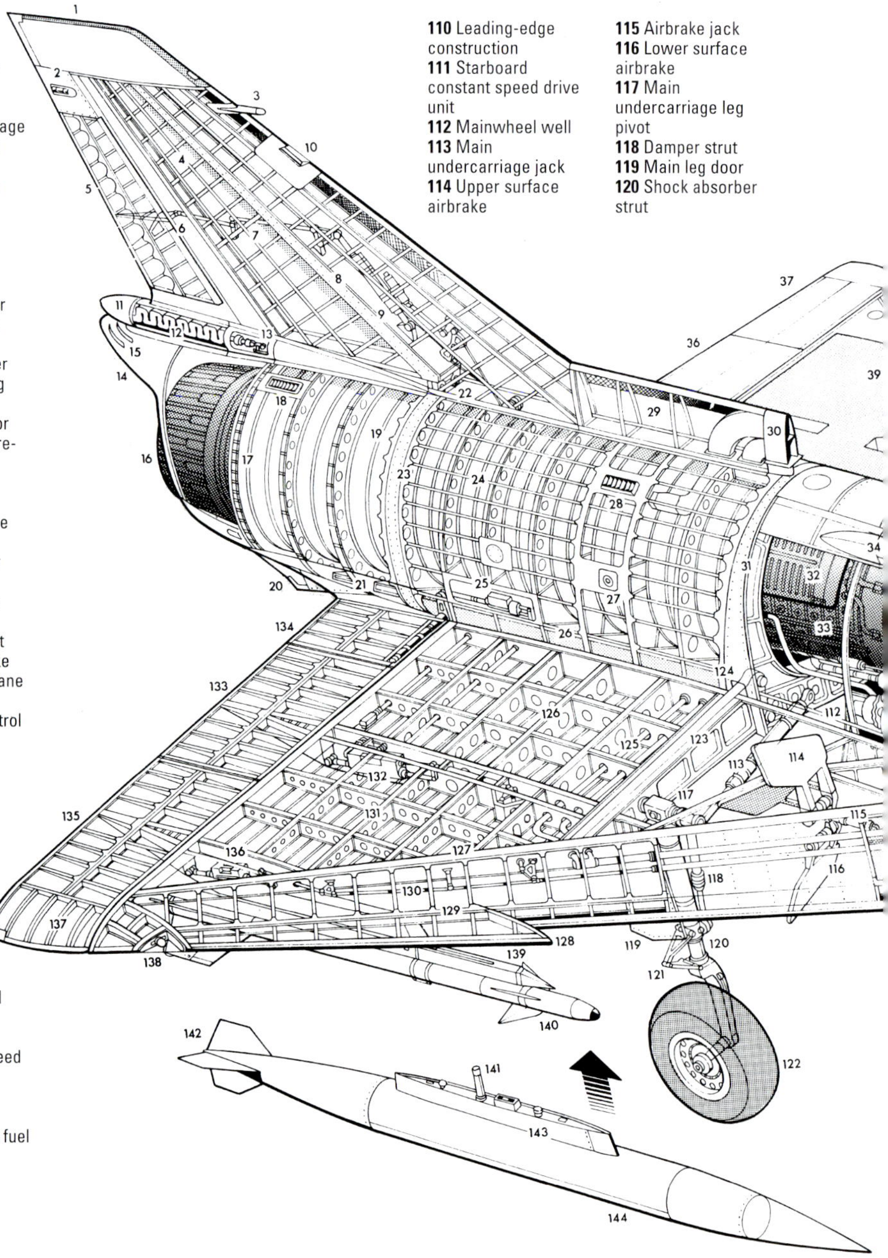

Kfir-C2

Cutaway key

1 Fintip UHF antenna
2 Rear navigation light
3 ECM antenna
4 Fin construction
5 Rudder construction
6 Rudder bellcrank
7 Rudder control rods
8 Fin spar
9 Rudderjack
10 Anti-collision beacon
11 Brake parachute fairing
12 Parachute
13 Release mechanism
14 Tail cone fairing
15 Airflow guide vanes
16 Variable exhaust nozzle
17 Tailcone attachment frame
18 Cooling air outlet
19 Jetpipe inner ducting
20 Tail bumper
21 Tail avionics boxes
22 Fin attachment
23 Fin attachment frame
24 Rear fuselage construction
25 Compensator jack
26 Belly fuel tank
27 Engine mounting attachment
28 Cooling air outlet
29 Finroot intake fairing
30 Cooling air intakes
31 Main fuselage frame
32 Oil tank
33 General Electric J79-GE-17 engine
34 Cooling air ducts
35 Engine front mounting cover
36 Port inboard elevon
37 Aileron
38 Port navigation light
39 Wing main fuel tank
40 Missile launch rail
41 Shafrir 2 air-to-air missile
42 Leading-edge fuel tank
43 Fuel supply piping
44 Fuselage fuel tanks
45 Port constant-speed drive unit
46 Engine starter
47 Port constant-speed drive unit
48 Intake ducting
49 Fuselage frame construction
50 Pressure sensor
51 Inverted flight accumulator
52 Dorsal fairing
53 Oxygen bottles
54 Forward fuselage fuel tank
55 Fuel filler
56 Canard foreplane construction
57 Canopy hinge attachment
58 Canopy external release handle
59 Ejection seat mounting
60 Avionics units
61 Martin Baker MJ6 ejection seat
62 Jettisonable canopy cover
63 Ejection seat firing handles
64 Pilot's control console
65 Instrument panel
66 Reflector sight
67 Windscreen
68 Instrument pitot
69 Nose construction
70 Radar ranging unit
71 Radome
72 Pitot boom
73 Nose strake
74 Yaw sensing vein
75 Autopilot controller
76 Radio and electronics equipment
77 Inertial platform
78 Static inverter
79 UHF aerial
80 Rudder pedal
81 Radar console
82 Control column
83 Ejection seat adjusting handle
84 Control rod linkage
85 Nosewheel leg doors
86 Nosewheel leg
87 Landing lights
88 Nosewheel suspension
89 Steerable nosewheel
90 Shimmy damper
91 Nosewheel leg pivot mounting
92 Locking cylinder
93 Air-conditioning plant
94 Nosewheel door
95 Air intake centre-body half-cone
96 Starboard air intake
97 Intake half-cone operating jack
98 Boundary layer duct
99 Cannon muzzle blast shield
100 Air intake duct
101 Auxiliary intake
102 Canard foreplane root fairing
103 Electrical control unit
104 Electrical servicing panel
105 Cannon barrel
106 DEFA 30-mm cannon
107 Ammunition feed chute
108 Front spar attachment
109 Leading-edge fuel tank
110 Leading-edge construction
111 Starboard constant speed drive unit
112 Mainwheel well
113 Main undercarriage jack
114 Upper surface airbrake
115 Airbrake jack
116 Lower surface airbrake
117 Main undercarriage leg pivot
118 Damper strut
119 Main leg door
120 Shock absorber strut

This Kfir-C7 has a typical air-to-air loadout consisting of four AIM-9 AAMs and a centreline fuel tank. The missiles, in combination with the DEFA cannon, give the Kfir a formidable close-in capability, although during the type's heyday in IDF/AF service the air-to-air role was mostly taken by the F-4 Phantom II and latterly by the even more capable F-15 and F-16. Also evident in this view is the truncated fairing beneath the jet pipe. At its rear, this houses an aft-facing strike camera.

SPECIFICATION

Kfir-C7

Dimensions

Length: 51 ft 4¼ in (15.65 m)
Wingspan: 26 ft 11⅓ in (8.22 m)
Wing aspect ratio: 1.94
Wing area: 374.60 sq ft (34.80 m²)
Canard foreplane span: 12 ft 3 in (3.73 m)
Canard foreplane area: 17.87 sq ft (1.66 m²)
Height: 14 ft 11¼ in (4.55 m)
Wheel track: 10 ft 6 in (3.20 m)
Wheel base: 15 ft 11⅓ in (4.87 m)

Powerplant

One IAI Bedek Division-built General Electric J79-J1E turbojet rated at 11,890 lb st (52.89 kN) dry and 18,750 lb st (83.41 kN) with afterburning

Weights

Empty: about 16,060 lb (7285 kg)
Normal take-off: 22,961 lb (10415 kg)
Maximum take-off: 36,376 lb (16500 kg)

Fuel and load

Internal fuel: 5,670 lb (2572 kg)
External fuel: up to 8,216 lb (3727 kg) in three 449-, 343-, 218-, 159- or 132-US gal (1700-, 1300-, 825-, 600- or 500-litre) drop tanks
Maximum ordnance: 13,415 lb (6085 kg)

Performance

Maximum level speed 'clean' at sea level: 750 kt (863 mph; 1389 km/h)
Maximum level speed 'clean' at 36,000 ft (10975 m): more than 1,317 kt (1,516 mph; 2440 km/h)
Maximum rate of climb at sea level: 45,930 ft (14000 m) per minute
Climb to 50,000 ft (15240 m) with full internal fuel and two Shafrir AAMs: 5 minutes 10 seconds
Zoom climb ceiling: 75,000 ft (22860 m)
Service ceiling: more than 50,000 ft (15240 m)
Stabilised supersonic ceiling: 58,000 ft (17680 m)
Take-off run at maximum take-off weight: 4,757 ft (1450 m)
Landing distance from 50 ft (15 m) at 25,500 lb (11566 kg): 5,102 ft (1555 m)
Landing run at 25,500 lb (11566 kg): 4,200 ft (1280 m)
Ferry range: 1,744 nm (2,000 miles; 3232 km) with one 343-US gal (1300-litre) and two 449-US gal (1700-litre) drop tanks
Combat radius on a hi-hi-hi interception mission with two Shafrir AAMs, one 218-US gal (825-litre) and two 343-US gal (1300-litre) drop tanks: 419 nm (482 miles; 776 km)
Combat radius on a 1-hour CAP with two Shafrir AAMs, one 449-US gal (1700-litre) and two 343-US gal (1300-litre) drop tanks: 476 nm (548 miles; 882 km)
Combat radius on a hi-lo-hi attack mission with two 800-lb (363-kg) and two 400-lb (181-kg) bombs, two Shafrir AAMs, and one 343-US gal (1300-litre) and two 449-US gal (1700-litre) drop tanks: 640 nm (737 miles; 1186 km)

g limits

+7.5

Armament

Two internal 30-mm IAI-built DEFA 552 cannon with 140 rounds per gun, plus a range of stores including M117, M118, Mk 82, Mk 83, Mk 84 and Israeli-designed/derived bombs; CBU-52/58 and Israeli-built TAL-1/2 cluster bombs, LAU-3A, LAU-10A and LAU-32A rocket pods, Python 3 and Shafrir 2 AAMs, and the Elta EL/L-8202 ECM pod

121 Undercarriage scissors link
122 Mainwheel
123 Main spar
124 Main spar attachment
125 Fuel system piping
126 Main wing fuel tank
127 Leading-edge spar
128 Leading-edge dogtooth
129 Leading-edge construction
134 Elevon compensator
135 Outboard elevon
136 Outboard elevon jack
137 Wingtip profile
138 Navigation light
139 Missile launcher
140 Shafrir 2 air to-air missile
141 Fuel tank pylon attachment
142 Fuel tank fins
143 Tank pylon
144 Fuel tank (110-Imp gal/500-litre capacity)

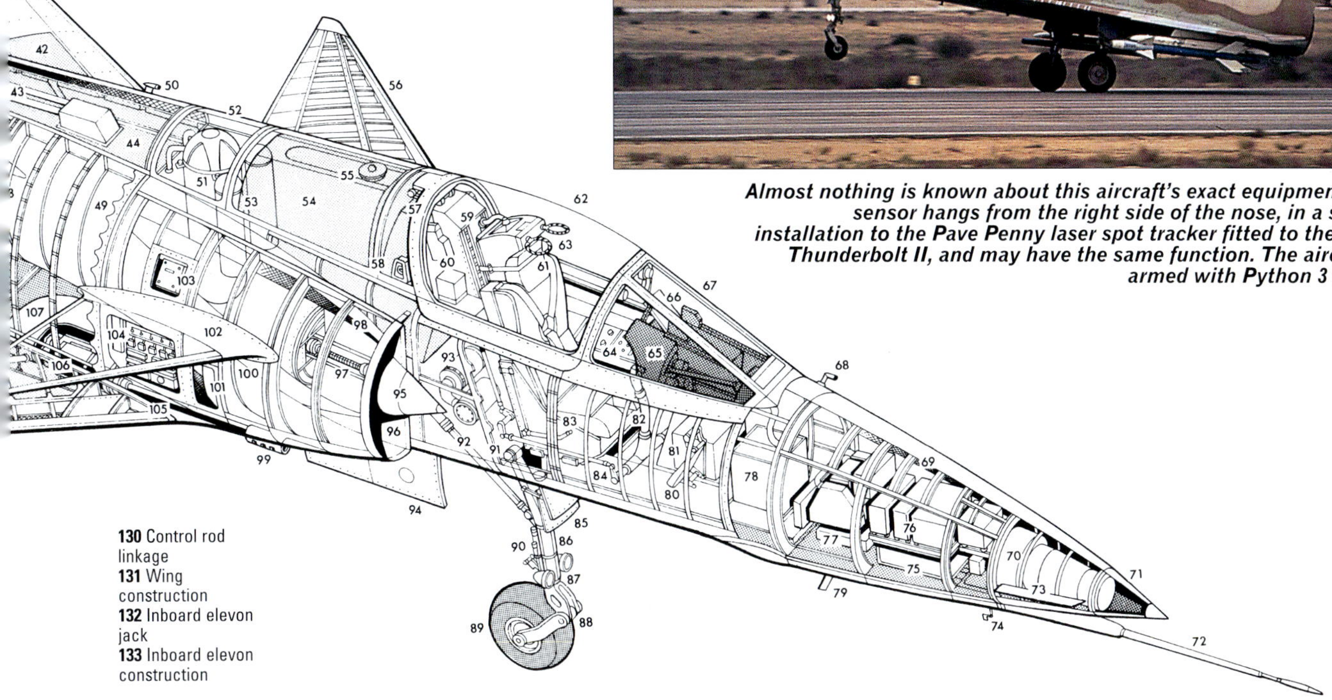

130 Control rod linkage
131 Wing construction
132 Inboard elevon jack
133 Inboard elevon construction

Almost nothing is known about this aircraft's exact equipment fit. A sensor hangs from the right side of the nose, in a similar installation to the Pave Penny laser spot tracker fitted to the A-10A Thunderbolt II, and may have the same function. The aircraft is armed with Python 3 AAMs.

Lockheed P-80 Shooting Star

Conceived in the shortest possible time, the Lockheed P-80 Shooting Star first flew in January 1944 and, in an enlarged and more powerful form, entered service in 1946. This weatherbeaten P-80A (later F-80) operated with the 56th Fighter Group at Selfridge AFB, Michigan. Of particular note is the fact that this aircraft carries Pearl Gray tip tanks and flaps. As this finish was applied only to very early aircraft, it seems likely that this aircraft's tip tanks and flaps were borrowed from another machine.

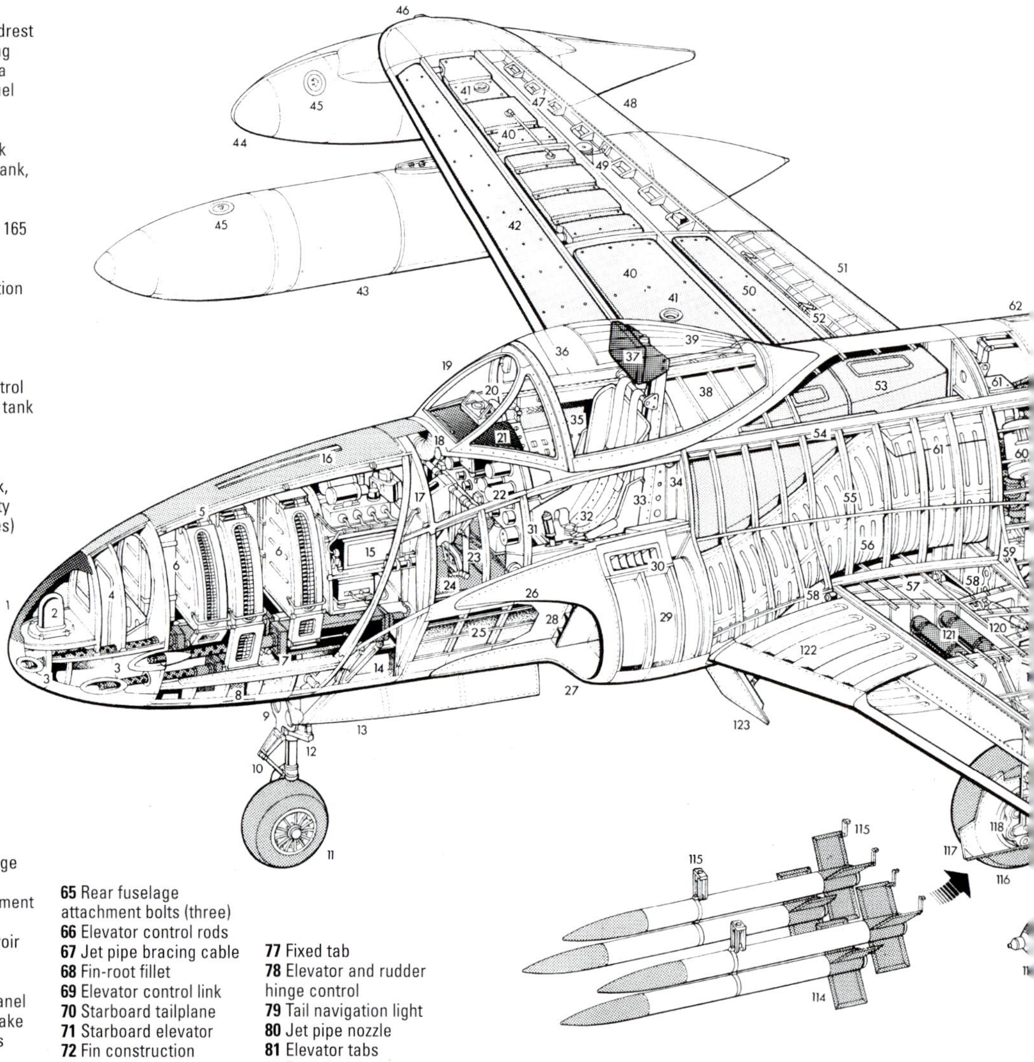

F-80C Shooting Star

Cutaway key
1 Nose antenna fairing
2 D/F loop aerial
3 Machine-gun muzzles
4 Oxygen tank
5 Nose compartment access panel
6 Port and starboard ammunition boxes, 300 rounds per gun
7 0.5-in (12.7-mm) machine-guns
8 Spent cartridge case and link ejector chutes
9 Landing and taxiing lamps
10 Nosewheel leg torque scissors
11 Nosewheel
12 Steering linkage
13 Nosewheel doors
14 Retraction strut
15 Radio and electrical equipment bay
16 External canopy-release handle
17 Cockpit front bulkhead
18 Windscreen heater duct
19 Bulletproof windscreen panel
20 Reflector gunsight
21 Instrument panel shroud
22 Instrument panel
23 Rudder pedals
24 Cockpit floor level
25 Nosewheel bay
26 Intake lip fairing
27 Port air intake
28 Boundary layer bleed air duct
29 Intake ducting
30 Boundary layer air exit louvres
31 Engine throttle control
32 Safety harness
33 Pilot's ejection seat
34 Cockpit rear bulkhead
35 Starboard side console panel
36 Sliding cockpit canopy cover
37 Ejection seat headrest
38 Canopy aft decking
39 D/F sense antenna
40 Starboard wing fuel tanks
41 Fuel filler caps
42 Leading-edge tank
43 Fletcher-type tip tank, capacity 265 US gal (1003 litres)
44 Tip tank, capacity 165 US gal (625 litres)
45 Tip tank filler cap
46 Starboard navigation light
47 Aileron balance weights
48 Starboard aileron
49 Aileron hinge control
50 Trailing-edge fuel tank
51 Starboard split trailing-edge flap
52 Flap control links
53 Fuselage fuel tank, total internal capacity 657 US gal (2487 litres)
54 Fuselage main longeron
55 Centre fuselage frames
56 Intake trunking
57 Main undercarriage wheel well
58 Wing spar attachment joints
59 Pneumatic reservoir
60 Hydraulic accumulator
61 Control access panel
62 Spring-loaded intake pressure relief doors
63 Allison J33-A-23 centrifugal-flow turbojet engine
64 Rear fuselage break point
65 Rear fuselage attachment bolts (three)
66 Elevator control rods
67 Jet pipe bracing cable
68 Fin-root fillet
69 Elevator control link
70 Starboard tailplane
71 Starboard elevator
72 Fin construction
73 Pitot tube
74 Fintip communications antenna fairing
75 Rudder mass balance
76 Rudder construction
77 Fixed tab
78 Elevator and rudder hinge control
79 Tail navigation light
80 Jet pipe nozzle
81 Elevator tabs
82 Port elevator construction
83 Elevator mass balance
84 Tailplane construction

SPECIFICATION

F-80C Shooting Star (early)

Dimensions

Length: 34 ft 5 in (10.49 m)
Height: 11 ft 3 in (3.42 m)
Wingspan: 38 ft 9 in (11.81 m)
Wing area: 237.5 sq ft (22.07 m^2)

Powerplant

One Allison J33-A-23/35 turbojet rated at 4,600 lb st (20.7 kN) dry, and 5,200 lb st (23.4 kN) with water injection

Weights

Empty: 8,420 lb (3819 kg)
Gross: 12,200 lb (5534 kg)
Maximum take-off: 16,856 lb (7646 kg)

Fuel

Fuel (normal): 425 US gal (1609 litres)
Fuel (maximum): 755 US gal (2858 litres) including drop tanks

Performance

Maximum level speed at sea level: 594 mph (956 km/h)
Maximum level speed at 25,000 ft (7620 m): 543 mph (874 km/h)
Cruising speed: 439 mph (707 km/h)
Landing speed: 122 mph (196 km/h)
Climb to altitude: climb to 25,000 ft (7620 m) in 7 minutes
Rate of climb: 6,870 ft (2094 m) per minute
Service ceiling: 46,800 ft (14265 m)

Range

Range: 825 miles (1328 km)
Maximum range: 1,380 miles (2221 km)

Armament

Four 0.50-in (12.7-mm) Colt-Browning M3 machine-guns each with 300 rounds, plus ten 5-in (127-mm) HVARs or two 1,000-lb (454-kg) bombs

This flight of P-80Cs from the 94th 'Hat-in-the-Ring' squadron was based at Ladd Field, Fairbanks, Alaska, for six months of training. During this time the aircraft had their wingtips and tails painted red so that they could be easily spotted from the air in the event of an emergency landing in the snow. Lockheed 'winterised' the aircraft by incorporating new greases and hydraulic units to allow the aircraft to operate below -65°C (-85°F).

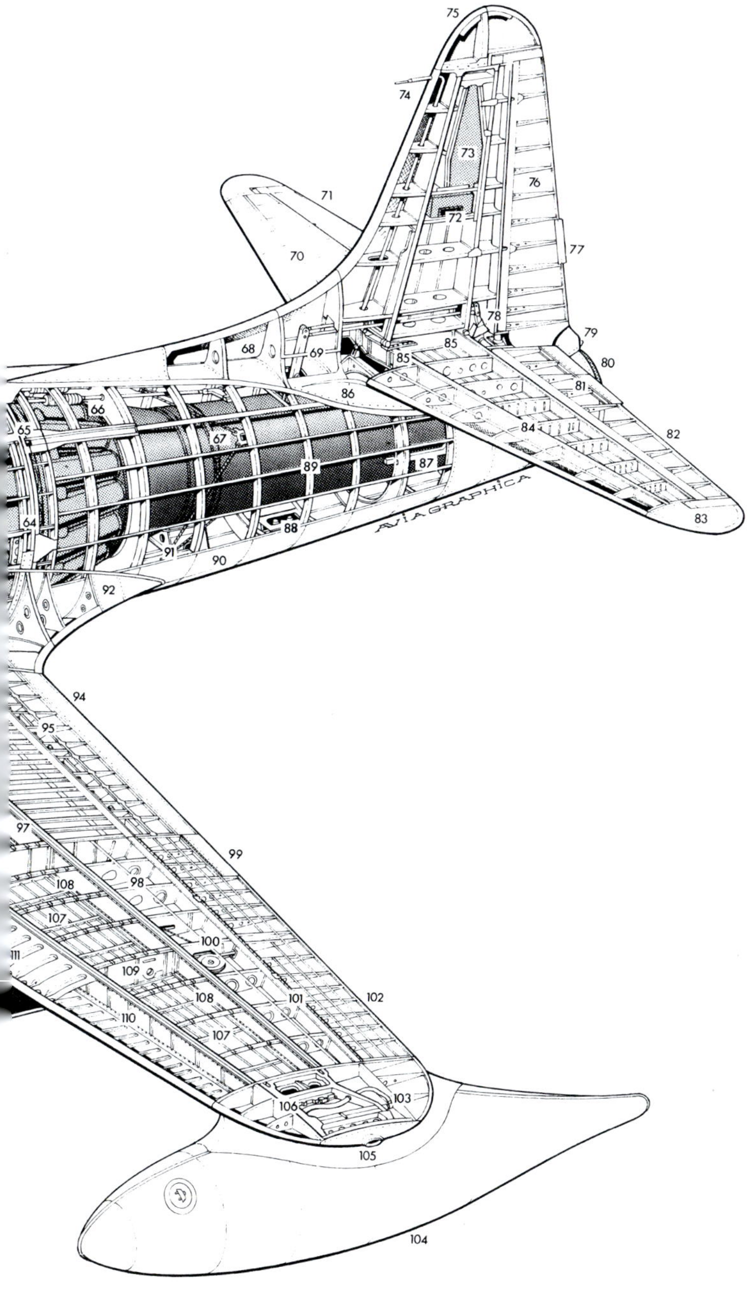

85 Fin/tailplane attachment joints
86 Tailplane fillet fairing
87 Jet pipe mounting rail
88 Rear fuselage frame and stringer construction
89 Gyrosyn radio compass flux valve
90 Fuselage skin plating
91 Jet pipe support frame
92 Trailing-edge wingroot fillet
93 Flap drive motor
94 Port split trailing-edge flap
95 Flap shroud ribs
96 Trailing-edge fuel tank bay
97 Rear spar
98 Trailing-edge ribs
99 Port aileron tab
100 Aileron hinge control
101 Upper-skin panel aileron hinge line
102 Aileron construction
103 Wingtip fairing construction
104 Tip tank
105 Port navigation light
106 Tip tank mounting and jettison control
107 Detachable lower wing skin panels
108 Port wing fuel tank bays
109 Inter tank bay ribs
110 Front spar
111 Corrugated leading-edge inner skin
112 Port stores pylon
113 1,000-lb (454-kg) HE bomb
114 5-in (127-mm) HVAR ground attack rockets (10 rockets maximum load)
115 HVAR mountings
116 Port mainwheel
117 Mainwheel doors
118 Wheel brake pad
119 Main undercarriage leg strut
120 Retraction jack
121 Upper skin panel wing stringers
122 Wingroot leading-edge extension
123 Port ventral airbrake

Having force-landed on a frozen lake 75 miles (120 km) from its base, this 65th FIS F-80C was resurrected over a three-week period, in which USAF ground crew fitted it with specially made aluminium skis. The take-off, the first successful jet-ski JATO take-off, was accomplished by Lt Col William F. Benedict. This method was used on a number of occasions to salvage stricken fighters throughout the area.

Lockheed AC-130 Hercules

An AC-130A from the 711th SOS lets fly with a 40-mm round during live-fire exercises. The gunship concept calls for the aircraft to fly in a smooth left-hand orbit with the guns trained on a central point at all times. Fine-tuning the aim is accomplished using roll and yaw inputs.

AC-130U

Cutaway key
1 Air data boom
2 Radome
3 AN/APG-80 radar scanner
4 Radar equipment racks
5 Front pressure bulkhead
6 Downward vision windows
7 Pilot's head-up-display
8 Cockpit radar repeater
9 Windscreen panels and wipers
10 Radar warning receivers
11 Overhead systems switch panels
12 Inflight-refuelling receptacle
13 Dual flight engineer's stations
14 Two-pilot cockpit with observer's seat
15 Control column
16 Rudder pedals
17 Side console panel
18 Pitot heads
19 Battery compartment
20 AN/AAQ-117 FLIR turret
21 Twin nosewheel undercarriage, forward-retracting
22 Nosewheel door
23 Avionics equipment racks
24 Crew entry door and airstairs
25 Access ladder to flight deck
26 Galley unit
27 Crew closet
28 Escape hatch access ladder
29 Cockpit roof escape hatch
30 Aerial cable lead-in
31 Crew rest compartment
32 Radar warning antennas and ECM transmitter fairing
33 Flexible gun seal
34 GAU-12U 25-mm five-barrelled rotary cannon
35 Ball Aerospace ALLTV turret
36 Cannon ammunition magazine, 3,000 rounds
37 Port-side equipment racks
38 Spare crew seats
39 Starboard side observer's seat
40 Observation hatch
41 UHF aerial
42 ADF loop antennas
43 Battle management centre
44 IR, TV, fire control, radar/nav and EW operators' seats and consoles
45 Hydraulic system equipment
46 Conditioned air delivery ducting
47 Wing spar attachment fuselage main frame
48 Engine fire extinguisher bottles
49 Wing centre-section integral fuel tank
50 Sigint antenna
51 Centre wing panel rib construction
52 Wing stringers
53 Starboard engine nacelles
54 External fuel tank
55 Hamilton Standard four-bladed constant-speed propellers
56 Starboard outer wing panel
57 Leading-edge flush aerial panel
58 Starboard navigation light
59 Starboard aileron
60 Single-slotted flaps
61 Flap shroud ribs
62 Life raft stowage
63 Satcom antenna
64 Aileron hydraulic booster
65 Mid-cabin escape hatch
66 Emergency equipment stowage
67 Rear fuselage air ducting
68 VHF/UHF aerial
69 Fin-root fillet construction
70 ELT antenna
71 AN/ALQ-172 EW equipment packs
72 Starboard tailplane
73 Fin spar box construction
74 VOR aerial
75 Anti-collision light
76 Rudder
77 Rudder tab
78 Tail radar warning and ECM transmitter fairing
79 Rudder and elevator hinge controls
80 Elevator tab
81 Port elevator
82 Tailplane rib construction
83 Elevator hydraulic booster unit
84 Rear ramp door hydraulic jack
85 Rear escape hatch
86 Ventral observation hatch
87 Observer's prone position
88 Ramp hydraulic jack
89 Rear loading ramp
90 Central flap drive hydraulic motor
91 105-mm Howitzer
92 Port single-slotted flaps
93 Flap rib construction
94 Aileron mass balance weights
95 Aileron tab
96 Port aileron
97 Static dischargers
98 Fuel jettison
99 Port navigation light
100 Leading-edge flush aerial panel
101 Outer wing panel integral fuel tank
102 Leading-edge ribs
103 Exhaust infra-red suppression mixing air duct
104 Allison T56-A-15 turboprop engine

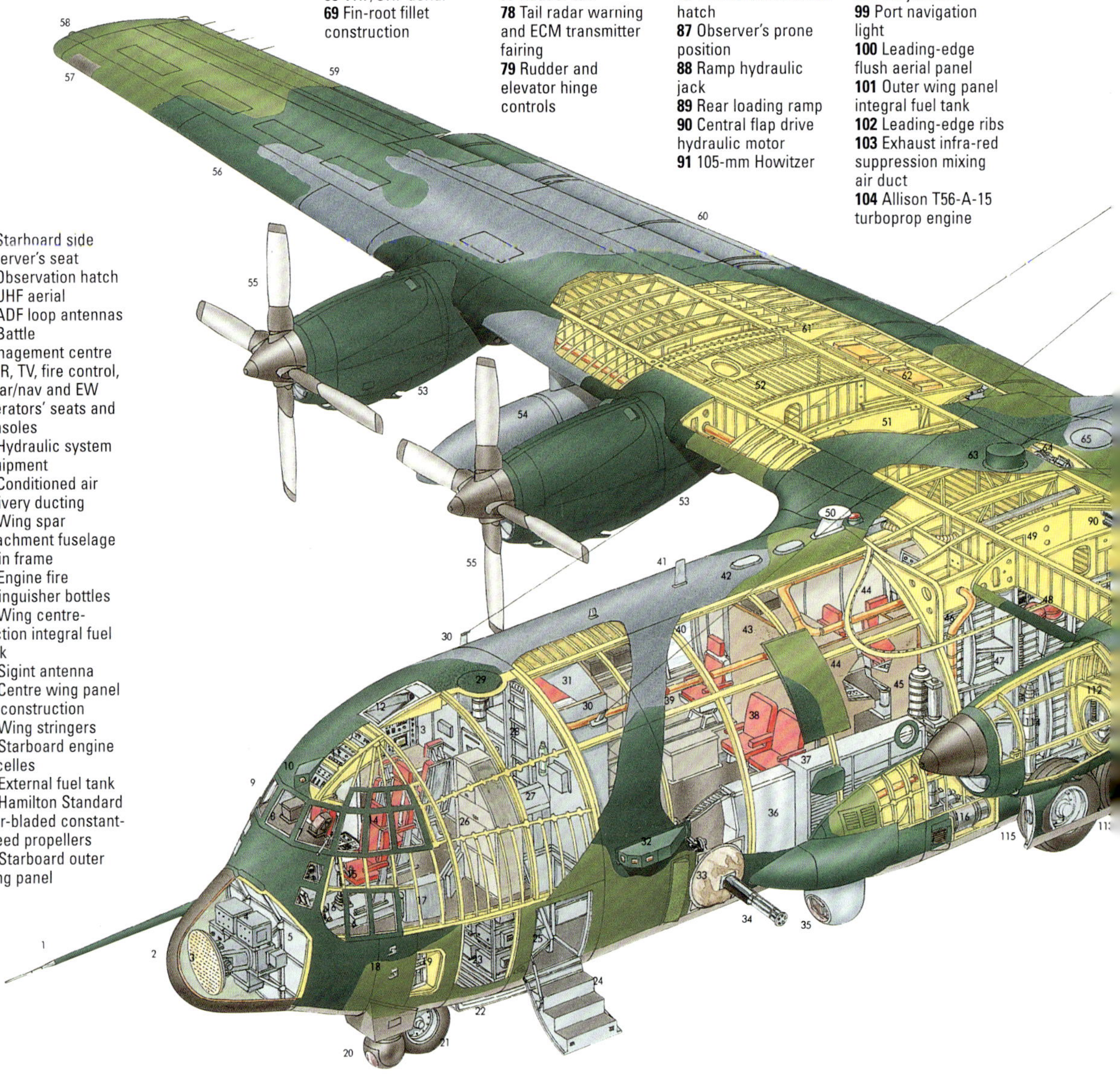

The gunship concept was developed for the war in Southeast Asia. AC-130s were the most powerful and best-equipped of the gunship types, and were used principally on night interdiction sorties over the Ho Chi Minh Trail.

SPECIFICATION

C-130H (AC-130H/U similar)

Dimensions

Wingspan: 132 ft 7 in (40.41 m)
Wing aspect ratio: 10.07
Wing area: 1,745 sq ft (162.12 m^2)
Tailplane span: 52 ft 8 in (16.05 m)
Tailplane area: 381 sq ft (35.40 m^2)
Length overall: 97 ft 9 in (29.79 m)
Height overall: 38 ft 3 in (11.66 m)
Wheelbase: 32 ft 1 in (9.77 m)
Wheel track: 14 ft 3 in (4.35 m)

Powerplant

Four Allison T56-A-15 turboprops, each rated at 4,508 shp (3362 kW), driving a Hamilton Standard 54H60 four-bladed constant-speed, reversible-pitch, fully-feathering propeller

Weights

Operating empty (C-130H): 76,469 lb (34686 kg)
Maximum take-off: 155,000 lb (70305 kg)
Maximum overload: 175,000 lb (79380 kg)

Fuel and load

Maximum internal fuel capacity: 6,820 US gal (5,679 Imp gal; 25816 litres)
Underwing tank capacity (combined): 2,800 US gal (2,332 Imp gal; 10600 litres)
Maximum payload (C-130H): 49,818 lb (22597 kg)

Performance

Maximum cruising speed: 315 kt (362 mph; 583 km/h)
Economical cruising speed: 300 kt (345 mph; 556 km/h)
Maximum rate of climb: 1,800 ft (548 m) per minute
Time to 20,000 ft (6100 m): 22 minutes
Service ceiling: 18,000 ft (5485 m)
Take-off run: 4,000 ft (1220 m)
Landing run: 1,500 ft (457 m)

Range

2,238 miles (3600 km) with 40,000-lb (18144-kg) payload

Armament

One single-round loaded M102 105-mm Howitzer and one clip-fed Bofors L-60 40-mm cannon (AC-130H/U); two belt-fed 20-mm M61 Vulcan cannon (AC-130H) or one belt-fed 25-mm GAU-12 rotary cannon (AC-130U)

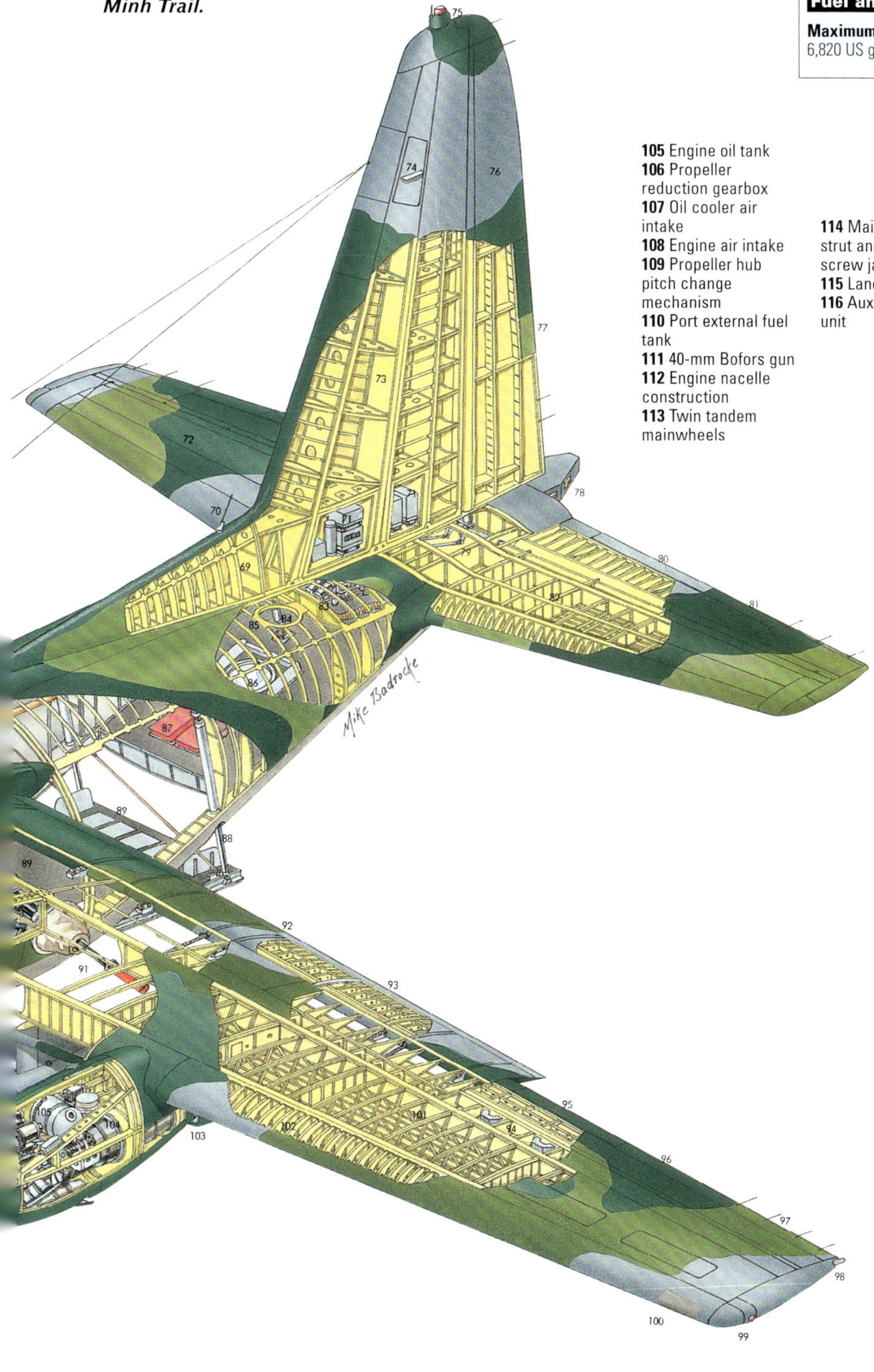

105 Engine oil tank
106 Propeller reduction gearbox
107 Oil cooler air intake
108 Engine air intake
109 Propeller hub pitch change mechanism
110 Port external fuel tank
111 40-mm Bofors gun
112 Engine nacelle construction
113 Twin tandem mainwheels
114 Mainwheel leg strut and retraction screw jack
115 Landing lamp
116 Auxiliary power unit

Lockheed F-117 Nighthawk

An extraordinary shape, revolutionary radar-defeating features and a top secret, yet high glamorous development have combined wit a star combat appearance in Desert Storm t make the Lockheed F-117 Nighthawk the best-known warplane in the world. Able to penetrate hostile airspace without being see by radars or infra-red sensors, the F-117 can use its sophisticated target acquisition and designation system to score strikes against vital targets with pinpoint accuracy.

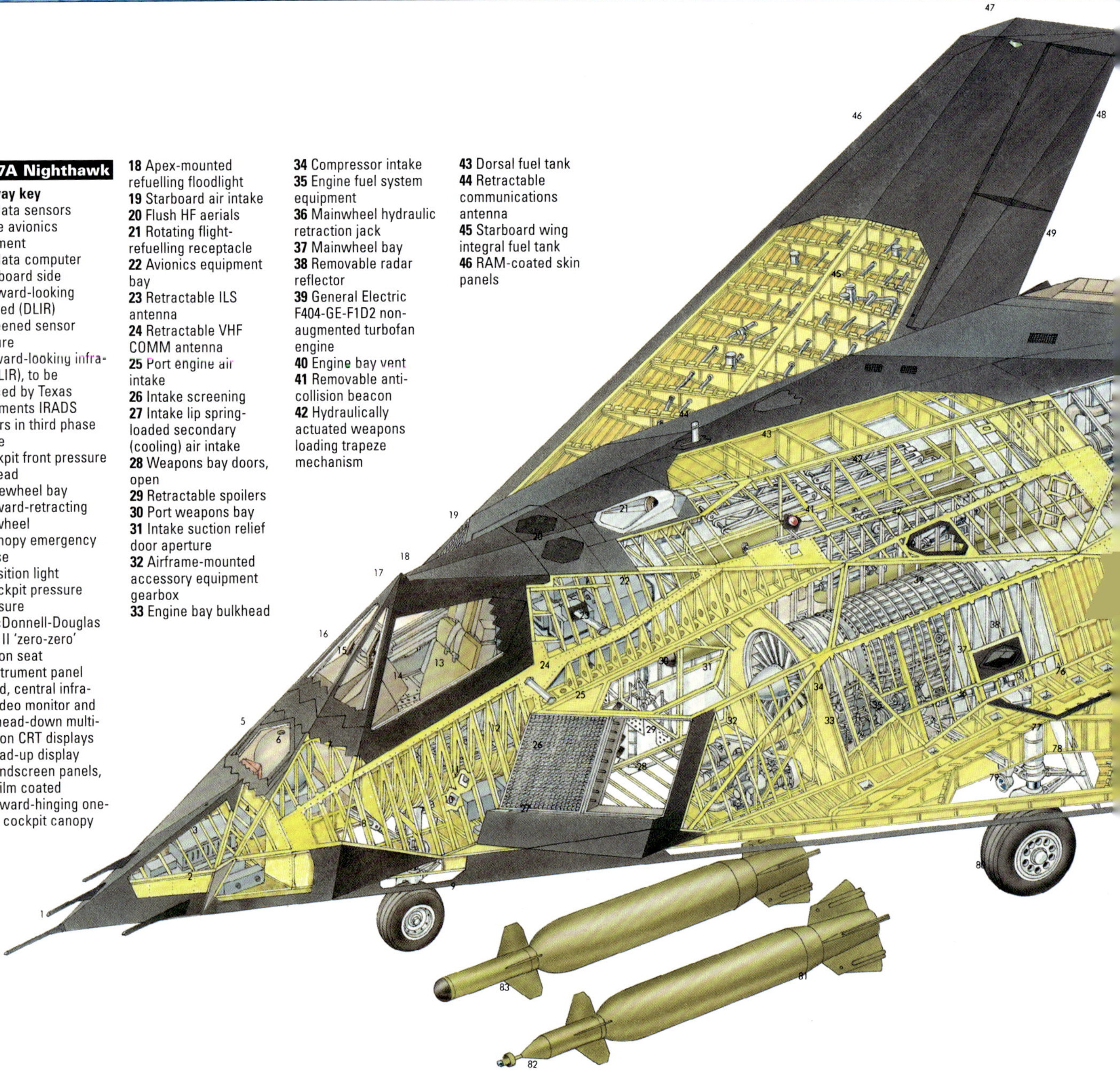

F-117A Nighthawk

Cutaway key

1 Air data sensors
2 Nose avionics equipment
3 Air data computer
4 Starboard side downward-looking infra-red (DLIR)
5 Screened sensor aperture
6 Forward-looking infra-red (FLIR), to be replaced by Texas Instruments IRADS sensors in third phase update
7 Cockpit front pressure bulkhead
8 Nosewheel bay
9 Forward-retracting nosewheel
10 Canopy emergency release
11 Position light
12 Cockpit pressure enclosure
13 McDonnell-Douglas ACES II 'zero-zero' ejection seat
14 Instrument panel shroud, central infra-red video monitor and dual head-down multi-function CRT displays
15 Head-up display
16 Windscreen panels, gold film coated
17 Upward-hinging one-piece cockpit canopy
18 Apex-mounted refuelling floodlight
19 Starboard air intake
20 Flush HF aerials
21 Rotating flight-refuelling receptacle
22 Avionics equipment bay
23 Retractable ILS antenna
24 Retractable VHF COMM antenna
25 Port engine air intake
26 Intake screening
27 Intake lip spring-loaded secondary (cooling) air intake
28 Weapons bay doors, open
29 Retractable spoilers
30 Port weapons bay
31 Intake suction relief door aperture
32 Airframe-mounted accessory equipment gearbox
33 Engine bay bulkhead
34 Compressor intake
35 Engine fuel system equipment
36 Mainwheel hydraulic retraction jack
37 Mainwheel bay
38 Removable radar reflector
39 General Electric F404-GE-F1D2 non-augmented turbofan engine
40 Engine bay vent
41 Removable anti-collision beacon
42 Hydraulically actuated weapons loading trapeze mechanism
43 Dorsal fuel tank
44 Retractable communications antenna
45 Starboard wing integral fuel tank
46 RAM-coated skin panels

Above: Nighthawk pilots form an elite within an elite. They have come from a variety of backgrounds, mainly from the USAF's fighter and attack fast jet communities, but have also included former SAC bomber pilots. Undergoing a selection procedure which is almost as rigorous as that for SR-71 aircrew, F-117 pilots are chosen as much for their flying skills as for such intangible qualities as stability of temperament.

SPECIFICATION

F-117A Nighthawk

Dimensions

Length overall: 65 ft 11 in (20.08 m)
Wing span: 43 ft 4 in (13.20 m)
Wing aspect ratio: 1.65
Wing area (estimated): 1,140 sq ft (105.9 m²)
Wing sweepback angle: 67° 30'
Ruddervator sweep angle: about 65°
Overall height: 12 ft 5 in (3.78 m)
Wheel track (estimated): 13 ft 5½ in (4.10 m)
Wheel base (estimated): 18 ft 7 in (5.66 m)
Maximum wing loading: 46.0 lb/sq ft (225 kg/m²)
Frontal radar cross section (estimated): 0.1 sq ft (0.009 m²)

Powerplant

Two General Electric F404-GE-F1D2 non-afterburning turbofans each rated at 10,800 lb (48.04 kN) maximum thrust

Weights

Empty operating: 29,000 lb (13154 kg)
Normal maximum take-off: 52,000 lb (23814 kg)
Maximum overload take-off: 54,000 lb (24494 kg)

Fuel and load

Total fuel load: 18,000 lb (8165 kg) of JP-4 fuel
Total fuel capacity: 2,769 US gal, (10483 litres)

Performance

(Applies to ISA conditions unless otherwise stated)
Maximum level speed: 561 kt (646 mph; 1040 km/h)
Normal maximum operating speed: Mach 0.9 at optimum altitude
Normal cruising speed: Mach 0.81, 488 kts (562 mph; 904 km/h) at 30,000 ft (9144 m)
Undercarriage limiting speed: 300 kts (345 mph; 556 km/h)
Service ceiling: 38,600 ft (11765 m)
Rotation speed: 152-195 kts (175-225 mph; 282-361 km/h) at 38,000-54,000 lb (17237-24494 kg) all-up weight (AUW)
Lift-off speed: 173-208 kts (199-240 mph; 321-385 km/h) at 38,000-54,000 lb (17237-24494 kg) AUW
Minimum take-off distance: 2,500-2,800 ft (762-853 m)
Minimum take-off distance at maximum take-off weight: 5,400-6,200 ft (1645-1890 m)
Approach speeds: 143-185 kts (165-213 mph; 265-343 km/h) at 30,000-50,000 lb (13608-22680 kg) landing weight
Approach angle of attack 9.5°
Minimum landing distance (from 50-ft/15-m) with brake parachute: 4,850 ft (1478 m)
Minimum landing roll with brake parachute: 2,790 ft (850 m)

Range

Unrefuelled combat radius with 4,000-lb (1814-kg) weapons load, normal diversion fuel and reserves: 465 nm (535 miles; 862 km) at 30,000 ft (9144 m)

g limit

Maximum: +7.0

Armament

Maximum weapons load: 5,000 lb (2268 kg)
Weapon stations: internal weapons bay with two stations each stressed for a 2,000-lb (907-kg) weapon
Internal weapons bay length: 15 ft 5 in (4.70 m)
Internal weapons bay width: 5 ft 9 in (1.75 m)
Operational weapons (primary): up to two laser-guided bombs (LGBs) comprising 496-lb (225-kg) GBU-12, 1,984-lb (900-kg) (approximate weight) GBU-10 or 2,169-lb (984-kg) GBU-27A/B; latter two LGBs fitted with either Mk 84 Paveway II or BLU-109B (Lockheed designation I-2000) Paveway III penetration warheads; other stores may include AGM-88 HARM anti-radar missiles
Probable nuclear strike weapon: B61 free-fall nuclear bomb weighing 719-765 lb (326-347 kg); F-117A has known nuclear capability and nuclear strike role
Training/utility stores: 500-lb (227-kg) Mk 82 free-fall general-purpose bombs dropped during early development; SUU-20 practice bomb and rocket dispenser with six BDU-33 bomb training shapes (no rockets carried)

47 Starboard flush-mounted navigation light, above and below
48 Outboard elevon
49 Inboard elevon
50 Starboard 'platypus' exhaust nozzle
51 Fixed lower portion of fin
52 Ruddervator torque shaft
53 Starboard ruddervator, thermoplastic graphite composite construction replacing earlier all-metal structure
54 Port ruddervator
55 Port engine 'platypus' exhaust nozzle
56 Exhaust nozzle lip heat shielding tiles
57 Hydraulic ruddervator actuator
58 Brake parachute housing
59 Rear equipment bay
60 Ventral emergency arrester hook
61 Auxiliary power unit (APU)
62 Venting air grille
63 Nickel alloy honeycomb exhaust duct with internal support posts
64 Inboard elevon hydraulic actuator
65 Elevon rib construction
66 Port inboard elevon
67 Composite trailing-edge structure
68 Port outboard elevon
69 Port flush-mounted navigation light, above and below
70 Outboard elevon hydraulic actuator, GEC Astronics quadruplex fly-by-wire flight control system
71 Port wing integral fuel tank
72 Composite leading-edge construction
73 Three-spar wing torsion box structure
74 Fuel system piping and contents capacitors
75 Wingroot rib
76 Multi-bolt wingroot attachment joints
77 Main undercarriage leg strut
78 Torque scissor links
79 Landing lamp
80 Forward-retracting mainwheel
81 GBU-10 2,000-lb (907-kg) laser-guided bomb
82 Paveway II laser seeker head
83 GBU-27A/B 2,000-lb (907-kg) laser-guided bomb with BLU-109 Paveway III penetrator warhead

McDonnell Douglas/BAe Harrier II

The AV-8B Harrier II Plus has provided the US Marine Corps and the Italian and Spanish navies with a formidable, radar-equipped fighter and ground-attack aircraft of unparalleled capability.

AV-8B Harrier II

Cutaway key
1 Glass-fibre radome
2 Planar radar scanner
3 Scanner tracking mechanism
4 Radar mounting bulkhead
5 Forward-Looking Infra-Red (FLIR)
6 APG-65 radar equipment module
7 Forward pitch control reaction air nozzle
8 Pitot head, port and starboard
9 Cockpit front pressure bulkhead
10 Pitch feel unit and trim actuator
11 Yaw vane
12 Single-piece wrap-round windscreen
13 Instrument panel shroud
14 Rudder pedals
15 Underfloor avionics bay, air data computer and inertial navigation equipment
16 Electro-luminescent and covert night vision goggle (NVG) formation lighting strips
17 Control column
18 Engine throttle and nozzle angle control levers
19 Instrument panel with full-colour multi-function CRT displays
20 Pilot's head-up display (HUD)
21 Sliding cockpit canopy with miniature detonating cord (MDC) emergency breaker
22 UPC/Stencel I lightweight ejection seat
23 Cockpit section framing, all-composite forward fuselage structure
24 Sloping seat mounting rear pressure bulkhead
25 Intake boundary layer separator
26 Port air intake
27 Landing/taxiing light
28 Levered suspension nosewheel, shortens on retraction
29 Intake suction relief doors, free-floating
30 Hydraulic nosewheel retraction jack
31 Hydraulic system accumulator
32 Demountable flight-refuelling probe
33 Cockpit air-conditioning pack
34 Intake boundary layer air spill duct
35 Heat exchanger ram air intakes
36 Rolls-Royce F402-RR-408A Pegasus 11-61 turbofan engine
37 Full-authority digital engine control (FADEC) unit
38 Upper formation lighting strips
39 Accessory equipment gearbox
40 Alternator
41 Engine oil tank
42 Forward fuselage fuel tank
43 Hydraulic system ground connectors and engine monitoring and recording equipment
44 Fuselage lift-improvement device (LID), lateral strake
45 Forward zero-scarf (fan air) swivelling exhaust nozzle
46 Centre fuselage fuel tank
47 Nozzle bearing
48 Gas turbine starter/auxiliary power unit
49 Leading-edge root extension (LERX)
50 Engine bay venting air intake
51 Wing centre-section integral fuel tank
52 Starboard wing integral tank
53 Fuel feed and vent piping
54 Starboard weapons pylons
55 RWR antenna
56 Starboard navigation light
57 Roll control reaction air valve, upper and lower surface vents
58 Wingtip formation lights
59 Fuel jettison
60 Starboard aileron
61 Outrigger wheel fairing
62 Starboard outrigger wheel, retracted position
63 Slotted flap
64 Articulated flap vane
65 VHF/UHF antenna
66 Anti-collision beacon
67 De-mineralised water tank
68 Engine fire suppression bottle
69 Water filler
70 Rear fuselage fuel tank
71 Electrical system distribution panels, port and starboard
72 Chaff/flare launchers
73 Heat exchanger ram air intake
74 Rudder hydraulic actuator
75 Starboard all-moving tailplane
76 Formation lighting strip
77 Fin conventional light alloy structure
78 MAD compensator
79 Temperature probe
80 Broad-band communications antenna

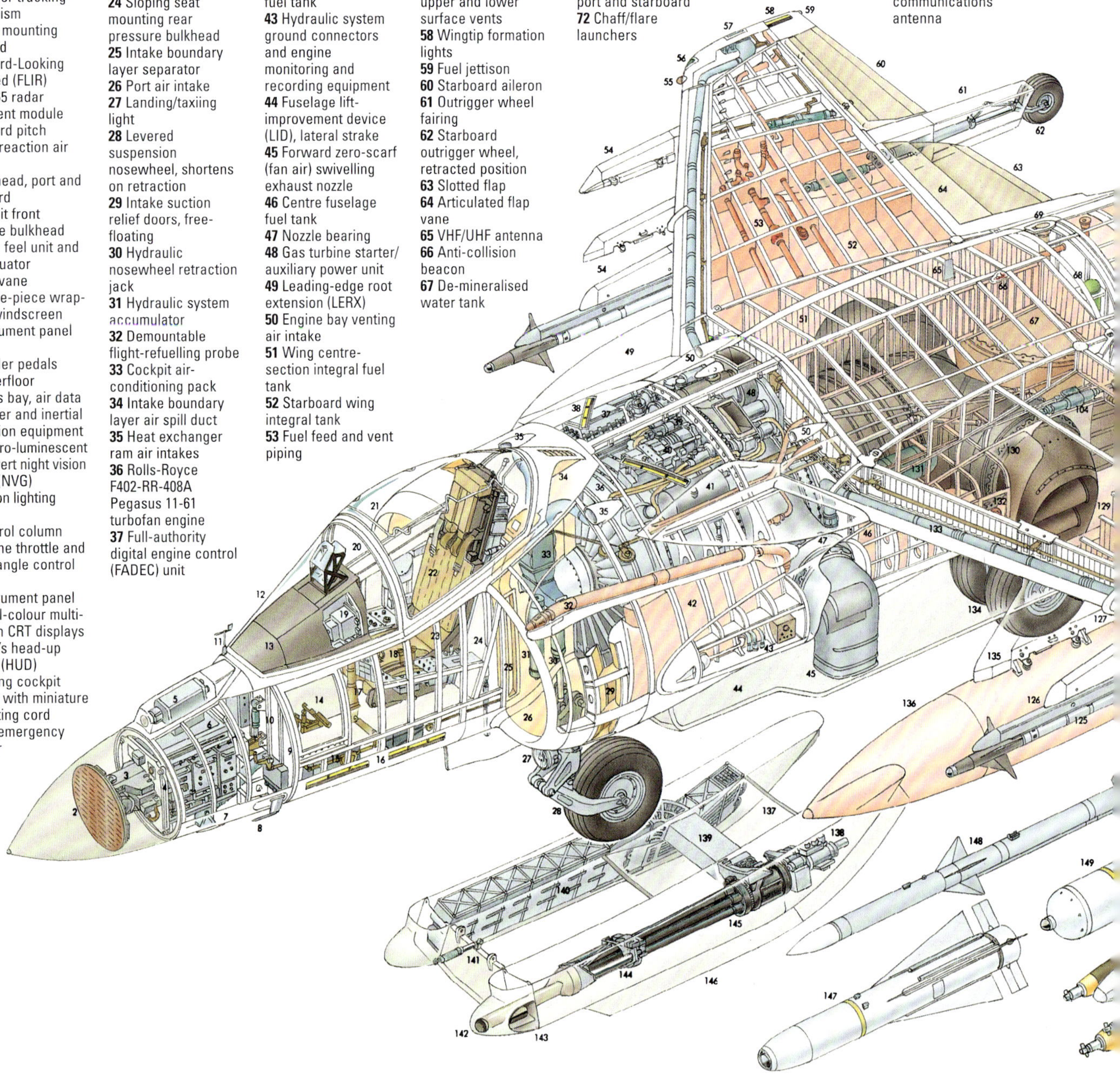

The APG-65 radar added a 1,000-lb (454-kg) weight penalty to the basic AV-8B, but since all Harrier II Pluses were powered by the more powerful F402-RR-408A engine, this was of little consequence. Other changes to the II Plus included the adoption of the RAF-type wing, with a 100 per cent LERX and, most significantly, four pylons per side, including the dedicated AAM pylons on the undercarriage outriggers.

SPECIFICATION

AV-8B Harrier II Plus

Dimensions

Length: 47 ft 9 in (14.55 m)
Wingspan: 30 ft 4 in (9.25 m)
Wing aspect ratio: 4.0
Tailplane span: 13 ft 11 in (4.24 m)
Wing area: 243.40 sq ft (22.61 m^2) including two 6.70-sq ft (0.62-m^2) LERXes
Height: 11 ft 7¾ in (3.55 m)
Outrigger wheel track: 17 ft (5.18 m)

Powerplant

One Rolls-Royce F402-RR-408A (Pegasus 11-61) vectored thrust turbofan engine rated at 23,800 lb st (106 kN)

Weights

Empty operating: 14,860 lb (6740 kg)
Normal take-off: 22,950 lb (10410 kg) for 7-*g* operation
Maximum take-off for 1,427-ft (435-m) STO under standard conditions: 31,000 lb (14061 kg)
Maximum take-off for VTO: 18,950 lb (8596 kg)

Fuel and load

Internal fuel: 7,759 lb (3519 kg)
External fuel: up to 8,070 lb (3661 kg) in four 300-US gal (1136-litre) drop tanks
Maximum ordnance: 13,235 lb (6003 kg)

Performance

Maximum level speed 'clean' at sea level: 575 kt (662 mph; 1065 km/h)
Maximum rate of climb at sea level: 14,715 ft (4485 m) per minute
Service ceiling: more than 50,000 ft (15240 m)
STO run at maximum take-off weight at 90°F (32°C): 1,700 ft (518 m)
Landing run at up to 19,937 lb (9043 kg): 0 ft (0 m)
Anti-ship combat radius with two AGM-84, two AIM-9 and two 300-US gal (1136-litre) drop tanks after a 450-ft (137-m) take-off run with a 6.5° ski-jump: 609 nm (701 miles; 1128 km)
Time on station for a CAP (including 2-minute combat) with four AIM-120 and two 300-US gal (1136-litre) drop tanks after a 450-ft (137-m) take-off run with a 6.5° ski-jump: 2 hours 42 minutes at 100-nm (115-mile, 185-km) radius or 2 hours 6 minutes at 200-nm (230-mile, 370-km) radius
Sea surveillance combat radius (including 50-nm (57-mile; 92-km) dash at sea level) with two AIM-9 and two 300-US gal (1136-litre) drop tanks after a 450-ft (137-m) take-off run with a 6.5° ski-jump: 608 nm (700 miles; 1127 km)

g limits

-3 to +8

Armament

AV-8B Harrier II Plus has four pylons under each wing for the carriage of AIM-9 Sidewinders, AIM-120 AMRAAM, Mk 7 cluster bomb dispensers, Mk 82/83 bombs, LAU-10/68/69 rocket pods, AGM-65 Maverick, AGM-84 Harpoon, CBU-55/72 fuel-air explosive, Mk 77 fire bombs and GBU-12/16 LGBs, the latter requiring designation from another source. The centreline hardpoint is used for an ALQ-167 ECM pod. Two fuselage packs contain a five-barrelled 25-mm GAU-12 cannon (port) and ammunition tank for 300 rounds (starboard)

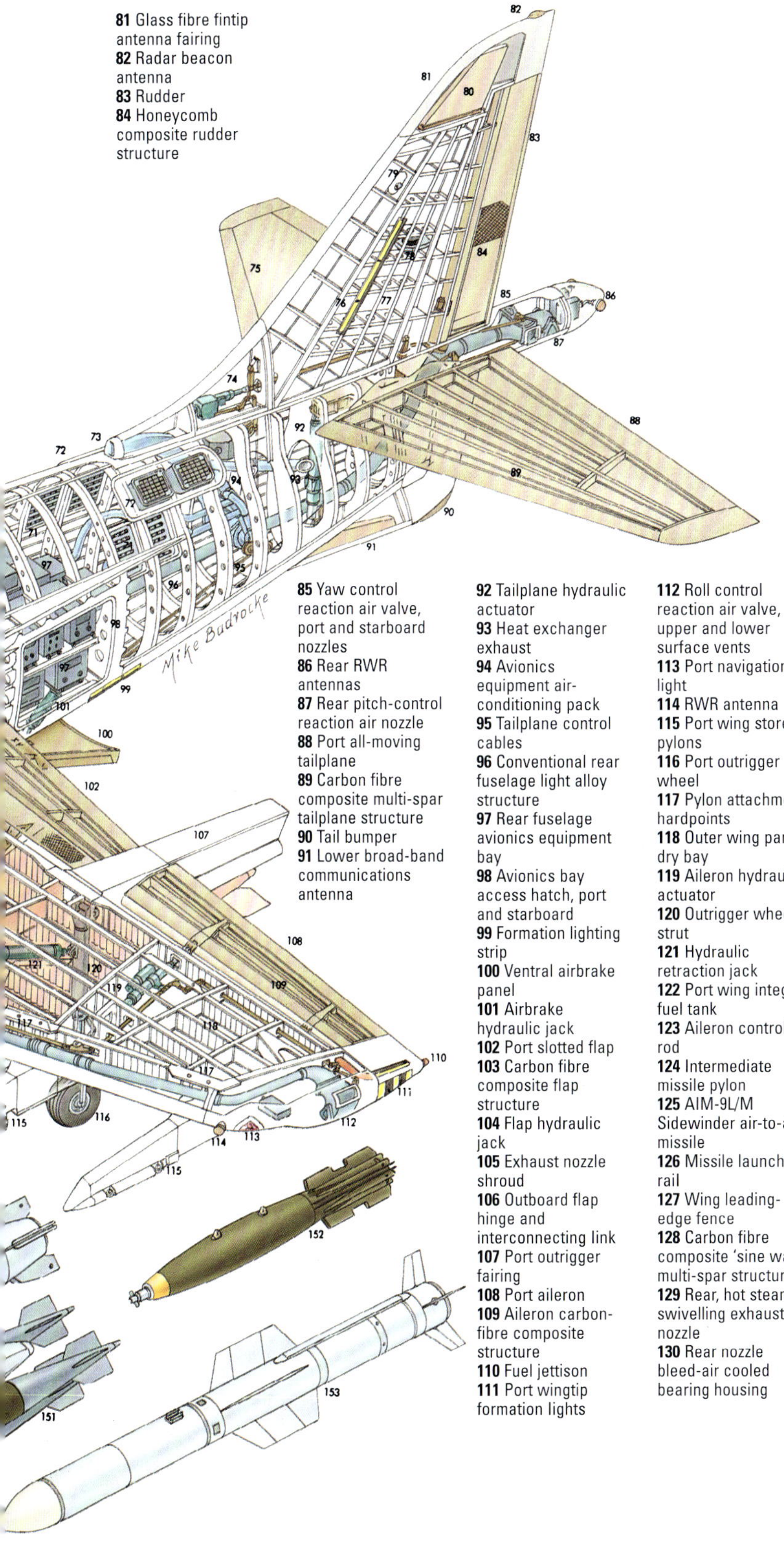

81 Glass fibre fintip antenna fairing
82 Radar beacon antenna
83 Rudder
84 Honeycomb composite rudder structure
85 Yaw control reaction air valve, port and starboard nozzles
86 Rear RWR antennas
87 Rear pitch-control reaction air nozzle
88 Port all-moving tailplane
89 Carbon fibre composite multi-spar tailplane structure
90 Tail bumper
91 Lower broad-band communications antenna
92 Tailplane hydraulic actuator
93 Heat exchanger exhaust
94 Avionics equipment air-conditioning pack
95 Tailplane control cables
96 Conventional rear fuselage light alloy structure
97 Rear fuselage avionics equipment bay
98 Avionics bay access hatch, port and starboard
99 Formation lighting strip
100 Ventral airbrake panel
101 Airbrake hydraulic jack
102 Port slotted flap
103 Carbon fibre composite flap structure
104 Flap hydraulic jack
105 Exhaust nozzle shroud
106 Outboard flap hinge and interconnecting link
107 Port outrigger fairing
108 Port aileron
109 Aileron carbon-fibre composite structure
110 Fuel jettison
111 Port wingtip formation lights
112 Roll control reaction air valve, upper and lower surface vents
113 Port navigation light
114 RWR antenna
115 Port wing stores pylons
116 Port outrigger wheel
117 Pylon attachment hardpoints
118 Outer wing panel dry bay
119 Aileron hydraulic actuator
120 Outrigger wheel strut
121 Hydraulic retraction jack
122 Port wing integral fuel tank
123 Aileron control rod
124 Intermediate missile pylon
125 AIM-9L/M Sidewinder air-to-air missile
126 Missile launch rail
127 Wing leading-edge fence
128 Carbon fibre composite 'sine wave' multi-spar structure
129 Rear, hot steam, swivelling exhaust nozzle
130 Rear nozzle bleed-air cooled bearing housing
131 Hydraulic reservoir, dual system, port and starboard
132 Pressure refuelling connection and control panel
133 Reaction-control air ducting
134 Aft retracting twin-wheel main undercarriage
135 Inboard 'wet' stores pylon
136 External fuel tank
137 Ventral gun pack, replaces fuselage LID strakes
138 Gun pneumatic drive unit
139 Ammunition cross-feed and link return chute
140 Ammunition magazine, 300 rounds
141 Retractable LID cross dam and hydraulic jack
142 Cannon muzzle aperture
143 Gun gas vent
144 Forward recoil mounting
145 GAU-12/U 25-mm five-barrelled rotary cannon
146 Gun pack LID strake
147 AGM-65A Maverick, laser-guided air-to-surface missile
148 AIM-120 AMRAAM, air-to-air missile
149 CBU-89B Gator, submunition dispenser
150 Triple ejector rack
151 Mk 82 LDGP 500-lb (227-kg) bomb
152 Mk 82SE Snakeye, retarded bomb
153 AGM-84A-D Harpoon, air-to-surface anti ship missile

McDonnell Douglas/BAe Harrier

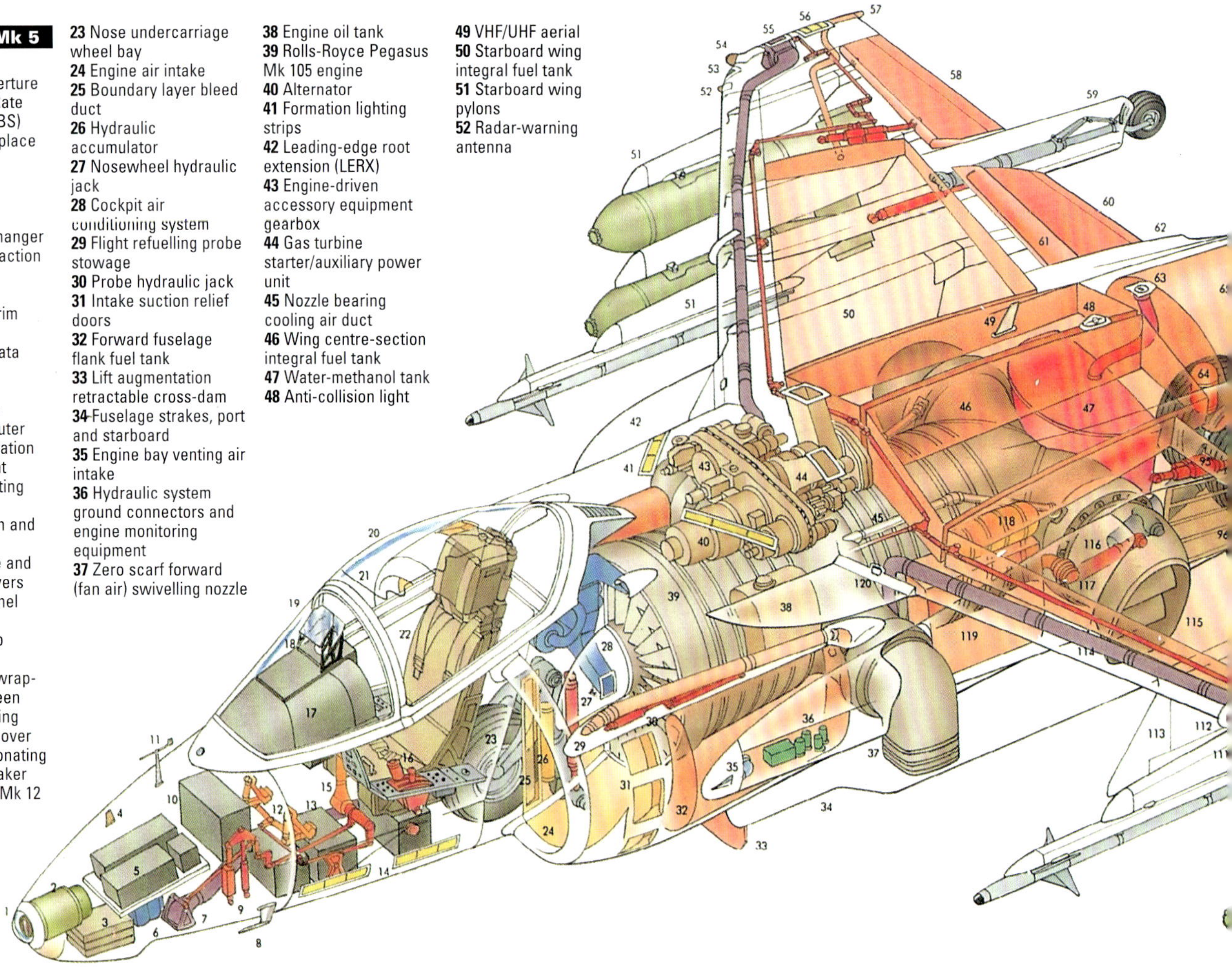

Harrier GR.Mk 5

Cutaway key
1 Glazed nose aperture
2 Hughes Angle Rate Bombing Set (ARBS)
3 Nose ballast in place of MIRLS
4 IFF aerial
5 Nose avionics equipment
6 ARBS heat exchanger
7 Pitch control reaction air valve
8 Pitot head
9 Pitch feel and trim actuators
10 ARBS signal data converter
11 Yaw vane
12 Rudder pedals
13 Air data computer and inertial navigation system equipment
14 Formation lighting strips
15 Control column and linkages
16 Engine throttle and nozzle control levers
17 Instrument panel shroud
18 Pilot's head-up display
19 Single piece 'wrap-around' windscreen
20 Rearward sliding cockpit canopy cover
21 Miniature detonating cord canopy breaker
22 Martin-Baker Mk 12 ejection seat
23 Nose undercarriage wheel bay
24 Engine air intake
25 Boundary layer bleed duct
26 Hydraulic accumulator
27 Nosewheel hydraulic jack
28 Cockpit air conditioning system
29 Flight refuelling probe stowage
30 Probe hydraulic jack
31 Intake suction relief doors
32 Forward fuselage flank fuel tank
33 Lift augmentation retractable cross-dam
34-Fuselage strakes, port and starboard
35 Engine bay venting air intake
36 Hydraulic system ground connectors and engine monitoring equipment
37 Zero scarf forward (fan air) swivelling nozzle
38 Engine oil tank
39 Rolls-Royce Pegasus Mk 105 engine
40 Alternator
41 Formation lighting strips
42 Leading-edge root extension (LERX)
43 Engine-driven accessory equipment gearbox
44 Gas turbine starter/auxiliary power unit
45 Nozzle bearing cooling air duct
46 Wing centre-section integral fuel tank
47 Water-methanol tank
48 Anti-collision light
49 VHF/UHF aerial
50 Starboard wing integral fuel tank
51 Starboard wing pylons
52 Radar-warning antenna

Above and left: The first-generation Hawker Siddeley Harriers entered service with the RAF in 1969 and brought new capabilities to RAF close support operations with their VTOL performance. However, they were hampered by short range, poor payload-carrying abilities and basic day-attack avionics. These shortcomings have been addressed by the second-generation Harriers which have overcome early teething troubles, to mature into some of the most effective attack aircraft in service today.

53 Starboard navigation light
54 Starboard/forward missile-warning antenna
55 Roll control reaction air valve
56 Wingtip formation light
57 Fuel jettison
58 Starboard aileron
59 Outrigger wheel fairing
60 Starboard slotted flap
61 Drooping flap vane
62 Wing root fairing
63 Water/methanol tank filler
64 Engine fire suppression bottle
65 Rear fuselage fuel tank
66 Aft avionics equipment bay
67 Electrical distribution panels
68 Heat exchanger ram air intake
69 Rudder hydraulic actuator
70 Starboard all-moving tailplane
71 Formation lighting strip
72 MAD compensator

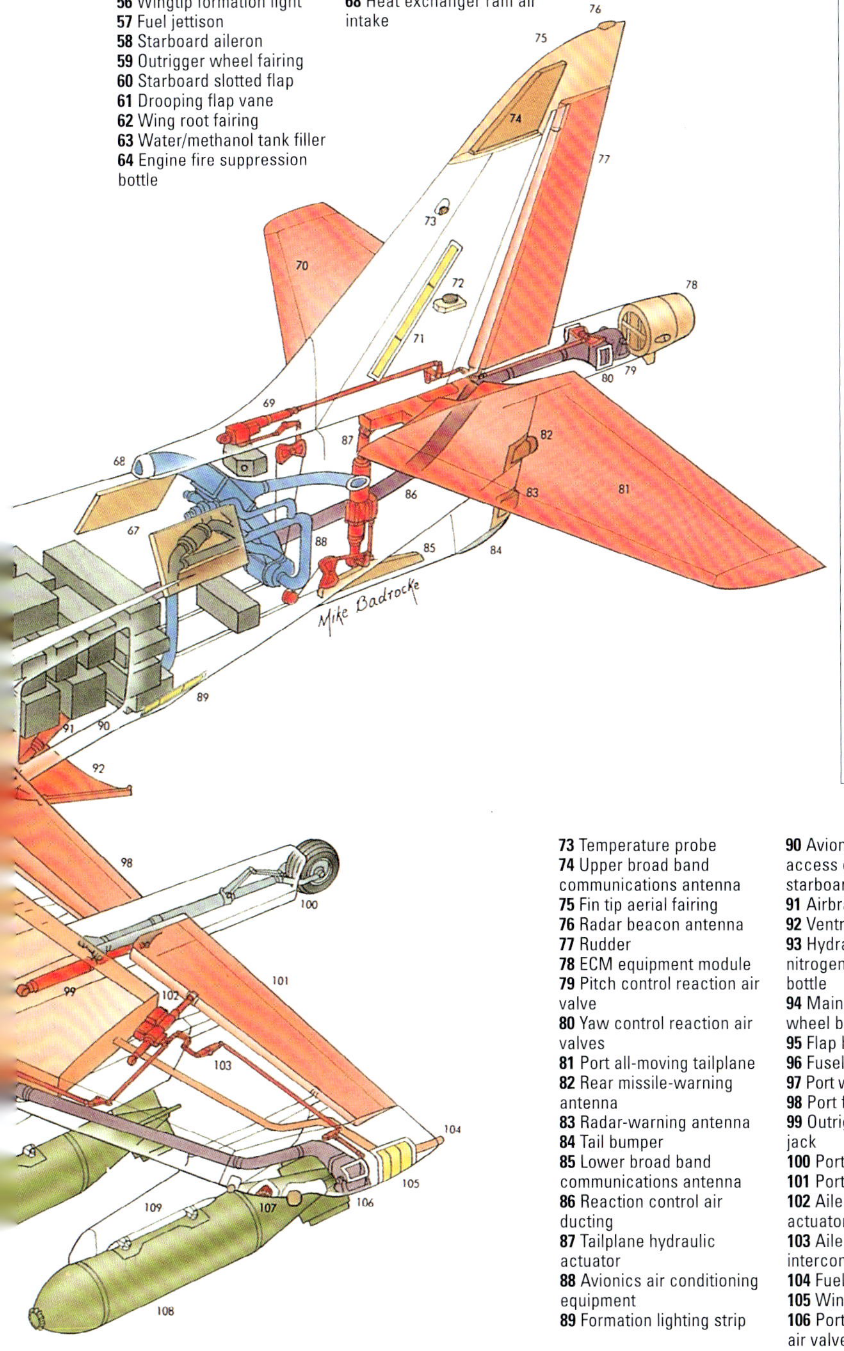

SPECIFICATION

Harrier GR.Mk 7

Dimensions

Length overall (flying attitude): 47 ft 8 in (14.53 m)
Wing span: 30 ft 4 in (9.25 m)
Wing aspect ratio: 4.0
Tailplane span: 13 ft 11 in (4.24 m)
Wing area (gross, excl. LERX): 230.0 sq ft (21.37 m²)
Total LERX area ('100%' LERX): 15.00 sq ft (1.39 m²)
Total trailing-edge flaps area: 31.00 sq ft (2.88 m²)
Total ailerons area: 12.40 sq ft (1.15 m²)
Total ventral fixed strakes area: 5.50 sq ft (0.51 m²)
Total retractable fence (LIDs) area: 2.60 sq ft (0.24 m²)
Total fin area: 26.60 sq ft (2.47m²)
Total rudder area (excl. tab): 5.30 sq ft (0.49 m²)
Tailplane area: 48.50 sq ft (4.51 m²)
Overall height: 11 ft 7¾ in (3.55 m)
Outrigger wheel track: 17 ft 0 in (5.18 m)
Maximum wing loading: 134.78 lb/sq ft (658.1 kg/m²)

Powerplant

GR Mk.7: one Rolls-Royce Pegasus Mk 105 vectored-thrust turbofan rated at 21,500 lb st (95.6 kN)

Weights

Empty operating (including pilot and unused fuel): 15,542 lb (7050 kg)
Basic flight design gross weight for 7-*g* operation: 22,950 lb (10410 kg)
Maximum take-off (after 1,430-ft (435-m) short take-off): 31,000 lb (14061 kg); (sea level vertical take-off, ISA conditions) 19,180 lb (8700 kg); (sea level vertical take-off, 32°C) 17,950 lb (8143 kg)
Design maximum landing weight: 25,000 lb (11340 kg)
Maximum vertical landing weight: 19,937 lb (9043 kg)

Fuel and load

Total fuel capacity: 1948 Imp gal (8858 litres)
Total usable internal fuel: 950 Imp gal (4318 litres)/7,759 lb (3519 kg), comprising integral wing fuel tanks with usable total 604 Imp gal (2746 litres), plus four fuselage tanks; front and rear 134 Imp gal (609 litres) each, plus two centre tanks 39 Imp gal (177 litres) each
Total usable internal fuel (Harrier T.Mk 10): 913 Imp gal (4150 litres)
Water injection tank capacity: 495 lb (225 kg)
External fuel: up to four 250-Imp gal (1135-litre) auxiliary fuel tanks on four inner underwing stations; single-point refuelling plus optional retractable bolt-on flight refuelling probe above port air intake
Maximum useful load: (including fuel, water injection for engine, stores, guns and ammunition) approx. 6,750 lb (3062 kg) with vertical take-off and 17,000 lb (7,710 kg) with short take-off

Performance

Maximum level speed at altitude: 645 mph (1041 km/h)
Maximum level speed at low level: 661 mph (1065 km/h)
Limiting Mach numbers: 0.98 at altitude; 0.87 at sea level
Maximum rate of climb at sea level: 14,715 ft (4485 m) per minute
Service ceiling above: 50,000 ft (15240 m)
STOL take-off run: (maximum weight, ISA conditions) 1,430 ft (435 m); (maximum weight, 32°C) 1,700 ft (518 m)

Range

Operational radius: (after short take-off with 12 500-lb (227-kg) bombs, internal fuel and 1-hr loiter) 103 miles (167 km); (hi-lo-hi profile after short take-off with seven 500-lb (227-kg) bombs, two external fuel tanks, no loiter) 684 miles (1101 km)
Ferry range: (unrefuelled with four underwing tanks, tanks retained) 1886 miles (3035 km)
Combat air patrol endurance: (at 115 miles (185 km) from base) three hours

g limits

Maximum/minimum: +8/-3, reduced manoeuvring limits apply when auxiliary fuel tanks carried

Armament

Fixed: two underfuselage pods each housing a single 25-mm Aden 25 cannon with 100 rounds
Weapon stations: centreline station stressed for 1,000 lb (454 kg), four stations under each wing stressed for loads up to 2,000 lb (907 kg) inboard, 1,000 lb (454 kg) intermediate and 630 lb (286 kg) outboard, additional weapon station ahead of outrigger wheel fairing for air-to-air missile
Current weapons: 540-lb (245-kg) and 1,000-lb (454-kg) general-purpose free-fall bombs, BL755 and CBU-87 cluster bombs, GBU-16, CPU-123/B, GBU-24/B 1,000-lb (454-kg) laser-guided bombs, Matra 155 rocket pods (each with 18 68-mm SNEB rockets) or CRV-7 rocket pods, AIM-9L/M Sidewinder air-to-air missiles for self-defence
Proposed stores and weapons : Brimstone anti-armour missiles, ALARM anti-radiation missile, CASOM, TIALD laser designator pod

73 Temperature probe
74 Upper broad band communications antenna
75 Fin tip aerial fairing
76 Radar beacon antenna
77 Rudder
78 ECM equipment module
79 Pitch control reaction air valve
80 Yaw control reaction air valves
81 Port all-moving tailplane
82 Rear missile-warning antenna
83 Radar-warning antenna
84 Tail bumper
85 Lower broad band communications antenna
86 Reaction control air ducting
87 Tailplane hydraulic actuator
88 Avionics air conditioning equipment
89 Formation lighting strip
90 Avionics equipment bay access door, port and starboard
91 Airbrake hydraulic jack
92 Ventral airbrake panel
93 Hydraulic system nitrogen pressurisation bottle
94 Main undercarriage wheel bay
95 Flap hydraulic jack
96 Fuselage heat shield
97 Port wing integral fuel tank
98 Port flap
99 Outrigger wheel hydraulic jack
100 Port outrigger wheel
101 Port aileron
102 Aileron hydraulic actuator
103 Aileron/air valve interconnection
104 Fuel jettison
105 Wingtip formation light
106 Port roll control reaction air valve
107 Port navigation light
108 BL755 cluster bombs
109 Outboard weapons pylons
110 AIM-9L/M Sidewinder air-to-air missile
111 Intermediate missile pylon
112 Wing fence
113 Inboard weapon/fuel tank pylon
114 Reaction control air ducting
115 Rear (hot-stream) swivelling exhaust nozzle
116 Main undercarriage hydraulic jack
117 Pressure refuelling connection
118 Hydraulic reservoir
119 Centre fuselage flank fuel tank
120 Engine bay venting air intake

North American F-86 Sabre

In recent years the Sabre has become extremely popular with the warbird fraternity, with about 50 potentially airworthy aircraft in the US and more under restoration. The fliers include single examples of the A, D, E, H and L, and 11 Fs. The remainder are CL-13s. This restored F-86F wears the markings of Maj. John Glenn, USMC who flew with the 25th Fighter Interceptor Squadron during the Korean War. Glenn was on Temporary Deployment (TDY) to the USAF and, during this period, he scored three kills over North Korean MiGs. Later in his career Glenn was to gain greater fame as America's first man in space and, more recently, as the world's oldest astronaut

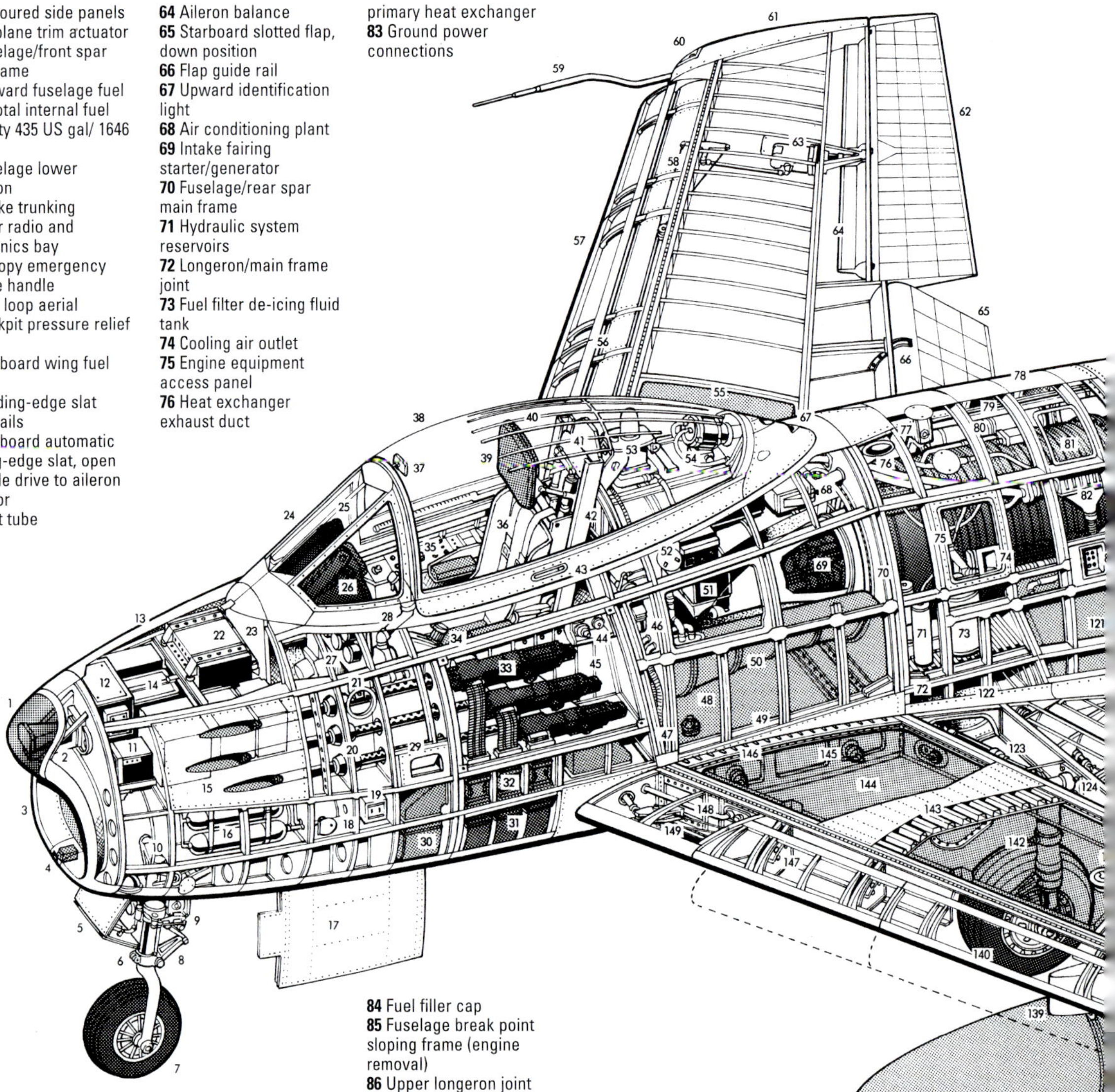

F-86E Sabre

Cutaway key
1 Radome
2 Radar antenna
3 Engine air intake
4 Gun camera
5 Nosewheel leg doors
6 Nose undercarriage leg strut
7 Nosewheel
8 Torque scissor links
9 Steering control valve
10 Nose undercarriage pivot fixing
11 Sight amplifier
12 Radio and electronics equipment bay
13 Electronics bay access panel
14 Battery
15 Gun muzzle blast troughs
16 Oxygen bottles
17 Nosewheel bay doors
18 Oxygen servicing point
19 Canopy switches
20 Machine-gun barrel mountings
21 Hydraulic system test connections
22 Radio transmitter
23 Cockpit armoured bulkhead
24 Windscreen panels
25 A-1CM radar gunsight
26 Instrument panel shroud
27 Instrument panel
28 Control column
29 Kick-in boarding step
30 Used cartridge case collector box
31 Ammunition boxes (267 rounds per gun)
32 Ammunition feed chutes
33 0.5-in (12.7-mm) Colt Browning machine-guns
34 Engine throttle
35 Starboard side console panel
36 North American ejection seat
37 Rear view mirror
38 Sliding cockpit canopy cover
39 Ejection seat headrest
40 ADF sense aerials
41 Pilot's back armour
42 Ejection seat guide rails
43 Canopy handle
44 Cockpit pressure valves
45 Armoured side panels
46 Tailplane trim actuator
47 Fuselage/front spar main frame
48 Forward fuselage fuel tank (total internal fuel capacity 435 US gal/ 1646 litres)
49 Fuselage lower longeron
50 Intake trunking
51 Rear radio and electronics bay
52 Canopy emergency release handle
53 ADF loop aerial
54 Cockpit pressure relief valve
55 Starboard wing fuel tank
56 Leading-edge slat guide rails
57 Starboard automatic leading-edge slat, open
58 Cable drive to aileron actuator
59 Pitot tube
60 Starboard navigation light
61 Wingtip fairing
62 Starboard aileron
63 Aileron hydraulic control unit
64 Aileron balance
65 Starboard slotted flap, down position
66 Flap guide rail
67 Upward identification light
68 Air conditioning plant
69 Intake fairing starter/generator
70 Fuselage/rear spar main frame
71 Hydraulic system reservoirs
72 Longeron/main frame joint
73 Fuel filter de-icing fluid tank
74 Cooling air outlet
75 Engine equipment access panel
76 Heat exchanger exhaust duct
77 Engine suspension links
78 Fuselage skin plating
79 Engine withdrawal rail
80 Starboard side oil tank
81 General Electric J47-GE-27 turbojet
82 Bleed air system primary heat exchanger
83 Ground power connections
84 Fuel filler cap
85 Fuselage break point sloping frame (engine removal)
86 Upper longeron joint
87 Engine bay air cooling duct
88 Cooling air outlet
89 Engine firewall bulkhead
90 Engine flame cans
91 Rear fuselage framing
92 Fuel jettison pipe

On 14 December 1953 NAA flew its TF-86F prototype 52-5016, which had been developed in response to a USAF requirement for a high-performance trainer. This aircraft was lost after a few flights, but was replaced by a second aircraft 52-1228 (illustrated), which differed in being fitted with a '6-3' wing. Although no orders were forthcoming, the second TF-86F had a long and productive career as a trainer and test/chase platform.

SPECIFICATION

F-86E Sabre

Dimensions

Length: 37 ft (11.27 m)
Height: 14 ft (4.26 m)
Wingspan: 37 ft (11.27 m)
Wing area: 288 sq ft (26.75 m²)

Powerplant

One General Electric J47-GE-13 turbojet rated at 5,450 lb (24.24 kN) static thrust

Weights

Empty: 10,555 lb (4788 kg)
Normal loaded: 16,346 lb (7414 kg)
Maximum take-off: 17,806 lb (8077 kg)

Fuel

Normal internal fuel: 435 US gal (1646 litres)
Maximum internal fuel: 675 US gal (2555 litres)

Performance

Maximum level speed clean at 35,000 ft (10688 m): 601 mph (967 km/h)
Maximum level speed clean at low level: 679 mph (1093 km/h)
Cruising speed: 537 mph (864 km/h)
Stalling speed: 123 mph (198 km/h)
Climb to 30,000 ft (9144 m): 6 minutes 18 seconds
Service ceiling: 47,200 ft (14387 m)
Range: 848 miles (1365 km)
Maximum range: 1,022 miles (1645 km)

Armament

Primary armament consisted of six 0.5-in (12.7-mm) Colt Browning machine-guns mounted in the forward fuselage with a total of 1,800 rounds; provision for either two 1,000-lb (454-kg) bombs or 16 0.5-in (12.7-mm) rocket projectiles mounted on underwing racks, in place of two 120-US gal (454-litre) drop tanks, a variety of other ordnance loads could also be carried

93 Fuselage top longeron
94 Fin/tailplane root fillet fairing
95 Control cable duct
96 Fin spar attachment joint
97 Tailplane/rudder control cables
98 All-moving tailplane hydraulic jack
99 Tailfin construction
100 Flush HF aerial panel
101 Starboard tailplane
102 Fintip dielectric aerial fairing
103 ADF aerial
104 Rudder construction
105 Rudder trim tab

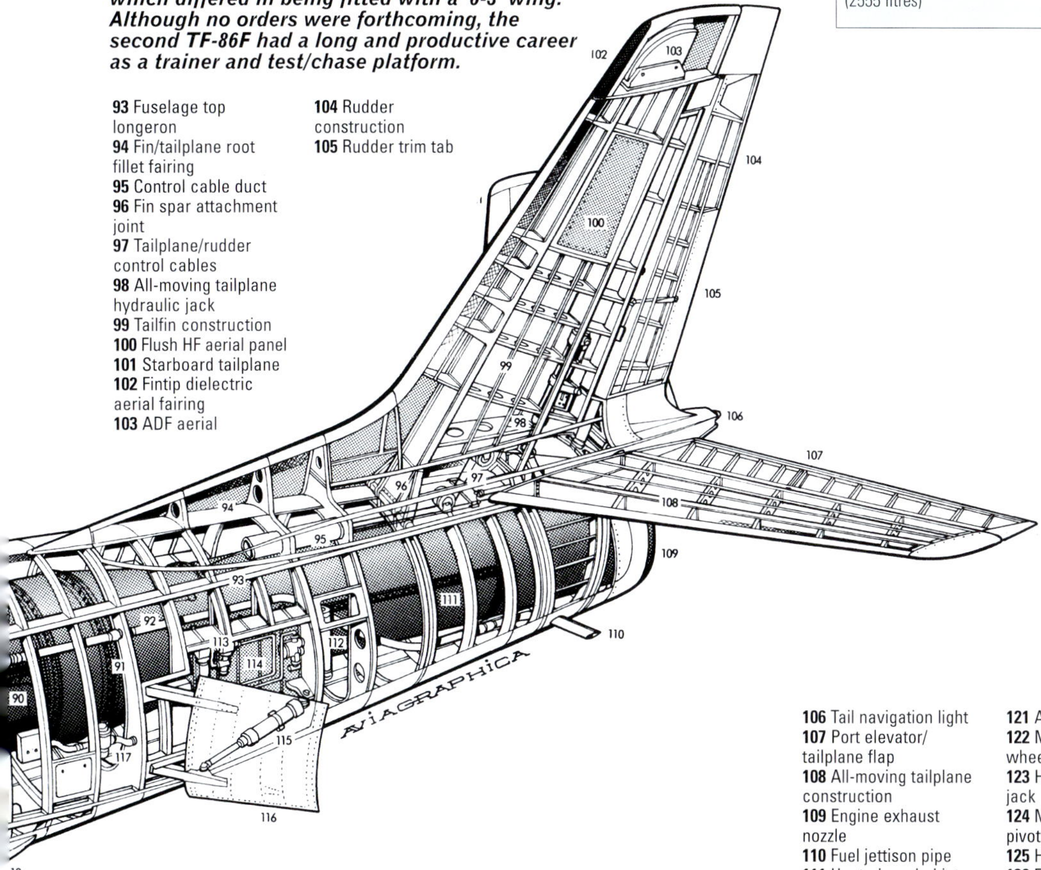

106 Tail navigation light
107 Port elevator/ tailplane flap
108 All-moving tailplane construction
109 Engine exhaust nozzle
110 Fuel jettison pipe
111 Heat-shrouded jet pipe
112 Power control compensator
113 Emergency hydraulic valves
114 Airbrake housing
115 Airbrake hydraulic jack
116 Port airbrake (open)
117 Hydraulic system emergency pump
118 Cooling air intake
119 Lower longeron joint
120 Trailing-edge root fillet
121 Aft main fuel tank
122 Main undercarriage wheel bay
123 Hydraulic retraction jack
124 Main undercarriage pivot fixing
125 Hydraulic flap jack
126 Flap shroud ribs
127 Port slotted flaps
128 Port aileron construction
129 Aileron hydraulic power control unit
130 Gyro compass remote transmitter
131 Wingtip fairing
132 Port navigation light
133 Port automatic leading-edge slat, open position
134 Leading-edge slat rib construction
135 Front spar
136 Wing rib and stringer construction
137 Wing skin/leading-edge piano hinge attachment joint
138 120-US gal (454-litre) drop tank
139 Drop tank pylon
140 Port main wheel
141 Fuel filler cap
142 Main undercarriage leg strut
143 Fuel tank bay corrugated double skin
144 Port wing fuel tank
145 Tank interconnectors
146 Skin panel attachment joint strap
147 Slat guide rails
148 Fuel feeders
149 Aileron cable drive

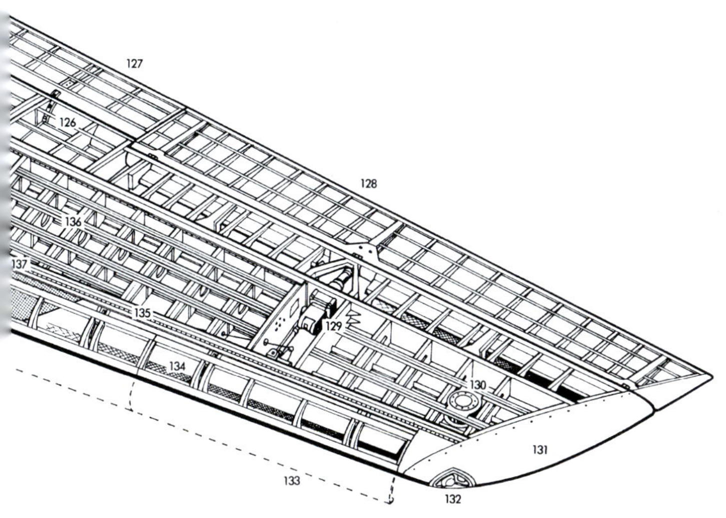

Panavia Tornado ADV

Demonstrating the adaptability o[f] Panavia's original Tornado/MRC[A] design, the ADV has a stretche[d] fuselage, different radar and avionics plus provision for eight air-to-a[ir] missiles. As one of the fastest aircraft a[t] low level, it is ideal for 'running dow[n]' bombers attacking down lo[w]

Tornado F.Mk 3

Cutaway key

1 Starboard taileron construction
2 Honeycomb trailing-edge panels
3 Compound sweep taileron leading edge
4 Taileron pivot fixing
5 Afterburner ducting, extended 14 in (36 cm)
6 Thrust-reverser bucket door actuator
7 Afterburner nozzle jack
8 Starboard fully-variable engine exhaust nozzle
9 Thrust-reverser bucket doors, open
10 Dorsal spine and fairing
11 Rudder hydraulic actuator
12 Honeycomb rudder construction
13 Rudder
14 Fuel jettison pipes
15 Tail navigation light
16 Aft passive ECM housing/radar warning antenna
17 Dielectric fin tip antenna housing
18 VHF aerial
19 Fuel jettison and vent valve
20 ILS aerial, port and starboard
21 Underside view showing semi-recessed missile positions
22 Extended fuselage section
23 Extended radar equipment bay
24 Radome
25 Secondary heat exchanger intake
26 Wing pylon-mounted missile rails
27 External fuel tanks
28 Port taileron
29 Fin leading edge
30 Fin integral fuel tank
31 Tailfin construction
32 Vortex generators
33 Heat shroud
34 Fin spar root attachment joints
35 Engine bay central firewall
36 Starboard airbrake, open
37 Airbrake hydraulic jack
38 Taileron actuator, fly-by-wire control system
39 Turbo-Union RB.199-34R Mk 104 three-spool afterburning turbofan
40 Hydraulic reservoir
41 Hydraulic system filters
42 Engine bay bulkhead
43 Bleed air duct
44 Heat exchanger exhaust duct
45 Primary heat exchanger
46 Ram air intake
47 HF aerial fairing
48 Engine compressor faces
49 Rear fuselage bag-type fuel tank
50 Intake trunking
51 Wing root pneumatic seal
52 KHD/Microtecnica/Lucas T312 APU
53 Engine-driven auxiliary gearbox
54 APU exhaust
55 Flap drive shaft
56 Starboard full-span double-slotted flaps, extended
57 Spoiler housings
58 Starboard wing fully-swept position
59 Flap guide rails
60 Flap screw jacks
61 Wingtip fairing
62 Starboard navigation light
63 Structural provision for outboard pylon attachment
64 Full-span leading-edge slats, extended
65 Starboard external fuel tank, capacity 495 Imp gal (2250 litres)
66 Fuel tank-stabilising fins
67 Swivelling wing stores pylon
68 Missile launching rail
69 AIM-9L Sidewinder air-to-air missiles
70 Leading-edge slat screw jacks
71 Slat guide rails
72 Wing rib construction
73 Two-spar wing torsion box construction
74 Swivelling pylon mounting
75 Starboard wing integral fuel tank
76 Main undercarriage leg strut
77 Starboard mainwheel
78 Mainwheel door
79 Undercarriage breaker strut
80 Wing pivot sealing fairing
81 Telescopic control linkages
82 Pylon swivelling link
83 Main undercarriage hydraulic retraction jack
84 Wing sweep actuator attachment joint
85 Starboard wing pivot bearing
86 Flexible wing seals
87 Wing pivot carry-through (Electron beam welded titanium box construction)
88 Wing pivot box integral fuel tank
89 Pitch and roll control non-linear gearing mechanism
90 Air-conditioning supply ducting (Normalair Garrett system)
91 Dorsal spine fairing
92 Anti-collision light
93 UHF aerials
94 Port wing pivot bearing
95 Flexible trailing-edge seals
96 Spoiler actuators
97 Port spoilers, open

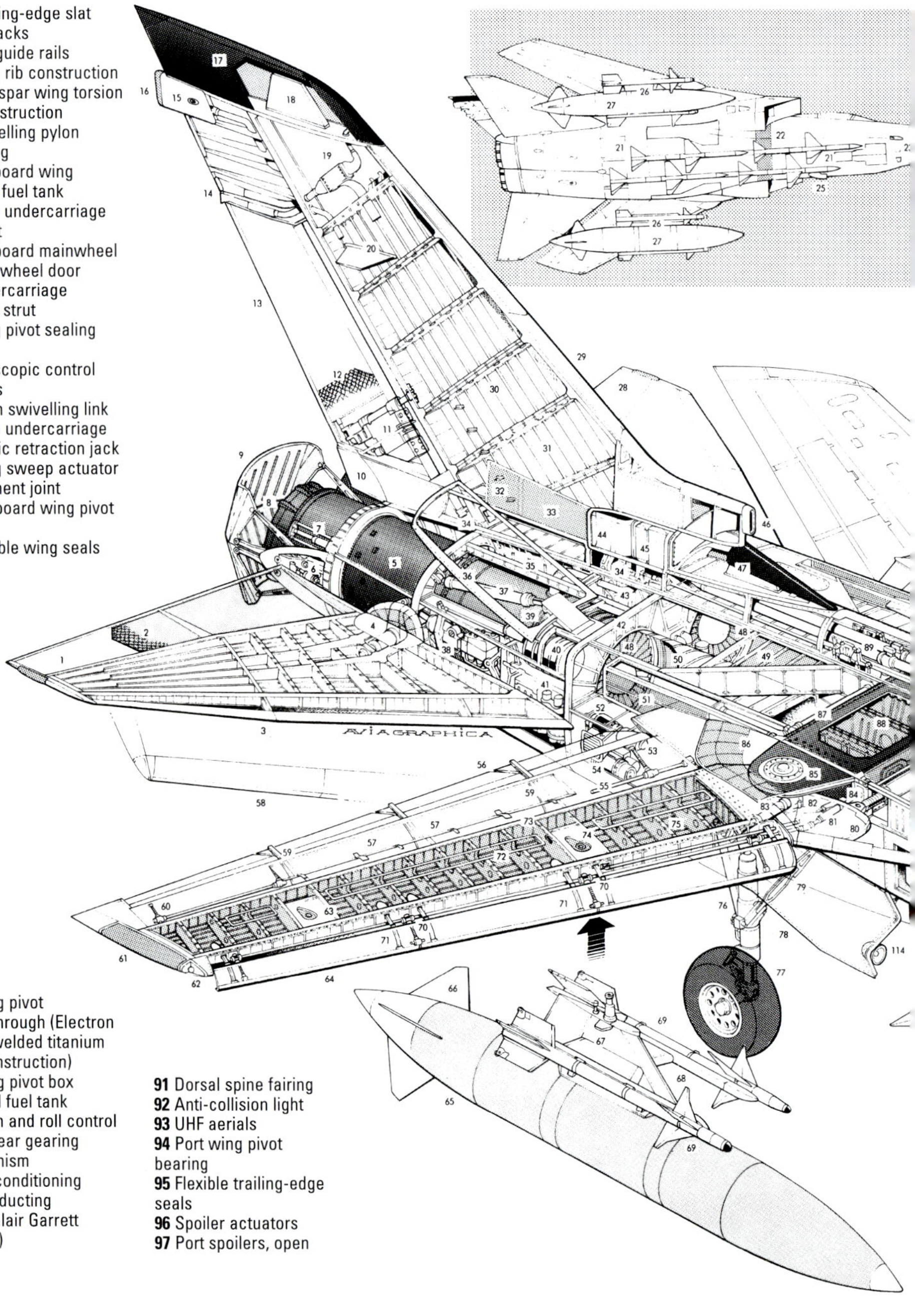

RAF Tornado F.Mk 3s are occasional visitors to the Red Flag exercises in the USA. These give participating aircrews the opportunity to fly missions in the vast desert ranges of southern Nevada against simulated enemy aggressor units. F.Mk 3s operate with Blue Force (friendly force) and are usually tasked with strike support, as part of the offensive counter-air force.

SPECIFICATION

Tornado F.Mk 3

Dimensions

Length: 61 ft 3½ in (18.68 m)
Height: 19 ft 6¼ in (5.95 m)
Wingspan spread: 45 ft 7½ in (13.91 m)
Wingspan swept: 28 ft 2½ in (8.60 m)
Aspect ratio spread: 7.73
Aspect ratio swept: 2.96
Wing area: 286.33 sq ft (26.60 m²)
Tailplane span: 22 ft 3½ in (6.80 m)
Wheel track: 10 ft 2 in (3.10 m)

Powerplant

Two Turbo-Union RB.199-34R Mk 104 turbofans each rated at 9,100 lb st (40.48 kN) dry and 16,520 lb st (73.48 kN) with afterburning

Fuel and load

Internal fuel: 12,544 lb (5690 kg)
External fuel: up to 12,800 lb (5806 kg) in two 495-Imp gal (2250-litre) and two 396-Imp gal (1500-litre) or four 330-Imp gal (1500-litre) drop tanks
Maximum ordnance: 18,740 lb (8500 kg)

Weights

Empty operating: 31,970 lb (14502 kg)
Maximum take-off: 61,700 lb (27986 kg)

Performance

Maximum level speed 'clean' at 36,000 ft (10975 m): 1,262 kt (1,453 mph; 2238 km/h)
Operational ceiling: 70,000 ft (21335 m)
Combat radius: more than 300 nm (345 miles; 556 km) supersonic or more than 1000 nm (1,151 miles; 1852 km) subsonic

Armament

One 27-mm IWKA Mauser cannon fitted to the starboard side. Main armament is four BAe SkyFlash semi-active radar-homing or four AIM-120 AMRAAM active radar AAMs with a range of 31 miles (50 km), carried semi-recessed under the fuselage. Four AIM-9L Sidewinders or four ASRAAMS are also carried for short-range combat. Self-defence is provided by a Bofors Phimat chaff dispenser carried on starboard outer Sidewinder pylon or Celsius Tech BOL integral chaff/flare dispenser in Sidewinder rail. Vicon 78 chaff/flare dispensers under rear fuselage. GEC-Marconi Ariel towed-radar decoy can be carried on outboard wing pylon

98 Port wing fully swept back position
99 Full-span double-slotted flaps, extended
100 Port wing fully forward position
101 Wing tip fairing
102 Port navigation light
103 Full-span leading-edge slats, extended
104 Port wing integral fuel tank
105 Swivelling pylon mounting
106 Pylon angle control link
107 Port wing sweep actuator
108 Wing flap and leading-edge slat drive motors
109 Starboard wing sweep actuator
110 Hydraulic drive motor and gearbox
111 Extended wingroot glove fairing
112 Forward radar-warning receiver, port and starboard
113 Supplementary 'blow-in' intake doors
114 Landing lamp, port and starboard
115 Starboard fully-variable engine air intake
116 Navigation light
117 Variable-intake ramp
118 Ramp control linkage
119 Ramp hydraulic jack
120 Bleed air exit louvres
121 Boundary layer spill duct
122 Enlarged forward fuselage bag-type fuel tank
123 Cockpit canopy pivot mounting
124 Air and fuel system ducting
125 Port intake bleed air outlet fairing
126 AIM-9L Sidewinder air-to-air missiles
127 Missile launching rail
128 Port external fuel tank
129 Intake lip
130 Navigator's cockpit enclosure
131 Navigator's ejection seat (Martin-Baker Mk 10A 'zero-zero' seat)
132 Canopy jack strut
133 Cockpit rear pressure bulkhead
134 Engine air intake curved inboard sidewall
135 Starboard avionics equipment and flight control system equipment bay
136 BAe SkyFlash air-to-air missile
137 Ventral semi-recessed missile housing
138 Cartridge case and link collector box
139 Navigator's side console

The RAF's acquisition of the Sentry AEW.Mk 1, which entered service in 1991, finally allowed the Tornado force to exploit and expand a modern integrated air defence AEW&C system. The belated addition of improved datalink systems and BVR missiles for the F.Mk 3 in the form of JTIDS and AMRAAM has enhanced the Tornado's defence capability still further.

140 Canopy centre arch
141 Navigator's instrument console
142 One-piece cockpit canopy cover
143 Ejection seat headrest
144 Pilot's ejection seat
145 Side console panel
146 Ammunition feed chute
147 Mauser 27-mm cannon, starboard only
148 Instrument pressure sensor
149 Cannon barrel
150 Radome, open position
151 Nosewheel leg strut
152 Twin nosewheels
153 Torque scissor links
154 Taxiing lamp
155 Nosewheel doors
156 Cannon muzzle blast tube
157 Electrical system equipment and ground test panels
158 Cockpit pressure floor
159 Rudder pedals
160 Control column
161 Instrument panel shroud
162 Pilot's head-up-display
163 Windscreen panels
164 Windscreen rain dispersal duct
165 Cockpit front pressure bulkhead
166 Avionics equipment, communications and navigation systems
167 Angle of attack transmitter
168 Blade antenna
169 In-flight refuelling probe, extended
170 Marconi-Elliot 'Foxhunter' airborne interception radar
171 Scanner tracking mechanism
172 Cassegrain radar antenna
173 Radar unit hinged to starboard for replacement of line replaceable units (LRUs)
174 Extended radome
175 Pitot tube

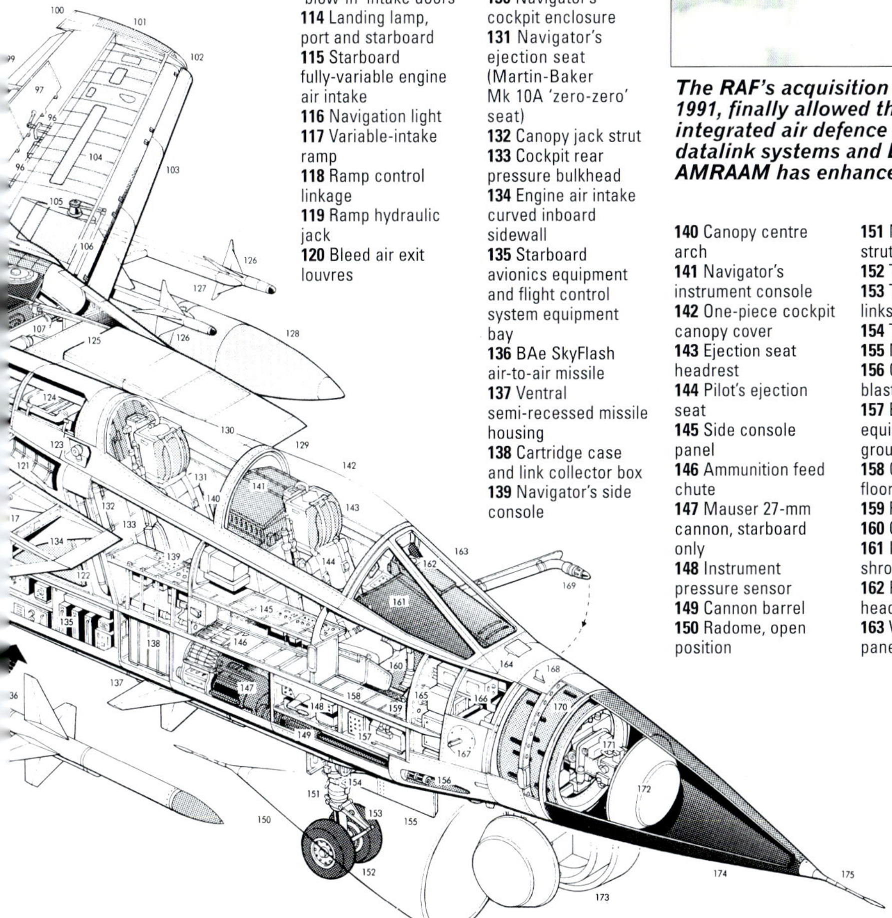

Republic F-84F Thunderstreak

F-84F Thunderstreak

Cutaway key
1 Engine air intake
2 Gun tracking radar antenna
3 Machine gun muzzles
4 Pitot tube
5 Nose undercarriage hydraulic retraction jack
6 Leg compression link
7 Nosewheel leg strut
8 Nosewheel
9 Mudguard
10 Steering jack
11 Taxiing lamp
12 Nose undercarriage leg rear strut
13 Nosewheel doors
14 Intake duct framing
15 Nose compartment 0.5-in (12.7-mm) 50 Colt-Browning M3 machine-guns (four)
16 Radar electronics equipment
17 Ammunition tanks, total of 1,800 rounds
18 Forward avionics bay including LABS bombing computer
19 Static ports
20 Battery
21 Intake ducting
22 Cockpit front pressure bulkhead
23 Rudder pedals
24 Instrument panel shroud
25 Windscreen panels
26 A-4 radar gunsight
27 Standby compass
28 Instrument panel
29 Ejection seat footrests
30 Aileron hydraulic booster
31 Intake suction relief door
32 Duct screen clearance access panel
33 Wingroot machine-gun muzzle
34 Intake duct screen
35 Port side console panel
36 Engine throttle
37 External canopy release handle
38 Pilot's ejection seat
39 Safety harness
40 Headrest
41 Ejection seat guide rails
42 Cockpit canopy cover
43 Starboard automatic leading edge slat, open
44 Slat guide rails
45 Starboard navigation light
46 Wingtip fairing
47 Starboard aileron
48 Aileron control rods
49 Starboard wing integral fuel tank; normal maximum fuel capacity 1,475 US gal (5583 litres)
50 Aileron fixed tab
51 Starboard flap
52 Spoiler
53 Starboard main undercarriage leg (retracted position)
54 Canopy rear hinge arm
55 Cockpit aft glazing
56 Ammunition feed chute
57 Cockpit rear pressure bulkhead
58 Wingroot 0.50-in (12.7-mm) Colt Browning M3 machine-gun
59 Engine electric starter/generator
60 Intake compressor face
61 Wing/fuselage spar lug attachment bolts (total four)
62 Fuselage main fuel tanks
63 Engine oil tank
64 Oxygen bottle
65 Air-conditioning and pressurisation pack
66 Fuselage fuel tank filler cap
67 Tail section attachment bolts (four)
68 Fuselage double main frames
69 Fuselage break point
70 Main engine mounting
71 Engine bay cooling intake
72 Wright J65-W-3 turbojet engine
73 Radio-compass antenna
74 Flush aerial fairing
75 Aft radio equipment bay
76 Engine firewall
77 Engine turbine section
78 Fuselage frame and stringer construction
79 Dorsal spine fairing
80 Anti-collision light
81 VHF aerial
82 Tailpipe cooling intake
83 Rear fuselage framing
84 Jet pipe
85 Rudder control rod, elevator on starboard side
86 Finroot fillet

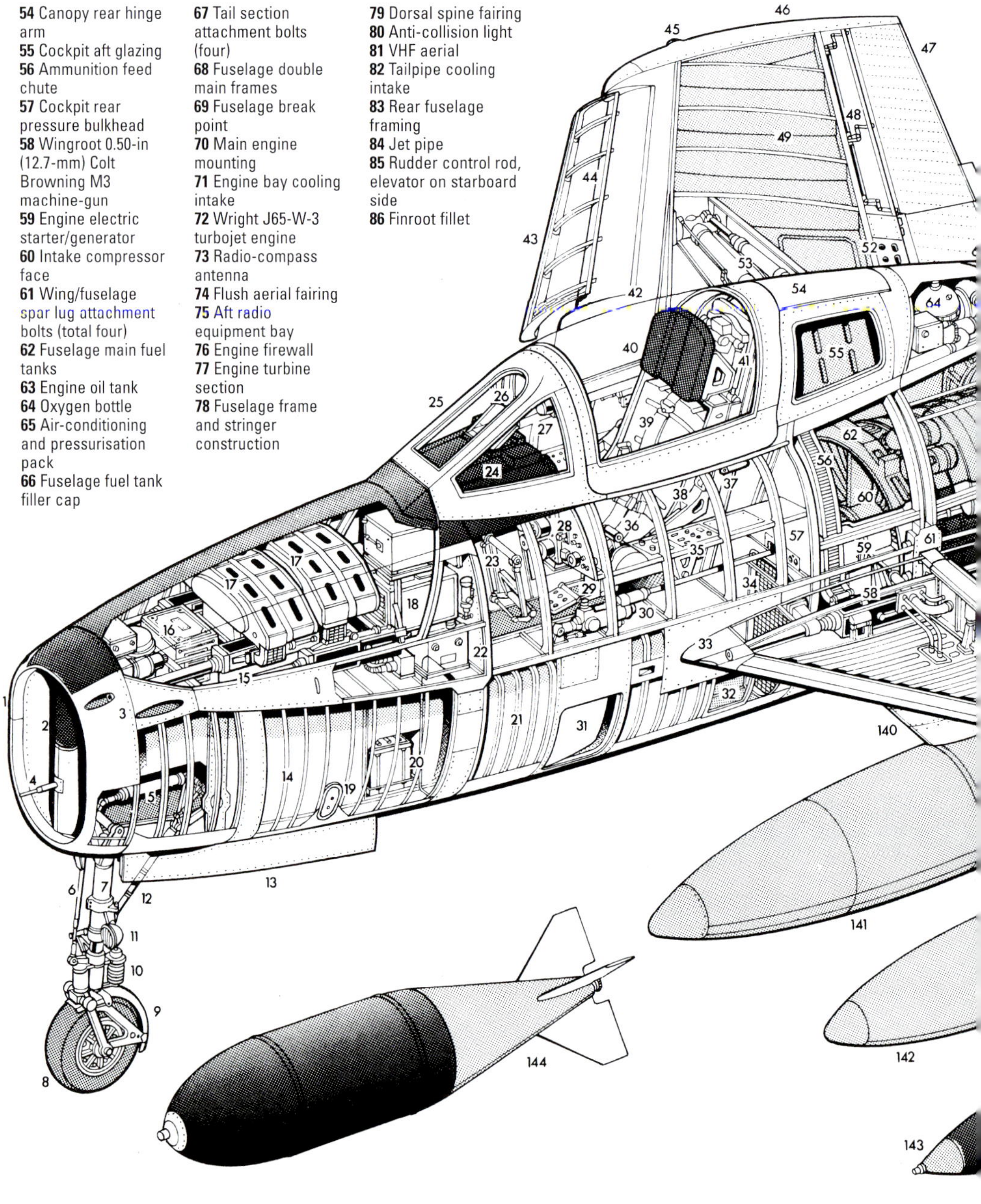

In 1949, Republic flew the first of its two XF-91 Thunderceptors. Similar to its F-84 forebears only in terms of its fuselage, the XF-91 had butterfly-like wings which were of inverse taper, becoming ever-broader towards the tips, and were mounted on pivots so as to have variable incidence. At the back were not only an afterburning J47 engine (the next engine after the J35), but also a battery of four rocket motors which, on 14 December, thrust the XF-91 beyond the speed of sound (the first time that this had been achieved by a non-Russian fighter). Many other features were also tested on the XF-91s, including a butterfly (V-type) tail, but there was no question of production.

Left: Over a 18-year period, F-84Fs served with 24 Air National Guard units in the interceptor (limited use only), special delivery (nuclear) and tactical fighter roles. The last ANG F-84Fs were phased out by Illinois's 170th TDFS and Ohio's 164th TFS (illustrated) in the last quarter of 1972.

SPECIFICATION

F-84F Thunderstreak

Dimensions

Length: 43 ft 4 in (13.23 m)
Wingspan: 33 ft 7 in (10.27 m)
Height: 14 ft 4½ in (4.39 m)
Wing area: 324.70 sq ft (98.90 m²)

Powerplant

One Wright J65-W-3 turbojet delivering 7,220 lb (32.50 kN) of thrust

Weights

Empty: 13,380 lb (6273 kg)
Maximum take-off: 28,000 lb (12700 kg)

Performance

Maximum speed at 20,000 ft (6095 m): 658 mph (1059 km/h)
Maximum speed at sea level: 695 mph (1118 km/h)
Initial climb rate: 8,200 ft (2500 m) per minute
Service ceiling: 46,000 ft (14020 m)
Combat radius (high with two drop tanks): 810 miles (1304 km)

Armament

Six 0.5-in (12.7-mm) Browning M3 machine-guns, and up to 6,000 lb (2722 kg) of external stores, originally including US tactical nuclear weapons

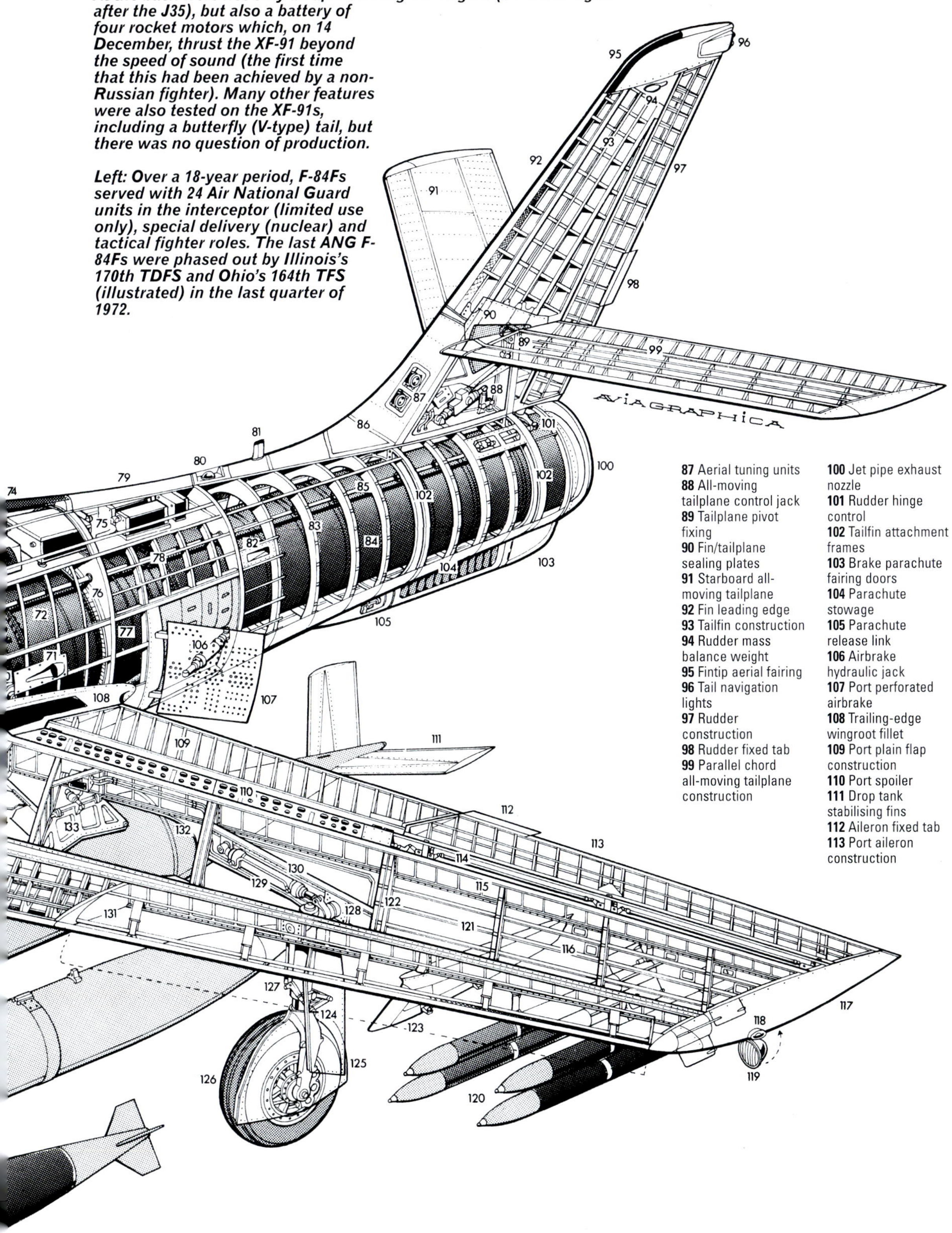

87 Aerial tuning units
88 All-moving tailplane control jack
89 Tailplane pivot fixing
90 Fin/tailplane sealing plates
91 Starboard all-moving tailplane
92 Fin leading edge
93 Tailfin construction
94 Rudder mass balance weight
95 Fintip aerial fairing
96 Tail navigation lights
97 Rudder construction
98 Rudder fixed tab
99 Parallel chord all-moving tailplane construction
100 Jet pipe exhaust nozzle
101 Rudder hinge control
102 Tailfin attachment frames
103 Brake parachute fairing doors
104 Parachute stowage
105 Parachute release link
106 Airbrake hydraulic jack
107 Port perforated airbrake
108 Trailing-edge wingroot fillet
109 Port plain flap construction
110 Port spoiler
111 Drop tank stabilising fins
112 Aileron fixed tab
113 Port aileron construction
114 Aileron control rods
115 Rear spar
116 Port wing integral fuel tank
117 Wingtip fairing
118 Port navigation light
119 Retractable landing lamp
120 5-in (12.7-cm) HVAR ground attack rockets
121 Wing stringers
122 Wing rib construction
123 Port outer wing pylon
124 Undercarriage leg torque scissors
125 Mainwheel doors
126 Port mainwheel
127 Main undercarriage leg strut
128 Undercarriage leg pivot fixing
129 Hydraulic retraction jack
130 Main undercarriage mounting beam
131 Port automatic leading-edge slat (closed position)
132 Mainwheel well
133 Inner wheel door
134 Flying boom in-flight refuelling adaptor (open)
135 Wing tank fuel filler cap
136 Ventral engine access doors
137 Front spar
138 Pylon fixing
139 Fixed leading-edge construction
140 Inboard pylon
141 375-US gal (1420-litre) ferry tank
142 191.5-US gal (725-litre) drop tank
143 Mk 84 1,000-lb (454-kg) HE bomb
144 2,000-lb (907-kg) free fall nuclear weapon

SEPECAT Jaguar

Jaguar

Cutaway key
1 Nose profile (Maritime Strike Variant)
2 Thomson-CSF Agave dual-role (air-air, air-ground) radar
3 Ferranti Type 105 Laser Ranger
4 Pitot tube
5 'Wedge-profile' optical sighting windows
6 Ferranti Laser Ranging and Marked Target Seeker
7 Total pressure probe (both sides)
8 Electronics cooling air duct
9 Air-data computer
10 Radio altimeter
11 Power amplifier
12 Avionics access doors
13 Waveform generator
14 Cooling air intake
15 Marconi Avionics nav/attack system equipment
16 Landing/taxiing lamps
17 Nosewheel leg door
18 Towing lug
19 Nosewheel forks
20 Nosewheel
21 Steering jacks
22 Nose undercarriage leg strut
23 Artificial feel control units
24 Rudder pedals
25 Instrument panel shroud
26 Retractable, in-flight refuelling probe
27 Windscreen panels
28 Smiths electronics head-up-display
29 Instrument panel
30 Smiths FS68 head-down navigational display
31 Control column
32 Engine throttles
33 Pilot's side console panel
34 Martin-Baker Mk 9 'zero-zero' ejection seat
35 Seat and parachute combined safety harness
36 Honeycomb cockpit side panel
37 Plexiglass cockpit canopy cover (upward-opening)
38 Ejection seat headrest
39 Canopy struts
40 Cockpit pressurisation valve
41 Rear pressure bulkhead
42 Gun muzzle blast trough
43 Battery and electrical equipment bay
44 Port engine intake
45 Gun gas vents
46 Spring-loaded secondary air intake doors
47 Boundary layer bleed duct
48 Forward fuselage fuel tank (total system capacity 924 Imp gal; 1200 litres)
49 Air conditioning unit
50 Secondary heat exchanger
51 Starboard engine air intake
52 VHF homing aerials
53 Heat exchanger intake/exhaust duct
54 Cable and hydraulic pipe ducting
55 Intake/fuselage attachment joint
56 Duct frames
57 Integrally-stiffened machined fuselage frames
58 Ammunition tank
59 30-mm ADEN cannon
60 Ground power supply socket
61 Mainwheel stowed position
62 Main undercarriage hydraulic lock strut
63 Leading-edge slat drive motors and gearboxes
64 Fuel system piping
65 Wing panel centreline joint
66 Anti-collision light
67 IFF aerial
68 Wing/fuselage forward attachment joint
69 Starboard wing integral fuel tank
70 Fuel piping provision for pylon-mounted tank
71 Overwing missile pylon
72 Missile launch rail
73 Matra 550 Magic air-to-air missile
74 Starboard leading-edge slat
75 Slat guide rails
76 Starboard navigation light
77 Tacan aerial
78 Flap guide rails and underwing fairings
79 Outboard double-slotted flap
80 Starboard spoilers
81 Inboard double slotted flap
82 Flap honeycomb construction
83 Flap drive shaft and screwjacks
84 Spoiler control links
85 Wing/fuselage aft attachment joint
86 Heat exchanger air scoop

French SEPECAT Jaguar two-seaters line up on the ramp awaiting another training hop. Although the Jaguar is being increasingly replaced by the Mirage 2000, three squadrons still operate this venerable aircraft.

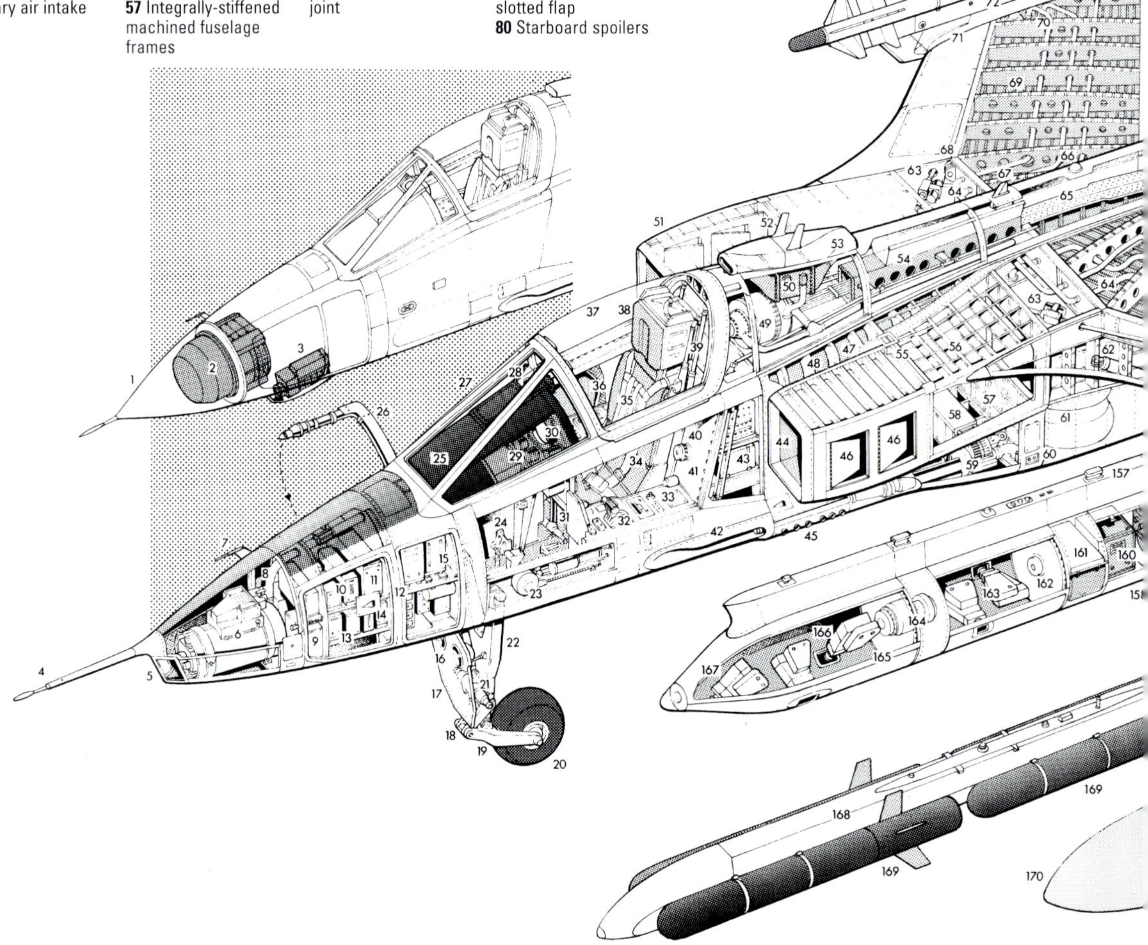

*The RAF used this picture as part of its recruitment campaign in the 1970s. Although such a close formation would be tantamount to suicide, this picture displays the power and agility of the **SEPECAT** Jaguar as the formation thunders over a Scottish loch.*

87 Control runs
88 Air conditioning supply ducting
89 Fuselage fuel tank access panels
90 Honeycomb intake duct construction
91 Engine intake frame
92 Hydraulic accumulator
93 Flap hydraulic motor and drive shaft
94 No. 2 system hydraulic reservoir
95 Primary heat exchanger
96 No. 1 system hydraulic reservoir
97 Heat exchanger exhaust ducts
98 Rear fuselage integral fuel tank
99 Inward/outward fuel vent valve
100 Dorsal spine fairing

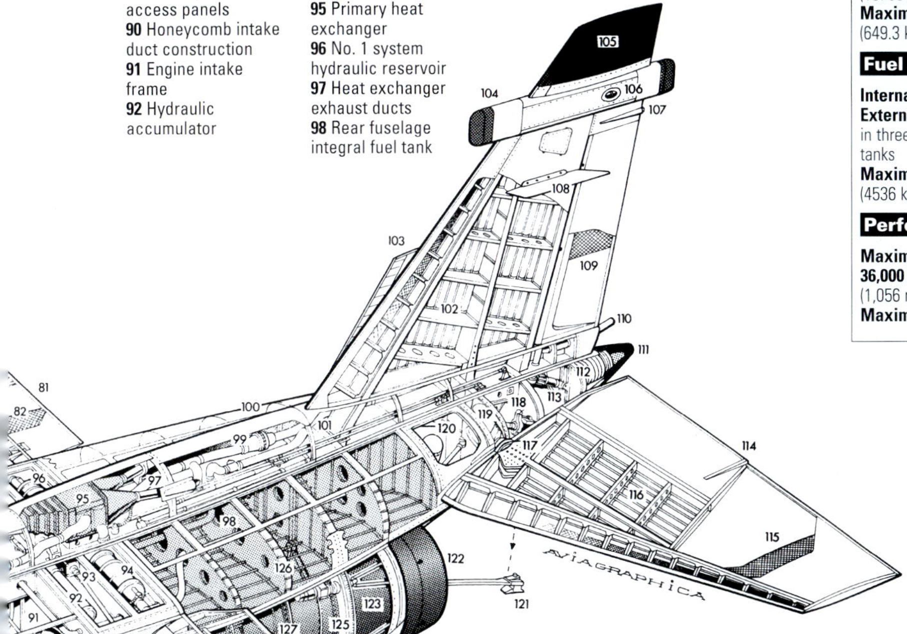

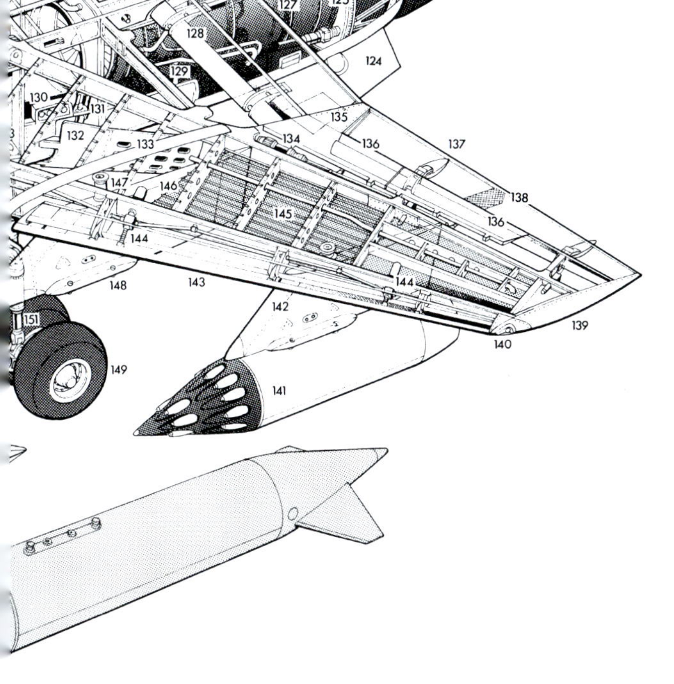

101 Fin spar attachment joint
102 Tailfin construction
103 Starboard tailplane
104 Fin tip ECM fairing
105 VHF/UHF antenna fairing
106 Recognition light
107 Tail navigation light
108 VOR aerial
109 Rudder honeycomb construction
110 Fuel jettison pipe
111 Tailcone
112 Brake parachute housing
113 Rudder hydraulic jack
114 Tailplane trailing-edge discontinuity
115 Honeycomb panel construction
116 Tailplane rib construction
117 Tailplane spar pivot joint
118 Differential all-moving tailplane hydraulic jack
119 Tailplane mounting frames
120 Fire extinguisher bottle
121 Arrester hook (extended)
122 Variable-area shrouded exhaust nozzle
123 Afterburner duct
124 Port ventral fin
125 Firewall
126 Engine rear suspension joint
127 Rolls-Royce/Turboméca Adour 804 (-26) turbofan
128 Port inboard double-slotted flap
129 Engine accessories
130 Hydraulic system ground servicing connectors
131 Airbrake hydraulic jack
132 Port airbrake (extended)
133 Wing fence (in place of missile pylon)
134 Spoiler hydraulic jack
135 Fixed portion of trailing edge
136 Port spoilers
137 Port outer double-slotted flap
138 Flap honeycomb construction
139 Wingtip fairing
140 Port navigation light
141 Matra Type 155 rocket launcher (18 SNEB rockets)
142 Outboard stores pylon
143 Port leading-edge slat
144 Slat screw jacks
145 Port wing integral fuel tank
146 Machined wing skin/stringer panel
147 Pylon fixing
148 Inboard stores pylon
149 Twin mainwheels
150 Pivoted axle beam
151 Shock absorber strut
152 Main undercarriage leg strut
153 Undercarriage pivot mounting
154 Fuselage sidewall construction
155 Main undercarriage leg door
156 Mainwheel doors
157 Fuselage centreline pylon
158 Reconnaissance pod
159 Infra-red linescan
160 Data converter
161 Air conditioning pack
162 Rear rotating camera drum (role interchangeable)
163 Twin Vinten F95 Mk 10 high oblique cameras
164 Drum rotating electric motor and gearbox
165 Forward-rotating camera drum
166 Twin Vinten F95 Mk 10 low oblique cameras
167 Forward-looking Vinten F95 Mk 7 reconnaissance camera
168 Multiple bomb carrier 430 lb (195 kg)
169 Four Matra Durandal penetration bombs
170 264 Imp-gal (1200-litre) auxiliary fuel tank

SPECIFICATION

Jaguar GR.Mk 1A (unless noted otherwise)

Dimensions

Length overall: 55 ft 2½ in (16.83 m)
T.Mk 2A: 57 ft 6¼ in (17.53 m)
Wingspan: 28 ft 6 in (8.69 m)
Wing area: 260.27 sq ft (24.18 m²)
Wing aspect ratio: 3.12
Wing chord at root: 11 ft 9 in (3.58 m)
Wing chord at tip: 3 ft 8½ in (1.13 m)
Height overall: 16 ft ½ in (4.89 m)
Tailplan span: 14 ft 10¼ in (4.53 m)
Wheel track: 7 ft 11 in (2.41 m)
T.Mk 2A: 7 ft 10½ in (2.40 m)
Wheelbase: 18 ft 8 in (5.69 m)
T.Mk 2A: 18 ft 7¼ in (5.67 m)

Powerplant

Two Rolls Royce/Turboméca Adour Mk 104 turbofans, each rated at 5,270 lb st (23.4 kN) dry and 7,900 lb st (35.1 kN) with afterburning

Weights

Typical empty: 15,342 lb (7000 kg)
Normal take-off: 24,149 lb (10964 kg)
Maximum take-off: 34,612 lb (15700 kg)
Maximum wing loading: 133 lb/sq ft (649.3 kg/m²)

Fuel and load

Internal fuel: 7,357 lb (3337 kg)
External fuel: up to 6,270 lb (2844 kg) in three 264-Imp gal (1200-litre) drop tanks
Maximum ordnance: 10,000 lb (4536 kg)

Performance

Maximum level speed 'clean' at 36,000 ft (10975 m): Mach 1.6 or 917 kt (1,056 mph; 1699 km/h)
Maximum level speed 'clean' at sea level: Mach 1.1 or 729 kt (840 mph; 1350 km/h)
Landing speed: 115 kt (132 mph; 213 km/h)
Time to 30,000 ft (9145 m): 1 minute 30 seconds
Service ceiling: 45,930 ft (14000 m)
Take-off run (clean): 1,854 ft (565 m)
Take-off run (four 1,000-lb/454-kg bombs): 4,101 ft (1250 m)
Landing distance from 50 ft (15 m): 2,575 ft (785 m)

Range

Ferry range with drop tanks: 1,902 nm (2,190 miles; 3524 km)
Combat radius on hi-lo-hi attack mission: 460 nm (530 miles; 852 km)
Combat radius on lo-lo-lo attack mission: 290 nm (334 miles; 537 km)

g-limits

+8.6 at typical weight or +12 ultimate

Armament

Fixed: Two 30-mm ADEN cannon in lower fuselage aft of cockpit with 150 rounds each
Weapon stations: One stores attachment on fuselage centreline and two under each wing. Centreline and inboard wing points can each carry up to 2,500 lb (1134 kg) of weapons, and the outboard underwing points up to 1,250 lb (567 kg)
Weapons carried: 1,000-lb (454-kg) bombs, various combinations of free-fall and retarded bombs, BL755 or CBU-87 cluster bombs, CPU-123/B Paveway II LGBs and CRV-7 rockets carried in 19-round LAU 5003 launchers. AIM-9L Sidewinder missiles can be carried for self-defence on overwing pylons. The Jaguar has relinquished its nuclear strike role, but could carry the AN 52 tactical nuclear weapon.

Sukhoi Su-24 'Fencer'

In addition to its standard nuclear strike (now no longer undertaken) and conventional attack roles, the Su-24M is a potent SEAD platform, highlighting the type's versatility. A number of anti-radiation missiles are available, including the Kh-58 (AS-11 'Kilter') seen here under test. Three pylons are available for the carriage of heavy missiles and pods, with another six available in total.

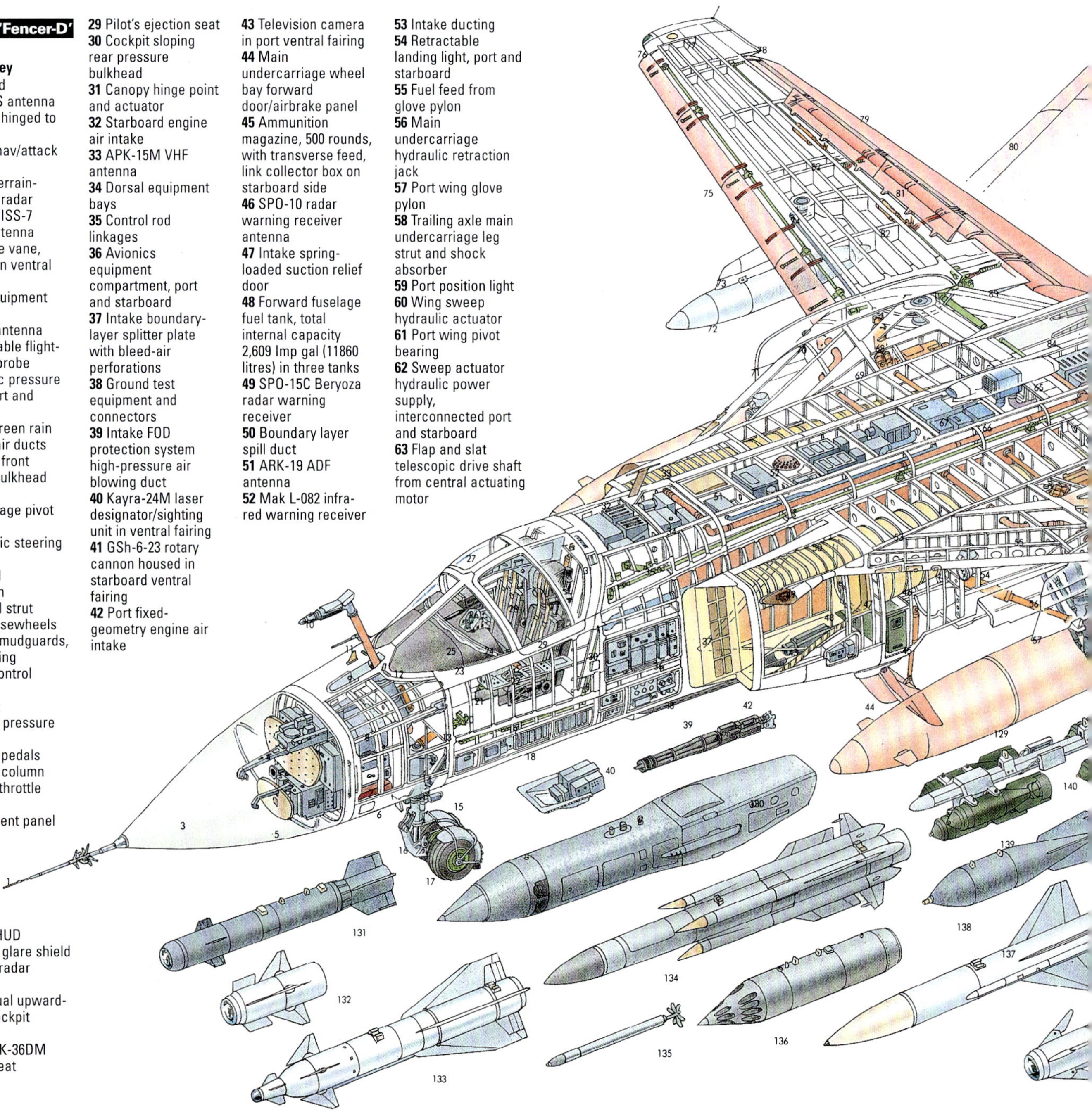

Su-24M 'Fencer-D'

Cutaway key
1 Pitot head
2 RSBN ILS antenna
3 Radome, hinged to port
4 Orion-A nav/attack radar
5 'Relyef' terrain-avoidance radar
6 Ventral DISS-7 Doppler antenna
7 Incidence vane, yaw vane in ventral position
8 Radar equipment modules
9 SRZ IFF antenna
10 Retractable flight-refuelling probe
11 Dynamic pressure probes, port and starboard
12 Windscreen rain dispersal air ducts
13 Cockpit front pressure bulkhead
14 Nose undercarriage pivot mounting
15 Hydraulic steering jacks
16 Levered suspension nosewheel strut
17 Twin nosewheels with fixed mudguards, aft-retracting
18 Flight control system equipment
19 Cockpit pressure floor
20 Rudder pedals
21 Control column
22 Engine throttle levers
23 Instrument panel shroud
24 Pilot's HUD
25 Folding glare shield
26 WSO's radar display
27 Individual upward-hinging cockpit canopies
28 WSO's K-36DM ejection seat
29 Pilot's ejection seat
30 Cockpit sloping rear pressure bulkhead
31 Canopy hinge point and actuator
32 Starboard engine air intake
33 APK-15M VHF antenna
34 Dorsal equipment bays
35 Control rod linkages
36 Avionics equipment compartment, port and starboard
37 Intake boundary-layer splitter plate with bleed-air perforations
38 Ground test equipment and connectors
39 Intake FOD protection system high-pressure air blowing duct
40 Kayra-24M laser designator/sighting unit in ventral fairing
41 GSh-6-23 rotary cannon housed in starboard ventral fairing
42 Port fixed-geometry engine air intake
43 Television camera in port ventral fairing
44 Main undercarriage wheel bay forward door/airbrake panel
45 Ammunition magazine, 500 rounds, with transverse feed, link collector box on starboard side
46 SPO-10 radar warning receiver antenna
47 Intake spring-loaded suction relief door
48 Forward fuselage fuel tank, total internal capacity 2,609 Imp gal (11860 litres) in three tanks
49 SPO-15C Beryoza radar warning receiver
50 Boundary layer spill duct
51 ARK-19 ADF antenna
52 Mak L-082 infra-red warning receiver
53 Intake ducting
54 Retractable landing light, port and starboard
55 Fuel feed from glove pylon
56 Main undercarriage hydraulic retraction jack
57 Port wing glove pylon
58 Trailing axle main undercarriage leg strut and shock absorber
59 Port position light
60 Wing sweep hydraulic actuator
61 Port wing pivot bearing
62 Sweep actuator hydraulic power supply, interconnected port and starboard
63 Flap and slat telescopic drive shaft from central actuating motor

Inflight refuelling is an important secondary task for the Su-24, for which it employs the UPAZ-1A Sakhalin pod (as also used by the Il-78 'Midas') on the centreline pylon. PTB-3000 tanks are normally carried to increase offloadable fuel. Tanking is usually performed to extend the range of other Su-24s, or to allow them to take off with a small fuel load but a heavy weaponload. The 'Fencer' has also tanked many other aircraft types. Here, the receiver is an Su-30 from the 'Test Pilots' display team, demonstrating the system at an air show.

SPECIFICATION

Su-24M 'Fencer-D'

Dimensions

Length: 74 ft 1½ in (22.59 m)
Height: 20 ft 4 in (6.19 m)
Maximum wingspan: 57 ft 10½ in (17.64 m)
Minimum wingspan: 34 ft (10.37 m)
Maximum wing area: 594.00 sq ft (55.166 m²)
Minimum wing area: 549.00 sq ft (51.024 m²)
Wheel base: 27 ft 11 in (8.51 m)
Wheel track: 10 ft 10 in (3.31 m)

Powerplant

Two Lyul'ka AL-21F-3 turbojets rated at 17,230 lb st (76.67) in dry thrust and 24,800 lb st (110.29 kN) with full afterburner activated

Weights

Empty: 49,162 lb (22300 kg)
Normal take-off: 79,365 lb (36000 kg)
Maximum take-off: 87,522 lb (39700 kg)
Maximum landing: 54,012 lb (24500 kg)

Fuel and load

Maximum internal fuel load: 21,715 lb (9850 kg)
Maximum external fuel load: 14,528 lb (6590 kg)
Maximum external weaponload: 19,841 lb (9000 kg)

Performance

Maximum speed at sea level: 870 mph (1400 km/h)
Maximum speed at sea level with external tanks: 823 mph (1325 km/h)
Maximum speed at 13000 m (theoretical): 1,317 mph (2120 km/h)
Minimum speed: 193 mph (310 km/h)
Service ceiling: 55,775 ft (17000 m)
Take-off run: 4,265 ft (1300 m)

Range

Maximum range: 1,770 miles (2850 km)
Operating radius at sea level: 348 miles (560 km)
Operating radius at sea level with external tanks: 777 miles (1250 km)
Landing run with parabrake: 3,120 ft (950 m)
Landing run without parabrake: 4,265 ft (1300 m)

Armament

'Fencers' can carry an immense array of weaponry. Conventional attacks can be carried out with air-to-surface missiles, unguided rockets, unguided bombs, cluster bombs and gun pods. For SAM hunting Kh-31T and Kh-58U missiles are used. Tactical nuclear bombs (now officially removed from the Russian inventory) carried include the 6U-57, 8U-49, 8U-63, RN-28, 244N and the 10-KT RN-24. Air defence is made possible by two R-60 IR missiles and a GsH-6-23 cannon is carried internally

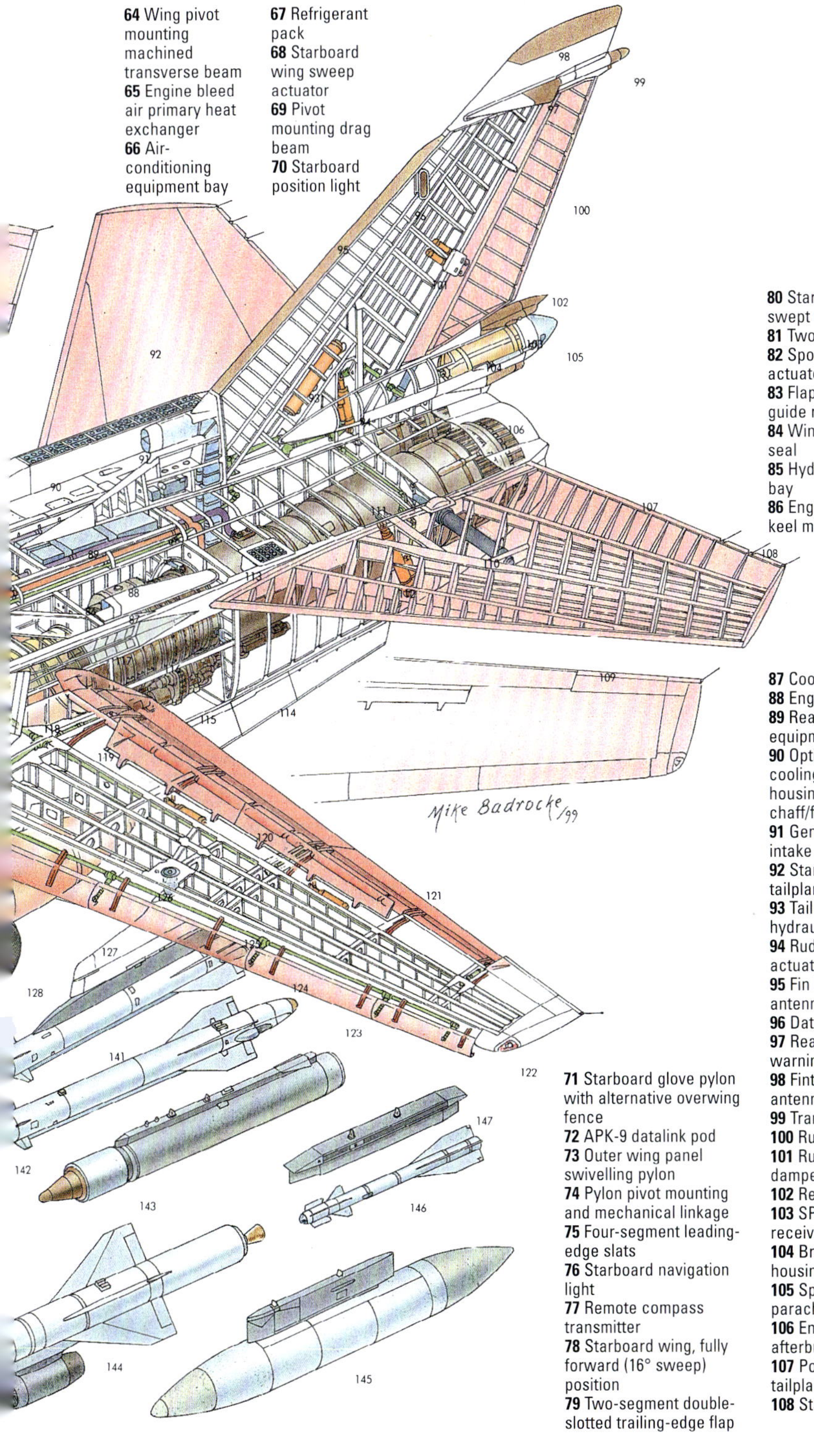

64 Wing pivot mounting machined transverse beam
65 Engine bleed air primary heat exchanger
66 Air-conditioning equipment bay
67 Refrigerant pack
68 Starboard wing sweep actuator
69 Pivot mounting drag beam
70 Starboard position light
71 Starboard glove pylon with alternative overwing fence
72 APK-9 datalink pod
73 Outer wing panel swivelling pylon
74 Pylon pivot mounting and mechanical linkage
75 Four-segment leading-edge slats
76 Starboard navigation light
77 Remote compass transmitter
78 Starboard wing, fully forward (16° sweep) position
79 Two-segment double-slotted trailing-edge flap
80 Starboard wing, fully swept (69°) position
81 Two-segment spoilers
82 Spoiler hydraulic actuators
83 Flap screw-jack and guide rails
84 Wingroot pneumatic seal
85 Hydraulic equipment bay
86 Engine bay central keel member/firewall
87 Cooling air intake
88 Engine oil tank
89 Rear fuselage dorsal equipment bay
90 Optional enlarged cooling intake fairing housing APP-50A chaff/flare launchers
91 Generator cooling air intake
92 Starboard all-moving tailplane
93 Tail control system hydraulic accumulator
94 Rudder hydraulic actuator
95 Fin leading-edge HF antenna
96 Datalink antenna
97 Rear SPO-15C radar warning antenna
98 Fintip UKB P-862 UHF antenna
99 Transponder antenna
100 Rudder
101 Rudder hydraulic dampers
102 Rear ILS antenna
103 SPO-10 radar warning receiver antenna
104 Brake parachute housing
105 Split conical fairing parachute doors
106 Engine variable-area afterburner nozzle
107 Port all-moving tailplane
108 Static dischargers
109 Port wing, fully-swept position
110 Tailplane pivot mounting
111 Afterburner duct
112 Tailplane hydraulic actuator
113 APP-50 12-round flare launcher
114 Ventral fin, port and starboard
115 Engine accessory equipment gearbox
116 Saturn (Lyul'ka) AL-21F-3 afterburning engine
117 Port wingroot pneumatic seal
118 Rear fuselage integral fuel tank
119 Inboard flap guide rail
120 Port two-segment spoiler panels
121 Two-segment double-slotted trailing-edge flap
122 Port navigation light
123 Port four-segment leading-edge slat
124 Slat guide rails
125 Slat drive shaft and screw jack actuators
126 Outer wing pylon pivot mounting
127 Port outer wing swivelling pylon
128 Twin mainwheels with pneumatic brakes, forward retracting
129 PTB-3000 external fuel tank (3000-litre/ 660-Imp gal) capacity
130 UPAZ-1A Sakhalin 'buddy' refuelling pack
131 KAB-500KR (500-kg/ 1,102-lb) television-guided bomb
132 Kh-29T television-guided missile nose section
133 Kh-29L (AS-14 'Kedge') laser-guided air-to-surface missile
134 Kh-31P (AS-17 'Krypton') radar-guided air-to-surface anti-shipping missile
135 S-8, 80-mm (3.15-in) FFAR
136 B-8M 20-round rocket pack
137 Kh-58 (AS-11 'Kilter') air-to-surface missile
138 FAB-500M-62 500-kg (1,102-lb) HE bomb
139 FAB-100 100-kg (220-lb) HE bomb
140 Multiple ejector rack
141 Kh-25ML (AS-10 'Karen') laser-guided air-to-surface missile
142 Kh-25MP radar-guided variant
143 S-25-OF, 250-mm (9.84-in) unguided rocket in launcher tube
144 Kh-59M (AS-18 'Kazoo') television-guided air-to-surface missile
145 'Fantasmagoria' passive radar pod
146 R-60 (AA-8 'Aphid') air-to-air 'self-defence' missile
147 R-60 twin missile launch rail, carried on outer wing pylon

Sukhoi Su-25 'Frogfoot'

With its wingtip airbrakes deployed, this Su-25TM from the Akhtubinsk Scientific and Technical Institute (NII) Flying Centre represents the latest combat variant of the 'Frogfoot' series. Based on the airframe of the two-seat Su-25UB, the Su-25TM introduces a large dorsal hump containing extra fuel and avionics, and a TV sensor in the newly-widened nose.

Su-25 'Frogfoot-A'

Cutaway key

1 Instrument boom
2 Fire-control system transducers
3 Pitot head
4 Glazed nose compartment
5 Laser ranger and marked target seeker
6 Strike camera
7 Localiser aerial port and starboard
8 'Odd-Rods' IFF antennas
9 Nose avionics equipment bays
10 Ventral Doppler aerial (starboard side)
11 30-mm six-barrelled cannon (port side)
12 Gun gas venting ducts
13 Cockpit floor level
14 Rudder pedals
15 Front pressure bulkhead
16 Control column
17 Instrument panel shroud
18 Armoured windscreen panels
19 Pilot's head-up display
20 Canopy open position
21 Rearview mirrors
22 Ejection seat headrest
23 Pilot's ejector seat
24 Canopy latch
25 Seat pan firing handles
26 Engine throttle levers
27 Side console panel
28 Nose undercarriage pivot fixing
29 Levered suspension nosewheel forks
30 Steerable nosewheel (forward-retracting)
31 Mudguard
32 Nosewheel leg door
33 Retractable boarding ladder
34 Fold-out step
35 Armoured cockpit enclosure
36 Rear pressure bulkhead
37 Ejection seat blast screen (retracted)
38 Grab handles
39 Forward fuselage equipment compartment
40 Port air intake
41 Ground power (intercom and telemetry sockets)
42 Intake ducting
43 Mainwheel retracted position
44 Forward fuselage fuel tanks
45 Wing spar centre section carry-through
46 Wing panel root attachment bolted joint
47 Centre-section fuel tank
48 VHF aerial
49 Starboard wingroot attachment joint
50 Inboard leading-edge slat segments (down position)
51 Starboard wing stores pylons
52 Leading-edge dogtooth
53 Starboard missile pylons
54 AA-8 'Aphid' self-defence air-to-air missile
55 Outboard leading-edge slat segments (down position)
56 Retractable landing/taxiing lamp
57 Dielectric ECM aerial fairing
58 Starboard navigation light
59 Wingtip pod fairing
60 Split trailing-edge airbrakes (open)
61 Static dischargers
62 Starboard aileron
63 Aileron tabs
64 Aileron hydraulic actuator
65 Starboard double-slotted tracked flaps (down position)
66 Flap guide rails
67 Flap jacks
68 Engine bay venting air intake
69 Starboard engine installation
70 Dorsal access panels (controls and systems ducting)
71 Rear fuselage fuel tanks
72 Fuel system venting air intake
73 Environmental system ram air intake
74 Tailfin
75 Starboard trimming tailplane
76 Starboard elevator
77 Fintip UHF aerial fairing
78 Tail navigation and position lights
79 Upper rudder segment
80 Rudder tabs
81 Lower rudder segment
82 Rudder hydraulic actuator
83 Radar warning antennas
84 Brake parachute housing
85 Ventral chaff/flare dispenser
86 Elevator tab
87 Port elevator
88 Static discharger
89 Port trimming tailplane

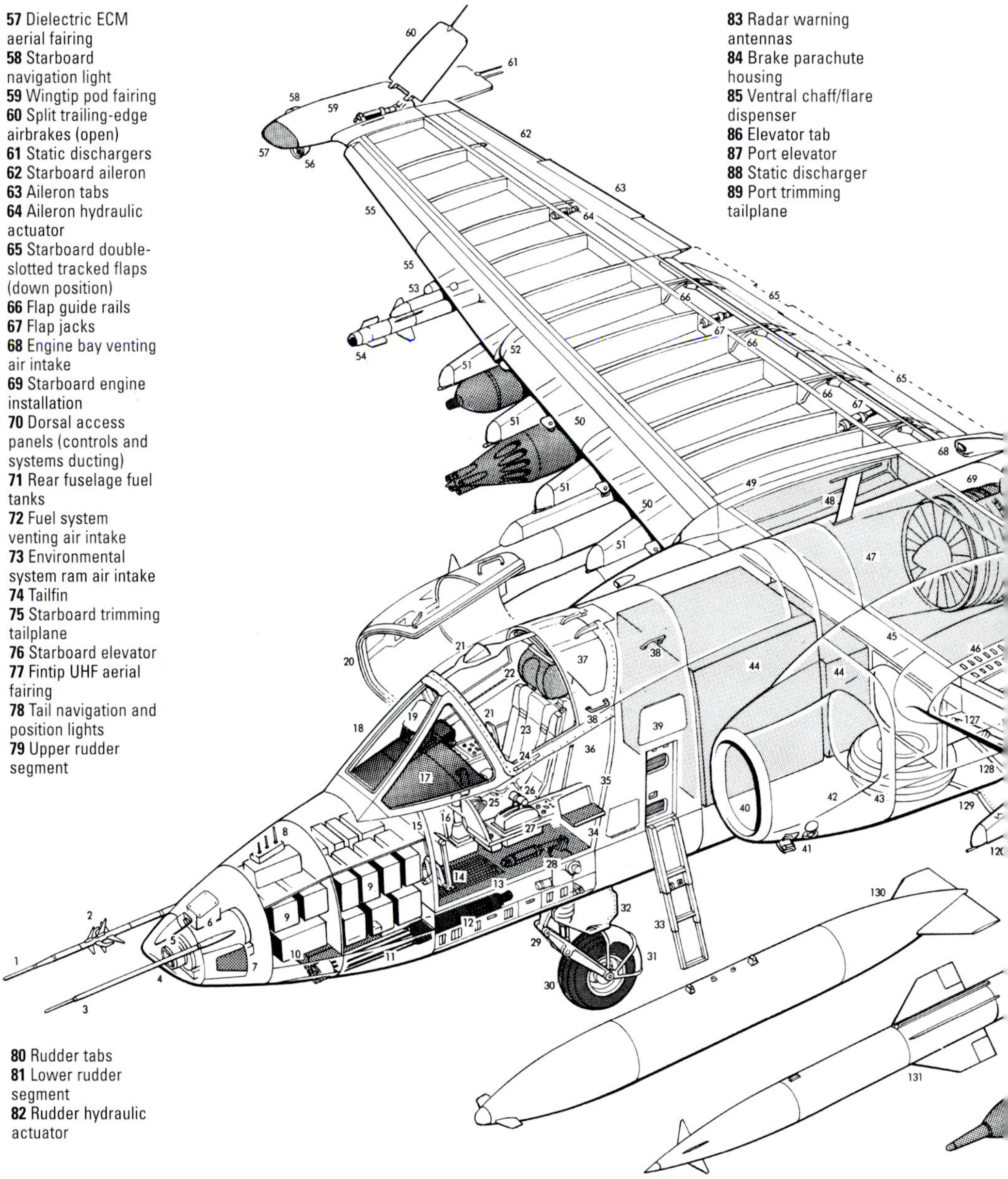

Developed partly as a result of VVS 'Frogfoot' experience in Chechnya, when bad weather grounded the Su-25 fleet, the Su-25T is an all-weather capable, ground-attack aircraft. Incorporated in the redesigned tail fairing is accommodation for enhanced countermeasures while the UV-26 dispenser is capable of carrying a mixture of PPI-26 infra-red decoys and PPR-26 chaff cartridges. 'Blue 09' carries a weaponload of R-60 (AA-8 'Aphid') AAMs outboard, with a B-13L 122-mm five-round rocket pod inboard.

90 Tailplane pivot fixing
91 Tailplane incidence control
92 Ventral 'Odd Rods' IFF antennas
93 Rear fuselage communications and ECM equipment
94 Ventral towel rail HF aerial
95 Environmental control system equipment
96 Engine exhaust nozzle
97 Soyuz/Gavrilov R-95Sh non-afterburning turbojet
98 Engine bay venting air intake
99 Accessory equipment gearbox
100 Flap guide rails
101 Flap control jacks
102 Two-segment, double-slotted flaps
103 Aileron tabs
104 Aileron hydraulic actuator
105 Port aileron
106 Split trailing-edge airbrake (open)
107 Static dischargers
108 Airbrake hydraulic jack
109 Port navigation light
110 Retractable landing/taxiing lamp
111 Dielectric ECM aerial fairing
112 Leading-edge slat segments
113 AA-8 'Aphid' self-defence air to-air missile
114 Missile launch rail
115 Outboard missile pylon
116 FAB-500 1,102-lb (500-kg) HE bomb
117 UV-16-57 rocket pack (16 x 57-mm rockets)
118 57-mm high velocity aircraft rocket (HVAR)
119 Leading-edge dog tooth
120 Port wing stores pylons
121 Wingrib construction (dry-bay, no fuel in wings)
122 Inboard leading-edge slat segments
123 Port mainwheel
124 Levered-suspension axle beam
125 Shock absorber strut
126 Main undercarriage leg pivot fixing
127 Hydraulic retraction jack
128 Leg rotating link
129 Mainwheel doors
130 132-Imp gal (600-litre) external fuel tank
131 AS-7 'Kerry' air-to-surface missile
132 BETA B-250 551-lb (250-kg) retarded bomb

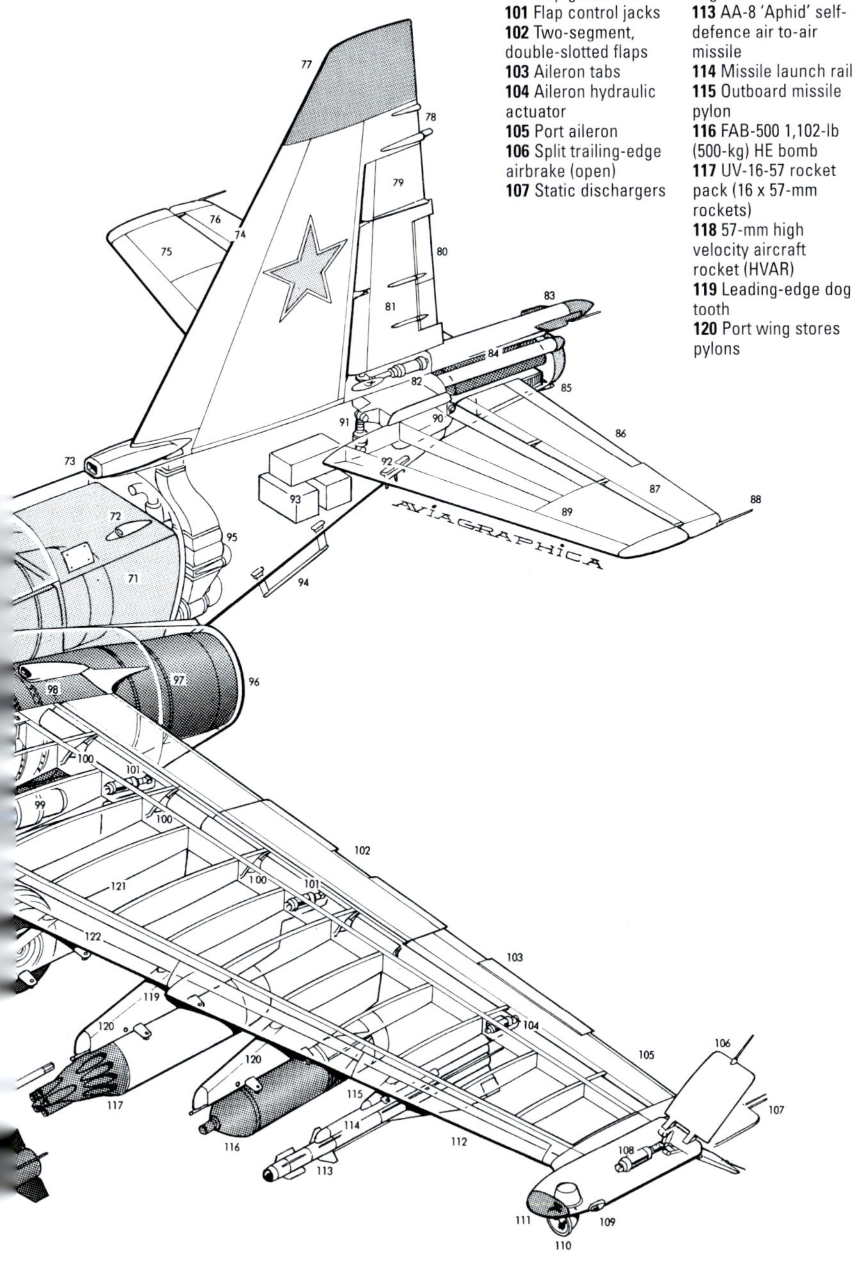

SPECIFICATION

Sukhoi Su-25 'Frogfoot-A'

Dimensions

Overall length including probe: 50 ft 11½ in (15.53 m)
Fuselage length excluding probe: 47 ft 10½ in (14.59 m)
Wingspan: 47 ft 1 in (14.36 m)
Wing area: 324 sq ft (30.1 m²)
Tailplane span: 15 ft 3 in (4.65 m)
Tailplane area: 69.7 sq ft (6.47 m²)
Overall height: 15 ft 9 in (4.8 m)
Wheel track: 8 ft 2 in (2.5 m)
Wheelbase: 11 ft 9 in (3.58 m)

Powerplant

Two Soyuz/Gavrilov R-95Sh turbojets each rated at 9,039-lb st (40.21-kN) thrust

Weights

Empty operating: 20,250 lb (9185 kg)
Normal take-off (early): 31,415 lb (14250 kg)
Normal take-off (late): 32,025 lb (14530 kg)
Maximum take-off (early): 38,250 lb (17350 kg)
Maximum take-off (late): 38,645 lb (17530 kg)
Normal landing: 23,810 lb (10800 kg)
Maximum landing: 29,320 lb (13300 kg)

Fuel load

Internal fuel: 6,615 lb (3000 kg)

Performance

Maximum level speed clean: 620 mph (1000 km/h)
Limiting Mach number (early): 0.71
Limiting Mach number (late): 0.82
Service ceiling: 22,950 ft (7000 m)
Take-off run: 1,640-2,953 ft (500-900 m)
Landing roll: 1,969-2,625 ft (600-800 m)
Take-off speed: 149-155 mph (240-270 km/h)
Landing speed: 140-162 mph (225-260 km/h)
***g* limits:** +6.5 to -3 at basic gross weight

Range

With underwing tanks (early): 1,150 miles (1850 km)
With underwing tanks (late): 1,212 miles (1950 km)
With 6,615 lb (3000 kg) of fuel: 310 miles (500 km)

Armament

One GSh-30-2 twin-barrelled 30-mm cannon with 250 rounds; 10 wing pylons to carry a normal weaponload of 2,954 lb (1340 kg), or a maximum weaponload of 9,568 lb (4340 kg)

As a result of the GSh-30-2 twin-barrelled 30-mm cannon mounted in the starboard side of the front fuselage, the Su-25 features an offset nose gear undercarriage. A mudguard is incorporated to reduce the risk of debris ingestion during operations from unprepared airstrips.

Bombers

Avro Vulcan B Mk.2

Between September 1963 and December 1970 the Vulcan B.Mk 2As of the Scampton Wing were equipped with the Avro Blue Steel missile. This used a Bristol Siddeley Stentor rocket motor (consisting of a high-thrust booster and less powerful sustainer) and had a Red Snow thermonuclear warhead. When launched from high altitude in standard mode the Blue Steel had a range of about 115 miles (185 km). In the terminal phase the weapon dived from around 70,000 ft (21336 m) at more than Mach 1.5. After modification for low-level launch, which employed both rocket motors together, range fell dramatically, restricting its use to peripheral targets.

Vulcan B.Mk 2

Cutaway key

1 Wingtip antennas
2 Starboard navigation light
3 Starboard wingtip construction
4 Outboard aileron
5 Inboard aileron
6 Rear spar
7 Outboard wing panel ribs
8 Front spar
9 Leading-edge ribs
10 Cranked leading edge
11 Corrugated leading-edge inner skin
12 Retractable landing and taxiing lamp
13 Fuel tank fire extinguisher bottles
14 Outer wing panel joint rib
15 Honeycomb skin panel
16 Outboard elevator
17 Inboard elevator
18 Elevator hydraulic jacks
19 No. 7 starboard fuel tank
20 No. 5 starboard fuel tank
21 Diagonal rib
22 Leading-edge de-icing air duct
23 Wing stringer construction
24 Parallel chord wing skin panels
25 No. 6 starboard fuel tank
26 No. 4 starboard fuel tank
27 No. 3 starboard fuel tank
28 Main undercarriage leg
29 Eight-wheel bogie
30 Mainwheel well door
31 Fuel tank fire extinguishers
32 Inboard leading-edge construction
33 De-icing air supply pipe
34 Fuel collectors and pumps
35 Main undercarriage wheel bay
36 Retracting mechanism
37 Airborne auxiliary power plant (AAPP)
38 Electrical equipment bay
39 Starboard engine bays
40 Rolls-Royce (Bristol Siddeley Olympus 301)
41 Air system piping
42 Engine bay dividing rib
43 Fire extinguishers
44 Jetpipes
45 Fixed trailing-edge construction
46 Jetpipe nozzles
47 Rear equipment bay
48 Oxygen bottles
49 Batteries
50 Rudder power control unit
51 Rear electronics bay
52 Electronic countermeasures system equipment
53 Cooling air intake
54 Tail warning radar scanner
55 Tail radome
56 Twin brake parachute housing
57 Brake parachute door
58 Rudder construction
59 Rudder balance weights and seals
60 Fin de-icing air outlet
61 Dielectric fin-tip fairing
62 Passive electronic countermeasures (ECM) antennas
63 Fin construction
64 Fin leading edge
65 Corrugated inner skin
66 Communications aerial
67 Fin de-icing air supply
68 Bomb bay rear bulkhead
69 Bomb bay roof arch construction
70 Flush air intake
71 Communications aerial
72 Port Olympus 301 engines
73 Engine bay top panel construction
74 Port jet pipe fairing
75 Electrical equipment bay
76 Chaff dispenser
77 'Green Satin' navigational radar bay
78 Elevator balance weights and seals
79 Elevator hydraulic jacks
80 Inboard elevator
81 Outboard elevator
82 Inboard aileron
83 Aileron balance weights
84 Control rods
85 Aileron power control jacks
86 Jack fairings
87 Outboard aileron
88 Port wingtip antennas
89 Retractable landing and taxiing lamp
90 Cranked leading edge
91 Fuel tank fire extinguishers
92 Cambered leading-edge profile
93 No. 7 port fuel tank
94 No. 5 port fuel tank
95 Leading-edge de-icing air duct
96 No. 6 port fuel tank
97 No. 4 port fuel tank
98 No. 3 port fuel tank
99 Port main undercarriage bay
100 Wing stringer construction
101 Port airbrakes
102 Airbrake drive mechanism
103 Intake ducts
104 Front wing spar attachment joints
105 Centre-section front spar frame
106 Suppressed aerial
107 Anti-collision light
108 Bomb bay longerons
109 Forward limit of bomb bay
110 Starboard airbrake housings
111 Boundary layer bleed air duct
112 Starboard intake ducts
113 No. 2 fuselage fuel tanks
114 Communications aerials
115 Port engine intake
116 No. 1 fuselage fuel tanks
117 Fuselage frame and stringer construction

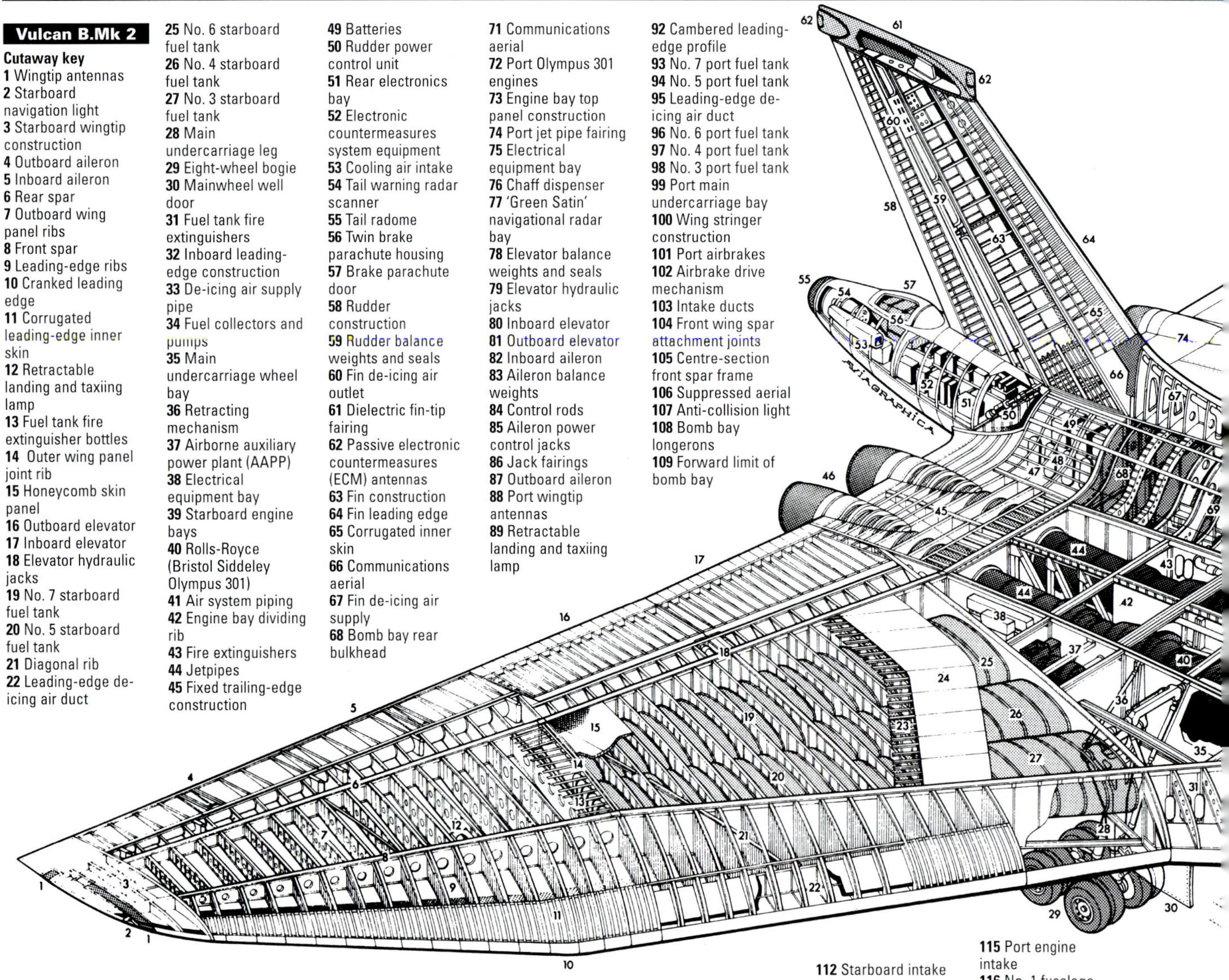

The 'schoolhouse' for the Vulcan fleet was No. 230 Operational Conversion Unit, which became the first service operator of both the B.Mk 1 and B.Mk 2 (illustrated). Initially based at Waddington, the OCU moved to Finningley in 1961, its aircraft acquiring the white rose of Yorkshire on the fin to signify their new base. In 1969 the OCU relocated to Scampton, from where it flew until August 1981. In addition to its Vulcans, between 1974 and 1977 the OCU also parented '1066 Flight', which was equipped with Hastings T.Mk 5 transports, modified with H_2S Mk 9 radar, to train Vulcan navigators.

SPECIFICATION

Vulcan B.Mk 2

Dimensions

Wingspan: 111 ft (33.83 m)
Length: 100 ft 1 in (30.50 m)
Height: 27 ft 2 in (8.29 m)
Wing area: 3,964 sq ft (368.3 m²)

Powerplant

Four Rolls-Royce (Bristol) B.O.16 Olympus Mk 201 turbojets, each rated at 17,000-lb (75.66-kN) thrust, or B.O.121 Olympus Mk 301 turbojets, each rated at 20,000-lb (89-kN) thrust

Weights

Maximum take-off: about 250,000 lb (113400 kg)

Performance

Maximum speed: 645 mph (1038 km/h)
Normal cruise: Mach 0.84
Service ceiling: 65,000 ft (19812 m)
Take-off run: 3,500 ft (1067 m) fully loaded with Mk 301 engines
Time to 40,000 ft (12192 m): 9 minutes
Range: 3,450 miles (5550 km) at low level; 4,600 miles (7400 km) at high altitude with bombload

Armament

Conventional – capability in all variants for 21 1,000-lb (454-kg) bombs carried internally in three clips of seven
B.Mk 1/1A nuclear – initially one 11,000-lb (4990-kg) Blue Danube strategic freefall weapon, followed by 9,000-lb (4082-kg) Violet Club. These were replaced by 6,000-lb (2722-kg) US-owned Mk 5 (Waddington Wing aircraft only 1958-62), 7,000-lb (3175-kg) Yellow Sun Mk 1 (1960-63) and by 7,250-lb (3289-kg) Yellow Sun Mk 2 (1962-66);. 2,000-lb (907-kg) Red Beard tactical nuclear bomb also available
B.Mk 2 nuclear – initially one Yellow Sun Mk 1/2, but from 1962-1970 aircraft so-modified (B.Mk 2A) could carry one Avro W.105 Blue Steel stand-off missile. From 1966 to 1982 multiple carriage of the 950-lb (431-kg) WE 177B strategic laydown weapon was the principal armament

118 Intake lip construction
119 Corrugated inner skin
120 Intake divider
121 Starboard intake
122 Boundary layer splitter plate
123 Nose section joint frame
124 Rear pressure bulkhead
125 Nosewheel doors
126 Nosewheel leg
127 Steering jack
128 Twin nosewheels
129 Radio and electronics equipment bay
130 Rearward facing crew members' stations: tactical navigator, radar operator and air electronics operator
131 Cabin side window
132 Chart table
133 Assisted exit seats
134 Jettisonable cockpit canopy
135 Pilot's ejection seat
136 Windscreen panels
142 Ladder
143 Pitot tube
144 Ventral bomb aiming blister fairing (not used on B.Mk 2 aircraft)
145 Destructor
146 Refuelling supply pipe
147 Cockpit pressure dome
148 Radar mounting
149 H_2S radar unit
150 Rotating radar scanner, 80-in (2.03-m) diameter

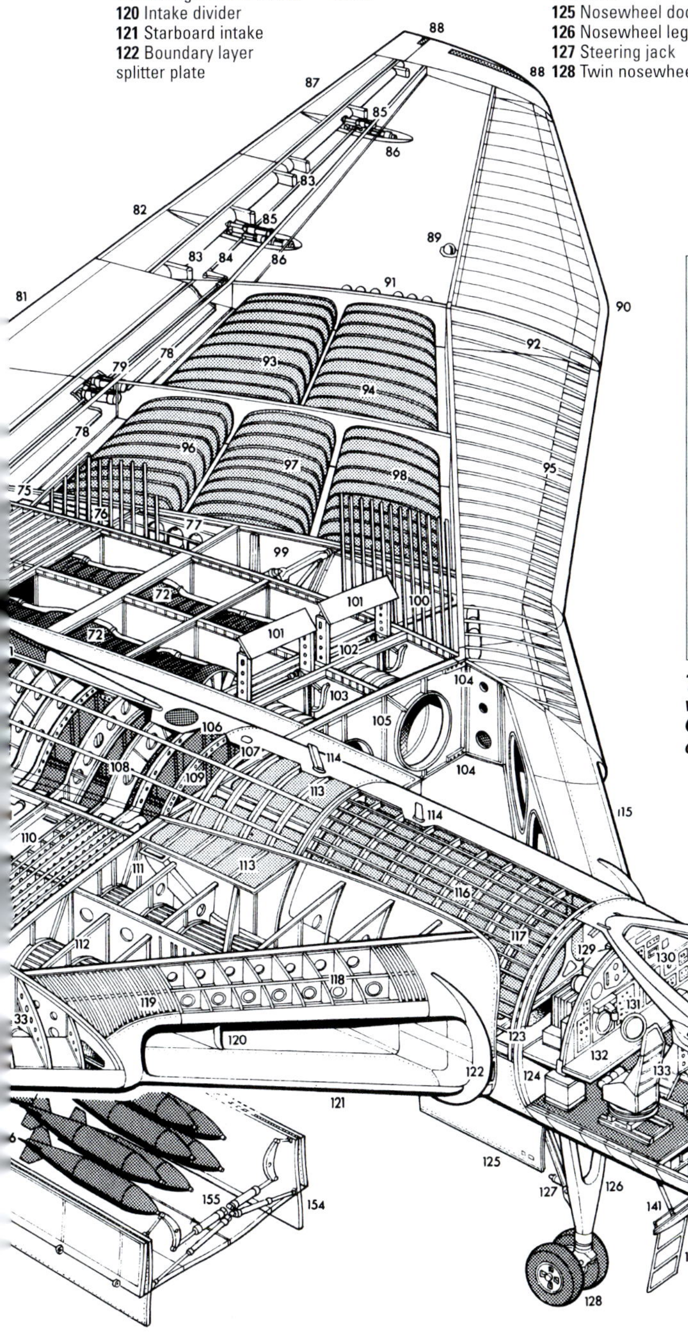

The Vulcan was used for several engine test programmes. Seen here fitted with an icing rig under the front fuselage, B.Mk 1 XA903 tests Concorde's Olympus 593 engine. The aircraft later flew with Tornado's RB199 engine, complete with representative intake and working cannon.

137 Instrument panel shroud
138 Windscreen wipers
139 Co-pilot's seat
140 Raised pilot's cockpit floor
141 Entry hatch
151 Radome
152 Terrain-following radar antenna
153 Flight refuelling probe
154 Bomb bay doors
155 Bomb door opening jacks
156 1,000-lb (454-kg) bombs, three groups of seven bombs each

Blackburn Buccaneer

Front-line service for the Buccaneer ended on 31 March
when Tornados took over the anti-ship role. Many 'B
veterans lamented the aircraft's passing, considering it
a better tool in this role than its replacement. Shortly be
withdrawal, a new camouflage scheme was devised fo
Buccaneer, comprising an overall light/medium- grey p
together with pink/light- blue national insignia and w
serials. It is widely believed that this scheme had more
with the ongoing 'fashion' for grey paint schemes than
any real operational v

Buccaneer S.Mk 2B

Cutaway drawing

1 Inflight-refuelling probe
2 Radar scanner
3 Multi-mode search and fire control radar
4 Weapon recorder
5 Radome (folded)
6 Radome hinge
7 Weapon-release computer
8 Windscreen rain dispersal duct
9 Windscreen wiper
10 Birdproof windscreen
11 Pilot's head-up display
12 Instrument panel shroud
13 Rudder pedals
14 Nosewheel leg hinge point
15 Landing and taxi lamp
16 Shock absorber strut
17 Nosewheel forks
18 Aft-retracting nosewheel
19 Avionics equipment
20 Engine throttles
21 Canopy side rail
22 Pilot's ejection seat
23 Seat firing handle
24 Aft-sliding canopy
25 Navigator's blast shield
26 Navigator's instrument display
27 Starboard engine air intake
28 Navigator's ejection seat
29 Cockpit floor structure
30 Head-up display symbol generator
31 Port engine air intake
32 Anti-icing air line
33 Air intake duct
34 Cockpit aft pressure bulkhead
35 Forward main fuselage fuel tank
36 Canopy motor
37 Canopy top rail
38 Rolls-Royce RB.168-1A Spey Mk 101 turbofan
39 Bleed air ducting
40 Detachable bottom cowling
41 Engine front mounting
42 Firewall frame
43 Engine aft mounting
44 Forward fuselage structure
45 Bleed air crossover duct
46 Canopy hand-winding shuttle
47 Detachable engine top cowling
48 Starboard slipper tank
49 Datalink acquisition pod
50 Datalink inboard pylon
51 Martel air-to-surface missile
52 Wing-fold hinge line
53 Leading-edge blowing air duct
54 UHF antenna
55 Dorsal spine structure
56 Anti-collision light
57 Wing-fold actuator
58 Wing-fold operating link
59 Starboard outer pylon
60 ARI 18218 aerial housing
61 Blown leading edge
62 Starboard navigation light
63 Formation light
64 Starboard blown aileron
65 Aileron actuator
66 Starboard wingtip (folded)
67 Aileron and flap blowing ducts
68 Starboard blown flap
69 Port wingtip (folded)
70 Centre fuselage fuel tank
71 Machined spar ring frames
72 Ring frame bolted attachment
73 Aft fuselage fuel tank
74 Electrical cable ducting in dorsal spine
75 Avionics equipment bay
76 Air data computer
77 HF notch aerial
78 Equipment bay cooling air intake
79 Fin spar attachment
80 Fin structure
81 Tailplane actuator
82 Tailplane operating rod
83 Tailplane blowing air duct
84 Bullet fairing
85 Forward passive warning system antenna
86 Blown tailplane leading edge
87 All-moving tailplane structure
88 Tailplane flap
89 Tailplane flap actuator
90 Hinge attachment point
91 Top fairing
92 Rear navigation light
93 Formation light
94 Aft passive warning system antenna
95 Port tailplane flap
96 Rudder structure
97 Rudder operating link
98 Rudder actuator
99 Airbrake jack
100 Drag-link hinge attachment
101 Airbrake operating slide
102 Split tailcone airbrake

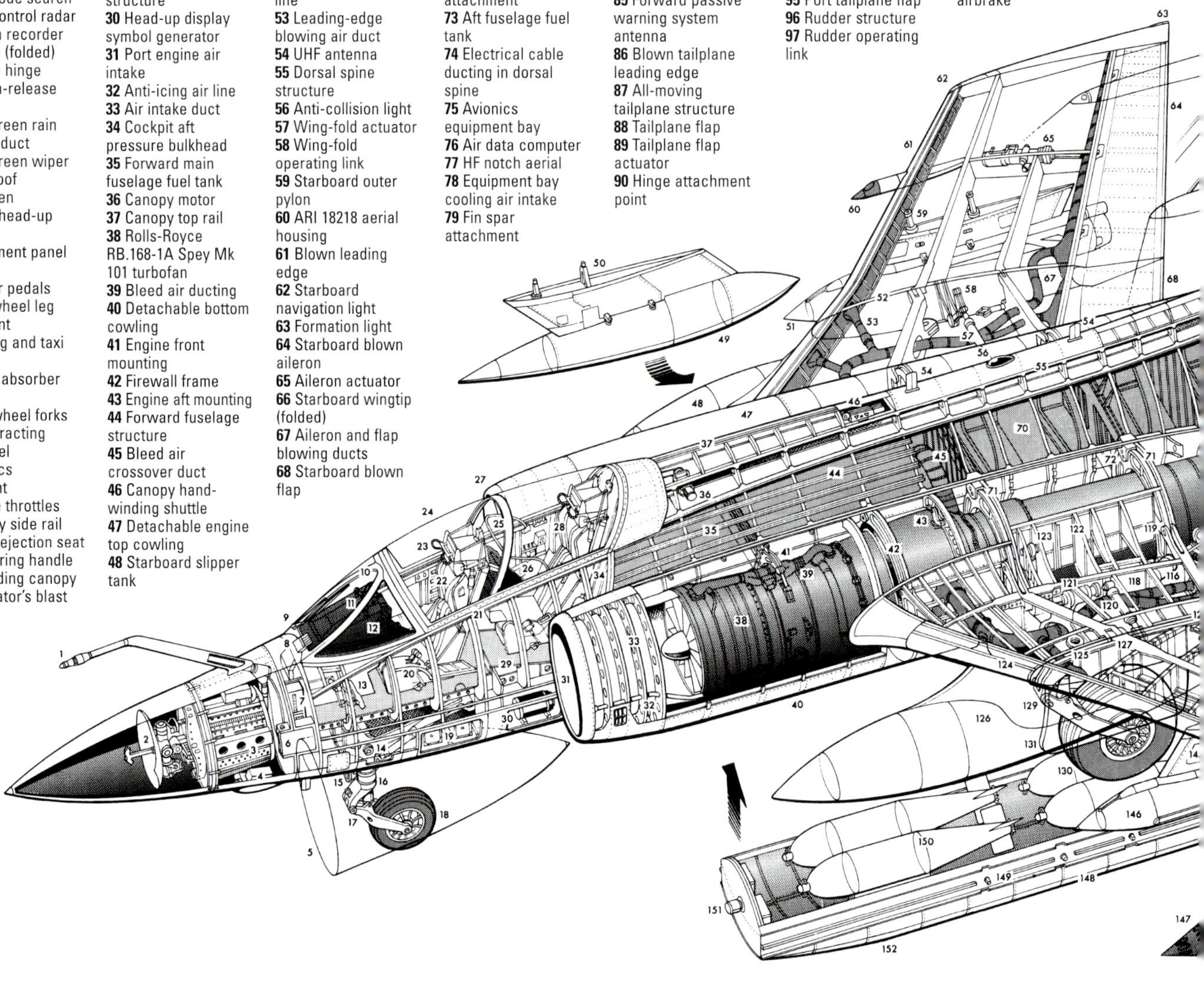

SPECIFICATION

Buccaneer S.Mk 2

Dimensions

Length: 63 ft 5 in (19.33 m)
Height: 16 ft 3 in (4.95 m)
Wingspan: 44 ft (13.41 m)
Wing area: 514.7 sq ft (47.82 m²)

Powerplant

Two Rolls-Royce RB.168-1A Spey Mk 101 turbofans each rated at 11,030-lb (49.08-kN) thrust plus (S.Mk 50 only) two Bristol-Siddeley/Rolls-Royce BS.605 rocket engines each rated at 8,000-lb (36-kN) thrust for 30 seconds

Weights

Typical landing: 35,000 lb (15876 kg)
Normal take-off: 46,000-56,000 lb (20866-25402 kg)
Maximum take-off: 62,000 lb (28132 lb)

Performance

Maximum speed at sea level: 691 mph (1112 km/h) or Mach 0.91
Attack speed at sea level: 621 mph (1000 km/h)
Maximum strike range with external fuel load and normal weapons load: 2,300 miles (3701 km)
Combat radius with full warload, hi-lo-hi: 600 miles (966 km)
Take-off distance at maximum take-off weight: 3,400 ft (1036 m)
Landing run: 3,150 ft (960 m)

Armament

(Buccaneer S.Mk 2B for maritime strike) four Martel ASMs (later Sea Eagle ASMs) or up to 16 1,000-lb (454-kg) bombs or (Buccaneer S.Mk 50) four AS30 ASMs or four 68-mm rocket pods, plus internal ordnance

One of No. 801 Squadron's Buccaneers is seen on the flight deck of HMS* Ark Royal *during the type's first carrier cruise. This was the work-up prior to the squadron embarking in HMS* Victorious. *No. 801 Squadron, formed from a nucleus of crews from No. 700Z Sqn, formally commissioned on the Buccaneer in July 1962. Clearly illustrated are the aircraft's folding wings and open speed brake which allowed it to utilise the smaller deck lifts of the British aircraft-carriers.

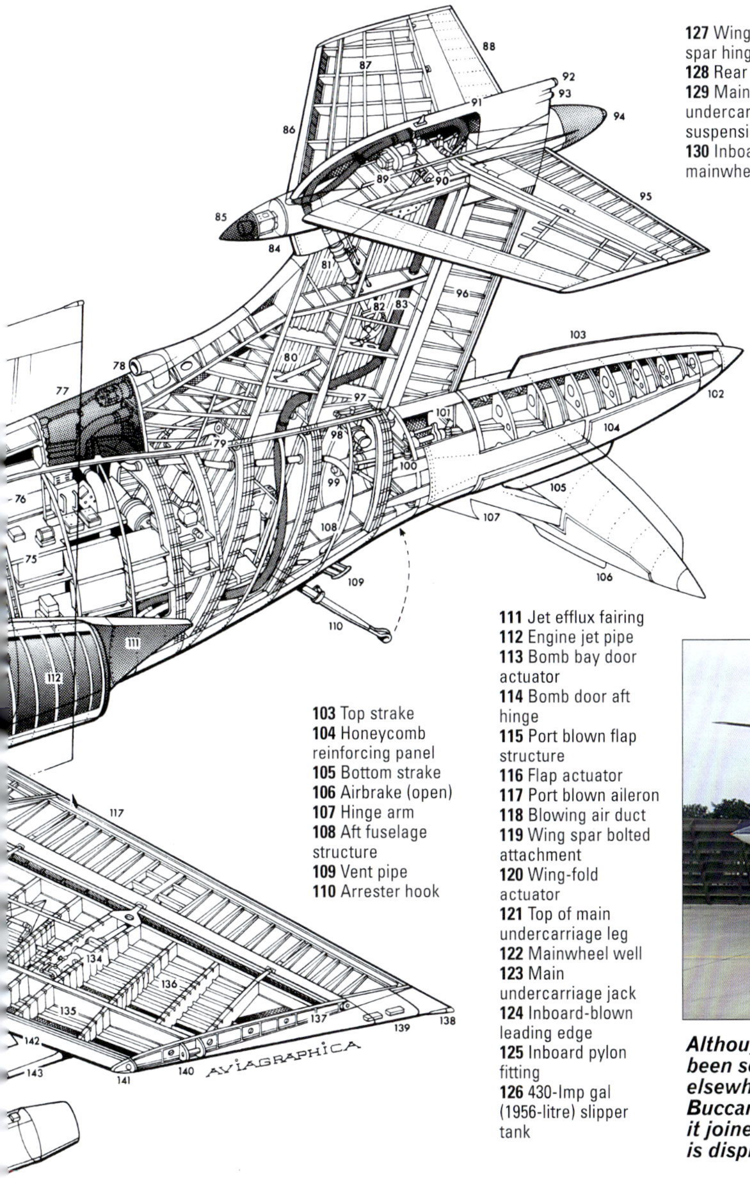

103 Top strake
104 Honeycomb reinforcing panel
105 Bottom strake
106 Airbrake (open)
107 Hinge arm
108 Aft fuselage structure
109 Vent pipe
110 Arrester hook
111 Jet efflux fairing
112 Engine jet pipe
113 Bomb bay door actuator
114 Bomb door aft hinge
115 Port blown flap structure
116 Flap actuator
117 Port blown aileron
118 Blowing air duct
119 Wing spar bolted attachment
120 Wing-fold actuator
121 Top of main undercarriage leg
122 Mainwheel well
123 Main undercarriage jack
124 Inboard-blown leading edge
125 Inboard pylon fitting
126 430-Imp gal (1956-litre) slipper tank
127 Wing-fold main spar hinge
128 Rear spar hinge
129 Main undercarriage levered suspension
130 Inboard retracting mainwheel
131 Mainwheel door
132 Outboard pylon fitting
133 Aileron operating rod
134 Port aileron actuator
135 Outer wing structure
136 Machined skin panels
137 Port wingtip
138 Formation light
139 Crash trip switches
140 Wing lifting lug
141 Port navigation light
142 Blown outboard leading edge
143 Pitot head
144 Port ARI 18218 aerial housing
145 Outboard pylon
146 Port Martel air-to-surface missile
147 36-tube rocket pod
148 Rotary bomb-bay door
149 Bomb door locks
150 1,000-lb (454-kg) bomb
151 Forward hinge point
152 425-Imp gal (1932-litre) bomb door auxiliary tank

Although most of the 193 British and 16 South African Buccaneers have now been scrapped, finding their way to yards at Elgin, Macclesfield and elsewhere, a few have been kept for display purposes. The first such Buccaneer (an S.Mk 1) went to the FAA Museum at RNAS Yeovilton, where it joined an NA.39. Having formerly flown with both Nos 801 and 809 NAS, it is displayed here in an early gloss finish of Extra Dark Sea Grey and white.

Boeing B-47 Stratojet

The 171st example of the B-47E to be manufactured b Lockheed under the Brickbat scheme, No. 233 features the standard Stratojet finish, bei predominantly natural metal with white anti-flas areas on the undersides. Rather unusually, no tail gu armament is carried by this aircra

B-47E-II Stratojet

Cutaway key

1 Inflight-refuelling receptacle, open
2 Bomb sight periscope aperture
3 Navigator's instrument panel
4 Bomb sight periscope
5 K-bombing system radar equipment
6 Nose compartment floor level
7 Ventral ejection hatch
8 Navigator/ bombardier's downward ejection seat
9 Nose compartment ditching hatch
10 Drift indicator
11 Radio and electronics equipment racks
12 Radar scanner
13 Ventral radome
14 K4A navigation and bombing radar equipment
15 Pressurised crew compartment inner skin
16 Internal walkway
17 Flight deck floor level
18 Rudder pedals
19 Control column
20 Instrument panel shroud
21 Windscreen wiper
22 Windscreen panels
23 Jettisonable cockpit canopy cover
24 Cockpit sunblinds
25 Starboard side console panels
26 Pilot's ejection seat
27 Co-pilot's instrument panel
28 Sextant aperture
29 Aerial mast
30 Co-pilot's ejection seat
31 Tail gunsight and firing controls
32 Oxygen bottles
33 Swivelling seat mounting
34 Pressurised section internal entry door
35 Forward auxiliary fuel tank
36 Maintenance access hatch
37 Crew entry hatch
38 Retractable boarding ladder
39 Cockpit air-conditioning plant
40 Forward main undercarriage hydraulic retraction jack
41 Nose compartment rear bulkhead
42 Forward mainwheel doors
43 Steerable twin wheel forward landing gear
44 Multi-plate disc brakes
45 Steering control unit
46 Main undercarriage leg pivot fixing
47 Bomb bay anti-buffet deflector door, open
48 Bomb door hydraulic jack
49 Hydraulic reservoir
50 Hydraulic equipment bay
51 Forward main fuel cell; total internal capacity 14,610 US gal (55305 litres)
52 Wing spar attachment bulkhead
53 Spar/fuselage attachment joint
54 Wing panel root attachment bolted joint
55 Centre-section fuel tank
56 Wing centre-section carry through
57 Dorsal control and cable ducting
58 Starboard wing panel root joint
59 Starboard fuel injection water/alcohol tanks, total capacity 600 US gal (2271 litres)
60 Fuel and air system piping
61 Starboard outrigger wheel
62 Inboard twin-engine nacelle
63 Detachable engine cowlings
64 Nacelle pylon
65 Inboard engine oil tanks, capacity 9.4 US gal (35.6 litres) each
66 Oil filler caps
67 Fuel piping to external tank
68 Starboard external fuel tank, capacity 1,780 US gal (6738 litres)
69 Fuel tank pylon
70 Outboard engine fuel and air system ducting
71 Vortex generators
72 Outboard engine nacelle
73 Outboard engine oil tank, capacity 9.4 US gal (35.6 litres)
74 Starboard navigation light
75 Wingtip fairing
76 Outboard aileron segment
77 Aileron hydraulic actuators
78 Nacelle tail fairing
79 Inboard aileron segment
80 Aileron tab
81 Outboard Fowler-type flap segment, down position
82 Inboard Fowler-type flap segment, down position

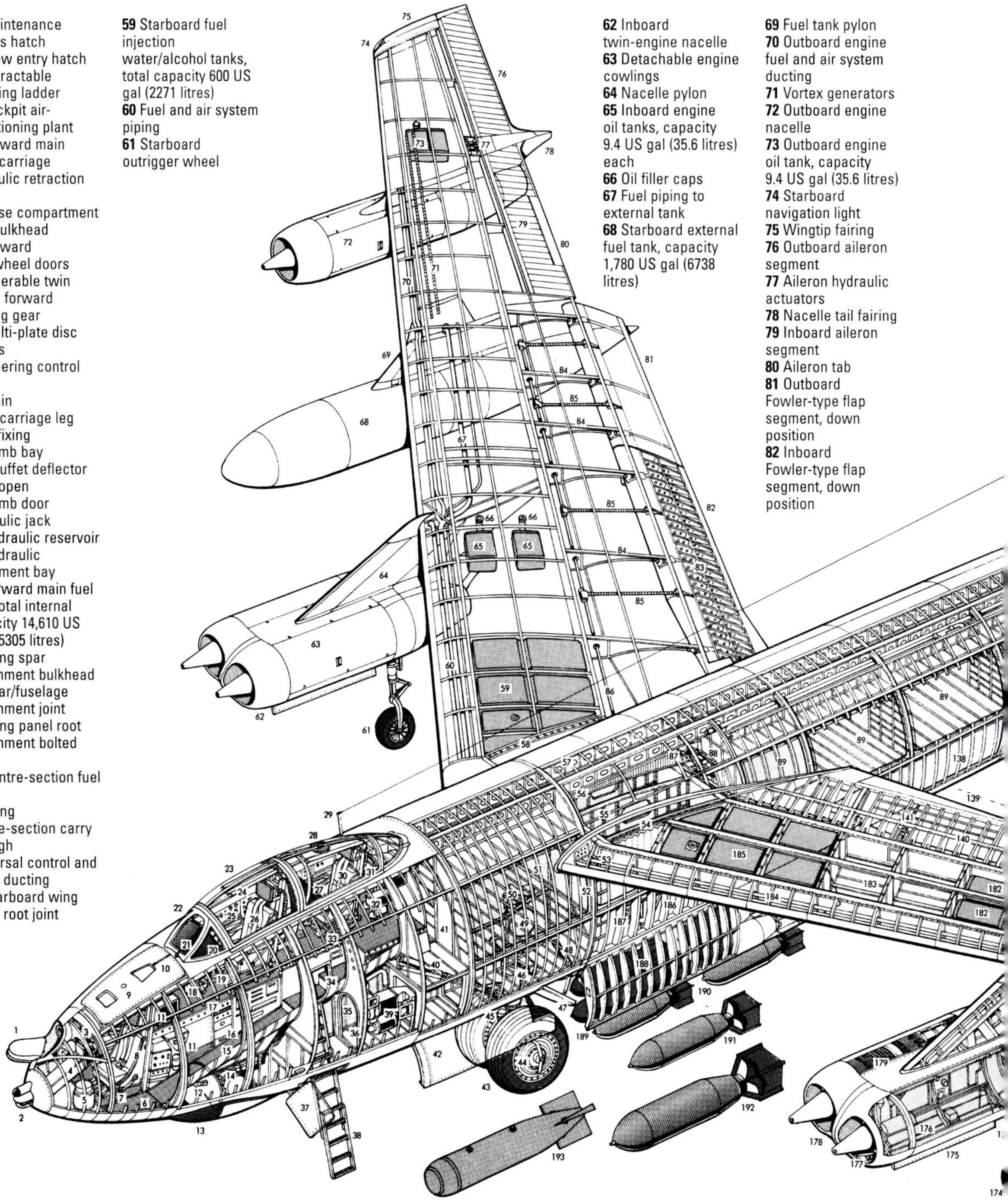

One of the most radically modified B-47s was aircraft 53-4296, an RB-47H which was given an extended lease of life when it was used for F-111 radar tests and fitted with an F-111 nose. The B-47 had a close association with the F-111, most notably when the Royal Australian Air Force was offered the loan of two squadrons of B/RB-47s, pending the delivery of its Aardvarks.

SPECIFICATION

B-47E-II Stratojet

Dimensions

Length: 109 ft 10 in (33.48 m)
Wingspan: 116 ft (35.36 m)
Height: 27 ft 11 in (8.51 m)
Wing area: 1,428 sq ft (132.66 m^2)
Aspect ratio: 11

Powerplant

Six General Electric J47-GE-25 or 25A turbojets, each developing 7,200 lb (32 kN) of thrust with water injection

Weights

Empty: 80,756 lb (36630 kg)
Maximum take-off with RATO pods: 198,180 lb (89893 kg)

Performance

Maximum speed at 16,300 ft (4970 m): 606 mph (975 km/h)
Cruising speed: 557 mph (896 km/h)
Service ceiling: 40,500 ft (12345 m)
Range: 4,000 miles (6437 km)
Climb: 2,430 ft (740 m) per minute with normal power, 4,660 ft (1420 m) per minute with water injection

Armament

Two 20-mm cannon (though the B model carried twin 0.50-in/12.7-mm guns) with 350 rounds in a remotely-controlled tail turret, plus up to 20,000 lb (9072 kg) of conventional or nuclear bombs such as the Mk 5, 6, 15, 18, 28, 36, 42, 53 and B 43 carried internally

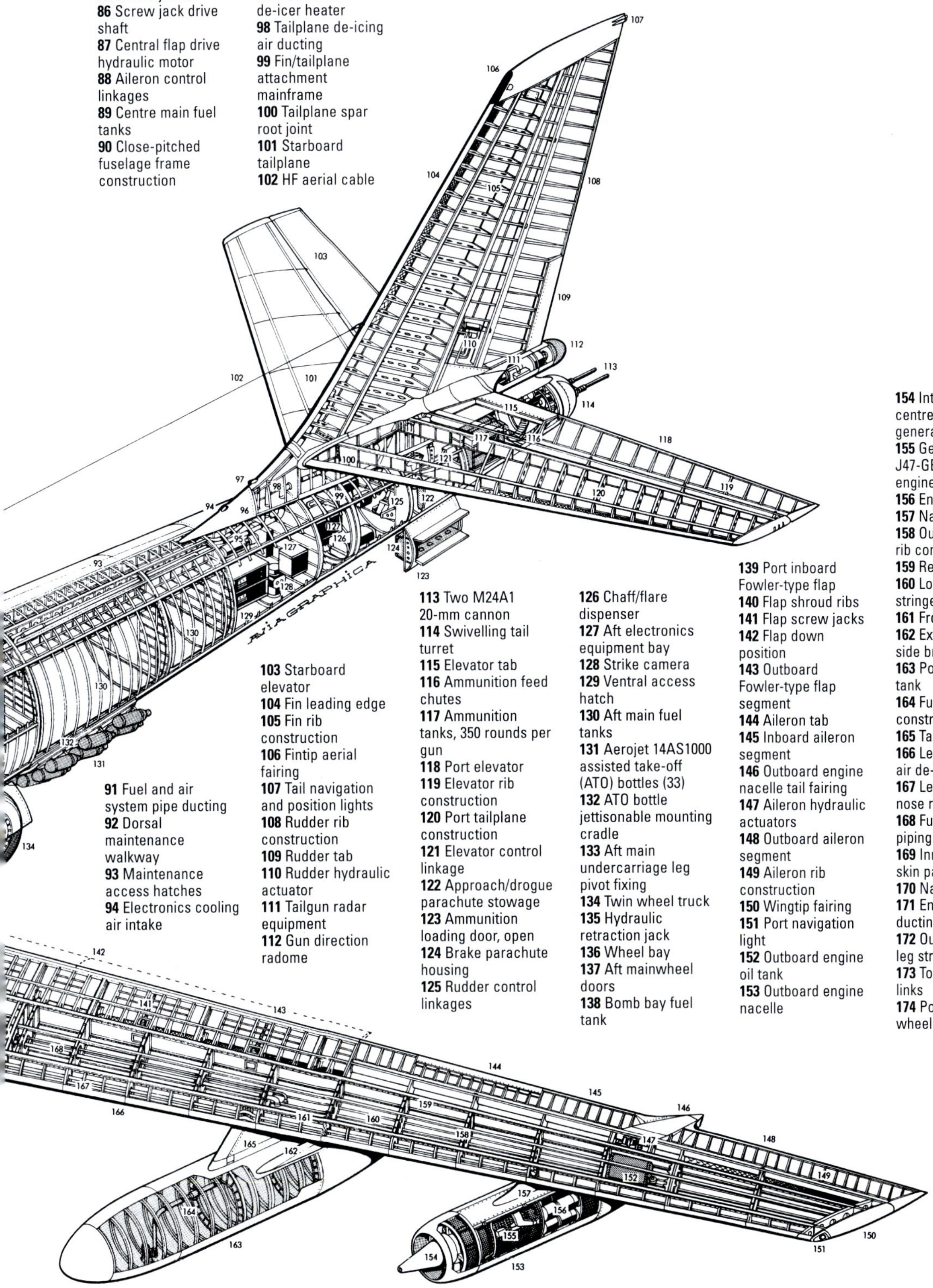

83 Flap rib construction
84 Flap guide rails
85 Screw jacks
86 Screw jack drive shaft
87 Central flap drive hydraulic motor
88 Aileron control linkages
89 Centre main fuel tanks
90 Close-pitched fuselage frame construction
91 Fuel and air system pipe ducting
92 Dorsal maintenance walkway
93 Maintenance access hatches
94 Electronics cooling air intake
95 Equipment air-conditioning plant
96 Finroot fillet
97 Air intake to de-icer heater
98 Tailplane de-icing air ducting
99 Fin/tailplane attachment mainframe
100 Tailplane spar root joint
101 Starboard tailplane
102 HF aerial cable
103 Starboard elevator
104 Fin leading edge
105 Fin rib construction
106 Fintip aerial fairing
107 Tail navigation and position lights
108 Rudder rib construction
109 Rudder tab
110 Rudder hydraulic actuator
111 Tailgun radar equipment
112 Gun direction radome
113 Two M24A1 20-mm cannon
114 Swivelling tail turret
115 Elevator tab
116 Ammunition feed chutes
117 Ammunition tanks, 350 rounds per gun
118 Port elevator
119 Elevator rib construction
120 Port tailplane construction
121 Elevator control linkage
122 Approach/drogue parachute stowage
123 Ammunition loading door, open
124 Brake parachute housing
125 Rudder control linkages
126 Chaff/flare dispenser
127 Aft electronics equipment bay
128 Strike camera
129 Ventral access hatch
130 Aft main fuel tanks
131 Aerojet 14AS1000 assisted take-off (ATO) bottles (33)
132 ATO bottle jettisonable mounting cradle
133 Aft main undercarriage leg pivot fixing
134 Twin wheel truck
135 Hydraulic retraction jack
136 Wheel bay
137 Aft mainwheel doors
138 Bomb bay fuel tank
139 Port inboard Fowler-type flap
140 Flap shroud ribs
141 Flap screw jacks
142 Flap down position
143 Outboard Fowler-type flap segment
144 Aileron tab
145 Inboard aileron segment
146 Outboard engine nacelle tail fairing
147 Aileron hydraulic actuators
148 Outboard aileron segment
149 Aileron rib construction
150 Wingtip fairing
151 Port navigation light
152 Outboard engine oil tank
153 Outboard engine nacelle
154 Intake centre-body/ starter generator housing
155 General Electric J47-GE-25A turbojet engine
156 Engine flame cans
157 Nacelle pylon
158 Outerwing panel rib construction
159 Rear spar
160 Lower wing skin/ stringer panel
161 Front spar
162 External fuel tank side brace
163 Port external fuel tank
164 Fuel tank rib construction
165 Tank pylon
166 Leading-edge hot air de-icing
167 Leading-edge nose ribs
168 Fuel system piping
169 Inner/outer wing skin panel joint rib
170 Nacelle tail fairing
171 Engine exhaust ducting
172 Outrigger wheel leg strut
173 Torque scissor links
174 Port outrigger wheel
175 Outrigger wheel doors
176 Engine nacelle construction
177 Ventral landing/ taxiing lamp
178 Engine air intakes
179 Inboard General Electric J47 engine
180 Nacelle pylon construction
181 Pylon attachment joint
182 Inboard engine oil tanks
183 Inner wing panel rib construction
184 Leading-edge fuel and air system piping
185 Port wing water/ alcohol injection tanks
186 Internal bomb bay
187 Bomb mounting racks
188 Bomb-door rib construction
189 Bomb-door, open
190 1,000-lb (464-kg) HE bombs; maximum bombload 20,000 lb (9072 kg)
191 2,000-lb (907-kg) HE bomb
192 4,000-lb (1814-kg) HE bomb
193 Mk 28 (B-28IN) free-fall 20-megaton nuclear weapon

Boeing B-52 Stratofortress

A heavily-laden Boeing B-52G Stratofortress lifts off from RAF Fairford in England, on its way to contribute to the round-the-clock punishment being inflicted on Iraq's Republican Guard during the Gulf War. B-52s flew 1,624 missions, dropping 25,700 tons of bombs (29 per cent of the total tonnage delivered during the war).

B-52G Stratofortress

Cutaway key
1 Nose radome
2 ALT28 ECM antenna
3 Electronic countermeasures (ECM)
4 Front pressure bulkhead
5 Electronic cooling air intake
6 Bombing radar
7 Low-light television scanner turret (EVS system), infra-red on starboard side
8 Television camera unit
9 ALQ-117 radar warning antenna
10 Underfloor control runs
11 Control column
12 Rudder pedals
13 Windscreen wipers
14 Instrument panel shroud
15 Windscreen panels
16 Cockpit eyebrow windows
17 Cockpit roof escape/ ejection hatches
18 Co-pilot's ejection seat
19 Drogue chute container
20 Pilot's ejection seat
21 Flight deck floor level
22 Navigator's instrument console
23 Ventral escape/ejection hatch, port and starboard
24 Radar navigator's downward ejection seat, navigator to starboard
25 Access ladder and hatch to flight deck
26 EWO instructor's folding seat
27 Electronics equipment rack
28 In-flight refuelling receptacle, open
29 Refuelling delivery line
30 Electronic warfare officer's (EWO) ejection seat
31 Rear crew members' escape/ejection hatches
32 EWO's instrument panel
33 Gunner's remote control
34 Gunner's ejection seat
35 Navigation instructor's folding seat
36 Radio and electronics racks
37 Ventral entry hatch and ladder
38 Lower deck rear pressure bulkhead
39 ECM aerials
40 ECM equipment bay
41 Cooling air ducting
42 Upper deck rear pressure bulkhead
43 Water injection tank capacity 1,200 US gal (4542 litres)
44 Fuselage upper longeron
45 Astro-navigation antenna
46 Tank access hatches
47 Leading edge 'strakelets' fitted to identify cruise missile carriers
48 Forward fuselage fuel tank
49 Air-conditioning plant
50 Forward starboard main undercarriage bogie
51 Landing lamp
52 Forward port main undercarriage bogie
53 Torque scissor links
54 Steering jacks
55 Main undercarriage door
56 Main undercarriage leg strut
57 Wing front spar/fuselage/main undercarriage attachment frame
58 Main undercarriage wheel bay
59 Doppler aerial
60 Central electronic equipment bay
61 Air-conditioning intake duct
62 Front spar attachment joint
63 Wingroot rib
64 Wing panel bolted attachment joint
65 Centre-section fuel tank bay
66 Wing centre-section carry through
67 Starboard wing attachment joint
68 Vortex generators
69 Starboard wing integral fuel tank bays, total fuel system capacity (includes external tanks) 48,030 US gal (181813 litres)
70 Engine ignition control unit
71 Bleed air ducting

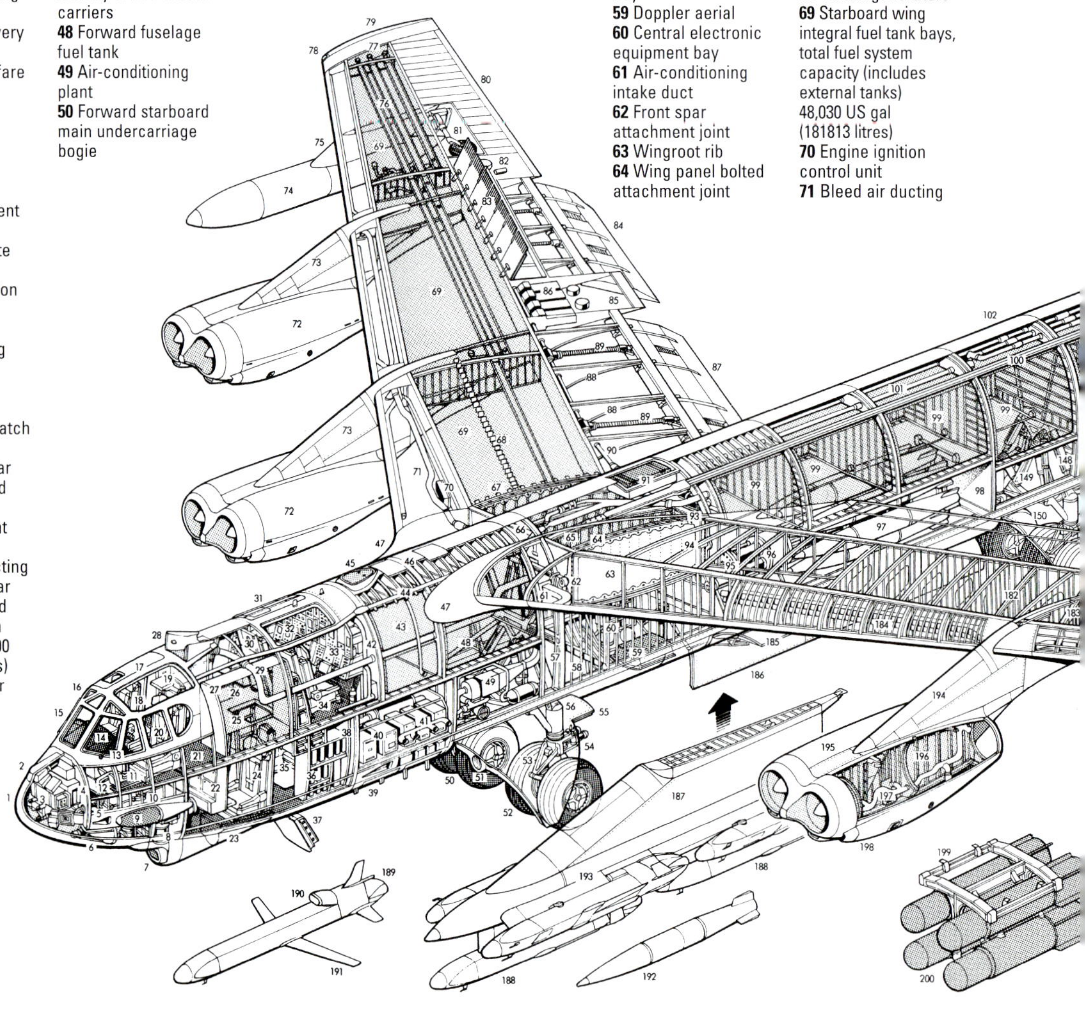

First delivered to Strategic Air Command (SAC) in June 1955 – to the 93rd Bombardment Wing (Heavy) at Castle AFB, California – B-52s ushered in a new phase in the Cold War. SAC boss General Curtis E. LeMay adopted the procedure of operating the aircraft under wartime conditions in a peacetime era. Across the US, B-52 crews were held on a 15-minute alert, ready to attack the Soviet Union and its allies. Here, the crew of a B-52B serving with the 22nd Bombardment Wing at March AFB, California runs to its aircraft during a simulated alert.

SPECIFICATION

B-52H Stratofortress

Dimensions

Length overall: 160 ft 10⅞ in (49.05 m)
Wingspan: 185 ft 9 in (56.39 m)
Wing area: 4,000.00 sq ft (371.60 m²)
Height: 40 ft 8 in (12.40 m)
Tailplane span: 52 ft (15.85 m)
Tailplane area: 900 sq ft (83.61 m²)
Wheel track: 8 ft 3 in (2.51 m)
Wheelbase: 50 ft 3 in (15.48 m)

Powerplant

Eight Pratt & Whitney TF33-P-3 (JT3D-2) low-bypass turbofans each rated at 17,000 lb st (75.62 kN)

Weights

Operating empty:138,385 lb (62771 kg)
Maximum payload: 51,615 lb (23413 kg)
Maximum take-off: 327,000 lb (148325 kg)
Maximum ramp weight: 328,000 lb (148780 kg)

Fuel and load

Fuel: 299,434 lb (135821 kg) plus provision for 9,114 lb (4134 kg) in two 700-US gal (2650-litre) non-jettisonable underwing tanks
Maximum ordnance: about 50,000 lb (22680 kg)

Performance

Maximum speed at high altitude: 595 mph (957 km/h)
Cruising speed at high altitude: 509 mph (819 km/h)
Penetration speed at low altitude: between 405 and 420 mph (652 and 676 km/h)
Range: more than 10,000 miles (16093 km)
Service ceiling: 55,000 ft (16765 m)
Take-off run: 9,500 ft (2896 m) at maximum take-off weight

Armament

In the nuclear role, the B-52H can carry 20 cruise missiles (eight internally on a rotary launcher and six under each wing). These can either be AGM-86B ALCMs or AGM-129 Advanced Cruise Missiles. Free-fall nuclear weapons such as B61 or B83, remain an option, but this capability is at present entrusted mainly to the B-2A Spirit fleet. Conventional weapons include AGM-86C cruise missiles, the AGM-142 Have Nap and AGM-84 Harpoon. Free-fall bombs can be carried on HSABs (heavy stores adaptor beams) or Hound Dog pylons, to a maximum of 51 750-lb (340-kg) class weapons. Alternatives to general-purpose bombs include cluster munitions and mines. The M61A1 Vulcan 20-mm cannon in the tail turret is no longer used

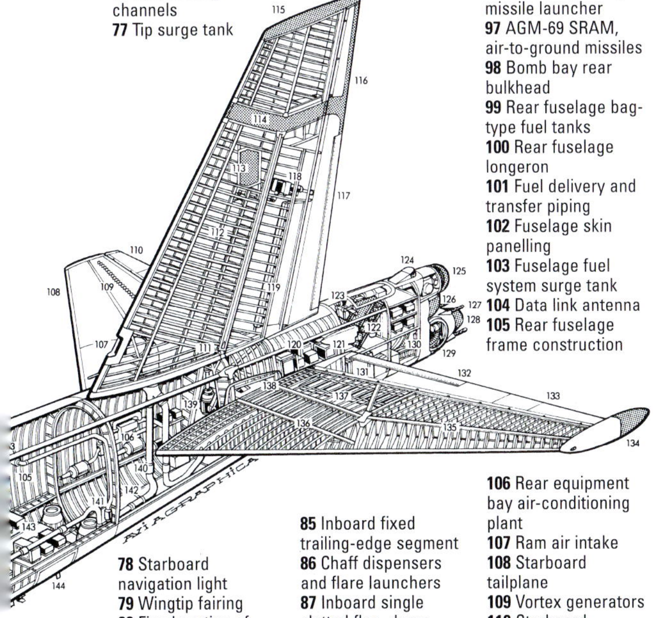

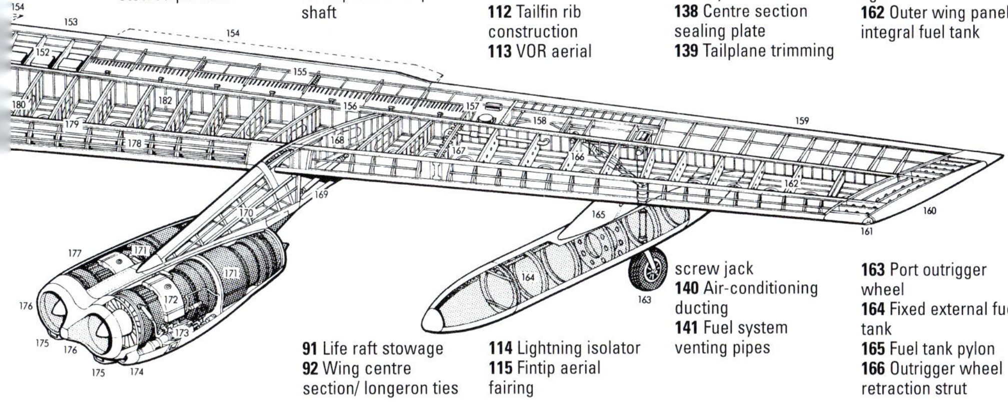

72 Starboard engine nacelles
73 Nacelle pylons
74 Fixed external fuel tank, capacity 700 US gal (2650 litres)
75 Tank pylon
76 Fuel venting channels
77 Tip surge tank
78 Starboard navigation light
79 Wingtip fairing
80 Fixed portion of trailing edge
81 Starboard outrigger wheel, stowed position
82 Hydraulic equipment bay
83 Roll control spoiler panels – open
84 Outboard single-slotted, Fowler-type flap, down position
85 Inboard fixed trailing-edge segment
86 Chaff dispensers and flare launchers
87 Inboard single slotted flap, down position
88 Flap guide rails
89 Flap screw jacks
90 Flap drive torque shaft
91 Life raft stowage
92 Wing centre section/ longeron ties
93 Central flap drive motor
94 Rear spar attachment joint
95 AGM-69 missile environmental control unit
96 Bomb bay rotary missile launcher
97 AGM-69 SRAM, air-to-ground missiles
98 Bomb bay rear bulkhead
99 Rear fuselage bag-type fuel tanks
100 Rear fuselage longeron
101 Fuel delivery and transfer piping
102 Fuselage skin panelling
103 Fuselage fuel system surge tank
104 Data link antenna
105 Rear fuselage frame construction
106 Rear equipment bay air-conditioning plant
107 Ram air intake
108 Starboard tailplane
109 Vortex generators
110 Starboard elevator
111 Fin spar attachment joint: fin folds to starboard
112 Tailfin rib construction
113 VOR aerial
114 Lightning isolator
115 Fintip aerial fairing
116 Rudder
117 Rudder tab
118 Hydraulic rudder control jack
119 Rudder aerodynamic balance
120 Rear ECM and fire control electronics pack
121 ECM aerial fairing
122 Brake parachute stowage
123 Parachute and door-release mechanism
124 ALQ-117 retractable aerial fairing
125 AN/ASG-15 search radome
126 ALQ-117 and APR-25 ECM radome
127 Four 0.5-in (12.7-mm) machine-guns
128 AN/ASG-15 tracking radome
129 Remote control gun turret
130 Ammunition feed chutes
131 Ammunition tanks, 600 rounds per gun
132 Elevator tab
133 Port elevator
134 ALQ-153 tail warning radar
135 All-moving tailplane construction
136 Tailplane carry-through box section spar
137 Elevator aerodynamic balance
138 Centre section sealing plate
139 Tailplane trimming screw jack
140 Air-conditioning ducting
141 Fuel system venting pipes
142 Ventral access hatch
143 Rear-fuselage ECM equipment bay
144 ECM aerials
145 Strike camera compartment
146 Rear main undercarriage wheel bay
147 Bomb/wheel bay box section longeron
148 Main undercarriage mounting frame
149 Hydraulic retraction jack
150 Rear main undercarriage bogie units
151 Flap shroud ribs
152 ECM dispensers
153 Fixed portion of trailing edge
154 Port flaps, down position
155 Outboard single slotted flap
156 Port roll control spoiler panels
157 Hydraulic reservoir
158 Outrigger wheel bay
159 Fixed portion of trailing edge
160 Glass-fibre wingtip fairing
161 Port navigation light
162 Outer wing panel integral fuel tank
163 Port outrigger wheel
164 Fixed external fuel tank
165 Fuel tank pylon
166 Outrigger wheel retraction strut
167 Outer wing panel attachment joint
168 Engine pylon mounting rib
169 Pylon rear attachment strut
170 Engine pylon construction
171 Pratt & Whitney J57-P-43WB turbojet engine
172 Engine oil tank, capacity 8.5 US gal (32 litres)
173 Accessory equipment gearbox
174 Generator cooling air duct
175 Oil cooler ram air Intakes
176 Engine air intakes
177 Detachable cowling panels
178 Leading-edge rib construction
179 Front spar
180 Wingrib construction
181 Rear spar
182 Port wing integral fuel tank bays
183 Inboard pylon mounting rib
184 Leading-edge bleed air and engine control runs
185 Weapons bay doors, open (loading) position
186 Bomb doors, open
187 Wing-mounted cruise missile pylon
188 Boeing AGM-86B Air-Launched Cruise Missiles (ALCMs), six per wing pylon, stowed configuration
189 AGM-86B missile in flight configuration
190 Retractable engine air intake
191 Folding wings
192 AGM-69 SRAM, alternative load
193 Missile adaptors
194 Nacelle pylon
195 Port inboard engine nacelles
196 Central engine mounting bulkhead/ firewall
197 Bleed air ducting
198 Generator cooling air ducting
199 Fuselage bomb-mounting cradle
200 Free-fall 25-megaton nuclear weapons (four)

Convair B-36 Peacemaker

In its B-36A form (illustrated), the Peacemaker was far from being an operational bomber, since it lacked a full armament fit and many other key systems were either not installed or were non-operational. BM-004 was the first production B-36A and, like many early B-36s, carried the large 'buzz number' on its forward fuselage to aid civilians when calling the USAF to complain about low flying.

B-36J

Cutaway key
1 Twin 20-mm nose cannon
2 ILS glideslope aerial
3 Nose-sighting station, hemispherical sight
4 Nose compartment glazing
5 Optically-flat sighting panel
6 Navigator's station
7 Nose turret mounting platform
8 Turret-actuating mechanism
9 Ammunition tanks, 400 rounds per gun
10 Windscreen panels
11 Instrument console
12 Rudder pedals
13 Entry hatch from wheel bay
14 Radar bombardier's station
15 K-system radar bombing equipment
16 ECM aerials
17 Radio compass housing
18 Retractable boarding ladder
19 Observer/radar technician's seat
20 Nosewheel doors
21 Marker beacon aerial
22 Pitot head
23 Electrical power distribution panels
24 Flight deck floor level
25 Pilot's seat
26 Centre instrument console
27 Co-pilot's seat
28 Overhead jet engine control panel
29 Cockpit canopy cover
30 Astrodome observation hatch
31 Flight engineer's seats (2)
32 Engineer's control panels
33 Canopy window panels
34 Lower deck radio operator's station
35 Forward cabin escape hatch
36 Nose undercarriage pivot mounting
37 Forward-retracting twin nosewheels
38 K-system radar antenna housing
39 Food locker
40 Access hatch to communications tunnel
41 Forward cabin pressure bulkhead
42 Port sighting station
43 Cabin pressurisation valve
44 Electrical system equipment
45 Starboard sighting station
46 VHF aerial
47 Life raft stowage
48 Retractable, remotely-controlled dorsal gun turrets, two 20-mm cannon
49 Ammunition feed chutes
50 Cannon bay sliding door, open
51 Ammunition magazines, 600 rounds per gun
52 Fire control equipment
53 Communicating tunnel, forward to rear pressurised crew compartments
54 ECM aerials
55 Bomb bay door frame construction
56 Forward bomb bay doors, open
57 Fuselage lower longitudinal beam
58 Communications tunnel railed personnel cart
59 Bomb bay girder frame construction
60 Oxygen bottles
61 Fuselage upper longitudinal beam
62 Interchangeable bomb racks
63 Frame-stiffened fuselage skin panels
64 No. 1 forward bomb bay
65 Fuselage maintenance walkway
66 No. 2 forward bomb bay
67 Starboard side crawlway
68 Transformer rectifier units
69 Fuselage girder frame/wing spar attachment joint
70 Wing panel centre line joint
71 Inboard integral fuel tank, capacity 4,212 US gal (15944 litres)
72 Fuel system piping
73 Centre integral fuel tank
74 Engine oil tanks, capacity 200 US gal (757 litres) each
75 Engine air intake ducting
76 Leading-edge pitot intakes
77 Outboard integral fuel tank, 2,262 US gal (8563 litres)
78 Leading-edge de-icing air supply ducting
79 Outer wing panel joint strap
80 Wing stringers
81 Wing skin panelling
82 Starboard jet engine nacelle
83 Nacelle pylon
84 Leading-edge thermal de-icing
85 Outer wing panel integral jet fuel tank, capacity 1432 US gal (5421 litres)
86 Wingtip fairing
87 Starboard navigation light
88 Starboard aileron
89 Blue formation lights
90 Aileron geared tab
91 Tab operating linkage
92 Aileron aerodynamic seal and balance
93 Aileron control linkage
94 Hydraulic lock
95 Outboard reciprocating engine nacelle
96 Curtiss-Wright three-bladed variable-pitch propellers, 19-ft (5.79-m) diameter
97 Propeller spinners
98 Outboard single-slotted flap segment, lowered

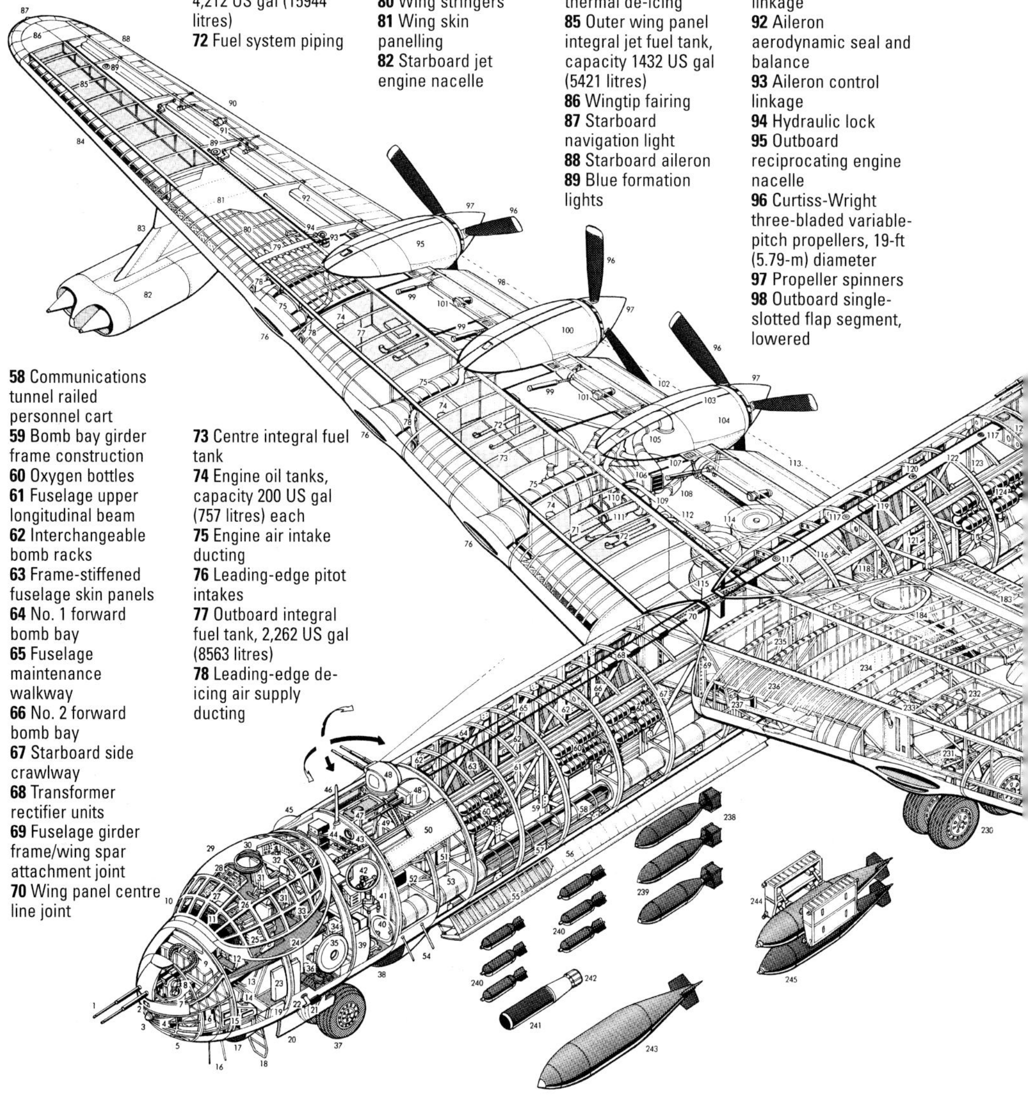

At the end of World War II, the USAAF closed seven gunnery schools and discharged almost all of its gunners. The few that remained were mostly B-29 crew, but there were insufficient numbers of them to man the guns of the few B-29s involved in the Korean War. Inevitably, when the B-36, with its demand for six gunners per aircraft, entered service this deficit was even more keenly felt and several months had passed before every B-36 could fly with a fully-competent gunnery crew.

SPECIFICATION

B-36J Peacemaker

Dimensions

Length: 162 ft 1 in (49.40 m)
Wingspan: 230 ft (70.10 m)
Wing area: 4,772 sq ft (443.32 m^2)
Height: 46 ft 8 in (14.22 m)

Powerplant

Six 3,800-hp (2834-kW) Pratt & Whitney R-4360-53 radial piston engines and four 5,200-lb st (23.12-kN) General Electric J47-GE-19 turbojet engines
B-36B: Six 3,500-hp (2610-kW) Pratt & Whitney R-4360-41 radial piston engines
B-36D: Six 3,500-hp (2610-kW) Pratt & Whitney R-4360-41 radial piston engines and four 5,200-lb st (23.12-kN) General Electric J47-GE-19 turbojet engines

Weights

Empty: 171,035 lb (77580 kg)
Maximum take-off: 410,000 lb (185973 kg)
Empty, B-36B: 140,640 lb (63794 kg)
Maximum take-off, B-36B: 328,000 lb (148780 kg)
Empty, B-36D: 158,843 lb (72051 kg)
Maximum take-off, B-36D: 357,500 lb (162162 kg)

Load

Maximum ordnance: 86,000 lb (39010 kg) with weight restrictions or normally up to 72,000 lb (32659 kg)

Performance

Maximum speed at 36,400 ft (11095 m): 411 mph (661 km/h)
Cruising speed: 391 mph (629 km/h)
Maximum rate of climb: 1,920 ft (585 m) per minute
Service ceiling: 39,900 ft (12160 m)
Maximum speed at 34,500 ft (10516 m), B-36B: 381 mph (613 km/h)
Cruising speed, B-36B: 202 mph (325 km/h)
Maximum rate of climb, B-36B: 1,510 ft (460 m) per minute
Service ceiling, B-36B: 42,500 ft (12954 m)
Maximum speed at 32,120 ft (9790 m), B-36D: 439 mph (706 km/h)
Cruising speed, B-36D: 225 mph (362 km/h)
Maximum rate of climb, B-36D: 1,740 ft (530 m) per minute
Service ceiling, B-36D: 45,200 ft (13777 m)

Range

With 10,000-lb (4536-kg) bombload: 6,800 miles (4536 km)
B-36B: 8,175 miles (13156 km)
B-36D: 7,500 miles (12070 km)

Armament

Six retractable and remotely-controlled fuselage turrets, each with twin 20-mm M24A1 cannon and similar weapons in nose and tail turrets. Maximum ammunition load 9,200 rounds. A wide range of conventional and primarily nuclear bombs could be carried

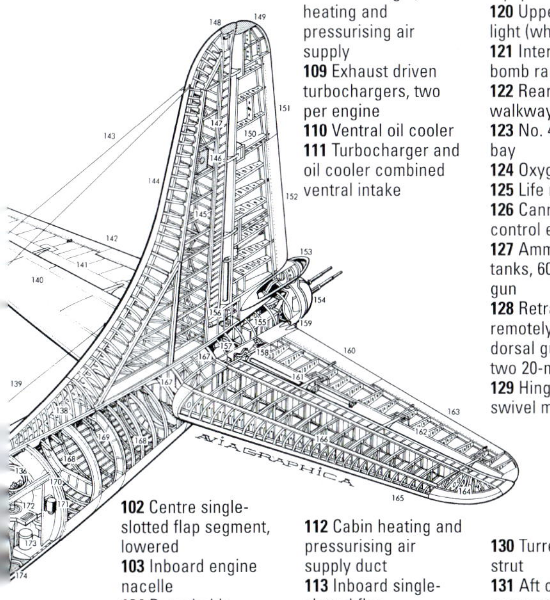

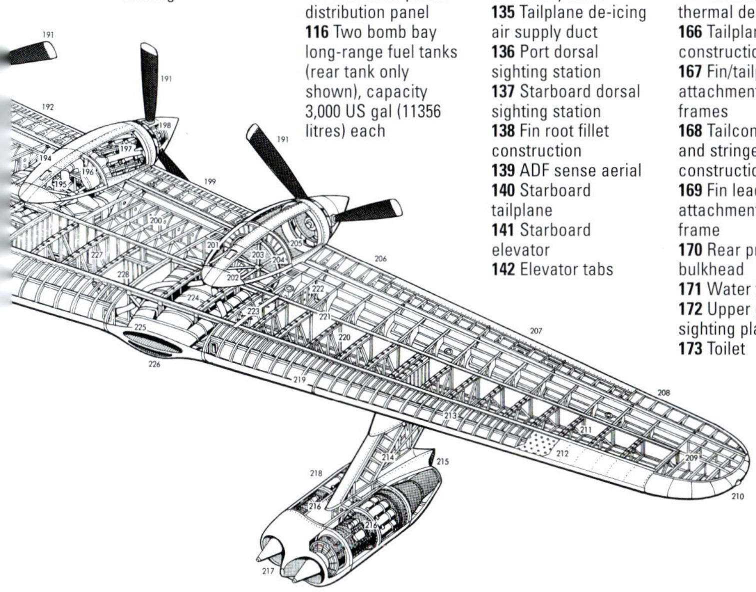

99 Flap screw jacks
100 Centre engine nacelle
101 Flap drive motor and torque shaft
102 Centre single-slotted flap segment, lowered
103 Inboard engine nacelle
104 Detachable engine cowling panels
105 Induction air ducting
106 Intercooler, two per engine
107 Intercooler exhaust
108 Exhaust primary heat exchanger, heating and pressurising air supply
109 Exhaust driven turbochargers, two per engine
110 Ventral oil cooler
111 Turbocharger and oil cooler combined ventral intake
112 Cabin heating and pressurising air supply duct
113 Inboard single-slotted flap
114 Starboard main undercarriage wheel bay
115 Electrical power distribution panel
116 Two bomb bay long-range fuel tanks (rear tank only shown), capacity 3,000 US gal (11356 litres) each
117 Fuselage formation lights (blue)
118 No. 3 aft bomb bay
119 Electrical system equipment
120 Upper navigation light (white)
121 Interchangeable bomb racks
122 Rear fuselage walkway
123 No. 4 aft bomb bay
124 Oxygen bottles
125 Life raft stowage
126 Cannon fire control equipment
127 Ammunition tanks, 600 rounds per gun
128 Retractable, remotely-controlled, dorsal gun turrets, two 20-mm cannon
129 Hinged turret swivel mounting
130 Turret retraction strut
131 Aft crew compartment pressure bulkhead
132 Crew rest bunks
133 D/F loop aerial
134 Galley units
135 Tailplane de-icing air supply duct
136 Port dorsal sighting station
137 Starboard dorsal sighting station
138 Fin root fillet construction
139 ADF sense aerial
140 Starboard tailplane
141 Starboard elevator
142 Elevator tabs
143 HF aerial cables
144 Fin leading-edge thermal de-icing
145 Tail fin rib construction
146 VOR aerial
147 Sternpost
148 Fintip VHF aerial
149 Rudder tip LORAN aerial
150 Rudder construction
151 Upper trim tab
152 Lower geared tab
153 Tail gun radar scanner
154 Remotely controlled tail gun turret, two 20-mm cannon
155 Ammunition tanks, 600 rounds per gun
156 Rudder hinge control
157 Tail radar modular unit
158 Elevator hinge control
159 Tail navigation lights
160 Elevator geared tab
161 Tab control torque shaft
162 Elevator rib construction
163 Outboard trim tab
164 Tailplane tip fairing construction
165 Leading-edge thermal de-icing
166 Tailplane rib construction
167 Fin/tailplane attachment main frames
168 Tailcone frame and stringer construction
169 Fin leading-edge attachment main frame
170 Rear pressure bulkhead
171 Water tank
172 Upper gunner's sighting platform
173 Toilet
174 Ventral entry hatch
175 Boarding ladder, stowed
176 Lateral/ventral sighting station, port and starboard
177 Lateral gunner's seat
178 Ventral strike camera
179 Communications tunnel aft section
180 Two ventral retractable gun turrets, two 20-mm cannon
181 Rear bomb bay doors, open
182 Inboard single-slotted flap segment
183 Flap rib construction
184 Port main undercarriage wheel bay
185 Inboard reciprocating engine nacelle
186 Engine-driven alternator
187 Engine accessory equipment gearbox
188 Pratt & Whitney R-4360-53 28-cylinder, four-row radial engine
189 Exhaust cooling air outlet ducts
190 Propeller spinner
191 Curtiss-Wright three-bladed propellers
192 Centre single-slotted flap segment
193 Trailing-edge rib construction
194 Centre engine nacelle
195 Induction air ducting
196 Water tank
197 Centre Pratt & Whitney R-4360-53 engine
198 Propeller hub pitch change mechanism
199 Outboard single-slotted flap segment
200 Flap shroud ribs
201 Outboard engine nacelle
202 Engine-driven cooling air fan
203 Engine bearer struts
204 Fireproof bulkhead
205 Engine bay cooling air variable outlet plug
206 Port aileron
207 Aileron geared tab
208 Aileron rib construction
209 Wingtip fairing construction
210 Port navigation light
211 Port jet engine integral fuel tank
212 Leading-edge double skin panelling
213 Leading-edge nose ribs
214 Nacelle pylon construction
215 Exhaust tail fairing
216 General Electric J47-GE-19 turbojet engines
217 Intake fairing
218 Detachable engine cowling panels
219 Leading-edge de-icing air ducting
220 Outer wing panel rib construction
221 Rear spar
222 Aileron shroud ribs
223 Outer wing panel joint strap
224 Reciprocating engine air ducting
225 Engine cooling and induction air upper pitot intake
226 Oil cooler and turbocharger lower pitot intake
227 Port wing integral fuel tank bays
228 Engine oil tank bays
229 Front spar
230 Inward-retracting four-wheel main undercarriage bogie
231 Main undercarriage leg
232 Mainwheel leg pivot mounting
233 Hydraulic retraction jack
234 Inboard fuel tank bay
235 Inboard wing panel rib construction
236 Leading-edge ribs
237 Central hydraulic equipment bay
238 Internal bombload, normal 72,000 lb (32660 kg); maximum 86,000 lb (39000 kg) at restricted gross weight
239 4,000-lb (1814-kg) HE bombs
240 500-lb (227-kg) bombs
241 Aerial mine
242 Mine parachute housing
243 43,000-lb (19505-kg) 'special' bomb
244 Large bomb adaptor racks
245 22,000-lb (9980-kg) bombs

Convair B-58 Hustler

The original Hustler, XB-58 55-660, first flew on 11 November 1956 and accumulated 150 flights (lasting 257 hours and 30 minutes). It was the first B-58 to reach Mach 1 and Mach 2. Nicknamed **Old Grandpappy**, *this aircraft was used for ALBM tests before being scrapped.*

B-58A Hustler

Cutaway key

1 Pitot tube
2 Nose probe
3 AN/ARC-10 long-range communications system antenna
4 Radome
5 Antenna coupler
6 Search radar scanner
7 Radar tracking mechanism
8 Air refuelling ramp door, open
9 Air refuelling receptacle
10 Angle of attack transmitter
11 Radar and nosewheel mounting bulkhead
12 Taxiing lamp
13 Landing lamp
14 Nose undercarriage leg strut
15 ILS glideslope aerial
16 Twin nosewheels
17 Torque scissor links
18 Steering jacks
19 Nose undercarriage hinge links
20 Nosewheel doors
21 Temperature probe
22 Hydraulic retraction jack
23 Secondary pitot tube
24 Search radar modulator unit
25 Liquid oxygen bottles
26 Cockpit front pressure bulkhead
27 Windscreen panels
28 Instrument panel shroud
29 Rudder pedals
30 Cockpit air-conditioning ducting
31 Cockpit pressure floor
32 Engine throttle levers
33 Side console panel
34 Pilot's ejection seat/escape capsule
35 Angle of attack indexer
36 Overhead windows
37 Pilot's escape capsule, deployed configuration
38 Integral control column
39 Folding blast shield
40 Canopy cover/entry hatch
41 Entry hatch, open position
42 Canopy actuator
43 Escape capsule, stowed
44 Ejection seat launch rails
45 Sloping cockpit bulkhead
46 Air data computer
47 Radio and electronics equipment racks
48 Modular navigation instrumentation unit
49 UHF command antenna
50 Navigation unit access/removal hatch
51 Starboard inner engine nacelle
52 Navigator's entry/escape hatch
53 Air-conditioning ducting
54 Hatch actuator
55 Navigator's ejection seat/escape capsule
56 Window panel
57 Side console panel
58 Defence equipment control module, DECM
59 TACAN aerial
60 Defensive systems operator's entry hatch
61 Side window panel
62 Defensive systems operator's ejection seat/escape capsule
63 Cockpit rear pressure bulkhead
64 Ventral air-conditioning pack
65 Fuel vent
66 Forward fuselage fuel tank/reservoir tank; total internal fuel capacity 10,924 US gal (41352 litres)
67 Main fuselage longeron
68 Air system water separators
69 Fuel tank bulkhead
70 Stabilisation amplifier
71 Astrotracker unit
72 Chaff dispenser pneumatic air bottle
73 Starboard main undercarriage, stowed position
74 Hydraulic retraction jack
75 Hydraulic reservoir
76 Fuel feed piping
77 Starboard wing forward main fuel tank
78 Inboard nacelle pylon
79 Main undercarriage wheel well fairing
80 Bleed air system ducting
81 Air/water heat exchanger
82 Water tank
83 Pneumatic starter air ducting to outboard engine
84 Port outboard engine nacelle
85 Engine air intake
86 BLU-2/B-2 fuel pod
87 Fuel and electrical disconnect to upper component pod
88 Upper component pod housing
89 Pod separation thruster
90 Pod release unit
91 Fuel pod frame construction
92 BLU-2/B-3 combined fuel and weapons pod
93 Forward fuel tank
94 Pod forward latch
95 Electrical disconnect
96 Integral munitions bay
97 Munitions bay access panel
98 Pylon mounting
99 Rear fuel tank
100 Pylon aft latches
101 Pod stabilising fins
102 Outboard nacelle pylon fairing
103 Starboard wing aft main fuel tanks
104 Fuel feed pipe to outer engine

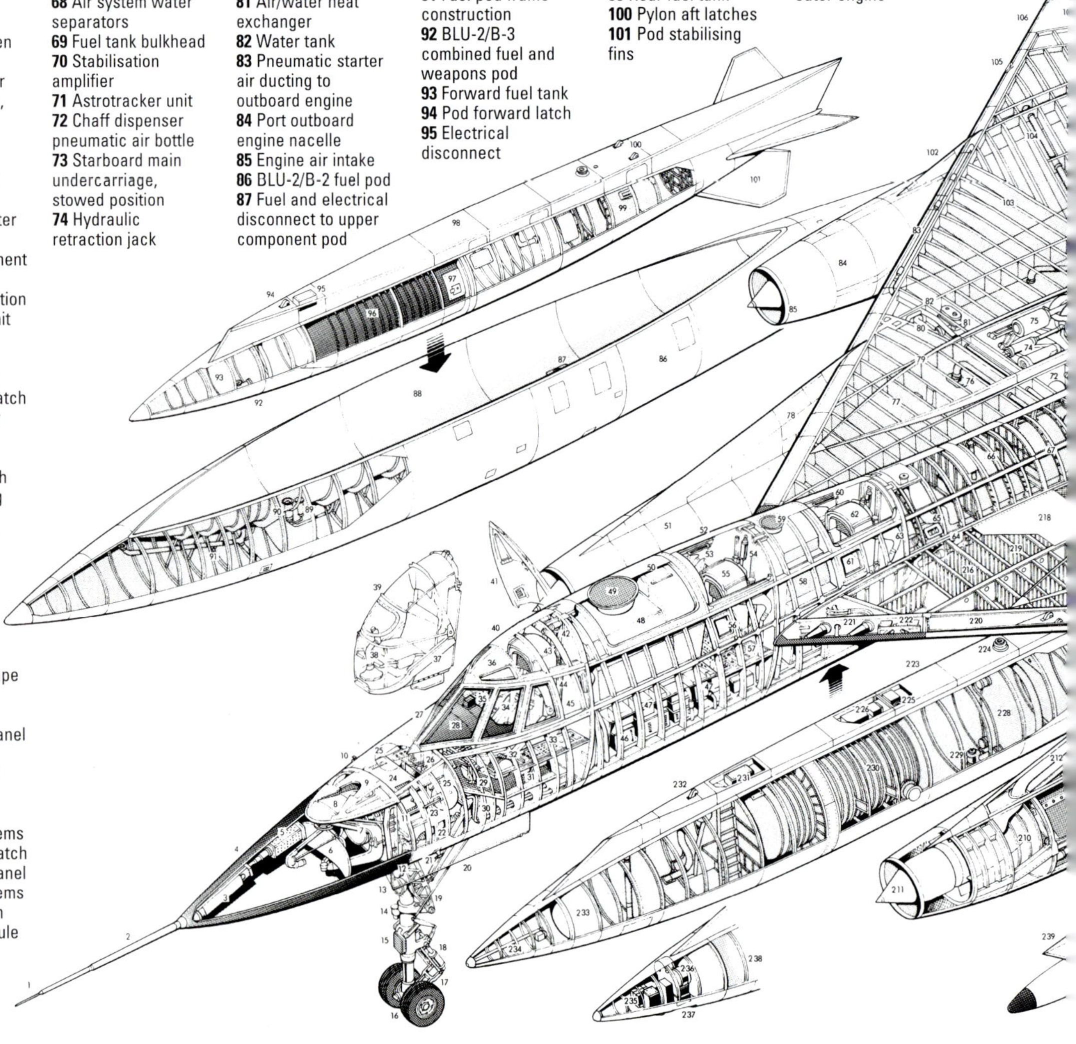

While the pilot of the B-58 had good visibility, the navigator and defensive systems operator just had two small windows for daylight, making conditions somewhat cramped and claustrophobic.

SPECIFICATION

B-58A Hustler

Dimensions

Overall length: 96 ft 9 in (29.5 m)
Wingspan: 56 ft 9 in (17.3 m)
Wing aspect ratio: 2.096
Wing area: 1,542 sq ft (143.25 m²)
Height: 29 ft 11 in (8.87 m)
Vertical fin area: 160 sq ft (48.7 m²)
Vertical fin height: 14 ft 6in (4.45 m)

Powerplant

Four General Electric J79-GE-5A turbojets, each rated at 15,600 lb (69.3 kN) thrust

Weights

Empty (without pod): 55,560 lb (25201 kg)
Basic (without pod): 57,916 lb (26270 kg)
Empty (with MB-1C pod): 64,115 lb (29081 kg)
Basic (with MB-1C pod): 66,471 lb (30150 kg)
Maximum gross weight (in flight): 176,890 lb (80235 kg)

Fuel

Total internal fuel: 10,924 US gal (41352 litres)
Total external fuel (in TCP pod): 657 US gal (2487 litres)

Performance

Maximum speed below 25,000 ft (7620 m): Mach 0.91
Maximum speed at 40,000 ft (12192 m): Mach 2.1
Cruise speed: 531 kt (611.4 mph; 983 km/h)
Initial climb rate: 17,400 ft (5310 m) per minute
Normal cruise altitude: 38,450 ft (11719 m)
Service ceiling: 63,400 ft (19324 m)
Time to 30,000 ft (9144 m): 11.2 min

Range

Range with internal fuel: 1,738 nm (2,000 miles; 3219 km)
Ferry range: 4,100 nm (4,718 miles; 7592 km)

Armament

Maximum weaponload: 19,450 lb (8823 kg)
One 20-mm T-171 rotary barrel cannon was situated in the tail to defend against aircraft attacking from the rear.
A number of weapon pod configurations were designed for the B-58, although few actually reached the production stage.
The MA-1C, with its warhead, was a rocket-propelled version of the MB-1, giving the B-58 stand-off capability. The MB-1C was a standard free-falling weapon. The TCP 'two-component pod' was similar to the MB-1, but was capable of retaining the warhead while ejecting the fuel cell when empty. B-58s were also tested with chemical and conventional weapons.

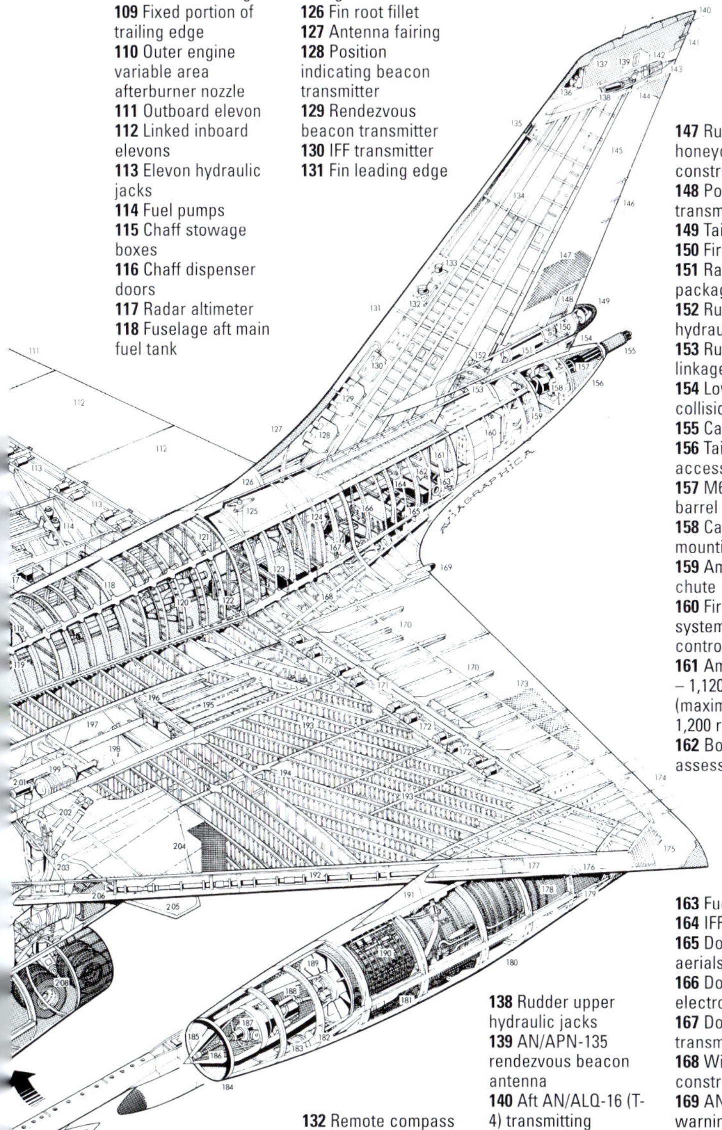

105 Starboard navigation light
106 Cambered leading edge
107 Wingtip fairing
108 Static dischargers
109 Fixed portion of trailing edge
110 Outer engine variable area afterburner nozzle
111 Outboard elevon
112 Linked inboard elevons
113 Elevon hydraulic jacks
114 Fuel pumps
115 Chaff stowage boxes
116 Chaff dispenser doors
117 Radar altimeter
118 Fuselage aft main fuel tank
119 Wing panel centreline joint
120 Fuel system piping
121 Fuel tank bulkhead
122 Power control unit linkage assembly
123 Radar track breaker (T-4) package
124 Aft fuselage fuel tank/balance tank
125 Rudder control linkage
126 Fin root fillet
127 Antenna fairing
128 Position indicating beacon transmitter
129 Rendezvous beacon transmitter
130 IFF transmitter
131 Fin leading edge
132 Remote compass transmitter
133 J-4 compass unit
134 Multi-spar tailfin construction
135 Anti collision light
136 AN/APX-47 IFF aerial
137 Fin tip radome
138 Rudder upper hydraulic jacks
139 AN/APN-135 rendezvous beacon antenna
140 Aft AN/ALQ-16 (T-4) transmitting antenna
141 Position indicating beacon antenna, AN/APN-136
142 AN/ALR-12 radar warning antenna
143 Tail navigation lights
144 VOR localiser aerial
145 Rudder
146 Static discharge wicks
147 Rudder honeycomb construction
148 Position indicator transmitter beacon
149 Tail radome
150 Fire control radar
151 Radar modulator package
152 Rudder lower hydraulic jacks
153 Rudder control linkage
154 Lower anti-collision light
155 Cannon barrels
156 Tailcone/cannon access
157 M61 Vulcan, 6-barrel rotary cannon
158 Cannon gimbal mounting
159 Ammunition feed chute
160 Fire control system tracking control unit
161 Ammunition tank – 1,120 rounds (maximum capacity 1,200 rounds)
162 Bomb damage assessment camera
163 Fuel jettison
164 IFF transponder
165 Doppler receiver aerials
166 Doppler electronics unit
167 Doppler radar transmitter aerials
168 Wing root fillet construction
169 AN/ALQ-16 radar warning receiver aerial
170 Port elevon construction
171 Elevon hinge rib
172 Elevon hydraulic jacks
173 Honeycomb trailing-edge panels
174 Static dischargers
175 Honeycomb wingtip fairing
176 Port navigation light
177 Honeycomb leading-edge panel
178 Engine afterburner duct
179 Variable area exhaust nozzle control jacks
180 Engine cowlings/access panels
181 Accessory equipment compartment
182 Pneumatic starter unit
183 Hydraulic fluid cooler
184 Engine variable air intake
185 Intake conical centre-body
186 Centre-body screw jack
187 Alternator
188 Constant speed drive unit
189 Inlet guide vanes
190 General Electric J79-GE-5B afterburning turbojet engine
191 Outboard nacelle pylon fairing
192 Leading-edge pneumatic air ducting
193 Port wing aft main fuel tanks
194 Diagonal multi-spar aft panel construction
195 Chaff stowage boxes
196 Chaff dispenser doors
197 Port main undercarriage wheel bay
198 Mainwheel doors
199 Hydraulic reservoir
200 Wheel brake hydraulic accumulator
201 Main undercarriage retraction jack
202 Undercarriage leg hinge links
203 Main undercarriage leg strut
204 Honeycomb wing skin panels
205 Fuel pod tail fins
206 Air system heat exchanger
207 Heat exchanger air ducting
208 Eight-wheel main undercarriage bogie
209 Inboard engine nacelle construction
210 Intake ducting
211 Moveable intake conical centre-body
212 Fresh air intake ducting
213 Pneumatic system piping
214 Inboard engine pylon construction
215 Pylon attachment joint
216 Port wing forward main fuel tanks
217 Wheel bay fairing
218 Honeycomb skin panel
219 Forward wing panel multi-spar construction
220 Front spar
221 Forward AN/ALQ-16 radar warning aerials
222 Aerial modulator unit
223 Fuselage centreline pylon
224 Aircraft system fuel connector
225 Ready/safe switch disconnect
226 Control unit disconnect.
227 MB-1C combined fuel and weapons pod
228 Pod integral fuel tank, total capacity 4856 US gal (15732 litres)
229 Integral munitions bay
230 Electrical system disconnect
231 Forward pylon latch
232 Fuel tank bulkhead
233 Retractable pitot tube
234 Camera electronic control units
235 KA-56 reconnaissance camera
236 Camera aperture
237 LA-331A reconnaissance pod, nose section
238 Wing root pylon
239 Ejector release unit
240 B43 free-fall nuclear weapon (four)
241 Ready/safe ground arming switch

Douglas A3D/A-3 Skywarrior

Douglas's A3D Skywarrior was built for strategic nuclear bombing of Russia, but until the 1991 Gulf War its only combat action was in Southeast Asia, where conventional ordnance was expended against tactical targets. More importantly, the 'Whale' proved adaptable to other roles, including air refuelling and electronic warfare.

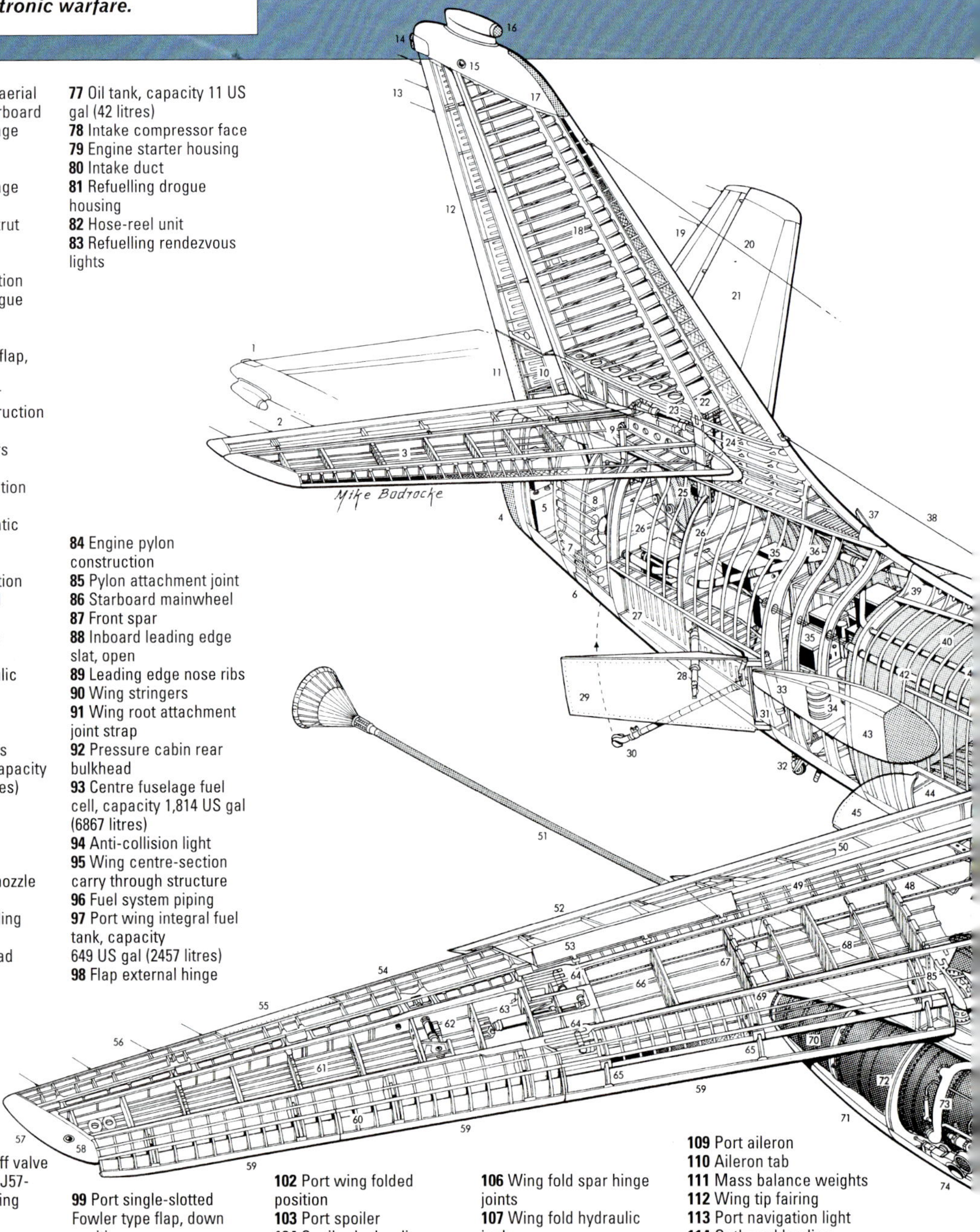

EKA-3B

Cutaway key

1 Tailfin folded position
2 Starboard elevator construction
3 Tailplane rib construction
4 Aft ECM aerial fairing
5 ECM transmission equipment
6 Brake parachute door
7 Parachute stowage
8 Aft fuselage framing
9 Tailplane pivot joint
10 Lower rudder segment
11 Rudder trim tab
12 Upper rudder segment construction
13 Static dischargers
14 Tail warning radar antenna
15 Tail navigation light
16 Forward radar warning receiver
17 Fin tip VHF aerial fairing
18 Tailfin rib construction
19 Port elevator
20 HF aerial cable
21 Port trimming tailplane
22 Fin fold hinge joint
23 Elevator hydraulic actuator
24 Fin fold jack
25 Tailplane trim screw jack
26 Fin/tailplane attachment bulkheads
27 Starboard airbrake housing
28 Airbrake hydraulic jack
29 Starboard airbrake, open
30 Arrester hook lowered
31 Airbrake hinge joint
32 Retractable tail bumper
33 Tail bumper hydraulic jack
34 Hydraulic reservoirs (two)
35 Avionics equipment racks
36 Fuel jettison pipe
37 Port airbrake, open
38 HF aerial cable
39 Fin root fillet construction
40 Rear fuselage fuel cell capacity 1,300 US gal (4921 litres)
41 Fuel vent piping
42 Control cable duct
43 AN/ALQ-86 ECM aerial fairing, port and starboard
44 Main undercarriage wheel well
45 Mainwheel door
46 Main undercarriage rear breaker strut
47 Mainwheel leg strut
48 Rear spar
49 Flap shroud ribs
50 Flap rib construction
51 Air refuelling drogue line
52 Starboard single slotted Fowler-type flap, down position
53 Starboard spoiler
54 Aileron rib construction
55 Aileron tab
56 Static dischargers
57 Wing-tip fairing
58 Starboard navigation light
59 Outboard automatic leading-edge slat segments, open
60 Slat rib construction
61 Outer wing panel construction
62 Aileron hydraulic actuator
63 Wing fold hydraulic jack
64 Spar hinge joints
65 Slat guide rails
66 Inner wing panels integral fuel tank, capacity 649 US gal (2457 litres)
67 Wing panel rib construction
68 Lower wing skin/stringer panel
69 Engine exhaust nozzle
70 Jet pipe
71 Detachable cowling panels
72 Fireproof bulkhead
73 Engine bleed air ducting
74 Bleed air blow-off valve
75 Pratt & Whitney J57-P-10 non after burning turbojet engine
76 Engine accessory equipment
77 Oil tank, capacity 11 US gal (42 litres)
78 Intake compressor face
79 Engine starter housing
80 Intake duct
81 Refuelling drogue housing
82 Hose-reel unit
83 Refuelling rendezvous lights
84 Engine pylon construction
85 Pylon attachment joint
86 Starboard mainwheel
87 Front spar
88 Inboard leading edge slat, open
89 Leading edge nose ribs
90 Wing stringers
91 Wing root attachment joint strap
92 Pressure cabin rear bulkhead
93 Centre fuselage fuel cell, capacity 1,814 US gal (6867 litres)
94 Anti-collision light
95 Wing centre-section carry through structure
96 Fuel system piping
97 Port wing integral fuel tank, capacity 649 US gal (2457 litres)
98 Flap external hinge
99 Port single-slotted Fowler type flap, down position
100 Nacelle tail fairing
101 Flap hydraulic jack
102 Port wing folded position
103 Port spoiler
104 Spoiler hydraulic actuator
105 Fuel pumps
106 Wing fold spar hinge joints
107 Wing fold hydraulic jack
108 Aileron hydraulic actuator
109 Port aileron
110 Aileron tab
111 Mass balance weights
112 Wing tip fairing
113 Port navigation light
114 Outboard leading edge slat segments, open
115 Nacelle pylon

Flight deck crewmen push an EA-3B belonging to VQ-1 into position aboard the USS* Nimitz*. The 'Whale' spent 35 years in service with the Navy, and the last ERA-3Bs of VAQ-33 and EA-3Bs of VQ-2 retired as recently as 1991. The longevity of the Skywarrior is of particular note since it was produced only in limited numbers (the last of 283 aircraft being accepted in January 1961). What is more, the aircraft have had busy lives, spending much of their time at sea and flying combat sorties throughout the Vietnam War and Operation Desert Storm.

SPECIFICATION

A-3B Skywarrior

Dimensions

Length with probe: 78 ft (23.77 m)
Length without probe: 74.5 ft (22.71 m)
Wing span: 72.5 ft (22.10 m)
Span with wings folded: 49.4 ft (15.06 m)
Height: 22.8 ft (6.95 m)
Height with fin folded: 15.9 ft (4.85 m)
Wing area: 812 sq ft (75.44 m^2)

Powerplant

Two Pratt & Whitney J57-P-10 non-afterburning turbojets with maximum rating of 10,500 lb st (46.7 kN) and normal rating of 9,000 lb st (40 kN)

Weights

Empty: 39,620 lb (17971 kg)
Design: 55,942 lb (25375 kg)
Combat: 62,089 lb (28163 kg)
Maximum carrier take-off: 73,000 lb (33112 kg)
Maximum field take-off: 78,000 lb (35380 kg)
Overload field ops: 83,259 lb (37765 kg)
Maximum carrier landing: 49,000 lb (22226 kg)
Maximum field landing: 56,000 lb (25401 kg)

Fuel and load

Standard internal fuel: 4338 US gal (16421 litres) though provisions were made for carrying a 748-US gal (2831-litre) auxiliary tank in the upper bomb bay.
Tanker operations: one 1,224-US gal (4633-litre) tank fitted to the lower bomb bay brought maximum fuel up to 40,514 lb (18367 kg)
Maximum load: 12,800 lb (5806 kg)

Performance

Maximum speed at sea level: 556 kt (640 mph; 1030 km/h)
Maximum speed at 35,000 ft (10670 m): 508 kt (585 mph; 941 km/h)
Cruising speed at 35,000 ft (10670m): 436 kt (501 mph; 807 km/h)
Maximum climb rate: 6,510 ft/min (33 m/sec)
Service ceiling: 41,000 ft (12495 m)
Combat ceiling: 39,900 to 42,300 ft (12160 to 12895 m) depending on weight
Unrefuelled combat radius with three Mk 27 nuclear bombs and upper bomb bay auxiliary tank: 1410 nm (1622 miles; 2610 km)
Tanker radius while transferring 19,200 lb (8710 kg) of fuel on a hi-hi-hi sortie: 900 nm (1019 miles; 1640 km)

Armament

After the Aero-21B turret fitted to the initial production aircraft was removed, the A-3B was left without defensive armament. Although a maximum of 12,800 lb (5806 kg) could be carried internally, load was normally limited to 6,300 lb (2858 kg). Typical load included conventional bombs (either 12 x 500-lb/227-kg GPs, 6 x 1,000-lb/454-kg GPs, 8 x 1,600-lb/726-kg APs or 4 x 2,000-lb/908-kg GPs), mines (two Mk 10s or XA-4As, four Mk 25s, six Mk 36s or Mk 52s, or 12 Mk 50s or Mk 53s), or special stores (such as the Mk 27 thermonuclear free-fall bomb with a yield of a few megatons and the Mk 28 thermonuclear bomb with a yield of 350-1100 kT

116 Port engine nacelle
117 Inboard leading edge slat segment, open
118 Pressure cabin section on roof framing
119 Cabin roof escape hatch
120 Electronics equipment racks
121 ECM operators display panels
122 ECM operators swivelling seats (four)
123 Forward fuselage frame and stringer construction
124 Retractable catapult strop hook
125 ALQ-100 sideways-looking radar (SLAR) fairing
126 Ventral access panel
127 ECM operators cabin floor level
128 Forward AN/ALQ-86 ECM aerial fairing, port and starboard
129 Forward radio and electronics equipment
130 Circuit breaker panels
131 Canopy aft glazing
132 Dinghy stowage
133 Rear window sunblinds
134 Signal pistol aperture
135 Cockpit roof ditching hatch
136 Rearward-facing navigator 's seat
137 Co-pilot/Air Refuelling Officer's seat
138 Side console panel
139 Heat exchanger air exhaust
140 Emergency ram air turbine
141 Ventral entry/escape hatch open
142 UHF aerial
143 Nose undercarriage leg strut
144 Forward retracting nosewheel
145 Torque scissor links
146 Cabin air conditioning plant
147 Cockpit floor level
148 Instrument panel
149 Control column handwheel
150 Pilot's seat
151 Instrument panel shroud
152 Windscreen panels
153 Windscreen wiper
154 Fresh air NACA intake
155 Front pressure bulkhead
156 Nosewheel door mounted landing/taxiing lamp
157 ALQ -126 ECM aerial
158 ASB radar scanner
159 Scanner tracking mechanism
160 Radar scanner mounting frame
161 Pitot tube
162 Radome
163 ILS aerial
164 Fixed inflight-refuelling probe

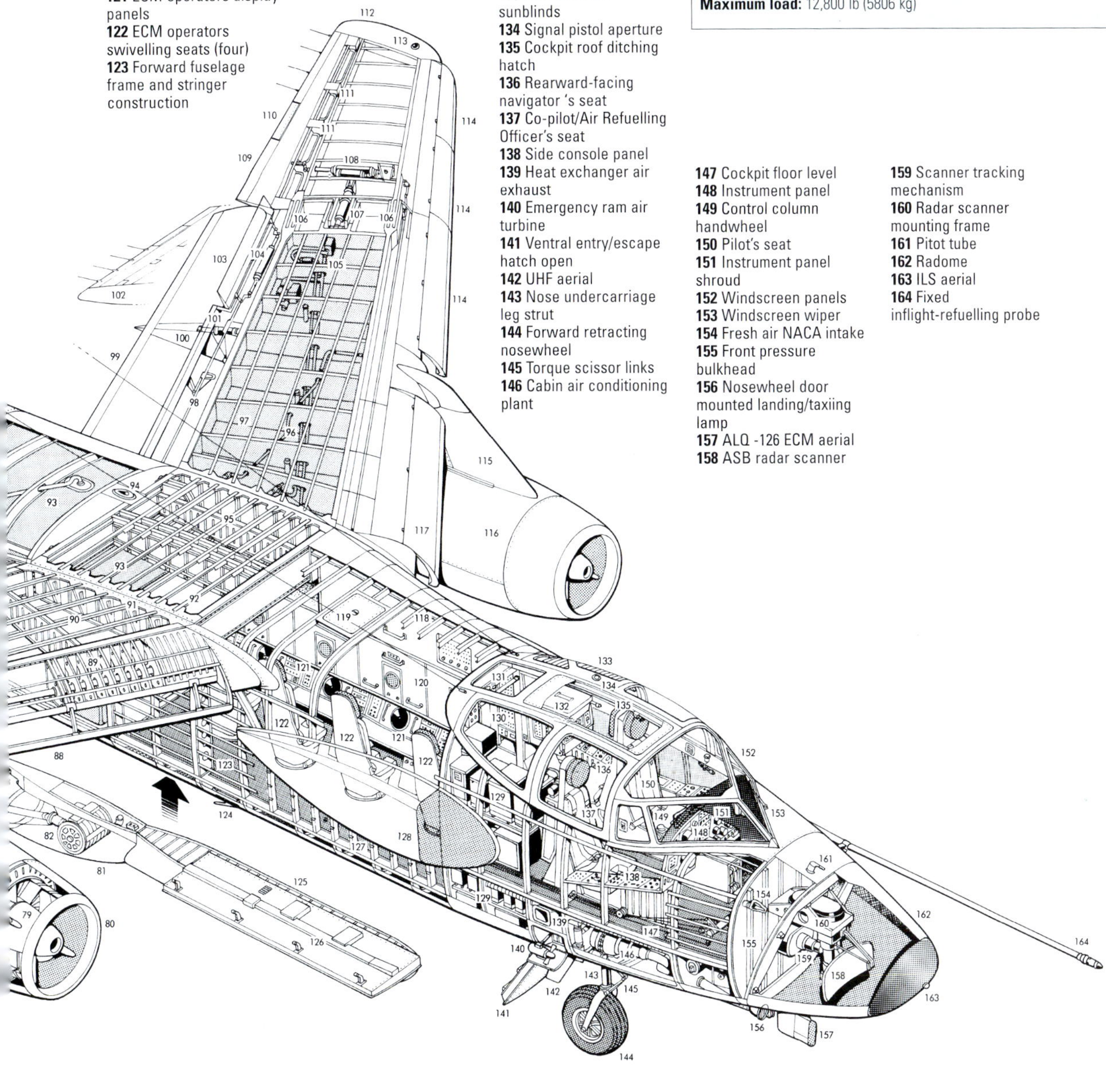

Handley Page Victor

The Victor could carry a theoretical maximum of 48 1,000-lb (454-kg) bombs in its capacious bays, although the normal maximum load was 35 – 14 more than the Vulcan. This remarkable picture captures a full load being dropped by a No. 57 Squadron B.Mk 1, showing the bombs falling as they were stowed: in five clips of seven.

Above right: XA922 was the sixth production B.Mk 1 and the second aircraft to be finished in anti-flash white, instead of the aluminium (silver) scheme adopted by the first four. Later, the serials and national insignia were made much paler.

Victor K.Mk 2

Cutaway key
1 Nose probe
2 Control feel system pressure intake
3 Nose compartment windows
4 Nose construction
5 Inflight-refuelling probe
6 Windscreen
7 Refuelling searchlights
8 Jettisonable roof hatch
9 Co-pilot's Martin Baker ejection seat
10 Pilot's roof hatch windows
11 Pilot's Martin Baker ejection seat
12 Control column
13 Instrument panel
14 Rudder pedals
15 Air intake to air-conditioning system
16 Radome
17 Throttles
18 Pilot's side console
19 Cockpit floor
20 External door handle
21 H_2S radar scanner
22 Radar mounting and equipment
23 Rearward facing crew members' seats
24 Cockpit door
25 Entry steps
26 Front fuselage construction
27 Rear view periscope
28 Rearward facing crew members' work table
29 Cabin side window
30 Air conditioning system
31 Nose freight compartment
32 Instrument panels
33 Radio and electronics racks
34 Pressure bulkhead
35 Air conditioning intake
36 Starboard emergency life-raft hatch
37 Port life-raft pack
38 Forward fuselage connecting construction
39 Wing spar bulkhead
40 Wing centre-section fuel tank
41 Overwing fuel tank
42 Starboard engine intake
43 Intake ducts
44 De-icing air system
45 Underwing fuel tank
46 Starboard wing fuel tanks
47 Fuel flow proportioner
48 De-icing connector to outer wing
49 Vortex generators
50 Starboard Flight Refuelling FR.20B refuelling pod
51 Power turbine propeller
52 Pylon mounting
53 Pitot head
54 Starboard wingtip
55 Starboard aileron
56 Trim tab
57 Refuelling hose
58 Trailing-edge fairing
59 Starboard flap
60 Flap track fairing
61 Flap mechanism
62 Starboard main undercarriage
63 Starboard engine bays
64 Exhaust pipe fairing
65 Bomb bay roof forward fuel tank
66 Forward refuelling bomb bay tank
67 Tank mountings
68 Fuel flow proportioner
69 Fuselage double frames
70 Fuselage stringer construction
71 Bomb-bay roof aft fuel tanks
72 Aft refuelling bomb-bay tank
73 Bomb-bay roof structure
74 Flight Refuelling FR.17B hose reel unit
75 Hose reel jack
76 Reel drive motor
77 Air system piping
78 Bomb-bay aft bulkhead
79 Retractable ram air turbine intakes
80 Rear fuselage fuel tank
81 Air system intake
82 Heat exchanger
83 Fin root fairing
84 Ram air turbine
85 Air exhaust louvres
86 Turbine intake jack
87 Tailplane de-icing air system
88 Fin root fixing
89 Fin leading edge
90 Corrugated inner skin
91 Fin construction
92 Rudder control unit
93 Tailplane fairing
94 Starboard refuelling drogue
95 Starboard tailplane construction
96 Elevator power control unit
97 Elevator construction

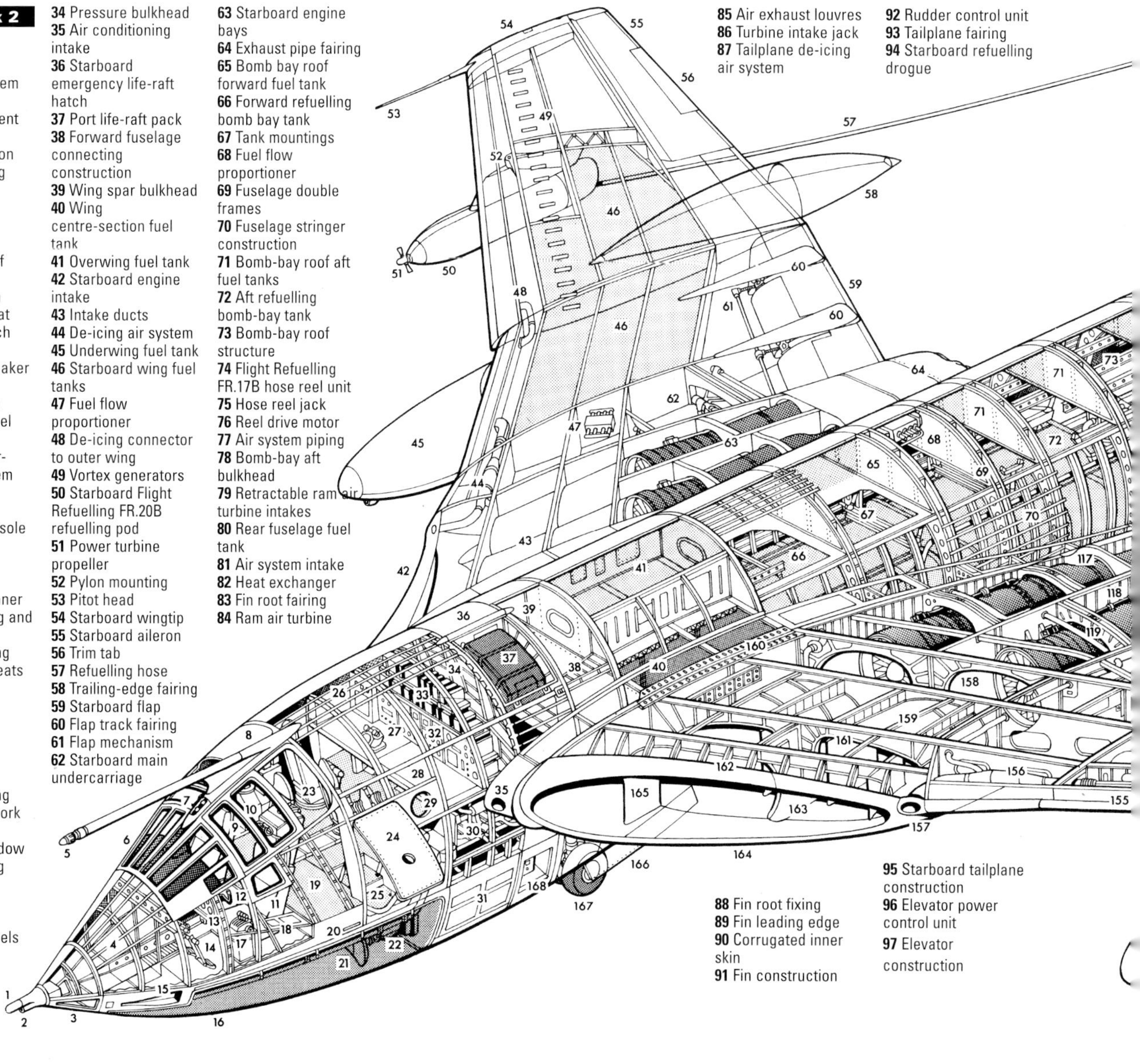

SPECIFICATION

Victor B.Mk 1

Dimensions

Length: 114 ft 11 in (35.02 m)
Height: 26 ft 9 in (8.15 m)
Wingspan: 110 ft (33.53 m)
Wing area: 2,406 sq ft (223.52 m²)

Powerplant

Four Armstrong Siddeley Sapphire 202 or 207 turbojets, each rated at 11,050 lb (49.17 kN) thrust

Weights

Empty: 79,000 lb (35834 kg)
Maximum take-off: 205,000 lb (92988 kg)
Maximum bombload: 35,000 lb (15876 kg)

Performance

Maximum speed at 40,000 ft (12192 m): 627 mph (1009 km/h)
Service ceiling: 56,000 ft (17069 m)
Range: 6,000 miles (9656 km)

Victor B.Mk 2R (as for B.Mk 1, except:)

Dimensions

Wingspan: 120 ft (36.58 m) (113 ft/34.44 m in K.Mk 2)
Wing area: 2,597 sq ft (241.26 m²)

Powerplant

Four Rolls-Royce RCo.17 Conway 201 turbofans, each rated at 20,600 lb (91.67 kN) thrust

Weights

Empty: 114,240 lb (51819 kg)
Maximum take-off: 223,000 lb (101153 kg)

Performance

Maximum speed at 40,000 ft (12192 m): 647 mph (1041 km/h)
Service ceiling: 60,000 ft (18288 m)

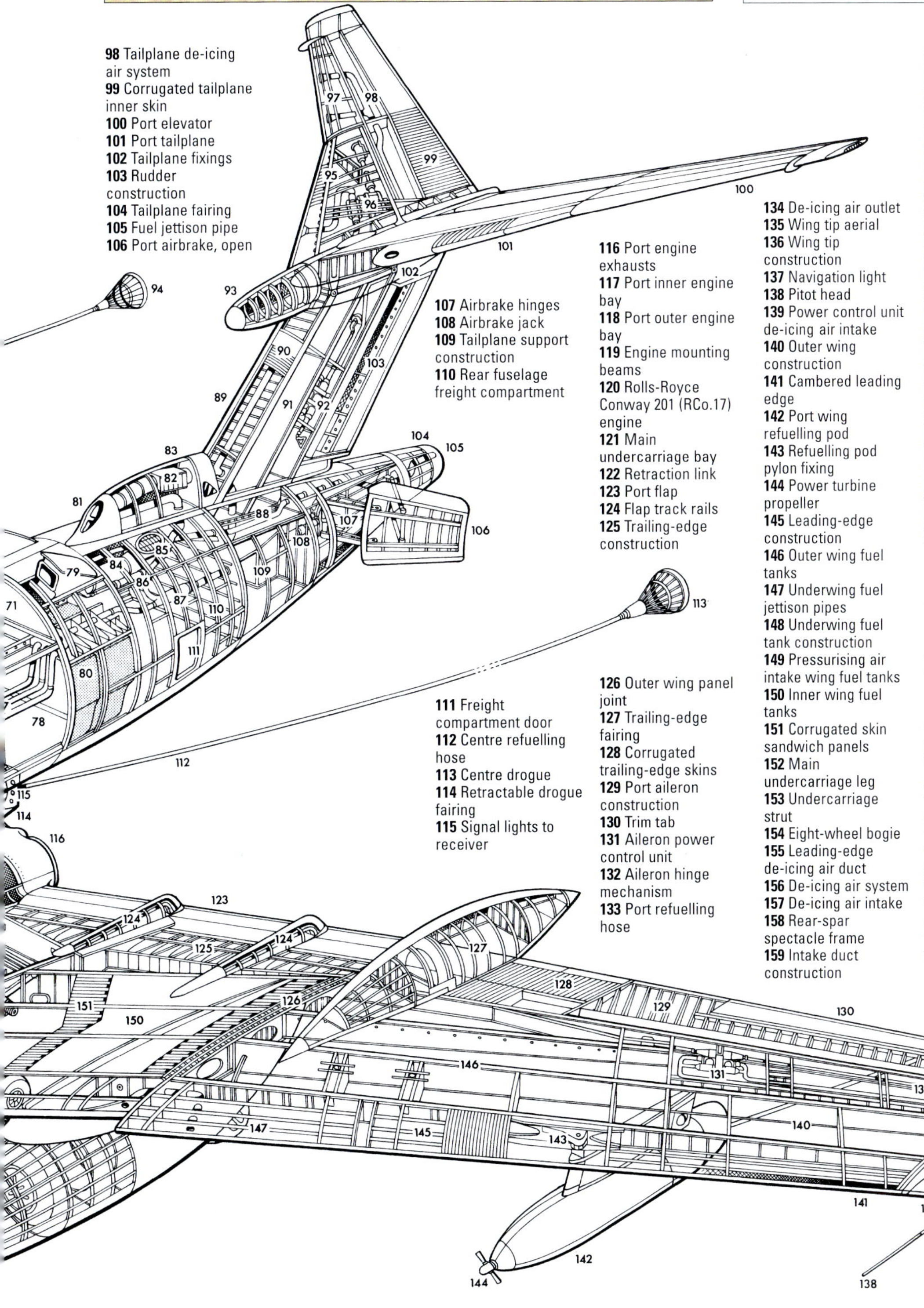

98 Tailplane de-icing air system
99 Corrugated tailplane inner skin
100 Port elevator
101 Port tailplane
102 Tailplane fixings
103 Rudder construction
104 Tailplane fairing
105 Fuel jettison pipe
106 Port airbrake, open
107 Airbrake hinges
108 Airbrake jack
109 Tailplane support construction
110 Rear fuselage freight compartment
111 Freight compartment door
112 Centre refuelling hose
113 Centre drogue
114 Retractable drogue fairing
115 Signal lights to receiver
116 Port engine exhausts
117 Port inner engine bay
118 Port outer engine bay
119 Engine mounting beams
120 Rolls-Royce Conway 201 (RCo.17) engine
121 Main undercarriage bay
122 Retraction link
123 Port flap
124 Flap track rails
125 Trailing-edge construction
126 Outer wing panel joint
127 Trailing-edge fairing
128 Corrugated trailing-edge skins
129 Port aileron construction
130 Trim tab
131 Aileron power control unit
132 Aileron hinge mechanism
133 Port refuelling hose
134 De-icing air outlet
135 Wing tip aerial
136 Wing tip construction
137 Navigation light
138 Pitot head
139 Power control unit de-icing air intake
140 Outer wing construction
141 Cambered leading edge
142 Port wing refuelling pod
143 Refuelling pod pylon fixing
144 Power turbine propeller
145 Leading-edge construction
146 Outer wing fuel tanks
147 Underwing fuel jettison pipes
148 Underwing fuel tank construction
149 Pressurising air intake wing fuel tanks
150 Inner wing fuel tanks
151 Corrugated skin sandwich panels
152 Main undercarriage leg
153 Undercarriage strut
154 Eight-wheel bogie
155 Leading-edge de-icing air duct
156 De-icing air system
157 De-icing air intake
158 Rear-spar spectacle frame
159 Intake duct construction
160 Wing attachment joint
161 Front-spar spectacle frame
162 Intake lip construction
163 Intake duct divider
164 Port engine air intake
165 Intake guide vane
166 Nosewheel bay door
167 Twin nosewheels
168 Battery bay

The badges on the noses of these Victors represent the three reconnaissance squadrons based at Wyton in the early 1970s: No. 39 (Canberra PR.Mk 9), No. 51 (Comet R.Mk 2/Nimrod R.Mk 1/Canberra) and No. 543 (Victor SR.Mk 2).

Ilyushin Il-28 'Beagle'

Romania's H-5s, exemplified here by an HJ-5 trainer, remain among the 494 H-5 aircraf remaining in service early in 2001. Al examples of the original Il-28 have passed ou of service, but China, North Korea an Romania all still fly H-5s

Il-28 'Beagle'

Cutaway key

1 Nose compartment glazing
2 Optically flat sighting panel
3 OPB-5 bombsight
4 Folding chart table
5 Nose undercarriage pivot fixing
6 Cannon muzzle
7 Nosewheel leg door
8 Twin nosewheels, aft retracting
9 Nosewheel steering unit
10 Retraction strut
11 Fixed NR-23 23-mm cannon, port and starboard
12 Ammunition tank, 100 rounds
13 Navigation equipment racks
14 Navigator's instrument panels
15 Nose compartment entry/escape hatch, offset to starboard
16 Navigator/ bombardie's ejection seat
17 Cockpit dividing bulkhead
18 Rudder pedals
19 Pilot's instrument panel
20 Engine throttle levers
21 Control column handwheel
22 Direct vision opening side window panel
23 Instrument panel shroud
24 Windscreen panels
25 Cockpit canopy cover, hinged to starboard
26 Ejection seat headrest
27 Safety harness
28 Canopy latch
29 Pilot's ejection seat
30 Side console panel
31 Underfloor control runs
32 Nosewheel leg doors
33 Oxygen bottles, air conditioning equipment on starboard side
34 Cockpit rear pressure bulkhead
35 Radar transmitter and receiver
36 Navigation and bombing radar scanner
37 Ventral radome
38 Forward fuselage fuel tanks, total internal capacity 9,921 lb (4500 kg)
39 Control duct access panel
40 Fuselage upper longeron
41 Cockpit aft fairing
42 Aerial mast
43 DF aerial fairing
44 Fuselage skin panelling
45 Fuel filler caps
46 Internal bomb bay; maximum load 6,614 lb (3000 kg)
47 Ventral strike/ reconnaissance camera
48 Bomb bay doors
49 5,511-lb (2500-kg) aerial mine
50 Fuselage control and cable ducting
51 Forward fuselage frame and stringer construction
52 Dorsal fuel tank
53 Wing spar attachment fuselage main frame
54 Wing centre-section carry through
55 Bolted wing root attachment joint
56 Wing stringers
57 Inboard leading-edge nose ribs
58 Starboard engine nacelle
59 Starboard mainwheel, stored position
60 Engine bay heat shield above wheel bay
61 Engine bearer struts
62 Starboard engine bay
63 Intake centre body fairing
64 Detachable nose cowling
65 Starboard engine intake
66 Il-28U 'Mascot' trainer variant, nose section profile
67 Leading edge hot air de-icing
68 Starboard outerwing panel
69 Fuel filler cap
70 Detachable long range wing tip tank; overload fuel capacity 14,550 lb (6600 kg)
71 Starboard aileron
72 Aileron hinge control

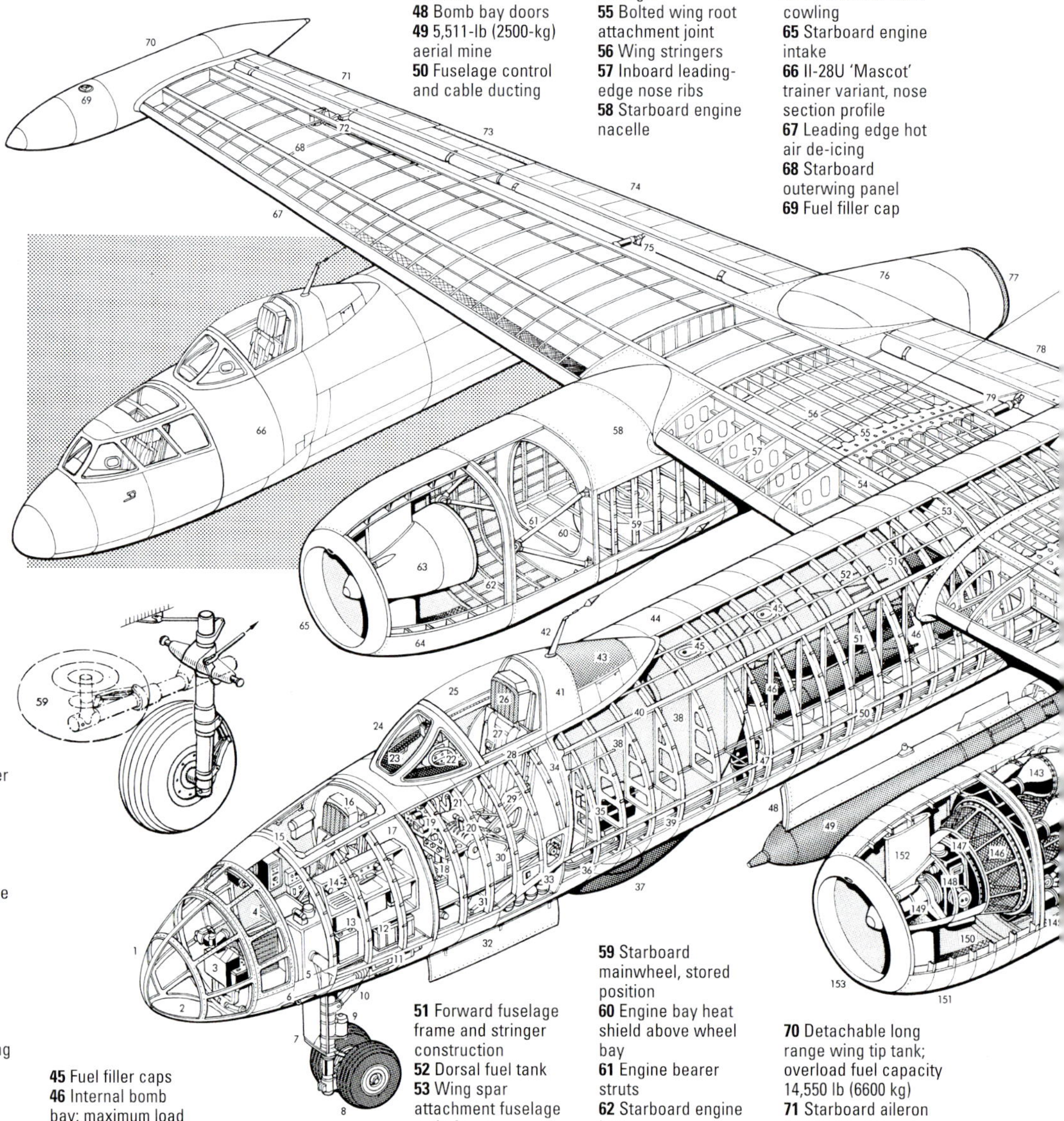

Above: Groundcrew at work on a Polish Il-28R. The recce 'Beagle' typically carried four vertical or oblique cameras in its modified bomb bay.

SPECIFICATION

Il-28 'Beagle'

Dimensions

Fuselage length (excluding cannon): 57 ft 11 in (17.65 m)
Height: 21 ft 11¾ in (6.70 m)
Wingspan (without tip tanks): 70 ft 4½ in (21.45 m) without tip tanks
Wing area: 654.47 sq ft (60.80 m²)

Powerplant

two Klimov VK-1A turbojet engines each rated at 5,952 lb st (26.48 kN)

Weights

Empty: 28,417 lb (11890 kg)
Normal take-off with 8,377 lb (3800 kg) of fuel, and a 2,200-lb (1000-kg) bomb load: 40,564 lb (1840 kg)
Maximum take-off: 46,738 lb (21200 kg)

Performance

Maximum speed at 14,765 ft (4500 m): 560 mph (902 km/h)
Climb to 32,810 ft (10000 m): 18 minutes
Maximum range at 32,810 ft (10000 m): 1,491 miles
Service ceiling: 40,350 ft (12300 m)

Armament

two 23-mm Nudel'man-Rikhter NR-23 fixed forward-firing cannon in the lower nose and two 23-mm Nudel'man-Rikhter NR-23 trainable rearward-firing cannon in the tail turret, plus up to 6,614 lb (3000 kg) of disposable stores carried in a lower-fuselage weapons bay, and generally comprising one 6,614-lb (3000-kg) FAB-3000 free-fall bomb, or four 1,102-lb (500-kg) FAB-500 free-fall bombs or eight 551-lb (250-kg) FAB-250 free-fall bombs

.Below right: This photograph of the first Il-28 prototype shows the aircraft after some degree of modification. However, it still retains the ventral strakes and rear-mounted radar location of its original form.

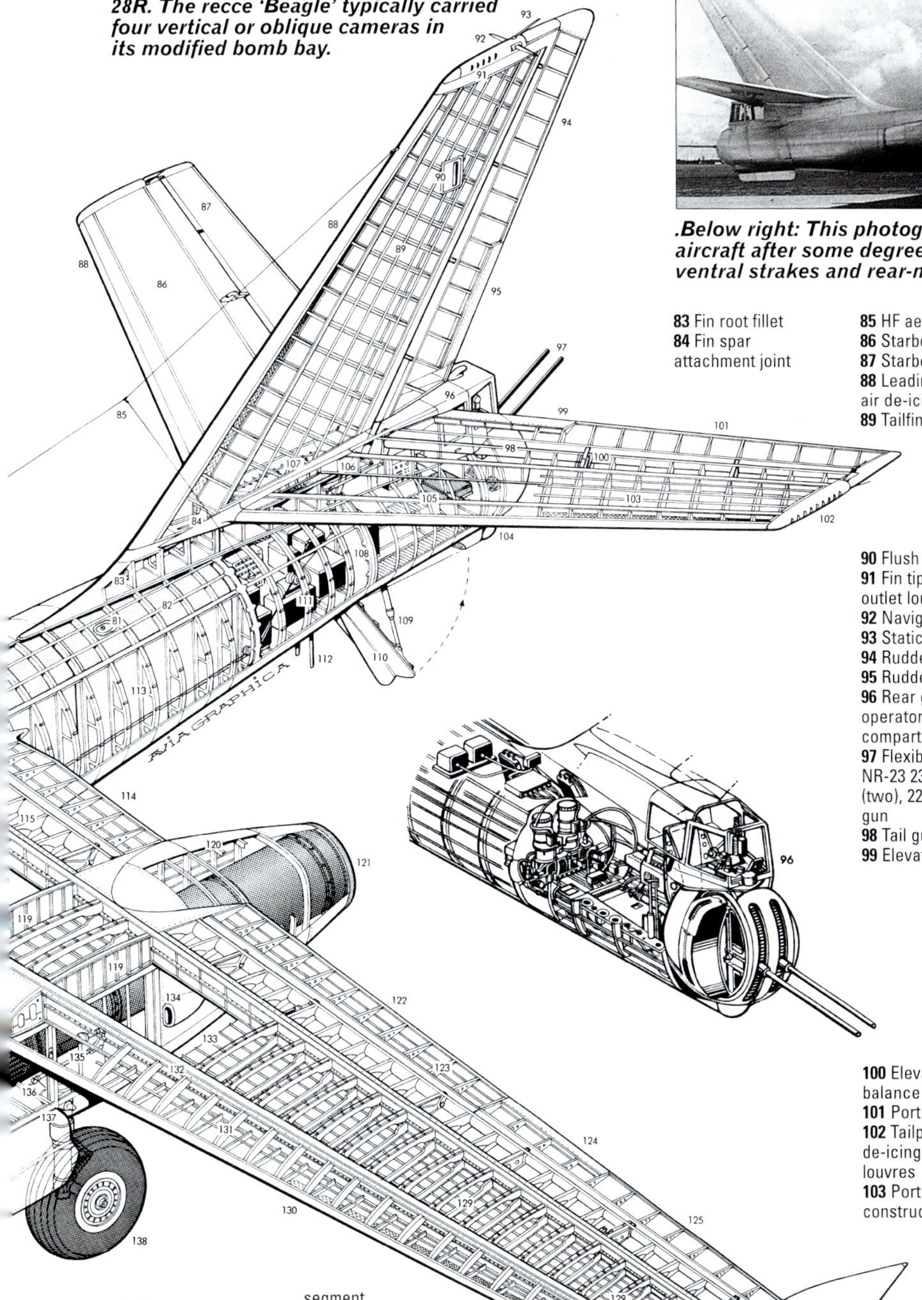

73 Aileron tab
74 Outboard plain flap segment
75 Flap hydraulic jack
76 Nacelle tail fairing
77 Engine exhaust nozzle
78 Inboard plain flap segment
79 Inboard flap hydraulic jack
80 Rear spar attachment fuselage main frame
81 Fuel filler caps
82 Rear fuselage fuel tanks
83 Fin root fillet
84 Fin spar attachment joint
85 HF aerial cable
86 Starboard tailplane
87 Starboard elevator
88 Leading edge hot air de-icing
89 Tailfin construction
90 Flush aerial
91 Fin tip de-icing air outlet louvres
92 Navigation aerial
93 Static discharger
94 Rudder
95 Rudder tab
96 Rear gunner/radio operator's compartment
97 Flexible mounted NR-23 23-mm cannon (two), 225 rounds per gun
98 Tail gun turret
99 Elevator tab
100 Elevator mass balance
101 Port elevator
102 Tailplane tip de-icing air outlet louvres
103 Port tailplane construction
104 Tail gun ranging antenna
105 Tail gunner/radio operator's seat
106 Rudder and elevator hinge controls
107 Tailplane centre-section bolted joint
108 Tail crew compartment pressure bulkhead
109 Door hydraulic jack
110 Ventral entry door/escape hatch
111 Radio and electronics equipment bay
112 Ventral aerials
113 Rear fuselage frame construction
114 Inboard plain flap segment
115 Flap shroud ribs
116 Rear spar
117 Bolted wing root attachment joint
118 Jettisonable rocket assisted take-off bottle (RATO)
119 Engine nacelle attachment main ribs
120 Tailpipe fairing
121 Port exhaust nozzle
122 Outboard plain flap segment
123 Flap rib construction
124 Aileron tab
125 Port aileron rib construction
126 Detachable wing-tip tank
127 Port navigation light
128 Fuel filler cap
129 Wing upper and lower panel half-rib construction
130 Leading-edge hot air de-icing (inoperable with tip tanks fitted)
131 Leading-edge nose ribs
132 Front spar
133 Lower wing skin/ stringer panel
134 Jet pipe cooling air scoop
135 Main undercarriage hydraulic retraction jack
136 Leg swivelling link
137 Mainwheel leg pivot fixing
138 Port mainwheel, forward retracting
139 Mainwheel doors
140 Heat shrouded jet pipe
141 Engine mounting bulkhead
142 Engine bearer struts
143 Klimov VK-1 centrifugal flow turbojet engine
144 Engine flame cans
145 Retractable landing/ taxiing lamp
146 Compressor intake filter screens
147 Gearbox mounting ring frame
148 Engine accessory equipment gearbox
149 Generator
150 Intake plenum chamber
151 Detachable front cowling
152 Intake centre fairing support struts
153 Port engine intake

Martin B-57 Canberra

Conspicuous due to its Dayglo orange and silver colour scheme, the B-57E was configured for target-towing duties. Combat equipment was replaced by four 1,500-ft (457-m) tow reels mounted in the bomb-bay door, though normally only two were carried, with operating equipment fitted in the rear cockpit.

RB/WB-57F Canberra

Cutaway key

1 Starboard navigation light
2 Leading edge anti-erosion sheathing
3 Glassfibre wingtip fairing
4 Outboard fuel tank
5 Spoiler hydraulic actuator
6 Starboard roll control spoiler, open
7 Outboard fixed portion of trailing edge
8 Static dischargers
9 Fuel boost pumps
10 Air sampling pod
11 Fuel system piping
12 Aileron/tab hinge control
13 Starboard aileron
14 Aileron geared tab
15 Fixed tab
16 Inboard fixed trailing-edge segment
17 Ventral fuel jettison pipe
18 Inboard three-bay main fuel tank
19 Fuel filler caps
20 Fuel tank or boost engine-mounting pylon
21 Starboard demountable auxiliary boost engine pod
22 Main engine fan section, detachable cowling panels
23 Fan air intake
24 Intake centre-body
25 Ventral oil cooler
26 Cockpit canopy sun blinds
27 Navigational sextant
28 Cockpit canopy, upward-hinged
29 Pitot head
30 Instrument panel shroud
31 Windscreen panels
32 Windscreen de-icing air duct
33 Alternative nose section (gust response measurement experiments)
34 Nose compartment hinged for access
35 Radome
36 Gas-sampling probes
37 ILS glideslope aerial
38 Temperature probe
39 Radar scanner
40 Scanner tracking mechanism
41 Ventral periscope housing
42 Alternative nose section (X-band sideways-looking radar)
43 Equipment cooling air scoop
44 Forward special mission equipment bay
45 Cockpit front pressure bulkhead
46 Rudder pedals
47 Control column
48 Cockpit floor level
49 Underfloor control linkages
50 Lower UHF aerial
51 Side console panel
52 Engine throttle levers
53 Pilot's McDonnell Escapac IC-6 ejection seat
54 Canopy hydraulic actuator
55 Folding chart board
56 Cockpit section framing
57 Nose wheel leg pivot fixing
58 Twin nose wheels
59 Nose wheel doors
60 Hydraulic retraction jack
61 Special Equipment Operator's (SEO) Escapac IC-6 ejection seat
62 SEO's instrument panel
63 Liquid oxygen converter
64 Cockpit rear pressure bulkhead
65 Radio and electronics equipment bay
66 Battery compartment
67 Ventral equipment pallet
68 Upper pressurised special equipment compartment
69 Avionics and experimental equipment racks
70 Front spar attachment double frame
71 Fuselage keel member
72 Wing main spar attachment bulkhead
73 Upper anti-collision light
74 Starboard main undercarriage wheel bay
75 Starboard mainwheel
76 Landing/taxing lamp
77 Leading-edge honeycomb skin panel
78 Hydraulic reservoir
79 Starboard main engine bay
80 Engine bay cooling air outlet duct
81 Turbine section venting air intakes
82 Main wing spar ring frames
83 Main undercarriage leg pivot fixing
84 Hydraulic retraction jack and side strut
85 Wingroot section construction
86 Fuselage skin panelling
87 VHF/communications aerial
88 Centre fuselage frame and stringer construction
89 Aileron and spoiler differential linkage
90 Wing rear spar attachment bulkhead
91 Aft equipment bay
92 Ground locks and tail stand stowage
93 Radio compass sense aerial
94 Upper UHF/ communications aerial
95 Fin leading edge
96 Starboard tailplane
97 HF aerial cable
98 Starboard elevator
99 Two-spar fin construction
100 Fin and rudder honeycomb skin panelling
101 VHF/nav aerial
102 Rudder horn balance
103 Static dischargers
104 Rudder
105 Rudder tab control
106 Rudder trim tab
107 Tailcone
108 Aft air data sensors
109 Tail navigation light
110 Elevator geared tab
111 Port elevator rib construction
112 Elevator horn balance
113 Tailplane rib construction
114 Vortex generators, above and below
115 Tab/elevator control linkage
116 Trimming tailplane incidence control screw jack
117 Tailplane rear-bracing strut
118 Fin-tailplane attachment mainframe
119 Tailplane spar attachment joint
120 Tailbumper
121 Rudder hydraulic actuator
122 Tailplane push-pull control rods
123 Rear fuselage frame and stringer construction
124 Starter kit stowage
125 Lower anti-collision light
126 Fixed trailing-edge panel
127 Jet pipe
128 Engine exhaust nozzle
129 'Eyeshade' exhaust nozzle fairing
130 Fixed trailing-edge honeycomb skin panels
131 Static dischargers
132 Fuel jettison pipe
133 Port aileron
134 Fixed tab
135 Aileron geared tab

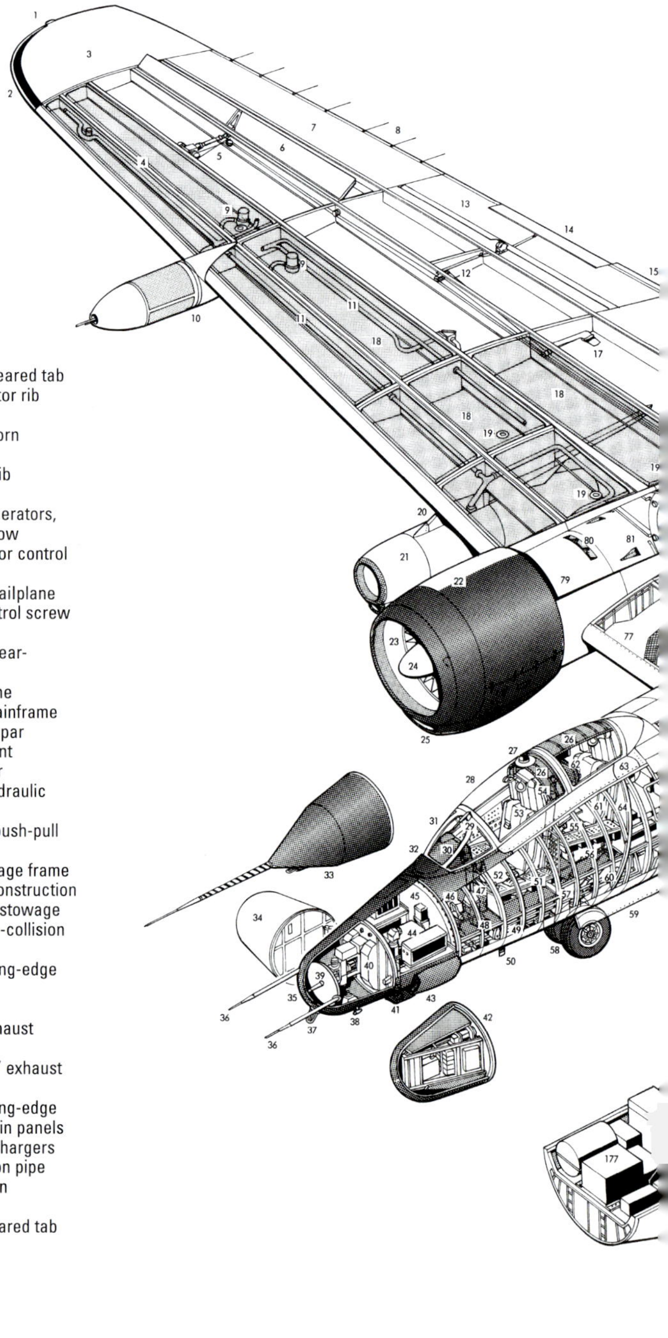

Natural metal finish was characteristic of the early production examples of the B-57A, as on this aircraft photographed on a test flight near the company's Middle River production plant. Externally similar to the Canberra B.Mk 2, the B-57A was powered by Wright-built Sapphire turbojets.

SPECIFICATION

RB-57E Canberra

Dimensions

Wingspan: 64 ft (19.51 m)
Length: 65 ft 6 in (19.96 m)
Height: 15 ft 7 in (4.75 m)
Wing area: 960.0 sq ft (89.19 m²)

Powerplant

Two Wright J65-W-1 or W-5 (Sapphire) turbojets each rated at 7,200 lb (32.4 kN) thrust

Weights

Empty: 27,000 lb (12247 kg)
Normal take-off: 45,000 lb (20412 kg)
Maximum take-off: 55,000 lb (24948 kg)

Performance

Maximum speed without tiptanks up to 5,000 ft (1524 m): 575 mph (925 km/h)
Maximum speed without tiptanks between 5,000 ft (1524 m) and 20,000 ft (6096 m): Mach 0.82
Maximum speed without tiptanks above 20,000 ft (6096 m): Mach 0.83 to 0.85
Service ceiling: 53,000 ft (16154 m)
Range with reserve fuel tanks: 2,300 miles (3701 km)

Armament

The bomber version of the B-57 could house four 750-lb (340-kg), nine 500-lb (227-kg), or 21 260-lb (118-kg) bombs, or two Mk 9 special weapons; four wing pylons could carry four 750-lb (340-kg) bombs, eight 5-in (127-mm) HVAR or 28 2.75-in (70-mm) FFAR rockets (double if eight pylons were used); eight M3 0.5-in (12.7-mm) machine-guns with 300 rounds, or four M39 20-mm cannon with 200 rounds

136 Aileron honeycomb core construction
137 Outboard fixed trailing-edge section
138 Static dischargers
139 Glassfibre/ honeycomb wingtip fairing construction
140 Port navigation light
141 Leading-edge anti erosion sheathing
142 Honeycomb sandwich wing skin panelling
143 Port spoiler
144 Spoiler hydraulic actuator
145 Spoiler/aileron interconnecting cables
146 Aileron tab hinge control
147 Lower wing skin access panels
148 Port wing outboard fuel tank bays
149 Fuel filler cap
150 Air sampling pod
151 Leading-edge nose ribs
152 Multi-spar outer wing panel construction
153 Tank bay dividing wing ribs
154 Port auxiliary boost engine pod
162 Inboard wing fuel tank bays
163 Fuel filler caps
164 Main sparring frames
165 Pratt & Whitney TF33-P11A turbofan engine
166 Engine bay fire wall
167 Port mainwheel, retracted position
168 Air-conditioning plant
169 Leading-edge heat exchanger air intake

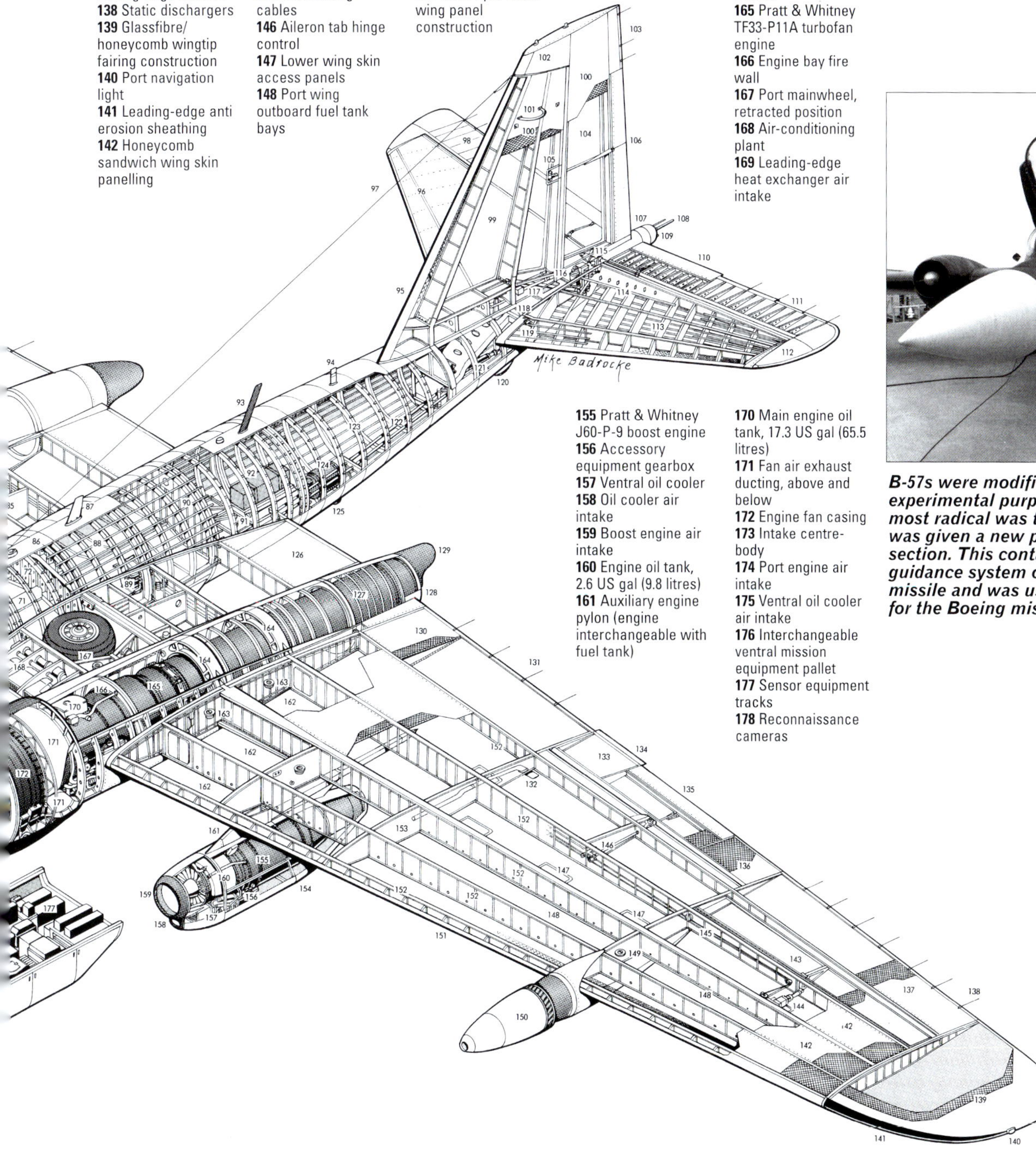

155 Pratt & Whitney J60-P-9 boost engine
156 Accessory equipment gearbox
157 Ventral oil cooler
158 Oil cooler air intake
159 Boost engine air intake
160 Engine oil tank, 2.6 US gal (9.8 litres)
161 Auxiliary engine pylon (engine interchangeable with fuel tank)
170 Main engine oil tank, 17.3 US gal (65.5 litres)
171 Fan air exhaust ducting, above and below
172 Engine fan casing
173 Intake centre-body
174 Port engine air intake
175 Ventral oil cooler air intake
176 Interchangeable ventral mission equipment pallet
177 Sensor equipment tracks
178 Reconnaissance cameras

B-57s were modified for a number of experimental purposes. One of the most radical was this aircraft, which was given a new pointed nose section. This contained the guidance system of a Bomarc missile and was used during tests for the Boeing missile's programme.

Northrop Grumman B-2A Spirit

AV-6 is seen here on a humid day over Edwards in March 1994 with strong laminal condensation flow building up as the aircraft approaches its critical Mach number. B-2s operating in hostile conditions would have to be careful to avoid tell-tale effects like this as they would be highly visible.

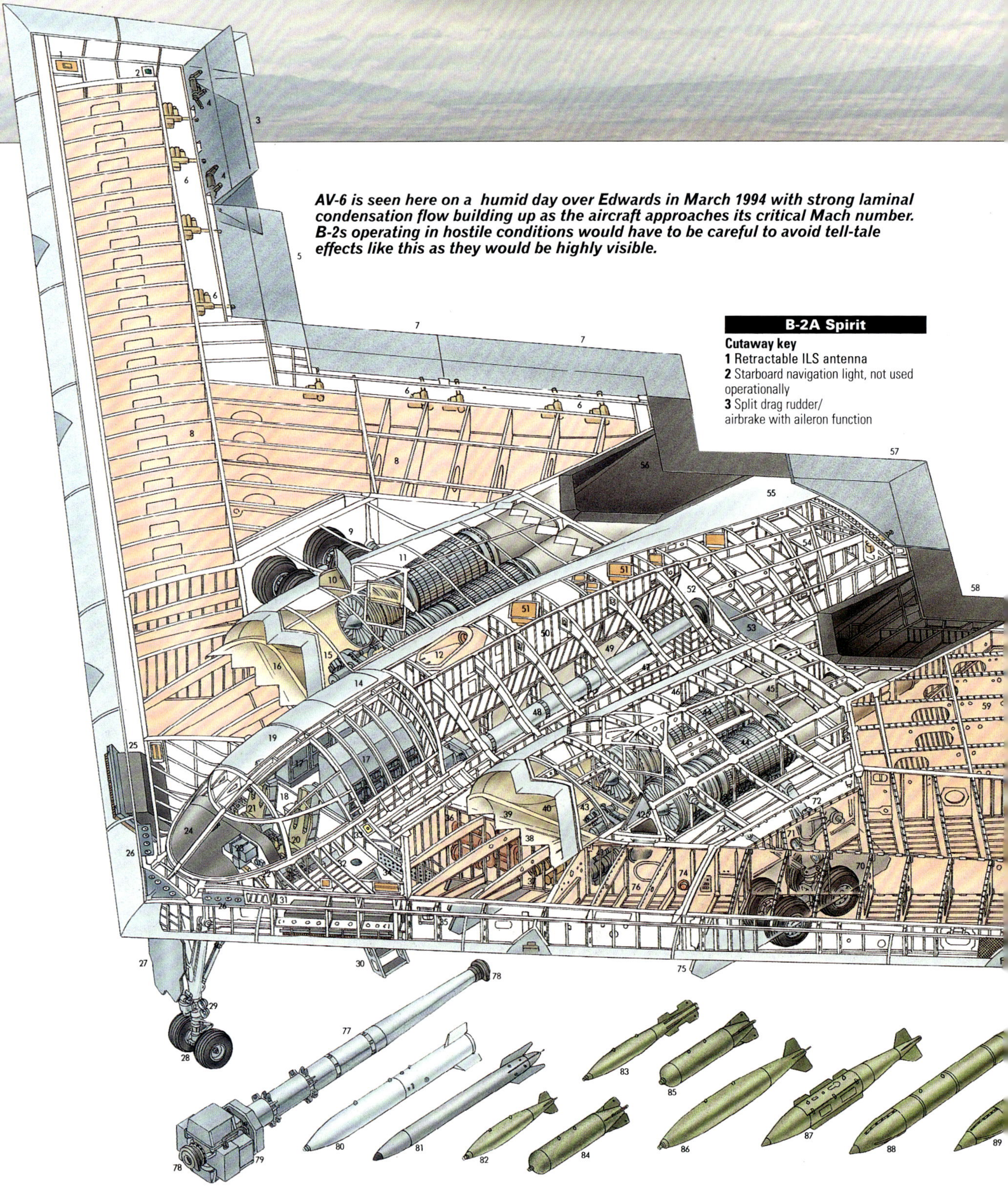

B-2A Spirit

Cutaway key

1 Retractable ILS antenna
2 Starboard navigation light, not used operationally
3 Split drag rudder/ airbrake with aileron function

SPECIFICATION

B-2A Spirit

Dimensions

Length: 69 ft (21.03 m)
Height: 17 ft (5.18 m)
Wingspan: 172 ft (52.43 m)
Wing aspect ratio: 5.92
Wing area: 5000 sq ft (464.5 m²) plus
Wheel track: 40 ft (12.2 m)

Powerplant

Four General Electric F118-GE-100 non-afterburning turbofans each rated at 19,000 lb (84.5 kN)

Weights

Empty: 153,700 lb (69717 kg)
Normal take-off: 336,500 lb (152635 kg)
Maximum take-off: 376,000 lb (170550 kg)
Maximum wing loading: 75.20 lb/sq ft (367.2 kg/m²)
Maximum power loading: 5.43 lb/lb st (554 kg/kN)

Fuel and load

Internal fuel capacity: 180,000 lb (81635 kg)
Maximum weapon load: 40,000 lb (22700 kg)

Performance

Cruising speed at high altitude: Mach 0.85 or 416 kt (475 mph; 900 km/h)
Cruising speed at low altitude: Mach 0.8 or 530 kt (610 mph; 980 km/h)
Approach speed: 140 kt (161 mph, 259 km/h)
Service ceiling: 50,000 ft plus (15240 m)
Range with one flight refuelling: 10,000 nm (11,515 miles; 18532 km)
Range unrefuelled with 8 SRAM and 8 B83: 4,410 nm (5075 miles; 8167 km)

Armament

Two side-by-side weapons bays in the lower centrebody house Boeing rotary launcher assemblies (RLA) which are detachable and each house eight large stores. Total capacity is 16 AGM-129 ACMs or 16 AGM-69 SRAM IIs. Designed with a nuclear mission in mind, the B-2 can carry 16 B83 or 16 B61 freefall nuclear bombs. For conventional missions, 80 Mk 82 500-lb (227-kg), 16 Mk 84 2,000-lb (908-kg) bombs, 36 CBU-87, -89, -97 and -98 cluster bombs, 80 Mk 36 560-lb (254-kg) or Mk 62 sea mines, 36 M117 750-lb (340-kg) fire bombs, 8 GAM-11 deep penetration bombs, 16 GAM-84s, or 16 JDAMs, can be carried

AV-1 and AV-2 prepare for an early morning launch at Edwards South Base. Testing for the B-2 was centred at South Base where a secure and purpose-built home for the aircraft was set up. It is now named the Birk Flight Test facility in honour of Colonel Frank Birk, the former B-2 director who was killed in an air crash.

4 Drag rudder rotary actuators
5 Outboard elevon
6 Elevon hydraulic actuators
7 Inboard elevons
8 Starboard integral fuel tankage, total capacity approximately 130,000 lb (58967 kg)
9 Starboard main undercarriage, stowed position
10 Intake suction relief doors
11 Starboard engine bays
12 Rotating inflight-refuelling receptacle
13 Cockpit rear pressure bulkhead
14 Structural cut-out, provision for third crew member ejection seat
15 Engine-driven auxiliary equipment gearbox
16 Starboard engine combined air intake
17 Avionics equipment racks
18 Two-crew flight deck
19 Cockpit roof escape hatches, port and starboard
20 Pilot's Aces II ejection seat
21 Mission commander's ejection seat
22 Conventional stick and rudder flight controls, quadruplex digital flight control system
23 Instrument panel with multi-function full-colour CRT displays
24 Instrument panel shroud
25 Starboard AN/APQ-181 Low Probability of Intercept (LPI) electronically scanned multi-function J-band radar unit
26 Airflow data sensors, above and below
27 Nosewheel leg door
28 Twin-wheel nose undercarriage, aft retracting
29 Taxiing lights
30 Crew door-mounted boarding ladder
31 Port AN/APQ-181 radar unit
32 Astro navigation sensor port
33 Cockpit hatch emergency release
34 Weapons bay retractable spoiler panels
35 Port weapons bay outer door
36 Environmental control system equipment bay, port and starboard
37 Hydraulic equipment bay
38 Boundary layer splitter
39 Boundary layer secondary intake duct to air systems and APU
40 Intake S-duct, common to both engines, bypass air to engine bay and exhaust cooling
41 Allied Signal auxiliary power unit (APU)
42 APU exhaust
43 Airframe-mounted auxiliary gearbox
44 Port General Electric F118-GE-100 non-afterburning turbofan engines
45 Jet pipes flattened towards aft exhausts
46 Engine/weapons bay fireproof structural bulkhead
47 Port weapons bay
48 Weapons bay rotary launcher
49 Starboard weapons bay
50 Centre fuselage keel structure
51 Flush antenna panels
52 Weapons bay rear bulkhead
53 Aft equipment bay
54 Rear fuselage frame structure
55 All-composite skin panelling
56 Exhaust duct thermal protection lining
57 Gust Load Alleviation System (GLAS) tail surface
58 Port engine exhaust duct
59 Rear integral fuel tank bay
60 Elevon hydraulic actuators
61 Port inboard elevons
62 Outboard elevon
63 Port split drag rudder/airbrake
64 Port navigation light, not used operationally
65 Port retractable ILS antenna
66 Outer wing panel rib structure
67 Port outboard integral fuel tankage
68 All-composite leading-edge structure
69 Leading-edge flush EW antennas
70 Four-wheel main undercarriage bogie
71 Landing lights
72 Main undercarriage leg strut
73 Retraction breaker strut
74 Port mainwheel bay pressure refuelling connection
75 Single-piece mainwheel door
76 Inboard integral fuel tankage
77 Weapons bay advanced rotary launcher (ARL)
78 Launcher mounting adaptors
79 Launcher drive unit and sequenced release mechanism
80 B83 nuclear weapon
81 B61 nuclear weapon
82 Mk 82 500-lb bomb
83 Mk 62 mine
84 CBU-87 CEM sub-munition dispenser
85 GBU-89 Gator sub-munition dispenser
86 Mk 84 2,000-lb (907-kg) bomb
87 GBU-31 2,000-lb (907-kg) Joint Direct Attack Munition (JDAM)
88 GQM-113
89 GAM-84
90 Joint Stand-Off Weapon (JSOW), sub-munitions or 500-lb (227-kg) HE unitary warhead
91 B61-11 penetrating nuclear weapon

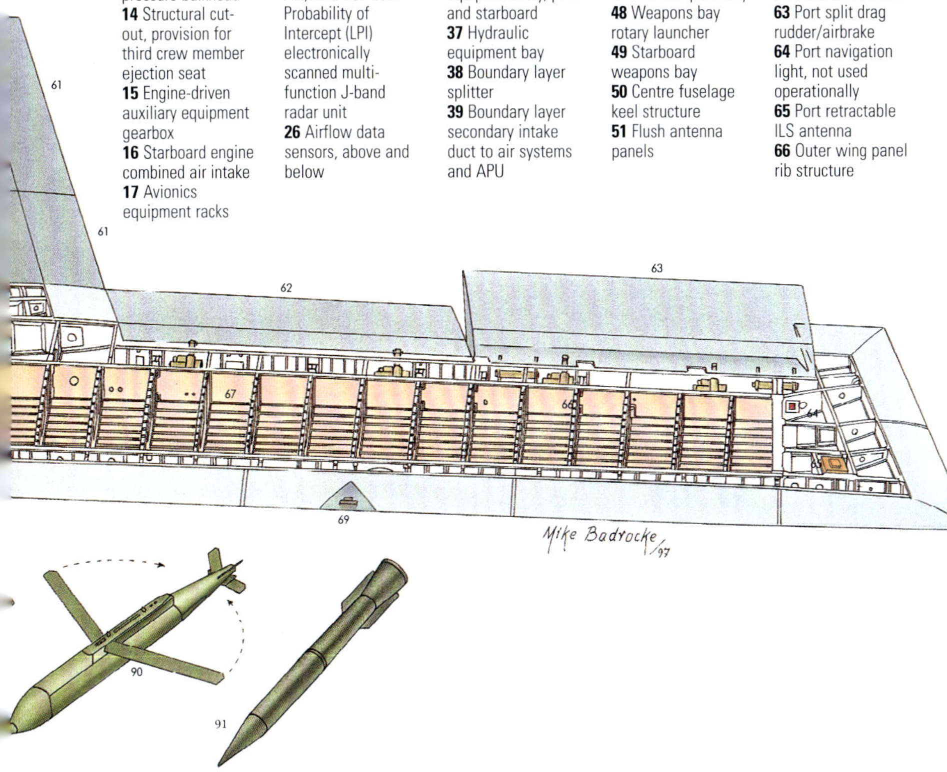

Panavia Tornado GR.Mk 1/4

The end of the Cold War led to a reduction in the number of Tornados committed to the overland strike role, with two four-squadron wings in RAF Germany being reduced to a single four-squadron wing. Three squadrons (Nos. XV, 16 and 20) were simply disbanded, while a fourth (No. II) returned to the UK. Some of the surplus aircraft were used to re-equip the two remaining maritime strike Buccaneer squadrons, being allocated the designation GR.Mk 1B. This Sea Eagle armed example is from No. 617 Sqn and carries two of the anti-ship missiles on dedicated launchers under the fuselage.

Tornado GR.Mk 4

1 Air data probe
2 Radome
3 Lightning conductor strip
4 Terrain following radar antenna
5 Ground mapping radar antenna
6 Radar equipment bay hinged position
7 Radome hinged position
8 IFF aerial
9 Radar antenna tracking mechanism
10 Radar equipment bay
11 UHF/TACAN aerial
12 Laser Rangefinder and Marked Target Seeker (LRMTS), starboard side
13 FLIR housing
14 Ventral Doppler aerial
15 Angle of attack transmitter
16 Canopy emergency release
17 Avionics equipment bay
18 Front pressure bulkhead
19 Windscreen rain dispersal airducts
20 Windscreen (Lucas-Rotax)
21 Retractable, telescopic, inflight refuelling probe
22 Probe retraction link
23 Windscreen open position, instrument access
24 Wide angle HUD
25 Instrument panel
26 Radar 'head down' display
35 Torque scissor links
36 Twin forward-retracting nosewheels (Dunlop)
37 Nosewheel steering unit
38 Nosewheel leg door
39 Electrical equipment bay
40 Ejection seat rocket pack
41 Engine throttle levers
42 Wing sweep control lever
43 Radar hand controller
44 Side console panel
45 Pilot's Martin-Baker Mk 10 ejection seat
46 Safety harness
47 Ejection seat head rest
48 Cockpit canopy cover (Kopperschmidt)
49 Canopy centre arch
50 Navigator's radar displays
51 Navigator's instrument panel and weapons control panels
52 Foot rests
53 Canopy external latch
54 Pitot head
55 Mauser 27-mm cannon, starboard side only
65 Navigator's Martin-Baker Mk 10 ejection seat
66 Starboard engine air intake
67 Intake spill duct
68 Canopy jack
69 Canopy hinge point
70 Rear pressure bulkhead
71 Intake ramp actuator linkage
72 Navigation light
73 Two dimensional variable area intake ramp doors
74 Intake suction relief doors
75 Wing glove Krüger flap
76 Intake bypass air spill ducts
77 Intake ramp hydraulic actuator
78 Forward fuselage fuel tank
79 Wing-sweep control screw jack (Microtecnica)
80 Flap and slat control drive shafts
81 Wing-sweep, flap and slat central control unit and motor (Microtecnica)
82 Wing pivot box integral fuel tank
83 Air system ducting
84 Anti collision light
85 UHF aerials

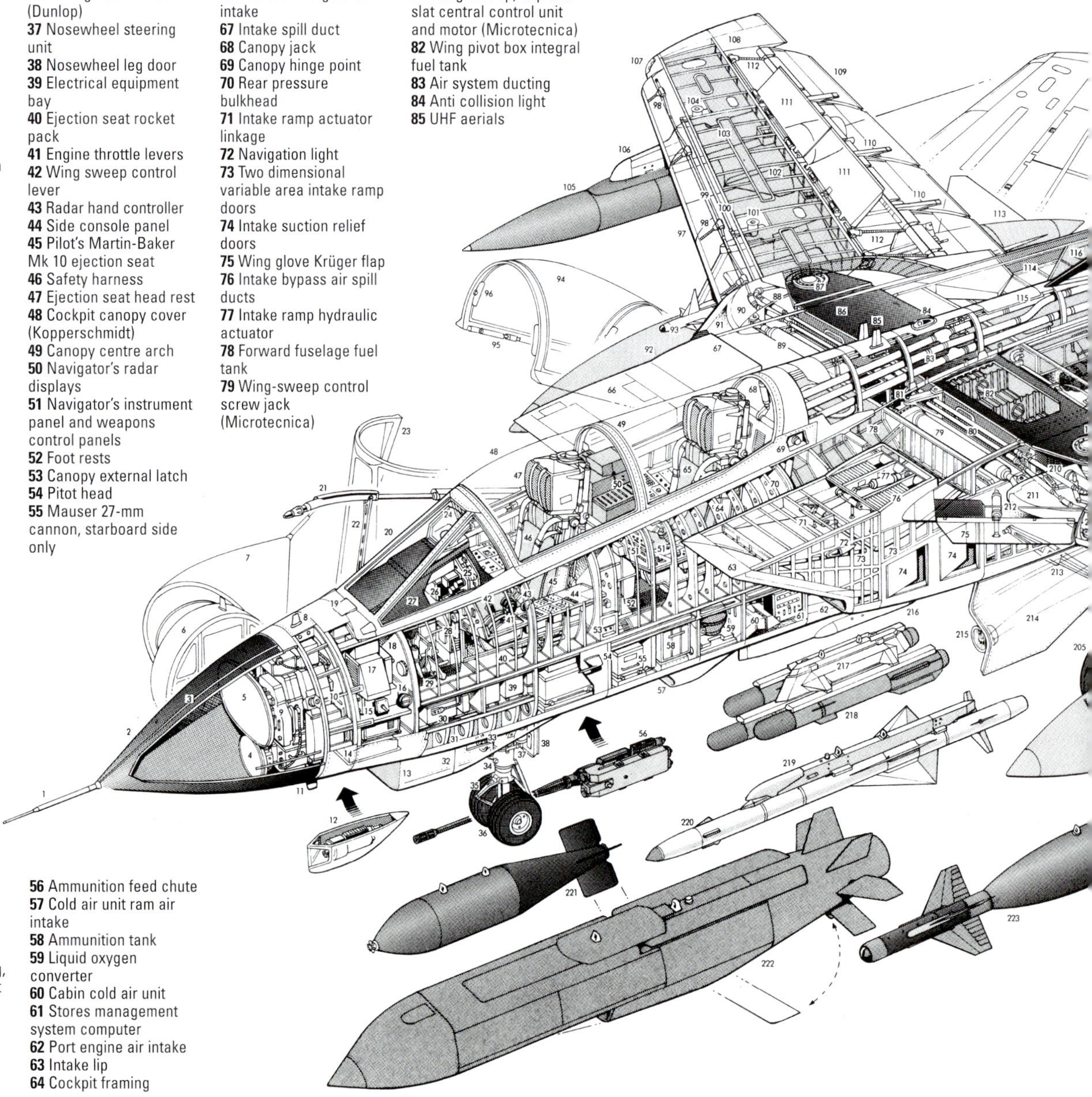

27 Instrument panel shroud
28 Control column
29 Rudder pedals
30 Battery
31 Cannon barrel housing, cannon deleted from port side
32 Nosewheel doors
33 Landing/taxiing lamp
34 Nose undercarriage leg strut (DowtyRotol)
56 Ammunition feed chute
57 Cold air unit ram air intake
58 Ammunition tank
59 Liquid oxygen converter
60 Cabin cold air unit
61 Stores management system computer
62 Port engine air intake
63 Intake lip
64 Cockpit framing

Saudi Arabia is the only export customer for the Tornado, equipping four squadrons with various IDS versions. The RSAF has adopted the RAF's equipment and weapons, including JP233, ALARM and the GR.Mk 1A reconnaissance system.

SPECIFICATION

Tornado GR.Mk 1

Dimensions:

Fuselage length: 54 ft 10¼ in (16.72 m)
Wing span: 45 ft 7½ in (13.91 m) minimum sweep (25°) and 28 ft 2½ in (8.60 m) maximum sweep (67°)
Wing aspect ratio: 7.73
Tailplane span: 22 ft 3½ in (6.80 m)
Tail height: 19 ft 6¼ in (5.95 m)
Wing area: 286.33 sq ft (26.60 m²)
Wheel track: 10 ft 2 in (3.10 m)
Wheel base: 20 ft 4 in (6.20 m)

Powerplant

Two Turbo-Union R.B.199-34R Mk 101 turbofans each rated at 8,475 lb st (37.70 kN) dry and 14,840 lb st (66.01 kN) with afterburning or, in later aircraft, Turbo-Union R.B.199-34R Mk 103 turbofans each rated at 8,650 lb st (38.48 kN) dry and 16,075 lb st (71.50 kN) with afterburning.

Weights

Empty operating: 31,065 lb (14091 kg)
Normal take-off: 45,000 lb (20411 kg)
Maximum take-off: 61,620 lb (27951 kg)

Fuel and load

Internal fuel: 11,221 lb (5090 kg)
External fuel: up to 13,200 lb (5988 kg) in two 495-Imp gal (2250-litre) and two 330-Imp gal (1500-litre), or four 330 Imp gal (1500-litre) drop tanks.
Maximum theoretical weapon load: more than 19,841 lb (9000 kg)

g limits

+7.5 at basic design gross weight

Performance

Maximum level speed 'clean' at altitude: 1,262 kt (1,453 mph; 2338 km/h)
Limiting Mach No: 1.4 with LRMTS
Service ceiling: 50,000 ft (15240 m)
Take-off run: less than 2,953 ft (900 m)
Landing run: 1,214 ft (370 m)

Range

Ferry range: 2,100 nm (2,420 miles; 3890 km) with four drop tanks
Combat radius: 750 nm (863 miles; 1390 km) on a typical hi-lo-hi attack mission with a heavy warload.

Armament

Maximum ordnance: 19,841 lb (9000 kg) carried on four underwing and three underfuselage hardpoints.

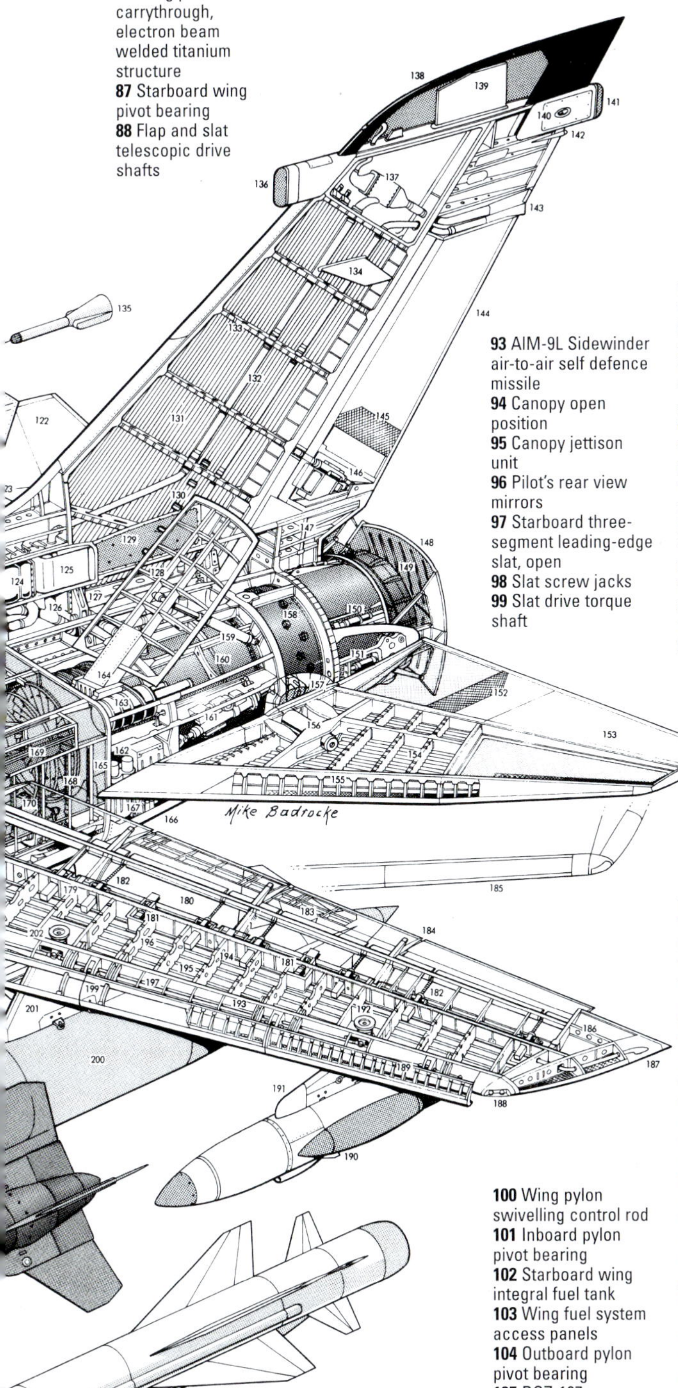

86 Wing pivot box carrythrough, electron beam welded titanium structure
87 Starboard wing pivot bearing
88 Flap and slat telescopic drive shafts
89 Starboard wing-sweep control screw jack
90 Leading edge sealing fairing
91 Wing root glove fairing
92 495-Imp gal (2250-litre) 'Hindenburger' fuel tank
93 AIM-9L Sidewinder air-to-air self defence missile
94 Canopy open position
95 Canopy jettison unit
96 Pilot's rear view mirrors
97 Starboard three-segment leading-edge slat, open
98 Slat screw jacks
99 Slat drive torque shaft
100 Wing pylon swivelling control rod
101 Inboard pylon pivot bearing
102 Starboard wing integral fuel tank
103 Wing fuel system access panels
104 Outboard pylon pivot bearing
105 BOZ-107 chaff/flare launcher
106 Outboard wing swivelling pylon
107 Starboard navigation and strobe lights
108 Wing tip fairing
109 Double-slotted Fowler type flaps, down position
110 Flap guide rails
111 Starboard spoilers, open
112 Flap screw jacks
113 External fuel tank tail fins
114 Wing-swept position trailing-edge housing
115 Dorsal spine fairing
116 Aft fuselage fuel tank
117 Fin root antenna fairing
118 HF aerial
119 Heat exchanger ram air intake
120 Starboard wing fully swept-back position
121 Air-brake, open
122 Starboard all-moving tailplane (taileron)
123 Air brake hydraulic jack
124 Primary heat exchanger
125 Heat exchanger exhaust duct
126 Engine bleed air ducting
127 Fin attachment joint
123 Port air brake rib construction
129 Fin heat shield
130 Vortex generators
131 Fin integral fuel tank
132 Fuel system vent piping
133 Tail fin structure
134 ILS aerial
135 GEC-Marconi Ariel towed radar decoy (TRD)
136 Forward passive ECM
138 Fin tip antenna fairing
139 VHF aerial
140 Tail navigation light
141 Aft passive ECM housing
142 Obstruction light
143 Fuel jettison
144 Rudder
145 Rudder honeycomb construction
146 Rudder hydraulic actuator (Fairey Hydraulics)
147 Dorsal spine tail fairing
148 Thrust reverser bucket doors, open
149 Variable area afterburner nozzle
150 Nozzle control jacks (four)
151 Thrust reverser door actuator
152 Honeycomb trailing edge construction
153 Port all-moving tailplane (taileron)
154 Tailplane rib construction
155 Leading-edge nose ribs
156 Tailplane pivot bearing
157 Tailplane bearing sealing plates
158 Afterburner duct
159 Airbrake hydraulic jack
160 Turbo-Union R.B.199-34R Mk 103 afterburning turbofan engine
161 Tailplane hydraulic actuator
162 Hydraulic system filters
163 Hydraulic reservoir (Dowty)
164 Air brake hinge point
165 Intake frame/ production joint
166 Engine bay ventral access panels
167 Engine oil tank
168 Rear fuselage fuel tank
169 Wing root pneumatic seal
170 Engine driven accessory gear boxes, port and starboard (KHD), airframe mounted
171 Integrated drive generator (two)
172 Hydraulic pump (two)
173 Gearbox interconnecting shaft
174 Starboard side Auxiliary Power Unit (KHD)
175 Telescopic fuel pipes
176 Port wing pivot bearing
177 Flexible wing sealing plates
178 Wing skin panelling
179 Rear spar
180 Port spoiler housings
181 Spoiler hydraulic actuators
182 Flap screw jacks
183 Flap rib construction
184 Port Fowler-type double slotted flaps, down position
185 Port wing fully swept-back position
186 Wing tip construction
187 Fuel vent
188 Port navigation and strobe lights
189 Leading-edge slat rib construction
190 Marconi Sky Shadow ECM pod
191 Outboard swivelling pylon
192 Pylon pivot bearing
193 Front spar
194 Port wing integral fuel tank
195 Machined wing skin/ stringer panel
196 Wing rib construction
197 Swivelling pylon control rod
198 Port leading-edge slat segments, open
199 Slat guide rails
200 Port 'Hindenburger' external fuel tank
201 Inboard swivelling pylon
202 Inboard pylon pivot bearing
203 Missile launch rail
204 AIM-9L Sidewinder air-to-air self defence missile
205 Port mainwheel (Dunlop), forward-retracting
206 Main undercarriage leg strut (DowtyRotol)
207 Undercarriage leg pivot bearing
208 Hydraulic retraction jack
209 Leg swivelling control link
210 Telescopic flap and slat drive torque shafts
211 Leading-edge sealing fairing
212 Krüger flap hydraulic jack
213 Main undercarriage leg breaker strut
214 Main wheel door
215 Landing lamp
216 Port fuselage pylon
217 Triple launch unit
218 GEC-Marconi Dynamics Brimstone anti-armour missile
219 ALARM launch rail
220 Matra-BAe Dynamics ALARM anti-radar missile
221 1000-lb (454-kg) HE bomb
222 Matra-BAe Dynamics Storm Shadow stand-off weapons dispenser
223 GBU-24/B 2000-lb (908-kg) Laser Guided Bomb
224 BAe Dynamics Sea Eagle air-to-surface anti-ship missile

Rockwell B-1B

The history of the B-1B has been a long and convoluted process, with it emerging as the B-1A, only to be cancelled, and then resurfacing as the B-1B. Having overcome a series of teething troubles, the 'Bone' is now the key element of the USA's strategic bomber force (despite still regularly receiving adverse publicity), being greater in number than the B-2 and more survivable than the B-52.

B-1B Lancer

Cutaway key
1 Radome
2 Multi-mode phased array radar antenna
3 Low observable shrouded scanner mounting and tracking mechanism
4 AN/APQ-146 offensive radar system equipment bays
5 Pitot heads
6 Foreplane hydraulic actuator
7 Structural mode control system (SMCS) foreplane
8 Nose undercarriage stowed position
9 Control column
10 Rudder pedals
11 Air refuelling receptacle
12 Windscreen panels
13 Fully-shrouded instrument panel
14 Cockpit roof escape hatches
15 Co-pilot's station
16 Weber ACES II zero-zero ejection seats, all positions
17 Pilot's station
18 Crew toilet
19 Conditioned air supply ducting
20 Ventral boarding ladder
21 Defensive Systems Operator's (DSO) station
22 Systems operator's display and control consoles
23 Observer's folding seat
24 Offensive systems operator's station (OSO)
25 Cockpit roof ejection hatches
26 Glideslope antenna
27 SATCOM antenna
28 Avionics racks, port and starboard, flight control and communications equipment
29 Avionics cooling air ground connection
30 Electrical equipment bay
31 Defensive avionics systems transmitting antennas
32 Forward fuselage integral fuel tanks, total system capacity 202,254 lb (91742 kg)
33 Weapons bay movable bulkhead
34 Weapons bay fuel tank, typical, various sizes up to 2,903 US gal (10989 litres)
35 Chaff/flare launchers
36 RFS antenna, port and starboard
37 Starboard leading-edge RFS/ECMS equipment bay
38 UHF/IFF antenna
39 Upper fuselage cable and systems ducting
40 Weapons bay door actuating mechanism
41 Retractable spoiler panel
42 Port leading-edge antenna panel
43 RFS/ECMS transmitting antennas
44 Defensive avionics system equipment
45 Weapons bay doors
46 Rotary weapons carrier/launcher
47 Wing pivot box integral fuel tank
48 Electron beam-welded titanium wing pivot box carry-through structure
49 Blended sidebody integral fuel tank
50 Wing sweep actuator hydraulic drive unit
51 Wing sweep control screw jack
52 Starboard wing integral fuel tank
53 Fuel system vent and feed piping
54 Seven-segment leading-edge slats
55 Starboard navigation and strobe lights
56 Fuel jettison vent
57 Spoiler panels/lift dumpers
58 Starboard single-slotted flap
59 Wing fully-swept position
60 Wing shroud panels
61 Main undercarriage wheel bay
62 Flight control system avionics equipment
63 Undercarriage bay roof fuel tank
64 Air supply ducting
65 Rear weapons bay rotary launcher
66 Weapons bay surround integral fuel tank
67 Starboard engine exhausts
68 Tailplane automatic flight control system equipment
69 Fin/tailplane support structure
70 All-moving tailplane hydraulic actuators
71 Starboard tailplane panel
72 Fin leading-edge HF antenna
73 Fin-tip ECMS and rendezvous beacon antennas
74 Tail navigation and strobe lights
75 Rudder
76 Rudder rotary actuators
77 Rudder SCAS unit
78 RFS/ECMS antenna
79 Port all-moving tailplane
80 Static dischargers
81 Tail radome
82 RFS/ECMS transmitting antennas
83 Defensive system avionics equipment racks
84 Rear fuselage fuel tank
85 Fuel tank pressurant nitrogen bottle
86 Rear weapons bay
87 Main engine mounting beam

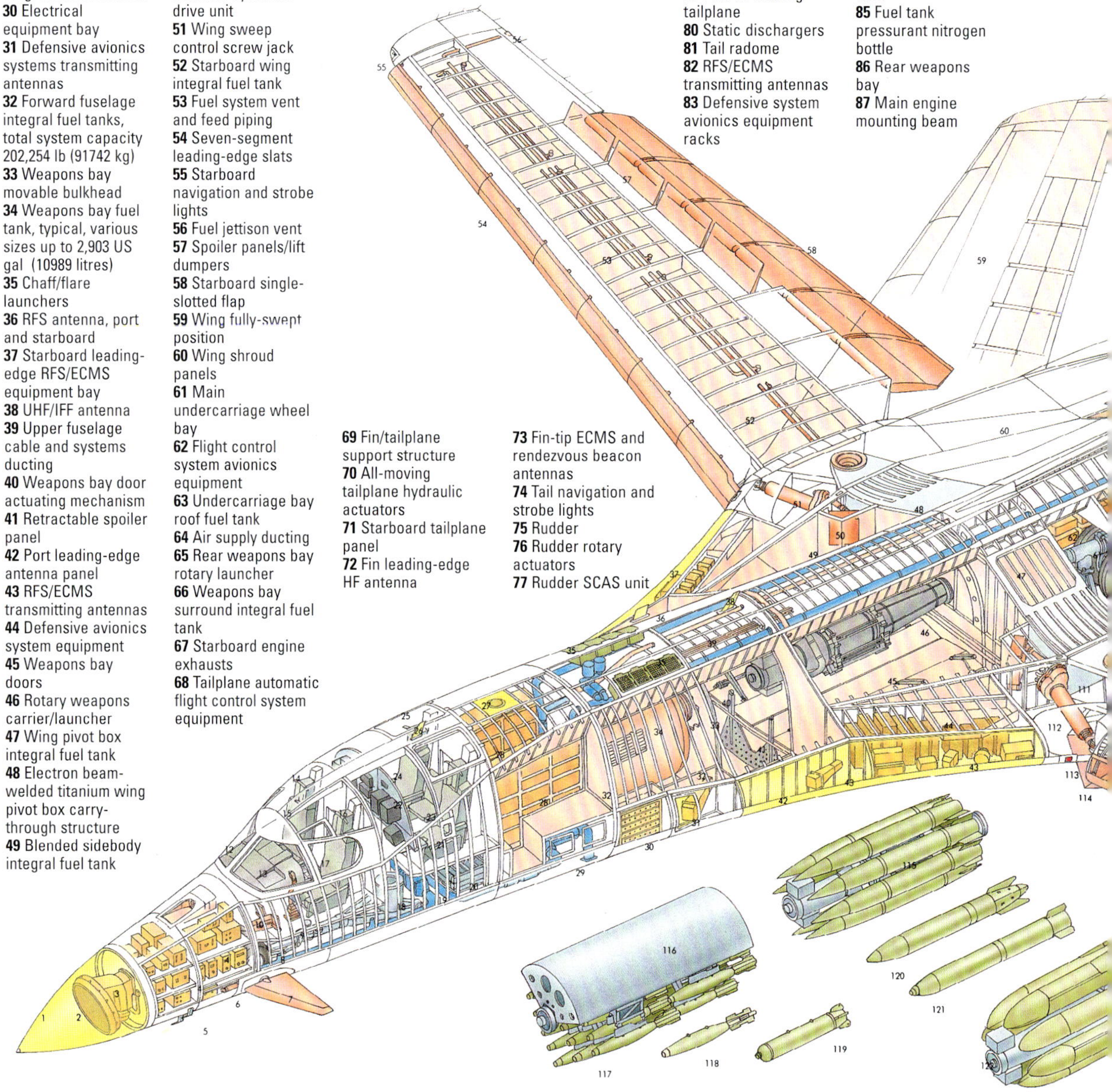

B-1B nose art, as with that of all US aircraft types, has moved away from the traditional glamourous female image following a recent ruling that all such art should be non-gender specific. Hence, decoration is now more carefully chosen, as was with case with* The Reluctant Dragon's *(86-0103) nose art, which originally appeared on a 96th Bomb Group B-17G and was applied when this particular 'Bone' took part in a B-1B publicity programme. UK visits were made to RAF Alconbury and Fairford in 1993 and 1996, with 86-0103 acting as the main display machine.

SPECIFICATION

B-1B Lancer

Dimensions

Fuselage length: 143 ft 3½ in (43.68 m)
Height: 34 ft (10.3 m)
Wingspan (maximum sweep): 78 ft 2½ in (23.84 m)
Wingspan (minimum sweep): 136 ft 8½ in (41.67 m)
Area: 1,950 sq ft (181.16 m²)
Aspect ratio (fully-spread): 9.58
Aspect ratio (fully-swept): 3.14
Tailplane span: 44 ft 10 in (13.67 m)
Wheel track: 14 ft 6 in (4.42 m)
Wheelbase: 57 ft 6 in (17.53 m)

Powerplant

Four General Electric F101-GE-102 turbofans each rated at 14,600 lb st (64.94 kN) dry and 30,780 lb st (136.92 kN) with afterburning

Weights

Basic empty: 182,360 lb (82840 kg)
Empty equipped: 192,000 lb (87091 kg)
Maximum take-off weight: 477,000 lb (216365 kg)
Low-altitude operating weight: 422,000 lb (191419 kg)
Nominal payload: 294,500 lb (133585 kg)

Fuel and load

Internal fuel: 193,405 lb (87728 kg)
Internal fuel with SIS/SEF FCS software: 206,160 lb (93514 kg)
Maximum internal ordnance: 75,000 lb (34019 kg)
Maximum external ordnance: 59,000 lb (26762 kg)

Performance

Maximum level speed 'clean' at high altitude: Mach 1.25 or 715 kt (823 mph; 1324 km/h)
Penetration speed at approximately 200 ft (61 m): Mach 0.92 or 521 kt (600 mph; 965 km/h)
Range with typical weaponload: 3000 nm (3,444 miles; 5542 km)
Service ceiling: 50,000 ft (15240 m)

Armament

The B-1B is fitted with three internal weapon bays though external stores are also used. In the nuclear role, up to 12 B28 or 14 B23 or 24 B61 or 24 B83 free-fall bombs or 24 AGM-69 SRAM or eight AGM-86B ALCMs can be carried. Conventional weapons may total 84 Mk 82 or 24 Mk 84 GP bombs or 8 AGM-86C ALCMs internally or 44 Mk 82 or 14 Mk 84 or 14 AGM-86C ALCMs externally

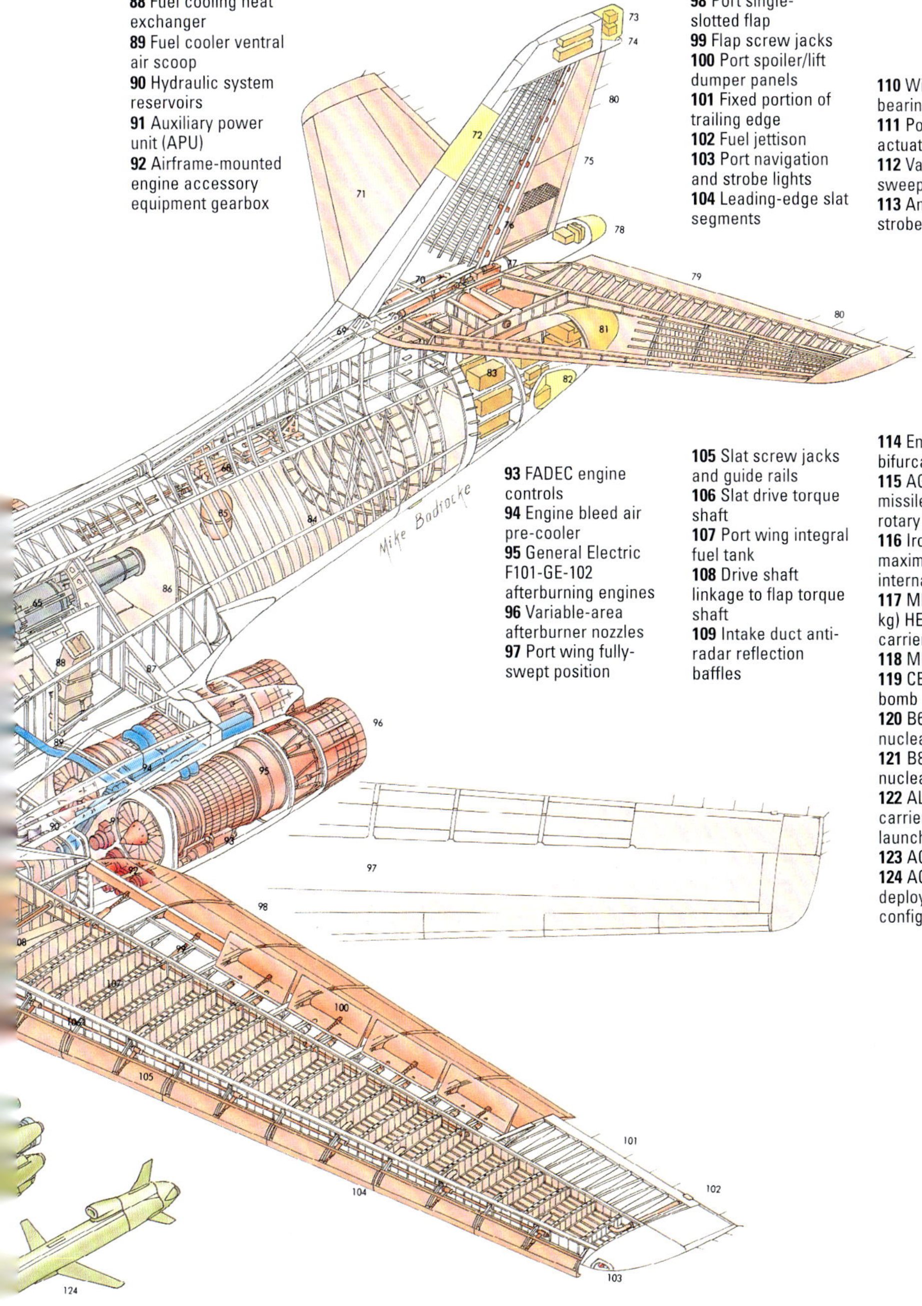

88 Fuel cooling heat exchanger
89 Fuel cooler ventral air scoop
90 Hydraulic system reservoirs
91 Auxiliary power unit (APU)
92 Airframe-mounted engine accessory equipment gearbox
93 FADEC engine controls
94 Engine bleed air pre-cooler
95 General Electric F101-GE-102 afterburning engines
96 Variable-area afterburner nozzles
97 Port wing fully-swept position
98 Port single-slotted flap
99 Flap screw jacks
100 Port spoiler/lift dumper panels
101 Fixed portion of trailing edge
102 Fuel jettison
103 Port navigation and strobe lights
104 Leading-edge slat segments
105 Slat screw jacks and guide rails
106 Slat drive torque shaft
107 Port wing integral fuel tank
108 Drive shaft linkage to flap torque shaft
109 Intake duct anti-radar reflection baffles
110 Wing pivot bearing
111 Port wing sweep actuator
112 Variable-wing sweep sealing horn
113 Anti-collision strobe light
114 Engine nacelle bifurcated air intake
115 AGM-69 SRAM missiles, eight on rotary launcher
116 Iron bomb carrier, maximum of three internally
117 Mk 82 500-lb (227-kg) HE bombs, 28 per carrier
118 Mk 36 mine
119 CBU-89B cluster bomb
120 B61 free-fall nuclear weapon
121 B83 free-fall nuclear weapon
122 ALCM eight-round carrier/rotary launcher
123 AGM-86B ALCM
124 AGM-86B ALCM deployed configuration

Tupolev Tu-16 'Badger'

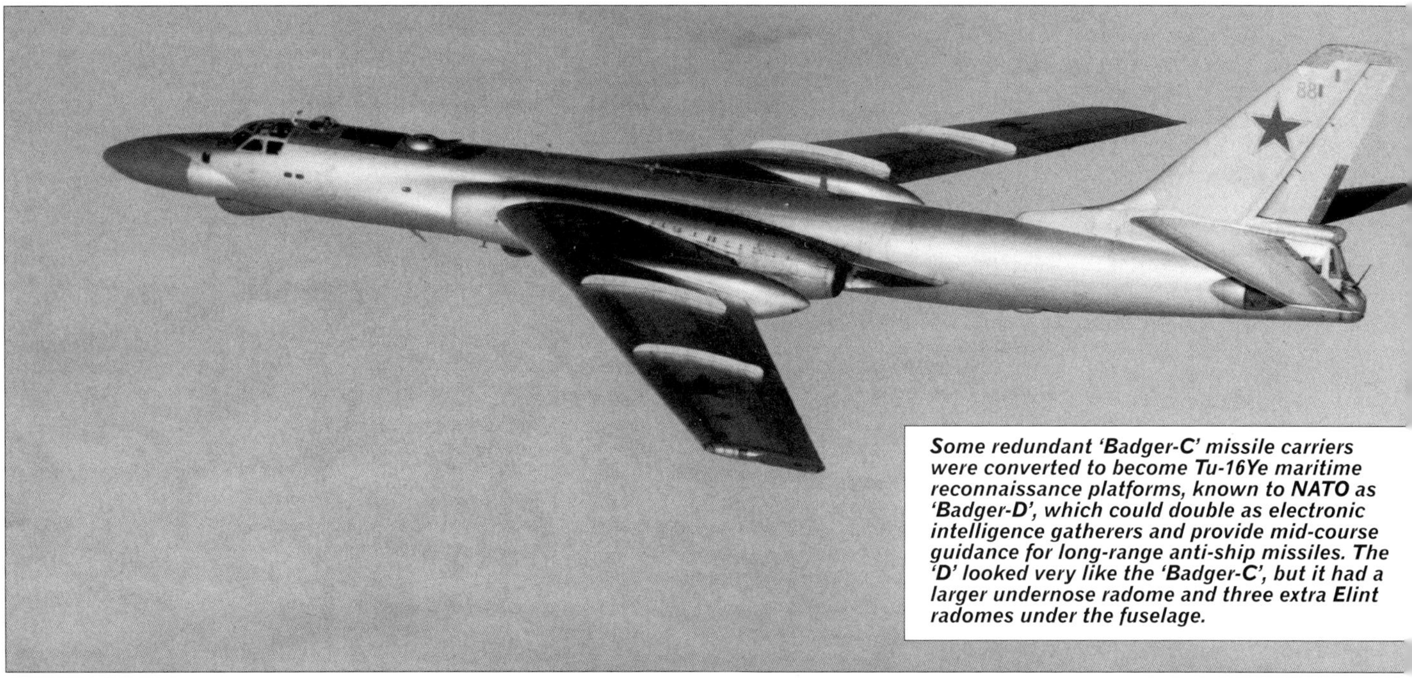

Some redundant 'Badger-C' missile carriers were converted to become Tu-16Ye maritime reconnaissance platforms, known to NATO as 'Badger-D', which could double as electronic intelligence gatherers and provide mid-course guidance for long-range anti-ship missiles. The 'D' looked very like the 'Badger-C', but it had a larger undernose radome and three extra Elint radomes under the fuselage.

Tu-16 'Badger'

Cutaway key

1 Radome, 'Badger-C' & 'D'
2 Weapons ranging and search radar scanner
3 Radar navigator/ bombardier's seat
4 Windscreen wipers
5 Pitot head
6 Windscreen panels
7 Cockpit eyebrow windows
8 Instrument panel shroud
9 Navigation radome
10 Pilot's seat
11 Co-pilot's seat
12 Cockpit roof escape hatches
13 Glazed nose section, all variants except 'C' & 'D'
14 Optically flat sighting window
15 Fixed forward firing NR-23, 23-mm cannon on starboard side only
16 Navigator/ bombardier's seat
17 Navigation radome
18 'Towel-rail' aerial
19 Astrodome observation hatch
20 Forward gunner's swivelling seat
21 Cabin side window panels
22 Ventral entry/exit hatch
23 Extending boarding ladder
24 Retractable landing/ taxiing lamps, port and starboard
25 Nose landing gear leg strut
26 Twin nosewheels, aft retracting
27 Nosewheel doors
28 Nose landing gear hydraulic jack
29 Electronics equipment racks, port and starboard
30 HF blade antenna
31 Remotely controlled dorsal gun turret
32 Twin NR-23, 23-mm cannon
33 Communications aerials, port and starboard
34 Port engine air intake
35 Radar altimeter aerial
36 Intake duct divided around front spar
37 Forward fuselage fuel cells, maximum capacity approximately 10,009 Imp gal (45500 litres)
38 Starboard engine air intake
39 Aerial mast
40 Starboard inboard wing panel
41 Outer wing panel joint
42 Starboard missile pylon
43 AS-6 'Kingfish' air-to-surface missile, 'Badger-G' 'modified'
44 Starboard wing integral fuel tanks
45 Inboard wing fence
46 Outboard wing fence
47 Outer wing panel
48 Starboard navigation light
49 Wing tip faring
50 Fuel jettison pipe
51 Starboard aileron
52 Aileron tab
53 Flap guide rails
54 Flap screw jacks
55 Starboard single-slotted track-mounted flap, down position
56 Starboard main landing gear fairing
57 Inboard flap segment
58 Starboard engine bay
59 Centre section internal weapons bay, capacity 19,842 lb (9000 kg)
60 Rear fuselage fuel cells
61 Blade antenna
62 'Badger-D' electronic reconnaissance variant, ventral view
63 Ventral radomes
64 'Badger-C' maritime strike variant
65 Semi-recessed missile housing, AS-2 'Kipper' air-to-surface missile
66 'Badger-A' bomber, ventral view
67 Weapons bay doors, open
68 Starboard trimming tailplane
69 Starboard elevator
70 HF aerial cable
71 Tailfin
72 Fin tip aerial fairing
73 Rudder
74 Rudder tab
75 Gun ranging radar antenna
76 Rear gunner's station
77 Twin NR-23, 23-mm cannon
78 Elevator tab
79 Port elevator
80 Port tailplane construction
81 Rear pressurised compartment ventral entry/exit hatches
82 Observation blister, port and starboard
83 Radio operator/ observer's station
84 Retractable tail bumper
85 'Odd-Rods' IFF aerials
86 Ventral remotely controlled gun turret, two NR-23, 23mm cannon
87 Exhaust fairing
88 Engine exhaust nozzle
89 Ventral strike camera
90 Outboard canted jet pipe
91 Engine bay access doors, above and below
92 Mikulin RD-3M (AM-3M) single shaft turbojet engine
93 Engine accessory equipment compartment
94 Port inboard flap
95 Port main landing gear housing
96 Main landing gear, stowed position
97 Flap cut-out for missile tail fin
98 Port single-slotted flap

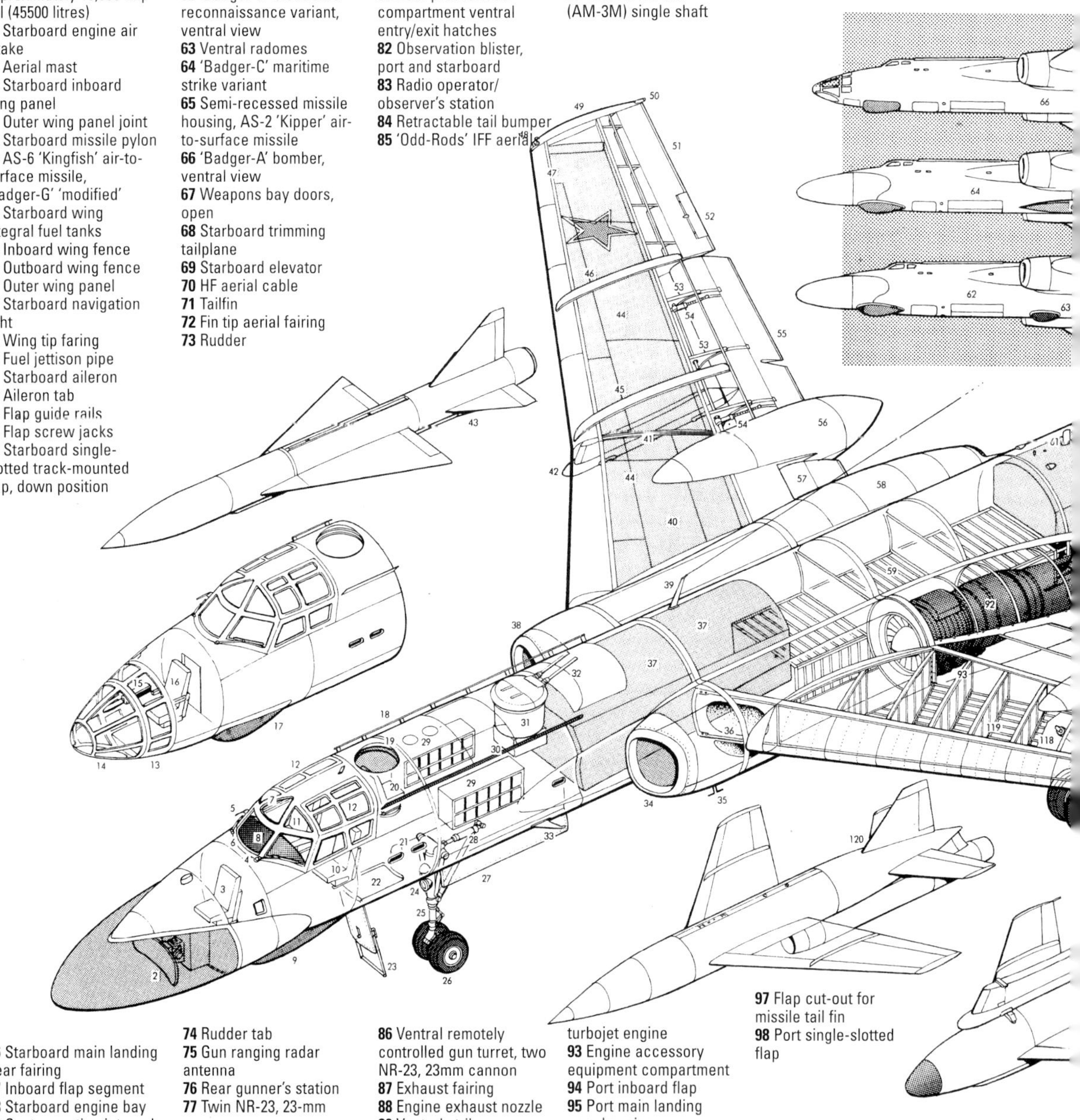

The 'Badger K' is an electronic reconnaissance variant of the Tu-16 first identified in the late 1980s. Probably based on the original anti-shipping 'Badger-B', it is identifiable by the two underfuselage radomes at each end of the weapons bay – much closer together than on the 'Badger-F'. According to some analysts, the 'Badger-K' was probably designed with highly automated systems for the precision collection of Elint in a dense signal environment.

SPECIFICATION

Tu-16K-11-16 'Badger-G'

Dimensions:

Wingspan: 108 ft ½ in (32.93 m)
Wing area: 1,772 sq ft (164.65 m²)
Length: 118 ft 11¼ in (36.25 m)
Height: 45 ft 11¼ in (14.00 m)
Wheel track: 32 ft ¾ in (9.77 m)
Wheel base: 34 ft 8 in (10.57 m)

Powerplant:

Two Mikulin KB Type RD-3M (AM-3M-500) single-shaft turbojets, each delivering 20,944-lb (93.16 kN) static thrust

Weights:

Empty equipped: 82,012 lb (37200 kg)
Max take-off: 167,110 lb (75800 kg)
Max landing: 110,250 lb (50000 kg)
Max landing (rough field): 105,840 lb (48000 kg)

Fuel and load

Internal fuel: 75,750 lb (34360 kg)
Maximum ordnance: 19,845 lb (9000 kg); normal bomb load 6,600 lb (3000 kg); naval variants carry one or two winged ASMs under wing (AS-5 'Kelt' 6,614 lb/3000 kg each; AS-6 'Kingfish' 11,023 lb/5000 kg each)

Performance

Maximum level speed at 19,700 ft (6000 m): 566 kt (654 mph; 1050 km/h)
Maximum level speed with combat load: 495 kt (571 mph; 918 km/h)
Maximum cruising speed with combat load: 445 kt (514 mph; 825 km/h)
Range: 3,885 nm (4,474 miles; 7200 km)
Range with 3000 kg warload: 3,130 nm (3,604 miles; 5800 km)
Range with 2 missiles underwing: 2,618 nm (3,013 miles; 4850 km)
Service ceiling: 49,215 ft (15000 m)
Take-off run: 4,620 ft (1250 m)
Landing speed: 120 kt (139 mph; 223 km/h)

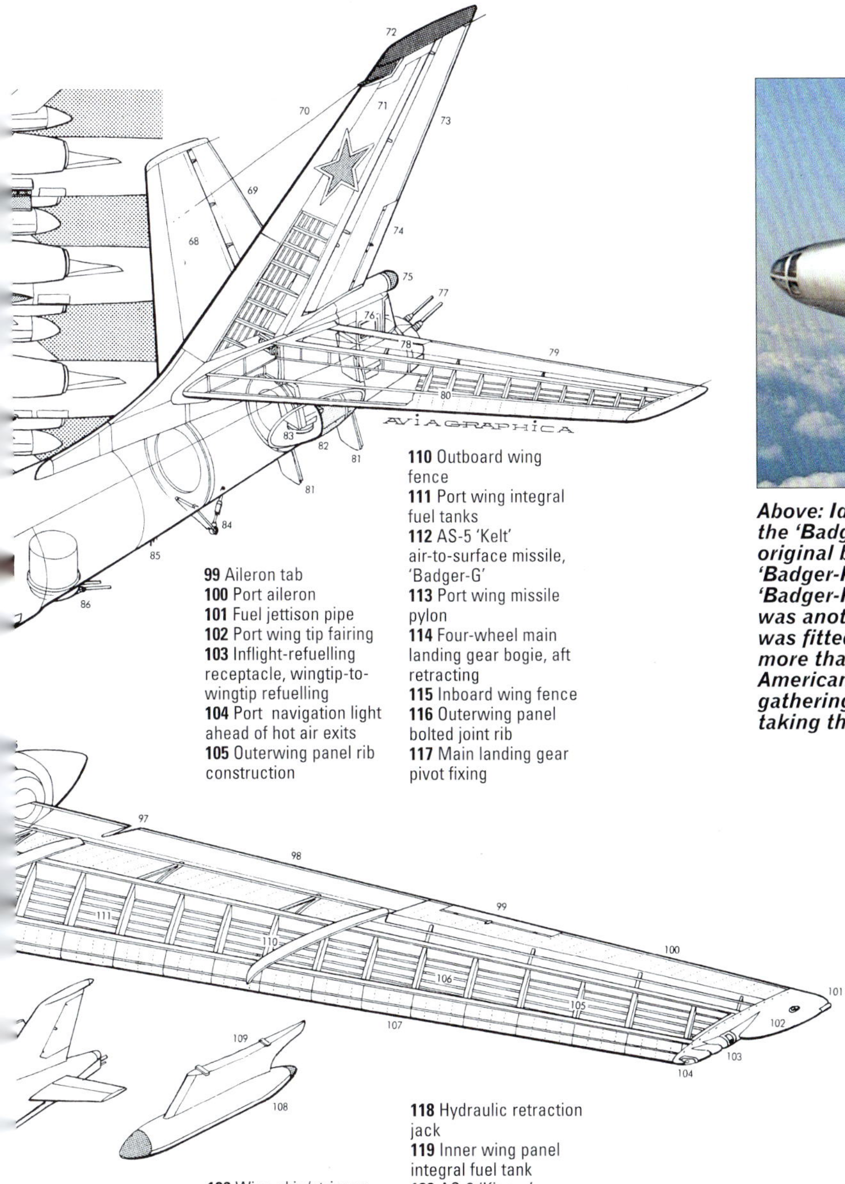

99 Aileron tab
100 Port aileron
101 Fuel jettison pipe
102 Port wing tip fairing
103 Inflight-refuelling receptacle, wingtip-to-wingtip refuelling
104 Port navigation light ahead of hot air exits
105 Outerwing panel rib construction
106 Wing skin/stringer panel
107 Thermally de-iced leading edge
108 Electronic intelligence gathering pod, 'Badger-F'
109 Electronic pod pylon
110 Outboard wing fence
111 Port wing integral fuel tanks
112 AS-5 'Kelt' air-to-surface missile, 'Badger-G'
113 Port wing missile pylon
114 Four-wheel main landing gear bogie, aft retracting
115 Inboard wing fence
116 Outerwing panel bolted joint rib
117 Main landing gear pivot fixing
118 Hydraulic retraction jack
119 Inner wing panel integral fuel tank
120 AS-2 'Kipper' air-to-surface missile, 'Badger-C'

Above: Identifying maritime electronic reconnaissance variants of the 'Badger' can be difficult, since most are conversions from the original bomber variant and look very similar. However, the 'Badger-K' seen here, can be distinguished from the very similar 'Badger-F' by its lack of underwing electronic pods. The 'Badger-L' was another similar variant, which had pods like the 'F' but which was fitted with an extended tailcone in place of the rear guns. For more than thirty years, aircraft like these shadowed NATO and American exercises all over the North Atlantic and northern Pacific, gathering intelligence on Western radar and communications and taking thousands of photographs of warships.

Tupolev Tu-22M 'Backfire'

The 'Backfire' would probably qualify as the greatest of the West's Cold War bugbears. Few aircraft have been as feared or as mocked, as overestimated or as dismissed.* Soviet Military Power, *the US chronicle of the USSR in the 1980s, considered the Tu-22M to be a long-range strategic aircraft. However, both it and the 'Blinder' have only ever been used in the medium-range theatre role.

Tu-22M2

Cutaway key
1 Detachable in-flight refuelling probe
2 Radome
3 'Down-Beat' bombing and navigation radar scanner
4 Radar equipment bay
5 Cockpit front pressure bulkhead
6 Pitot head (port and starboard)
7 Forward ECM aerial (port and starboard)
8 Rudder pedals
9 Control column
10 Instrument panel shroud
11 Windscreen panels
12 Cockpit roof escape hatches
13 Co-pilot's ejector seat
14 Pilot's ejector seat
15 Instrument consoles
16 Blade aerial
17 Navigator/bombardier's and electronic systems officer's seats
18 Rear cockpit side window panel
19 Cockpit floor level
20 Ventral observation and visual bomb-aiming cupola
21 Blade aerial
22 Nose landing gear leg strut
23 Twin nosewheels
24 Nosewheel doors
25 Nose landing gear retraction mechanism
26 Cockpit rear pressure bulkhead
27 Avionics equipment racks
28 Equipment cooling air spill duct
29 'Backfire-C' nose profile
30 Wedge-type engine air intakes
31 Starboard engine air inlet
32 Forward fuselage fuel tank
33 Boundary layer bleed air spill duct
34 Boundary layer splitter plate
35 Variable-area inlet ramp doors
36 Port engine air inlet
37 Inlet ducting
38 Retractable landing/taxiing lamp (port and starboard)
39 Multiple weapons racks (port and starboard)
40 Air system equipment
41 Wing pivot box centre section carry-through
42 Centre fuselage fuel tank
43 Auxiliary air intakes (open)
44 Wing pivot bearing
45 Glove section leading-edge slat (open)
46 Wing fence
47 Starboard wing integral fuel tank
48 Leading-edge slat segments (open)
49 Starboard wing, fully forward (20° sweep) position
50 Starboard navigation lights
51 Wingtip fairing
52 Fixed portion of trailing edge
53 Starboard slotted flap-down position
54 Two-segment spoilers/lift dumpers (open)
55 Wing glove sealing plate
56 Starboard wing, fully-swept (65°) position
57 Flush aerial fairing
58 Centre fuselage weapons bay
59 Extended fin root fillet
60 Flush aerial
61 Fin root fuel tank
62 Starboard all-moving tailplane (taileron)
63 Tailfin
64 Navigational aerials
65 Fin tip fairing
66 ATC transponders and Sirena 3 radar warning receiver
67 Tail navigation light
68 Rudder
69 ECM aerial
70 'Bar-Tail' tail gun control radar
71 Rear ECM aerial (port and starboard)
72 Ammunition magazines
73 Remotely-controlled gun
74 Two twin-barrelled 23-mm cannon
75 Afterburner duct cooling air scoop
76 Variable-area afterburner nozzle
77 Afterburner nozzle control jacks
78 Tailplane/fuselage fillet fairing
79 Port all-moving tailplane (taileron)
80 Tailplane pivot fixing
81 Tailplane hydraulic actuator
82 Afterburner ducting
83 Kuznetsov afterburning turbofan engine
84 Engine accessory equipment gearbox
85 Rear fuselage integral fuel tank
86 Ventral weapons bay doors
87 Main landing gear wheel bay
88 Hydraulic retraction mechanism
89 Mainwheel leg pivot fixing
90 Wing glove section slotted trailing-edge flap
91 Flap hydraulic actuator
92 Ventral landing gear leg pivot fairing

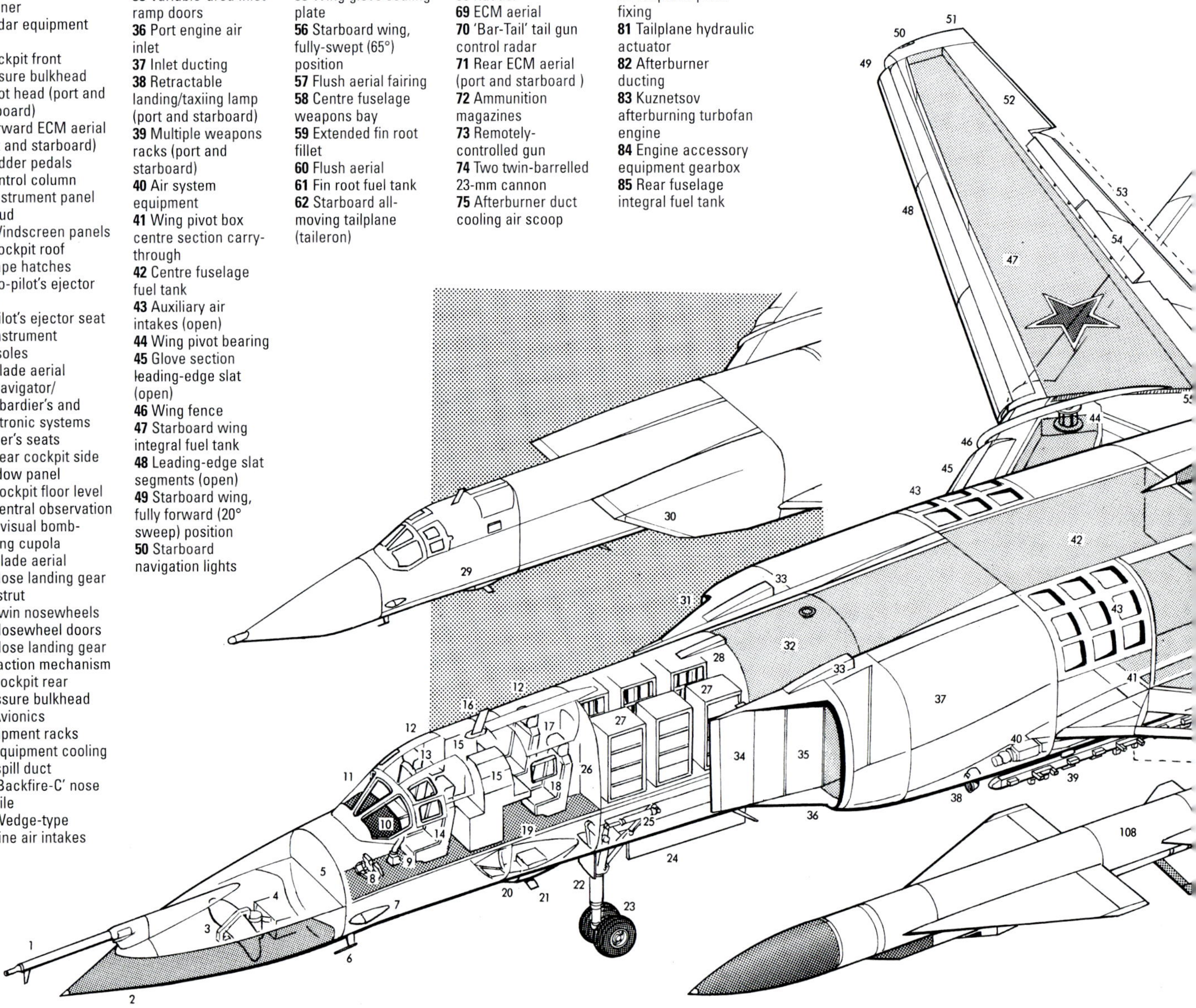

The dragon badge of the 341st TBAP (Heavy Bomber Regiment) was first worn in Afghanistan, but was retained on its Tu-22Ps in later years. As the Soviet Union dissolved, most of the 'Blinders' were stationed in Belarus. However, the Belarussians allowed the Russians to withdraw the regiments and return them to Engels air base for disbandment. As of 1997, 92 such aircraft were waiting for 'demilitarisation'.

SPECIFICATION

Tu-22M3 'Backfire-C'

Dimensions

Length: 139 ft 4 in (42.46 m)
Height: 36 ft 3 in (11.05 m)
Wingspan (fully-spread): 112 ft 7 in (34.30 m)
Wingspan (fully-swept): 76 ft 11 in (23.40 m)
Weapons bay length: 22 ft 11 in (7.00 m)
Weapons bay width: 5 ft 10 in (1.80 m)

Powerplant

Two side-by-side Kuznetsov/KKBM NK-25 turbofans in rear fuselage, each rated at 55,115 lb st (245.2 kN) with afterburning

Weights and loads

Maximum weaponload: 52,910 lb (24000 kg)
Maximum take-off weight: 273,370 lb (124000 kg)
Maximum landing weight: 194,000 lb (88000 kg)
Fuel load: 110,230 lb (50000 kg)

Performance

Maximum level speed at high altitude: Mach 1.88 (1,080 kt; 1,242 mph; 2000 km/h)
Maximum level speed at low altitude: Mach 0.86 (567 kt; 652 mph; 1050 km/h)
Normal cruising speed at height: 485 kt (560 mph; 900 km/h)
Take-off speed: 200 kt (230 mph; 370 km/h)
Take-off run: 6,560-6,890 ft (2000-2100 m)
Normal landing run: 3,940-4,265 ft (1200-1300 m)

Armament

Maximum offensive load is three Kh-22 air-to-surface missiles or 52,910 lb (24000 kg) of conventional bombs and mines, half carried internally and half on racks under the wings. Internal bombs can be replaced by rotary launcher for six Kh-15P short-range missiles. Loads can include Kh-31 or Kh-35 missiles and FAB-3000, FAB-1500, FAB-500 or FAB-250 bombs.

93 Wing glove sealing plate
94 Port three-segment slotted flap
95 Two-segment spoilers
96 Port wing fully-swept position
97 Fixed portion of trailing edge
98 Port wing, fully forward position
99 Port navigation lights
100 Leading-edge slat segments
101 Port wing integral fuel tank
102 Wing pivot bearing

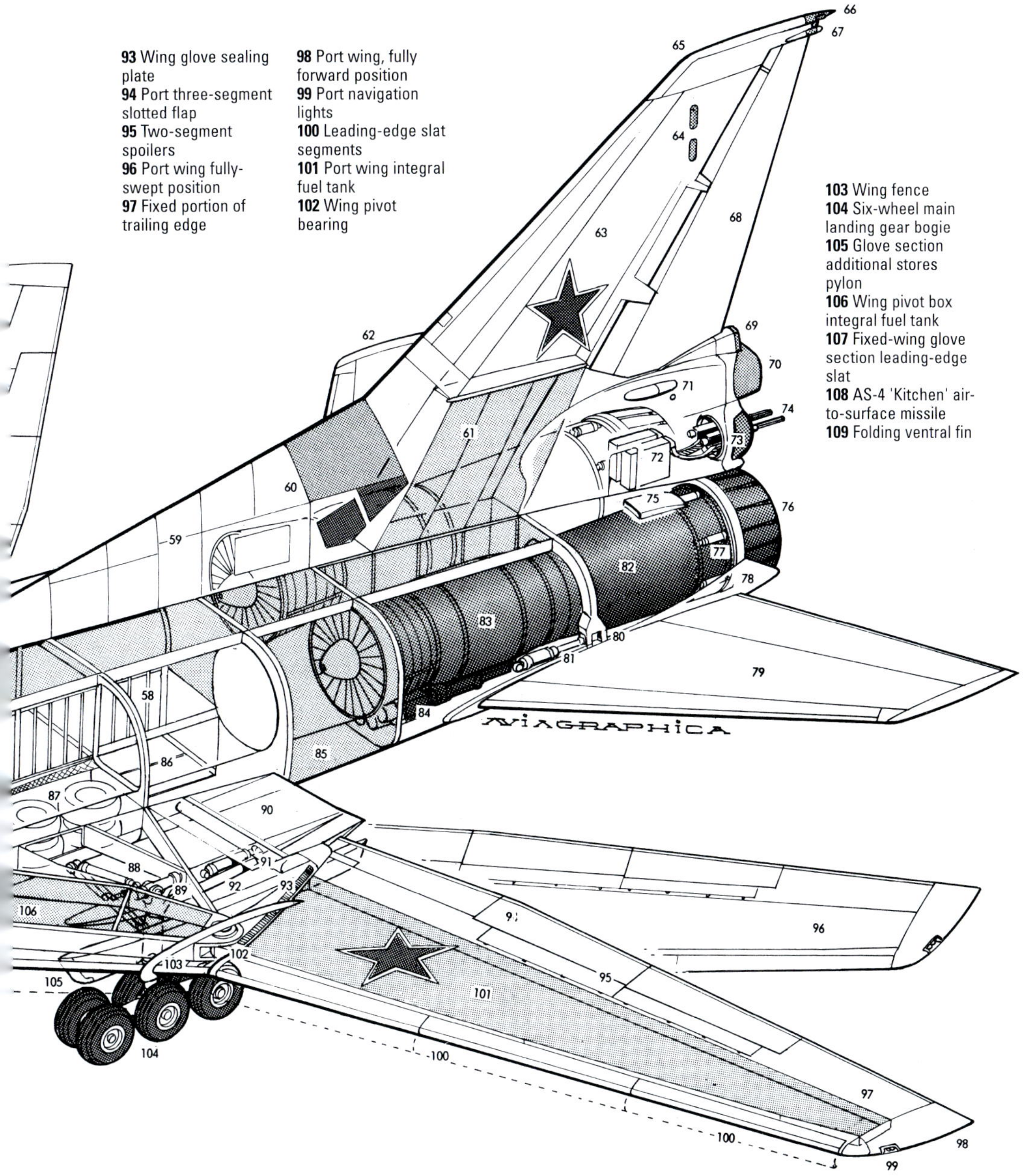

103 Wing fence
104 Six-wheel main landing gear bogie
105 Glove section additional stores pylon
106 Wing pivot box integral fuel tank
107 Fixed-wing glove section leading-edge slat
108 AS-4 'Kitchen' air-to-surface missile
109 Folding ventral fin

Vickers Valiant

A Valiant tanker of No. 214 Squadron is towed past a Bloodhound SAM batter at RAF Marham, Norfolk. This location became the main Valiant operating bas in the final years of the type's career, housing two tanker units (Nos 90 and 21 Sqns) and three NATO-assigned bomber units (Nos 49, 148 and 207 Sqns

Valiant B.Mk 1

Cutaway key
1 Flight refuelling probe, detachable
2 Refuelling probe blanking adaptor
3 De-icing fluid tank
4 Refuelling floodlight
5 H_2S amplifier
6 Gyro unit
7 Radar scanner mounting frame
8 Radome
9 H_2S radar scanner
10 Oxygen and nitrogen bottles
11 Sloping front pressure bulkhead
12 Instrument panel
13 Side mounted control column
14 Rudder pedals
15 Pilot's floor level
16 Bomb sight
17 Ventral bomb aimers windows
18 Bombardier's prone position couch
19 IFF aerial
20 Crew entry door
21 Main crew compartment
22 Air Electronics Officer's seat
23 Door opening bail-out blast shield
24 Dual Navigator's seats to starboard
25 Refuelling line fairing (outside pressurised compartment)
26 Direct vision opening side window panel
27 Curved windscreen panels
28 Jettisonable cockpit roof hatch
29 Second Pilot's Martin-Baker Mk.3A ejection seat
30 Cockpit eyebrow window
31 First Pilot's ejection seat
32 Sextant dome, jettisonable
33 Sextant
34 Cabin roof ditching hatch
35 UHF aerials
36 Radio equipment racks
37 Rear pressure bulkhead
38 Lower crew compartment window, escape hatch on starboard side
39 Nosewheel leg pivot mounting
40 Twin nosewheels, aft retracting
41 Position of HF aerial rail on starboard nosewheel door
42 Mudguard
43 Nosewheel bay
44 Air conditioning equipment
45 Flood-flow cooling air intake
46 Heat exchanger
47 Radar equipment racks
48 Oxygen bottles
49 Dinghy stowage
50 Dinghy inflation bottles
51 ADF aerial
52 Electrical and electronics equipment bay
53 Fuel tank bay bulkhead
54 Fire extinguisher bottles
55 Port engine air intakes
56 Battery compartment
57 Radar altimeter aerials
58 Ground power connection
59 Retractable bomb bay deflectors
60 Intake Spraymat de-icing
61 Inward and upward opening bomb doors
62 Bomb door rack-and-pinion drive mechanism
63 Bomb bay roof
64 Front spar/fuselage frame attachment
65 Reserve tank
66 No 1 fuel tank, total fuel capacity including wing external tanks, 9,972 Imp gal (45,332 litre)
67 Wing front spar, joined on aircraft centreline
68 No 2 fuel tank
69 Fuel recupurators
70 Fuel system piping
71 Upper fuselage central keel member
72 Starboard engine bays
73 Intake ducts
74 Engine bay firewalls
75 De-icing air heat exchanger intake
76 Starboard main undercarriage, stowed position
77 Leading edge de-icing air duct
78 External fuel tank
79 Tank pylon
80 Starboard wing fuel tanks
81 Wing fence
82 Wing stringers
83 Vortex generators
84 Wing skin panelling
85 Pitot head
86 Starboard navigation light
87 ILS glideslope aerial
88 Fuel vent
89 Starboard two-

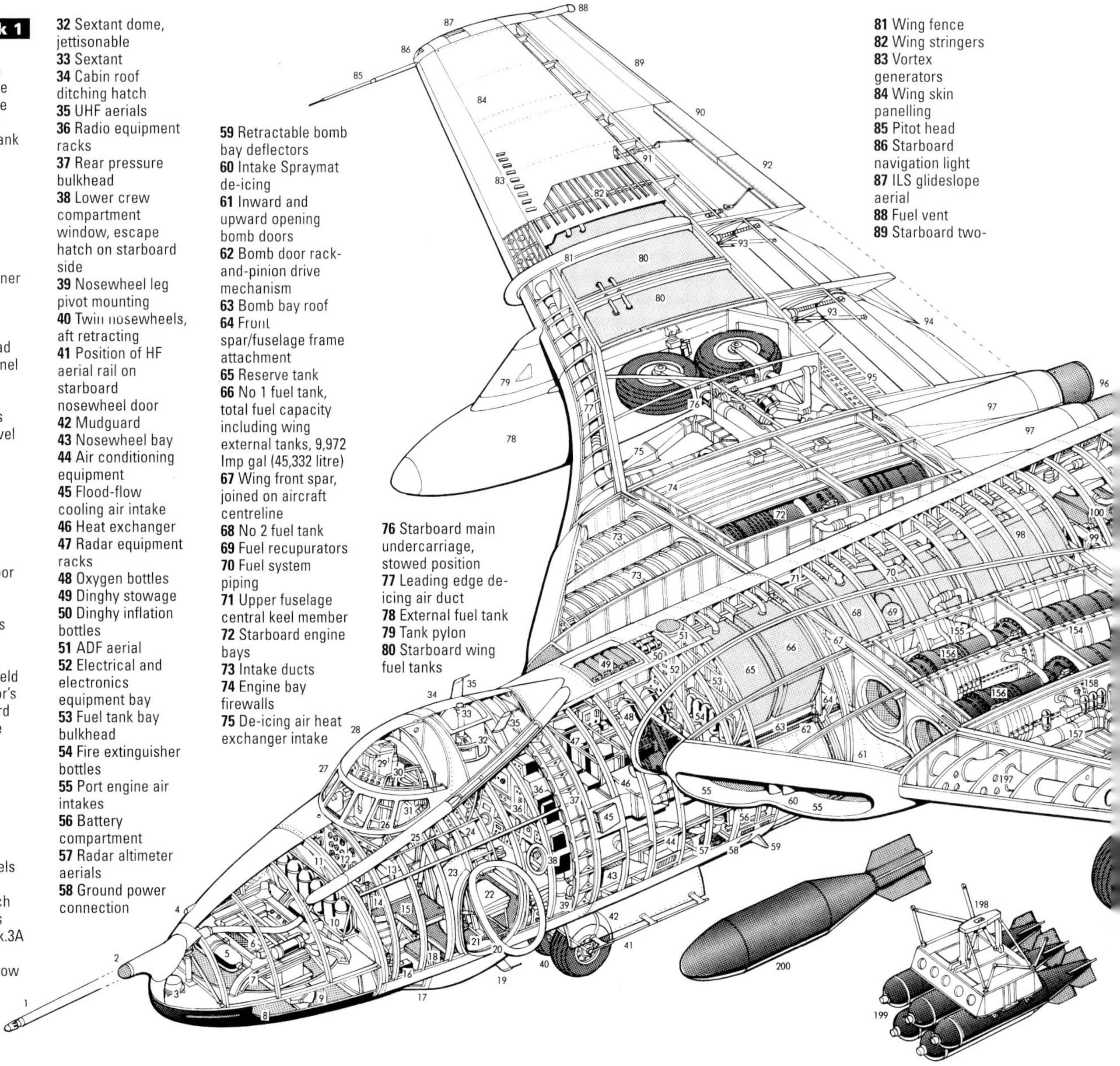

Throughout the Valiant's career crew training was performed at Gaydon by 232 Operational Conversion Unit. From 1957 the unit also trained Victor crews, the Valiant element becoming 'B' Squadron. Although formation flying had no operational relevance to the V-force, it was occasionally practised.

SPECIFICATION

Valiant B(K).Mk 1

Dimensions

Length: 114 ft 4 in (34.85 m)
Height: 32 ft 2 in (9.80 m)
Wing area: 2,362 sq ft (219.43 m²)

Powerplant

Four Rolls-Royce Avon RA.28 (200 series), each rated at 10,000 lb (44.48 kN) thrust

Weights

Empty: 75,881 lb (34419 kg)
Maximum all-up: 140,000 lb (63503 kg)

Performance

Maximum speed at 30,000 ft (9144 m): 567 mph (912 km/h)
Service ceiling: 54,000 ft (16459 m)
Range: 4,500 miles (7242 km) with underwing tanks

Bomb load

No gun armament. Bomb load comprised either one 10,000-lb (4536-kg) nuclear weapon (Blue Danube) or 21 1,000-lb (454-kg) iron bombs. In the tanking role a pair of 1,615-Imp gal (7342-litre) ful tanks could be carried in the bomb bay.

Only one Valiant was saved from the scrapper's torch, though fortunately it was an airframe with an important and interesting history. XD818 was one of the eight Grapple aircraft and dropped Britain's first thermonuclear weapon on 15 May 1957. After several years at the mercy of the elements, on the gate at RAF Marham, this unique machine was saved by the RAF Museum and resides in the Bomber Command Hall.

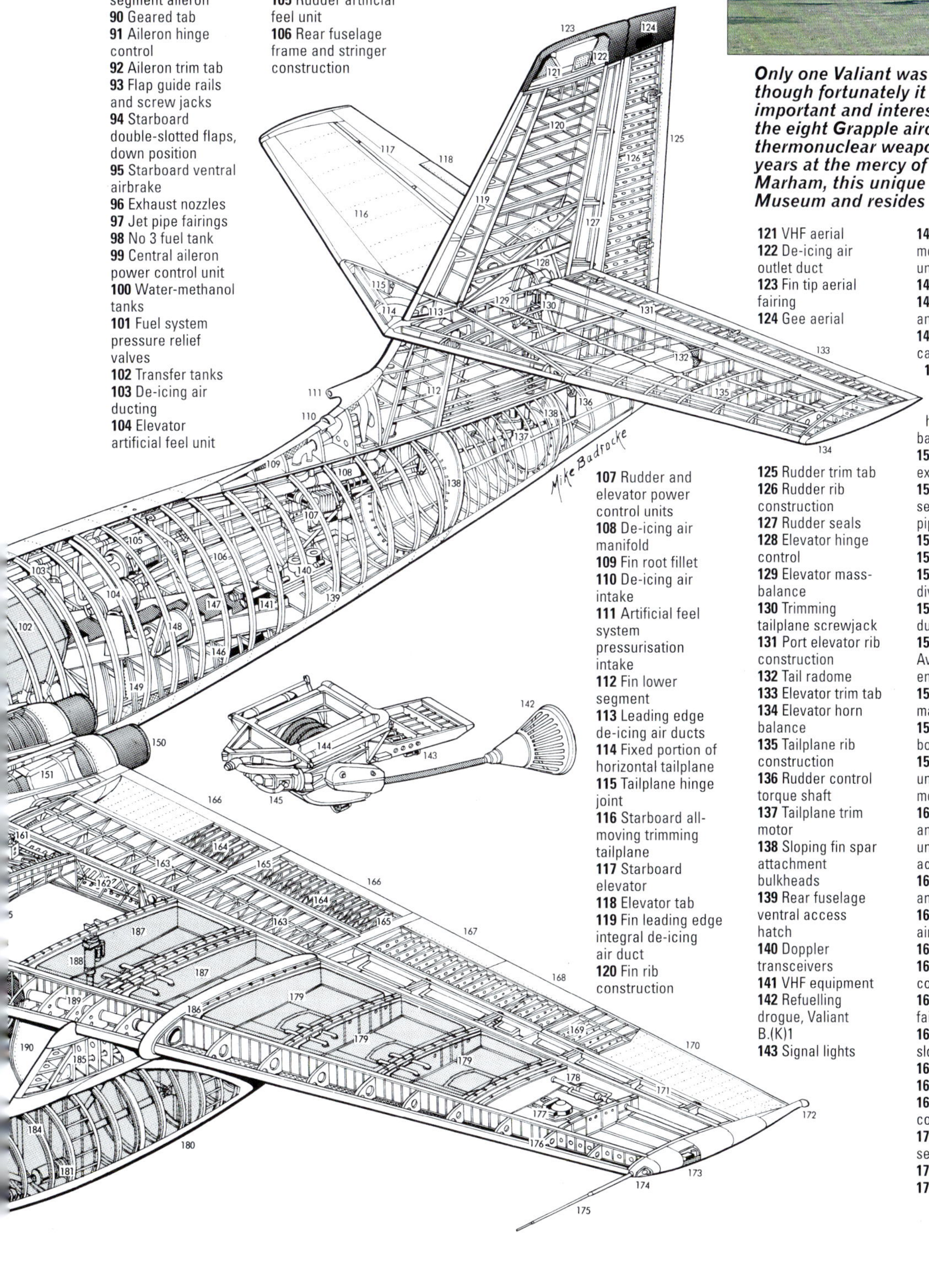

segment aileron
90 Geared tab
91 Aileron hinge control
92 Aileron trim tab
93 Flap guide rails and screw jacks
94 Starboard double-slotted flaps, down position
95 Starboard ventral airbrake
96 Exhaust nozzles
97 Jet pipe fairings
98 No 3 fuel tank
99 Central aileron power control unit
100 Water-methanol tanks
101 Fuel system pressure relief valves
102 Transfer tanks
103 De-icing air ducting
104 Elevator artificial feel unit
105 Rudder artificial feel unit
106 Rear fuselage frame and stringer construction
107 Rudder and elevator power control units
108 De-icing air manifold
109 Fin root fillet
110 De-icing air intake
111 Artificial feel system pressurisation intake
112 Fin lower segment
113 Leading edge de-icing air ducts
114 Fixed portion of horizontal tailplane
115 Tailplane hinge joint
116 Starboard all-moving trimming tailplane
117 Starboard elevator
118 Elevator tab
119 Fin leading edge integral de-icing air duct
120 Fin rib construction
121 VHF aerial
122 De-icing air outlet duct
123 Fin tip aerial fairing
124 Gee aerial
125 Rudder trim tab
126 Rudder rib construction
127 Rudder seals
128 Elevator hinge control
129 Elevator mass-balance
130 Trimming tailplane screwjack
131 Port elevator rib construction
132 Tail radome
133 Elevator trim tab
134 Elevator horn balance
135 Tailplane rib construction
136 Rudder control torque shaft
137 Tailplane trim motor
138 Sloping fin spar attachment bulkheads
139 Rear fuselage ventral access hatch
140 Doppler transceivers
141 VHF equipment
142 Refuelling drogue, Valiant B.(K)1
143 Signal lights
144 Bomb bay mounted hose drum unit
145 Fuel pumps
146 Ventral Doppler antenna
147 Rear fuselage catwalk
148 Aileron artificial feel unit
149 Upward hinged rear bomb bay deflector
150 Port engine exhaust nozzles
151 Inboard flap segment beneath jet pipes
152 Port jet pipes
153 Rearspar
154 Engine bay dividing firewall
155 Bleed air ducting
156 Rolls-Royce Avon R.A.28 Mk.204 engines
157 De-icing air manifold
158 Fire extinguisher bottles
159 Main undercarriage pivot mounting
160 Electric main and emergency undercarriage leg actuators
161 Flap drive shaft and screw jacks
162 Port ventral airbrake
163 Flap shroud ribs
164 Flap rib construction
165 Flap track fairings
166 Port double-slotted flaps
167 Aileron trim tab
168 Geared tab
169 Aileron rib construction
170 Port two-segment aileron
171 Aileron seals
172 Fuel vent
173 VOR localiser aerial
174 Port navigation light
175 Pitot head
176 Leading edge nose ribs
177 Retractable landing lamp
178 Fuel vent valve
179 Wing outboard fuel tanks
180 External fuel tank
181 Fuel pump
182 Nitrogen pressurising bottles
183 Ground nitrogen and air charging points
184 Tank rib construction
185 Tank pylon construction
186 Port wing fence
187 Wing inboard fuel tanks
188 Mainwheel door electric actuator
189 Leading edge de-icing air duct
190 Mainwheel door
191 Twin tandem mainwheels
192 Disc brakes and anti-skid units
193 Horizontal telescopic torque link
194 Shock absorber leg struts
195 Side stay and breaker strut
196 Balance strut
197 Bifurcated de-icing air duct
198 Bomb bay weapons carrier
199 1,000-lb (454-kg) HE bombs, maximum load 21 in groups of 5 and 3
200 10,000-lb (4536-kg) free-fall nuclear weapon

Fighters

04
FARA
39

Convair F-102 Delta Dagger

509th FIS F-102As pass over defoliated and bomb-blasted terrain in South Vietnam. With little air threat, the F-102 was used as an escort for the B-52, and in a limited ground attack capacity (for which it was credited with the destruction of 106 buildings, 16 sampans and a bridge during 199 sorties). Vietnam involvement began in March 1962 for the F-102, when four examples were deployed to Tan Son Nhut. Further examples were sent out by President Johnson following the Gulf of Tonkin incident in August 1964 under the codename Water Glass. By the end of 1968, six aircraft were on alert at Bien Hoa, six at Da Nang, and a further 10 were in Thailand.

F-102A

Cutaway key
1 Pitot head
2 Radome
3 Radar scanner
4 Scanner tracking mechanism
5 ILS glideslope aerial
6 Radar mounting bulkhead
7 Radar pulse generator and modulator units
8 Nose compartment access doors
9 Static port
10 Lower IFF aerial
11 Angle of attack transmitter
12 TACAN aerial
13 MG-10 fire control system electronics
14 Nose compartment longeron
15 Infra-red detector
16 Electronics cooling air duct
17 Windscreen panels
18 Central vision splitter
19 Instrument panel shroud
20 Rudder pedals and linkages
21 Cockpit front pressure bulkhead
22 Air-conditioning system ram air intake
23 Boundary layer splitter plate
24 Electrical system equipment
25 Port air intake
26 Nosewheel door
27 Taxiing lamp
28 Nosewheel, forward-retracting
29 Nose undercarriage leg strut
30 Torque scissor links
31 Intake duct framing
32 Nose undercarriage pivot mounting
33 Cockpit pressure floor
34 Port side console panel
35 Engine throttle lever
36 Two-handed control grip, radar and flight controls
37 Pilot's ejection seat
38 Canopy handle
39 Starboard side console panel
40 Radar display
41 Optical sight
42 Cockpit canopy cover, upward-hinging
43 Ejection seat headrest
44 Boundary layer spill duct
45 Sloping cockpit rear pressure bulkhead
46 Air-conditioning plant
47 Canopy external release
48 Canopy jack
49 Air exit louvres
50 Equipment bay access hatches, port and starboard
51 Canopy hinge
52 Radio and electronics equipment bay
53 Forward position light
54 Intake trunking
55 Missile bay cooling air duct
56 Missile bay door pneumatic jacks
57 Canopy emergency release
58 Liquid oxygen converter
59 Electrical system equipment bay
60 Fuselage upper longeron
61 Upper IFF aerial
62 Wing front spar attachment bulkhead
63 Pneumatic system air bottles
64 Bifurcated intake duct
65 Close-pitched fuselage frame construction
66 Engine bleed air duct
67 Anti-collision light
68 Starboard wing forward main fuel tank, total internal capacity 1,085 US gal (4107 litres)
69 Inboard wing fence
70 Fuel system piping
71 Centre-section wing dry bay
72 Wing pylon mountings and connectors
73 Starboard main undercarriage pivot mounting
74 Dorsal spine fairing
75 Intake duct mixing chamber
76 Engine intake centre-body fairing
77 Wing main spar attachment bulkheads
78 Intake compressor face
79 Forward engine mounting
80 Pratt & Whitney J57-P-23A afterburning turbojet engine
81 Engine oil tank, capacity 5.5 US gal (21 litres)
82 Oil filler cap
83 Starboard wing aft main fuel tanks
84 Fuel feed and vent piping
85 Ventral actuator fairing
86 Outboard wing fence
87 Cambered leading edge
88 Wing tip camber wash-out
89 Starboard navigation light
90 Fixed portion of trailing edge
91 Starboard outer elevon
92 Elevon hydraulic actuator
93 Trailing-edge dry bay
94 Fin leading-edge rib construction
95 Aerial tuning units
96 Fin attachment joints
97 Tailfin construction
98 Artificial feel system pitot intakes
99 Sloping front spar
100 Upper fin multi-spar construction
101 Fintip aerial fairing
102 UHF aerials
103 VOR localiser aerial
104 Rudder
105 Honeycomb core rudder construction
106 Split airbrake panels
107 Airbrake pneumatic jacks
108 Airbrake, open position
109 Variable-area afterburner exhaust nozzle
110 Aft fuselage aerodynamic (area-rule) fairing
111 Exhaust nozzle control jacks (eight)
112 Tailcone attachment joint frame (engine removal)
113 Rear position lights
114 Afterburner duct
115 Engine bay internal heat shield
116 Brake parachute housing
117 Rudder hydraulic actuator
118 Rudder trim and feel force control units
119 Afterburner fuel manifold
120 Rear engine mounting
121 Inboard elevon hydraulic actuator
122 Engine turbine section
123 Bleed air connections
124 Bleed air blow-off valve
125 Engine accessory equipment gearbox
126 Wing spar/fuselage frame pin joints
127 Wingroot rib
128 Port wing aft integral fuel tanks

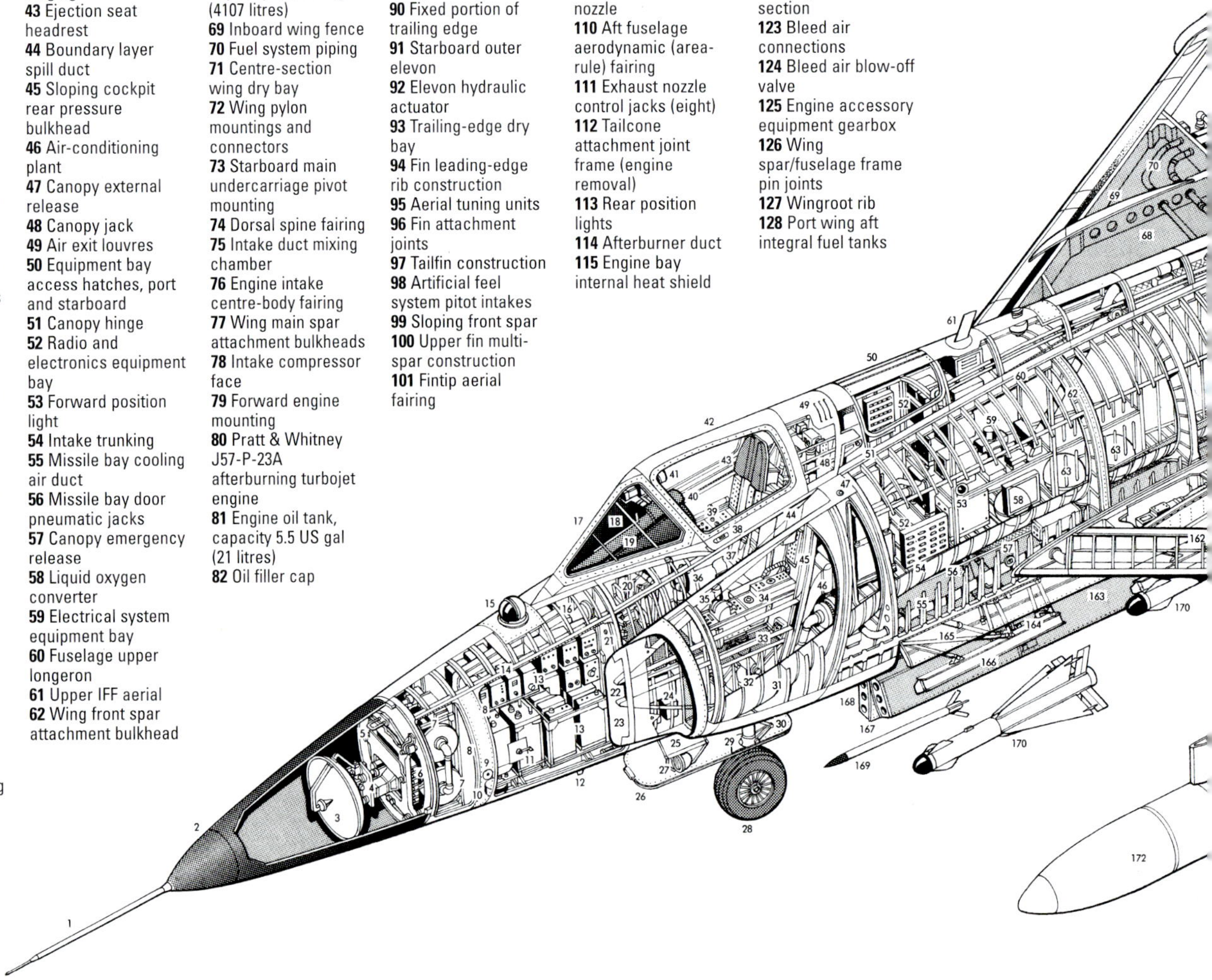

*The first **YF-102** (Model 8-80), 52-7994, is seen on the dry lake at Edwards **AFB** test facility. After only six flights the aircraft was lost after an engine flame-out during take-off. First flown by test pilot **Dick Johnson** on 24 October 1953, the **YF-102** suffered from severe instability and buffeting in the transonic region, and the second aircraft refused to accelerate past Mach 1. The failure of the initial **YF-102** necessitated a complete redesign, which finally resulted in the **YF-102A** (8-90).*

SPECIFICATION

F-102A Delta Dagger

Dimensions

Length: 68 ft 4½ in (20.82 m)
Wingspan: 38 ft 1½ in (11.60 m)
Height: 21 ft 2½ in (6.45 m)
Wing area: 695 sq ft (64.56 m²)

Powerplant

One Pratt & Whitney J57-P-23 turbojet developing 11,700 lb (53 kN) of thrust dry and 17,200 lb (77 kN) thrust with afterburning

Weights

Empty: 19,350 lb (8777 kg)
Normal loaded, 'clean': 27,700 lb (12565 kg)
Normal loaded, point interception: 28,150 lb (12769 kg)
Maximum take-off: 31,500 lb (14288 kg)

Fuel

Internal: 1,085 US gal (4107 litre)
Maximum, with two 215-US gal (814-litre) drop tanks: 1,515 US gal (5735 litre)

Performance

Maximum speed 'clean' at 40,000 ft (12190 m): 825 mph (1328 km/h)
Normal cruising speed at 35,000 ft (10670 m): 540 mph (869 km/h)
Stalling speed: 154 mph (248 km/h)
Service ceiling: 54,000 ft (16460 m)
Tactical radius with two 215-US gal (814-litre) drop tanks and full armament: 500 miles (805 km)
Maximum range: 1,350 miles (2173 km)
Initial climb rate: 17,400 ft (5304 m) per minute

Armament

Three AIM-4C Falcon infra-red homing AAMs and one AIM-26A Nuclear Falcon AAM, or three AIM-4A/E radar-guided and three AIM-4C/F infra-red homing AAMs, or up to 24 unguided 2.75-in (70-mm) folding fin aircraft rockets in early aircraft (this latter facility was later deleted in service)

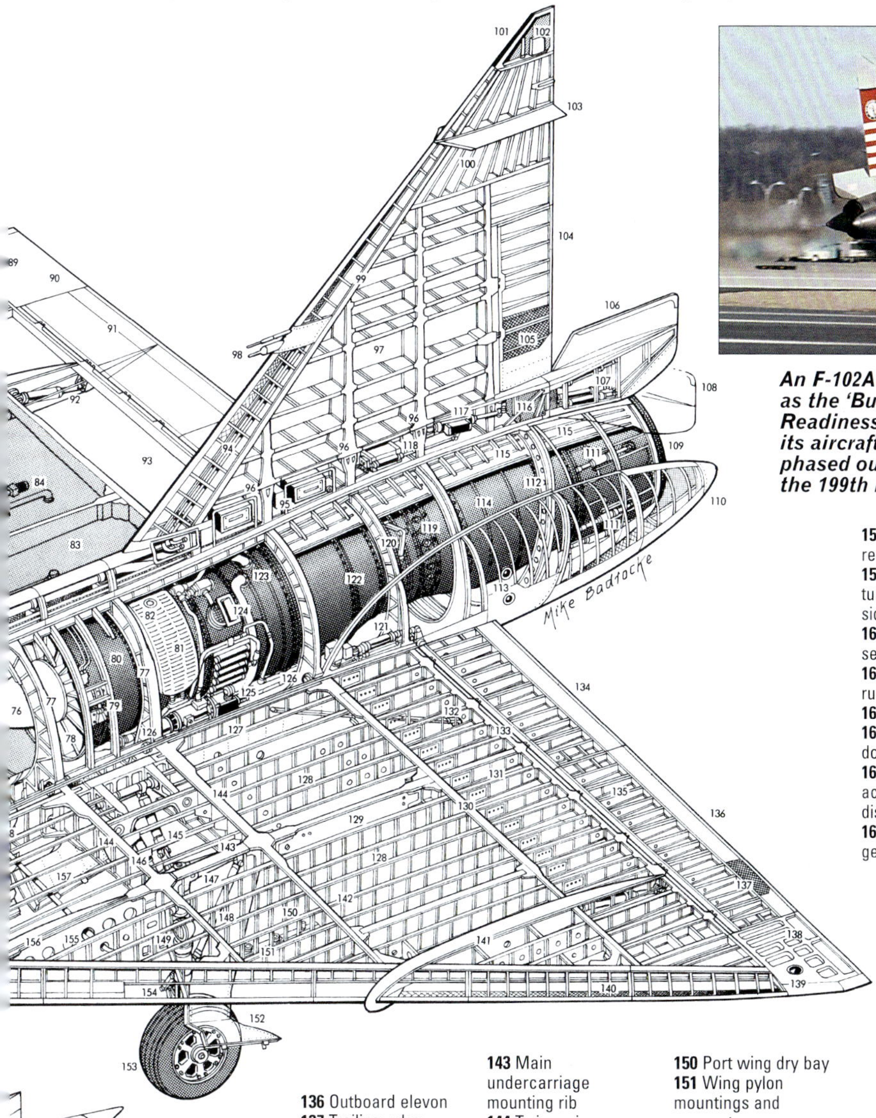

*An **F-102A** of the Wisconsin **ANG**'s 176th **FIS** (sometimes known as the 'Bushy Badgers') takes off during an Operational Readiness Inspection (**ORI**). The first **F-102** squadron gave up its aircraft in 1964, but the bulk of the 23 **ANG** units so-equipped phased out the 'Deuce' in the 1969-71 period. The final user was the 199th **FIS** at Hickam **AFB**, retiring the **F-102** in October 1976.*

129 Fuel tank dividing rib
130 Rear spar
131 Trailing-edge ribs
132 Runway emergency arrester hook, lowered
133 Elevon spar
134 Inboard elevon
135 Elevon rib construction
136 Outboard elevon
137 Trailing-edge honeycomb
138 Wingtip fairing construction
139 Port navigation light
140 Cambered leading-edge rib construction
141 Outboard wing fence
142 Wing rib construction
143 Main undercarriage mounting rib
144 Twin main spars
145 Main undercarriage side strut
146 Hydraulic retraction jack
147 Main undercarriage leg pivot mounting
148 Drag strut and pneumatic brake reservoir
149 Landing lamp
150 Port wing dry bay
151 Wing pylon mountings and connectors
152 Main undercarriage leg door
153 Port mainwheel
154 Torque scissor links
155 Port wing forward integral fuel tank
156 Inboard wing fence
157 Mainwheel door
158 Hydraulic reservoirs
159 Position of ram air turbine on starboard side
160 Missile bay aft section doors
161 Retractable over-run barrier probe
162 Wing front spar
163 Port missile bay doors
164 Pantographic action missile displacement gear
165 Displacement gear hydraulic jack
166 Missile launch rail
167 Missile bay door integral rocket launch tubes
168 Centre missile bay door
169 2.75-in (70-mm) FFAR folding-fin rockets (24)
170 AIM-4D Falcon air-to-air missile (6)
171 Port wing fuel tank pylon
172 215-US gal (814-litre) external fuel tank

Convair F-106 Delta Dart

'Spittin' Kittens' F-106s originally wore a large tail chevron as seen on this aircraft – the F-106 belonging to the 5th FIS's commanding officer – in May 1974. The more famous lightning bolt and stars appeared several years later.

F-106A Delta Dart

Cutaway key

1 Pitot head
2 Radome
3 Radar scanner dish
4 Radar tracking mechanism
5 Hughes MA-1 weapons system radar unit
6 Radar mounting bulkhead
7 Pulse generator units
8 TACAN aerial
9 Angle of attack transmitter
10 MA-1 weapons system electronics units
11 Electronics bay access door
12 Infra-red detector fairing
13 Retractable infra-red detector
14 Knife-edged windscreen panels
15 Central vision splitter
16 Instrument panel shroud
17 'Head down' tactical display panel
18 Canopy external release
19 Rudder pedals
20 Cockpit front pressure bulkhead
21 Electrical relay panel
22 Nose undercarriage wheel bay
23 Nosewheel door
24 Taxiing lamp
25 Twin nosewheels
26 Torque scissor links
27 UHF aerial
28 Nose undercarriage leg strut
29 Oxygen filler point and gauge
30 Nosewheel leg pivot fixing
31 Liquid oxygen converter
32 Cockpit air-conditioning ducting
33 Cockpit pressure floor
34 Control column
35 Two-handed control grip, radar and flight controls
36 Engine throttle lever
37 Pilot's ejection seat
38 Radar display
39 Optical sight
40 Cockpit canopy cover
41 Ejection seat headrest
42 Ejection seat launch rails
43 Cockpit rear pressure bulkhead
44 Side console panel
45 Ground power supply connections
46 Doppler navigation unit
47 Aft lower electronics compartment
48 Aft upper electronics equipment bays, port and starboard
49 Electronics bay door
50 Cockpit rear decking
51 Over pressurisation relief valve
52 Canopy pneumatic jack
53 Canopy hinge
54 Air exit louvres
55 Starboard engine air intake
56 Fuel tank access panel
57 Upper longeron
58 Fuselage fuel tank, total internal capacity, 1,514 US gal (5731 litres)
59 Fuselage frame construction
60 Ventral weapons bay
61 Missile pallet hinge arms
62 Bottom longeron
63 Boundary layer splitter plate
64 Port engine air intake
65 Variable area intake ramp
66 Ramp bleed air louvres
67 Air-conditioning system intake duct
68 Intake duct framing
69 Starboard side pressure refuelling connection
70 Forward missile pallet pneumatic jack
71 Air-conditioning plant
72 De-icing fluid reservoir
73 Heat exchanger air exit duct
74 Air refuelling ramp door, open
75 Pneumatic system air bottles
76 Bifurcated intake ducting
77 Aft missile pylon pneumatic jacks
78 AIR-2 Genie air-to-air missile housing
79 Hydraulic accumulators
80 Hydraulic reservoirs, duplex systems
81 Intake trunking
82 Wing spar attachment fuselage main frames
83 Oil cooler air duct
84 Intake centre body fairing
85 Engine intake compressor face
86 Bleed air ducting
87 Dorsal spine fairing
88 Fuel boost pump
89 Starboard main undercarriage pivot fixing
90 Wing forward fuel tank
91 Dry bay
92 Wing pylon mountings and connectors
93 Fuel system piping
94 Starboard wing main fuel tank
95 Leading-edge slot
96 Cambered leading edge
97 Wingtip fairing
98 Starboard navigation light
99 Outboard elevon
100 Elevon hydraulic jack
101 Elevon jack ventral fairing
102 Inboard elevon
103 Starboard wing aft fuel tank
104 Fuel system vent piping
105 Engine oil tank, 45 US gal (17 litres)
106 Pratt & Whitney J75-P-17 turbojet engine
107 Forward engine mounting
108 Ventral accessory equipment compartment
109 Cooling air ducting
110 Wing and fin spar attachment main frame
111 Inboard elevon hydraulic jack
112 Engine turbine section
113 Exhaust pipe heat shroud
114 Rear engine mounting
115 Aerial tuning units
116 Artificial feel system pitot intakes
117 Fin leading edge
118 Tailfin construction
119 Air-to-air identification (AAI) aerial
120 Fin tip aerial fairing
121 UHF/TACAN aerial
122 Tail navigation light
123 Rudder
124 Rudder honeycomb construction
125 Split air brake panels

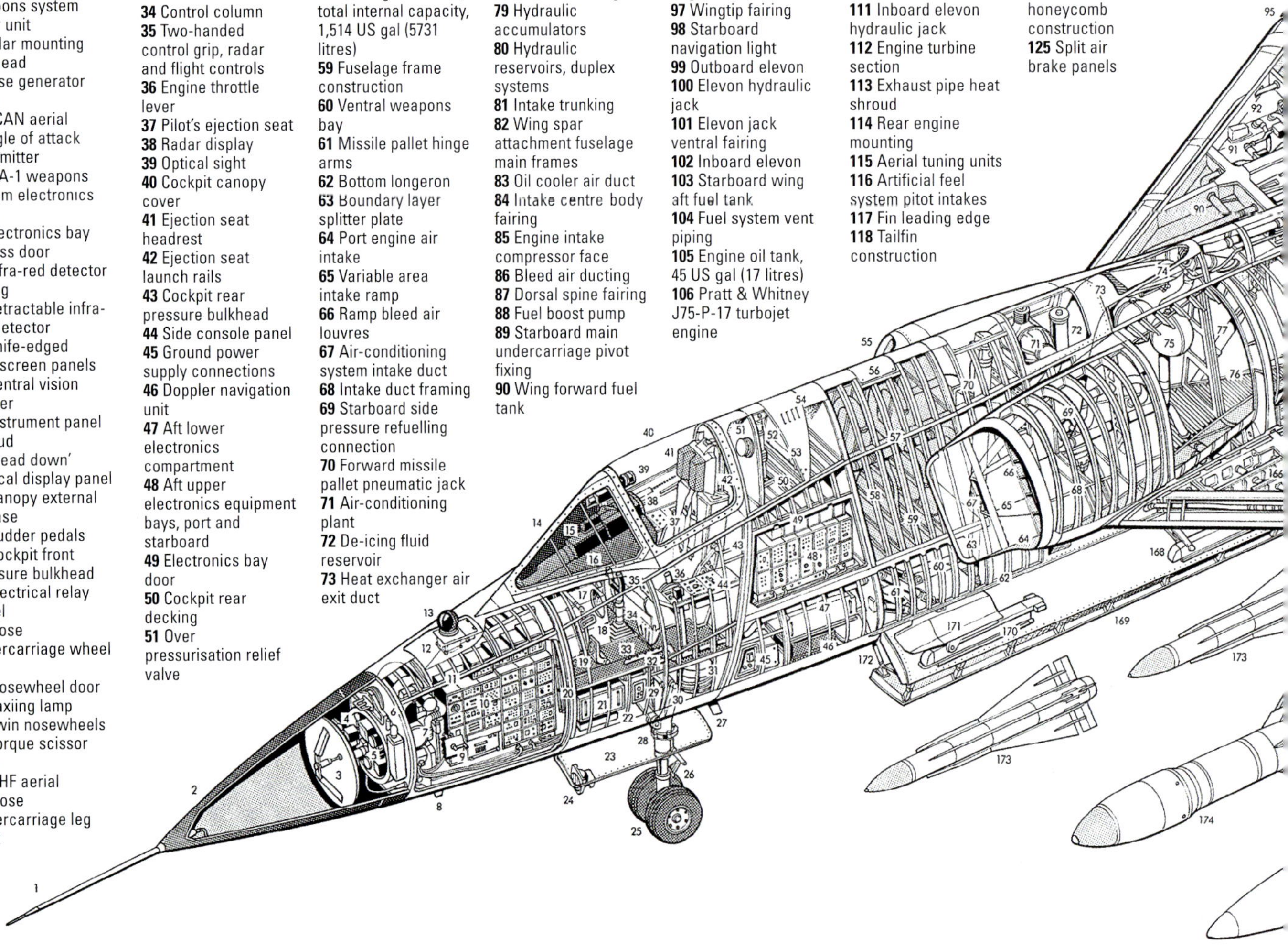

*Before the first **F-106** even flew, authorisation was given for the construction of a two-seat variant of the **Delta Dart**. Unlike the **TF-102A**, with its side-by-side seating and inferior performance to the single-seater, the **F-106B** was designed to be as close to the **F-106A** as possible. Due to the fact that it was more than a trainer and carried all the mission equipment of a basic interceptor, it was not given the **TF-106A** designation. This particular aircraft **(57-2540)** was, in common with many **F-106s**, converted into a drone (**QF-106**) and shot down with an **AIM-120 AMRAAM**.*

SPECIFICATION

F-106A Delta Dart

Dimensions

Length (with probe): 70 ft 8 in (21.55 m)
Wingspan: 38 ft 3½ in (11.67 m)
Wing area: 697.5 sq ft (64.8 m²)
Height: 20 ft 3⅓ in (6.18 m)

Powerplant

One Pratt & Whitney J75-P-17 turbojet rated at 17,200 lb (77.4 kN) dry and 24,500 lb (110.25 kN) with full afterburner

Weights

Empty: 23,814 lb (10800 kg)
Normal loaded: 35,500 lb (16012 kg)
Maximum take-off: 38,250 lb (17350 kg)

Performance

Maximum speed (without tanks) at 40,000 ft: Mach 2.25 (1,487 mph/ 2393 km/h)
Sustained ceiling: 58,000 ft (17680 m)
Combat radius (with internal fuel): 575 miles (925 km)
Combat radius (with external tanks and aerial refuelling): 1,950 miles (3138 km)
Time to 57,000 ft (17374 m): 4.5 min

Armament

Most aircraft had an M61A-1 20-mm cannon. Standard missile armament was four AIM-4E/F and/or AIM-4G Falcon AAMs (Super Falcons) plus one AIR-2A Genie nuclear rocket. The F-106 also tested the XAIM-97A ASAT weapon

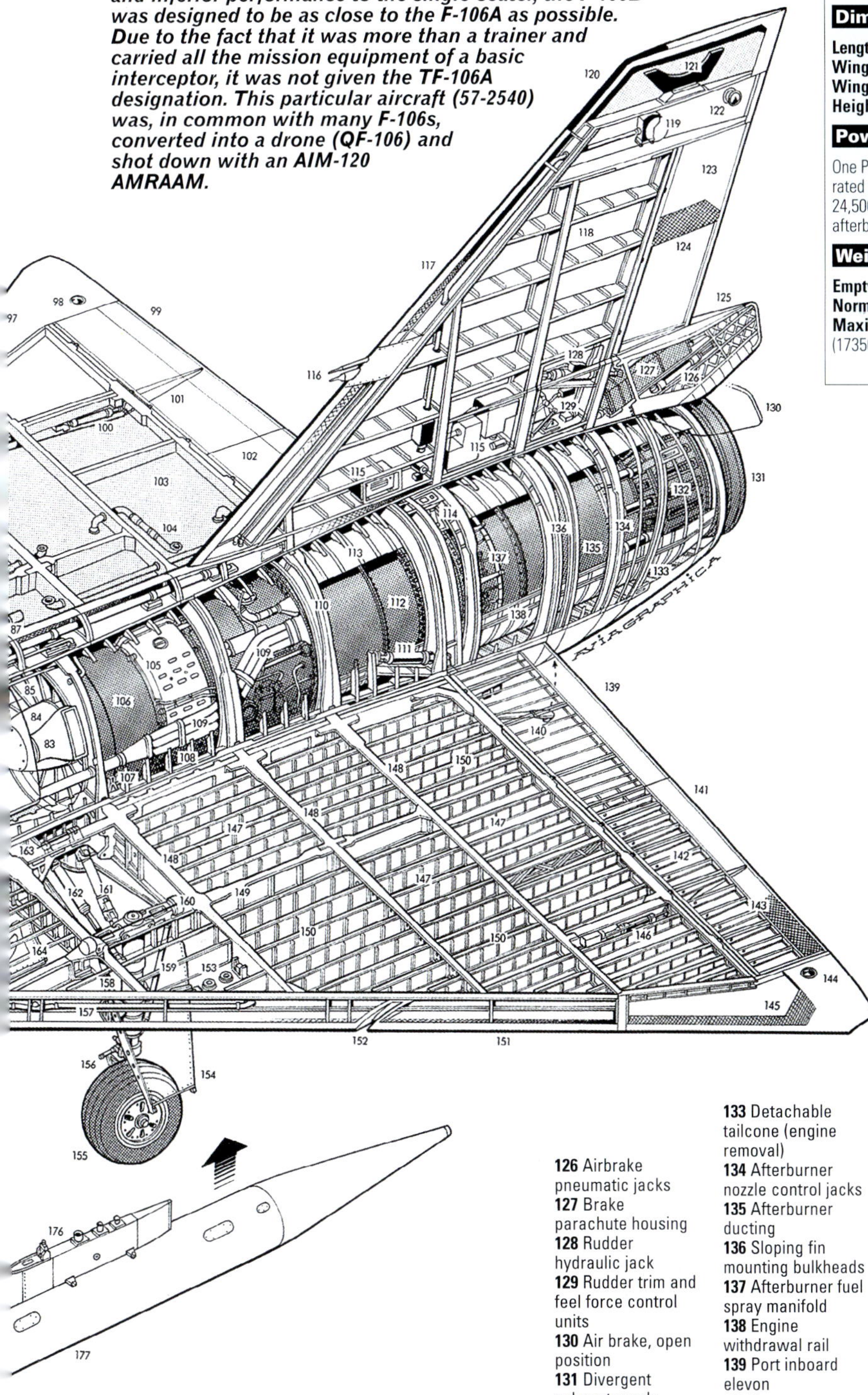

126 Airbrake pneumatic jacks
127 Brake parachute housing
128 Rudder hydraulic jack
129 Rudder trim and feel force control units
130 Air brake, open position
131 Divergent exhaust nozzle
132 Variable-area afterburner exhaust nozzle
133 Detachable tailcone (engine removal)
134 Afterburner nozzle control jacks
135 Afterburner ducting
136 Sloping fin mounting bulkheads
137 Afterburner fuel spray manifold
138 Engine withdrawal rail
139 Port inboard elevon
140 Runway emergency arresting hook, lowered
141 Port outboard elevon
142 Elevon rib construction
143 Honeycomb trailing-edge panels
144 Port navigation light
145 Honeycomb wingtip fairing
146 Outboard elevon hydraulic jack
147 Port wing integral fuel tank
148 Machined wing spars
149 Machined main undercarriage mounting rib
150 Wing rib construction
151 Cambered leading edge
152 Leading-edge slot
153 Port wing pylon connectors
154 Main wheel leg door
155 Port mainwheel
156 Torque scissor links
157 Landing lamp
158 Main undercarriage leg strut
159 Drag brace and pneumatic brake reservoir
160 Main undercarriage leg pivot fixing
161 Breaker strut
162 Hydraulic retraction jack
163 Main undercarriage wheel bay
164 Mainwheel doors
165 Emergency ram air turbine
166 Port wing forward fuel tank bay
167 Fuel system vent pipe
168 Aft single missile pylon, port and starboard, lowered position
169 Weapons bay doors, open
170 Missile launch rail
171 Forward twin missile pallet
172 Weapons bay door pneumatic jack
173 AIM-4F Falcon air-to-air missile (4)
174 Single AIR-2A Genie air-to-air nuclear missile
175 Missile folding fins, deployed position
176 Port wing pylon
177 227-US gal (859-litre) external fuel tank

Gloster Meteor

As the operational life of the Meteor began to draw to a close, a large number of Meteor F.Mk 8s was modified to become target-towing tugs and advanced trainers. The target tugs were unofficially designated 'F(TT).Mk 8' and the trainers 'T.Mk 8'. Both types received high-visibility markings and a 'T.Mk 8' of No. 85 Sqn, with bright orange Dayglo markings, is seen here accompanying a Lightning of No. 5 Sqn.

Meteor F.Mk III

Cutaway key

1 Starboard detachable wingtip
2 Starboard navigation light
3 Starboard recognition light
4 Starboard aileron
5 Aileron balance tab
6 Aileron mass balance weights
7 Aileron control coupling
8 Aileron torque shaft
9 Chain sprocket
10 Cross-over control runs
11 Front spar
12 Rear spar
13 Aileron (inboard) mass balance
14 Nacelle detachable tail section
15 Jet pipe exhaust
16 Internal stabilising struts
17 Rear spar 'spectacle' frame
18 Fire extinguisher spray ring
19 Main engine mounting frame
20 Engine access panel(s)
21 Nacelle nose structure
22 Intake internal leading-edge shroud
23 Starboard engine intake
24 Windscreen de-icing spray tube
25 Reflector gunsight
26 Cellular glass bulletproof windscreen
27 Aft-sliding cockpit canopy
28 Demolition incendiary (cockpit starboard wall)
29 RPM indicators (left and right of gunsight)
30 Pilot's seat
31 Forward fuselage top deflector skin
32 Gun wobble button
33 Control column grip
34 Main instrument panel
35 Nosewheel armoured bulkhead
36 Nose release catches (10)
37 Nosewheel jack bulkhead
38 Nose ballast weight location
39 Nosewheel mounting frames
40 Radius rod (link and jack omitted)
41 Nosewheel pivot bearings
42 Shimmy-damper/ self-centring strut
43 Gun camera
44 Camera access
45 Aperture
46 Nose cone
47 Cabin cold-air intake
48 Nosewheel leg door
49 Picketing rings
50 Tension shock absorber
51 Pivot bracket
52 Mudguard
53 Torque strut
54 Doorhoop
55 Wheel fork
56 Retractable nosewheel
57 Nosewheel doors
58 Port cannon trough fairings
59 Nosewheel cover
60 Intermediate diaphragm
61 Blast tubes
62 Gun front mount rails
63 Pilot's seat pan
64 Emergency crowbar
65 Canopy de-misting silica gel cylinder
66 Bulletproof glass rear view cut-outs
67 Canopy track
68 Sea bulkhead
69 Entry step
70 Link ejection chutes
71 Case ejection chutes
72 20-mm Hispano Mk III cannon
73 Belt feed mechanism
74 Ammunition feed necks
75 Ammunition tanks
76 Aft glazing (magazine bay top door)
77 Leading ramp
78 Front spar bulkhead
79 Oxygen bottles (2)
80 Front spar carry-through
81 Tank bearer frames
82 Rear spar carry-through
83 Self-sealing (twin compartment) main fuel tank, capacity 165 Imp gal (750 litres) in each half
84 Fuel connector pipe
85 Return pipe
86 Drain pipes
87 Fuel filler caps
88 Tank doors (2)
89 T.R.1143 aerial mast
90 Rear spar bulkhead (plywood face)
91 Aerial support frame
92 R.3121 (or B.C.966M IFF installation
93 Tab control cables
94 Amplifier
95 Fire extinguisher bottles (2)
96 Elevator torque shaft
97 T.R.1143 transmitter/ receiver radio installation
98 Pneumatic system filler
99 Pneumatic system (compressed) air cylinders
100 Tab cable fairlead
101 Elevator control cable
102 Top longeron
103 Fuselage frame
104 IFF aerial
105 DR compass master unit
106 Rudder cables
107 Starboard lower longeron
108 Cable access panels (port and starboard)
109 Tail section joint
110 Rudder linkage
111 Tail ballast weight location
112 Fin spar/fuselage frame
113 Rudder tab control
114 Fin structure
115 Torpedo fairing
116 Tailplane spar/upper fin attachment plates
117 Upper fin section
118 Starboard tailplane
119 Elevator horn and mass balance
120 Starboard elevator

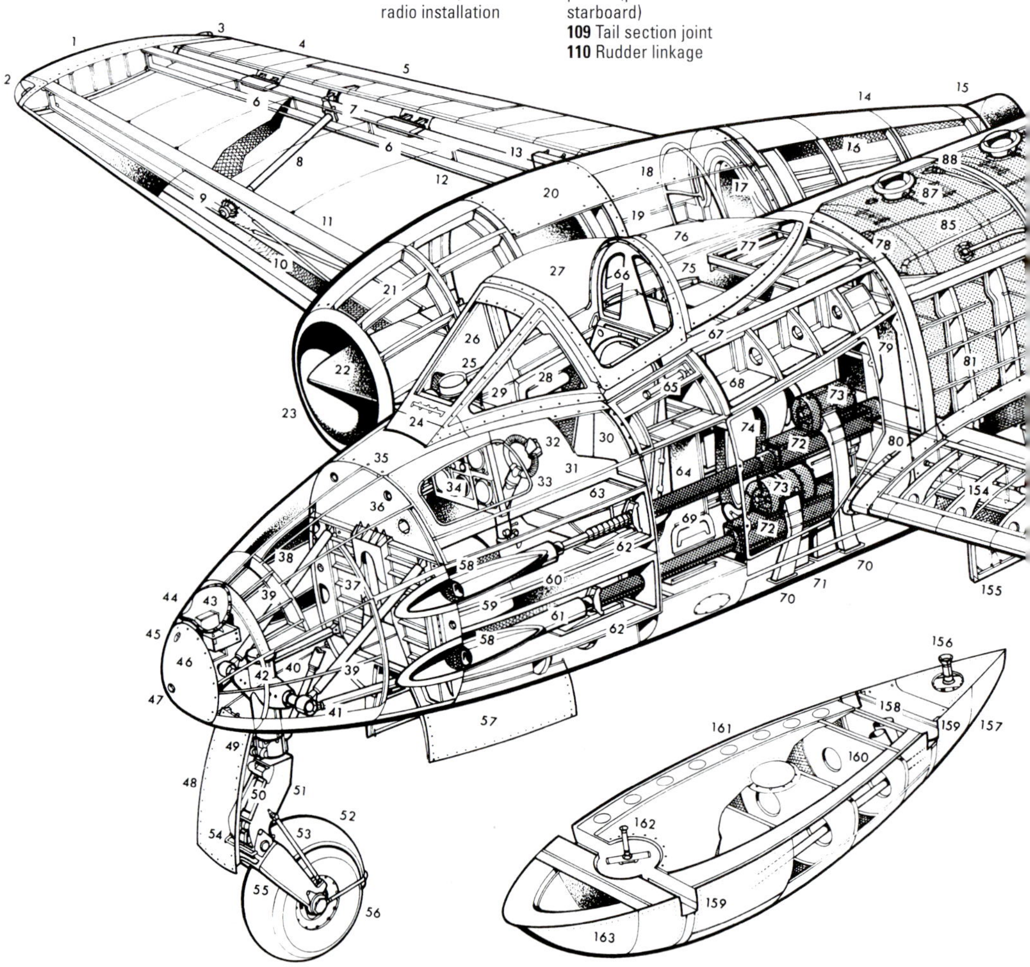

EE389 was the first Meteor involved in inflight-refuelling trials, in April 1949. Here it is seen, with its airbrakes deployed and probe clearly evident, about to refuel from a Lancaster Mk III tanker. The neat probe installation in the nose of the aircraft served as the basis of a similar fitting applied to a number of Mk 4s and Mk 8s.

SPECIFICATION

Meteor F.Mk 8

Dimensions

Length: 44 ft 7 in (13.59 m)
Wingspan: 37 ft 2 in (11.33 m)
Height: 13 ft (3.96 m)
Wing area: 350 sq ft (32.515 m²)
Aspect ratio: 3.9
Root chord: 11 ft 9 in (3.6 m)

Powerplant

Two 3,500-lb (15.5-kN) thrust Rolls-Royce Derwent 8 turbojets

Weights

Empty: 10,684 lb (4846 kg)
Maximum overload: 15,700 lb (7122 kg)

Performance

Maximum speed at sea level: 592 mph (953 km/h)
Maximum speed at 30,000 ft (9144 m): 550 mph (885 km/h)
Climb to 30,000 ft (9144 m): 6 minutes 30 seconds
Service ceiling: 44,000 ft (13410 m)
Range without wing drop tanks: 690 miles (1111 km)
Endurance at 40,000 ft (12192 m) with 420 Imp gal (1909 litres) of fuel: 592 mph (953 km/h)

Armament

Four fixed 20-mm British Hispano cannon in the nose with 195 rounds per gun

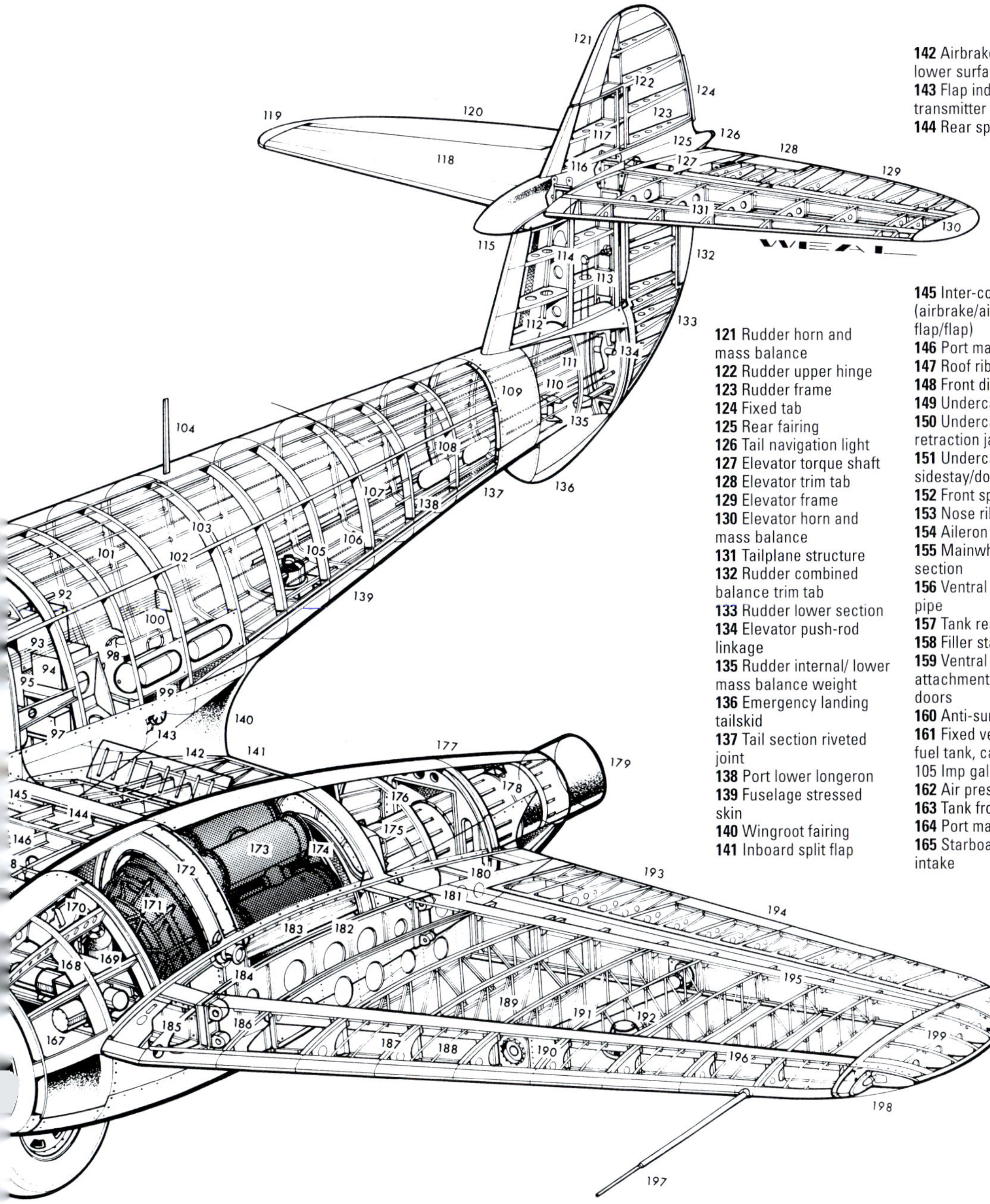

121 Rudder horn and mass balance
122 Rudder upper hinge
123 Rudder frame
124 Fixed tab
125 Rear fairing
126 Tail navigation light
127 Elevator torque shaft
128 Elevator trim tab
129 Elevator frame
130 Elevator horn and mass balance
131 Tailplane structure
132 Rudder combined balance trim tab
133 Rudder lower section
134 Elevator push-rod linkage
135 Rudder internal/ lower mass balance weight
136 Emergency landing tailskid
137 Tail section riveted joint
138 Port lower longeron
139 Fuselage stressed skin
140 Wingroot fairing
141 Inboard split flap
142 Airbrake (upper and lower surfaces)
143 Flap indicator transmitter
144 Rear spar
145 Inter-coupler cables (airbrake/airbrake and flap/flap)
146 Port mainwheel well
147 Roof rib station
148 Front diaphragm
149 Undercarriage beam
150 Undercarriage retraction jack
151 Undercarriage sidestay/downlock
152 Front spar
153 Nose ribs
154 Aileron control runs
155 Mainwheel door inner section
156 Ventral tank transfer pipe
157 Tank rear fairing
158 Filler stack pipes
159 Ventral tank attachment strap access doors
160 Anti-surge baffles
161 Fixed ventral fuel tank, capacity 105 Imp gal (477 litres)
162 Air pressure inlet
163 Tank front fairing
164 Port mainwheel
165 Starboard engine intake
166 Intake internal leading edge shroud
167 Auxiliary gearbox drives (vacuum pump/generator)
168 Nacelle nose structure
169 Starter motor
170 Oil tank
171 Rolls-Royce W.2B/23C Derwent I
172 Main engine mounting frame
173 Combustion chambers
174 Rear spar spectacle frame
175 Jet pipe thermo-coupling
176 Nacelle aft frames
177 Nacelle detachable tail section
178 Jet pipe suspension link
179 Jet pipe exhaust
180 Gap fairing tail section
181 Rear-spar outer wing fixing
182 Outer wing rib No. 1
183 Engine end rib
184 Engine mounting/ removal trunnion
185 Gap fairing nose section
186 Front-spar outer wing fixing
187 Nose ribs
188 Intermediate riblets
189 Wing ribs
190 Aileron drive chain sprocket
191 Aileron torque shaft
192 Retractable landing lamp
193 Port aileron
194 Aileron balance tab
195 Rear spar
196 Front spar
197 Pitot head
198 Port navigation light
199 Outer wing rib No. 10/wingtip attachment
200 Port recognition light

Lockheed F-94 Starfire

The 15th production Starfire, 50-970, poses for one of a series of publicity photographs. This aircraft was delivered to the Air Proving Ground Command at Eglin AFB on 3 October 1952 where it was used as a test aircraft in the programme to develop wing rocket pods for the F-94C.

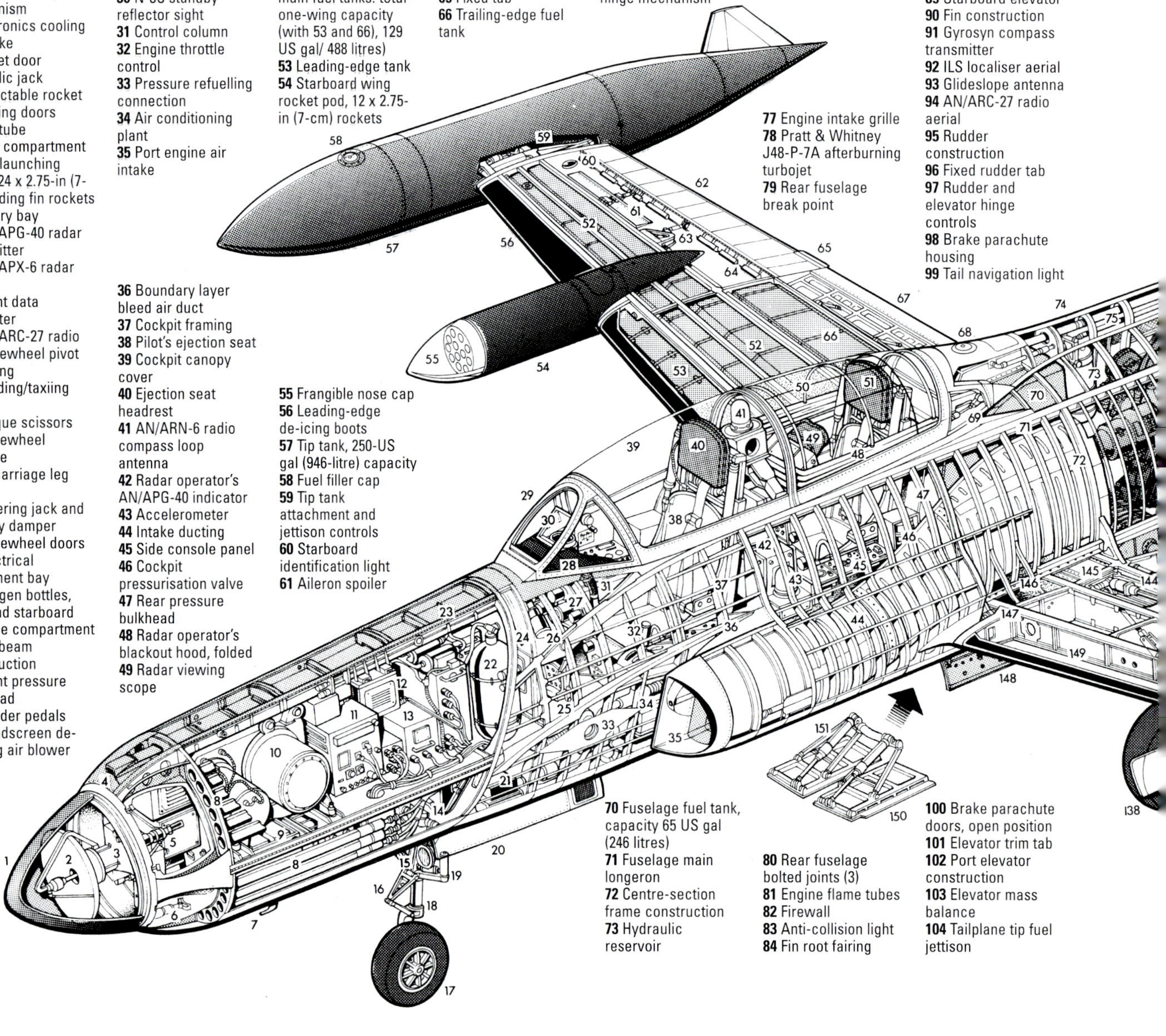

F-94C Starfire

Cutaway key

1 Radome
2 Radar scanner
3 Radar tracking mechanism
4 Electronics cooling air intake
5 Rocket door hydraulic jack
6 Retractable rocket launching doors
7 Pitot tube
8 Nose compartment rocket launching tubes, 24 x 2.75-in (7-cm) folding fin rockets
9 Battery bay
10 AN/APG-40 radar transmitter
11 AN/APX-6 radar unit
12 Flight data computer
13 AN/ARC-27 radio
14 Nosewheel pivot mounting
15 Landing/taxiing lamps
16 Torque scissors
17 Nosewheel
18 Nose undercarriage leg strut
19 Steering jack and shimmy damper
20 Nosewheel doors
21 Electrical equipment bay
22 Oxygen bottles, port and starboard
23 Nose compartment upper beam construction
24 Front pressure bulkhead
25 Rudder pedals
26 Windscreen de-misting air blower
27 Instrument panel
28 Instrument panel shroud
29 Windscreen framing
30 N-3C standby reflector sight
31 Control column
32 Engine throttle control
33 Pressure refuelling connection
34 Air conditioning plant
35 Port engine air intake
36 Boundary layer bleed air duct
37 Cockpit framing
38 Pilot's ejection seat
39 Cockpit canopy cover
40 Ejection seat headrest
41 AN/ARN-6 radio compass loop antenna
42 Radar operator's AN/APG-40 indicator
43 Accelerometer
44 Intake ducting
45 Side console panel
46 Cockpit pressurisation valve
47 Rear pressure bulkhead
48 Radar operator's blackout hood, folded
49 Radar viewing scope
50 Canopy-mounted ADF sense aerial
51 Radar operator's ejection seat
52 Starboard wing main fuel tanks: total one-wing capacity (with 53 and 66), 129 US gal/ 488 litres)
53 Leading-edge tank
54 Starboard wing rocket pod, 12 x 2.75-in (7-cm) rockets
55 Frangible nose cap
56 Leading-edge de-icing boots
57 Tip tank, 250-US gal (946-litre) capacity
58 Fuel filler cap
59 Tip tank attachment and jettison controls
60 Starboard identification light
61 Aileron spoiler
62 Starboard aileron
63 Aileron hinge control
64 Aileron balance weights
65 Fixed tab
66 Trailing-edge fuel tank
67 Starboard split trailing-edge flap
68 Fuselage fuel tank filler cap
69 Cockpit canopy hinge mechanism
70 Fuselage fuel tank, capacity 65 US gal (246 litres)
71 Fuselage main longeron
72 Centre-section frame construction
73 Hydraulic reservoir
74 Dorsal spine fairing
75 Fuel system vent pipe
76 Engine accessory equipment
77 Engine intake grille
78 Pratt & Whitney J48-P-7A afterburning turbojet
79 Rear fuselage break point
80 Rear fuselage bolted joints (3)
81 Engine flame tubes
82 Firewall
83 Anti-collision light
84 Fin root fairing
85 Fuel jettison valves
86 Tailplane leading-edge de-icing boots
87 Starboard tailplane
88 Fuel jettison pipe
89 Starboard elevator
90 Fin construction
91 Gyrosyn compass transmitter
92 ILS localiser aerial
93 Glideslope antenna
94 AN/ARC-27 radio aerial
95 Rudder construction
96 Fixed rudder tab
97 Rudder and elevator hinge controls
98 Brake parachute housing
99 Tail navigation light
100 Brake parachute doors, open position
101 Elevator trim tab
102 Port elevator construction
103 Elevator mass balance
104 Tailplane tip fuel jettison

Based at Great Falls (later Malmstrom) AFB, Montana, the 29th FIS was the only USAF squadron to fly the F-94C from the 'Northern Tier' states. These three aircraft were photographed in the mid-1950s, possibly during a gunnery meet: their missing rocket pod nosecones suggest that their FFARs have been fired.

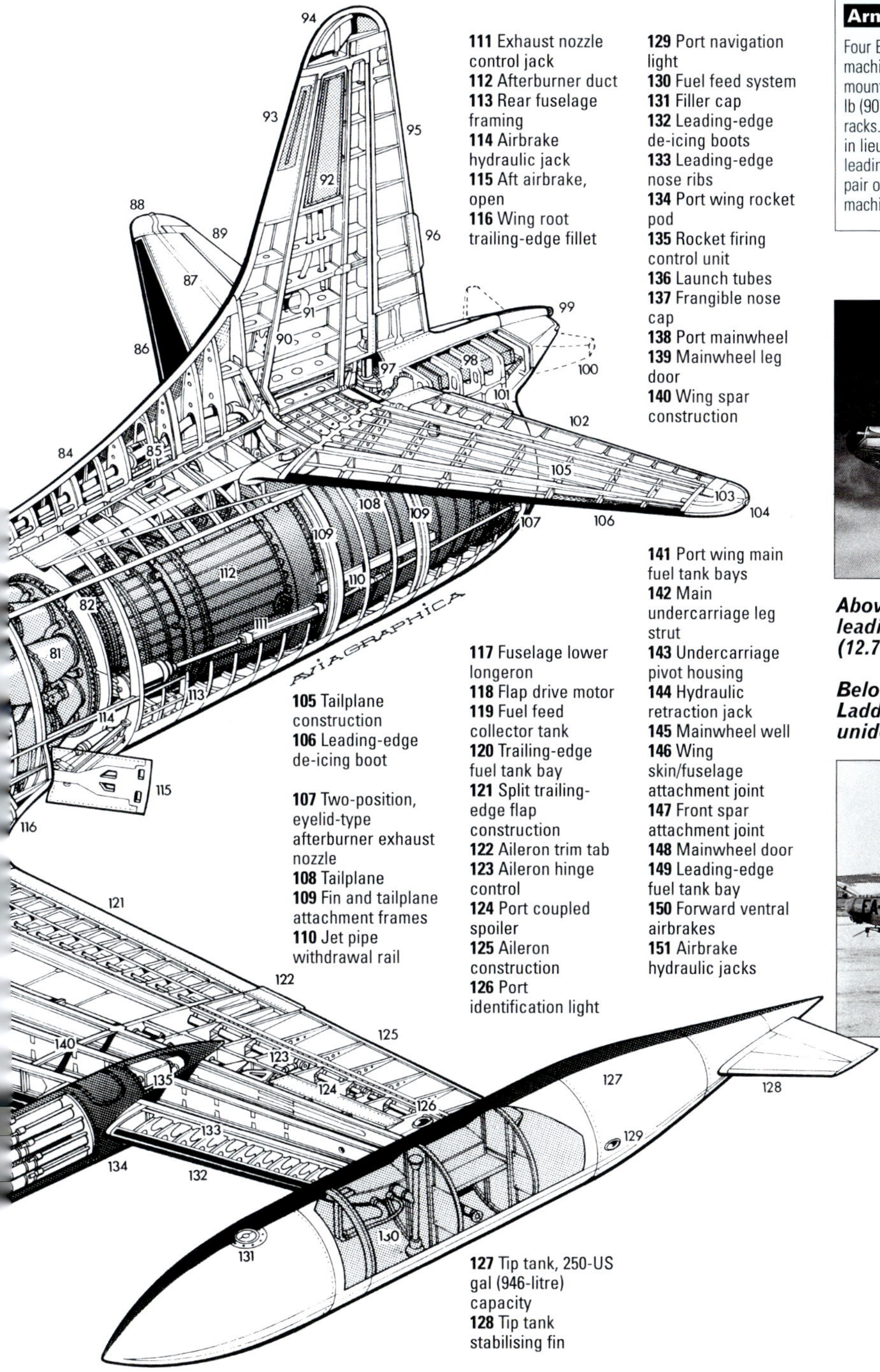

111 Exhaust nozzle control jack
112 Afterburner duct
113 Rear fuselage framing
114 Airbrake hydraulic jack
115 Aft airbrake, open
116 Wing root trailing-edge fillet
129 Port navigation light
130 Fuel feed system
131 Filler cap
132 Leading-edge de-icing boots
133 Leading-edge nose ribs
134 Port wing rocket pod
135 Rocket firing control unit
136 Launch tubes
137 Frangible nose cap
138 Port mainwheel
139 Mainwheel leg door
140 Wing spar construction
141 Port wing main fuel tank bays
142 Main undercarriage leg strut
143 Undercarriage pivot housing
144 Hydraulic retraction jack
145 Mainwheel well
146 Wing skin/fuselage attachment joint
147 Front spar attachment joint
148 Mainwheel door
149 Leading-edge fuel tank bay
150 Forward ventral airbrakes
151 Airbrake hydraulic jacks
105 Tailplane construction
106 Leading-edge de-icing boot
107 Two-position, eyelid-type afterburner exhaust nozzle
108 Tailplane
109 Fin and tailplane attachment frames
110 Jet pipe withdrawal rail
117 Fuselage lower longeron
118 Flap drive motor
119 Fuel feed collector tank
120 Trailing-edge fuel tank bay
121 Split trailing-edge flap construction
122 Aileron trim tab
123 Aileron hinge control
124 Port coupled spoiler
125 Aileron construction
126 Port identification light
127 Tip tank, 250-US gal (946-litre) capacity
128 Tip tank stabilising fin

SPECIFICATION

F-94B

Dimensions

Length: 40 ft 1 in (12.22 m)
Height: 12 ft 8 in (3.86 m)
Wingspan: 37 ft 6 in (11.43 m)
Wingspan (with tip tanks): 38 ft 11 in (11.86 m)
Wing area: 234.8 sq ft (21.813 m²)

Powerplant

One Allison J33-A-33 or -33A turbojet, rated at 5,200 lb (23.13 kN) thrust

Weights

Empty: 10,064 lb (4565 kg)
Loaded: 13,474 lb (6112 kg)
Maximum take-off: 16,844 lb (7640 kg)

Performance

Maximum speed at sea level: 606 mph (975 km/h)
Cruising speed: 452 mph (727 km/h)
Initial rate of climb: 6,850 ft/min (2088 m/min)
Service ceiling: 48,000 ft (14630 m)
Normal range: 665 miles (1070 km)
Maximum range: 905 miles (1455 km)

Armament

Four Browning M3 0.5-in (12.7-mm) machine-guns, each with 300 rounds, mounted in the nose, plus up to 2,000 lb (907 kg) of bombs on underwing racks. Some aircraft modified to carry, in lieu of bombs, a gun pod on the leading edge of each wing, containing a pair of Browning M3 0.5-in (12.7-mm) machine-guns and 265 rounds per gun.

F-94C Starfire

Dimensions

Length: 44 ft 6 in (13.56 m)
Height: 14 ft 11 in (4.55 m)
Wingspan: 37 ft 4 in (11.38 m)
Wing area: 232.8 sq ft (21.628 m²)

Powerplant

One Pratt & Whitney J48-P-5, -5A or -7A turbojet, rated at 8,750 lb (38.91 kN) thrust with afterburner

Weights

Empty: 12,708 lb (5764 kg)
Loaded: 18,300 lb (8301 kg)
Maximum take-off: 24,184 lb (10970 kg)

Performance

Maximum speed at sea level: 640 mph (1030 km/h)
Cruising speed: 493 mph (793 km/h)
Initial rate of climb: 7,980 ft/min (2432 m/min)
Service ceiling: 51,400 ft (15665 m)
Normal range: 805 miles (1295 km)
Maximum range: 1275 miles (2050 km)

Armament

Twenty-four 2.75-in (6.99-cm) folding-fin aircraft rockets (FFARs) in four clusters of six arranged around the nose, plus (from the 100th aircraft) 12 FFARs in a pod on the leading edge of each wing.

Above: This 175th FIS, South Dakota ANG F-94A carries leading-edge gun pods, each with a pair of 0.5-in (12.7-mm) Brownings.

Below: Aircrew of the 449th FIS stand by in their F-94As at Ladd AFB, Alaska, ready to scramble and intercept unidentified aircraft reported by AEW radar.

Lockheed F-104 Starfighter

Italy's Starfighters represent the ultimate expression of the F-104 in the interceptor role. Thanks to local upgrades, the aircraft are able to launch the AIM-7 Sparrow AAM or the indigenous Aspide medium-range AAM, as well as the all-aspect AIM-9L Sidewinder.

F-104S Starfighter

1 Pitot tube
2 Radome
3 Radar scanner dish
4 R21G/H multi-mode radar equipment
5 Radome withdrawal rails
6 Communications aerial
7 Cockpit front bulkhead
8 Infra-red sight
9 Windscreen panels
10 Reflector gunsight
11 Instrument panel shroud
12 Rudder pedals
13 Control column
14 Nose section frame construction
15 Control cable runs
16 Pilot's side console panel
17 Throttle control
18 Safety harness
19 Martin-Baker IQ-7A ejection seat
20 Face blind seat firing handle
21 Cockpit canopy cover
22 Canopy bracing struts
23 Seat rail support box
24 Angle of attack probe
25 Cockpit rear bulkhead
26 Temperature probe
27 Nosewheel doors
28 Taxiing lamp
29 Nosewheel leg strut
30 Nosewheel
31 Steering linkage
32 AIM-7 Sparrow avionics (replacing M61 gun installation of strike model)
33 Inertial platform
34 Avionics compartment
35 Avionics compartment shroud cover
36 Cockpit aft glazing
37 Ram air turbine
38 Emergency generator
39 Avionics compartment access cover
40 Fuselage frame construction
41 Pressure bulkhead
42 Ammunition compartment auxiliary fuel tank (101.5-Imp gal/ 462-litre capacity)
43 Fuel feed pipes
44 Flush-fitting UHF aerial panel
45 Anti-collision light
46 Starboard intake
47 Engine bleed air supply to air-conditioning
48 Gravity fuel fillers
49 Fuselage main fuel tanks (total internal capacity 746 Imp gal/ 3391 litres)
50 Pressure refuelling adaptor
51 Intake shock cone centre body
52 De-iced intake lip
53 Port intake
54 Shock cone boundary layer bleed
55 Boundary layer bleed air duct
56 Auxiliary intake
57 Hinged auxiliary intake door
58 Navigation light
59 Leading-edge flap jack
60 Intake trunking
61 Fuselage main longeron
62 Wingroot attaching members
63 Intake flank fuel tanks
64 Wing-mounting fuselage mainframes
65 Control cable runs
66 Electrical junction box
67 Dorsal spine fairing
68 Starboard inboard pylon
69 Leading-edge flap (lowered)
70 AIM-7 Sparrow AAM
71 Missile launch rail
72 Starboard outer pylon
73 Tip tank vane
74 Tip tank latching unit
75 Starboard wingtip tank
76 Fuel filler caps
77 Starboard aileron
78 Aileron power control jacks
79 Power control servo valves
80 Fuel lines to auxiliary tanks
81 Flap blowing duct
82 Starboard blown flap (lowered)
83 Engine intake compressor face
84 Intake spill flaps
85 Aileron torque shaft
86 Hydraulic reservoir
87 Air-conditioning bleed air supply pipe
88 General Electric J79-GE-19 turbojet
89 Engine withdrawal rail
90 Starboard airbrake (open)
91 Fin root fillet
92 Elevator servo controls
93 Elevator/all-moving tailplane hydraulic jacks
94 Push-pull control rods
95 Tailfin construction
96 Fin tip fairing
97 Tailplane rocking control arm
98 Starboard tailplane
99 One-piece tailplane construction
100 Tailplane spar
101 Tailplane spar central pivot
102 Fin trailing-edge construction

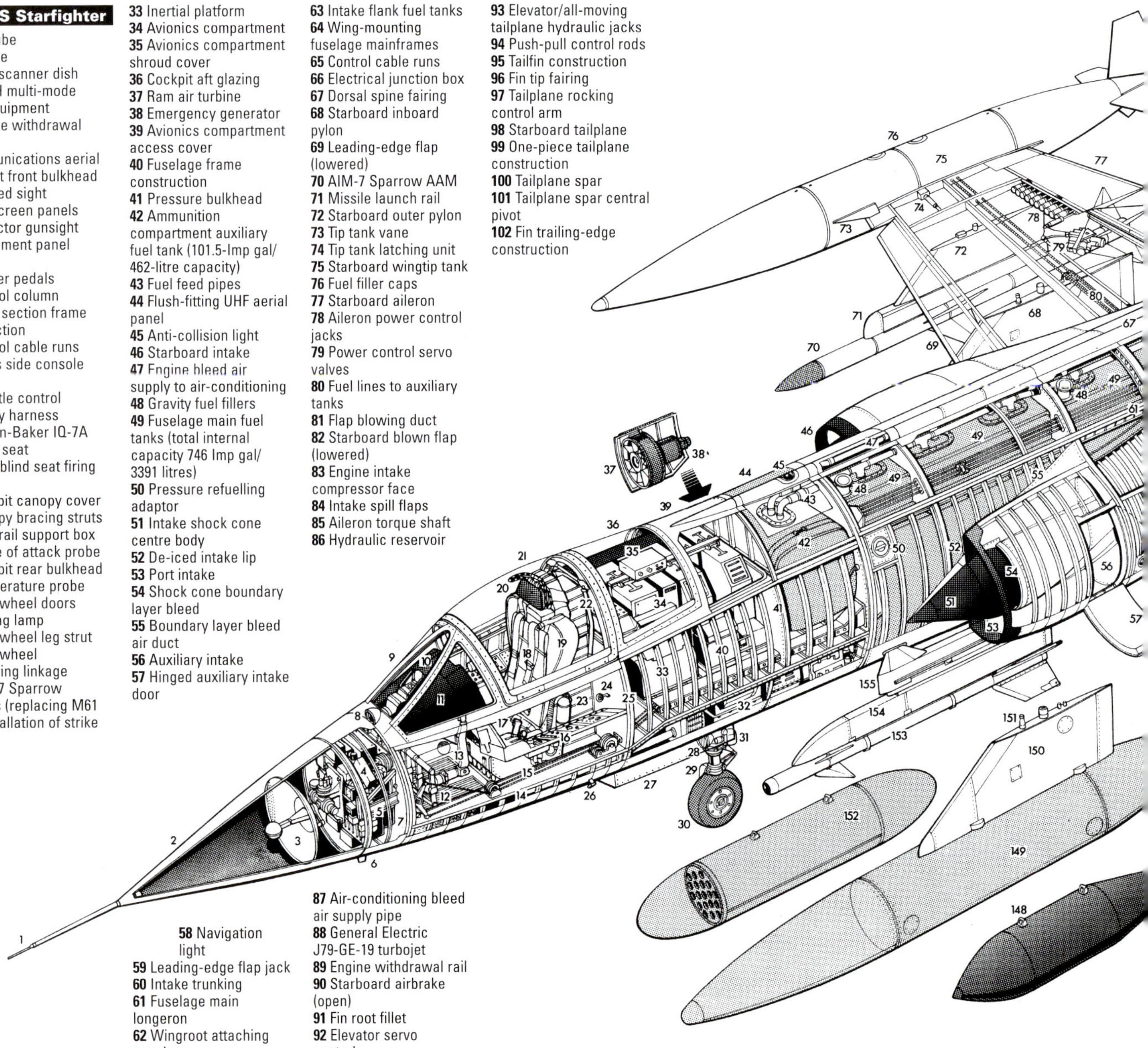

A willing customer during the original 'sale of the century', the Netherlands was to follow several of its NATO allies in replacing its F-104s with the F-16 Fighting Falcon in the second 'sale of the century'. As it had in the US, the F-104 initially suffered an appallingly high accident rate in European service, the Dutch losing some 35.8 per cent of their Starfighter fleet.

SPECIFICATION

F-104G Starfighter

Dimensions

Length: 54 ft 9 in (16.69 m)
Wingspan (without tip-mounted AAMs): 21 ft 11 in (6.68 m)
Wing area: 196.10 sq ft (18.22 m²)
Wing aspect ratio: 2.45
Height 13 ft 6 in (4.11 m)
Tailplane span: 11 ft 11 in (3.63 m)
Wheel track: 9 ft (2.74 m)
Wheelbase: 15 ft ½ in (4.59 m)

Powerplant

One General Electric J79-GE-11A turbojet rated at 10,000 lb st (44.48 kN) dry and 15,800 lb st (70.28 kN) with afterburning

Weights

Empty equipped: 14,082 lb (6387 kg)
Normal take-off: 21,639 lb (9840 kg)
Maximum take-off: 28,779 lb (13054 kg)

Fuel and load

Internal fuel: 896 US gal (3392 litres)
External fuel: up to 955 US gal (3615 litres) in one 225-US gal (852-litre) and two 195-US gal (740-litre) drop tanks and two 170-US gal (645-litre) tip tanks
Maximum ordnance: 4,310 lb (1955 kg)

Performance

Maximum level speed 'clean' at 36,000 ft (10975 m): 1,262 kt (1,453 mph; 2338 km/h)
Cruising speed at 36,000 ft (10975 m): 530 kt (610 mph; 981 km/h)
Maximum rate of climb at sea level: 55,000 ft (16765 m) per minute
Service ceiling: 58,000 ft (17680 m)
Take-off distance to 50 ft (15 m): 4,600 ft (1402 m)
Landing distance from 50 ft (15 m): 3,250 ft (990 m)

Range

Ferry range with four drop tanks: 1,893 nm (2,180 miles; 3510 km)
Combat radius: 648 nm (746 miles; 1200 km)
Combat radius on hi-lo-hi attack mission with maximum warload: 261 nm (300 miles; 483 km)

Armament

Fixed: One General Electric 20-mm M61A-1 Vulcan six-barrelled rotary cannon with 725 rounds
Weapon stations: One underfuselage, four underwing and two wingtip hardpoints for AIM-9 AAMs and a variety of bombs, pods and rockets

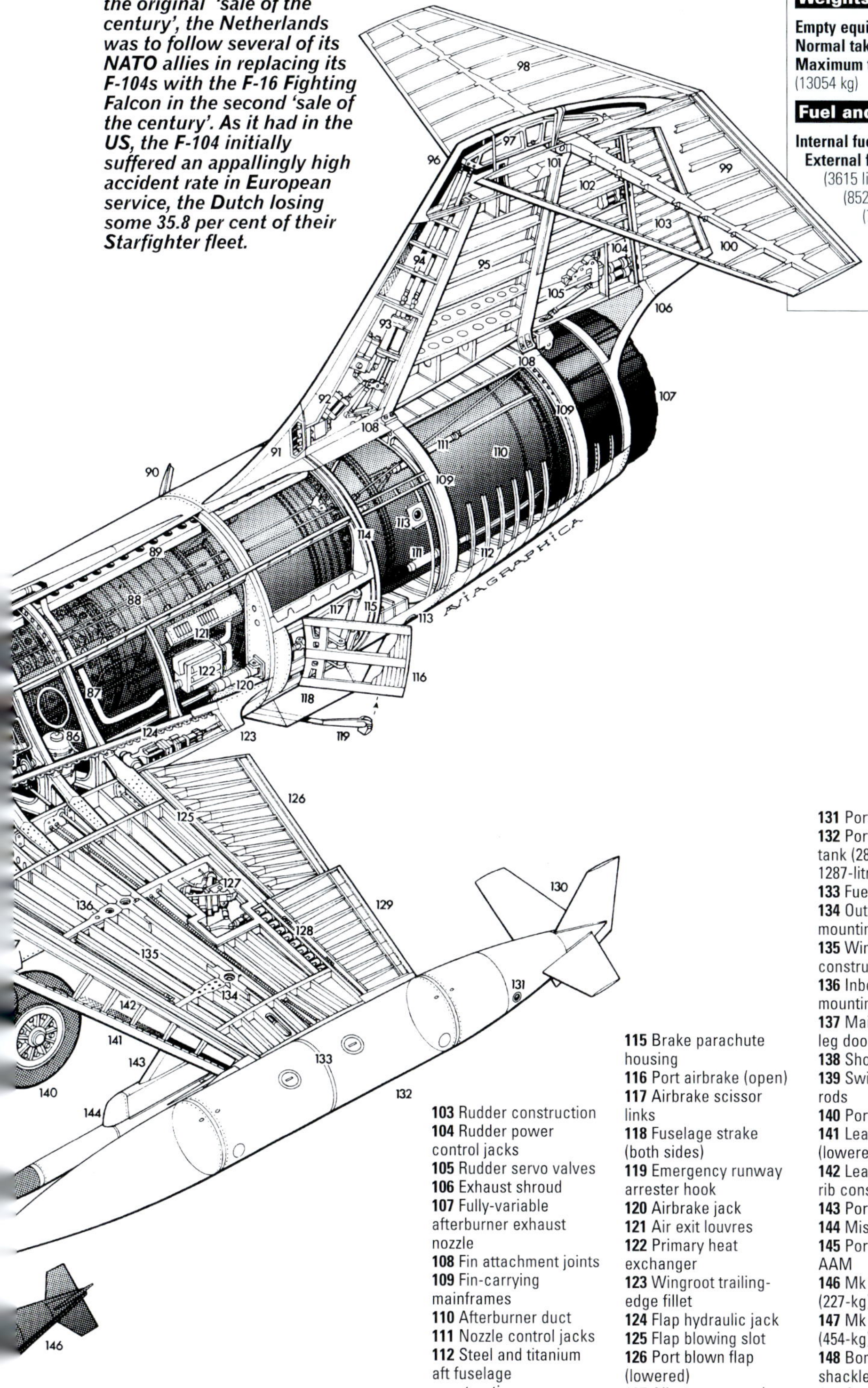

103 Rudder construction
104 Rudder power control jacks
105 Rudder servo valves
106 Exhaust shroud
107 Fully-variable afterburner exhaust nozzle
108 Fin attachment joints
109 Fin-carrying mainframes
110 Afterburner duct
111 Nozzle control jacks
112 Steel and titanium aft fuselage construction
113 Rear navigation lights
114 Aft fuselage attachment joint
115 Brake parachute housing
116 Port airbrake (open)
117 Airbrake scissor links
118 Fuselage strake (both sides)
119 Emergency runway arrester hook
120 Airbrake jack
121 Air exit louvres
122 Primary heat exchanger
123 Wingroot trailing-edge fillet
124 Flap hydraulic jack
125 Flap blowing slot
126 Port blown flap (lowered)
127 Aileron servo valves
128 Aileron power control jacks
129 Port aileron
130 Tip tank fins
131 Port navigation light
132 Port wingtip fuel tank (283-Imp gal/ 1287-litre) capacity
133 Fuel filler caps
134 Outboard pylon mounting rib
135 Wing multi-spar construction
136 Inboard pylon mounting rib
137 Main undercarriage leg door
138 Shock absorber strut
139 Swivel axle control rods
140 Port mainwheel
141 Leading-edge flap (lowered)
142 Leading-edge flap rib construction
143 Port outboard pylon
144 Missile launch rail
145 Port AIM-7 Sparrow AAM
146 Mk 82 500-lb (227-kg) bomb
147 Mk 83 1,000-lb (454-kg) bomb
148 Bomb mounting shackles
149 Auxiliary fuel tank (163-Imp gal/740-litre) capacity
150 Port inboard wing pylon
151 Pylon attachments
152 LAU-3A 2.75-in (70-mm) FFAR pod (19 rockets)
153 AIM-9 Sidewinder AAM
154 Missile launch rail
155 Fuselage stores pylon adaptor

Lockheed Martin F-16 Fighting Falcon

Conceived as a lightweight no-frills air combat fighter, the F-16 has matured into a sophisticated multi-role warplane. This Dutch F-16A is equipped with a centrally mounted Orpheus reconnaissance pod.

F-16C Block 50/52

Cutaway key

1 Pitot head/air data probe
2 Glass-fibre radome
3 Lightning conducting strips
4 Planar radar scanner
5 Radome hinge point, opens to starboard
6 Scanner tracking mechanism
7 ILS glideslope antenna
8 Radar mounting bulkhead
9 Incidence vane, port and starboard
10 IFF antenna
11 GBU-12B laser-guided bomb
12 AN/APG-68 digital pulse-Doppler, multi-mode radar equipment bay
13 Forward oblique radar warning antennas, port and starboard
14 Front pressure bulkhead
15 Static ports
16 Fuselage forebody strake fairing
17 Forward avionics equipment bay
18 Canopy jettison charge
19 Instrument panel shroud
20 Instrument panel, multi-function CRT head-down displays
21 Sidestick controller, fly-by-wire control system
22 Video recorder
23 GEC wide-angle head-up display
24 CBU-52/58/71 submunition dispenser
25 LAU-3A 19-round rocket launcher
26 2.75-in (68-mm) FFAR
27 CBU-87/89 Gator submunition dispenser
28 Starboard intake flank (No. 5R) stores pylon adaptor
29 LANTIRN (FLIR) targeting pod
30 One-piece frameless cockpit canopy
31 Ejection seat headrest
32 McDonnell-Douglas ACES II zero-zero ejection seat
33 Side console panel
34 Canopy frame fairing
35 Canopy external emergency release
36 Engine throttle lever incorporating HOTAS (hands-on throttle-and-stick) radar controls
37 Canopy jettison handle
38 Cockpit section frame structure
39 Boundary layer splitter plate
40 Fixed-geometry engine air intake
41 Nosewheel, aft retracting
42 LANTIRN (FLIR/TFR) navigation pod
43 Port intake flank (No. 5L) stores pylon adaptor
44 Port position light
45 Intake duct framing
46 Intake ducting
47 Gun gas suppression muzzle aperture
48 Aft avionics equipment bay
49 Cockpit rear pressure bulkhead
50 Canopy hinge point
51 Ejection seat launch rails
52 Canopy rotary actuator
53 Conditioned air delivery duct
54 Canopy sealing frame
55 Canopy aft glazing
56 600-US gal (500-Imp gal; 2271-litre) external fuel tank
57 Garrett hydrazine turbine emergency power unit (EPU)
58 Hydrazine fuel tank
59 Fuel tank bay access panel
60 Forward fuselage bag-type fuel tank, total internal capacity 6972 lb (3162 kg)
61 Fuselage upper longeron
62 Conditioned air ducting
63 Cannon barrels
64 Forebody frame construction
65 Air system ground connection
66 Ventral air conditioning system equipment bay
67 Centreline 300-US gal (250-Imp gal; 1136-litre) fuel tank
68 Mainwheel door hydraulic actuator
69 Mainwheel door
70 Hydraulic system ground connectors
71 Gun bay ventral gas vent
72 GE M61A1 Vulcan 20-mm rotary cannon
73 Ammunition feed chute
74 Hydraulic gun drive motor
75 Port hydraulic reservoir
76 Centre fuselage integral fuel tank
77 Leading-edge flap drive hydraulic motor
78 Ammunition drum with 511 rounds
79 Upper position light/refuelling floodlight
80 TACAN antenna
81 Hydraulic accumulator
82 Starboard hydraulic reservoir
83 Leading-edge flap drive shaft
84 Inboard, No. 6 stores station 4,500-lb (2041-kg) capacity
85 Pylon attachment hardpoint

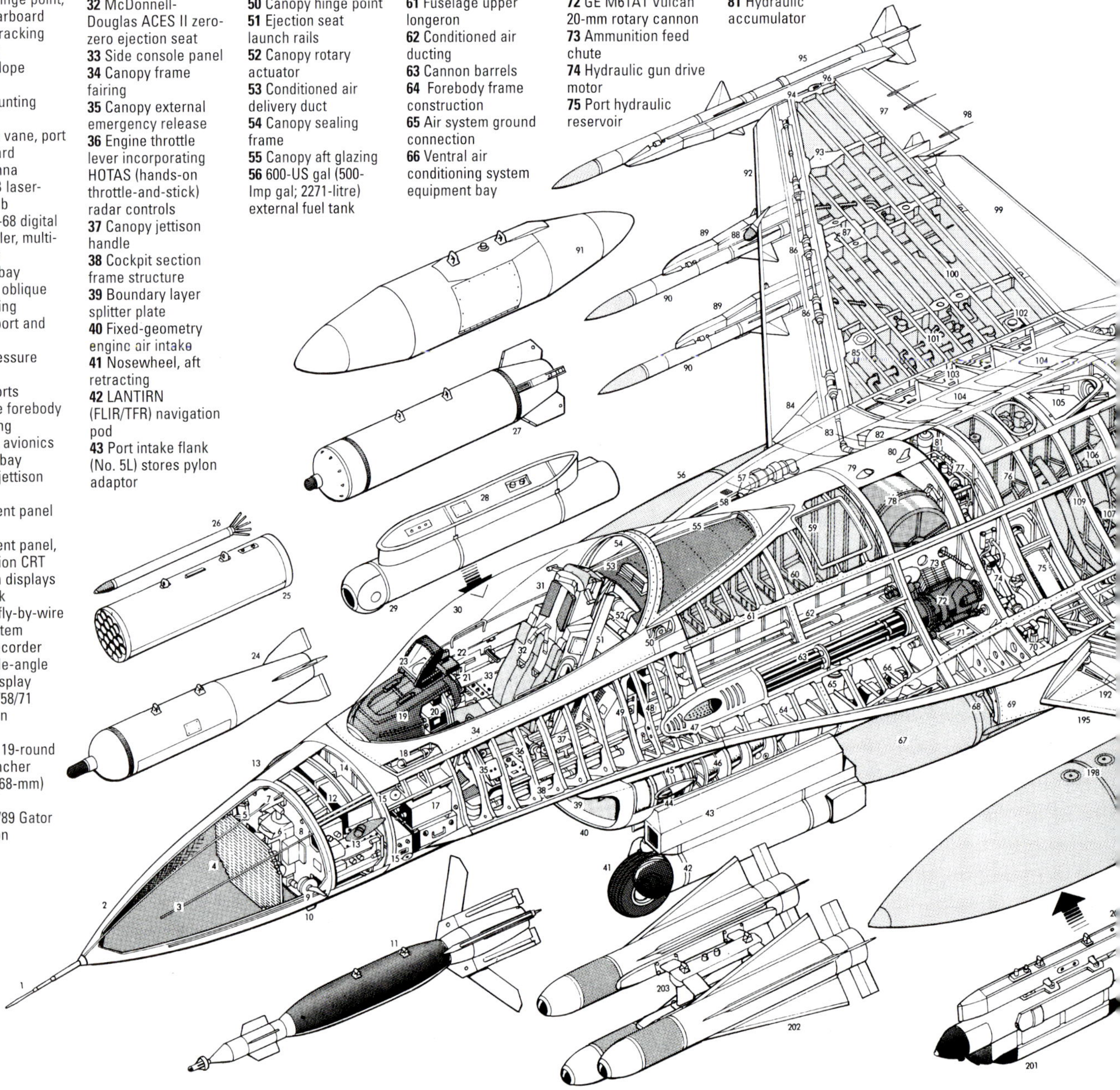

86 Leading-edge flap drive shaft and rotary actuators
87 No. 7 stores hardpoint, capacity 3,500 lb (1588 kg)
88 Starboard forward radar warning antenna
89 Missile launch rails
90 AIM-120 AMRAAM medium-range AAMs
91 MXU-648 baggage pod, carriage of essential ground equipment and personal effects for off-base deployment
92 Starboard leading-edge manoeuvre flap, down position
93 Outboard, No. 8 stores station, capacity 700 lb (318 kg)
94 Wingtip, No. 9 stores station, capacity 425 lb (193 kg)
95 Wingtip AMRAAM
96 Starboard navigation light
97 Fixed portion of trailing edge
98 Static dischargers
99 Starboard flaperon
100 Starboard wing integral fuel tank
101 Fuel system piping
102 Fuel pump
103 Starboard wingroot attachment fishplates
104 Fuel tank access panels
105 Universal air refuelling receptacle (UARSSI), open
106 Engine intake centrebody fairing
107 Airframe mounted accessory equipment gearbox
108 Jet fuel starter
109 Machined wing attachment bulkheads
110 Engine fuel management equipment
111 Pressure refuelling receptacle ventral adaptor
112 Pratt & Whitney F100-PW-229 afterburning turbofan engine
113 VHF/IFF antenna
114 Starboard flaperon hydraulic actuator
115 Fuel tank tail fins
116 Sidebody fairing integral fuel tank
117 Position light
118 Cooling air ram air intake
119 Finroot fairing
120 Forward engine support link
121 Rear fuselage integral fuel tank
122 Thermally insulated tank inner skin
123 Tank access panels
124 Radar warning system power amplifier
125 Finroot attachment fittings
126 Flight control system hydraulic accumulators

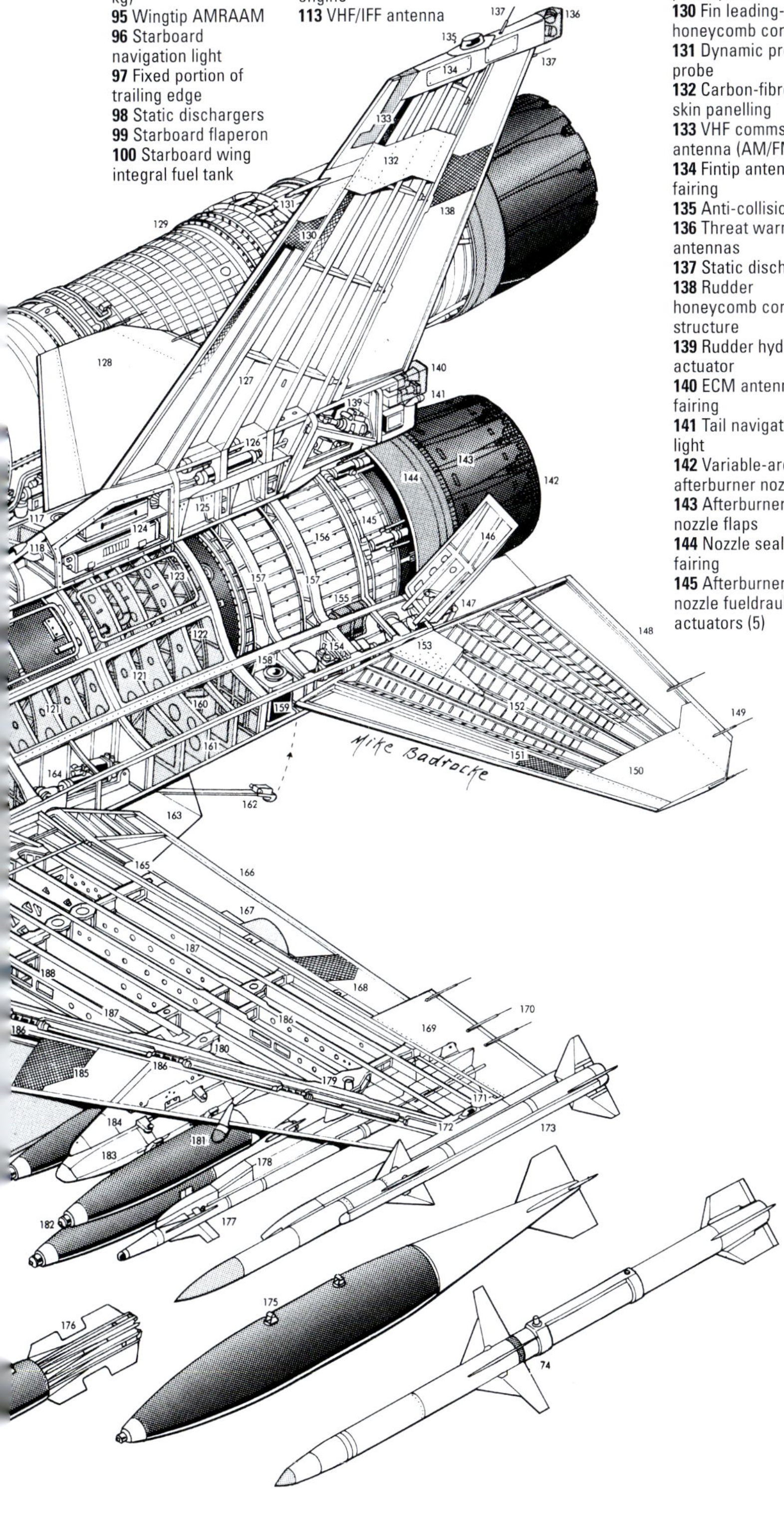

The F-16 takes on fuel from tankers via a receptacle in the upper fuselage. Like all USAF tactical combat aircraft, it refuels from 'flying boom'-equipped tankers.

127 Multi-spar fin torsion box structure
128 Starboard all-moving tailplane (tailplane panels interchangeable)
129 General Electric F110-GE-129 alternative powerplant
130 Fin leading-edge honeycomb core
131 Dynamic pressure probe
132 Carbon-fibre fin skin panelling
133 VHF comms antenna (AM/FM)
134 Fintip antenna fairing
135 Anti-collision light
136 Threat warning antennas
137 Static dischargers
138 Rudder honeycomb core structure
139 Rudder hydraulic actuator
140 ECM antenna fairing
141 Tail navigation light
142 Variable-area afterburner nozzle
143 Afterburner nozzle flaps
144 Nozzle sealing fairing
145 Afterburner nozzle fueldraulic actuators (5)
146 Port split trailing-edge airbrake panel, open, upper and lower surfaces
147 Airbrake actuating linkage
148 Port all-moving tailplane
149 Static dischargers
150 Graphite-epoxy tailplane skin panels
151 Leading-edge honeycomb construction
152 Corrugated aluminium sub-structure
153 Tailplane pivot mounting
154 Tailplane hydraulic actuator
155 Fuel jettison chamber, port and starboard
156 Afterburner ducting
157 Rear fuselage machined bulkheads
158 Port navigation light
159 AN/ALE-40(VO-4) chaff/flare launcher, port and starboard
160 Main engine thrust mounting, port and starboard
161 Sidebody fairing frame structure
162 Runway arrester hook
163 Composite ventral fin, port and starboard
164 Port flaperon hydraulic actuator
165 Flaperon hinges
166 Port flaperon, lowered
167 External fuel tank tail fairing
168 Flaperon honeycomb core structure
169 Fixed portion of trailing edge
170 Static dischargers
171 Port navigation light
172 Wingtip, No. 1 stores station, capacity 425 lb (193 kg)
173 Port wingtip AMRAAM
174 AGM-88 HARM (High-speed Anti-Radiation Missile}
175 Mk 84 low-drag 2,000-lb (907-kg) HE bomb
176 Mk 83 Snakeye retarded bomb
177 AIM-9L Sidewinder air-to-air missile
178 Missile launch rails
179 No. 2 stores station, capacity 700 lb (318 Kg)
180 No. 3 stores station, capacity 3,500 lb (1588 Kg)
181 Port forward radar warning antenna
182 Mk 82 500 lb (227 kg) HE bombs
183 Triple ejector rack
184 Intermediate wing pylon
185 Leading-edge manoeuvre flap honeycomb core structure
186 Flap drive shaft and rotary actuators
187 Multi-spar wing torsion box structure
188 Port wing integral fuel tankage
189 No. 4 stores station hardpoint, capacity 4,500 lb (2041 kg)
190 Wing panel root attachment fishplates
191 Undercarriage leg mounted landing light
192 Articulated retraction/drag link
193 Main undercarriage leg strut
194 Shock absorber strut
195 Port leading-edge manoeuvre flap, down position
196 Inboard wing pylon
197 Port mainwheel, forward retracting
198 Fuel filler caps
199 Port 370-US gal (308-Imp gal; 1400-litre) external tank
200 Centreline, No. 5 stores pylon, capacity 2,200 lb (998 kg)
201 AN/ALQ-184(V)-2 (short) ECM pod
202 AGM-65 Maverick air-to-surface missiles
203 LAU-88 triple missile carrier/launcher

SPECIFICATION

F-16C Fighting Falcon Block 30

Dimensions:

Fuselage length: 49 ft 4 in (15.03 m)
Wing span with tip-mounted AAMs: 32 ft 9¾ in (10.00 m)
Wing area: 300.00 sq ft (27.87 m²)
Wing aspect ratio: 3.0
Tailplane span: 18 ft 3¾ in (5.58 m)
Vertical tail surfaces: 54.75 sq ft (5.09 m²)
Height: 16 ft 8½ in (5.09 m)
Wheel track: 7 ft 9 in (2.36 m)
Wheelbase: 13 ft 1½ in (4.00 m)

Powerplant:

One General Electric F110-GE-100 turbofan rated at 28,984 lb st (128.9 kN) with afterburning, or a Pratt & Whitney F100-PW-220 23,770 lb st (105.7 kN) in Blocks 40/42

Weights:

Empty operating: 19,100 lb (8663 kg)
Typical combat take-off: 21,585 lb (9791 kg)
Maximum take-off for air-to-air mission without drop tanks: 25,071 lb (11372 kg)
Maximum take-off, with maximum external load: 42,300 lb (19187 kg)

g limits

Max symmetrical design *g* limit with full internal fuel load ±9

Performance:

Maximum level speed 'clean' at altitude : 1,146 kt (1,320 mph; 2124 km/h)
Maximum level speed at sea level: 795 kt (915 mph; 1472 km/h)
Maximum rate of climb at sea level: 50,000 ft (15240 m) per minute
Service ceiling: more than 50,000 ft (15240 m)
Combat radius: 295 nm (340 miles; 547 km) on a hi-lo-hi mission with six 1,000-lb (454-kg) bombs

Armament:

One internal M61 Vulcan 20-mm cannon; maximum ordnance of 15,200 lb (6894 kg) on one fuselage pylon and six underwing pylons.

McDonnell Douglas F-4 Phantom II

The F-4G was the definitive defence suppression Phantom, and the last in front-line USAF service. This pair illustrates the primary weapons employed in this role in later years, the furthest aircraft carrying the AGM-45 Shrike and cluster bombs, while that in the foreground is armed with AGM-65 Maverick and AGM-88 HARM. F-4Gs routinely carried weapons such as Maverick and CBUs to attack missile installations once the anti-radiation missiles had taken out the radars.

RF-4C Phantom II

Cutaway key

1 Pitot head
2 Radome
3 Radar scanner dish
4 Radar dish tracking mechanism
5 Texas Instruments AN/APQ-99 forward-looking radar unit
6 Nose compartment construction
7 No. 1 camera station
8 KS-87 forward oblique camera
9 Forward radar warning antennas, port and starboard
10 Camera bay access hatches
11 Ventral camera aperture
12 KA-57 low-altitude panoramic camera
13 Lateral camera aperture (alternative KS-87 installation)
14 No. 2 camera station
15 ADF sense aerial
16 Windscreen rain dispersal air duct
17 Camera viewfinder periscope
18 Nose undercarriage emergency air bottles
19 Recording unit
20 No. 3 camera station
21 KA-91 high-altitude panoramic camera
22 Air-conditioning ram air intake
23 Landing/taxiing lamp (2)
24 Lower UHF/VHF aerial
25 Nosewheel leg door
26 Torque scissor links
27 Twin nosewheels, aft retracting
28 Nosewheel steering mechanism
29 AN/AVQ-26 'Pave Tack' laser designator pod
30 Swivelling optical package
31 Fuselage centreline pylon adaptor
32 Sideways-looking radar antenna (SLAR)
33 Electro-luminescent formation lighting strip
34 Canopy emergency release handle
35 Air-conditioning plant, port and starboard
36 Cockpit floor level
37 Front pressure bulkhead
38 Rudder pedals
39 Control column
40 Instrument panel
41 Radar display
42 Instrument panel shroud
43 LA-313A optical viewfinder
44 Windscreen panels
45 Forward cockpit canopy cover
46 Face blind seat firing handle
47 Pilot's Martin-Baker Mk.H7 ejection seat
48 External canopy latches
49 Engine throttle levers
50 Side console panel
51 Intake boundary layer splitter plate
52 APQ-102R/T SLAR equipment
53 AAS-18A infra-red reconnaissance package
54 Intake front ramp
55 Port engine air intake
56 Intake ramp bleed air holes
57 Rear canopy external latches
58 Rear instrument console
59 Canopy centre arch
60 Starboard engine air intake
61 Starboard external fuel tank, capacity 308 Imp gal (1400 litres)
62 Rear view mirrors
63 Rear cockpit canopy cover
64 Navigator/Sensor Operator's Martin-Baker ejection seat
65 Intake ramp bleed air spill louvres
66 Avionics equipment racks
67 Rear pressure bulkhead
68 Liquid oxygen converter
69 Variable intake ramp jack
70 Intake rear ramp door
71 Fuselage centreline external fuel tank, capacity 500 Imp gal (2271 litres)
72 Position of pressure refuelling connection on starboard side
73 ASQ-90B data annotation system equipment
74 Cockpit voice recorder
75 Pneumatic system air bottle
76 Bleed air ducting
77 Fuselage No. 1 fuel cell, capacity 179 Imp gal (814 litres)
78 Intake duct framing
79 Boundary layer spill duct
80 Control cable runs
81 Aft avionics equipment bay
82 IFF aerial
83 Upper fuselage light
84 Fuselage No. 2 fuel cell, capacity 154 Imp gal (700 litres)
85 Centre fuselage frame construction
86 Electro-luminescent formation lighting strip
87 Engine intake centre-body fairing
88 Intake duct rotary spill valve
89 Wing spar attachment fuselage main frames
90 Control cable ducting

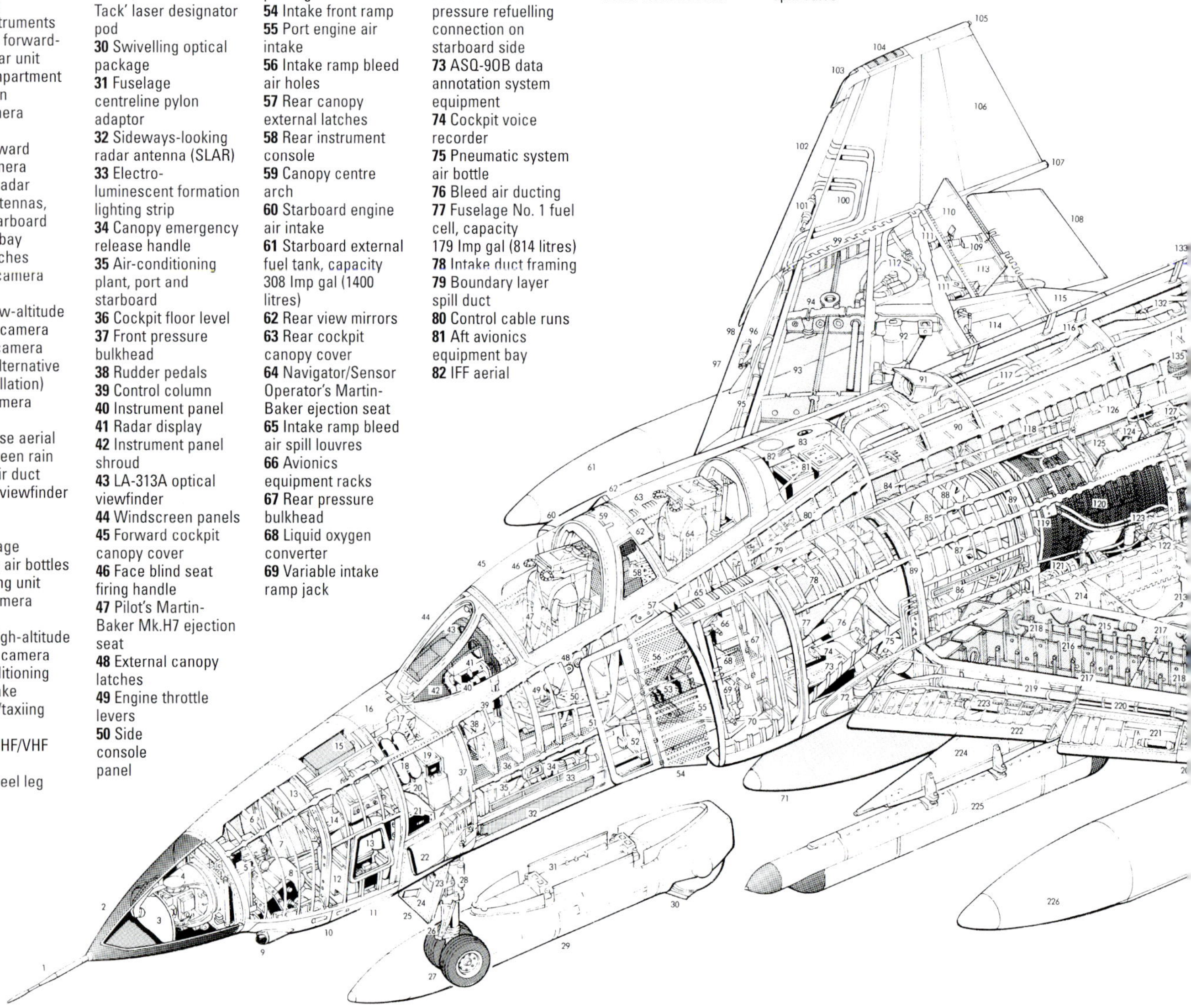

USAF Phantoms first went to war in a peacetime grey/white scheme, but soon acquired tactical camouflage. This F-4C carries a mixed load of ground attack weapons, comprising three M117 bombs, four rocket pods and two napalm canisters.

SPECIFICATION

F-4E Phantom II

Dimensions

Wingspan: 38 ft 7½ in (11.77 m)
Wingspan (folded): 27 ft 7 in (8.41 m)
Wing aspect ratio: 2.82
Wing area: 530 sq ft (49.2 m^2)
Length: 63 ft (19.20 m)
Wheel track: 17 ft 10½ in (5.45 m)
Height: 16 ft 5½ in (5.02 m)

Powerplant

Two General Electric J79-GE-17A turbojets, each rated at 17,900 lb (80 kN) thrust with afterburning

Weights

Empty: 30,328 lb (13757 kg)
Operating empty: 31,853 lb (14448 kg)
Combat take-off: 41,487 lb (18818 kg)
Maximum take-off: 61,795 lb (28030 kg)

Fuel and load

Internal fuel capacity: 1,855 US gal (1,545 Imp gal; 7022 litres)
Provision for one 600-US gal (500-Imp gal; 2271-litre) tank on centreline and two 370-US gal (308-Imp gal; 1400-litre) tanks under the wings
Maximum weaponload: 16,000 lb (7250 kg)

Performance

Maximum speed: approximately Mach 2.2
Maximum rate of climb: 61,400 ft (18715 m) per minute
Service ceiling: 62,250 ft (18975 m)
Take-off run at maximum take-off weight: 4,390 ft (1338 m)
Landing run at maximum landing weight: 3,780 ft (1152 m)

Range

Ferry range: 1,978 miles (3184 km)
Area intercept combat radius: 786 miles (1266 km)
Defensive counter-air combat radius: 494 miles (795 km)
Interdiction combat radius: 712 miles (1145 km)

Armament

Fixed internal M61A1 Vulcan 20-mm six-barrelled cannon; standard intercept load of four AIM-7 Sparrow missiles in fuselage recesses and four AIM-9 Sidewinders on wing pylon shoulder stations; four wing pylons and one centreline station available for carriage of wide range of air-to-ground ordnance, including M117 and Mk 80 series bombs, cluster weapons, laser-guided bombs, gun pods, napalm, fuel-air explosives and rocket pods; nuclear weapon options included B28, B43, B57 and B61; various ECM pods, training targets and laser designator pods available; air-to-surface missiles included AGM-12 Bullpup, AGM-45 Shrike, AGM-65 Maverick and AGM-78 Standard

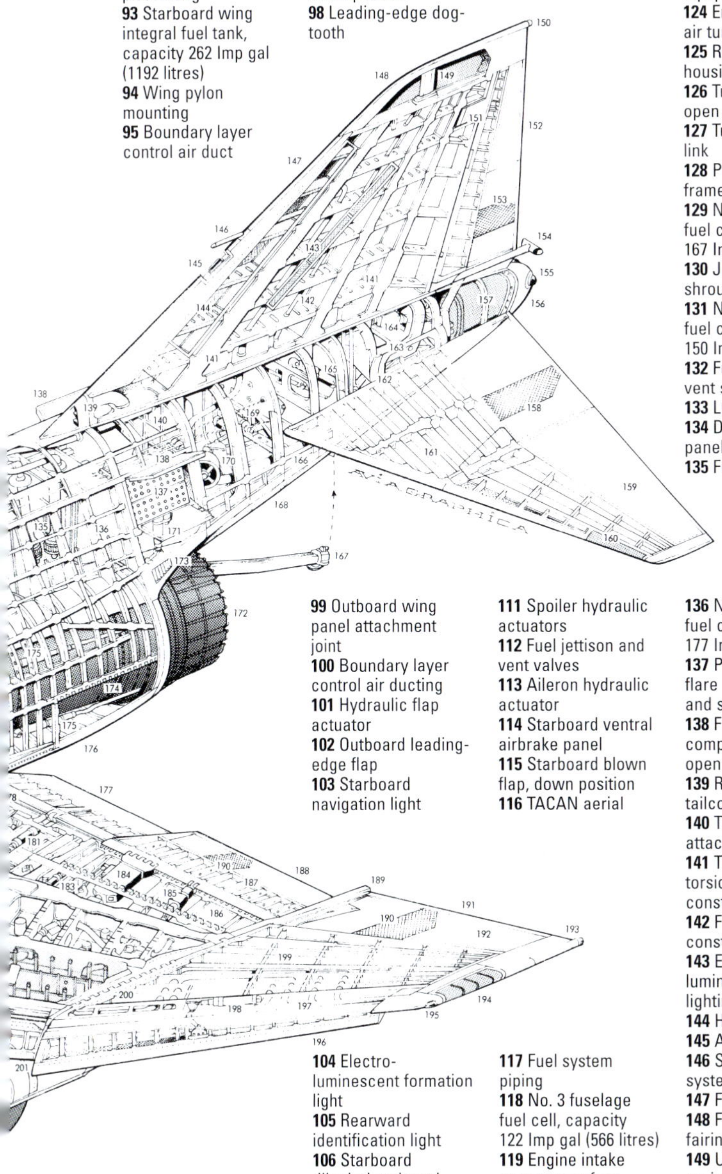

91 In-flight refuelling receptacle, open
92 Starboard main undercarriage leg pivot fixing
93 Starboard wing integral fuel tank, capacity 262 Imp gal (1192 litres)
94 Wing pylon mounting
95 Boundary layer control air duct
96 Leading-edge flap hydraulic actuator
97 Inboard leading-edge flap segment, down position
98 Leading-edge dog-tooth
99 Outboard wing panel attachment joint
100 Boundary layer control air ducting
101 Hydraulic flap actuator
102 Outboard leading-edge flap
103 Starboard navigation light
104 Electro-luminescent formation light
105 Rearward identification light
106 Starboard dihedral outboard wing panel
107 Wing fuel tank vent pipe
108 Starboard drooping aileron, down position
109 Aileron flutter damper
110 Starboard spoilers, open
111 Spoiler hydraulic actuators
112 Fuel jettison and vent valves
113 Aileron hydraulic actuator
114 Starboard ventral airbrake panel
115 Starboard blown flap, down position
116 TACAN aerial
117 Fuel system piping
118 No. 3 fuselage fuel cell, capacity 122 Imp gal (566 litres)
119 Engine intake compressor face
120 General Electric J79 GE-15 afterburning turbojet engine
121 Ventral engine accessory equipment gearbox
122 Wing rear spar attachment joint
123 Engine and afterburner control equipment
124 Emergency ram air turbine
125 Ram air turbine housing
126 Turbine doors, open
127 Turbine actuating link
128 Port engine bay frame construction
129 No. 4 fuselage fuel cell, capacity 167 Imp gal (759 litres)
130 Jet pipe heat shroud
131 No. 5 fuselage fuel cell, capacity 150 Imp gal (681 litres)
132 Fuel feed and vent system piping
133 LORAN aerial
134 Dorsal access panels
135 Fuel pumps
136 No. 6 fuselage fuel cell, capacity 177 Imp gal (806 litres)
137 Photographic flare dispenser, port and starboard
138 Flare compartment doors, open
139 Ram air intake, tailcone venting
140 Tailcone attachment bulkhead
141 Three-spar fin torsion box construction
142 Fin rib construction
143 Electro-luminescent formation lighting strip
144 HF aerial panel
145 Anti-collision light
146 Stabilator feel system pressure head
147 Fin leading edge
148 Fin tip aerial fairing
149 Upper UHF/VHF aerial
150 Tail navigation light
151 Rudder horn balance
152 Rudder
153 Honeycomb trailing-edge panels
154 Fuselage fuel cell jettison pipe
155 Rear radar warning antennas
156 Tailcone/brake parachute hinged door
157 Brake parachute housing
158 Honeycomb trailing-edge panel
159 Port all-moving tailplane/stabilator
160 Stabilator mass balance weight
161 Stabilator multi-spar construction
162 Pivot sealing plate
163 All-moving tailplane hinge mounting
164 Rudder hydraulic actuator
165 Tailplane hydraulic actuator
166 Heat-resistant tailcone skinning
167 Arrester hook, lowered
168 Arrester hook stowage
169 Stabilator feel system balance mechanism
170 Artificial feel system pneumatic bellows
171 Arrester hook jack and shock absorber
172 Variable-area afterburner exhaust nozzle
173 Engine bay cooling exit louvres
174 Afterburner duct
175 Exhaust nozzle actuators
176 Hinged engine cowling panels
177 Port blown flap, down position
178 Boundary layer control air blowing slot
179 Lateral autopilot servo
180 Airbrake jack
181 Flap hydraulic jack
182 Rear spar
183 Port spoiler hydraulic jack
184 Aileron hydraulic actuator
185 Aileron flutter damper
186 Port spoiler housing
187 Aileron rib construction
188 Port drooping aileron, down position
189 Wing fuel tank jettison pipe
190 Honeycomb trailing-edge panels
191 Port dihedral outer wing panel
192 Fixed portion of trailing edge
193 Rearward identification light
194 Electro-luminescent formation light
195 Port navigation light
196 Outboard leading-edge flap, lowered
197 Boundary layer control air blowing slot
198 Leading-edge flap actuator
199 Outer wing panel multi-spar construction
200 Outer wing panel attachment joint
201 Leading-edge dog-tooth
202 Port mainwheel
203 Mainwheel multi-plate disc brake
204 Mainwheel leg door
205 Outboard wing pylon
206 Inner wing panel outboard leading-edge flap, down position
207 Leading-edge flap rib construction
208 Wing pylon mounting
209 Main undercarriage leg pivot fixing
210 Hydraulic retraction jack
211 Undercarriage uplock
212 Port ventral airbrake panel, open
213 Main undercarriage wheel bay
214 Hydraulic reservoir
215 Hydraulic system accumulator
216 Port wing integral fuel tank, capacity 262 Imp gal (1192 litres)
217 Two-spar torsion box fuel tank construction
218 Wing skin support posts
219 Leading-edge boundary layer control air duct
220 Bleed air blowing slot
221 Outboard flap actuator
222 Inboard leading-edge flap, lowered
223 Hydraulic flap actuator
224 Inboard wing pylon
225 AN/ALQ-101 ECM pod
226 Port external fuel tank, capacity 308 Imp gal (1400 litres)

McDonnell Douglas F-15 Eagle

F-15Es serve with two overseas commands (USAFE and PACAF). Like USAFE's 48th FW, PACAF's 3rd Wing (previously the 21st Composite Wing) operates a mix of F-15Cs and F-15Es, the latter flying with the 90th FS 'Pair o'Dice'. Here the wing commander leads aircraft from the three Eagle squadrons through the Alaskan mountains from the unit's base at Elmendorf AFB.

F-15E Eagle

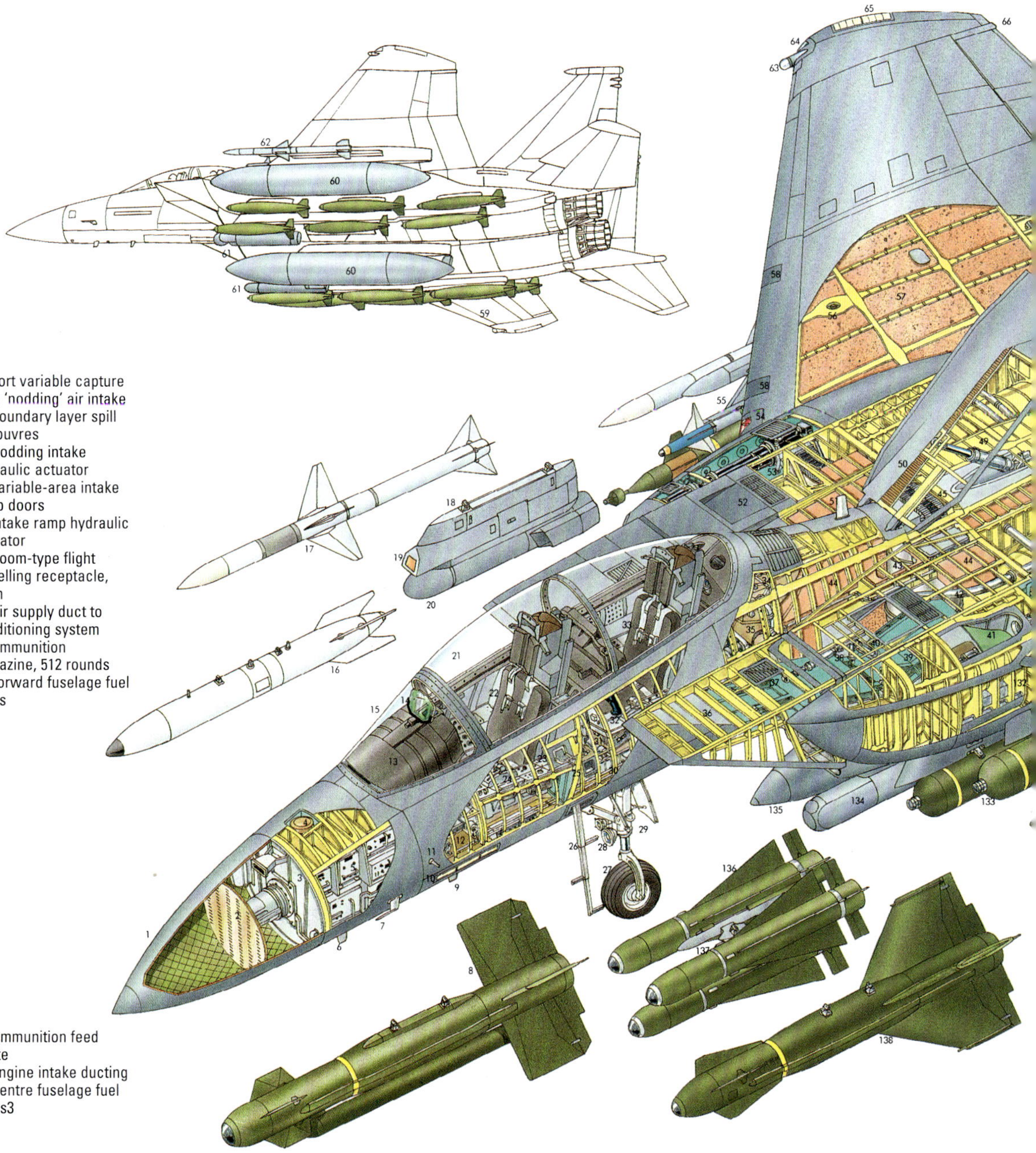

Cutaway key

1 Glass-fibre radome
2 Hughes AN/APG-70 I-band pulse-Doppler radar scanner
3 Radar mounting bulkhead
4 ADF sense antenna
5 Avionics equipment bay, port and starboard
6 UHF antenna
7 Pitot head
8 AGM-130 TV-guided air-to-surface weapon
9 TACAN antenna
10 Formation lighting strip
11 Incidence probe
12 Rudder pedals
13 Instrument panel shroud
14 Pilot's head-up display
15 Frameless windscreen panel
16 B61 tactical nuclear weapon
17 AIM-7F Sparrow air-to-air missile
18 LANTIRN navigation pod, mounted beneath starboard intake
19 FLIR aperture
20 Terrain-following radar
21 Upward-hinging cockpit canopy
22 Pilot's ACES II ejection seat
23 Side console panel
24 Engine throttle levers
25 Boarding steps
26 Extended boarding ladder
27 Forward-retracting nosewheel
28 Landing/taxiing lights
29 Nosewheel leg shock absorber strut
30 Underfloor control runs
31 Flying controls duplicated in rear cockpit
32 Radar hand controller
33 Weapons Systems Officer's ACES II ejection seat
34 Canopy hinge point
35 Cockpit air-conditioning pack
36 Port variable capture area 'nodding' air intake
37 Boundary layer spill air louvres
38 Nodding intake hydraulic actuator
39 Variable-area intake ramp doors
40 Intake ramp hydraulic actuator
41 Boom-type flight refuelling receptacle, open
42 Air supply duct to conditioning system
43 Ammunition magazine, 512 rounds
44 Forward fuselage fuel tanks
45 Ammunition feed chute
46 Engine intake ducting
47 Centre fuselage fuel tanks3

An F-15A of the now defunct 48th Fighter Interceptor Squadron (FIS) fires an AIM-7. Despite being at the end of its development life, the Sparrow remains an important F-15 weapon even though the AIM-120 is now well established in service. During the Gulf War, most of the kills made by F-15s were achieved with the AIM-7.

48 Fuel tank bay access panel
49 Airbrake hydraulic jack
50 Dorsal airbrake honeycomb construction
51 Upper UHF antenna
52 Starboard intake by-pass air spill duct
53 M61A-1 Vulcan 20-mm cannon
54 Anti-collision light
55 Starboard wing pylon carrying GBU-10, AIM-7M and AIM-120
56 Pylon mounting hardpoint
57 Starboard wing integral fuel tank, fire suppressant foam filled
58 Leading edge flush HF antenna panels
59 Ventral view showing carriage of 12 Mk 82 500-lb (227-kg) bombs
60 610-US gal (2309-litre) external fuel tanks (3)
61 LANTIRN navigation and targeting pods
62 Wing pylon mounted AIM-9M and AIM-120 air-to-air missiles
63 Forward ECM transmitting antenna
64 Starboard navigation light

SPECIFICATION

F-15E

Dimensions

Length: 63 ft 9 in (19.43 m)
Height: 18 ft 5½ in (5.63 m)
Wingspan: 42 ft 9¾ in (13.05 m)
Wing aspect ratio: 3.01
Tailplane span: 28 ft 3 in (8.61 m)
Wheel track: 9 ft ¼ in (2.75 m)
Wheelbase: 17 ft 9½ in (5.42 m)

Powerplant

Original F-15E had powerplant of F-15C/D, but with option of General Electric F110-GE-129. Aircraft from 135 onwards (90-0233), built from August 1991, have two Pratt and Whitney F100-PW-229s rated at 29,000 lb st (129 kN)

Weights

Empty operating: 31,700 lb (14379 kg)
Normal take-off: 44,823 lb (20331 kg)
Maximum take-off: 36,741 lb (81000 kg)

Fuel and load

Internal fuel: 13,123 lb (5952 kg)
External fuel: 21,645 lb (9818 kg)
Maximum weaponload: 24,500 lb (11113 kg)

Performance

Maximum level speed: Mach 2.5
Maximum combat radius: 685 nm (790 miles; 1270 km)
Maximum combat range: 2,400 nm (2,765 miles; 4445 km)

Armament

One 20-mm M61A-1 six-barrel gun, with 512 rounds, in starboard wingroot. Wing pylons for AIM-9 Sidewinder, AIM-120 AMRAAM, and AIM-7 Sparrow. Single or triple rail launchers for AGM-65 Maverick on wing pylons. A wide range of guided and unguided weapons including Mk 20 Rockeye, Mk 82, MK 84, BSU-49 and -50, GBU-10, -12, -15, and -24; CBU-52, -58, -71, -87, -89, -90, -92 and -93; LAU-3A rockets; B57 and B61 nuclear weapons

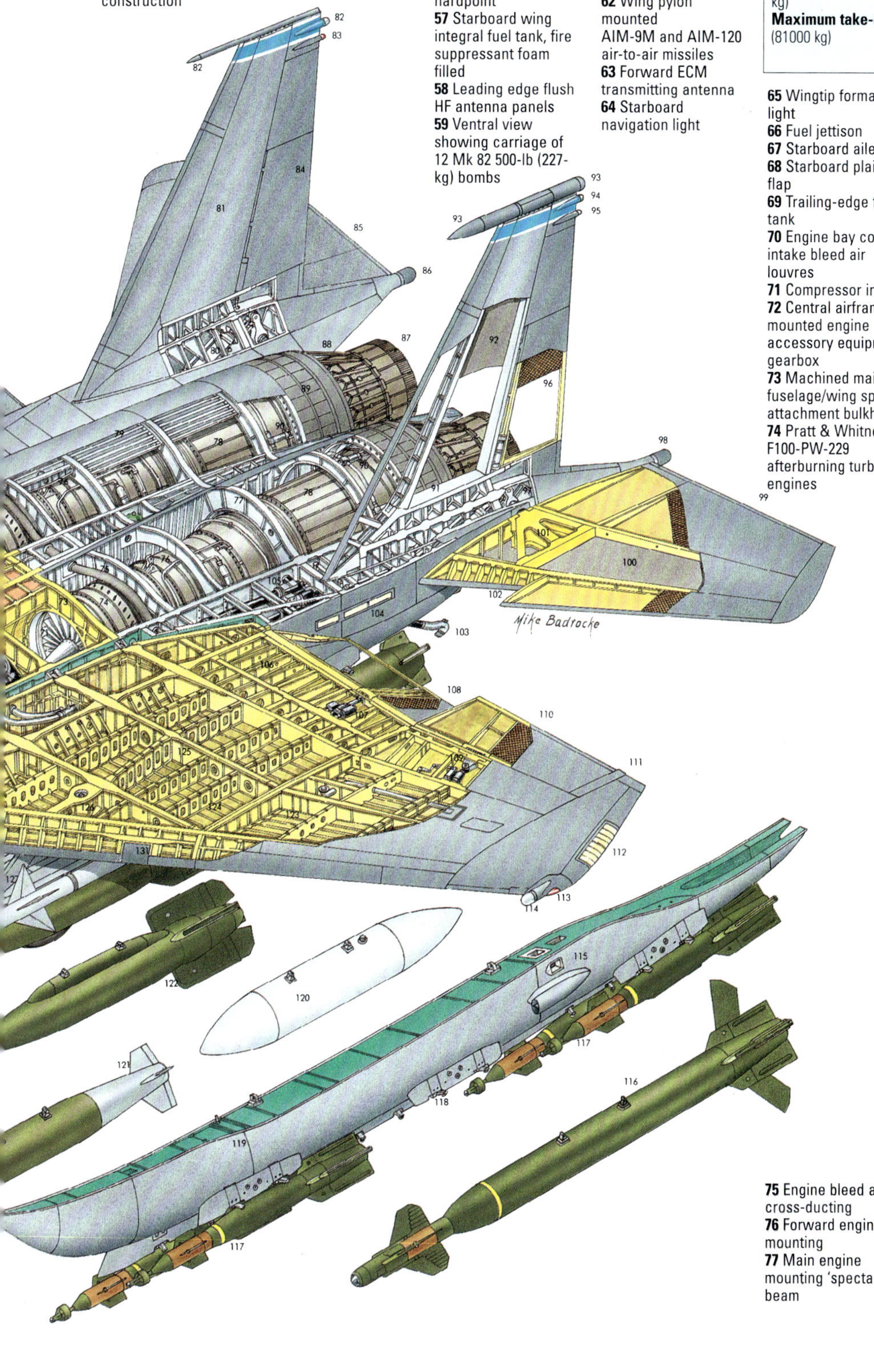

65 Wingtip formation light
66 Fuel jettison
67 Starboard aileron
68 Starboard plain flap
69 Trailing-edge fuel tank
70 Engine bay cooling intake bleed air louvres
71 Compressor intake
72 Central airframe-mounted engine accessory equipment gearbox
73 Machined main fuselage/wing spar attachment bulkheads
74 Pratt & Whitney F100-PW-229 afterburning turbofan engines
75 Engine bleed air cross-ducting
76 Forward engine mounting
77 Main engine mounting 'spectacle' beam
78 Afterburner ducting
79 Rear fuselage/engine bay diffusion-bonded all-titanium structure
80 Tailplane hydraulic actuator
81 Starboard fin
82 Fintip ECM antenna
83 Anti-collision light
84 Starboard rudder
85 Starboard all-moving tailplane
86 Aft ECM transmitting antenna
87 Variable area afterburner nozzle
88 Nozzle actuating linkage
89 Nozzle shroud panels
90 Fueldraulic afterburner nozzle actuators
91 Two-spar fin torsion box structure
92 Boron-fibre fin skin panelling
93 Radar warning antenna
94 Port rear ECM antenna
95 White strobe light
96 Port rudder honeycomb core construction
97 Tailplane pivot mounting
98 Port aft ECM transmitting antenna
99 Port all-moving tailplane
100 Boron-fibre tailplane skin panelling
101 Machined tailplane trunion mounting fitting
102 Leading edge dog-tooth
103 Runway emergency arrester hook, lowered
104 Formation lighting strip
105 Engine bleed air primary heat exchangers, port and starboard
106 Port trailing-edge fuel tank bay
107 Flap hydraulic jack
108 Port plain flap
109 Aileron hydraulic actuator
110 Port aileron honeycomb core construction
111 Fuel jettison
112 Port formation light
113 Port navigation light
114 Forward ECM transmitting antenna
115 Engine bleed air primary heat exchanger air intake and exhaust ducts
116 GBU-28 'Deep-Throat' laser-guided bomb
117 GBU-12 laser-guided bombs
118 CFT pylons
119 Port conformal fuel tank (CFT)
120 AXQ-14 datalink pod
121 Mk 84 2,000-lb (907-kg) HE bomb
122 GBU-24 laser-guided bomb
123 Outer wing panel dry bay
124 Port wing integral fuel tankage
125 Multi-spar wing panel structure
126 Port pylon hardpoint
127 Wing stores pylon
128 Missile launch rails
129 AIM-120 AMRAAM
130 AIM-9M Sidewinder air-to-air missile
131 Leading-edge flush HF antenna
132 Stores management system equipment
133 CBU-87 sub-munition dispensers
134 Port LANTIRN targeting pod
135 Centreline external tank
136 AGM-65 Maverick air-to-surface missiles
137 Triple missile carrier/launch rail
138 GBU-15 electro-optical guided glide bomb

McDonnell F-101 Voodoo

Conceived as a long-range fighter to accompany SAC bombers, the F-88 never reached production status, but it did provide the basis for the later F-101 Voodoo. The Voodoo was also initially designed as a strategic fighter, yet it saw service in three other major roles: nuclear bomber, tactical reconnaissance platform and defender of North American airspace. In the latter tasking, it served with Canada into the late 198

F-101C

Cutaway key
1 Radome
2 Scanner dish
3 Radar tracking mechanism
4 Radar mounting bulkhead
5 Refuelling probe doors
6 Radar modulating units
7 Refuelling probe hydraulic jack
8 Flight refuelling probe, extended
9 Forward avionics equipment bay, radar and weapons system equipment
10 Nose compartment access panels
11 Angle of attack transducer
12 Pitot head
13 Nosewheel doors
14 Emergency brake reservoir
15 Cannon muzzles
16 Cockpit pressure floor
17 Cockpit air-conditioning ducting
18 Front pressure bulkhead
19 Rudder pedals
20 Control column
21 Instrument panel
22 Instrument panel shroud
23 K-19 (Mk 7) gunsight
24 Armoured glass windscreen panel
25 MA-7 flight indicator radar scope
26 Canopy cover
27 Canopy mounted flush aerial
28 Headrest
29 Safety harness
30 Canopy external release
31 Pilot's ejection seat
32 Throttle levers
33 Side console panel
34 Cockpit pressurisation valve
35 Cannon barrel seals
36 Nose undercarriage pivot fixing
37 Cannon barrel fairings
38 Nose undercarriage leg strut
39 Landing and taxiing lamps
40 Twin nosewheels
41 Torque scissor links
42 Ventral AW aerial
43 Cannon barrels
44 Control rod runs
45 Anti-'g' valve
46 Rear pressure bulkhead
47 Canopy hydraulic jack
48 Canopy aft fairing
49 Rear avionics equipment bay, navigation and communications systems
50 Canopy hinge
51 Ammunition access door
52 Ammunition magazine, 375 rounds per gun
53 Feed chutes
54 M39 20-mm cannon (single cannon on starboard side, fourth weapon replaced by transponder equipment)
55 Heat exchanger flush air intake
56 Circuit breaker panel
57 Air-conditioning plant
58 Autopilot rate gyros
59 Control linkages
60 Hydraulic accumulators
61 Boundary layer splitter plate
62 Port engine air intake
63 Intake duct framing
64 Port hydraulic system reservoir
65 Boundary layer bleed air spill duct
66 Wing spar attachment main bulkhead
67 Forward fuselage fuel tanks; total system capacity 2,146 US gal (8123 litres)
68 Fuel filler cap, pressure refuelling connector on starboard side
69 Fuel system piping
70 Anti-collision light
71 Starboard wing panel
72 Wing fence
73 Starboard navigation light
74 Fixed portion of trailing edge
75 Starboard aileron
76 Aileron mass-balance weights
77 Aileron hydraulic actuator
78 Main undercarriage pivot fixing
79 Starboard split trailing-edge flap
80 Boom type refuelling receptacle, open
81 Wing spar and engine mounting main bulkheads
82 Centre fuselage fuel tank
83 Fuel vent piping
84 Fuselage upper access panels
85 Fuselage top longeron
86 Aft fuselage fuel tanks
87 Control cable duct
88 Fuel filler cap
89 Finroot fillet
90 Tailcone joint frame
91 Artificial feel system bellows
92 Starboard airbrake, open
93 Tailfin construction
94 Remote compass transmitter
95 Artificial feel system ram air intake
96 VHF aerial

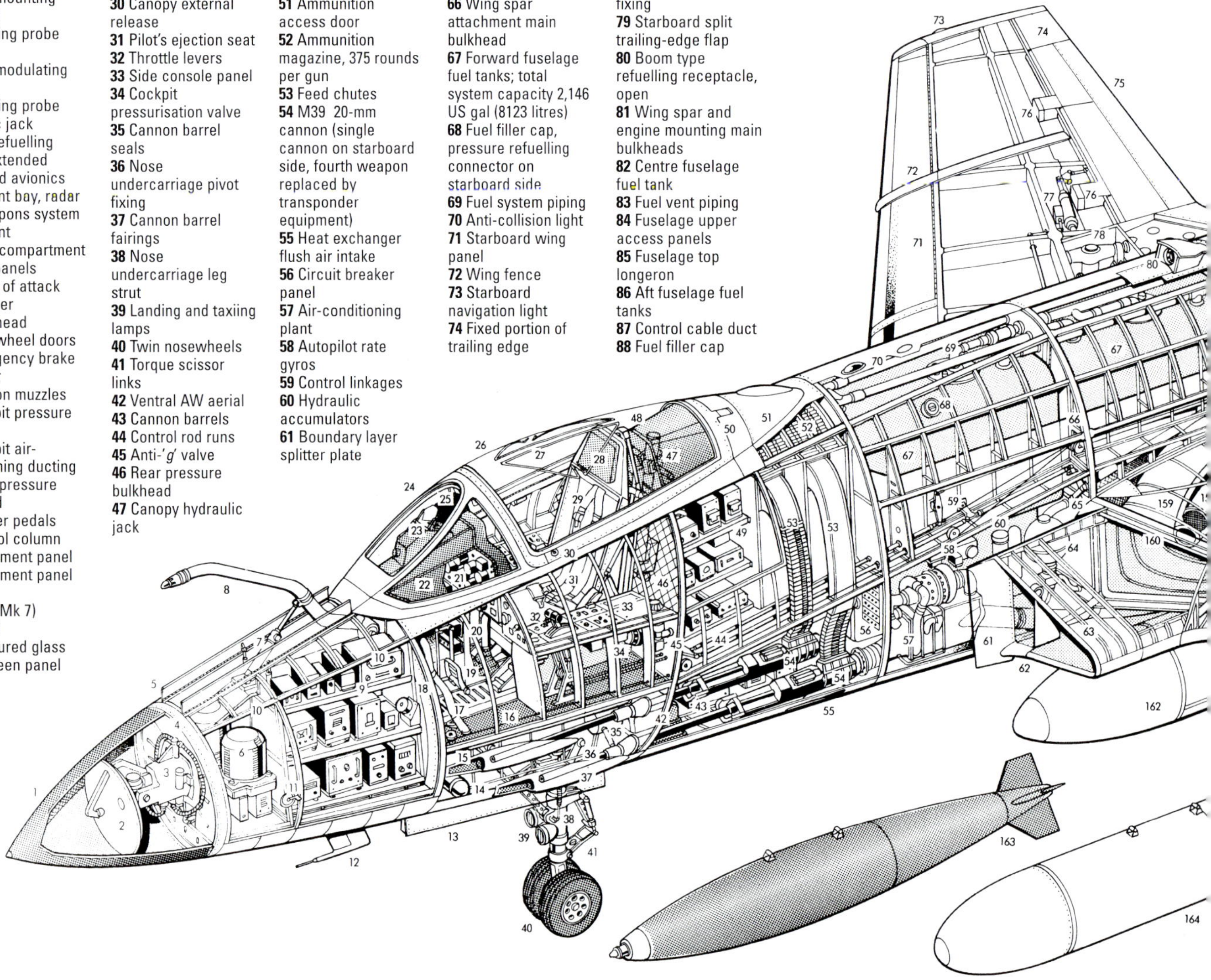

F-101 Voodoo 60236 was the fifth F-101B built and was used for the testing of onboard systems, in this case the MB-1, which would later be named the AIR-2A Genie. The Hughes MG-13 fire-control system on the Voodoo handled both nuclear and non-nuclear air-to-air rocket missiles and projectiles, and it was in 1961 that the Voodoo underwent a modification programme to let it fire the awesome Genie. Designed to scatter incoming Soviet bomb formations, the first 'live' Genie was tested on 19 July 1957 when it was fired from a Northrop F-89 Scorpion over Yucca Flat, Nevada, in a 1.5-kT blast.

SPECIFICATION

RF-101C Voodoo

Dimensions

Length: 69 ft 3 in (21.1 m)
Height: 18 ft (5.49 m)
Wing span: 39 ft 8 in (12.09 m)
Wing area 368 sq ft (34.19 m²)

Powerplant

Two Pratt & Whitney J57-13 turbojets each rated at 14,880 lb (66.2 kN) with maximum afterburner

Weights

Empty: 25,610 lb (11617 kg)
Loaded (clean): 42,550 lb (19300 kg)
Maximum (with two tanks): 48,720 lb (22099 kg)

Wing loading: 130.8 lb/sq ft (638.6 kg/sq m)
Power loading: 1.6 lb/lb st (1.6 kg/kgp)

Performance

Maximum speed (clean, at height): Mach 1.7 (1,120 mph;1802 km/h)
Service ceiling: 52,000 ft (15850 m)
Range (with internal tanks at high altitude): 1,890 miles (3040 km)
Range (with two 375-US gal (1705-litre) drop tanks: 2,400 miles (3862 km)

Armament

No weapons were carried though a single nuclear weapon could be mounted on the centreline hardpoint in the event of nuclear war

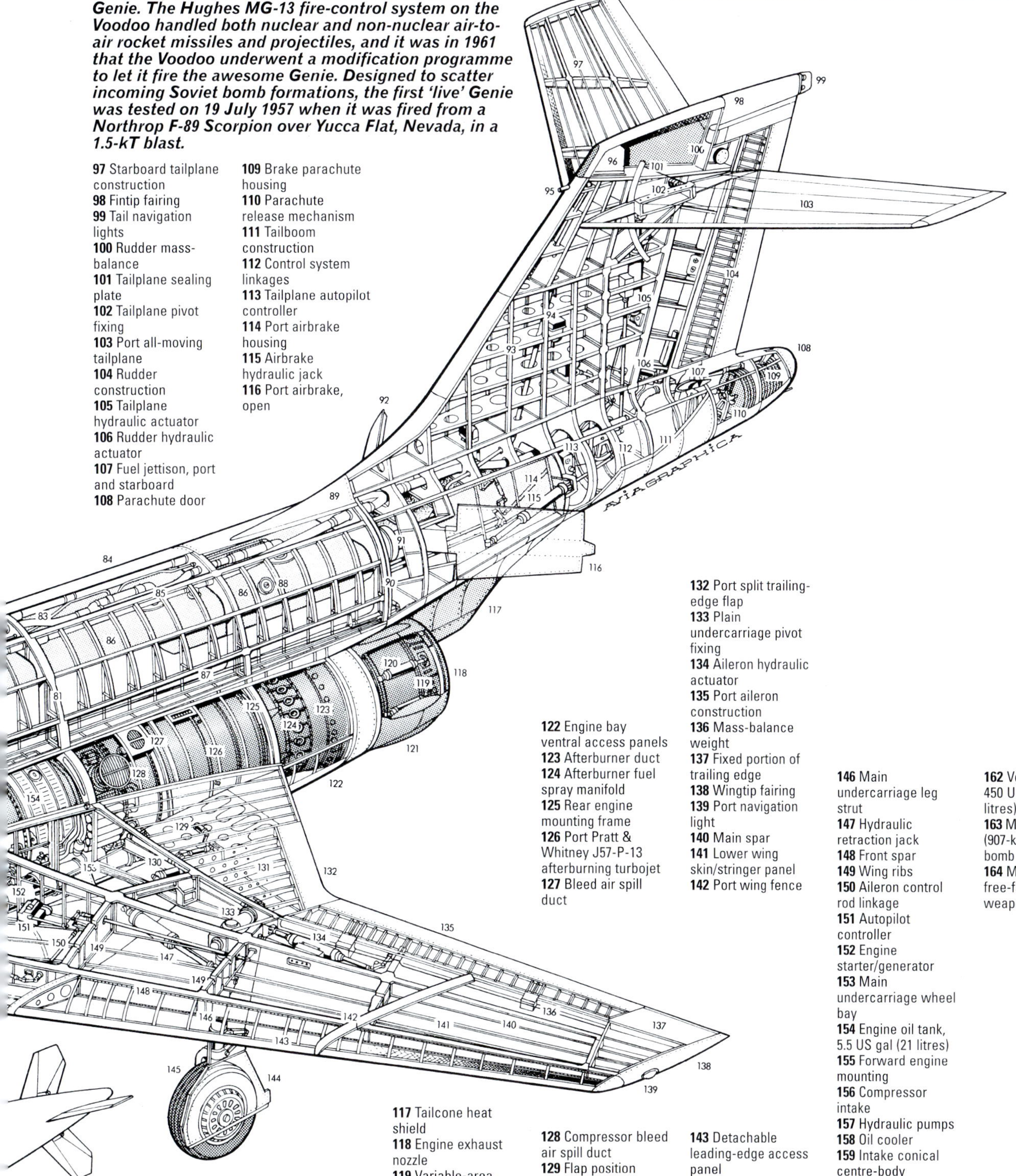

97 Starboard tailplane construction
98 Fintip fairing
99 Tail navigation lights
100 Rudder mass-balance
101 Tailplane sealing plate
102 Tailplane pivot fixing
103 Port all-moving tailplane
104 Rudder construction
105 Tailplane hydraulic actuator
106 Rudder hydraulic actuator
107 Fuel jettison, port and starboard
108 Parachute door
109 Brake parachute housing
110 Parachute release mechanism
111 Tailboom construction
112 Control system linkages
113 Tailplane autopilot controller
114 Port airbrake housing
115 Airbrake hydraulic jack
116 Port airbrake, open
117 Tailcone heat shield
118 Engine exhaust nozzle
119 Variable-area afterburner nozzle
120 Nozzle control jacks
121 Nozzle shroud
122 Engine bay ventral access panels
123 Afterburner duct
124 Afterburner fuel spray manifold
125 Rear engine mounting frame
126 Port Pratt & Whitney J57-P-13 afterburning turbojet
127 Bleed air spill duct
128 Compressor bleed air spill duct
129 Flap position transmitter
130 Flap hydraulic jack
131 Flap shroud ribs
132 Port split trailing-edge flap
133 Plain undercarriage pivot fixing
134 Aileron hydraulic actuator
135 Port aileron construction
136 Mass-balance weight
137 Fixed portion of trailing edge
138 Wingtip fairing
139 Port navigation light
140 Main spar
141 Lower wing skin/stringer panel
142 Port wing fence
143 Detachable leading-edge access panel
144 Mainwheel doors
145 Port mainwheel
146 Main undercarriage leg strut
147 Hydraulic retraction jack
148 Front spar
149 Wing ribs
150 Aileron control rod linkage
151 Autopilot controller
152 Engine starter/generator
153 Main undercarriage wheel bay
154 Engine oil tank, 5.5 US gal (21 litres)
155 Forward engine mounting
156 Compressor intake
157 Hydraulic pumps
158 Oil cooler
159 Intake conical centre-body
160 Intake duct main frame
161 Wing spar attachment joints
162 Ventral fuel tank 450 US gal (1700 litres)
163 Mk 84 2,000-lb (907-kg) low-drag HE bomb
164 Mk 7 1-megaton free-fall nuclear weapon

Mikoyan-Gurevich MiG-17 'Fresco'

China's locally-developed two-seat version of the MiG-17, the Chengdu JJ-5/FT-5 was believed to be still in use in China, North Korea, Pakistan and Sudan. These are Pakistani examples.

This former Moroccan air force early-productic aircraft was acquired by the Champlin Fight Museum and is displayed at their Mesa, Arizo facility. A large number of ex-Polish Lim-5s and - have been purchased by warbird enthusiasts the US and Europ

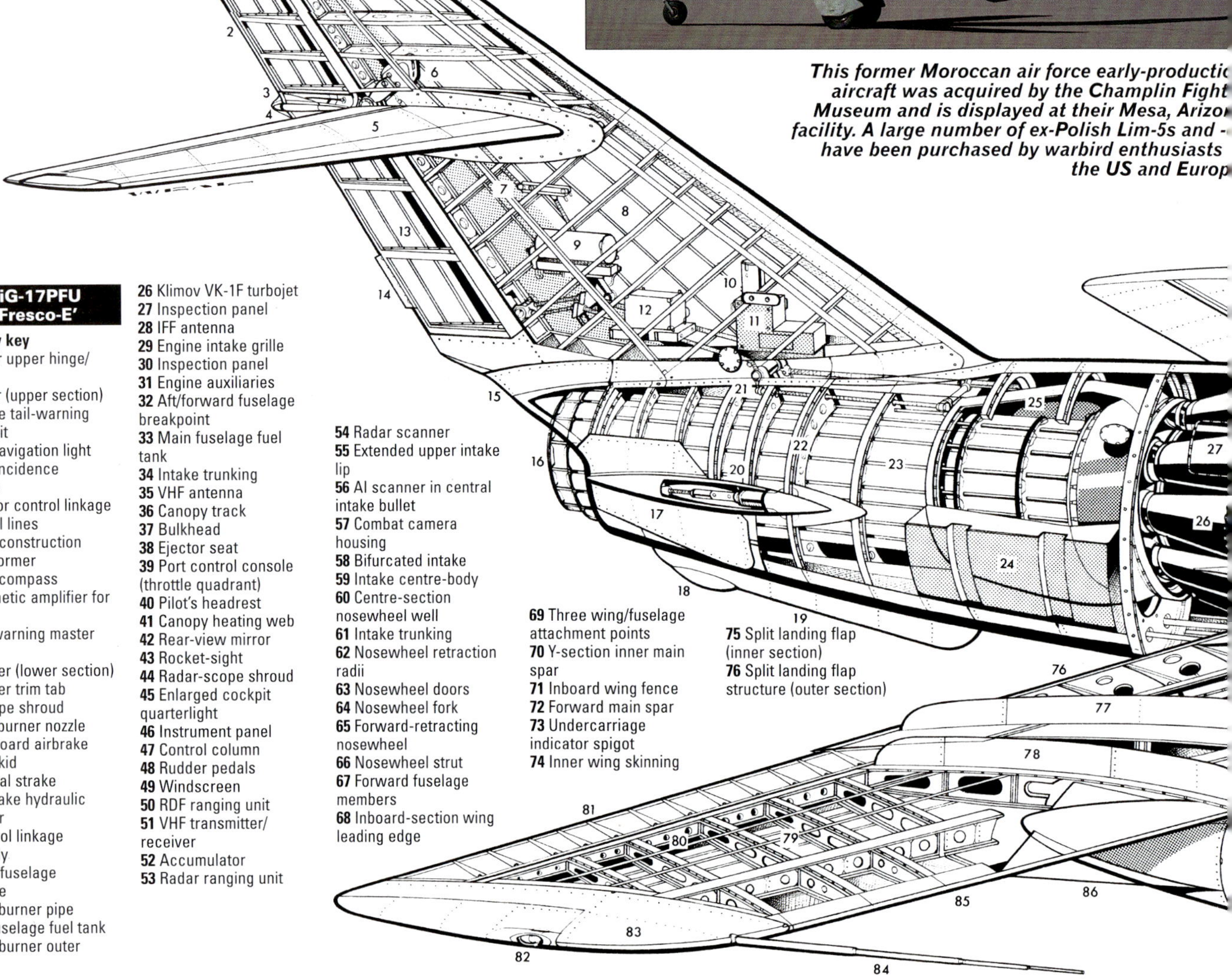

MiG-17PFU 'Fresco-E'

Cutaway key
1 Rudder upper hinge/balance
2 Rudder (upper section)
3 Passive tail-warning radar unit
4 Rear navigation light
5 Fixed incidence tailplane
6 Elevator control linkage
7 Control lines
8 Tailfin construction
9 Transformer
10 Gyro compass
11 Magnetic amplifier for gyro
12 Tail-warning master unit
13 Rudder (lower section)
14 Rudder trim tab
15 Tailpipe shroud
16 Afterburner nozzle
17 Starboard airbrake
18 Tail skid
19 Ventral strake
20 Airbrake hydraulic activator
21 Control linkage assembly
22 Rear fuselage structure
23 Afterburner pipe
24 Aft fuselage fuel tank
25 Afterburner outer casing
26 Klimov VK-1F turbojet
27 Inspection panel
28 IFF antenna
29 Engine intake grille
30 Inspection panel
31 Engine auxiliaries
32 Aft/forward fuselage breakpoint
33 Main fuselage fuel tank
34 Intake trunking
35 VHF antenna
36 Canopy track
37 Bulkhead
38 Ejector seat
39 Port control console (throttle quadrant)
40 Pilot's headrest
41 Canopy heating web
42 Rear-view mirror
43 Rocket-sight
44 Radar-scope shroud
45 Enlarged cockpit quarterlight
46 Instrument panel
47 Control column
48 Rudder pedals
49 Windscreen
50 RDF ranging unit
51 VHF transmitter/receiver
52 Accumulator
53 Radar ranging unit
54 Radar scanner
55 Extended upper intake lip
56 AI scanner in central intake bullet
57 Combat camera housing
58 Bifurcated intake
59 Intake centre-body
60 Centre-section nosewheel well
61 Intake trunking
62 Nosewheel retraction radii
63 Nosewheel doors
64 Nosewheel fork
65 Forward-retracting nosewheel
66 Nosewheel strut
67 Forward fuselage members
68 Inboard-section wing leading edge
69 Three wing/fuselage attachment points
70 Y-section inner main spar
71 Inboard wing fence
72 Forward main spar
73 Undercarriage indicator spigot
74 Inner wing skinning
75 Split landing flap (inner section)
76 Split landing flap structure (outer section)

Above: This worm's eye view of a MiG-17PF, with its flaps deployed for landing, (note the afterburner nozzle and radar) shows well the design of the type's wing, with its virtually untapered planform and 45° sweep angle.

SPECIFICATION

MiG-17F 'Fresco-C'

Dimensions:

Wingspan: 31 ft 7 in (9.628 m)
Wing area: 243.27 sq ft (22.60 m²)
Length: 36 ft 11½ in (11.264 m)
Height: 12 ft 5½ in (3.80 m)
Wheel track: 12 ft 7½ in (3.849 m)
Wheel base: 3.368 m (11 ft ½ in)

Powerplant:

one Klimov VK-1F turbojet rated at 5,732 lb st (29.50 kN) dry and 7,451 lb st (33.14 kN) with afterburning

Weights:

Empty equipped: 8,664 lb (3930 kg)
Maximum take-off: 13,380 lb (6069 kg)

Fuel and load

Internal fuel: 2,579 lb (1170 kg)
External fuel: up to 1,444 lb (655 kg) in two 106- or 63-US gal (400- or 240-litre) drop tanks;
Maximum ordnance: 1,102 lb (500 kg)

Performance

Limiting Mach number: 1.03
Maximum level speed 'clean' at 9,845 ft (3000 m): 594 kt (684 mph; 1100 km/h)
Maximum level speed 'clean' at 32,810 ft (10000 m): 578 kt (666 mph; (1071 km/h)
Speed limit with drop tanks: 486 kt (559 mph; 900 km/h)
Ferry range: 1,091 nm (1,255 miles; 2020 km) with drop tanks
Combat radius: 378 nm (435 miles; 700 km) on a hi-lo-hi attack mission with two 551-lb (250-kg) bombs and two drop tanks
Maximum rate of climb at sea level: 12,795 ft (3900 m) per minute;
Service ceiling: 49,215 ft (15000 m) at dry thrust and 54,460 ft (16600 m) at afterburning thrust
Take-off run: 1,936 ft (590 m) at normal take-off weight
Landing run: 2,789 ft (850 m) at normal landing weight

Right: After testing in MiG-17 prototype SP-6, a modified radar set and missile launch boxes were fitted to all surviving MiG-17PFs, bringing them up to MiG-17PFU standard. Known to NATO as 'Fresco-E', this was the first production fighter in Europe to feature AAM armament, this comprising a quartet of K-5 (RS-2US) beam-riding missiles.

77 Centre wing fence
78 Outboard wing fence
79 Wing construction
80 Rearspar
81 Aileron construction
82 Starboard navigation light
83 Wingtip
84 Starboard pitot head
85 Outboard-section wing leading edge
86 Auxiliary-tank fin assembly
87 Triple-strut auxiliary-tank bracing
88 Mainwheel leg
89 Starboard mainwheel
90 Mainwheel door
91 Auxiliary tank (88 Impgal /400 litre capacity)
92 Mainwheel retraction rod
93 AA-1 'Alkali'-type beam-riding air-to-air missiles
94 Weapon pylons
95 Altimeter radio dipole (port outboard/ starboard inboard)

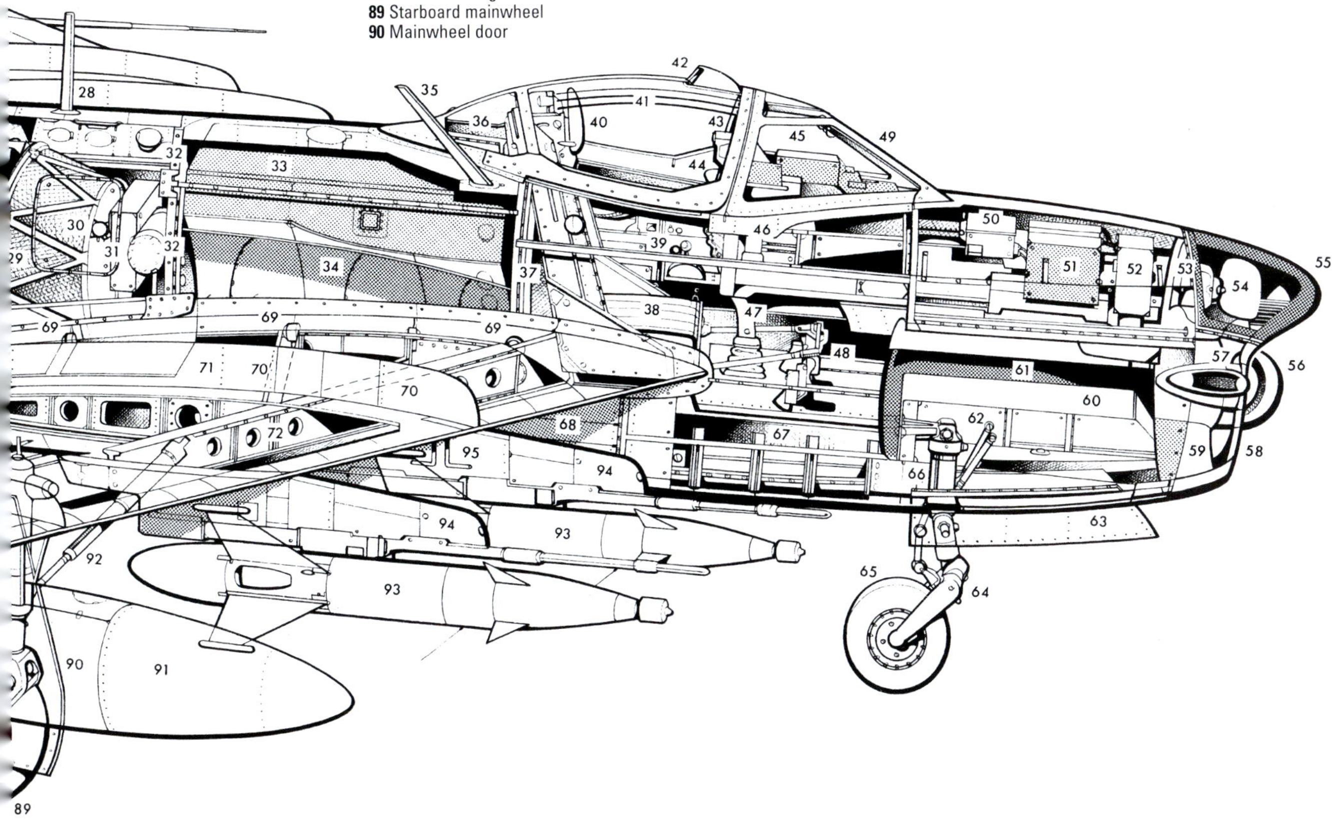

Mikoyan-Gurevich MiG-21 'Fishbed'

Finland operated the MiG-21 between 1963 and 1998, when the final aircraft were retired. Along the way, four variants – the MiG-21F-13, U, UM a BIS – were all operated. The picture above shows the very last MiG-21 to fly, MiG-21BIS MG-138 piloted by Captain Yrjö Rantamäki, preparing t at Rissala air base on 7 March 1998. Finland has now replaced its MiG-21s with Boeing F-18s.

MiG-21MF 'Fishbed-J'

Cutaway key
1 Pitot static boom
2 Pitch vanes
3 Yaw vanes
4 Conical three-position intake centre body
5 'Spin Scan' search-and-track radar antenna
6 Boundary layer slot
7 Engine air intake
8 'Spin Scan' radar
9 Lower boundary layer exit
10 IFF antennas
11 Nosewheel doors
12 Nosewheel leg and shock absorbers
13 Castoring nosewheel
14 Anti-shimmy damper
15 Avionics bay access
16 Attitude sensor
17 Nosewheel well
18 Spill door
19 Nosewheel retraction pivot
20 Bifurcated intake trunking
21 Avionics bay
22 Electronics equipment
23 Intake trunking
24 Upper boundary layer exit
25 Dynamic pressure probe for q-feel
26 Semi-elliptical armour glass windscreen
27 Gunsight mounting
28 Fixed quarterlight
29 Radar scope
30 Control column (with tailplane trim switch and two firing buttons)
31 Rudder pedals
32 Underfloor control runs
33 KM-1 two-position zero-level ejection seat
34 Port instrument console
35 Undercarriage handle
36 Seat harness
37 Canopy release/lock
38 Starboard wall switch pane
39 Rear-view mirror fairing
40 Starboard hinged canopy
41 Ejection seat headrest
42 Avonics bay
43 Control rods
44 Air-conditioning plant
45 Suction relief door
46 Intake trunking
47 Wingroot attachment fairing
48 Wing/fuselage spar-lug attachment points (four)
49 Fuselage ring frames
50 Intermediary frames
51 Main fuselage fuel tank
52 RSIU radio bay
53 Auxiliary intake
54 Leading-edge integral fuel tank
55 Starboard outer weapons pylon
56 Outboard wing construction
57 Starboard navigation light
58 Leading-edge suppressed aerial
59 Wing fence
60 Aileron control jack
61 Starboard aileron
62 Flap actuator fairing
63 Starboard blown flap SPS (*sduva pogranichnovo slova*)
64 Multi-spar wing structure
65 Main integral wing fuel tank
66 Undercarriage mounting/pivot point
67 Starboard mainwheel leg
68 Auxiliaries compartment
69 Fuselage fuel tanks Nos 2 and 3
70 Mainwheel well external fairing
71 Mainwheel (retracted)
72 Trunking contours
73 Control rods in dorsal spine
74 Compressor face
75 Oil tank
76 Avionics pack
77 Engine accessories
78 Tumanskii R-13 turbojet
79 Fuselage break/transport joint
80 Intake
81 Tail surface control linkage
82 Artificial feel unit
83 Tailplane jack
84 Hydraulic accumulator
85 Tailplane trim motor
86 Fin spar attachment plate
87 Rudder jack
88 Rudder control linkage
89 Fin structure
90 Leading-edge panel
91 Radio cable access
92 Magnetic detector
93 Fin mainspar
94 RSIU *(radio-stantsiya istrebitelnaya ultrakorotkykh vol'n* – very short-wave fighter radio) antenna plate
95 VHF/UHF aerials
96 IFF antennas
97 Formation light
98 Tail warning radar
99 Rear navigation light
100 Fuel vent
101 Rudder construction
102 Rudder hinge
103 Braking parachute hinged bullet fairing
104 Braking parachute stowage
105 Tailpipe (variable convergent nozzle)
106 Afterburner installation
107 Afterburner bay cooling intake
108 Tail plane linkage fairing
109 Nozzle actuating cylinders
110 Tailplane torque tube
111 All-moving tailplane
112 Anti flutter weight

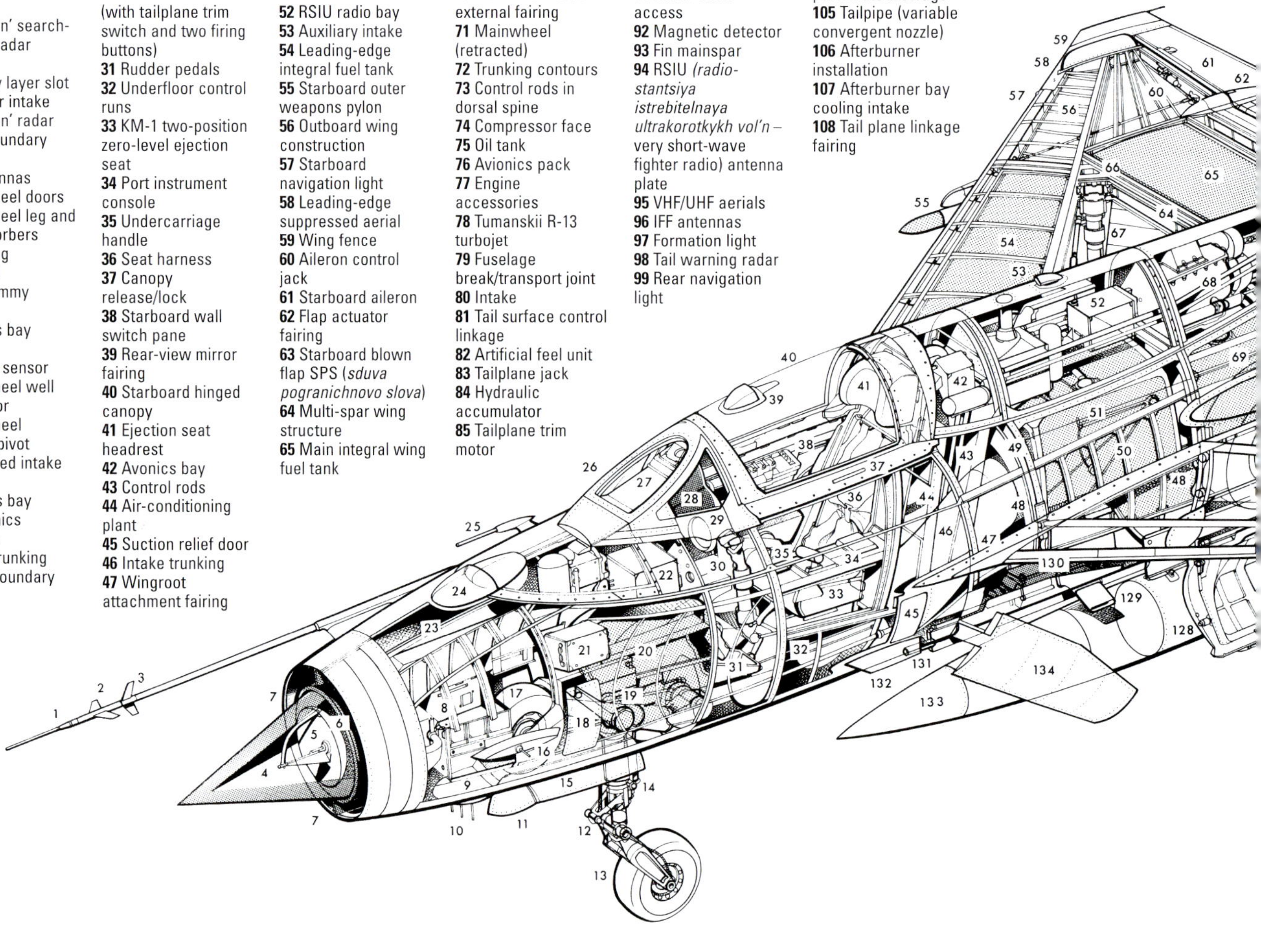

A handful of MiG-21s was delivered to the West during the Cold War by means of defections from Communist countries or by other more covert methods. This example, a Soviet-built MiG-21, is seen flying over the Groom Lake/Area 51 complex and was part of the secret 4477th Test & Evaluation Squadron or 'Red Eagles'. Aircraft like these were flown, in Operation Have Doughnut, against the latest Western types to evaluate their performance, and so enable planners to formulate tactics. The obvious intense secrecy that surrounded these test flights has helped to shape the legend of Area 51 and it is entirely plausible that the supposed 'alien' aircraft are simply foreign aircraft.

SPECIFICATION

MiG-21MF 'Fishbed-J'

Dimensions

Length with probe: 51 ft 8½ in (15.76 m)
Length excluding probe: 40 ft 4 in (12.29 m)
Height: 13 ft 6 in (4.13 m)
Span: 23 ft 6 in (7.15 m)
Wing area: 247.5 sq ft (23 m²)
Wing aspect ratio: 2.23
Wheel track: 9 ft 1¾ in (2.79 m)
Wheel base: 15 ft 5½ in (4.71 m)

Powerplant

One MNPK 'Soyuz' (Tumanskii/Gavrilov) R-13-300 turbojet rated at 8,972 lb st (39.92 kN) dry and 14,037 lb st (63.66 kN) with afterburning

Weights

Empty: 11,795 lb (5350 kg)
Normal take-off with four AAMs and three 129-US gal (490-litre) drop tanks: 17,967 lb (8150 kg)
Maximum take-off: 20,723 lb (9400 kg)

Fuel and load

Internal fuel: 687 US gal (2600 litres)
External fuel: up to 387 US gal (1470 litres) in three drop tanks
Maximum ordnance: 4,409 lb (2000 kg)

Performance

Maximum rate of climb at sea level: 23,622 ft (7200 m) per minute
Service ceiling: 59,711 ft (18200 m)
Take off run: 2,625 ft (800 m)

Range

Ferry range: 971 nm (1,118 miles; 1800 km) with three drop tanks
Combat radius: 200 nm (230 miles; 370 km) on a hi-lo-hi attack mission with four 551-lb (250-kg) bombs, or 400 nm (460 miles;740 km) on a hi-lo-hi mission with two 551-lb (250-kg) bombs and drop tanks

Armament

Standard gun is the GSh-23L which has a calibre of 23 mm and can fire AP or HE ammunition, with 420 rounds being carried. The only guided missiles normally carried are for air-to-air use. The MF is capable of firing the K-13A (AA-2 'Atoll') and the AA-2-2 Advanced 'Atoll'. As with other MiG-21s, up to eight R-60 (AA-8 'Aphid') infra-red missiles can also be carried. There is provision for various FABs (free-fall general-purpose bombs), up to 1,102 lb (500 kg) in weight. A wide range of fragmentation, chemical, cluster bombs and rocket-boosted penetrators for use against concrete can be carried, as well as 57-mm or 240-mm calibre rockets.

113 Intake
114 Afterburner mounting
115 Fixed tailplane root fairing
116 Longitudinal lap joint
117 External duct (nozzle hydraulics)
118 Ventral fin
119 Engine guide rail
120 ATO assembly canted nozzle
121 ATO assembly thrust plate forks (rear-mounting)
122 ATO assembly pack

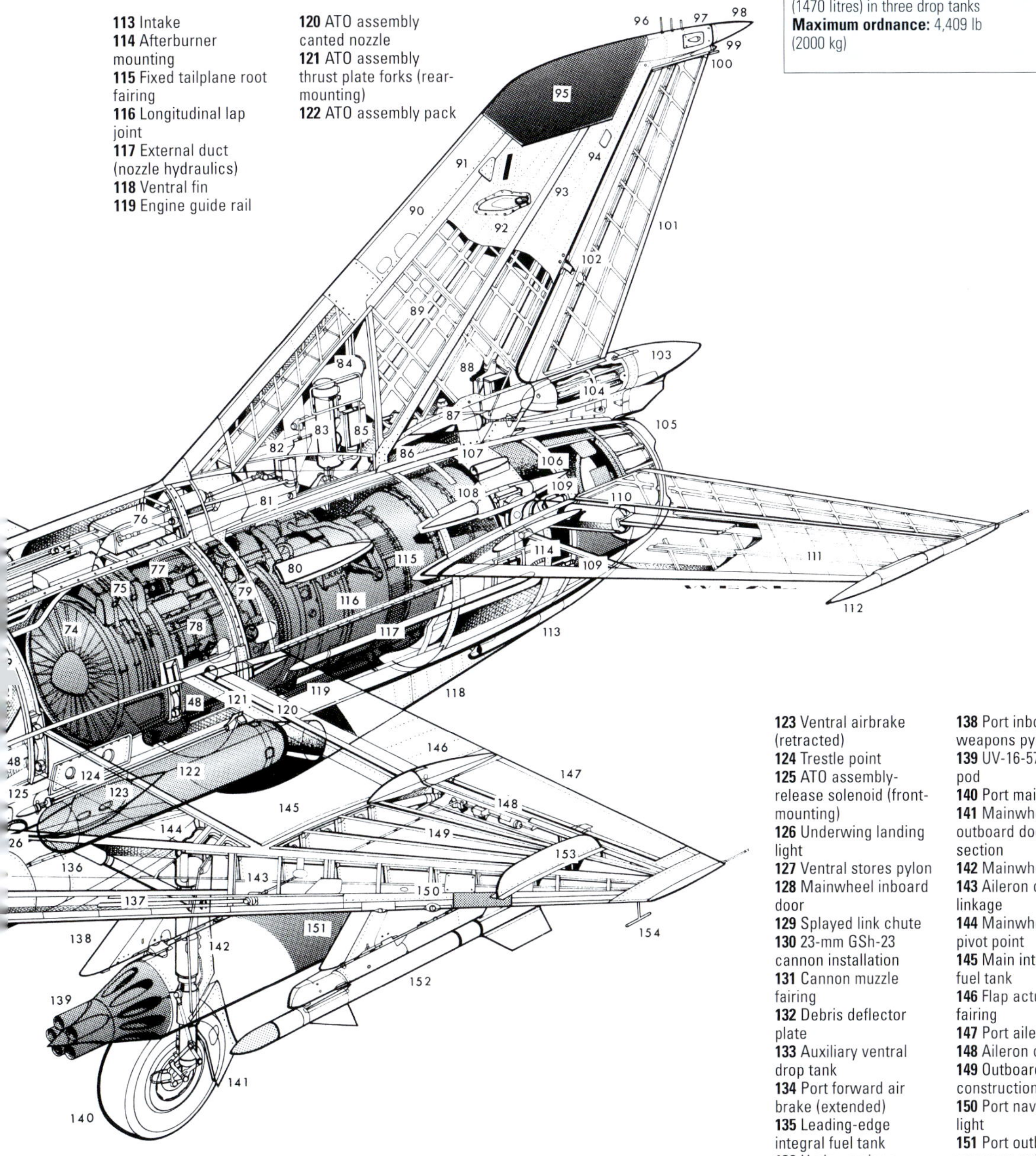

123 Ventral airbrake (retracted)
124 Trestle point
125 ATO assembly-release solenoid (front-mounting)
126 Underwing landing light
127 Ventral stores pylon
128 Mainwheel inboard door
129 Splayed link chute
130 23-mm GSh-23 cannon installation
131 Cannon muzzle fairing
132 Debris deflector plate
133 Auxiliary ventral drop tank
134 Port forward air brake (extended)
135 Leading-edge integral fuel tank
136 Undercarriage retraction strut
137 Aileron control rods in leading edge
138 Port inboard weapons pylon
139 UV-16-57 rocket pod
140 Port mainwheel
141 Mainwheel outboard door section
142 Mainwheel leg
143 Aileron control linkage
144 Mainwheel leg pivot point
145 Main integral wing fuel tank
146 Flap actuator fairing
147 Port aileron
148 Aileron control jack
149 Outboard wing construction
150 Port navigation light
151 Port outboard weapons pylon
152 'Advanced Atoll' IR-homing AAM
153 Wing fence
154 Radio altimeter antenna

Mikoyan-Gurevich MiG-25 'Foxbat'

Developed to meet a bomber threat never materialised, Mikoyan-Gurevi MiG-25 'Foxbat' may have lacked technology of its NATO adversaries, more than made up for this wit incredible performance. In the MiG the Soviet Union and now, am others, Russia, Azerbaijan, Bela India, Iraq, Libya and Syria, have excellent reconnaissance platform bomber destro

MiG-25 'Foxbat-A'

Cutaway key

1 Ventral airbrake
2 Starboard tailplane (aluminium alloy trailing edge)
3 Steel tailplane spar
4 Titanium leading edge
5 Tail bumper
6 Fully-variable engine exhaust nozzle
7 Exhaust nozzle actuator
8 Starboard rudder
9 Static dischargers
10 Sirena 3 tail warning radar and ECM transmitter
11 Transponder aerial
12 Twin brake parachute housing
13 Port engine exhaust nozzle
14 Port rudder
15 Static dischargers
16 VHF aerial
17 HF leading-edge aerial
18 Port tailfin (steel primary structure)
19 Rudder actuator
20 Titanium rear fuselage skins
21 Dorsal spine fairing
22 Fireproof bulkhead between engine bays
23 Engine afterburner duct
24 Cooling air intake
25 Tailplane hydraulic actuator
26 Starboard ventral fin
27 VHF and ECM aerial housing
28 Aileron actuator
29 Starboard aileron
30 Static discharger
31 All-steel wing construction
32 Wingtip fairing
33 Sirena 3 radar warning receiver and ECM transmitter
34 Continuous-wave target-illuminating radar
35 AA-6 'Acrid' semi-active radar guided-air-to air missile
36 Missile-launching rail
37 Outboard missile pylon
38 Pylon attachments
39 Wing titanium leading edge
40 Inboard pylon
41 Wing fence
42 Engine access panels
43 Engine accessory gearbox
44 Tumanskii R-31 single-shaft afterburning turbojet engine
45 Port flap
46 Aileron hydraulic actuator
47 Port aileron
48 Fixed portion of trailing edge
49 Sirena 3 radar warning receiver and ECM transmitter
50 Continuous-wave target-illuminating radar
51 Titanium leading edge
52 Port wing fences
53 AA-6 'Acrid' semi-active radar-guided air-to-air missile
54 Infra-red-guided AA-6 'Acrid' missile
55 Stainless steel wing skins
56 Intake flank fuel tanks
57 Controls and systems ducting
58 Main fuel tanks (welded steel integral construction), total system capacity 31,575 lb (14322 kg), nitrogen-pressurised
59 Intake bleed air ducts engine bay cooling
60 Engine compressor face
61 Wing spar attachments
62 Main undercarriage leg strut
63 Starboard mainwheel
64 Mainwheel doors
65 Mainwheel stowed position
66 Starboard infra-red guided AA-6 'Acrid' missile
67 Retractable landing/taxiing lamp
68 Intake duct control vanes
69 Steel fuselage primary structure
70 Intake bleed air outlet ducts
71 UHF communications aerials
72 Variable-intake ramp doors
73 Ramp jacks
74 Intake water/methanol injection duct
75 Electric intake tip actuator
76 Variable lower intake lip
77 Nose wheel door/mudguard
78 Twin nose wheels
79 Nose wheel leg doors
80 Starboard navigation light
81 Curved intake inboard sidewall
82 Rear avionics bay, communications and ECM equipment
83 Cockpit canopy cover, hinges to starboard
84 Pilot's ejection seat
85 Cockpit rear pressure bulkhead
86 UHF communications aerial
87 Radar altimeter
88 Pilot's side console panel
89 Control column
90 Instrument panel shroud
91 Stand-by visual sighting system for infra-red missiles
92 Windscreen panels
93 'Odd Rods' IFF aerials
94 Pitot tube
95 Forward avionics compartment, radar and navigation equipment
96 'Fox Fire' fire control radar system
97 Angle-of-attack probe
98 Scanner tracking mechanism
99 Radar scanner dish, 2ft 9½-in (85-cm) diameter
100 Radome
101 'Swift Rod' ILS antenna
102 Pitot tube

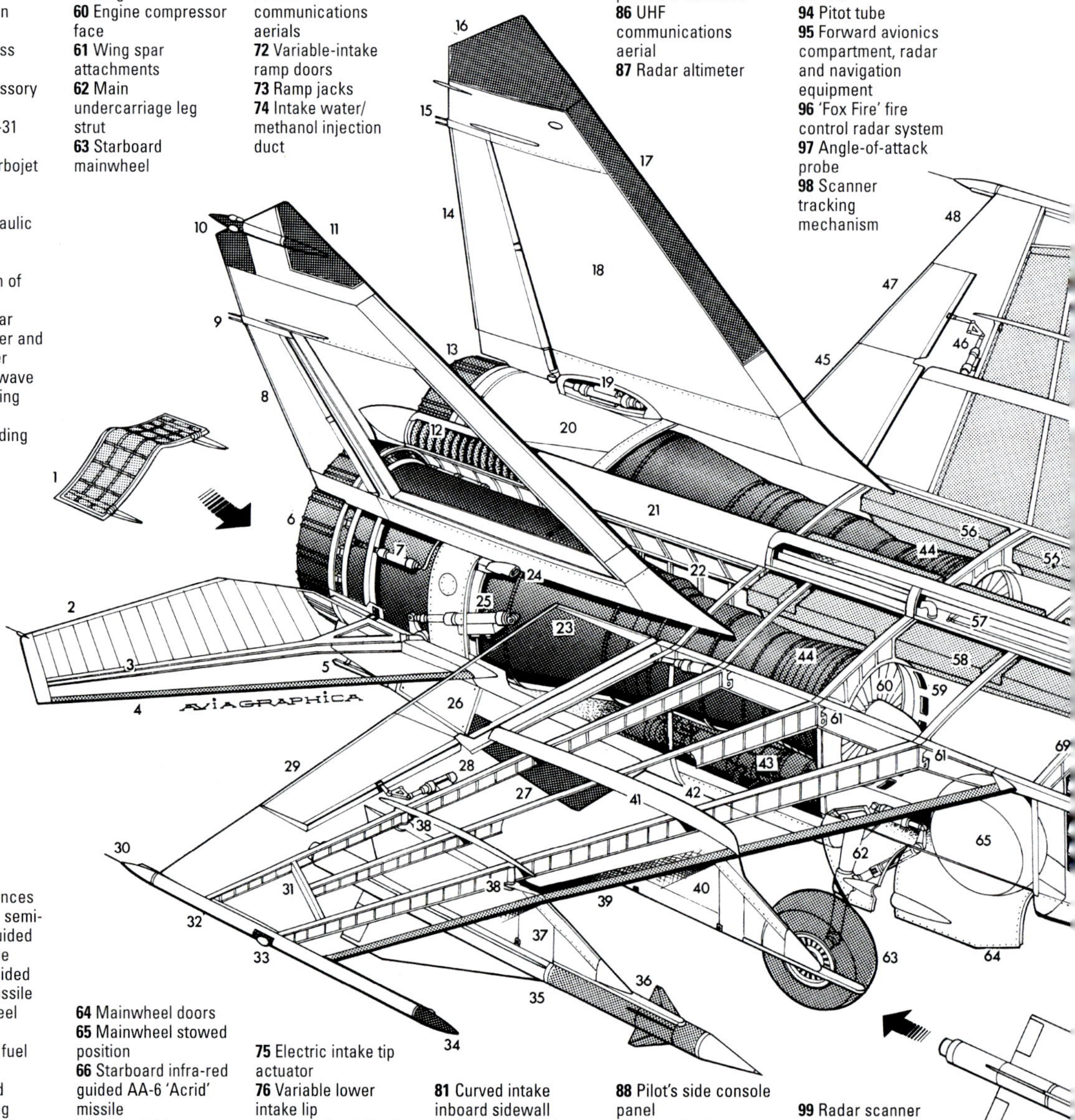

A refuelling capability was considered an important asset for the MiG-25 and a single MiG-25PD (coded 'Blue 45') was modified as an inflight-refuelling receiver testbed. The aircraft was fitted with an inverted-L-shaped probe immediately in front of the windscreen. Ultimately, the PVO and VVS remained pitifully short of tankers, and priority was given to equipping newer aircraft types with probes, so nothing came of the proposal to retro-fit in-service MiG-25PDs for inflight refuelling.

SPECIFICATION

MiG-25PDS 'Foxbat-E'

Dimensions

Length: 78 ft 1¾ in (23.82 m)
Length for aircraft modified with IFR capability: 78 ft 11½ in (24.07 m)
Height: 20 ft ¼ in (6.10 m)
Wingspan: 45 ft 11¾ in (14.02 m)
Wing aspect ratio: 3.2
Wing area: 660.93 sq ft (61.40 m²)
Wheel track: 12 ft 7½ in (3.85 m)
Wheel base: 16 ft 10¼ in (5.14 m)

Powerplant

Two MNPK 'Soyuz' (Tumanskii) R-15BD-300 turbojets each rated at 24,691 lb st (109.83 kN) with afterburning

Weights

Normal take-off weight with four R-40s (AA-6s) and 100 per cent internal fuel: 80,952 lb (36720 kg)

Fuel and loads

Internal fuel: 32,121 lb (14570 kg)
External fuel in an underbelly tank: 9,634 lb (4370 kg)
Maximum ordnance: 8,818 lb (4000 kg)

Performance

Maximum level speed 'clean' at 42,650 ft (13000 m): Mach 2.8 or 1,619 kt (1,864 mph; 3000 km/h)
Maximum level speed 'clean' at sea level: Mach 2.8 or 647 kt (745 mph; 1200 km/h)
Climb to 65,615 ft (20000 m): 8 minutes 54 seconds
Service ceiling: 67,915 ft (20700 m)
Take-off run at normal take-off weight: 4,101 ft (1250 m)
Landing run at normal landing weight with brake chutes: 2,624 ft (800 m)
***g* limits:** + 4.5 supersonic

Range

With internal fuel at subsonic speed: 933 nm (1,075 miles; 1730 km)
With internal fuel at supersonic speed: 675 nm (776 miles;1730 km)
Endurance: 2 hours 5 minutes

Armament

Standard intercept fit is two or four R-40 (AA-6 'Acrid') missiles. MiG-25PDS aircraft are armed with two R-40s and four R-60 (AA-8 'Aphid') AAMs

103 MiG-25U 'Foxbat-C' two-seat operational training variant
104 Student pilot's cockpit enclosure
105 Instructor's cockpit
106 MiG-25R 'Foxbat-B' reconnaissance variant
107 Reconnaissance cameras, one vertical and four oblique
108 Sideways-looking airborne radar (SLAR) aperture
109 Ground mapping and Doppler radar antenna
110 'Jay-Bird' forward-looking radar

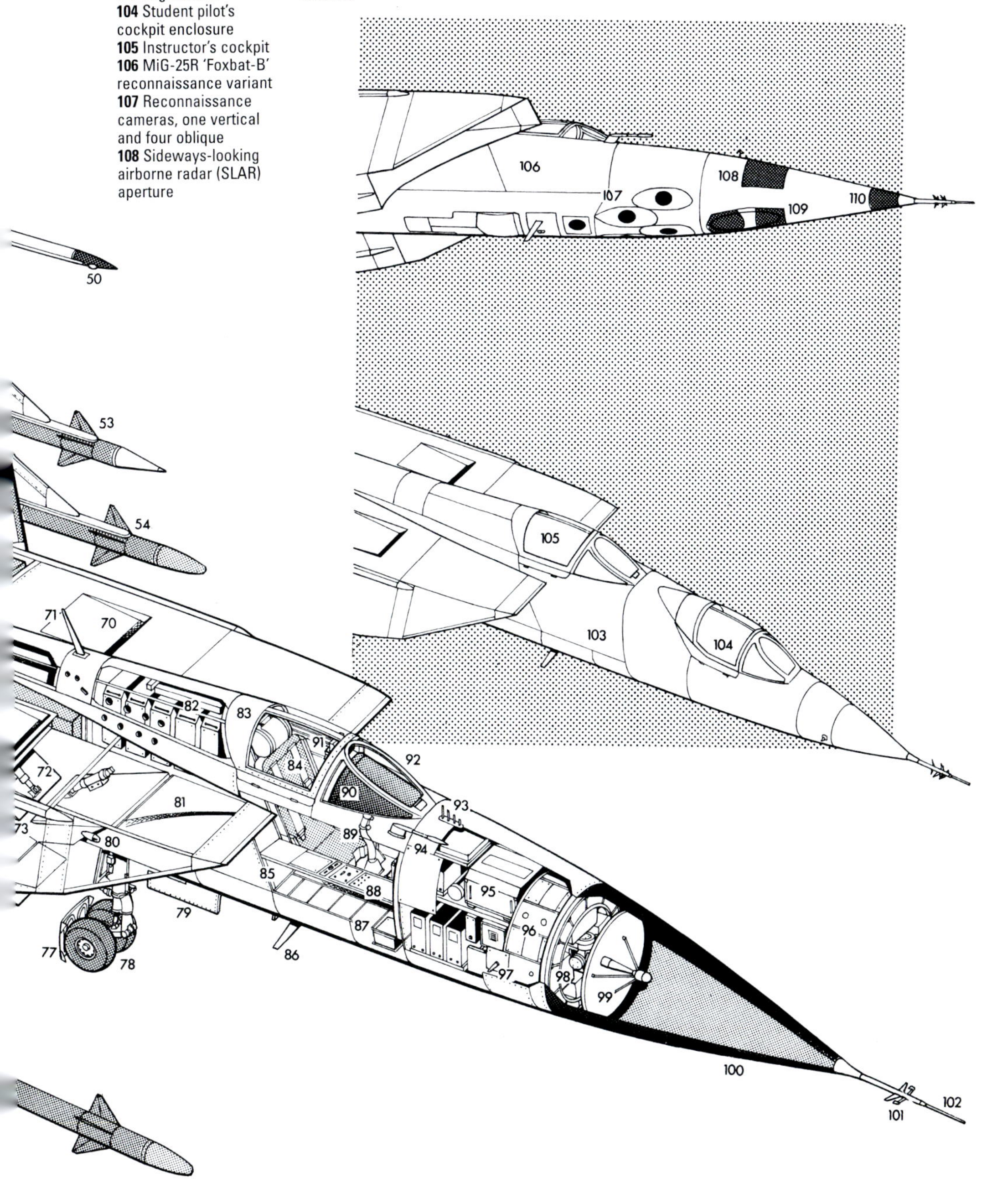

Mikoyan-Gurevich MiG-29 'Fulcrum'

The first squadron of Poland's No. 1 PLM (Pulk Lotnictwa Mysliwskiego/Air Fighter Regiment) operates some 19 'Fulcrum-As', nine of which were obtained from the Czech Republic. Four 'Fulcrum UBs' are also used and one of these was also obtained from the Czechs. Poland still hopes to replace its remaining MiG-21Ms with new 'Fulcrums'.

MiG-29 'Fulcrum-A'

Cutaway Key
1 Pitot head
2 Vortex generating nose strake
3 Glass-fibre radome
4 Pulse-Doppler radar scanner
5 Scanner tracking mechanism
6 N-019 (NATO: 'Slot Back') radar equipment module
7 Angle-of-attack transmitter
8 ILS aerial fairing
9 SRO-2 (NATO: 'Odd Rods') IFF aerial
10 UHF antenna
11 Forward avionics equipment bay
12 Infra-red search and track sensor and laser ranger
13 Dynamic pressure probe
14 Frameless windscreen panel
15 Pilot's head-up-display
16 Instrument panel shroud
17 Rudder pedals and control column
18 Fuselage blended chine fairing
19 Cannon muzzle aperture
20 Cannon barrel
21 Slide mounted engine throttle levers
22 Canopy latch
23 K-36D 'zero-zero' ejection seat
24 Upward-hingeing cockpit canopy cover
25 Electrical distribution centre
26 Cockpit rear pressure bulkhead
27 Cannon bay venting air louvres
28 Nosewheel retraction jack
29 Levered suspension nosewheel leg
30 Twin nosewheels, aft retracting
31 Mudguard
32 ECM aerial panels
33 Cartridge case and link collector box
34 Ammunition magazine
35 Centre avionics equipment bay
36 Canopy hinge point
37 Canopy hydraulic jack
38 HF aerial
39 Mechanical control rods
40 Rear avionics equipment bay
41 Air intake louvres/ blow-in doors
42 Variable area intake ramp doors
43 Ramp hydraulic actuator
44 Port engine air intake
45 Weapons interlock access
46 Landing lamp
47 Mainwheel door
48 Forward fuselage integral fuel tank
49 Port main undercarriage wheel bay
50 Flight control system hydraulic equipment module
51 ADF aerial
52 Starboard main undercarriage wheel bay
53 Chaff/flare cartridge housing
54 Starboard wing integral fuel tank
55 Starboard wing missile carriage
56 Leading edge manoeuvre flap
57 Starboard navigation light
58 Radar warning antenna
59 Starboard aileron
60 Plain flap
61 Flap hydraulic jack
62 Centre fuselage integral fuel tank
63 Engine compressor face
64 Cooling air scoop
65 Top-mounted engine accessory equipment gear boxes
66 Central gas turbine starter/APU
67 Engine bay/tailplane attachment machined main frames
68 Airbrake hydraulic jack
69 RD-33D afterburning turbofan engine
70 Fin rib construction
71 Carbon fibre fin skin panelling
72 Fin tip VHF aerial fairing
73 Radar warning antenna
74 'Swift Rod' ILS aerial
75 Starboard rudder
76 Rudder hydraulic actuator
77 Tailplane hydraulic actuator
78 Starboard all-moving tailplane
79 Airbrake, upper and lower split surfaces
80 Brake parachute housing
81 Variable area afterburner nozzle
82 Port tailfin
83 Tail navigation light
84 Sirena-3 ECM aerial fairing
85 Static discharger
86 Port rudder composite construction
87 Port all-moving tailplane
88 Static dischargers
89 Carbon fibre trailing edge skin panelling
90 Tailplane spar box construction
91 Tailplane pivot point
92 Fuselage side-body fairing construction
93 Artificial feel system pitot heads and control valves
94 Port plain flap composite construction
95 Main undercarriage hydraulic retraction jack
96 Port chaff/flare cartridge
97 Main undercarriage leg pivot fixing
98 Pylon attachment hardpoints
99 Flap hydraulic jack
100 Port wing integral fuel tank
101 Aileron hydraulic actuator
102 Port aileron composite construction

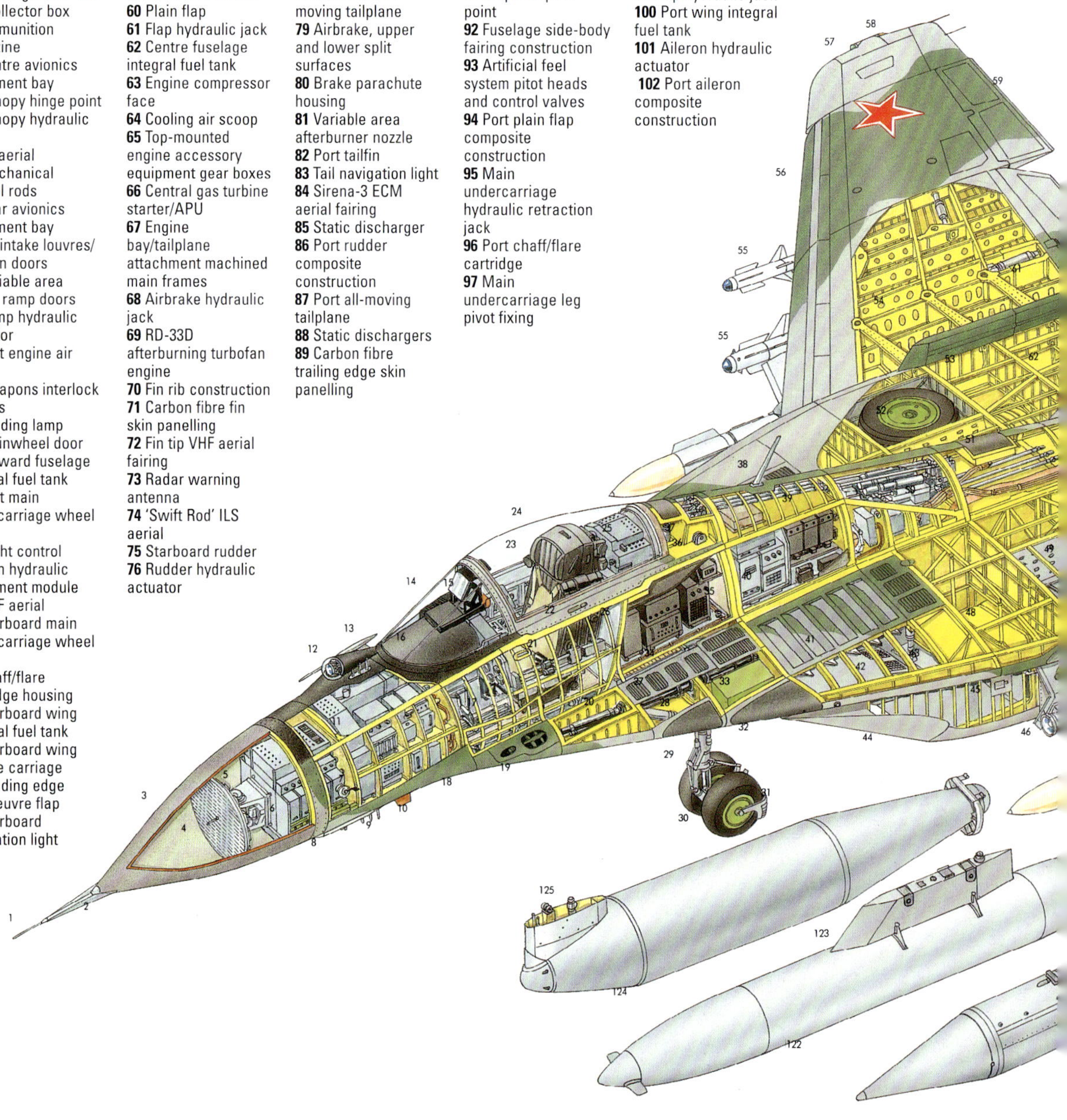

Despite being conceived in the early 1970s, the MiG-29 remains in service around the world today. The MiG-MAPO design and production facility has also decided that the MiG-29 will remain its basic export product until 2005, by which time a new fighter will have been designed. It is a tribute to the 'Fulcrum' and its designers that it is still in service and, with upgrades, should remain there for many years.

103 Carbon fibre skin panelling
104 Static dischargers
107 Downward identification light and remote compass housing
108 Outer wing panel rib construction
109 Port leading edge manoeuvre flap
110 Port wing missile pylons
111 Leading edge flap hydraulic jacks
112 Port mainwheel
113 Main undercarriage leg strut
114 Three-spar wing torsion box construction
115 Spar root attachment joints
116 Undercarriage bay pressure refuelling connection
117 AA-10 'Alamo' long-range air-to-air missile
118 AA-11 'Archer' intermediate-range air-to-air missile
119 AA-8 'Aphid' short-range air-to-air missile
120 57-mm rocket pack
121 Cluster bomb
122 Wing-mounted external fuel tank
123 Tank pylon
124 Centre fuselage 'tunnel' fuel tank
125 Tank pylon attachment

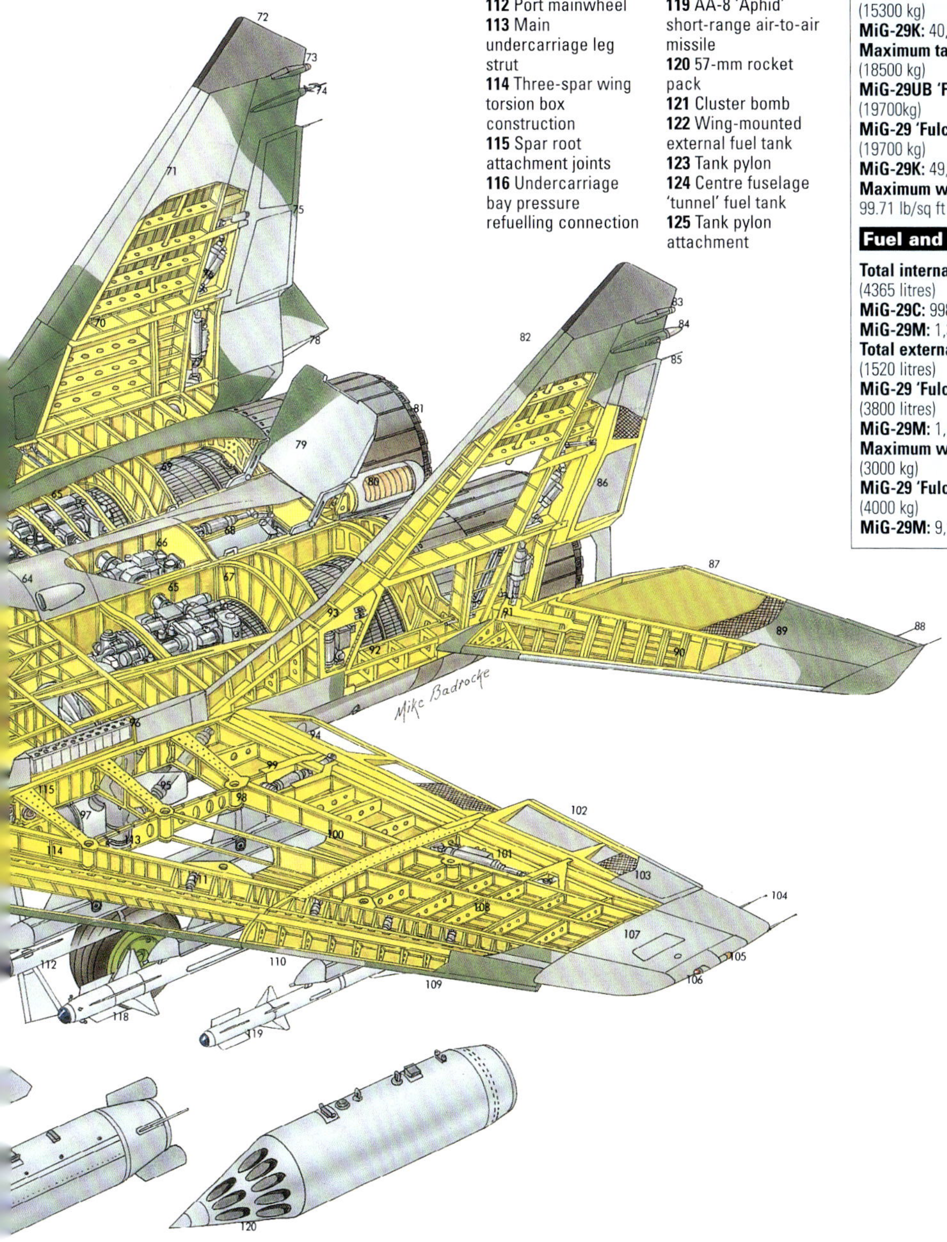

SPECIFICATION

MiG-29 'Fulcrum-A'
(unless otherwise noted)

Dimensions

Fuselage length (including probe): 56 ft 10 in (17.32 m)
Wing span: span 37 ft 3¼ in (11.36m)
Wing aspect ratio: 3.4
Wing area: 409.04 sq ft (38 m²)
Tailplane span: 25 ft 6¼ in (7.78m)
Wheel track: 10 ft 2 in (3.10m)
Wheel base: 12 ft 0.5 in (3.67m)
Height: 15 ft 6.⅕ in (4.73m)

Powerplant

Two Klimov/Leningrad (Isotov/Sarkisov) RD-33 augmented turbofans each rated at 11,111 lb st (49.42 kN) dry and 18,298 lb st (81.39 kN) with afterburning.
MiG-29M: Two Klimov/Leningrad (Isotov/Sarkisov) RD-33K turbofans rated at 12,125 lb st (53.95 kN) dry. 19,400 lb st (86.33 kN) with afterburning, and with an 'emergency regime' rating of 20,725 lb st (92.22 kN)

Weights

Empty operating: 24,030 lb (10900 kg)
Normal take-off interceptor: 33,600 lb (15240 kg)
MiG-29UB 'Fulcrum-B': 33,730 lb (15300 kg)
MiG-29 'Fulcrum-C': 33,730 lb (15300 kg)
MiG-29K: 40,705 lb (18480 kg)
Maximum take-off: 40,785 lb (18500 kg)
MiG-29UB 'Fulcrum-B': 43,430 lb (19700kg)
MiG-29 'Fulcrum-C': 42,680 lb (19700 kg)
MiG-29K: 49,340 lb (22400 kg)
Maximum wing loading: 99.71 lb/sq ft (486.8 kg/m²)

Fuel and load

Total internal fuel: 960 Imp gal (4365 litres)
MiG-29C: 998 Imp gal (4540 litres)
MiG-29M: 1,375 Imp gal (6250 litres)
Total external fuel: 334 Imp gal (1520 litres)
MiG-29 'Fulcrum-C': 836 Imp gal (3800 litres)
MiG-29M: 1,100 Imp gal (5000 litres)
Maximum weapon load: 6,614 lb (3000 kg)
MiG-29 'Fulcrum-C': 8,818 lb (4000 kg)
MiG-29M: 9,921lb (4500 kg)

Performance

Maximum level speed 'clean' at 36,090 ft (11000 m): 1,320 kt (1,520 mph; 2445 km/h)
MiG-29K: 1,242 kt (1,430 mph; 2300 km/h)
Maximum level speed at low level: 810 kt (932 mph; 1500 km/h)
Limiting Mach numbers: 2.3 at 36,090 ft (11000 m); 1.22 at sea level
Take-off speed: 119 kt (137 mph; 220 km/h)
Take-off run: 820 ft (250 m)
Approach speed: 140 kt (162 mph; 260 km/h)
Landing speed: 127 kt (146 mph; 235 km/h)
Landing run with brake chute: 1,970-2,300 ft (600-700 m)
Acceleration at 3,280 ft (1000 m): 325-595 kt (373-683 mph; 600-1100 km/h) in 13.5 s
595-700 kt (683-805 mph; 1100-1300 km/h) in 8.7 s
***G*-Limits:** +9 below Mach 0.85 and +7 above Mach 0.85
Service ceiling: 55,780 ft (17000 m)

Range

With maximum internal fuel: 810 nm (932 miles; 1500 km)
MiG-29M 1,080 nm (1,243 miles; 2000 km)
Ferry range with three external tanks: 1,134 nm (1,305 miles; 2100 km)
MiG-29M 1,728 nm (1,988 miles; 3200 km)

Armaments

Maximum weapon load: 4,410 lb (2000 kg)
MiG-29 'Fulcrum-C': 6,614 lb (3000 kg) of ordnance on six underwing hardpoints, with provision for two 253-Imp gal (1150-litre) drop tanks underwing and an optional centreline hardpoint for a 330 Imp gal (1500 litre) fuel tank
Cannon: one 30-mm GSh-301 cannon in port wingroot leading edge with 150 rounds
Air-to Air missiles: R-60 (AA-8 'Aphid'), R-27 (AA-10 'Alamo'), R-73 (AA-11 'Archer')
Ground attack weapons: FAB-250 and -500 bombs, KMGU-2 submunitions dispenser, ZB-500 napalm tank, 20 round 80-mm rocket pods, 130-mm and 240-mm rockets. One 30-kT RN-40 nuclear bomb on port inboard pylon

North American F-100 Super Sabre

Although it suffered from more than its fair share of problems, the F-100 represented the absolute cutting-edge of US fast-jet technology when the F-100A entered operational service in 1954. As a combat aircraft the F-100A lacked capability, however, and was soon replaced by the more capable -100C. These F-100Cs demonstrate the original straight-wing trailing edge, which was 'cranked' on the F-100D and F thanks to their increased-area flaps.

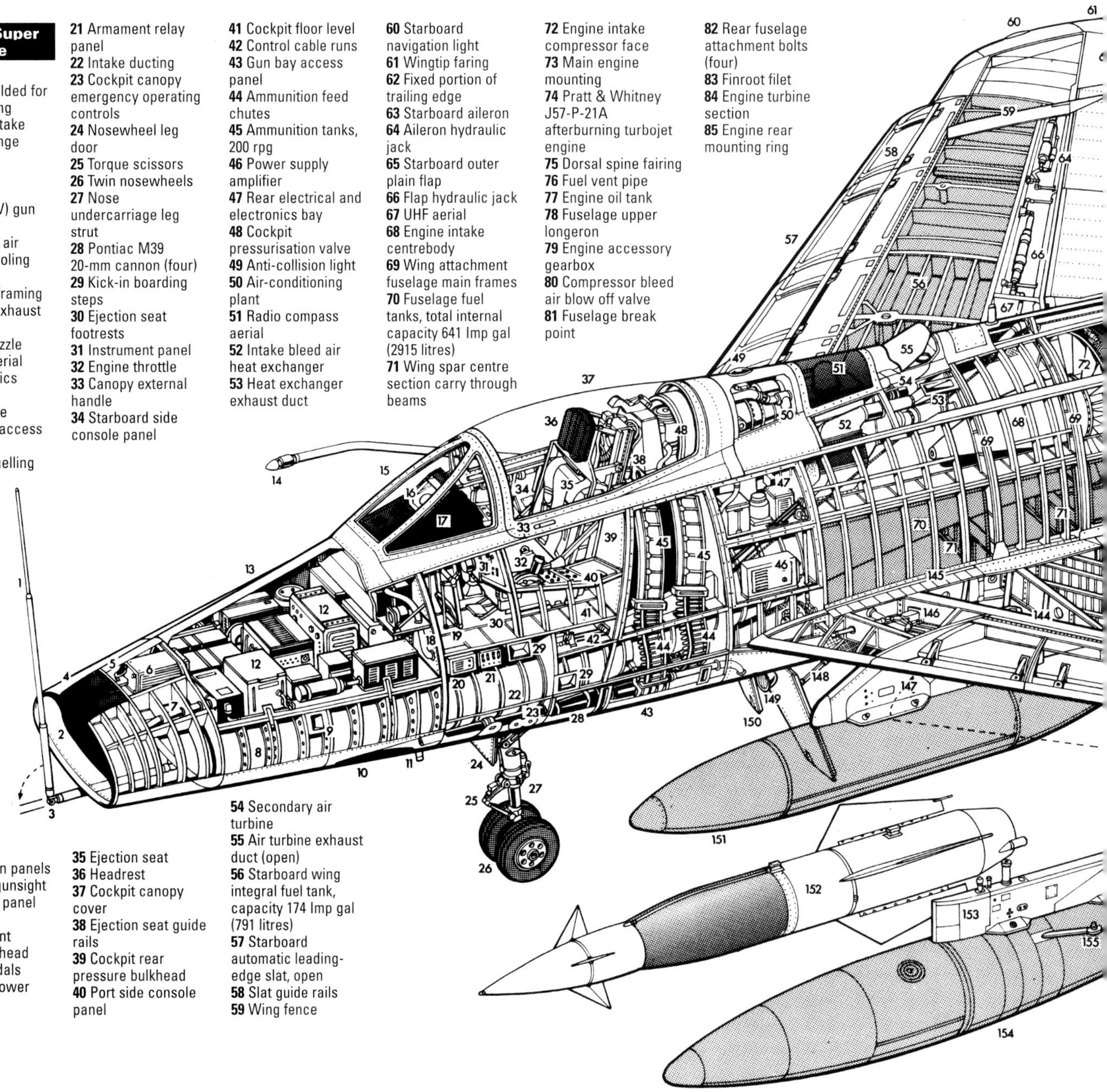

F-100D Super Sabre

Cutaway key

1 Pitot tube, folded for ground handling
2 Engine air intake
3 Pitot tube hinge point
4 Radome
5 IFF aerial
6 AN/APR-25(V) gun tracking radar
7 Intake bleed air electronics cooling duct
8 Intake duct framing
9 Cooling air exhaust duct
10 Cannon muzzle port **11** UHF aerial
12 Nose avionics compartment
13 Hinged nose compartment access door
14 Inflight refuelling probe
15 Windscreen panels
16 A-4 radar gunsight
17 Instrument panel shroud
18 Cockpit front pressure bulkhead
19 Rudder pedals
20 Gunsight power supply
21 Armament relay panel
22 Intake ducting
23 Cockpit canopy emergency operating controls
24 Nosewheel leg door
25 Torque scissors
26 Twin nosewheels
27 Nose undercarriage leg strut
28 Pontiac M39 20-mm cannon (four)
29 Kick-in boarding steps
30 Ejection seat footrests
31 Instrument panel
32 Engine throttle
33 Canopy external handle
34 Starboard side console panel
35 Ejection seat
36 Headrest
37 Cockpit canopy cover
38 Ejection seat guide rails
39 Cockpit rear pressure bulkhead
40 Port side console panel
41 Cockpit floor level
42 Control cable runs
43 Gun bay access panel
44 Ammunition feed chutes
45 Ammunition tanks, 200 rpg
46 Power supply amplifier
47 Rear electrical and electronics bay
48 Cockpit pressurisation valve
49 Anti-collision light
50 Air-conditioning plant
51 Radio compass aerial
52 Intake bleed air heat exchanger
53 Heat exchanger exhaust duct
54 Secondary air turbine
55 Air turbine exhaust duct (open)
56 Starboard wing integral fuel tank, capacity 174 Imp gal (791 litres)
57 Starboard automatic leading-edge slat, open
58 Slat guide rails
59 Wing fence
60 Starboard navigation light
61 Wingtip faring
62 Fixed portion of trailing edge
63 Starboard aileron
64 Aileron hydraulic jack
65 Starboard outer plain flap
66 Flap hydraulic jack
67 UHF aerial
68 Engine intake centrebody
69 Wing attachment fuselage main frames
70 Fuselage fuel tanks, total internal capacity 641 Imp gal (2915 litres)
71 Wing spar centre section carry through beams
72 Engine intake compressor face
73 Main engine mounting
74 Pratt & Whitney J57-P-21A afterburning turbojet engine
75 Dorsal spine fairing
76 Fuel vent pipe
77 Engine oil tank
78 Fuselage upper longeron
79 Engine accessory gearbox
80 Compressor bleed air blow off valve
81 Fuselage break point
82 Rear fuselage attachment bolts (four)
83 Finroot filet
84 Engine turbine section
85 Engine rear mounting ring

*As the F-100's front-line service career began drawing to a close, airframes became available for conversion as target drones. The first **YQF-100** prototype was converted and flown by Sperry Flight Systems in 1979 and was followed by a second **YQF-100**, three **QF-100s** for the USAF, three for the US Army and a single **QF-100F**. A major programme then got underway, with Tracor producing a further 72 **QF-100Ds** and Tracor/Flight Systems another 209 **QF-100D/F** drones. A **QF-100D** is illustrated.*

SPECIFICATION

F-100D Super Sabre

Dimensions

Length: 49 ft 6 in (15.09 m)
Height: 16 ft 2⅔ in (4.95 m)
Wingspan: 38 ft 9 in (11.81 m)
Wing area: 385.20 sq ft (35.79 m^2)

Powerplant

One Pratt & Whitney J57-P-21A turbojet rated at 11,700 lb st (52.02 kN) dry and 16,950 lb st (75.40 kN) with afterburning

Weights

Empty: 21,000 lb (9525 kg)
Loaded: 29,762 lb (13500 kg)
Combat: 30,061 lb (13633 kg)
Maximum take-off: 38,048 lb (17256 kg)

Fuel and load

Internal fuel: 1,189 US gal (4500 litres)
External fuel: up to 1,070 US gal (4050 litres)
Drop tanks: normally flown with two 450-US gal (1703-litre) tanks, although 200-US gal (757-litre), 275-US gal (1041-litre) and 335-US gal (1268-litre) tanks were also available
Maximum external load: 7,500 lb (3402 kg)

Performance

Maximum speed at 35,000 ft (10670 m): 864 mph (1390 km/h)
Rate of climb: 16,000 ft (4875 m) per minute
Climb to 35,000 ft (10670 m) at combat weight with maximum power: 3 minutes 30 seconds

Range

Combat: 1,500 miles (2415 km)
Ferry: 1,973 miles (3176 km)

Armament

Four 20-mm Pontiac M39E 20-mm cannon with 200 rounds per gun, plus provision for tactical nuclear weapons and a wide range of conventional stores including Mk 80 series and M117 bombs, cluster munitions, practice bomb dispensers, rocket pods, fire bombs and napalm tanks, AGM-12A/B/C ASMs and AIM-9B AAMs

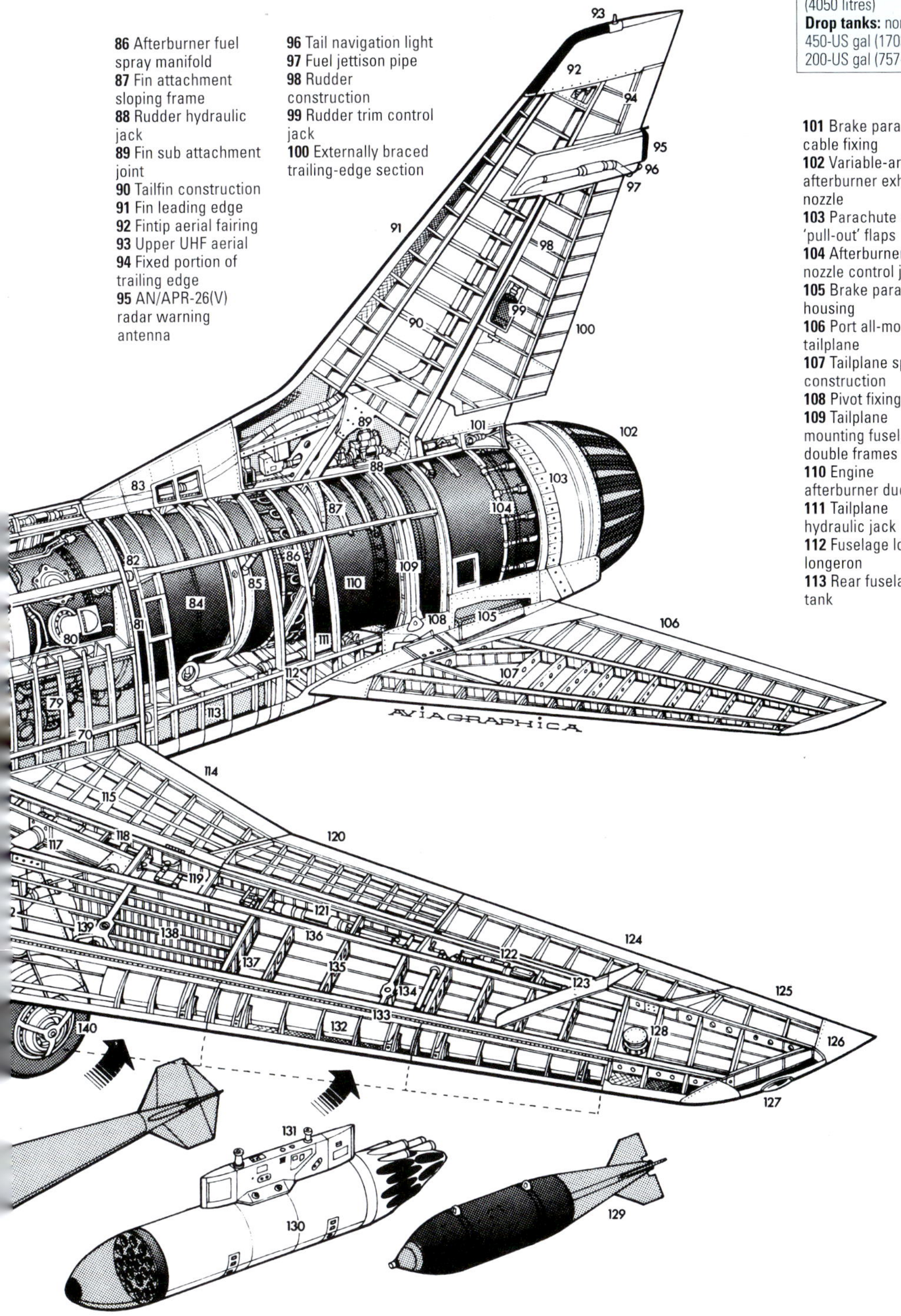

86 Afterburner fuel spray manifold
87 Fin attachment sloping frame
88 Rudder hydraulic jack
89 Fin sub attachment joint
90 Tailfin construction
91 Fin leading edge
92 Fintip aerial fairing
93 Upper UHF aerial
94 Fixed portion of trailing edge
95 AN/APR-26(V) radar warning antenna
96 Tail navigation light
97 Fuel jettison pipe
98 Rudder construction
99 Rudder trim control jack
100 Externally braced trailing-edge section
101 Brake parachute cable fixing
102 Variable-area afterburner exhaust nozzle
103 Parachute cable 'pull-out' flaps
104 Afterburner nozzle control jacks
105 Brake parachute housing
106 Port all-moving tailplane
107 Tailplane spar box construction
108 Pivot fixing
109 Tailplane mounting fuselage double frames
110 Engine afterburner duct
111 Tailplane hydraulic jack
112 Fuselage lower longeron
113 Rear fuselage fuel tank
114 Port inner plain flap
115 Flap rib construction
116 Main undercarriage wheel bay
117 Undercarriage leg pivot fixing
118 Flap hydraulic jack
119 Flap interconnecting linkage
120 Port outer flap
121 Flap hydraulic jack
122 Aileron jack
123 Wing fence
124 Port aileron
125 Fixed portion of trailing edge
126 Wingtip fairing
127 Port navigation light
128 Compass master transmitter
129 750-lb (340-kg) M117 HE bomb
130 SUU-7A CBU 19-round rocket pod and bomblet dispenser
131 Outboard wing pylon
132 Leading-edge slat rib construction
133 Hinged leading-edge attachment joint
134 Outboard pylon fixing
135 Wing rib construction
136 Rear spar
137 Port wing integral fuel tank, 174 Imp gal (791 litres)
138 Multi-spar inner wing panel construction
139 Centre pylon fixing
140 Multi-plate disc brake
141 Port mainwheel
142 Main undercarriage leg strut
143 Undercarriage mounting rib
144 Front spar
145 Wing/fuselage attachment skin joint
146 Aileron cable control run
147 Inboard pylon
148 Airbrake hydraulic jacks (two)
149 Retractable landing/taxiing lamps, port and starboard
150 Ventral airbrake
151 166.5-Imp gal (757-litre) napalm container
152 AGM-12C Bullpup B tactical missile
153 Centre wing pylon
154 279-Imp gal (1268-litre) air refuellable supersonic fuel tank
155 Tank side bracing strut

Northrop F-5

Around 45 Brazilian F-5E/Fs are to undergo an upgrade programme by Elbit Systems. The aircraft are to receive a new multi-mode fire-control radar, HOTAS controls and large-format multi-function LCD displays. Brazil's F-5Es are used mainly for ground-attack and this pre-upgrade F-5E is carrying four unguided bombs on its underwing hardpoints, MAA-1 missiles on the wingtip launchers and an auxiliary fuel tank on the centreline pylon.

RF-5E 'Tigereye'

Cutaway key
1 Pilot head
2 Forward radar-warning antennas
3 KS-87B forward camera, station 1
4 Forward camera compartment
5 Pallet 3, HIAC-1 Long-Range Oblique Photography (LOROP) camera, requires reconfigured window aperture panel
6 LOROP camera rotary drive
7 Main camera compartment
8 Pallet 1 option, KA-95B medium-altitude panoramic camera, station 2
9 KA-56E low-altitude panoramic camera, station 3
10 RS-710 infra-red linescanner, station 4
11 Pallet 2 option, KA-93B high-altitude panoramic camera, stations 2 and 3
12 KA-56E low-altitude panoramic camera, station 4
13 Camera-mounting pallet
14 Optical viewing panel, hinged to starboard
15 Optional vertical KS-87B camera replacing IR linescanner at station 4 of pallet 1
16 Forward-retracting nosewheel
17 Temperature probe
18 Gun gas venting retractable air scoop
19 Ammunition magazine, 280 rounds
20 Ammunition feed chute
21 Single M39A2 20-mm cannon
22 Central avionics equipment compartment
23 Avionics equipment relocated to starboard cannon bay
24 Television sighting camera, located at base of starboard cannon bay
25 Windscreen de-icing fluid tank
26 Gun gas venting air ducts
27 Cartridge case ejector chute
28 Static ports
29 UHF/IFF antenna
30 Rudder pedals
31 Canopy emergency release
32 Position of angle-of attack transmitter on starboard side
33 Control column
34 Instrument panel shroud
35 Frameless windscreen panel
36 AN/ASG-31 lead computing gunsight
37 Upward-hinged canopy
38 Pilot's lightweight rocket-powered ejection seat
39 External canopy handle
40 Engine throttle levers
41 Fold-out boarding steps
42 275-US gal (1041-litre) centreline fuel tank
43 Port engine air intake
44 Liquid oxygen converter
45 Cabin air-conditioning plant
46 Rear avionics equipment bay, port and starboard access
47 Electro-luminescent formation lighting strip
48 Canopy hinge arms and hydraulic actuator
49 Engine bleed air duct to air conditioning heat exchanger
50 Forward fuel cell, bag-type tanks, total internal capacity 677 US gal (2563 litres)
51 Inverted flight reservoir
52 Pressure refuelling connection
53 Port navigation light
54 Ventral retractable landing light
55 Missile control relay boxes
56 Wing leading-edge root extension
57 Leading-edge flap actuator
58 Ventral airbrake panel, port and starboard
59 Airbrake hydraulic jack
60 Intake ducting
61 Centre fuselage fuel cell
62 Gravity fuel fillers
63 Starboard wing tank pylon
64 Leading-edge manoeuvring flap
65 Wingtip missile installation
66 Starboard position light
67 Aileron control linkage
68 Starboard aileron
69 Starboard plain flap
70 Fuel feed pipes
71 Rear fuselage fuel cell
72 Fuel jettison pipe
73 Starboard all-moving tailplane
74 Anti-collision flashing beacon
75 Pressure head
76 Tail position light
77 Fintip antenna fairing
78 UHF antenna
79 Trailing-edge communications antennas
80 Fuel jettison
81 Rudder
82 Rudder and hydraulic actuators
83 Parachute anchorage and release link
84 Brake parachute housing
85 Exhaust nozzle shrouds
86 Variable-area afterburner nozzle
87 Rear radar warning antenna, port and starboard
88 Afterburner ducting
89 Port all-moving tailplane
90 Tailplane pivot mounting and hydraulic actuator
91 General Electric J85-GE-21 afterburning engine
92 Engine accessory equipment
93 Runway emergency arrester hook
94 Engine auxiliary air intake doors
95 Hydraulic reservoir, dual systems port and starboard

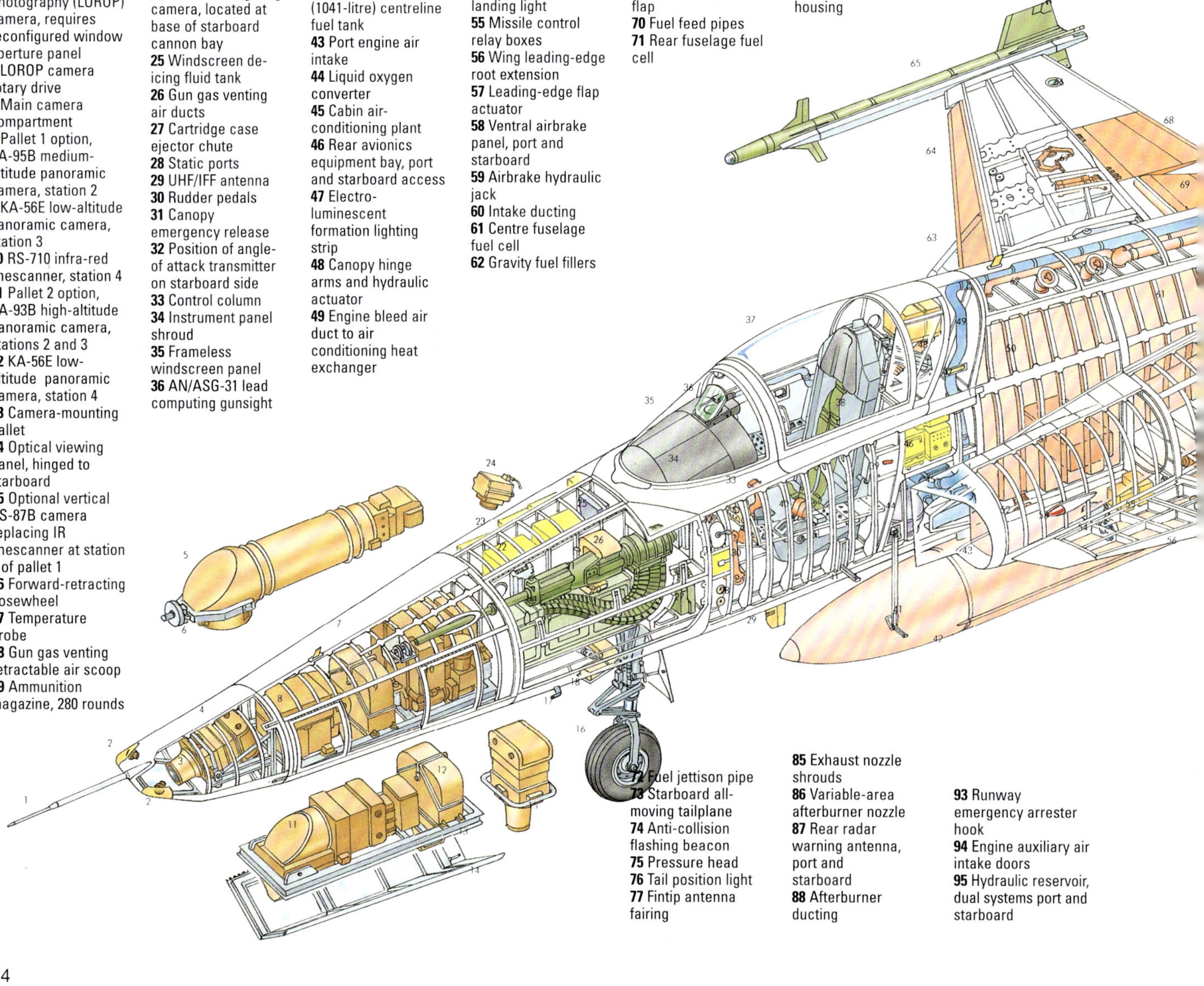

Thirty years after their introduction, Norway's F-5s are still soldiering on in service. This F-5A(G) is seen cruising over Norwegian fjords in the 1970s prior to the aircraft receiving a new light-grey colour scheme in the 1980s to help prevent excessive corrosion of the airframe. During the 1980s the F-5s also underwent the associated Service Life Extension Programme (SLEP), and in the 1990s underwent a further upgrade with the Programme for Avionics and Weapon System (PAWS). The F-5s are likely to remain in service until at least 2003.

SPECIFICATION

F-5E Tiger II

Dimensions

Length including probe: 47 ft 4¾ in (14.45 m)
Height: 13 ft 4½ in (4.08 m)
Wingspan without wingtip AAMs: 26 ft 8 in (8.13 m)
Wingspan with wingtip AAMs: 28 ft (8.53 m)
Wing area: 186.00 sq ft (17.28 m^2)
Wing aspect ratio: 3.82
Tailplane span: 14 ft 1½ in (4.31 m)
Wheel track: 12 ft 5½ in (3.80 m)
Wheel base: 16 ft 11½ in (5.17 m)

Powerplant

Two General Electric J85-GE-21B turbojets each rated at 3,500 lb st (15.5 kN) dry and 5,000 lb st (22.2 kN) with afterburning

Weights

Empty: 9,558 lb (4349 kg)
Maximum take-off: 24,664 lb (11187 kg)

Fuel and load

Maximum internal fuel: 677 US gal (2563 litres)
Maximum external fuel: up to three 275-US gal (1040-litre) auxiliary drop tanks
Maximum ordnance: 7,000 lb (3175 kg)

Performance

Maximum level speed 'clean' at 36,000 ft (10975 m): 917 kt (1,056 mph; 1700 km/h)
Cruising speed at 36,000 ft (10975 m): 562 kt (647 mph; 1041 km/h)
Maximum rate of climb at sea level: 34,300 ft (10455 m) per minute
Service ceiling: 51,800 ft (15590 m)
Take-off run: 2,000 ft (610 m) at 15,745 lb (7142 kg)
Take-off distance to 50 ft (15 m): 2,800 ft (853 m) at 15,745 lb (7142 kg)
Landing run: 2,450 ft (747 m) at 11,340 lb (5143 kg) with brake parachute

Range

Ferry range: 2,010 nm (2,314 miles; 3720 km) with empty auxiliary tanks dropped
Combat radius: 760 nm (875 miles; 1405 km) with two AIM-9 Sidewinder air-to-air missiles

Armament

Two 20-mm Pontiac (Colt-Browning) M39A2 cannon in fuselage nose with 280 rounds per gun; two AIM-9 Sidewinder air-to-air missiles on wingtip launchers and up to 7,000 lb (3175 kg) of mixed ordnance on one underfuselage and four wing hardpoints, including M129 leaflet bombs, 500-lb (227-kg) Mk 82 and Snakeye bombs, 2000-lb (907-kg) Mk 84 bomb, various air-launched rockets, CBU-24, -49, -52 or -58 cluster bomb units and SUU-20 bomb and rocket packs. Can also be adapted to carry AGM-65 Maverick, a centreline multiple ejector rack and laser guided bombs

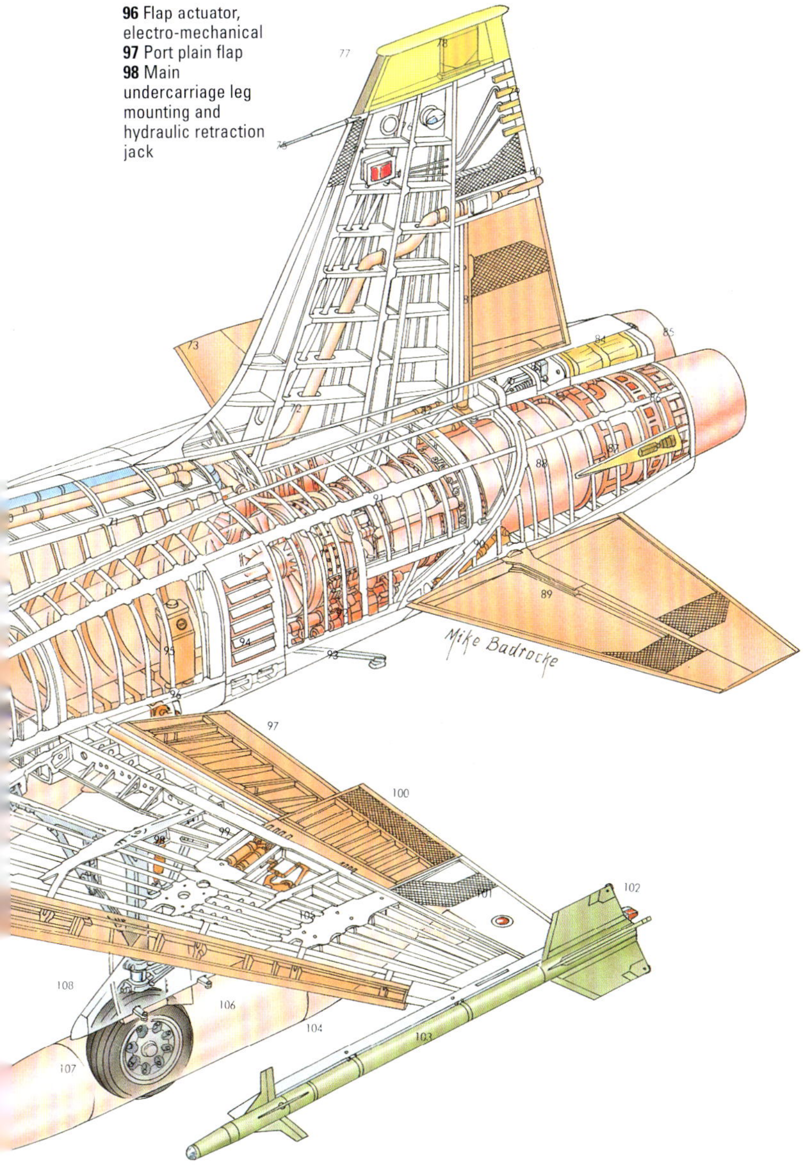

96 Flap actuator, electro-mechanical
97 Port plain flap
98 Main undercarriage leg mounting and hydraulic retraction jack
99 Aileron tandem hydraulic actuators
100 Port aileron
101 Port position light
102 Navigation light repeater
103 AIM-9L Sidewinder air-to-air missile
104 Missile launch rail
105 Outboard pylon hardpoint (unused)
106 150-US gal (568-litre) external fuel tank
107 Port mainwheel
108 External tank pylon
109 Port leading-edge manoeuvring flap

The 425th Tactical Fighter Training Squadron (formerly the 4441st CCTS) was formed in 1969 and, during its 20-year existence, operated the F-5A, B, E and F. Its primary role was the training of foreign F-5 pilots, although from 1979 the parent 405th TTW also acted as the replacement training unit for the F-15.

Republic F-84 Thunderjet

***F-84G-26-RE 51-16719* displays the wel known colours of the USAF Thunderbirds display team. The F-84 was the team's first type upon i formation in 1953. Five aircraft wer assigned to the team and the F-84G wa employed for two years. In 1955 they wer replaced by F-84F Thunderstreak**

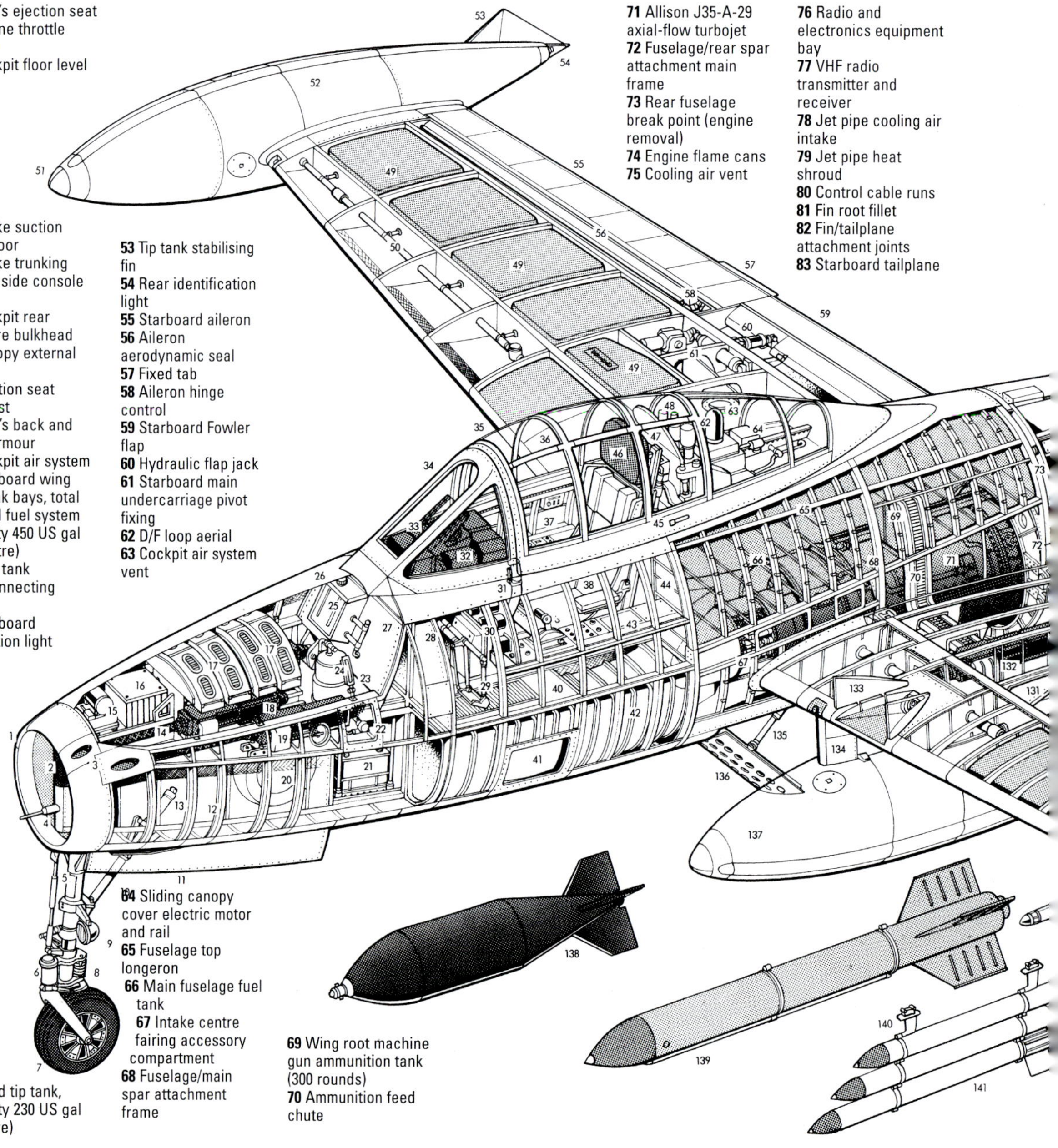

F-84G Thunderjet

Cutaway key
1 Engine air intake
2 Gun laying radar seeker
3 Machine gun muzzles
4 Pitot tube
5 Main undercarriage leg strut
6 Steering control
7 Nosewheel
8 Shimmy damper
9 Taxiing lamp
10 Nosewheel retraction strut
11 Nosewheel doors
12 Bifurcated intake ducting
13 Nosewheel hydraulic retraction jack
14 Machine gun barrels
15 Gyro compass unit
16 Ballast weights
17 Ammunition tanks (300 rounds per gun)
18 M3 0.5-in (12.7-mm) machine guns
19 Spent cartridge case collector chute
20 Nosewheel bay between intake ducts
21 Battery
22 Servicing access panels
23 Gun bay access panel latch
24 Oxygen converter
25 Hydraulic system header tank
26 Gun bay access panel
27 Armoured bulkhead
28 Cockpit front pressure bulkhead
29 Rudder pedals
30 Instrument panel
31 Control column
32 Instrument panel shroud
33 Sperry radar gunsight
34 Bullet-proof windscreen
35 Cockpit canopy cover
36 Canopy framing
37. Starboard side console panel
38 Pilot's ejection seat
39 Engine throttle control
40 Cockpit floor level
41 Intake suction relief door
42 Intake trunking
43 Port side console panel
44 Cockpit rear pressure bulkhead
45 Canopy external latch
46 Ejection seat headrest
47 Pilot's back and head armour
48 Cockpit air system
49 Starboard wing fuel tank bays, total internal fuel system capacity 450 US gal (1709 litre)
50 Fuel tank interconnecting piping
51 Starboard navigation light
52 Fixed tip tank, capacity 230 US gal (870 litre)
53 Tip tank stabilising fin
54 Rear identification light
55 Starboard aileron
56 Aileron aerodynamic seal
57 Fixed tab
58 Aileron hinge control
59 Starboard Fowler flap
60 Hydraulic flap jack
61 Starboard main undercarriage pivot fixing
62 D/F loop aerial
63 Cockpit air system vent
64 Sliding canopy cover electric motor and rail
65 Fuselage top longeron
66 Main fuselage fuel tank
67 Intake centre fairing accessory compartment
68 Fuselage/main spar attachment frame
69 Wing root machine gun ammunition tank (300 rounds)
70 Ammunition feed chute
71 Allison J35-A-29 axial-flow turbojet
72 Fuselage/rear spar attachment main frame
73 Rear fuselage break point (engine removal)
74 Engine flame cans
75 Cooling air vent
76 Radio and electronics equipment bay
77 VHF radio transmitter and receiver
78 Jet pipe cooling air intake
79 Jet pipe heat shroud
80 Control cable runs
81 Fin root fillet
82 Fin/tailplane attachment joints
83 Starboard tailplane

Large numbers of F-84s were supplied to NATO and other allied air forces under the Mutual Defense Assistance Program, including that of the Netherlands (pictured). The Dutch received 21 F-84Es in 1951-52 (most of which were later converted as photo-reconnaissance aircraft) and 166 F-84Gs. Other Thunderjet operators have included France (below right), Greece, Portugal, Taiwan, Turkey and Yugoslavia.

SPECIFICATION

F-84G Thunderjet

Dimensions

Wingspan: 36 ft 5 in (11.09 m)
Length: 38 ft 1 in (11.60 m)
Height: 12 ft 7 in (3.83 m)
Wing area: 260.00 sq ft (24.25 m²)

Powerplant

one Allison J35-A-29 axial-flow turbojet rated at 5,600 lb st (24.91 kN)

Weights

Empty: 11,095 lb (5033 kg)
Normal loaded: 18,645 lb (8457 kg)
Maximum loaded: 23,525 lb (10670 kg)

Fuel load

Internal fuel: 451 US gal (1709 litres)
External fuel: two wingtip fuel tanks and two underwing drop tanks, each of 230-US gal (870-litre) capacity

Performance

Maximum speed at sea level: 622 mph (1001 km/h)
Maximum speed at 20,000 ft (6095 m): 575 mph (925 km/h)
Maximum speed at 36,000 ft (10970 m): 540 mph (869 km/h)
Cruising speed at 35,000 ft (10670 m): 483 mph (777 km/h)
Time to 35,000 ft (10670 m): 7 minutes 54 seconds; 9 minutes 24 seconds with external tanks
Service ceiling: 40,500 ft (12345 m)
Range with internal fuel: 670 miles (1078 km)
Range with tiptanks: 1,330 miles (2140 km)
Range with maximum external fuel: 2,000 miles (3217 km)

Armament

Six 0.5-in (12.7-mm) Colt-Browning M3 machine-guns, with 300 rounds per gun, and provision for up to 4,000 lb (1184 kg) of external ordnance, including 100-, 500- and 1,000-lb (45-, 227-, 454-kg) GP bombs, napalm and 'Tiny Tim' 12-in (30-cm) and 5-in (12.7-cm) HVARs

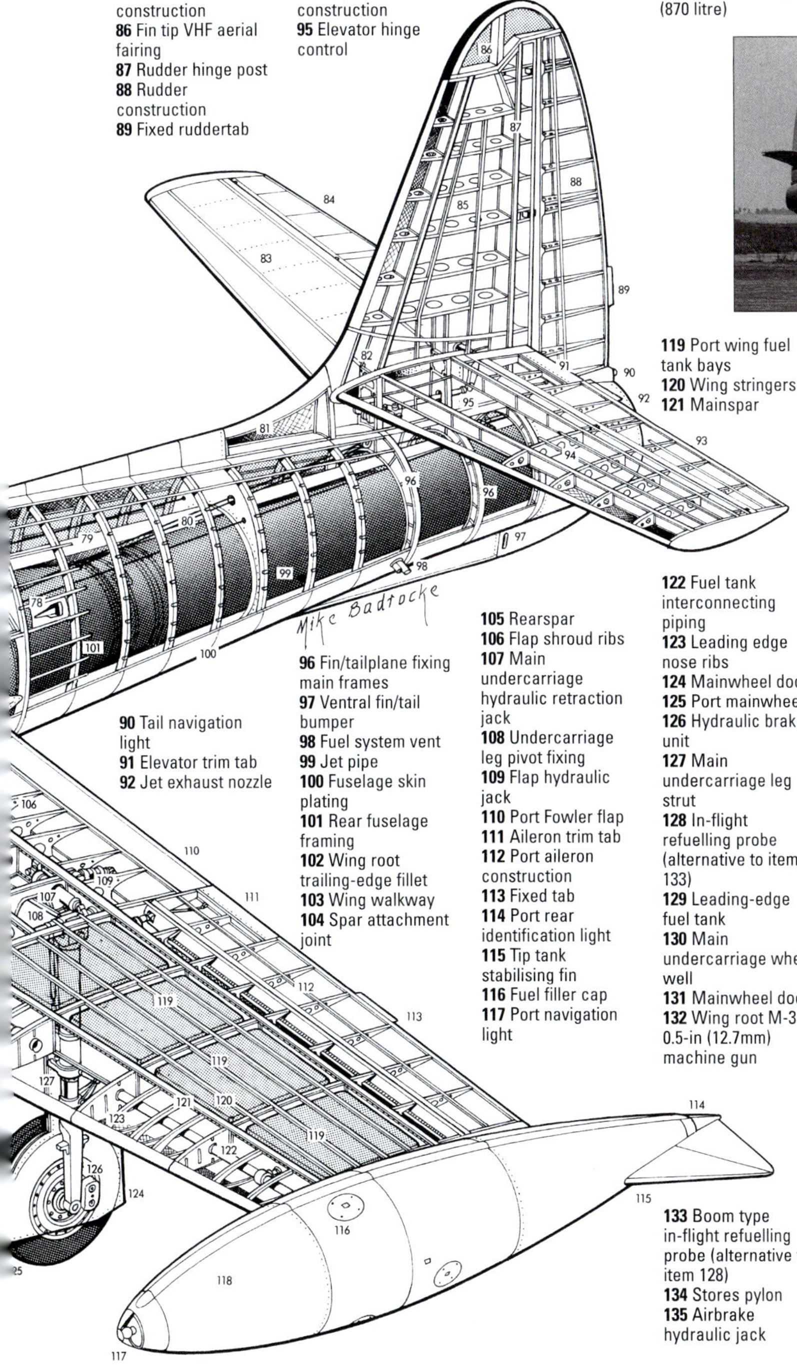

84 Starboard elevator
85 Tailfin construction
86 Fin tip VHF aerial fairing
87 Rudder hinge post
88 Rudder construction
89 Fixed ruddertab
90 Tail navigation light
91 Elevator trim tab
92 Jet exhaust nozzle
93 Port elevator
94 Tailplane construction
95 Elevator hinge control
96 Fin/tailplane fixing main frames
97 Ventral fin/tail bumper
98 Fuel system vent
99 Jet pipe
100 Fuselage skin plating
101 Rear fuselage framing
102 Wing root trailing-edge fillet
103 Wing walkway
104 Spar attachment joint
105 Rearspar
106 Flap shroud ribs
107 Main undercarriage hydraulic retraction jack
108 Undercarriage leg pivot fixing
109 Flap hydraulic jack
110 Port Fowler flap
111 Aileron trim tab
112 Port aileron construction
113 Fixed tab
114 Port rear identification light
115 Tip tank stabilising fin
116 Fuel filler cap
117 Port navigation light
118 Fixed tip tank, capacity 230 US gal (870 litre)
119 Port wing fuel tank bays
120 Wing stringers
121 Mainspar
122 Fuel tank interconnecting piping
123 Leading edge nose ribs
124 Mainwheel doors
125 Port mainwheel
126 Hydraulic brake unit
127 Main undercarriage leg strut
128 In-flight refuelling probe (alternative to item 133)
129 Leading-edge fuel tank
130 Main undercarriage wheel well
131 Mainwheel door
132 Wing root M-3 0.5-in (12.7mm) machine gun
133 Boom type in-flight refuelling probe (alternative to item 128)
134 Stores pylon
135 Airbrake hydraulic jack
136 Perforated ventral airbrake
137 Drop tank, capacity 230 US gal (870 litre)
138 500-lb (227-kg) HE bomb
139 'Tiny Tim' 30-cm air-to-ground rocket
140 Rocket fixing shackles
141 HVAR ground attack rockets

USAF experiments with the zero-length launcher (ZELL) concept, which culminated in test flights employing F-100s and F-104s, began with this F-84. Equipped with a booster bottle and mounted on a mobile launcher developed by Martin Aircraft of Baltimore, the ZELL system was mooted as the future for air defence aircraft, rendering vulnerable conventional airstrips obsolete. However, the concept did not find sufficient favour in official circles and was ultimately abandoned.

Republic F-105 Thunderchief

A flight of F-105D Thunderchiefs of 335th TFS, Tactical Air Command is seen during a practice mission over North Carolina in 1964. In May of that year, the squadron had the distinction of being the first 'Thunderchief' unit to temporarily deploy to Europe to take part in Operation Long Thrust. This exercise saw TAC units operating alongside NATO forces under simulated wartime conditions. The Thunderchief in the foreground (FH-161) was one of the 30 'Thuds' later converted to the T-Stick II configuration.

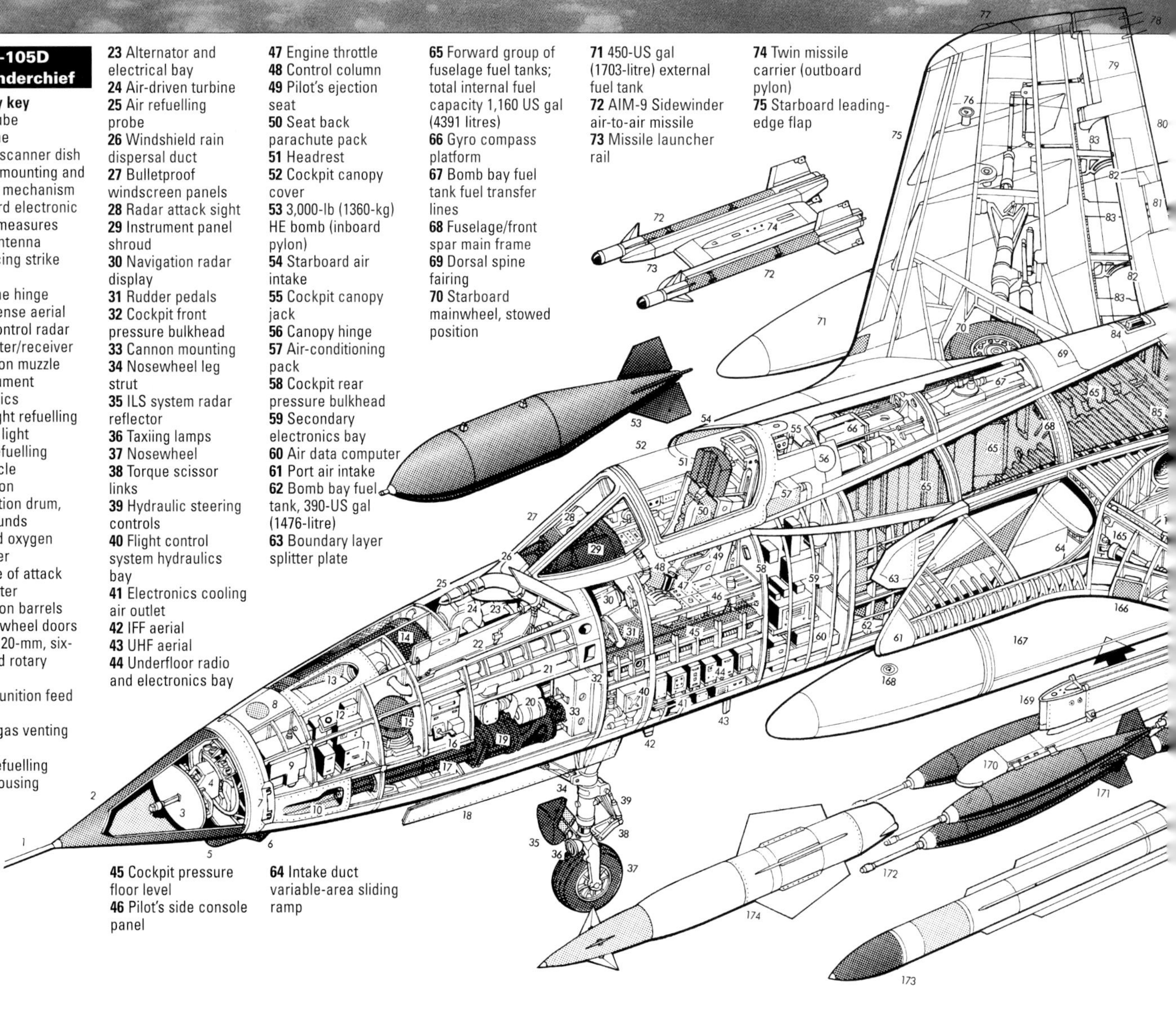

F-105D Thunderchief

Cutaway key

1 Pitot tube
2 Radome
3 Radar scanner dish
4 Radar mounting and tracking mechanism
5 Forward electronic countermeasures (ECM) antenna
6 Aft-facing strike camera
7 Radome hinge
8 ADF sense aerial
9 Fire-control radar transmitter/receiver
10 Cannon muzzle
11 Instrument electronics
12 In-flight refuelling position light
13 Air refuelling receptacle
14 Cannon ammunition drum, 1,028 rounds
15 Liquid oxygen converter
16 Angle of attack transmitter
17 Cannon barrels
18 Nosewheel doors
19 M61, 20-mm, six-barrelled rotary cannon
20 Ammunition feed chute
21 Gun gas venting pipe
22 Air refuelling probe housing
23 Alternator and electrical bay
24 Air-driven turbine
25 Air refuelling probe
26 Windshield rain dispersal duct
27 Bulletproof windscreen panels
28 Radar attack sight
29 Instrument panel shroud
30 Navigation radar display
31 Rudder pedals
32 Cockpit front pressure bulkhead
33 Cannon mounting
34 Nosewheel leg strut
35 ILS system radar reflector
36 Taxiing lamps
37 Nosewheel
38 Torque scissor links
39 Hydraulic steering controls
40 Flight control system hydraulics bay
41 Electronics cooling air outlet
42 IFF aerial
43 UHF aerial
44 Underfloor radio and electronics bay
45 Cockpit pressure floor level
46 Pilot's side console panel
47 Engine throttle
48 Control column
49 Pilot's ejection seat
50 Seat back parachute pack
51 Headrest
52 Cockpit canopy cover
53 3,000-lb (1360-kg) HE bomb (inboard pylon)
54 Starboard air intake
55 Cockpit canopy jack
56 Canopy hinge
57 Air-conditioning pack
58 Cockpit rear pressure bulkhead
59 Secondary electronics bay
60 Air data computer
61 Port air intake
62 Bomb bay fuel tank, 390-US gal (1476-litre)
63 Boundary layer splitter plate
64 Intake duct variable-area sliding ramp
65 Forward group of fuselage fuel tanks; total internal fuel capacity 1,160 US gal (4391 litres)
66 Gyro compass platform
67 Bomb bay fuel tank fuel transfer lines
68 Fuselage/front spar main frame
69 Dorsal spine fairing
70 Starboard mainwheel, stowed position
71 450-US gal (1703-litre) external fuel tank
72 AIM-9 Sidewinder air-to-air missile
73 Missile launcher rail
74 Twin missile carrier (outboard pylon)
75 Starboard leading-edge flap

The USAF's* Thunderbirds *flight demonstration team usually flew the 'hottest' aircraft on the inventory, but the Thunderchief seemed to be too much for them. The team adopted the F-105B Thunderchief for the 1964 season. The transition was fairly smooth and the first display was at Norfolk, Virginia in April 1964. Modifications were necessary to convert the Thunderchiefs to the aerobatic team role, with changes in the rudder, flap and fuel systems, for extended inverted flight. The life of the F-105B with the Thunderbirds was short-lived, only six displays being given before a fatal accident, in May 1964, resulted in a decision to revert to the F-100 Super Sabre for the remainder of the season.

SPECIFICATION

F-105D Thunderchief

Dimensions

Length: 64 ft 4 in (19.61 m)
Height: 19 ft 7 in (5.97 m)
Wingspan: 34 ft 9 in (10.59 m)
Wing area: 385 sq ft (35.77 m²)

Powerplant

One Pratt & Whitney J75-P-19W turbojet rated at 17,200 lb (76.0 kN) thrust dry and 24,500 lb (110.25 kN) thrust with afterburning; water injection permitted 60-second rating of 26,500 lb (117.7 kN) thrust in afterburner mode

Weights

Empty: 27,500 lb (12474 kg)
Maximum overload take-off: 52,838 lb (23967 kg)

Fuel

Normal internal fuel: 435 US gal (1646 litres)
Maximum internal fuel: 675 US gal (2555 litres)

Performance

Maximum level speed clean at 36,000 ft (10970 m): 1,390 mph (2237 km/h)
Initial climb rate: 34,400 ft (10485 m) per minute in clean configuration
Service ceiling: 41,200 ft (12560 m)
Range: 920 miles (1480 km) with two 450-US gal (1703 litre) drop-tanks underwing, one 650-US gal (2461-litre) drop-tank on centreline and two AGM-12 Bullpup ASMs; ferry range 2,390 miles (3846 km) with maximum external fuel at 584 mph (940 km/h)

Armament

Combination of 750-lb (340-kg) M117 bombs, 1,000-lb (454-kg) Mk 83 bombs, 3,000-lb (1361-kg) M118 bombs, AGM-12 Bullpup ASMs, AIM-9 Sidewinder AAMs, 2.75-in (70-mm) rocket pods, napalm containers, Mk 28/43 special weapons, chemical bombs, leaflet bombs, 5-in (127-mm) rocket pods and MLU-10/B mines; also one M61 Vulcan 20-mm cannon with 1,028 rounds of ammunition

Big Sal, ***flown by Capt. John Hoffman, is seen on the way back from a bombing mission over North Vietnam. Hoffman's F-105D (61-0086) survived the war to serve with the Virginia Air National Guard until 1979.***

76 Outboard pylon fixing/drop tank filler cap
77 Starboard navigation light
78 Static dischargers
79 Starboard aileron
80 Starboard fowler flap
81 Trim tab, starboard only
82 Flap guide rails
83 Roll control spoilers
84 Anti-collision light
85 Air intake ducting
86 Ground running secondary air intake
87 Wing spar attachment joint
88 Fuselage/rear spar main frame
89 Engine compressor face
90 Forward engine mounting frame
91 Rear fuselage group of fuel tanks
92 Fuel pipe ducting
93 Drop tank tail fins
94 Afterburner duct cooling ram air intake
95 Starboard all-moving tailplane
96 Tailfin construction
97 Fin tip ECM aerials
98 Tail position light
99 Static dischargers
100 Rudder mass balance
101 Rudder
102 Formation light
103 Water injection tank 36-US gal (136-litre) capacity
104 Rudder-power control unit
105 Brake parachute housing
106 Parachute door
107 Petal-type airbrakes, open position
108 Republic convergent/ divergent ram air ejector nozzle flaps
109 Airbrake/nozzle flap jacks
110 Internal variable-area afterburner nozzle

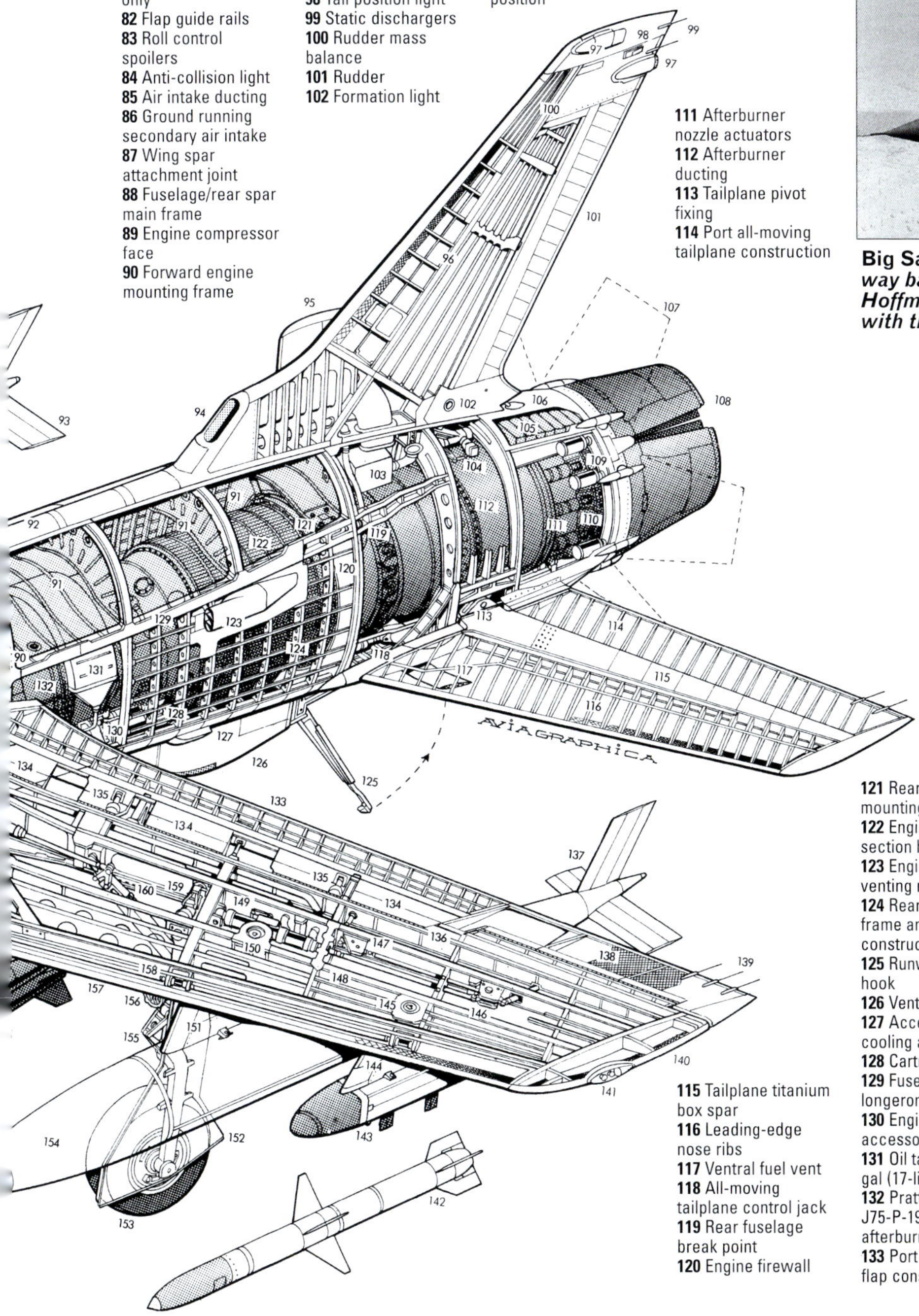

111 Afterburner nozzle actuators
112 Afterburner ducting
113 Tailplane pivot fixing
114 Port all-moving tailplane construction
115 Tailplane titanium box spar
116 Leading-edge nose ribs
117 Ventral fuel vent
118 All-moving tailplane control jack
119 Rear fuselage break point
120 Engine firewall
121 Rear engine mounting
122 Engine turbine section heat shroud
123 Engine bay venting ram air intake
124 Rear fuselage frame and stringer construction
125 Runway arrester hook
126 Ventral fin
127 Accessory cooling air duct
128 Cartridge starter
129 Fuselage top longeron
130 Engine-driven accessory gearbox
131 Oil tank, 4.5-US gal (17-litre) capacity
132 Pratt & Whitney J75-P-19W afterburning turbojet
133 Port Fowler-type flap construction
134 Five-section roll control spoilers
135 Flap screw jacks
136 Aileron mass balance
137 Port drop tank tail fins
138 Honeycomb aileron construction
139 Static dischargers
140 Wingtip fairing
141 Port navigation light
142 AGM-45 Strike anti-radar missile
143 ECM pod
144 Outboard stores pylon
145 Pylon fixing/fuel filler cap
146 Aileron hinge control
147 Aileron/spoiler mixer linkage
148 Multi-spar wing construction
149 Aileron power control unit
150 Inboard pylon fixing
151 Inboard stores pylon
152 Mainwheel leg door
153 Port mainwheel
154 450-US gal (1703-litre) drop tank
155 Main undercarriage leg torque scissor links
156 Landing lamp
157 Port leading-edge flap
158 Leading-edge flap rotary actuators
159 Main undercarriage pivot mounting
160 Undercarriage side breaker strut
161 Hydraulic retraction jack
162 Diagonal wing spar
163 Mainwheel housing
164 Inner mainwheel door
165 Leading-edge flap actuator
166 Leading-edge flush aerial
167 650-US gal (2461-litre) centreline fuel tank
168 Fuel tank filler cap
169 Centreline stores pylon
170 Triple ejection rack
171 Six M117, 750-lb (340-kg) HE bombs
172 Anti-personnel extended bomb fuse
173 AGM-78 Standard anti-radar missile
174 AGM-12C Bullpup air-to-ground missile

Saab J-35 Draken

As the final J 35J Drakens underwent major overhaul in the early 1990s, they were repainted in this highly effective two-tone air superiority grey camouflage. This was the final colour scheme worn by Swedish air force Drakens before the last front-line squadron was eventually retired at the end of 1998.

Saab J 35F-2 Draken

Cutaway key
1 Nose probe
2 Glass-fibre nose cone
3 Radar scanner
4 Scanner mounting frame
5 Radar pack
6 Saab S7-collision-course fire control
7 L. M. Ericsson (Hughes licence) infra-red seeker
8 Electronics pack
9 Front pressure bulkhead
10 Data-handling unit
11 Rudder pedal assembly
12 Port instrument console
13 Side panel
14 Instrument panel/ radar scope shroud
15 Windscreen frame
16 Weapons sight
17 Windscreen
18 Starboard intake
19 Glassfibre intake lip
20 Aft-hinged cockpit canopy
21 Cockpit sill
22 Control panel
23 Control column
24 Throttle quadrant
25 Pilot's Saab RS 35 ejection seat
26 Canopy hinge mechanism
27 Seat support frame
28 Rear pressure bulkhead
29 Navigation computer
30 Forward avionics equipment bay
31 Gyro unit
32 TACAN transmitter-receiver
33 Auxiliary air intake
34 Starboard intake trunk
35 Starboard fuel tanks
36 Dorsal spine
37 Starboard forward bag-type fuel tank,
38 30-mm ADEN cannon
39 Ammunition magazine (100 rounds)
40 Dorsal antenna
41 Electrical wiring
42 Mid-fuselage production break line
43 Intake trunking
44 Oil-cooler air intake
45 Volvo Flygmotor RM6C (Rolls-Royce Avon 300 series) turbojet
46 Louvres
47 Access panels
48 Fuselage frames
49 Engine firewall
50 Cooling air inlet scoop
51 Finroot fairing
52 Fuel transfer
53 Starboard mainwheel door
54 Door actuating rod
55 Inner/outer wing joint strap
56 Starboard navigation light
57 Wing skinning
58 Starboard outer elevon
59 Hinge point
60 Actuating jack access
61 Control hinge
62 Access panels
63 Starboard aft integral fuel tank
64 Starboard aft bag-type fuel tanks (3)
65 Intake grille
66 Jet pipe
67 Engine aft mounting ring
68 Access
69 Tailfin main spar attachment
70 Control stick angle indicator unit
71 Computer amplifier
72 Synchroniser pack
73 Tailfin structure
74 Pitot tube
75 Rudder mass balance
76 Rudder structure
77 Rudder post
78 Tailfin rear spar
79 Rudder servo mechanism and actuator
80 Attachment point
81 Speed brake
82 Fuselage structure
83 Detachable tail cone (engine removal)
84 Access panel
85 Brake parachute housing
86 Aft fairing
87 Afterburner assembly
88 Exhaust
89 Air intake (afterburner housing)
90 Control surface blunt trailing edge
91 Port inner elevon
92 Hinge points
93 Elevon actuator
94 Rear spar
95 Twin (retractable) tailwheels
96 Port aft
97 Inner/outer wing joint
98 Wing outer structure
99 Rib stations
100 Port outer elevon
101 Elevon actuator
102 Hinge points
103 Port wingtip
104 Anti-buffet underwing fences (6)
105 Stores pylons (maximum 8)
106 Nose ribs
107 Forward spar
108 Wheel door
109 Port navigation light
110 Port main wheel
111 Door inboard section
112 Port main wheel well
113 Fuel transfer
114 Wing join strap
115 Port aft bag-type fuel tanks (3)
116 Fuel collector
117 Mainwheel retraction mechanism
118 Mainwheel oleo leg mounting
119 Engine accessory gearbox
120 Port cannon ammunition magazine
121 Port 30-mm ADEN cannon (Saab 35F has starboard gun only, earlier intercept and export 35X versions retaining port gun as illustrated)
122 Port forward bag-type fuel tank
123 Port forward integral fuel tanks
124 Cannon port
125 Inner wing/ fuselage integral structure
126 Angled frame member
127 Emergency ventral ram-air turbine
128 Trunking formers
129 Gyro amplifiers
130 Intake trunking
131 Nosewheel leg
132 Glassfibre intake lip
133 Forward retracting nosewheel
134 Steering mechanism
135 Possible stores (including jettisonable tanks)
136 Pod containing 19 x 3-in (75-mm) rockets
137 Rb 28 (Sidewinder) IR-homing missile
138 5.3-in (13.5-cm) rocket
139 Rb 27 (Falcon) radar-homing missile
140 1,102-lb (500-kg) bomb

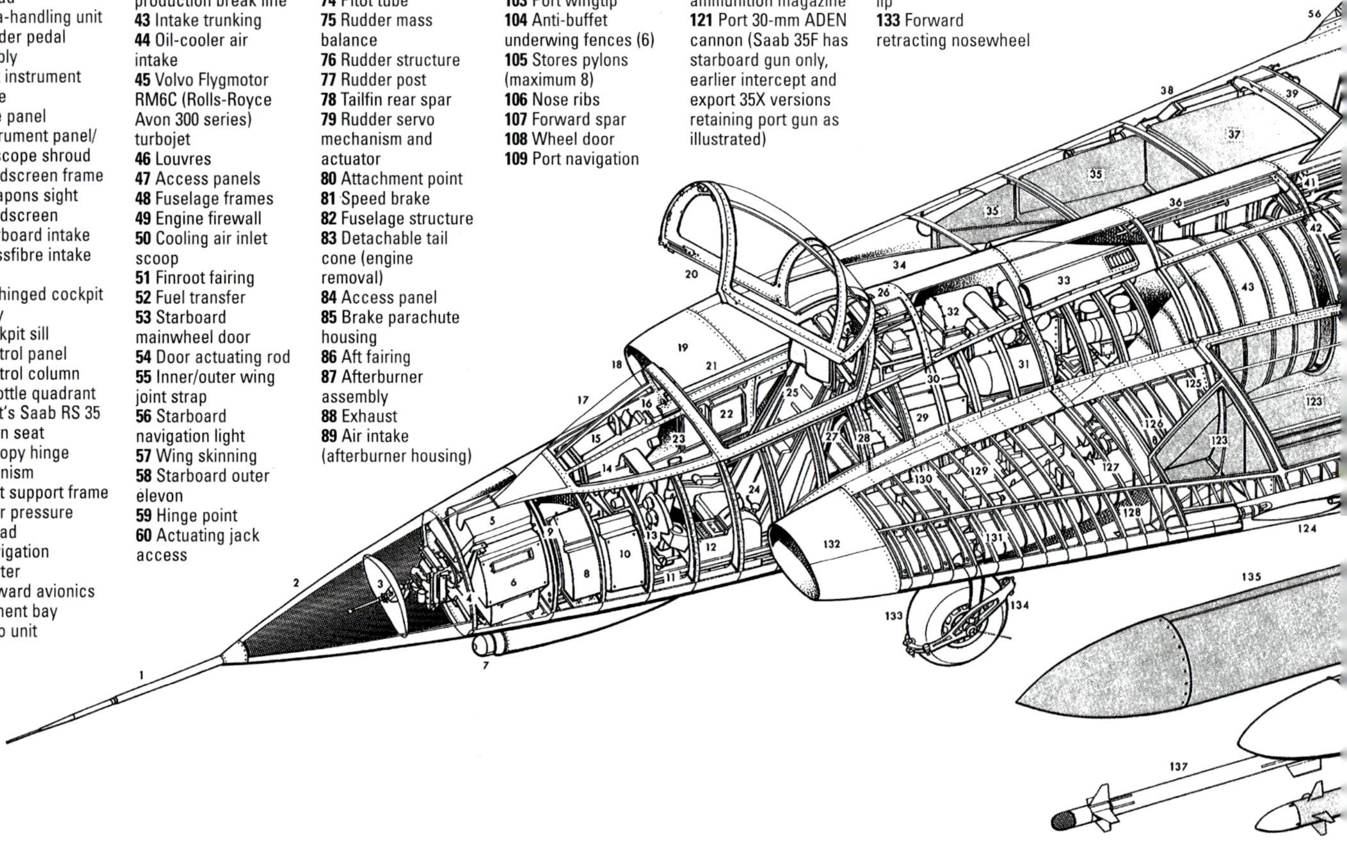

Above: HavLlv 21 was one of two squadrons to operate the Draken in Finnish air force service. The unit operated the type for 25 years from 1972, before becoming the first operational unit in the Finnish air force to fly the F-18C/D Hornet. The remaining Draken operator, HavLlv 11, will operate the type into the next century.

Above right: At its introduction into service, the Draken was one of the best-performing warplanes in the world, thanks in part to its excellent powerplant. The large flame emitted in reheat was courtesy of a locally-produced afterburner added to the licence-built Rolls-Royce Avon engine.

SPECIFICATION

J 35J Draken

Dimensions

Length: 50 ft 4 in (15.35 m)
Wingspan: 30 ft 10 in (9.40 m)
Height: 12 ft 9 in (3.89 m)
Wing area: 529.60 sq ft (49.20 m^2)
Wing aspect ratio: 1.77
Wheel track: 8 ft 10½ in (2.70 m)

Powerplant

One 12,790-lb st (56.89-kN) dry or 17,650-lb st (78.5-kN) with afterburning Volvo Flygmotor RM6C turbojet (licence-built Rolls-Royce Avon Series 300 turbojet fitted with a Swedish-designed afterburner)

Weights

Empty: 18,188 lb (8250 kg)
Normal take-off: 25,132 lb (11400 kg)
Maximum take-off: 27,050 lb (12270 kg) for interceptor mission or 33,069 lb (17650 kg) for attack mission

Fuel

Internal fuel: 1,057 US gal (4000 litres)
External fuel: provision for up to 1,321 US gal (5000 litres) in external drop tanks

Performance

Maximum level speed 'clean' at 36,000 ft (10975 m): more than 1,147 kt (1,317 mph; 2119 km/h)
Maximum speed at 300 ft (90 m): 793 kt (910 mph; 1465 km/h)
Maximum climb rate at sea level: 34,450 ft (10500 m) per minute with afterburning
Service ceiling: 65,600 ft (19995 m)
Take-off run: 2,133 ft (650 m) at normal take-off weight
Take-off distance to 50 ft (15 m): 3,150 ft (960 m) at normal take-off weight

Range

Ferry range: 1,533 nm (1,763 miles; 2837 km)
Combat radius: 304 nm (350 miles; 564 km) on a hi-lo-hi attack mission with internal fuel only

Armament

Usual air-to-air armament of 2 x AIM-9J Sidewinder air-to-air missiles on centre-section pylons, 2 x Hughes Falcon air-to-air missiles on wing pylons and one 30-mm ADEN cannon with 90 rounds in starboard wing. Maximum ordnance of 6,393 lb (2900 kg)

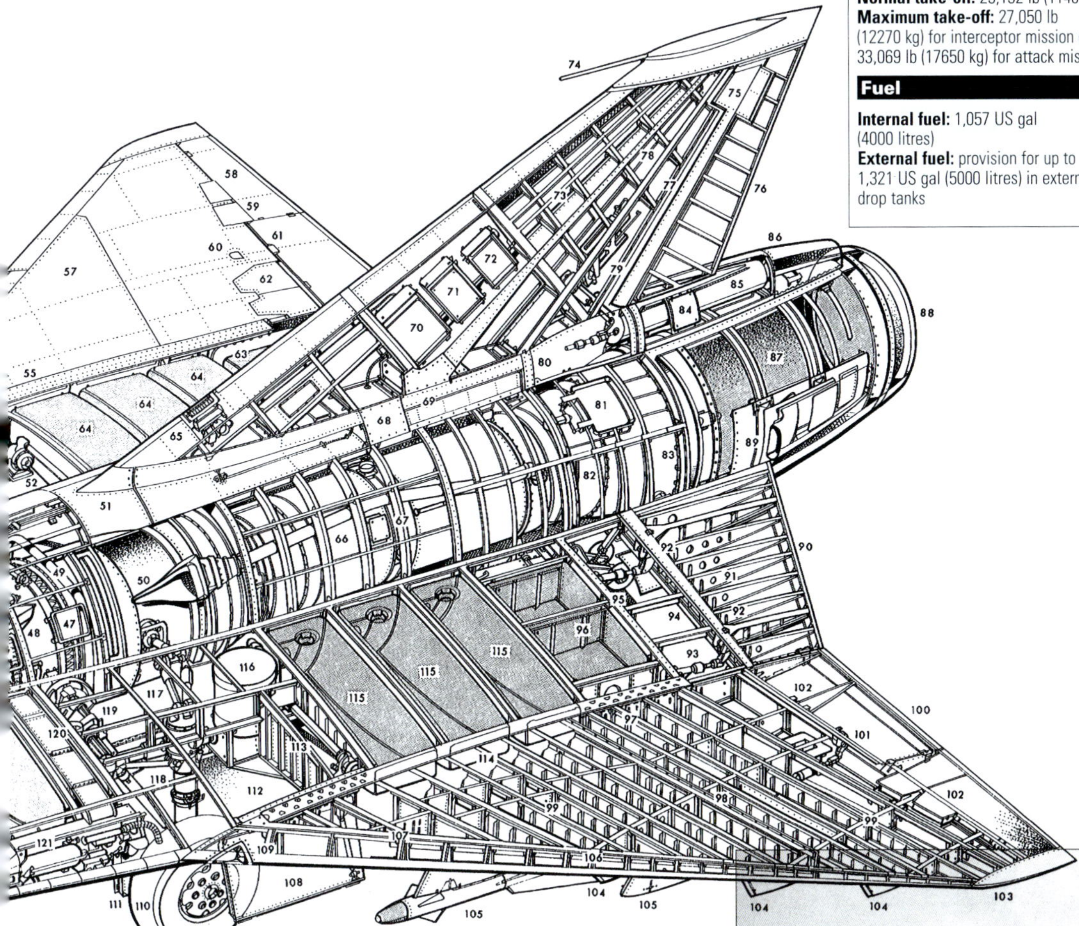

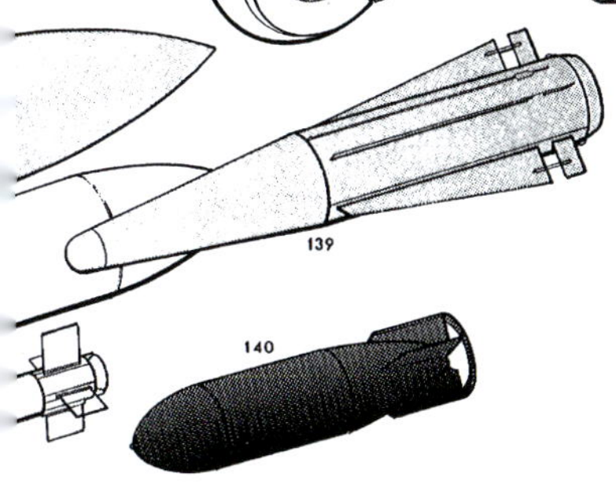

Below: The Saab 210 Draken (later Lilldraken) was fundamentally a seven-tenths scale prototype of the definitive Draken. It was the first double-delta wing type to fly and played a major role in the subsequent success of the Draken project.

Saab Viggen

Displaying the Viggen's distinctive wing planform, this JA 37 is loaded with inert training BAe Sky Flash and AIM-9 Sidewinder missiles. This load is representative of the Viggen's intercept mission although a fuel tank would normally be carried on the fuselage centreline to help overcome the aircraft's poor range.

SH 37 Viggen

Cutaway key
1 Pitot head
2 Glass-fibre radome
3 Radar scanner housing
4 LM Ericsson PS-37/A radar equipment module
5 Incidence probe
6 Cockpit pressure bulkhead
7 Forward avionics equipment bay
8 Rudder pedals
9 Instrument panel shroud
10 One piece frameless windscreen panel
11 Pilot's head-up display
12 Upward-hinging cockpit canopy
13 Ejection seat arming lever
14 Saab rocket-powered ejection seat
15 Engine throttle lever
16 Boundary layer splitter plate
17 Port air intake
18 Landing/taxiing lamp
19 Twin nosewheels, forward retracting
20 Hydraulic steering control
21 Red Baron multi-sensor reconnaissance pod
22 Centreline external fuel tank
23 Electro-luminescent formation lighting strip
24 Central avionics equipment bay
25 Intake ducting
26 Boundary layer spill duct
27 Forward fuselage integral fuel tank
28 Dorsal avionics equipment bay
29 Starboard canard foreplane
30 Canard flap
31 SATT AQ31 ECM jamming pod
32 SSR transponder aerial
33 Anti-collision light
34 Air-conditioning equipment bay
35 Heat exchanger air exhaust
36 Intake flank fuel tankage
37 Engine compressor face
38 Accessory equipment gearbox
39 Foreplane spar attachment joint
40 Fuselage flank avionics equipment bays, port and starboard
41 Emergency ram air turbine
42 Port canard foreplane flap honeycomb panel
43 Hydraulic reservoirs
44 Formation lighting strip
45 Centre fuselage integral fuel tankage
46 Main engine mounting
47 Volvo Flygmoto RM8A afterburning turbofan engine
48 Engine bleed air pre-cooler
49 Fuel-cooled engine oil cooler
50 Wing spar attachment fuselage main frame
51 Fuel system recuperators
52 ADF aerial
53 Starboard wing panel
54 Outboard missile pylon
55 ECM antenna fairing
56 Extended-chord outboard leading edge
57 Starboard navigation light
58 Starboard elevon panels
59 Artificial feel system pressure head
60 Fin tip aerial fairing
61 Multi spar fin construction
62 Rudder hydraulic actuator
63 Fin spar attachment joints
64 Hydraulic hand pump for hangaring fin folding
65 Port lateral airbrake
66 Airbrake hydraulic jack
67 Afterburner ducting
68 Variable area afterburner nozzle control jack
69 Exhaust duct ejector seal (closed at speeds above Mach 1)
70 Ejector seal screw jack
71 Thrust reverser door pneumatic actuator
72 Radar warning antennas
73 Engine/afterburner exhaust nozzle
74 Thrust reverser blocker doors
75 Tail navigation light
76 Lower thrust reverser door pneumatic actuator
77 Port inboard elevon
78 Elevon hydraulic actuators
79 Elevon honeycomb construction
80 Port outboard elevon
81 Port navigation lights
82 Outboard elevon hydraulic actuator
83 Saab Bofors Rb 24 (licence-built Sidewinder) air-to-air self-defence missile
84 Missile launch rail
85 ECM antenna fairing
86 Bofors BOZ-9 flare launcher pod
87 Wing stores pylon
88 Honeycomb wing skin panels
89 Multispar wing panel construction
90 Wing panel integral fuel tank
91 Main spar
92 Main undercarriage wheel bay
93 Side breaker strut
94 Hydraulic retraction jack
95 Main undercarriage mounting rib
96 Mainwheel leg strut
97 Torque scissor links
98 Tandem mainwheels
99 Starboard fuselage pylon
100 Long-range camera pod
101 Rb 05A air-to-surface missile
102 Rb 04E air-to-surface anti-ship missile
103 Missile launch adaptor

Viggen reconnaissance pods

There are two Viggen reconnaissance variants: the SH 37 is used for maritime surveillance, while the SF 37 is employed in the overland role. The SH 37 is equipped with an RKA 40 recording camera and can carry a night-photography pod on its port or starboard sides. There is also provision for a single, forward-facing long-range optical pod (LOROP) on the starboard shoulder pylon containing an SKA 24D camera with 600-mm lens (top left).

Mounted on the shoulder pylons of the SF 37 are reconnaissance pods (top right). To the port side is a pod (Mörkerspaningskapsel) which contains three SKA 34 cameras with 75-mm lenses, each having a 120° field of view, and situated at the rear are two flash windows. The starboard pod (Blixtkapsel) contains capacitors which are charged from the aircraft's electrical generators to release power to the flashguns when required for illumination from heights of up to approximately 1,640 ft (500 m).

SPECIFICATION

JA 37 Viggen

Dimensions

Length: 53 ft 9¾ in (16.40 m)
Height: 19 ft 4¼ in (5.90 m)
Wingspan: 34 ft 9¼ in (10.60 m)
Wing area: 495.16 sq ft (46.00 m^2)
Canard foreplane span: 17 ft 10½ in (5.45 m)
Canard foreplane area: 66.74 sq ft (6.20 m^2)
Wheel base: 18 ft 8 in (5.69 m)
Wheel track: 15 ft 7½ in (4.76 m)

Powerplant

One Volvo Flygmotor RM8B turbofan (Pratt & Whitney JT8D-22 with Swedish-designed afterburner and thrust reverse) rated at 16,600 lb st (73.84 kN) maximum military dry and 28,109 lb st (125 kN) with afterburning

Weights

Normal take-off: 33,069 lb (15000 kg)
Maximum take-off interceptor: 37,478 lb (17000 kg)
Maximum take-off attack: 45,194 lb (20500 kg)

Fuel and load

Internal fuel: 1,506 US gal (5700 litres)

Performance

Maximum level speed clean at 36,000 ft (10975 m): More than 1,147 kt (1,321 mph; 2126 km/h)
Climb to 32,800 ft (10000 m): Less than 1 minute 40 seconds from brakes off with afterburning
Service ceiling: 60,000 ft (18290 m)
Take-off run at typical take-off weight: 1,312 ft (400 m)
Landing run: 1,640 ft (500 m) at normal landing weight
Combat radius on hi-lo-hi mission: 539 nm (621 miles; 1000 km)
Combat radius on lo-lo-lo mission: 270 nm (311 miles; 500 km)

Armament

Primary armament consists of six AAMs. The standard BVR weapon is the medium-range, semi-active radar-guided, all-weather BAeD Rb 71 Sky Flash. Rb 74 (AIM-9L) IR-homing Sidewinders are fielded for short-range work. The JA 37 also has an integral 30-mm Oerlikon KCA revolver cannon with 150 rounds.Seven to nine pylons accommodate up to 13,000 lb (5987 kg) of external stores. These include four pods each containing six Bofors 5.3-in (13.5-cm) rockets for air-to-surface use.

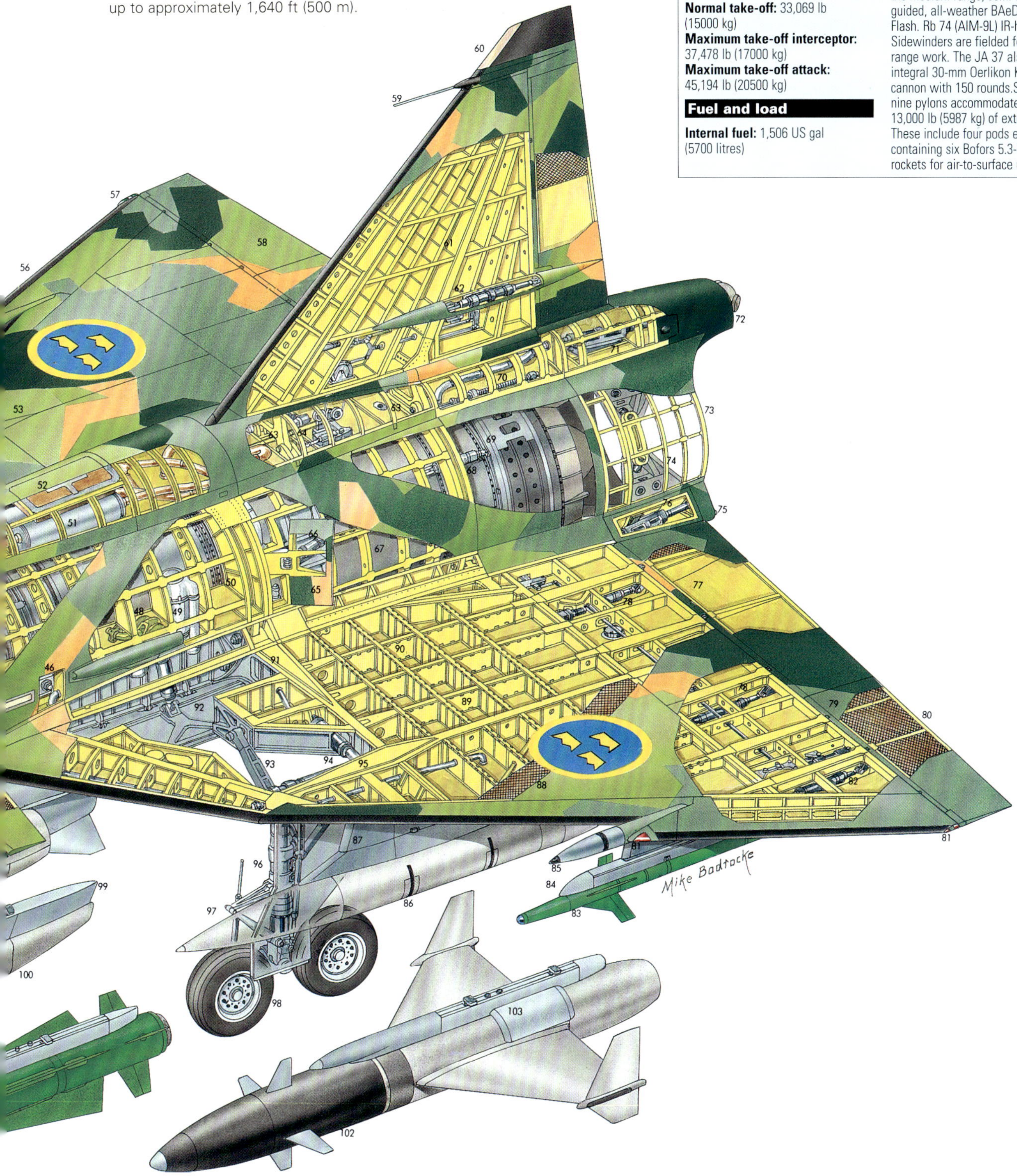

Sukhoi Su-7 'Fitter'

In Indian Air Force service, the Su-7 gained a reputation as a formidable combat asset during the confrontation with Pakistan in 1971. Approximately 150 Su-7BMs were operational during the war, equipping seven units, five of which were based on the Western Front. 221 Sqn was particularly active: raiding the sole PAF airbase at Tezgaon outside Dacca and the airfield at Jessore. Some 32 'Fitter-As' were lost.

Su-7BMK 'Fitter-A'

Cutaway key
1 Pitot tube
2 Pitch vanes
3 Yaw vanes
4 Engine air intake
5 Fixed intake centre-body
6 Radome
7 Ranging radar scanner
8 ILS aerial
9 Radar controller
10 Weapon release ballistic computer
11 Retractable taxiing lamp
12 SRO-2M 'Odd-rods' IFF aerials
13 Intake suction relief doors
14 Intake duct divider
15 Instrument access panel
16 Su-7UM 'Moujik' two-seat operational training variant
17 Armoured glass windscreen
18 Reflector sight
19 Instrument panel shroud
20 Control column
21 Rudder pedals
22 Control linkages
23 Nose undercarriage wheel well
24 Nosewheel doors
25 Torque scissor links
26 Steerable nosewheel
27 Low pressure 'rough-field' tyre
28 Hydraulic retraction jack
29 Cockpit pressure floor
30 Engine throttle
31 Pilot's side console panel
32 Ejection seat
33 Canopy release handle
34 Parachute pack headrest
35 Rear view mirror
36 Sliding cockpit canopy cover
37 Instrument venturi
38 Radio and electronics equipment bay
39 Intake ducting
40 Air conditioning plant
41 Electrical and pneumatic systems ground connections
42 Cannon muzzle
43 Skin doubler/blast shield
44 Fuel system components access
45 Main fuel pumps
46 Fuel system accumulator
47 Filler cap
48 External piping ducts
49 Starboard main undercarriage leg pivot fixing
50 Shock absorber pressurisation charging valve
51 Gun camera
52 Starboard wing integral fuel tank
53 Starboard wing fence
54 Outer wing panel dry bay
55 Wing tip fence
56 Static discharger
57 Starboard aileron
58 Flap guide rail
59 Starboard fowler flap
60 Flap jack
61 Fuselage skin plating
62 Fuselage fuel tank
63 Wing/fuselage attachment double frame
64 Engine compressor face
65 Ram air intake
66 Engine oil tank
67 Bleed air system 'blow-off' valve
68 Fuselage break point, engine removal
69 Lyulka AL-71F-1 turbojet
70 Afterburner duct
71 Fin root fillet
72 Autopilot controller
73 Starboard upper airbrake, open
74 Rudder power control unit
75 Artificial feel unit
76 Tailfin construction
77 VHF/UHF aerial fairing
78 RSIU (very short wave fighter radio) aerial
79 Tail navigation light
80 Sirena-3 tail warning radar
81 Rudder
82 Brake parachute release tank
83 Brake parachute housing
84 Parachute doors
85 Engine exhaust nozzle
86 Port all-moving tailplane
87 Static discharger
88 Tailplant anti-flutter weight
89 Tailplane construction
90 Pivot mounting
91 Tailplane limit stops
92 Variable area exhaust nozzle flaps
93 Nozzle control jacks
94 Fin/tailplane attachment fuselage frame
95 Afterburner cooling air intake
96 Rear fuselage frame and stringer construction
97 Insulated tailplane
98 Airbrake housing
99 Hydraulic jack
100 Tailplane power control unit
101 'Odd rods' IFF aerials
102 Port lower airbrake, open
103 Engine accessories
104 Jettisonable RATO bottle
105 Port fowler flap
106 Port wing integral fuel tanks
107 Aileron control rod
108 Port aileron construction
109 Static discharger
110 Wing tip fairing
111 Port navigation light
112 Wing tip fence
113 Pitot tube
114 Wing rib and stringer construction
115 Port outer stores pylon

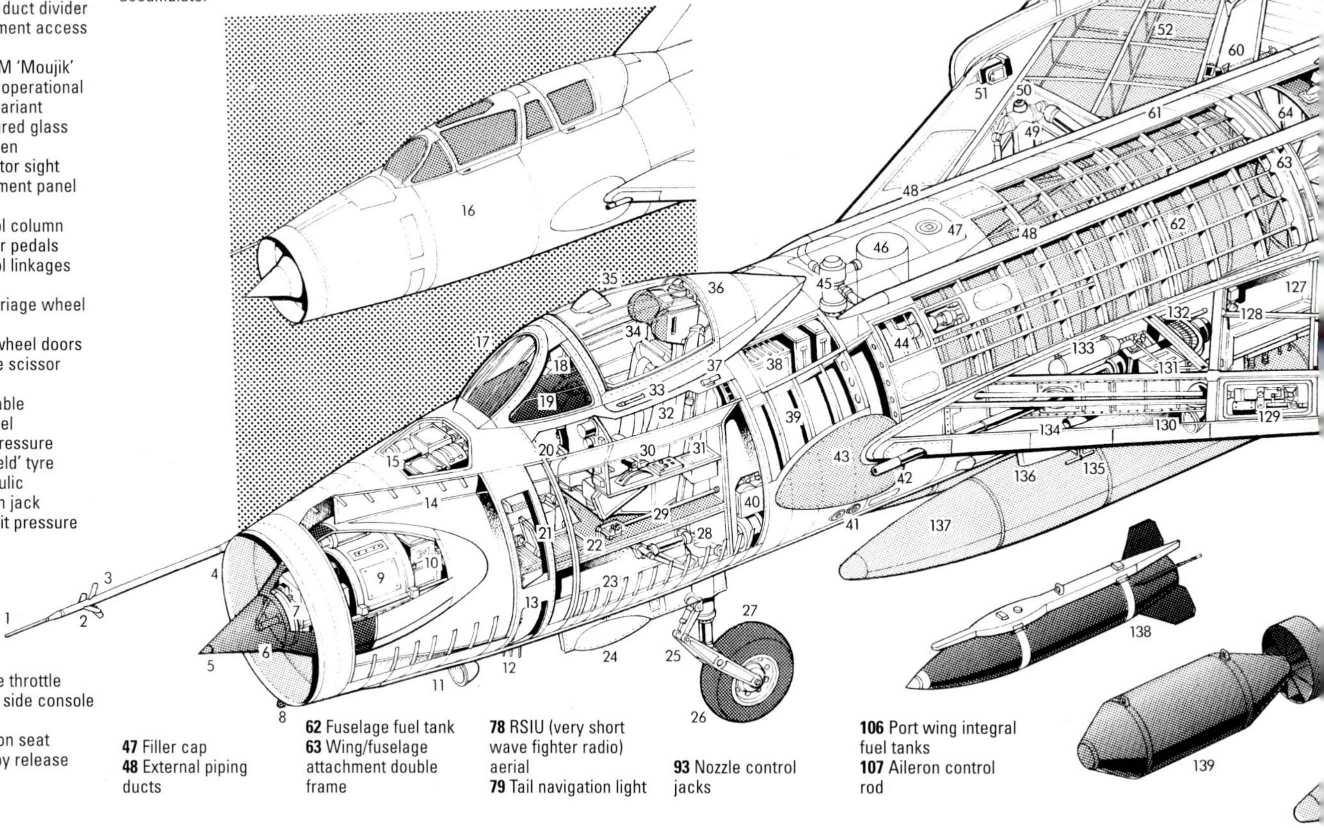

SPECIFICATION

Su-7BMK 'Fitter-A'

Dimensions

Wingspan: 29 ft 3½ in (8.93 m)
Length: 57 ft (17.37 m) including probe
Height: 15 ft (4.57 m)
Wing area: 297.09 sq ft (27.60 m²)
Aspect ratio: 2.89

Powerplant

One NPO Saturn (Lyul'ka) AL-7F-1 turbojet rated at 15,432 lb st (68.65 kN) dry and 22,282 lb st (99.12 kN) with afterburning

Weights

Empty equipped: 19,004 lb (8620 kg)
Normal take-off: 26,455 lb (12000 kg)
Maximum take-off: 29,762 lb (13500 kg)

Fuel load

Internal fuel: 5,181 lb (2350 kg)
External fuel: up to two 159-US gal (600-litre) and two 458- or 238-US gal (1800- or 900-litre) drop tanks; maximum ordnance 5,511 lb (2500 kg)

Performance

Maximum level speed 'clean' at 36,090 ft (11000 m): 1,055 mph (1700 km/h)
Maximum level speed at sea level: 840 mph (1350 km/h)
Maximum rate of climb at sea level: about 29,920 ft (9120 m) per minute
Service ceiling: 49,705 ft (15150 m)
Take-off run: 2,887 ft (880 m)
Ferry range with drop tanks: 901 miles (1450 km)
Combat radius on a hi-lo-hi attack mission with a 2,205-lb (1000-kg) warload and two drop tanks: 214 miles (345 km)

Armament

Two 30-mm Nudelmann Richter NR-30 cannon in the wing roots, plus a wide range of bombs, rockets and unguided missiles

These Polish Su-7BM 'Fitter-As' appear to have been updated to Su-7BMK standard with the enlarged fairing – obscured here by the technicians who appear to be repacking the 'chutes – for the twin brake 'chutes of this variant. All single-seat Su-7s were designated 'Fitter-A' by ASCC/NATO, while the two-seat Su-7UM was known as 'Moujik'. Experimental Su-7s included the 100LDU testbed of 1968, which was derived from n Su-7UM, and as a remotely-piloted aircraft was used to investigate flight stability for the T-4.

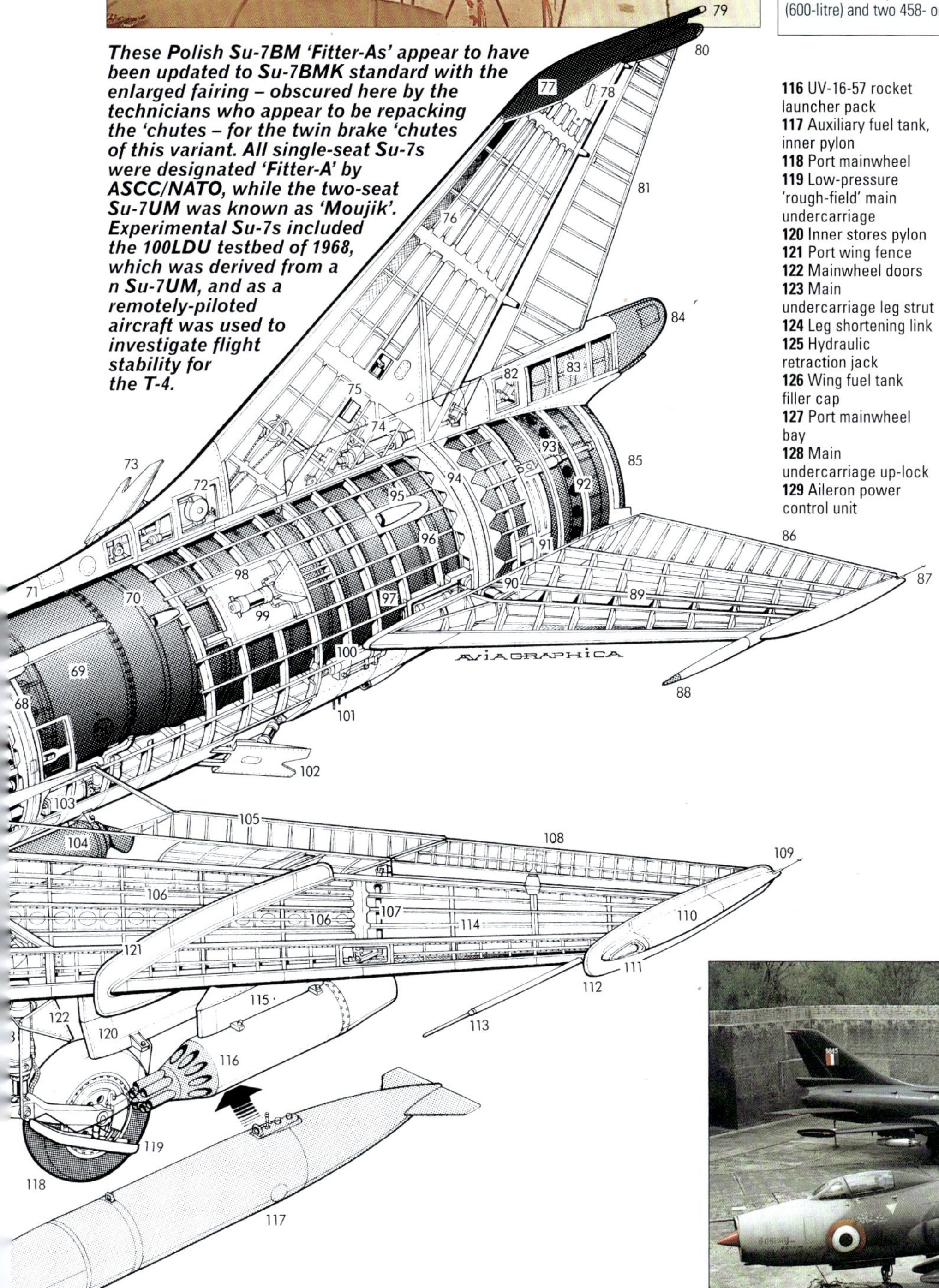

116 UV-16-57 rocket launcher pack
117 Auxiliary fuel tank, inner pylon
118 Port mainwheel
119 Low-pressure 'rough-field' main undercarriage
120 Inner stores pylon
121 Port wing fence
122 Mainwheel doors
123 Main undercarriage leg strut
124 Leg shortening link
125 Hydraulic retraction jack
126 Wing fuel tank filler cap
127 Port mainwheel bay
128 Main undercarriage up-lock
129 Aileron power control unit
130 Retractable landing lamp
131 Ammunition tank (80 rounds per gun)
132 30-mm NR-30 cannon
133 Cannon pressurisation bottle
134 Ventral gun gas venting intake
135 Radar altimeter
136 Fuselage pylon, port and starboard
137 Twin fuselage mounted auxiliary fuel tanks
138 551-lb (250-kg) concrete piercing bomb
139 1,102-lb (500-kg) HE bomb

This pair of specially-marked Su-7s was photographed in a poor condition at an Indian air force base. Both aircraft have UB-16 rockets pods under their outer wings and both are moored to concrete blocks and in a poor state of finish.

Sukhoi Su-17/20/22 'Fitter'

*Formerly the prime operator of the **Su-17** family, the **'Fitter'** is now rapidly disappearing from the air forces of the former Soviet Union .The intended replacement for the **Su-17M** (and **MiG-27**) in Soviet service was the now aborted **MiG-29M**, specifically developed to meet the requirements of the **Group of Soviet Forces in Germany**. The **Su-17M-4** (illustrated) has now left the front-line of the Russian inventory in response to post-Cold War CFE guidelines.*

Su-22M-4 'Fitter-K'

Cutaway key

1 Instrumentation data probe
2 Yaw and pitch vanes
3 Fire control system computer transducers
4 Pitot head
5 Conical intake centre-body shock cone/ radome
6 Engine air intake
7 'High Fix' 1-band ranging radar
8 Laser marked target designator
9 Radar altimete r
10 Ventral doppier navigation aerial
11 Angle of attack transmitter
12 Radar equipment module
13 Bifurcated intake ducting
14 Spring loaded intake suctton relief doors, open
15 Temperature probe
16 Nose avionics equipment compartment, ASP-5ND fire control system
17Su-17'Fitter G' two-seat tandem trainer variant, nose profile
18 Student pilot's cockpit
19 Retractable forward vision periscope
20 Instructor's cockpit
21 Armoured glass windscreen panels
22 Pilot's head-up display and attack sight
23 Instrument panel shroud
24 Control column
25 Rudder pedals
26 Nose undercarriage wheel bay
27Retractablelanding/ taxiing lamp, port and starboard
28 Nosewheel doors
29 Nosewheel forks
30 Steerable nosewheel forward retracting
31 SRO-21VI 'Odd Rods'IFF aerials
32 Nosewheel leg pivot fixing
33 Hydraulic retraction jack
34 Cockpit floor level
35 Close pitched fuselage frames
36 Port console panel
37 Engine throttle lever
38 Canopy latch
39 Pilot's 'zero-zero' ejection seat
40 Ejection seat headrest
41 Rear view mirror
42 Cockpit canopy cover, upward tingeing
43 Canopy jack
44 Cockpit pressurisation valve
45 Rear pressure bulkhead
46 Air conditioning plant
47 intake duct framing
48 Ground power and intercom sockets
49 Cannon muzzle blast- shield/skin doubler
50 Avionics equipment racks
51 Cockpit aft fairing additional avionics equipment
52 Fuel system access panel
53 Intake trunking
54 Main fuel pumps
55 Inverted fight accumulator
56 Front wing spar attachment main frame
57 Centre fuselage frame and stringer construction
58 Fuselage fuel tanks
59 Fuel system components
60 Dorsal spine fairing
61 ADF aerial
62 Starboard wing fixed root section
63 Strike camera
64 Wing pivot bearing
65 Outboard wing fence
66 Fuselage centreline reconnaissance pod
67 GSh-231-cannon pod
68 Leading edge slats, down position
69 Slat hydraulic actuator
70 Starboard wing integral fuel tank
71 Aileron hydraulic actuator
72 Slat guide rails
73 Starboard navigation light
74 Wing tip fairing
75 Static discharger
76 Starboard aileron
77 Starboard wing fully swept position
78 Outboard single-slotted flap, down postion
79 Wing glove section
80 Spine fairing access panels
81 Fuselage skin panelling
82Wing main spar attachment double frame
83 Engine compressor intake
84 Engine oil tank
85 Lyulka AL-21 F-3 afterburning engine (Tumanskii R-29B alternative instailation)
86 Rear fuselage break point (engine removal)
87 Engine turbine section
88 Cooling air intakes
89 Forward 'Sirena-3' radar warning and ECIA aerial (repositioned from leading edge of three-pylon wing)
90 Starboard upper airbrake, open
91 Autopilot controller
92 HF aerial
93 Starboard tailplane anti flutterweight
94 Rudder control linkages
95 Rudder hydraulic actuator
96 Tailfin construction
97 PSI U (very short wave fighter control) aerial
98 Fin tip UHF aerial fairing
99 Tail navigation light
100 Rudder
101 Rear'Sirena-3' radar warning and ECM aerial
102 Parachute release link
103 Brake parachute housing
104 Tailcone/parachute door conic fairing
105 Transponder aerial
106 Engine exhaust nozzle
107 Port all-moving tailplane
108 Static discharger
109 Tailplane tip anti-flutter weight
110 Tailplane rib construction
111 Tailplane spar
112 Pivot mounting
113 Tailplane limit stops
114 Variable area afterburner nozzle
115 Nozzle control jacks
116 Fin and tailplane attachment fuselage double frame
117 Afterburner ducting
118 Tailplane hydraulic actuator
119 RO-2M'Odd Rods'IFF aerial
120 Airbrake housing
121 Airbrake hydraulic jack
122 Rear fuselage frame and stringer construction

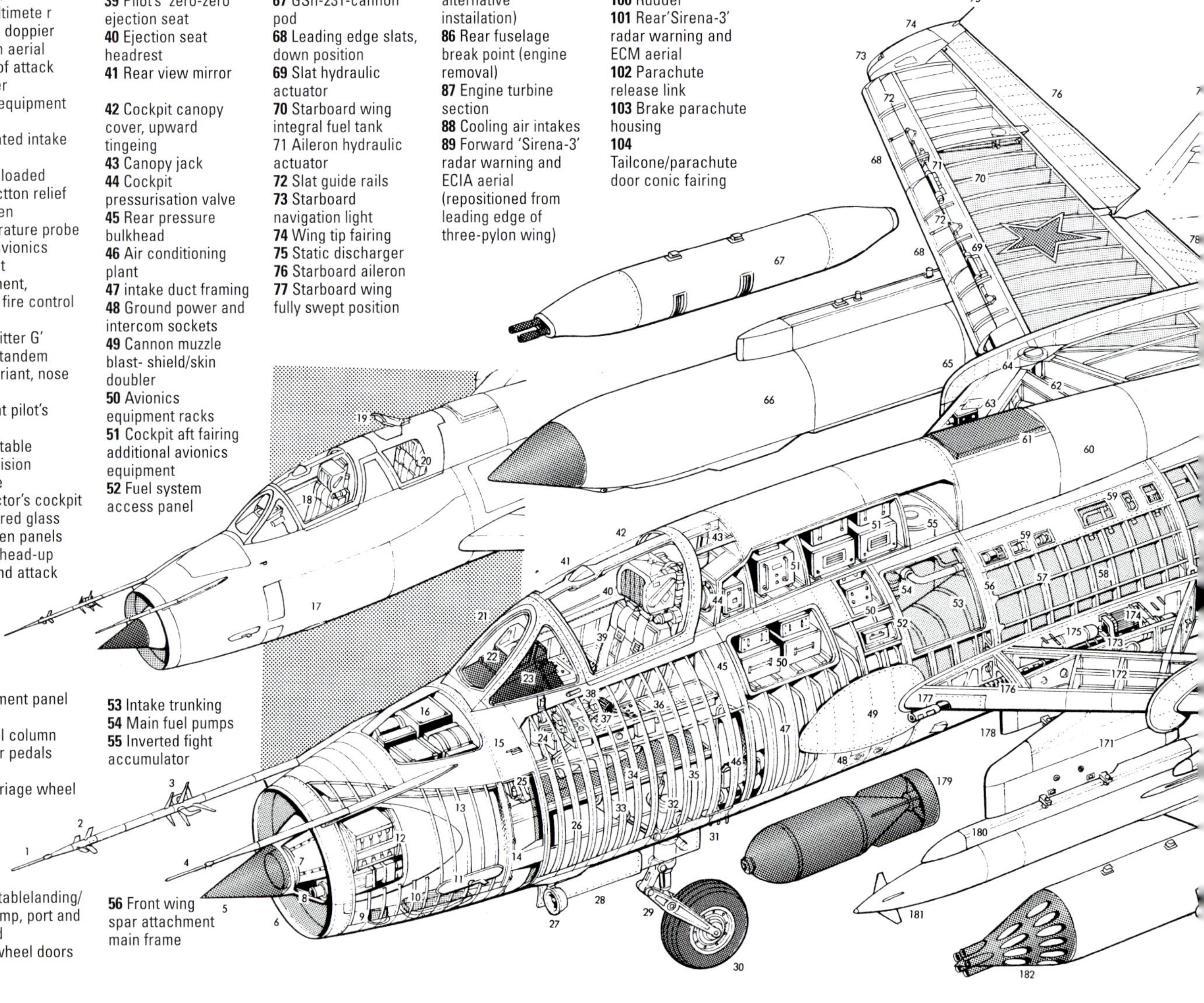

Above: The definitive two-seat 'Fitter' was exported to allied air forces as the R-29BS-300-engined Su-22UM-3, delivered from 1982. This was quickly superceded by the Su-22UM-3K delivered from 1983, and incorporating the AL-21F-3 powerplant carried as standard by Soviet Su-17Ms. East Germany operated eight examples of the Su-22UM-3K within a total 'Fitter' fleet comprising 53 aircraft. During 1979-80 all Soviet Su-17UM two-seaters were updated to Su-17UM-3 standard.

SPECIFICATION

Su-17M-4 'Fitter-K'

Dimensions

Length (including probes): 61 ft 6¼ in (18.75 m)
Height: 16 ft 5 in (5.00 m)
Span (spread): 45 ft 3 in (13.80 m)
Span (swept): 32 ft 10 in (10.00 m)
Wing area (spread): 430.57 sq ft (40.00 m²)
Wing area (swept): 398.28 sq ft (37.00 m²)

Powerplant

One NPO Saturn (Lyul'ka) AL-21F-3 turbojet rated at 17,196 lb st (76.49 kN) dry and 24,802 lb st (110.32 kN) with afterburning, plus provision for two RATO units

Weights

Normal take-off: 36,155 lb (16400 kg)
Maximum take-off: 42,989 lb (19500 kg)

Performance

Maximum level speed 'clean' at sea level: 870 mph (1400 km/h)
Maximum rate of climb at sea level: 45,276 ft (13800m) per minute
Service ceiling: 49,870 ft (15200 m)
Combat radius: 621 nm (715 miles; 1150 km) on a hi-lo-hi mission with a 4,409 lb (2000-kg) warload, or 378 nm (435 miles; 700 km) on a lo-lo-lo mission with a 4,409 lb (2000-kg) warload

Armament

A wide range of freefall bombs and podded and unpodded unguided rocket projectiles ranging in calibre from 57 mm to 330 mm; for precision attacks a variety of air-to-surface missiles, including Kh-25 (AS-10 'Karen' and AS-12 'Kegler') and Kh-29, and the Kh-58E (AS-11 'Kilter') anti-radar missile. To further improve self-defence capability, four 32-round upward-firing ASO chaff/flare dispensers can be scabbed on to either side of the tailfin, augmenting the two six-tube KDS-23 dispensers mounted flush with the dorsal spine.For strafe attacks, the 'Fitter-K's wingroot-mounted NR-30 30-mm cannon (each with 80 rounds) can be augmented by gun pods carried under the wings or fuselage. .The 'Fitter-K' can be used in the tactical reconnaissance role, carrying the same KKR reconnaissance pod as has been applied to the 'Fitter-C' and 'Fitter-H'. This contains three optical cameras, flares and Elint modules, and is usually carried in association with the SPS ECM pod

123 Ventral fin
124 Port lowerairbrake, open
125 Engine accessory equipment access panel
126 Accessory equipment gearbox compartment
127 Port inboard single-slotted flap
128 Flap actuator
129 Auxiliary rear spar
130 Wing sweep control hydraulic jack
131 Rear spar guide rails
132 Wing glove section external stiffeners
133 Outboard wing fence
134 Outboard single-slotted flap
135 Flap rib construction
136 Flap actuator
137 Aileron construction
138 Port wing fully swept (63 deg) position
139 Port aileron
140 Aileron hinge control linkage
141 Static discharger
142 Wing tip fairing
143 Port wing fully forward (28 deg) position
144 Port navigation light
145 Three-segment leading edge slats, down position
146 BETA-B 250-kg (551-1b) retarded concrete piercing bomb
147 S-24 240-mm aircraft rocket
148AA-2 (K-13A)'Atoli' air-to air self-defence missile
149 Missile launch rail
150 Leading edge slat rib construction
151 Aileron hydraulic actuator

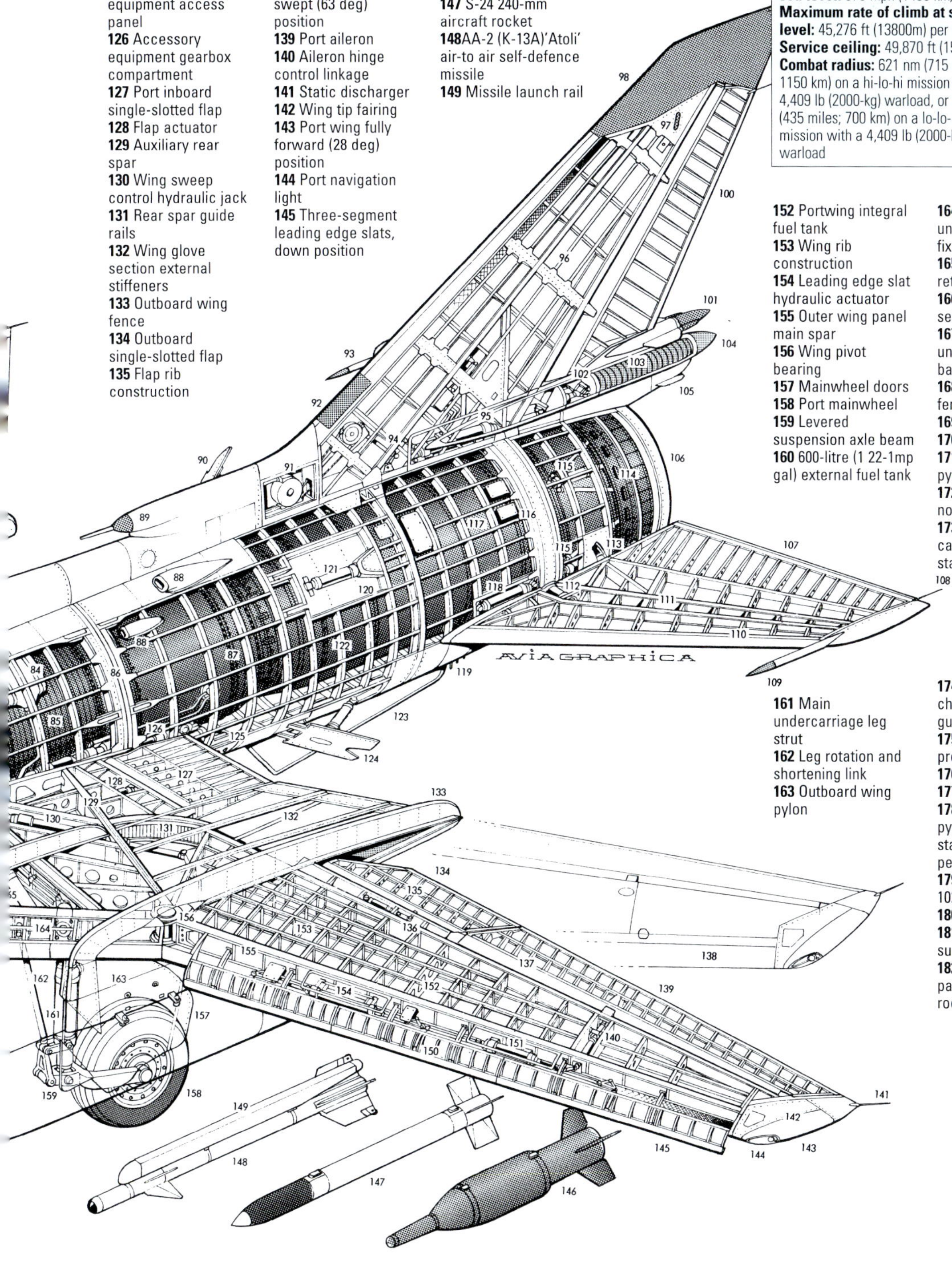

152 Portwing integral fuel tank
153 Wing rib construction
154 Leading edge slat hydraulic actuator
155 Outer wing panel main spar
156 Wing pivot bearing
157 Mainwheel doors
158 Port mainwheel
159 Levered suspension axle beam
160 600-litre (1 22-1mp gal) external fuel tank
161 Main undercarriage leg strut
162 Leg rotation and shortening link
163 Outboard wing pylon
164 Main undercarriage pivot fixing
165 Hydraulic retraction jack
166 Fixed wing section main spar
167 Main undercarriage wheel bay
168 Inboard wing fence
169 Front spar
170 Centre wing pylon
171 Inboard wing pylon
172 Leading edge nose ribs
173 N R 30-mm cannon, port and starboard
174 Ammunition feed chute, 70-rounds per gun
175 Cannon pressurisation bottle
176 Recoil mounting
177 Cannon muzzle
178 Fuselage stores pylons, port and starboard, one or two per side
179 FAB 500, 500-kg (1 102-1b) HE bomb
180 Missile launch rail
181AS-7 'Kerry'air-to-surface missile
182 UV-32-57 rocket pack 32x57-mm rockets

Sukhoi Su-27 'Flanker'

The formidable Sukhoi Su-27 'Flanker' family of fighter aircraft is one of the most successful ever to emerge from the former Soviet Union. Supremely manoeuvrable, heavily armed and blessed with exceptional range, it has been compared favourably with the best Western fighters.

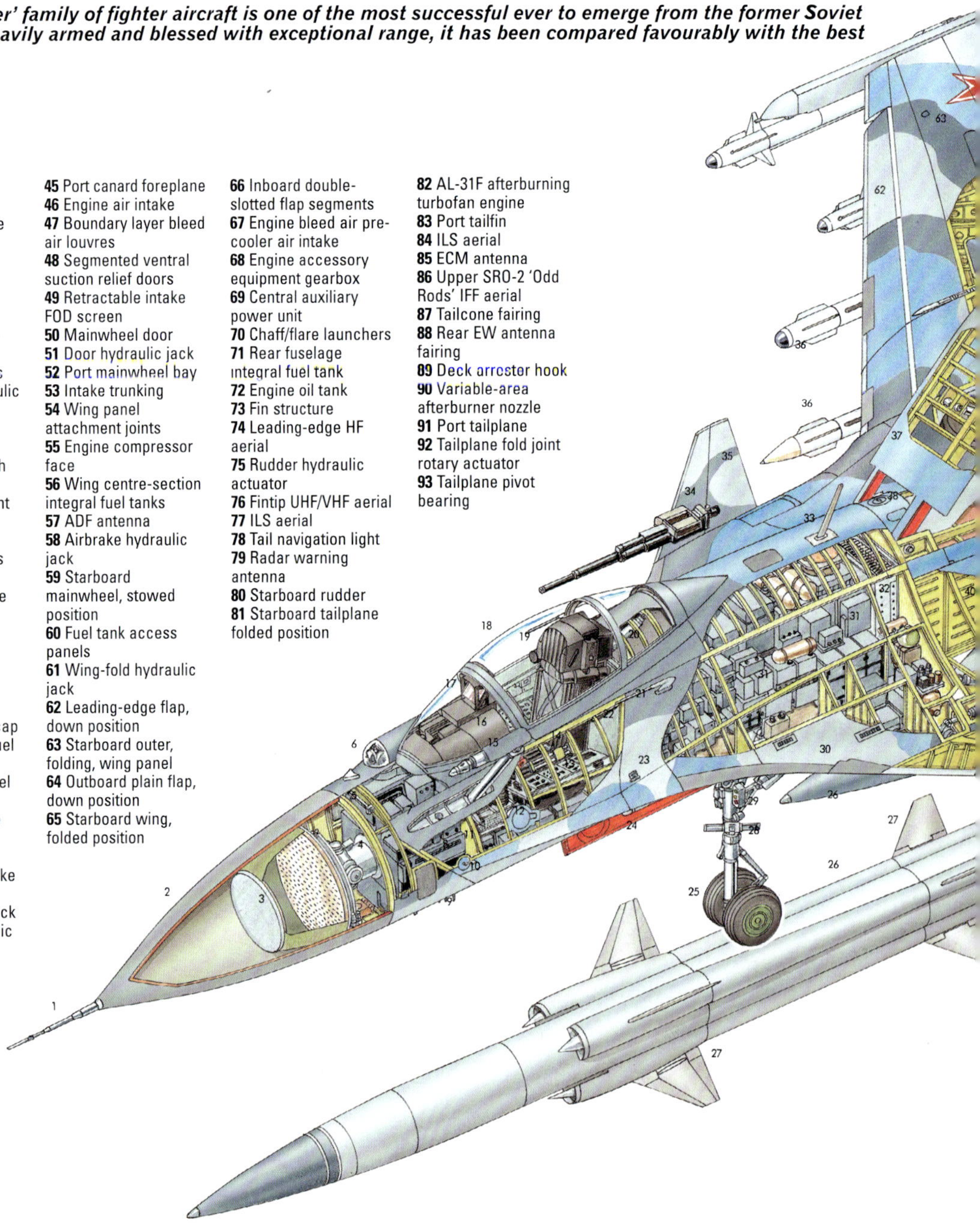

Su-27K (Su-33)

Cutaway key
1 Pitot head
2 Upward-hinging radome
3 Radar scanner
4 Scanner mounting
5 Radome hinge point
6 Infra-red search and tracking scanner
7 Refuelling probe housing
8 Radar equipment module; tilts down for access
9 Lower SRO-2 'Odd-Rods' IFF aerial
10 Incidence transmitter
11 Cockpit front pressure bulkhead
12 Retractable spotlight, port and starboard
13 Cockpit side console panel
14 Slide-mounted throttle levers
15 Flight-refuelling probe, extended
16 Instrument panel shroud
17 Pilot's head-up display
18 Upward-hinging cockpit canopy
19 K-36MD 'zero-zero' ejection seat
20 Canopy hydraulic jack
21 Dynamic pressure probe, port and starboard
22 Cockpit rear pressure bulkhead
23 Temperature probe
24 Nosewheel door
25 Twin nosewheels, forward-retracting
26 ASM-MSS long-range ramjet and rocket-powered anti-shipping missile
27 Missile folding fins
28 Nosewheel hydraulic steering jacks
29 Deck approach 'traffic-lights'
30 Leading-edge flush EW aerial
31 Avionics equipment bay
32 Ammunition magazine, 149 rounds
33 HF aerial
34 Starboard fuselage GSh-30-1 30-mm cannon
35 Canard foreplane
36 Starboard wing missile armament
37 Dorsal airbrake
38 Gravity fuel filler cap
39 Centre fuselage fuel tank
40 Forward lateral fuel tanks
41 ASM-MSS missile carrier on fuselage centreline station
42 Variable-area intake ramp doors
43 Ramp hydraulic jack
44 Foreplane hydraulic actuator
45 Port canard foreplane
46 Engine air intake
47 Boundary layer bleed air louvres
48 Segmented ventral suction relief doors
49 Retractable intake FOD screen
50 Mainwheel door
51 Door hydraulic jack
52 Port mainwheel bay
53 Intake trunking
54 Wing panel attachment joints
55 Engine compressor face
56 Wing centre-section integral fuel tanks
57 ADF antenna
58 Airbrake hydraulic jack
59 Starboard mainwheel, stowed position
60 Fuel tank access panels
61 Wing-fold hydraulic jack
62 Leading-edge flap, down position
63 Starboard outer, folding, wing panel
64 Outboard plain flap, down position
65 Starboard wing, folded position
66 Inboard double-slotted flap segments
67 Engine bleed air pre-cooler air intake
68 Engine accessory equipment gearbox
69 Central auxiliary power unit
70 Chaff/flare launchers
71 Rear fuselage integral fuel tank
72 Engine oil tank
73 Fin structure
74 Leading-edge HF aerial
75 Rudder hydraulic actuator
76 Fintip UHF/VHF aerial
77 ILS aerial
78 Tail navigation light
79 Radar warning antenna
80 Starboard rudder
81 Starboard tailplane folded position
82 AL-31F afterburning turbofan engine
83 Port tailfin
84 ILS aerial
85 ECM antenna
86 Upper SRO-2 'Odd Rods' IFF aerial
87 Tailcone fairing
88 Rear EW antenna fairing
89 Deck arrester hook
90 Variable-area afterburner nozzle
91 Port tailplane
92 Tailplane fold joint rotary actuator
93 Tailplane pivot bearing

***Above:** The sheer size of the 'Flanker' is evident in this view of two Su-27Ks of the 1st Squadron, Severomorsk Regiment, AV-MF (Russian naval aviation). The Su-27K (Su-33) is a naval variant of the 'Flanker-B' interceptor. The large lifting area makes the Su-27 suitable for operations from aircraft-carriers, while its size enables a huge internal fuel load to be carried, conferring excellent long-range performance. Clearly visible are the Su-27K's canards and double-slotted trailing-edge flaps.*

SPECIFICATION

Su-27P 'Flanker-B'

Dimensions

Fuselage length (including probe): 72 ft 0 in (21.94 m)
Wing span over tip missile launch rails: 48 ft 3 in (14.70 m)
Wing aspect ratio: 7.76
Tailplane span: 32 ft 5 in (9.88 m)
Wing area: 667.8 sq ft (62.04 m²)
Horizontal tail area: 131.75 sq ft (12.24 m²)
Total fin area: 165.76 sq ft (15.40 m²)
Distance between fin tips: 14 ft 1¼ in (4.30 m)
Overall height: 19 ft 6 in (5.93 m)
Wheel track: 14 ft 3 in (4.34 m)
Wheelbase: 19 ft 4 in (5.88 m)
Maximum wing loading: 93.4 lb/sq ft (456.2 kg/m²)

Powerplant

Two Saturn Lyul'ka AL-31F afterburning turbofans each rated at 16,755 lb st (74.53 kN) dry and 27,558 lb st (122.59 kN) with afterburning

Weights

Empty operating: 36,112 lb (16380 kg)
Normal take-off: 50,705 lb (23000 kg)
Maximum take-off: 62,391 lb (28300 kg)

Fuel and load

Internal fuel: (normal) 11,620 lb (5270 kg), (maximum) 20,723 lb (9400 kg) or 2,640 Imp gal (12000 litres) in three main fuselage tanks, with additional tanks in outer wing panels; the basic Su-27 has no provision for inflight refuelling or for the carriage of external fuel tanks (but see under individual variant briefings for exceptions)
Maximum theoretical weapon load: 17,636 lb (8000 kg)
Normal weapon load: 8,818 lb (4000 kg)

g limits

8-9 at basic design gross weight

Performance

Maximum level speed at sea level (estimated): 743 kt (850 mph; 1370 km/h)
Maximum level speed 'clean' at altitude: 1,236 kt (1,418 mph; 2280 km/h)
Limiting Mach No.: 2.35
Absolute ceiling: 60,700 ft (18500 m)
Practical service ceiling (estimated): 58,070 ft (17700 m)
Take-off run: 1,640 ft (500 m) or 1,476 ft (450 m)
Landing roll: 1,968 ft (600 m) or 2,297 ft (700 m)
Landing speed: 121-124 kt (140-143 mph; 225-230 km/h)

Range

Maximum range: 1,987 nm (2,285 miles; 3680 km) at altitude, 740 nm (851 miles; 1370 km) at low level
Radius of action (high-altitude): 590 nm (677 miles; 1090 km)
Radius of action (low-altitude): 227 nm (261 miles; 420 km)

Armament

Note: This specification should be treated with some caution. Sukhoi has released widely differing performance figures on different occasions (and even releases different dimensions for the same aircraft), while rarely specifying the loads carried for particular range or performance figures.

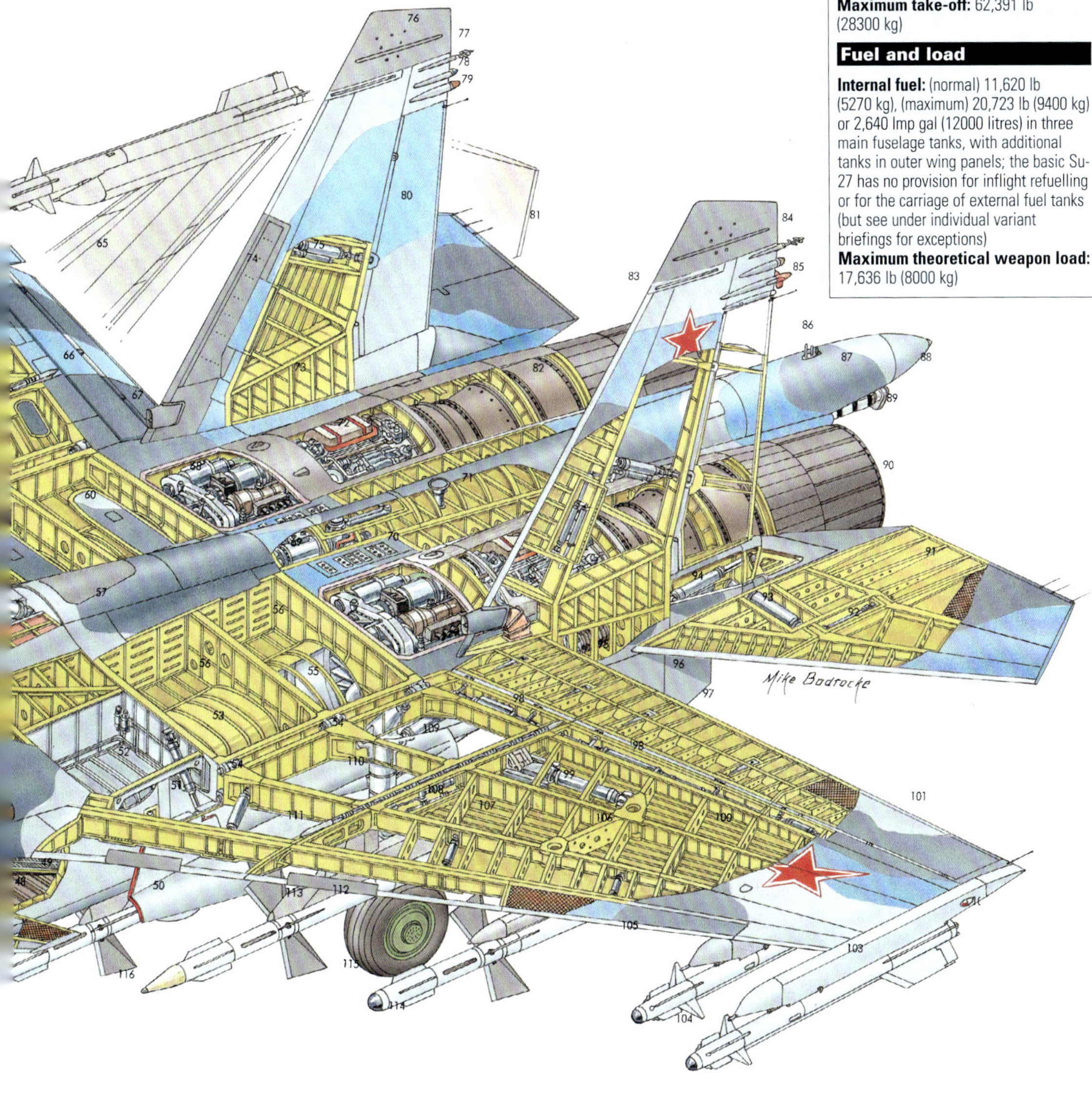

94 Hydraulic actuator
95 Hydraulic accumulator
96 Ventral fin
97 Port inboard double-slotted flap segments
98 Flap hydraulic actuators
99 Wing-fold hydraulic jack
100 Outer wing panel structure
101 Outboard plain flap segment
102 Port navigation light
103 Wingtip missile launch rail
104 Vympel R-73 (AA-11 'Archer') air-to-air missiles
105 Leading-edge flap
106 Pylon attachment hardpoints
107 Port wing integral fuel tank
108 Wing-fold locking mechanism jack
109 Main undercarriage hydraulic retraction jack
110 Mainwheel leg strut
111 Wing-fold hinge joint
112 Leading-edge flush EW aerial panels
113 Missile pylon
114 Vympel R-27 (AA-10 'Alamo-B') IR-homing air-to-air missile
115 Port mainwheel
116 Vympel R-27 (AA-10 'Alamo-C') RHAAM

Helicopters

Aérospatiale Alouette III

Aérospatiale's SA.316B is one of the world's t
selling light helicopters – nearly 1,500 examp
have been built and the helicopter has be
exported to 190 civil and military operators in
countries. A further 400 examples have been b
by HAL in India (the Chetak) and by ICA-Brasov
Romania (where they were designated
IAR 316B). This example is from the South Afric
Air Force and is being escorted by a Pur

SA.316B Alouette III

Cutaway key

1 FM homing antennas, port and starboard
2 Pitot head
3 Instrument access panel
4 Cockpit ventilating air intake
5 Antenna mounting
6 Downward-view windows
7 Curved windscreen panels
8 Standby compass
9 Instrument panel shroud
10 Pilot's instrument console
11 Weapons system control panel
12 Centre control pedestal
13 Yaw control rudder pedals
14 Landing lamp
15 Floor beam construction
16 Levered suspension nose landing gear leg strut
17 Non-retracting castoring nosewheel
18 Port navigation light
19 Door jettison linkage
20 Cyclic pitch control column
21 Central power and engine condition levers
22 Collective pitch control lever
23 Control column handgrip
24 Missile hand-controller
25 Safety harness
26 Starboard jettisonable cockpit door
27 Outside air temperature gauge
28 Starboard sliding cabin door
29 Cabin roof glazing
30 APX-Bezu 260 gyro-stabilised sight
31 Retractable sight controller and binocular viewer
32 Pilot's seat
33 Co-pilot/weapons officer's seat
34 Sliding side window panel
35 Port jettisonable cockpit door
36 Collective pitch-control lever
37 Seat mounting rails (three-abreast front seat row)
38 Boarding step
39 Lower fuselage 'raft' section skin panelling
40 Port sliding cabin door
41 Passenger/cargo loading
42 Door hatches/ stretcher handle apertures
43 Door latches
44 Troop-carrying folding seats (four)
45 Fixed backrest
46 Control rod linkages
47 Sliding doortop rail
48 Cabin rear-sloping bulkhead
49 Trim/insulating panelling
50 First-aid kit
51 Anti-collision light
52 VHF aerial
53 Cabin roof skin panelling
54 Rotor head control rods
55 Main transmission gearbox
56 Swash plate mechanism
57 Blade pitch angle control rods
58 Torque scissor links
59 Rotor head hinge fitting
60 Lifting fitting
61 Hydraulic drag hinge dampers
62 Bracing cables

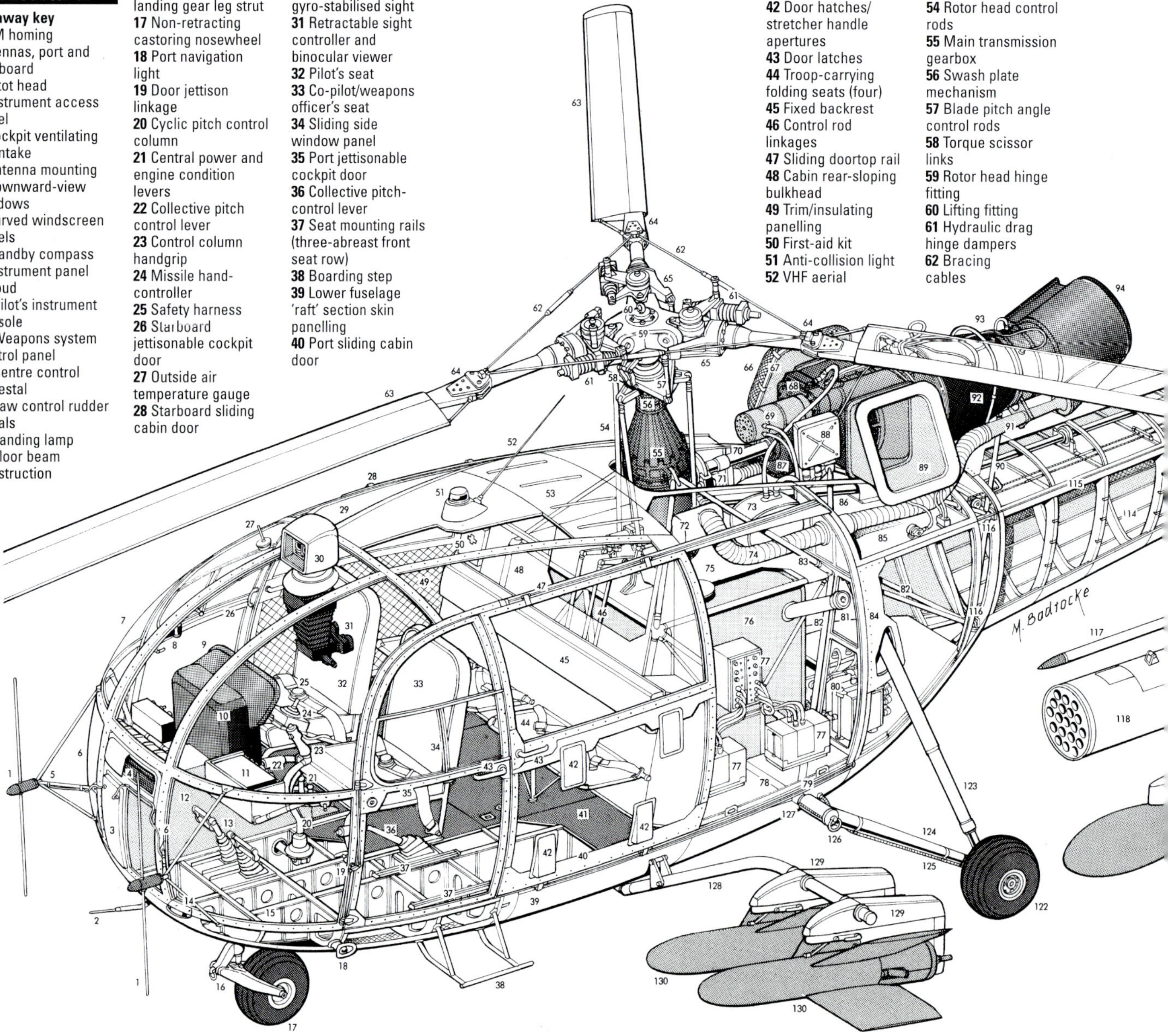

France still retains a significant Alouette capability and the Aéronavale is the major operator. These ageing aircraft remain in service due to the delays over the entry into service of the NH 90. About 70 examples of the Alouette II and the Alouette III are operated by 20, 22 and 23 Escadrille de Servitude, performing in the general utility and training roles.

SPECIFICATION

SA.319 Alouette III Astazou

Dimensions

Fuselage length: 32 ft 10¾ in (10.03 m)
Length with rotors turning: 42 ft 1½ in (12.84 m)
Height: 9 ft 10 in (3 m)
Wheeltrack: 8 ft 6¼ in (2.60 m)
Main rotor diameter: 36 ft 1¾ in (11.02 m)
Tail rotor diameter: 6 ft 3¼ in (1.90 m)
Main rotor disk area: 1,026.68 sq ft (95.38 m^2)
Tail rotor disk area: 30.84 sq ft (2.87 m^2)

Powerplant

One 870-shp (649 kW) Turboméca Astazou XIV derated to 600 shp (447 kW)

Weights

Empty: 2,513 lb (1140 kg)
Maximum take-off: 4,960 lb (2250 kg)

Fuel and load

Internal fuel: 126.5 Imp gal (575 litres)
External fuel: None
Maximum payload: 1,653 lb (750 kg)

Performance

Maximum level speed 'clean' at sea level: 118 kt (136 mph; 220 km/h)
Maximum cruising speed at sea level: 106 kt (136 mph; 220 km/h)
Maximum rate of climb at sea level: 885 ft (270 m) per minute
Hovering ceiling: 10,170 ft (3100 m)
Range: 326 nm (375 miles; 605 km)

Armament

A wide range of light weaponry can be carried by the Alouette when it is engaged in combat duties. A 0.3-in (7.62-mm) machine-gun can be fired through the port door, while a 20-mm cannon in a fixed axial fairing can be mounted on the port side of the cabin. One or two MATRA 155H rocket pods firing 2⅝ in (68-mm) unguided rockets are useful against soft or dispersed targets. For anti-armour missions, the Euromissile HOT, the AS11 or the FN ETNA TMP-5-twin 0.3-in (7.62-mm) machine gun pod is used. Naval Alouettes carry a pair of Mk 44 torpedoes or a MAD bird and single torpedo or search radar.

63 Three-bladed main rotor
64 Blade root attachment joints
65 Blade pitch angle control horn
66 Engine inlet filter screen
67 Starboard engine inlet
68 Accessory equipment gearbox
69 Generator
70 Engine transmission shaft
71 Rotor brake
72 Transmission oil cooler
73 Oil tank
74 Oil cooler airduct
75 Gearbox mounting deck
76 Fuel tank, capacity 126.5 Imp gal (575 litres)
77 Electrical system equipment
78 Equipment loading deck
79 Sliding cabin door bottom rail
80 Missile system avionics equipment
81 Position of fuel filter on starboard side
82 Welded steel tube centre fuselage framework
83 Gearbox mounting struts
84 Non-structural skin panelling
85 Fireproof engine mounting deck
86 Angled tail rotor transmission shaft
87 Engine reduction gearbox
88 Ignition control unit
89 Port engine bellmouth air inlet
90 Rear engine mounting strut
91 Tailpipe negative pressure cooling air duct
92 Engine combustion section
93 Turboméca Artouste IIIB turboshaft engine
94 Engine exhaust duct
95 Tailboom top decking/access panel
96 Tail rotor transmission shaft
97 Transmission shaft bearings
98 Tail rotor control cables
99 Starboard fixed tailplane
100 Endplate tailfin
101 Three-bladed tail rotor
102 All-metal tail rotor blades
103 Blade pitch control mechanism
104 Right-angle final drive gearbox
105 Tail navigation light
106 Steel tube tailskid/ rotor protector
107 Port fixed tailplane
108 Port endplate tailfin
109 Tailplane bracing struts
110 Main rotor blade balance weights
111 Aluminium alloy blade spar
112 Moltoprene foam trailing-edge filler
113 Bonded aluminium alloy rotor blade skin panels
114 Tailboom frame and stringer construction
115 Upper longeron
116 Tailboom attachment joints
117 68-mm folding fin aircraft rocket (FFAR)
118 MATRA rocket launcher pack
119 Missile pylon adaptor
120 Missile launch rail
121 AS12 wire-guided air-to-surface missiles (two)
122 Port mainwheel
123 Shock absorber leg strut
124 Trailing axle beam
125 Hydraulic brake pipe
126 Tie-down point
127 Axle beam pivot fixing
128 Weapons pylon mount
129 Detachable missile pylons
130 AS11 wire guided air-to-surface missiles (four)

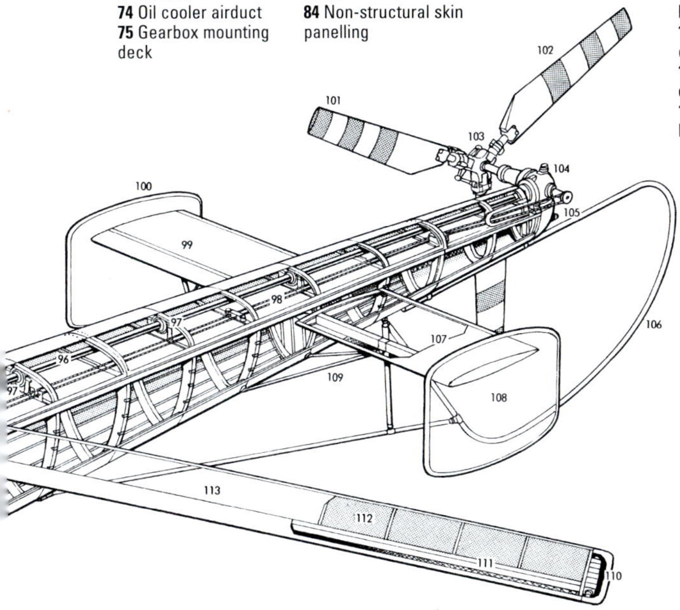

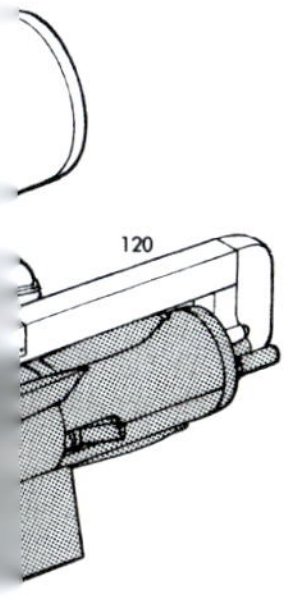

Still remaining in service with several Alouette III operators, the AS12 air-to-surface missile entered service in 1960. Four of these missiles can be carried by the SA.316B and are generally used in the anti-tank role. The AS12 has three different warhead fits: semi-armour piercing, shaped charge and fragmentation. Detonation of the missile is delayed until the missile has passed through ¾ in (20 mm) of armour; it will then explode 6½ feet (2 m) beyond the entry point. The most famous use of the AS12 was by British Wasp helicopters, which used the weapon to damage the Argentine submarine, Santa Fé. Most Alouette III operators now use the HOT missile rather than the AS12.

Aérospatiale SA321 Super Frelon

One of France's few remaining Super Frelons is seen wearing the new two-tone colour scheme, which has replaced the all-over sea-grey finish of old. Aéronavale SA 321s lost their ASW role to the Westland Lynx and now see limited operational use.

SA 321GM Super Frelon

Cutaway key

1 Nose radome (Exocet armed anti-shipping role)
2 Target designation radar scanner
3 Omera Segid ORB 32 radar equipment
4 Radome mounting fairing
5 FM homing aerials, port and starboard
6 Pitot heads
7 Windscreen wipers
8 Windscreen panels
9 Pilot's instrument console
10 Instrument panel shroud
11 Centre control pedestal
12 Cyclic pitch control column
13 Rudder pedals
14 Downward view window
15 Lower IFF aerial
16 Anti-collision light
17 Fixed steerable twin nosewheels
18 Boat hull chine member
19 Collective pitch control lever
20 Door jettison handle
21 Jettisonable cockpit doors
22 Safety harness
23 Co-pilot's seat
24 Observer's folding seat
25 Cockpit doorway
26 Sliding side window panel
27 Pilot's seat
28 Engine power levers
29 Folding sun visors
30 Overhead systems switch panel
31 Circuit breaker panel
32 Starboard side maintenance access ladder
33 Rescue hoist/winch
34 Flight control rod linkages
35 Engine control runs
36 Cabin heater
37 Heater intake grille
38 Starboard side main cabin doorway (sliding door)
39 Cockpit rear bulkhead
40 Avionics and electrical equipment racks, port and starboard
41 Radar altimeter equipment
42 Tie-down ring
43 Boat hull waterproof bulkheads
44 Position of pressure refuelling connection on starboard side
45 Main cabin loading floor
46 Cargo tie down rings
47 Forward group of underfloor, bag type fuel tanks total fuel capacity 874 Imp gal (3975 litres)
48 Honeycomb floorpanels
49 Cabin window panels
50 HF aeriel rail
51 Aerial support masts
52 Fuselage frame and stringer construction
53 Cabin wall soundproof lining
54 Engine mounting deck
55 Forward engine air intakes
56 Cooling air intake
57 Starboard engine bay hinged cowling/ work platform
58 Main rotor blade hollow D-section aluminium alloy main spar
59 Starboard engine exhaust fairing
60 Engine bay dividing fireproof bulkhead
61 Generator
62 Engine oil tank
63 Ignition control unit
64 Turboméca Turmo 111 C6 turboshaft engine
65 Engine transmission gearbox
66 Bifurcated exhaust ducts
67 Oil cooler fan
68 Forward engine power take-off shafts
69 Combining gearbox
70 Main reduction gearbox
71 Rotor head swash plate mechanism
72 Blade pitch control rods
73 Main rotor hub
74 Hydraulic fluid reservoirs
75 Folding rotor blade hinge joints
76 Six-bladed main rotor
77 Blade root attachment joints
78 Pressurised blade crack detection indicator

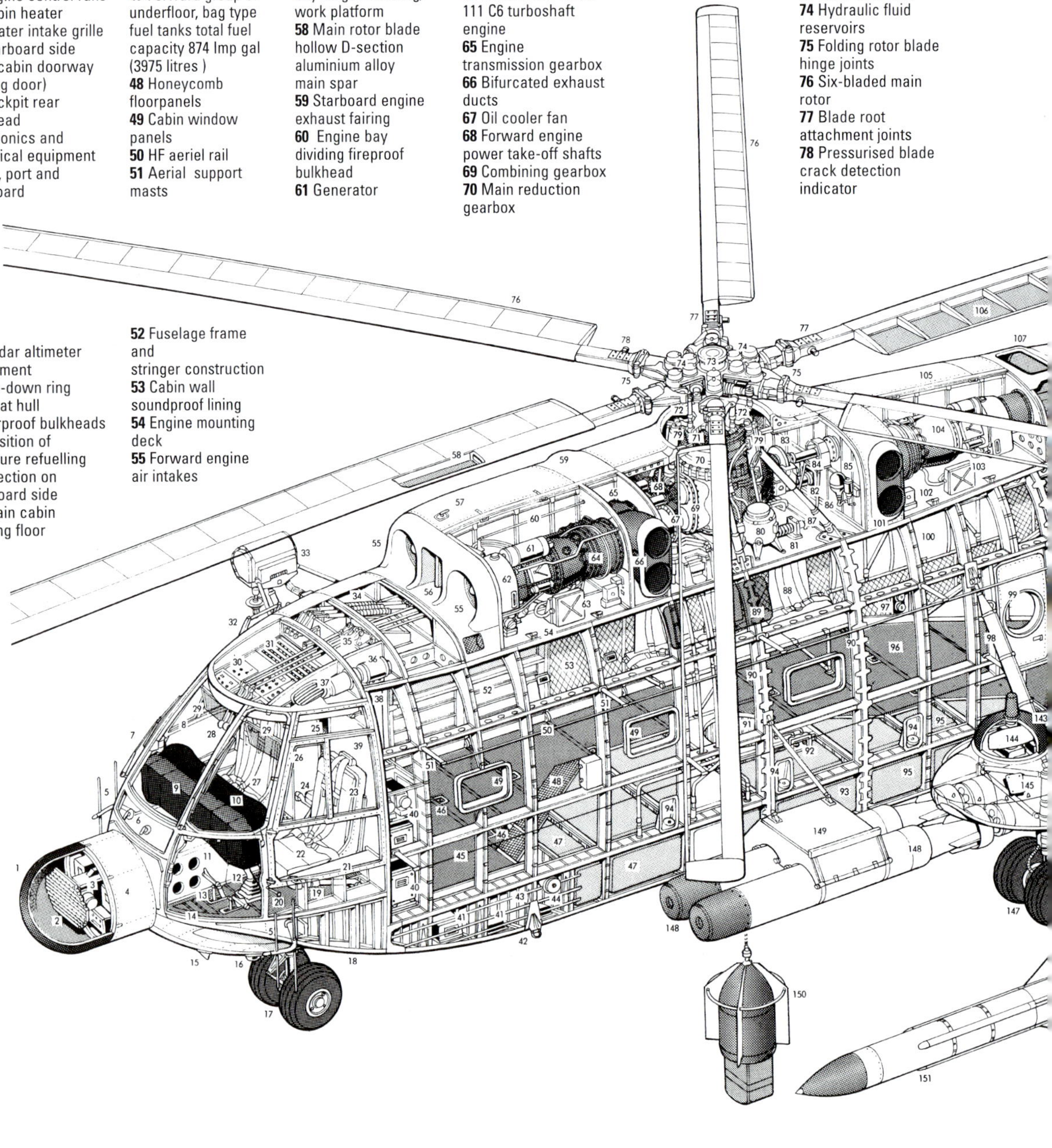

The SA 321F was the unsuccessful civil version of the SA 321 Super Frelon. It was designed to transport 34-37 passengers and had a soundproofed cabin that was claimed to be 'as comfortable as that of any modern airliner'. It had a cruise speed of 130 kt (150 mph; 241 km/h) and it was designed to be used for feeder and short haul shuttle services over medium distance routes.

SPECIFICATION

SA 321G Super Frelon

Dimensions

Main rotor diameter: 62 ft (18.90 m)
Tail rotor diameter: 13 ft 1½ in (4 m)
Length overall (rotors turning): 75 ft 6⅞ in (23.03 m)
Height overall: 22 ft 2¼ in (6.76 m)
Wheel track: 14 ft 1 in (4.30 m)

Powerplant

Three Turboméca Turmo IIIC3 turboshafts each rated at 1,475 hp (1100 kW) or, in later helicopters, three Turboméca Turmo IIIC7 turboshafts each rated at 1,610 hp (1201 kW)

Weight

Empty: 15,130 lb (6863 kg)
Maximum take-off in early versions: 27,557 lb (12500 kg)
Maximum take-off in late versions: 28,660 lb (13000 kg)

Performance

Maximum rate of climb at sea level: 984 ft (300 m) per minute
Service ceiling: 10,170 ft (3100 m)
Hovering ceiling (in ground effect): 6,400 ft (1950 m)
Never-exceed speed at sea level: 149 kt (171 mph; 275 km/h)
Maximum cruising speed at sea level: 134 kt (154 mph; 248 km/h)
Range with a 7,716-lb (3500-kg) payload: 550 nm (633 miles; 1020 km)
Endurance: Up to four hours

Fuel and load

Internal fuel: 1,050 US gal (3975 litres) plus provision for 262 US gal (1191 litres) of auxiliary fuel in two cabin tanks
External fuel: Up to two 132-US gal (600-litre) auxiliary tanks
Maximum payload: 11,023 lb (5000 kg)

Armament

Two AM39 Exocet missiles or four Mk 46 homing torpedoes

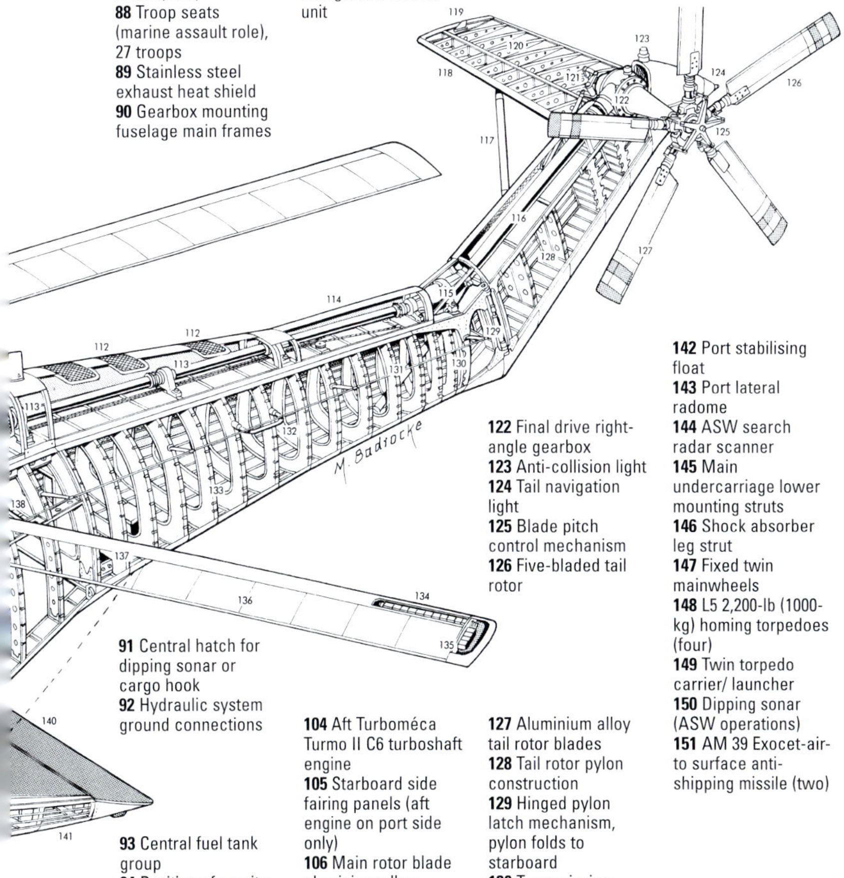

79 Rotor head hydraulic control actuator (three)
80 Hydraulic reservoir
81 Gearbox mounting deck
82 Tail rotor transmission shaft
83 Rotor brake
84 Rear engine power take-off shaft
85 Fireproof bulkhead
86 Engine fire extinguisher bottle
87 Gearbox mounting struts (four)
88 Troop seats (marine assault role), 27 troops
89 Stainless steel exhaust heat shield
90 Gearbox mounting fuselage main frames
91 Central hatch for dipping sonar or cargo hook
92 Hydraulic system ground connections
93 Central fuel tank group
94 Position of gravity fuel fillers on starboard side
95 Rear underfloor fuel tank group
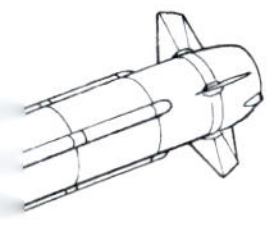

96 Rear cabin sloping floor section
97 Fuel pump
98 Main undercarriage upper energy absorbing strut
99 Emergency escape hatch, port and starboard
100 Troop seats folded against cabin wall
101 Aft engine bifurcated exhaust duct
102 Fireproof engine bay decking
103 Ignition control unit
104 Aft Turboméca Turmo II C6 turboshaft engine
105 Starboard side fairing panels (aft engine on port side only)
106 Main rotor blade aluminium alloy trailing-edge pocket construction
107 Aft engine air intake
108 Intake plenum
109 Tail rotor transmssion shaft
110 Tail rotor control cables
111 VHF aerial
112 Venting air grilles
113 Transmission shaft bearings
114 Dorsal spine fairing
115 Bevel drive gearbox
116 Tailrotor angled drive shaft
117 Tailplane bracing strut
118 Fixed tailplane
119 Upper IFF aerial
120 Tailplane rib construction
121 Tail rotor control linkage
122 Final drive right-angle gearbox
123 Anti-collision light
124 Tail navigation light
125 Blade pitch control mechanism
126 Five-bladed tail rotor
127 Aluminium alloy tail rotor blades
128 Tail rotor pylon construction
129 Hinged pylon latch mechanism, pylon folds to starboard
130 Transmission gearbox mounting main frame
131 Internal bracing strut
132 VOR aerial
133 Tailboom frame and stringer construction
134 Main rotor blade balance weights
135 Tracking weights
136 Aluminium alloy blade skin panelling
137 IFF transceiver
138 Tailboom access hatch
139 Ramp hydraulic jacks, port and starboard
140 Cargo loading ramp (flight openable), down position
141 Ramp frame and stringer construction
142 Port stabilising float
143 Port lateral radome
144 ASW search radar scanner
145 Main undercarriage lower mounting struts
146 Shock absorber leg strut
147 Fixed twin mainwheels
148 L5 2,200-lb (1000-kg) homing torpedoes (four)
149 Twin torpedo carrier/ launcher
150 Dipping sonar (ASW operations)
151 AM 39 Exocet-air-to surface anti-shipping missile (two)

An SA 321G development aircraft fires an Aérospatiale Exocet anti-shipping missile. The Exocet/Super Frelon combination proved deadly during the Gulf War between Iraq and Iran.

Super Frelon at war

First to see combat were the assault transport SA 321Ks of the Israeli Defence Force/Air Force (IDF/AF). The initial few from an order for 12 had only just been delivered when they were involved in the Six-Day War of June 1967. They transported troops for a daring assault on the airfield at Sharm el Sheikh at the southern tip of the Sinai peninsula, and literally carried off an Egyptian missile guidance radar station so that electronics experts could reveal its secrets at their leisure. Over a decade later, in 1978, plans were announced to re-engine Israeli Super Frelons with 1,896-hp (1413-kW) General Electric T58-16 turboshafts to improve hot-and-high performance and to give powerplant commonality with the IDF's other transport helicopter, the Sikorsky CH-53. Eventually, the CH-53, with its greater lifting capacity and power, replaced the Super Frelon, though one heavy lift squadron; 114 Tayeset, has retained the name 'The Super Frelon Squadron'. During Desert Storm, at least one of Iraq's Super Frelons was destroyed by the Allies.

Puma/Super Puma/Cougar

Due to its relatively large size, the Puma is vulnerable to return fire if employed in the attack role. However, under favourable circumstances, it may be equipped with 72 rocket projectiles with blast, hollow-charge, fragmentation or smoke warheads, as demonstrated by this Aérospatiale-owned SA 330L.

AS 332 Super Puma

Cutaway key
1 Radome
2 Weather radar scanner
3 Retractable landing/ taxiing lamp
4 Communications and navigation system electronic equipment
5 Pitot tubes
6 Cockpit front bulkhead
7 Fresh air scoop
8 Windscreen wipers
9 Windscreen panels
10 Instrument panel shroud
11 Instrument panel
12 Centre control console
13 Cyclic pitch control column
14 Yaw control rudder pedals
15 Brake pedal connections
16 Nose undercarriage leg strut
17 Downward-vision window
18 Twin nosewheels
19 Nose undercarriage wheel bay
20 Cockpit step
21 Flight-deck floor level
22 Collective pitch control lever
23 Hydraulic system hand pump
24 Cockpit door
25 Co-pilot's seat
26 Safety harness
27 Opening side-window panel
28 Cockpit bulkhead
29 Fire extinguisher bottle
30 Control rod linkages
31 Pilot's seat
32 Starboard side cockpit door
33 Engine power levers and fuel cocks
34 Overhead control panel
35 Cabin air intake
36 Cabin heater unit
37 Intake ice and debris shield
38 Engine air intake
39 Starboard engine cowling panels
40 Engine bay central firewall
41 Accessory equipment
42 Engine oil tank
43 Engine mounting deck

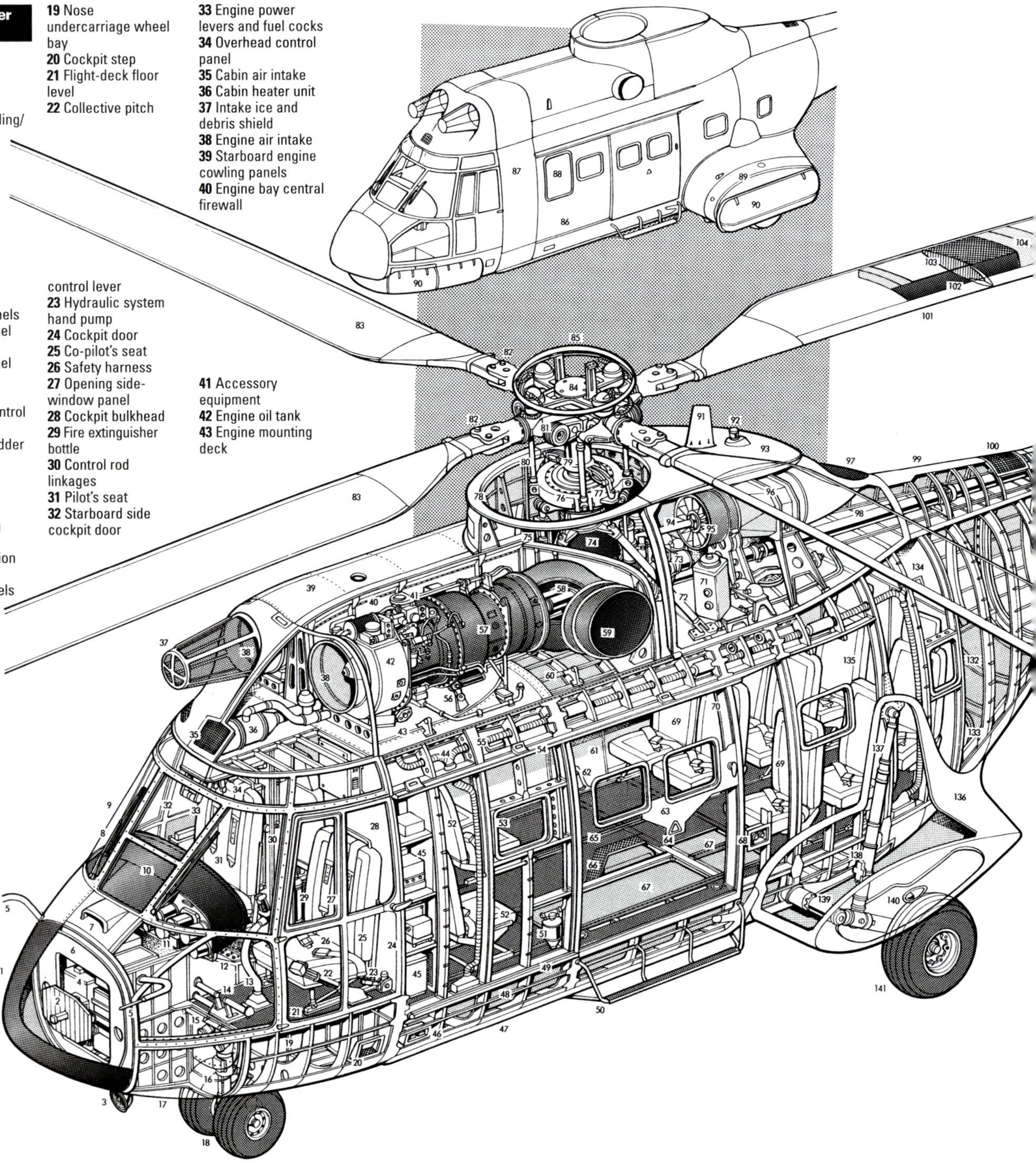

The French deployed the Puma early in Operation Desert Shield as part of the French Army Light Aviation (ALAT) advanced party. By the outbreak of Desert Storm, a total of 38 French Pumas was in theatre; during the conflict, they were mainly used for army support, carrying 16 troops or 2.5 tonnes of underslung load, such as ammunition.

SPECIFICATION

SA 330L Puma

Dimensions

Length (rotors turning): 59 ft 6½ in (18.15 m)
Fuselage length: 46 ft 1½ in (14.06 m)
Main rotor diameter: 49 ft 2½ in (15.00 m)
Tail rotor diameter: 9 ft 11½ in (3.04 m)
Main rotor disc area: 1,902.20 sq ft (176.71 m²)
Tail rotor disc area: 78.13 sq ft (7.26 m²)
Height overall: 16 ft 10½ in (5.14 m)
Height to top of rotor head: 14 ft 4½ in (4.38 m)
Wheel base: 13 ft 3 in (4.05 m)
Wheel track: 7 ft 10¾ in (2.38 m)

Powerplant

Two Turboméca Turmo IVC turboshafts each rated at 1,575 shp (1175 kW)

Weights

Empty: 7,970 lb (3615 kg)
Maximum take-off: 16,534 lb (7500 kg)

Fuel and load

Internal fuel load: 408 US gal (1544 litres) plus provision for 502 US gal (1900 litres) of auxiliary fuel in four cabin tanks
External fuel load: up to two 93-US gal (350-litre) auxiliary tanks
Maximum payload: 7,055 lb (3200 kg)

Performance

Never-exceed speed: 158 kt (182 mph; 293 km/h)
Maximum cruising speed 'clean' at optimum altitude: 146 kt (168 mph; 270 km/h)
Maximum rate of climb at sea level: 1,810 ft (552 m) per minute
Service ceiling: 19,685 ft (6000 m)
Hover ceiling (IGE): 14,435 ft (4400 m)
Hover ceiling (OGE): 13,940 ft (4250 m)
Range: 308 nm (355 miles; 572 km)

Armament

The Puma can carry a wide variety of weaponry for several different roles, including one or two 0.3-in (7.62-mm) general-purpose machine-guns for self-defence, 72 SNEB 2.68-in (68-mm) rocket projectiles for close support, two Aérospatiale AM39 Exocets for the long-range anti-shipping role or six Aérospatiale AS15TT missiles for the medium-range anti-shipping role

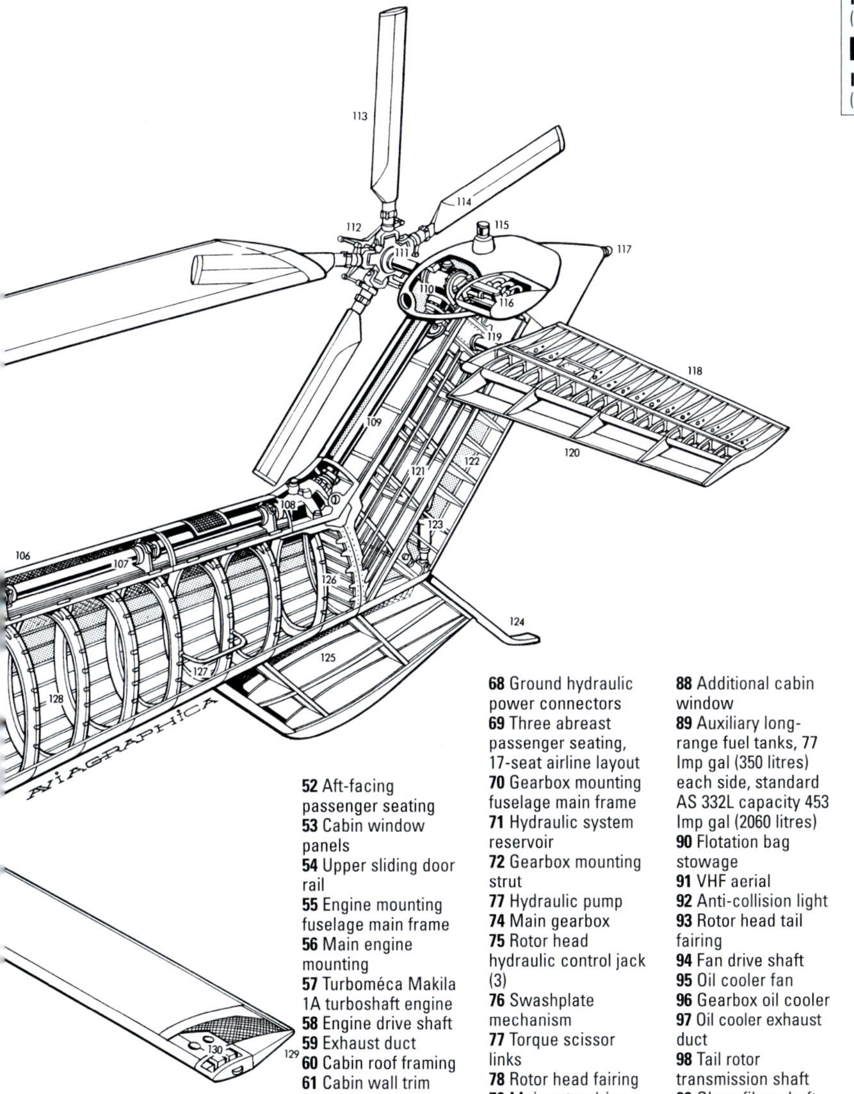

44 Cabin heater air ducting
45 Radio and electronics equipment rack
46 Ground electrical power socket
47 External cable and pipe duct
48 Forward underfloor fuel cell
49 Lower sliding door rail
50 Boarding step
51 Fuel system filter
52 Aft-facing passenger seating
53 Cabin window panels
54 Upper sliding door rail
55 Engine mounting fuselage main frame
56 Main engine mounting
57 Turboméca Makila 1A turboshaft engine
58 Engine drive shaft
59 Exhaust duct
60 Cabin roof framing
61 Cabin wall trim panelling
62 Starboard side refuelling pipe
63 Sliding main cabin door, port and starboard
64 Door emergency release handle
65 Main cabin flooring
66 Honeycomb floor panels
67 Underfloor fuel tanks; total fuel capacity 343 Imp gal (1560 litres)
68 Ground hydraulic power connectors
69 Three abreast passenger seating, 17-seat airline layout
70 Gearbox mounting fuselage main frame
71 Hydraulic system reservoir
72 Gearbox mounting strut
77 Hydraulic pump
74 Main gearbox
75 Rotor head hydraulic control jack (3)
76 Swashplate mechanism
77 Torque scissor links
78 Rotor head fairing
79 Main rotor drive shaft
80 Blade pitch control rods
81 Vibration damper
82 Blade root fixing
83 Composite construction main rotor blades
84 Main rotor hub mechanism
85 Rotor head fairing
86 AS 332L stretched variant
87 AS 332L 2-ft 6-in (0.76-m) fuselage plug
88 Additional cabin window
89 Auxiliary long-range fuel tanks, 77 Imp gal (350 litres) each side, standard AS 332L capacity 453 Imp gal (2060 litres)
90 Flotation bag stowage
91 VHF aerial
92 Anti-collision light
93 Rotor head tail fairing
94 Fan drive shaft
95 Oil cooler fan
96 Gearbox oil cooler
97 Oil cooler exhaust duct
98 Tail rotor transmission shaft
99 Glass-fibre shaft fairing
100 Cooling air grille
101 Rotor blade titanium leading-edge capping strip
102 Glass-fibre roving blade spar
103 Honeycomb blade core
104 Carbon fibre inner skins
105 Glass-fibre outer skin covering
106 Dorsal spine fairing
107 Drive shaft bearings
108 Bevel drive gearbox
109 Tail rotor angled transmission shaft
110 Right-angled final drive gearbox
111 Tail rotor hub fixing
112 Blade pitch control mechanism
113 Five-bladed tail rotor
114 Composite construction tail rotor blades
115 Anti-collision light
116 Tail rotor hydraulic control jack
117 Tail navigation light
118 Fixed tailplane construction
119 Tailplane mounting shaft
120 Fixed leading-edge slat
121 Tail pylon construction
122 Glass-fibre trailing-edge section
123 Skid shock absorber
124 Tail skid
125 Ventral fin construction
126 Gearbox/tail pylon mounting bulkhead
127 VOR aerial
128 Tailboom frame and stringer construction
129 Main rotor blade tip fairing
130 Blade tracking weights
131 Fuselage/tailcone production joint
132 Ventral hatch/ emergency exit
133 Downward vision window
134 Cabin rear bulkhead
135 Aft cabin seating
136 Glass-fibre main undercarriage fairing
137 Hydraulic retraction jack
138 Shock absorber leg strut
139 Pivoted suspension main axle beam
140 Port navigation light
141 Port mainwheel

This atmospheric shot of ALAT Pumas was taken during their deployment to the Gulf after the invasion of Kuwait in 1990. As well as the support role, the Puma carried out communications duties, one example – using a simplified version of the Orchidée system – being used for battlefield surveillance.

Bell AH-1 Huey Cobra

Hunting for targets at tree-top height, this production is seen on a practice mission on one of the vast test ra the United States. The widespread introduction of the Apache has resulted in the HueyCobra being withdraw front-line US Army operations, but for many foreign ope the Cobra remains their most important attack heli

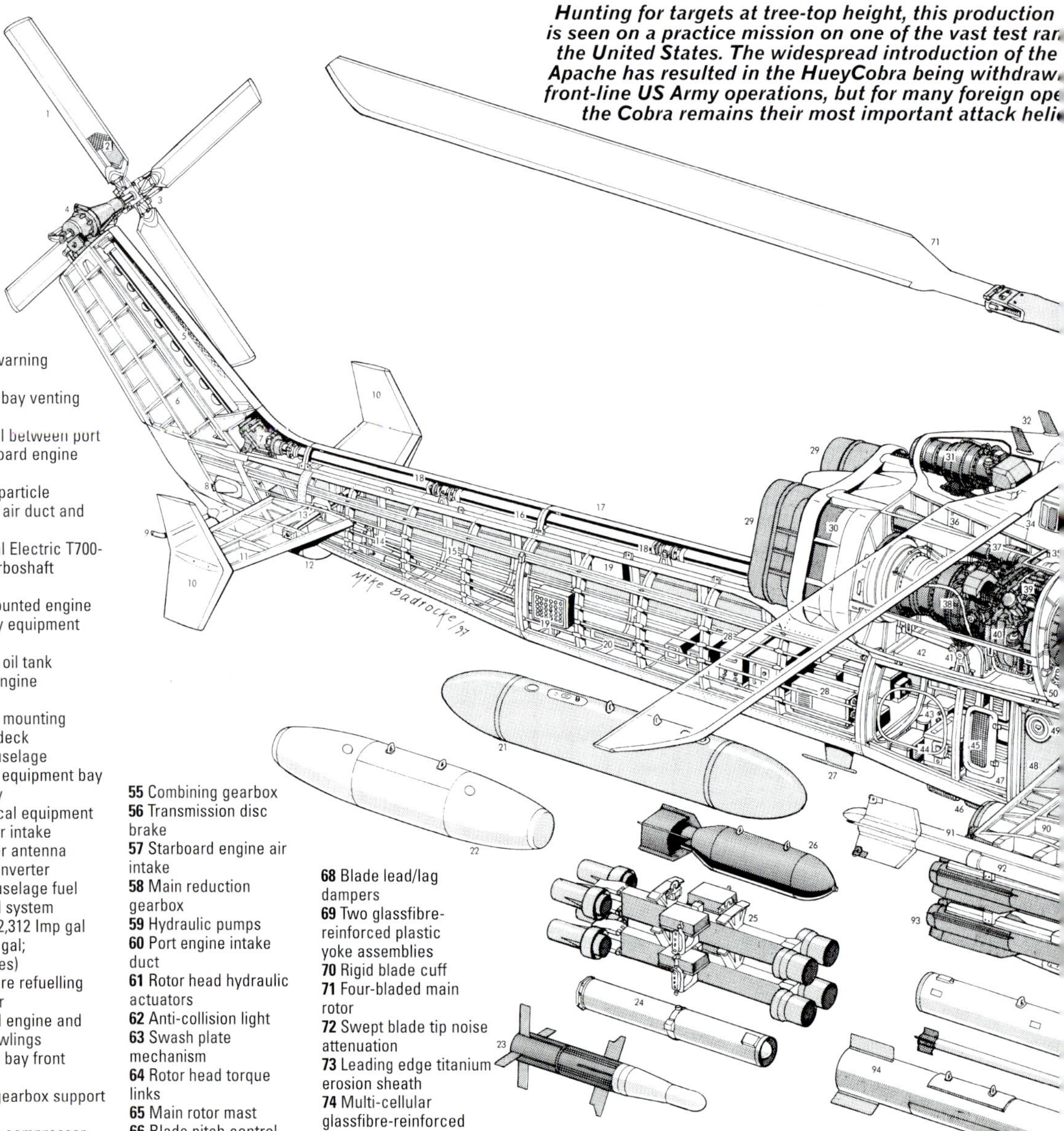

AH-1(4B)W

Cutaway key
1 Four-bladed tail rotor
2 All-composite tail rotor blades
3 Pitch control links
4 Final drive right-angle gearbox
5 Angled drive shaft
6 Tail pylon structure
7 Angle-drive intermediate gearbox
8 Rear radar warning antenna
9 Tail protection skid
10 Endplate fins
11 Elevator
12 Ventral fin
13 Elevator hinge mounting
14 Control rod linkages
15 Tail boom frame and stringer structure
16 Tail rotor drive shaft
17 Detachable shaft housing
18 Shaft bearings and couplings
19 Rear chaff/flare launcher, port and starboard
20 Electro-luminescent external lighting strip
21 83.3 Imp gal (100 US gal; 379 litre) external fuel tank
22 83.3 Imp gal (100 US gal; 379 litre) napalm tank
23 BGM-71 TOW air-to-surface anti-armour missile
24 TOW missile container
25 Four-round TOW missile carrier/launcher
26 250-lb (113-kg) fragmentation bomb
27 Lower UHF/IFF antenna
28 Tail boom avionics equipment bay, access on port side
29 Infra-red suppression engine exhaust ducts
30 Exhaust cooling air intake
31 Auxiliary power unit (APU)
32 VHF homing antennae
33 Infra-red jammer
34 Laser warning detector
35 Engine bay venting air intake
36 Firewall between port and starboard engine bays
37 Intake particle separator air duct and blower
38 General Electric T700-GE-401 turboshaft engine
39 Top-mounted engine accessory equipment gearbox
40 Engine oil tank
41 Main engine mounting
42 Engine mounting fireproof deck
43 Rear fuselage electrical equipment bay
44 Battery
45 Electrical equipment cooling air intake
46 Doppler antenna
47 Static inverter
48 Rear fuselage fuel tank, total system capacity 2,312 Imp gal (2,776 US gal; 10,508 litres)
49 Pressure refuelling connector
50 Hinged engine and intake cowlings
51 Engine bay front firewall
52 Main gearbox support structure
53 Engine compressor intake
54 Anti-vibration gearbox mounting
55 Combining gearbox
56 Transmission disc brake
57 Starboard engine air intake
58 Main reduction gearbox
59 Hydraulic pumps
60 Port engine intake duct
61 Rotor head hydraulic actuators
62 Anti-collision light
63 Swash plate mechanism
64 Rotor head torque links
65 Main rotor mast
66 Blade pitch control links
67 Rotor hub
68 Blade lead/lag dampers
69 Two glassfibre-reinforced plastic yoke assemblies
70 Rigid blade cuff
71 Four-bladed main rotor
72 Swept blade tip noise attenuation
73 Leading edge titanium erosion sheath
74 Multi-cellular glassfibre-reinforced blade spar
75 Laminated glassfibre blade skin

The Bell company endeared itself to the US Army with a succession of highly successful utility/transport helicopters. The AH-1G, seen in the foreground flying with a UH-1B (centre) and a UH-1D (rear), was radically different, yet used many components of the earlier UH-1 models.

SPECIFICATION

AH-1F HueyCobra

Dimensions

Length overall, rotors turning: 53 ft 1 in (16.18 m)
Fuselage length: 44 ft 7 in (13.59 m)
Main rotor diameter: 44 ft (13.41 m)
Tail rotor diameter: 8 ft 6 in (2.59 m)
Height to top of rotor head: 13 ft 5 in (4.09 m)
Main rotor disc area: 1,520.23 sq ft (141.26 m²)
Tail rotor disc area: 56.75 sq ft (5.27 m²)
Stabiliser span: 6 ft 11 in (2.11 m)
Skid track: 7 ft (2.13 m)

Powerplant

One 1,800-shp (1342-kW) Textron Lycoming T53-L-703 turboshaft, transmission-limited to 1,290 shp (962 kW) for take-off and 1,134 shp (845 kW) for continuous running

Weights

Operating weight: 6,598 lb (2993 kg)
Normal take-off: 9,975 lb (4524 kg)
Maximum take-off: 10,000 lb (4536 kg)

Performance

Never exceed speed in TOW configuration: 195 mph (315 km/h)
Maximum level speed at optimum altitude in TOW configuration: 141 mph (227 km/h)
Service ceiling: 12,200 ft (3720 m)
Range: 315 miles (507 km)

Armament

Primary armament is the TOW missile, eight of which can be carried along with 2.75-in (70-mm) FFARs with a variety of warheads. Secondary armament is the M197 triple-barrelled 20-mm cannon in a chin turret

AH-1W SuperCobra

Dimensions

Length overall, rotors turning: 58 ft (17.68 m)
Fuselage length: 45 ft 6 in (13.87 m)
Main rotor diameter: 48 ft (14.63 m)
Tail rotor diameter: 9 ft 9 in (2.97 m)
Height to top of rotor head: 13 ft 6 in (4.11 m)
Main rotor disc area: 1,809.56 sq ft (168.11 m²)
Tail rotor disc area: 74.7 sq ft (6.94 m²)
Stabiliser span: 6 ft 11 in (2.11 m)
Skid track: 7 ft (2.13 m)

Powerplant

Two 1,625-shp (1212-kW) General Electric T700-GE-401 turboshafts, transmission-limited to a total of 2,032 shp (1515 kW) for take-off and 1,725 shp (1286 kW) continuous running

Weights

Empty weight: 10,200 lb (4627 kg)
Maximum take-off: 14,750 lb (6691 kg)

Performance

Never exceed speed: 219 mph (352 km/h)
Maximum level speed 'clean' at sea level: 175 mph (282 km/h)
Service ceiling: more than 12,200 ft (3720 m)
Range: 395 miles (635 km)

Armament

Uniquely qualified to carry both TOW and Hellfire, plus up to four seven-round LAU-69A rocket pods containing 2.75-in (70-mm) Hydra 70 rockets. Self-defence is provided by a single AIM-9L Sidewinder AAM mounted on a hardpoint above each wingtip

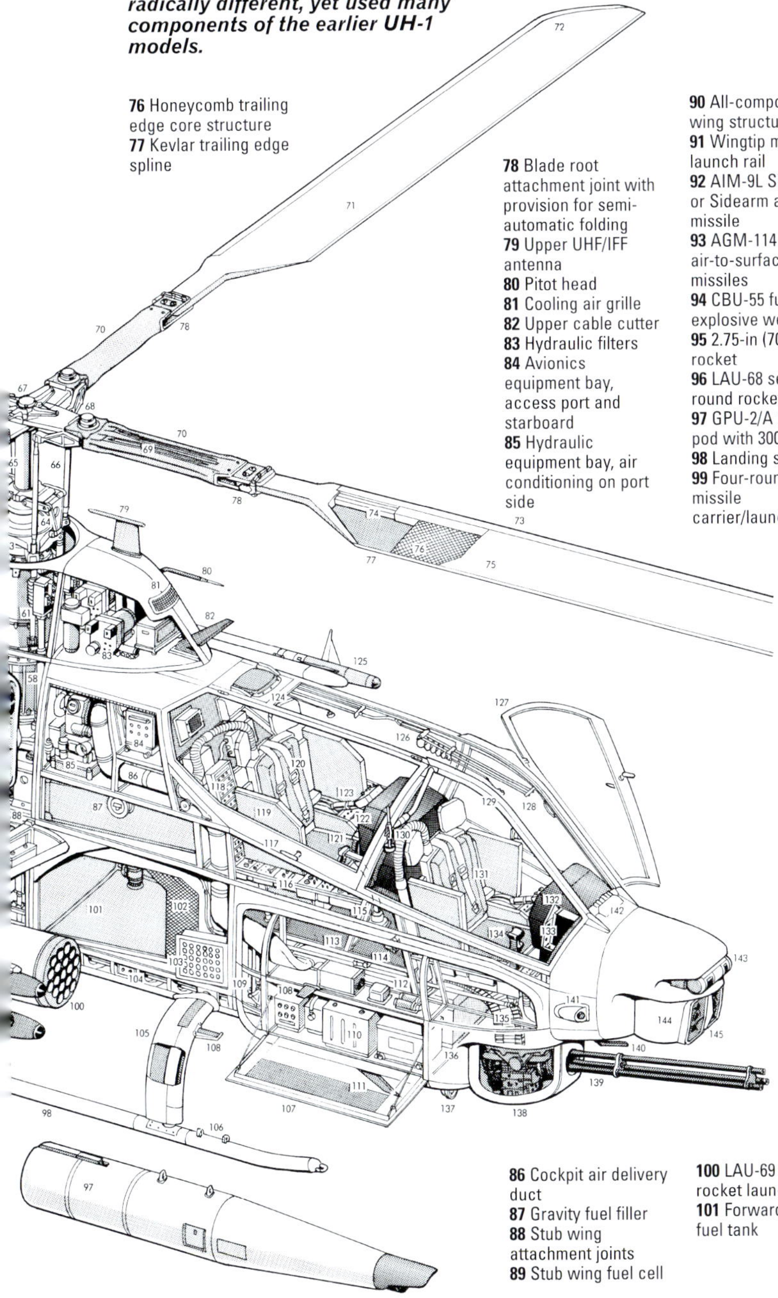

76 Honeycomb trailing edge core structure
77 Kevlar trailing edge spline
78 Blade root attachment joint with provision for semi-automatic folding
79 Upper UHF/IFF antenna
80 Pitot head
81 Cooling air grille
82 Upper cable cutter
83 Hydraulic filters
84 Avionics equipment bay, access port and starboard
85 Hydraulic equipment bay, air conditioning on port side
86 Cockpit air delivery duct
87 Gravity fuel filler
88 Stub wing attachment joints
89 Stub wing fuel cell
90 All-composite stub wing structure
91 Wingtip missile launch rail
92 AIM-9L Sidewinder or Sidearm air-to-air missile
93 AGM-114 Hellfire air-to-surface missiles
94 CBU-55 fuel/air explosive weapon
95 2.75-in (70-mm) rocket
96 LAU-68 seven-round rocket launcher
97 GPU-2/A 20mm gun pod with 300-rounds
98 Landing skid
99 Four-round Hellfire missile carrier/launcher
100 LAU-69 19-round rocket launcher
101 Forward fuselage fuel tank
102 Fuel tank bay kevlar composite armoured structure
103 Forward chaff/flare launcher, port and starboard
104 Control stability augmentation system actuators
105 Front skid strut with aerodynamic fairing
106 Ground handling wheel attachment points
107 Forward fuselage equipment bay access door, port and starboard
108 Boarding steps
109 Flight control linkages
110 TOW missile system avionics equipment
111 Lower cable cutter
112 Forward fuselage lateral equipment bay, port and starboard
113 Pilot's cockpit armoured floor
114 Yaw control rudder pedals
115 Collective pitch control column
116 Side console panel
117 Starboard side pilot's entry hatch
118 Circuit breaker panel
119 Detachable shoulder armour
120 Pilot's seat
121 Armament and weapons display control handgrips; flight controls and weapons system can be operated from either cockpit
122 Pilot's instrument console, full-colour CRT displays
123 Cyclic pitch control lever
124 ADF antenna
125 Port wingtip missile installation
126 Helmet-mounted sight provision for possible future integration
127 Co-pilot/gunner's entry hatch open
128 Single-piece curved windscreen panel
129 Windscreen wire deflecting strakes
130 Low-speed air data sensor
131 Co-pilot/gunner's armoured seat
132 Front cockpit instrument console
133 CRT cockpit displays
134 Front cockpit sidestick controller
135 Ammunition feed chute
136 Cannon ammunition magazine, 750 rounds
137 Retractable landing light
138 GE universal gun turret
139 M197 three-barrel 20-mm rotary cannon
140 Turret protection cable cutter
141 Forward radar warning antenna, port and starboard
142 Windscreen rain dispersal air duct
143 Forward-looking infra-red (FLIR)
144 Night targeting system turret
145 High resolution FLIR dual field-of-view low-light television camera and laser rangefinder designator

Boeing CH-47 Chinook

The CH-47JA has radar, AAQ-16 FLIR and long-range tanks to increase greatly the operational capability of the type in JGSDF service. In addition to the Kyoiku Sien Hiko-tai at Akeno, CH-47JAs serve with the JGSDF's Dai 1 Konsei-Dan and Dai 1 Herikoputa-dan. The Seibu Homen Herikoputa-tai (Western Army Helicopter Squadron) has, in 1999, begun to receive this advanced variant to replace the KV-107-II.

CH-47D Chinook

Cutaway key
1 Pitot tubes
2 Forward lighting
3 Nose compartment access hatch
4 Vibration absorber
5 IFF aerial
6 Windscreen panels
7 Windscreen wipers
8 Instrument panel shroud
9 Rudder pedals
10 Yaw sensing ports
11 Downward vision window
12 Pilot's footboards
13 Collective pitch control
14 Cyclic pitch control column
15 Co pilot's seat
16 Centre instrument console
17 Pilot's seat
18 Glideslope indicator
19 Forward transmission housing fairing
20 Cockpit overhead window
21 Doorway from main cabin
22 Cockpit emergency exit doors
23 Sliding side window panel
24 Cockpit bulkhead
25 Vibration absorber
26 Cockpit door release handle
27 Radio and electronics racks
28 Sloping bulkhead
29 Stick boost actuators
30 Stability augmentation system actuators
31 Forward transmission mounting structure
32 Windscreen washer bottle**33** Rotor control hydraulic jack
34 Forward transmission gearbox
35 Rotor head fairing
36 Forward rotor head mechanism
37 Pitch change control levers
38 Blade drag dampers
39 Glassfibre rotor blades
40 Titanium leading-edge capping with de-icing provision
41 Rescue hoist/winch
42 Forward transmission aft fairing
43 Hydraulic system modules
44 Control levers
45 Front fuselage frame and stringer construction
46 Emergency exit window, main entry door on starboard side
47 Forward end of cargo floor
48 Fuel tank fuselage side fairing
49 Battery
50 Electrical system equipment bay
51 Aerial cable
52 Stre tcher rack (up to 24 stretchers)
53 Cabin window panel
54 Cabin heater duct outlet
55 Troop seats stowed against cabin wall
56 Cabin roof transmission and control run tunnel
57 Formation-keeping lights
58 Rotor blade cross section
59 Static dischargers
60 Blade balance and tracking weights pocket
61 Leading-edge anti-erosion strip
62 Fixed tab
63 Fuselage skin plating
64 Maintenance walkway
65 Transmission tunnel access doors
66 Troop seating, up to 44 troops
67 Cargo hook access hatch
68 VOR aerial
69 Cabin lining panels
70 Control runs
71 Main transmission shaft
72 Shaft couplings
73 Centre fuselage construction
74 Centre aisle seating (optional)
75 Main cargo floor, 1,440-cu ft (40.78-m^3) cargo volume
76 Ramp-down 'dam' for waterborne operations
77 Ramp hydraulic jack
78 Engine bevel drive gearbox
79 Transmission combining gearbox
80 Rotor brake
81 Transmission oil tank
82 Oil cooler
83 Engine drive shaft fairing
84 Engine screen
85 Starboard engine nacelle
86 Cooling air grilles
87 Tail rotor pylon construction
88 Hydraulic equipment
89 Access door
90 Maintenance step
91 Tail rotor drive shaft
92 Tail rotor bearing mounting
93 Rotor head fairing

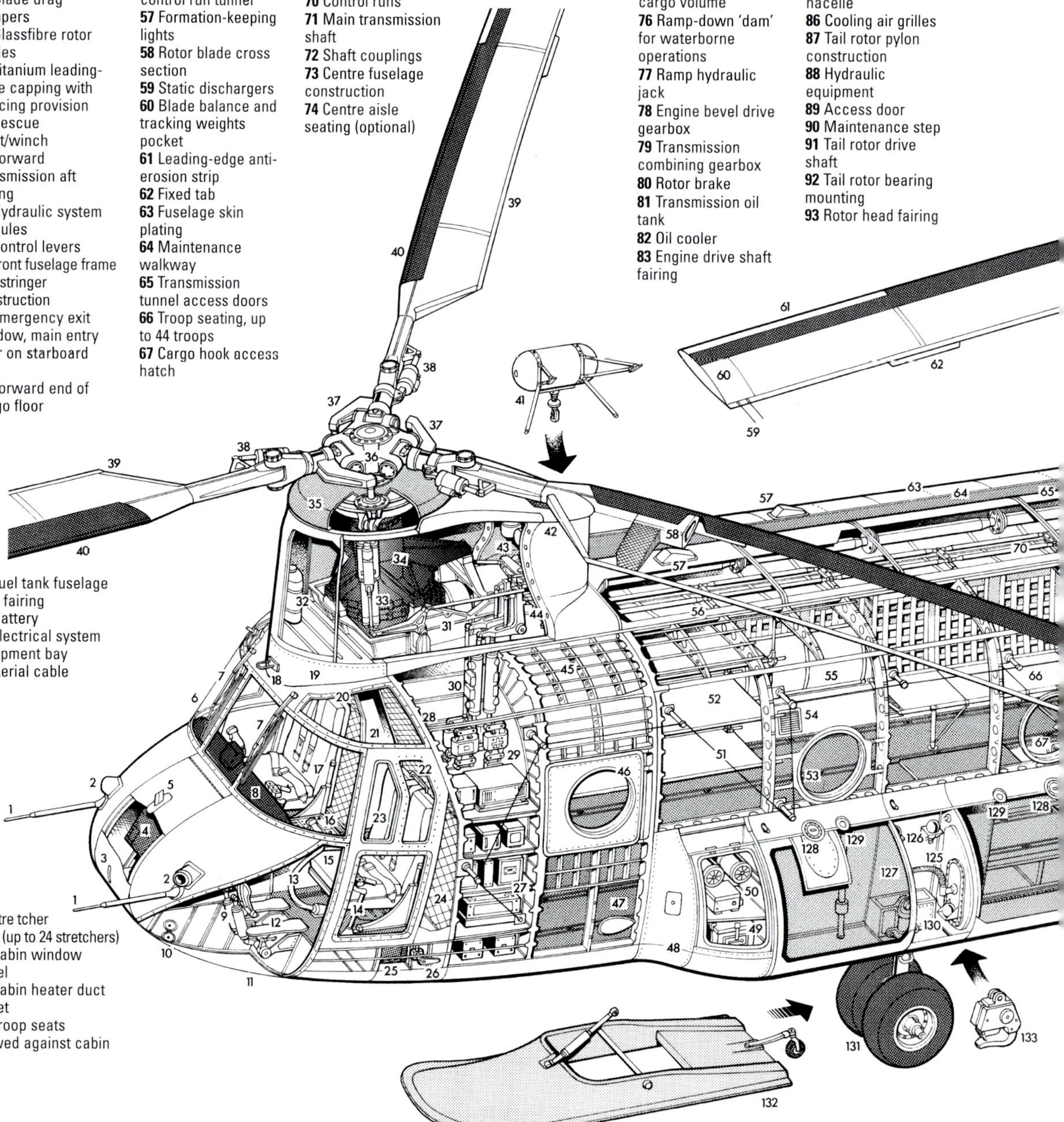

The ultimate expression of the Special Forces Chinook is the MH-47E, a purpose-built version for the US Army. Dripping with defensive avionics and low-level night penetration aids, the MH-47E also introduced a 'glass' cockpit, and bulged 'saddle' tanks based on those developed for the Model 234LR Commercial Chinook.

SPECIFICATION

CH-47D Chinook

Dimensions

Length overall, rotors turning: 98 ft 10¾ in (30.14 m)
Fuselage: 51 ft (15.54 m)
Height to top of rear rotor head: 18 ft 11 in (5.77 m)
Wheel track: 10 ft 6 in (3.20 m)
Wheel base: 22 ft 6 in (6.86 m)
Rotor diameter: 60 ft (18.29 m)
Rotor disc area: 5,654.86 sq ft (525.34 m²)

Powerplant

Two Textron Lycoming T55-L-712 turboshafts each rated at 3,750 shp (2796 kW) for take-off and 3,000 shp (2237 kW) for continuous running, or two Textron Lycoming T55-L-712 SSB turboshafts each rated at 4,378 shp (3264 kW) for take-off and 3,137 shp (2339 kW) for continuous running, in both cases driving a transmission rated at 7,500 shp (5593 kW) on two engines and 4,600 shp (3430 kW) on one engine

Weights

Empty: 22,379 lb (10151 kg)
Normal take-off: 46,000 lb (20866 kg)
Maximum take-off: 50,000 lb (22679 kg)

Fuel and load

Internal fuel: 1,030 US gal (3899 litres)
External fuel: None
Maximum payload: 22,798 lb (10341 kg)

Range

Ferry range: 1,093 nm (1,259 miles; 2026 km)
Operational radius with maximum internal and maximum external payloads respectively: Between 100 and 30 nm (115 and 35 miles; 185 and 56 km)

Performance

Maximum level speed at sea level: 161 kt (185 mph; 298 km/h)
Maximum cruising speed at optimum altitude: 138 kt (159 mph; 256 km/h)
Maximum rate of climb at sea level: 2,195 ft (669 m) per minute
Service ceiling: 22,100 ft (6735 m)
Hovering ceiling: 10,550 ft (3215 m)

94 Tail rotor head mechanism
95 Main rotor blades, glassfibre construction
96 Rotor control hydraulic jack
97 Vibration absorber
98 Pylon aft fairing construction
99 Rear lighting
100 Solar T62T-2B auxiliary power unit
101 APU-driven generators
102 Maintenance walkways
103 Engine exhaust duct
104 Avco Lycoming T55-L-712 turboshaft engine
105 Detachable engine cowlings
106 Aft fuselage frame and stringer construction
107 Rear cargo doorway
108 Ramp extensions
109 Cargo ramp, lowered
110 Ramp ventral strake
111 Fuselage side fairing aft extension
112 Ramp control lever
113 Ramp hydraulic jack
114 Rear landing gear shock absorber
115 Landing gear leg strut
116 Single rear wheels
117 Rear wheel optional ski fitting
118 Maintenance steps
119 Rear fuel tank
120 Fuel tank interconnections
121 Ventral strake
122 Main fuel tank; total system capacity 1,030 US gal (3899 litres)
123 Floor beam construction
124 Fuel tank attachment joint
125 Fuel system piping
126 Fire extinguishers
127 Forward fuel tank
128 Fuel filler caps
129 Fuel capacity transmitters
130 Front landing-gear mounting
131 Twin forward wheels
132 Forward wheels optional ski-fitting
133 Triple cargo hook system; forward and rear hooks 20,000-lb (9072-kg) capacity
134 Main cargo hook, 28,000-lb (12701-kg) capacity

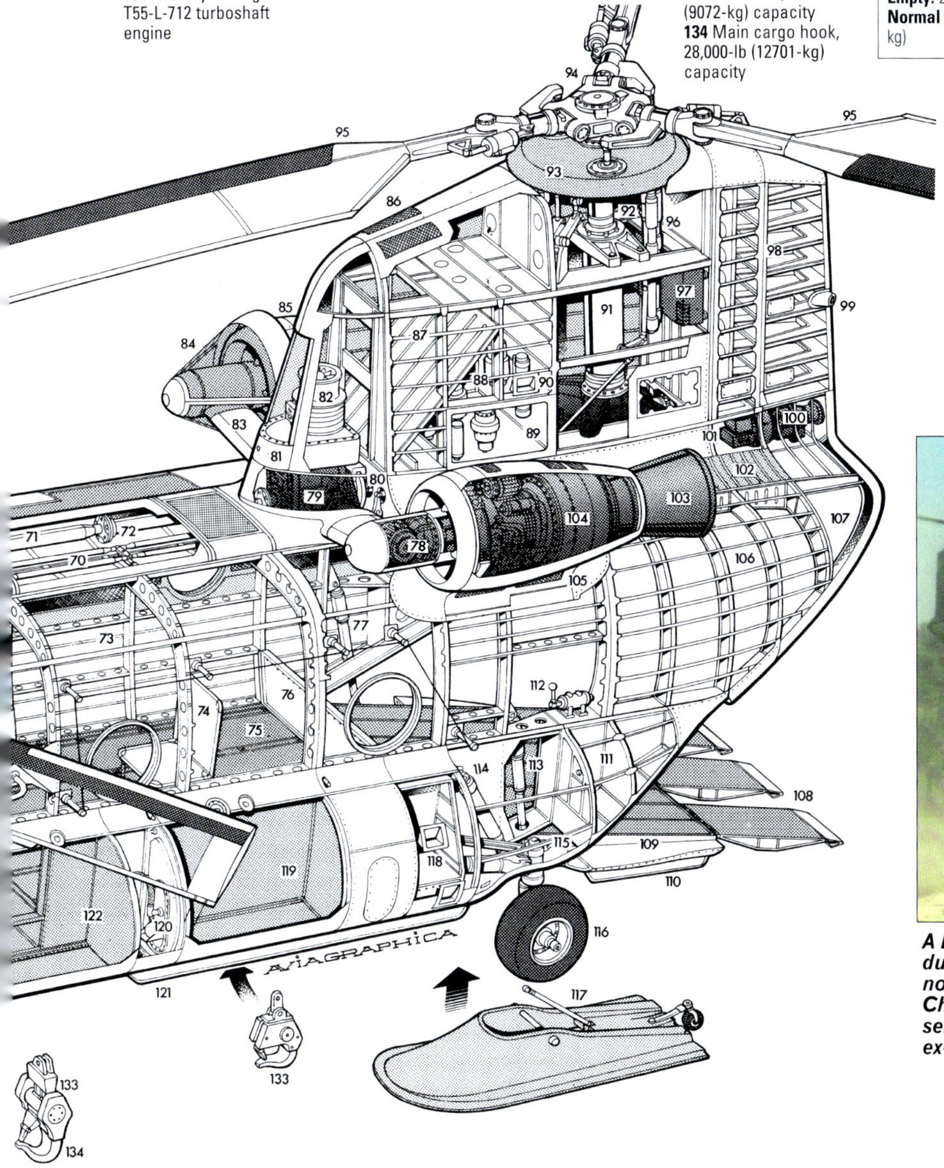

A Dutch CH-47D repositions a tactical vehicle during manoeuvres. Equipped with EFIS cockpit, nose radar and T55-L-714 engines, the Dutch Chinooks are among the most advanced in service. The first seven were converted from ex-Canadian CH-147s.

Kamov Ka-25/Ka-27

Civil versions of the 'Helix' are designated Ka-32 by Kamov. The models are often simplified military versions lacking any offensive equipment and radar. This particular example is a Ka-32T (wearing Aeroflot airline markings), a variant which has been heavily marketed in the US. Despite its low purchase price of US$3 million however, the Ka-32 series has yet to make a real impact on the civil market.

Ka-29TB 'Helix-B'

Cutaway key
1 Pitch and yaw vanes
2 Dual pitot heads
3 Air data instrumentation boom
4 Ventral periscope sighting unit and FLIR
5 Retractable landing lamps
6 Instrumentation and autostabilisation equipment racks
7 ECM antenna
8 Gun compartment hinged door, open
9 Flexibly-mounted 7.62-mm four-barrel Gatling gun
10 Rear view mirror, port and starboard
11 Armoured windscreen panels
12 Weapons Systems Officer sighting unit
13 Windscreen wipers
14 Outside temperature gauge
15 Weapons control panel
16 Gunner's periscope sight
17 Remote gun controller
18 Pilot's head-up display
19 Cyclic pitch control column
20 Instrument panel
21 Radar warning antenna
22 Yaw control rudder pedals
23 Radar director unit for AT-6 Spiral missile
24 Cockpit armour panelling
25 Nose undercarriage torque scissor links
26 Nose wheel rebound position
27 Steerable nose wheel
28 Shock absorber leg strut
29 Cockpit floor level
30 Collective pitch control lever and throttle
31 Pilot's seat
32 Sliding cockpit door
33 Centre control pedestal
34 Gunner's folding seat
35 Cockpit bulkhead doorway
36 IFF aerial
37 Overhead switch panel
38 Starboard split cabin door, upper segment, open
39 Engine air intakes, bleed air de-iced
40 Control rod linkages
41 Cockpit door rail
42 Pilot's shoulder armour
43 ECM antenna
44 Avionics equipment racks
45 Battery compartment
46 Step
47 Lower fuselage strake
48 Fuselage chine member
49 Underfloor fuel cells; maximum capacity 8,113 lb (3680 kg)
50 Stub wing pylon assembly support struts
51 Seat mounting rails
52 Cabin window panel
53 Engine bay armoured panel
54 Isotov TV3-117BK turboshaft engine
55 Engine accessory equipment gearbox
56 Generator
57 Engine bay firewall
58 Oil tank access panels
59 Generator cooling air scoop
60 Glass/carbon fibre D-section rotor blade spar
61 Honeycomb core trailing-edge pockets
62 Composite skin panelling
63 Hinged engine cowling/work platform
64 Starboard engine exhaust duct
65 Rotor head hydraulic actuators
66 Lower rotor swashplate mechanism
67 Articulated rotor hub
68 Blade pitch control rods
69 Upper/lower rotor (interconnecting)
70 Upper rotor swash plate mechanism
71 Blade pitch control links and rods
72 Upper rotor articulated hub
73 Blade-folding hinge joints, manual
74 Three-bladed, coaxial, contra-rotating main rotors
75 Electro-thermal leading-edge de-icing
76 Vibration-damping pendulum weights, lower rotor only
77 Blade root attachment fittings
78 Radar warning antenna
79 Maintenance handgrips
80 Cooling air louvres
81 Oil cooler fan
82 Rotor brake
83 Gearbox mounting struts
84 Port engine lateral exhaust
85 Engine/gearbox coupling shaft
86 Central combining and main reduction gearbox
87 Cabin fresh air scoop
88 Gearbox sump fairing
89 Folding troop seats along cabin wall (12)
90 Gearbox mounting fuselage double main frames
91 Cabin floor panelling
92 Main undercarriage leg shock absorber strut
93 Gravity fuel filler
94 Central underfloor internal weapons bay ([torpedoes, depth charges, etc.)
95 Split main cabin door, upper segment
96 Air intake grilles
97 Auxiliary Power Unit (APU)
98 Cabin heater
99 ESM 'Flower Pot' antenna
100 Anti-collision light
101 Rear avionics equipment racks
102 Cabin rear bulkhead
103 Tailcone joint frame
104 'Hot Brick' infra-red jammer
105 Radar warning antenna
106 Tailcone frame and stringer construction
107 Communications aerial
108 Starboard tailplane
109 Radar warning antenna
110 Starboard elevator
111 Rudder control linkage
112 Tailfin fixed leading-edge slat
113 Starboard tailfin
114 Starboard rudder
115 Upper rotor blades [folded position]
116 Blade locking struts
117 Lower rotor blades [folded position]
118 IFF aerial
119 ESM antenna fairing
120 Aft radome
121 Port rudder
122 Rudder composite construction
123 Fin/tailplane joint rib
124 Port navigation light
125 UB-32 rocket pod
126 2¼-in (57-mm) rocket projectiles
127 Port fin leading-edge slat

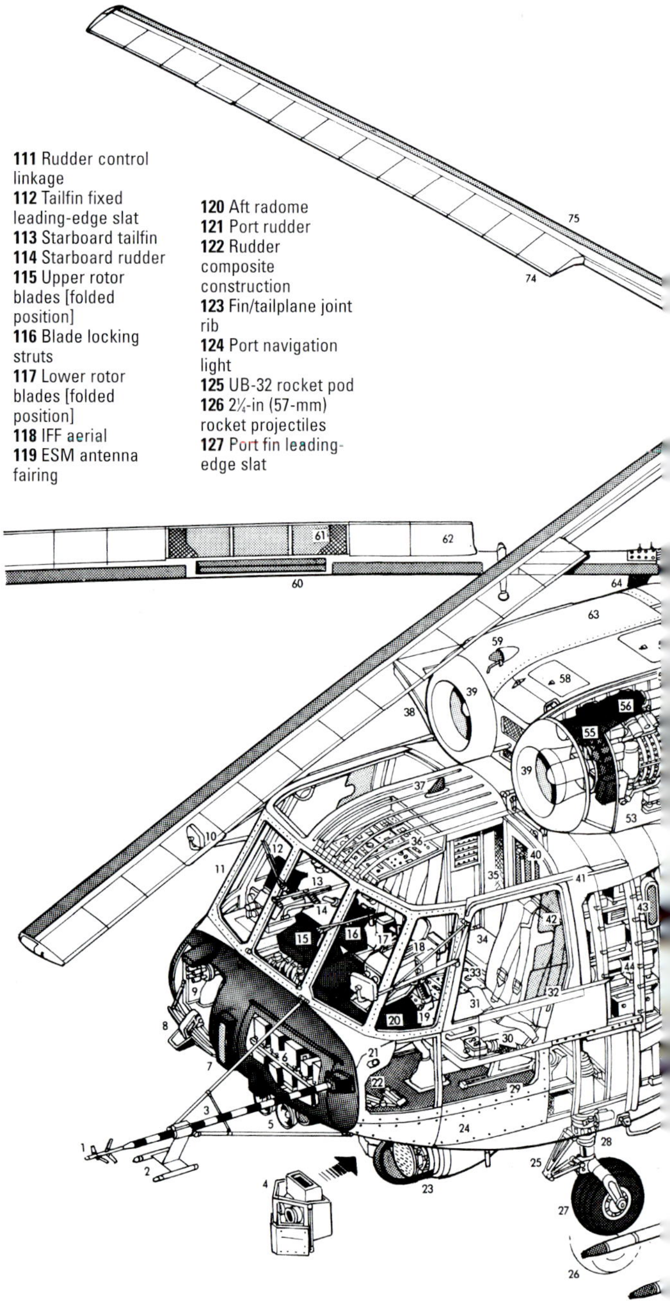

This Ka-25BSh 'Hormone-A' (displaying the flag of the Soviet navy) is bereft of flotation gear, fuel tanks and all the usual ASW equipment. In this configuration, the Ka-25 can carry a useful load of freight or 12 passengers, enabling it to perform an important secondary ship-to-shore transport role. Throughout the Cold War years, 'Hormones' were often seen shadowing NATO warships for intelligence-gathering purposes. For this role, a photographer would hang out of the main cabin door to record images of the warship of interest.

SPECIFICATION

Ka-25BSh 'Hormone-A'

Dimensions

Fuselage length: 32 ft (9.75 m)
Height overall: 17 ft 5 in (5.37 m)
Stabiliser span including endplate surfaces: 12 ft 4 in (3.76 m)
Wheel track: 4 ft 7½ in (1.41 m) for the front unit and 11 ft 6¼ in (3.52 m) for the rear unit
Rotor diameter (each): 52 ft 7¾ in (15.74 m)
Rotor disc area: 4,188.83 sq ft (389.15 m²)

Powerplant

Two OMKB Isotov (Glushenkov) GTD-3F turboshafts each rated at 898 shp (671 kW) in early helicopters, or two OMKB Isotov (Glushenkov) GTD-3BM turboshafts each rated at 900 shp (738 kW) in late helicopters

Performance

Maximum speed 'clean' at optimum altitude: 130 mph (209 km/h)
Normal cruising speed at optimum altitude: 120 mph (193 km/h)
Ferry range with auxiliary fuel: 404 miles (650 km)
Range with standard fuel: 249 miles (400 km)

Weights

Empty: 10,505 lb (4765 kg)
Maximum take-off: 16,534 lb (7500 kg)
Maximum payload: 2,866 lb (1300 kg)

Armament

Armament is not normally carried although the helicopter can be fitted with a long 'coffin-like' weapons bay which runs along the belly from the radome back to the tailboom, and small bombs or depth charges can be carried on tiny pylons just aft of the nosewheels. The underfuselage weapons bay can carry a variety of weapons, including nuclear depth charges. When wire-guided torpedoes are carried, a wire reel is mounted on the port side of the forward fuselage

128 Tailplane bracing strut
129 Horizontal tailplane construction
130 Tailplane mounting bulkhead
131 Rear navigation avionics equipment bay
132 Twin gyromagnetic compass unit
133 Pressure refuelling connection
134 Ground power socket
135 Ventral weapons bay loading doors, open
136 Chaff/flare dispenser housing, port and starboard
137 Split cabin door lower segment
138 Integral boarding steps
139 External fuel tank
140 Fixed blade tab
141 Spiral missile launch tubes
142 Missile pylon adaptor
143 UPK-23 twin barrel, 23-mm cannon pod
144 AT-6 'Spiral' command guidance air-to-surface missile
145 Mainwheel rebound position
146 Port fixed mainwheel
147 Mainwheel forks
148 Main undercarriage leg strut
149 Upper and lower wishbone links
150 Pylon stub wing
151 Fixed stores pylons
152 UB-20 rocket pods
153 3.15-in (80-mm) rocket projectiles

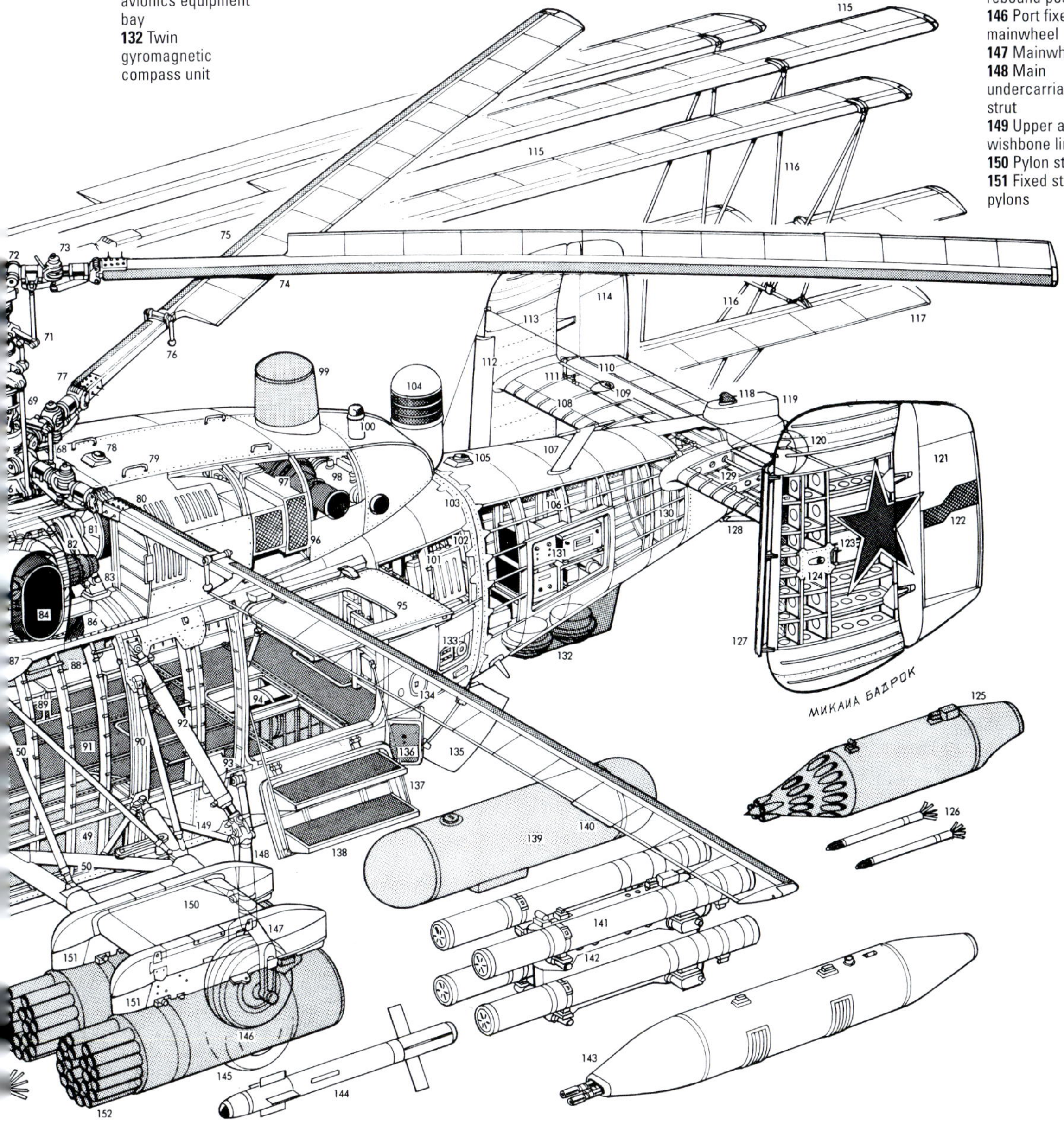

McDonnell Douglas AH-64 Apache

This Apache is one of those operated by the US Army's 1st Aviation Training Brigade, located at Ft Rucker, Alabama – the home of US Army aviation. All training helicopters based at Ft Rucker wear large white identification codes. Students who have graduated onto the AH-64 from basic flying training join the 1-14 AVN for the 12 week conversion course. They first learn all about basic aircraft systems before progressing onto advanced flying training and weapons familiarisation. Increasing numbers of foreign air crew are now travelling to Ft Rucker for their Apache training.

AH-64A Apache

Cutaway key

1 Night systems sensor scanner
2 Pilot's Night Vision System (PNVS)
3 Electro-optical target designation and night sensor systems turret
4 Target Acquisition and Designation System (TADS) daylight scanner
5 Azimuth motor housing
6 TADS/PNVS swivelling turret
7 Turret drive motor housing
8 Sensor turret mounting
9 Rear-view mirror
10 Nose compartment access hatches
11 Remote terminal unit
12 Signal data converter
13 Co-pilot/gunner's yaw control rudder pedals
14 Forward radar warning antenna
15 M230E1 Chain Gun barrel
16 Fuselage sponson fairing
17 Avionics cooling air ducting
18 Boron armoured cockpit flooring
19 Co-pilot/gunner's 'fold down' control column
20 Weapons control panel
21 Instrument panel shroud
22 Windscreen wiper
23 Co-pilot/gunner's armoured windscreen
24 Head-down sighting system viewfinder
25 Pilot's armoured windscreen panel
26 Windscreen wiper
27 Co-pilot/gunner's Kevlar armoured seat
28 Safety harness
29 Side console panel
30 Engine power levels
31 Avionics equipment bays, port and starboard
32 Avionics bay access door
33 Collective pitch control lever
34 Adjustable crash-resistant seat mouldings
35 Pilot's rudder pedals
36 Cockpit side window panel
37 Pilot's instrument console
38 Inter-cockpit acrylic blast shield
39 Starboard side window entry hatches
40 Rocket launcher pack
41 Starboard wing stores pylon
42 Cockpit roof glazing
43 Instrument panel shroud
44 Pilot's Kevlar armoured seat
45 Collective pitch control lever
46 Side console panel
47 Engine power levers
48 Rear cockpit floor level
49 Main landing gear shock absorber mounting
50 Linkless ammunition feed chute
51 Forward fuel tank: total fuel capacity 312 Imp gal (1419 litres)
52 Control rod linkages
53 Cockpit ventilating air louvres
54 Display adjustment panel
55 Grab handles/ maintenance steps
56 Control system hydraulic actuators (three)
57 Ventilating air intake
58 UHF aerial
59 Starboard stub wing
60 Main rotor blade
61 Laminated blade-root attachment joints
62 Vibration absorbers
63 Blade pitch bearing housing
64 Air data sensor mast
65 Rotor hub unit
66 Offset flapping hinges
67 Elastomeric lead/lag dampers
68 Blade pitch control rod
69 Pitch control swashplate
70 Main rotor mast
71 Airturbine starter/ auxiliary power unit (APU) input shaft
72 Rotor head control mixing linkages
73 Gearbox mounting plate
74 Transmission oil coolers, port and starboard
75 Rotor brake
76 Main gearbox

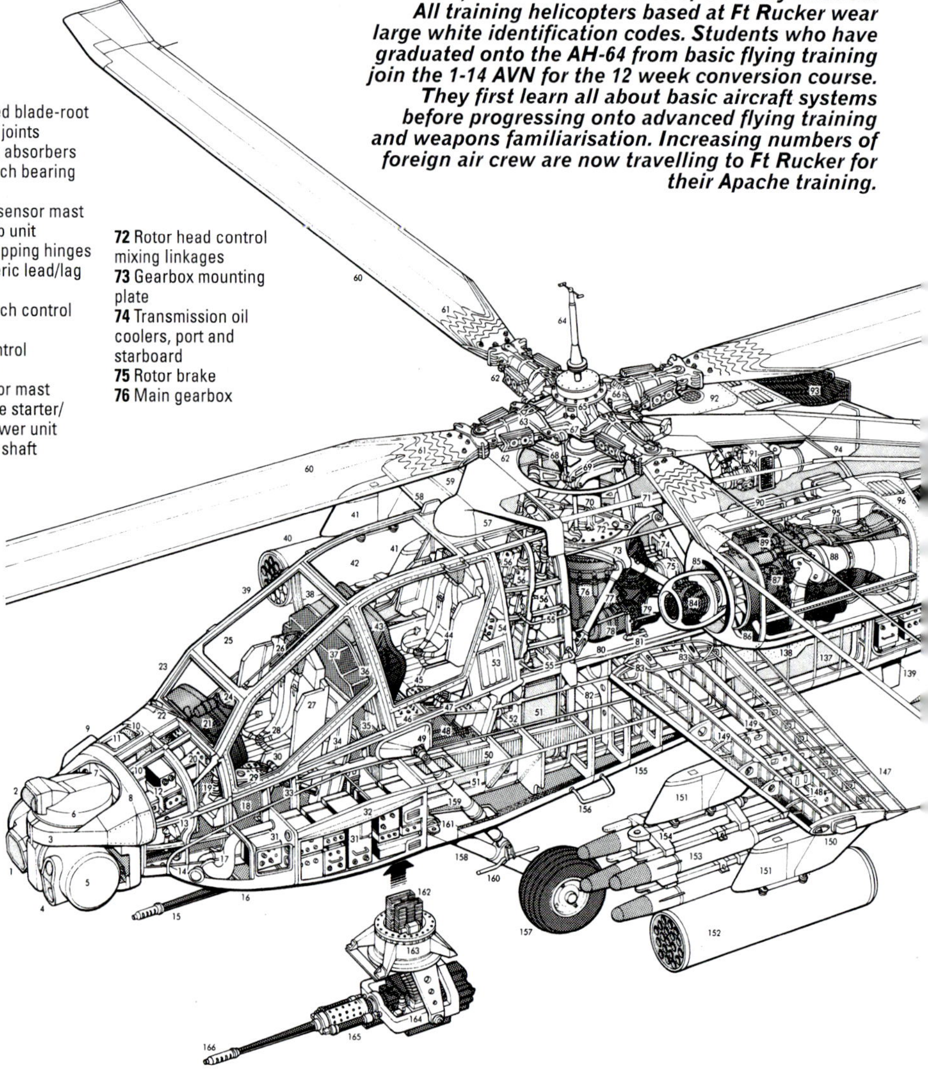

The Apache's performance during Operation Desert Storm persuaded several nations, such as Greece, the UK and the Netherlands, to finally accelerate their search for a new attack helicopter – while existing customers (such as Saudi Arabia and the UAE) came back for more. Over 200 Apaches have now been exported.

SPECIFICATION

AH-64A Apache

(unless otherwise noted)

Dimensions

Fuselage length including both rotors turning: 58 ft 3⅛ in (17.76 m)
Main rotor diameter: 48 ft (14.63 m)
Tail rotor diameter: 9 ft 2 in (2.79 m)
Height over tail rotor: 14 ft 1¼ in (4.30 m)
Total height AH-64D: 16 ft 3 in (4.95 m)
Main rotor disc area: 1,809.5 sq ft (168.11 m^2)
Tail rotor disc area: 66 sq ft (6.13 m^2)
Wingspan: 17 ft 2 in (5.23 m)
Tailplane span: 11 ft 2 in (3.4 m)
Wheelbase: 34 ft 9 in (10.59 m)
Wheel track: 6 ft 8 in (2.03 m)
Main rotor ground clearance (turning): 11 ft 9¼ in (3.59 m^2)

Powerplant

Two 1,696-shp (1265-kW) General Electric T700-GE-701 turboshafts, each derated for normal operations or, from 604th helicopter, two General Electric T700-GE-701C turboshafts, each rated at 1,890 shp (1409 kW)
AH-64D: two General Electric T700-GE-701C turboshafts, each rated at 1,800 shp (1342 kW)

Weights

Empty: 11,387 lb (5165 kg)
AH-64D: 11,800 lb (5352 kg)
Normal take-off for primary mission: 14,445 lb (6552 kg)
Design mission weight: 17,650 lb (8006 kg)
Maximum external stores: 1,700 lb (772 kg)
Maximum take-off: 21,000 lb (9525 kg)

Fuel and load

Internal fuel: 2,550 lb (1157 kg)
External fuel (four Brunswick tanks): 2,712 lb (5980 kg)

Performance

Maximum level and cruising speed: 158 kt (182 mph; 293 km/h)
AH-64D: 141 kt (162 mph; 261 km/h)
Never-exceed speed: 197 kt (227 mph; 365 km/h)
Maximum rate of climb at sea level: 3,240 ft (990 m) per minute
AH-64D: 3,090 ft (942 m) per minute
Maximum vertical rate of climb at sea level: 2,500 ft (762 m) per minute
AH-64D: 1,555 ft (474 m) per minute
Service ceiling: 21,000 ft (6400 m)
Service ceiling, one engine out: 10,800 ft (3290 m)
Hovering ceiling: 15,000 ft (4570 m)
AH-64D: 13,500 ft (4115 m)

Range

Maximum range, internal fuel only: 260 nm (300 miles; 482 km)
AH-64D: 220 nm (253 miles; 407 km)
Ferry range, max internal and external fuel in still air: 1,024 nm (1,180 miles; 1899 km)

Armament

Maximum ordnance: some 1,700 lb (771 kg) of ordnance can be carried by the Apache. One McDonnell Douglas M230 30-mm Chain Gun is located between the mainwheel legs in an underfuselage mounting. Normal rate of fire is 625 rds/min of HE (high-explosive) or HEDP (high-explosive, dual-purpose) ammunition, with a maximum load of 1,200 rounds. There are four underwing hardpoints, upon which can be carried 16 Rockwell AGM-114A Hellfire anti-tank missiles or up to 77 2.75-in (70-mm) FFAR (folding fin aircraft rockets) in their launchers, or a combination of Hellfires and FFAR. Planned modifications include two extra hardpoints for four Stinger, four Mistral or two Sidewinder missiles for an air-defence role. The co-pilot has responsibility for firing the gun and missiles, but the pilot can override his controls in the event of an emergency.

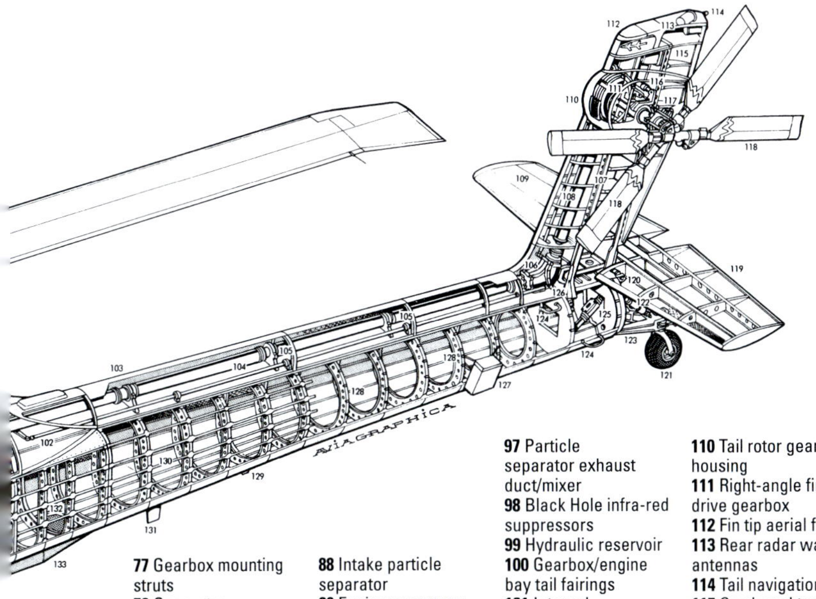

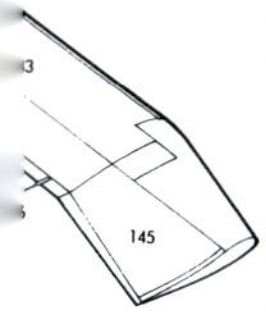

77 Gearbox mounting struts
78 Generator
79 Input shaft from port engine
80 Gearbox mounting deck
81 Tail rotor control rod linkage
82 Ammunition magazine
83 Stub wing attachment joints
84 Engine transmission gearbox
85 Air intake
86 Engine integral oil tank
87 General Electric T-700-GE-701 turboshaft
88 Intake particle separator
89 Engine accessory equipment gearbox
90 Oil cooler plenum
91 Gas turbine starter/auxiliary power unit
92 Starboard engine cowling panels/fold-down maintenance platform
93 Starboard engine exhaust ducts
94 APU exhaust
95 Pneumatic system and environmental control equipment
96 Cooling air exhaust louvres
97 Particle separator exhaust duct/mixer
98 Black Hole infra-red suppressors
99 Hydraulic reservoir
100 Gearbox/engine bay tail fairings
101 Internal maintenance platform
102 Tail rotor control rod
103 Spine shaft housing
104 Tail rotor transmission shaft
105 Shaft bearings and couplings
106 Bevel drive intermediate gearbox
107 Fin/rotor pylon construction
108 Tail rotor drive shafts
109 All-moving tailplane
110 Tail rotor gearbox housing
111 Right-angle final drive gearbox
112 Fin tip aerial fairing
113 Rear radar warning antennas
114 Tail navigation light
115 Cambered trailing-edge section (directional stability)
116 Tail rotor pitch actuator
117 Tail rotor hub mechanism
118 Asymmetric (noise attenuation) tail rotor blades
119 Tailplane construction
120 Tailplane pivot bearing
121 Castoring tailwheel
122 Tailwheel shock absorber
123 Tailwheel yoke attachment
124 Handgrips/ maintenance steps
125 Tailplane control hydraulic jack
126 Fin/rotor pylon attachment joint
127 Chaff and flare dispenser
128 Tailboom ring frames
129 Ventral radar warning aerial
130 Tailcone frame and stringer construction
131 UHF aerial
132 ADF loop aerial
133 ADF sense aerial
134 Access hatch
135 Handgrips/ maintenance steps
136 Radio and electronics equipment bay
137 Rear fuel tank
138 Reticulated foam fire suppressant tank bay linings
139 VHF aerial
140 Main rotor blade stainless steel spars (five)
141 Glassfibre sparlinings
142 Honeycomb trailing-edge panel
143 Glassfibre blade skins
144 Trailing-edge fixed tab
145 Swept-blade tip fairing
146 Static discharger
147 Stub wing trailing-edge flap
148 Stub wing rib construction
149 Twin spar booms
150 Port navigation and strobe lights
151 Port wing stores pylons
152 Rocket pack: 19 x 2.75-in (70-mm) FFAR rockets
153 Rockwell Hellfire AGM-114 anti-tank missiles
154 Missile launch rails
155 Fuselage sponson aft fairing
156 Boarding step
157 Port mainwheel
158 Main landing gear leg strut
159 Shock absorber strut
160 Boarding steps
161 Main landing gear leg pivot fixing
162 Ammunition feed and cartridge case return chutes
163 Gun swivelling mounting
164 Azimuth control mounting frame
165 Hughes M230E1 Chain Gun 30-mm cannon
166 Blast suppression cannon muzzle

Mil Mi-8/9/17 'Hip'

Fitted with a Doppler navigation box beneath th tailboom, this Mi-8T belongs to the Karinės Oro Pajėgo (Lithuanian air force) and was one of three acquire shortly after the air arm was established in 1993. Or example was destroyed in an accident and the remainir two examples serve with 13 Squadron at Zoknia

Mi-8TB 'Hip-E'

Cutaway key

1 0.5-in (12.7-mm) machine-gun barrel
2 Flexible gun mounting
3 Downward vision windows
4 Pitot heads
5 Yaw control rudder pedals
6 Gunsight
7 Windscreen wipers
8 Windscreen panels
9 Weapons Systems Officer's sighting unit
10 Overhead switch panels
11 'Odd-Rods' IFF aerials
12 Cockpit roof escape hatch
13 Main cabin doorway
14 Radio and electrical equipment racks
15 Co-pilot/Weapons Systems Officer's armoured seat
16 Gunner's folding seat
17 Instrument consoles
18 Stand-by compass
19 Cyclic pitch control column
20 Cockpit floor level
21 Twin nosewheels
22 Collective pitch control lever
23 Safety harness
24 Pilot's armoured seat
25 Adjustable seat mounting
26 Sliding cockpit side-window panel
27 Ground power and intercom sockets
28 Batteries (2)
29 Cockpit rear bulkhead
30 Control rod ducting
31 Engine air intake
32 Anti-ingestion intake guard
33 Internal particle separator
34 Main rotor blade hollow steel spar boom
35 Honeycomb trailing-edge panels
36 Starboard missile launch rails (2)
37 AT-2 'Swatter' air-to-surface missile
38 Hinged engine cowling panel/work platform
39 Engine bay fireproof bulkhead
40 Accessory equipment gearbox
41 Generator
42 Generator cooling air duct
43 Isotov TV2-117A turboshaft engine
44 Engine mounting deck
45 Starboard side-folding troop seats, maximum 19 troops
46 Rescue hoist/winch
47 Main cabin loading deck
48 Port folding troop seats
49 Cabin wall-mounted heating duct
50 Entry doorway
51 Folding entry steps
52 Sliding main entry door
53 Door latch
54 Port external fuel tank, total fuel capacity 411 Imp gal (1870 litres)
55 Fuel filler cap
56 Centre-section underfloor fuel cells (2)
57 Sliding door rail
58 Engine exhaust duct
59 Engine/gearbox drive shaft
60 Oil cooler
61 Cooler air fan
62 Oil cooler air intake
63 5-bladed main rotor
64 Mi-8 'Hip-C' commercial transport variant

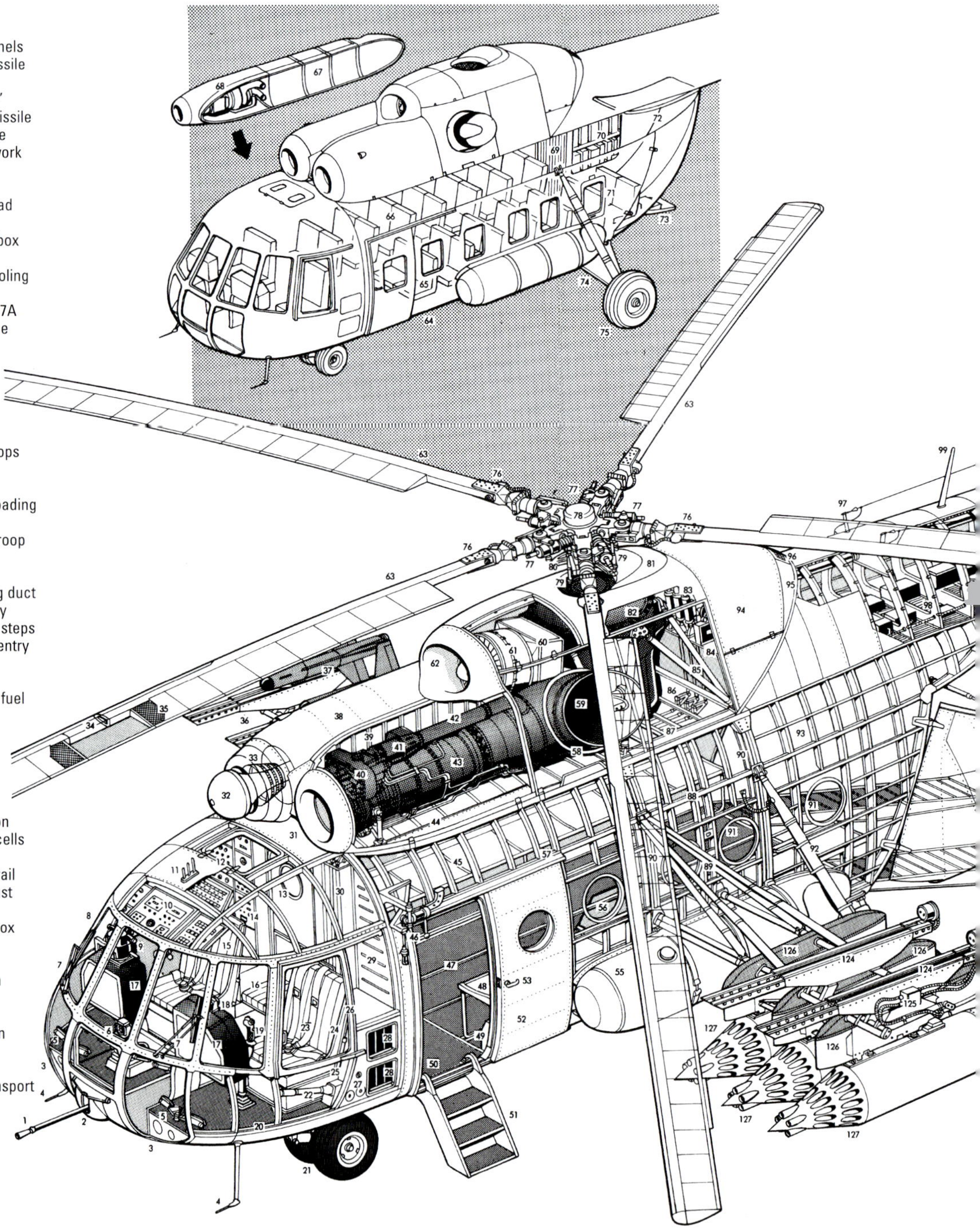

Above: The Eesti Piirivalve (Estonian Border Guard) acquired four Mi-8s (three ex-Luftwaffe and one ex-East German civil) in November 1995. This Mi-8T is one of two which have been updated with new radar, IR sensor and hoist to make them more effective in the SAR role; the other two are in storage, awaiting funds for upgrade.

SPECIFICATION

Mi-8T 'Hip-C'

Dimensions

Length (rotors turning): 82 ft 9¾ in (25.24 m)
Fuselage length: 59 ft 7⅓ in (18.17 m)
Main rotor diameter: 69 ft 10¼ in (21.29 m)
Tail rotor diameter: 12 ft 10 in (3.91 m)
Main rotor disc area: 3,832.08 sq ft (356.00 m²)
Tail rotor disc area: 129.25 sq ft (12.01 m²)
Height overall: 18 ft 6½ in (5.65 m)
Wheel base: 13 ft 11¾ in (4.26 m)
Wheel track: 14 ft 9 in (4.50 m)

Powerplant

Two Klimov (Isotov) TV2-117A turboshafts each rated at 1,481 shp (1104 kW)

Weights

Typical empty: 15,784 lb (7160 kg)
Normal take-off: 24,471 lb (11100 kg)
Maximum take-off: 26,455 lb (12000 kg)

Fuel load

Standard fuel load: 494 US gal (1870 litres)
External fuel load: 258 US gal (980 litres)

Performance

Maximum level speed at sea level: 134 kt (155 mph; 250 km/h)
Maximum rate-of-climb at sea level: 14 ft 8⅘ in (4.50 m) per second
Service ceiling: 14,760 ft (4500 m)
Hover ceiling (IGE): 6,235 ft (1900 m)
Hover ceiling (OGE): 2,625 ft (800 m)
Ferry range: 501 nm (577 miles; 930 km)
Radius of action: 188 nm (217 miles; 350 km)

Armament

Outriggers carry four pylons each capable of carrying a UV-16-57, a UV-32-57 rocket pod or alternatively each pylon can carry a bomb of up to 551 lb (250 kg)

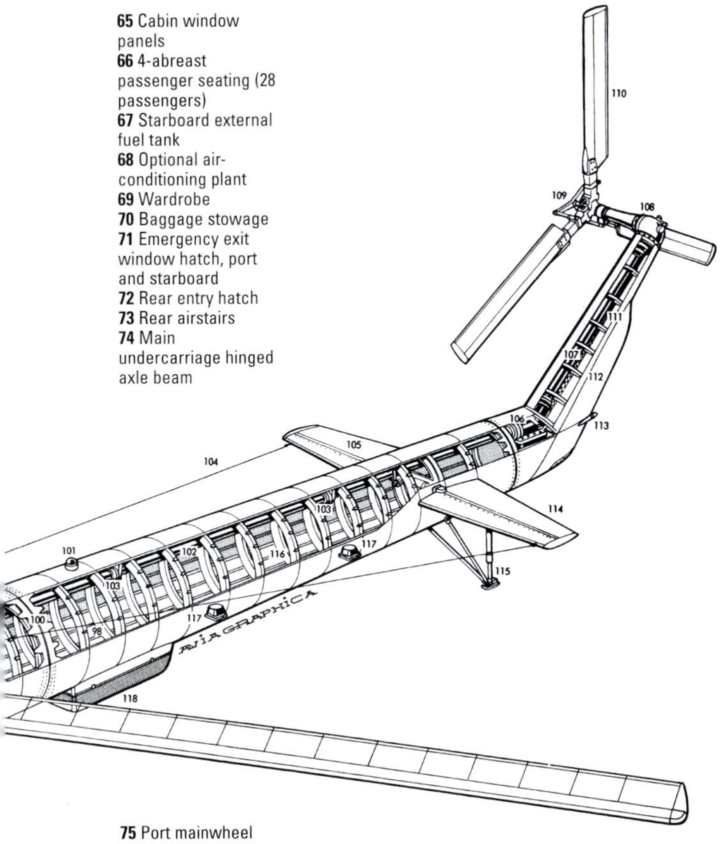

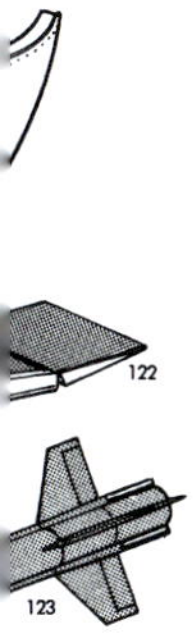

65 Cabin window panels
66 4-abreast passenger seating (28 passengers)
67 Starboard external fuel tank
68 Optional air-conditioning plant
69 Wardrobe
70 Baggage stowage
71 Emergency exit window hatch, port and starboard
72 Rear entry hatch
73 Rear airstairs
74 Main undercarriage hinged axle beam
75 Port mainwheel
76 Blade root attachment joints
77 Hydraulic drag dampers
78 Rotor head hydraulic reservoir
79 Blade pitch control rods
80 Swash plate mechanism
81 Rotor head tail fairing
82 Main reduction gearbox
83 Rotor head hydraulic control jacks (3)
84 Fuel system collector and feed tank
85 Gearbox mounting struts
86 Control rod linkages
87 Gearbox mounting deck
88 Fuselage upper longeron
89 External stores pylon mounting struts
90 Gearbox mounting fuselage main frames
91 Cabin window panels
92 Main undercarriage shock absorber leg strut
93 Fuselage frame and stringer construction
94 Control system access hatch
95 Engine/gearbox aft fairing
96 Cooling air exit louvres
97 Aerial mast
98 Radio communications and navigation equipment
99 HVF aerial
100 Tailboom attachment joint ring frame
101 Anti-collision light
102 Tail rotor transmission shaft
103 Shaft bearings
104 HF aerial cable
105 Starboard variable-incidence tailplane
106 Bevel drive gearbox
107 Tail rotor drive shaft
108 Final drive right-angle gearbox
109 Tail rotor pitch control mechanism
110 3-bladed tail rotor
111 Tail rotor pylon
112 Pylon tail fairing
113 Tail navigation light
114 Port variable-incidence tailplane
115 Fixed tail bumper
116 Tailboom frame and stringer construction
117 Radio altimeter aerials
118 Doppler aerial fairing
119 Aft clamshell doors, open
120 Door hydraulic jacks
121 Vehicle loading ramps
122 Ramp toe-plate
123 AT-2 'Swatter' air-to-surface missile
124 Missile launch rails
125 Missile fire control unit
126 Port stores pylons (3)
127 UV-32-57 rocket launcher, 32 x 57-mm folding fin rockets

East Germany, like most Warsaw Pact states, obtained Mi-8s/17s. This Mi-8TB was operated by that nation's Volksmarine and was fitted with three pylons on each side of the fuselage. After German re-unification, the Marineflieger continued to operate the type until the mid-1990s.

Mil Mi-24 'Hind'

In addition to its normal underwing load of rocket pods and guided-missiles, the 'Hind' is capable of carrying a variety of stores including bombs, napalm tanks and fuel tanks, as illustrated by these Czech 'Hind-Es'.

Mi-24RCh & -24K 'Hind-G1 & G-2'

Cutaway key

1 Remotely-controlled camera (Mi-24K) mounted beneath nose cannon barbette
2 Hinged lens cover
3 Low-speed precision airflow sensors
4 Air data sensor boom
5 IFF antenna
6 Armoured windscreen panel
7 Windscreen wiper
8 Gunsight
9 Pitot heads
10 Turret mechanism access doors
11 Ammunition feed mechanism
12 Four-barrelled 9-A-624 12.7-mm rotary machine-gun with 1,470 rounds
13 Gun turret; ±20° traverse and +20°/-60° elevation/depression
14 Radar warning antenna
15 Retractable landing lamp
16 Boarding step
17 Kick-in steps
18 Oxygen bottle
19 Weapons Systems Officer (WSO) hatch
20 Collective pitch control column, flying controls duplicate in front cockpit
21 WSO's seat
22 Safety harness
23 Side console panel
24 WSO's hatch, open position
25 Cabin-mounted camera pallet (Mi-24K)
26 Hinged lens aperture, replaces starboard cabin door
27 Film magazine
28 Cabin-mounted data-link console (Mi-24RCh)
29 Pilot's entry door, open position
30 Armoured windscreen panel
31 Windscreen wiper
32 Head-up-display
33 Instrument panel shroud
34 Cyclic pitch control column
35 Yaw control rudder pedals
36 Underfloor control linkages
37 Search light
38 Nosewheel leg door and indicator light
39 Levered suspension axle beam
40 Aft-retracting twin nosewheels
41 Conditioned air ducting (ammunition magazine on starboard side)
42 Nosewheel bay (semi-retracted housing)
43 Cockpit section armoured skin panelling
44 Cyclic pitch control lever
45 Fuel cocks
46 Circuit-breaker panel
47 Oxygen bottle
48 Rear view mirror
49 Pilot's seat
50 Engine air intake vortex-type dust/debris extractors
51 Debris ejection chute
52 Intake cowling
53 Generator cooling air intake
54 Starboard engine cowling/hinged work platform

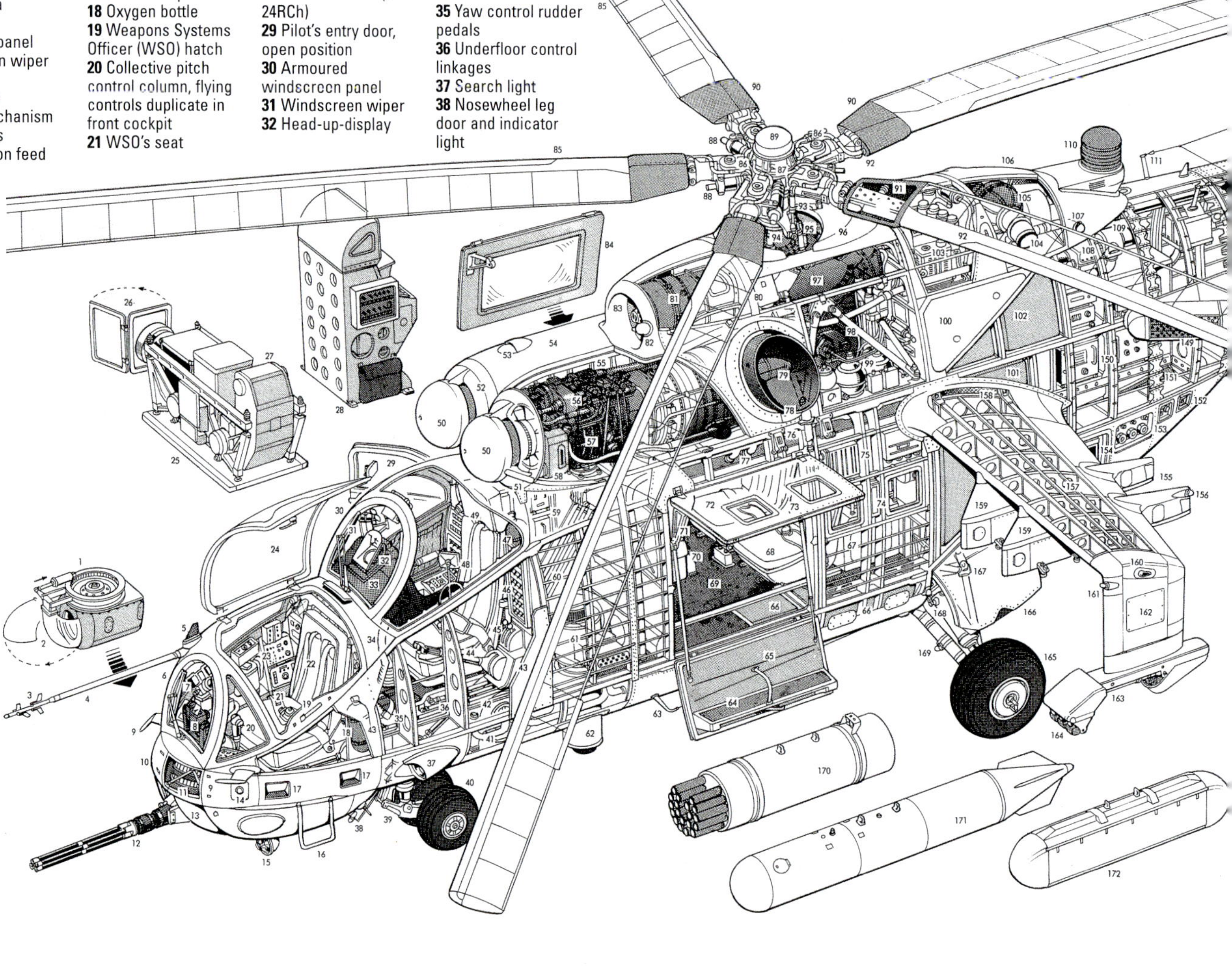

The Mi-24DU 'Hind-D' trainer saw extensive use in the Soviet Union, for training both Soviet pilots and those of client nations, almost all of whom sent their students to the USSR. These helicopters belonged to the Syzran Air Force Academy.

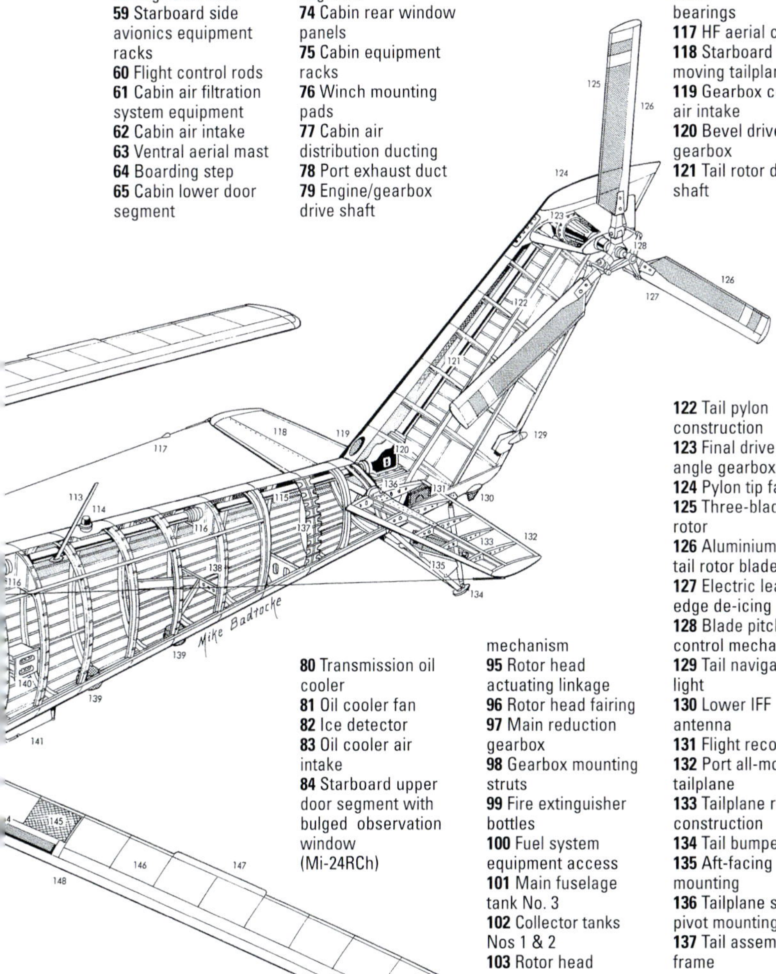

55 Engine bay dividing firewall
56 Accessory equipment gearbox
57 Klimov TV3-117 turboshaft engine
58 Engine oil tank
59 Starboard side avionics equipment racks
60 Flight control rods
61 Cabin air filtration system equipment
62 Cabin air intake
63 Ventral aerial mast
64 Boarding step
65 Cabin lower door segment
66 Underfloor fuel tanks Nos 4 and 5, total internal fuel capacity 427 Imp gal (1940 litres)
67 Cabin equipment consoles
68 Observer's seat
69 Main cabin floor panelling
70 Starboard side data-link console (Mi-24RCh)
71 Door interconnecting linkage
72 Tactical navigator's seat
73 Upper door segment
74 Cabin rear window panels
75 Cabin equipment racks
76 Winch mounting pads
77 Cabin air distribution ducting
78 Port exhaust duct
79 Engine/gearbox drive shaft
80 Transmission oil cooler
81 Oil cooler fan
82 Ice detector
83 Oil cooler air intake
84 Starboard upper door segment with bulged observation window (Mi-24RCh)
85 Five-bladed main rotor
86 Blade root hinge joints
87 Titanium rotor head
88 Hydraulic drag dampers
89 Hydraulic reservoir
90 Blade root cuffs
91 Blade spar crack indicator (pressurised nitrogen-filled)
92 Electric leading-edge de-icing
93 Blade pitch control rods
94 Swash plate
mechanism
95 Rotor head actuating linkage
96 Rotor head fairing
97 Main reduction gearbox
98 Gearbox mounting struts
99 Fire extinguisher bottles
100 Fuel system equipment access
101 Main fuselage tank No. 3
102 Collector tanks Nos 1 & 2
103 Rotor head control hydraulic module
104 APU exhaust
105 AI-9V auxiliary power unit
106 Hinged access panels
107 Air system vent
108 Auxiliary equipment gearbox
109 Generator
110 L-166V-11E Ispanka, microwave-pulse infra-red jammer
111 Aerial lead-in
112 UHF aerial mast
113 VHF aerial
114 Anti-collision light
115 Tail rotor transmission shaft
116 Transmission shaft bearings
117 HF aerial cable
118 Starboard all-moving tailplane
119 Gearbox cooling air intake
120 Bevel drive gearbox
121 Tail rotor drive shaft
122 Tail pylon construction
123 Final drive right-angle gearbox
124 Pylon tip fairing
125 Three-bladed tail rotor
126 Aluminium alloy tail rotor blades
127 Electric leading-edge de-icing
128 Blade pitch control mechanism
129 Tail navigation light
130 Lower IFF antenna
131 Flight recorder
132 Port all-moving tailplane
133 Tailplane rib construction
134 Tail bumper
135 Aft-facing camera mounting
136 Tailplane spar pivot mounting
137 Tail assembly joint frame
138 Tailcone frame and stringer construction
139 Radar altimeter antennas
140 Signal cartridge firing unit
141 Short-wave aerial cable
142 DISS-15D gyromagnetic compass units
143 Tailcone joint frame
144 Hollow D-section titanium blade spar
145 Honeycomb trailing-edge panels
146 Glass-fibre skin panelling
147 Fixed blade tab
148 Leading-edge anti-erosion sheath
149 Chaff-flare dispensers, port and starboard
150 Rear avionics equipment bay
151 Ventral access hatch
152 Ground power socket
153 Pneumatic system around connectors
154 Battery bay, both sides
155 Wing pylon tail fairings
156 Radar warning antenna
157 Stub wing rib and spar structure
158 Stub wing attachment joints
159 Port stores pylons
160 Port navigation light
161 Radar warning antenna
162 Stub wing endplate pylon fairing
163 'Clutching-hand' ground sample collector
164 Hydraulically-operated scoops, three per side
165 Port mainwheel
166 Mainwheel leg door
167 Mainwheel leg aft pivot mounting
168 Shock absorber strut
169 Undercarriage indicator light
170 UV-32A-24 rocket launcher
171 PTB-450 110-Imp gal (500-litre) auxiliary fuel tank
172 Container store

SPECIFICATION

Mi-24D 'Hind-D'

Dimensions

Length overall, rotors turning: 64 ft 11 in (19.79 m)
Fuselage length (excluding rotors and gun): 57 ft 5½ in (17.51 m)
Main rotor diameter: 56 ft 9 in (17.30 m)
Tail rotor diameter: 12 ft 10 in (3.91 m)
Height overall, rotors turning: 21 ft 4 in (6.50 m)
Height to top of rotor head: 14 ft 6¾ in (4.44 m)
Main rotor disc area: 2,529.52 sq ft (235 m²)
Tail rotor disc area: 129.12 sq ft (11.99 m²)
Wingspan: 21 ft 5½ in (6.54 m)
Tailplane span: 10 ft 9 in (3.27 m)
Wheel base: 14 ft 5 in (4.39 m)
Wheel track: 9 ft 11½ in (3.03 m)
Maximum rotor disc loading: 10.46 lb/sq ft (51.05 kg/m²)

Powerplant

Two 2,200-shp (1640-kW) Klimov (Isotov) TV3-117 Series III turboshafts

Weights

Empty: 18,519 lb (8400 kg)
Normal take-off: 24,250 lb (11000 kg)
Maximum take-off: 27,557 lb (12500 kg)

Fuel and load

Internal fuel: 3,307 lb (1500 kg) or 469 Imp gal (2130 litres) plus provision for 2,205 lb (1000 kg) or 187 Imp gal (850 litres) of auxiliary fuel in an optional cabin tank
External fuel (with auxiliary internal tank removed): 2,646 lb (1200 kg) in four 110-Imp gal (500-litre) drop tanks
Maximum ordnance: 5,291 lb (2400 kg)

Performance

Maximum level speed 'clean' at optimum altitude: 168 kt (192 mph; 310 km/h)
Maximum cruising speed at optimum altitude: 140 kt (162 mph; 260 km/h)
Maximum rate of climb at sea level: 2,461 ft (750 m) per minute
Service ceiling: 14,765 ft (4500 m)
Hovering ceiling (out-of-ground effect): 7,220 ft (2200 m)

Range

Maximum range, internal fuel only: 405 nm (466 miles; 750 km)
Combat radius with maximum military load: 86 nm (99 miles; 160 km)
Combat radius with two drop tanks: 135 nm (155 miles; 250 km)
Combat radius with four drop tanks: 155 nm (179 miles; 288 km)

Armament

One four-barrelled 12.7-mm (0.5-in) YakB-12.7 Gatling-type machine-gun in remotely-controlled undernose USPU-24 turret, plus four 9M17P Skorpion (AT-2 'Swatter') radio-guided anti-tank missiles, and four UV-32 rocket pods each containing 32 S-5 57-mm (2.24-in) unguided rockets; or four twenty-round B-8V-20 80-mm (3.15-in) S-8 rocket pods; or four five-round B-13L 130-mm (5.12-in) S-13 rocket pods; or four 240-mm (9.45-in) S-24B rockets. Possible underwing stores also include UPK-23-250 gun pods containing one GSh-23L twin-barrelled 23-mm cannon; GUV gun pods containing one four-barrelled 12.7-mm (0.5-in) machine-gun or two four-barrelled 7.62-mm (0.3-in) 9-A-622 machine-guns or one 30-mm AGS-17 Plamia grenade launcher; bombs, mine dispensers and napalm tanks

Sikorsky H-3/Westland Sea King

Westland achieved considerable export success with the Sea King, including the supply of three Commando Mk 2As to Qatar (two of which are illustrated). The Commando was originally conceived as a dedicated assault version of the Sea King but, through a number of variants, its roles now include AEW, anti-ship and VIP transport operations.

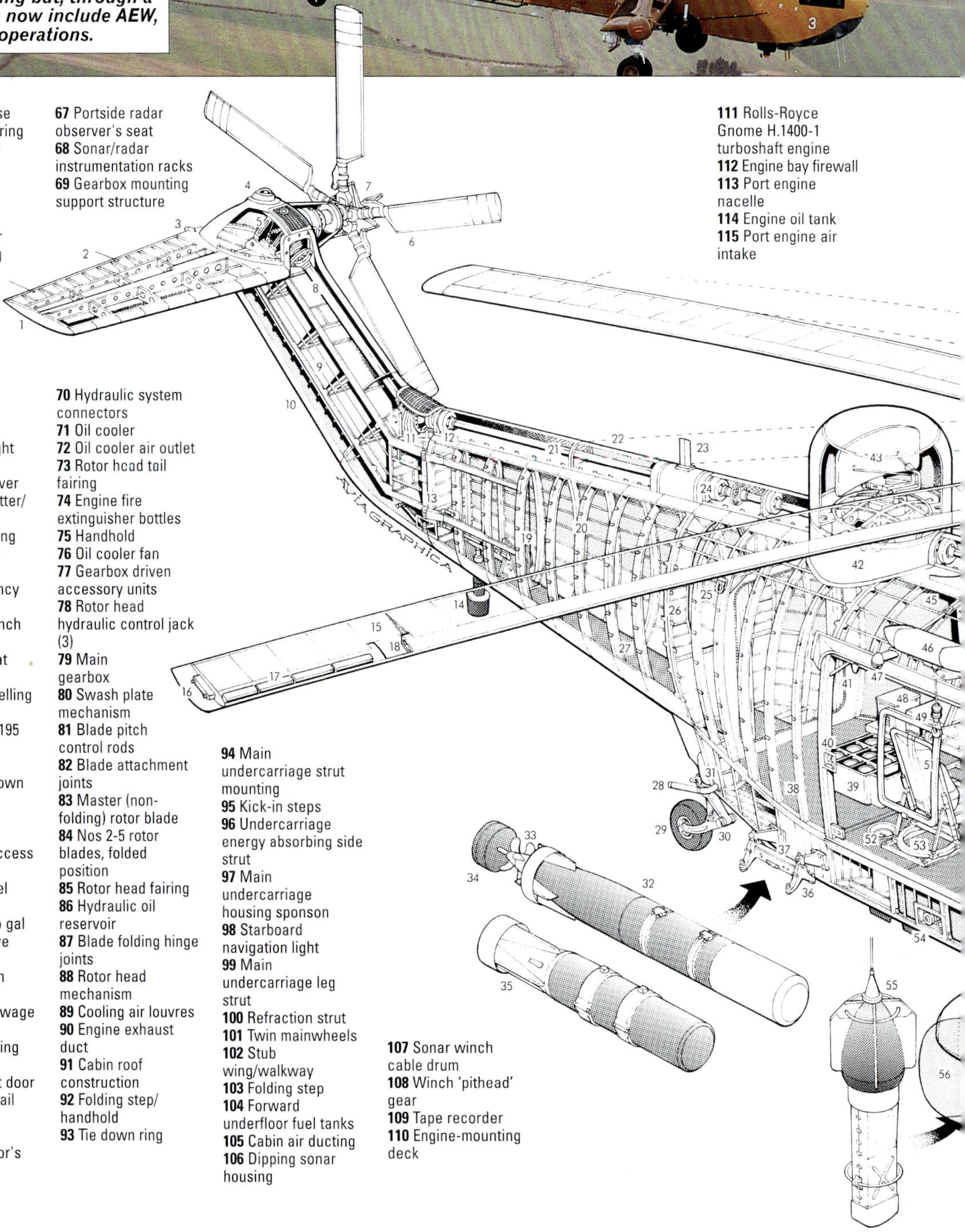

Sea King HAS. Mk 5

Cutaway key
1 Fixed-tailplane construction
2 Static dischargers
3 Tail navigation light
4 Anti-collision light
5 Tail rotor gear box
6 Six-bladed tail rotor
7 Blade pitch change mechanism
8 Tail rotor drive shaft
9 Tail pylon construction
10 Glassfibre trailing-edge panel
11 Intermediate shaft gearbox
12 Shaft coupling
13 Folding tail pylon hinges
14 Transponder aerial
15 Rotor blade cross section
16 Blade tracking weight
17 Blade balance weights
18 D-section aluminium spar
19 Tail rotor control gear
20 Tailcone frame and stringer construction
21 Tail rotor transmission shaft
22 Dorsal spine fairing
23 UHF aerial
24 Shaft bearings
25 Tie-down ring
26 Fuselage/tailcone production joint
27 Maintenance walkway
28 Fuel jettison pipe
29 Non-retracting tailwheel
30 Tailwheel levered suspension leg strut
31 Tailwheel castoring leg fixing
32 Mk 46 torpedo
33 Torpedo propellers
34 Parachute launch pack
35 Mk 11 depth charge
36 Weapon pylon shackles
37 Weapon release unit **38** Cabin flooring
39 Smoke marker container
40 Door latch
41 Cabin rear bulkhead
42 Radar scanner support mounting
43 MEL Sea Searcher radar scanner
44 HF aerial cable
45 Cabin wall soundproofing panels
46 Rescue hoist/winch
47 Winch floodlight
48 Transponder transmitter/receiver
49 Radar transmitter/receiver
50 Data-processing station (Marconi LAPADS)
51 Crew emergency exit window
52 Sonobuoy launch tube
53 Swivelling seat mounting
54 Pressure refuelling connection
55 Plessey Type 195 dipping sonar
56 Emergency flotation bag (shown inflated)
57 Flotation bag inflation bottles
58 Bilge pump access covers
59 Underfloor fuel tanks, total fuel capacity 704 Imp gal (3200 litres) in five cells
60 Fuselage main longeron
61 Sonobuoy stowage racks
62 Winch operating control lever
63 Sliding freight door
64 Freight door rail
65 Data display panels
66 Sonar operator's seat
67 Portside radar observer's seat
68 Sonar/radar instrumentation racks
69 Gearbox mounting support structure
70 Hydraulic system connectors
71 Oil cooler
72 Oil cooler air outlet
73 Rotor head tail fairing
74 Engine fire extinguisher bottles
75 Handhold
76 Oil cooler fan
77 Gearbox driven accessory units
78 Rotor head hydraulic control jack (3)
79 Main gearbox
80 Swash plate mechanism
81 Blade pitch control rods
82 Blade attachment joints
83 Master (non-folding) rotor blade
84 Nos 2-5 rotor blades, folded position
85 Rotor head fairing
86 Hydraulic oil reservoir
87 Blade folding hinge joints
88 Rotor head mechanism
89 Cooling air louvres
90 Engine exhaust duct
91 Cabin roof construction
92 Folding step/handhold
93 Tie down ring
94 Main undercarriage strut mounting
95 Kick-in steps
96 Undercarriage energy absorbing side strut
97 Main undercarriage housing sponson
98 Starboard navigation light
99 Main undercarriage leg strut
100 Refraction strut
101 Twin mainwheels
102 Stub wing/walkway
103 Folding step
104 Forward underfloor fuel tanks
105 Cabin air ducting
106 Dipping sonar housing
107 Sonar winch cable drum
108 Winch 'pithead' gear
109 Tape recorder
110 Engine-mounting deck
111 Rolls-Royce Gnome H.1400-1 turboshaft engine
112 Engine bay firewall
113 Port engine nacelle
114 Engine oil tank
115 Port engine air intake

Since the late 1970s, the Sea King HAR.Mk 3 has been the RAF's standard SAR helicopter. An initial batch of 16 had been delivered by 1979, with a further three delivered in 1985. Six examples of the advanced HAR.Mk 3A are also now in service.

116 Engine starter housing
117 Starboard engine air intake
118 Engine mounting strut
119 Intake foreign object deflector
120 Pitot tube
121 Control rod linkages
122 Heating/ventilation system air intake
123 Fuel control computers
124 Cabin heater/blower
125 Boat hull chine longeron
126 Electrical equipment bay
127 Cockpit floor level
128 Fire extinguisher
129 Sliding side window pane
130 Pilot's seat
131 Cockpit bulkhead
132 Radio rack
133 Port entry doorway
134 Rotor brake lever
135 Overhead switch panels
136 Engine control cables
137 Entry door upper segment (open)
138 Pitot tube
139 Cockpit eyebrow windows
140 Windscreen panels
141 Windscreen wipers
142 Air temperature probe
143 Co-pilot's seat
144 Instrument panel shroud
145 Centre control console
146 Cyclic pitch control column
147 Back of instrument panel
148 Yaw control rudder pedals
149 Downward vision window
150 Radio and electronics equipment
151 Anti-collision light
152 Homing aerials
153 Retractable landing/taxiing lamps
154 Vertical landing lamps
155 Bow compartment hinged access door
156 Battery compartment
157 Fresh air intakes
158 VHF aerial

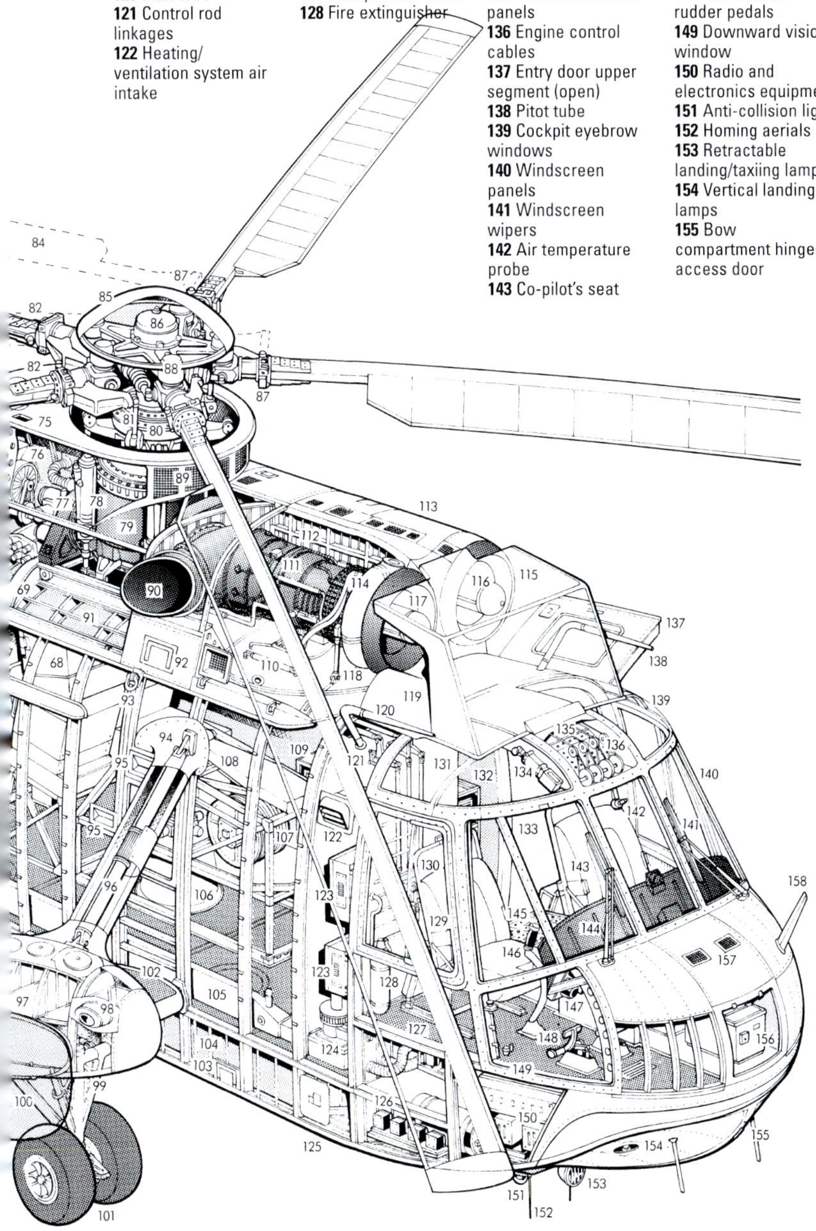

SPECIFICATION

Advanced Sea King

Dimensions

Length overall, rotors turning: 72 ft 8 in (22.15 m)
Fuselage length: 55 ft 10 in (17.02 m)
Length with main rotor blades folded: 57 ft 2 in (17.42 m)
Length with main rotor blades and tail pylon folded: 47 ft 3 in (14.40 m)
Main rotor diameter: 62 ft (18.90 m)
Tail rotor diameter: 110 ft 4 in (3.16 m)
Height overall, rotors turning: 16 ft 10 in (5.13 m)
Height overall, rotors stationary: 15 ft 11 in (4.85 m)
Height to top of rotor head: 15 ft 6 in (4.72 m)
Main rotor disc area: 3,019.07 sq ft (280.47 m²)
Tail rotor disc area: 83.86 sq ft (7.79 m²)
Wheel base: 23 ft 5 in (7.14 m)
Wheel track: 13 ft (3.96 m)

Powerplant

Two Rolls-Royce Gnome H.1400-1T turboshafts each rated at 1,660 shp (1238 kW) for take-off and 1,465 shp (1092 kW) for continuous running

Weights

Basic empty with sponsons: 11,891 lb (5393 kg)
Basic empty without sponsons: 11,845 lb (5373 kg)
Empty equipped (ASW role): 16,377 lb (7428 kg)
Empty equipped (ASV role): 16,689 lb (7570 kg)
Empty equipped (AEW role): 17,143 lb (7776 kg)
Empty equipped (SAR role): 13,760 lb (6241 kg)
Empty equipped (troop transport role): 12,594 lb (5712 kg)
Empty equipped (freight role): 12,536 lb (5686 kg)
Empty equipped (VIP role): 15,917 lb (7220 kg)
Maximum take-off: 21,500 lb (9752 kg)

Fuel and load

Internal fuel capacity: 817 Imp gal (3714 litres)
Auxiliary fuel: 190 Imp gal (863 litres) in an auxiliary fuselage tank
Maximum ordnance: 2,500 lb (1134 kg)

Performance

Never-exceed speed at sea level: 140 mph (226 km/h)
Maximum cruising speed at sea level: 126 mph (204 km/h)
Maximum rate of climb at sea level: 2,030 ft (619 m) per minute
Service ceiling (one engine out): 4,000 ft (1220 m)
Hovering ceiling (in ground effect): 6,500 ft (1980 m)
Hovering ceiling (out of ground effect): 4,700 ft (1435 m)

Range

Ferry range with auxiliary fuel: 1,082 miles (1742 km)
Range with standard fuel: 921 miles (1482 km)

Armament

Up to four Mk 46, A244S or Sting Ray torpedoes can be carried externally, or four Mk 11 depth charges. Sea Eagle or Exocet anti-ship missiles are anti-ship options. Pintle-mounted machine-guns can be fitted to starboard door.

Sikorsky H-19/Westland Whirlwind

Operated by the US Marines test unit HMX-6, this HRS-3 could be fitted with rocket-assisted rotors to boost performance. Two of these helicopters were capable of transporting a whole 'geodetic dome' field hospital unit as shown in the photograph.

S-55

Cutaway key
1 Two-bladed tail rotor
2 Tail rotor pitch change mechanism
3 Feathering counterweights
4 Right-angle final drive gearbox
5 Tail rotor drive shaft
6 Tail rotor pylon
7 Rear navigation lights
8 Tailcone fairing
9 Tail bumper/rotor guard
10 Bevel drive intermediate gearbox
11 Pylon attachment ring frame
12 Anhedral stabiliser, port and starboard
13 VHF aerial
14 Tailboom
15 Tail rotor transmission shaft
16 Dorsal spine fairing
17 Shaft bearings
18 Tail rotor control cables
19 Tailboom frame and stringer construction
20 Ventral fairing construction
21 Main rotor blade trailing-edge construction
22 Extruded aluminium blade spar
23 Tailcone joint ring frame
24 Bolted tailcone joint
25 Cabin heater intake
26 Combustion heater unit
27 Winch cable hydraulic motor
28 Baggage/equipment stowage space
29 Radio and electronics equipment bay
30 Equipment bay access door
31 Main undercarriage shock absorber strut
32 Starboard mainwheel
33 Float mainwheel
34 Pivoted axle beam
35 Hydraulic brake pipe
36 Sliding cabin door lower rail
37 Rear fuel tank group filler cap
38 3-abreast passenger seating, 8-seat all-passenger layout
39 Cabin window panel
40 Heater air ducting
41 Cabin rear bulkhead
42 Baggage compartment door
43 Sliding door top rail
44 Winch cable emergency cutter
45 Gearbox support deck double frame
46 Rear underfloor fuel tanks, total system capacity 150 Imp gal (180 US gal; 682 litres)
47 Axle beam pivot fixing
48 Floor beam construction
49 Jettisonable window panel
50 Cabin wall trim panelling
51 Folding maintenance platform, open
52 Rescue hoist floodlight
53 Cockpit sliding entry hatch rail
54 Internal maintenance platform
55 Cockpit rearward vision window panels
56 Gearbox mounting deck
57 Gearbox oil cooler
58 Oil cooler air outlet fairing
59 Oil cooler belt drive
60 Gearbox mounting strut
61 Main gearbox
62 Rotor head control jacks (3)
63 Swash plate assembly
64 Blade pitch angle control rods
65 Torque scissor links
66 Pitch control arm
67 Drag damper
68 Blade root attachment joints
69 Main rotor head
70 3-bladed main rotor
71 Cockpit eyebrow windows
72 Co-pilot's seat
73 Control linkages
74 Rotor brake
75 Pilot's seat
76 Sliding side window/entry hatch
77 Rescue hoist/winch
78 Main cabin sliding door
79 Door latch
80 Forward group of underfloor fuel tanks
81 Cabin floor panelling
82 Winch hook
83 External load sling, 2000-lb (907-kg) capacity
84 Forward float mounting strut
85 Grab handle/step
86 Forward fuel tank filler cap
87 Aft-facing seat row
88 Boarding steps
89 Oil tank
90 Oil filler cap
91 Ventilating air grille
92 Cockpit floor level
93 Cockpit heater duct
94 Cyclic pitch control column
95 Collective pitch control lever
96 Centre control console, engine controls
97 Instrument panel shroud
98 Temperature probe
99 Windscreen panels
100 Windscreen de-misting air duct
101 Yaw control rudder pedals
102 Engine cooling air intake grille
103 Sloping engine bay fireproof bulkhead
104 Cooling air fan
105 Engine bay door hinges
106 Engine oil cooler
107 Nose undercarriage leg strut
108 Torque scissor links
109 Starboard castoring nosewheel
110 Cooling air ventral outlet
111 Carburettor intake duct
112 Main engine mounting
113 Carburettor heater intake duct
114 Exhaust pipe heater muff
115 Exhaust collector ring, exhaust on port side
116 Pratt & Whitney R-1340-40 nine-cylinder radial engine
117 Engine bay clamshell doors
118 Clamshell door latch
119 Engine accessory equipment housing
120 Generator
121 Carburettor
122 Port castoring nosewheel
123 Fixed float, H-19B amphibious rescue variant
124 Bilge pump access panels
125 Mooring attachment
126 Float castoring nosewheels

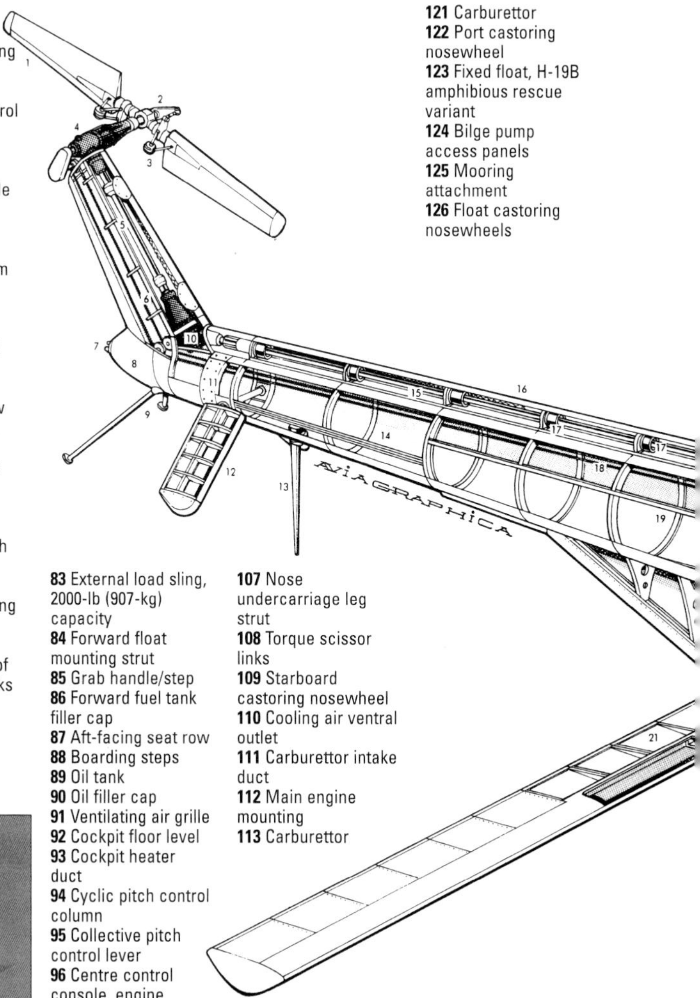

Four specially-painted H-19Ds were employed by the US Army as a 'square dance display team' operating at air shows and other demonstrations around the USA. The 'female' H-19 in the foreground is seen wearing a 'skirt' whereas the 'male' H-19 behind is 'smoking a pipe'.

SPECIFICATION

Whirlwind HAR.Mk 2

Dimensions

Length overall: 62 ft 4 in (19.00 m)
Nose-to-tail rotor distance: 44 ft 2 in (13.50 m)
Main rotor diameter: 53 ft (16.15 m)
Height overall: 15 ft 7½ in (4.76 m)
Wheel track: 11 ft 3 in (3.45 m)
Main rotor disc area: 2,205 sq ft (204.84 m^2)

Powerplant

One Pratt & Whitney R-1340-40 nine-cylinder radial air cooled geared and supercharged engine rated at 600 hp (447 kW)

Performance

Maximum level speed: 99 mph (159 km/h)
Maximum cruising speed: 85 mph (137 km/h)
Rate of climb at sea level: 600 ft (183 m) per minute
Service ceiling at normal weight: 8,600 ft (2621 m)
Standard range at normal operating weight: 320 miles (515 km)

Weights

Empty: 5,327 lb (2416 kg)
Normal loaded: 7,200 lb (3266 kg)

Armament

Up to four Nord SS.11 ASMs

In the late 1950s and early 1960s the Royal Canadian Air Force not only operated the Sikorsky H-19 (bottom two), but also the Sikorsky H-34 (second from top, first ordered in 1955) and the Piasecki (later Vertol) H-21 (top). All three types were heavily used during the construction of the Mid-Canada Line for NORAD and were mainly operated by civilian contractors.

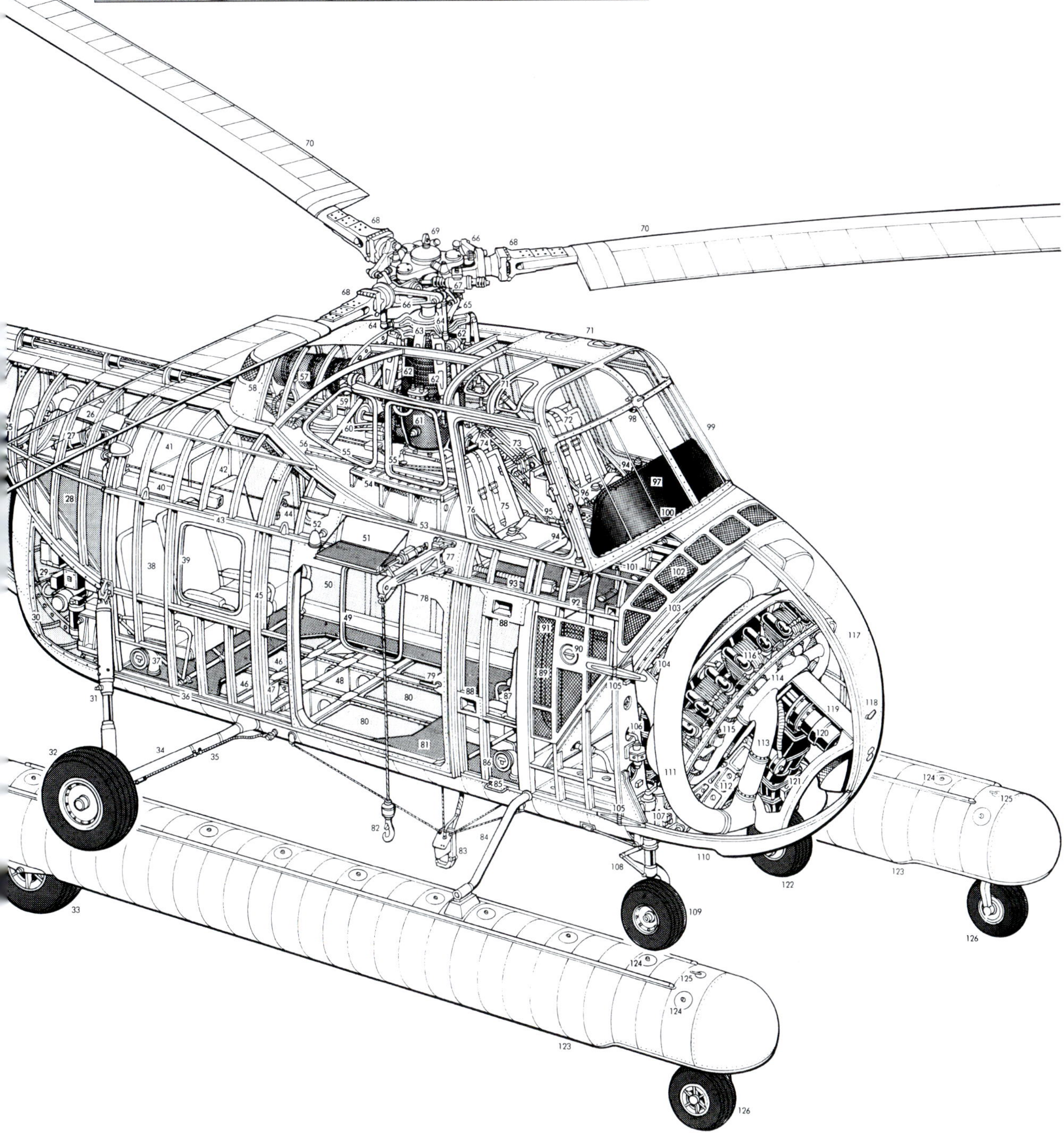

Sikorsky H-53

Bulky loads are carried underneath a central cargo hook. This CH-53 is delivering a field g to troops. Such capability allows rapid mobili around the battlefield for heavy equipment, o early deployment so that artillery is in place soon after the first troops land.

CH-53E

Cutaway key

1 Retractable-in-flight refuelling boom
2 Refuelling boom fairing
3 Instrument compartment access door
4 Glideslope aerial
5 Fresh air intakes
6 Yaw control rudder pedals
7 Landing lamp
8 Downward vision windows
9 Nose undercarriage leg strut
10 Twin nosewheels
11 Radio and electronics bay, port and starboard
12 Cockpit floor level
13 Collective pitch control lever
14 Cyclic pitch control column
15 Co-pilot's armoured seat
16 Instrument panel shroud
17 Windscreen wipers
18 Windscreen panels
19 Rescue hoist/winch
20 Pitot tube
21 UHF aerial
22 Overhead control panel
23 Pilot's armoured seat
24 Cockpit eyebrow window
25 Flight leader's folding jump seat
26 Cockpit bulkhead
27 Jettisonable side window panel
28 Starboard side crew entry door
29 Fuselage and stringer construction
30 Emergency exit window
31 Engine air intake particle separator
32 Bevel drive gearbox
33 Engine oil cooler
34 Auxiliary power unit (APU)
35 Cabin heater unit
36 Starboard engine intake particle separator
37 Engine cowlings armoured on lower surface
38 Auxiliary gearbox
39 Hydraulic reservoirs
40 Gearbox drive shaft
41 Port engine transmission shaft
42 Folding troop seats maximum 37 troops
43 Cargo loading floor
44 Roller conveyor
45 Cargo hook support links
46 General Electric T64-GE-416 turboshaft engine
47 Gearbox mounting fuselage main frame
48 Engine exhaust duct
49 Centre engine intake
50 Main transmission gearbox
51 Blade pitch control rotating swashplate
52 Rotor head mechanism
53 Blade pitch control links
54 Blade folding hinge points
55 Rotor head fairing
56 Seven-bladed main rotor 79-ft (24.08-m) diameter
57 Centre engine oil cooler
58 Maintenance handrail
59 Engine compartment firewall
60 Centre General Electric T64-GE-416 turboshaft engine
61 Cabin wall soundproofing trim panel
62 Rear troop seats
63 Fuselage/main undercarriage main frame
64 Cargo ramp hydraulic jack
65 Production break double frame
66 Centre engine exhaust duct
67 Oil cooler exhaust
68 Rotor blade cross-section
69 D-section titanium spar
70 Honeycomb trailing edge panel
71 Glass-fibre blade skin
72 Leading edge anti-erosion strip
73 Dorsal spine fairing
74 Tail rotor transmission shaft
75 TACAN aerial

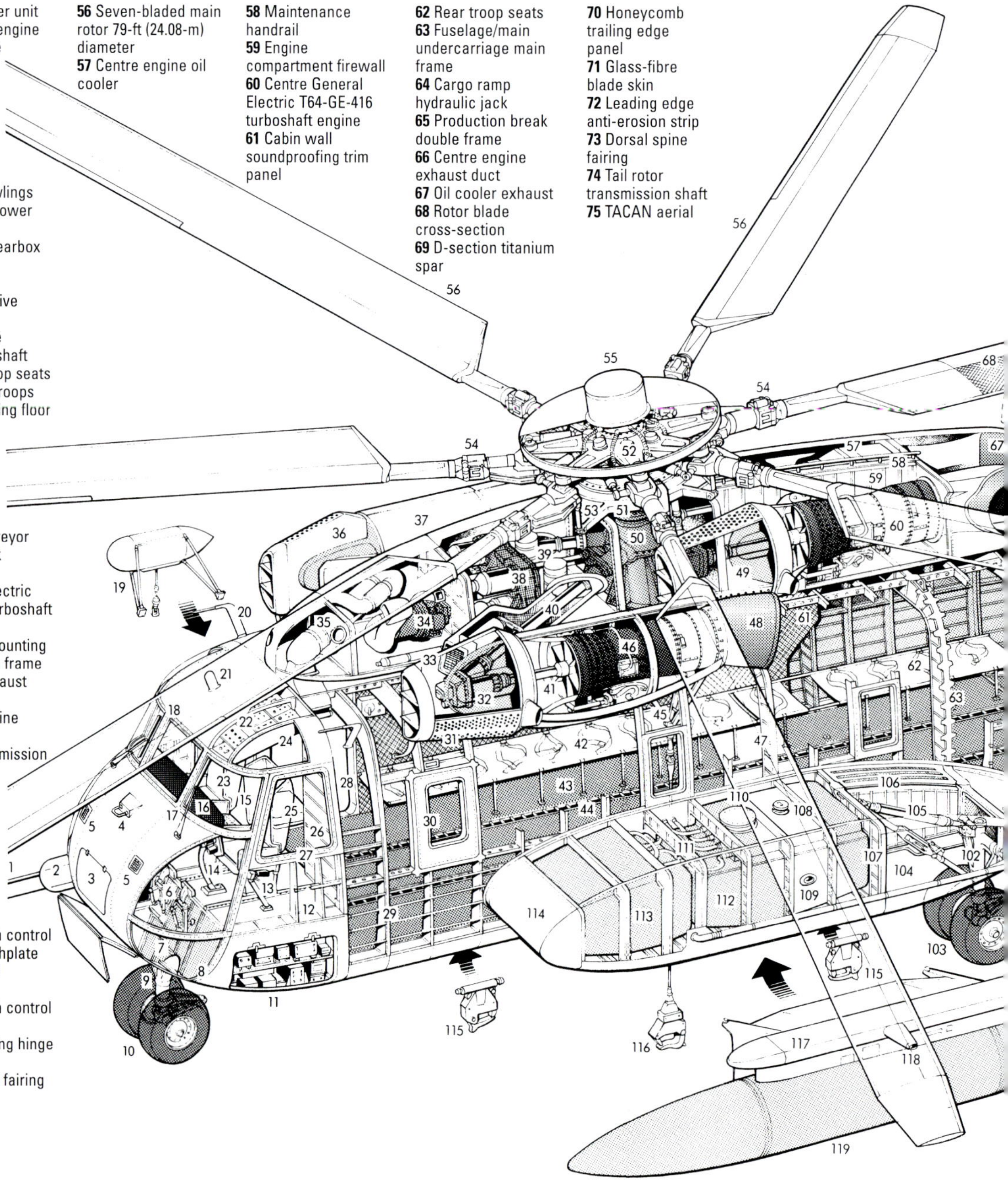

H-53 in Vietnam

It was the shortcomings of other helicopters in Vietnam that provided the impetus for the construction of the H-53. First entering service in 1967, the CH-53A, and soon after the HH-53B, proved popular for transporting supplies and rescuing downed aircrew all over the region. Two years later, in September 1969, the HH-53C made its combat debut. Outwardly similar but possessing many upgraded capabilities the Super Jolly was equipped with an external cargo hook which enabled it to carry loads of up to 20,000 lb (9072 kg), while for the rescue mission it was fitted with a hoist complete with a 250-ft (76-m) cable to penetrate the tallest jungle canopy. For defence up to three 0.3-in (7.62-mm) Miniguns were carried, and these could suppress enemy forces intent on reaching the downed crews before the rescue helicopter.

SPECIFICATION

CH-53E Super Stallion

Dimensions

Length overall, rotors turning: 99 ft ½ in (30.19 m)
Fuselage length: 73 ft 4 in (22.35 m)
Length with rotor and tail folded: 60 ft 6 in (18.44 m)
Height overall: 29 ft 5 in (8.97 m)
Main rotor diameter: 79 ft (24.08 m)
Tail rotor diameter: 20 ft (6.10 m)
Main rotor disc area: 4,901.7 sq ft (455.38 m²)
Tail rotor disc area: 314.2 sq ft (29.19 m²)

Powerplant

Three General Electric T64-GE-416 engines rated at 4,380 shp (3266 kW) for ten minutes, 4,145 shp (3091 kW) for 30 minutes and 3,696 shp (2756 kW) for continuous running

Weights

Empty: 33,228 lb (15072 kg)
Maximum take-off with an internal payload: 69,750 lb (31640 kg)
Maximum take-off with an external payload: 73,500 lb (33340 kg)

Fuel and load

Internal fuel: 1,017 US gal (3849 litres)
External fuel: up to two 650 US gal (2461-litre) drop tanks
Maximum payload internally over 115-mile (185-km) radius: 36,000 lb (16330 kg)
Maximum payload externally over 57.5-mile (92.5-km) radius: 32,000 lb (14515 kg)

Performance

Maximum level speed 'clean' at sea level: 196 mph (315 km/h)
Cruising speed at sea level: 173 mph (278 km/h)
Maximum rate of climb at sea level with a 25,000-lb (11340-kg) payload: 2,500 ft (762 m) per minute
Service ceiling: 18,500 ft (5640 m)
Hovering ceiling: 11,500 ft (3520 m)

Range

Ferry range without aerial refuelling: 1,290 miles (2075 km)
Radius with a 20,000-lb (9072-kg) external payload: 575 miles (925 km)
Radius with a 32,000-lb (14515-kg) external payload: 57.5 miles (92.5 km)

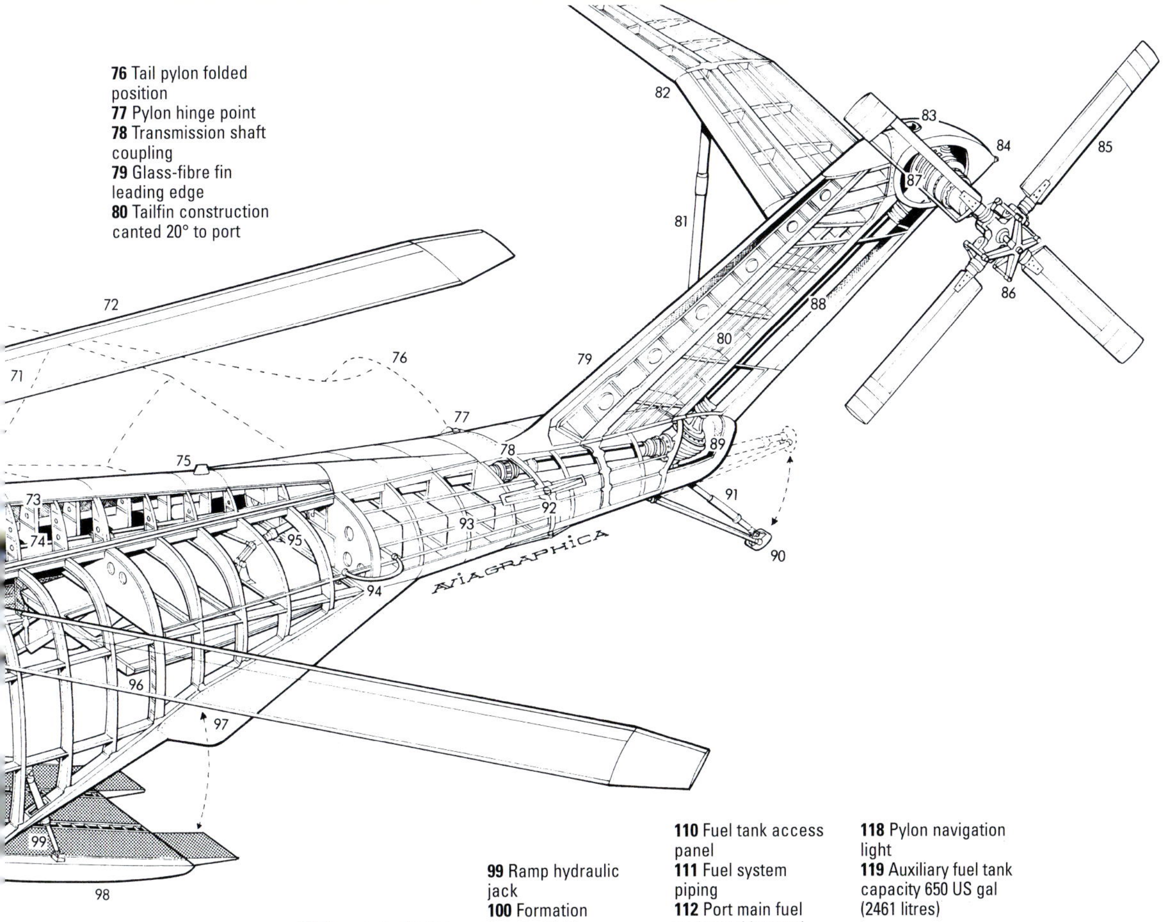

76 Tail pylon folded position
77 Pylon hinge point
78 Transmission shaft coupling
79 Glass-fibre fin leading edge
80 Tailfin construction canted 20° to port
81 Stabiliser bracing strut
82 Gull-wing horizontal stabiliser
83 Anti-collision light
84 Tail navigation light
85 Four-bladed tail rotor, 20-ft (6.1-m) diameter
86 Tail rotor pitch control mechanism
87 Tail rotor gearbox
88 Final drive shaft
89 Bevel gear box
90 Retractable tail bumper
91 Bumper hydraulic jack
92 Folding tail pylon latches
93 Tail boom construction
94 VOR/localiser aerial
95 Upper cargo door hydraulic jack
96 Upper cargo door, open position
97 Doorway side strakes
98 Cargo loading ramp down position
99 Ramp hydraulic jack
100 Formation keeping light
101 Fuel jettison pipe
102 Main undercarriage leg strut
103 Twin mainwheels
104 Mainwheel bay
105 Hydraulic retraction jack
106 Maintenance platform walkway
107 Fuselage sponson main frame
108 Fuel filler cap
109 Port navigation light
110 Fuel tank access panel
111 Fuel system piping
112 Port main fuel tank; total internal capacity 1,017 US gal (3850 litres)
113 Secondary fuel tank
114 Sponson nose fairing
115 Two-point suspension cargo hooks
116 Single-point cargo hook; maximum external slung load 32,200 lb (14606 kg)
117 Auxiliary fuel tank pylon
118 Pylon navigation light
119 Auxiliary fuel tank capacity 650 US gal (2461 litres)

Sikorsky H-60 Black Hawk

Black Hawks were designed from the outset to have an offensive capability on their external pylons. Up to 10,000 lb (4536 kg) of armament can be carried, including Hellfire missiles, 20-mm cannon, and 2.75-in (70-mm) FFARs (Folding Fin Aerial Rockets), which can be seen here being fired.

UH-60A Black Hawk

Cutaway key

1 Nose radio and electronics bay
2 Nose glazing
3 Radio compartment access door
4 Air grille
5 Windscreen wipers
6 Windscreen panels
7 Instrument panel shroud
8 Rudder pedals
9 Downward vision window
10 Boarding step
11 Cockpit door
12 Cyclic pitch control lever
13 Sliding side window
14 Co-pilot's seat
15 Control column
16 Centre instrument console
17 Armoured seat backs
18 Pilot's seat
19 Sliding side armour panels
20 Cockpit eyebrow windows
21 Armoured headrests
22 Electrical fuse panels
23 Pitot tubes
24 Control equipment sliding access cover
25 Cooling intake grille
26 Sliding fairing guide rails
27 Control runs
28 Handrail/aerial bar
29 Gunner's sliding side windows, open
30 Cockpit step/main axle fairing
31 Port navigation light
32 Main undercarriage pivoted axle
33 Ground power supply
34 Swivelling gun mounting
35 Gunner's side-facing seat
36 Port M-23D 0.3-in (7.62-mm) machine-gun
37 Undercarriage shock-absorber strut
38 Port mainwheel
39 Cabin floor level
40 Folding maintenance step
41 Undercarriage mounting main frames
42 Hydraulic pump
43 Flight control mixer unit
44 Cabin heater
45 Engine-driven accessory units
46 Rotor blade titanium main spar
47 Bevel gearboxes
48 Starboard engine intake
49 Central main reduction gearbox
50 Rotor control swash plate
51 Rotor mast
52 Main rotor head (elastomeric non-lubricated bearings)
53 Bifilar vibration damper
54 Rotor head fairing
55 Blade root attachments
56 Composite titanium spar/glass-fibre main rotor blades
57 Rotor blade drooped leading edge
58 Fixed trailing-edge tabs
59 Blade pitch control rods
60 Port engine intake
61 Engine drive shaft
62 Aft-sliding cabin door
63 Emergency door release handle
64 Cargo hook, 8,000-lb (3650-kg) maximum capacity
65 Cabin accommodation, 13 troops plus up to two crew chief/gunners or four stretchers or internal cargo
66 Cabin rear bulkhead
67 Troop seats
68 Sliding door rail fairings

The UH-60L superseded the A model on the production line. It introduces more powerful engines, uprated transmission and a refined gearbox to give a much improved performance, especially in 'hot-and-high' conditions or with external loads. The Hover Infra-Red Suppressor Subsystem (HIRSS – pronounced 'herz') was fitted as standard on all production UH-60s from 1987. This system, which redirects engine exhaust and mixes cool air into the exhaust stream, is effective in all flight profiles.

SPECIFICATION

UH-60A Black Hawk

Dimensions

Length overall with rotors turning: 64 ft 10 in (19.76 m)
Height overall: 16 ft 10 in (5.13 m)
Main rotor diameter: 53 ft 8 in (16.36 m)
Tail rotor diameter: 11 ft (3.35 m)
Wheel track: 8 ft 10½ in (2.71 m)
Wheel base: 28 ft 11¾ in (8.83 m)

Powerplant

Two General Electric T700-GE-700 turboshafts each rated at 1,560 shp (1151 kW) or in export S-70s, two General Electric T700-GE-701A turboshafts each rated at 1,723 shp (1285 kW)

Performance

Maximum level 'clean' speed at sea level: 160 kt (184 mph; 296 km/h)
Maximum cruising speed at 4,000 ft (1220 m): 145 kt (167 mph; 268 km/h)
Ferry range with four external auxiliary tanks: 1200 nm (1,382 miles; 2224 km)
Standard range: 319 nm (368 miles; 592 km)

Fuel and load

Internal fuel: 360 US gal (1361 litres) plus provision for 370 US gal (1400 litres) of auxiliary fuel in two fuselage tanks
External fuel: Up to two 230-US gal (870-litre) and/or two 450-US gal (1703-litre) tanks
Maximum payload: 2,640 lb (1197 kg) carried internally or 8,000 lb (3629 kg) carried externally

Weights

Empty: 11,284 lb (5118 kg)
Normal take-off: 16,994 lb (7708 kg)
Maximum take-off: 20,250 lb

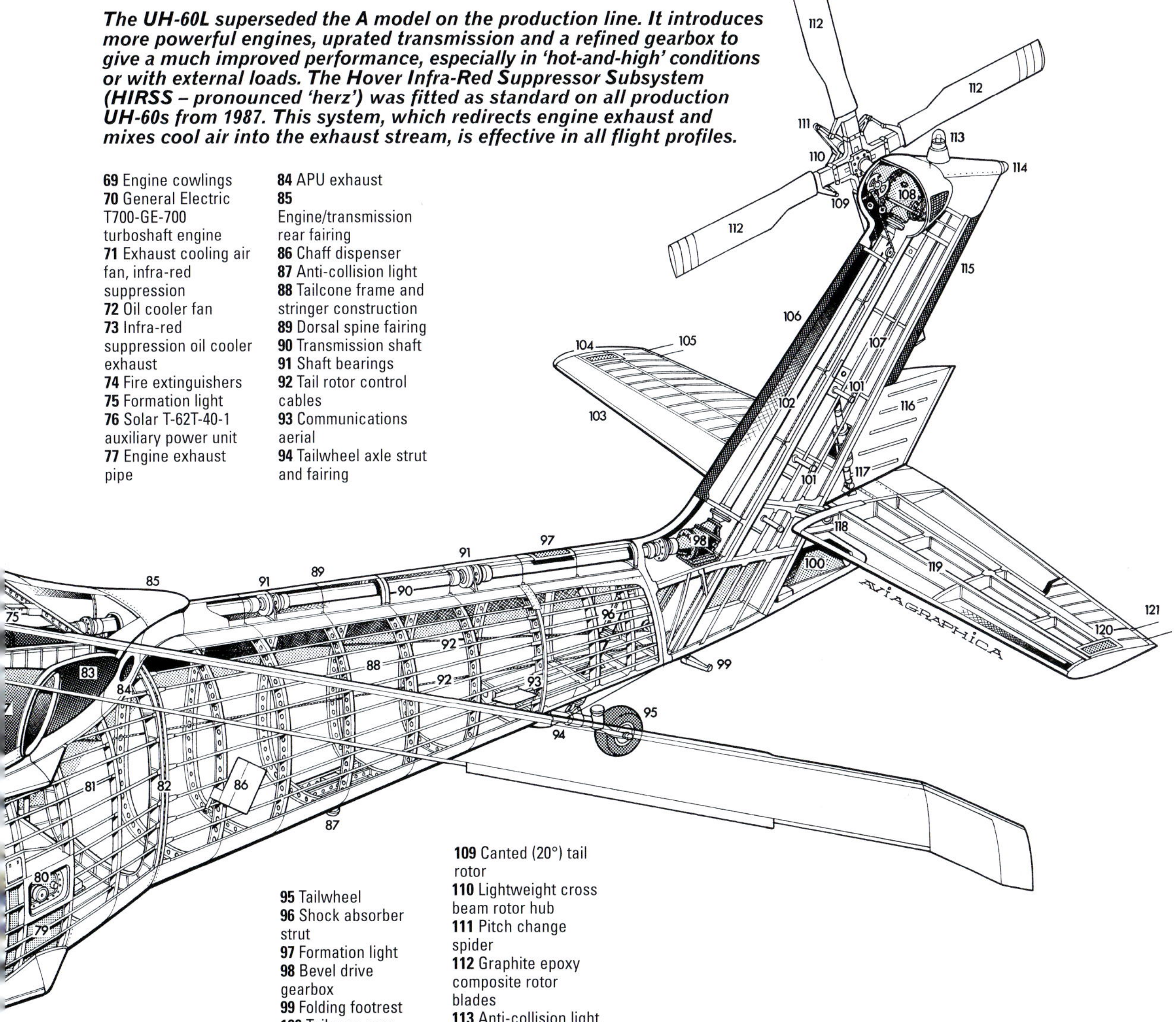

69 Engine cowlings
70 General Electric T700-GE-700 turboshaft engine
71 Exhaust cooling air fan, infra-red suppression
72 Oil cooler fan
73 Infra-red suppression oil cooler exhaust
74 Fire extinguishers
75 Formation light
76 Solar T-62T-40-1 auxiliary power unit
77 Engine exhaust pipe
78 Spring-loaded maintenance steps
79 Fireproof main fuel tanks, port and starboard, capacity 157 US gal (594 litres) each
80 Pressure refuelling connection
81 Fuselage frame and stringer construction
82 Tailcone joint frame
83 Engine exhaust shroud
84 APU exhaust
85 Engine/transmission rear fairing
86 Chaff dispenser
87 Anti-collision light
88 Tailcone frame and stringer construction
89 Dorsal spine fairing
90 Transmission shaft
91 Shaft bearings
92 Tail rotor control cables
93 Communications aerial
94 Tailwheel axle strut and fairing
95 Tailwheel
96 Shock absorber strut
97 Formation light
98 Bevel drive gearbox
99 Folding footrest
100 Tailcone rear fairing
101 Pull-out maintenance steps
102 Tail rotor drive shaft
103 Starboard tailplane
104 Formation light
105 Static discharge wicks
106 Fin leading-edge suppressed aerial
107 Fin construction
108 Tail rotor drive gearbox
109 Canted (20°) tail rotor
110 Lightweight cross beam rotor hub
111 Pitch change spider
112 Graphite epoxy composite rotor blades
113 Anti-collision light
114 Tail navigation light
115 Troop commander's communications aerial
116 Cambered section fin
117 Tailplane hydraulic jack
118 Tailplane pivot fixing
119 Port tailplane construction
120 Formation light
121 Static discharge wicks

Westland Wessex

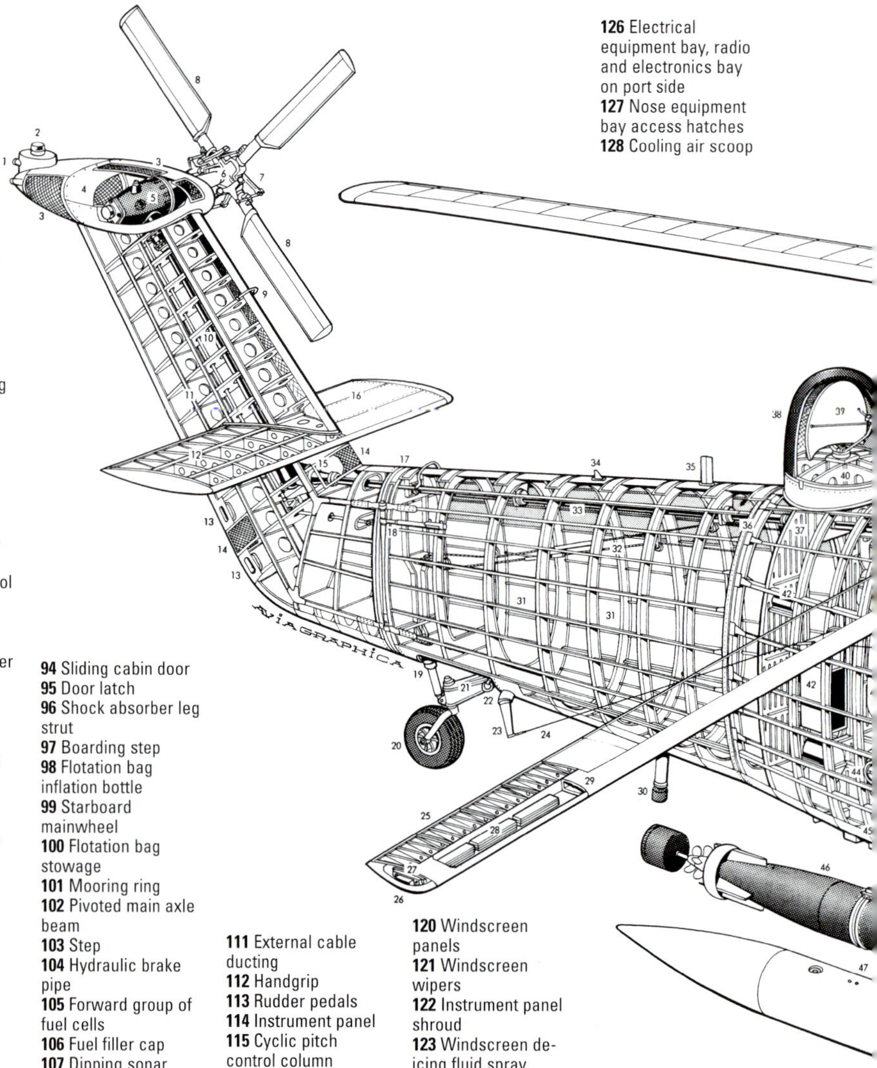

Wessex HAS.Mk 3

Cutaway key

1 Tail navigation lights
2 Anti-collision light
3 Cooling air grilles
4 Tail rotor gearbox fairing
5 Final drive right-angle gearbox
6 Tail rotor hub mechanism
7 Blade pitch angle control linkage
8 Four-bladed tail rotor
9 Handgrip
10 Tail rotor drive shaft
11 Tail pylon construction
12 Fixed horizontal tailplane construction
13 Ground handling grips
14 Cooling air grilles
15 Bevel drive gearbox
16 Port tailplane
17 Folding tail pylon hinge
18 Tail pylon latching mechanism
19 Tailwheel shock absorber strut
20 Castoring tailwheel
21 Hinged axle beam
22 Mooring ring
23 Aerial mast
24 HF aerial cable
25 Rotor blade trailing edge rib construction
26 Tip fairing
27 Blade tracking weight
28 Blade balance weights
29 'D'-section aluminium blade spar
30 Transponder aerial
31 Tailcone frame and stringer construction
32 Tail rotor control cables
33 Tail rotor transmission shaft
34 Upper IFF aerial
35 UHF aerial
36 Tailcone/fuselage joint frame
37 Equipment bay bulkhead
38 Dorsal radome
39 Search radar scanner
40 Radome mounting structure
41 Port side cabin heater
42 Electrical system equipment
43 Fuel delivery piping
44 Rear fuel tank group filler cap
45 Pressure refuelling connection
46 Mk 46 torpedo
47 External fuel tank, capacity 100 Imp gal (454 litres)
48 External cable ducting
49 Aft crashproof fuel cells; total fuel capacity 266 Imp gal (1209 litres)
50 Cabin window/ escape hatch
51 Cabin rear bulkhead
52 Curtained aperture to equipment bay
53 Vent piping
54 Oil cooler air exit louvres
55 Rotor head rear aerodynamic fairing
56 Oil cooler
57 Rear fairing access panels
58 Cabin heating ducting
59 Smoke marker stowage
60 Marker launch tube cover
61 Cabin floor panelling
62 External stores carrier
63 Stores pylon fixing
64 Dipping sonar
65 Floor beam construction
66 Cabin door
67 Seat mounting rails
68 Tactical navigator and sonar operator seats
69 Instrument consoles
70 Rescue hoist/winch
71 Gearbox mounting deck
72 Gearbox support struts
73 Rotor brake
74 Oil cooler fan
75 Gearbox deck access panels
76 Rotor head servo control units
77 Blade pitch control linkage
78 Torque scissor links
79 Blade drag damper
80 Hydraulic oil reservoir
81 Rotor head mechanism
82 Four-bladed main rotor
83 Blade root attachment joints
84 Cooling air grilles
85 Cockpit roof glazing
86 Overhead switch panel
87 Chart case
88 Servomotor switching control panel
89 Cockpit rear bulkhead
90 Pilot's seat
91 Sliding side window/ entry hatch
92 Main undercarriage leg strut attachment
93 Cabin door jettison lever
94 Sliding cabin door
95 Door latch
96 Shock absorber leg strut
97 Boarding step
98 Flotation bag inflation bottle
99 Starboard mainwheel
100 Flotation bag stowage
101 Mooring ring
102 Pivoted main axle beam
103 Step
104 Hydraulic brake pipe
105 Forward group of fuel cells
106 Fuel filler cap
107 Dipping sonar winch mechanism
108 Cockpit access steps
109 Bifurcated engine exhaust pipes, port and starboard
110 Cockpit floor level
111 External cable ducting
112 Handgrip
113 Rudder pedals
114 Instrument panel
115 Cyclic pitch control column
116 Collective pitch lever
117 Co-pilot's seat
118 Rotor brake control lever
119 Temperature gauge
120 Windscreen panels
121 Windscreen wipers
122 Instrument panel shroud
123 Windscreen de-icing fluid spray nozzle
124 Sloping cockpit front bulkhead
125 Engine/gearbox transmission shaft
126 Electrical equipment bay, radio and electronics bay on port side
127 Nose equipment bay access hatches
128 Cooling air scoop

SPECIFICATION

Wessex HC.Mk 2

Dimensions

Overall length with rotors turning: 65 ft 9 in (20.04 m)
Fuselage length: 48 ft 4½ in (14.74 m)
Length with main rotor blades and tail folded: 38 ft 6 in (11.73 m)
Height to top of rotor head: 14 ft 5 in (4.39 m)
Height overall: 16 ft 2 in (4.93 m)
Main rotor diameter: 56 ft (17.07 m)
Tail rotor diameter: 9 ft 6 in (2.90 m)
Main rotor disc area: 2,643.01 sq ft (228.81 m²)
Tail rotor disc area: 70.88 sq ft (6.58 m²)
Wheel track: 12 ft (3.66 m)

Powerplant

two Rolls-Royce (Bristol Siddeley) Gnome Mk 110/111 coupled turboshafts each rated at 1,350 shp (1007 kW)

Weights

Empty operating: 8,304 lb (3767 kg)
Maximum take-off: 13,500 lb (6123 kg)

Fuel and load

Internal fuel: 300 Imp gal (1364 litres)
Auxiliary fuel: 200 Imp gal (909 litres)
Maximum payload: 4,000 lb (1814 kg)

Performance

Maximum level speed at sea level: 132 mph (212 km/h)
Maximum cruising speed at optimum altitude: 121 mph (195 km/h)
Ferry range with standard fuel: 478 miles (769 km)
Ferry range with auxiliary fuel: 645 miles (1040 km)
Maximum rate of climb at sea level: 1,650 ft (503 m) per minute
Service ceiling: 12,000 ft (3658 m)
Hovering ceiling: 4,000 ft (1220 m)

Above: The Wessex HC.Mk 2 saw widespread service with the RAF and as late as 1994, when this photograph was taken, Nos 28, 60, 72 and 84 Sqns were still operational with the type.

Left: This Wessex HU.Mk 5 was photographed on operations in Malaysia during 1965 while serving with No. 848 Sqn Fleet Air Arm aboard HMS Albion. No. 848 reformed on the Wessex in May 1964.

129 Batteries (two)
130 Engine oil tank
131 Engine turbine section
132 Exhaust compartment firewalls
133 Ground power socket
134 Ventilating air intake
135 Starboard navigation light
136 Main axle beam mounting
137 Nose compartment framing
138 Engine bay access door
139 Throttle control linkage
140 Engine withdrawal rail
141 Engine mounting struts
142 Rolls-Royce (Napier) Gazelle 22 turboshaft engine
143 Engine bay ventilating air intake
144 Starter cartridge magazine
145 Hydraulic pump
146 Fire extinguisher bottles
147 Engine air intake
148 Hinged nose cone access panel
149 Engine accessory equipment gearbox
150 Generator
151 Intake plenum
152 Retractable landing lamp
153 Lower IFF aerial

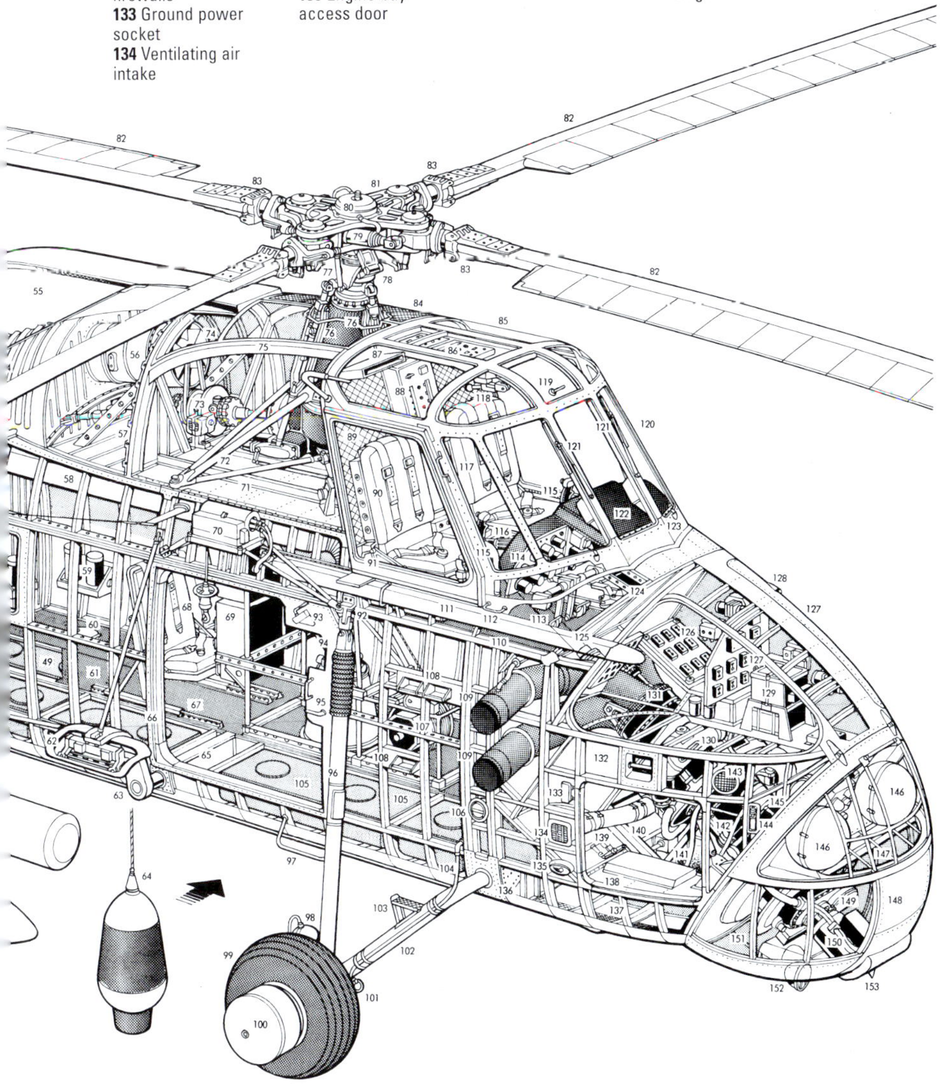

For hangar stowage aboard ship, the Wessex has a folding main rotor, as demonstrated by this FAA Wessex HAS.Mk 1. Not illustrated in this photograph is the folding tail pylon, which served to reduce the overall length of the aircraft by 27 ft (8.23 m).

Sikorsky SH-60 Sea Hawk

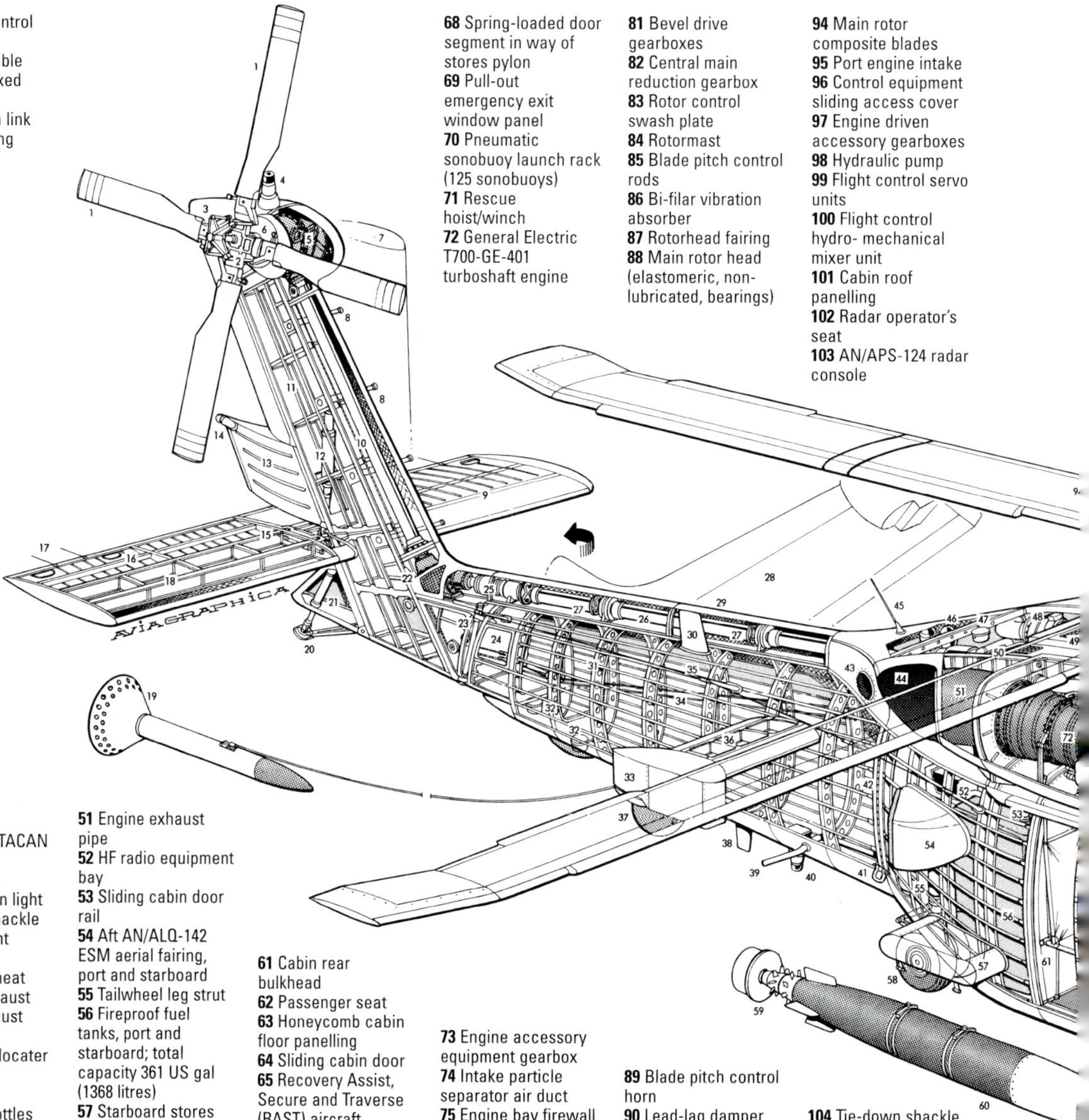

SH-60B Seahawk

Cutaway key

1 Graphite epoxy composite tail rotor blades
2 Lightweight cross beam rotor hub
3 Blade pitch change spider
4 Anti-collision light
5 Tail rotor final drive bevel gearbox
6 Rotor hub canted 20°
7 Horizontal tailplane folded position
8 Pull-out maintenance steps
9 Port tailplane
10 Tail rotor drive shaft
11 Fin pylon construction
12 Tailplane hydraulic jack
13 Cambered trailing edge section
14 Tail navigation light
15 Tailplane hinge joint (manual folding)
16 Handgrips
17 Static dischargers
18 Starboard tailplane construction
19 Towed magnetic anomaly-detector (MAD) bird
20 Tail bumper
21 Shock absorber strut
22 Bevel drive gearbox
23 Tail pylon latch joint
24 Tail pylon hinge frame (manual folding)
25 Transmission shaft disconnect
26 Tail rotor transmission shaft
27 Shaft bearings
28 Tail pylon folded position
29 Dorsal spine fairing
30 UHF aerial
31 Tailcone frame and stringer construction
32 Magnetic compass remote transmitters
33 MAD detector housing and reeling unit
34 Tail rotor control cables
35 HF aerial cable
36 MAD unit fixed pylon
37 Ventral data link antenna housing
38 Lower UHF/TACAN aerial
39 Fuel jettison
40 Anti-collision light
41 Tie-down shackle
42 Tailcone joint frame
43 Air system heat exchanger exhaust
44 Engine exhaust shroud
45 Emergency locater aerial
46 Engine fire suppression bottles
47 IFF aerial
48 Port side auxiliary power unit (APU)
49 Oil cooler exhaust grille
50 Starboard side air conditioning plant
51 Engine exhaust pipe
52 HF radio equipment bay
53 Sliding cabin door rail
54 Aft AN/ALQ-142 ESM aerial fairing, port and starboard
55 Tailwheel leg strut
56 Fireproof fuel tanks, port and starboard; total capacity 361 US gal (1368 litres)
57 Starboard stores pylon
58 Castoring twin tail wheels
59 Torpedo parachute housing
60 Mk 46 lightweight torpedo
61 Cabin rear bulkhead
62 Passenger seat
63 Honeycomb cabin floor panelling
64 Sliding cabin door
65 Recovery Assist, Secure and Traverse (RAST) aircraft haul-down fitting
66 Ventral cargo hook, 6,000 lb (2722-kg) capacity
67 Floor beam construction
68 Spring-loaded door segment in way of stores pylon
69 Pull-out emergency exit window panel
70 Pneumatic sonobuoy launch rack (125 sonobuoys)
71 Rescue hoist/winch
72 General Electric T700-GE-401 turboshaft engine
73 Engine accessory equipment gearbox
74 Intake particle separator air duct
75 Engine bay firewall
76 Oil cooler fan
77 Rotor brake unit
78 Engine intake ducts
79 Maintenance step
80 Engine drive shafts
81 Bevel drive gearboxes
82 Central main reduction gearbox
83 Rotor control swash plate
84 Rotormast
85 Blade pitch control rods
86 Bi-filar vibration absorber
87 Rotorhead fairing
88 Main rotor head (elastomeric, non-lubricated, bearings)
89 Blade pitch control horn
90 Lead-lag damper
91 Individual blade folding joints, electrically actuated
92 Blade spar crack detectors
93 Blade root attachment joints
94 Main rotor composite blades
95 Port engine intake
96 Control equipment sliding access cover
97 Engine driven accessory gearboxes
98 Hydraulic pump
99 Flight control servo units
100 Flight control hydro- mechanical mixer unit
101 Cabin roof panelling
102 Radar operator's seat
103 AN/APS-124 radar console
104 Tie-down shackle
105 Gearbox and engine mounting main frames
106 Maintenance steps
107 Main undercarriage leg

Above: An SH-60F Ocean Hawk departs USS Nimitz. The Ocean Hawk was designed to replace the SH-3H Sea King, two examples of which can be seen in the background. Among the A-6s, A-7s, F-14s and SH-3s on this late-1980's carrier deck, the SH-60F plays an unexpectedly important role in the carrier's survival, performing as it does the close-in ASW mission and therefore acting as the ship's last line of defence against enemy submarines.

Left: The Seahawk has never been noted for its bright markings, this HSL-41 SH-60B proving that the red and yellow MAD bird is often the only thing to relieve the type's grey camouflage.

SPECIFICATION

SH-60B Seahawk

Dimensions

Overall length with rotors turning: 64 ft 10 in (19.76 m)
Fuselage length: 50 ft ¾ in (15.26 m)
Length with main rotor blades and tail folded: 40 ft 11 in (12.47 m)
Length with main rotor blades and tail folded (HH-60H): 41 ft ⅝ in (12.51 m)
Length with main rotor blades and tail folded (HH-60J): 43 ft ⅞ in (13.13 m)
Height to top of rotor head: 11 ft 11 in (3.63 m)
Height overall, rotors turning: 17 ft (5.18 m)
Height with tail folded: 13 ft 3¼ in (4.04 m)
Stabiliser span: 14 ft 4½ in (4.38 m)
Main rotor diameter: 53 ft 8 in (16.36 m)
Tail rotor diameter: 11 ft (3.35 m)
Main rotor disc area: 2,262.03 sq ft (210.05 m^2)
Tail rotor disc area: 95.03 sq ft (8.83 m^2)
Wheel track: 9 ft 2 in (2.79 m)
Wheel base: 15 ft 10 in (4.83 m)

Powerplant

Two General Electric T700-GE-401 turboshafts each rated at 1,690 shp (1260 kW) or, in helicopters delivered from 1988, two General Electric T700-GE-401C turboshafts each rated at 1,900 shp (1417 kW)

Weights

Empty (for the ASW mission): 13,648 lb (6191 kg)
Empty (HH-60H): 13,480 lb (6114 kg)
Empty (HH-60J): 13,417 lb (6086 kg)
Mission take-off (for the ASW mission): 20,244 lb (9182 kg)
Maximum take-off (HH-60H and SH-60B for the utility mission): 21,884 lb (9926 kg)
Maximum take-off (SH-60F and SH-60R): 23,500 lb (10659 kg)
Maximum take-off (HH-60J): 21,246 lb (9637 kg)

Fuel and load

Internal fuel: 590 US gal (2233 litres)
External fuel: up to two 120-US gal (455-litre) drop tanks
Internal payload (HH-60H): 4,100 lb (1860 kg)
Useful load (HH-60J): 7,829 lb (3551 kg)
Maximum payload: 8,000 lb (3629 kg)

Performance

Dash speed at 5,000 ft (1525 m): 145 mph (234 km/h)
Cruising speed at sea level (HH-60H): 169 mph (272 km/h)
Cruising speed at sea level (HH-60J): 168 mph (271 km/h)
Operational radius (for a 3-hour loiter): 58 miles (93 km)
Operational radius (for a 1-hour loiter): 173 miles (278 km)
Operational radius (HH-60H on a SAR mission): 288 miles (463 km)
Operational radius (HH-60H on a SEAL insertion/extraction mission): 230 miles (370 km)
Maximum vertical rate of climb at sea level: 700 ft (213 m) per minute

mounting
108 Shock absorber leg strut
109 Starboard mainwheel
110 Pivoted axle beam
111 Starboard navigation light
112 Cockpit step/main axle fairing
113 Forward cabin access panel
114 Collective and cyclic pitch control rods
115 Sliding fairing guide rails
116 Cooling air grille
117 Main rotor blade glass-fibre skins
118 Honeycomb trailing edge panel
119 Titanium tube blade spar
120 Rotor blade drooped leading edge
121 Leading edge anti-erosion sheathing
122 Fixed trailing edge tab
123 Cockpit eyebrow window
124 Rearview mirrors
125 Overhead engine throttle and fuel cock control levers
126 Circuit breaker panel
127 Pilot's seat
128 Safety harness
129 Crash resistant seat mounting
130 Pull-out emergency exit window panel
131 Flight deck floor level
132 Cockpit door
133 Boarding step
134 AN/APS-124 search radar antenna
135 Ventral radome
136 Retractable landing/hovering lamp
137 Downward vision window
138 Yaw control rudder pedals
139 Cyclic pitch control column
140 Instrument panel
141 Centre instrument console
142 Stand-by compass
143 ATO/co-pilot's seat
144 Outside air temperature gauge
145 Instrument panel shroud
146 Air data probes
147 Windscreen panels
148 Windscreen wipers
149 Hinged nose compartment access panel
150 Pitot tubes
151 Avionics equipment bay
152 Forward data link antenna
153 Forward AN/ALQ-142 ESM aerial housings

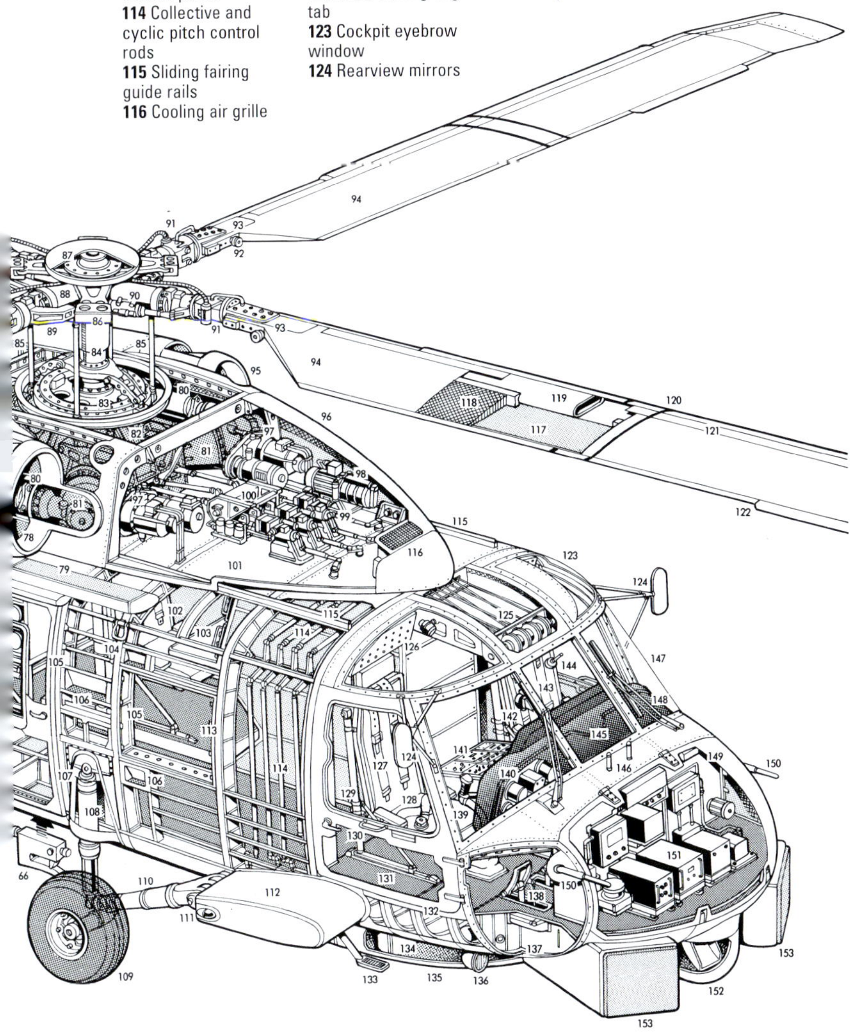

Early SH-60Bs used the lower-rated T700-GE-401 engine, which was later replaced in production by the T700-GE-401C of the HH-60H and HH-60J. By 2002, the US Navy expects its SH-60R to begin entering service. The SH-60R sees all the capabilities of the SH-60B combined with the dipping sonar of the SH-60F to produce the so-called LAMPS Block 2 helicopter. Up to 170 SH-60Bs, 77 SH-60Fs and 42 HH-60Hs are likely to be brought up to the new standard, which is optimised for littoral warfare.

Westland Lynx

The Brazilian navy received nine Lynx Mk 21s, based on the Royal Navy's HAS.Mk 2 with some minor changes in avionics equipment. The surviving seven airframes are being upgraded to Super Lynx standard as Mk 21As and are being joined by nine new-build examples from Westland.

Lynx AH.Mk 9

Cutaway key

1 Nose cone
2 Circuit-breaker panel
3 Forward radar warning antennas, port and starboard
4 Electrical system relays
5 Nose avionics equipment bay
6 Gyro platform
7 UHF aerials
8 Ventilating air intake
9 Flight-control computer
10 Avionics bay access door, port and starboard
11 Battery
12 Cockpit front bulkhead
13 Pitot head
14 Underfloor control linkages
15 Yaw control rudder pedals
16 Cyclic pitch control column
17 Downward-vision window
18 Instrument panel
19 Instrument panel shroud
20 Windscreen wipers
21 Curved windscreen panels
22 Outside air temperature probe
23 Composite rotor blade construction
24 Honeycomb core trailing edge
25 Glassfibre blade skins
26 Leading-edge anti-erosion sheath
27 Rescue hoist/winch, 600-lb (272-kg) capacity
28 Hoist pintle mounting, starboard cabin doorway
29 Cockpit roof glazing
30 Engine power and condition levers
31 Overhead switch panel
32 Safety harness
33 Pilot's seat
34 Centre control pedestal
35 Chart case
36 Boarding step
37 Port navigation light
38 Collective pitch control lever
39 Jettisonable cockpit door
40 Control rod linkages
41 Forward fuselage arch double main frame
42 Direct-vision sliding side window panel
43 Co-pilot/observer's seat
44 TOW missile sight periscope
45 Swivelling turret mounting
46 Hughes (British Aerospace licence-built) TOW sighting unit
47 Optical sighting aperture
48 Miniaturised thermal imaging system scanners
49 Control system ducting
50 Sliding cowling rail
51 Starboard aft-sliding main cabin door
52 Cabin wall trim/soundproofing panelling
53 Glassfibre honeycomb cabin roof construction
54 Lightweight detachable troop seats (nine fully-armed troops)
55 Side window panel
56 Port aft sliding cabin door
57 Sliding door latch
58 Emergency jettison handle
59 Forward underfloor fuel tank, total fuel capacity 1,616 lb (773 kg)
60 Lower sliding door rail
61 Floor beam construction
62 Lower fuselage access panels
63 Fuselage keel web member
64 Fuel collector tanks (two)
65 Forged gearbox mounting main frame
66 Forward gearbox mounting
67 Lateral control auto-stabilised servo-actuator
68 Hydraulic reservoir, dual system
69 Hydraulic equipment module, port and starboard
70 Alternators (two)
71 Alternator cooling air duct
72 Collective pitch servo-actuator
73 Longitudinal pitch control autostabilised servo-actuator
74 Forward-sliding equipment bay cowling
75 Hingeless main rotor hub, titanium
76 Blade pitch control horns
77 Pitch bearings
78 Flexible blade arm
79 Blade root attachment joints
80 Four-bladed main rotor
81 Drag hinge dampers
82 Blade pitch control rods
83 Main rotor mast
84 Main gearbox
85 Gearbox mounting deck
86 Sliding cabin door top rail fairing
87 Main cabin frame and stringer construction
88 Fuel tank access panel
89 Main fuel tanks, port and starboard
90 Fuel filler cap
91 Intake dust/debris ejector
92 Engine air intake filter screen
93 Engine/gearbox transmission shaft
94 Rotor brake
95 Compressor air intake
96 Starter/generator intake/exhaust fairing
97 Starboard engine cowling

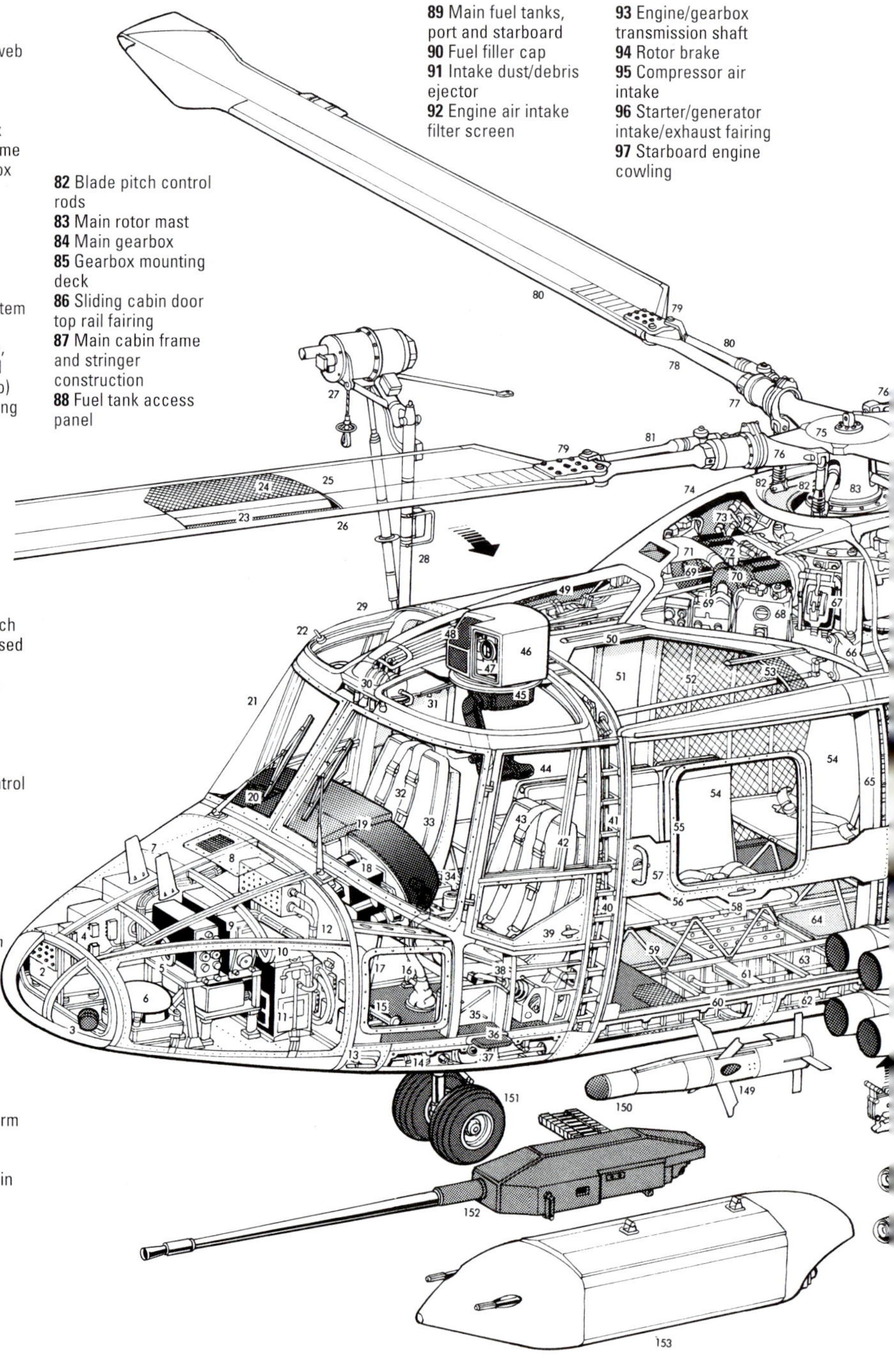

West Germany eventually received a total of 19 Lynx Mk 88s for the ASW role. The aircraft were delivered to a standard similar to that of the FAA's HAS.Mk 2, but with Gem 41-2 turboshafts and a non-folding tail boom.

SPECIFICATION

Lynx AH.Mk 7

Dimensions

Length overall, rotors turning: 49 ft 9 in (15.16 m)
Fuselage length: 39 ft 7 in (12.06 m)
Main rotor diameter: 40 ft (12.80 m)
Tail rotor diameter: 7 ft 3 in (2.21 m)
Height overall, rotors stationary: 12 ft (3.66 m)
Main rotor disc area: 1,385.44 sq ft (128.71 m^2)
Tail rotor disc area: 41.28 sq ft (3.84 m^2)
Stabiliser span: 5 ft 10 in (1.78 m)
Skid track: 6 ft 8 in (2.03 m)

Powerplant

Two Rolls-Royce Gem 41-1 turboshafts each rated at 1,120 shp (835 kW); from 1987 the engines were upgraded to Gem 42-1 standard, rated at 1,135 shp (846 kW)

Weights

Empty: 5,683 lb (2578 kg)
Maximum take-off: 10,750 lb (4876 kg)

Performance

Maximum continuous cruising speed: 140 kt (161 mph; 259 km/h)
Economical cruising speed: 70 kt (81 mph; 130 km/h)
Hover ceiling out-of-ground effect: 10,600 ft (3230 m)
Maximum rate of climb: 2,480 ft (756 m) per minute at sea level
Ferry range: 724 nm (835 miles; 1342 km) with auxiliary fuel
Standard range: 340 nm (392 miles; 630 km)
Typical range on troop-carrying mission: 292 nm (336 miles; 540 km)
Endurance: 3 hours

Fuel and load

Internal fuel: 214 Imp gal (973 litres) plus provision for 47 Imp gal (214 litres) in one fuselage tank or 192 Imp gal (873 litres) in two fuselage ferry tanks
Maximum ordnance: about 1,210 lb (549 kg)

Armament

One or two 20-mm cannon, 0.3-in (7.62-mm) Miniguns or rocket projectile pods; six AS11s, two Stinger or eight HOT, Hellfire, TOW or ATOW air-to-surface missiles

98 Engine bay dividing fireproof bulkhead, titanium
99 Starter/generator
100 Engine accessory equipment gearbox
101 Main engine mounting
102 Fireproof engine mounting deck, titanium
103 Rolls-Royce Gem 42-1 turboshaft engines
104 Engine bay rear fireproof bulkhead
105 Multi-lobe exhaust duct
106 Exhaust cooling air intakes
107 Infra-red suppressor fairings
108 Tail rotor transmission shaft
109 Shaft bearings
110 Dorsal spine fairings/access panels
111 Bevel drive intermediate gearbox
112 Tail pylon bracing struts
113 Tail rotor angled drive shaft
114 Gearbox cooling air intake
115 Fixed horizontal stabiliser
116 Stabiliser tubular spar
117 Final drive right-angled gearbox
118 Articulated tail rotor hub
119 Tail rotor power control unit
120 Blade pitch control spider
121 Anti-collision light
122 Tail navigation light
123 Four-bladed advanced tail rotor
124 Glassfibre pylon tail fairing/access panel
125 Tail rotor control linkage
126 Tail pylon construction
127 Tailboom ring frame construction
128 Access panels
129 Tail rotor control cables
130 UHF/IFF aerial
131 Main rotor blade balance weight pocket
132 BERP rotor blade tip profile
133 VHF aerial
134 Tailboom attachment joint frame
135 Fixed blade tab
136 Ferranti AWARE-3 rear radar-warning receiver
137 ALE-39 chaff/flare dispenser
138 Fuselage lower longeron
139 Rear fuselage frame construction
140 Top longeron
141 Aft fuselage equipment bay
142 Electrical system equipment racks
143 Cooling air scoop
144 HF homing aerial rail, port and starboard
145 Ventral access hatch
146 Main undercarriage shock absorber leg strut
147 TOW missile carrier/launch tubes
148 External cargo hook, 3000-lb (1361-kg) capacity
149 Missile folding fins
150 Hughes TOW air-to-surface anti-armour missile
151 Fixed castoring nosewheels
152 Oerlikon KAD 20-mm cannon (25-mm alternative)
153 FN ETNA gun pack, two 0.30-in (7.62-mm) machine-guns
154 Rockwell Hellfire laser-guided anti-armour missiles (8)
155 Missile-carrier/launch- rail
156 2.75-in (70-mm) folding-fin aircraft rockets (FFAR)
157 Rocket-launcher pack (19 FFARS)
158 SURA 3.1-in (80-mm) air-to-surface unguided rockets
159 Ammunition magazine
160 Emmerson Flexible Turret System (FTS) gun mounting
161 M134 Minigun, 0.3-in (7.62-mm) six-barrel rotary machine-gun

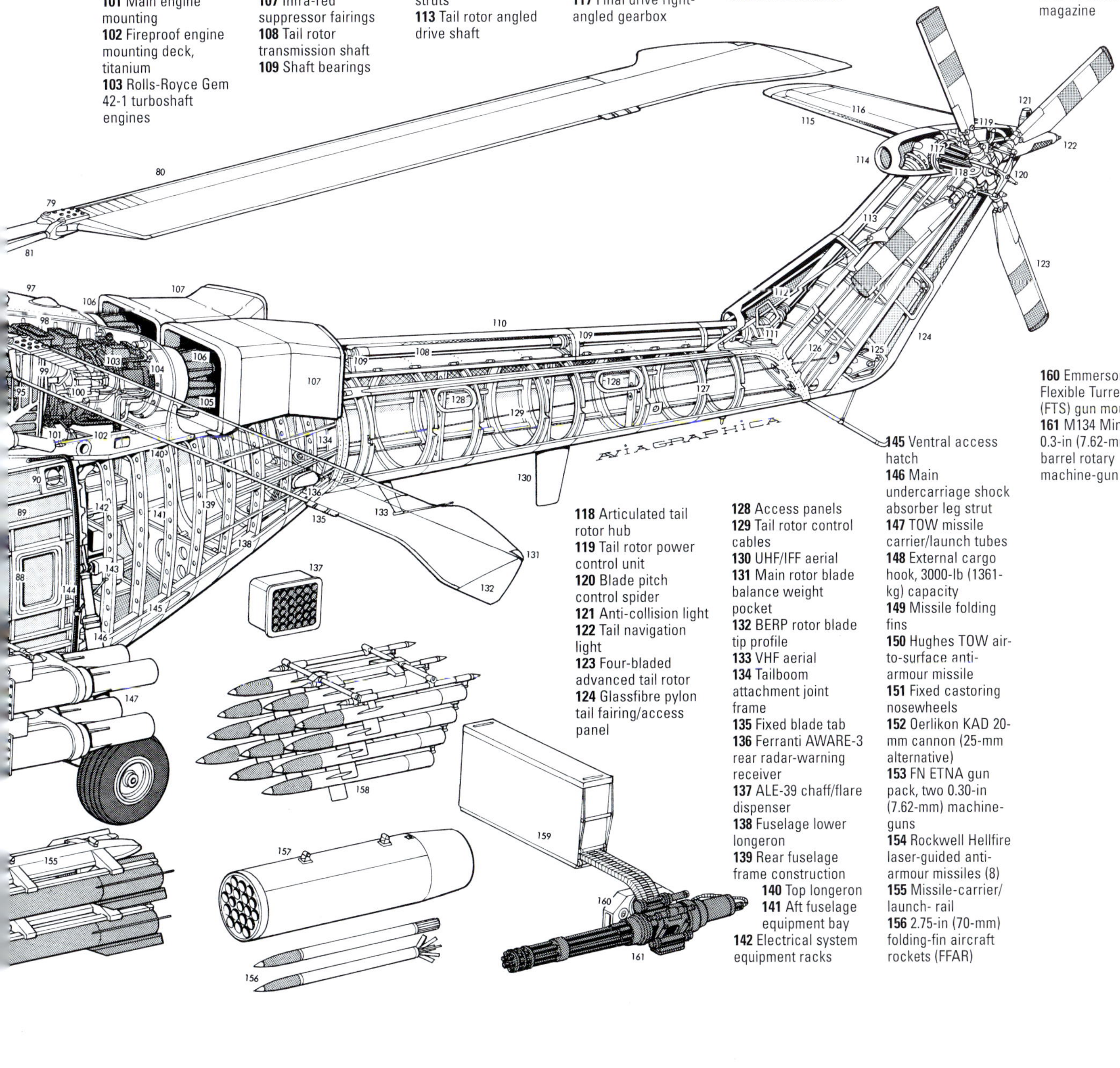

Naval Aircraft

Breguet Atlantic/Dassault Atlantique 2

The most noticeable differences between the Atlantique 2 (seen here) and the Atlantic are the former's wingtip ECM pods, nose-mounted Tango FLIR turret, the prominent cooling intakes below the cockpit and the re-shaped fin top, housing an ECM aerial.

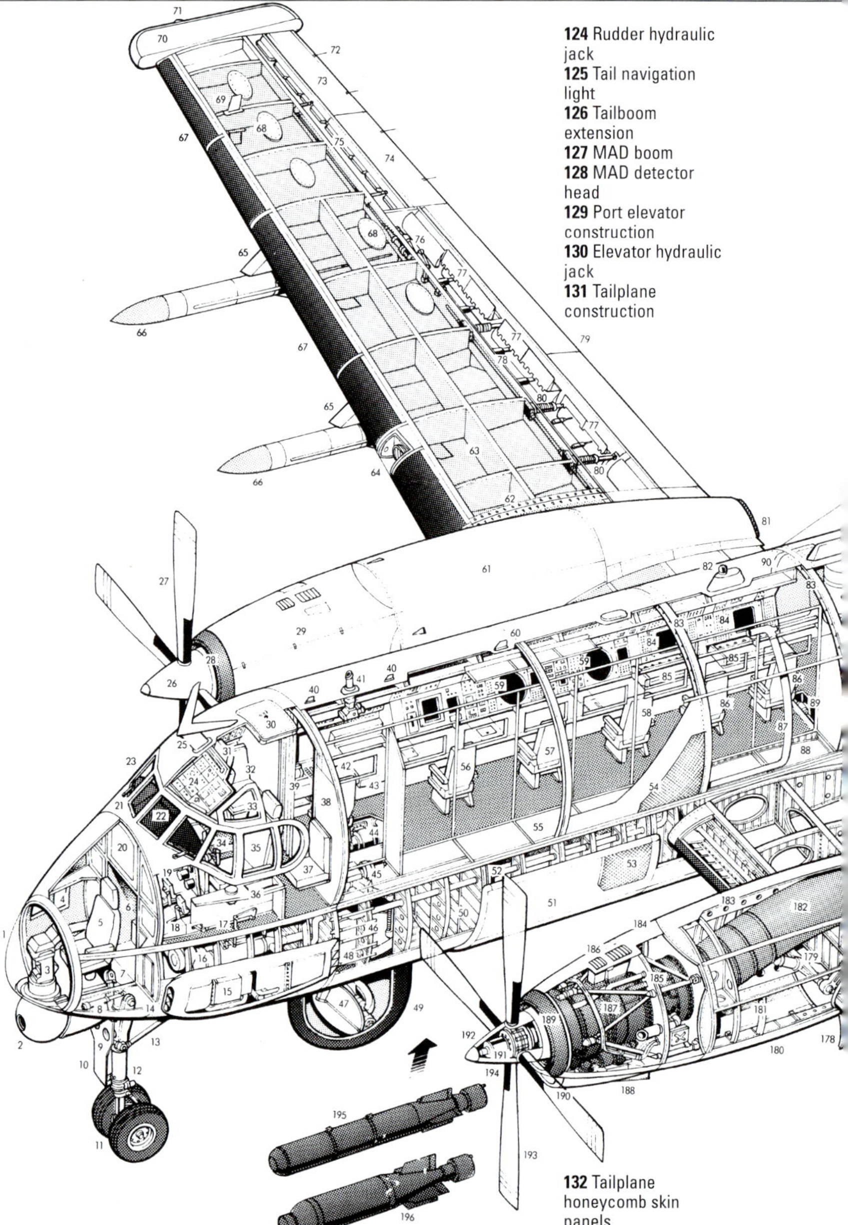

Atlantique 2

Cutaway key
1 Nose compartment glazing
2 Forward-looking infra-red sensor (FLIR)
3 Observer's sight
4 Side window
5 Forward observer's seat
6 Access doorway to flight deck
7 Nose landing gear pivot fixing
8 Nosewheel steering jacks
9 Taxiing lamps
10 Nosewheel leg door
11 Twin nosewheels
12 Nose landing-gear leg strut
13 Hydraulic retraction jack
14 Air-conditioning system ram air intake
15 Heat exchangers
16 Air-conditioning plant, electronics systems cooling air
17 Control rod linkages
18 Rudder pedals
19 Instrument panel
20 Flight deck bulkhead
21 Windscreen-wipers
22 Instrument panel shroud
23 Windscreen panels
24 Overhead switch panel
25 VHF aerial
26 Starboard propeller spinner
27 Four-bladed constant-speed propeller
28 Engine air inlet
29 Detachable engine cowling panels
30 Cockpit roof escape hatch
31 Aircraft commander's seat
32 Flight engineer's swivelling seat
33 Cockpit eyebrow window
34 Control column handwheel
35 Pilot's seat
36 Side console panel
37 Observer's folding seat
38 Main cabin bulkhead
39 Curtained doorway
40 TACAN aerials
41 Periscope sextant mounting
42 Radio navigator's station
43 Moving map display
44 Starboard underfloor APU bay
45 Radome raising and lowering hydraulic motor
46 Fuselage lower lobe frame construction
47 Thomson-CSF Iguane search radar
48 Air-conditioning system exhaust duct
49 Retractable radome
50 Weapons bay forward bulkhead
51 Externally sliding weapons bay doors
52 Door guide rails
53 Bomb door honeycomb construction
54 Fuselage pressurised section honeycomb skin panels
55 Port side radio and electronics racks
56 ESM, ECM and MAD systems operator's seat
57 Radar operator's seat
58 Tactical co-ordinator's seat
59 Display consoles
60 IFF aerial
61 Starboard engine nacelle fairing
62 Outer-wing panel joint
63 Starboard wing integral fuel tank; total system capacity 5,086 Imp gal (23120 litres)
64 Landing/search light
65 Wing stores pylons
66 AM39 Exocet air-to-surface missiles
67 Leading-edge pneumatic de-icing boots
68 Wing access panels
69 UHF aerial
70 Wingtip ECM pod
71 Starboard navigation light
72 Static dischargers
73 Starboard outer aileron
74 Starboard inner aileron
75 Aileron mass balance weights
76 Aileron hydraulic jack
77 Spoiler/airbrake panels, open
78 Spoiler hydraulic jacks
79 Outboard, two-segment double-slotted flaps
80 Flap screw jacks
81 Starboard engine exhaust nozzle
82 Anti-collision light
83 Wing/fuselage attachment main frames
84 Sonobuoy display consoles
85 Teleprinters
86 Sonobuoy operators' seats (two)
87 Electronics racks cooling air ducting
88 Wing centre-section carry through
89 Central flap hydraulic motor
90 Starboard escape hatch
91 DF aerial
92 Life raft stowage
93 Port escape hatch
94 Pressure floor beam construction
95 Bomb-bay door hydraulic motor
96 Crew rest area seating, port and starboard
97 Galley compartment
98 Dining table
99 Toilet compartment
100 Wardrobe
101 Curtained doorway
102 Rear observers' seats, port and starboard
103 Binocular mounting rail
104 Observation bubble window
105 Cabin doorway
106 Rear pressure bulkhead
107 Flare stowage rack
108 Sonobuoy stowage rack, maximum load 72 A or A3 sonobuoys
109 Rear fuselage frame and stringer construction
110 Tailplane mounting bulkhead
111 Finroot fillet
112 Tailplane leading-edge de-icing boots
113 Starboard HF aerial cable
114 Starboard tailplane
115 Starboard elevator
116 Static dischargers
117 Fin leading-edge de-icing boot
118 Fin construction
119 Fin honeycomb skin panels
120 Fintip ECM aerial housing
121 Static dischargers
122 Rudder mass balance weights
123 Rudder construction
124 Rudder hydraulic jack
125 Tail navigation light
126 Tailboom extension
127 MAD boom
128 MAD detector head
129 Port elevator construction
130 Elevator hydraulic jack
131 Tailplane construction
132 Tailplane honeycomb skin panels
133 Leading-edge de-icing boot
134 Port HF aerial cable
135 Rudder and elevator control rods
136 Rear entry hatch
137 Extending boarding ladder

The five crew positions on the starboard side of the Atlantique 2's cabin are used for monitoring the aircraft's various systems. The two closest positions are sonobuoy display consoles, which are the responsibility of the sonobuoy operators. Next is the tactical co-ordinator's screen which displays correlated information from all of the systems. To the left is the radar operator and, finally, the ESM, ECM and MAD systems operator's position.

SPECIFICATION

Atlantique 2

Dimensions

Length: 103 ft 9 in (31.62 m)
Height: 35 ft 8¾ in (10.89 m)
Wingspan: 122 ft 9¼ in (37.42 m)
Wing area: 1,295.37 sq ft (120.34 m²)
Wing aspect ratio: 10.94
Tailplane span: 40 ft 4½ in (12.31 m)
Wheel track: 29 ft 6¼ in (9.00 m)
Wheel base: 30 ft 10 in (9.40 m)

Powerplant

Two Rolls-Royce Tyne RTy.20 Mk 21 turboprops each rated at 6,100 ehp (4549 ekW)

Weights

Empty equipped: 56,437 lb (25600 kg)
Normal take-off: 97,443 lb (44200 kg) for ASW role or 99,206 lb (45000 kg) for combined ASW and ASV roles
Maximum take-off: 101,852 lb (46200 kg)

Fuel and load

Internal fuel: 40,785 lb (18500 kg)
External fuel: none
Maximum external ordnance: 7,716 lb (3500 kg)
Maximum internal ordnance: 5,512 lb (2500 kg)

Performance

Maximum level speed 'clean': 349 kt (402 mph; 648 km/h)
Maximum cruising speed at 23,620 ft (7200 m): 300 kt (345 mph; 556 km/h)
Normal patrol speed between sea level and 5,000 ft (1525 m): 170 kt (196 mph; 315 km/h)
Maximum rate of climb at sea level: 2,900 ft (884 m) per minute
Service ceiling: 30,000 ft (9145 m)
Take-off distance to 35 ft (10.50 m) at maximum take-off weight: 6,037 ft (1840 m)
Landing distance from 35 ft (10.50 m) at normal landing weight: 4,921 ft (1500 m)
Ferry range: 4,897 nm (5,639 miles; 9075 km)
Operational radius: 1,799 nm (2,071 miles; 3333 km) for a 2-hour ASV patrol or 599 nm (690 miles; 1110 km) for an 8-hour ASW patrol at low-level
Endurance: 18 hours

Armament

The weapons bay can accommodate all NATO-standard bombs, depth charges, two ASMs, up to eight Mk 46 torpedoes or seven Murène advanced torpedoes; the four wing pylons can carry additional stores, including future ASMs, AAMs and equipment pods

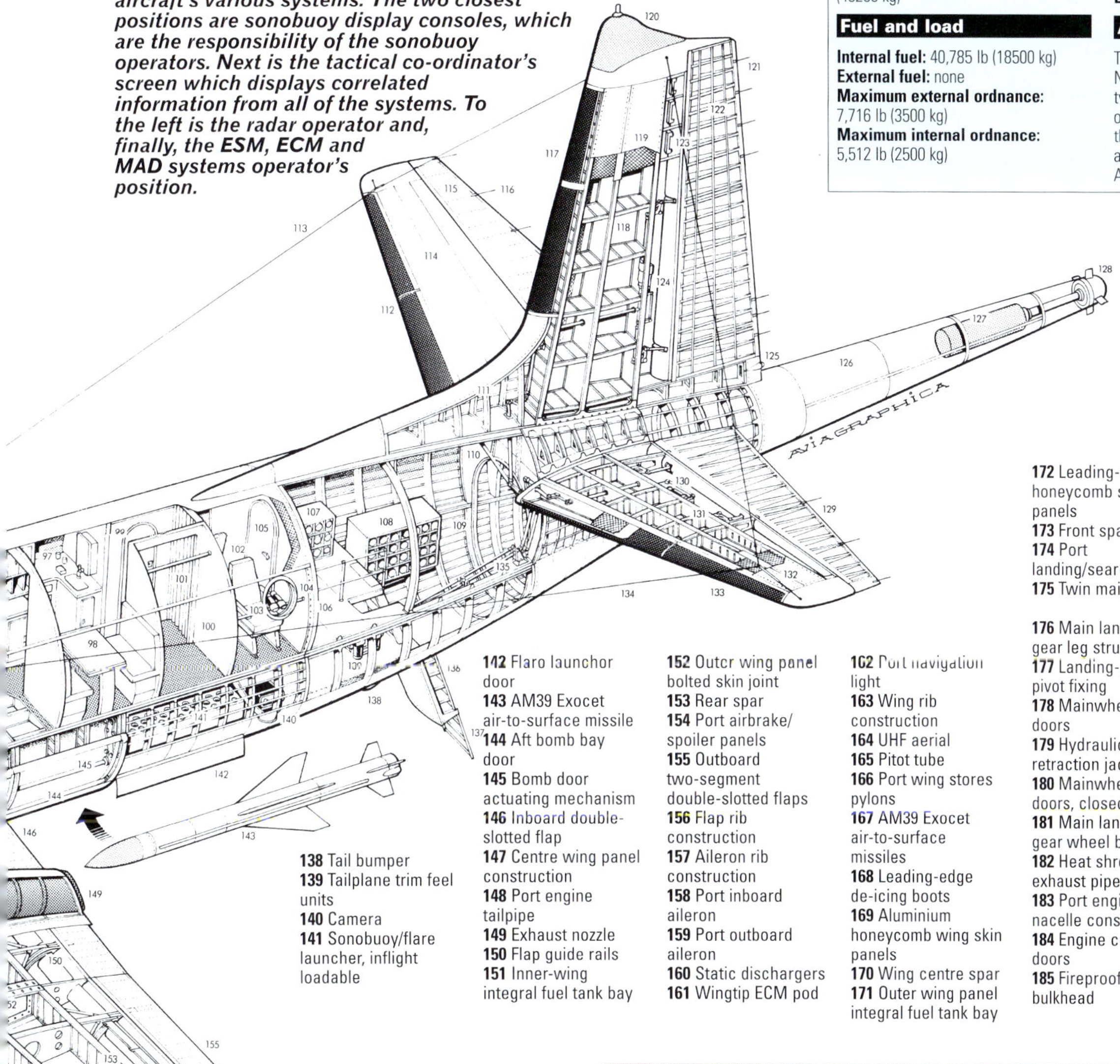

138 Tail bumper
139 Tailplane trim feel units
140 Camera
141 Sonobuoy/flare launcher, inflight loadable
142 Flare launcher door
143 AM39 Exocet air-to-surface missile
144 Aft bomb bay door
145 Bomb door actuating mechanism
146 Inboard double-slotted flap
147 Centre wing panel construction
148 Port engine tailpipe
149 Exhaust nozzle
150 Flap guide rails
151 Inner-wing integral fuel tank bay
152 Outer wing panel bolted skin joint
153 Rear spar
154 Port airbrake/spoiler panels
155 Outboard two-segment double-slotted flaps
156 Flap rib construction
157 Aileron rib construction
158 Port inboard aileron
159 Port outboard aileron
160 Static dischargers
161 Wingtip ECM pod
162 Port navigation light
163 Wing rib construction
164 UHF aerial
165 Pitot tube
166 Port wing stores pylons
167 AM39 Exocet air-to-surface missiles
168 Leading-edge de-icing boots
169 Aluminium honeycomb wing skin panels
170 Wing centre spar
171 Outer wing panel integral fuel tank bay
172 Leading-edge honeycomb skin panels
173 Front spar
174 Port landing/search light
175 Twin mainwheels
176 Main landing-gear leg strut
177 Landing-gear leg pivot fixing
178 Mainwheel leg doors
179 Hydraulic retraction jack
180 Mainwheel bay doors, closed
181 Main landing-gear wheel bay
182 Heat shrouded exhaust pipe
183 Port engine nacelle construction
184 Engine cowling doors
185 Fireproof bulkhead
186 Engine bleed air and pre-cooler exhaust louvres
187 Rolls-Royce Tyne RTy.20 Mk 21 turboprop engine
188 Ventral oil cooler duct
189 De-iced engine air inlet
190 Oil cooler ram air intake
191 Propeller hub pitch change mechanism
192 Spinner
193 Four-bladed constant-speed propeller
194 Propeller blade root de-icing cuffs
195 Mk 46 lightweight torpedo
196 Depth charge

This 88° Gruppo, 41° Stormo Atlantic was photographed in 1992 wearing a special paint scheme to celebrate the aircraft's 20 years in Italian service.

British Aerospace Sea Harrier

Following the success of the Sea Harrier/AIM-9L Sidewinder combination during the Falklands conflict, the Royal Navy's aircraft were modified to carry twin launcher rails under each wing, thus doubling their potential AAM capacity. The two aircraft illustrated wear the 'winged fist' insignia of No. 899 NAS, a squadron which formed at Yeovilton in April 1980 from the existing No. 700A Squadron. A No. 899 Squadron pilot scored the first Argentine fighter kill (a Mirage III) in the Falklands.

Sea Harrier FRS.Mk 1

Cutaway key
1 Pitot head
2 Radome
3 Ferranti Blue Fox radar scanner
4 Radar equipment module
5 Radome hinge
6 Nose pitch reaction control valve
7 Pitch feel and trim control mechanism
8 Rudder pedals
9 Starboard side oblique camera
10 Inertial platform
11 IFF aerial
12 Cockpit ram air intake
13 Yaw vane
14 Pressurisation spill valve
15 Windscreen wiper
16 Head-up display
17 Instrument panel shroud
18 Control column and linkages
19 Doppler antenna
20 TACAN aerial
21 UHF aerial
22 Nose undercarriage wheel bay
23 Radar hand controller
24 Throttle and nozzle angle control levers
25 Martin-Baker Mk 10H zero-zero ejection seat
26 Miniature detonating cord (MDC) canopy breaker
27 Boundary layer spill duct
28 Cockpit air-conditioning pack
29 Nose undercarriage hydraulic retraction jack
30 Hydraulic accumulator
31 Boundary layer bleed duct
32 Engine air intake
33 Intake suction relief doors (spring-loaded)
34 Forward fuselage flank fuel tank
35 Hydraulic system ground connectors
36 Engine monitoring and recording equipment
37 Engine oil tank
38 Rolls-Royce Pegasus Mk 104 turbofan engine
39 UHF homing aerials
40 Alternator
41 Accessory equipment gearbox
42 Gas turbine starter/auxiliary power unit (GTS/APU)
43 Starboard external fuel tank
44 Starboard wing integral fuel tank
45 Twin missile pylon
46 Starboard navigation light
47 Roll control reaction air valve
48 Outrigger wheel fairing
49 Starboard outrigger wheel
50 Starboard aileron
51 Aileron hydraulic actuator
52 Fuel jettison valve
53 Starboard plain flap
54 Anti-collision light
55 Water methanol tank
56 Engine-fire suppression bottle
57 Flap hydraulic actuator
58 Water-methanol filler cap
59 Rear fuselage fuel tank
60 Emergency ram air turbine extended
61 Ram air turbine actuator
62 Heat exchanger air intake
63 HF tuner
64 HF notch aerial
65 Rudder control linkage
66 Starboard all-moving tailplane
67 Temperature probe
68 Forward radar-warning antenna
69 VHF aerial
70 Rudder
71 Rudder trim tab
72 Rear rudder-warning antenna
73 Tail pitch control reaction air valve
74 Yaw control reaction air valves
75 Port all-moving tailplane
76 IFF notch aerial
77 Tail bumper
78 Radar altimeter aerials
79 Reaction control air duct
80 Tailplane hydraulic actuator
81 Rear equipment bay air conditioning pack
82 Chaff/flare dispensers
83 Avionics equipment bay
84 Airbrake hydraulic jack
85 Ventral airbrake
86 Liquid oxygen converter
87 Hydraulic system nitrogen pressurising bottle
88 Main undercarriage stowage

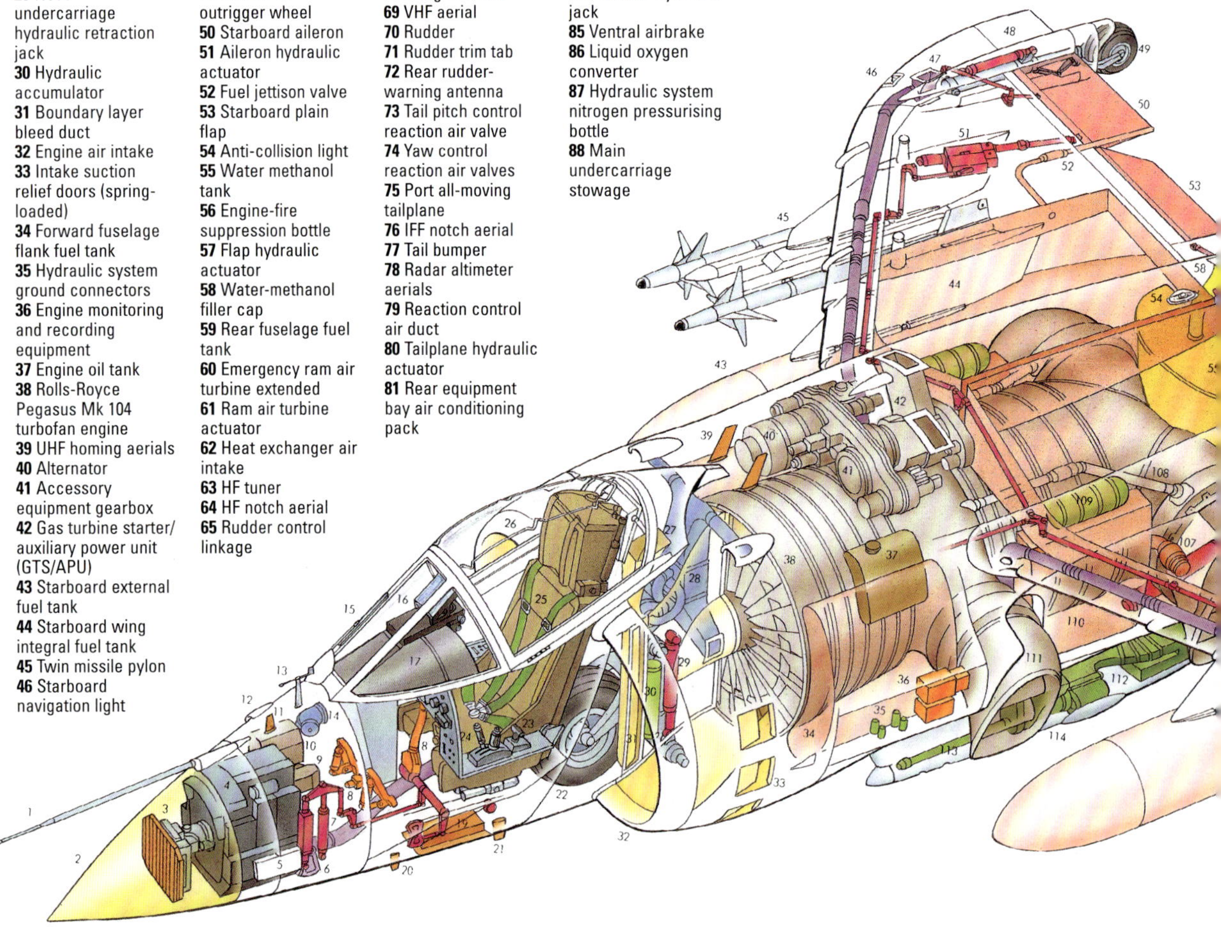

The new-generation Sea Harrier (initially known as the FRS.Mk 2) features a stretched fuselage (with a plug inserted behind the wing) and a refined wing, with new wingtips and a revised, kinked leading edge. A further advance was the introduction of the Ferranti Blue Vixen radar, providing a new multiple target-tracking and all-weather look-down/shoot-down capability.

SPECIFICATION

Sea Harrier FRS.Mk 1

Dimensions

Overall length: 47 ft 7 in (14.50 m)
Length with nose folded: 41 ft 9 in (12.73 m)
Wingspan: 25 ft 3 in (7.70 m)
Wingspan with ferry tips: 29 ft 8 in (9.04 m)
Wing area: 202.10 sq ft (18.68 m²)
Wing aspect ratio: 3.175

Powerplant

One Rolls-Royce Pegasus Mk 104 vectored thrust turbofan rated at 21,500 lb st (95.6 kN)

Weights

Basic empty: 13,000 lb (5897 kg)
Operating empty: 14,502 lb (6374 kg)
Maximum take-off weight: 26,200 lb (11884 kg)

Fuel load

Maximum internal fuel: 5,060 lb (2295 kg)
Maximum external fuel: 5,300 lb (2404 kg) in two 100-Imp gal (455-litre) drop tanks or two 330- or 190-Imp gal (1500- or 864-litre) ferry tanks; maximum ordnance 8,000 lb (3629 kg)

Performance

Maximum speed at high altitude: 825 mph (1328 km/h)
Maximum speed 'clean' at sea level: more than 736 mph (1185 km/h)
Cruising speed at 36,000 ft (10975 m): 528 mph (850 km/h)
Maximum rate of climb at sea level: 50,000 ft (15240 m) per minute
Service ceiling: 51,000 ft (15545 m)
Take-off run: 1,000 ft (305 m) at maximum take-off weight without 'ski jump'; landing run 0 ft (0 m) at normal landing weight

Range

Combat radius: 460 miles (750 km) on a hi-hi-hi interception mission with four AAMs, or 288 miles (463 km) on a hi-lo-hi attack mission.
***g* limit:** +7.8/-4.2

Armament

Underfuselage mounts for two 30-mm ADEN cannon, and four underwing hardpoints stressed for up to 8,000 lb (3629 kg). Standard carrying capabilities as follows: underfuselage and inboard wing hardpoints 2,000 lb (907 kg) each; outboard wing pylons 650 lb (295 kg) each. Cleared for carriage of standard British 1,000-lb (454-kg) free-fall and retarded HE bombs, BAe Sea Eagle ASM, AGM-84 Harpoon ASM, WE177 tactical nuclear free-fall bomb, Lepus flare units, CBLS 100 practice bomb dispenser and most NATO-standard bombs, rockets and flares. Air-to-air armament can comprise four AIM-9L Sidewinders on twin-rail launchers, or MATRA Magic missiles on Indian aircraft

89 Nozzle blast shield
90 Port wing integral fuel tank
91 Port plain flap
92 Fuel jettison
93 Port aileron
94 Outrigger wheel hydraulic retraction jack
95 Port outrigger wheel
96 Roll control reaction air valve
97 Port navigation light
98 AIM-9L Sidewinder air-to-air missiles
99 Twin missile carrier/launcher
100 Outboard stores pylon
101 Reaction control air duct
102 Port aileron hydraulic actuator
103 External fuel tank
104 Inboard wing pylon
105 Rear (hot-stream) swivelling exhaust nozzle
106 Main undercarriage hydraulic retraction jack
107 Pressure refuelling connection
108 Nozzle bearing cooling air duct
109 Hydraulic system reservoir, port and starboard
110 Centre fuselage flank fuel tank
111 Fan air (cold-stream) swivelling nozzle
112 Ammunition magazine
113 ADEN 30-mm cannon
114 Ventral gun pack, port and starboard

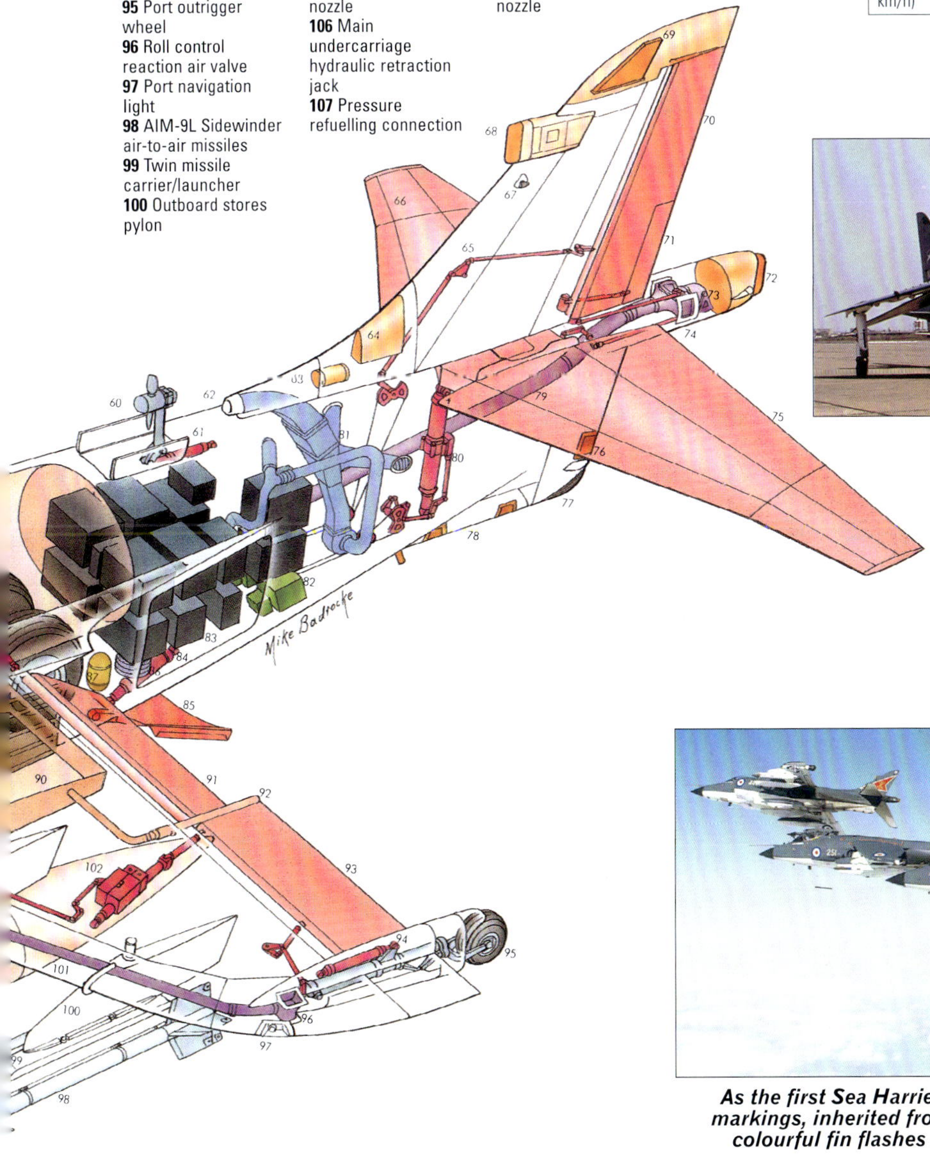

India acquired four two-seat Harrier T.Mk 60s to complement its squadron of FRS.Mk 51s. Indian Sea Harriers operate alongside Sea King helicopters on the nation's remaining carrier, INS* Viraat *(ex-HMS* Hermes*).

As the first Sea Harriers entered Royal Navy service, high-visibility unit markings, inherited from their operator's Phantoms, were* de rigeur. *The colourful fin flashes and gloss white bellies of these No. 800 Squadron aircraft were removed during the Falklands War.

De Havilland D.H.110 Sea Vixen

In 1961 two Sea Vixen FAW.Mk 1s fresh off the Christchurch production line – XN684 and XN685 – were flown to Hatfield and converted to an interim FAW.Mk 2 standard. Once procedures for the conversion programme were established both aircraft were employed in development work concerning the Red Top AAM. XN684 carries four in this view and has a special test camera fitted under its nose.

Sea Vixen FAW.Mk 1/2

Cutaway key

1 Glass-fibre radome
2 GEC A1 radar scanner
3 Scanner gimballing mechanism
4 Radome hinge point, opens to starboard
5 Wave form generator
6 Power pulse unit
7 Radar modulator
8 Radome latch
9 Radar transmitter/ receiver
10 Lower IFF aerial
11 UHF aerial
12 2-in folding fin rocket
13 Landing taxiing and carrier approach lights
14 Nosewheel leg door
15 Levered suspension nosewheel forks
16 Nosewheel, aft-retracting
17 Microcell 14-round rocket launcher, extended
18 Rocket launcher hydraulic jack
19 Nose undercarriage pivot mounting
20 Front pressure bulkhead
21 Rudder pedals
22 Control column
23 Nosewheel housing
24 Cockpit rain dispersal air duct fairing
25 Instrument panel shroud
26 Ferranti gun sight
27 Upper IFF aerial
28 Knife-edged windscreen panels
29 Observer's instrument panel
30 Sliding cockpit canopy cover
31 Canopy framing, FAW.Mk 1 (clear canopy on FAW.Mk 2)
32 Observer's instrument console
33 Pilot's Martin-Baker Mk 4 ejection seat
34 Engine throttle levers
35 Fuel cocks
36 Rudder Q-feel dashpot
37 Airbrake hinge point
38 Ventral airbrake, open
39 Airbrake strake
40 Airbrake hydraulic jack
41 Boundary layer splitter plate
42 Electronics equipment racks
43 Port engine air intake
44 Ventral air-conditioning system ram air intake
45 Boundary layer spill duct
46 Canopy release strut
47 Canopy de-misting air duct
48 Observer's ejection seat
49 Bulged hatch fairing (on modified FAW.Mk 2 aircraft)
50 Observer's entry hatch
51 Starboard engine air intake
52 Rear pressure bulkhead
53 Electrical and Low-Altitude Bombing System (LABS) equipment bays, port and starboard
54 Boundary layer spill air louvres
55 Centre section fuel tank
56 Fuel filler, port and starboard
57 Intake de-icing air duct
58 Engine intake ducting
59 Central air-conditioning equipment bay
60 Intake flank fuel tank
61 Ventral catapult strop hook
62 Mainwheel door actuator
63 Engine fire suppression bottle
64 Port main undercarriage wheel bay
65 Engine bleed air ducting
66 Rolls-Royce Avon 208 engine
67 Starter housing intake fairing
68 Bleed air spill duct
69 Air-conditioning controllers
70 VHF aerials
71 Starboard mainwheel, stowed position
72 Starboard intake flank fuel tank
73 Tailboom fuel tank
74 Extended tailboom fairing (FAW.Mk 2 only)
75 Outer wing panel integral fuel tank
76 Wing fence
77 Aileron trim actuator
78 Autopilot actuator
79 Extended chord outboard leading edge
80 Pitot head
81 Starboard navigation light
82 Aileron horn balance
83 Starboard aileron
84 Aileron hydraulic actuator
85 Chain and sprocket flap drive
86 Outboard Fowler-type flap
87 Extended tailboom aft fairing
88 Flap drive shafts
89 Wing fold hydraulic deck
90 Hydraulic reservoir, leader tank and accumulator
91 Inboard flap segment guide rails

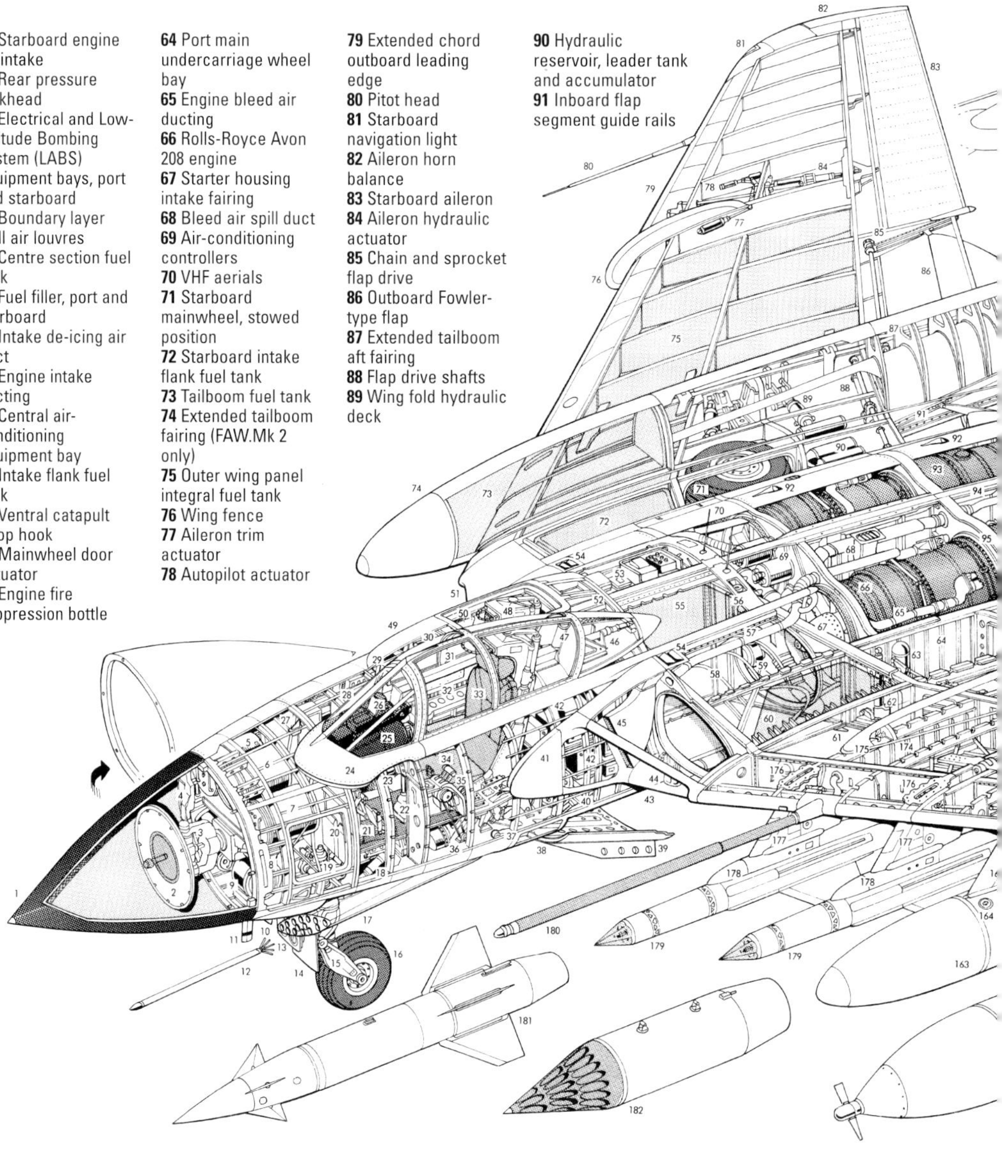

***A concession in the DH.110's design, made to improve the aircraft's suitability for deck operations, was the offset cockpit. This allowed the Sea Vixen's pilot to see forward over the large radome, forward visibility being an important consideration when landing an aircraft on a carrier deck. This aircraft, seen being catapulted from the deck of a carrier, has the 'V' tailcode of HMS* Victorious.**

SPECIFICATION

Sea Vixen FAW.Mk 1

Dimensions

Length: 55 ft 7 in (16.94 m)
Length (nose folded): 50 ft 2½ in (15.3 m)
Height: 10 ft 9 in (3.28 m)
Wingspan: 50 ft (15.24 m)
Wing area: 648 sq ft (60.20 m²)

Powerplant

Two Rolls-Royce Avon 208 turbojet engines each rated at 11,230 lb st (49.95 kN)

Weights

Empty: 27,952 lb (12679 kg)
All-up: 35,000 lb (15876 kg)
Maximum take-off: 41,575 lb (18858 kg)

Fuel and load

External fuel: Provision for two 150-Imp gal (682-litre) drop tanks
Maximum ordnance: up to 2,000 lb (907 kg) of ordnance on underwing pylons

Performance

Maximum speed at sea level: 690 mph (1110 km/h)
Maximum speed at 10,000 ft (3048 m): 645 mph (1038 km/h)
Time to 10,000 ft (3048 m): 1½ minutes
Time to 40,000 ft (12192 m): approximately 6½ minutes
Service ceiling: 48,000 ft (14630 m)

Armament

Up to four de Havilland Firestreak IR-homing air-to-air missiles on underwing pylons; two retractable Microcell rocket packs under the forward fuselage, each holding 14 2-in (5.08-cm) rockets; a 500-lb (227-kg) or 1,000-lb (454-kg) GP bomb, napalm tank, rocket pod (carrying either 2-in/5.08-cm or 3-in/7.62-cm rockets) or Martin Bullpup A air-to-surface missile on each of the inner pair of AAM pylons

92 Engine bay venting air intakes
93 Starboard engine bay
94 Fuselage centre keel member
95 Detachable firewall upper segment (for engine removal)
96 Flap drive cable drum
97 Flap hydraulic actuator
98 Arrester hook jack and damper
99 Engine jet pipe
100 Arrester hook stowage
101 Exhaust nozzle fairing
102 Emergency ram air turbine, extended
103 Static inverters
104 Tailboom electrical equipment bay
105 Hydraulic accumulators
106 Tail bumper
107 Starboard rudder
108 Tailplane rib and stringer construction

109 Wing panels, folded position
110 Rudder hydraulic actuator
111 All-moving horizontal tailplane
112 Tailplane rib construction
113 Fin/tailplane fairing **114** Tail navigation lights
116 Tab balance weights
117 Tailplane tip fairing
118 Tab control linkage
119 Tailplane pivot mounting
120 Tailplane hydraulic actuator
121 Rudder mass balance weight
122 Port rudder hydraulic actuator
123 Rudder rib construction
124 Tie-down point
125 Tail bumper
126 Bumper strut
127 Tailboom/fin joint frame
128 Tailboom frame construction
129 Telebriefing socket
130 Tailplane control cables
131 Port electrical equipment bay
132 Inverter cooling air intake
133 Tailboom joint ring frame
134 Air bottles
135 Port inboard flap segment
136 Autopilot control unit
137 Flap drive shafts and gearboxes
138 Wing fold hydraulic jack
139 Wing fold hinge fittings
140 Outboard flap segment guide rails
141 Flap rail fairings
142 Flap shroud ribs
143 Rear spar
144 Aileron hydraulic actuator
145 Port aileron rib construction
146 Refuelling drogue, extended
147 Port aileron horn balance
148 Tip fairing construction
149 Outer wing panel rib construction
150 Port navigation light
151 Leading-edge nose ribs
152 Red Top air-to-air missile (arming FAW.Mk 2 only)
153 Pitot head
154 Aileron control rod linkage
155 Leading-edge dog-tooth
156 Wing fence honeycomb construction
157 Fuel tank end rib
158 Semi-span centre spar
159 Port outer wing panel integral fuel tank
160 Front spar
161 Mainwheel leg door
162 'Buddy' refuelling pack
163 150-Imp gal (682-litre) external fuel tank
164 Fuel filler
165 Fuel tank pylon
166 Port mainwheel
167 Torque scissor links
168 Pylon hard point
169 Main undercarriage leg strut
170 Side breaker strut
171 Undercarriage leg pivot mounting
172 Hydraulic retraction jack
173 Wing fold hinge rib
174 Wing inboard segment integral fuel tank
175 Tailboom forward fairing (FAW.Mk 1)
176 Light stores pylon fittings
177 Missile pylons
178 Missile launch rails
179 Firestreak air-to-air missiles (arming FAW.Mk 1)
180 Flight refuelling boom
181 Bullpup A air-to-surface missile
182 36 x 2-in (50-mm) rocket pack

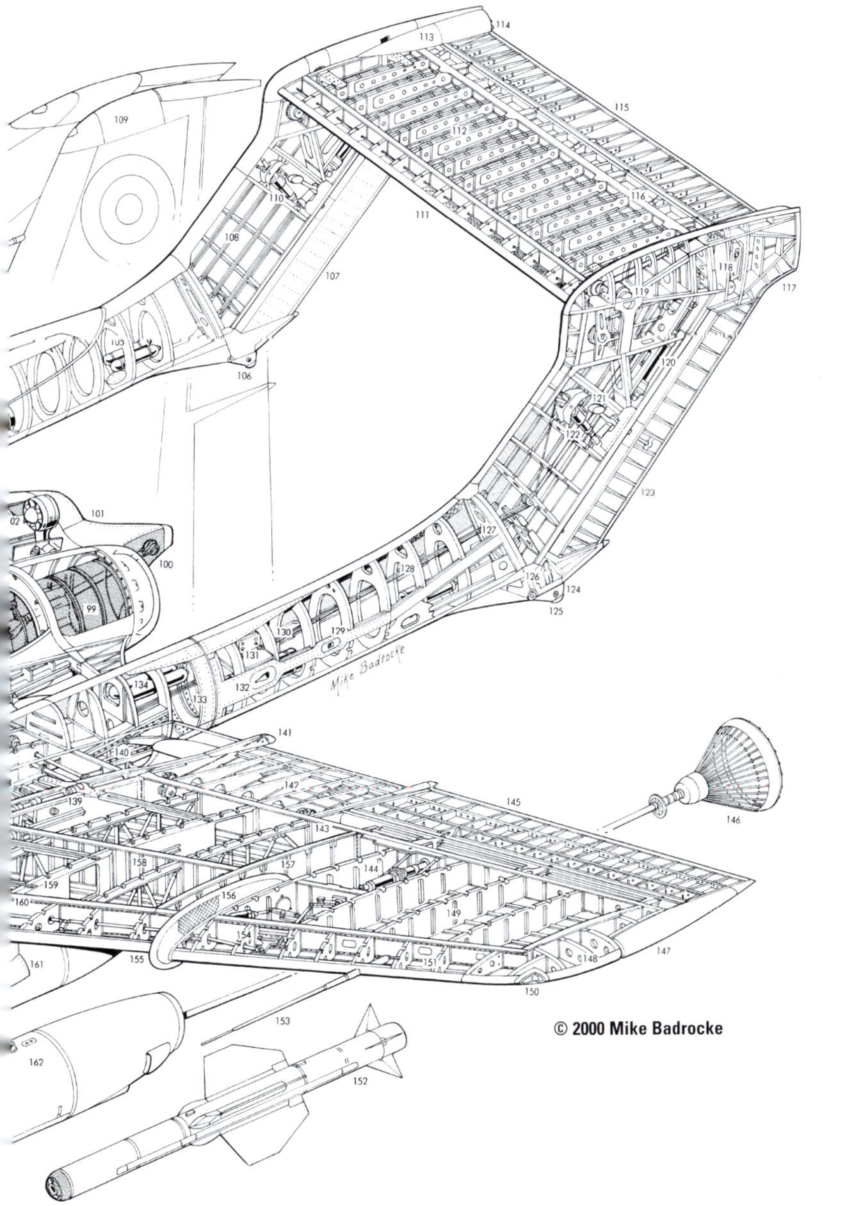

Douglas AD/A-1 SkyRaider

With its distinctive exhau staining around the engi cowling, this early unmark Skyraider is seen on a test flight. wears the second colour scher introduced on the **Skyraider** *fam consisting of gull-grey upp surfaces and white undersid*

A-1H Skyraider

Cutaway key

1 Aeroproducts four-bladed, variable-pitch propeller, 13 ft 6-in (4.1-m) diameter
2 Propeller hub pitch change mechanism
3 Gearbox cowling
4 Retractable cooling air baffles
5 Starboard underwing fuel tank
6 Cowling nose ring
7 Propeller reduction gearbox
8 Wright R-3350-26WA 18-cylinder, two-row radial engine
9 Detachable cowling panels
10 Starboard mainwheel
11 Mainwheel disc brake
12 Centre line auxiliary fuel tank, capacity 300 US gal (1136 litres)
13 Oil cooler intake
14 Exhaust stubs
15 Oil cooler
16 Engine bearer lower segment
17 Oil tank, capacity 38.5 US gal (146 litres)
18 Engine mounting ring
19 Engine accessory equipment
20 Port magneto
21 Carburettor
22 Carburettor intake
23 Cockpit air-ducting
24 Armoured front bulkhead
25 Engine bearer upper segment
26 Cowling air exit flaps
27 Exhaust shields
28 Rudder pedals
29 Cockpit floor level
30 Hydraulic reservoir
31 Oxygen bottle
32 Electrical system distribution box
33 Autopilot controller
34 Boarding step
35 Control linkages
36 Port side console panel
37 Engine throttle and propeller control levers
38 Control column
39 Circuit-breaker panel
40 Instrument panel
41 Windscreen de-misting air duct
42 Instrument panel shroud
43 Reflector gun sight
44 Armoured glass windscreen
45 Starboard mainwheel, retracted position
46 Ammunition tank, 200 rounds
47 Main undercarriage retraction jack
48 Gun camera
49 Starboard inboard pylon
50 Approach light
51 BLU-11B, 500-lb (227-kg) napalm tanks
52 Cannon barrels
53 M3 20-mm cannon
54 Front spar hinge joint
55 Ammunition feed drums
56 Outboard ammunition tank, 200 rounds
57 Starboard outer underwing pylons (six)
58 Ventral pitot tube
59 Radar warning antennae
60 Starboard navigation light
61 Wingtip fairing
62 Static dischargers
63 Starboard aileron
64 Starboard wing, folded position
65 Cannon bay access panels
66 Rear spar hinge joint
67 Wing fold hydraulic jack
68 Starboard Fowler-type flap
69 Sliding cockpit canopy cover
70 Armoured headrest
71 Pilot's seat
72 Canopy external handle
73 Safety harness
74 Adjustable seat mountings
75 Armoured cockpit rear bulkhead
76 Main fuel tank; internal fuel capacity 378 US gal (1431 litres)
77 Hand grip
78 Fuselage top longeron
79 Fuel vent pipe
80 Sliding canopy rail
81 Port wing, folded position
82 VHF aerial
83 Dorsal section frame and stringer construction
84 IFF aerial
85 ADF antenna
86 Fuselage dorsal skin panelling

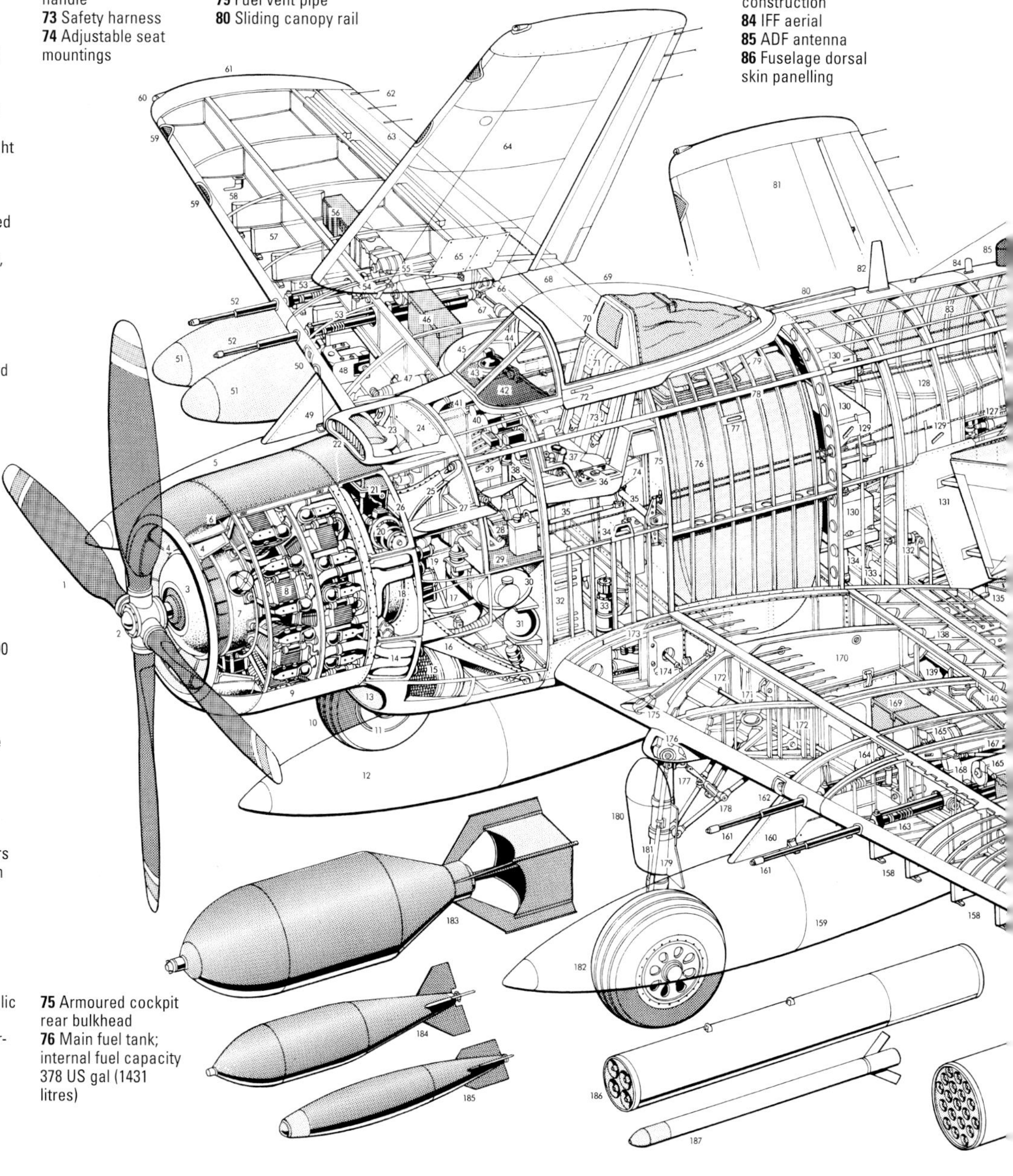

This diving AD-2 Skyraider illustrates just one of the proposed (but not adopted) weapons configurations available to the Skyraider. Mounted on the outer-wing hardpoints are 12 unguided High-Velocity Aircraft Rockets (HVARs), while the inner pylons carry two 11.75-in (298-mm) Tiny Tim air-to-ground rockets.

SPECIFICATION

A-1H Skyraider

Dimensions

Length: 38 ft 10 in (11.84 m)
Wingspan: 50 ft ¼ in (15.25 m)
Wing area: 400 sq ft (37.19 m²)
Height: 15 ft 8¼ in (4.78 m)

Powerplant

One 2,700-hp (2013-kW) Wright R-3350-26WA 18-cylinder, two-row radial piston engine driving a four-bladed constant-speed propeller.

Weights

Empty: 11,968 lb (5429 kg)
Normal take-off: 18,106 lb (8213 kg)
Maximum take-off: 25,000 lb (11340 kg)

Fuel and load

Total main internal fuel: 314.5 Imp gal (1431 litres)
Maximum bombload: 8,000 lb (3629 kg)

Performance

Maximum speed: 322 mph (518 km/h) at 18,000 ft (5485 m)
Cruising speed: 198 mph (319 km/h)
Initial climb rate: 2,850 ft (869 m) per minute
Service ceiling: 28,500 ft (8685 m)
Take-off run (to 50 ft/15 m): 4,649 ft (1417 m)
Landing run (from 50 ft/15 m): 3,002 ft (915 m)

Range

Operational range : 1,315 miles (2116 km)

Armament

Four wing-mounted 20-mm cannon, each with 200 rounds of ammunition plus up to 8,000 lb (3629 kg) of external stores

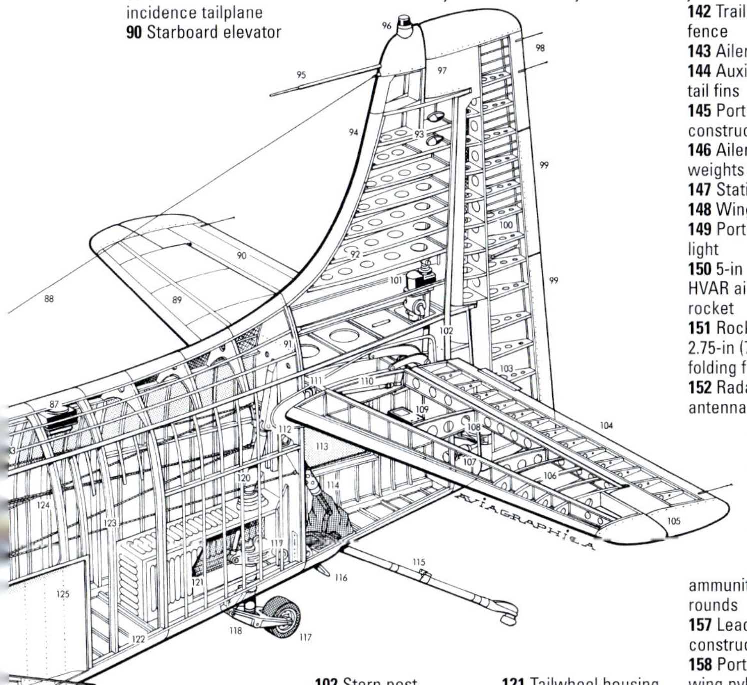

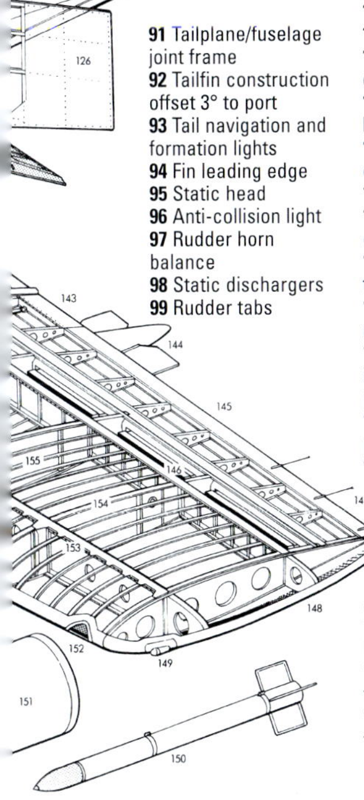

87 Remote compass transmitter
88 HF aerial
89 Starboard variable-incidence tailplane
90 Starboard elevator
91 Tailplane/fuselage joint frame
92 Tailfin construction offset 3° to port
93 Tail navigation and formation lights
94 Fin leading edge
95 Static head
96 Anti-collision light
97 Rudder horn balance
98 Static dischargers
99 Rudder tabs
100 Rudder construction
101 Tailplane incidence control jack
102 Stern post
103 Fixed elevator tab
104 Port elevator construction
105 Elevator horn balance
106 Port tail plane construction
107 Rudder tab jack
108 Rudder hinge control
109 Radar altimeter transmitter/receiver
110 Elevator control rod
111 Variable-incidence tailplane pivot fixing
112 Tailplane sealing plate
113 Starboard side access panel
114 Arrester hook jack damper
115 Deck arresting hook, down
116 Catapult hold-back link
117 Tailwheel solid tyre
118 Castoring tailwheel forks
119 Tailwheel pivot fixing
120 Hydraulic retraction jack
121 Tailwheel housing
122 Rear fuselage box section lower longeron
123 Fuselage side-panel frame construction
124 Tail control cables
125 Airbrake housing
126 Port lateral airbrake, open
127 Airbrake hydraulic actuators
128 Starboard lateral airbrake housing
129 Handgrips
130 Radio and electronics equipment racks
131 Airbrake-reinforced hinge panel
132 Central flap hydraulic jack
133 Flap torque shaft
134 Battery
135 Boarding step
136 Ventral airbrake, open
137 Port Fowler-type flap construction
138 Flap shroud ribs
139 Flap external hinge
140 Wing fold hydraulic jack
141 Rear spar hinge joint
142 Trailing-edge fence
143 Aileron tab
144 Auxiliary fuel tank tail fins
145 Port aileron construction
146 Aileron balance weights
147 Static dischargers
148 Wingtip fairing
149 Port navigation light
150 5-in (127-mm) HVAR air-to-ground rocket
151 Rocket pack, 19 x 2.75-in (70-mm) folding fin rockets
152 Radar warning antenna
153 Main spar
154 Port wing rib construction
155 Inter-rib stiffeners
156 Outboard ammunition tank, 200 rounds
157 Leading-edge rib construction
158 Port outboard wing pylons
159 Auxiliary fuel tank. capacity 300 US gal (1136 litres)
160 Inboard pylon
161 Cannon barrels
162 Approach light
163 Recoil spring
164 Front spar hinge joint
165 Ammunition feed drums
166 Aileron control rod
167 M3 20-mm cannon
168 Cartridge case ejector chute
169 Inboard ammunition tank, 200 rounds
170 Main undercarriage wheel bay
171 Hydraulic retraction jack
172 Main undercarriage mounting diagonal ribs
173 Wing spar/fuselage attachment joint
174 Aileron push/pull control rod
175 Catapult strop attachment link
176 Main undercarriage pivot fixing
177 Retraction linkage
178 Folding rear struts
179 Main undercarriage leg strut
180 Leg strut fairing
181 Wheel rotation push rod, wheel rotated 90° to lie flat in wing bay
182 Port mainwheel
183 AN-M66A2 2,000-lb (907-kg) HE bomb
184 Mk 82 500-lb (227-kg) bomb
185 Mk 81 250-lb (113-kg) low-drag bomb
186 SUU-11A rocket launcher, 4 x 5-in (12.7-cm) folding fin rockets
187 5-in (12.7-cm) folding fin air-to-ground rocket

Douglas Skyraider variants

XBT2D: first version with 2,300-hp (1716-kW) R-3350-24W, included prototypes of five other versions; total 25
AD-1: redesignation of BT2D-1, with 2,500-hp (1865-kW) R-3350-24W engine and strengthened structure; total 242
AD-1Q: ECM platform with jammer pod on left wing and ECM operator in fuselage cabin; total 35
AD-2: further strengthening, more fuel, 2,700-hp (2014-kW) R-3350-26W engine, mainwheel doors added; total 156
AD-2D: conversion to drone (RPV) directors
AD-2Q: ECM version; total 22
AD-2Q(U): further rebuild to tow Mk 22 target
AD-3: further strengthening, long-stroke main gears; Aeroproducts propeller and new canopy; total 124
AD-3E: conversion for ASW search
AD-3N: night-attack version; total 15
AD-3Q: ECM version; total 23
AD-3S: anti-submarine attack, partner to AD-3E; all conversions
AD-3W: AEW version with improved APS-20 surveillance radar and two operators in fuselage cabin, plus auxiliary fins; total 31
AD-4: refined structure, cleared for great increase in gross weight from 18,500 lb (8392 kg) to 24,000 lb (10886 kg); P-1 autopilot, modified windscreen, improved radar (APS-19A) option; total 344
AD-4B: four cannon, provision for nuclear bombs; total 194
AD-4L: conversion for winter (Arctic)
AD-4N: night attack version with APS-19A, total 248
AD-4NA: night version stripped for day attack; total 23 plus conversions; redesignated (A-1D) from 1962
AD-4NL: winterised night version; total 36
AD-4Q: ECM version; total 39
AD-4W: RN designation AEW.Mk 1; total 168
AD-5 (A-1E): redesigned multi-role model with wide forward and longer overall fuselage, side-by-side cockpit, taller fin, side dive brakes removed (leaving ventral brake), four guns standard, provision for quick role conversions; cleared to 25,000 lb (11340 kg); total 212
AD-5N (A-1G): night-attack version; total 239
AD-5Q (EA-1F): ECM conversions of 54 aircraft
AD-5S: (no 1962 designation) anti-submarine conversion
AD-5U (UA-1E): conversions as target tow/transport for 12 seats or 3,000 lb (1361 kg) of freight
AD-5W (EA-1E): AEW version; total 156
AD-6 (A-1H): new standard close-support single-seater, LABS toss-bombing avionics and reinforced wing as AD-4B; total 713
AD-7 (A-1J): further reinforced wing and main gear, 3,050-hp (2275-kW) R-3350-26WB engine; total 72

Douglas A-4 Skyhawk

*Two-seat Skyhawks provided carrier-cap training facilities for **USN/USMC** aircrew from until 1999, when the last training examples w replaced by **T-45A Goshawks**. Pictured is a **TA-4** **VT-7**, which finally retired the type from the US training establishment in September 1999, lea Puerto Rico-based target facilities squadron V as the only Navy unit with Skyhawks in 2*

A-4M Skyhawk

Cutaway key

1 Fixed inflight refuelling probe
2 Nose ECM recording and suppression aerials
3 Angle-Rate Bombing System (ARBS) laser seeker head
4 Hinged nose compartment access door
5 Laser seeker system electronics
6 Electronics cooling air inlet
7 Pitot tube
8 Avionics access panel
9 APN-153(V) navigation radar
10 Lower TACAN aerial
11 Communications electronics
12 Cockpit front pressure bulkhead
13 Pressurisation valve
14 Windshield rain dispersal air duct
15 Rudder pedals
16 Angle-of-attack sensor
17 Air conditioning refrigeration plant
18 Nosewheel door
19 Control system access
20 Cockpit floor level
21 Pilot's side console panel
22 Engine throttle
23 Control column
24 Instrument panel shroud
25 Head-up display (HUD)
26 Windscreen panels
27 AIM-9L Sidewinder air-to-air missile
28 Missile launch rail
29 D-704 flight refuelling pack containing 300 US gal (1135 litres) **30** Cockpit canopy cover
31 Face blind firing handle
32 Ejection seat headrest
33 Safety harness
34 McDonnell Douglas ESCAPAC IG-3 'zero-zero' ejection seat
35 Anti-*g* valve
36 Cockpit insulation and fragmentation blanket
37 Rear pressure bulkhead
38 Emergency canopy release handle
39 Nose undercarriage leg strut
40 Steering linkage
41 Nosewheel
42 Leg shortening link
43 Hydraulic retraction strut
44 Emergency wind-driven generator
45 Port cannon muzzle
46 Intake gun gas shield
47 Port air intake
48 Boundary layer splitter plate
49 Self-sealing fuselage fuel cell, capacity 240 US gal (908 litres)
50 Fuel system piping
51 Canopy hinge cover
52 Starboard air intake duct
53 Fuel system gravity filler cap
54 UHF aerial
55 Electronics cooling air inlet
56 Engine-driven generator
57 Constant-speed drive
58 Bifurcated intake duct
59 Reel-type ammunition magazine (200 rounds per gun)
60 Intake compressor face
61 Electrical system power amplifier
62 Engine accessory drive gearbox
63 Wing spar attachment fuselage double frame
64 Engine mounting trunnion
65 Engine fuel system access panel
66 Pratt & Whitney J52-P-408 turbojet
67 Dorsal avionics bays
68 Compressor bleed air exhaust duct
69 Upper TACAN aerial
70 Starboard wing integral fuel tank (total wing capacity 560 US gal/2120 litres)
71 Wing tank access panels
72 Slat guide rails
73 Starboard automatic leading-edge slat (open)
74 Wing fences
75 Vortex generators
76 Starboard navigation light
77 Wing tip communications aerial
78 Aileron horn balance
79 Starboard aileron
80 Split trailing-edge spoiler (open position)
81 Starboard split trailing-edge flap (down position)
82 Anti-collision light
83 Cooling air exit louvres

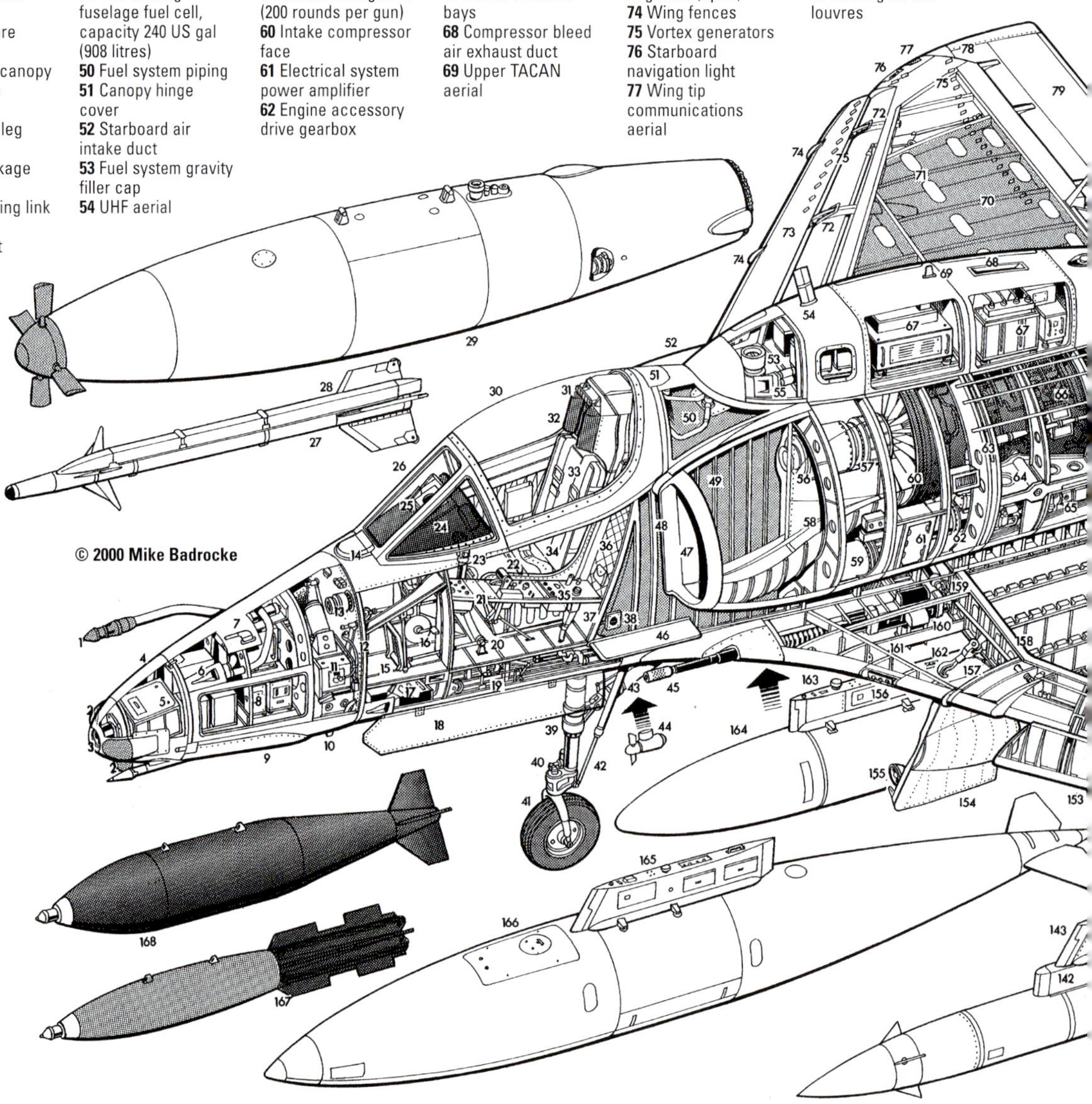

© 2000 Mike Badrocke

BuNo. 160264 was the 2,960th and last Skyhawk constructed, after a production run that had lasted no less than 27 years and finally ended in 1979. An A-4M for the USMC, the aircraft incorporated all the improvements introduced during the 'Mike's 10-year production life, including a redesigned cockpit, Elliott HUD, laser spot tracker, ECM equipment and a new GE generator. An angle-rate bombing system (ARBS) had been added in 1977. With so many changes to the A-4M consideration was given to redesignating these late-production aircraft as A-4Ys, but this was not proceeded with as most early A-4Ms were eventually upgraded.

SPECIFICATION

A4D-2N Skyhawk

Dimensions

Fuselage length: 39 ft 4¾ in (12.01 m)
Height: 15 ft (4.57 m)
Wingspan: 27 ft 6 in (8.38 m)
Wing area: 260.00 sq ft (24.15 m²)

Powerplant

One Wright J65-W-16A turbojet rated at 7,700 lb st (34.25 kN)

Weights

Empty: 9,559 lb (4336 kg)
Normal loaded: 17,295 lb (7845 kg)

Fuel and load

Internal fuel capacity: 770 US gal (2910 litres)
External fuel capacity: 500 US gal (1893 litres)

Performance

Maximum speed at sea level: 680 mph (1094 km/h)

Armament

Two Colt Mk 12 20-mm cannon, each with 200 rounds of ammunition, plus up to 5,000 lb (2268 kg) of weapons on three external hardpoints

A-4M Skyhawk

Dimensions

Fuselage length: 40 ft 3½ in (12.29 m)
Height: 15 ft (4.57 m)
Wingspan: 27 ft 6 in (8.38 m)
Wing area: 260.00 sq ft (24.15 m²)

Powerplant

One Pratt & Whitney J52-P-408A turbojet rated at 11,200 lb st (49.80 kN)

Weights

Empty: 10,465 lb (4747 kg)
Maximum take-off: 24,500 lb (11113 kg)

Fuel and load

Internal fuel capacity: 800 US gal (3028 litres)
External fuel capacity: up to 1,000 US gal (3786 litres)

Performance

Maximum speed at sea level: 685 mph (1102 km/h)
Maximum rate of climb at sea level: 10,300 ft (3140 m) per minute
Service ceiling: 38,700 ft (11795 m)
Combat radius: 345 miles (547 km) with a 4,000-lb (1814-kg) warload

Armament

Two Colt Mk 12 20-mm cannon, each with 200 rounds of ammunition, plus up to 9,155 lb (4153 kg) of weapons on five external hardpoints

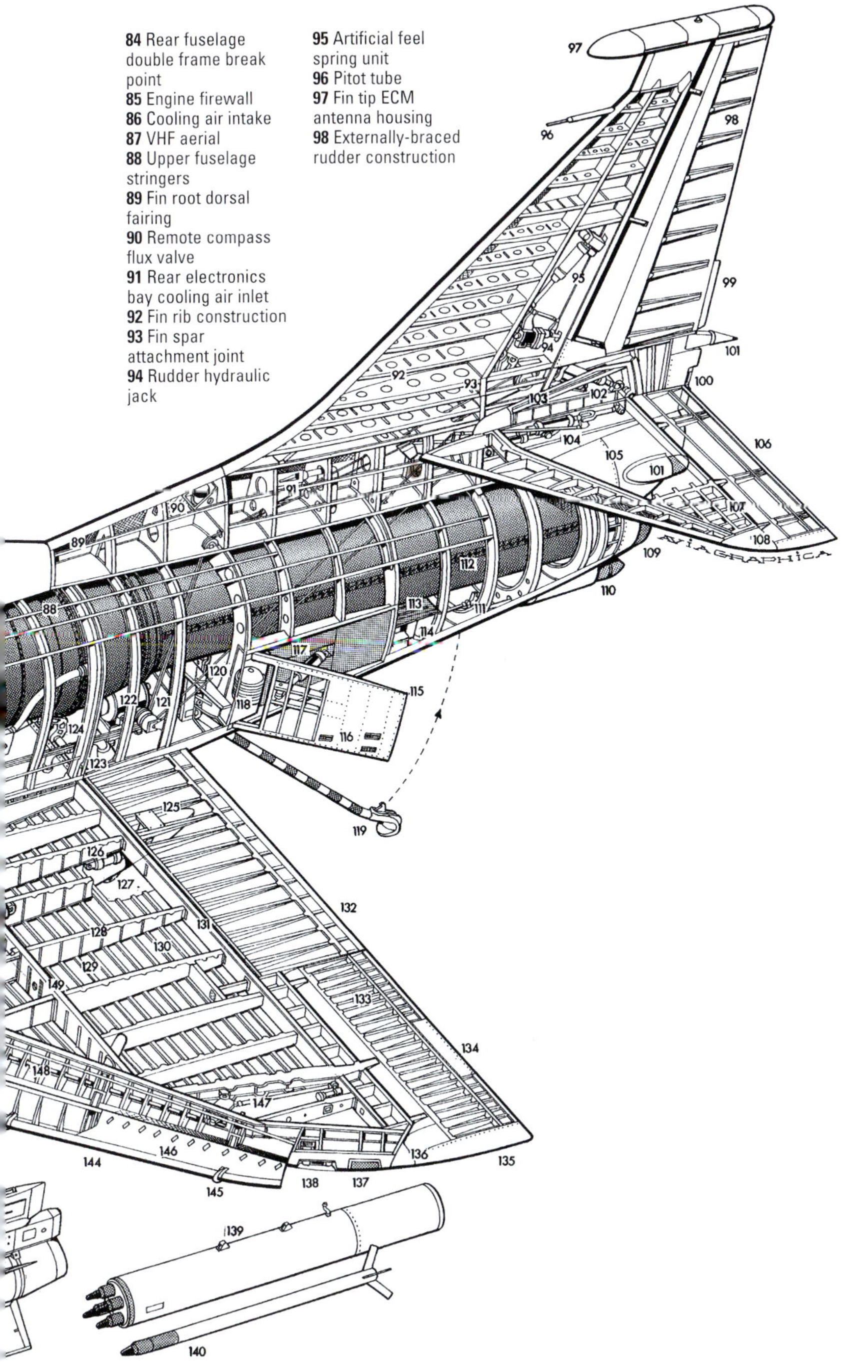

84 Rear fuselage double frame break point
85 Engine firewall
86 Cooling air intake
87 VHF aerial
88 Upper fuselage stringers
89 Fin root dorsal fairing
90 Remote compass flux valve
91 Rear electronics bay cooling air inlet
92 Fin rib construction
93 Fin spar attachment joint
94 Rudder hydraulic jack
95 Artificial feel spring unit
96 Pitot tube
97 Fin tip ECM antenna housing
98 Externally-braced rudder construction
99 Fixed rudder tab
100 Tail navigation light
101 ECM antennas
102 Tailplane trim jack
103 Tailplane seal plate
104 Elevator hydraulic jack
105 Tailpipe fairing
106 Port elevator
107 All-moving tailplane construction
108 Elevator horn balance
109 Jet pipe exhaust nozzle
110 Drake parachute housing for 16-ft (4.88-m) diameter, ribbon type 'chute
111 Brake parachute release linkage
112 Insulated jet pipe
113 Electronics bay heat shield
114 Rear electronics bay, automatic flight control system (AFCS)
115 Port airbrake (open)
116 JATO bottle attachment hardpoints
117 Airbrake hydraulic jack
118 2.65-US gal (10-litre) liquid oxygen (LOX) converter
119 Arrester hook (down position)
120 Arrester hook hydraulic jack
121 Control cable runs
122 Inertial platform
123 Ventral pressure refuelling connection
124 Central hydraulic flap drive linkage
125 Port upper surface spoiler
126 Spoiler hydraulic jack
127 Ventral anti-collision light
128 Wing rib construction
129 Stringer construction
130 Port wing integral fuel tank (single tank tip-to-tip)
131 Rear spar
132 Port split trailing-edge flap
133 Port aileron construction
134 Aileron trim tab
135 Tip fairing
136 Aileron horn balance
137 Wing tip antenna fairing
138 Port navigation light
139 LAU-10A Zuni rocket launcher
140 5-in (12.7-cm) folding fin rocket
141 AGM-12 Bullpup air-to-ground missile
142 Missile launcher rail
143 Outboard wing pylon (1,000-lb/454-kg capacity)
144 Port automatic leading-edge slat (open)
145 Wing fences
146 Vortex generators
147 Aileron control rod linkage
148 Leading-edge ribs
149 Wing centre spar
150 Main undercarriage hydraulic retraction jack
151 Undercarriage leg pivot mounting
152 Slat guide rail fuel sealing can
153 Port mainwheel
154 Mainwheel door
155 Position of landing lamp on starboard mainwheel door
156 Approach lights
157 Retractable catapult hook
158 Cranked wing front spar
159 Aileron servo control
160 Colt Mk 12 20-mm cannon
161 Spent cartridge case and link ejector chutes
162 Mainwheel well
163 Centreline pylon (3,575-lb/1622-kg capacity)
164 150-US gal (568-litre) fuel tank
165 Inboard wing pylon (2,240-lb/1016-kg capacity)
166 400-US gal (1514-litre) long-range fuel tank
167 Snakeye 500-lb (227-kg) retarded bomb
168 Mk 83 1,000-lb (454-kg) HE bomb

Grumman A-6E Intruder

An A-6E from VA-34, carrying one ATM-84E Stand-c Land Attack Missile, receives fuel from a Kansas A National Guard KC-135E during the SLAM-O-RAM exercise over the Naval Air Weapons Station, Patuxe River, MD. During Desert Storm, an A-6E from the US Kennedy made the first-ever combat launch of a SLAI

A-6E (TRAM) Intruder

Cutaway key
1 Radome
2 Radome open position
3 Norden AN/APQ-148 multi-mode radar scanner
4 Scanner-tracking mechanism
5 Intermediate frequency unit
6 ILS aerials
7 TRAM rotating turret mounting
8 Target Recognition and Attack Multisensor turret (TRAM)
9 Taxiing lamp
10 Deck approach lights
11 Nosewheel leg door
12 Hydraulic nosewheel steering unit
13 Catapult launch strap
14 Twin nosewheels (aft-retracting)
15 Retraction/breaker strut
16 Shock-absorber leg strut
17 Torque scissor links
18 Radome latch
19 Hinged avionics equipment pallet (port and starboard)
20 Radar scanner mounting
21 Radome hydraulic jack
22 Flight-refuelling probe
23 ALQ-165 ECM system forward spiral antenna
24 Refuelling probe spotlight
25 Windscreen rain-repellent air duct
26 Front-pressure bulkhead
27 Nosewheel bay-mounted pressure refuelling connection
28 Boundary layer splitter plate
29 Port engine air intake
30 Nosewheel bay electronic equipment racks
31 VHF aerial
32 UHF aerial
33 Intake duct framing
34 Temperature probe
35 Canopy emergency release handle
36 TACAN aerial
37 Folding boarding ladder
38 Integral boarding steps
39 Angle-of-attack transmitter
40 Boundary layer spill duct
41 Cockpit floor level
42 Rudder pedals
43 Engine throttle levers
44 Control column
45 Instrument panel shroud
46 Pilot's optical sighting unit/head-up display
47 Windscreen panels
48 Aft-sliding cockpit canopy cover
49 Forward-Looking Infra-Red (FLIR) viewing scope
50 Navigator/ Bombardier's Martin Baker GRU-7 ejection seat
51 Ejection seat headrests
52 Seat-reclining mechanism
53 Centre console
54 Pilot's GRU-7 ejection seat
55 Safety/parachute harness
56 Port side console panel
57 Electrical system equipment
58 Destruct initiator
59 Leading-edge stall warning buffet strip
60 Engine intake compressor face
61 Engine bay venting air scoop
62 Accessory equipment gearbox
63 Pratt & Whitney J52-P-8B non-afterburning turbofan
64 Mainwheel door
65 Leading-edge antenna fairing (port and starboard)
66 ILQ-165 high-, mid- and low-band ECM aerials
67 Mainwheel well
68 Hydraulic system reservoir
69 Cockpit rear-pressure bulkhead
70 Cooling air spill louvres
71 Electrical equipment bay
72 Electronics and avionics equipment bay
73 Forward fuselage bag-type fuel tank
74 Weapons-monitoring module
75 Sliding canopy rail
76 Canopy hydraulic jack
77 Canopy aft fairing
78 Starboard wing inboard integral fuel tank; total fuel capacity 1,951 Imp gal (2344 US gal/8870 litres)
79 Fuel system piping
80 Inboard wing fence
81 Leading-edge slat drive shaft
82 Slat guide rails
83 Slat screw jacks
84 AGM-65 Maverick air-to-surface missiles
85 Triple missile carrier/launcher
86 Starboard wing stores pylons
87 AIM-9P Sidewinder 'self-defence' air-to-air missile
88 Wing fold twin hydraulic jacks
89 Spar hydraulic latch pins
90 Wing fold hinge joint

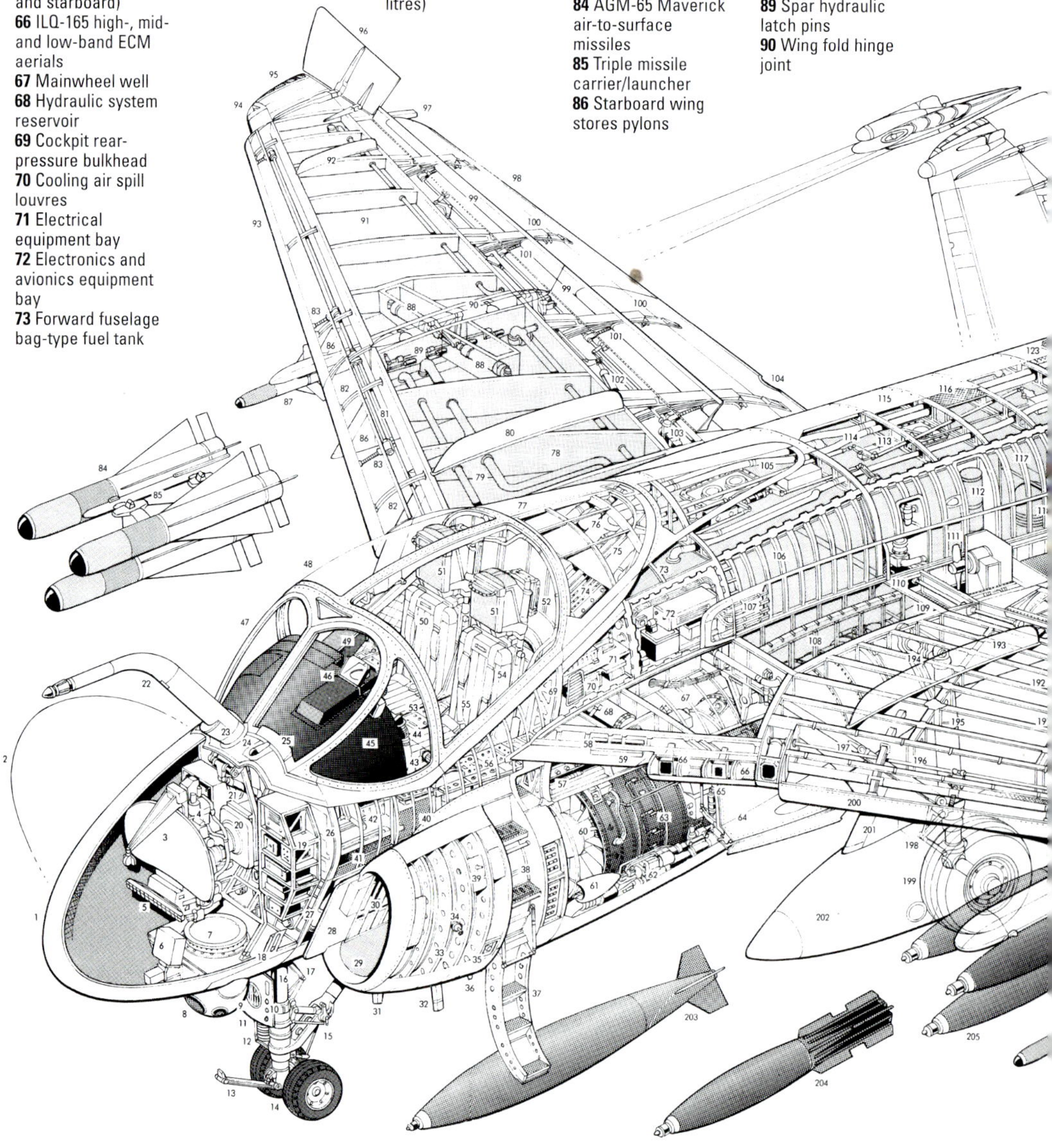

EA-6B Prowlers performed sterling service throughout the Gulf War, providing stand-off (nicknamed 'white snow') jamming on every strike. Desert Storm also gave the Prowler the chance to demonstrate its ability as a HARM-shooter on strike escort missions.

SPECIFICATION

A-6E Intruder

Dimensions

Length: 54 ft 9 in (16.69 m)
Height: 16 ft 2 in (4.93 m)
Wingspan: 53 ft (16.15 m)
Width, folded: 25 ft 4 in (7.72 m)
Wing area: 528.90 sq ft (49.13 m²)
Wing aspect ratio: 5.31
Wheel track: 10 ft 10½ in (3.32 m)

Powerplant

Two Pratt & Whitney J52-P-8B turbojets, each rated at 9,300 lb st (41.4 kN) dry

Weights

Empty: 27,613 lb (12525 kg)
Maximum take-off for catapult launch: 58,600 lb (26580 kg)
Maximum take-off for field launch: 60,400 lb (27397 kg)

Fuel and load

Internal fuel: 15,939 lb (7230 kg)
External fuel: up to 10,050 lb (4558 kg) in five 400-US gal (1514-litre) drop tanks
Maximum ordnance: 18,000 lb (8165 kg)

Performance

Never-exceed speed: 700 kt (806 mph; 1297 km/h)
Maximum level speed 'clean' at sea level: 560 kt (644 mph; 1037 km/h)
Cruising speed at optimum altitude: 412 kt (474 mph; 763 km/h)
Service ceiling: 42,400 ft (12925 m)

Range

Ferry range: 2,818 nm (3,245 miles; 5222 km) with empty tanks dropped
Ferry range: 2,380 nm (2,740 miles; 4410 km) with empty tanks retained
Range with maximum military load: 878 nm (1,011 miles; 1627 km)

Armament

The A-6 carried its stores on one centreline and four wing pylons. Virtually all the US Navy/Marine Corps inventory of stores could be carried. Stand-off weapons included the AGM-62 Walleye, AGM-84 Harpoon, AGM-84E SLAM, AGM-88 HARM and AGM-123 Skipper. Mk 82 and 83 general-purpose bombs, B57 and B61 nuclear bombs, ADM-141 decoy drones and AIM-9s were also compatible.

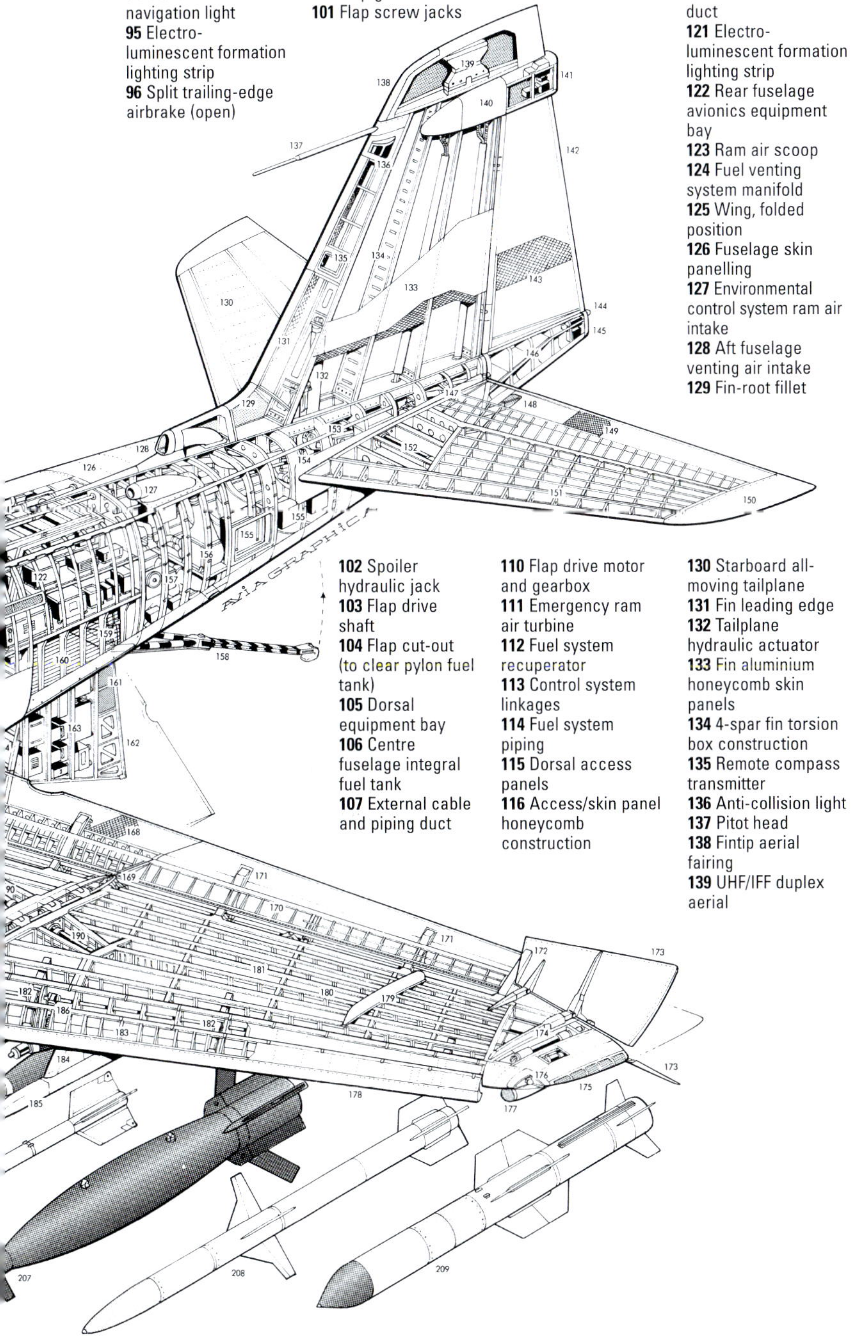

91 Outer wing panel integral fuel tank
92 Outboard wing fence
93 Starboard leading-edge slat (open)
94 Starboard navigation light
95 Electro-luminescent formation lighting strip
96 Split trailing-edge airbrake (open)
97 Fuel jettison
98 Single-slotted Fowler-type flap (down position)
99 Roll control spoiler/lift dumper
100 Flap guide rails
101 Flap screw jacks
102 Spoiler hydraulic jack
103 Flap drive shaft
104 Flap cut-out (to clear pylon fuel tank)
105 Dorsal equipment bay
106 Centre fuselage integral fuel tank
107 External cable and piping duct
108 Wing centre-section integral fuel tank
109 Wing centre-section carry-through spar box
110 Flap drive motor and gearbox
111 Emergency ram air turbine
112 Fuel system recuperator
113 Control system linkages
114 Fuel system piping
115 Dorsal access panels
116 Access/skin panel honeycomb construction
117 Aft fuselage bag-type fuel tank
118 Liquid oxygen converters (2)
119 Outboard canted jet pipe
120 External cable duct
121 Electro-luminescent formation lighting strip
122 Rear fuselage avionics equipment bay
123 Ram air scoop
124 Fuel venting system manifold
125 Wing, folded position
126 Fuselage skin panelling
127 Environmental control system ram air intake
128 Aft fuselage venting air intake
129 Fin-root fillet
130 Starboard all-moving tailplane
131 Fin leading edge
132 Tailplane hydraulic actuator
133 Fin aluminium honeycomb skin panels
134 4-spar fin torsion box construction
135 Remote compass transmitter
136 Anti-collision light
137 Pitot head
138 Fintip aerial fairing
139 UHF/IFF duplex aerial
140 ECM aerial fairing
141 ALQ-165 high-, mid-and low-band ECM transmitting aerials
142 Rudder
143 Rudder honeycomb construction
144 Tail navigation light
145 ALQ-165 ECM receiving aerials
146 Tailcone rudder segment
147 Rudder hydraulic actuator
148 Fuel jettison
149 Tailplane honeycomb trailing-edge section
150 Tailplane tip fairing
151 Multi-spar tailplane construction
152 All-moving tailplane pivot fixing
153 Tailplane hinge control arm
154 Tailplane sealing plate
155 ECM transmitting and receiving equipment
156 Avionics equipment environmental control system
157 Static discharge port
158 Deck arrester hook (lowered)
159 Arrester hook hydraulic jack and damper
160 Structural provision for fuselage air brake (deleted)
161 ALE 45 chaff and flare dispensers
162 Avionics bay ventral access door, open
163 'Birdcage' avionics equipment rack (lowered)
164 Telescopic access ladder
165 Port engine exhaust nozzle
166 Flap rib construction
167 Spoiler hydraulic actuator
168 Flap honeycomb trailing-edge segment
169 Wing fold control linkage breakers
170 Port spoiler rib construction
171 Flap track fairings
172 Fuel jettison
173 Port split trailing-edge airbrake (open)
174 Airbrake hydraulic jack
175 Wingtip electro-luminescent formation lighting strip
176 Port navigation light
177 ALR 45 radar warning receiver
178 Port leading-edge slat (open)
179 Outboard wing fence
180 Multi-spar outer wing panel construction
181 Port wing integral fuel tank
182 Slat guide rails
183 Slat rib construction
184 Additional outer wing missile pylon
185 Missile launch rail
186 Slat screw jack
187 Multiple ejector rack
188 Port outer stores pylon
189 Wing fold hinge joint
190 Wing fold hydraulic jacks
191 Inboard integral fuel tank
192 Inner wing panel multi-spar construction
193 Inboard wing fence
194 Main undercarriage leg pivot fixing
195 Main undercarriage leg strut
196 Leading-edge slat drive shaft
197 Undercarriage leg retraction/breaker strut
198 Torque scissor links
199 Port mainwheel
200 Inboard leading-edge slat segment
201 Inboard stores pylon
202 External fuel tank, capacity 250 or 330 Imp gal (300 or 400 US gal/1135 or 1514 litres)
203 2000-lb (907-kg) low-drag HE bomb
204 Snakeye Mk 92 retarded bomb
205 Mk 83 500-lb (227-kg) HE bombs (6 per rack)
206 AIM-9P Sidewinder 'self-defence' air-to-air missile
207 GBU-10 Paveway (2000-lb/907-kg) laser-guided bomb
208 AGM-88 Harm air-to-surface anti-radar missile
209 AGM-84A Harpoon air-to-surface anti-shipping missile

Grumman EA-6B Prowler

This EA-6B Prowler is an ICAP (Improved Capability) II Block 82 aircraft with just two dorsal antennas and none under the nose. ICAP II Prowlers were retrospectively dubbed 'Block 82' after the introduction of the so-called 'Block 86' avionics upgrade in 1986.

EA-6B Prowler

1 Inflight-refuelling probe
2 Radome, upward-hinging
3 APQ-92 radar antenna
4 Pitot head, port and starboard
5 TACAN aerial
6 IFF L-band antenna
7 Forward avionics equipment bay
8 Rudder pedals
9 Pilot's head-down radar display
10 Control column
11 Instrument panel shroud
12 Refuelling probe spotlight
13 Windscreen rain dispersal air duct
14 Upward-hinging cockpit canopy
15 Electronic countermeasures officer No. 1 ejection seat (ECMO-1)
16 Pilot's Martin-Baker GRUEA-7 ejection seat
17 Engine throttle levers
18 Fold-down boarding step
19 Boundary layer splitter plate
20 Nosewheel leg door-mounted landing and approach lights
21 Port engine air intake
22 Temperature probe
23 Canopy emergency release
24 Hinged boarding ladder
25 Nose undercarriage, stowed position
26 Hydraulic retraction jack
27 ECMO display and control consoles
28 Forward canopy actuator
29 Tactical jamming system pod
30 Rear upward-hinging canopy
31 ECMO-2 ejection seat
32 Rear canopy actuator
33 ECMO-3 ejection seat
34 Leading-edge stall-warning buffet strip
35 Stand-by hydraulic pump and selector valves
36 Generator cooling air intake
37 Engine accessory equipment gearbox
38 Pratt & Whitney J52-P-408 turbojet engine
39 Leading-edge radome panel
40 Forward ECM transmitting antennas
41 Main undercarriage stowed position
42 Hydraulic reservoir
43 Central electrical and avionics equipment bay
44 Anti-collision light
45 Starboard wing inboard integral fuel tank
46 Inboard wing fence
47 Wing-fold hydraulic jacks
48 Wing-fold hinge joint
49 Leading-edge slat
50 Outboard integral fuel tank
51 Outboard wing fence
52 Radar warning antenna
53 Starboard navigation light
54 Formation light
55 Split trailing-edge airbrake
56 Fuel jettison
57 Port single-slotted Fowler-type flap
58 Port spoiler/lift dumper
59 UHF/TACAN aerial
60 Fuselage fuel tanks
61 Fuel recuperator
62 ADF aerial
63 Flight control system mechanical linkages
64 Avionics cooling air intakes
65 Cooling system air cycle machine
66 Downward-hinging avionics equipment pallet
67 HF antenna
68 Starboard all-moving tailplane
69 Tailplane hydraulic actuator
70 Band 1 and 2 transmitting antennas
71 Fintip radome

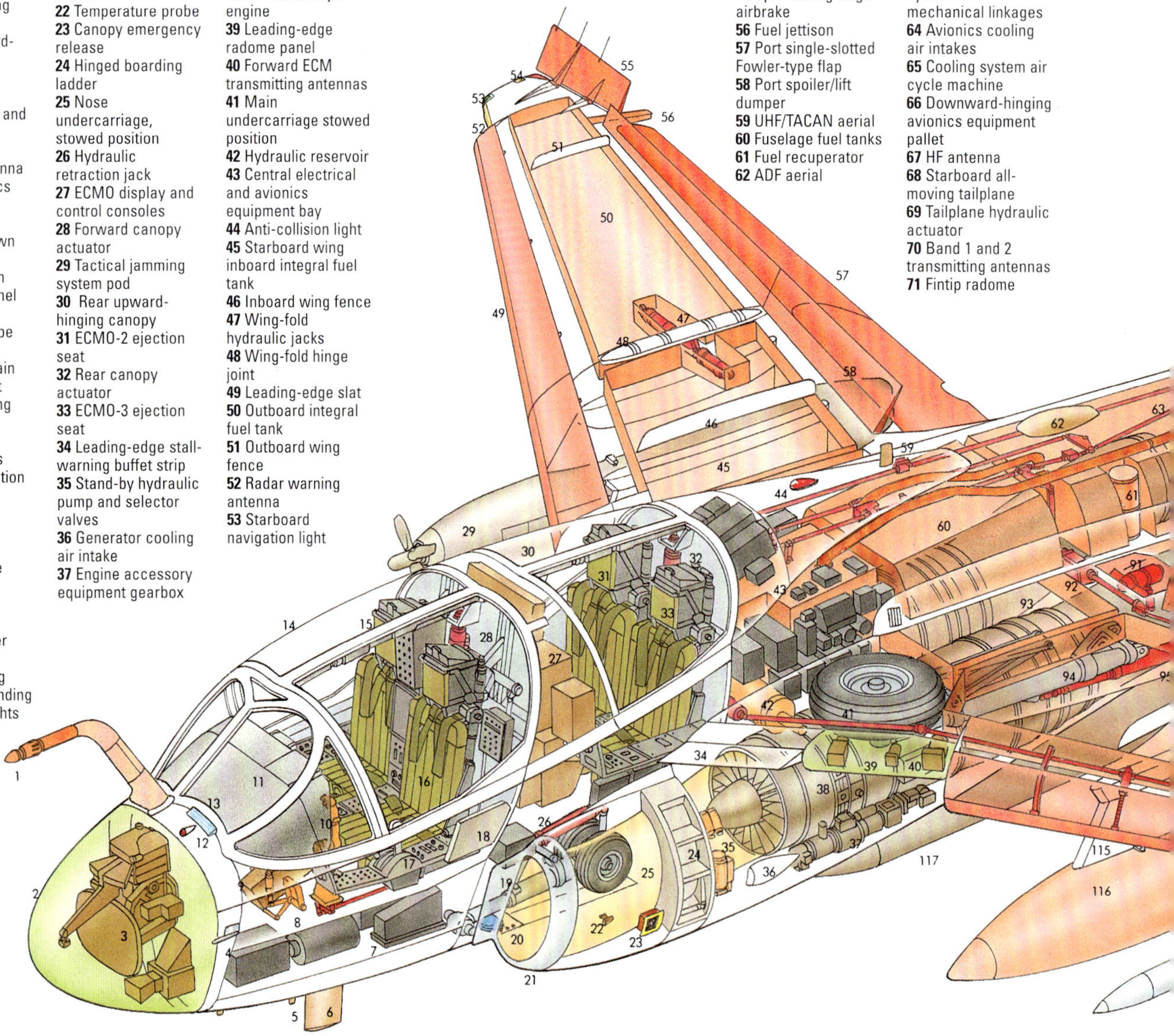

*The key to the EA-6B's success as a jamming system – and one of the reasons it survived and the **EF-111A** did not – is its four-man crew. Having three dedicated **ECMOs**, plus a pilot, allows a far greater division of duties between the crew than in comparable aircraft. Here, the ground crew prepare a **VAQ-137** aircraft before a mission from **USS America**.*

SPECIFICATION

EA-6B Prowler

Dimensions

Length: 59 ft 10 in (18.24 m)
Wingspan: 53 ft (16.15 m)
Wingspan, folded: 25 ft 10 in (7.87 m)
Wing aspect ratio: 5.31
Wing area: 528.90 sq ft (49.13 m²)
Height: 16 ft 3 in (4.95 m)
Tailplane span: 20 ft 4½ in (6.21 m)
Wheel track: 10 ft 10½ in (3.32 m)
Wheel base: 17 ft 2 in (5.23 m)

Powerplant

Two Pratt & Whitney J52-P-408 turbojets each rated at 11,200 lb st (49.80 kN) dry

Weights

Empty: 31,572 lb (14321 kg)
Normal take-off (equipped): 54,461 lb (24703 kg)
Normal take-off (equipped) with maximum fuel: 60,610 lb (27493 kg)
Maximum take-off: 65,000 lb (29484 kg)

Fuel

Internal fuel: 15,422 lb (6995 kg)
External fuel: up to 10,025 lb (4547 kg) in five 400-US gal (1514-litre) drop tanks

Performance

Never-exceed speed: 710 kt (817 mph; 1315 km/h)
Maximum level speed 'clean' at sea level: 566 kt (651 mph; 1048 km/h)
Maximum level speed carrying five jammer pods at sea level: 530 kt (610 mph; 982 km/h)
Cruising speed at optimum altitude: 418 kt (481 mph; 774 km/h)
Maximum rate of climb 'clean' at sea level: 12,900 ft (3932 m) per minute
Maximum rate of climb with five jammer pods at sea level: 10,030 ft (3057 m) per minute
Service ceiling 'clean': 41,200 ft (12550 m)
Service ceiling with five jammer pods: 38,000 ft (11580 m)
Take-off run with five jammer pods: 2,670 ft (814 m)
Take-off distance to 50 ft (15 m) with five jammer pods: 3,495 ft (1065 m)
Landing distance from 50 ft (15 m) at maximum landing weight: 2,700 ft (823 m)
Landing run with five jammer pods: 2,150 ft (655 m)
Ferry range: 2,085 nm (2,399 miles; 3861 km) with empty tanks dropped
Range: 955 nm (1,099 miles; 1769 km) with maximum external load

*Air-to-air refuelling with drogue-equipped USAF aircraft, such as this **KC-135R**, has become routine for Prowler crews. However, they must now adapt to a new era of completely integrated operations.*

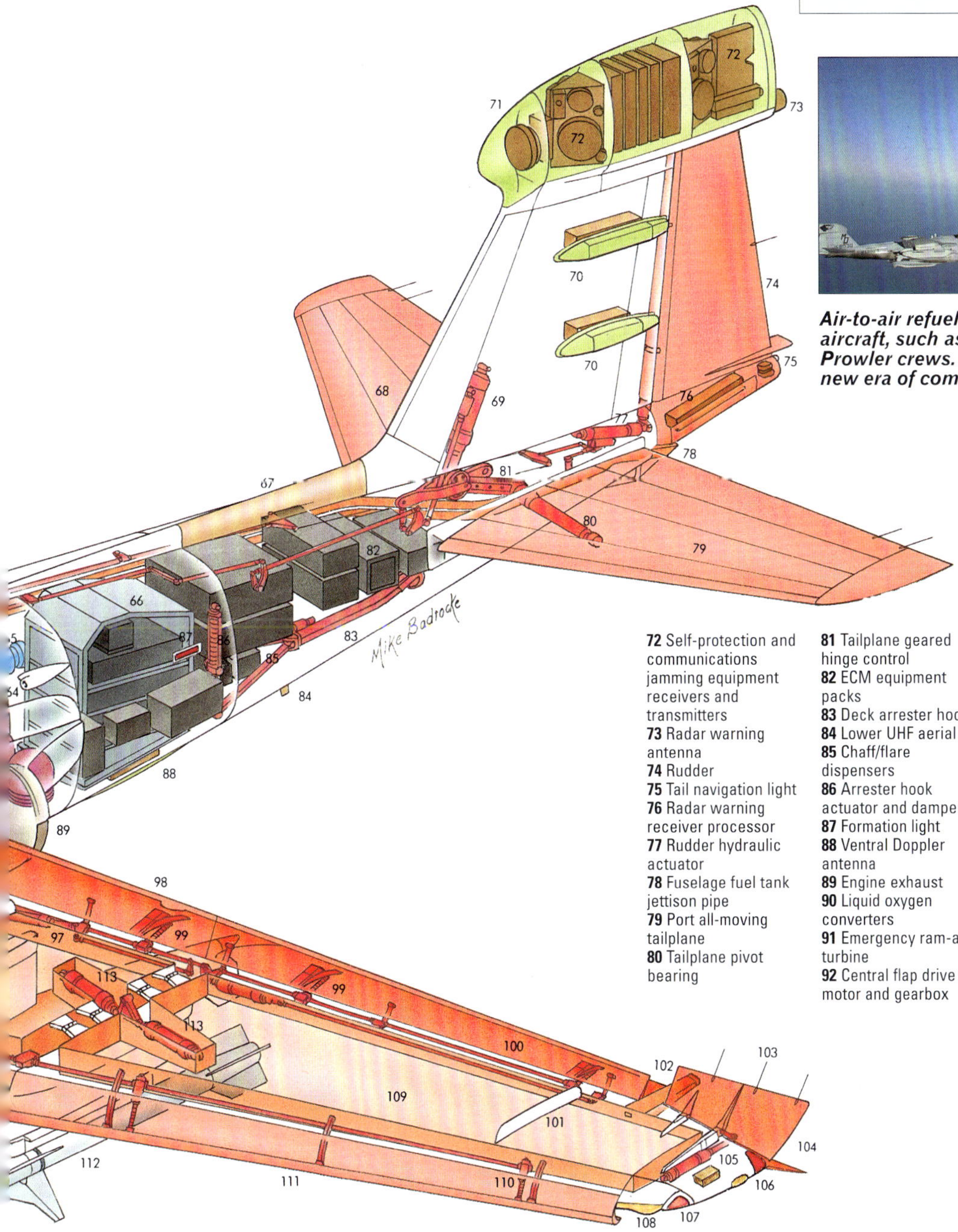

72 Self-protection and communications jamming equipment receivers and transmitters
73 Radar warning antenna
74 Rudder
75 Tail navigation light
76 Radar warning receiver processor
77 Rudder hydraulic actuator
78 Fuselage fuel tank jettison pipe
79 Port all-moving tailplane
80 Tailplane pivot bearing
81 Tailplane geared hinge control
82 ECM equipment packs
83 Deck arrester hook
84 Lower UHF aerial
85 Chaff/flare dispensers
86 Arrester hook actuator and damper
87 Formation light
88 Ventral Doppler antenna
89 Engine exhaust
90 Liquid oxygen converters
91 Emergency ram-air turbine
92 Central flap drive motor and gearbox
93 Wing centre-section integral tuel tank
94 Main undercarriage leg strut
95 Hydraulic retraction jack
96 Port inboard wing fence
97 Spoiler hydraulic actuator
98 Port single-slotted flap
99 Flap screw jacks and guide rails
100 Port spoiler/lift dumper
101 Outboard wing fence
102 Fuel jettlson
103 Static dischargers
104 Port split trailing-edge airbrakes
105 Airbrake hydraulic jack
106 Formation light
107 Port navigation light
108 Radar warning antenna
109 Wing outboard integral fuel tank
110 Slat screw jacks and guide
111 Port leading-edge slat
112 AGM-88A HARM anti-radar missile
113 Wing-fold hydraulic jack
114 Inboard integral fuel tank
115 Wing store pylons
116 External fuel tank
117 Centreline-mounted tactical jamming system pod

Grumman F-14 Tomcat

Symbolising the might of the US Navy's carrierborne air power, the Grumman F-14 Tomcat has been the world's premier long-range interceptor for over 20 years. This F-14A Tomcat of VF-2 'Bounty Hunters' climbs with afterburners blazing at full power.

F-14D Tomcat

Cutaway key
1 Pitot head
2 Glass-fibre radome
3 IFF aerial array
4 AN/APG-71 flat plate radar scanner
5 Scanner tracking mechanism
6 Infra-red search and track sensor (IRST) and television camera housing
7 Cannon port
8 Weapons system avionics equipment bay
9 Angle of attack transmitter
10 ADF aerial
11 Flight refuelling probe
12 Pilot's head-up display
13 Instrument panel shroud
14 Temperature probe
15 Rudder pedals
16 Control column
17 Electro-luminescent formation lighting strip
18 Nosewheel doors
19 Catapult strop link
20 Twin nosewheels, forward-retracting
21 Boarding ladder, extended
22 M61A1 Vulcan cannon
23 Ammunition drum
24 Pull-out steps
25 Pitot static head
26 Engine throttle levers
27 Pilot's Martin-Baker Mk 14 Navy Aircrew Common Ejection Seat (NACES)
28 Upward-hinged cockpit canopy cover
29 Naval flight officer's instrument console
30 Kick-in step
31 Tactical information display hand controller
32 NFO's ejection seat
33 Rear avionics equipment bay
34 Air data computer
35 Electrical system relays
36 Fuselage missile pallet
37 AIM-54A Phoenix air-to-air missile
38 Port engine air intake
39 Port navigation light
40 Variable-area intake control ramps
41 Intake ramp hydraulic actuators
42 Air conditioning pack
43 Forward fuselage fuel tanks
44 Canopy hinge point
45 UHF/TACAN aerial
46 Starboard navigation light
47 Mainwheel stowed position
48 Starboard intake duct spill door
49 Dorsal control and cable duct
50 Central flap and slat drive hydraulic motor
51 Emergency hydraulic generator
52 Intake by-pass door
53 Electron-beam welded titanium wing pivot box
54 Port wing pivot bearing
55 Pivot box beam integral fuel tank
56 UHF datalink/IFF aerial
57 Honeycomb skin panels
58 Wing glove stiffeners
59 Starboard wing pivot bearing
60 Flap and slat drive shaft and gearbox
61 Starboard leading-edge slat
62 Wing panel fully forward position
63 Navigation light
64 Wingtip formation light
65 Roll control spoilers
66 Outboard manoeuvre flaps
67 Inboard high-lift flap
68 Flap sealing vane
69 Mainwheel leg hinge fitting
70 Variable wing-sweep screw jack
71 Wing glove sealing plates
72 Wing glove pneumatic seal
73 Starboard wing fully swept position
74 Starboard all-moving tailplane
75 Fin-tip aerial fairing
76 Tail navigation light
77 Starboard rudder
78 Rudder hydraulic actuator
79 Variable-area afterburner nozzle control jack
80 Dorsal airbrake (split ventral surfaces)
81 Chaff/flare dispensers
82 Fuel jettison
83 ECM antenna
84 Aluminium honeycomb fin skin panels
85 Anti-collision light
86 Formation lighting strip
87 ECM aerial
88 Port rudder
89 Variable-area afterburner nozzle
90 Port all-moving tailplane
91 Tailplane boron-fibre skin panels
92 Tailplane pivot bearing
93 Afterburner ducting
94 Tailplane hydraulic actuator
95 Ventral fin
96 Formation lighting strip
97 Hydraulic equipment bay
98 Hydraulic reservoir
99 General Electric F110-GE-400 afterburning turbofan engine
100 Rear fuselage fuel tank bays
101 Flight control system linkages

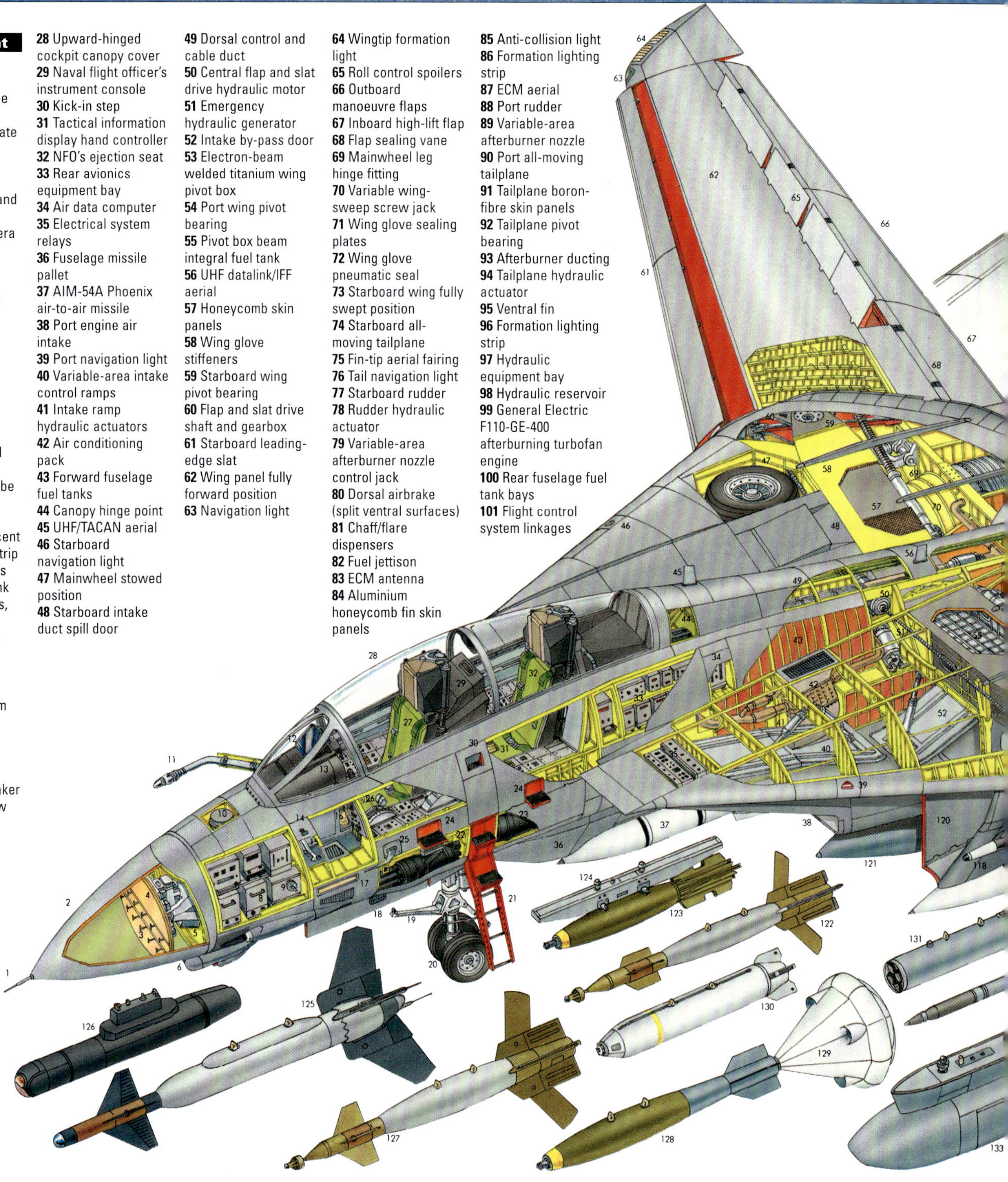

Left: Exploitation of the Tomcat's air-to-ground capability has turned it into a true multi-role combat aircraft, able to provide a formidable self-escort capability. An F-14A of VF-51 'Screaming Eagles' banks to show a heavy ordnance load of four 2,000-lb (907-kg) 'slick' Mk 84 free-fall bombs.

102 Engine bleed air ducting
103 Port wing-sweep crew jack
104 Inboard high-lift flap hydraulic jack
105 Flap hinge links
106 Flap honeycomb construction
107 Port wing fully swept position
108 Port manoeuvre flaps
109 Wingtip formation light
110 Navigation light
111 Port leading-edge slat
112 Slat guide rails
113 Wing integral fuel tank
114 Machined wing rib construction
115 Main undercarriage leg strut
116 Port mainwheel, forward-retracting
117 Wing glove mounted AIM-54A Phoenix air-to-air missile
118 AIM-9L Sidewinder air-to-air missile
119 Wing glove pylon
120 Mainwheel door
121 External fuel tank
122 GBU-12D/B Paveway II, 500-lb (227-kg) laser-guided bomb
123 Mk 82 Snakeye, 500-lb (227-kg) retarded bomb
124 Phoenix pallet weapons adapter
125 GBU-24A/B Paveway III, 2,000-lb (907-kg) laser-guided bomb
126 AN/AAQ-14 LANTIRN navigation and targeting pod, carried on starboard glove pylon
127 GBU-16 Paveway II, 1,000-lb (454-kg) laser-guided bomb
128 Mk 83 AIR, 1,000-lb (454-kg) retarded bomb
129 Mk 83 AIR inflated ballute
130 Mk 7 submunition dispenser
131 LAU-97, 4-round rocket launcher
132 5-in (127-mm) Zuni FFAR (Folding-Fin Air Rocket)
133 TARPS reconnaissance pod, carried in centreline tunnel
134 ALQ-167 counter-measures pod, carried on forward fuselage Phoenix pallet station

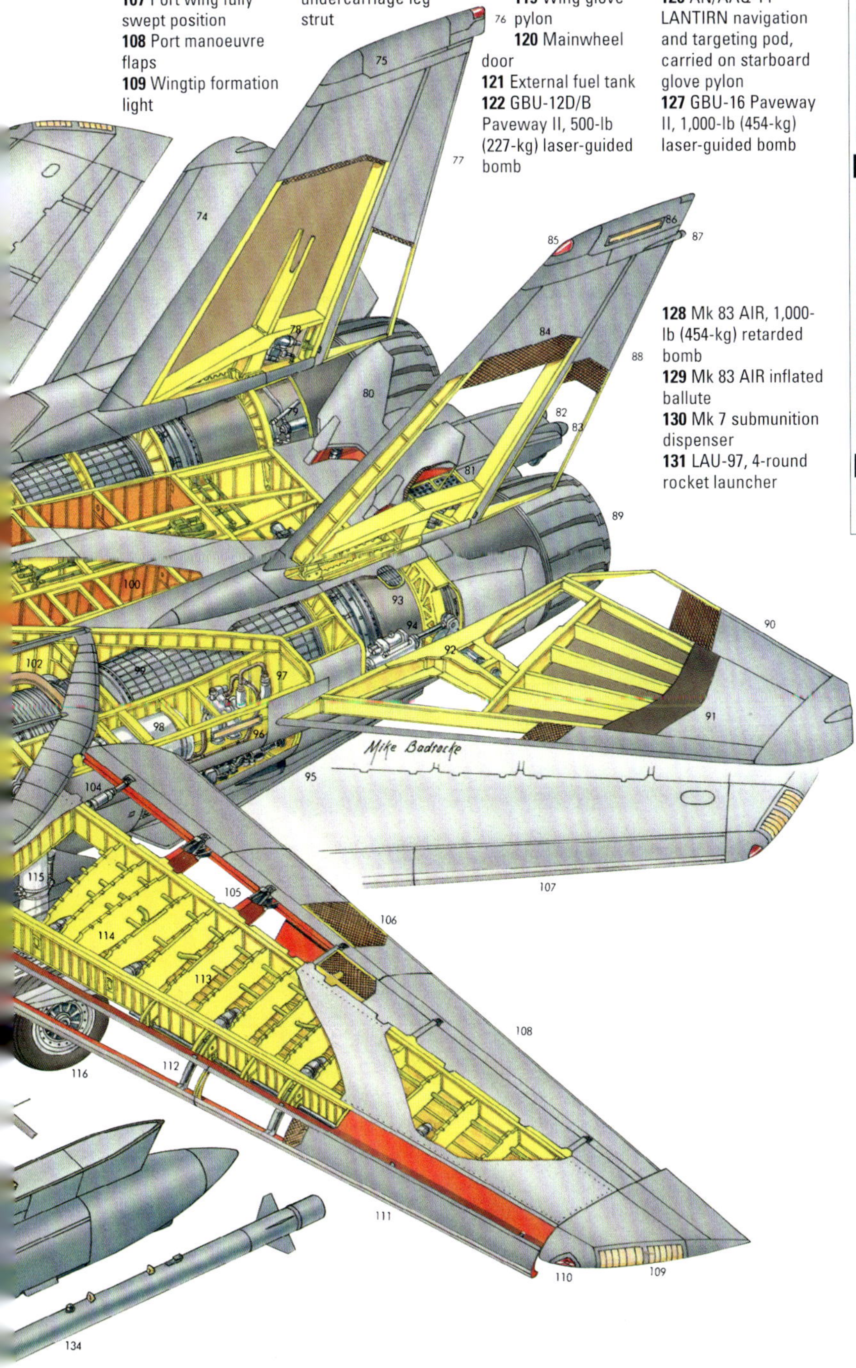

SPECIFICATION

F-14A Tomcat (unless otherwise noted)

Dimensions

Fuselage length (including probe): 62 ft 8 in (19.10 m)
Wing span: (unswept) 64 ft 1½ in (19.54 m); (swept) 38 ft 2½ in (11.65 m); (overswept) 33 ft 3½ in (10.15 m)
Wing aspect ratio: 7.28
Tailplane span: 32 ft 8½ in (9.97 m)
Wing area: 565 sq ft (52.49 m^2)
Total slat area: 46.2 sq ft (4.29 m^2)
Total flap area: 106.3 sq ft (9.87 m^2)
Total spoiler area: 21.2 sq ft (1.97 m^2)
Horizontal tail area: 140 sq ft (13.01 m^2)
Total fin area: 85 sq ft (7.90 m^2)
Total rudder area: 33 sq ft (3.06 m^2)
Tailplane area: 140 sq ft (13.01 m^2)
Distance between fin-tips: 10 ft 8 in (3.25 m)
Overall height: 16 ft (4.88 m)
Wheel track: 16 ft 5 in (5.00 m)
Wheelbase: 23 ft 0½ in (7.02 m)
Wing loading: 90 lb/sq ft (439 kg/m^2)
Wing/fuselage loading: 55 lb/sq ft (269 kg/m^2)

Powerplant

F-14A: two Pratt & Whitney TF30-P-412A/414A turbofans each rated at 20,900 lb st (92.97 kN) with afterburning
F-14B & F-14D: two General Electric F110-GE-400 turbofans each rated at 14,000 lb st (62.27 kN) dry and 27,600 lb st (122.8 kN) with afterburning

Weights

Empty operating: (F-14A) 40,104 lb (18191 kg); (F-14B) 41,780 lb (18951 kg); (F-14D) 43,735 lb (19838 kg)
Normal take-off, four Sparrow: 59,714 lb (27086 kg)
Normal take-off, six Phoenix: 70,764 lb (32098 kg)
Maximum take-off (all versions, service limit): 72,000 lb (32659 kg)
Maximum take-off (manufacturer's limit): 74,349 lb (33725 kg)
Design landing weight: 51,830 lb (23510 kg)

Fuel load

Total internal fuel: 2385 US gal (9030 litres) [approx 16,200 lb/ 7348 kg] in six main tanks, comprising: forward fuselage tank 691 US gal (2616 litres); rear fuselage tank 648 US gal (2453 litres); combined left and right feed tanks 456 US gal (1727 litres); wing tanks 295 US gal (1117 litres) each
External fuel: two 267-US gal (1011-litre) under-intake fuel tanks; single-point pressure refuelling via port on starboard lower fuselage, below inflight refuelling probe
Normal weapon load: (air-to-ground and tactical reconnaissance) up to 14,500 lb (6577 kg) of stores

Performance

Maximum level speed at altitude: 1,342 kt (1,544 mph; 2485 km/h)
Maximum level speed at low level: 792 kt (912 mph; 1468 km/h)
Limiting Mach numbers: 2.38 at altitude; 2.4 attained, but initially limited to 2.25 in service; 1.2 at low level
Maximum cruising speed: 550 kt (633 mph; 1019 km/h)
Maximum rate of climb at sea level: 30,000 ft (9140 m) per minute
Absolute ceiling: 56,000 ft (17069 m)
Service ceiling: 50,000 ft (15240 m) (F-14A); 53,000 ft (16154 m) (F-14B/D)
Normal carrier approach speed: 134 kt (154 mph; 248 km/h)
Stalling speed, landing configuration: 115 kt (132 mph; 213 km/h)
Take-off run: (with full fuel, and four AIM-7s) 1,400 ft (427 m)
Landing run: 2,900 ft (884 m)

Range

Combat air patrol endurance: (with four AIM-54s, two AIM-7, two AIM-9s and external fuel) 90 min at 150 nm (173 miles; 278 km); one hour at 253 nm (292 miles; 470 km)
Radius: (deck-launched intercept with four AIM-54s, two AIM-7s, two AIM-9s and external fuel) 171 nm (197 miles; 317 km) at Mach 1.3; 134 nm (154 miles; 248 km) at Mach 1.5
Ferry range: (F-14A with two tanks) 1,730 nm (2,000 miles; 3200 km); (F-14B with two tanks) 2,050 nm (2,369 miles; 3799 km)

Grumman F-14A+/B/D Tomcat

Applying the superb F110 engine to the Tomcat made it the aircraft it had always promised to be. As a dogfighter, the F-14 now possessed carefree engine handling, while as a long-range air defence machine it could fly further or stay on station longer. In addition, it could carry a hefty weight of air-to-ground ordnance, while burning less fuel for greater performance and, on top of all this, it offered safer carrier operations.

F-14D Tomcat

Cutaway key
1 Pitot head
2 Radar target horn
3 Upward-hinging glass-fibre radome
4 Radome hinge point
5 AN/APG-71 radar scanner
6 Articulated scanner mounting
7 Undernose IRST/TCS sensor pod
8 Infra-Red Search and Track (IRST)
9 Television camera set (TCS)
10 Anti-collision light
11 Cannon barrel aperture
12 Incidence transmitter
13 Weapons system avionics equipment
14 ADF antenna
15 Retractable inflight-refuelling probe
16 Windscreen panels
17 Pilot's head-up display
18 Instrument panel shroud
19 Temperature probe
20 Rudder pedals
21 Avionics cooling air exhaust
22 Electro-luminescent formation lighting strips
23 Gun gas purging intake
24 Nosewheel doors
25 Canopy emergency release
26 Dynamic pressure probe
27 Engine throttle levers
28 Control column
29 Pilot's instrument panel with dual multi-function displays
30 Cockpit canopy, open position
31 Rear-view mirrors
32 Ejection seat headrest with canopy breakers
33 Martin-Baker Mk 14 NACES ejection seat
34 Boarding step
35 M61 Vulcan six-barrelled rotary cannon
36 Catapult strop link
37 Fold-out boarding ladder
38 Steerable twin nosewheels, forward retracting
39 Nosewheel undercarriage leg strut
40 Hydraulic retraction jack
41 Ammunition magazine, 675 rounds
42 Ammunition feed and cartridge case return chutes
43 Tactical information display hand controller
44 Radar Intercept Officer's display console
45 RIO's ejection seat
46 Canopy hydraulic jack
47 Electrical system controller
48 Electrical relays
49 Engine intake lip
50 Ventral missile pallet
51 EW antenna
52 Port engine air intake
53 Intake sidewall honeycomb core structure
54 Port navigation light
55 Conditioned air ducting
56 Rear avionics equipment bay
57 Canopy hinge point
58 UHF/TACAN antenna
59 Starboard wing glove fairing
60 Starboard navigation light
61 Dorsal control and cable ducting
62 Forward fuselage fuel tank bays, total internal fuel capacity 1,986 Imp gal (9029 litres)
63 Air-conditioning system heat exchanger, port and starboard dual system for crew and avionics
64 Variable-area intake ramp doors
65 Intake ramp hydraulic actuators
66 Main undercarriage wheel bay
67 Mainwheel door
68 Rear intake ramp
69 Wing glove sealing horn fairing
70 Telescopic flap/slat drive shaft
71 Port wing pivot bearing
72 Electron beam welded titanium wing pivot box
73 Intake bypass air spill duct
74 Emergency hydraulic generator
75 Central flap/slat drive motor
76 UHF datalink/IFF antenna
77 Fuselage upper longeron/pivot box attachment links
78 Wing pivot box integral fuel tank
79 Telescopic fuel feed pipes
80 Variable wing sweep control screw jacks
81 Centre-section fuel tankage
82 Intake ducting
83 Honeycomb skin panels
84 Starboard mainwheel, stowed position
85 Starboard wing

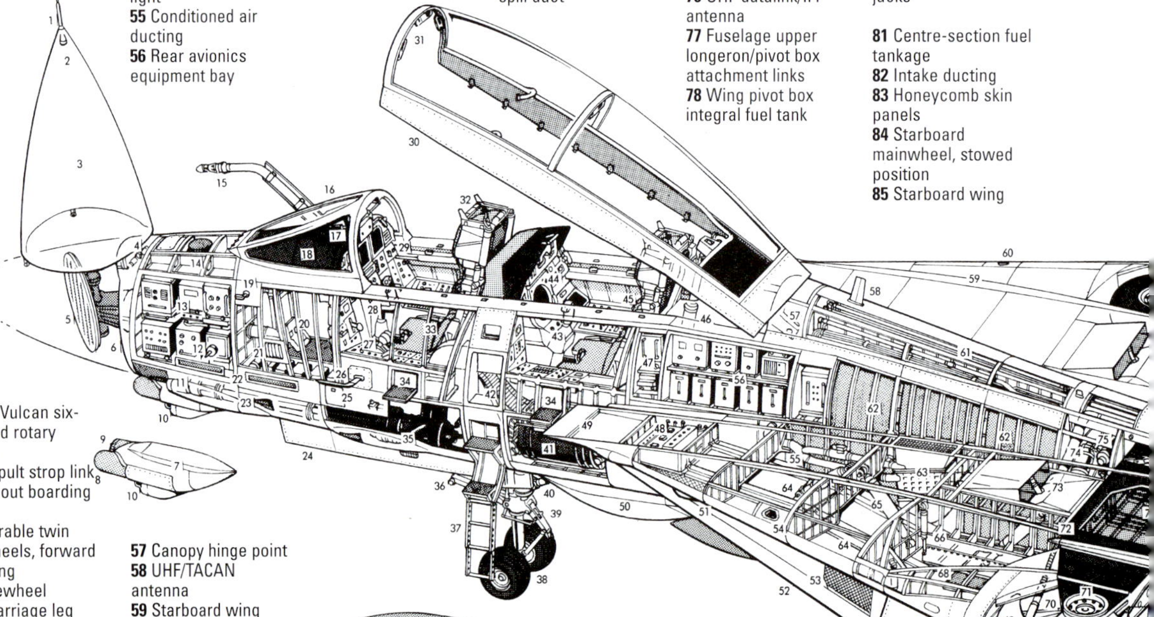

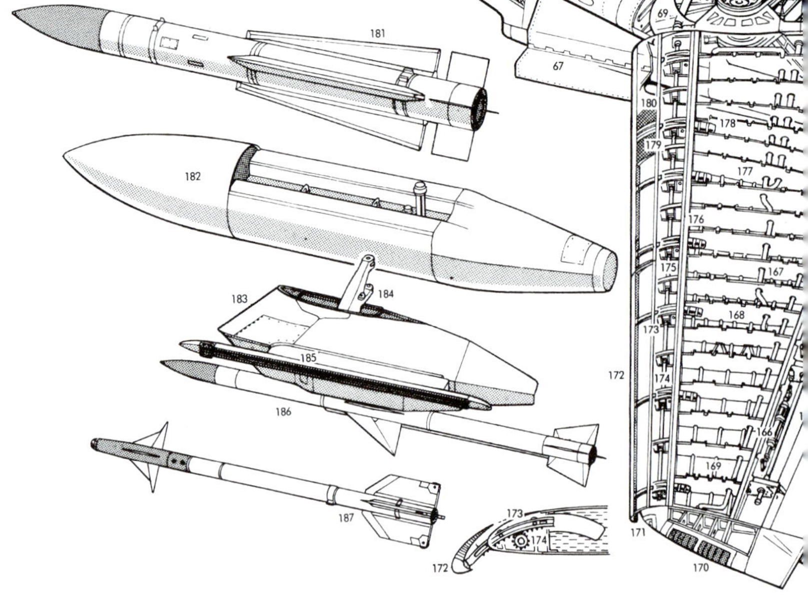

VF-74 was just completing its transition to the F-14A+ when it left for the Red Sea on 7 August 1990 aboard USS* Saratoga*. The squadron returned on 28 March 1991, making a further cruise on* Saratoga *in 1992. A subsequent stint aboard* Constellation *was followed by a rearrangement of units, which left VF-74 without a front-line role. It then flew in the adversary role from NAS Oceana, using specially painted aircraft such as this F-14B, until disbanded on 28 April 1994.

SPECIFICATION

F-14D Tomcat

Dimensions

As F-14A, see File 133 Sheet 02

Powerplant

Two General Electric F110-GE-400 turbofans each rated at 14,000 lb st (62.27 kN) dry and 23,100 lb st (102.75 kN) with afterburning

Weights

Empty: 41,780 lb (18951 kg)
Normal take-off, fighter/escort mission: 64,093 lb (29072 kg)
Normal take-off, fleet air defence mission: 73,098 lb (33157 kg)
Maximum take-off: 74,349 lb (33724 kg)

Fuel

Internal fuel: 16,200 lb (7348 kg)
External fuel: up to 3,800 lb (1724 kg) in two 267-US gal (1011-litre) drop tanks

Performance

Maximum level speed 'clean' at high altitude: 1,078 kt (1,241 mph; 1997 km/h)
Cruising speed at optimum altitude: 413 kt (475 mph; 764 km/h)
Maximum climb rate at sea level: more than 30,000 ft (9145 m) per minute
Service ceiling: more than 53,000 ft (16150 m)
Take-off run, on land: 2,500 ft (762 m) at maximum take-off weight
Landing run, on land: 2,400 ft (732 m)

Range

Maximum range with internal and external fuel: about 1,600 nm (1,842 miles; 2965 km)

pivot bearing
86 Flap/slat interconnecting drive shaft
87 Starboard wing integral fuel tank
88 Starboard two-segment leading-edge slats
89 Wing forward (20° sweep) position
90 Starboard navigation/ strobe light
91 Wingtip formation lighting panels
92 Two-segment slotted flaps, down position
93 Starboard spoiler panels
94 Inboard auxiliary flap
95 Wing glove flexible sealing plates
96 External glove stiffeners/dorsal fences
97 Forward/rear fuselage longeron joint
98 Flight control system artificial feel units
99 Control rods and linkages
100 Rear fuselage fuel tank bays
101 Starboard engine bay
102 Finroot fairing
103 Pneumatic wingroot glove seat
104 Starboard wing fully swept (68°) position, may be 'overswept' to 72° for carrier deck stowage
105 Starboard fin
106 Fin honeycomb core skin panels
107 Fintip antenna fairing
108 Tail navigation light
109 Starboard rudder, honeycomb core structure
110 Port fintip antenna fairing
111 Anti-collision light
112 Formation lighting strip
113 ECM antenna
114 Starboard all-moving tailplane
115 Variable-area afterburner nozzle
116 Carbon-fibre composite nozzle shroud
117 Flexible sealing plates
118 Flight control system back-up hydraulic module
119 Dorsal airbrake panel, split lower surfaces
120 Airbrake hydraulic jacks
121 Airbrake housing
122 Ventral AN/ALE-29 chaff/flare launchers
123 Fuel jettison
124 ECM antenna
125 Deck arrester hook, stowed
126 Port engine exhaust nozzle
127 Afterburner duct outer sealing plate
128 Variable-area nozzle actuator
129 Afterburner duct
130 Rudder hydraulic actuator
131 Fin/tailplane main mounting frame
132 Tailplane pivot bearing
133 Multi-spar tailplane structure
134 Arrester hook, down position
135 Honeycomb trailing-edge panel
136 AN/ALR-45 (V) radar warning antenna
137 Boron fibre tailplane skin panels
138 Wing rib (typical), machined on inner face
139 Port wing, fully-swept position
140 Ventral fin
141 Afterburner duct cooling air intake
142 Tailplane hydraulic actuator
143 Rear fuselage sponson fairing structure
144 Port General Electric F110-GE-400 afterburning turbofan
145 Hydraulic system filters
146 Formation lighting strip
147 Hydraulic reservoir, port and starboard
148 Engine bay access panel
149 Engine accessory equipment bay
150 Port auxiliary flap
151 Main undercarriage hydraulic retraction jack
152 Auxiliary flap hydraulic jack
153 Main undercarriage leg pivot mounting
154 Retraction breaker/ drag strut
155 Shock absorber leg strut
156 Torque scissor links
157 Port mainwheel
158 Trailing-edge flap section, cruise condition
159 Flap eyebrow fairing
160 Flap 10° down, manoeuvre position
161 Flap slotted, 35° down landing position
162 Port outboard flap segments
163 Flap honeycomb core structure
164 Port spoiler panels
165 Flap drive torque shaft
166 Spoiler hydraulic actuators
167 Fuel system piping
168 Machined wing ribs
169 Bottom wing skin/ stringer panel
170 Wingtip formation light
171 Port navigation/ strobe light
172 Two-segment leading-edge slat, extended
173 Slat guide rails
174 Slat drive torque shaft
175 Leading-edge ribs
176 Two-spar wing torsion box structure
177 Port wing integral fuel tank
178 Slat guide rail fuel sealing cans
179 Leading-edge slat honeycomb core construction
180 Tank pylon beneath intake trunk
181 AIM-54A Phoenix long-range air-to-air missile
182 267-US gal (1011-litre) external fuel tank
183 Glove pylon
184 Pylon attachment link
185 Shoulder-mounted Sidewinder launch rail
186 AIM-120 AMRAAM medium-range air-to-air missile
187 AIM-9L Sidewinder short range air-to-air missile

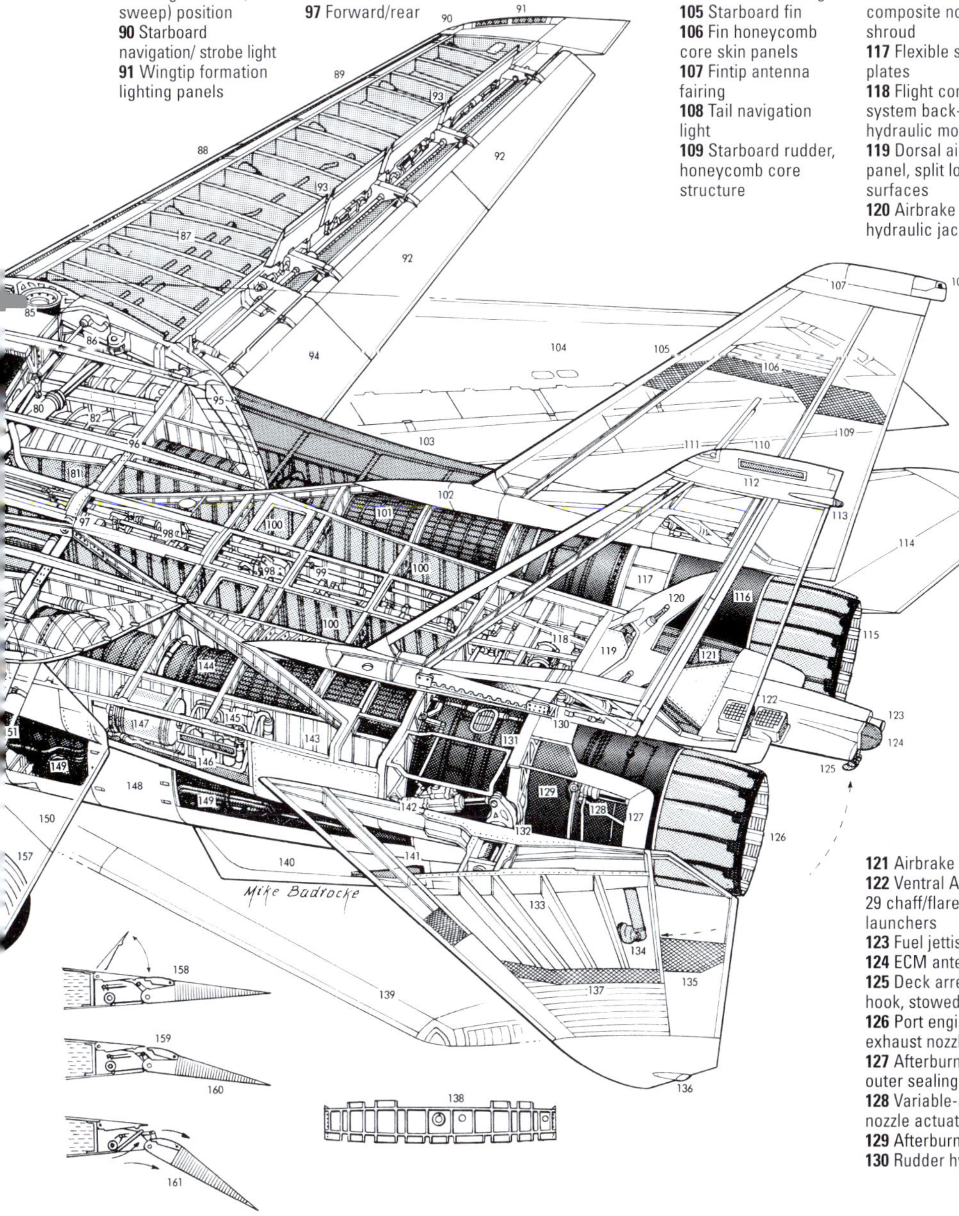

Grumman F9F Panther/Cougar

The last examples of the final F9F fighter variant – the F9F-8 – were delivered to the USN in March 1957. BuNo. 141140 is a late example with many of the improvements introduced during the course of production of this variant, including an in-flight refuelling probe, an undernose UHF homing antenna and the capability to launch AAM-N-7 (later AIM-9) Sidewinder AAMs.

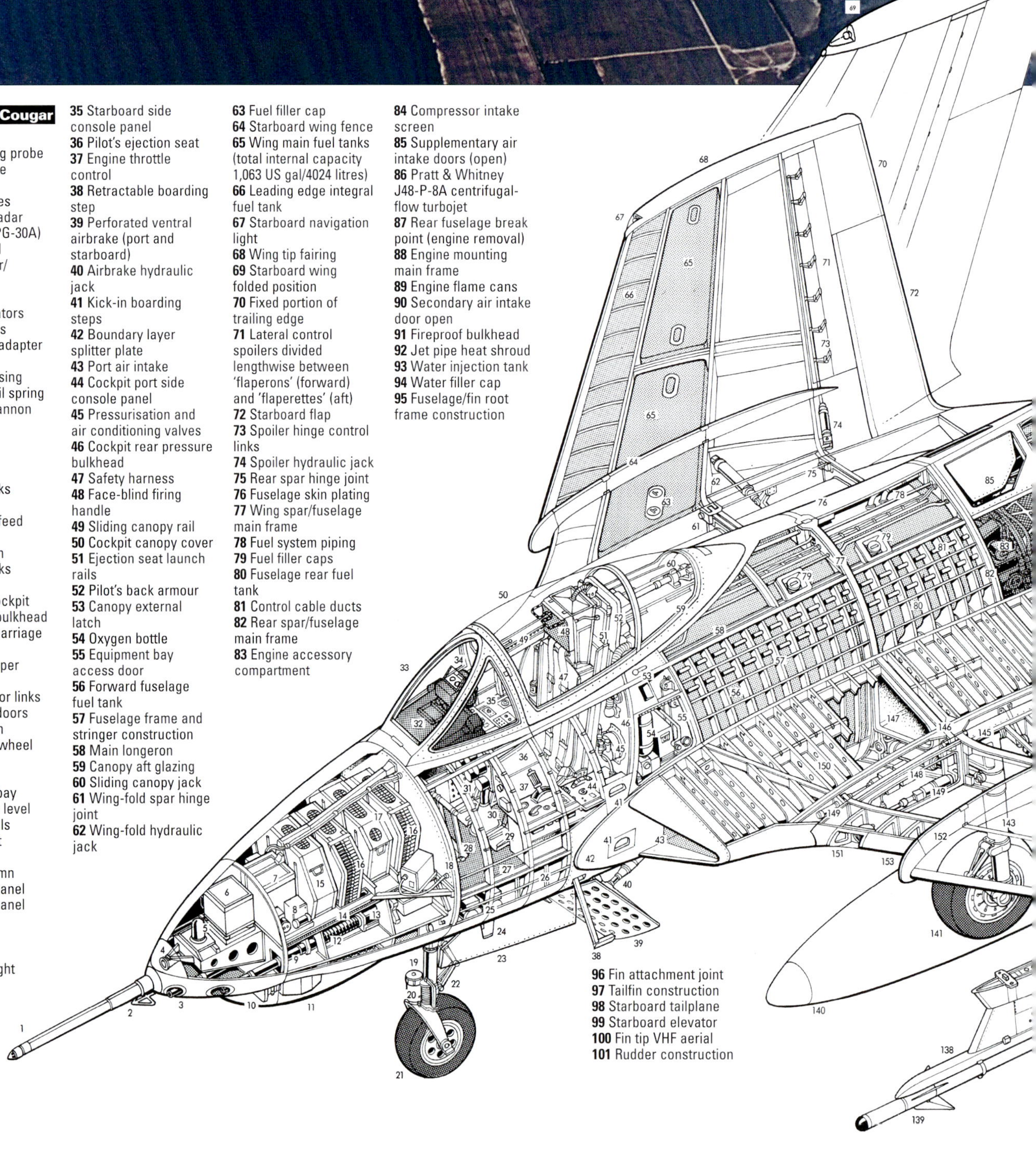

F9F-8 (F-9J) Cougar

Cutaway key
1 Flight refuelling probe
2 Deck barricade deflector
3 Cannon muzzles
4 Gun ranging radar antenna (AN/APG-30A)
5 D/F loop aerial
6 D/F transmitter/ receiver
7 Battery
8 Voltage regulators
9 Cannon barrels
10 UHF homing adapter antenna
11 Antenna housing
12 Cannon recoil spring
13 M3 20-mm cannon (four)
14 Nose cone withdrawal rail
15 Inboard gun ammunition tanks (190 rpg)
16 Ammunition feed chutes
17 Outboard gun ammunition tanks (190 rpg)
18 Armoured cockpit front pressure bulkhead
19 Nose undercarriage leg strut
20 Shimmy damper
21 Nosewheel
22 Torque scissor links
23 Nosewheel doors
24 VHF aerial on starboard nosewheel door
25 Alternators
26 Nosewheel bay
27 Cockpit floor level
28 Rudder pedals
29 Ejection seat footrests
30 Control column
31 Instrument panel
32 Instrument panel shroud
33 Bullet proof windscreen
34 Radar gunsight (Aero 5D-1)
35 Starboard side console panel
36 Pilot's ejection seat
37 Engine throttle control
38 Retractable boarding step
39 Perforated ventral airbrake (port and starboard)
40 Airbrake hydraulic jack
41 Kick-in boarding steps
42 Boundary layer splitter plate
43 Port air intake
44 Cockpit port side console panel
45 Pressurisation and air conditioning valves
46 Cockpit rear pressure bulkhead
47 Safety harness
48 Face-blind firing handle
49 Sliding canopy rail
50 Cockpit canopy cover
51 Ejection seat launch rails
52 Pilot's back armour
53 Canopy external latch
54 Oxygen bottle
55 Equipment bay access door
56 Forward fuselage fuel tank
57 Fuselage frame and stringer construction
58 Main longeron
59 Canopy aft glazing
60 Sliding canopy jack
61 Wing-fold spar hinge joint
62 Wing-fold hydraulic jack
63 Fuel filler cap
64 Starboard wing fence
65 Wing main fuel tanks (total internal capacity 1,063 US gal/4024 litres)
66 Leading edge integral fuel tank
67 Starboard navigation light
68 Wing tip fairing
69 Starboard wing folded position
70 Fixed portion of trailing edge
71 Lateral control spoilers divided lengthwise between 'flaperons' (forward) and 'flaperettes' (aft)
72 Starboard flap
73 Spoiler hinge control links
74 Spoiler hydraulic jack
75 Rear spar hinge joint
76 Fuselage skin plating
77 Wing spar/fuselage main frame
78 Fuel system piping
79 Fuel filler caps
80 Fuselage rear fuel tank
81 Control cable ducts
82 Rear spar/fuselage main frame
83 Engine accessory compartment
84 Compressor intake screen
85 Supplementary air intake doors (open)
86 Pratt & Whitney J48-P-8A centrifugal-flow turbojet
87 Rear fuselage break point (engine removal)
88 Engine mounting main frame
89 Engine flame cans
90 Secondary air intake door open
91 Fireproof bulkhead
92 Jet pipe heat shroud
93 Water injection tank
94 Water filler cap
95 Fuselage/fin root frame construction
96 Fin attachment joint
97 Tailfin construction
98 Starboard tailplane
99 Starboard elevator
100 Fin tip VHF aerial
101 Rudder construction

Blue Angels

The first F9F-2s to enter US Navy service were those assigned to the *Blue Angels* demonstration team on 20 August 1949 (top). The team's first jet aircraft, the jet-age Panthers replaced F8F Bearcats and were employed until July 1950, when the team was stood-down to form the nucleus of VF-191 in anticipation of the war in Korea. After a 14-month stand-down and a 1952 season spent with the F7U-1 Cutlass, the *Angels* were reunited with members of the F9F family the following year. It had been intended that F9F-6 Cougars be issued to the team at this point, but the grounding of the type in 1953 forced the team to use overhauled F9F-5 Panthers instead. These remained in use until 1955, when the Cougar was finally made available. The first F9F-8s (above) were debuted on the air show circuit in 1955 and stayed with the team until 1957, when the first F11F-1 Tigers were introduced. This was not the last word however; the team acquired an F9F-8T when it transitioned to the F11F (replacing a Lockheed TV-2) and this was retained until 1969, when the team received McDonnell Phantom IIs.

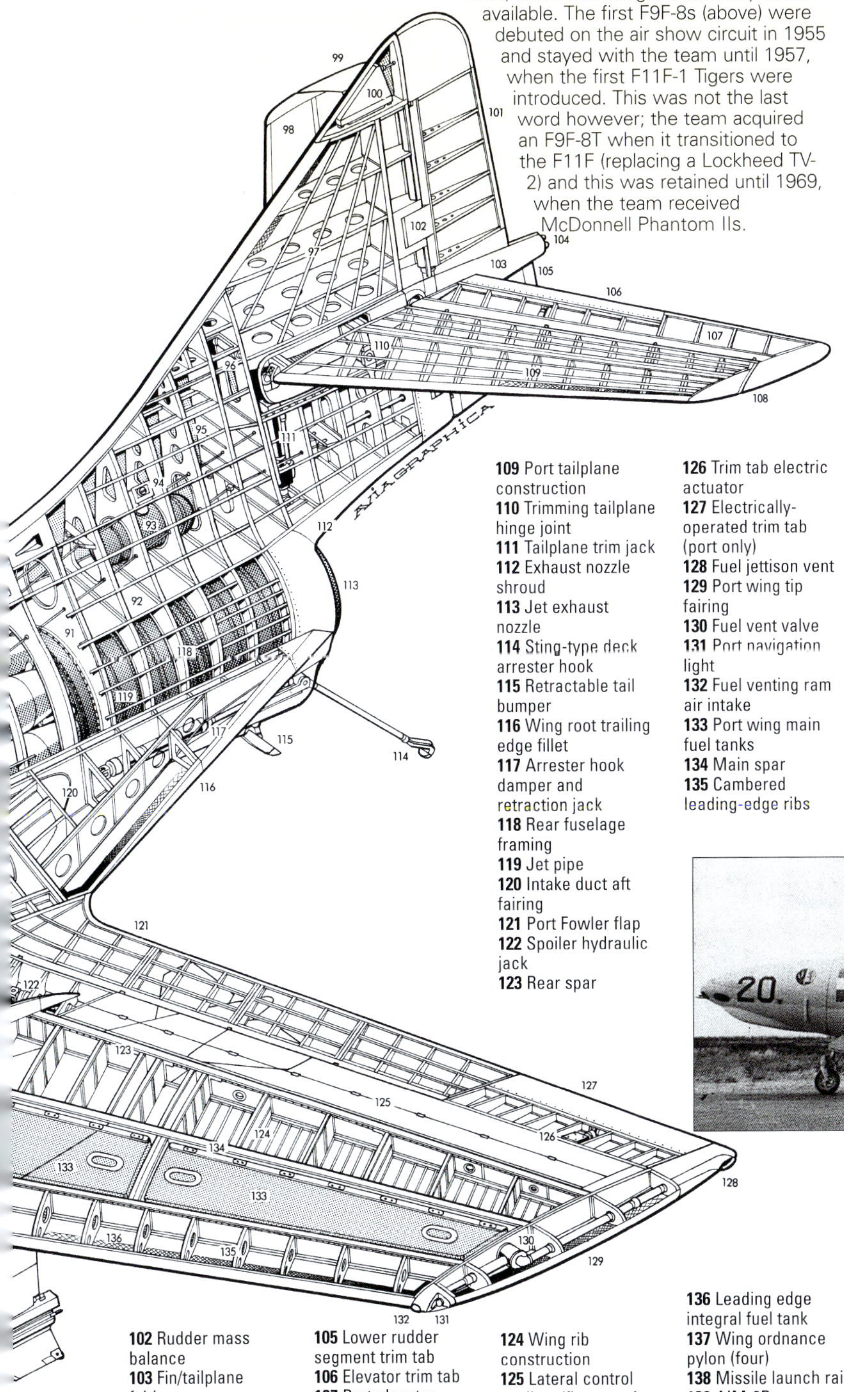

109 Port tailplane construction
110 Trimming tailplane hinge joint
111 Tailplane trim jack
112 Exhaust nozzle shroud
113 Jet exhaust nozzle
114 Sting-type deck arrester hook
115 Retractable tail bumper
116 Wing root trailing edge fillet
117 Arrester hook damper and retraction jack
118 Rear fuselage framing
119 Jet pipe
120 Intake duct aft fairing
121 Port Fowler flap
122 Spoiler hydraulic jack
123 Rear spar
126 Trim tab electric actuator
127 Electrically-operated trim tab (port only)
128 Fuel jettison vent
129 Port wing tip fairing
130 Fuel vent valve
131 Port navigation light
132 Fuel venting ram air intake
133 Port wing main fuel tanks
134 Main spar
135 Cambered leading-edge ribs

102 Rudder mass balance
103 Fin/tailplane fairing
104 Tail navigation lights
105 Lower rudder segment trim tab
106 Elevator trim tab
107 Port elevator
108 Elevator horn balance
124 Wing rib construction
125 Lateral control spoilers 'flaperons' (forward) and 'flaperettes' (aft)
136 Leading edge integral fuel tank
137 Wing ordnance pylon (four)
138 Missile launch rail
139 AIM-9B Sidewinder air-to-air missile
140 150 US gal (568 litre) auxiliary fuel tank
141 Port mainwheel
142 Fuel tank bay corrugated double skin
143 Main undercarriage leg strut
144 Wing fold hydraulic jack
145 Main undercarriage pivot housing
146 Main spar hinge joint
147 Intake duct
148 Undercarriage hydraulic retraction jack
149 Wing-fold locking cylinders
150 Intake duct framing
151 Landing/taxiing lamp
152 Port wing fence
153 Leading-edge dog-tooth

SPECIFICATION

F9F-2 Panther	F9F-8 Cougar
Dimensions	**Dimensions**
Length: 37 ft 5⅜ in (11.41 m) **Height:** 11 ft 4 in (3.45 m) **Span:** 38 ft (11.58 m) **Span (folded):** 23 ft 5 in (7.14 m) **Wing area:** 250 sq ft (23.23 m²)	**Length:** 42 ft 2 in (12.85 m) **Height:** 12 ft 3 in (3.73 m) **Span:** 34 ft 6 in (10.52 m) **Span (folded):** 14 ft 2 in (4.32 m) **Wing area:** 337 sq ft (31.31 m²)
Powerplant	**Powerplant**
One Pratt & Whitney J42-P-4, P-6 or P-8 turbojet rated at 5,000 lb (22.24 kN) thrust dry, or 5,750 lb (25.58 kN) thrust with water injection	One Pratt & Whitney J48-P-8A or P-8C turbojet rated at 6,250 lb (27.80 kN) thrust dry, or 7,250 lb (32.25 kN) thrust with water injection
Weights	**Weights**
Empty: 9,303 lb (4220 kg) **Loaded:** 16,450 lb (7462 kg) **Maximum take-off:** 19,494 lb (8842 kg)	**Empty:** 11,866 lb (5382 kg) **Loaded:** 20,098 lb (9116 kg) **Maximum take-off:** 24,763 lb (11232 kg)
Performance	**Performance**
Maximum speed: 575 mph (925 km/h) at sea level **Cruising speed:** 487 mph (784 km/h) **Climb rate:** 6,000 ft (1829 m) per minute **Service ceiling:** 44,600 ft (13594 m) **Normal range:** 1,353 miles (2177 km)	**Maximum speed:** 647 mph (1041 km/h) at 2,000 ft (607 m) **Cruising speed:** 516 mph (830 km/h) **Climb rate:** 5,750 ft (1753 m) per minute **Service ceiling:** 42,000 ft (12802 m) **Normal range:** 1,208 miles (1944 km) **Maximum range:** 1,312 miles (2111 km)
Armament	**Armament**
Four 20-mm cannon, each with 190 rounds per gun. Most F9F-2s were later modified with four racks under each wing; the inner pair was stressed to carry a 1,000-lb (454-kg) bomb or 150-US gal (568-litre) drop tank, while the outer racks could each carry a 250-lb (114-kg) bomb or a 5-in (127-mm) HVAR. Total external load was 3,000 lb (1361-kg)	Four 20-mm cannon, plus two bomb racks each able to carry one 1,000-lb (454-kg) bomb or 150-US gal (568-litre) drop tank. Late production aircraft were equipped with four additional pylons, each wired to carry an AAM-N-7 Sidewinder air-to-air missile

The only other nation to operate F9Fs took delivery of exampes of both the Panther and Cougar. Argentina's Servicio de Aviación Naval received 24 refurbished F9F-2s (below) in 1958. Issued to 1ª Escadrilla Aeronavale de Ataque, 2ª Escuadra Aeronavale at BAN Comandante Espora the aircraft were confined to operations from land bases as the catapults aboard the carrier ARA Independencia were not sufficiently powerful to launch a jet aircraft of this type at normal operating weights. In April 1963 four of the Panthers were destroyed on the ground and a fifth collided with a Corsair during a three-day coup in which navy personnel fought with those of the army and air force. Two years later, by which time the 1ª Escadrilla Aeronavale de Ataque, had been transferred to the 3ª Escuadra Aeronavale, the Panthers were again in action, flying border patrols during a dispute with neighbouring Chile. Surviving aircraft were grounded in 1969 by a lack of spare parts. Meanwhile, in 1962 a pair of F9F-8Ts had been acquired for use by the 1ª Escadrilla Aeronavale de Ataque. Both had been withdrawn by 1971.

Grumman S2F/S-2 Tracker

When it first appeared, Grumman's stubby Tracker (commonly known as the 'Stoof' due to its S2F designation) was the first aircraft to combine submarine hunter/killer operations in one carrierborne airframe. A stalwart of US Navy operations, it was widely exported and spawned several vastly differing variants

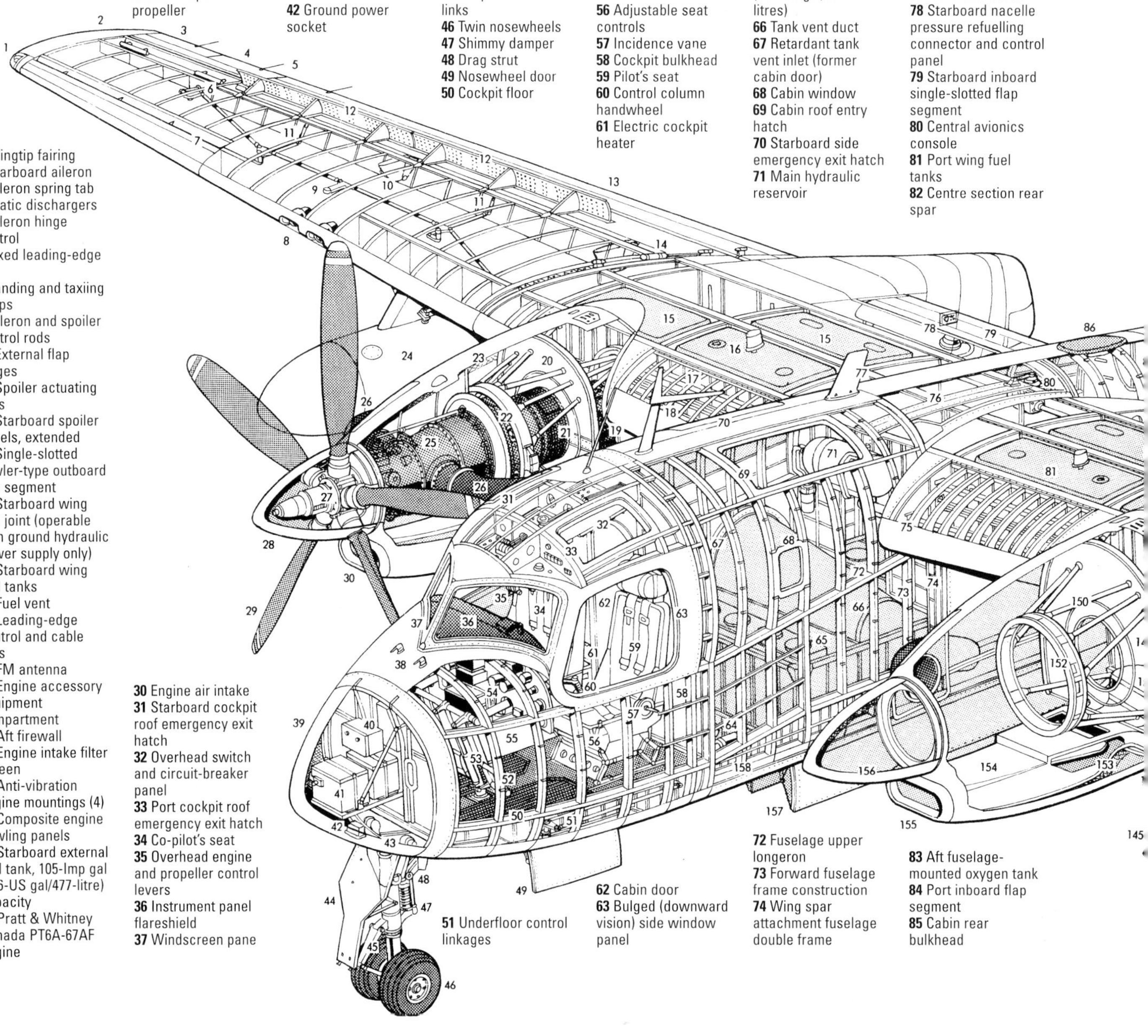

Conair Turbo Firecat

Cutaway key

1 Starboard navigation light
2 Wingtip fairing
3 Starboard aileron
4 Aileron spring tab
5 Static dischargers
6 Aileron hinge control
7 Fixed leading-edge slot
8 Landing and taxiing lamps
9 Aileron and spoiler control rods
10 External flap hinges
11 Spoiler actuating links
12 Starboard spoiler panels, extended
13 Single-slotted Fowler-type outboard flap segment
14 Starboard wing fold joint (operable with ground hydraulic power supply only)
15 Starboard wing fuel tanks
16 Fuel vent
17 Leading-edge control and cable runs
18 FM antenna
19 Engine accessory equipment compartment
20 Aft firewall
21 Engine intake filter screen
22 Anti-vibration engine mountings (4)
23 Composite engine cowling panels
24 Starboard external fuel tank, 105-Imp gal (126-US gal/477-litre) capacity
25 Pratt & Whitney Canada PT6A-67AF engine
26 Exhaust stubs
27 Propeller hub pitch change mechanism
28 Spinner
29 Hartzell five-bladed constant-speed propeller
30 Engine air intake
31 Starboard cockpit roof emergency exit hatch
32 Overhead switch and circuit-breaker panel
33 Port cockpit roof emergency exit hatch
34 Co-pilot's seat
35 Overhead engine and propeller control levers
36 Instrument panel flareshield
37 Windscreen pane
38 Dual pitot heads
39 Composite nosecone
40 Radio junction box
41 Twin NiCd battery installation
42 Ground power socket
43 Nose undercarriage trunion fittings
44 Nosewheel leg door
45 Torque scissor links
46 Twin nosewheels
47 Shimmy damper
48 Drag strut
49 Nosewheel door
50 Cockpit floor
51 Underfloor control linkages
52 Rudder pedals
53 Brake master cylinders
54 Instrument panel
55 Central avionics console
56 Adjustable seat controls
57 Incidence vane
58 Cockpit bulkhead
59 Pilot's seat
60 Control column handwheel
61 Electric cockpit heater
62 Cabin door
63 Bulged (downward vision) side window panel
64 Retardant tank dump door rotary actuators (4)
65 Retardant tank, capacity 735 Imp gal (882 US gal/3341 litres)
66 Tank vent duct
67 Retardant tank vent inlet (former cabin door)
68 Cabin window
69 Cabin roof entry hatch
70 Starboard side emergency exit hatch
71 Main hydraulic reservoir
72 Fuselage upper longeron
73 Forward fuselage frame construction
74 Wing spar attachment fuselage double frame
75 Hinged inboard leading-edge segment (control access)
76 Wing panel centreline joint
77 No. 1 VHF aerial
78 Starboard nacelle pressure refuelling connector and control panel
79 Starboard inboard single-slotted flap segment
80 Central avionics console
81 Port wing fuel tanks
82 Centre section rear spar
83 Aft fuselage-mounted oxygen tank
84 Port inboard flap segment
85 Cabin rear bulkhead

Utilising the fuselage of a TF-1 Trader, replacing the original vertical tail with twin fins and rudders and adding a 17-ft 6-in (5.33-m) diameter radome, Grumman created the E-1B Tracer. A total of 88 aircraft was built and the first operational example was delivered on 20 January 1969. Typical sorties for this AEW machine saw Tracers operating some 150 nm (172 miles; 278 km) from carriers on missions lasting up to seven hours. In the anti-submarine warfare support role, Tracers were used to control S-2s and to ensure that they were flying precise search patterns.

SPECIFICATION

S-2E Tracker

Dimensions

Length: 43 ft 6 in (13.26 m)
Height: 16 ft 7 in (5.06 m)
Wingspan: 72 ft 7 in (22.13 m)
Width, wings folded: 27 ft 4 in (8.33 m)
Wing area: 496.00 sq ft (46.08 m²)
Aspect ratio: 10.63
Wheel track 18 ft 6 in (5.64 m)

Powerplant

Two Wright R-1820-82 WA Cyclones each rated at 1,525 hp (1137 kW)

Weight

Empty: 18,750 lb (8505 kg)
Normal take-off: 24,413 lb (11074 kg)
Maximum take-off: 29,150 lb (13222 kg)

Fuel and load

Internal fuel: 4,368 lb (1981 kg)
Maximum ordnance: 4,810 lb (2182 kg)

Performance

Maximum level speed 'clean' at sea level: 230 kt (265 mph; 426 km/h)
Cruising speed at optimum altitude: 180 kt (207 mph; 333 km/h)
Patrol speed at 1,500 ft (457 m): 130 kt (150 mph; 241 km/h)
Ferry range: 1,130 nm (1,301 miles; 2094 km)
Range: 1,000 nm (1,152 miles; 1853 km)
Endurance: 9 hours
Maximum rate of climb at sea level: 1,390 ft (425 m) per minute
Service ceiling: 21,000 ft (6400 m)
Take-off run at maximum take-off weight: 1,300 ft (396 m)

Armament

Early Trackers could carry either one Mk 34 or one Mk 41 or two Mk 43 torpedoes or one Mk 24 mine in the ventral bomb bay and either four Mk 19 mines or four Mk 43 torpedoes or four Mk 54 depth charges or six HVAR rockets beneath the wings. The S-2F-2 version had an enlarged bomb bay which enabled carriage of the Mk 90 nuclear depth charge; this weapon was later replaced by the smaller Mk 101 'Lulu'. Later models could also carry rocket pods, Mk 44 or Mk 46 torpedoes or the Mk 57 nuclear weapon. The varying export operators have each added weapons packages of their own, such as gun pods or ASMs to meet individual needs

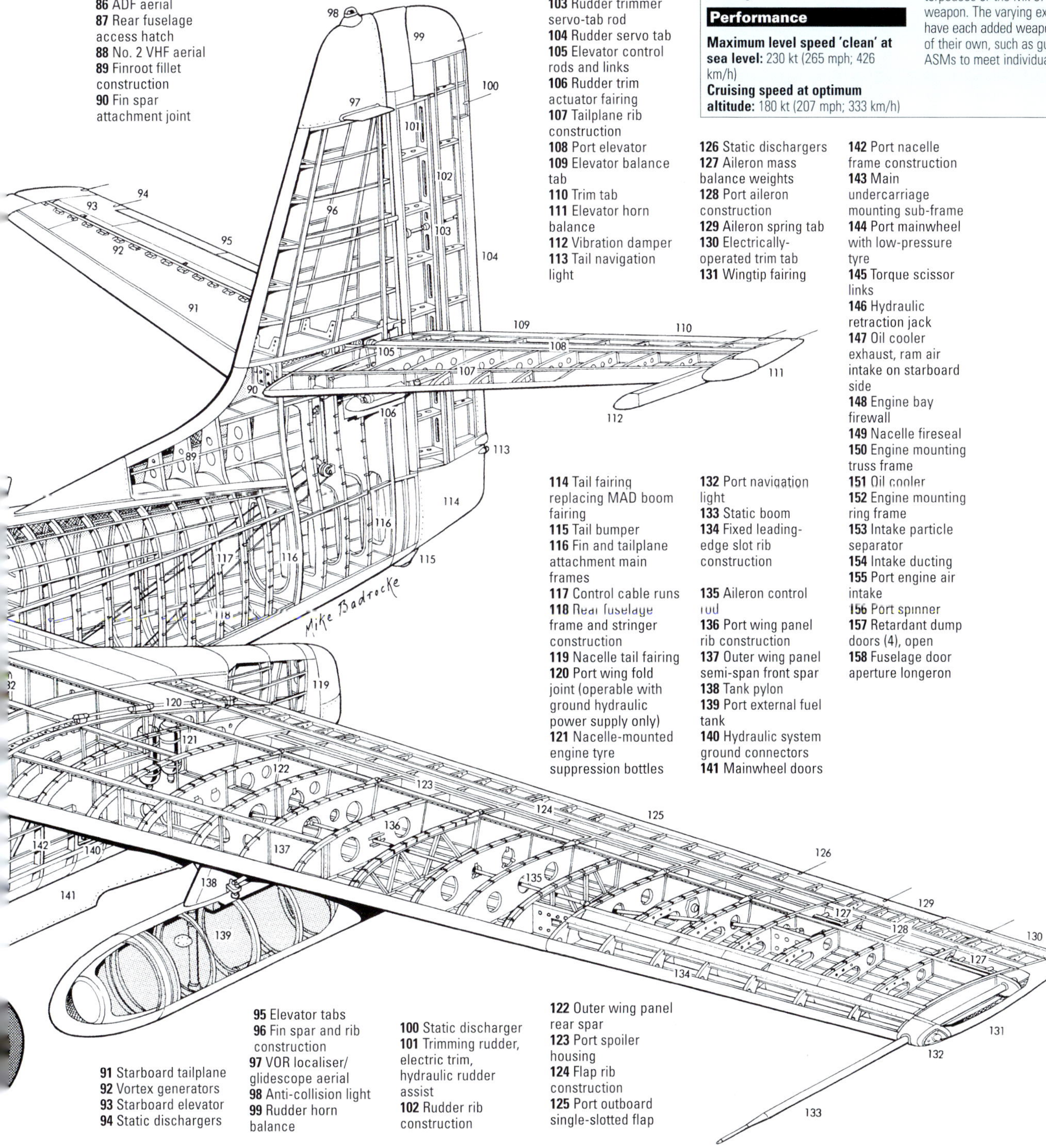

86 ADF aerial
87 Rear fuselage access hatch
88 No. 2 VHF aerial
89 Finroot fillet construction
90 Fin spar attachment joint
91 Starboard tailplane
92 Vortex generators
93 Starboard elevator
94 Static dischargers
95 Elevator tabs
96 Fin spar and rib construction
97 VOR localiser/ glidescope aerial
98 Anti-collision light
99 Rudder horn balance
100 Static discharger
101 Trimming rudder, electric trim, hydraulic rudder assist
102 Rudder rib construction
103 Rudder trimmer servo-tab rod
104 Rudder servo tab
105 Elevator control rods and links
106 Rudder trim actuator fairing
107 Tailplane rib construction
108 Port elevator
109 Elevator balance tab
110 Trim tab
111 Elevator horn balance
112 Vibration damper
113 Tail navigation light
114 Tail fairing replacing MAD boom fairing
115 Tail bumper
116 Fin and tailplane attachment main frames
117 Control cable runs
118 Rear fuselage frame and stringer construction
119 Nacelle tail fairing
120 Port wing fold joint (operable with ground hydraulic power supply only)
121 Nacelle-mounted engine tyre suppression bottles
122 Outer wing panel rear spar
123 Port spoiler housing
124 Flap rib construction
125 Port outboard single-slotted flap
126 Static dischargers
127 Aileron mass balance weights
128 Port aileron construction
129 Aileron spring tab
130 Electrically-operated trim tab
131 Wingtip fairing
132 Port navigation light
133 Static boom
134 Fixed leading-edge slot rib construction
135 Aileron control rod
136 Port wing panel rib construction
137 Outer wing panel semi-span front spar
138 Tank pylon
139 Port external fuel tank
140 Hydraulic system ground connectors
141 Mainwheel doors
142 Port nacelle frame construction
143 Main undercarriage mounting sub-frame
144 Port mainwheel with low-pressure tyre
145 Torque scissor links
146 Hydraulic retraction jack
147 Oil cooler exhaust, ram air intake on starboard side
148 Engine bay firewall
149 Nacelle fireseal
150 Engine mounting truss frame
151 Oil cooler
152 Engine mounting ring frame
153 Intake particle separator
154 Intake ducting
155 Port engine air intake
156 Port spinner
157 Retardant dump doors (4), open
158 Fuselage door aperture longeron

Hawker Sea Fury

Nos 803 and 883 Sqns, Fleet Air Arm were disbanded 1946 and 1948 respectively, upon becoming Roy Canadian Navy units. Equipped with Sea Furies th were later renumbered Nos 870 and 871 Sqns, RC

Sea Fury FB.Mk 11

Cutaway key
1 Spinner
2 Rotor five -bladed constant speed propeller of 12 ft 9 in (3.89 m) diameter
3 Propeller hub pitch change mechanism
4 Spinner backplate
5 Engine cowling ring
6 Cooling air intake
7 Propeller reduction gear casing
8 Detachable engine cowlings
9 Bristol Centaurus Mk 18 18-cylinder two-row radial sleeve valve engine
10 Exhaust stubs
11 Carburettor intake ducting
12 Starboard British Hispano Mk 5 20-mm cannon
13 Recoil springs
14 Cannon muzzles
15 60-lb (27-kg) ground attack rocket projectiles
16 Zero-length rocket launcher rails
17 Wing folding jack
18 Wing fold latching mechanism
19 Starboard outer wing panel
20 Starboard navigation light
21 Wingtip fairing
22 Starboard aileron
23 Aileron hinge control
24 Push-button control rod
25 Aileron spring tab
26 Retractable landing/taxiing lamp
27 Ammunition box (290 rounds port and starboard)
28 Starboard wing folded position
29 Outer split trailing edge flap
30 Ammunition feed drum blister fairings
31 Cannon breeches
32 Oil tank (14 Imp gal/ 63.65 litre capacity)
33 Engine cartridge starter
34 Engine bearer struts
35 Hydraulic reservoir
36 Accessory drive gearbox
37 Engine cooling air outlet
38 Wing front spar attachment joint
39 Fireproof engine compartment bulkhead
40 Fuselage double frame
41 Main fuel tank (97 Imp gal/441 litre)
42 Fuel tank vent
43 Filler cap
44 Fuselage top longeron
45 Rudder pedals
46 Auxiliary fuselage fuel tank (30 Imp gal/ 136 litre)
47 Fuselage bottom longeron
48 Rear wing spar attachment joint
49 Oxygen bottle
50 Control column
51 Instrument panel
52 Bullet proof windscreen
53 Mk 4B reflector sight
54 Windscreen framing
55 Pilot's starboard side console
56 Pilot's seat
57 Engine throttle and propeller controls
58 Radio equipment
59 Port side console
60 Seat back armour plate
61 Safety harness
62 Headrest
63 Armoured headrest support
64 Sliding cockpit canopy cover
65 Canopy rails
66 Tailplane control rod
67 Rear fuselage joint frame
68 Whip aerial
69 Fuselage skin plating
70 Elevator push-pull control rod
71 Tailplane attachment joint frame
72 Fin root fillet
73 Starboard tailplane
74 Starboard elevator
75 Tailfin construction
76 Curved fin leading edge
77 Stempost
78 Rudder construction
79 Mass balance weight
80 Rudder tab
81 Deck arrester hook
82 Elevator trim tab
83 Port elevator
84 Tailplane construction
85 Tailplane spar joints
86 Rudder hinge control
87 Tail navigation light
88 Arresting hook attachment link
89 Tailwheel hydraulic retraction jack
90 Tailwheel
91 Tailwheel doors
92 Rear fuselage double bulkhead
93 Tailwheel bay
94 Tailwheel bay bulkhead

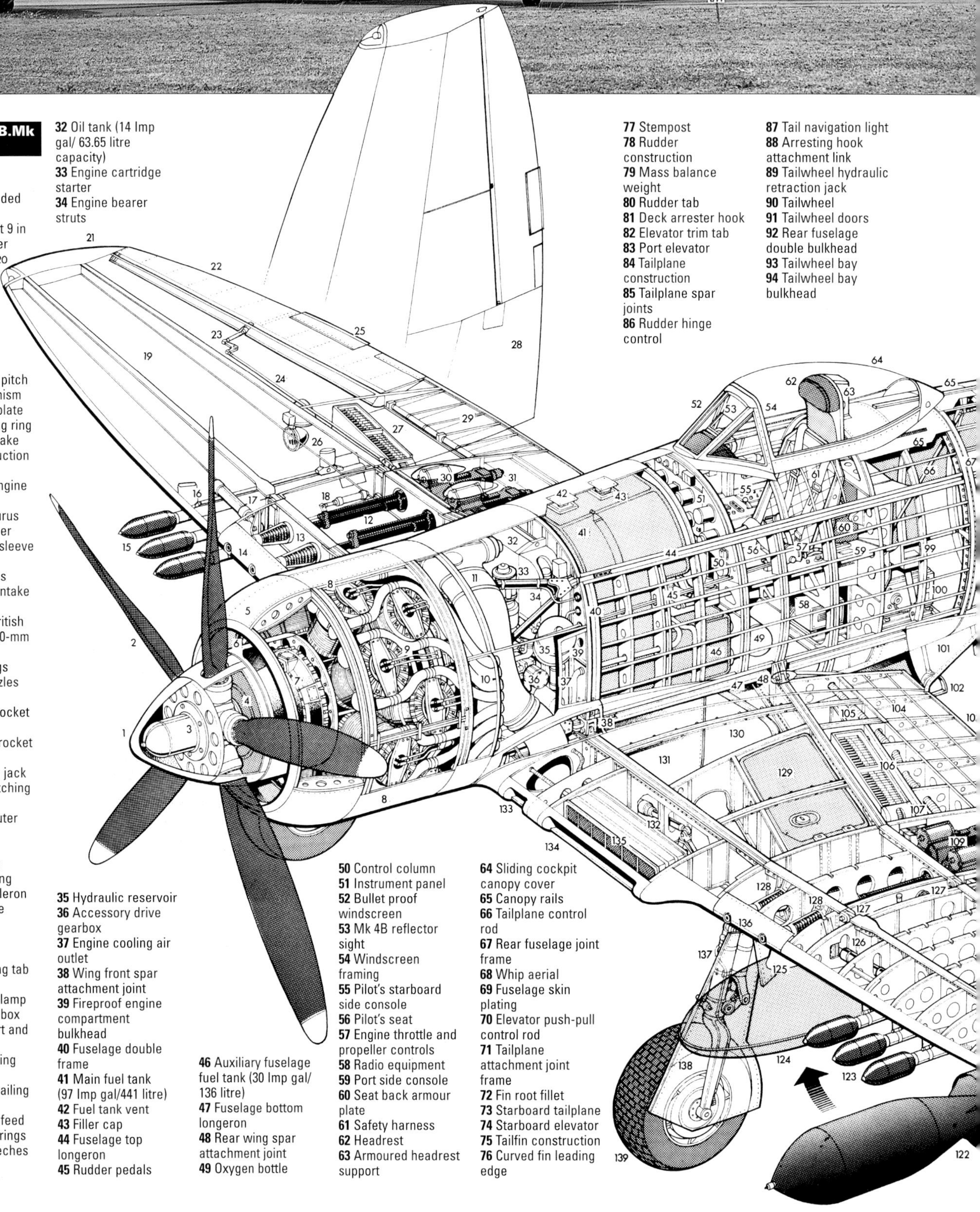

Canada's Sea Furies flew from HMCSs Warrior and Magnificent, ex-RN carriers completed after the end of World War II and loaned to the RCN. Here a Sea Fury cruises over Magnificent.

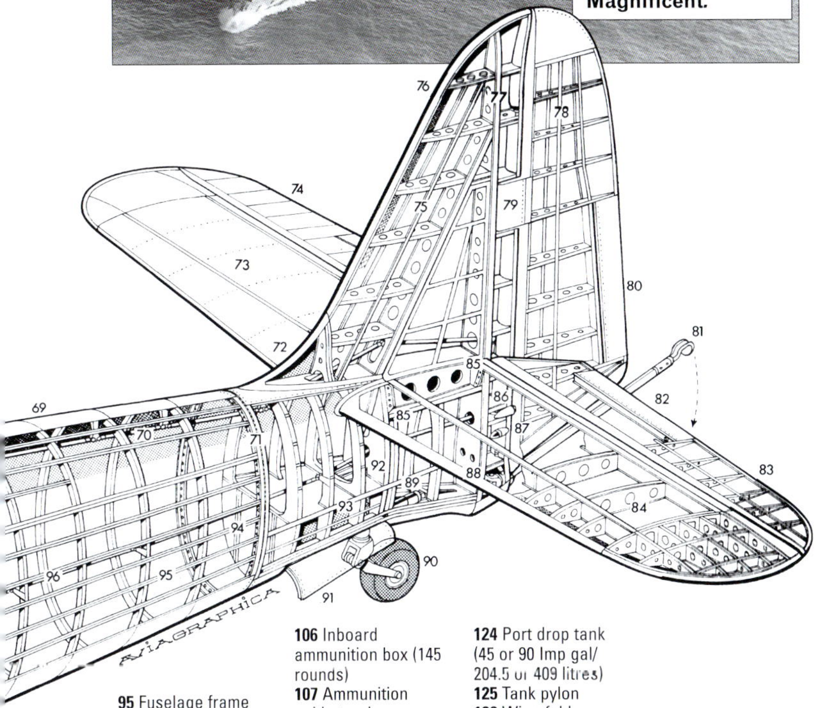

SPECIFICATION

Sea Fury F.Mk X	Sea Fury FB.Mk 11
Dimensions	**Dimensions**
Length: 34 ft 3 in (10.44 m) **Wingspan:** 38 ft 4½ in (11.70 m) **Wingspan (folded):** 16 ft 1 in (4.90 m) **Wing area:** 280 sq ft (26.01 m²)	**Length:** 34 ft 3 in (10.44 m) **Wingspan:** 38 ft 4½ in (11.70 m) **Wingspan (folded):** 18 ft 2 in (5.54 m) **Wing area:** 280 sq ft (26.01 m²)
Powerplant	**Powerplant**
One Bristol Centaurus Mk 18 18-cylinder, air-cooled radial rated at 2,480 hp (1849 kW) and driving a Rotol four-bladed propeller	One Bristol Centaurus Mk 18 18-cylinder, air-cooled radial rated at 2,480 hp (1849 kW) and driving a Rotol five-bladed propeller
Weights	**Weights**
Empty: 9,070 lb (4114 kg) **Loaded:** 10,660 lb (4835 kg)	**Empty:** 9,240 lb (4191 kg) **Loaded:** 12,300 lb (5579 kg)
Performance	**Performance**
Maximum speed at 18,000 ft (5486 m): 465 mph (748 km/h) **Climb to 30,000 ft (9144 m):** 9 minutes 48 seconds **Range (clean):** 710 miles (1142 km) **Service ceiling:** 36,180 ft (11028 m)	**Maximum speed at 18,000 ft (5486 m):** 450 mph (724 km/h) **Climb to 30,000 ft (9144 m):** 10 minutes 48 seconds **Range (clean):** 810 miles (1304 km) **Service ceiling:** 37,800 ft (11521 m)
Armament	**Armament**
Four 20-mm Hispano Mk 5 cannon	Four 20-mm Hispano Mk 5 cannon, plus an underwing stores load comprising a pair of 500-lb or 1,000-lb (227-kg or 454-kg) bombs (or equivalent weights in napalm or sea mines), or up to 12 25-lb (11-kg) rocket projectiles

95 Fuselage frame and stringer construction
96 Rudder push-pull control rod
97 Remote compass transmitter
98 Ventral aerial
99 Hand grip
100 Radio transmitter/receiver
101 Trailing edge wing root fillet
102 Retractable 'stirrup-type' step
103 Inboard split trailing edge flap
104 Flap shroud structure
105 Gun heater duct
106 Inboard ammunition box (145 rounds)
107 Ammunition guide track
108 Port British Hispano Mk 5 20-mm cannon
109 Ammunition feed drums
110 Outer ammunition box (145 rounds)
111 Outer split trailing edge flap
112 Port retractable landing/taxiing lamp
113 Aileron spring tab
114 Aileron construction
115 Wing tip fairing
116 Port navigation light
117 Pitot tube
118 Rear spar
119 Wing rib construction
120 Main spar
121 Leading-edge nose ribs
122 1,000-lb (454-kg) HE bomb
123 60-lb (27-kg) ground attack rockets
124 Port drop tank (45 or 90 Imp gal/ 204.5 or 409 litres)
125 Tank pylon
126 Wing fold hydraulic jack
127 Wing fold hinge joints
128 Cannon barrel mountings
129 Port interspar fuel tank (28 Imp gal/ 127 litres)
130 Main undercarriage wheel bay
131 Mainwheel door
132 Hydraulic retraction jack
133 Port carburettor air intake
134 Oil cooler ram air intake
135 Oil radiator (starboard leading edge has 17 Imp gal/77 litre fuel tank)
136 Port cannon muzzles
137 Pivoted main undercarriage shock absorber leg strut
138 Undercarriage leg fairing door
139 Port mainwheel

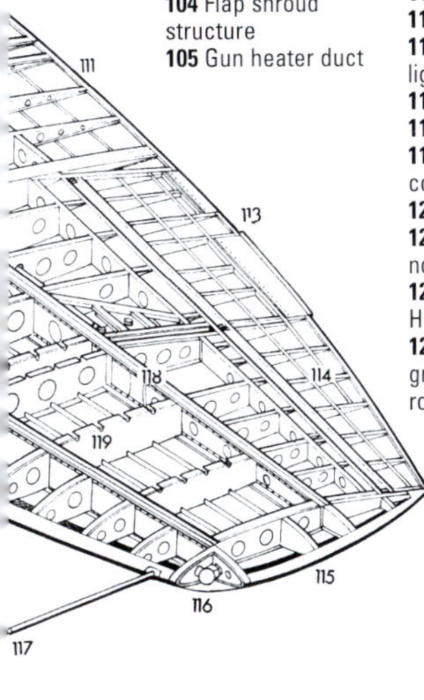

With Sea Furies of No. 802 Sqn and Fireflies of No. 825 Sqn ranged on her deck, HMS Ocean steams from Japan to the waters off Korea, July 1952. Ocean made two deployments to the region, in 1952 and 1953, including the last by a Royal Navy carrier during the Korean conflict.

Vought A-7 Corsair II

Pictured are A-7Ds of the 57th Fighter Weapons Wing of the Tactical Fighter Weapons Center (TFWC), Nellis AFB, Nevada, a unit which more recently flew the A-10A, F-15C/D and F-16A/B/C/D. The TFWC was established in 1966 and the 57th FWW (formerly 57th FIW) became part of the unit in 1969. The TFWC has established small detachments at other USAF bases, assisting the service entry of new types; for the A-7D, this included an establishment at Luke AFB, Arizona.

A-7D

Cutaway key

1 Radome
2 Radar scanner dish
3 AN/APQ-126 radar equipment module
4 Scanner tracking mechanism
5 Radar-mounting bulkhead
6 Pitot tubes
7 Windscreen rain dispersal air ducts
8 Cooling air louvres
9 Radar transmitter/receiver equipment
10 Engine air intake
11 ILS aerial
12 Forward radar warning antenna
13 Pave Penny laser ranger and marked target seeker
14 Pave Penny avionics pack
15 Intake duct framing
16 Cockpit floor level
17 Boron carbide (HCF) cockpit armour panelling
18 Cockpit pressurisation valve
19 Armoured front pressure bulkhead
20 Rudder pedals
21 Control column
22 Instrument panel
23 Head-down projected map display
24 Instrument panel shroud
25 Armoured windscreen panels
26 AN/AVQ-7(V) head-up display (HUD)
27 Pilot's rear view mirrors
28 Cockpit canopy cover, upward-hinged
29 Ejection seat headrest
30 Seat arming/safety lever
31 Safety harness
32 McDonnell Douglas Escapac 1-C2 rocket-powered 'zero-zero' ejection seat
33 Starboard side console panel
34 External canopy latch
35 Engine throttle lever
36 Port side console panel
37 Static ports
38 Boarding steps
39 Cannon muzzle
40 Retractable boarding ladder
41 Taxiing lamp
42 Nose undercarriage leg strut
43 Levered suspension axle beam
44 Twin nosewheels, aft retracting
45 Nosewheel doors
46 Cannon barrels
47 Intake trunking
48 Cockpit rear-pressure bulkhead
49 Angle-of-attack transmitter
50 Electrical system equipment bay
51 Ejection seat launch rails
52 Canopy aft framing
53 Hydraulic canopy jack
54 Canopy hinge point
55 TACAN antenna
56 Ammunition feed drive mechanism
57 Canopy emergency release
58 Ammunition feed and link return chutes
59 M61A1 Vulcan, 20-mm rotary cannon
60 Gun gas spill duct
61 Rotary cannon/ammunition drive, flexible interconnection
62 Cannon bay, air-conditioning plant on starboard side
63 Liquid oxygen converter
64 Emergency hydraulic accumulator
65 Electronics systems built-in test equipment panel (BITE)
66 Ground power socket
67 Ventral Doppler navigation aerial
68 Port avionics equipment bay
69 Cooling air extractor fan
70 Forward fuselage fuel cell; total internal fuel capacity 1,249 US gal (4728 litres)
71 Fuselage stores pylon, capacity 500 lb (227 kg)
72 Wing front spar/fuselage attachment joint
73 Control rod runs
74 Ammunition drum, 1,000 rounds
75 Universal air refuelling receptacle, open
76 Centre-section integral fuel tank
77 Wing panel centre-section carry-through structure
78 Wing skin panel centreline joint strap
79 Upper anti-collision light
80 Starboard wing integral fuel tank
81 Fuel system piping
82 Pylon attachment hardpoints
83 Inboard leading-edge flap
84 Flap hydraulic actuators
85 Centre wing pylon, capacity 3,500 lb (1558 kg)
86 AIM-9 Sidewinder air-to-air missile
87 Missile launch rail
88 Fuselage missile pylon
89 Mk 82 HE bombs, 500 lb (227 kg)
90 Multiple ejector rack
91 Outboard wing pylon, 3,500-lb (1558-kg) capacity
92 Leading-edge dog-tooth
93 Wing fold hydraulic jack
94 Outer wing panel hinge joint
95 Leading-edge flap hydraulic actuators
96 Snakeye, retarded bomb, 500 lb (227 kg)
97 Outboard leading-edge flap segment
98 Starboard navigation light
99 Wingtip fairing
100 Formation light
101 Outer wing panel, folded position
102 Starboard aileron
103 Aileron hydraulic actuator
104 Fuel jettison pipe
105 Starboard single-slotted trailing-edge flap (down)
106 Flap hydraulic jack
107 Starboard spoiler
108 Spoiler hydraulic actuator
109 Dorsal spine fairing
110 Control rod linkages
111 Rear spar/fuselage attachment joint
112 Gravity fuel filler cap
113 Rear fuselage fuel cell

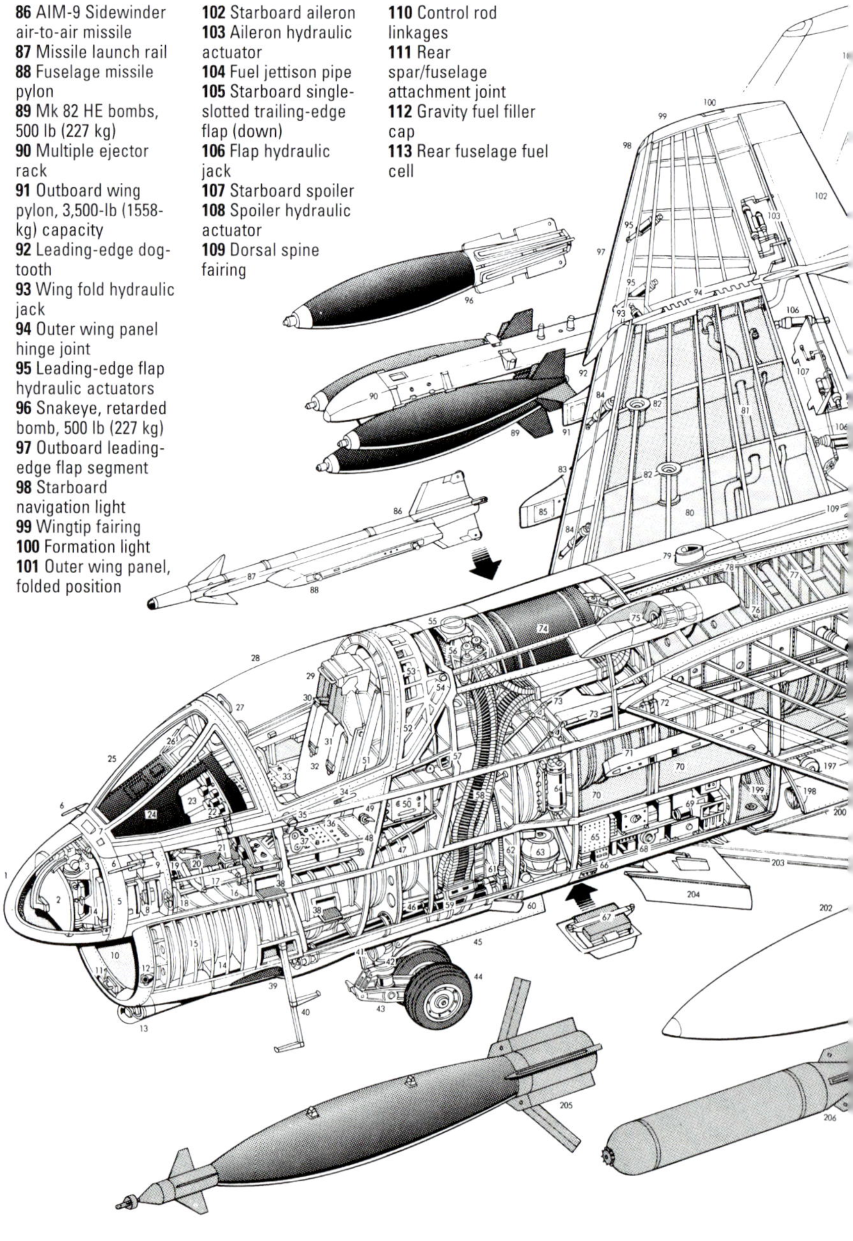

The first two-seat TA-7C made its maiden flight in mid-1975. Used for pilot familiarisation by the USN, the TA-7C was initially delivered with the TF30 engine, and from March 1983 six aircraft were redesignated EA-7L and assigned to VAQ-34 for electronic warfare training and simulation. The USN began taking delivery of re-engined (TF41) and upgraded TA-7Cs in January 1985. The TA-7Cs pictured served with VA-174 'Hell Razors' of the east coast Replacement Air Group (RAG), based at NAS Cecil Field, Florida.

SPECIFICATION

A-7D

Dimensions

Wingspan: 38 ft 9 in (11.81 m)
Length: 46 ft 1½ in (14.06 m)
Height: 16 ft ¾ in (4.90 m)
Wing area: 375 sq ft (34.84 m²)

Powerplant

One 14,500-lb (64.5-kN) Allison TF41-A-1 (licence-built Rolls-Royce Spey) turbojet, without afterburning

Weights

Empty: 19,781 lb (8973 kg)
Maximum take-off: 42,000 lb (19051 kg)

Fuel and load

Internal fuel 9,263 lb (4202 kg); external fuel up to four 300-US gal (1136-litre) drop tanks; maximum ordnance 20,000 lb (9072 kg) theoretical, 15,000 lb (6804 kg) practical with reduced internal fuel and 9,500 lb (4309 kg) with maximum internal fuel

Performance

Maximum speed (clean): 662 mph (1065 km/h) at 2,000 ft (610 m)
Maximum speed with 6,000-lb (2722-kg) of stores: 647 mph (1041 km/h) at 5,000 ft (1525 m)
Radius of action: ferry range 2,485 nm (2,861 miles; 4604 km) with maximum internal and external fuel or 1,981 nm (2,281 miles; 3671 km) with internal fuel; combat radius 620 nm (714 miles; 1149 km) on a hi-lo-hi mission
Ferry range: 2,870 miles (4619 km) with four 250-Imp gal (1137-litre) tanks

Armament

One internal 20-mm M61 Vulcan cannon with up to 1,000 rounds, plus up to 15,000 lb (6804 kg) of ordnance on six underwing pylons, and two fuselage hardpoints

114 Control rod spring damper
115 Engine compressor intake
116 Intake centre-body fairing
117 Fuselage upper longeron
118 Engine bleed air ducting
119 Rear fuselage main frames
120 Hydraulic system reservoir
121 Vertical tail control rod
122 Finroot fillet
123 Vertical tail trim feel unit
124 Vertical tail autopilot controller
125 Rudder feel control unit
126 Tailfin construction
127 Flush VHF aerial
128 Starboard all-moving tailplane
129 Fin leading-edge ribs
130 Dielectric fintip aerial fairing
131 UHF/IFF aerial
132 VOR aerial
133 Tail navigation light
134 Tail radar warning antenna (electronic countermeasures, ECM)
135 Rudder
136 Rudder rib construction
137 Rudder hydraulic actuator
138 Fin attachment post
139 Detachable tail cone
140 Jet pipe
141 Engine exhaust nozzle
142 Port all-moving tailplane
143 Tailplane rib construction
144 Tailplane spar box
145 Leading-edge nose ribs
146 Tailplane pivot fixing
147 Tailplane control lever arm
148 Hydraulic actuator
149 Back-up tailplane control interconnecting yoke
150 Rear engine mounting
151 Allison TF41-A-1 turbofan
152 Fuselage lower longeron
153 Ventral chaff dispenser
154 Engine bay access panels
155 Boron carbide (HCF) engine bay armour
156 Emergency runway arrester hook
157 Hook hydraulic actuator/damper
158 Engine accessory equipment gearbox
159 Main engine mounting trunion
160 Hydraulic accumulators
161 Position of strike camera on starboard side
162 Fuel vent mast
163 Port spoiler
164 Flap hinge arms
165 Flap hydraulic jack
166 Flap rib construction
167 Port single-slotted trailing-edge flap
168 Fuel jettison pipe
169 External fuel tank tail fins
170 Aileron hydraulic actuator
171 Port aileron
172 Fixed portion of trailing edge
173 Port formation light
174 Wingtip fairing
175 Port navigation light
176 Outboard leading-edge flap
177 Leading-edge flap (down)
178 Leading-edge flap rib construction
179 Flap hydraulic jacks
180 Outer wing panel multi-spar construction
181 Wing hinge rib
182 Wing fold hydraulic actuator
183 Port outer stores pylon
184 Leading-edge dog-tooth
185 Port mainwheel
186 Inner wing panel multi-spar construction
187 Port wing integral fuel tank
188 Centre pylon attachment hardpoint
189 Main undercarriage leg strut
190 Hydraulic retraction jack
191 Shock/absorber strut
192 Main undercarriage leg pivot fixing
193 Aileron feel trim control unit
194 Centre fuselage fuel cell
195 Inner wing pylon hardpoint, capacity 2,500 lb (1134 kg)
196 Hydraulic reservoir
197 Undercarriage bay pressure refuelling connection
198 Position of landing lamp in starboard wheel bay
199 Fuel sump cell
200 Main wheel doors
201 Port centre wing pylon
202 250-Imp gal (1137-litre) external fuel tank
203 Ventral airbrake
204 Airbrake retractable side flap
205 GBU-10 Paveway II laser-guided bomb (2,000-lb/907-kg Mk 84)
206 Rockeye cluster bomb
207 AGM-65A Maverick TV or laser-guided air-to-ground missile
208 LAU-37 air-to-ground rocket launcher

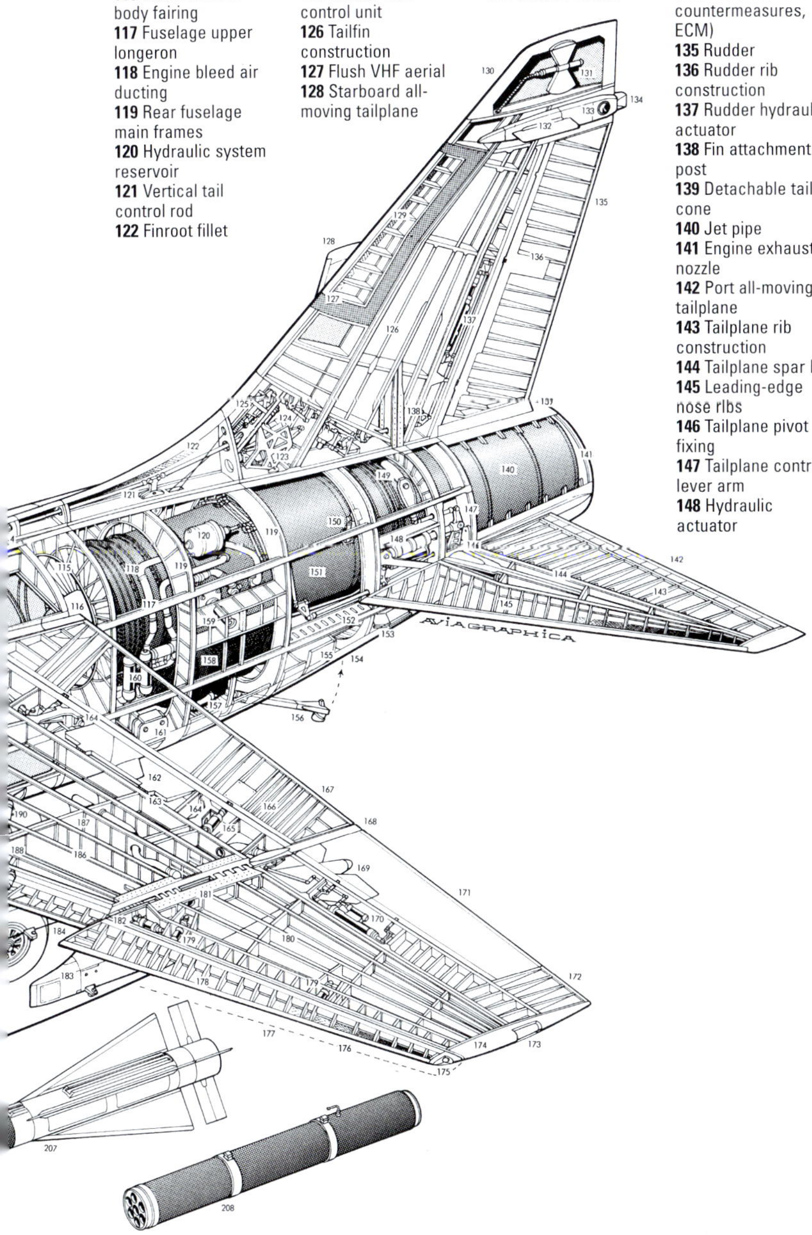

Vought F-8 Crusader

Fighter Squadron One Hundred and Three (VF-103) transitioned to the F8U in 1957, then to the F-8E in 1963. Known as the 'Sluggers', the squadron was aboard USS Foresstal *in the eastern Mediterranean at the time of the Lebanon crisis (1958), during which US Marines undertook an amphibious landing in an effort to restore peace to the area.*

F-8E Crusader

Cutaway key
1 Fintip VHF aerial fairing
2 Tail warning radar
3 Tail navigation light
4 Rudder construction
5 Rudder hydraulic jack
6 Engine exhaust nozzle
7 Variable-area nozzle flaps
8 Afterburner cooling air duct
9 Nozzle control jacks
10 Starboard all-moving tailplane construction
11 Tailplane spar box
12 Leading-edge ribs
13 Tailplane pivot fixing
14 Tailplane hydraulic control jack
15 Tailpipe cooling air vents
16 Fin attachment main frame
17 Afterburner duct
18 Rudder control linkages
19 Fin leading-edge construction
20 Port all-moving tailplane
21 Fin-root fillet construction
22 Rear engine mounting
23 Fuselage break point double frame (engine removal)
24 Afterburner fuel spray manifold
25 Tailplane autopilot control system
26 Deck arrester hook
27 Starboard ventral fin
28 Rear fuselage fuel tank
29 Pratt & Whitney J57-P-20A afterburning turbojet
30 Engine bay cooling air louvres
31 Wingroot trailing-edge fillet
32 Bleed air system piping
33 Engine oil tank (85 US gal/322 litres)
34 Wing spar pivot fixing
35 Hydraulic flap jack
36 Starboard flap
37 Control rod linkages
38 Rear spar
39 Engine accessory gearbox compartment
40 Inboard wing panel multi-spar construction
41 Starboard wing integral fuel tank, total fuel system capacity 1,348 US gal (5103 litres)
42 Aileron power control unit
43 Starboard drooping aileron construction
44 Hydraulic wing fold jack
45 Trailing-edge ribs
46 Fixed portion of trailing edge
47 Wingtip fairing
48 Starboard navigation light
49 Leading-edge flap, lowered position
50 Leading-edge flap rib construction
51 Outer wing panel spar construction
52 Leading-edge flap hydraulic jack
53 Wingfold hinge
54 Front spar
55 Leading-edge flap inboard section
56 Leading-edge dog-tooth
57 Wing pylon
58 AGM-12B Bullpup A air-to ground missile
59 Starboard mainwheel
60 Main undercarriage leg strut
61 Shock absorber strut
62 Hydraulic retraction jack
63 Landing lamp
64 Wheel bay doors
65 Main undercarriage pivot fixing
66 Wing spar/front engine mounting main bulkhead
67 Engine compressor intake
68 Wingroot rib
69 Centre-section fuel tank
70 Wing spar carry-through structure
71 Dorsal fairing
72 Port flap jack
73 Port plain flap, lowered position
74 Port drooped aileron, lowered position
75 Aileron power control unit
76 Fuel system piping
77 Wing-fold hydraulic jack
78 Fixed portion of trailing edge
79 Port wing folded position
80 Wingtip fairing
81 Port navigation light
82 Port outboard leading-edge flap, lowered
83 Outboard flap hydraulic jack
84 Leading-edge dog-tooth
85 Wing-fold hinge
86 Inboard leading-edge flap hydraulic jacks
87 Port wing integral fuel tank
88 Anti-collision light
89 Missile system avionics
90 Two-position variable-incidence wing, raised position
91 Intake trunking
92 Wing incidence hydraulic jack
93 Fuselage upper longeron
94 Air system exhaust heat shield
95 Main fuselage fuel tank
96 Airbrake hydraulic jack
97 Airbrake housing
98 Ventral airbrake, lowered
99 Rocket launch tubes
100 Rocket launcher pylon adaptor
101 Zuni folding-fin ground attack rockets (8)
102 Emergency air-driven generator
103 Liquid oxygen bottle (LOX)
104 Fuselage stores pylon

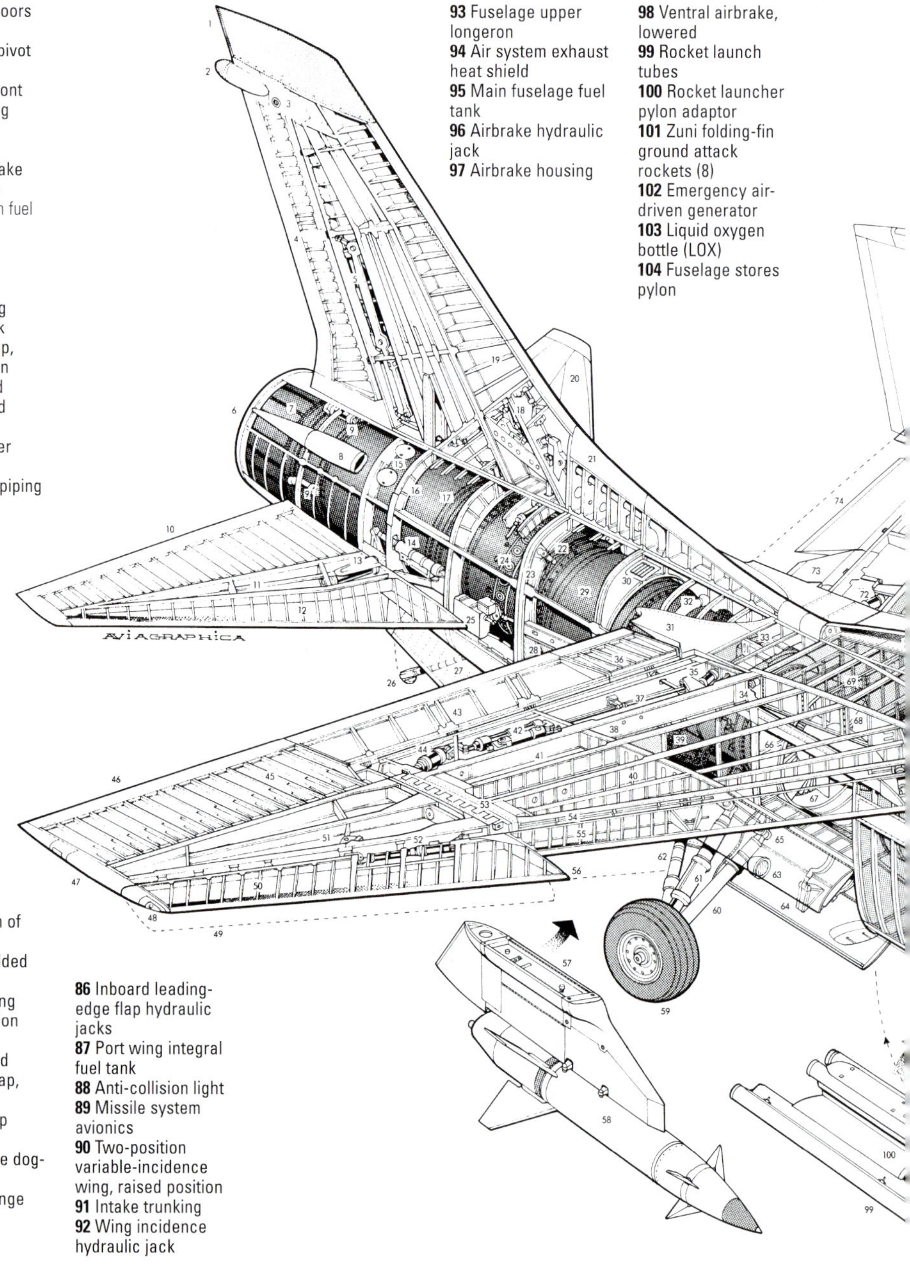

SPECIFICATION

F-8E Crusader

Dimensions

Length: 54 ft 6 in (16.61 m)
Height: 15 ft 9 in (4.80 m)
Wingspan: 35 ft 2 in (10.72 m)
Wing area: 350 sq ft (35.52 m²)

Powerplant

One Pratt & Whitney J57-P-20A turbojet rated at 10,700 lb (48.15 kN) static thrust or 18,000 lb st (81 kN) with afterburner

Weights

Empty: 17,541 lb (7957 kg)
Gross weight: 28,765 lb (13048 kg)
Combat weight: 25,098 lb (11304 kg)
Maximum take-off: 34,000 lb (15422 kg)

Performance

Maximum level speed at sea level: 764 mph (1230 km/h)
Maximum level speed at 40,000 ft (12192 m): 1,120 mph (1802 km/h)
Cruising speed: 570 mph (917 km/h)
Stalling speed: 162 mph (261 km/h)
Rate of climb in one minute: 31,950 ft (9738 m)
Service ceiling: 58,000 ft (17678 m)
Combat ceiling: 53,400 ft (16276 m)

Range

Range: 453 miles (729 km)
Maximum range: 1,737 miles (2795 km)

Armament

Four Colt-Browning (20-mm) Mk 12 cannon with 144 rounds per gun; plus up to four AIM-9 Sidewinder AAMs; or 12 250-lb (113-kg bombs); or eight 500-lb (227-kg) bombs; or eight Zuni rockets; or two AGM-12A or AGM-12B Bullpup A AGMs

The exceptional speed of the Crusader meant that the aircraft was used to accomplish a number of high-profile speed records. By far the best-known was the record coast-to coast flight of 16 July 1957, known as 'Project Bullet'. This was carried out in an F8U-1P flown by Major John Glenn (illustrated) and an F8U-1 flown by LCdr Charles Demmler. Taking off from Los Angeles, the pair flew supersonically to the first refuelling, where Demmler's probe was damaged, forcing him to retire. Glenn carried on, landing at New York 3 hours and 23 minutes after take-off, having photographed the entire strip of land beneath his aircraft during the run.

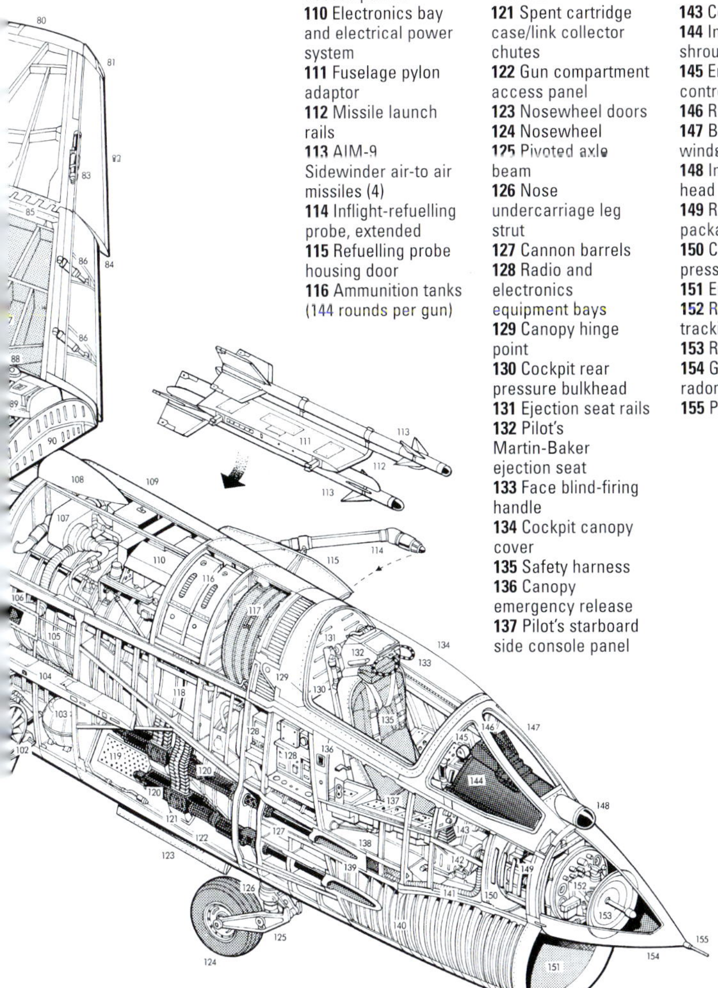

105 Intake duct
106 Heat exchanger air exhaust
107 Air-conditioning plant
108 Dorsal fairing
109 Upper fuselage access panels
110 Electronics bay and electrical power system
111 Fuselage pylon adaptor
112 Missile launch rails
113 AIM-9 Sidewinder air-to air missiles (4)
114 Inflight-refuelling probe, extended
115 Refuelling probe housing door
116 Ammunition tanks (144 rounds per gun)
117 Avionics system inertial platform
118 Ammunition feed chutes
119 Gun bay gas vent panel
120 Mk 12 20-mm cannon
121 Spent cartridge case/link collector chutes
122 Gun compartment access panel
123 Nosewheel doors
124 Nosewheel
125 Pivoted axle beam
126 Nose undercarriage leg strut
127 Cannon barrels
128 Radio and electronics equipment bays
129 Canopy hinge point
130 Cockpit rear pressure bulkhead
131 Ejection seat rails
132 Pilot's Martin-Baker ejection seat
133 Face blind-firing handle
134 Cockpit canopy cover
135 Safety harness
136 Canopy emergency release
137 Pilot's starboard side console panel
138 Cockpit floor level
139 Cannon muzzle blast troughs
140 Intake duct framing
141 Radar cooling air piping
142 Rudder pedals
143 Control column
144 Instrument panel shroud
145 Engine throttle control
146 Radar gunsight
147 Bullet-proof windshield
148 Infra-red seeker head
149 Radar electronics package
150 Cockpit front pressure bulkhead
151 Engine air intake
152 Radar scanner tracking mechanism
153 Radar antenna
154 Glass-fibre radome
155 Pitot tube

Vought modified the 74th F-8A into a dual (tandem)-seat aircraft, designated F8U-1T (later TF-8A). As a one-of-a-kind prototype, the lone TF-8A performed extremely well. After serving with the US Navy (illustrated), the 'Two-sader' served with NASA/Vought at Edwards AFB, where one of its roles was as a chase-plane. The TF-8A was lost on 28 July 1978 while on a training mission.

Reconnaissance Aircraft

Boeing E-3 Sentry

The size of the AN/APY-1 radar held within rotodome is clearly evident in this view USAF E-3B. The radar's sophistication allow to operate in several modes, provic comprehensive surveillance tailored to E-3B's tactical needs over a wide ra

Boeing E-3C/D Sentry (AWACS)

Cutaway key

1 Fixed flight-refuelling probe, British and French aircraft only
2 Refuelling lights
3 Refuelling receptacle, open
4 Flight deck, two pilots, navigator and flight engineer
5 Forward 'Quick-Look' ESM system antennas, retrofitted to USAF and NATO aircraft
6 Forward entry door
7 Forward toilet
8 Safety equipment locker
9 Communications console with Joint Tactical Information Distribution System (JTIDS)
10 Communications equipment racks
11 Lateral 'Quick-Look' antennas, port and starboard
12 Underfloor flight-essential avionics racks
13 Bale-out chute
14 Computer racks
15 Communications technician's console
16 Data display controller
17 Underfloor power distribution centre
18 Operator consoles (14)
19 GPS antenna
20 Escape hatch with automatic life raft
21 Standard overwing escape hatches, port and starboard
22 CFM56-2A-3 engines, British, French and Saudi aircraft, USAF and NATO aircraft have Pratt & Whitney TF33-P-100/-100A
23 Wingtip Loral 10171 ('Yellow Gate') ESM pod, British aircraft only
24 HF antenna
25 Rotodome, rotates at 6 rpm operationally
26 IFF antenna
27 TADIL-C array
28 Radar-cooling air duct
29 Antenna ancillary equipment
30 Westinghouse APY-2 phased-array antenna
31 Radar maintenance station
32 Radar receiving and signal processing racks
33 Supernumerary and relief crew seats
34 JTIDS terminal
35 Underfloor radar equipment bay, regulators, filters and pulser/Klystron equipment
36 Radar-cooling system ram air intake to heat exchanger
37 Auxiliary power unit (APU) bay
38 ESM equipment rack
39 Relief crew seating
40 Bunks, port and starboard
41 Starboard-side service door
42 Galley
43 Rear entry door
44 Toilet compartments, port and starboard
45 Rear 'Quick-Look' ESM antenna

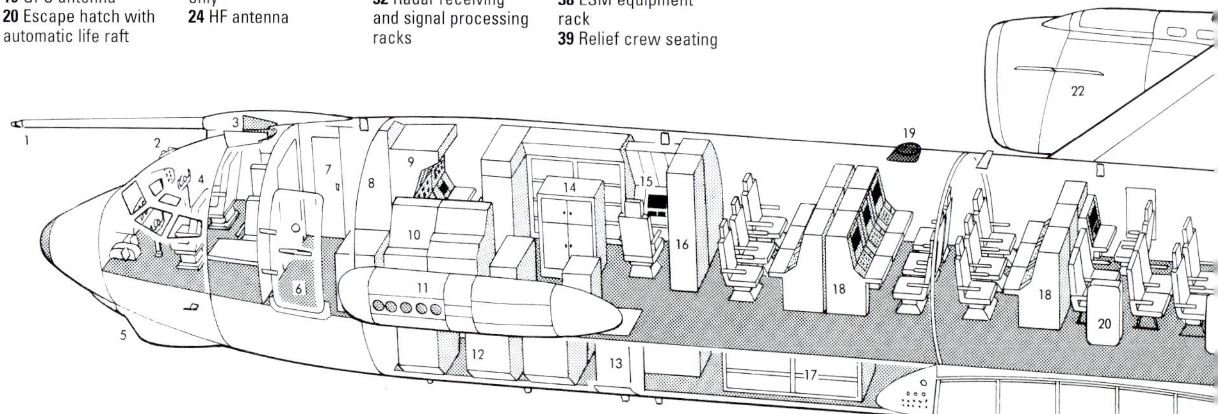

Boeing E-767

In December 1991 Boeing announced the project for an AWACS version of the Boeing 767-200ER airliner, fitted with the Northrop Grumman AN/APY-2 radar. Japan expressed an immediate interest and in November 1993 two examples were ordered for the JASDF. Two further examples were ordered the following year and in March 1998 the first two examples were delivered. The aircraft has a flight crew of two and up to 19 mission crew, although this number can vary depending on the mission profile. Substantial structural modifications were needed, including two additional bulkheads and reinforced floor beams as well as the rotodome to hold the rotating radar. The aircraft offers twice the floor space and three times the internal volume of the Boeing E-3. When operational, the four aircraft will be based at Hamamatsu Air Base.

Left: To mark the 50th anniversary of NATO in 1999, NE-3A Sentry LX-N-90442 was repainted in this spectacular colour scheme. The flags of all 19 member states of NATO are depicted on the side of the aircraft.

Below left: During the 1990s, NATO's E-3 fleet underwent a major upgrade. This included the installation of an AN/AYR-1 ESM system, identified by the additional bulges beneath the nose and on the side of the front fuselage, and an upgraded computer system.

SPECIFICATION

Sentry AEW.Mk 1 (E-3D)

Dimensions

Length: 152 ft 11 in (46.61 m)
Wingspan: 147 ft 7 in (44.98 m)
Wingspan (E-3A/B/C): 145 ft 9 in (44.42 m)
Wing area: 3,050 sq ft (283.35 m²)
Height: 41 ft 9 in (12.73 m)

Powerplant

Four CFM56-2A-3 turbofan engines each rated at 24,000 lb st (106.8 kN) for take-off or 23,405 lb st (104.1 kN) maximum continuous power
E-3A/B/C: Four Pratt & Whitney TF33-P-100/-100A turbofan engines each rated at 21,000 lb st (93.41 kN)

Weights

Normal take-off: 325,000 lb (147417 kg)
Maximum take-off: 332,500 lb (150820 kg)
Maximum ramp: 335,000 lb (151953 kg)

Fuel load

Total fuel weight (JP4): 155,448 lb (70,510 kg)

Performance

Maximum level speed at operating altitude: 460 kt (530 mph; 853 km/h)
Service ceiling: more than 35,000 ft (10670 m)
Maximum unrefuelled range: more than 5,000 nm (5,758 miles; 9266 km)
Operational radius: 870 nm (1,000 miles; 1610 km) with six hours on allocated station
Maximum unrefuelled endurance: more than 11 hours

Accommodation

Flight crew: 4
Mission crew and operators: 17

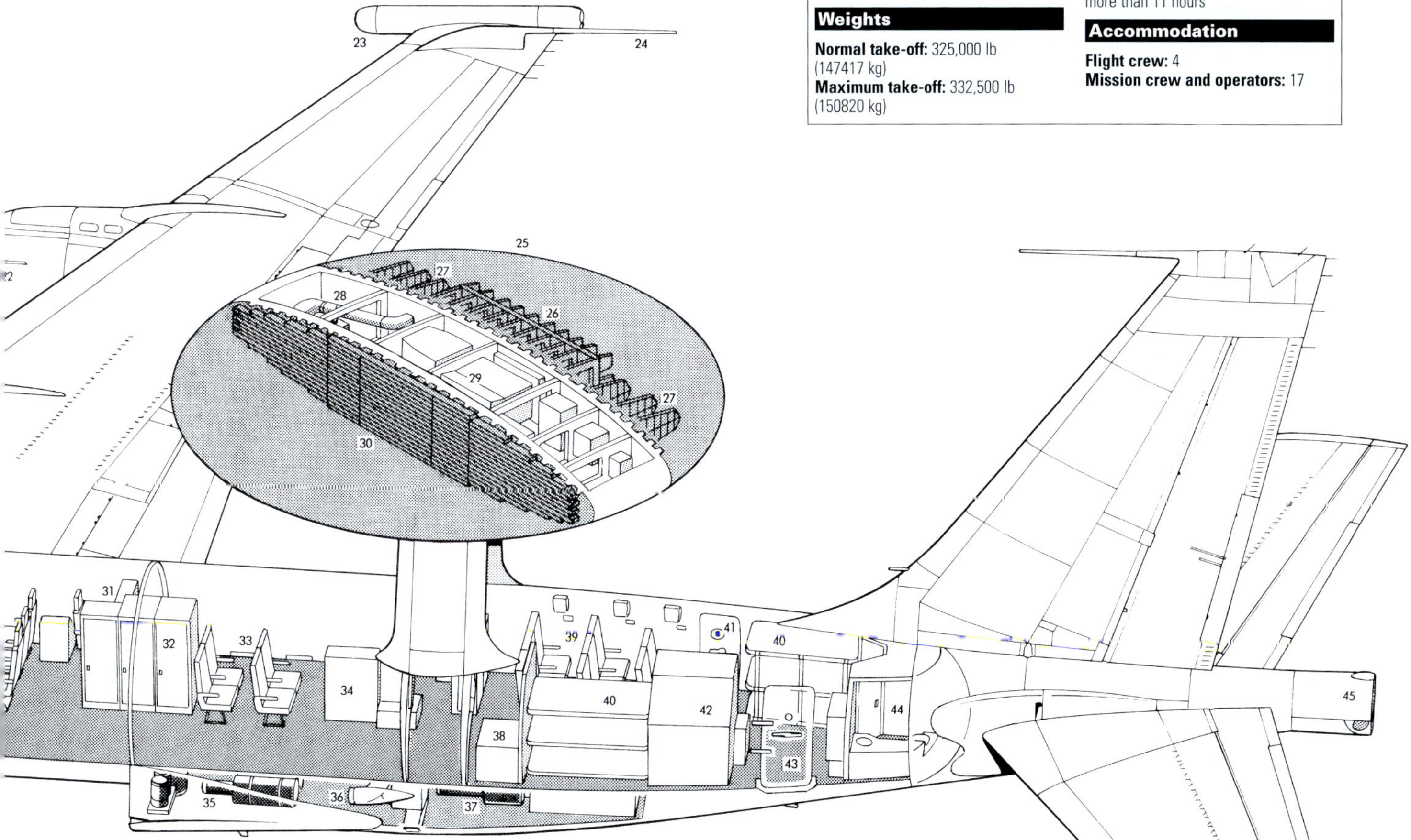

As the UK component of the NATO Airborne Early Warning Force (AEWF), the RAF's E-3Ds are equipped with a refuelling probe to allow air-to-air refuelling from the RAF's VC10 and TriStar tanker assets. Each of the seven aircraft is named after characters in the story, **Snow White and the Seven Dwarfs.**

Boeing RC-135

Despite the collapse of the Soviet threat, the disbandment of Strategic Air Command in 1992 and the introduction of modern spy satellites and J-STARS, the RC-135 remains an important type in the USAF's reconnaissance inventory. During Operation Allied Force in 1999, elements of the RC-135V/W Rivet Joint fleet operated from Mildenhall as a vital component of the electronic warfare/surveillance effort over the former Yugoslavia.

RC-135W

Cutaway key
1 Radome
2 Forward radar antenna
3 Front pressure bulkhead
4 Ventral antennas
5 Extended nose radome fairing
6 Nose compartment framing
7 Cockpit floor level
8 Pilot's side console panel
9 Rudder pedals
10 Control column
11 Instrument panel
12 Windscreen wipers
13 Windscreen panels
14 Cockpit eyebrow windows
15 Overhead systems switch panels
16 Co-pilot's seat
17 Direct vision opening side window panel
18 Pilot's seat
19 Safety equipment stowage
20 Chart-plotting table
21 Navigator's instrument console
22 Boom type inflight refuelling receptacle, open
23 Dual navigators' seats
24 Retractable escape spoiler, stowed position
25 Entry hatch floor grille
26 Nose landing gearwheel bay
27 Crew entry hatch, open
28 Retractable boarding ladder
29 Twin nosewheels, forward retracting
30 Nose landing gear pivot fixing
31 Underfloor avionics equipment racks
32 Electrical equipment racking
33 Super numerary crew seal
34 Star tracking windows, celestial navigation system
35 Cockpit doorway
36 Circuit breaker panels
37 Overhead air distribution ducting
38 No. 1 UHF/VHF aerial
39 Starboard side avionics equipment racks
40 Toilet compartment
41 Water heater
42 Wash basin
43 Water storage tanks
44 Toilet
45 Side-Looking Airborne Radar (SLAR) antenna panels
46 SLAR equipment fairing
47 Cargo doorway, electronic equipment loading
48 Main cabin flooring
49 Modular equipment package
50 Cargo door hydraulic jacks and hinges
51 Cargo door, open
52 ADF aerials
53 Electronics equipment racks
54 Air conditioning ducting
55 Aerial lead-in
56 Front spar attachment fuselage main frame
57 Centre section fuel tanks, 27656 litres (7,306 US gal) capacity
58 Overwing emergency exit hatch, port only
59 Floor beam construction
60 AN/ASD-1 avionics equipment racks
61 Tacan aerial
62 No. 1 satellite navigation system aerial
63 Inboard wing fuel tank, 8612 litres (2,275 US gal) capacity
64 Fuel filler cap
65 Detachable engine cowling panels
66 No. 3 starboard inboard engine nacelle
67 Intake cowling
68 Nacelle pylon
69 Pylon strut access panels
70 Wing centre main fuel tank, 7805 litres (2,062 US gal) capacity
71 Fuel venting channels
72 Leading edge flap hydraulic jacks
73 Krueger-type leading edge flap, down position
74 No. 4 starboard outboard engine nacelle
75 Outboard nacelle pylon
76 Outer wing panel joint rib
77 Wing outboard fuel tank, 1643 litres (434 US gal) capacity
78 HF antenna tuner
79 Lightning arrester panel
80 HF aerial mast
81 Pitot static boom
82 Starboard navigation light
83 Static dischargers
84 Outboard, low speed, aileron
85 Aileron internal balance panels
86 Spoiler interconnection linkage
87 Aileron hinge control mechanism
88 Aileron tab
89 Outboard double-slotted Fowler-type flap, down position
90 Outboard spoilers, open
91 Spoiler hydraulic jacks
92 Flap guide rails
93 Flap screw jacks
94 Aileron control and trim tab
95 Inboard, high speed, aileron
96 Gust damper
97 Aileron hinge control linkage
98 Inboard spoilers, open
99 Spoiler hydraulic jacks
100 Inboard double-slotted Fowler-type flap, down position
101 Rear spar attachment fuselage main frame
102 Pressure floor above wheel bay
103 ECM operator's seats
104 AN/ASD-1 Elint system control console
105 No. 2 UHF/VHF aerial
106 Cabin divider
107 Production break fuselage mainframe
108 Main cabin floor beams
109 Rear underfloor fuel cells, not used on Sigint aircraft
110 Signals Intelligence (Sigint) operator's seats
111 Sigint instrument and control consoles
112 No. 2 satellite navigation system aerial
113 QRC-259 superheterodyne receiver system console
114 Rear cabin emergency exit and service hatch, starboard only
115 QRC-259 operator's seat
116 Avionics equipment racks
117 Equipment modules
118 Table
119 Crew rest area seating
120 Hatches to underfloor radar equipment bay
121 Recorder unit
122 Rear fuselage close pitched frame construction
123 Galley unit
124 Aft toilet compartment

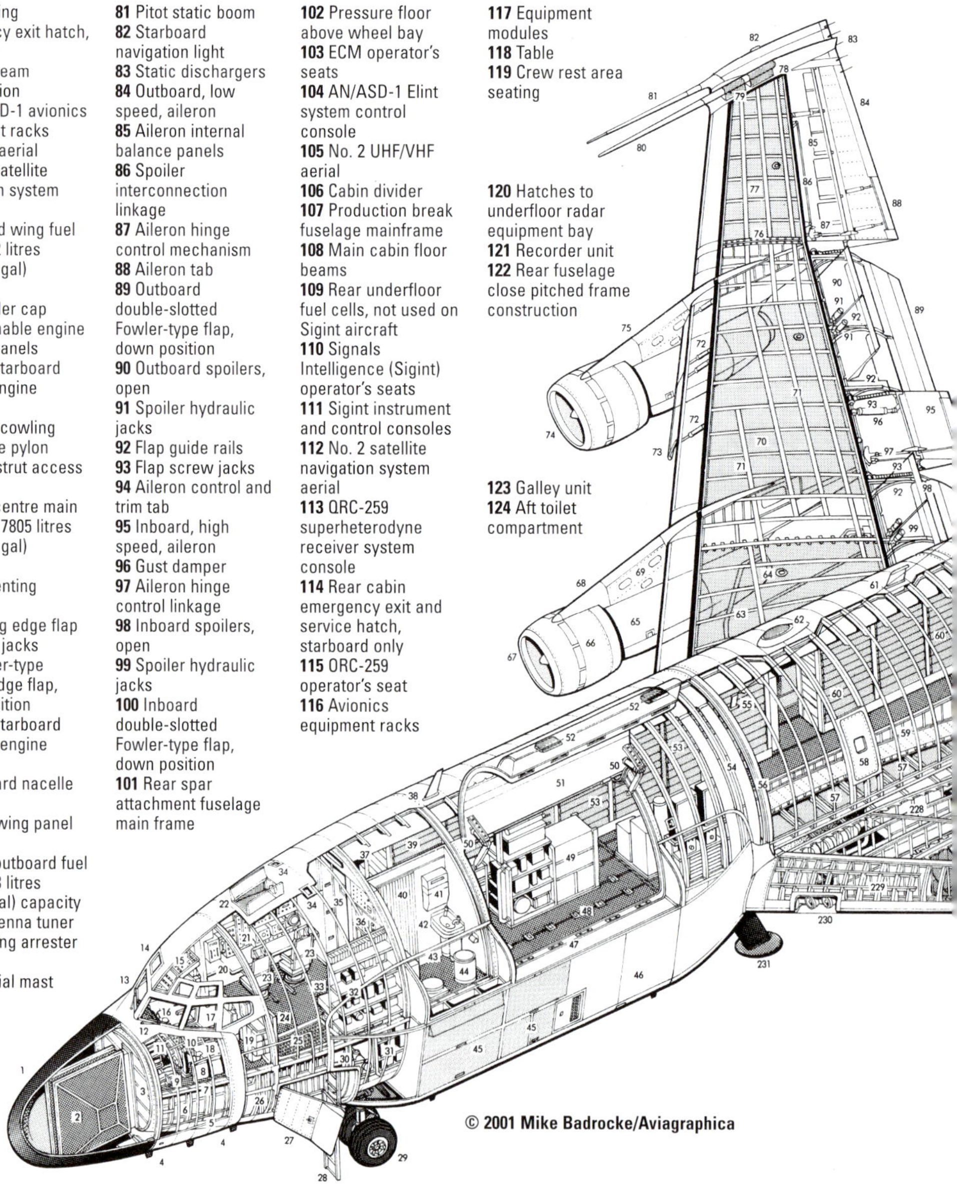

SPECIFICATION

RC-135V Rivet Joint

Dimensions:

Wingspan: 130 ft 10 in (39.88 m)
Wing area (less ailerons): 2,313.4 sq ft (214.9 m²)
Wing area (flaps extended): 2,754.4 sq ft (255.9 m²)
Length: 135 ft 1 in (41.17 m)
Height: 41 ft 9 in (12.73 m)

Powerplant:

Four Pratt & Whitney TF33-P-9 turbofans, each rated at 18,000 lb st (80.07 kN) dry; re-engined aircraft powered by four CFM International F108-CF-100 turbofans, each rated at 22,000 lb st (97.86 kN)

Weights:

Maximum gross (taxi): 301,600 lb (136803 kg)

Performance

Generally similar to KC-135E Stratotanker

The basic flight crew of the RC-135 comprises two pilots and two navigators, with the rear part of the cabin containing seven operator stations for Sigint missions. Today's Rivet Joint crew includes members of several 55th Wing squadrons. Pilots, navigators and maintainers are assigned to the 38th Reconnaissance Sqn, EWOs ('Ravens') and maintainers are drawn from the 343rd RS while the remaining crew are provided by the 97th Intelligence Sqn.

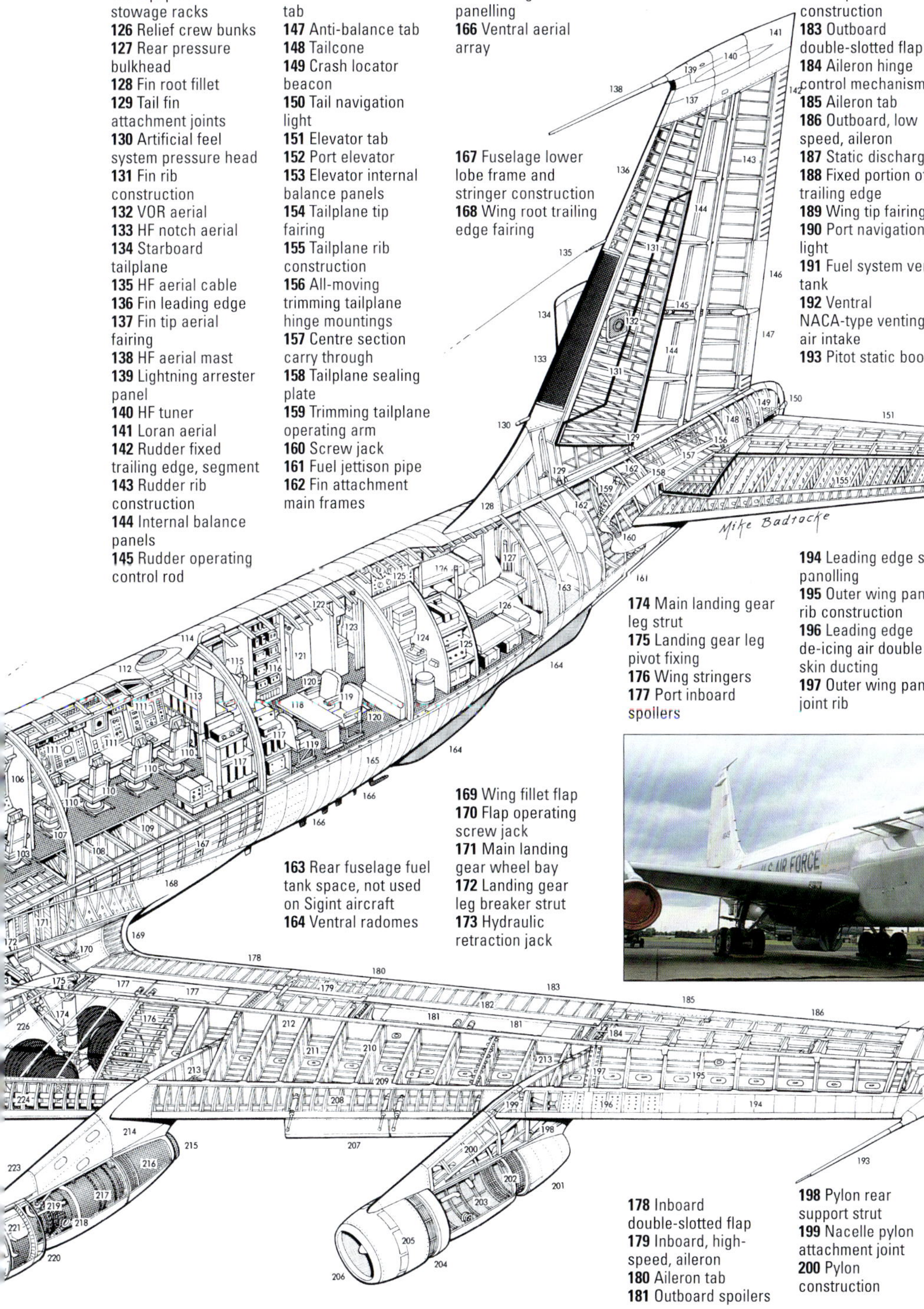

125 Equipment stowage racks
126 Relief crew bunks
127 Rear pressure bulkhead
128 Fin root fillet
129 Tail fin attachment joints
130 Artificial feel system pressure head
131 Fin rib construction
132 VOR aerial
133 HF notch aerial
134 Starboard tailplane
135 HF aerial cable
136 Fin leading edge
137 Fin tip aerial fairing
138 HF aerial mast
139 Lightning arrester panel
140 HF tuner
141 Loran aerial
142 Rudder fixed trailing edge, segment
143 Rudder rib construction
144 Internal balance panels
145 Rudder operating control rod
146 Rudder control tab
147 Anti-balance tab
148 Tailcone
149 Crash locator beacon
150 Tail navigation light
151 Elevator tab
152 Port elevator
153 Elevator internal balance panels
154 Tailplane tip fairing
155 Tailplane rib construction
156 All-moving trimming tailplane hinge mountings
157 Centre section carry through
158 Tailplane sealing plate
159 Trimming tailplane operating arm
160 Screw jack
161 Fuel jettison pipe
162 Fin attachment main frames
163 Rear fuselage fuel tank space, not used on Sigint aircraft
164 Ventral radomes
165 Fuselage skin panelling
166 Ventral aerial array
167 Fuselage lower lobe frame and stringer construction
168 Wing root trailing edge fairing
169 Wing fillet flap
170 Flap operating screw jack
171 Main landing gear wheel bay
172 Landing gear leg breaker strut
173 Hydraulic retraction jack
174 Main landing gear leg strut
175 Landing gear leg pivot fixing
176 Wing stringers
177 Port inboard spoilers
178 Inboard double-slotted flap
179 Inboard, high-speed, aileron
180 Aileron tab
181 Outboard spoilers
182 Flap rib construction
183 Outboard double-slotted flap
184 Aileron hinge control mechanism
185 Aileron tab
186 Outboard, low speed, aileron
187 Static dischargers
188 Fixed portion of trailing edge
189 Wing tip fairing
190 Port navigation light
191 Fuel system vent tank
192 Ventral NACA-type venting air intake
193 Pitot static boom
194 Leading edge skin panelling
195 Outer wing panel rib construction
196 Leading edge de-icing air double skin ducting
197 Outer wing panel joint rib
198 Pylon rear support strut
199 Nacelle pylon attachment joint
200 Pylon construction
201 Aft translating exhaust cowling, open
202 Thrust reverser cascades
203 Engine cowling panels
204 Fan air reverser, open
205 Spring loaded intake pressure relief doors
206 No. 1 outboard engine cowling
207 Port leading edge Krueger flap, down position
208 Leading edge nose ribs
209 Front spar
210 Wing rib construction
211 Port wing integral fuel tankage
212 Rear spar
213 Diagonal pylon mounting ribs
214 No. 2 inboard engine mounting pylon
215 Engine hot stream exhaust nozzle
216 Tailpipe
217 Pratt & Whitney TF33-9 turbofan engine
218 Engine accessory equipment gearbox
219 Main engine mounting
220 Fan air, cold stream, exhaust duct
221 Engine oil tank
222 Compressor intake face
223 Inboard nacelle pylon
224 Bleed air ducting
225 Four-wheel main landing gear bogie
226 Wing skin panelling
227 Inboard integral fuel tanks
228 Ventral air conditioning pack, port and starboard
229 Leading edge rib construction
230 Landing taxiing lamps
231 Sigint antennae

RC-135U* Bark like a Dawg *operated out of Mildenhall during the late 1980s. Initially a minor modification program,* Combat Sent *later added a chin radome, a second radome below the forward fuselage, two 'rabbit's ear' fairings above the cheek fairings, and an angular fairing over the boom position, to produce the most radically-altered sub-type in service.

British Aerospace Nimrod

*A number of Nimrods was modified for the carriage of **BOZ** chaff/flare pods during the Gulf War. These aircraft also received an underwing **FLIR** turret to starboard and a fin-mounted **TRD** (Towed Radar Decoy).*

Nimrod MR.Mk 2

Cutaway key
1 Radome
2 Taxiing lamp
3 EMI Searchwater radar scanner
4 Windscreen panels and wipers
5 Flight refuelling probe
6 Two-pilot flight deck
7 Pitot head
8 Interrogator unit
9 Doppler antenna
10 Forward radio crate
11 Flight engineer's station
12 Escape hatch (inoperable with refuelling boom fitted)
13 Starboard electrical equipment crate
14 Scarbe aerials
15 Upper IFF aerial
16 Periscope sextant aperture
17 Starboard side forward entry door
18 Systems equipment crate
19 Toilet compartment
20 Nose undercarriage wheel bay
21 Weapons bay door ground control panel
22 AGM-86A Harpoon air-to-surface anti-shipping missile
23 Weapons bay doors
24 Underfloor weapons bay
25 Armament distribution panel
26 Black-out curtain
27 Domed observation window, port and starboard
28 Port beam observer's station
29 Starboard beam observer's station
30 Dual sonic homers
31 ADF loop aerial
32 Navigation console
33 Route navigator's station
34 Tactical navigator's station
35 Observation window, starboard only
36 AEO's seat
37 Communications officer's seat
38 Searchwater radar equipment crates
39 Sonics operators' stations (2)
40 UHF/VHF aerials
41 Loran aerial
42 Avionics equipment cooling air ducts
43 ASV operator's seat
44 AC electrical equipment crate, port and starboard
45 Emergency exit window hatches, port and starboard

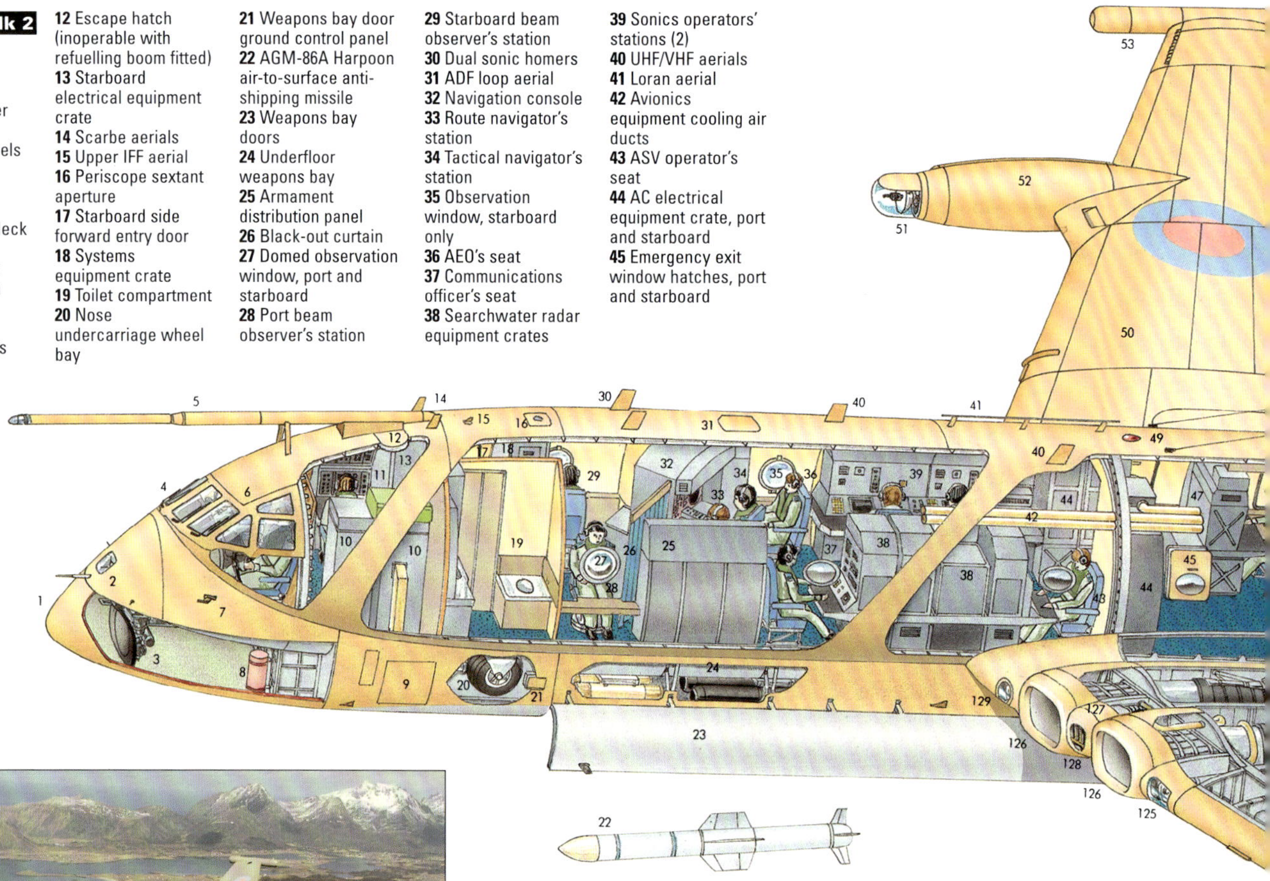

46 Document stowage
47 ESM/MAD console
48 ESM/MAD operator's station
49 Anti-collision light
50 Starboard wing panel
51 Steerable searchlight, 70-million candlepower
52 External fuel tank
53 Wingtip ECM pod
54 Starboard aileron
55 Airbrake panel, upper and lower surfaces
56 Starboard plain flaps
57 Galley compartment
58 Crew rest area
59 Curtained bulkheads
60 Acoustic equipment crates (2)
61 Sonobuoy stowage racks, two-port, six-starboard
62 Stores loader's station
63 Pressurised launcher
64 Rotary launchers (2)

For onboard crew training in the ASW role, the Nimrod can utilise the Airborne Crew Trainer Mk 1 (ACT-1). This consists of a single exercise control unit driven by a reel of magnetic tape containing a software programme which simulates submarine threats on the AQS-901 acoustic detection and display system. This allows the crew to train in authentic conditions, without expending sonobuoys.

SPECIFICATION

Nimrod MR.Mk 2

Dimensions

Length: 126 ft 9 in (38.63 m)
Height: 29 ft 8½ in (8.60 m)
Wing span: 114 ft 10 in (35.00 m)
Wing area: 2,121 sq ft (197.04 m²)
Tailplane span: 47 ft 7¼ in (14.51 m)
Wheel track: 28 ft 2½ in (8.60 m)
Wheel base: 46 ft 8½ in (14.24 m)

Powerplant

Four Rolls-Royce RB.168-20 Spey Mk 250 turbofans each rated at 12,140 lb st (54.00 kN)

Weights

Typical empty: 86,000 lb (39010 kg)
Maximum normal take-off: 177,500 lb (80514 kg)
Maximum overload take-off: 192,000 lb (87091 kg)

Fuel and load

Internal fuel: 85,840 lb (38937 kg) plus provision for 15,100 lb (6849 kg) of auxiliary fuel in six weapon-bay tanks
Maximum ordnance: 13,500 lb (6124 kg)

Performance

Maximum speed at optimum altitude: 500 kt (575 mph; 926 km/h)
Maximum cruising speed at optimum altitude: 475 kt (547 mph; 880 km/h)
Economical cruising speed at optimum altitude: 425 kt (490 mph; 787 km/h)
Typical patrol speed at low level: 200 kt (230 mph; 370 km/h) on two engines
Service ceiling: 42,000 ft (12800 m)
Take-off run at normal maximum take-off weight: 4,800 ft (1463 m)
Landing run at normal landing weight: 5,300 ft (1615 m)
Ferry range: 5,000 nm (5,758 miles; 9266 km)
Endurance: 12 hours typical, 15 hours maximum and 19 hours with one refuelling

Armament

Weapons bay can hold up to six lateral rows of ASW weapons including up to nine torpedoes as well as bombs or depth charges. Bay in rear fuselage for storing and launching active and passive sonobuoys and marine markers through two rotary and two single-barrel launchers. Hardpoints beneath each wing can carry two AIM-9 Sidewinder AAMs, a Harpoon air-surface missile, rockets, cannon pods, torpedoes or ECM equipment

65 Starboard side emergency door
66 UHF aerial
67 Intercom panel
68 Retro-launcher
69 Spare camera magazines
70 Hat rack
71 ECM pre-amplifier
72 Conditioning air intake, port only
73 Extended finroot fillet
74 HF aerial cables
75 Starboard tailplane
76 Fintip ECM aerial fairing
77 Rudder
78 MAD sensor boom
79 Elevator trim tab
80 Port elevator
81 Auxiliary fins, above and below
82 Port tailplane
83 Ventral fin
84 Tail bumper
85 Lower sonics aerial
86 Systems equipment air-conditioning pack
87 Auxiliary Power Unit (APU)
88 Internal walkway and tailcone access
89 Rear pressure dome
90 Cooling air blowers
96 Wingroot fillet
97 Dinghy stowage compartment
98 Engine exhaust nozzles
99 Thrust reverser cascades, outboard engines only
100 Port two-segment plain flaps
101 Fuel vent pipe
102 Dual fuel jettison pipes
103 Port airbrake panel
111 Aileron actuator linkage
112 Wing bumper
113 Port external fuel tank
128 Heat exchanger air intake
129 Taxiing lamp

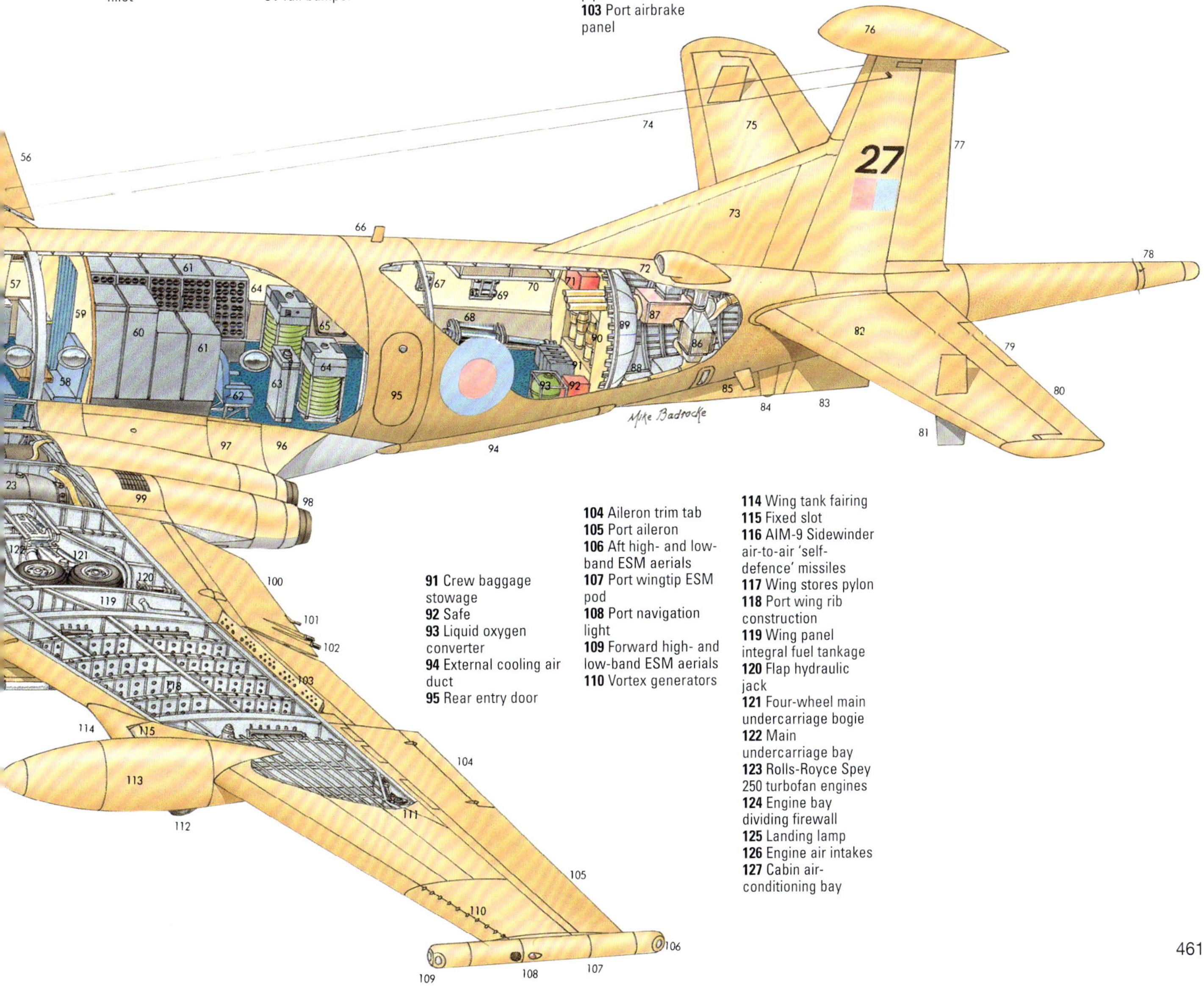

104 Aileron trim tab
105 Port aileron
106 Aft high- and low-band ESM aerials
107 Port wingtip ESM pod
108 Port navigation light
109 Forward high- and low-band ESM aerials
110 Vortex generators
91 Crew baggage stowage
92 Safe
93 Liquid oxygen converter
94 External cooling air duct
95 Rear entry door
114 Wing tank fairing
115 Fixed slot
116 AIM-9 Sidewinder air-to-air 'self-defence' missiles
117 Wing stores pylon
118 Port wing rib construction
119 Wing panel integral fuel tankage
120 Flap hydraulic jack
121 Four-wheel main undercarriage bogie
122 Main undercarriage bay
123 Rolls-Royce Spey 250 turbofan engines
124 Engine bay dividing firewall
125 Landing lamp
126 Engine air intakes
127 Cabin air-conditioning bay

Grumman E-2 Hawkeye

A VAW-116 E-2 serenely cruises over the Pacific. All five Hawke squadrons of the Pacific Fleet are equipped with the Group II version, was VAW-114 'Hormel Hawgs" before its disbandment in March 19

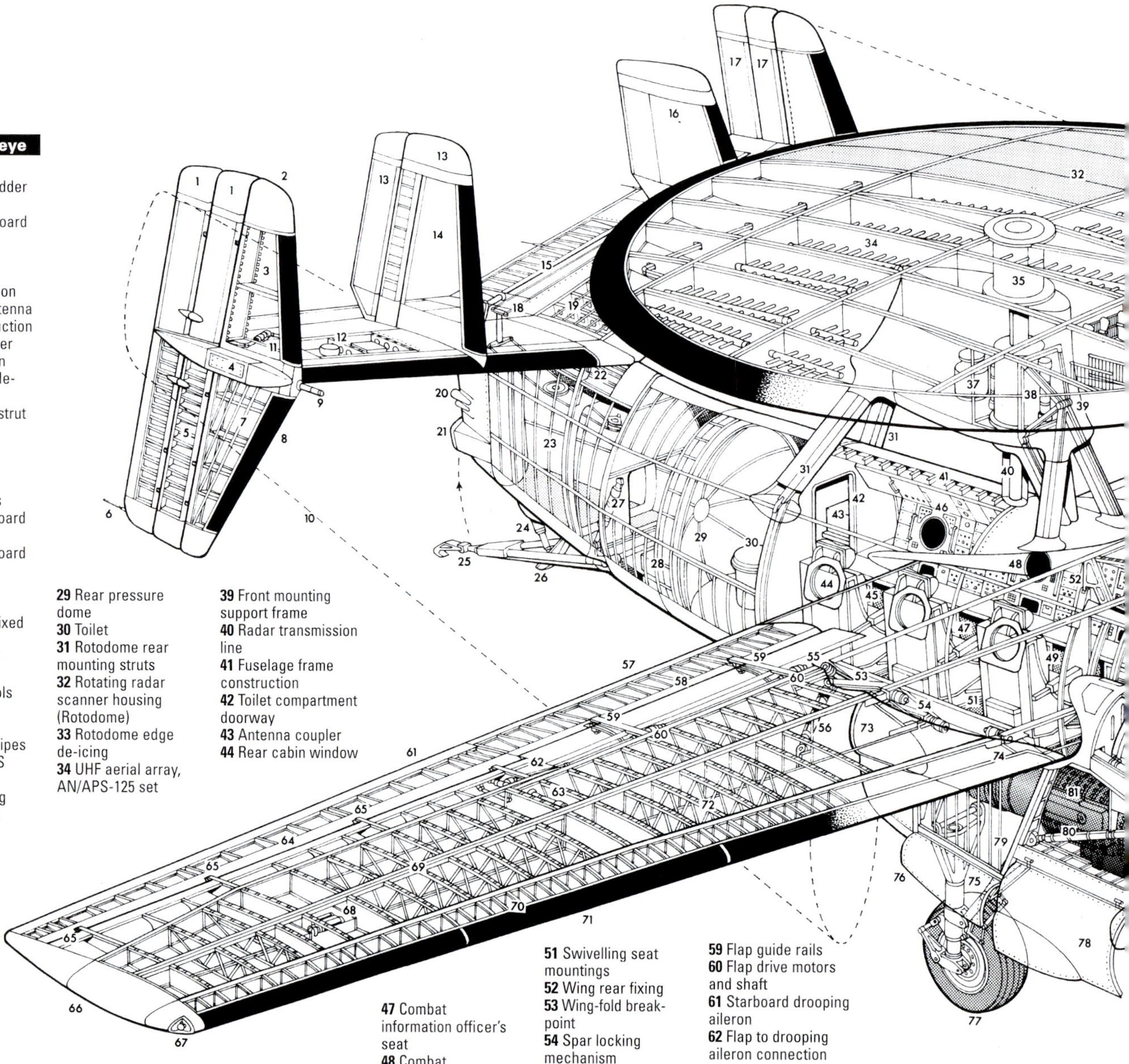

E-2C Hawkeye

Cutaway key

1 Two-section rudder panels
2 Starboard outboard fin
3 Glassfibre fin construction
4 Passive detection system (PDS) antenna
5 Rudder construction
6 Static discharger
7 Fin construction
8 Leading-edge de-icing
9 Wing-fold jury strut lock
10 Wing folded position
11 Rudder jack
12 PDS receivers
13 Starboard inboard rudder sections
14 Starboard inboard glassfibre fin
15 Port elevator construction
16 Port inboard fixed fin
17 Port outboard rudder sections
18 Rudder controls
19 Tailplane construction
20 Fuel jettison pipes
21 Rearward PDS antenna
22 Tailplane fixing
23 Rear fuselage construction
24 Tailskid jack
25 Arrester hook
26 Tailskid
27 Arrester hook jack
28 Lower PDS receiver and antenna
29 Rear pressure dome
30 Toilet
31 Rotodome rear mounting struts
32 Rotating radar scanner housing (Rotodome)
33 Rotodome edge de-icing
34 UHF aerial array, AN/APS-125 set
35 Pivot bearing housing
36 IFF aerial array
37 Rotodome motor
38 Hydraulic lifting jack
39 Front mounting support frame
40 Radar transmission line
41 Fuselage frame construction
42 Toilet compartment doorway
43 Antenna coupler
44 Rear cabin window
45 Air controller's seat
46 Radar and instrument panels
47 Combat information officer's seat
48 Combat information radar panel
49 Radar operator
50 Radar panel and instruments
51 Swivelling seat mountings
52 Wing rear fixing
53 Wing-fold break-point
54 Spar locking mechanism
55 Wing-fold hinge jack
56 Wing-folding hydraulic jack
57 Starboard outboard flap
58 Flap construction
59 Flap guide rails
60 Flap drive motors and shaft
61 Starboard drooping aileron
62 Flap to drooping aileron connection
63 Aileron jack
64 Aileron construction
65 Aileron hinges
66 Starboard wingtip construction
67 Navigation light
68 Jury strut locking mechanism
69 Outer wing construction
70 Leading-edge construction

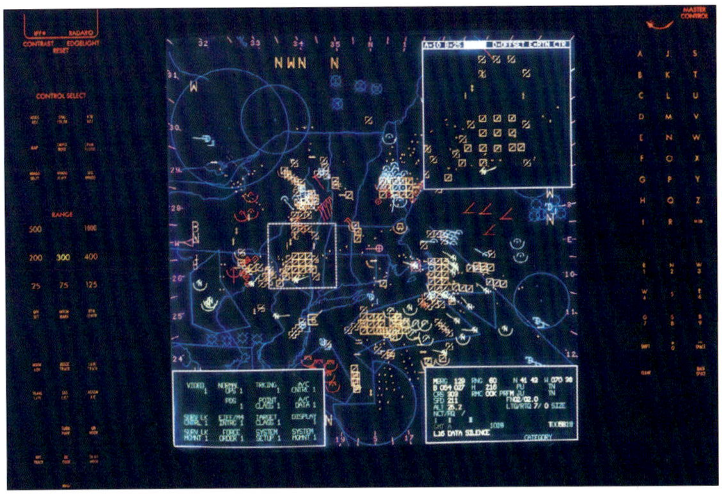

Hawkeye WSOs view tactical displays like the one above. The 11-in (27.94-cm) screen displays a background map, with colour-coded symbols detailing the origin, status, vector and intentions of the objects causing the radar returns.

SPECIFICATION

E-2C Hawkeye

Dimensions

Length: 57 ft 6¾ in (17.54 m)
Height: 18 ft 3¾ in (5.58 m)
Wingspan: 80 ft 7 in (24.56 m)
Wing, folded width: 29 ft 4 in (8.94 m)
Wing, aspect ratio: 9.3
Wing area: 700.00 sq ft (65.03 m²)
Tailplane span: 26 ft 2½ in (7.99 m)
Wheel track: 19 ft 5¾ in (5.93 m)
Wheel base: 23 ft 2 in (7.06 m)

Powerplant

Two Allison T56-A-425 turboprop engines, each rated at 4,910 ehp (3661 kW)

Empty: 38,063 lb (17265 kg)
Maximum take-off: 51,933 lb (23556 kg)
Internal fuel: 12,400 lb (5624 kg)

Performance

Maximum level speed: 323 kt (372 mph; 598 km/h)
Maximum cruising speed at optimum altitude: 311 kt (358 mph; 576 km/h)
Ferry cruising speed at optimum altitude: 268 kt (308 mph; 496 km/h)
Ferry range: 1,394 nm (1,605 miles; 2583 km)
Operational radius: 175 nm (200 miles; 320 km) for a patrol of 3 to 4 hours
Endurance with maximum fuel: 6 hours 6 minutes
Maximum rate of climb at sea level: 2,515 ft (767 m) per minute
Service ceiling: 30,800 ft (9390 m)
Minimum take-off run: 2,000 ft (610 m)

71 Leading-edge de-icing
72 Lattice rib construction
73 Engine exhaust pipe fairing
74 Front spar locking mechanism
75 Main undercarriage leg
76 Undercarriage leg door
77 Single mainwheel
78 Mainwheel door
79 Engine pylon construction
80 Engine mounting strut
81 Allison T56-A-425 engine
82 Oil cooler
83 Oil cooler intake
84 Engine intake
85 Hamilton Standard four-bladed propeller
86 Gearbox drive shaft
87 Propeller mechanism
88 Cooling air intake
89 Engine-to-propeller gearbox
90 Oil tank, usable capacity 9.25 US gal (35 litres) each nacelle
91 Bleed air supply duct
92 Vapour cycle air-conditioning plant
93 Wing front fixing
94 Computer bank
95 Wing centre rib joint
96 Inboard wing section fuel tank, capacity 912 US gal (3452 litres) each wing
97 Lattice rib construction
98 Port inboard flap
99 Wing-fold hinge
100 Wing-fold joint line
101 Sloping hinge rib
102 Port outboard flap
103 Aileron jack
104 Port aileron
105 Port outer wing panel
106 Port wingtip
107 Navigation light
108 Leading-edge de-icing
109 Aileron control cable mechanism
110 Engine mounting strut attachment
111 Engine-to-propeller gearbox
112 Propeller spinner fairing
113 Hamilton Standard four-bladed propeller
114 Engine intake
115 Gearbox drive shaft
116 Port engine
117 Fuel system piping
118 Cooling air intake
119 Vapour cycle system radiator
120 Cooling air outlet duct
121 Radar processor
122 IFF processor
123 Radar transmission line
124 Range finder amplifier
125 Port side entry doorway
126 Equipment cooling air duct
127 Port side equipment rack
128 Starboard side radio and electronics racks
129 Radar duplexer
130 Electronics boxes
131 Forward fuselage frame construction
132 Lower electronics racks
133 Scrambler boxes
134 Navigation equipment
135 Cockpit air-conditioning duct
136 Cockpit doorway
137 Electrical system junction box
138 Air-conditioning diffuser
139 Signal equipment
140 Cockpit floor level
141 Co-pilot's seat
142 Parachute stowage
143 Pilot's seat
144 Headrest
145 Cockpit roof window
146 Cockpit roof construction
147 Instrument panel shroud
148 Windscreen wiper
149 Bulged cockpit side window
150 Instrument panel
151 Control column
152 Nose undercarriage strut
153 Nose undercarriage door
154 Rudder pedals
155 Nose construction
156 Pilot head
157 Sloping front bulkhead
158 Navigation code box
159 Nose electrical junction box
160 Rudder pedal linking mechanism
161 Windscreen heater unit
162 Nose undercarriage leg
163 Steering mechanism
164 Twin nosewheels
165 Catapult strop attachment arm
166 Nosewheel leg door
167 Nosewheel emergency air bottle
168 Nose PDS receivers
169 Oxygen tank
170 Landing lamp
171 Landing and taxi light window
172 Nose PDS antenna array
173 Nose aerial fairing

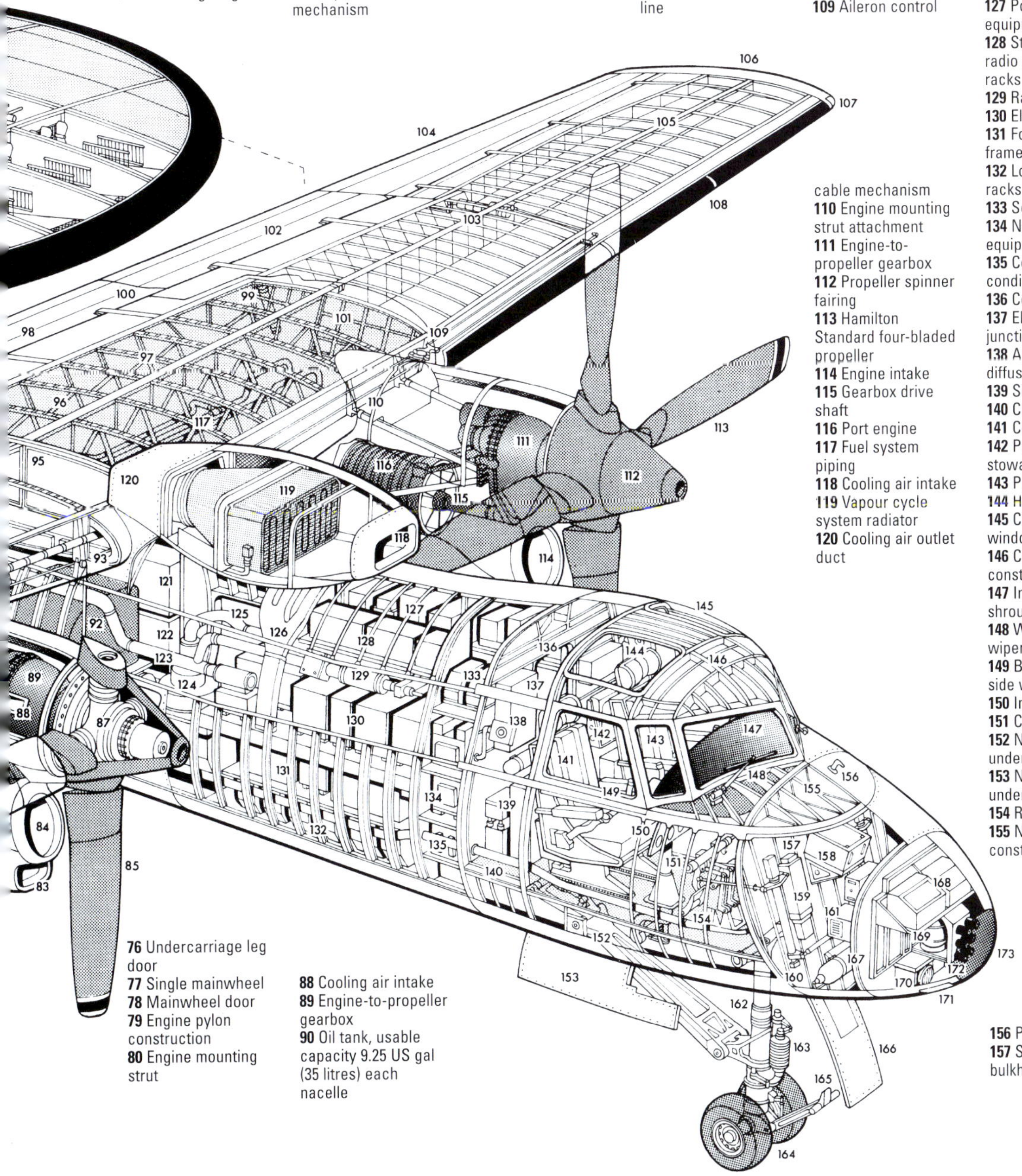

Grumman OV-1/RV-1 Mohawk

The first Mohawk variant equipped with a SLAR pod was the AO-1B (later OV-1B), 101 of which were built. To offset the weight and drag of the AN/APS-94B pod, the AO-1B had longer span wings of 48 ft (14.6 m), compared to the 42 ft (12.8 m) of the AO-1A). The third AO-1B, 59-2623, has been pictured shortly after completion. Note the large defensive flare dispensers atop the aircraft's wing, at its junction with the fuselage.

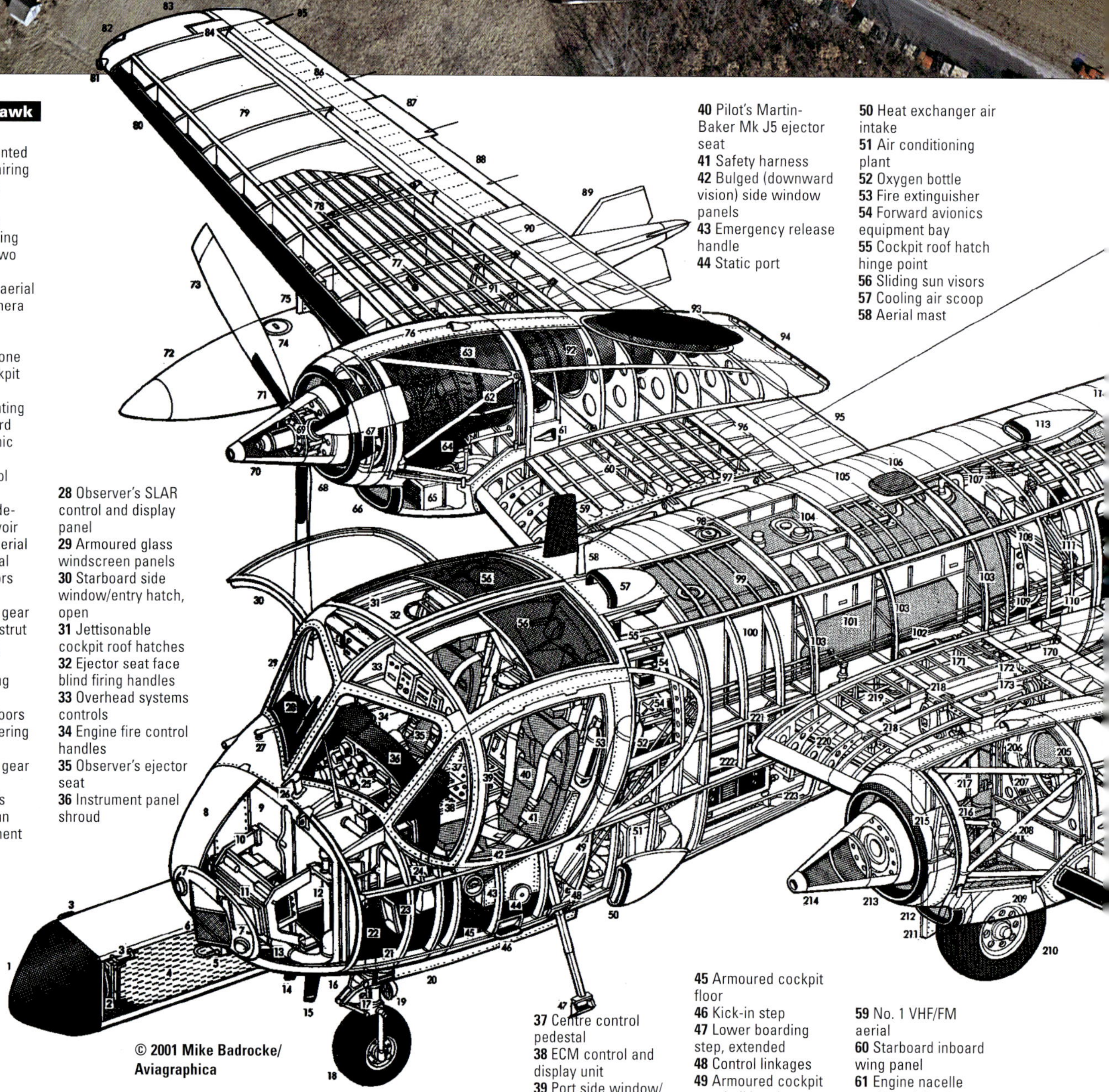

OV-1D Mohawk

Cutaway key

1 Externally mounted SLAR antenna fairing
2 Antenna tilting mechanism
3 Lifting handles
4 Sideways-looking radar antenna, two back-to-back
5 ILS glideslope aerial
6 'Flat Plate' camera aperture
7 ECM aerials
8 Hinged nose cone
9 Armoured cockpit front bulkhead
10 Camera mounting
11 KA 60c forward oblique panoramic camera
12 Rudder control torque shaft
13 Windscreen de-icing fluid reservoir
14 Forward IFF aerial
15 Data link aerial
16 Torque scissors links
17 Nose landing gear shock absorber strut
18 Aft retracting nosewheel
19 Landing taxiing lamp
20 Nosewheel doors
21 Hydraulic steering control unit
22 Nose landing gear leg pivot
23 Rudder pedals
24 Control column
25 Pilot's instrument panel
26 Pitot head
27 Windscreen wipers
28 Observer's SLAR control and display panel
29 Armoured glass windscreen panels
30 Starboard side window/entry hatch, open
31 Jettisonable cockpit roof hatches
32 Ejector seat face blind firing handles
33 Overhead systems controls
34 Engine fire control handles
35 Observer's ejector seat
36 Instrument panel shroud
37 Centre control pedestal
38 ECM control and display unit
39 Port side window/entry hatch
40 Pilot's Martin-Baker Mk J5 ejector seat
41 Safety harness
42 Bulged (downward vision) side window panels
43 Emergency release handle
44 Static port
45 Armoured cockpit floor
46 Kick-in step
47 Lower boarding step, extended
48 Control linkages
49 Armoured cockpit rear bulkhead
50 Heat exchanger air intake
51 Air conditioning plant
52 Oxygen bottle
53 Fire extinguisher
54 Forward avionics equipment bay
55 Cockpit roof hatch hinge point
56 Sliding sun visors
57 Cooling air scoop
58 Aerial mast
59 No. 1 VHF/FM aerial
60 Starboard inboard wing panel
61 Engine nacelle cooling air intake

© 2001 Mike Badrocke/
Aviagraphica

In order to evaluate the suitability of the Mohawk in the armed reconnaissance/ ground support role, two OV-1As were modified as JOV-1As, equipped with an extra 500-lb (227-kg) stores station on each wing. A cockpit gunsight, gun firing and stores release equipment and armour plating were also added. Stores cleared for carriage by the aircraft included 0.50-in (12.7-mm) machine-gun pods, 2.75-in (7-cm) and 5-in (12.7-cm) FFARs, 5-in (12.7-cm) HVARs, 250-, 500- and 1,000-lb (114-, 227 and 454-kg) low-drag bombs, fire bombs and Sidewinder air-to-air missiles.

SPECIFICATION

OV-1D Mohawk

Dimensions

Wingspan: 48 ft (14.63 m)
Wing area: 360 sq ft (33.45 m²)
Length (including SLAR pod): 44 ft 11 in (13.69 m)
Height: 13 ft (3.96 m)

Powerplant

Two Lycoming T53-L-701 turboprops each rated at 1,400 hp (1044 kW)

Weights

Empty: 11,757 lb (5333 kg)
Loaded: 15,741 lb (7140 kg)
Maximum: 18,109 lb (8214 kg)
Wing loading*: 43.7 lb/sq ft (213.5 kg/m²)

Power loading*: 5.6 lb/shp (2.6 kg/shp)

Performance

Maximum speed: 305 mph (491 km/h) at 5,000 ft (1525 m)
Cruising speed: 207 mph (333 km/h) **Climb rate:** 3,618 ft/min (18 m/sec)
Service ceiling: 25,000 ft (7620 m)
Maximum range: 1,010 miles (1625 m)

*Wing and power loadings are calculated at normal loaded weight and maximum take-off power

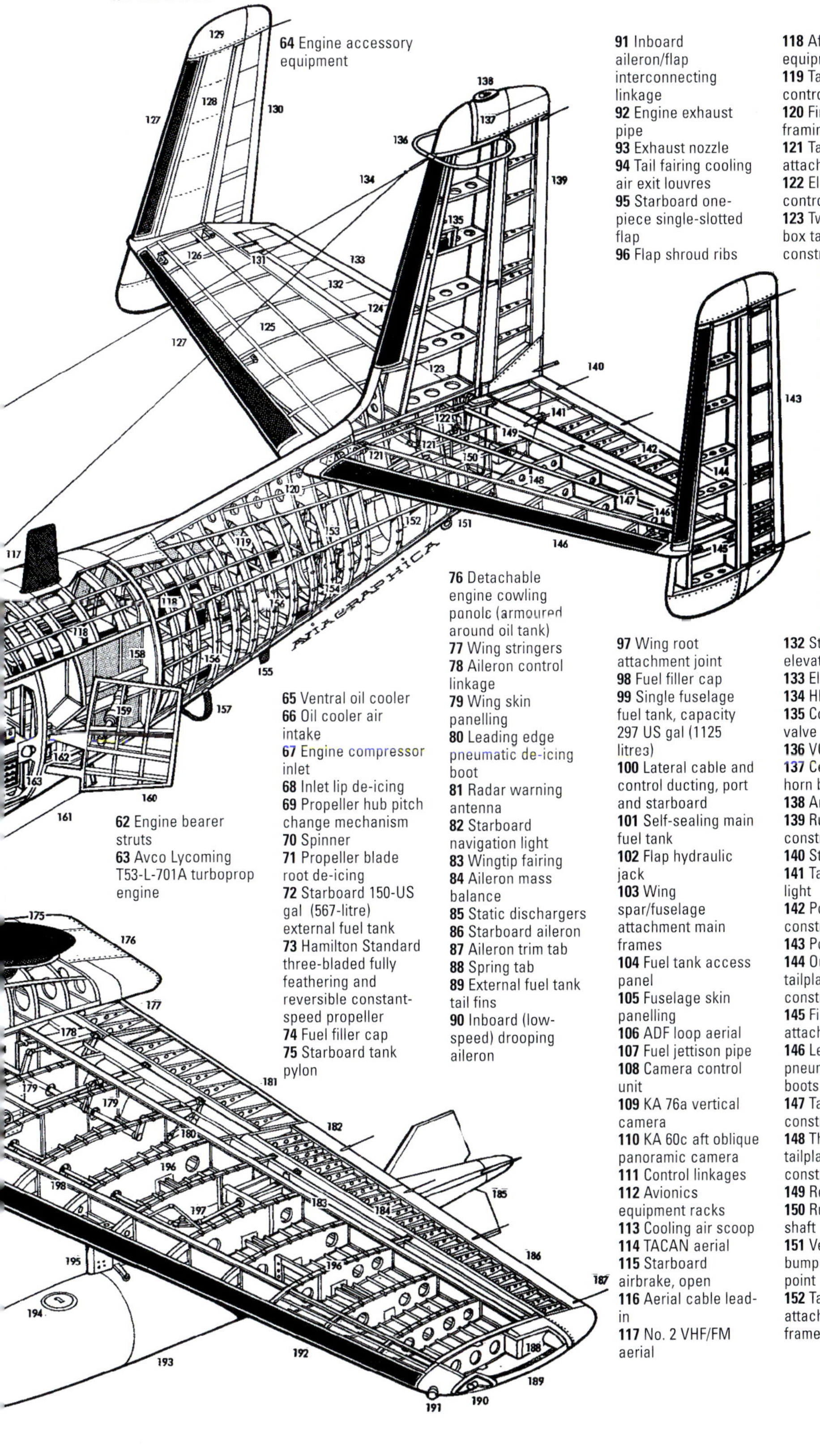

62 Engine bearer struts
63 Avco Lycoming T53-L-701A turboprop engine
64 Engine accessory equipment
65 Ventral oil cooler
66 Oil cooler air intake
67 Engine compressor inlet
68 Inlet lip de-icing
69 Propeller hub pitch change mechanism
70 Spinner
71 Propeller blade root de-icing
72 Starboard 150-US gal (567-litre) external fuel tank
73 Hamilton Standard three-bladed fully feathering and reversible constant-speed propeller
74 Fuel filler cap
75 Starboard tank pylon
76 Detachable engine cowling panels (armoured around oil tank)
77 Wing stringers
78 Aileron control linkage
79 Wing skin panelling
80 Leading edge pneumatic de-icing boot
81 Radar warning antenna
82 Starboard navigation light
83 Wingtip fairing
84 Aileron mass balance
85 Static dischargers
86 Starboard aileron
87 Aileron trim tab
88 Spring tab
89 External fuel tank tail fins
90 Inboard (low-speed) drooping aileron
91 Inboard aileron/flap interconnecting linkage
92 Engine exhaust pipe
93 Exhaust nozzle
94 Tail fairing cooling air exit louvres
95 Starboard one-piece single-slotted flap
96 Flap shroud ribs
97 Wing root attachment joint
98 Fuel filler cap
99 Single fuselage fuel tank, capacity 297 US gal (1125 litres)
100 Lateral cable and control ducting, port and starboard
101 Self-sealing main fuel tank
102 Flap hydraulic jack
103 Wing spar/fuselage attachment main frames
104 Fuel tank access panel
105 Fuselage skin panelling
106 ADF loop aerial
107 Fuel jettison pipe
108 Camera control unit
109 KA 76a vertical camera
110 KA 60c aft oblique panoramic camera
111 Control linkages
112 Avionics equipment racks
113 Cooling air scoop
114 TACAN aerial
115 Starboard airbrake, open
116 Aerial cable lead-in
117 No. 2 VHF/FM aerial
118 Aft avionics equipment racks
119 Tailplane autopilot controllers
120 Fin root fillet framing
121 Tailplane attachment joints
122 Elevator hinge control links
123 Two-spar torsion box tailfin construction
124 Fin leading edge pneumatic de-icing boot
125 Starboard tailplane
126 Outboard rudder interconnecting linkage
127 Leading edge pneumatic de-icing boots
128 Starboard tailfin
129 Rudder horn balance
130 Starboard rudder
131 Elevator mass balance
132 Starboard elevator
133 Elevator tab
134 HF aerial cable
135 Compass flux valve
136 VOR aerial
137 Centre rudder horn balance
138 Anti-collision light
139 Rudder rib construction
140 Static dischargers
141 Tail navigation light
142 Port elevator rib construction
143 Port rudder
144 Outboard tailplane rib construction
145 Fin/tailplane attachment joints
146 Leading edge pneumatic de-icing boots
147 Tailplane rib construction
148 Three-spar tailplane torsion box construction
149 Rear IFF aerial
150 Rudder torque shaft
151 Ventral tail bumper/ tie-down point
152 Tailplane attachment main frames
153 Rear fuselage frame and stringer construction
154 Fuselage lower longeron
155 Lower TACAN aerial
156 Radar altimeter aerials
157 FM homing aerial
158 Port airbrake housing
159 Hydraulic jack
160 Port airbrake construction
161 ADF sense aerial
162 Airbrake hinge point
163 Equipment bay access door, port and starboard
164 Electrical system equipment
165 Ground power socket
166 Battery
167 Ventral VHF/UHF aerial
168 Camera equipment light sensor
169 Marker beacon antenna
170 Port flap operating rod
171 Stub wing construction
172 Rear spar bolted joint
173 Main landing gear leg pivot fixing
174 Port single-slotted flap
175 Port engine exhaust nozzle
176 Nacelle tail fairing
177 Flap rib construction
178 Outboard flap operating rod
179 Flap/drooping aileron interconnection
180 Swinging link flap/aileron hinge
181 Port low-speed drooping aileron
182 Aileron geared tab
183 Rear spar
184 Aileron rib construction
185 External tank tail fins
186 Port aileron
187 Static dischargers
188 Aileron mass balance
189 Wingtip fairing
190 Port navigation light
191 Radar warning antenna
192 Leading edge de-icing boot
193 Port 150-US gal (567-litre) external fuel tank
194 Fuel filler cap
195 Port fuel tank pylon
196 Wing rib construction
197 Aileron control linkage
198 Front spar
199 Aileron interconnecting link
200 Auxiliary centre spar
201 Main landing gear wheel bay
202 Engine nacelle framing
203 Port engine exhaust duct
204 Nacelle venting air intake
205 Rear engine mounting mainframe
206 Main landing gear hydraulic retraction jack
207 Side breaker strut
208 Engine bearer struts
209 Lower hinged engine cowling panels
210 Port main wheel
211 Main wheel doors
212 Oil cooler air intake
213 Engine air inlet
214 Port spinner
215 Cowling nose ring
216 Forward engine mounting ring frame
217 Engine oil tank, capacity 2½ US gal (9.50 litres)
218 Forward and centre spar bolted joints
219 Aileron autopilot controller
220 Leading edge engine control runs
221 SLAR signal receiver (interchangeable with IR receiver)
222 SLAR signal processor (interchangeable with IR recorder)
223 Ventral equipment bay access doors

Lockheed S-3 Viking

In all 187 of 199 planned Vikings were buil about 119 of these eventually being reworke to S-3B standard. Eight of the original aircra were pre-production prototypes and tes airframes. Production ended in 197

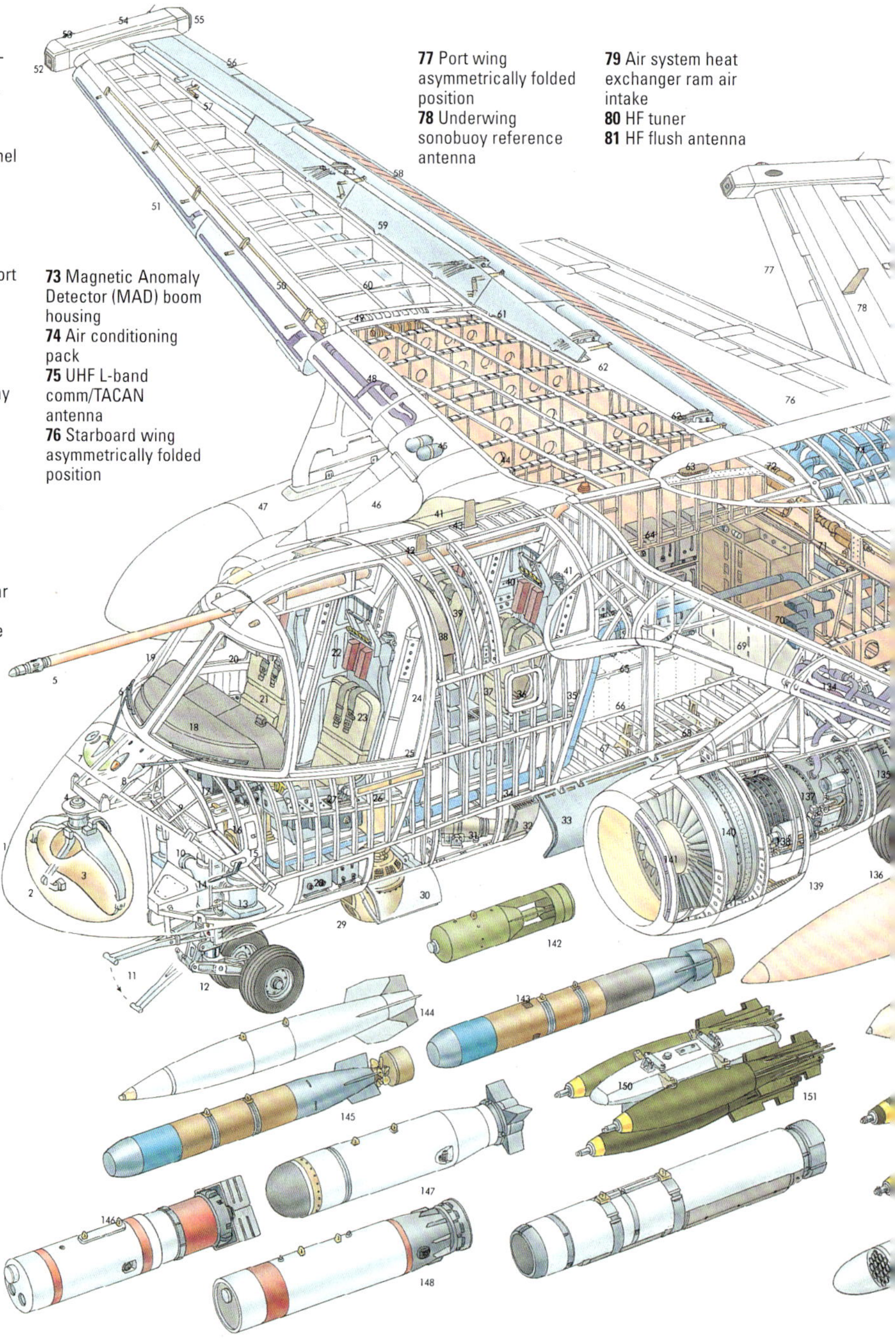

S-3B Viking

Cutaway key

1 Upward-hinging glass-fibre radome
2 Scanner protective housing
3 AN/APS(V)1 radar scanner
4 Rotating scanner mounting
5 Retractable flight refuelling probe
6 Windscreen wipers
7 Windscreen de-icing fluid reservoir
8 Forward identification light
9 Cockpit front pressure bulkhead
10 Nose undercarriage leg pivot mounting
11 Catapult strop link
12 Trailing link nosewheel suspension
13 Cabin conditioning and pressurisation outflow valves
14 Pitot head
15 Canopy external release
16 Rudder pedals
17 Instrument panel
18 Instrument panel shroud
19 Electrically heated windscreen panels
20 Overhead switch panels
21 Second pilot's seat
22 Tactical co-ordinator's (TACCO's) console
23 Pilot's Escapac 1-E ejection seat
24 Seat mounting/ejection rails
25 Jettisonable side window hatch
26 Electro-luminescent formation lighting strip
27 Engine throttle levers
28 OR-89AA infra-red equipment bay, radar equipment to starboard
29 Retractable Forward-Looking Infra-Red (FLIR) turret
30 FLIR turret doors
31 Auxiliary Power Unit (APU) bay, crew entry hatch on starboard side
32 APU exhaust duct
33 Port weapons bay door
34 Cabin conditioning air ducting
35 Sloping seat mounting bulkhead
36 Side window with rotating Polaroid blind
37 Sensor operator's (SENSO's) seat
38 SENSO's instrument console
39 TACCO's seat
40 Circuit breaker panels
41 Rear crew compartment ejection/escape hatch
42 UHF L-band UHF/IFF antenna
43 VHF antenna
44 Fixed inboard wing panel integral fuel tank, total internal capacity 1900 US gal (7192 litres)
45 Engine fire suppression bottles
46 Starboard engine pylon
47 CNU-264 cargo pod
48 De-icing air ducting
49 Starboard wing-fold hinge joint and rotary actuator
50 Leading-edge torque shaft and actuating links
51 Starboard drooped leading edge
52 Forward and forward oblique ECM antennas
53 Starboard navigation light
54 Wing-tip ECM equipment pod
55 Aft and aft oblique ECM antennas
56 Starboard aileron
57 Aileron hinge link
58 Starboard single-slotted flap
59 Outboard spoiler panels
60 Ventral airbrake/spoiler panel
61 Inboard spoiler panel
62 Flap guide rails
63 ADF antenna
64 Avionics equipment racks, port and starboard
65 Starboard weapons bay
66 Equipment bay centre aisle
67 Port weapons bay
68 BRU-14A bomb racks
69 Univac main computer
70 'Cold-plate' avionics cooling air ducts
71 Control surface actuators, ailerons and spoilers, on rear face of rear spar
72 Central flap drive unit
73 Magnetic Anomaly Detector (MAD) boom housing
74 Air conditioning pack
75 UHF L-band comm/TACAN antenna
76 Starboard wing asymmetrically folded position
77 Port wing asymmetrically folded position
78 Underwing sonobuoy reference antenna
79 Air system heat exchanger ram air intake
80 HF tuner
81 HF flush antenna

Sharing ramp space at Palmdale with the first, third, and seventh prototype and first production aircraft, the fifth S-3A prototype (BuNo. 157996) is run-up prior to another test flight.

SPECIFICATION

S-3A Viking

Dimensions

Wingspan: 68 ft 8 in (20.93 m)
Wingspan (folded): 29 ft 6 in (8.99 m)
Wing area: 598.00 sq ft (55.56 m²)
Length overall: 53 ft 4 in (16.26 m)
Length (tail folded): 49 ft 5 in (15.06 m)
Height overall: 22 ft 9 in (6.93 m)
Height (tail folded): 15 ft 3 in (4.65 m)

Powerplant

Two General Electric TF34-GE-2 turbofans each rated at 9,275 lb st (41.26 kN) dry

Weights

Empty: 26,650 lb (12088 kg)
Normal take-off: 42,500 lb (19277 kg)
Maximum take-off: 52,540 lb (23832 kg)

Fuel and load

Internal fuel: 12,863 lb (5753 kg)
External fuel: up to two 300-US gal (1136-litre) drop tanks
Maximum ordnance: 7,000 lb (3175 kg) including 4,000 lb (1814 kg) carried internally

Performance

Maximum level speed 'clean' at sea level: 439 kt (506 mph; 814 km/h)
Maximum cruising speed at optimum altitude: more than 350 kt (403 mph; 649 km/h)
Patrol speed at optimum altitude: 160 kt (184 mph; 296 km/h)
Ferry range: more than 3,000 nm (3,454 miles; 5558 km)
Operational radius: more than 945 nm (1,088 miles; 1751 km)
Endurance: 7 hours 30 minutes
Maximum rate of climb at sea level: more than 4,200 ft (1280 m) per minute
Service ceiling: more than 35,000 ft (10670 m)

Armament (S-3B)

The S-3B has two internal weapon bays on the 'corners' of its fuselage, able to carry four Mk 46 or 50 torpedoes, four Mk 36, 62 or 82 bomb/destructors, or two B57 nuclear depth charges (no longer carried on US carriers). A wing pylon outboard of each nacelle is able to carry two Mk 52, 55, 56 or 60 mines, six Mk 36, 62 or 82 destructor/bombs, six Mk 7 cluster dispensers, six ADM-141 decoys, six rocket pods, six flare dispensers or two AGM-84 Harpoons or AGM-84E SLAM. In refuelling role the D-704 pod is carried under the port wing, with a 300-US gal (1135-litre) fuel tank to starboard

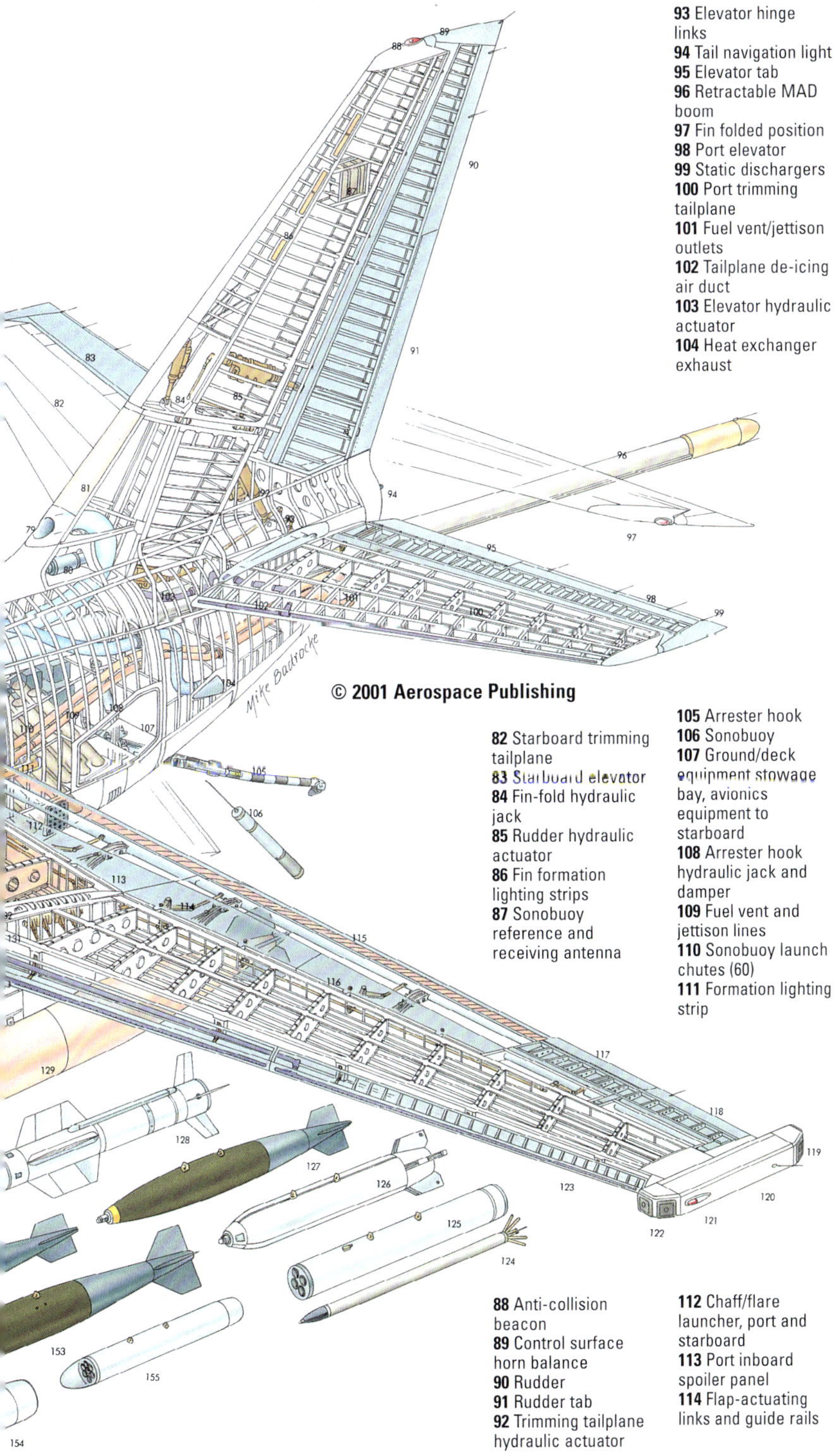

Mike Badrocke

82 Starboard trimming tailplane
83 Starboard elevator
84 Fin-fold hydraulic jack
85 Rudder hydraulic actuator
86 Fin formation lighting strips
87 Sonobuoy reference and receiving antenna
88 Anti-collision beacon
89 Control surface horn balance
90 Rudder
91 Rudder tab
92 Trimming tailplane hydraulic actuator
93 Elevator hinge links
94 Tail navigation light
95 Elevator tab
96 Retractable MAD boom
97 Fin folded position
98 Port elevator
99 Static dischargers
100 Port trimming tailplane
101 Fuel vent/jettison outlets
102 Tailplane de-icing air duct
103 Elevator hydraulic actuator
104 Heat exchanger exhaust
105 Arrester hook
106 Sonobuoy
107 Ground/deck equipment stowage bay, avionics equipment to starboard
108 Arrester hook hydraulic jack and damper
109 Fuel vent and jettison lines
110 Sonobuoy launch chutes (60)
111 Formation lighting strip
112 Chaff/flare launcher, port and starboard
113 Port inboard spoiler panel
114 Flap-actuating links and guide rails
115 Port single-slotted flap
116 Outboard spoiler panels
117 Aileron tab
118 Port aileron
119 Aft/aft oblique ECM antennas
120 Port wingtip ECM pod
121 Port navigation light
122 Forward and forward oblique ECM antennas
123 Port drooped leading edge
124 4.5-in FFAR
125 LAU-10 Zuni rocket launcher
126 Mk 7 dispenser, CBU-59 APAM or Mk 20 Rockeye II
127 Mk 83, 1000-lb HE bomb
128 AGM 84A Harpoon
129 Acro 1D 300-US gal (1136-litre) external tank
130 Port wing stores pylon
131 Wing-fold rotary actuator, hydraulically powered
132 Wing-fold hinge joint
133 Port wing integral fuel tank
134 Engine bleed air ducting
135 Mainwheel leg strut
136 Port mainwheel
137 General Electric TF34-GE-400
138 Engine accessory equipment
139 Fan air exhaust duct
140 Fan casing
141 Intake lip de-icing air duct
142 Mk 54 350-lb depth bomb
143 Mk 50 Barracuda torpedo
144 Mk 57 'Special' weapon
145 Mk 46 torpedo
146 Mk 52 mine
147 Mk 55 moored mine
148 Mk 56 mine
149 Mk 60 Captor mine
150 Triple ejector rack (TER)
151 Mk 36 destructor mines
152 Mk 82 500-lb HE bomb
153 Mk 84 2000-lb HE bomb
154 LAU-69, 19-round 2.75-in rocket launcher
155 LAU-68, 7 round 2.75-in rocket launcher

In 2002 the Viking is very much in the twilight of its career, relegated to anti-surface and land attack offensive taskings and an increasingly important tanking role. Some aircraft will be converted to a permanent IFR role, possible redesignated as KS-3Bs.

Lockheed SR-71 Blackbird

Undoubtedly the most impressive military jet ever built, Lockheed's SR-71 is still officially the world's fastest air-breathing aircraft twelve years after its premature retirement from service in 1990. Tasked with performing highly classified reconnaissance missions, the aircraft's sensor systems collected material that helped to formulate US foreign policy for more than 20 years.

SR-71 Blackbird

Cutaway key
1 Pitot tube
2 Air data probe
3 Radar warning antennas
4 Nose mission equipment bay
5 Panoramic camera aperture
6 Detachable nosecone joint frame
7 Cockpit front pressure bulkhead
8 Rudder pedals
9 Control column
10 Instrument panel
11 Instrument panel shroud
12 Knife-edged windscreen panels
13 Upward-hinged cockpit canopy covers
14 Ejection seat headrest
15 Canopy actuator
16 Pilot's Lockheed F-1 'zero-zero' ejection seat
17 Engine throttle levers
18 Side console panel
19 Fuselage chine close-pitched frame construction
20 Liquid oxygen converters (2)
21 Side console panel
22 Reconnaissance Systems Officer's (RSO) instrument display
23 Cockpit rear pressure bulkhead
24 RSO's Lockheed F-1 'zero-zero' ejection seat
25 Canopy hinge point
26 SR-71B dual-control trainer variant, nose profile
27 Raised instructor's rear cockpit
28 Astro-inertial navigation star tracker
29 Navigation and communications systems electronic equipment
30 Nosewheel bay
31 Nose undercarriage pivot fixing
32 Landing and taxiing lamps
33 Twin nosewheels, forward-retracting
34 Hydraulic retraction jack
35 Cockpit environmental system equipment bay
36 Air refuelling receptacle, open
37 Fuselage upper longeron
38 Forward fuselage frame construction
39 Forward fuselage integral fuel tanks
40 Palletised, interchangeable reconnaissance equipment packs
41 Fuselage chine member
42 Forward/centre fuselage joint ring frame
43 Centre fuselage integral fuel tanks; total system capacity 12,219 US gal (46254 litres)
44 Beta B.120 titanium alloy skin panelling
45 Corrugated wing skin panelling

The inability of surveillance satellites to provide the specific, targetted reconnaissance required by US military commanders saw the brief resurrection of the SR-71, although the type has now once more been retired by the USAF.

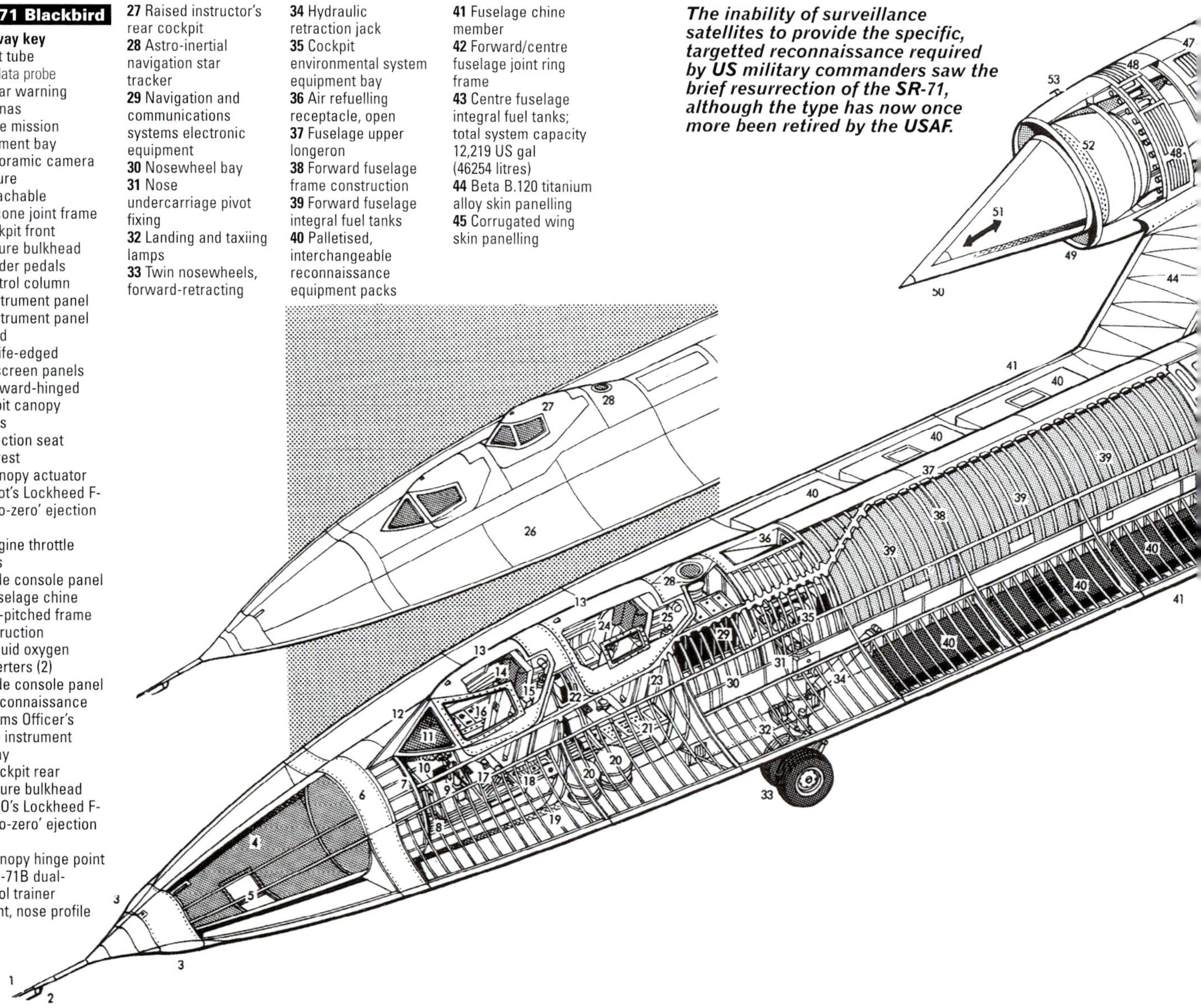

46 Starboard main undercarriage, stowed position
47 Intake centre-body bleed air louvres
48 Bypass duct suction relief louvres
49 Starboard engine air intake
50 Moveable intake conical centre-body
51 Centre-body retracted (high-speed position)
52 Boundary layer bleed air holes
53 Automatic intake control system air data probe
54 Diffuser chamber
55 Variable inlet guide vanes
56 Hinged engine cowling/outer wing panel
57 Pratt & Whitney JT11D-20B (J58) single-spool bleed-bypass engine
58 Engine accessory equipment
59 Bypass duct suction relief doors
60 Compressor bleed air bypass doors
61 Afterburner fuel manifold
62 Tailfin fixed root section
63 Starboard outer wing panel
64 Under-cambered leading edge
65 Outboard roll control elevon
66 All-moving starboard fin
67 Continuously-operating afterburner duct
68 Afterburner nozzle
69 Engine bay tertiary air flaps
70 Exhaust nozzle ejector flaps
71 Variable-area exhaust nozzle
72 Starboard wing integral fuel tank bays
73 Brake parachute doors, open
74 Ribbon parachute stowage
75 Aft fuselage integral fuel tanks
76 Skin doubler
77 Aft fuselage frame construction
78 Elevon mixer unit
79 Inboard elevon torque control unit
80 Tailcone
81 Fuel vent
82 Port all-moving fin
83 Fin rib construction

SPECIFICATION

SR-71A Blackbird

Dimensions

Length overall: 103 ft 10 in (31.65 m)
Length overall (including probe): 107 ft 5 in (32.74 m)
Wing span: 55 ft 7 in (16.94 m)
Wing area: 1,605 sq ft (149.10 m²)
Moving vertical tail area: 70.2 sq ft (6.52 m²)
Height: 18 ft 6 in (5.64 m)
Wheel track: 16 ft 8 in (5.08 m)
Wheel base: 37 ft 10 in (11.53 m)

Powerplant

Two Pratt & Whitney J58 afterburning bleed turbojets, each rated at 32,500 lb (144.57 kN) of thrust with afterburning

Weights

Empty: 67,500 lb (30617 kg)
Maximum take-off: 172,000 lb (78017 kg)

Fuel and load

Total fuel capacity: 12,219 US gal (46254 litres)
Internal sensor payload (approximate): 2,770 lb (1256 kg)

Performance

Design maximum speed: Mach 3.2 - 3.5 at 80,000 ft (24385 m) (limited by structural integrity of windscreen)
Maximum speed: Mach 3.35 at 80,000 ft (24385 m)
Maximum cruising speed: Mach 3.35 at 80,000 ft (24385 m)
Maximum sustained cruising speed: Mach 3.2 or approximately 2,100 mph (3380 km/h) at 80,000 ft (24385 m)
Maximum altitude (approximate): 100,000 ft (30480 m)
Operational ceiling: 85,000 ft (25908 m)
Take-off run at 140,000-lb (63503-kg) gross weight: 5,400 ft (1646 m)
Landing run at maximum landing weight: 3,600 ft (1097 m)

Range

Maximum unrefuelled range at Mach 3.0: 3,250 miles (5230 km)
Operational radius (typical): 1,200 miles (1931 km)
Maximum unrefuelled endurance at Mach 3.0: 1 hour 30 minutes

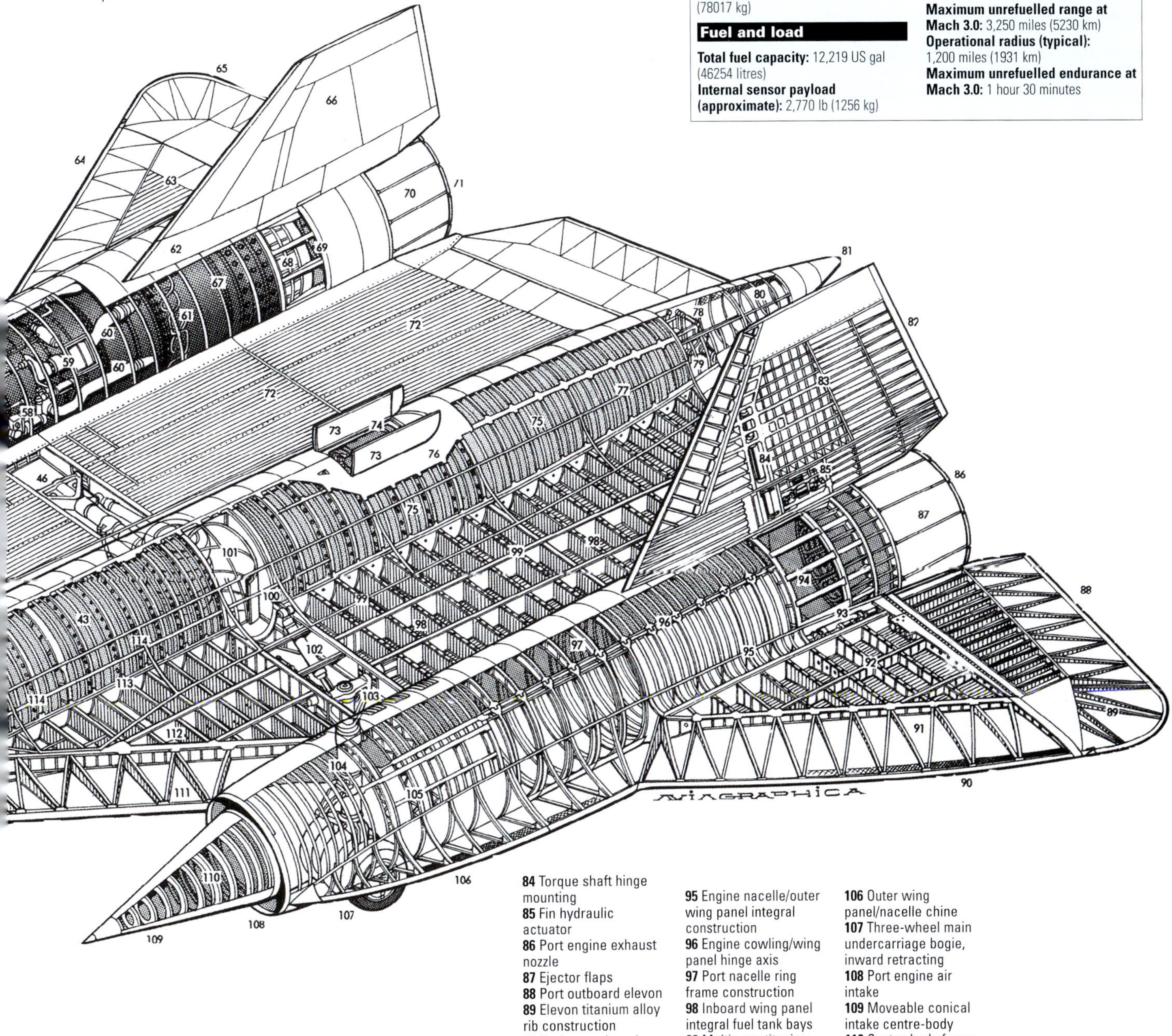

84 Torque shaft hinge mounting
85 Fin hydraulic actuator
86 Port engine exhaust nozzle
87 Ejector flaps
88 Port outboard elevon
89 Elevon titanium alloy rib construction
90 Under-cambered leading edge
91 Leading-edge diagonal rib construction
92 Outer wing panel titanium alloy construction
93 Outboard elevon hydraulic actuator
94 Engine bay tertiary air flaps
95 Engine nacelle/outer wing panel integral construction
96 Engine cowling/wing panel hinge axis
97 Port nacelle ring frame construction
98 Inboard wing panel integral fuel tank bays
99 Multi-spar titanium alloy wing construction
100 Main undercarriage wheel bay
101 Wheel bay heat shield
102 Hydraulic retraction jack
103 Main undercarriage pivot fixing
104 Mainwheel leg strut
105 Intake duct framing
106 Outer wing panel/nacelle chine
107 Three-wheel main undercarriage bogie, inward retracting
108 Port engine air intake
109 Moveable conical intake centre-body
110 Centre-body frame construction
111 Inboard leading-edge diagonal rib construction
112 Inner wing panel integral fuel tank
113 Wingroot/fuselage attachment root rib
114 Close-pitched fuselage titanium alloy frames
115 Wing/fuselage chine blended fairing panels

Lockheed U-2

***Above:* NASA flies the ER-2 from Moffett Field on high-altitude experiment work, having earlier operated the U-2C. The aircraft in the foreground is configured with high-altitude atmospheric sampling equipment.**

U-2R/TR-1A

Cutaway key

1 Nose radome
2 Radar cooling air intake
3 Hughes Advanced Synthetic Aperture Radar System (ASARS-2) antennas
4 Radar system equipment module
5 Interchangeable nose-section mounting bulkhead
6 Avionics equipment bay
7 Pitot head
8 Downward-vision periscope/driftsight
9 Front pressure bulkhead
10 Instrument panel
11 Windscreen panels
12 Cockpit canopy, hinged to port
13 Canopy ultra-violet shield
14 Rear view mirror
15 Canopy emergency release
16 Pilot's zero-zero ejection seat
17 Sloping rear pressure bulkhead
18 Photint system
19 Itec panoramic (horizon-to-horizon) optical bar camera
20 Equipment conditioning airducts
21 Q-bay mission equipment compartment
22 Astro-inertial navigation system equipment package
23 Satellite antenna
24 E-bay avionics equipment compartment
25 Port engine air intake
26 Intake air spill duct
27 Mainwheel doors
28 Twin mainwheels, forward retracting
29 Landing/taxiing lamps
30 Main undercarriage wheel bay
31 Ventral antenna 'farm' – Senior Spear Comint package
32 Engine bay bulkhead
33 Engine compressor intake
34 Hydraulic pumps
35 Liquid oxygen converter
36 Air-conditioning equipment bay
37 Dorsal UHF communications aerial
38 Starboard interchangeable mission equipment superpod
39 Leading-edge stall strip
40 Wingtip skid
41 Starboard navigation light
42 Wingtip threat warning receiver pod
43 Starboard aileron
44 IRCM dummy pod
45 Starboard plain flap, inboard and outboard segments
46 Equipment pod tail fairing
47 Anti-collision light
48 Engine oil tank
49 Wing panel attachment joints
50 Machined wing support mainframes
51 Port wing integral fuel tank
52 Fuel filler cap
53 Wing rib construction
54 Pod support machined ribs
55 Flap shroud ribs
56 Inboard plain flap segment
57 Pratt & Whitney J75-P-13B non-afterburning turbojet
58 Rear fuselage break point, engine removal
59 Extended fin root fillet fairing
60 Communications equipment compartment
61 Starboard trimming tailplane
62 Starboard elevator
63 Fin leading-edge HF-aerial
64 Tail navigation light
65 Fuel vent
66 ECM antenna
67 Rudder
68 Fixed rudder tab

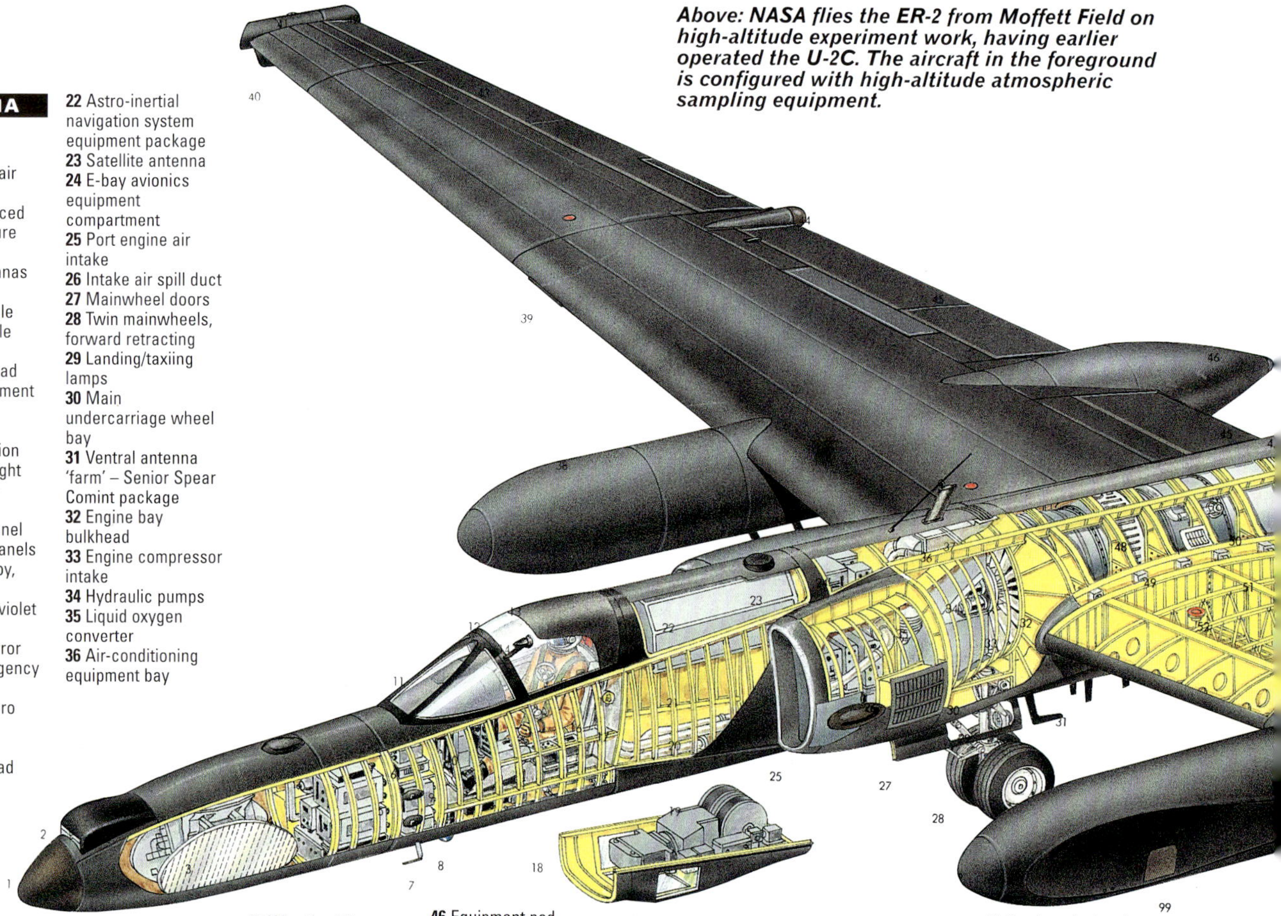

When the USAF retired its final Convair F-106 Delta Dart, the U-2R became the last type in the inventory powered by the Pratt & Whitney J75. Moreover, just as with the U-2A in the 1950s, the U-2R was becoming airframe-limited. Re-engining with the General Electric F118 (similar to the B-2's engine) was the answer, a test aircraft flying for the first time on 23 May 1989. Despite highly encouraging test results, it was not until 1994 that the first production U-2S took to the air. The F118 engine is mounted centrally in the airframe, exhausting through a long duct to the jet pipe. The engine produces much less thrust at its operating altitude, where the U-2 coasts along at a cruise speed not far from its stall speed.

69 Rear threat warning radar receiver
70 Trimming tailplane incidence control jack
71 Elevator tab
72 Port elevator
73 Tailplane leading-edge skin stiffeners
74 Convergent-divergent thrust augmentor nozzle
75 Trimming tailplane pivot point
76 Heat-shrouded jet pipe
77 Ventral mission equipment bay
78 Datalink antenna
79 Tailwheel doors
80 Solid-tyre twin tailwheels
81 Port airbrake
82 Airbrake hydraulic jack
83 Port superpod tail fairing
84 Spoiler/lift dump panels
85 Outboard plain flap segment
86 Fuel jettison
87 Port aileron
88 Wingtip threat warning receiver pod
89 Port navigation light
90 Abradable wingtip skid
91 Manually-folding wingtip hinge joint
92 Port jettisonable outrigger wheel
93 Wing panel outboard integral fuel tank
94 Fuel filler cap

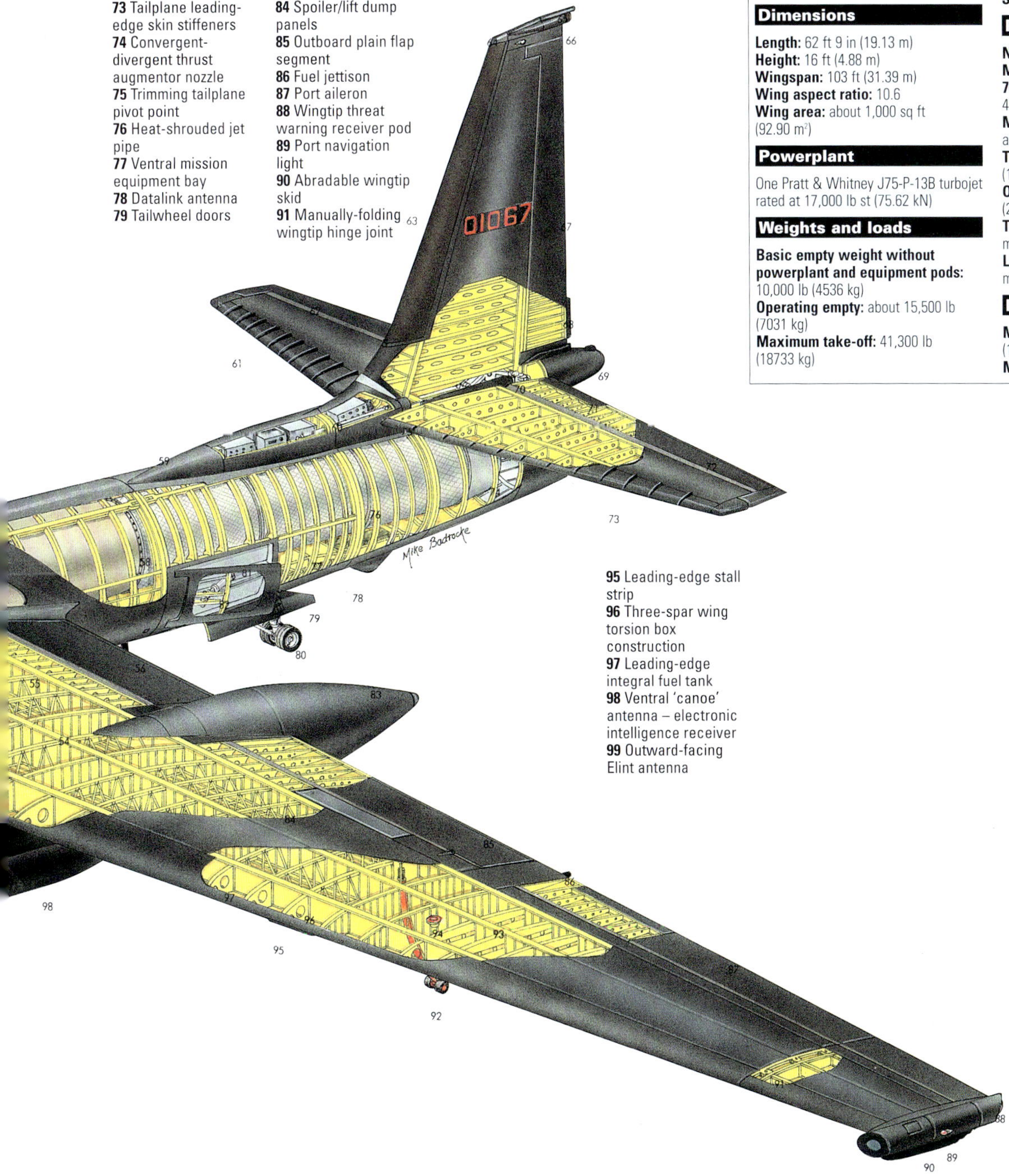

95 Leading-edge stall strip
96 Three-spar wing torsion box construction
97 Leading-edge integral fuel tank
98 Ventral 'canoe' antenna – electronic intelligence receiver
99 Outward-facing Elint antenna

SPECIFICATION

U-2R

Dimensions
Length: 62 ft 9 in (19.13 m)
Height: 16 ft (4.88 m)
Wingspan: 103 ft (31.39 m)
Wing aspect ratio: 10.6
Wing area: about 1,000 sq ft (92.90 m²)

Powerplant
One Pratt & Whitney J75-P-13B turbojet rated at 17,000 lb st (75.62 kN)

Weights and loads
Basic empty weight without powerplant and equipment pods: 10,000 lb (4536 kg)
Operating empty: about 15,500 lb (7031 kg)
Maximum take-off: 41,300 lb (18733 kg)
Fuel load: 7,649 lb (3469 kg)
Sensor payload: 3,000 lb (1361 kg)

Performance
Never-exceed speed: Mach 0.8
Maximum cruising speed at 70,000 ft (21335 m): more than 430 mph (692 km/h)
Maximum rate of climb at sea level: about 5,000 ft (1525 m) per minute
Time to climb: climb to 65,000 ft (19812 m) in 35 minutes
Operational ceiling: 80,000 ft (24385 m)
Take-off run: about 650 ft (198 m) at maximum take-off weight
Landing run: about 2,500 ft (762 m) at maximum landing weight

Range
Maximum range: about 6,250 miles (10060 km)
Maximum endurance: 12 hours

North American RA-5C Vigilante

The large size of the Vigilante made it an unpopular aircraft with commanders and a dangerous one for inexperienced pilots. However, upon its conversion to a tactical reconnaissance platform, it proved peerless in its abilities and still today, 20 years after its retirement, a comparable successor has yet to be found.

RA-5C Vigilante

Cutaway key

1 Pitot static boom
2 Hinged radome
3 Search radar antenna
4 Hinged radar and AN/ASB-12 forward package (servicing position)
5 TV optical scanner
6 Inflight-refuelling line
7 Inflight-refuelling probe (stowed)
8 AN/ASB-12 bomb directing set
9 Radome actuator
10 Radome (folded)
11 LOX converter
12 Instrument panel shroud
13 One-piece acrylic windscreen
14 Radar-flight projected display indicator
15 Control column
16 Rudder controls
17 TACAN antenna
18 ADF antenna
19 AN/APR-27 antenna
20 Viewfinder
21 Pilot's ejection seat
22 Underseat high-pressure emergency oxygen bottle
23 Cockpit air supply
24 Canopy emergency air bottle
25 Headrest
26 Pilot's canopy
27 Emergency escape system ballistic charges
28 Pilot's canopy actuator
29 Indicating power supply
30 Bombing computer
31 UHF antenna
32 Radar altimeter
33 AN/ALQ-100 antenna
34 Navigator's side console
35 Underseat high-pressure emergency oxygen bottle
36 Navigator's ejection seat
37 Canopy emergency air bottle
38 Navigator's window
39 Headrest
40 Navigator's canopy actuator
41 LOX storage converters (2)
42 Master flight reference gyro
43 Nosewheel well
44 Pre-closing nosewheel doors
45 Nosewheel gear steering unit
46 Taxiing light
47 Forward-retracting nosewheel
48 Nosewheel centring unit
49 Nosewheel gear actuator
50 Flight control main electronics bay
51 Flight control relays
52 IFF antenna
53 Bulkhead
54 Forward fuselage fuel cell (455 US gal/ 1722 litres)
55 Inlet sidewall structure
56 Forward variable ramp
57 Nacelle inlet assembly
58 Port intake
59 Nacelle structure
60 Aft variable ramp
61 Ramp actuator
62 Intake duct
63 (Ventral) launch catapult hooks (2)
64 Main wing/fuselage frame forging
65 Wing forward attachment pick-up point
66 BLC ducting
67 Fuselage sump fuel cell (490 US gal/ 1855 litres)
68 Starboard wingroot fillet
69 Starboard auxiliary drop-tanks (400 US gal/ 1514 litres each)
70 AN/ALQ-41 and -100 forward transmit antennas
71 AN/APR-25 and AN/ALQ-41 and -100 forward receive antennas
72 Leading-edge wing-droop (inner section)
73 Droop actuator and torque rod
74 Conduit to wing fold (hydraulic and electrical)
75 Wing structure
76 Starboard wing integral fuel (715 US gal/2707 litres)
77 Span-wise corrugated stiffeners
78 Wing-fold line
79 Leading-edge wing-droop (outer section)
80 Starboard navigation lights
81 Starboard formation light
82 Wing outer section (folded)
83 Outboard spoiler deflector (downward airflow)
84 AN/ALQ-4 and-100 aft receiver antennas
85 Central (closed) and inboard spoiler deflectors (upward airflow)
86 Starboard flaps
87 BLC flap-blowing duct
88 Wing aft attachment pick-up point
89 Dorsal fairing
90 Overwing saddle tank (210 US gal/ 795 litres)
91 Wing centreline splice assembly
92 Starboard intake ducting
93 Bomb-bay forward fuel cell
94 Hydraulic reservoir air storage tank
95 Port intake ducting
96 Port mainwheel well
97 Retraction jack
98 Universal pivot
99 Mainwheel leg down-lock
100 Wing aft attachment pick-up point
101 Steel mainframe and firewall (canted)
102 General Electric J79-GE-10 turbojet
103 BLC cross-over ducting
104 Bomb-bay central fuel cell
105 No. 1 hydraulic system reservoir
106 No. 2 hydraulic system reservoir
107 Aft fuselage saddle tank (130 US gal/ 492 litres)
108 Anti-collision beacon
109 Starboard engine oil tank (6.10 US gal/ 23 litres)
110 Fuselage aft structure
111 Horizontal stabiliser
112 Horizontal stabiliser attachment frame
113 Horizontal stabiliser actuator
114 Vertical stabiliser actuator
115 Bomb-bay aft fuel cell (total internal capacity: 885 US gal/ 3350 litres)
116 Fuselage aft frame
117 Vertical stabiliser pivot
118 Vertical stabiliser lower section structure
119 Conduits (front to rear: electrical, hydraulic, tail-fold cable)
120 Leading-edge dielectric panel
121 Tail-fold hinge line
122 Tail-fold actuator
123 Vertical stabiliser upper-section structure
124 Front spar
125 Duplex UHF comm/ALQ-55 antenna
126 Electrical conduit
127 Vertical stabiliser (folded)
128 Rear formation light
129 DECM antenna, AN/APR-18, AN/APR-25(v) or AN/ALR-45(v)
130 Buddy tanker lights
131 Fuel vent
132 AN/APR-18 antenna, if fitted
133 Fuel vent line
134 Electrical conduits
135 Honeycomb structure

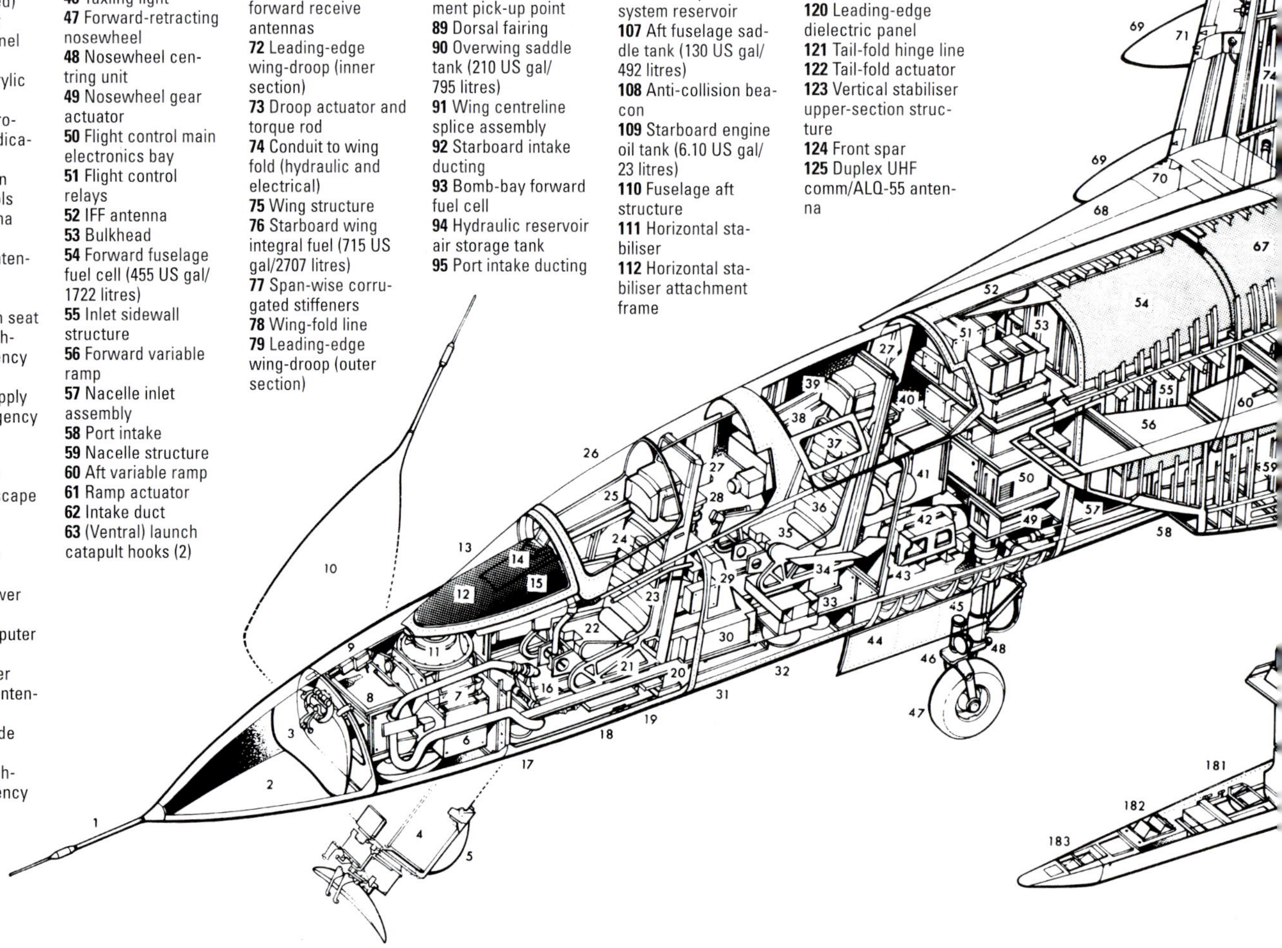

It was not unknown for Vigilantes to lose their linear bay fuel tanks during a 'cat shot', all three fuel cells landing unceremoniously on the flight deck. An explosion and fire on the deck would usually result although, more often than not, damage to the ship was minor and experienced pilots were able to continue their take-offs normally, as in this example, where Cdr John H. Huber of RVAH-12 lost his fuel cells, and the 885 US gal (3350 litres) of fuel they contained, aboard USS Independence on 4 September 1969. Others were not so lucky, including the crew of BuNo. 156609 from the same unit which, during spring 1973, lost one fuel tank and caught fire during a 'cat shot'. The crew ejected as the aircraft rolled out of control; both survived.

SPECIFICATION

RA-5C Vigilante

Dimensions

Length: 76 ft 6 in (23.35 m)
Length (with vertical fin and radome folded): 65 ft 4⅜ in (19.92 m)
Wingspan: 53 ft (16.17 m)
Wingspan (folded): 42 ft (12.8 m)
Wing area: 753.7 sq ft (70.02 m²)
Height: 19 ft 4¾ in (5.91 m)
Height (with vertical tail folded): 15 ft 6 in (4.72 m)

Powerplant

Two General Electric J79-GE-10 turbojets each rated at 17,900 lb (79.63 kN) thrust with maximum afterburner

Weights

Empty: 37,498 lb (17024 kg)
Basic: 38,219 lb (17336 kg)
Combat: 55,617 lb (25227 kg)
Maximum landing (field): 65,988 lb (29931 kg)
Maximum landing (arrested): 47,000 lb (21319 kg)

Performance

Maximum speed at sea level: 806 mph (1297 km/h)
Maximum speed at 40,000 ft (12192 m): 1,320 mph (2124 km/h)
Initial climb rate: 6,600 ft (2012 m) per minute at sea level
Service ceiling: 49,000 ft (14935 m)
Combat radius (attack): 1,284 miles (2066 km)
Combat radius: 1,508 miles (2427 km)
Ferry range: 2,050 miles (3299 km)

Armament

None normally carried, though two pylons could be fitted under each wing and a full weapons delivery capability was retained

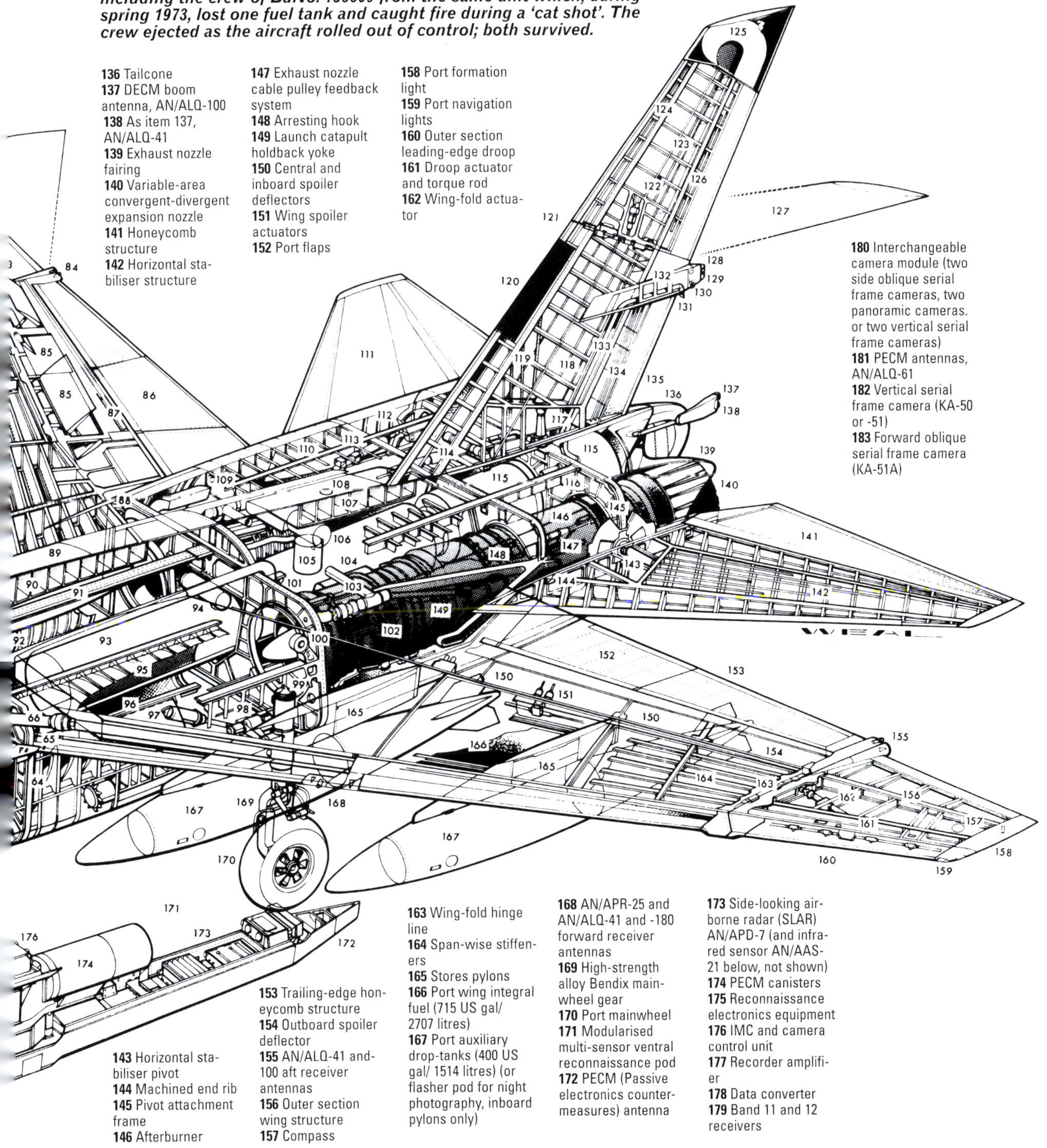

136 Tailcone
137 DECM boom antenna, AN/ALQ-100
138 As item 137, AN/ALQ-41
139 Exhaust nozzle fairing
140 Variable-area convergent-divergent expansion nozzle
141 Honeycomb structure
142 Horizontal stabiliser structure
143 Horizontal stabiliser pivot
144 Machined end rib
145 Pivot attachment frame
146 Afterburner
147 Exhaust nozzle cable pulley feedback system
148 Arresting hook
149 Launch catapult holdback yoke
150 Central and inboard spoiler deflectors
151 Wing spoiler actuators
152 Port flaps
153 Trailing-edge honeycomb structure
154 Outboard spoiler deflector
155 AN/ALQ-41 and -100 aft receiver antennas
156 Outer section wing structure
157 Compass
158 Port formation light
159 Port navigation lights
160 Outer section leading-edge droop
161 Droop actuator and torque rod
162 Wing-fold actuator
163 Wing-fold hinge line
164 Span-wise stiffeners
165 Stores pylons
166 Port wing integral fuel (715 US gal/ 2707 litres)
167 Port auxiliary drop-tanks (400 US gal/ 1514 litres) (or flasher pod for night photography, inboard pylons only)
168 AN/APR-25 and AN/ALQ-41 and -180 forward receiver antennas
169 High-strength alloy Bendix mainwheel gear
170 Port mainwheel
171 Modularised multi-sensor ventral reconnaissance pod
172 PECM (Passive electronics countermeasures) antenna
173 Side-looking airborne radar (SLAR) AN/APD-7 (and infrared sensor AN/AAS-21 below, not shown)
174 PECM canisters
175 Reconnaissance electronics equipment
176 IMC and camera control unit
177 Recorder amplifier
178 Data converter
179 Band 11 and 12 receivers
180 Interchangeable camera module (two side oblique serial frame cameras, two panoramic cameras, or two vertical serial frame cameras)
181 PECM antennas, AN/ALQ-61
182 Vertical serial frame camera (KA-50 or -51)
183 Forward oblique serial frame camera (KA-51A)

Trainers

01

Aermacchi MB326/MB339

*Although not as glamorous as the front-line fast jets to which their pilots graduate, Aermacchi's **MB-326** and **MB-339** have proved to be highly capable in both the training and attack roles. This photograph shows an **MB-339CB** of the **Royal New Zealand Air Force. RNZAF** aircraft are equipped to deliver **AGM-65** Maverick air-to-ground missiles and **AIM-9** Sidewinder **IR**-homing air-to-air missiles. In addition to gun pods, un-guided weapons include the Matra 155 rocket pod (illustrated), containing 18 Brandt 68mm spin-stabilised **SNEB** rocket projectiles. These may be fitted with different types of warhead according to the type of target to be attacked.*

MB-339A

Cutaway key

1 Starboard fixed wingtip tank (69.5-Imp gal/316-litre capacity)
2 Fuel filler point
3 Starboard servo-powered aileron
4 Aileron balance tab
5 Aileron balance weights
6 Wingtip tank fuel line
7 Starboard outer pylon (507-lb/230-kg capacity)
8 Missile launcher rail
9 Matra R550 Magic IR-guided air-to-air missile
10 Starboard centre pylon (750-lb/340-kg capacity)
11 Jettisonable auxiliary fuel tank (71-Imp gal/ 325-litre capacity)
12 Starboard wing fence
13 Aileron hinge mechanism
14 Aileron servo
15 Starboard single-slotted flap
16 Flap control linkage
17 Pod attachment spigot
18 Macchi 0.5-in (12.7-mm) AN/M-3 machine-gun pod
19 Ammunition feed chute
20 Ammunition box (350 rounds)
21 Cartridge case ejection chute
22 Starboard-hinged canopy
23 Rear Martin-Baker Mk 1T-10F zero-zero ejection seat
24 Headrest parachute container
25 Seat safety harness
26 Rear Aeritalia fixed reflector sight
27 Rear instrument panel shroud
28 Front Martin-Baker Mk 1T-10F zero-zero ejection seat
29 Headrest parachute container
30 Starboard instrument console
31 Front Aeritalia fixed reflector sight (Aeritalia Saab RGS2 computer gyroscopic sight or Thomson/CSF RD 21 simple gyroscopic sight optional)
32 Fully automatic OMERA-SEGID 110-3 gun camera
33 Front instrument panel shroud
34 Curved one-piece windshield
35 Pitot tubes
36 Antenna
37 Nosewheel retraction jack
38 Nosewheel well
39 Nosewheel bay front bulkhead
40 Nose cone
41 Landing/taxi light
42 Nosewheel shock absorber
43 Chined steerable nosewheel
44 Nose wheel doors
45 Front radio and electronics bay
46 Cockpit forward bulkhead
47 Rudder pedals
48 Control column
49 Underfloor control linkage
50 Throttle
51 C-section forward fuselage frames
52 Niche-type step
53 Ventral antenna
54 Control runs
55 Rear flight controls
56 Canopy emergency release handle
57 Access panel
58 Rear cockpit niche-type steps
59 Rear throttle
60 Port instrument side console
61 Ventral air brake
62 Air brake jack
63 Port engine air intake
64 Aft-sloping rear cockpit bulkhead
65 Intake trunk cut-out
66 Fuselage/main spar attachment
67 Main spar carry-through
68 Forward fuselage frame-and-stringer construction
69 Fuselage double frame
70 Top longeron
71 Rear radio and electronics bay
72 Antenna
73 Access panel
74 Engine air intake trunking
75 Cooling air louvres
76 Fuel filler
77 Fuel system piping
78 Batteries (2 x 24V)
79 Access panel
80 Engine intake
81 Hydraulic system reservoirs
82 Side access panel
83 Engine electrical accessories
84 Piaggio-built Rolls-Royce Viper 632-43 engine
85 Engine bay air louvres
86 Dorsal navigation light
87 Dorsal spine construction
88 Engine mounting frame
89 Fuselage break point
90 Firewall
91 Fuel jettison pipe
92 Jet pipe flush cooling air intakes
93 Jet pipe
94 Fuselage/fin spar attachment
95 Elevator linkage
96 Fin construction
97 Starboard tailplane
98 Starboard elevator
99 Fin leading edge
100 Dielectric fin tip
101 Anti-collision light
102 UHF antenna
103 Fin VOR antenna
104 Rudder balance
105 Rudder construction
106 Fixed tab
107 Trim tab jack
108 Rudder trim tab
109 Rudder control linkage
110 Tailplane centre section
111 Elevator hinge control
112 Tailcone
113 Jet pipe nozzle shroud
114 Tail navigation light
115 Elevator trim tab
116 Port elevator
117 Balance weight

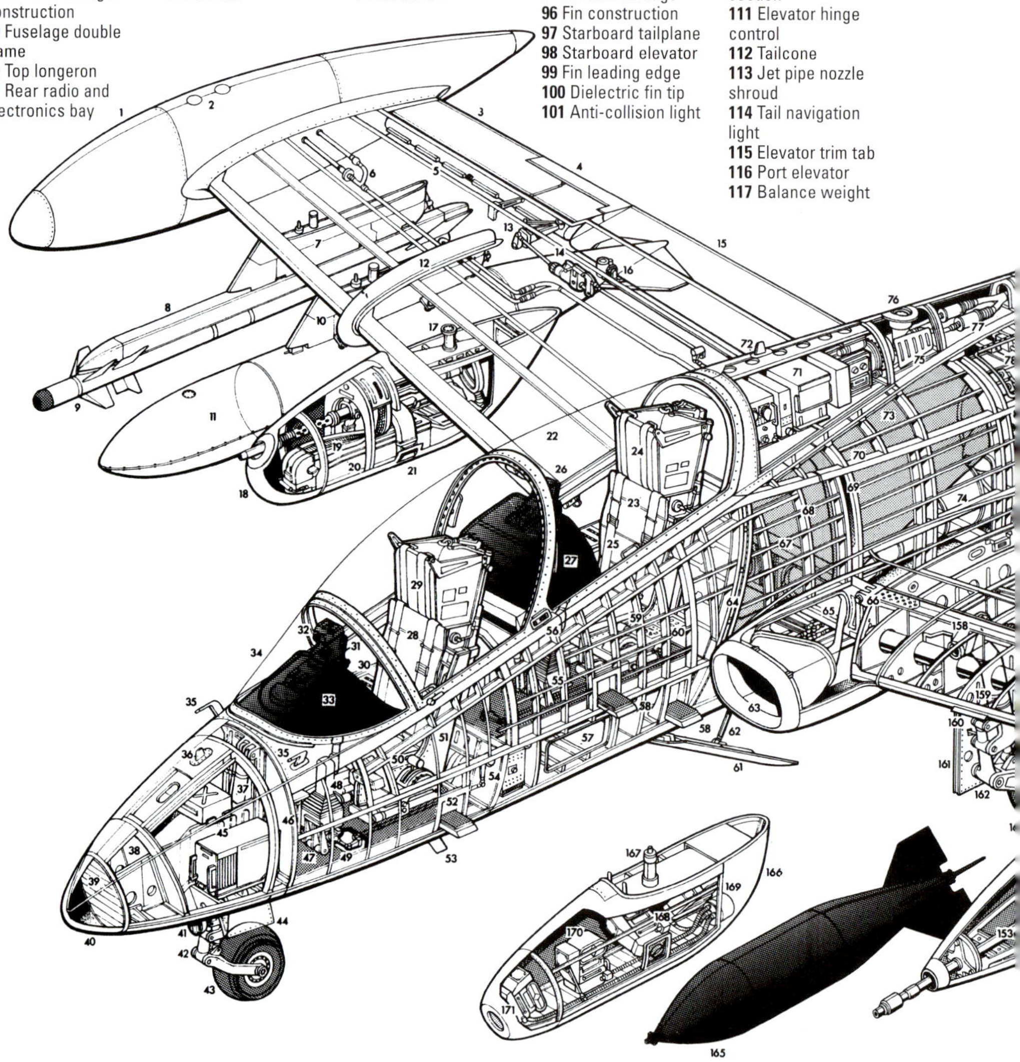

Left: For many years the MB-326 was the backbone of the AMI's jet training force. For this role, the aircraft wore a highly conspicuous orange colour scheme. The docile handling characteristics of the machine soon came to be appreciated by novice jet pilots, resulting in an accident rate of only 0.8 per 10,000 flying hours. Today, the type has given way to the MB-339, but a few examples still soldier on in the communications role and as squadron 'hacks'.

SPECIFICATION

MB-339A

Dimensions

Length: 36 ft (10.97 m)
Height: 13 ft 1¼ in (3.99 m)
Wingspan: 35 ft 7½ in (10.86 m) over tip tanks
Wing area: 207.74 sq ft (19.30 m²)
Wing aspect ratio: 6.1
Wheel track: 8 ft 1¾ in (2.48 m)
Wheel base: 14 ft 4 in (4.37 m)

Powerplant

One Piaggio-built Rolls-Royce (Bristol Siddeley) Viper Mk 632-43 rated at 4,000 lb st (17.79 kN) dry

Weights

Empty equipped: 6,889 lb (3125 kg)
Operating empty: 6,913 lb (3136 kg)
Normal take-off: 9,700 lb (4400 kg)
Maximum take-off: 12,996 lb (5895 kg)

Fuel and load

Internal fuel: 2,425 lb (1100 kg)
External fuel: up to two 71-Imp gal (325-litre) drop tanks
Maximum ordnance: 4,497 lb (2040 kg)

Performance

Never-exceed speed: 575 mph (926 km/h)
Maximum level speed 'clean' at 30,000 ft (9145 m): 508 mph (817 km/h)
Maximum speed at sea level: 558 mph (898 km/h)
Stalling speed: 93 mph (149 km/h)
Maximum rate of climb at sea level: 6,595 ft (2010 m) per minute
Climb to 30,000 ft (9145 m): in 7 minutes 6 seconds
Service ceiling: 48,000 ft (14630 m)

Range/Armamente 1

Ferry range: 1,311 miles (2110 km) with drop tanks; range 1,094 miles (1760 km); combat radius 369 miles (593 km) on a hi-lo-hi attack mission with four Mk 82 bombs and two drop tanks, or 244 miles (393 km) on a hi-lo-hi attack mission with six Mk 82 bombs, or 317 miles (510 km) on a hi-lo-hi attack mission with two 30-mm cannon pods, two rocket launchers and two drop tanks, or 351 miles (565 km) on a hi-lo-hi attack mission with four rocket launchers and two drop tanks, or 190 miles (306 km) on a hi-lo-hi attack mission with six rocket launchers, or 230 miles (371 km) on a lo-lo-lo attack mission with four Mk 82 bombs and two drop tanks, or 168 miles (271 km) on lo-lo-lo attack mission with six Mk 82 bombs, or 219 miles (352 km) on a lo-lo-lo attack mission with two 30-mm cannon pods, two rocket launchers and two drop tanks, or 222 miles (358 km) on a lo-lo-lo attack mission with four rocket launchers and two drop tanks, or 142 miles (228 km) on a lo-lo-lo attack mission with six rocket launchers; endurance 3 hours 45 minutes with drop tanks or 2 hours 50 minutes on internal fuel

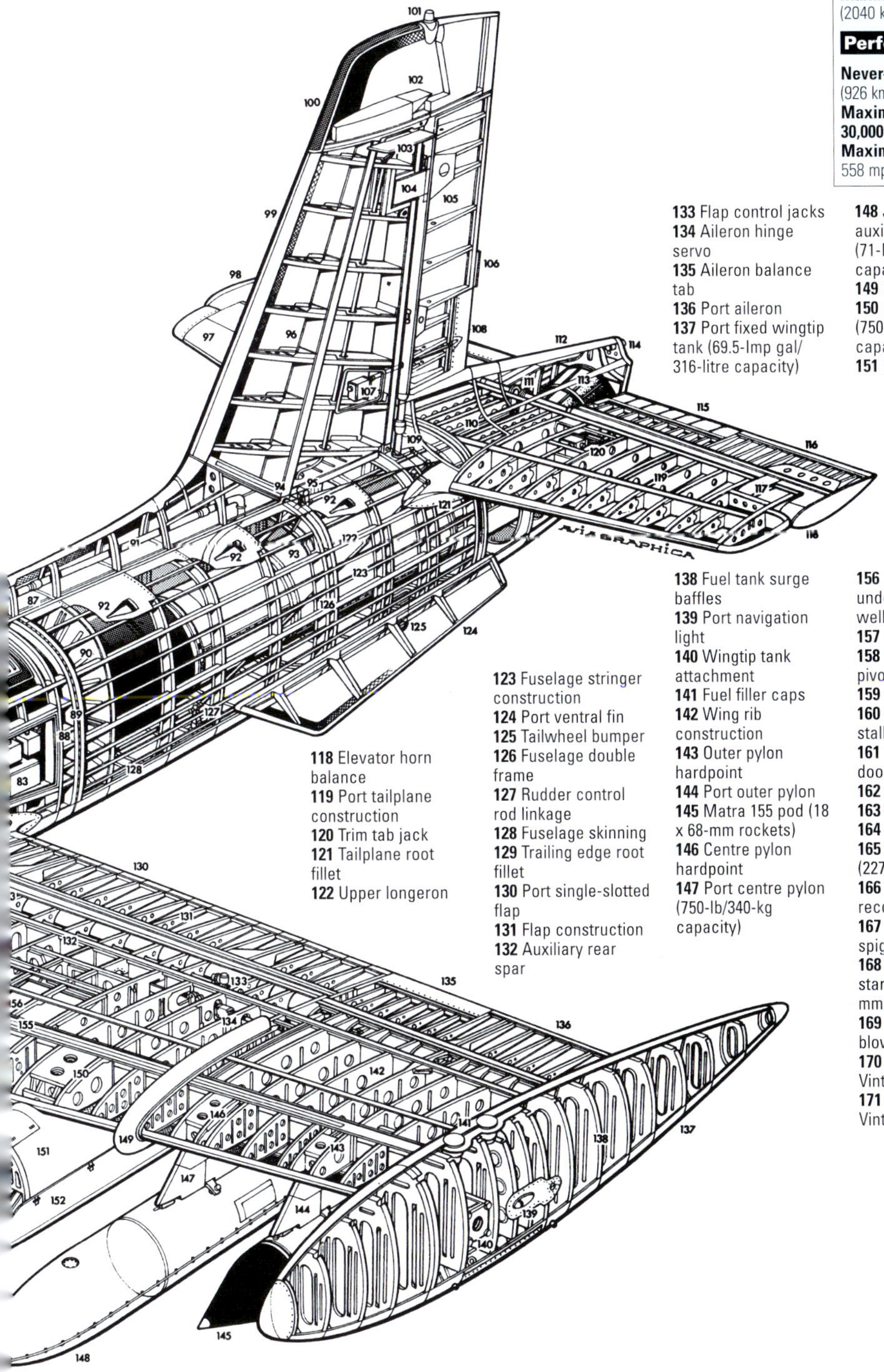

118 Elevator horn balance
119 Port tailplane construction
120 Trim tab jack
121 Tailplane root fillet
122 Upper longeron
123 Fuselage stringer construction
124 Port ventral fin
125 Tailwheel bumper
126 Fuselage double frame
127 Rudder control rod linkage
128 Fuselage skinning
129 Trailing edge root fillet
130 Port single-slotted flap
131 Flap construction
132 Auxiliary rear spar
133 Flap control jacks
134 Aileron hinge servo
135 Aileron balance tab
136 Port aileron
137 Port fixed wingtip tank (69.5-Imp gal/316-litre capacity)
138 Fuel tank surge baffles
139 Port navigation light
140 Wingtip tank attachment
141 Fuel filler caps
142 Wing rib construction
143 Outer pylon hardpoint
144 Port outer pylon
145 Matra 155 pod (18 x 68-mm rockets)
146 Centre pylon hardpoint
147 Port centre pylon (750-lb/340-kg capacity)
148 Jettisonable auxiliary fuel tank (71-Imp gal/325-litre capacity)
149 Port wing fence
150 Port inboard pylon (750-lb/340-kg capacity) hardpoint
151 DEFA 553 (30-mm) cannon pod
152 Ammunition magazine (120-round capacity)
153 Cannon barrel
154 Oxygen bottle (one in each wing)
155 Main spar
156 Main undercarriage wheel well
157 Retraction jack
158 Mainwheel leg pivot
159 Mainwheel leg
160 Leading-edge stall strip
161 Undercarriage leg door
162 Shock absorber
163 Mainwheel forks
164 Port mainwheel
165 Mk 82 500-lb (227-kg) HE bomb
166 Photo-reconnaissance pod
167 Pod mounting spigot
168 Port and starboard oblique 70-mm Vinten cameras
169 Heater and blower unit
170 Vertical 70-mm Vinten camera
171 Forward 70-mm Vinten camera

Aero L-39 Albatros

The L-39 has scored heavily in the 'real' expor market (as opposed to Warsaw Pact nations, notably in the Middle East, from where Egyp Iraq, Libya and Syria placed large orders fo 'first-generation' aircraft. Egypt followed up b buying the L-59E, this example being illustrate in the four-tank ferry fit used to deliver the jets

L-139 Albatros

Cutaway key

1 Glass-fibre nosecone
2 ILS antenna
3 Ground intercom socket
4 Navigational antenna
5 Nosewheel door, closed after cycling of undercarriage
6 Avionics equipment compartment
7 Nosewheel housing
8 Pitot head
9 Hinged access doors, port and starboard
10 On-Board Oxygen Generator System (OBOGS)
11 Front pressure bulkhead
12 Nosewheel pivot mounting
13 Nosewheel leg strut
14 Levered suspension shock absorber
15 Forward-retracting nose wheel
16 Shimmy damper
17 Ventral cannon pack
18 Rudder pedals
19 Incidence transmitter
20 Control column, fully duplicated controls
21 Instrument console
22 Undercarriage position visual indicator
23 Hot air de-iced one-piece windscreen
24 Front cockpit instrument panel with Electronic Flight Information System (EFIS) displays
25 Rear cockpit instrument panel with EFIS displays
26 Pilot's head-up display (HUD)
27 Stand-by horizon
28 Rear cockpit monitor screen
29 Stand-by compass
30 Individual cockpit canopies, hinged to starboard
31 Student pilot's VS-2R rocket-assisted ejection seat
32 Seat harness
33 Engine throttle lever
34 Side console panel
35 Front cockpit floor level
36 Boarding steps
37 Underfloor equipment bays
38 Rear cockpit floor level
39 Canopy external release
40 Canopy lifting handle
41 Rear instrument console
42 Canopy centre arch
43 Instructor's VS-2R ejection seat
44 Rear side console panel
45 Kick-in steps
46 Rear pressure bulkhead
47 Fuselage centre bag-type fuel tanks, total internal capacity (excluding tip tanks) 232 Imp gal (1055 litres)
48 Boundary layer splitter plate
49 Port air intake
50 Fuselage tank gravity filler
51 Tailplane control rods
52 Starboard air intake
53 Flap actuating linkage
54 Starboard outer wing pylon
55 Starboard wing panel
56 Pitot head
57 Landing/taxiing light
58 Starboard navigation light
59 22-Imp gal (100-litre) fixed wingtip fuel tank
60 Starboard aileron
61 Servo tab
62 Aileron operating linkage
63 Flap track fairing
64 Starboard double-slotted flap
65 Anti-collision

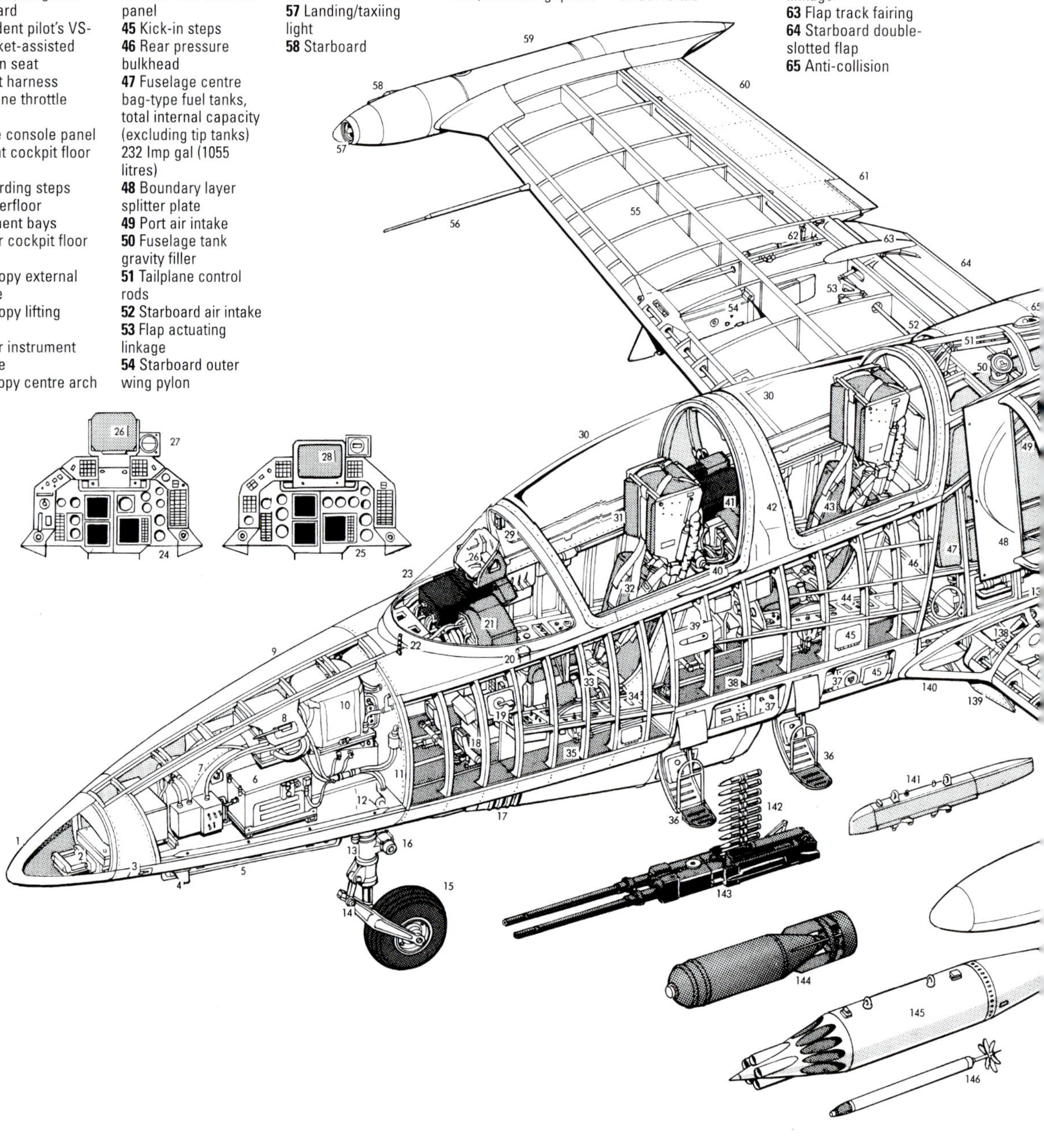

Flying from Korat with No. 1 Wing, Thailand's two L-39ZA/ART squadrons provide fighter lead-in training and a light attack capability. The black/yellow chequerboard markings on the tail denotes No. 101 Squadron.

SPECIFICATION

L-39ZO Albatros

Dimensions

Length: 39 ft 9½ in (12.13 m)
Height: 15 ft 7¾ in (4.77 m)
Wingspan: 31 ft ½ in (9.46 m)
Aspect ratio: 4.4 or 5.2 including tip tanks
Wing area: 202.37 sq ft (18.80 m²)
Tailplane span: 14 ft 5 in (4.40 m)
Wheel track: 8 ft (2.44 m)
Wheel base: 14 ft 4¾ in (4.39 m)

Powerplant

One ZMDB Progress (Ivchenko) AI-25TL turbofan engine rated at 3,792 lb (16.87 kN) dry

Weights

Empty equipped: 7,804 lb (3540 kg)
Normal take-off: 9,976 lb (4525 kg)
Maximum take-off: 10,362 lb (4700 kg)

Fuel and load

Internal fuel: 1,816 lb (824 kg) plus provision for 344 lb (156 kg) in two 48-US gal (180-litre) non-jettisonable tip tanks
External fuel: up to 1,199 lb (544 kg) in two 110-US gal (420-litre) drop tanks
Maximum ordnance: 2,200 lb (1000 kg)

Performance

Never-exceed speed at 36,090 ft (11000 m): 459 kt (528 mph; 850 km/h)
Maximum level speed 'clean' at 16,405 ft (5000 m): 407 kt (466 mph; 755 km/h)
Maximum speed at sea level: 388 kt (447 mph; 720 km/h)
Ferry range: 944 nm (1,087 miles; 1750 km) with drop tanks
Standard range: 593 nm (683 miles; 1100 km) with internal fuel
Endurance at 22,975 ft (7000 m): 3 hours 50 minutes with drop tanks or 2 hours 30 minutes with internal fuel
Maximum rate of climb at sea level: 4,134 ft (1260 m) per minute
Climb to 16,405 ft (5000 m): 5 minutes
Service ceiling: 36,090 ft (11000 m)
Take-off run: 1,740 ft (530 m) at normal take-off weight
Landing run: 2,135 ft (650 m) at normal landing weight
***g* limits:** -4 to +8 at operational weights or up to +12 at 9,259 lb (4200 kg)

Armament

Four underwing hardpoints, inboard pair each stressed for loads of up to 1,102 lb (500 kg) and the outer pair for loads of up to 551 lb (250 kg) each. Typical underwing stores can include various combinations of bombs (two 1,102-lb/500-kg, four 551-lb/250-kg or six 220-lb/100-kg) or four UB-16-57 M pods containing 16 S-5 2¼ (57-mm) air-to-surface rockets or infra-red air-to-air missiles (outer pylons only) or a five-camera day reconnaissance pod (port inboard pylon only)

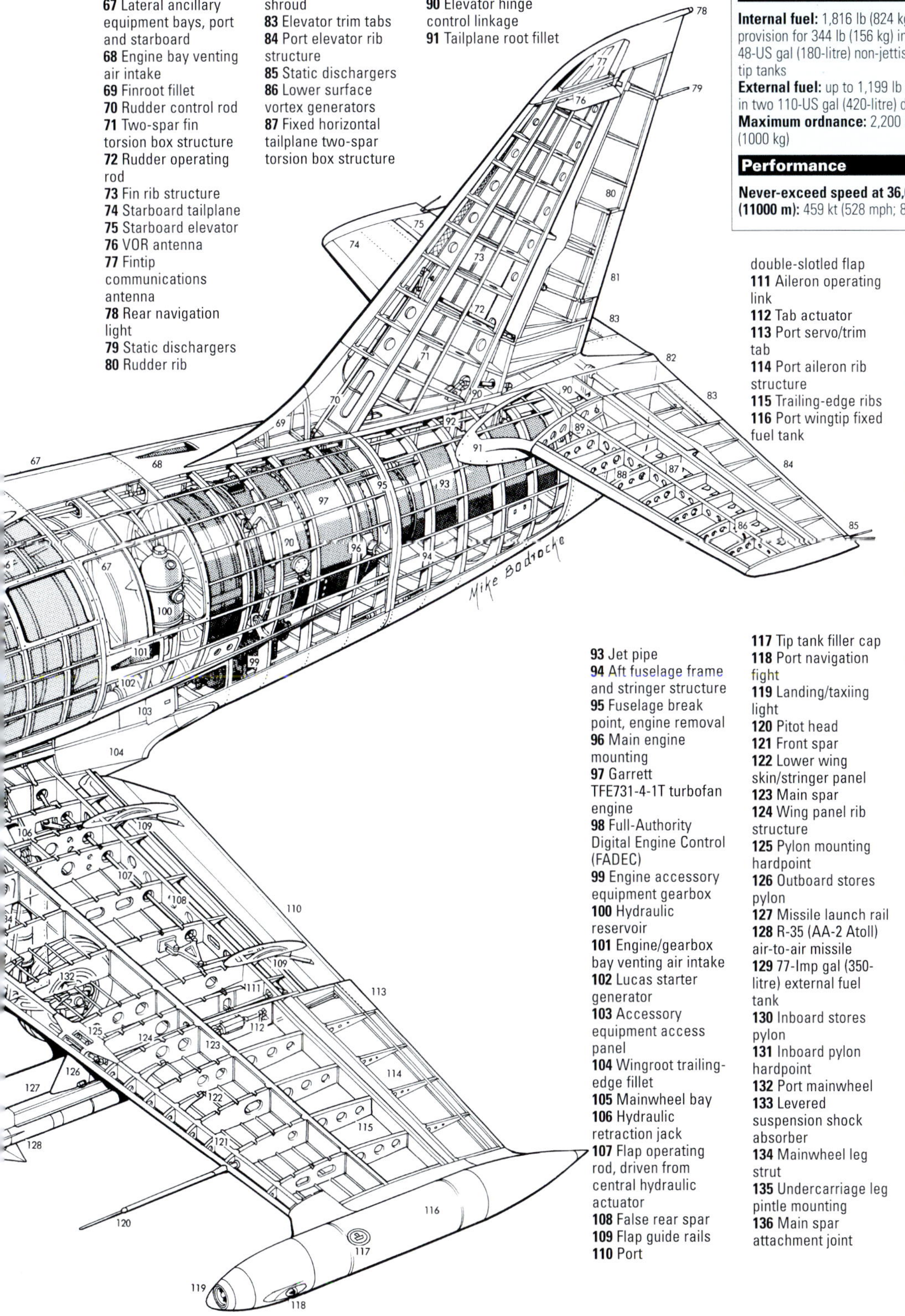

strobe light
66 Intake flank fuel tank
67 Lateral ancillary equipment bays, port and starboard
68 Engine bay venting air intake
69 Finroot fillet
70 Rudder control rod
71 Two-spar fin torsion box structure
72 Rudder operating rod
73 Fin rib structure
74 Starboard tailplane
75 Starboard elevator
76 VOR antenna
77 Fintip communications antenna
78 Rear navigation light
79 Static dischargers
80 Rudder rib structure
81 Rudder trim tab
82 Exhaust nozzle shroud
83 Elevator trim tabs
84 Port elevator rib structure
85 Static dischargers
86 Lower surface vortex generators
87 Fixed horizontal tailplane two-spar torsion box structure
88 Leading-edge ribs
89 Tailplane spar attachment joints
90 Elevator hinge control linkage
91 Tailplane root fillet fairing
92 Fin spar joint
93 Jet pipe
94 Aft fuselage frame and stringer structure
95 Fuselage break point, engine removal
96 Main engine mounting
97 Garrett TFE731-4-1T turbofan engine
98 Full-Authority Digital Engine Control (FADEC)
99 Engine accessory equipment gearbox
100 Hydraulic reservoir
101 Engine/gearbox bay venting air intake
102 Lucas starter generator
103 Accessory equipment access panel
104 Wingroot trailing-edge fillet
105 Mainwheel bay
106 Hydraulic retraction jack
107 Flap operating rod, driven from central hydraulic actuator
108 False rear spar
109 Flap guide rails
110 Port double-slotted flap
111 Aileron operating link
112 Tab actuator
113 Port servo/trim tab
114 Port aileron rib structure
115 Trailing-edge ribs
116 Port wingtip fixed fuel tank
117 Tip tank filler cap
118 Port navigation fight
119 Landing/taxiing light
120 Pitot head
121 Front spar
122 Lower wing skin/stringer panel
123 Main spar
124 Wing panel rib structure
125 Pylon mounting hardpoint
126 Outboard stores pylon
127 Missile launch rail
128 R-35 (AA-2 Atoll) air-to-air missile
129 77-Imp gal (350-litre) external fuel tank
130 Inboard stores pylon
131 Inboard pylon hardpoint
132 Port mainwheel
133 Levered suspension shock absorber
134 Mainwheel leg strut
135 Undercarriage leg pintle mounting
136 Main spar attachment joint
137 Fuselage lower main longeron
138 Airbrake hydraulic jack
139 Ventral airbrake panels (2)
140 Extended chord wingroot fairing
141 Light stores dispenser
142 Ammunition feed, 150 rounds housed beneath rear cockpit floor
143 GSh-23 twin-barrel 23-mm cannon
144 250-lb (113-kg) HE bomb
145 UV-16-57, 16-round rocket launcher
146 57-mm (2¼-in) rocket

British Aerospace Hawk

Originally intending to buy licence-built Aermacchi MB.326s, the Indonesian air force abandoned its plans in favour of becoming the third Hawk export customer in 1978 with an initial order for eight T.Mk 53s. Another 12 examples were subsequently ordered and the aircraft are used for both advanced training (in a red and white scheme) and for tactical weapons training, as is the case with this example, in two-tone camouflage.

Hawk 60 series

Cutaway key

1 Starboard all-moving tailplane
2 Tailplane multi-spar construction
3 Engine exhaust nozzle
4 Tail navigation light
5 Optional brake parachute housing
6 Tailplane pivot fixing
7 Tailplane sealing plate
8 Hydraulic tailplane actuator
9 Port all-moving tailplane
10 Rudder trim tab
11 Trim tab rotary actuator
12 Rudder hinge control
13 Mass balance weights
14 Rudder honeycomb construction
15 Rudder
16 Fintip aerial fairing
17 VHF aerial
18 Fin leading edge
19 Tailfin construction
20 Rudder control rod linkage
21 Fin spar attachment joints
22 Aft fuselage frame construction
23 Heat-shrouded exhaust pipe
24 Tail bumper
25 Lower IFF aerial
26 Ventral fin, port and starboard
27 Airbrake hydraulic jack
28 Ventral airbrake, lowered
29 Engine bay access panel
30 Fireproof bulkhead
31 Hydraulic reservoir, port and starboard
32 Finroot fillet
33 Ram air turbine actuator
34 Emergency ram air turbine, extended
35 Ram air turbine doors
36 Equipment bay decking
37 Engine combustion and turbine section
38 Rolls-Royce/ Turboméca Adour 861 turbofan engine
39 Engine oil tank
40 Wing trailing-edge root fillet
41 Engine accessory equipment gearbox
42 Engine bay framing
43 Bleed air ducting
44 Gas turbine starter
45 Starter exhaust
46 Starter turbine intake grille, port and starboard
47 Wing rear spar attachment frame
48 Engine compressor face
49 Rear spar attachment joint
50 Trailing-edge hinged access panels
51 Flap rib construction
52 Starboard double-slotted flap
53 Flap operating link
54 Rear spar
55 Reduced span flap vane
56 Aileron hydraulic actuator
57 Aileron honeycomb construction
58 Starboard aileron
59 Localiser aerial
60 Brandt 100-4 rocket launcher, 3.9-in (100-mm) rockets (four)
61 Starboard navigation light
62 Outer wing panel dry bay
63 Matra 155 18-tube rocket-launcher
64 2.75-in (70-mm) HVAR folding-fin rockets
65 Outboard stores pylon
66 Wing tank end rib
67 Wing fence
68 Stall strip
69 Leading-edge rib construction
70 Auxiliary wing fences
71 Inboard stores pylon
72 Two 550-lb (250-kg) low-drag bombs
73 Twin stores carrier
74 250-lb (113-kg) low-drag bomb
75 Starboard mainwheel
76 Mainwheel leg doors
77 Levered suspension axle beam
78 Mainwheel shock absorber leg strut
79 Auxiliary front spar
80 Main spar
81 Wingrib construction
82 Machined wing skin/ stringer panel
83 Starboard wing integral fuel tank; total internal fuel capacity 375 Imp gal (1705 litres)
84 Main undercarriage leg pivot fixing
85 Hydraulic retraction jack
86 Main spar attachment joint
87 Fuel system access panels
88 Fuselage bag-type tank
89 Bleed air control valve
90 Rudder control rod (elevator control on port side)

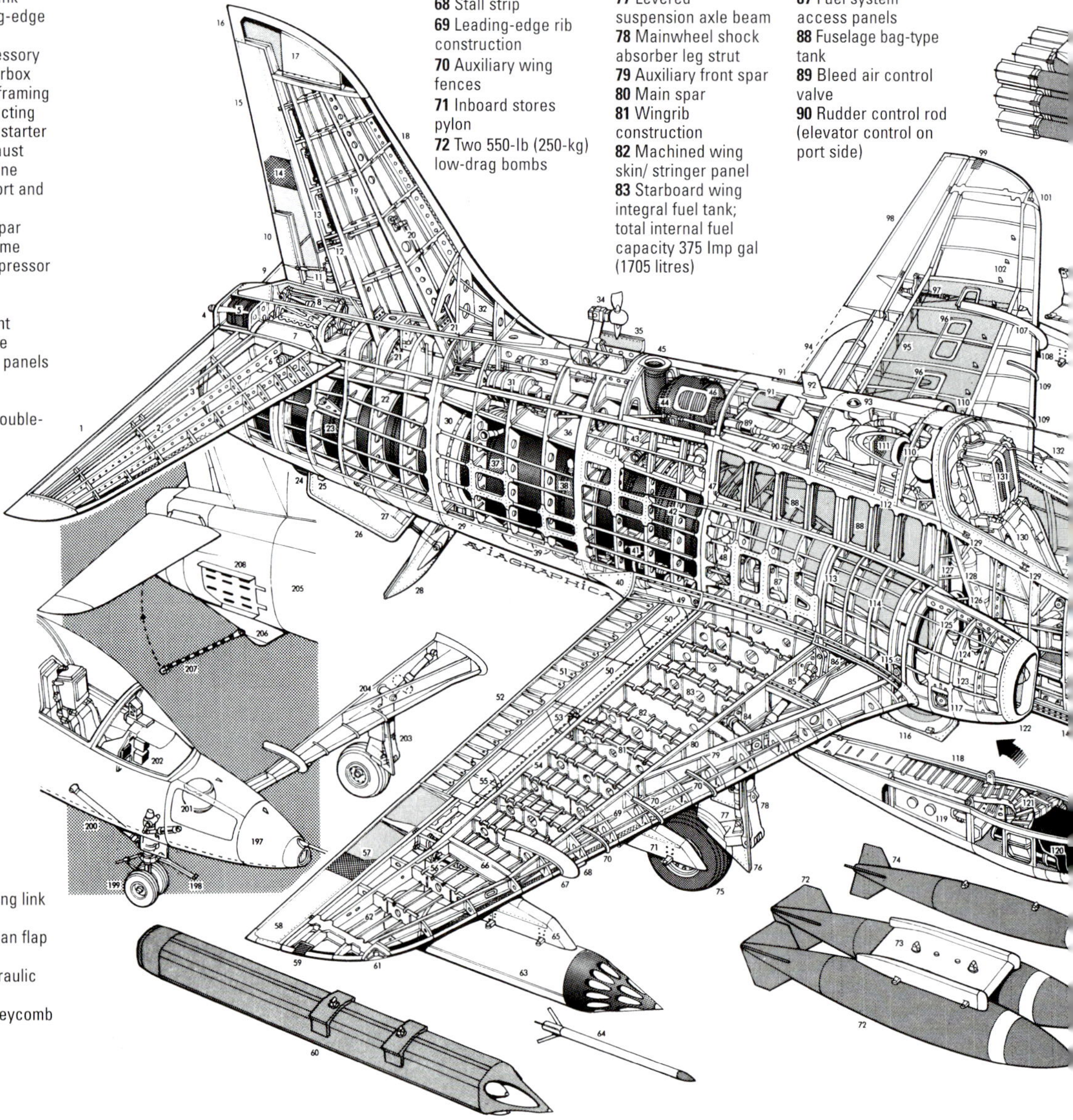

After rigorous evaluation in competition with the Aero L-39 Albatross, Aermacchi MB.339, Dassault/Dornier Alpha Jet and the Saab 105A, the Hawk gained its first export order, from the Finnish air force. Used for advanced training, Finnish Hawks originally wore this attractive three-tone camouflage, which has now been replaced by a flat grey scheme. The success of the aircraft in service led to follow-up orders and, to date, Finland has received 57 Hawk Mk 51/51As.

SPECIFICATION

Hawk T.Mk 1A

Dimensions

Wingspan: 30 ft 10 in (9.39 m)
Overall length: 38 ft 11 in (11.85 m)
Height: 13 ft 1 in (4.00 m)
Wing area: 180 sq ft (16.90 m^2)
Wheel track: 11 ft 5 in (3.47 m)
Wheel base: 14 ft 9 in (4.50 m)

Powerplant

One Rolls-Royce/Turboméca Adour Mk 151-01 turbofan rated at 5,200 lb st (23.13 kN), also quoted as 5,340 lb st (23.67 kN)

Weights

Empty equipped: 8,013 lb (3635 kg)
Maximum take-off: 18,390 lb (8340 kg)
Maximum landing: 17,000 lb (7650 kg)

Performance

Maximum level speed: 560 kt (646 mph; 1040 km/h), Mach 0.88 at sea level
Maximum speed (shallow dive): 572 kt (658 mph; 1060 km/h), Mach 1.2 at 3,000 ft (914 m)
Demonstrated Mach No.: 1.2 IMN
Service ceiling: 50,000 ft (15240 m)
Maximum rate of climb: 9,300 ft (2835 m) per minute
Time to 30,000 ft (9144 m): 6 minutes
Take-off run: 1,600 ft (488 m)
Maximum endurance: 5 hours 30 minutes
Ferry range: 1,300 nm (1,491 miles; 2400 km) with internal fuel; 1,700 nm (1,957 miles; 3150 km) with external fuel
Combat radius: 500 nm (578 miles; 930 km) hi-lo-hi with gun, four 1,000-lb (454-kg) bombs and two fuel tanks
***g* limit:** -4 to +7.5 *g* (service limit, cleared to 9 *g*); sustained turn 4.7 *g* at 1,600 ft (487 m)

Fuel and load

Internal fuel: 2,970 lb (1347 kg), 375 Imp gal (1705 litres)
Maximum external fuel: two 190-Imp gal (864-litre) fuel tanks
Maximum external load: tested up to 6,500 lb (2948 kg), weaponload up to 6,800 lb (3100 kg) on export versions

Armament

All Hawks can be fitted with a centreline gun pod, containing a single 30-mm ADEN Mk 4 cannon and 120 rounds of ammunition. Two underwing pylons are fitted as standard, but most Hawk series can be fitted with four underwing pylons. RAF Hawk T.Mk 1As have provision for underwing AIM-9L Sidewinder missiles. In service, they are usually limited to external loads of 1,500 lb (680 kg), carrying carrier, bomb. light store (CBLS) practice bomb carriers or SNEB rocket pods. The Hawk 60 and Hawk 100 can carry up to 6,614 lb (3000 kg), including a wide range of weapons plus wingtip AAM launch rails

91 Heat exchanger exhaust ducts
92 Upper UHF aerial
93 Anti-collision light
94 Port double-slotted flap, down position
95 Port wing integral fuel tank
96 Fuel tank access panels
97 Aileron hydraulic actuator
98 Port aileron
99 Localiser aerial
100 Thomson Brandt BAT 120 runway-cratering retarded bombs (nine)
101 Port navigation light
102 Vortex generators
103 AIM-9L Sidewinder air-to-air missiles
104 Missile launch rails
105 Twin missile carrier
106 Outboard stores pylon
107 Port wing fence
108 Leading-edge stall strip
109 Auxiliary wing fences
110 Air-conditioning system heat exchanger intakes
111 Heat exchangers (two)
112 Fuselage upper longeron
113 Mainspar attachment double frame
114 Intake duct framing
115 Auxiliary front spar attachment joint
116 Mainwheel door
117 Position of pressure refuelling connection on port side
118 Ventral gun pack (two)
119 Ammunition magazine
120 30-mm ADEN cannon
121 Ammunition feed and link return chutes
122 Starboard air inlet
123 Bifurcated inlet ducting
124 Fire extinguisher bottle
125 Nitrogen bottle, port and starboard (emergency pressurisation of hydraulic system)
126 Boundary layer spill duct
127 Oxygen bottle (two)
128 Cockpit rear pressure bulkhead
129 Canopy hinge joints
130 Instructor's Martin-Baker 'zero-zero' ejection seat
131 Ejection seat headrest
132 Port inboard stores pylon
133 Canopy, hinged to starboard
134 Rear cockpit internal windshield
135 Ferranti F.195 weapons sight
136 Rear instrument panel shroud
137 Side console panels
138 Rear control column
139 Rudder pedals
140 Rear cockpit pressurised floor panels
141 External fuse panel, port and starboard
142 Position of retractable boarding step on port side
143 Underfloor electrical equipment bay
144 Lower UHF aerial
145 Ventral gun pack attachment link
146 Cannon barrel

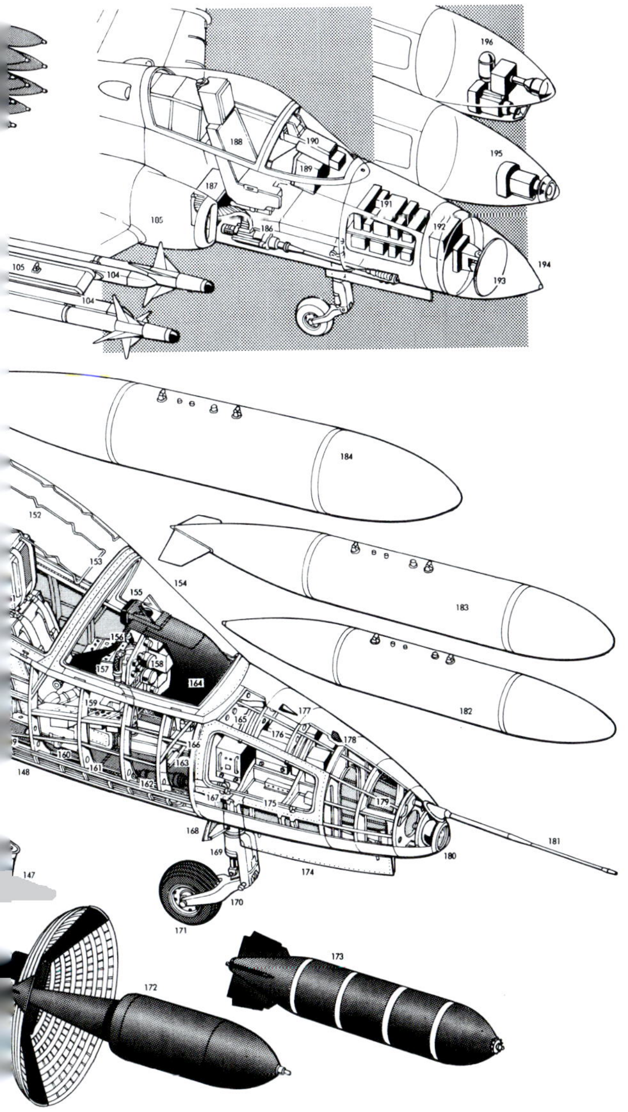

147 Cannon muzzle fairing
148 Electrical cable ducting, port and starboard
149 Fuselage lower longeron
150 Position of fold-out step handhold on port side
151 Student pilot's Martin Baker 'zero-zero' ejection seat
152 Canopy miniature detonating cord (MDC)
153 Canopy arch frame
154 Curved frameless windscreen panel
155 Student pilot's F.195 weapon sight and recording camera
156 Engine throttle lever
157 Front control column
158 Instrument panel
159 Starboard side console panel
160 Ejection seat rocket pack
161 Forward cockpit pressurised floor panels
162 Nose-landing gear hydraulic retraction jack
163 Rudder pedals
164 Instrument panel shroud
165 Avionics equipment bay
166 Cockpit front pressure bulkhead
167 Nose-landing gear pivot fixing
168 Nosewheel leg door
169 Shock absorber nosewheel leg strut
170 Levered suspension nosewheel forks
171 Nosewheel, forward-retracting
172 1,000-lb (454-kg) retarded bomb, deployed configuration
173 BL755 600-lb (272 kg) cluster bomb
174 Nosewheel doors
175 Avionics bay doors, port and starboard
176 Avionics equipment racks
177 Fresh air intake
178 Upper IFF aerial
179 Nosewheel bay construction
180 Landing/taxiing lamp
181 Pitot head
182 100-Imp gal (455-litre) auxiliary fuel tank
183 130-Imp gal (591-litre) auxiliary fuel tank
184 190-Imp gal (864-litre) auxiliary fuel tank
185 Hawk 200 single-seat conflguration
186 Built-in cannon (two) 25-mm, 27-mm or 30-mm
187 Ammunition magazine
188 Pilot's seat
189 Radar head-down display
190 Head-up display
191 Avionics equipment and radar processing equipment bay
192 Radar transmitter/ receiver
193 Radar scanner
194 Radome
195 Alternative nose configuration with Laser Ranger and Marked Target Seeker (LRMTS)
196 Alternative nose configuration with Laser Ranger and Infra-Red detector
197 Boeing/British Aerospace T-45A Goshawk, nose configuration
198 Catapult towbar
199 Twin-wheel nose landing gear
200 Lowered fuselage nose profile
201 Liquid oxygen converter
202 Advanced Cathode-Ray Tube (CRT) cockpit displays
203 T-45A Goshawk long-stroke main landing gear leg, carrier-compatible
204 T-45A main landing gear leg pivot, moved outboard
205 T-45A Goshawk rear fuselage configuration
206 Ventral fin and tail bumper
207 Deck arrester hook, lowered
208 Repositioned lateral airbrakes, port and starboard

Casa C.101 Aviojet

The C.101 has fulfilled the basic jet training role for the Spanish air force since 1980 and has proved to be safe and economical in this role. Sales of the aircraft never reached the 300 mark, which CASA had proposed in 1978, but this was due more to overenthusiastic predictions than to any shortcomings in the aircraft.

C.101CC

Cutaway key
1 Glassfibre nosecone
2 ILS glideslope antenna
3 Nosewheel doors
4 Nose undercarriage wheel bay
5 Oxygen bottles
6 Pitot head
7 TACAN antenna
8 Temperature probe
9 Cockpit fresh air intake
10 Avionics equipment bay
11 Nosewheel leg pivot mounting
12 Hydraulic retraction jack
13 Avionics bay access doors, port and starboard
14 Nosewheel leg strut
15 Forward-retracting nosewheel
16 Electro-luminescent formation lighting strip
17 Cockpit front pressure bulkhead
18 Static ports
19 Rudder pedals
20 Control column
21 Front instrument console
22 Avimo gunsight
23 Frameless windscreen panel
24 Canopy open position, individual canopy covers
25 Pilot's rear view mirrors
26 Student pilot's cockpit canopy
27 Ejection seat headrest
28 Martin-Baker Mk E10 'zero-zero' ejection seat
29 Canopy external latch
30 Engine throttle lever
31 Front seat mounting sloping bulkhead
32 Underfloor control runs
33 Cockpit pressure floor
34 DEFA 30-mm ventral cannon pack
35 Twin Browning 0.5-in (12.7-mm) machine-gun pack (alternative fit)
36 Machine-gun ammunition magazine, 220 rounds per gun
37 Cartridge case ejection chute
38 Battery
39 Cannon ammunition magazine, 130 rounds
40 Nitrogen bottle, emergency undercarriage lowering
41 Rear cockpit pressure floor
42 Rear canopy external latch
43 Rear instrument console
44 Internal windscreen between cockpits
45 Instructor's Martin-Baker Mk E10 ejection seat
46 Instructor's cockpit canopy
47 Starboard outer integral fuel tank, total internal capacity 531 Imp gal/638 US gal (2414 litres); 380 Imp gal/457 US gal (1730 litres) without outboard tanks
48 Starboard stores pylons (3)
49 Machined wing skin/ stringer panel over tank bay
50 Outer wing tank filler cap
51 Formation lighting strip
52 Starboard navigation light
53 Static dischargers
54 Starboard aileron
55 Aileron hydraulic actuator
56 Flap operating link and torque shaft
57 Starboard single-slotted flap
58 UHF antenna
59 Control rod linkages
60 Cockpit sloping rear pressure bulkhead
61 Port air intake
62 Boundary layer diverter
63 Airbrake hydraulic jack
64 Ventral airbrake panel
65 Pressure refuelling connection
66 Air-conditioning pack
67 Wing spar/fuselage bolted attachment joint
68 Front spar attachment fuselage mainframe
69 Fuselage bag-type fuel tank
70 Intake duct structure
71 Boundary layer spill duct
72 Centre/forward fuselage joint frame
73 Tank bay access panel
74 Fuel feed/vent piping

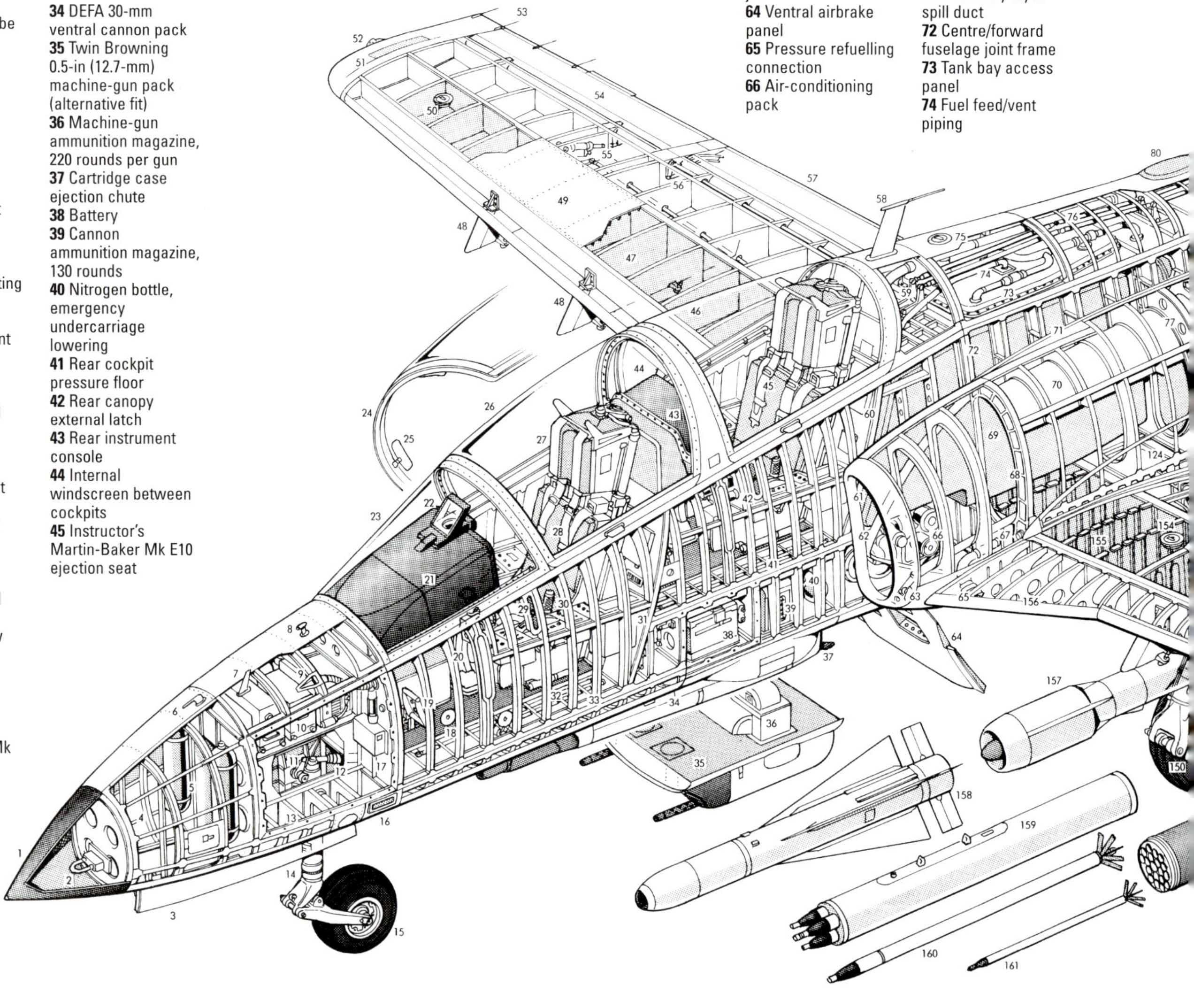

Although relatively docile in the jet trainer role, the C.101 can pack a punch in the light attack role. The aircraft can be configured with six underwing hardpoints, as illustrated by this CASA demonstrator, carrying a wide range of stores including bombs, missiles, rockets and guns. Armed variants are operated by the air arms of Chile, Honduras and Jordan.

C.101CC Aviojet

Dimensions

Length: 41 ft (12.50 m)
Height: 13 ft 11¼ in (4.25 m)
Wingspan: 34 ft 9½ in (10.60 m)
Aspect ratio: 5.6
Wing area: 215.29 sq ft (20.00 m²)
Tailplane span: 14 ft 2 in (4.32 m)
Wheel track: 10 ft 5¼ in (3.18 m)
Wheel base: 15 ft 7¾ in (4.77 m)

Powerplant

One Garrett TFE731-5-1J turbofan rated at 4,300 lb st (19.13 kN) dry normal and 4,700 lb st (20.91 kN) dry with military power reserve

Weights

Empty equipped: 7,716 lb (3500 kg)
Normal take-off: 11,023 lb (5000 kg)
Maximum take-off: 13,889 lb (6300 kg)

Fuel and load

Internal fuel: 4,017 lb (1822 kg)
External fuel: None
Maximum ordnance: 4,960 lb (2250 kg)

Performance

Never-exceed speed: 450 kt (518 mph; 834 km/h)
Maximum level speed 'clean' at 20,000 ft (6095 m): 435 kt (501 mph; 806 km/h)
Maximum speed at sea level: 415 kt (478 mph; 769 km/h)
Economical cruising speed at 30,000 ft (9145 m): 354 kt (407 mph; 656 km/h)
Ferry range: 2,000 nm (2,303 miles; (3706 km)
Combat radius: 280 nm (322 miles; (519 km) on a lo-lo-lo mission with cannon pod and four 551-lb (250-kg) bombs, or 520 nm (599 miles; 964 kg) on a hi-lo-hi photo-reconnaissance mission
Maximum rate of climb at sea level: 4,900 ft (1494 m) per minute at normal power and 6,100 ft (1859 m) per minute with military power reserve
Climb to 25,000 ft (7620 m): 6 minutes 30 seconds
Service ceiling: 42,000 ft (12800 m)
Take-off run: 1,835 ft (559 m) at 9,921 lb (4500 kg)
Take-off distance to 50 ft (15 m): 2,460 ft (750 m) at 9,921 lb (4500 kg)
Landing run: 1,575 ft (480 m) at 10,361 lb (4700 kg)
***g* limits:** -3.9 to +7.5 at 10,582 lb (4800 kg) or -1 to +5.5 at 13,889 lb (6300 kg)

Armament

One 30-mm DEFA fixed forward-firing cannon under the fuselage or two 0.5-in (12.7-mm) Colt-Browning fixed forward-firing machine-guns in the lower-fuselage bay, plus up to 4,056 lb (1840 kg) of disposable stores, including bombs, rockets or missiles, carried on six underwing hardpoints

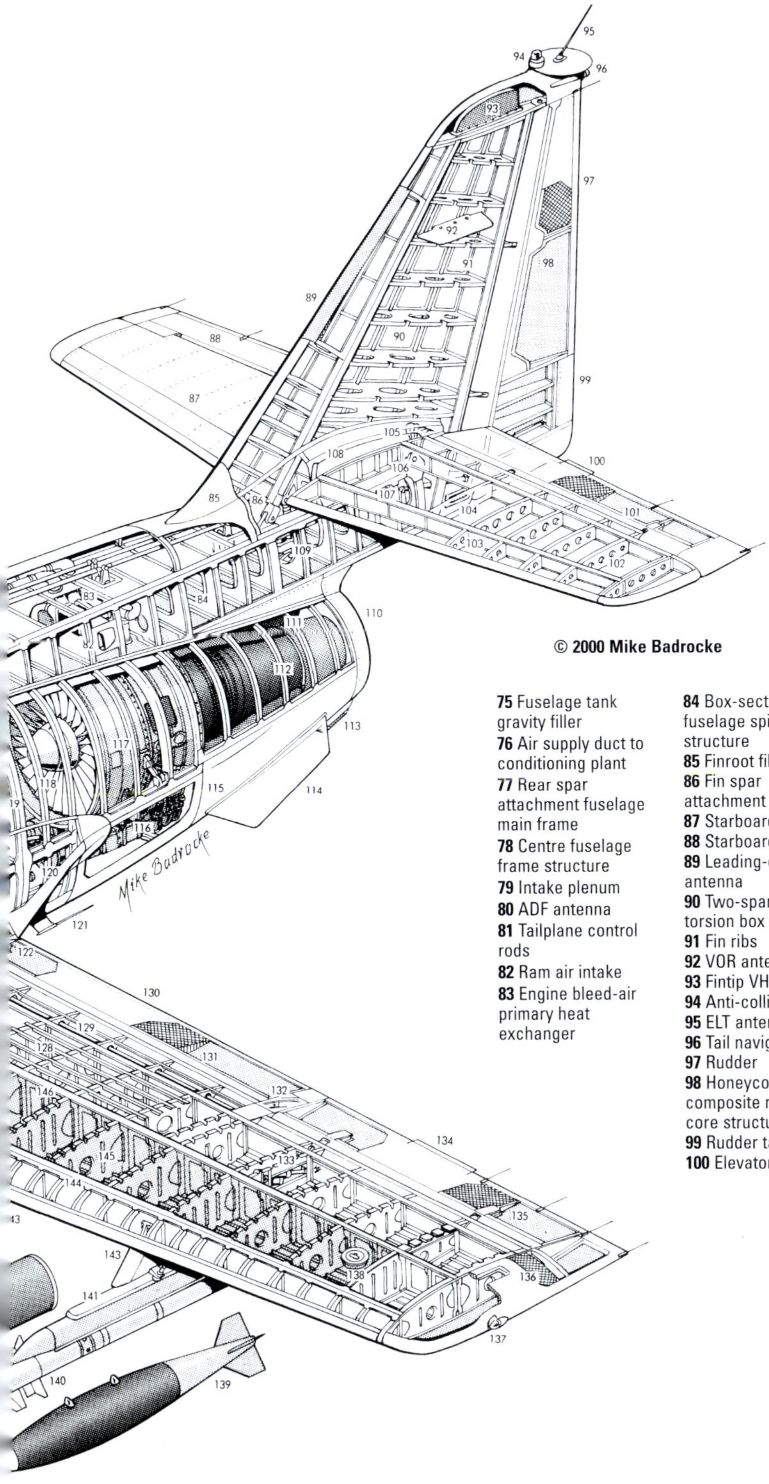

75 Fuselage tank gravity filler
76 Air supply duct to conditioning plant
77 Rear spar attachment fuselage main frame
78 Centre fuselage frame structure
79 Intake plenum
80 ADF antenna
81 Tailplane control rods
82 Ram air intake
83 Engine bleed-air primary heat exchanger
84 Box-section rear fuselage spine structure
85 Finroot fillet
86 Fin spar attachment joint
87 Starboard tailplane
88 Starboard elevator
89 Leading-edge HF antenna
90 Two-spar fin torsion box structure
91 Fin ribs
92 VOR antenna
93 Fintip VHF antenna
94 Anti-collision light
95 ELT antenna
96 Tail navigation light
97 Rudder
98 Honeycomb composite rudder core structure
99 Rudder tab
100 Elevator fixed tab
101 Port elevator honeycomb composite core structure
102 Tailplane ribs
103 Two-spar torsion box tailplane structure
104 Rear formation lighting strip
105 Elevator hinge control
106 Trimming tailplane hinge mounting
107 Rudder hinge control
108 Trimming tailplane sealing plate
109 Tailplane trim control screw jack, electrically operated
110 Engine exhaust nozzle
111 Fan air, cold stream, exhaust duct
112 Core engine, hot stream, exhaust
113 Tail bumper
114 Ventral fin, port and starboard
115 Engine bay cowling panels
116 Accessory equipment gearbox
117 Allied Signal TFE731-5-1J turbofan engine
118 Engine front fan
119 Rear fuselage joint frame
120 Wingroot trailing-edge fillet
121 Lower UHF antenna
122 Flap inboard guide rail
123 Hydraulic reservoir
124 Main undercarriage wheel bay
125 Hydraulic retraction jack
126 Undercarriage mounting auxiliary wing spar
127 Rear spar
128 Wing stringers
129 Flap shroud ribs
130 Port single-slotted flap
131 Flap honeycomb composite core structure
132 Flap outboard guide rail
133 Port aileron hydraulic actuator
134 Fixed aileron tab
135 Port aileron honeycomb composite core structure
136 Composite wingtip fairing
137 Port navigation light
138 Port outer wing tank gravity filler
139 Mk 82 500-lb (227-kg) HE bomb
140 MATRA Magic air-to-air 'self-defence' missile
141 Missile launch rail
142 LAU-3/A rocket launcher, 19 x 2.75-in (70-mm) FFAR
143 Port wing stores pylons (3)
144 Front spar
145 Port outer integral wing tank
146 Main spar
147 Wing rib structure
148 Leading-edge ribs
149 Port mainwheel
150 Trailing-axle suspension
151 Mainwheel leg doors
152 Retractable landing light
153 Main undercarriage leg pivot mounting
154 Wing centre tank gravity filler
155 Centre-section integral fuel tank
156 Extended chord wingroot section
157 Elettronica ELT/555 jamming pod
158 ASM-65A Maverick air-to-surface missile
159 LAU-10 rocket launcher, 4 x 5-in (127-mm) FFAR
160 5-in (127-mm) Zuni rocket
161 2.75-in (70-mm) rocket

Cessna T-37/A-37 Dragonfly

The T-37 has been the primary USAF trainer for almost four decades. First scheduled to be replaced by the Fairchild T-46A (cancelled in 1987), the 'Tweet' now awaits replacement by the Raytheon T-6A Texan II, procured jointly under the JPATS scheme with the US Navy.

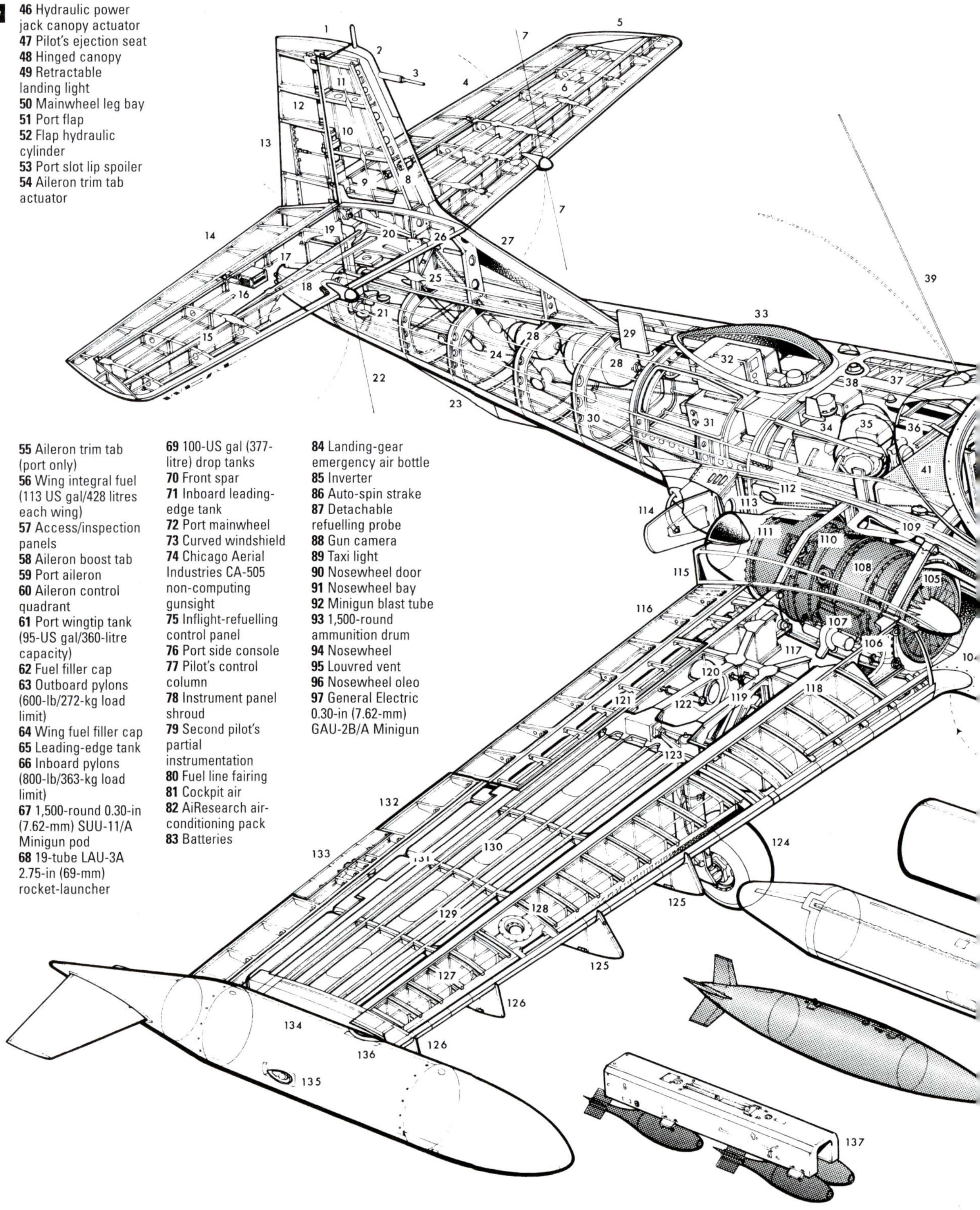

A-37B Dragonfly

Cutaway key

1 Rudder top hinge and aerodynamic balance
2 VHF communications antenna
3 Pitot head
4 Port elevator
5 Elevator aerodynamic balance
6 Tailplane structure
7 FM homing dipole
8 Elevator trim tab actuator
9 Elevator trim tab (port only)
10 Rudder trim tab actuator
11 Tailfin structure
12 Rudder
13 Rudder trim tab
14 Starboard elevator
15 Tailplane front spar
16 Magnetic detector
17 Rear navigation lights
18 Tail cone
19 Tailplane/fin rear spar attachment
20 Elevator push rod
21 Rudder control bell crank assembly
22 FM homing dipole
23 Tail bumper fairing
24 Rudder/elevator control cables
25 Elevator control quadrant assembly
26 Tailplane fin front spar attachment
27 Tailfin fairing
28 Oxygen cylinders
29 Dorsal UHF TACAN antenna
30 Fuselage frames
31 Avionics bay access door
32 Radio/avionics equipment
33 Low frequency ADI antenna
34 Dorsal IFF antenna
35 Hydraulic reservoir
36 Inverter
37 Fuselage construction
38 Dorsal position light
39 FM communications whip aerial
40 Canopy fairing
41 Fuselage fuel tank, 91 US gal (344 litres)
42 Seat ejection unit
43 Weber ejection seat
44 Harness
45 Headrest
46 Hydraulic power jack canopy actuator
47 Pilot's ejection seat
48 Hinged canopy
49 Retractable landing light
50 Mainwheel leg bay
51 Port flap
52 Flap hydraulic cylinder
53 Port slot lip spoiler
54 Aileron trim tab actuator
55 Aileron trim tab (port only)
56 Wing integral fuel (113 US gal/428 litres each wing)
57 Access/inspection panels
58 Aileron boost tab
59 Port aileron
60 Aileron control quadrant
61 Port wingtip tank (95-US gal/360-litre capacity)
62 Fuel filler cap
63 Outboard pylons (600-lb/272-kg load limit)
64 Wing fuel filler cap
65 Leading-edge tank
66 Inboard pylons (800-lb/363-kg load limit)
67 1,500-round 0.30-in (7.62-mm) SUU-11/A Minigun pod
68 19-tube LAU-3A 2.75-in (69-mm) rocket-launcher
69 100-US gal (377-litre) drop tanks
70 Front spar
71 Inboard leading-edge tank
72 Port mainwheel
73 Curved windshield
74 Chicago Aerial Industries CA-505 non-computing gunsight
75 Inflight-refuelling control panel
76 Port side console
77 Pilot's control column
78 Instrument panel shroud
79 Second pilot's partial instrumentation
80 Fuel line fairing
81 Cockpit air
82 AiResearch air-conditioning pack
83 Batteries
84 Landing-gear emergency air bottle
85 Inverter
86 Auto-spin strake
87 Detachable refuelling probe
88 Gun camera
89 Taxi light
90 Nosewheel door
91 Nosewheel bay
92 Minigun blast tube
93 1,500-round ammunition drum
94 Nosewheel
95 Louvred vent
96 Nosewheel oleo
97 General Electric 0.30-in (7.62-mm) GAU-2B/A Minigun

66-7978, a T-37B-CE, was typical of the bulk of 'Tweety Bird' production for the USAF Air Training Command (ATC). A total of 552 T-37Bs was built, making it the most numerous Model 318 training variant. The T-37 powerplant is the Continental/Teledyne J69 turbojet, a licence-built version of the Turboméca Marboré, the engine used by the Magister. The T-37B introduced the uprated (1,025-lb/4.61-kN static thrust) J69-T-25 in place of the T-37A's (920-lb/4.14-kN static thrust) J69-T-9.

SPECIFICATION

OA-37B Dragonfly

Dimensions

Length excluding probe: 29 ft 3¼ in (8.93 m)
Height: 8 ft 10½ in (2.70 m)
Wingspan with tip tanks: 35 ft 10½ in (10.93 m)
Wing area: 183.90 sq ft (17.09 m²)
Wing aspect ratio: 6.2
Tailplane span: 13 ft 11¼ in (4.25 m)
Wheel track: 14 ft ½ in (4.28 m)
Wheel track: 7 ft 10 in (2.39 m)

Powerplant

Two General Electric J85-GE-17A turbojets each rated at 2,850 lb st (12.68 kN)

Weights

Basic empty: 6,211 lb (2817 kg)
Empty equipped: 5,843 lb (2650 kg)
Maximum take-off: 14,000 lb (6350 kg)

Fuel and load

Internal fuel: 3,307 lb (1500 kg)
External fuel: up to four 100-US gal (378-litre) drop tanks
Maximum ordnance: 4,100 lb (1860 kg)

Performance

Never-exceed speed: 455 kt (524 mph; 843 km/h)
Maximum level speed at 16,000 ft (4875 m): 440 kt (507 mph; 816 km/h)
Maximum cruising speed at 25,000 ft (7620 m): 425 kt (489 mph; 787 km/h)
Maximum rate of climb at sea level: 6,990 ft (2130 m) per minute
Service ceiling: 41,765 ft (12730 m)
Take-off run at maximum take-off weight: 1,740 ft (531 m)
Take-off distance to 50 ft (15 m) at maximum take-off weight: 2,595 ft (791 m)
Landing run at normal landing weight: 1,710 ft (521 m)

Armament

Up to 4,100 lb (1860 kg) of stores mounted on eight underwing pylons, including unguided bombs and rockets

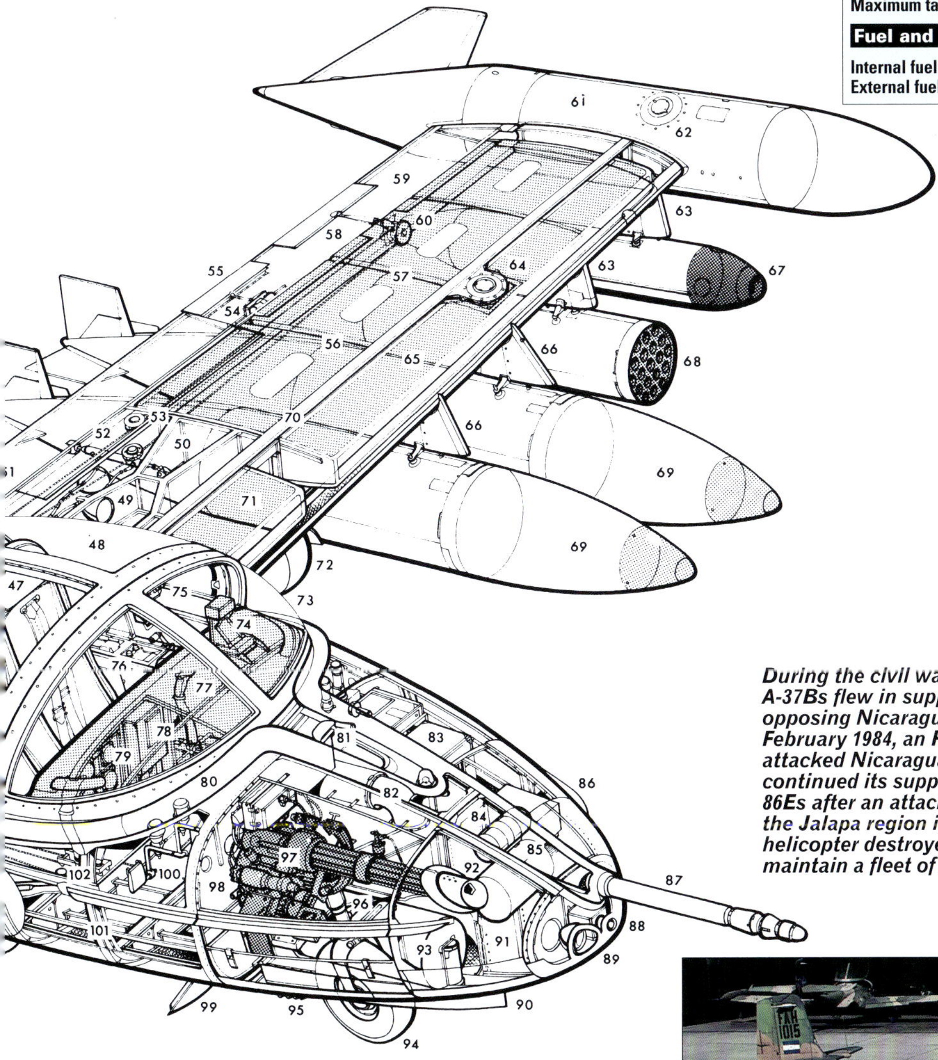

During the civil war in Nicaragua, Fuerza Aérea Hondureña (FAH) A-37Bs flew in support of the US-backed Nicaraguan Contras, opposing Nicaragua's left-wing Sandinista government. In February 1984, an FAH A-37B, supported by five helicopters, attacked Nicaraguan positions. The following year the FAH continued its support of the Contras, striking with A-37Bs and F-86Es after an attack by Fuerza Aérea Sandinista Mi-8s and Mi-25s in the Jalapa region in September, when the FAH claimed one Mil helicopter destroyed. In early 2000, Honduras continues to maintain a fleet of A-37B Dragonflies, principally based at La Ceiba.

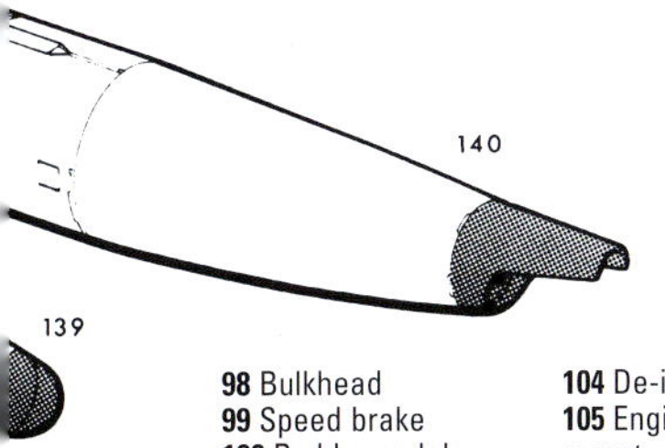

98 Bulkhead
99 Speed brake
100 Rudder pedals
101 Fuselage structure
102 Control column pushrod assembly
103 Automatically-actuated inlet screen
104 De-icing inlet lip
105 Engine forward mount
106 Front spar attachment points
107 Engine accessories
108 General Electric J85-GE-17A engine
109 Supplementary intake
110 Engine rear mount
111 Tailpipe
112 Elevator controls
113 Jet exhaust fairing
114 Hydraulically-operated thrust attenuator
115 Wingroot/engine nacelle fairing
116 Starboard flap
117 Wheelwell
118 Front spar
119 Mainwheel retraction strut
120 Retractable landing light
121 Starboard slot lip spoiler
122 Flap hydraulic cylinder
123 Mainwheel oleo
124 Starboard mainwheel
125 Inboard pylons (see item 66)
126 Outboard pylons (see item 63)
127 Leading-edge structure
128 Fuel filler cap
129 Wing construction
130 Access/inspection panels
131 Rear spar
132 Starboard aileron
133 Aileron boost tab
134 Starboard wingtip tank
135 Starboard navigation light
136 Fuel filler cap
137 Practice-bomb carrier (four 35-lb/16-kg BDU-33 bombs)
138 500-lb (227-kg) Mk 82 low-drag bomb
139 BLU-1/B napalm fire bomb
140 GPU-2/A 20-mm cannon pod

Dassault/Dornier Alphajet

Two French air force Alpha Jets from the advanced training unit at Tours formate with a pair of aircraft from the Moroccan air force equivalent. In addition to their primary role of providing advanced training, the 'Gadgets' of the Ecole de l'Aviation de Chasse 314 are used for fast-jet navigator training. Around 50 French Alpha Jets are being upgraded with 'glass' cockpits to make them more suitable as lead-in trainers for modern fighters such as the Mirage 2000-5F and Rafale.

Alpha Jet E

1 Nosecone
2 Nosewheel bay bulkhead
3 Fixed nose strake
4 Nose landing-gear wheel bay
5 Nosewheel door mechanism
6 Temperature probe
7 Fresh air intake
8 Nose landing-gear leg strut
9 Pivoted axle beam
10 Spray suppression nosewheel tyre
11 Pitot tube
12 Nosewheel leg door
13 Nosewheel pivot fixing
14 Oxygen filler point
15 Liquid oxygen container
16 Cockpit front pressure bulkhead
17 Rudder pedals
18 Rudder pedal access panel
19 Footboards
20 Control column linkage
21 Rudder cable run
22 Cockpit coaming
23 Instrument panel shroud
24 Windscreen panels
25 Pilot's head-up display (navigational data display on trainer version)
26 Starboard side console panel
27 Control column
28 Canopy latch
29 Engine throttle levers
30 Front seat boarding step
31 Aileron control rod run
32 Port side console panel
33 Seat/parachute harness
34 Student pilot's Martin Baker AJR1VIA-4 ejection seat
35 Ejection seat headrest
36 Face blind firing handle
37 Canopy breaker arms
38 Forward hinged canopy
39 Canopy operating jack
40 Rear cockpit blast shield
41 Ejection seat launch rails
42 Rear rudder pedal linkages
43 Canopy emergency release handle
44 SFIM 550 inertial platform
45 Rear cockpit elevated floor level
46 Boundary layer splitter plate
47 Port engine air intake
48 Forward avionics equipment bay
49 Intake duct framing
50 Splitter plate honeycomb construction
51 Boarding step
52 Rear throttle levers
53 Boundary layer spill duct
54 Rear canopy latch
55 Rear instrument panel shroud
56 Canopy centre section glazing
57 Fixed canopy arch
58 Rear pilot/instructor's ejector seat
59 Rear hinged canopy cover
60 Ejection seat headrest
61 Cockpit rear bulkhead
62 Canopy hinges
63 Control rod runs
64 Wing leading-edge fairing
65 Wing centre-section carry through
66 Skin panel bolted joint strap
67 Central flap hydraulic jack
68 Electrical cable ducting
69 Dorsal spine fairing
70 UHF aerial
71 Starboard wing panel bolted joint
72 Wing skin panel spanwise joint
73 Starboard wing integral fuel tank. Total internal fuel capacity 418 Imp gal (1900 litres)
74 Compound sweep leading-edge section
75 Inboard pylon hardpoint
76 68-Imp gal (310-litre) auxiliary fuel tank
77 Tank filler cap
78 Outboard tank pylon
79 Leading-edge dog-tooth
80 Fuel system access panels
81 Starboard navigation light
82 Wingtip fairing
83 Starboard aileron
84 Static dischargers
85 Aileron control linkage
86 Hydraulic operating jack
87 Aileron push-pull rod
88 Flap hinge fairings
89 Flap vane
90 Starboard double-slotted tracked flap
91 Inboard flap guide rail
92 Control system mechanical mixer unit
93 Flap operating mechanism
94 Rear spar
95 Rear fuselage fuel tank
96 Fixed trailing-edge fillet
97 Fuel filler cap
98 Air-conditioning plant
99 Dorsal spine access panels

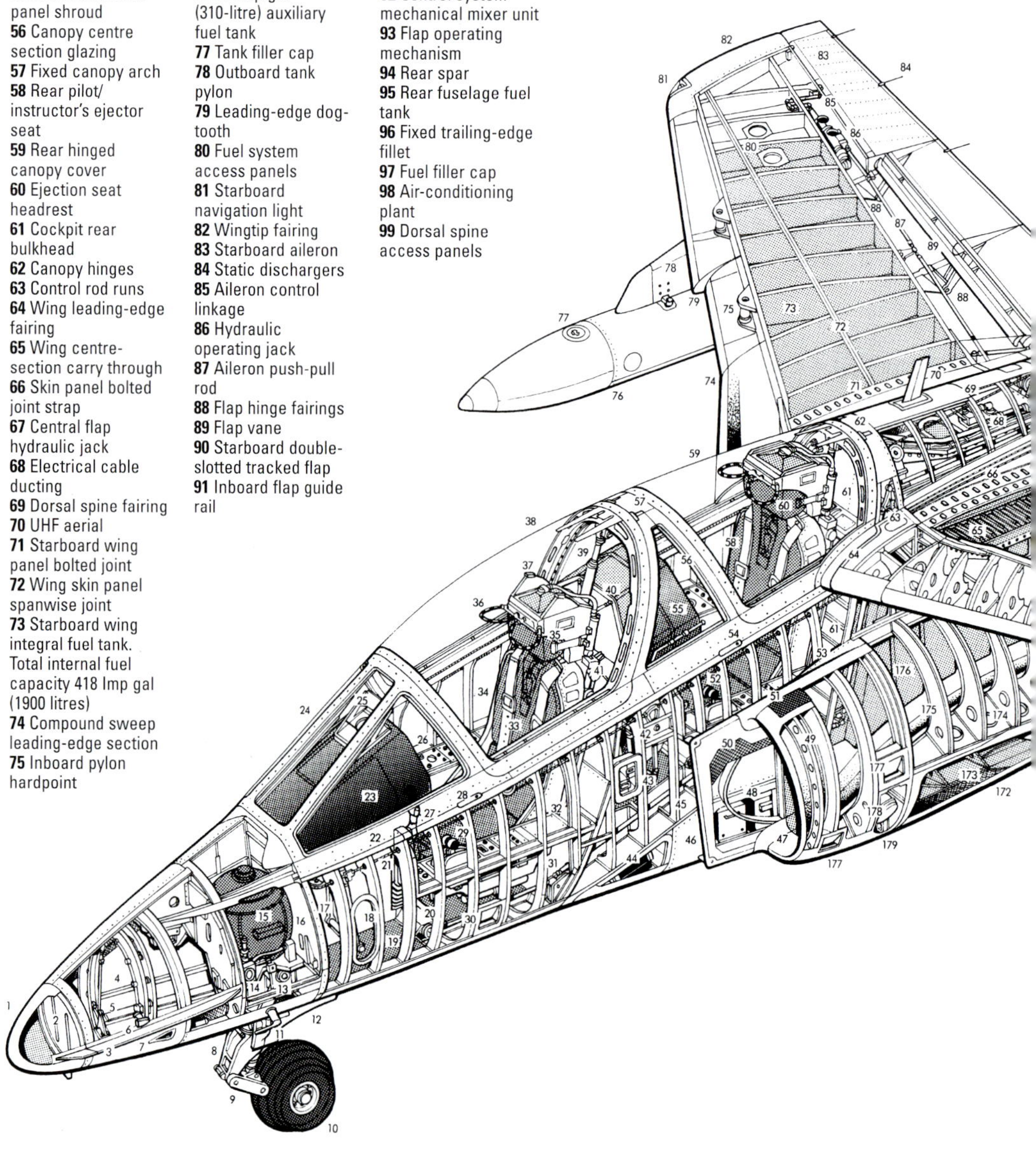

During the 1991 Gulf War the Alpha Jet received a call-to-alert from a surprising direction. While US and NATO aircraft were assigned directly to the war against Iraq, aircraft from other nations were drafted in to maintain NATO's commitment to defend Turkey. Among the aircraft deployed were the Alpha Jets of 2./JBG 49 (492 Squadron in NATO parlance), seen here at Erhac.

SPECIFICATION

Alpha Jet E

Dimensions

Length: 38 ft 6½ in (11.75 m)
Height: 13 ft 9 in (4.19 m)
Wingspan: 29 ft 10¾ in (9.11 m)
Wing area: 188.37 sq ft (17.50 m²)
Wing aspect ratio: 4.8
Tailplane span: 14 ft 2½ in (4.33 m)
Wheel track: 8 ft 10¾ in (2.71 m)
Wheelbase: 15 ft 5¾ in (4.72 m)

Powerplant

Two SNECMA/Turboméca Larzac 04-C6 turbofans, each rated at 2,976 lb (13.24 kN) thrust

Weights

Empty equipped: 7,374 lb (3345 kg)
Maximum take-off: 17,637 lb (8000 kg)

Fuel and load

Internal fuel: 2,976 lb (1350 kg)
External fuel: up to four 99-Imp gal (450-litre) or 68-Imp gal (310-litre) drop tanks
Maximum ordnance: more than 5,511 lb (2500 kg)

Performance

Maximum level speed at 32,810 ft (10000 m): 494 kt (569 mph; 916 km/h)
Maximum level speed at sea level: 539 kt (621 mph; 1000 km/h)
Maximum rate of climb at sea level: 12,008 ft (3660 m) per minute
Climb to 30,020 ft (9150 m): less than 7 minutes
Service ceiling: 48,000 ft (14630 m)
Take-off run at normal take-off weight: 1,215 ft (370 m)
Landing run at normal landing weight: 1,640 ft (500 m)
Ferry range: more than 2,159 nm (2,486 miles; 4000 km)
Operational radius: 361 nm (416 miles; 670 km) on a lo-lo-lo training mission with two drop tanks, or 664 nm (764 miles; 1230 km) on a hi-hi-hi mission
Endurance on internal fuel: more than 3 hours 30 minutes at high altitude or 2 hours 30 minutes at low level

Armament

Removable centreline pod for one 30-mm DEFA 553 cannon plus 150 rounds (27-mm IWKA-Mauser cannon on Alpha Jet A). Four underwing hardpoints for drop tanks and various unguided munitions, including general-purpose and cluster bombs, MATRA F4 or 155 rocket pods, or Brandt BAT120 retarded bombs

100 Heat exchanger fresh air scoop
101 Tailplane control runs
102 Starboard airbrake, open position
103 Anti-collision light
104 Rudder control cables
105 Finroot fairing
106 Starboard tailplane
107 Tailfin construction
108 Fin main spar

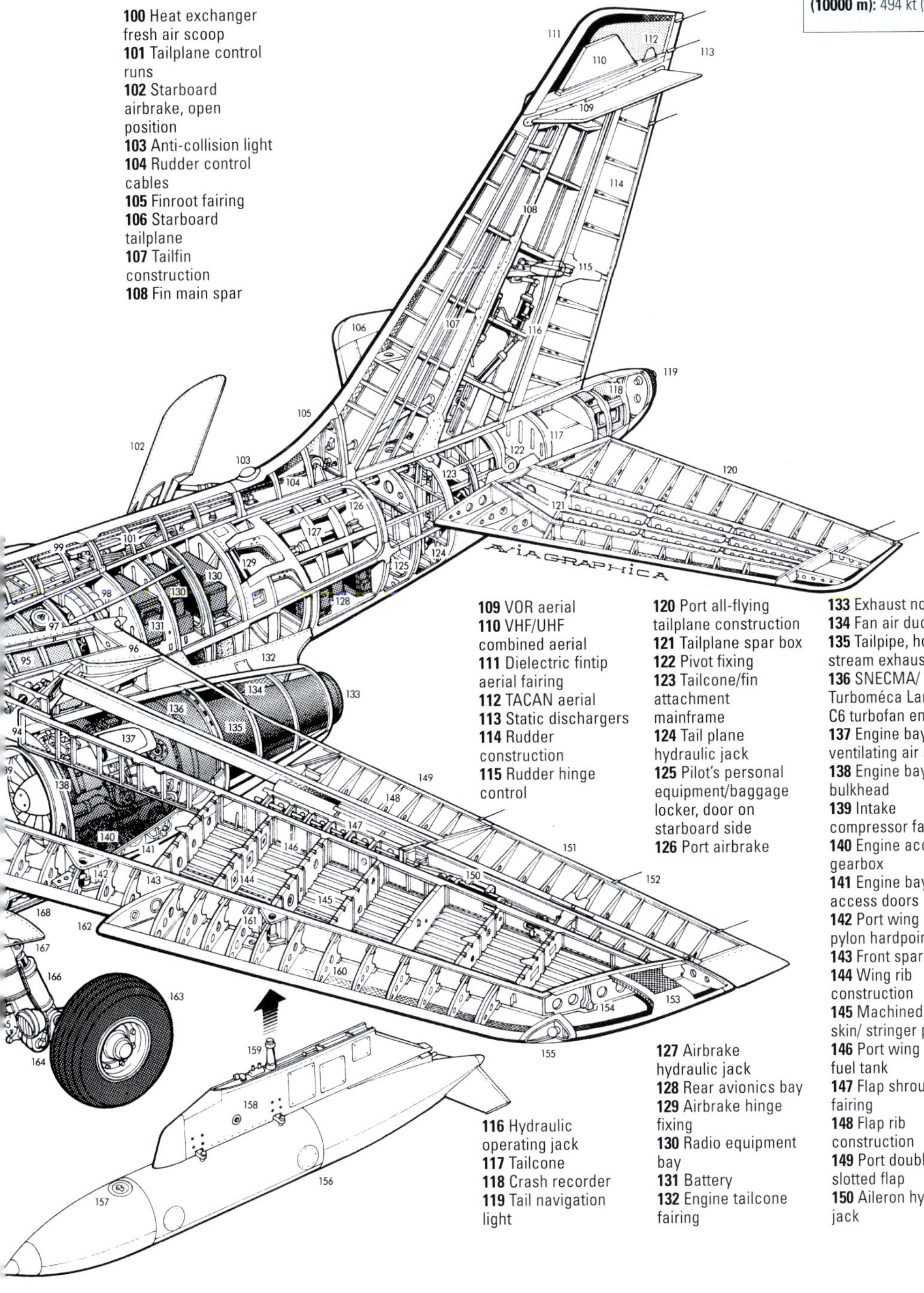

109 VOR aerial
110 VHF/UHF combined aerial
111 Dielectric fintip aerial fairing
112 TACAN aerial
113 Static dischargers
114 Rudder construction
115 Rudder hinge control
116 Hydraulic operating jack
117 Tailcone
118 Crash recorder
119 Tail navigation light
120 Port all-flying tailplane construction
121 Tailplane spar box
122 Pivot fixing
123 Tailcone/fin attachment mainframe
124 Tail plane hydraulic jack
125 Pilot's personal equipment/baggage locker, door on starboard side
126 Port airbrake
127 Airbrake hydraulic jack
128 Rear avionics bay
129 Airbrake hinge fixing
130 Radio equipment bay
131 Battery
132 Engine tailcone fairing
133 Exhaust nozzle
134 Fan air duct
135 Tailpipe, hot stream exhaust
136 SNECMA/ Turboméca Larzac 04-C6 turbofan engine
137 Engine bay ventilating air scoop
138 Engine bay bulkhead
139 Intake compressor face
140 Engine accessory gearbox
141 Engine bay access doors
142 Port wing inboard pylon hardpoint
143 Front spar
144 Wing rib construction
145 Machined wing skin/ stringer panel
146 Port wing integral fuel tank
147 Flap shroud fairing
148 Flap rib construction
149 Port double-slotted flap
150 Aileron hydraulic jack
151 Port aileron construction
152 Static dischargers
153 Glassfibre honeycomb wingtip fairing
154 Wingtip jacking point
155 Port navigation light
156 Port 68-Imp gal (310-litre) auxiliary fuel tank
157 Fuel filler cap
158 Tank pylon
159 Pylon attachment spigot
160 Wing leading-edge rib construction
161 Outboard pylon hardpoint
162 Leading-edge dog-tooth
163 Port mainwheel
164 Pivoted axle beam
165 Landing/taxiing lamp
166 Shock absorber strut
167 Mainwheel leg door
168 Ground connections panel, electrical, hydraulic and intercom
169 Main landing gear leg pivot fixing
170 Hydraulic retraction jack
171 Hydraulic downlock strut
172 Mainwheel door
173 Main landing wheel bay
174 Intake duct framing
175 Intake trunking
176 Centre fuselage bag-type fuel tank
177 Rear seat boarding steps
178 Position of pressure refuelling connection (actually located on starboard side)
179 Fuselage jacking point

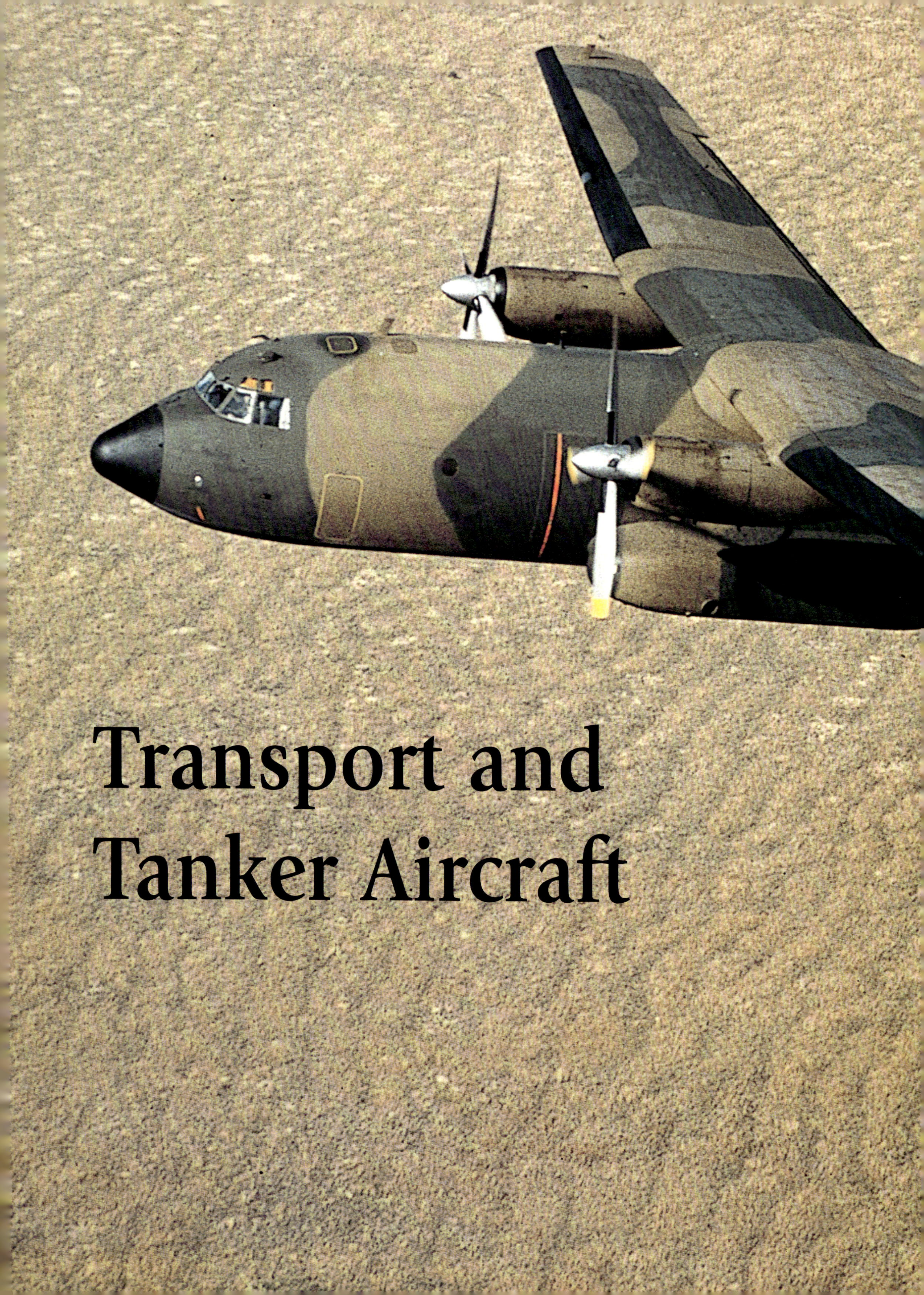

Transport and Tanker Aircraft

Antonov An-12 'Cub'

Having demonstrated all the versatility of the Hercules in both transport and 'specia roles, the An-12 is in the twilight of i career with Russian forces. Elsewhere remains an important type however and still the subject, in upgraded form, of limite production in Chin

An-12BP 'Cub-A'

Cutaway key
1 Nose compartment glazing
2 Optically flat lower viewing panel
3 Nose radome
4 Weather and navigational radar scanner
5 Chart table
6 Navigator's station
7 Nose compartment entry hatch
8 'Odd Rods' IFF aerials
9 Windscreen panels
10 Windscreen wipers
11 Instrument panel shroud
12 Pilot's instrument panel
13 Control column
14 Rudder pedals
15 Boarding ladder
16 Door mounted retractable taxiing lamp
17 Blade antennas
18 Crew entry door/ escape hatch, open
19 Avionics equipment racks
20 Flight deck floor level
21 Pilot's seat
22 Cockpit eyebrow windows
23 Co-pilot's seat
24 Overhead systems switch panel
25 Aerial lead-in
26 Cockpit roof escape hatch
27 Flight engineer's instrument panels
28 Engineer's swivelling seat
29 Flight deck doorway
30 Cockpit pressure bulkhead
31 Radio operator's station
32 Nose landing gear pivot fixing
33 Pitot head
34 Nosewheel hydraulic steering control unit
35 Twin nosewheels, aft retracting
36 Blade antenna
37 Cargo deck floor level
38 Ventral access hatch
39 Cabin window panels
40 Port side emergency exit window hatch
41 Paratroop seating, 100 troops maximum
42 Central 'back-to-back' seat rows, removable
43 Cabin wall removable troop seat
44 Cabin wall insulating and trim panelling
45 Starboard side emergency exit window hatch
46 D/F loop aerials
47 Fuselage frame and stringer construction
48 Cargo deck floor beams
49 Underfloor bulk stowage compartment
50 Main cargo loading deck
51 Crew/passenger entry door
52 Wing spar/fuselage attachment main frame
53 Engine floodlight
54 Wing root fillet fairings
55 Front spar centre-section carry-through
56 Wing panel bolted root joints
57 Centre section ribs
58 Inboard bag-type fuel tanks (three), total fuel capacity 3058 Imp gal (13901 litres), 3981 Imp gal (18100 litres) with overload tanks
59 Starboard inner engine nacelle
60 Hinged engine cowling panels
61 Ventral oil cooling intake
62 Propeller spinners
63 AV-68 four-bladed fully feathering and reversible, variable pitch propellers
64 Starboard outer engine nacelle
65 Intermediate bag-type fuel tanks (five)
66 Outboard bag-type fuel tanks (three)
67 Outerwing panel joint rib
68 Ventral navigational antennas
69 Anhedral outerwing panel
70 Starboard navigation lights
71 Static dischargers
72 Starboard two-segment aileron
73 Aileron trim tab
74 'Cub-B' electronic intelligence variant (Elint)
75 Variant with maritime surveillance radar
76 'Cub-C' Electronic countermeasures variant (ECM)
77 ASW version with magnetic anomaly detection equipment (MAD)
78 Starboard double-slotted flap, down position
79 Flap guide rails
80 Wing root trailing edge fillet
81 ADF sense aerial, port and starboard
82 Starboard emergency exit window hatch
83 Overhead travelling cargo handling crane
84 Rear cabin roof escape hatches
85 Starboard ramp door, open
86 Fin root fillet
87 Tailfin support structure
88 Two-spar torsion box tailfin construction
89 Starboard tailplane
90 Starboard elevator
91 Fin leading edge thermal de-icing
92 HF aerial cables
93 Short wave ground- control communications antennas
94 Anti-collision light
95 Static dischargers
96 Rudder
97 Rudder trim tabs
98 Tail navigation light
99 Tail warning radar antenna
100 Rear gunner's station
101 Gun turret, two 23-mm NR-23 cannon
102 Elevator tab
103 Port elevator
104 Static dischargers
105 Tailplane leading edge thermal de-icing

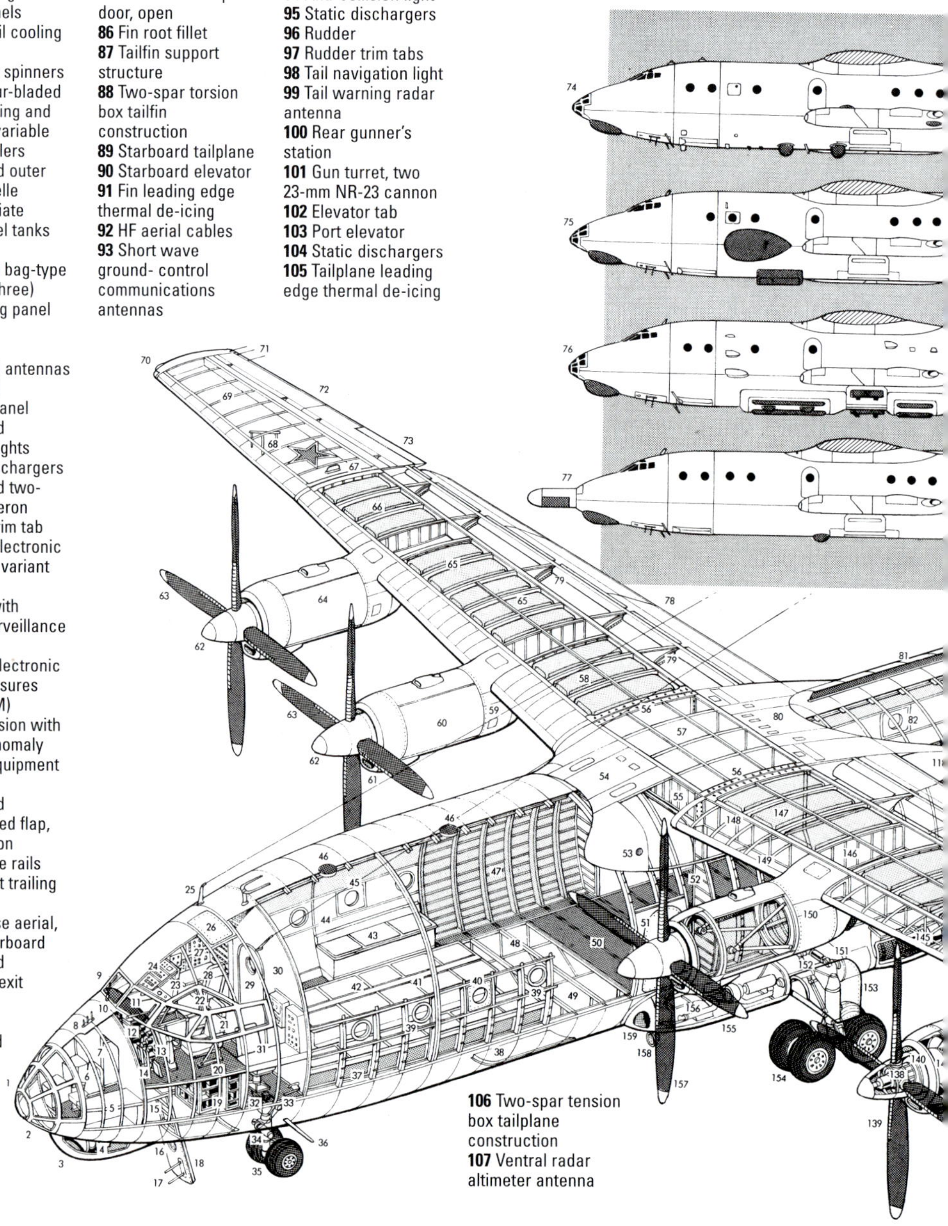

106 Two-spar tension box tailplane construction
107 Ventral radar altimeter antenna

This An-12PS was photographed at Saki in the Crimea during 2000, while on the strength of the Severomorsk Regiment and wearing an unusual polar bear marking. The An-12PS is based on the airframe of the basic An-12B and is believed to have been produced both by conversion and on the production line from 1969. The official line seems to be that the An-12PS was a search and rescue platform, with the initials PS standing for Poiskovo Spasatel'nii. It could reportedly carry and deploy a Type 03473 rescue boat, with a three-man crew. The deployment of the An-12PS suggests that the variant also fulfilled a vital Elint role, for it was frequently encountered shadowing NATO naval forces or monitoring NATO exercises.

SPECIFICATION

An-12BP 'Cub-A'

Dimensions

Wingspan: 124 ft 8 in (38 m)
Wing area: 1,310.01 sq ft (121.7 m²)
Length: 108 ft 7¼ in (33.10 m)
Height: 34 ft 6½ in (10.53 m)
Wing aspect ratio: 11.85
Tailplane span: 40 ft ¼ in (12.20 m)
Wheel track: 17 ft 9½ in (5.42 m)
Wheel base: 35 ft 6 in (10.82 m)

Powerplant

Four ZMDB Progress (Ivchyenko) AI-20K turboprops each rated at 4,000 ehp (2983 kW)

Typical weights

Empty: 61,728 lb (2800 kg)
Normal take-off: 121,473 lb (55100 kg)
Maximum take-off: 134,480 lb (61000 kg)

Fuel and load

Internal fuel: 4,781 US gal (18100 litres)
Maximum payload: 44,092 lb (20000 kg)

Performance

Maximum level speed 'clean' at optimum altitude: 482 mph (777 km/h)
Maximum cruising speed at optimum altitude: 416 mph (670 km/h)
Maximum rate of climb at sea level: 1,969 ft (600 m) per minute
Service ceiling: 33,465 ft (10200 m)
Take-off run at maximum take-off weight: 2,297 ft (700 m)
Landing run at normal landing weight: 1,640 ft (500 m)
Range with maximum fuel: 3,542 miles (5700 km)
Range with maximum payload: 2,237 miles (3600 km) with maximum payload

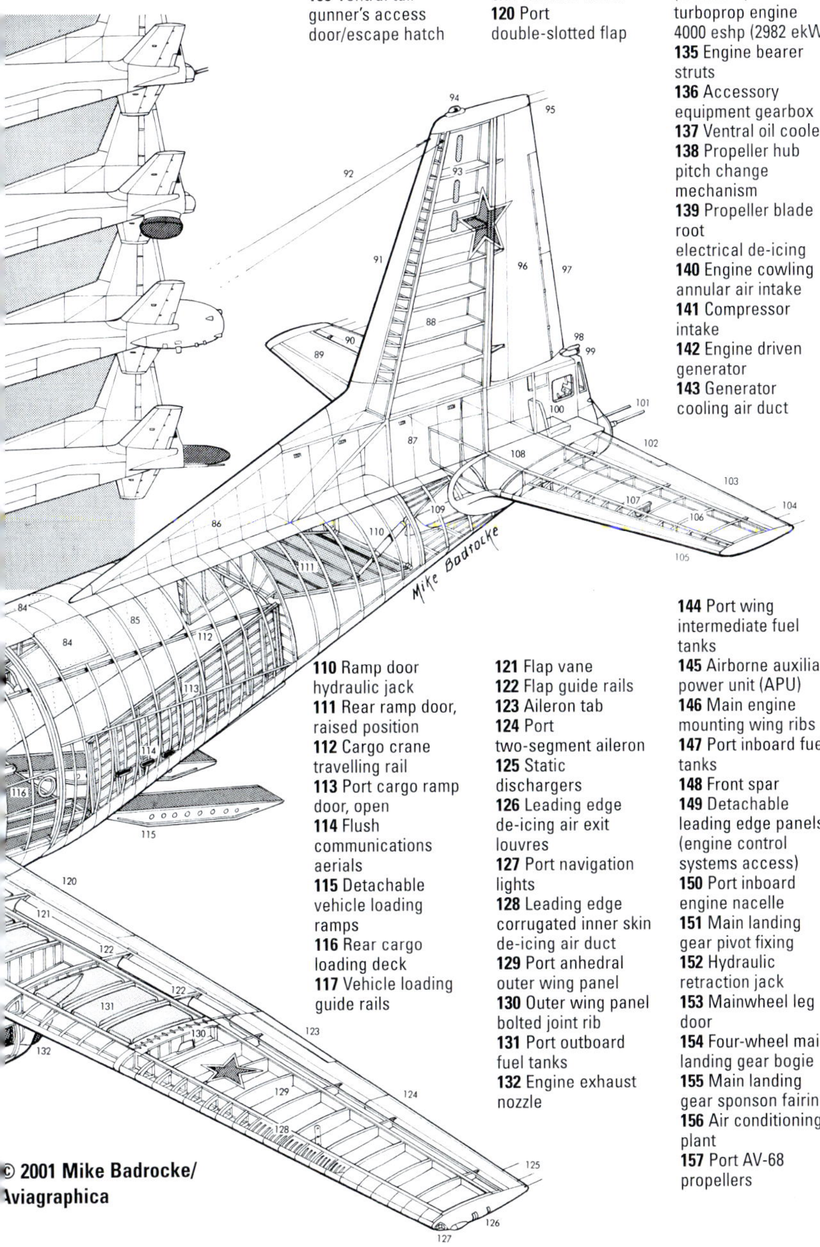

108 Tailplane centre-section carry-through
109 Ventral tail gunner's access door/escape hatch
110 Ramp door hydraulic jack
111 Rear ramp door, raised position
112 Cargo crane travelling rail
113 Port cargo ramp door, open
114 Flush communications aerials
115 Detachable vehicle loading ramps
116 Rear cargo loading deck
117 Vehicle loading guide rails
118 Wing root trailing edge fillet
119 Port emergency exit window hatch
120 Port double-slotted flap
121 Flap vane
122 Flap guide rails
123 Aileron tab
124 Port two-segment aileron
125 Static dischargers
126 Leading edge de-icing air exit louvres
127 Port navigation lights
128 Leading edge corrugated inner skin de-icing air duct
129 Port anhedral outer wing panel
130 Outer wing panel bolted joint rib
131 Port outboard fuel tanks
132 Engine exhaust nozzle
133 Port outer engine nacelle
134 ZMDB Progress (Ivchenko) AI-20K turboprop engine 4000 eshp (2982 ekW)
135 Engine bearer struts
136 Accessory equipment gearbox
137 Ventral oil cooler
138 Propeller hub pitch change mechanism
139 Propeller blade root electrical de-icing
140 Engine cowling annular air intake
141 Compressor intake
142 Engine driven generator
143 Generator cooling air duct
144 Port wing intermediate fuel tanks
145 Airborne auxiliary power unit (APU)
146 Main engine mounting wing ribs
147 Port inboard fuel tanks
148 Front spar
149 Detachable leading edge panels (engine control systems access)
150 Port inboard engine nacelle
151 Main landing gear pivot fixing
152 Hydraulic retraction jack
153 Mainwheel leg door
154 Four-wheel main landing gear bogie
155 Main landing gear sponson fairing
156 Air conditioning plant
157 Port AV-68 propellers
158 Retractable landing lamp, port and starboard
159 Air conditioning system cooling air ram intake

Replacing the An-12A on the production lines at Tashkent and Voronezh in 1963, the An-12B, as photographed here at Sperenberg in 1992, featured a TG-16 APU with a prominent exhaust in the port undercarriage fairing. This provided an autonomous self-start capability at airfields of up to 3,281 ft (1000 m) in elevation. The rudder trim tab was replaced by a pair of separate tabs, each with its own actuator fairing, which reached higher up the trailing edge of the rudder. The An-12B also lost the provision for the rear pair of external bomb racks.

Antonov An-24/26/30/32

Instantly recognisable by the enlarged nacelles for its Ivchenko AI-40DM engines perched high above the fuselage, this An-32 belongs to the Indian air force (where it is named Sutlej). The good 'hot-and-high' performance of the 'Cline' was well appreciated by the Indians, who chose the aircraft over the G222, DHC-5 and Andover, and the IAF operates six squadrons of the type in 2000.

An-24V Series II 'Coke'

Cutaway key
1 Radome
2 Weather radar scanner
3 Scanner tracking mechanism
4 Radome hinges
5 ILS glideslope aerial
6 VOR localiser aerial
7 Radar transmitters and receivers
8 Forward pressure bulkhead
9 Nose undercarriage wheel bay
10 Rudder pedals
11 Instrument panel shroud
12 Radar display
13 Curved windscreen panels
14 Windscreen wipers
15 Cockpit eyebrow windows
16 Overhead systems switch panel
17 Co-pilot/ navigator/radio operator's seat
18 Instrument panel
19 Control column
20 Cockpit floor level
21 Nose undercarriage pivot fixing
22 Twin steerable nosewheels, forward retracting
23 Lower electrical equipment bay port and starboard
24 Underfloor control runs
25 Space provision for radio operator
26 Side console panel
27 Pilot's seat
28 Opening (direct vision) side window panel
29 Space provision for flight engineer
30 Circuit breaker panels
31 Aerial lead-in
32 Cockpit roof escape hatch, interchangeable with jettisonable astrodome observation hatch
33 Cockpit doorway
34 Control linkages
35 Cockpit rear bulkhead
36 Radio and electronics equipment racks
37 Baggage compartment
38 Baggage loading shelving
39 Starboard side 'up-and-over' baggage door
40 Crew wardrobe
41 Curtained cabin doorway
42 Passenger cabin front bulkhead
43 Fuselage skin doubler in line with propellers
44 Four-abreast passenger seating, 50-seat all tourist-class layout
45 Cabin window panels
46 Passenger cabin floor panelling
47 VHF aerial
48 Seat mounting rails
49 Emergency exit window hatch
50 Floor beam construction
51 Cabin wall trim panelling
52 Curtained window panels
53 Centre fuselage frame and stringer construction
54 D/F loop aerial
55 Air supply ducting
56 Wingroot fillet
57 Leading-edge de-icing air duct
58 Cabin air supply duct
59 Fuel filler cap
60 Inboard bag-type fuel tanks
61 Leading-edge engine control runs
62 Starboard nacelle
63 Starboard main undercarriage, stowed position
64 Fireproof bulkhead
65 Air-conditioning system, hot air supply
66 Ivchenko AI-24A turboprop engine
67 Engine auxiliary equipment
68 Hot-air de-iced intake lip
69 Propeller hub pitch change mechanism
70 Spinner
71 Propeller blade root electric de-icing
72 AV-72 four-bladed, constant-speed propeller
73 Engine cowling panels
74 Exhaust duct, exhausts on outboard side of nacelle
75 Wing panel joint rib
76 Fuel vent
77 Fuel filler cap
78 Outer wing panel integral fuel tank; total system capacity 1,220 Imp gal (5550 litres)
79 Leading edge de-icing air duct
80 Retractable landing/taxiing lamp
81 Outer wing panel joint rib

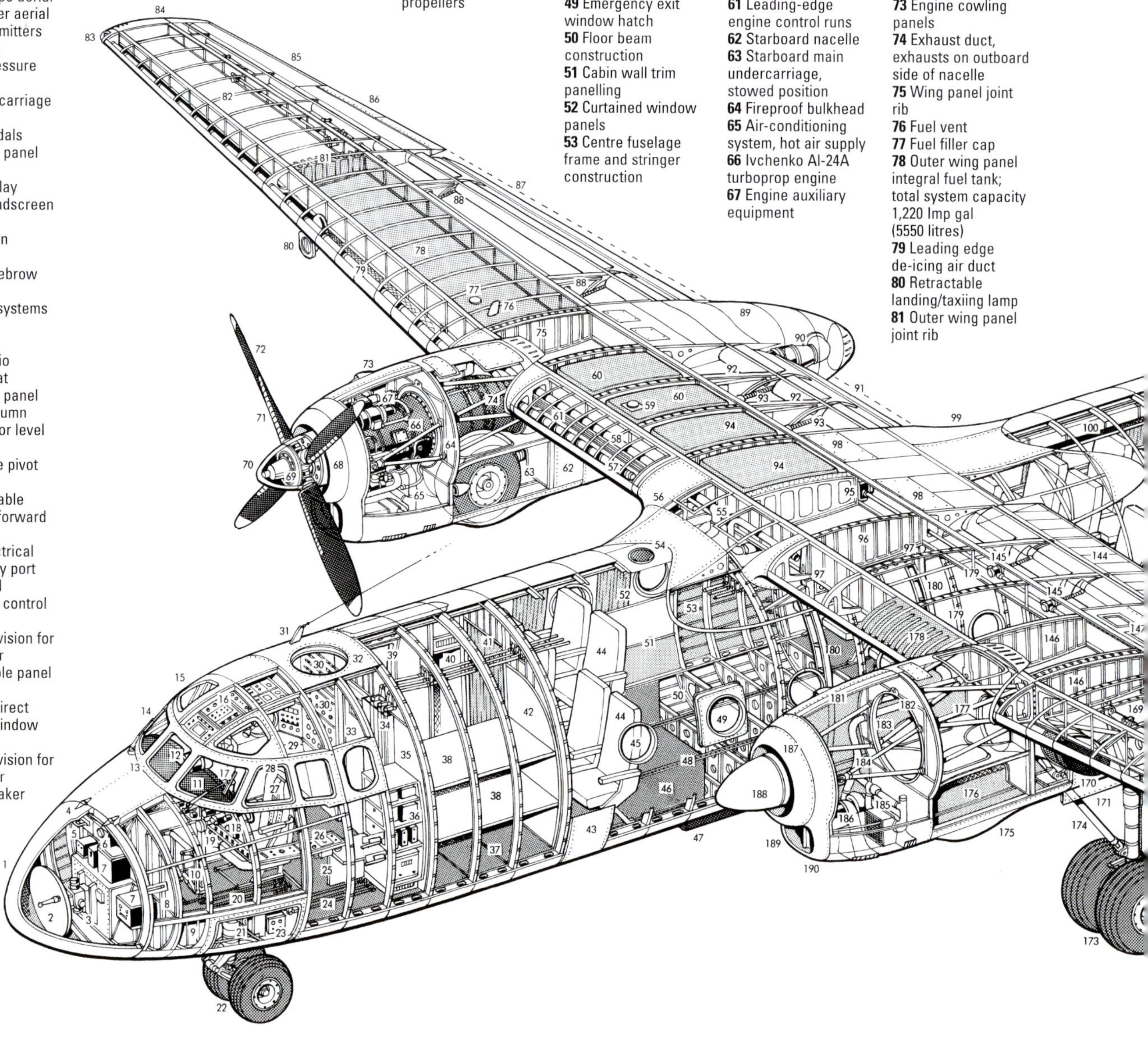

Russia retained about 120 An-26s after the break-up of the Soviet Union; they remain the principal light transport aircraft in service and there are no current plans to replace them. Variants of the 'Curl' also act as command posts, VIP transport and liaison and SAR/casualty evacuation aircraft.

SPECIFICATION

An-26B 'Curl-A'

Dimensions

Length: 78 ft 1 in (23.80 m)
Height: 28 ft 1½ in (8.575 m)
Wingspan: 95 ft 9½ in (29.20 m)
Aspect ratio: 11:7
Wing area: 807.10 sq ft (74.98 m²)
Tailplane span: 32 ft 8¾ in (9.97 m)
Wheel track: 25 ft 11 in (7.90 m)
Wheel base: 25 ft 1¼ in (7.65 m)

Powerplant

Two ZMDB Progress (Ivchenko) AI-24VT turboprops each rated at 2,820 ehp (2103 kW) and one Soyuz (Tumanskii) RU-19A-300 turbojet rated at 1,765 lb st (7.85 kN)

Weights

Empty: 33,950 lb (15400 kg)
Normal take-off: 50,705 lb (23000 kg)
Maximum take-off: 53,790 lb (24400 kg)
Maximum payload: 12,125 lb (5500 kg)

Fuel

Internal fuel: 12,125 lb (5500 kg)

Performance

Maximum level speed at 16,400 ft (5000 m): 292 kt (336 mph; 540 km/h)
Maximum level speed at sea level: 275 kt (317 mph; 510 km/h)
Cruising speed at 19,685 ft (6000 m): 237 kt (273 mph; 440 km/h)
Maximum rate of climb at sea level: 1,575 ft (480 m) per minute
Service ceiling: 24,605 ft (7500 m)
Take-off run at maximum take-off weight: 2,559 ft (780 m)
Landing run at normal landing weight: 2,395 ft (730 m)

Range

With maximum fuel: 1,376 nm (1,585 miles; 2550 km)
With maximum payload: 593 nm (683 miles; 1100 km)

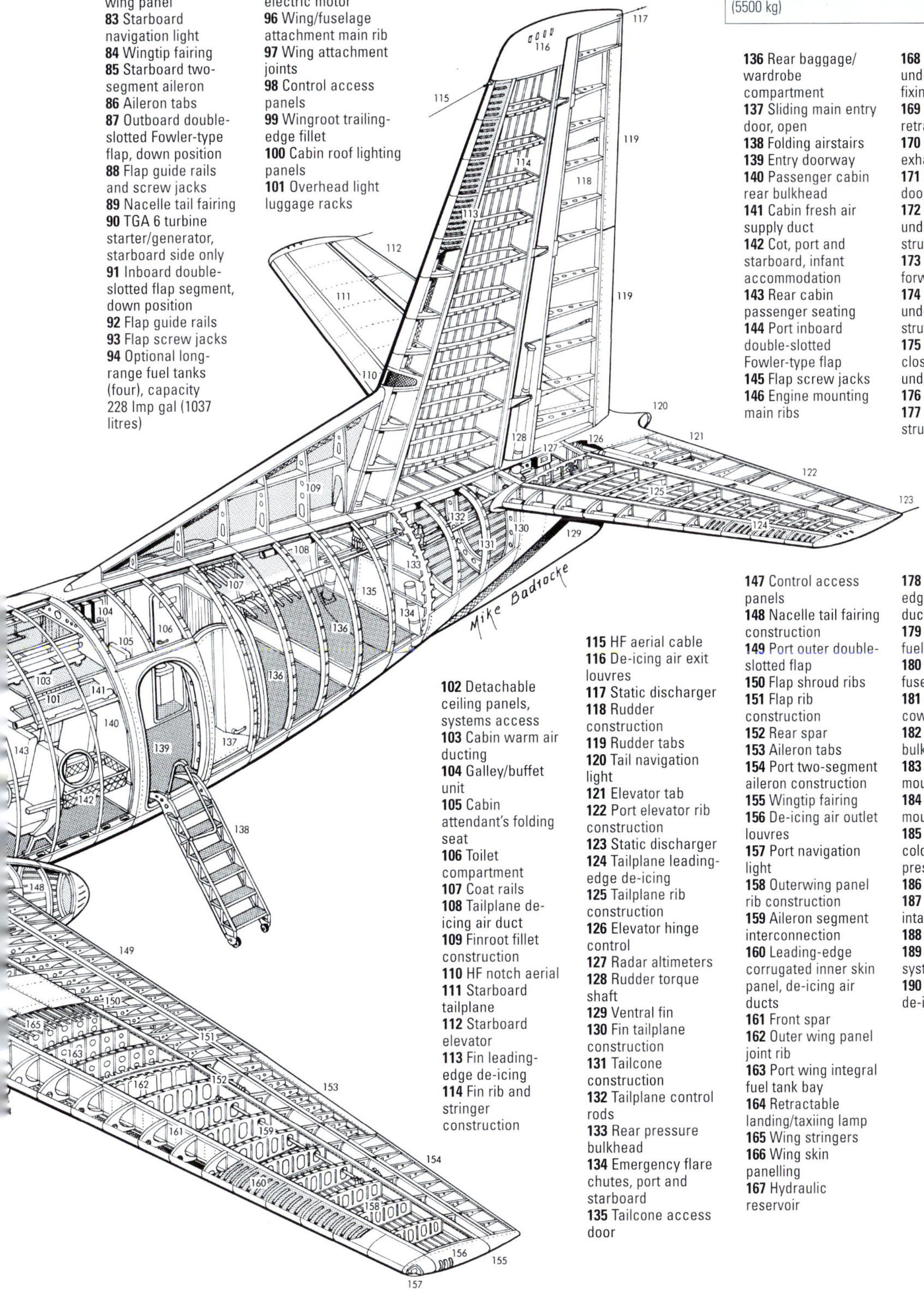

82 Anhedral outer wing panel
83 Starboard navigation light
84 Wingtip fairing
85 Starboard two-segment aileron
86 Aileron tabs
87 Outboard double-slotted Fowler-type flap, down position
88 Flap guide rails and screw jacks
89 Nacelle tail fairing
90 TGA 6 turbine starter/generator, starboard side only
91 Inboard double-slotted flap segment, down position
92 Flap guide rails
93 Flap screw jacks
94 Optional long-range fuel tanks (four), capacity 228 Imp gal (1037 litres)
95 Central flap drive electric motor
96 Wing/fuselage attachment main rib
97 Wing attachment joints
98 Control access panels
99 Wingroot trailing-edge fillet
100 Cabin roof lighting panels
101 Overhead light luggage racks
102 Detachable ceiling panels, systems access
103 Cabin warm air ducting
104 Galley/buffet unit
105 Cabin attendant's folding seat
106 Toilet compartment
107 Coat rails
108 Tailplane de-icing air duct
109 Finroot fillet construction
110 HF notch aerial
111 Starboard tailplane
112 Starboard elevator
113 Fin leading-edge de-icing
114 Fin rib and stringer construction
115 HF aerial cable
116 De-icing air exit louvres
117 Static discharger
118 Rudder construction
119 Rudder tabs
120 Tail navigation light
121 Elevator tab
122 Port elevator rib construction
123 Static discharger
124 Tailplane leading-edge de-icing
125 Tailplane rib construction
126 Elevator hinge control
127 Radar altimeters
128 Rudder torque shaft
129 Ventral fin
130 Fin tailplane construction
131 Tailcone construction
132 Tailplane control rods
133 Rear pressure bulkhead
134 Emergency flare chutes, port and starboard
135 Tailcone access door
136 Rear baggage/wardrobe compartment
137 Sliding main entry door, open
138 Folding airstairs
139 Entry doorway
140 Passenger cabin rear bulkhead
141 Cabin fresh air supply duct
142 Cot, port and starboard, infant accommodation
143 Rear cabin passenger seating
144 Port inboard double-slotted Fowler-type flap
145 Flap screw jacks
146 Engine mounting main ribs
147 Control access panels
148 Nacelle tail fairing construction
149 Port outer double-slotted flap
150 Flap shroud ribs
151 Flap rib construction
152 Rear spar
153 Aileron tabs
154 Port two-segment aileron construction
155 Wingtip fairing
156 De-icing air outlet louvres
157 Port navigation light
158 Outerwing panel rib construction
159 Aileron segment interconnection
160 Leading-edge corrugated inner skin panel, de-icing air ducts
161 Front spar
162 Outer wing panel joint rib
163 Port wing integral fuel tank bay
164 Retractable landing/taxiing lamp
165 Wing stringers
166 Wing skin panelling
167 Hydraulic reservoir
168 Main undercarriage pivot fixing
169 Hydraulic retraction jack
170 Port engine exhaust pipe
171 Mainwheel leg doors
172 Main undercarriage leg strut
173 Twin mainwheels, forward retracting
174 Main undercarriage front strut
175 Mainwheel doors, closed after cycling of undercarriage leg
176 Mainwheel bay
177 Engine bearer struts
178 Inboard leading-edge de-icing air ducting
179 Inner wing panel fuel tank bays
180 Wing attachment fuselage main frames
181 Port engine cowling panels
182 Fireproof bulkhead
183 Main engine mounting ring frame
184 Forward engine mounting struts
185 Cabin air system cold air and pressurising supply
186 Oil cooler
187 Engine annular air intake
188 Propeller spinner
189 Oil cooler and air system
190 Intake lip hot air de-icing

Beech (Raytheon) C-12

The US Navy operates two RC-12Fs (illustrated) and two RC-12Ms as RANSAC (Range Surveillance Aircraft), equipped with surface search radar under the belly. The RC-12Fs fly from Barking Sands in Hawaii, while the RC-12Ms serve at Point Mugu, California.

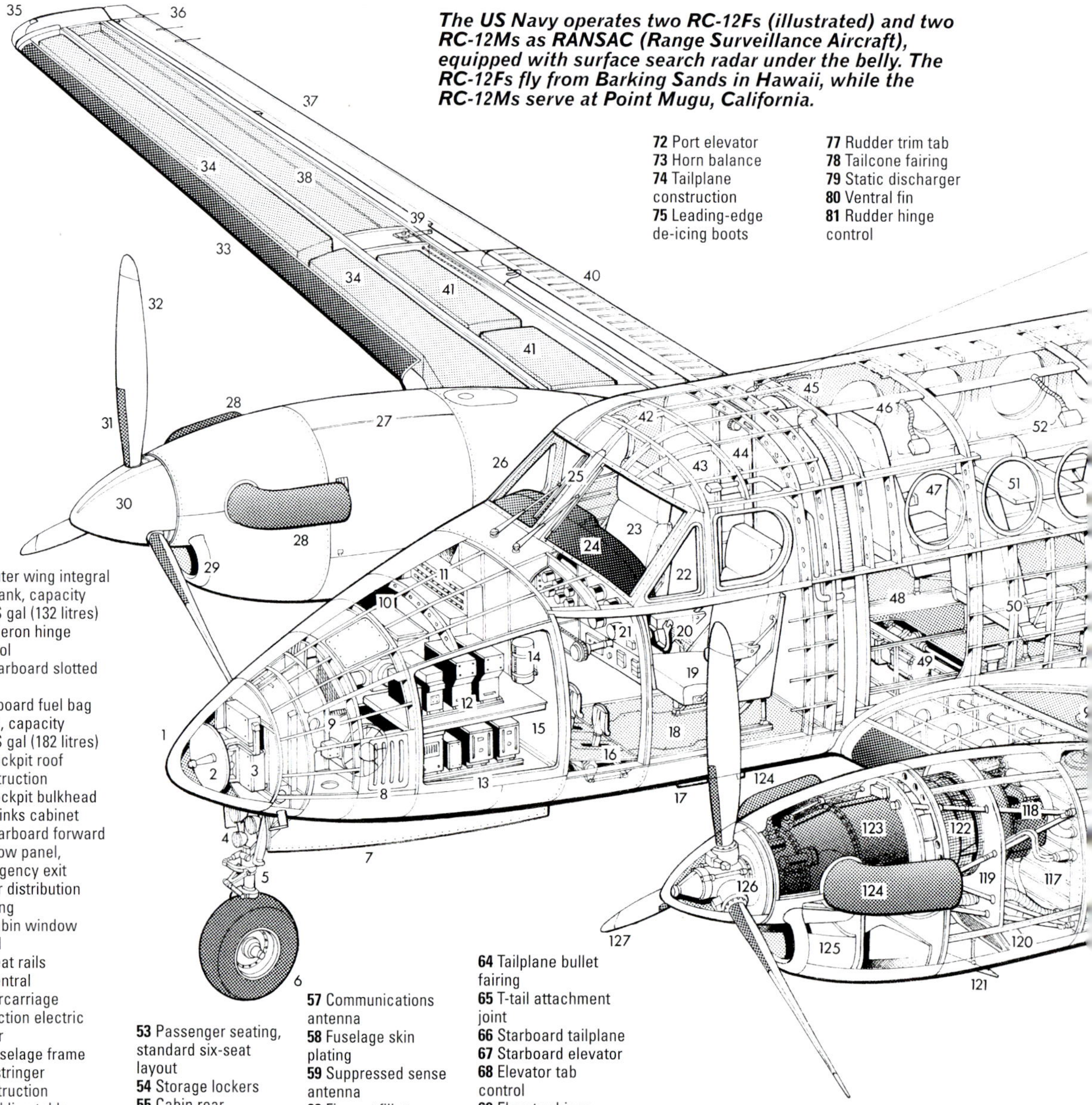

Super King Air 200

Cutaway key

1 Nosecone
2 Weather radar
3 Radar transmitter
4 Landing and taxiing lamps
5 Nose undercarriage leg strut
6 Nosewheel
7 Nosewheel doors
8 Air louvres
9 Air conditioning plant
10 Nose compartment construction
11 Electrical equipment bay
12 Radio and electronics bay
13 Access door
14 Brake hydraulic reservoir
15 Front pressure bulkhead
16 Rudder pedals
17 Ventral aerials
18 Cockpit floor level
19 Pilot's seat
20 Control column handwheel
21 Instrument panel
22 Opening side window panel
23 Co-pilot's seat
24 Instrument panel shroud
25 Windscreen wipers
26 Electrically heated windscreen panels
27 Starboard engine nacelle cowlings
28 Exhaust stubs
29 Engine intake
30 Propeller spinner
31 Blade root de-icing boots
32 Three-bladed variable-pitch reversible propeller
33 Leading-edge de-icing boots
34 Leading-edge fuel bag tanks, capacity 53 US gal (200 litres)
35 Starboard navigation lights
36 Static discharge wicks
37 Starboard aileron
38 Outer wing integral fuel tank, capacity 35 US gal (132 litres)
39 Aileron hinge control
40 Starboard slotted flap
41 Inboard fuel bag tanks, capacity 48 US gal (182 litres)
42 Cockpit roof construction
43 Cockpit bulkhead
44 Drinks cabinet
45 Starboard forward window panel, emergency exit
46 Air distribution ducting
47 Cabin window panel
48 Seat rails
49 Central undercarriage retraction electric motor
50 Fuselage frame and stringer construction
51 Folding table
52 Cabin trim panels
53 Passenger seating, standard six-seat layout
54 Storage lockers
55 Cabin rear bulkhead
56 Toilet compartment
57 Communications antenna
58 Fuselage skin plating
59 Suppressed sense antenna
60 Fin root fillet
61 VOR aerial
62 Fin construction
63 Antenna cables
64 Tailplane bullet fairing
65 T-tail attachment joint
66 Starboard tailplane
67 Starboard elevator
68 Elevator tab control
69 Elevator hinge control gear
70 Tail navigation light
71 Elevator tab
72 Port elevator
73 Horn balance
74 Tailplane construction
75 Leading-edge de-icing boots
76 Rudder construction
77 Rudder trim tab
78 Tailcone fairing
79 Static discharger
80 Ventral fin
81 Rudder hinge control
82 Sloping fin mounting frames

Civilian-model King Air 200s serve in small numbers with a wide variety of nations, mostly employed as light transport/utility types, although a few undertake maritime patrol or training duties. No. 42 Squadron, RNZAF, at Whenuapai flies three leased aircraft on multi-engine training, general transport and VIP duties formerly undertaken by Andovers and Cessna 421s.

SPECIFICATION

King Air B200

Dimensions

Length: 43 ft 10 in (13.36 m)
Height: 14 ft 10 in (4.52 m)
Wingspan: 54 ft 6 in (16.61 m)
Tailplane span: 18 ft 5 in (5.61 m)
Wing area: 303 sq ft (28.15 m²)
Wheel track: 17 ft 2 in (5.23 m)
Wheelbase: 14 ft 11½ in (4.56 m)
Passenger cabin volume: 31,285 cu ft (885.90 m³)

Powerplant

Two Pratt & Whitney Canada PT6A-42 turboprops, each rated at 850 shp (634 kW)

Weights

Empty: 8,192 lb (3716 kg)
Maximum take-off: 12,500 lb (5670 kg)
Maximum ramp weight: 12,590 lb (5710 kg)

Fuel and load

Fuel capacity: 544 US gal (453 Imp gal; 2059 litres)
Maximum fuel load: 3,645 lb (1653 kg)
Baggage load: 550 lb (249 kg)

Performance

Maximum level speed: 336 mph (541 km/h)
Service ceiling: 35,000 ft (10670 m)
Take-off run: 1,860 ft (567 m)
Maximum range: 2,139 miles (3442 km)

Accommodation

Flight deck crew of two, plus up to seven passengers in the cabin

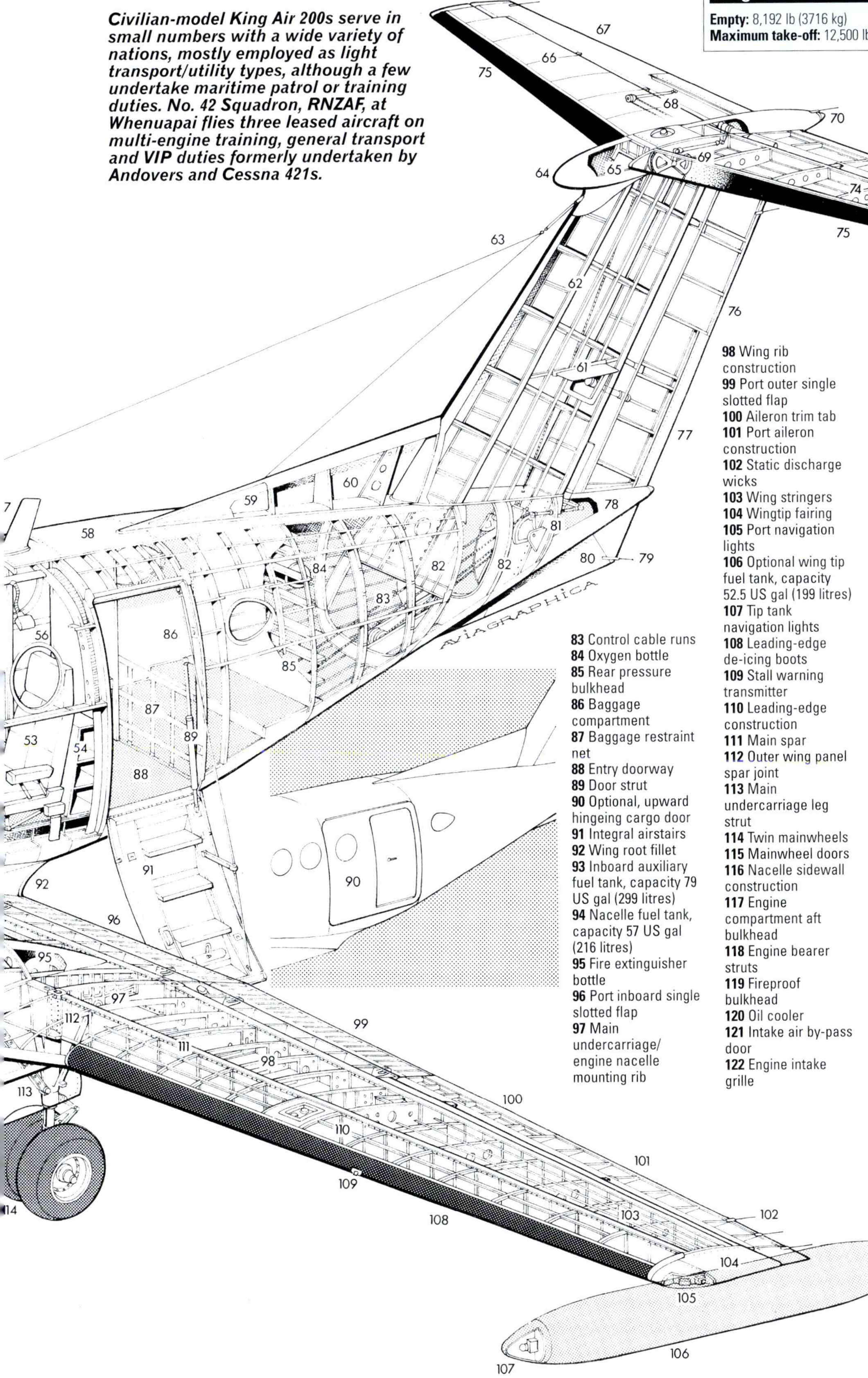

83 Control cable runs
84 Oxygen bottle
85 Rear pressure bulkhead
86 Baggage compartment
87 Baggage restraint net
88 Entry doorway
89 Door strut
90 Optional, upward hingeing cargo door
91 Integral airstairs
92 Wing root fillet
93 Inboard auxiliary fuel tank, capacity 79 US gal (299 litres)
94 Nacelle fuel tank, capacity 57 US gal (216 litres)
95 Fire extinguisher bottle
96 Port inboard single slotted flap
97 Main undercarriage/ engine nacelle mounting rib
98 Wing rib construction
99 Port outer single slotted flap
100 Aileron trim tab
101 Port aileron construction
102 Static discharge wicks
103 Wing stringers
104 Wingtip fairing
105 Port navigation lights
106 Optional wing tip fuel tank, capacity 52.5 US gal (199 litres)
107 Tip tank navigation lights
108 Leading-edge de-icing boots
109 Stall warning transmitter
110 Leading-edge construction
111 Main spar
112 Outer wing panel spar joint
113 Main undercarriage leg strut
114 Twin mainwheels
115 Mainwheel doors
116 Nacelle sidewall construction
117 Engine compartment aft bulkhead
118 Engine bearer struts
119 Fireproof bulkhead
120 Oil cooler
121 Intake air by-pass door
122 Engine intake grille
123 Pratt & Whitney Canada PT6A-41 turboprop engine
124 Engine exhaust stubs
125 Intake ducting
126 Propeller hub pitch change mechanism
127 Hartzell three-bladed propeller

USAF purchases of the King Air 200 reached 82 aircraft. Forty were C-12Fs, based on the B200C model, and used for general transport duties, although several were reassigned to the US Army in 1995 as C-12F-3s. This example carries the 'ZZ' tailcode of the 18th Wing at Kadena AFB, Okinawa

Boeing KC-135 Stratotanker

Playing a major part in the USAF refuelling effort are the 'Weekend Warriors' of the Air Force Reserve and the Air National Guard. AFRes has three squadrons of KC-135Rs and one of Es, while the ANG has 10 KC-135E (with an example from the 190th ARW at Forbes Field, Kansas illustrated) and 13 R units.

KC-135R Stratotanker

Cutaway key

1 Radome
2 Weather radar scanner
3 ILS glideslope antenna
4 Front pressure bulkhead
5 Underfloor equipment bay
6 Ventral access hatch
7 Rudder pedals
8 Instrument panel
9 Windscreen wipers
10 Instrument panel shroud
11 Windscreen panels
12 Overhead systems switch panel
13 Ditching handholds
14 Cockpit eyebrow windows
15 Co-pilot's seat
16 Pilot's seat
17 Pitot head
18 Nosewheel bay
19 Escape spoiler
20 Entry hatch
21 Twin nosewheels, forward-retracting
22 Boarding ladder
23 Entry/escape hatch
24 Instructor's seat
25 Navigator's station
26 Flight-refuelling receptacle
27 Star tracking windows, celestial navigation system
28 TACAN antenna
29 Cockpit doorway
30 Avionics equipment rack
31 Navigator's stool
32 Supernumerary crew seat
33 Electrical equipment rack
34 Flight deck air supply duct
35 Battery stowage
36 Wash basin
37 Crew toilet
38 Director lighting strip for receiving aircraft, port and starboard
39 Forward underfloor fuel cells (4), capacity 4,830 Imp gal/5,800 US gal (21955 litres)
40 Cargo door latches
41 Door aperture (9 ft 6 in x 6 ft 6 in/2.9 m x 1.98 m)
42 Cargo deck floor structure
43 Tie-down fittings
44 Cargo door hydraulic jacks and hinges
45 Upward-opening cargo door
46 VHF/UHF antenna
47 Door-mounted ADF antennas
48 Conditioned air risers to overhead distribution duct
49 Wing inspection light
50 Front spar attachment fuselage main frame
51 Centre-section fuel tanks (6), capacity 6,084 Imp gal/7,306 US gal (27656 litres)
52 Overwing escape hatches, port and starboard
53 Wing centre-section carry-through structure
54 Floor beam structure
55 Fuselage frame and stringer structure
56 Cabin overhead air distribution duct
57 Inboard integral wing tank, capacity 1,894 Imp gal/2,275 US gal (8612 litres)
58 Tank filler
59 No. 3, starboard inner engine nacelle
60 Nacelle pylon
61 Wing centre main integral tank, capacity 1,717 Imp gal/2,062 US gal (7805 litres)
62 Fuel venting channels
63 Leading-edge flap hydraulic jacks
64 Krüger-type leading-edge flap
65 No. 4, starboard outer engine nacelle
66 Outboard nacelle pylon
67 Outboard reserve integral fuel tank, capacity 361 Imp gal/434 US gal (1643 litres)
68 Optional drogue-type refuelling pod, carried by French C-135FR
69 Starboard navigation light
70 Outboard, low-speed aileron
71 Aileron internal balance panels
72 Spoiler interconnection linkage
73 Aileron hinge control linkage
74 Aileron tab
75 Outboard double-slotted Fowler-type flap segment, extended
79 Flap screw jacks
80 Aileron geared tab
81 Inboard, high-speed, aileron
82 Gust damper
83 Aileron actuating linkage
84 Inboard spoiler panels, open
85 Inboard double-slotted Fowler-type flap segment, extended
86 Anti-collision beacon
87 Pressure floor above wheel bay
88 Rear spar attachment fuselage main frame

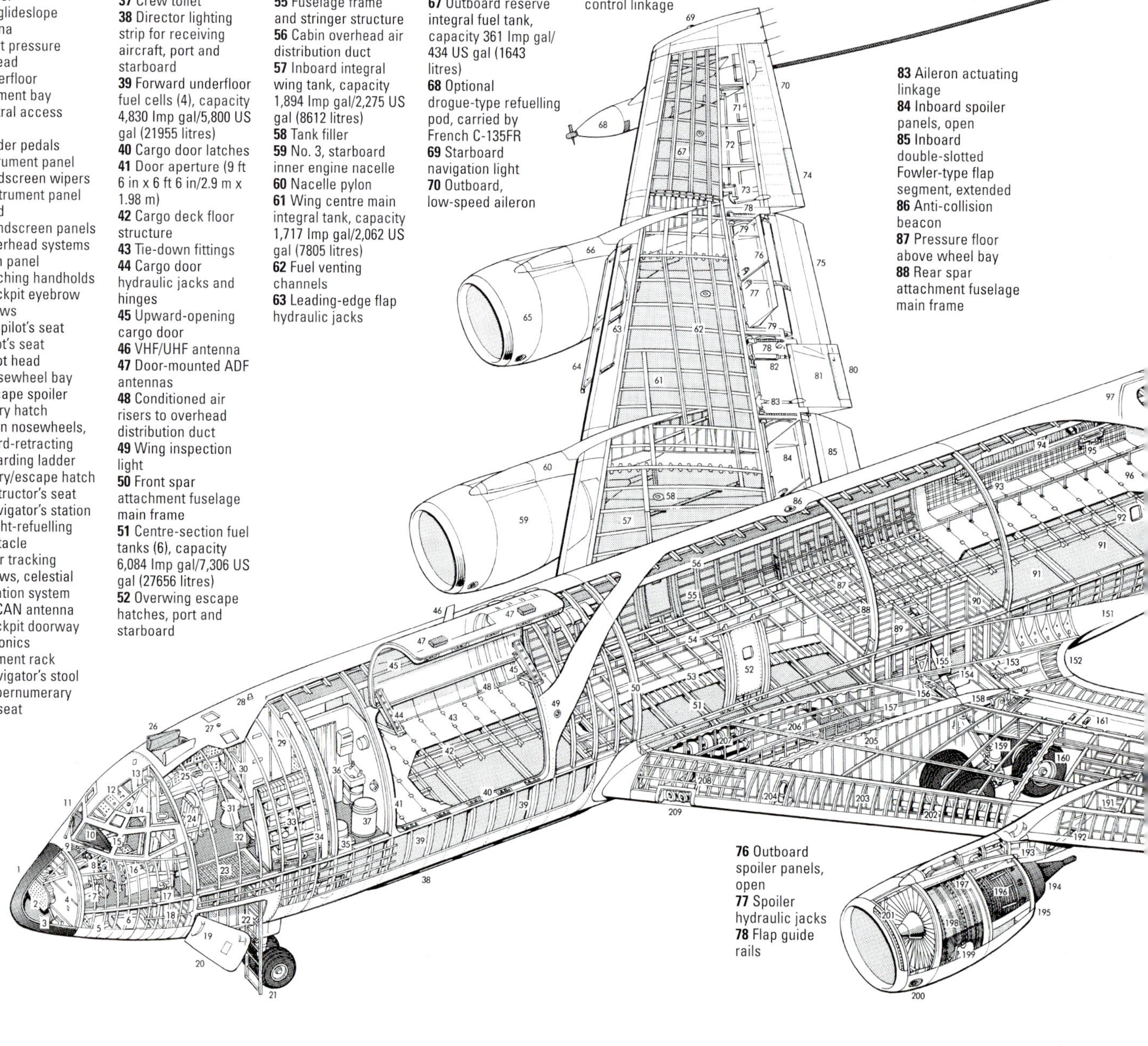

76 Outboard spoiler panels, open
77 Spoiler hydraulic jacks
78 Flap guide rails

During Desert Shield, KC-135As refuelled aircraft which were deploying from the USA and Europe to the Persian Gulf, before switching to the delivery of personnel and equipment – a task in which they assisted MAC airlifters. Approximately 200 KC-135s were directly assigned to the theatre, being operated by provisional air refuelling wings, while several hundred other tankers regularly operated between the USA (as seen above) and the Gulf region. Some 15,000 refuelling sorties were flown during the conflict itself, with almost 46,000 aircraft from the USAF, USN, USMC and the Coalition receiving fuel. Though the KC-135s were officially restricted to designated refuelling areas above northern Saudi Arabia, the aircraft sometimes ventured across the Iraqi border to provide fuel for aircraft with critical fuel states.

SPECIFICATION

KC-135A Stratotanker

Dimensions

Length: 136 ft 3 in (41.53 m)
Height: 41 ft 8 in (12.70 m)
Wing span: 130 ft 10 in (39.88 m)
Aspect ratio: 7.04
Wing area: 2,433.00 sq ft (226.03 m²)
Tailplane span: 40 ft 3 in (12.27 m)
Wheel base: 46 ft 7 in (14.20 m)

Powerplant

Four Pratt & Whitney J57-P-59W turbojets, each rated at 13,750 lb st (61.16 kN)

Weights

Operating empty: 106,306 lb (48220 kg)
Maximum take-off: 316,000 lb (143335 kg)

Fuel and load

Internal fuel: 189,702 lb (86047 kg)
Maximum payload: 83,000 lb (37650 kg)

Performance

Maximum level speed at high altitude: 530 kt (610 mph; 982 km/h)
Cruising speed at 35,000 ft (10670 m): 462 kt (532 mph; 856 km/h)
Operational radius to offload 120,000 lb (54432 kg) of fuel: 1000 nm (1,151 miles; 1854 km)
Service ceiling: 45,000 ft (13715 m)
Typical take-off run: 10,700 ft (3261 m) increasing to 14,000 ft (4267 m) under 'hot-and-high' conditions at maximum take-off weight
Maximum rate of climb at sea level: 1,290 ft (393 m) per minute

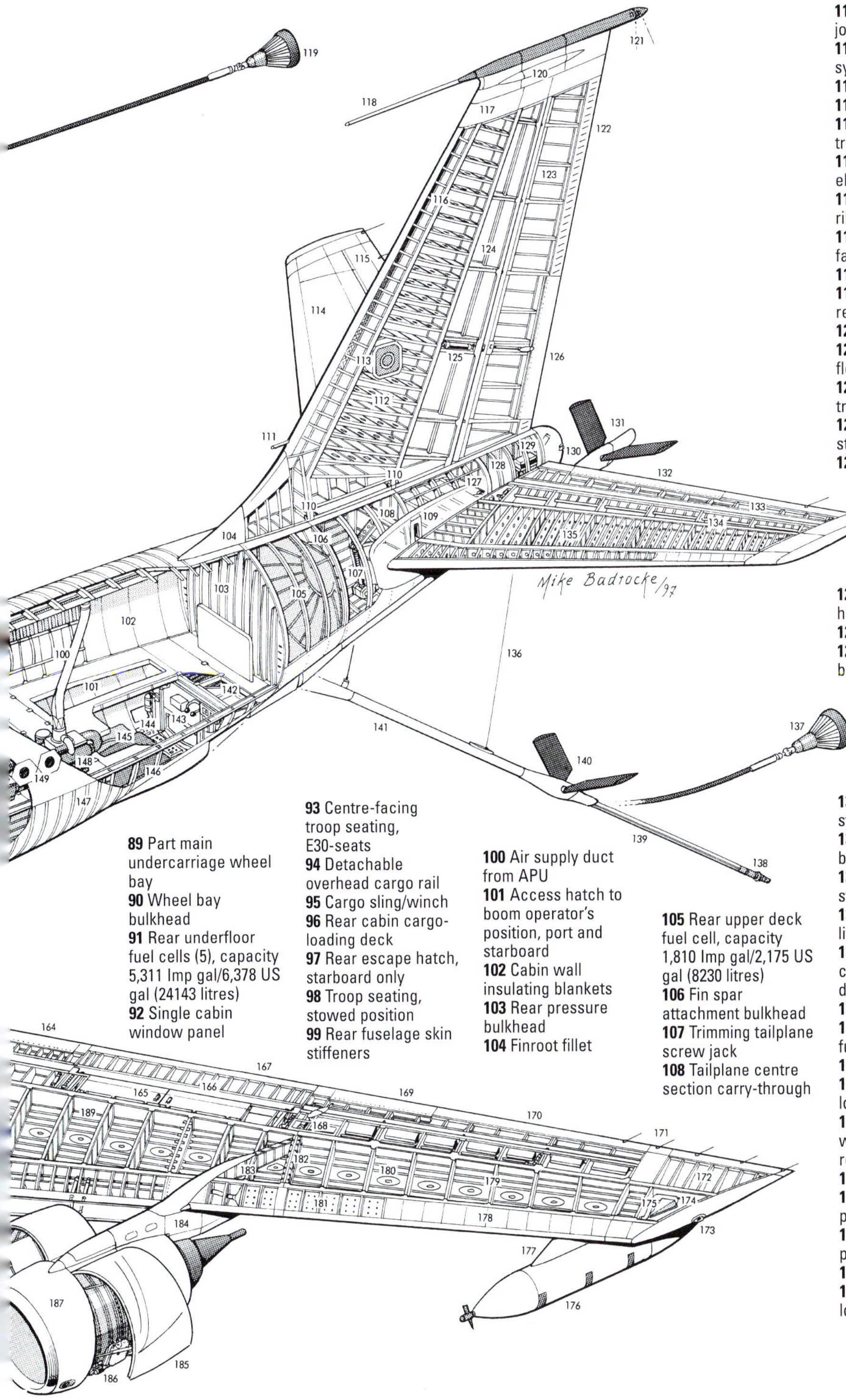

89 Part main undercarriage wheel bay
90 Wheel bay bulkhead
91 Rear underfloor fuel cells (5), capacity 5,311 Imp gal/6,378 US gal (24143 litres)
92 Single cabin window panel
93 Centre-facing troop seating, E30-seats
94 Detachable overhead cargo rail
95 Cargo sling/winch
96 Rear cabin cargo-loading deck
97 Rear escape hatch, starboard only
98 Troop seating, stowed position
99 Rear fuselage skin stiffeners
100 Air supply duct from APU
101 Access hatch to boom operator's position, port and starboard
102 Cabin wall insulating blankets
103 Rear pressure bulkhead
104 Finroot fillet
105 Rear upper deck fuel cell, capacity 1,810 Imp gal/2,175 US gal (8230 litres)
106 Fin spar attachment bulkhead
107 Trimming tailplane screw jack
108 Tailplane centre section carry-through
109 Trimming tailplane seal
110 Fin attachment joints
111 Artificial feel system pressure head
112 Fin rib structure
113 VOR antenna
114 Starboard trimming tailplane
115 Starboard elevator
116 Fin leading-edge ribs
117 Fintip antenna fairing
118 HF antenna
119 Starboard refuelling drogue
120 HF tuner
121 Refuelling floodlight
122 Rudder fixed trailing-edge segment
123 Rudder rib structure
124 Internal balance panels
125 Rudder hydraulic actuator
126 Rudder tab
127 Trimming tailplane hinge mounting
128 Tailcone structure
129 Crash locator beacon
130 Tail navigation and strobe lights
131 Refuelling boom, stowed position
132 Elevator tab
133 Port elevator structure
134 Elevator internal balance panels
135 Port tailplane rib structure
136 Refuelling boom lifting cable
137 Alternative central refuelling drogue
138 Refuelling adaptor
139 Refuelling boom, fully extended
140 Ruddervators
141 Refuelling boom, lowered position
142 Boom operator's window cover, retracted
143 Viewing window
144 Refuelling control panel
145 Boom operator's pallet
146 Instructor's pallet
147 Fuselage lower lobe skin stiffeners
148 Optional Auxiliary Power Unit (APU)
149 APU exhaust ducts
150 Fuselage lower lobe frame and stringer structure
151 Wingroot trailing-edge fillet
152 Fillet flap
153 Flap operating screw jack
154 Mainwheel doors
155 Mainwheel leg breaker strut
156 Hydraulic retraction jack
157 Wingroot integral fuel tank bay, capacity 1,895 Imp gal/2,275 US gal (8615 litres)
158 Main undercarriage leg pivot mounting
159 Shock absorber leg strut
160 Four-wheel main undercarriage bogie
161 Port inboard spoiler panels
162 Inboard double-slotted flap segment
163 Inboard, high-speed, aileron
164 Aileron tab
165 Outboard spoiler panels
166 Flap rib structure
167 Outboard double-slotted flap segment
168 Port aileron hinge control
169 Aileron tab
170 Port outboard, low-speed aileron
171 Static dischargers
172 Fixed trailing-edge segment rib structure
173 Port navigation light
174 Fuel system vent tank
175 Ventral NACA-type venting intake
176 Port optional drogue-type refuelling pod
177 Refuelling pod pylon
178 Leading-edge skin panelling
179 Outer-wing panel rib structure
180 Wing lower skin/stringer panel with access manholes
181 Leading-edge de-icing air double skin ducting
182 Outer wing panel joint rib
183 Engine pylon mounting rib
184 Port outboard nacelle pylon
185 Hinged cowling panels, engine access
186 Engine accessory equipment gearbox
187 No. 1, port outer engine nacelle
188 Port leading-edge Krüger-type flaps
189 Port wing integral fuel tankage
190 Port wing panel rib structure
191 Inboard nacelle mounting rib
192 Nacelle drag strut
193 Nacelle pylon structure
194 Core engine, hot-stream exhaust
195 Fan air, cold-stream exhaust
196 Engine turbine section
197 CFM international F108-CF-100 (CFM56-2A2) turbofan engine
198 Engine fan casing
199 Long-range oil tank
200 De-icing air exhaust
201 Intake lip bleed air de-icing
202 Engine bleed air ducting
203 Leading-edge rib structure
204 Pressure refuelling connection, port and starboard
205 Main undercarriage mounting rib
206 Air-conditioning system heat exchanger
207 Ventral air-conditioning pack
208 Heat exchanger ram air intake
209 Landing/taxiing lights

Lockheed C-141 Starlifter

Like the USAF's other long-range transport assets, the StarLifter has been expected to perform its appointed roles in all conditions. This Compass Grey C-141B was photographed from a tanker over inhospitable territory during a training mission.

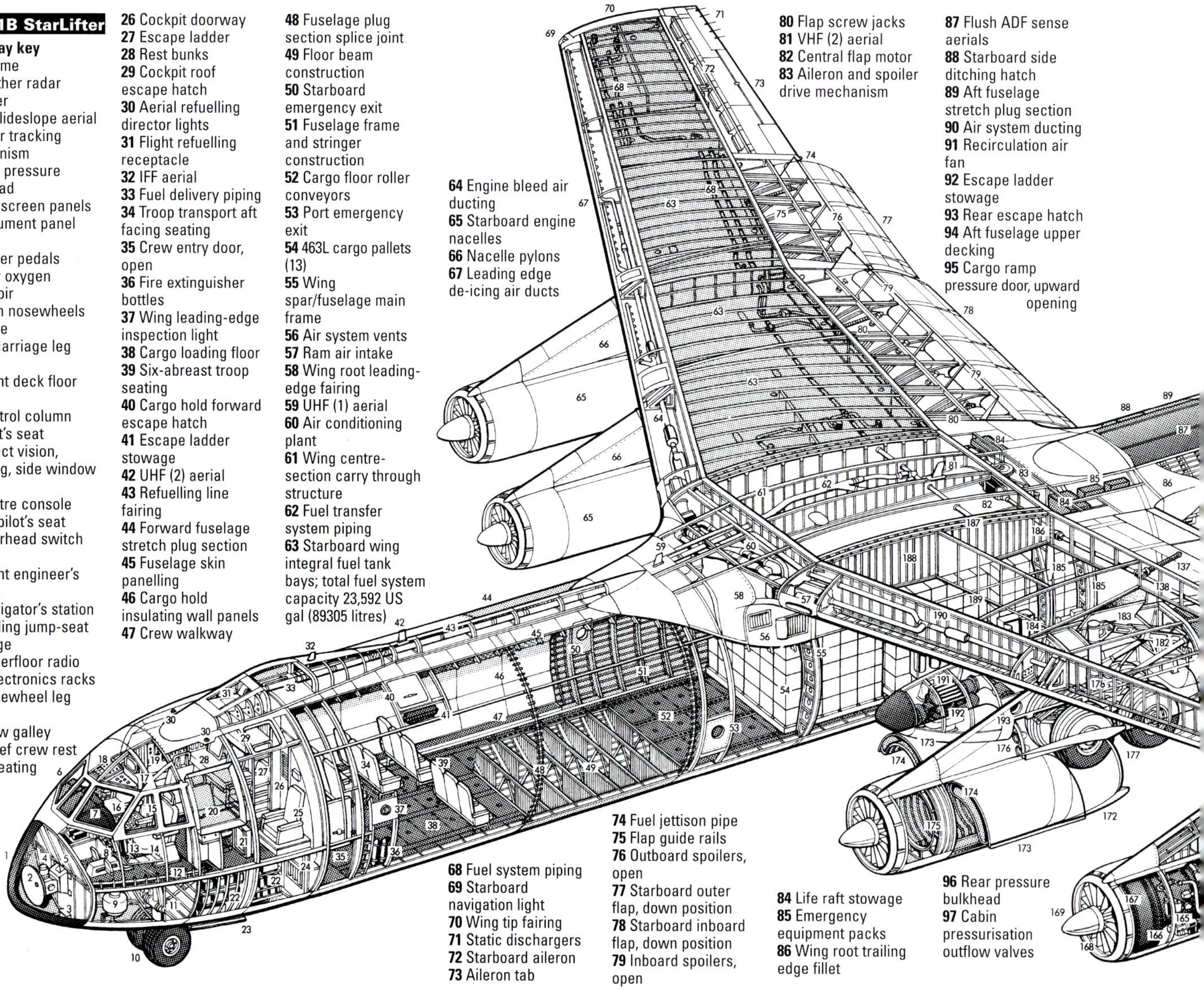

C-141B StarLifter

Cutaway key

1 Radome
2 Weather radar scanner
3 ILS glideslope aerial
4 Radar tracking mechanism
5 Front pressure bulkhead
6 Windscreen panels
7 Instrument panel shroud
8 Rudder pedals
9 Crew oxygen reservoir
10 Twin nosewheels
11 Nose undercarriage leg strut
12 Flight deck floor level
13 Control column
14 Pilot's seat
15 Direct vision, opening, side window panel
16 Centre console
17 Co-pilot's seat
18 Overhead switch panel
19 Flight engineer's station
20 Navigator's station
21 Folding jump-seat stowage
22 Underfloor radio and electronics racks
23 Nosewheel leg door
24 Crew galley
25 Relief crew rest area seating
26 Cockpit doorway
27 Escape ladder
28 Rest bunks
29 Cockpit roof escape hatch
30 Aerial refuelling director lights
31 Flight refuelling receptacle
32 IFF aerial
33 Fuel delivery piping
34 Troop transport aft facing seating
35 Crew entry door, open
36 Fire extinguisher bottles
37 Wing leading-edge inspection light
38 Cargo loading floor
39 Six-abreast troop seating
40 Cargo hold forward escape hatch
41 Escape ladder stowage
42 UHF (2) aerial
43 Refuelling line fairing
44 Forward fuselage stretch plug section
45 Fuselage skin panelling
46 Cargo hold insulating wall panels
47 Crew walkway
48 Fuselage plug section splice joint
49 Floor beam construction
50 Starboard emergency exit
51 Fuselage frame and stringer construction
52 Cargo floor roller conveyors
53 Port emergency exit
54 463L cargo pallets (13)
55 Wing spar/fuselage main frame
56 Air system vents
57 Ram air intake
58 Wing root leading-edge fairing
59 UHF (1) aerial
60 Air conditioning plant
61 Wing centre-section carry through structure
62 Fuel transfer system piping
63 Starboard wing integral fuel tank bays; total fuel system capacity 23,592 US gal (89305 litres)
64 Engine bleed air ducting
65 Starboard engine nacelles
66 Nacelle pylons
67 Leading edge de-icing air ducts
68 Fuel system piping
69 Starboard navigation light
70 Wing tip fairing
71 Static dischargers
72 Starboard aileron
73 Aileron tab
74 Fuel jettison pipe
75 Flap guide rails
76 Outboard spoilers, open
77 Starboard outer flap, down position
78 Starboard inboard flap, down position
79 Inboard spoilers, open
80 Flap screw jacks
81 VHF (2) aerial
82 Central flap motor
83 Aileron and spoiler drive mechanism
84 Life raft stowage
85 Emergency equipment packs
86 Wing root trailing edge fillet
87 Flush ADF sense aerials
88 Starboard side ditching hatch
89 Aft fuselage stretch plug section
90 Air system ducting
91 Recirculation air fan
92 Escape ladder stowage
93 Rear escape hatch
94 Aft fuselage upper decking
95 Cargo ramp pressure door, upward opening
96 Rear pressure bulkhead
97 Cabin pressurisation outflow valves

Above: The third StarLifter off the production line was assigned a permanent test role throughout its USAF career. Known as an NC-141A, it was latterly employed as an advanced radar testbed, allowing nose-mounted radar sets to be tested in an 'ECM environment'. The aircraft was retired to AMARC in 1997.

SPECIFICATION

C-141A StarLifter

Dimensions

Length: 145 ft (44.2 m)
Height: 39 ft 4 in (11.99 m)
Wingspan: 160 ft (48.77 m)
Wing area: 3,228.1 sq ft (299.901 m²)

Powerplant

Four Pratt & Whitney TF33-P-7A turbofans rated at 21,000 lb st (93.4 kN)

Weights

Empty: 136,900 lb (62097 kg)
Maximum take-off: 323,100 lb (146556 kg)

Performance

Maximum speed at 24,000 ft (7440 m): 565 mph (909 km/h)
Cruising speed: 478 mph (769 km/h)
Initial rate of climb: 7,925 ft (2416 m) per minute
Service ceiling: 51,700 ft (15760 m)
Range with maximum payload: 4,155 miles (6685 km)
Ferry range: 6,575 miles (10580 km)

Load

Five crew and either 138 troops in aft-facing seats, 124 paratroops on side-facing bucket seats, 80 litters and 23 attendents, or a maximum payload of 62,717 lb (28448 kg) of military cargo

C-141B StarLifter

as for C-141A, except:

Dimensions

Length: 168 ft 3½ in (51.29 m)

Weights

Empty: 153,350 lb (69558 kg)

Performance

Initial rate of climb: 2,990 ft/min (911 m/min)
Range with maximum payload: 3,200 miles (5150 km)
Maximum unrefuelled range without payload: 6,385 miles (10275-km)

Load

Max. payload of 89,152 lb (40439 kg)

The SOLL II Special Operations C-141Bs are likely to be among the last StarLifters to be replaced by C-17As when the type is finally retired in 2003.

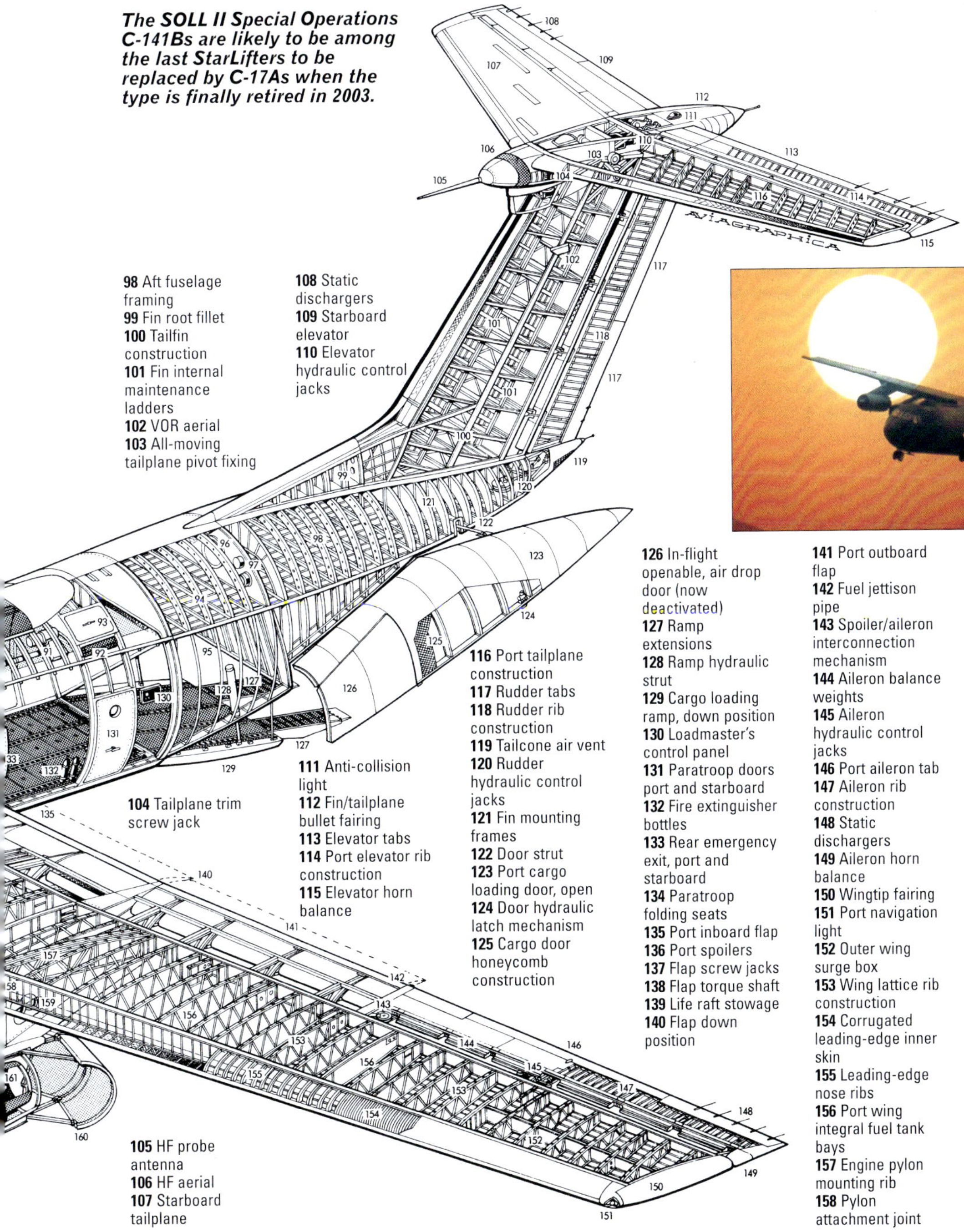

Now in the twilight of its career, the StarLifter is likely to have celebrated 40 years service by the time it is retired.

98 Aft fuselage framing
99 Fin root fillet
100 Tailfin construction
101 Fin internal maintenance ladders
102 VOR aerial
103 All-moving tailplane pivot fixing
104 Tailplane trim screw jack
105 HF probe antenna
106 HF aerial
107 Starboard tailplane
108 Static dischargers
109 Starboard elevator
110 Elevator hydraulic control jacks
111 Anti-collision light
112 Fin/tailplane bullet fairing
113 Elevator tabs
114 Port elevator rib construction
115 Elevator horn balance
116 Port tailplane construction
117 Rudder tabs
118 Rudder rib construction
119 Tailcone air vent
120 Rudder hydraulic control jacks
121 Fin mounting frames
122 Door strut
123 Port cargo loading door, open
124 Door hydraulic latch mechanism
125 Cargo door honeycomb construction
126 In-flight openable, air drop door (now deactivated)
127 Ramp extensions
128 Ramp hydraulic strut
129 Cargo loading ramp, down position
130 Loadmaster's control panel
131 Paratroop doors port and starboard
132 Fire extinguisher bottles
133 Rear emergency exit, port and starboard
134 Paratroop folding seats
135 Port inboard flap
136 Port spoilers
137 Flap screw jacks
138 Flap torque shaft
139 Life raft stowage
140 Flap down position
141 Port outboard flap
142 Fuel jettison pipe
143 Spoiler/aileron interconnection mechanism
144 Aileron balance weights
145 Aileron hydraulic control jacks
146 Port aileron tab
147 Aileron rib construction
148 Static dischargers
149 Aileron horn balance
150 Wingtip fairing
151 Port navigation light
152 Outer wing surge box
153 Wing lattice rib construction
154 Corrugated leading-edge inner skin
155 Leading-edge nose ribs
156 Port wing integral fuel tank bays
157 Engine pylon mounting rib
158 Pylon attachment joint
159 Engine fire extinguisher bottles
160 Thrust reverser bucket doors, open
161 Hot stream exhaust nozzle
162 Fan air exhaust duct
163 Pratt & Whitney TF33-P-7 turbofan engine
164 Nacelle firewall
165 Engine accessory equipment bay
166 Front fan casing
167 Suction relief doors
168 Inlet guide vanes
169 Intake centre-body fairing
170 Engine pylon construction
171 Cable and pipe ducting
172 Inboard engine nacelle
173 Side cowling panels, open
174 Cowling integral by-pass ducting
175 Bifurcated fan air duct
176 Landing/taxiing lamps
177 Four-wheel main undercarriage bogie
178 Main undercarriage leg pivot fitting
179 Mainwheel door
180 Position of refuelling adaptor on starboard side
181 Undercarriage side-body fairing construction
182 Main undercarriage retraction strut
183 Upper leg door
184 Hydraulic equipment service centre
185 Wing/fuselage main frames
186 Spar attachment joint
187 Wing panel/centre section bolted joint
188 Wing root rib
189 Inboard fuel tank bays
190 Front spar
191 APU intake grille
192 Auxiliary power unit (APU)
193 APU exhaust

Lockheed C-130 Hercules

With tensions with China at a high state, Taiwan built up its armed forces during the 1990s. The air force's tactical transport requirements are met by 24 C-130H Hercules aircraft, based at Pingtung North.

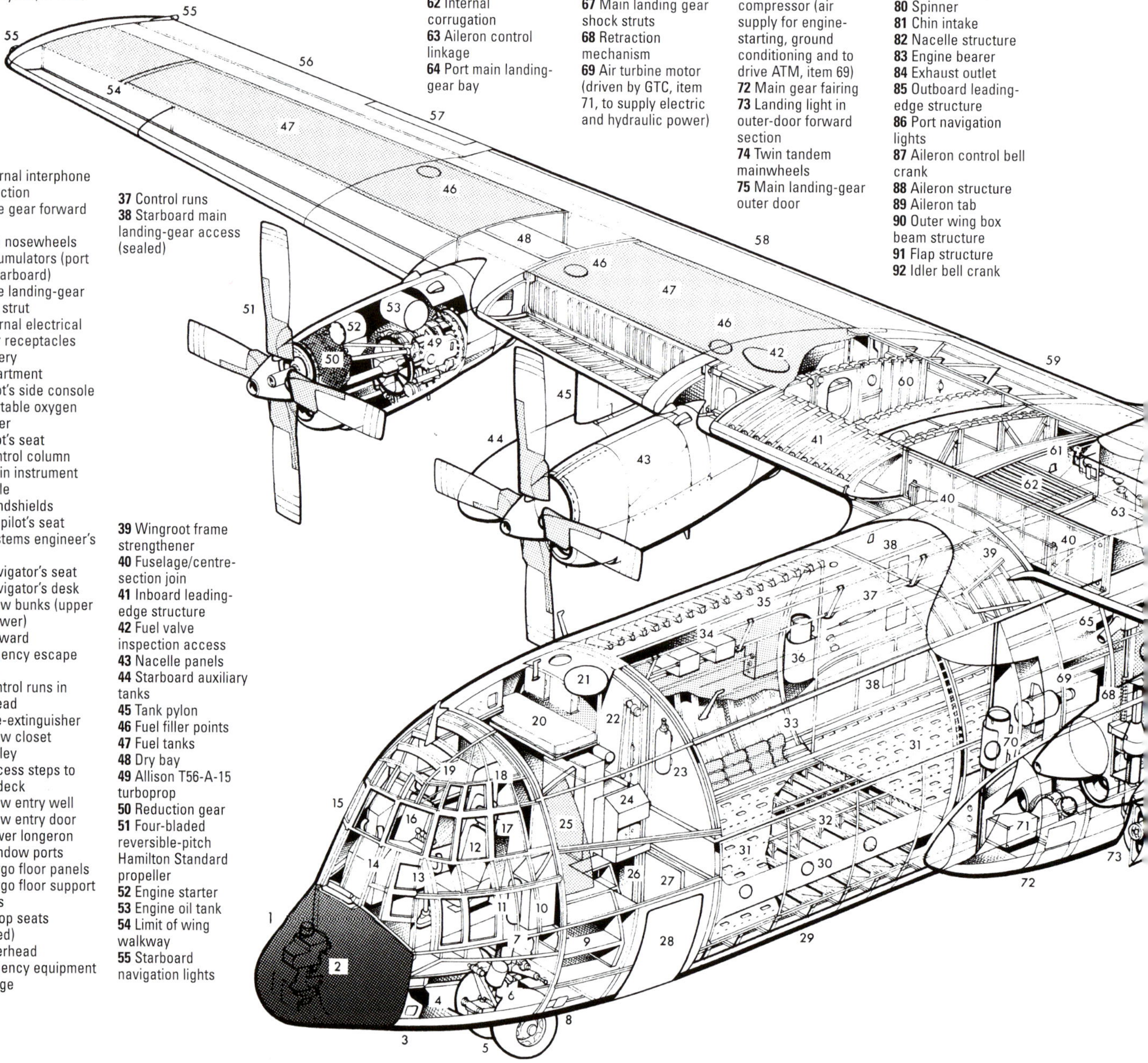

C-130H Hercules

Cutaway key

1 Radome
2 Sperry AN/APN-59 radar
3 External interphone connection
4 Nose gear forward door
5 Twin nosewheels
6 Accumulators (port and starboard)
7 Nose landing-gear shock strut
8 External electrical power receptacles
9 Battery compartment
10 Pilot's side console
11 Portable oxygen cylinder
12 Pilot's seat
13 Control column
14 Main instrument console
15 Windshields
16 Co-pilot's seat
17 Systems engineer's seat
18 Navigator's seat
19 Navigator's desk
20 Crew bunks (upper and lower)
21 Forward emergency escape hatch
22 Control runs in bulkhead
23 Fire-extinguisher
24 Crew closet
25 Galley
26 Access steps to flight deck
27 Crew entry well
28 Crew entry door
29 Lower longeron
30 Window ports
31 Cargo floor panels
32 Cargo floor support frames
33 Troop seats (stowed)
34 Overhead emergency equipment stowage
35 Fuselage frames
36 Booster hydraulic system reservoir and accumulator
37 Control runs
38 Starboard main landing-gear access (sealed)
39 Wingroot frame strengthener
40 Fuselage/centre-section join
41 Inboard leading-edge structure
42 Fuel valve inspection access
43 Nacelle panels
44 Starboard auxiliary tanks
45 Tank pylon
46 Fuel filler points
47 Fuel tanks
48 Dry bay
49 Allison T56-A-15 turboprop
50 Reduction gear
51 Four-bladed reversible-pitch Hamilton Standard propeller
52 Engine starter
53 Engine oil tank
54 Limit of wing walkway
55 Starboard navigation lights
56 Starboard aileron
57 Aileron tab
58 Outer wing flap
59 Centre-section flap
60 Centre-section wing box beam structure
61 Flap drive control
62 Internal corrugation
63 Aileron control linkage
64 Port main landing-gear bay
65 Hydraulic actuator motor
66 Fire-extinguisher bottles
67 Main landing gear shock struts
68 Retraction mechanism
69 Air turbine motor (driven by GTC, item 71, to supply electric and hydraulic power)
70 Utility hydraulic system reservoir and accumulator
71 Gas turbine compressor (air supply for engine-starting, ground conditioning and to drive ATM, item 69)
72 Main gear fairing
73 Landing light in outer-door forward section
74 Twin tandem mainwheels
75 Main landing-gear outer door
76 Inner-door section
77 Air deflector door
78 Tank pylon
79 Port auxiliary tank
80 Spinner
81 Chin intake
82 Nacelle structure
83 Engine bearer
84 Exhaust outlet
85 Outboard leading-edge structure
86 Port navigation lights
87 Aileron control bell crank
88 Aileron structure
89 Aileron tab
90 Outer wing box beam structure
91 Flap structure
92 Idler bell crank

From 1970 the US Navy's demonstration team – the Blue Angels – operated a specially marked C-130F to support the team during engagements, both in the US and sometimes abroad. The C-130F was replaced by a TC-130G in the early 1990s.

SPECIFICATION

C-130F Hercules

Dimensions

Length: 97 ft 9 in (29.79 m)
Height: 38 ft 3 in (11.66 m)
Wingspan: 132 ft 7 in (40.41 m)
Wing area: 1,745 sq ft (161.12 m²)
Wing aspect ratio: 10.09
Tailplane span: 52 ft 8 in (16.05 m)
Wheel track: 14 ft 3 in (4.35 m)
Wheel base: 32 ft 1 in (9.77 m)

Powerplant

Four Allison T56-A-7 turboprops each rated at 4,050 ehp (3020 ekW)

Weights

Empty equipped: 69,300 lb (31434 kg)
Maximum take-off: 135,000 lb (61236 kg)

Fuel and load

Internal fuel: 5,050 US gal (19116 litres)
External fuel: two 450-US gal (1703-litre) underwing tanks
Maximum payload: 35,700 lb (16194 kg)

Performance

Maximum speed at 30,000 ft (9145 m): 321 kt (370 mph; 595 km/h)
Maximum climb rate at sea level: 2,000 ft (610 m) per minute
Service ceiling: 34,000 ft (10365 m)
Take-off distance to 50 ft (15 m): 4,300 ft (1311 m) at maximum take-off weight

Accommodation

The cabin can hold 78 troops (92 in a high-density configuration) or 64 paratroops or up to 74 stretcher litters. In cargo configuration, vehicles, artillery, small helicopters and numerous other cargoes can be carried or up to six palletised loads

One of 12 C-130Es delivered to the Royal Australian Air Force from 1966, this No. 37 Squadron example is seen at Clark AFB, Philippines in 1981 while participating in Cope Thunder. The RAAF's C-130Es are currently being replaced by C-130Js.

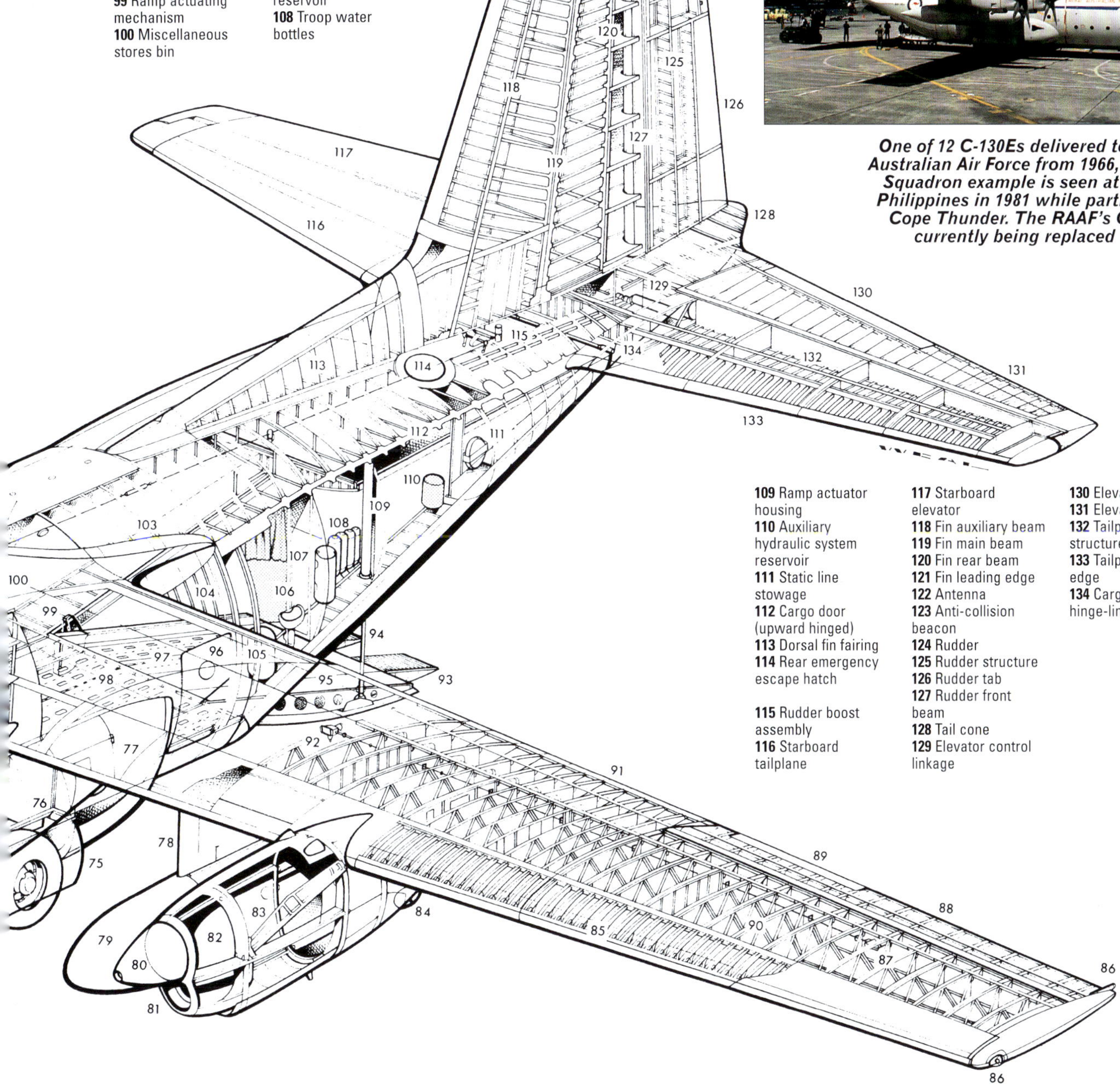

93 Auxiliary ground-loading ramp
94 Ramp actuating cylinder
95 Cargo ramp (lowered)
96 Port paratroop door
97 Cargo ramp floor panels
98 Ramp hinge line
99 Ramp actuating mechanism
100 Miscellaneous stores bin
101 Starboard paratroop door
102 Centre emergency escape hatch
103 Wingroot fairing
104 Fuselage frames
105 Toilet
106 Urinal
107 Ramp and auxiliary hydraulic reservoir
108 Troop water bottles
109 Ramp actuator housing
110 Auxiliary hydraulic system reservoir
111 Static line stowage
112 Cargo door (upward hinged)
113 Dorsal fin fairing
114 Rear emergency escape hatch
115 Rudder boost assembly
116 Starboard tailplane
117 Starboard elevator
118 Fin auxiliary beam
119 Fin main beam
120 Fin rear beam
121 Fin leading edge
122 Antenna
123 Anti-collision beacon
124 Rudder
125 Rudder structure
126 Rudder tab
127 Rudder front beam
128 Tail cone
129 Elevator control linkage
130 Elevator tab
131 Elevator structure
132 Tailplane box structure
133 Tailplane leading edge
134 Cargo door rear hinge-line

McDonnell Douglas KC-10 Extender

During the Allied Force operations over Kosovo in 1999, the USAF assigned 24 KC-10As to the Allied Force tanker fleet, split between Moron in Spain and Rhein-Main in Germany. The example shown here was operated from the latter, as part of the 60th Air Expeditionary Wing.

KC-10 Extender

Cutaway key
1 Radome
2 Weather radar scanner
3 Radar mounting
4 Front pressure bulkhead
5 Radome hinge panel
6 Windscreen wipers
7 Windscreen panels
8 Instrument panel shroud
9 Control column
10 Rudder pedals
11 Underfloor radio and electronics racks
12 Flight deck floor level
13 Pilot's seat
14 Overhead systems control panel
15 Flight engineer's control panel
16 Observer's seat
17 Cockpit doorway
18 Refuelling floodlights
19 Universal air refuelling receptacle (UARSSI)
20 Toilet compartment
21 Crew baggage locker
22 Galley
23 Air conditioning ram air intake duct
24 Entry doorway
25 Air conditioning system access panels
26 Nose landing gear strut
27 Twin nosewheels
28 Nosewheel leg doors
29 Air conditioning plant
30 Passenger seating, six crew and 14 support personnel layout
31 Forward cabin roof trim panels
32 Upper formation light
33 IFF aerial
34 Overhead air conditioning ducting
35 Crew rest bunks (four)
36 Environmental curtain
37 Cargo winch
38 Cargo safety net
39 Powered cargo handling system control box
40 Low voltage formation lighting strip
41 Underfloor oxygen bottle stowage
42 Powered roller cargo handling floor
43 Underfloor water tank
44 Door hydraulic jack
45 Cargo door, 102 in x 140 in (2.59 m x 3.56 m)
46 TACAN aerial
47 VHF aerial
48 Starboard engine nacelle
49 UHF SATCOM aerial, structural provision
50 USAF 463L cargo pallet, 25 pallets in configuration shown
51 Main cabin doorway
52 Forward underfloor fuel cell group, total underfloor cell capacity 18,075 US gal (68420 litres)
53 Fuselage frame and stringer construction
54 Director lights, port and starboard
55 Wing root fillet
56 Runway light
57 Electrical system distribution equipment centre
58 Access ladder to equipment bay
59 Central slat drive unit
60 Wing centre section carry through
61 Single cabin window, port and starboard
62 Centre section fuel tank, aircraft basic fuel system, capacity 238,565lb (108211 kg)
63 Floor beam construction
64 Wing spar/fuselage main frame
65 Overwing integral fuel tank
66 De-activated centre section doors
67 Anti-collision light
68 Starboard wing integral fuel tank
69 Inboard leading edge slat
70 Engine thrust reverser cascades, open
71 Starboard nacelle pylon
72 Outboard slat drive mechanism
73 Pressure refuelling connections
74 Fuel system piping
75 Slat guide rails
76 Outboard leading edge slat segments
77 Starboard navigation light
78 Wing tip formation lights
79 Starboard wing tip strobe light
80 Static dischargers
81 Aileron balance weights
82 Aileron hydraulic jack
83 Outboard low-speed aileron
84 Fuel jettison pipe
85 Outboard spoiler segments (four), open
86 Spoiler hydraulic jacks
87 Flap hydraulic jacks
88 Flap hinge fairings
89 Outboard double slotted flap, down position
90 High-speed aileron
91 Inboard spoiler
92 Inboard double slotted flap, down position
93 Fuselage skin plating
94 UHF aerial
95 Centre fuselage construction
96 Pressure floor above wheel bay
97 Centre landing gearwheel bay
98 Cargo loading floor
99 Roller conveyors
100 Cabin wall trim panels
101 Access ladder to lower deck refuelling station
102 Drogue refuelling hose reel unit
103 Drogue housing
104 Ground emergency exit doorway
105 Rear cabin air conditioning duct
106 HF aerial
107 Centre engine pylon construction
108 Centre engine intake
109 Intake duct construction
110 Intake duct ring frames
111 Tailfin attachment joint
112 Starboard tailplane
113 Starboard elevator
114 Tailfin construction
115 J-band and I-band beacon antennas
116 VOR localiser-1 aerial
117 Fin tip fairing
118 VOR localiser 2 aerial
119 Rudder mass balance
120 Two-segment rudder
121 Rudder hydraulic jacks
122 Fin low voltage formation lighting strip
123 Centre engine installation
124 Detachable engine cowlings
125 Bleed-air system pre-cooler
126 Engine mounting pylon

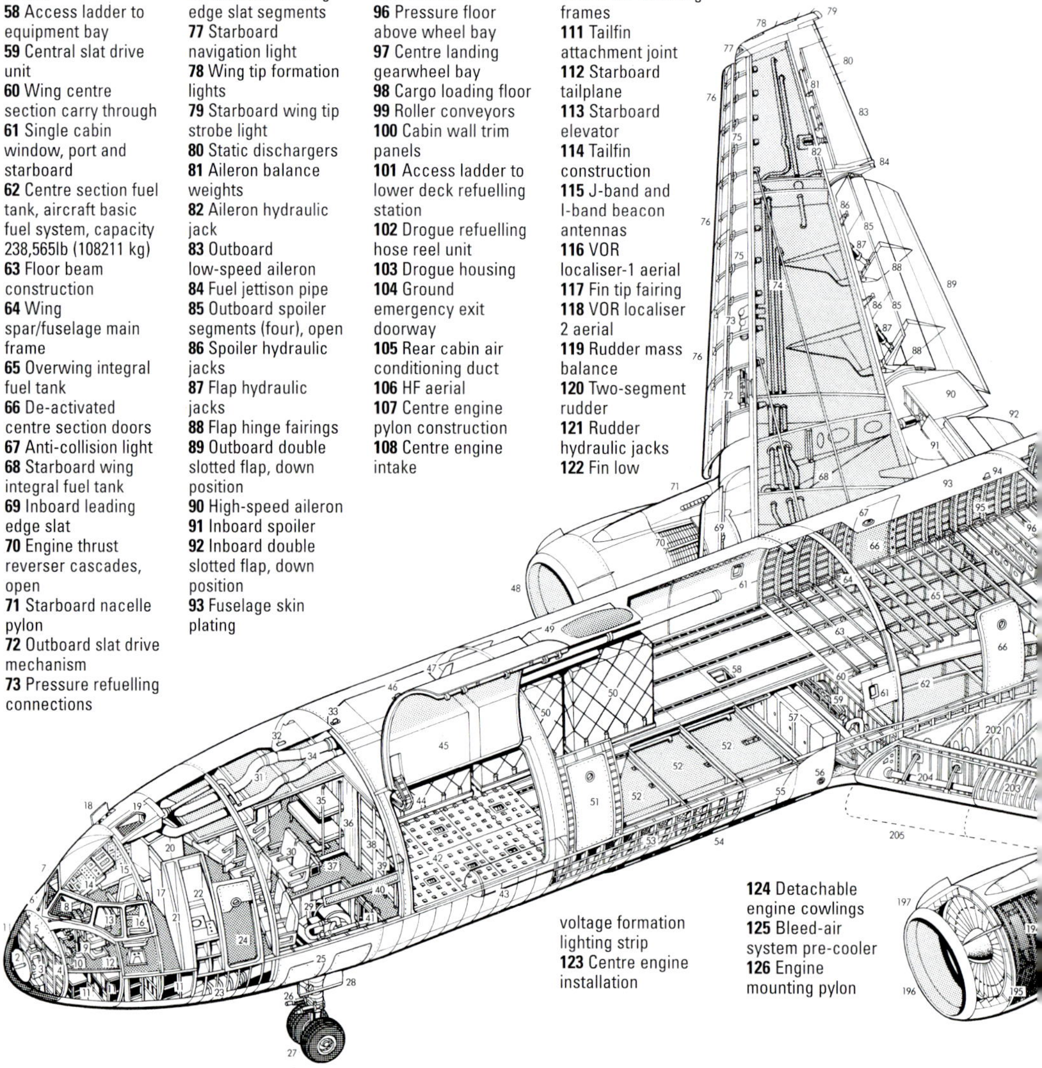

Above: Crew members from the 2nd Air Refuelling Squadron based at McGuire AFB in New Jersey load up a KC-10A using the upward-hinging cargo door in the forward port fuselage, at NSF Diego Garcia, British Indian Ocean Territory. A total of 60 Extenders was built for the USAF, most of which now carry the all-over charcoal grey colour scheme.

SPECIFICATION

KC-10A Extender

Dimensions

Wingspan: 155 ft 4 in (47.34 m)
Wing aspect ratio: 6.8
Wing area: 3,861.00 sq ft (358.69 m²)
Length: 181 ft 7 in (55.35 m)
Height: 58 ft 1 in (17.70 m)
Tailplane span: 71 ft 2 in (21.69 m)
Wheel track: 34 ft 8 in (10.57 m)
Wheel base: 72 ft 5 in (22.07 m)

Powerplant

Three General Electric CF6-50C2 turbofans each rated at 52,500 lb st (233.53 kN)

Weights

Operating empty as a tanker: 240,065 lb (108891 kg)
Operating empty as a cargo transport: 244,630 lb (110962 kg)
Maximum take-off: 590,000 lb (267620 kg)

Fuel and load

Basic aircraft fuel system: 238,236 lb (108062 kg)
Fuselage bladder fuel cells: 117,829 lb (53446 kg)
Total internal fuel: 356,065 lb (161508 kg)
Maximum cargo payload: 169,409 lb (76843 kg)

Performance

Never exceed speed: Mach 0.95
Maximum level speed 'clean' at 25,000 ft (7620 m): 610 mph (982 km/h)
Maximum cruising speed at 30,000 ft (9145 m): 564 mph (908 km/h)
Maximum rate of climb at sea level: 2,900 ft (884 m) per minute
Service ceiling: 33,400 ft (10180 m)
Nominal range with 100,000 lb (45400 kg) payload: 6,905 miles (11112 km)
Maximum range with maximum cargo: 4,370 miles (7032 km)
Ferry range: 11,500 miles (18507 km)
Take-off balanced field length at maximum take-off weight: 10,400 ft (3170 m)
Landing balanced field length at maximum landing weight: 6,130 ft (1868 m)

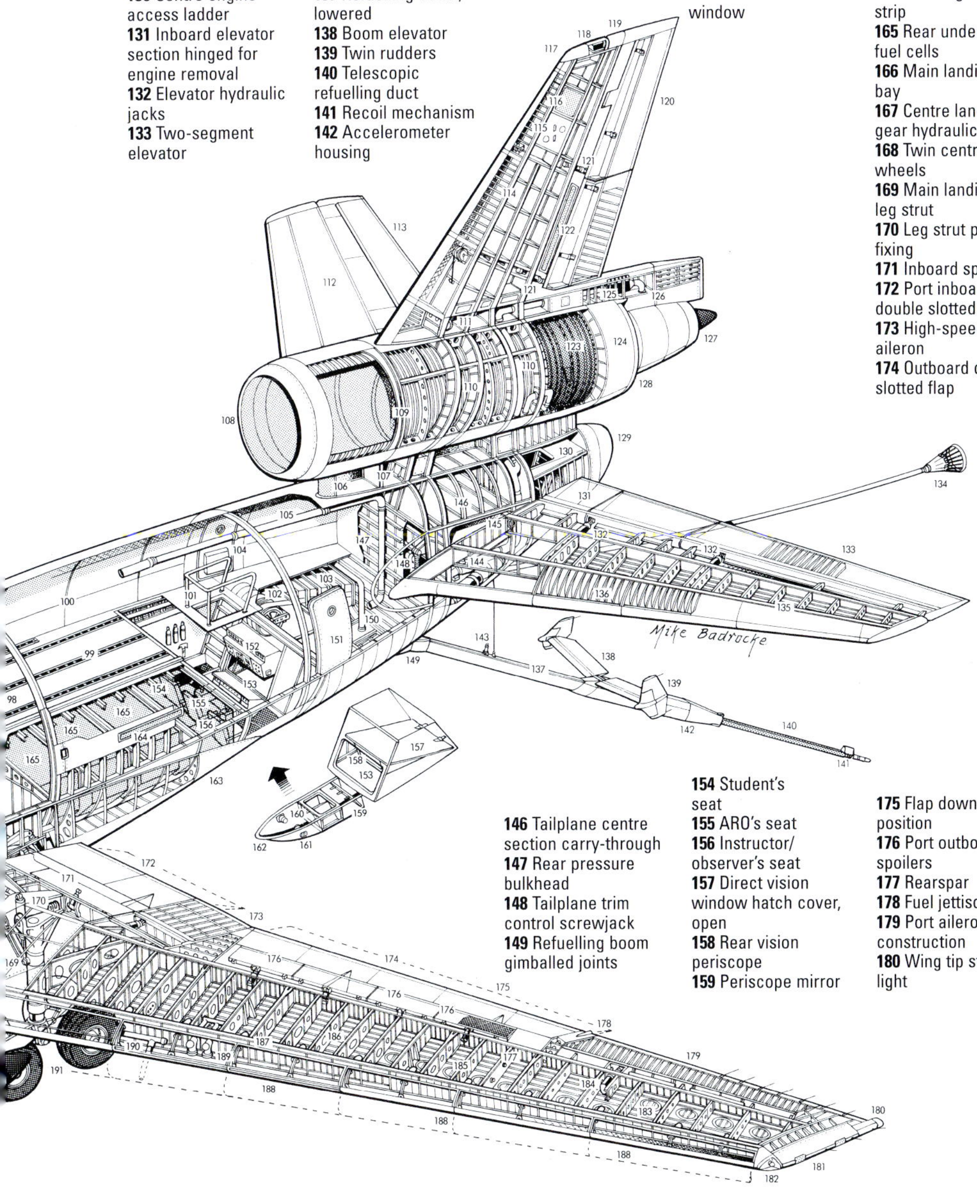

127 Hot stream exhaust nozzle
128 Fan air exhaust duct
129 Detachable tailcone fairing
130 Centre engine access ladder
131 Inboard elevator section hinged for engine removal
132 Elevator hydraulic jacks
133 Two-segment elevator
134 Flight-refuelling drogue, deployed
135 Port tailplane construction
136 Leading edge nose ribs
137 Refuelling boom, lowered
138 Boom elevator
139 Twin rudders
140 Telescopic refuelling duct
141 Recoil mechanism
142 Accelerometer housing
143 Boom hoist cable and up-lock
144 Auxiliary power unit (APU)
145 Tailplane pivot fixing
146 Tailplane centre section carry-through
147 Rear pressure bulkhead
148 Tailplane trim control screwjack
149 Refuelling boom gimballed joints
150 Fuel feed pipe
151 De-activated doorway
152 Air refuelling officer's (ARO's) control panel
153 Direct vision window
154 Student's seat
155 ARO's seat
156 Instructor/observer's seat
157 Direct vision window hatch cover, open
158 Rear vision periscope
159 Periscope mirror
160 Side view mirrors
161 Wing floodlights
162 Mirror fairing
163 Wing root trailing edge fillet
164 Low voltage formation lighting strip
165 Rear underfloor fuel cells
166 Main landing gear bay
167 Centre landing gear hydraulic jack
168 Twin centre wheels
169 Main landing gear leg strut
170 Leg strut pivot fixing
171 Inboard spoiler
172 Port inboard double slotted flap
173 High-speed aileron
174 Outboard double slotted flap
175 Flap down position
176 Port outboard spoilers
177 Rearspar
178 Fuel jettison pipe
179 Port aileron construction
180 Wing tip strobe light
181 Port wing tip formation lights
182 Port navigation light
183 Lower wing skin access panels
184 Aileron hydraulic jack housing
185 Wing ribconstruction
186 Port wing integral fuel tank
187 Front spar
188 Port leading edge slat segments
189 Pressure refuelling connections
190 Leading edge de-icing telescopic air duct
191 Four wheel main landing gear bogie
192 Port engine installation
193 Thrust reverser cascade, closed
194 General Electric CF6-50C2 turbofan engine
195 Fan easing mounted accessory gearbox
196 Engine air intake
197 Nacelle strakes
198 Nacelle pylon construction
199 Pylon attachment joint
200 Wing skin panelling
201 Wing stringers
202 Inboard wing ribs
203 Inboard leading edge slat rib construction
204 Bleed air ducting
205 Slat down position

Transall C-160

In 1970, the South African Air Force received the first of nine French-built Transalls, the Germans being politically unwilling to deal with the South African government. These were designated C.160Zs, and ordered largely because the US had embargoed further deliveries of Lockheed C-130 Hercules aircraft, which the SAAF had been using up until that time. However, the C.160 proved remarkably successful for the SAAF, and several were returned to service after being withdrawn and stored in 1993.

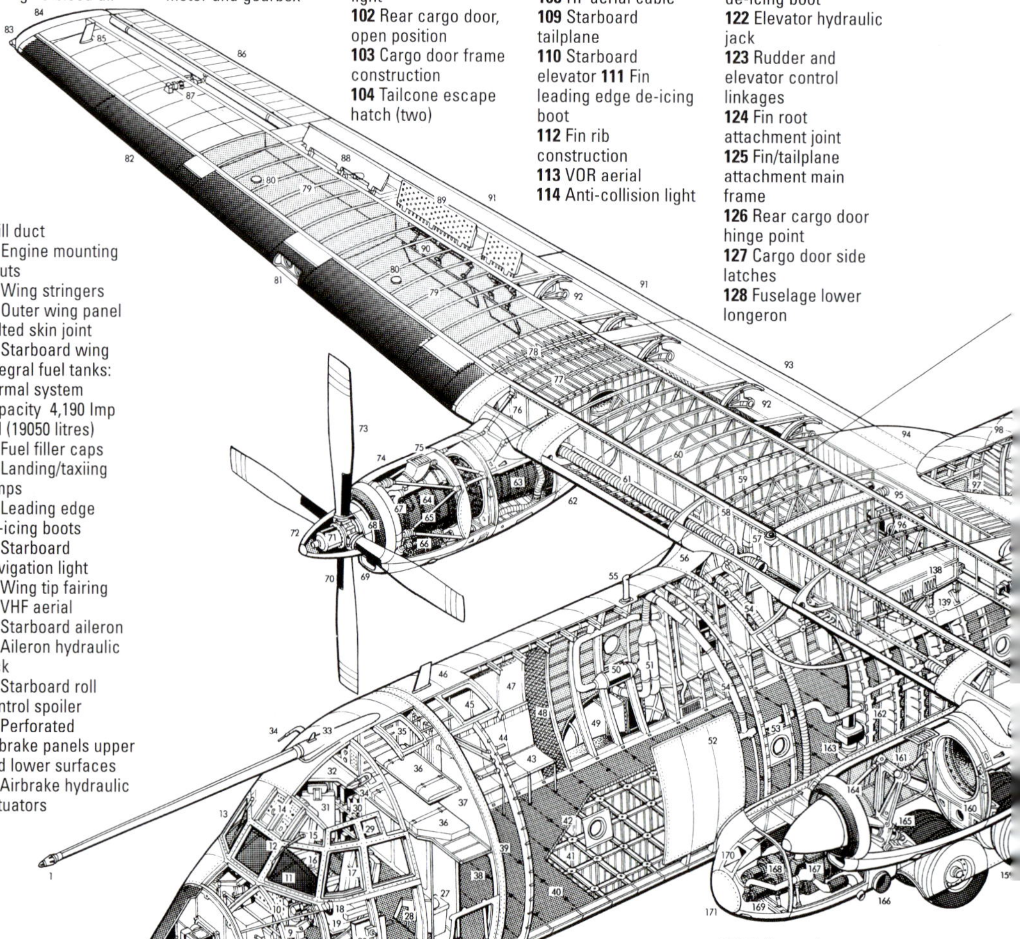

C.160

Cutaway key

1 Fixed inflight-refuelling probe
2 Radome
3 Radar tracking mechanism
4 Weather radar scanner
5 Cockpit front pressure bulkhead
6 Battery bay port and starboard
7 Cockpit floor level
8 Rudder pedals
9 Control column
10 Instrument panel
11 Instrument panel shroud
12 Windscreen wipers
13 Windscreen panels
14 Overhead switch panel
15 Co-pilot's seat
16 Centre control pedestal
17 Sliding side window panel
18 Control column handwheel
19 Pilot's seat
20 Chart case
21 Nosewheel bay
22 Twin nosewheels
23 Nosewheel door
24 Boarding steps
25 Door external latch
26 Crew entry door
27 Crew toilet
28 Cockpit steps
29 Cockpit eyebrow windows
30 Radio and electronics racks
31 Navigator's swivelling seat
32 Chart table
33 TACAN aerial
34 Pitot tubes
35 Crew escape hatch
36 Twin rest bunks
37 Cockpit rear bulkhead
38 Main cabin doorway
39 Centre fuselage/cockpit section production joint
40 Main cargo loading deck
41 Floor beam construction
42 Honeycomb floor panels
43 Folding troop seats: up to 93 troops
44 Control rod runs
45 Cabin escape hatch
46 VHF aerial
47 Cabin wall trim panels
48 Soundproofing insulation
49 Starboard sponson air system heat exchanger
50 Water extractor
51 Air conditioning plant
52 Foreign object damage propeller guard skin reinforcing plate
53 Cabin window panel
54 Air system piping
55 HF aerial mast
56 Wing fillet fairing
57 Anti-collision light
58 Wing front spar
59 Strengthened wing centre section construction provision for additional fuel tankage of 1,980 Imp gal (9000 litres)
60 Wing lattice ribs
61 Engine bleed air piping
62 Starboard engine nacelle
63 Engine exhaust duct
64 Rolls-Royce Tyne RTy.20 Mk 22 turboprop engine
65 Engine bearer struts
66 Engine accessory units
67 Oil tank 7½ Imp gal (34 litres)
68 Annular engine air intake
69 Oil cooler air intake
70 Propeller blade root de-icing boots
71 Propeller pitch change mechanism
72 Spinner
73 Four-bladed constant speed propeller
74 Detachable cowling panels
75 Engine bleed air spill duct
76 Engine mounting struts
77 Wing stringers
78 Outer wing panel bolted skin joint
79 Starboard wing integral fuel tanks: normal system capacity 4,190 Imp gal (19050 litres)
80 Fuel filler caps
81 Landing/taxiing lamps
82 Leading edge de-icing boots
83 Starboard navigation light
84 Wing tip fairing
85 VHF aerial
86 Starboard aileron
87 Aileron hydraulic jack
88 Starboard roll control spoiler
89 Perforated airbrake panels upper and lower surfaces
90 Airbrake hydraulic actuators
91 Outboard double-slotted flap segments
92 Flap guide rails
93 Inboard double slotted flap
94 Wing root trailing edge fillet
95 Cabin pressurisation valves
96 Central flap drive motor and gearbox
97 Starboard paratroop door
98 Fuselage skin plating
99 Control rod runs
100 Centre fuselage/tailcone production joint
101 Lower formation light
102 Rear cargo door, open position
103 Cargo door frame construction
104 Tailcone escape hatch (two)
105 Fin root fillet
106 Upper formation light
107 Tailplane leading edge de-icing boot
108 HF aerial cable
109 Starboard tailplane
110 Starboard elevator **111** Fin leading edge de-icing boot
112 Fin rib construction
113 VOR aerial
114 Anti-collision light
115 Rudder construction
116 Rudder hydraulic jack
117 Tailcone
118 Tail navigation light
119 Port elevator construction
120 Fixed tailplane construction
121 Leading edge de-icing boot
122 Elevator hydraulic jack
123 Rudder and elevator control linkages
124 Fin root attachment joint
125 Fin/tailplane attachment main frame
126 Rear cargo door hinge point
127 Cargo door side latches
128 Fuselage lower longeron
129 Tailcone frame construction
130 Detachable vehicle loading ramps
131 Forward cargo door/main loading ramp, lowered position
132 Rear toilet

In terms of their general layout the C-130 Hercules (background) and Transall C.160 are very similar. Both have a shoulder-mounted wing leaving the cabin clear of intrusion by the wing spars, a rugged undercarriage with its main units housed in external sponsons and an upswept beaver tail with an in-built loading ramp/door. While the Hercules employs four engines of around 4,050 shp (3020 kW) each, the Transall uses a pair of considerably more powerful Tyne engines. This results in a machine with similar payload capabilities to the US airlifter, but with generally inferior performance, a fact perhaps reflected by l'Armée de l'Air's decision to procure the Hercules as well as the Transall.

SPECIFICATION

C.160 (first generation)

Dimensions

Length: 106 ft 3½ in (32.40 m)
Height: 38 ft 5 in (11.65 m)
Wingspan: 131 ft 3 in (40.00 m)
Wing area: 1,723.36 sq ft (160.10 m²)
Wing aspect ratio: 10
Tailplane span: 47 ft 7 in (14.50 m)
Wheel track: 16 ft 9 in (5.10 m)
Wheel base: 34 ft 4½ in (10.48 m)

Powerplant

Two Rolls-Royce Tyne RTy.20 Mk 22 each rated at 6,100 ehp (4548 ekW)

Weights

Empty equipped: 63,400 lb (28758 kg)
Normal take-off: 97,443 lb (44200 kg)
Maximum take-off: 108,245 lb (49100 kg)

Fuel and load

Internal fuel: 4,359 US gal (16500 litres)
Maximum payload: 35,273 lb (16000 kg)

Performance

Maximum level speed 'clean' at 14,765 ft (4500 m): 333 mph (536 km/h)
Maximum cruising speed at 18,045 ft (5500 m): 319 mph (513 km/h)
Maximum cruising speed at 26,245 ft (8000 m): 308 mph (495 km/h)
Maximum rate of climb at sea level: 1,444 ft (440 m) per minute
Service ceiling: 27,885 ft (8500 m)
Take-off run at maximum take-off weight: 2,608 ft (795 m)
Take-off distance to 35 ft (10.70 m) at maximum take-off weight: 3,609 ft (1100 m)
Landing distance from 50 ft (15 m) at normal landing weight: 2,100 ft (640 m)
Landing run at normal landing weight: 1,181 ft (360 m)
Range with an 8000-kg (17,637-lb) payload: 2,796 miles (4500 km)
Range with a 35,273-lb (16000-kg) payload: 734 miles (1182 km)

Above: Turkey maintains a tactical transport force similar in constitution to that of France, with C.160Ds flying alongside c-130 Hercules and CN-235s. The Transalls were acquired as 19 secondhand aircraft from West Germany in 1971, and are based at Erkilet with 221 Filo of the Türk Hava Kuvvetleri.

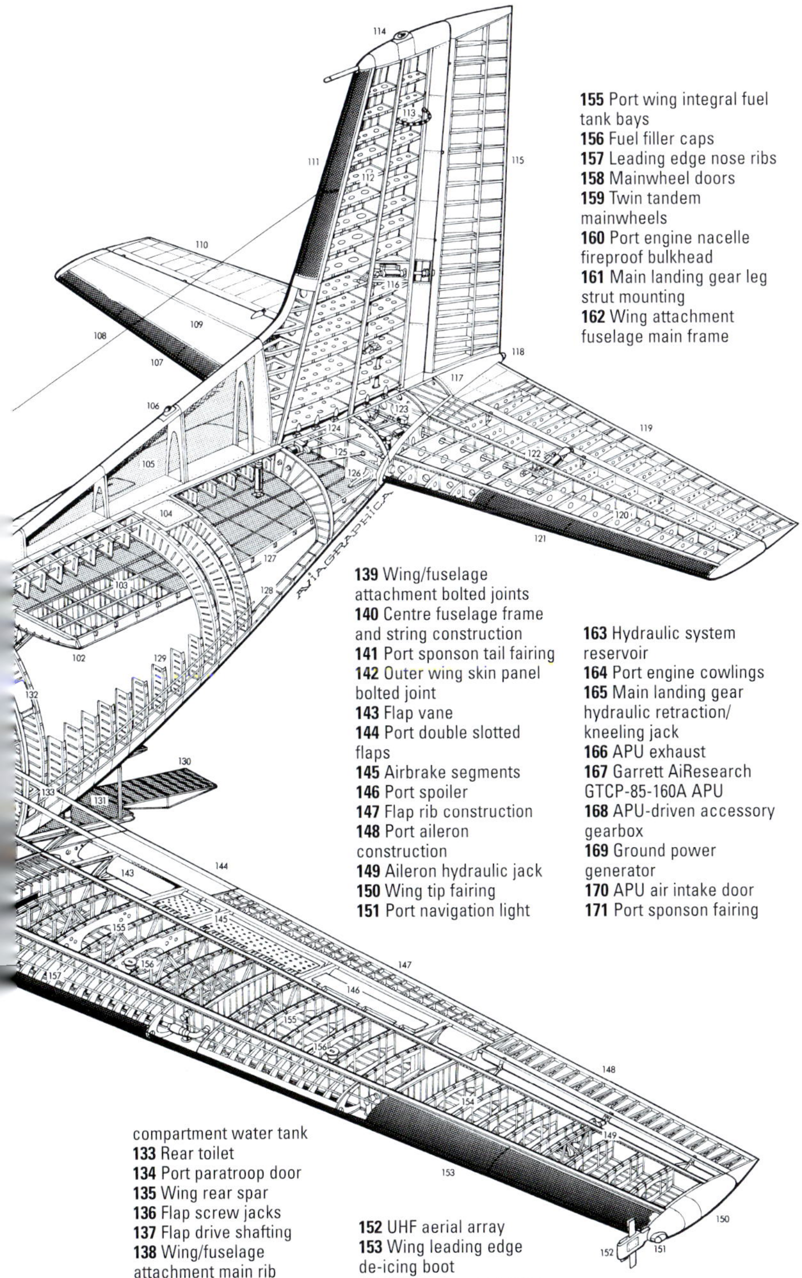

compartment water tank
133 Rear toilet
134 Port paratroop door
135 Wing rear spar
136 Flap screw jacks
137 Flap drive shafting
138 Wing/fuselage attachment main rib
139 Wing/fuselage attachment bolted joints
140 Centre fuselage frame and string construction
141 Port sponson tail fairing
142 Outer wing skin panel bolted joint
143 Flap vane
144 Port double slotted flaps
145 Airbrake segments
146 Port spoiler
147 Flap rib construction
148 Port aileron construction
149 Aileron hydraulic jack
150 Wing tip fairing
151 Port navigation light
152 UHF aerial array
153 Wing leading edge de-icing boot
154 Wing rib construction
155 Port wing integral fuel tank bays
156 Fuel filler caps
157 Leading edge nose ribs
158 Mainwheel doors
159 Twin tandem mainwheels
160 Port engine nacelle fireproof bulkhead
161 Main landing gear leg strut mounting
162 Wing attachment fuselage main frame
163 Hydraulic system reservoir
164 Port engine cowlings
165 Main landing gear hydraulic retraction/kneeling jack
166 APU exhaust
167 Garrett AiResearch GTCP-85-160A APU
168 APU-driven accessory gearbox
169 Ground power generator
170 APU air intake door
171 Port sponson fairing

The capability for off-airfield operations is a must for any tactical airlifter. Low-pressure tyres allow the C.160 to operate from a variety of unprepared surfaces, although operations from autobahns are more demanding of the crew than the aircraft. Provision was made in the Transall design for the installation of underwing booster jets in the 5,250-lb st (23.34-kN) class, for improved take-off and cruise performance, but these have never been used in service.

Vickers VC-10

Although the RAF's VC10 fleet is coming to the end of its useful service life, the type is still a vitally important element in the UK's front-line. The VC10s served with distinction during Desert Storm, refuelling US Navy aircraft in addition to RAF machines and have since been active over the Balkans and in support of various UN operations. One K.Mk 2 was initially painted in a grey/green camouflage, but otherwise the hemp scheme has been universal.

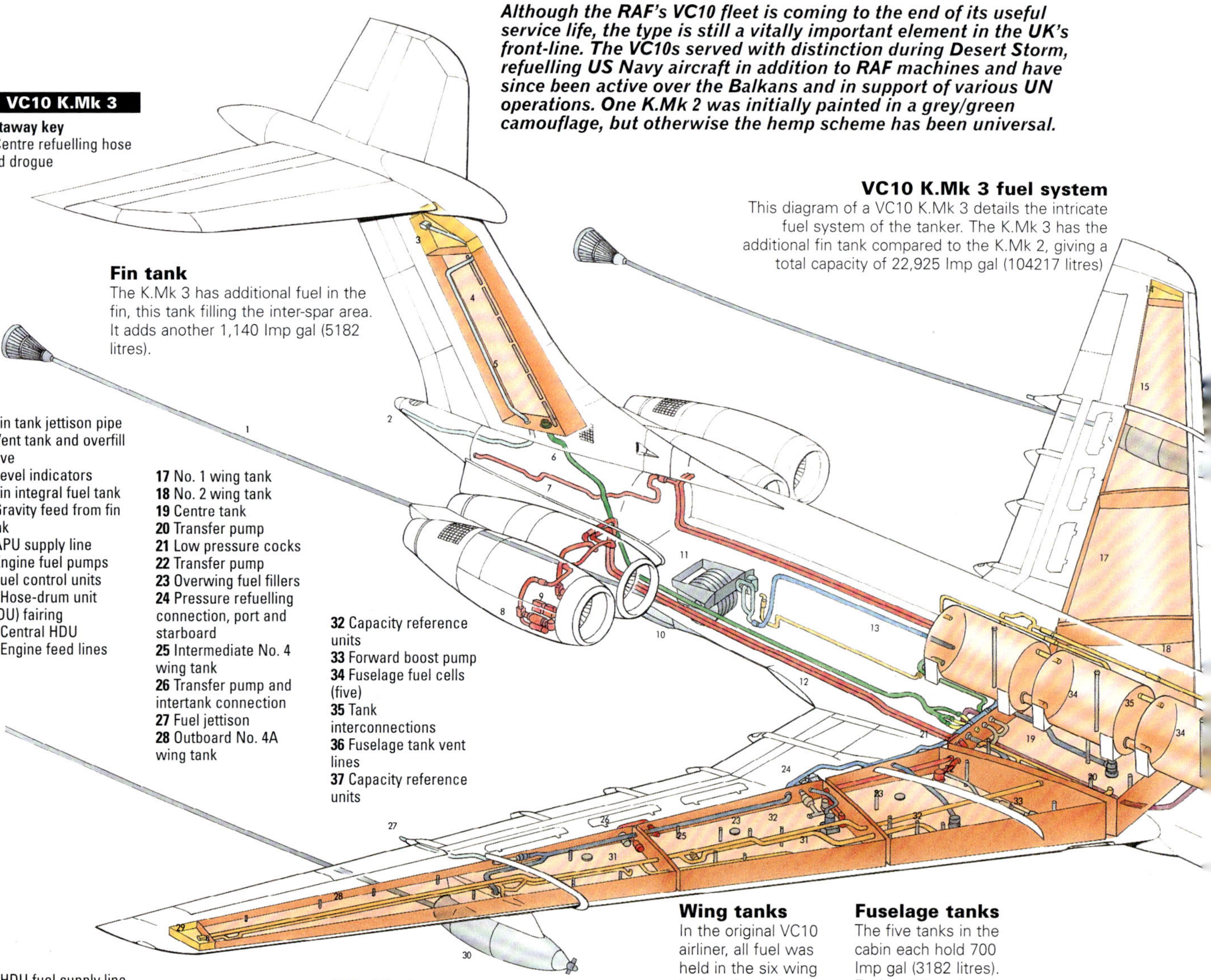

VC10 K.Mk 3

Cutaway key
1 Centre refuelling hose and drogue
2 Fin tank jettison pipe
3 Vent tank and overfill valve
4 Level indicators
5 Fin integral fuel tank
6 Gravity feed from fin tank
7 APU supply line
8 Engine fuel pumps
9 Fuel control units
10 Hose-drum unit (HDU) fairing
11 Central HDU
12 Engine feed lines
13 HDU fuel supply line
14 Wingtip vent surge tank
15 Outboard No. 1A wing tank
16 Flight Refuelling Mk 32 wing pod
17 No. 1 wing tank
18 No. 2 wing tank
19 Centre tank
20 Transfer pump
21 Low pressure cocks
22 Transfer pump
23 Overwing fuel fillers
24 Pressure refuelling connection, port and starboard
25 Intermediate No. 4 wing tank
26 Transfer pump and intertank connection
27 Fuel jettison
28 Outboard No. 4A wing tank
29 Port wingtip vent surge tank
30 Starboard Mk 32 refuelling pod
31 Tank vent lines
32 Capacity reference units
33 Forward boost pump
34 Fuselage fuel cells (five)
35 Tank interconnections
36 Fuselage tank vent lines
37 Capacity reference units
38 Feed line from refuelling probe
39 Refuelling control panel at flight engineer's station
40 Fixed flight-refuelling probe

Fin tank
The K.Mk 3 has additional fuel in the fin, this tank filling the inter-spar area. It adds another 1,140 Imp gal (5182 litres).

VC10 K.Mk 3 fuel system
This diagram of a VC10 K.Mk 3 details the intricate fuel system of the tanker. The K.Mk 3 has the additional fin tank compared to the K.Mk 2, giving a total capacity of 22,925 Imp gal (104217 litres)

Wing tanks
In the original VC10 airliner, all fuel was held in the six wing tanks, situated between the main spars. These offer a capacity of 17,925 Imp gal (81480 litres).

Fuselage tanks
The five tanks in the cabin each hold 700 Imp gal (3182 litres). Each unit is a double-walled metal cylinder with flexible inner bag, mounted on two floor beams and restrained by an A-frame structure.

Extra tanks

The K.Mk 2s and K.Mk 3s have an extra 3,500 Imp gal (15910 litres) of fuel contained in five equal-sized cells in the former passenger cabin. The tanks were installed via the freight doors of the 'Supers' before they were sealed, but the K.Mk 2s had to be cut in half during conversion. The six standard wing tanks in all VC10s hold a total of 17,925 Imp gal (81480 litres), in addition to the optional fin tank and some 20 Imp gal (91 litres) in each of the three refuelling units' reservoirs.

Hose-drum units

Two types of refueller are used on the VC10. In the lower rear fuselage is a Flight Refuelling Ltd Mk 17B hose-drum unit (HDU) deploying up to 70 ft (21 m) of hose and capable of delivering fuel at up to 4,000 lb (500 Imp gal; 2270 litres) per minute. Outboard, beneath the wings are two FRL Mk 32/2800 pods with a 48-ft (14.60-m) hose length and fuel flow of 2,800 lb (350 Imp gal; 1591 litres) per minute. Normal operating speeds for tanking are between 250-390 mph (400-630 km/h) with either one large aircraft on the centreline or two fighters refuelling from the wing pods. Both Mk 17B and Mk 32 (right) hose-drum units are equipped with signal lights for receiver aircraft and aligning marks. The drogues themselves are equipped with white lights to provide visual cues for night refuelling.

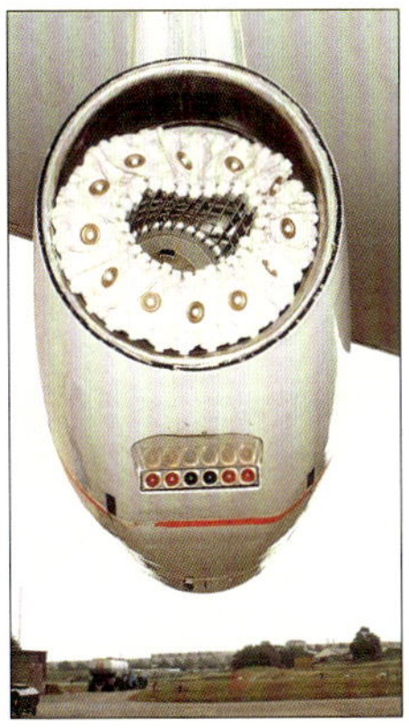

SPECIFICATION

VC10 C.Mk 1

Dimensions

Wingspan: 146 ft 2 in 144.55 m)
Wing area: 2,932 sq ft (272.38 m²)
Aspect ratio: 7.29
Length (excluding probe): 158 ft 8 in (48.38 m)
Height: 39 ft 6 in (12.04 m)
Tailplane span: 43 ft 10 in (13.36 m)
Wheel track: 21 ft 5 in (6.53 m)
Wheel base: 65 ft 10½ in (20.08 m)

Powerplant

Four Rolls-Royce Conway RCo.43 Mk 301 turbofans each rated at 21,800 lb st (96.97 kN)

Weights

Empty: 146,000 lb (66224 kg)
Maximum take-off: 323,000 lb (146510 kg)
Maximum take-off (K.Mk 2): 313,056 lb (142000 kg)
Maximum take-off (K.Mk 3): 334,882 lb (151900 kg)
Maximum take-off (K.Mk 4): 334,882 lb (151900 kg)

Fuel and load

Maximum payload: 57,400 lb (26037 kg)
Internal fuel: 19,365 Imp gal (88032 litres)
Internal fuel (K.Mk 2): 21,485 Imp gal (97671 litres)
Internal fuel (K.Mk 3): 22,925 Imp gal (104217 litres)
Internal fuel (K.Mk 4): 19,425 Imp gal (88306 litres)

Performance

Maximum cruising speed at 31,000 ft (9450 m): 581 mph (935 km/h)
Economical cruising speed at 30,000 ft (9145 m): 426 mph (684 km/h)
Range with maximum payload: 3,898 miles (6273 km)
Maximum rate of climb at sea level: 3,050 ft (930 m) per minute
Service ceiling: 42,000 ft (12800 m)
Take-off distance to 35 ft (10.70 m) at maximum take-off weight: 8,300 ft (2530 m)
Balanced landing field length at normal landing weight: 7,000 ft (2134 m)

VC10 C.Mk 1(K) and K.Mk 2 aircraft are being retired as their service life is deemed to expire. A major factor in the case of the C.Mk 1(K) is the cost of deep maintenance and as aircraft become due for their multimillion pound major overhauls, they are being retired. The VC10 tanker/transport fleets are therefore slowing running down, while the demands for their services remain constant – this situation is leading to low morale within the VC10 community. By 2007 the RAF envisages that the VC10's refuelling role will have been put out to a civilian contractor using either Boeing 767- or Airbus A310-based tankers. The practicality of sending civilian aircraft, possibly operated by civilian crews, into a combat environment while laden with fuel has yet to be decided.

Above: The VC10 tanker conversion added nose refuelling probes to the aircraft, along with a Turboméca Artouste 520 APU in the tailcone. The 'new' machines received the revised company designations V1112 VC10 and V1164 Super VC10. The latter also had their forward fuselage freight doors sealed.

Right: Painted with grey and green disruptive upper surface camouflage, the first K.Mk 2 conversion, ZA141, first flew at Filton on 22 June 1982, piloted by Roy Radford. A structural weakness in the fin resulted in it being fitted with the tail unit of an ex-civil VC10, XX914, which had recently been withdrawn from use as a testbed at RAE, Bedford and, as such, ZA141 was delivered to Boscombe Down on 9 June 1983 for trials by the A&AEE

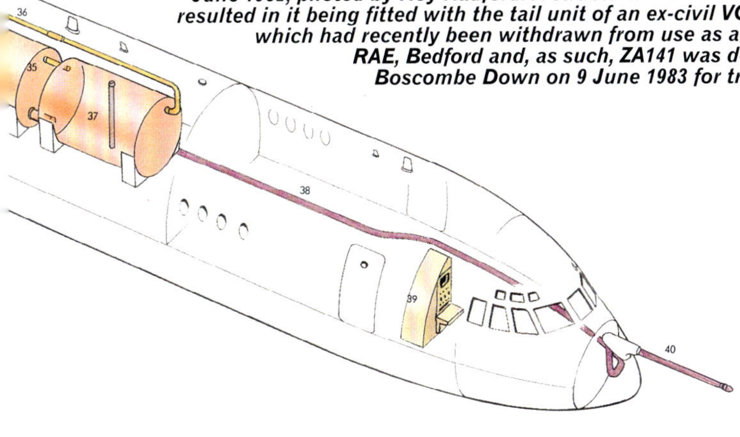

Index

Page numbers in *italics* refer to illustration captions